ISBN 978-0-266-99866-2
PIBN 10921069

CLASS-BOOK OF BOTANY,

DESIGNED FOR

COLLEGES, ACADEMIES AND OTHER SEMINARIES.

IN TWO PARTS.

PART I.
THE ELEMENTS OF BOTANICAL SCIENCE.

PART II.
THE NATURAL ORDERS.

ILLUSTRATED BY

A FLORA

OF THE NORTHERN, MIDDLE AND WESTERN STATES.

PARTICULARLY OF

THE UNITED STATES NORTH OF THE CAPITOL, LAT. 38°.

BY ALPHONSO WOOD, A. M.,

And he spake of trees, from the cedar in Lebanon even unto the hyssop that springeth out of the wall. 1 Kings 4 : 33
Consider the lilies of the field : *** even Solomon in all his glory was not arrayed like one of these.   Matt. 6 : 28, 29.

Forty-first Edition, Revised and Enlarged.

NEW YORK:
PUBLISHED BY A. S. BARNES & CO.,
111 & 113 WILLIAM STREET (COR. JOHN).
1869.

Entered according to Act of Congress, in the year 1846.

, BY ALPHONSO WOOD,

In the Clerk's Office of the District Court of the District of New Hampshire.

TO THE

REV. CHESTER DEWEY, M. D., D. D.,

PROFESSOR OF NATURAL SCIENCE IN THE BERKSHIRE AND OTHER MEDICAL
INSTITUTIONS, AUTHOR OF THE REPORT ON THE HERBACEOUS
PLANTS OF MASSACHUSETTS, MONOGRAPH ON
THE CARICES, ETC. ETC.

THIS VOLUME

IS RESPECTFULLY DEDICATED, BY

THE AUTHOR.

PREFACE.

THE science of botany is as eminently progressive as it is delightful and ennobling. By recent discoveries it has been established on the basis of inductive philosophy, and elevated almost to the rank of an exact science. The theory of the floral structure which refers each organ to the principle of the leaf, now enters into almost every department of botany, and gives a new aspect to the whole; revealing more clearly than any other discovery has ever done, the beauty and simplicity of the plan on which Creative Power is exerted in the production of the countless forms of vegetable existence.

The present treatise contains, first, the Elements of Botany, according to the latest authorities, written in the form of simple propositions, briefly illustrated, and broken into short paragraphs, with direct reference to the convenience of the learner. Brief as it is, it is hoped that it will be found to embody all the established principles of the science contained in former school treatises, together with those newly discovered principles in Organography and Physiology, by which botany has been really enriched and advanced.*

The Flora is adapted particularly to that section of the United States which lies north of the Capitol, that is, of the 39th parallel, including essentially the States lying north of the Ohio river and Maryland.† It comprehends all the Phænogamia, or flowering plants, with the Ferns, &c. which have hitherto been discovered and described as indigenous in these States, together with the naturalized exotics, and those which are more generally cultivated either as useful or ornamental. The descriptions are

---

* The student who aims at the highest attainments will by no means rest satisfied with mere outlines, such as our limits here admit of. It affords us pleasure to be able to recommend to all such as would advance beyond first principles, the full and elaborate "Text Book" of Dr. Asa Gray,—an American work of the highest merit.

† With some exceptions, therefore, this Flora will answer for the adjacent States of Delaware, Maryland, Virginia, Kentucky, Missouri, and the Canadas.

*1**

as extended and minute as appeared to the author necessary for the complete recognition of the plants, and for imparting a knowledge of whatever is peculiar or interesting in their habit, culture, or use.

With regard to the sources of information from which this part of the work has been prepared, it is proper to state, that I have for several years been engaged in the collection of materials, during which I have made extensive tours with this object in view, in nearly every section of country which this Flora represents, and have had access to numerous and extensive herbaria. By these means, I have been able to draw the description of about nine tenths of the species mentioned from the living or dried specimen. For additional information, I have availed myself of the best authorities within my reach, among which are the botanical works of Bigelow, Eaton, Wright, Pursh, Michaux, Smith, Nuttall, Torrey, Beck, Loudon, Elliott, Darlington, Dewey, Barton, Hooker, Decandolle, and Torrey and Gray.

With few exceptions, I have adopted, for our native and naturalized plants, the nomenclature of the "North American Flora" of the two last distinguished authors; and for our cultivated exotics, the nomenclature of the "Prodromus" of Decandolle, (that is, so far as these floras at present extend,) regarding these, as they truly are, standard works.

The present Flora is accompanied with numerous Analytical Tables, designed to facilitate the hitherto tedious process of botanical analysis. The object aimed at in their construction is to exhibit at one view the most striking characteristics of each group to which they respectively relate, so arranged as to conduct the mind from a single *radiating point* to any desired genus contained in the volume. That we have *fully* realized this plan, or that the tables are free from error, is not to be expected; yet we do hope that they will afford facilities for analysis greater than any system hitherto available.

To Dr. EDWARD E. PHELPS, Lecturer on Medical Botany, &c. in Dartmouth College, I am indebted for many highly valuable suggestions, particularly in regard to the tables above mentioned. In this department of the work, I would gratefully acknowledge his aid. A few tables under the Natural Orders were originally of his construction, although necessarily much altered in this edition, by the admission of additional genera. He has also granted me free access to his botanical books and specimens.

To the Rev. Professor CHESTER DEWEY, to whom I am permitted to dedicate this volume, I am indebted for that part of the Flora which relates to the difficult yet deeply interesting family of the Carices. He has not only granted me access to his former excellent monograph of that genus, but has prepared the article for the present work with his own *hand.*

PREFACE.

It gives me pleasure to be able also to introduce in this connection the name of Dr. JAMES W. ROBBINS, — a name long since enrolled among American botanists. Our Flora has been recently in part reviewed by him; and he has communicated to me, in relation to this work, the results of his extensive and accurate observations, as will be seen by the frequent recurrence of his name in the following pages.

Grateful acknowledgments are also due to Dr. JOSEPH BARRATT. He has kindly communicated his unequalled monograph of the willows, "Salices Americanæ;" also that of the "Eupatoria verticillata;" both of which, abridged, I have adopted in this work.

Among other friends and correspondents, from whom I have received specimens and highly valuable information, and whose kindness will be held by me in most grateful remembrance, are the following: viz. Dr. ALBERT G. SKINNER and Dr. JOHN PLUMMER, Indiana; Dr. SAMUEL B MEAD and Rev. E. JENNEY, Illinois; Mr. WM. S. SULLIVANT, Professor LOCKE, and Mr. JOSEPH CLARK, Ohio; Dr. I. A. LAPHAM, Wisconsin; Dr. WM. DARLINGTON, Pennsylvania; President HITCHCOCK, Professor ASA GRAY, and Dr. AUGUSTUS A. GOULD, Massachusetts; Mr. TRUMAN RICKARD, New Hampshire; Dr. SARTWELL and Mr. BROWNNE, New York; Mr. S. T. OLNEY, Rhode Island; &c. &c.

In the present edition, a chapter on the principles of Agricultural Chemistry has been added to the "Elements," and some important alterations introduced. The Flora has been to a great extent rewritten, and enlarged, not only by the addition of such plants as were necessary to adapt it to a wider section of country, as mentioned above, but by many plants recently discovered within our former limits. At the end is appended a synoptical view of the lower orders of the Cryptogamia, with their genera; and the whole is now illustrated with numerous engravings.

Finally, the present work is again submitted to the public with the assurance that, on the part of the author, it shall still be his constant care to detect and rectify its imperfections in future editions; and with the earnest hope that his labors may serve to awaken in the minds of others as deep an interest in this branch of the study of Nature as they have done in his own.

The author still solicits communications from his numerous readers concerning their difficulties or discoveries in this pursuit. and is always especially grateful for notices of new or rare species, or for *specimens* of such, wherever found. Please address him at *Ohio Female College, College Hill*, (near Cincinnati.) *O.*

THE GREEK ALPHABET, &c.

Large.	Small.	Rom. letters.	Names.		Numerals.
A	α	a	Alpha.		
B	β	b	Beta.		1. εἷς, μονας.
Γ	γ	g	Gamma.		2. δυο, δις.
Δ	δ	d	Delta.		3. τρεις.
E	ε	ĕ	Epsílon.		4. τεσσαρες, τετραϰ.
Z	ζ	z	Zeta.		5. πεντε.
H	η	ē	Eta.		6. ἑξ.
Θ	ϑ, θ	th	Theta.		7. ἑπτα.
I	ι	i	Iota.		8. οϰτω.
K	ϰ	c (k)	Kappa.		9. εννεα.
Λ	λ	l	Lambda.		10. δεϰα.
M	μ	m	Mu.		11. ἑνδεϰα.
N	ν	n	Nu.		12. δωδεϰα.
Ξ	ξ	x	Xi.		20. εικοσι.
O	ο	ŏ	Omíkron.		Many, πολις.
Π	π, ϖ	p	Pi.		
P	ρ	r	Rho.		
Σ	σ, ς	s	Sigma.		———
T	τ	t	Tau.		
Υ	υ	y	Üpsílon.		
Φ	φ	ph	Phi.		Upon, επι.
X	χ	ch	Chi.		Around, παρι.
Ψ	ψ	ps	Psi.		Under, ὑπο.
Ω	ω	ō	Ómega.		

PRONUNCIATION.

1. Every Latin word has as many syllables as it has separate vowels and diph thongs.

2. The penult (last syllable but one) is always accented in words of two sylla bles. In words of more than two syllables, the penult, if *long in quantity*, is ac cented; if short, the *antepenult* (last syllable but two) is accented. A word may have, also, a secondary accent, &c.

3. A vowel before another vowel, or the letter *h*, or marked with this (˘) char acter, is short in quantity. A diphthong, a vowel before two consonants, or a double consonant, or the letter *j*, or marked with this (¯) character, is *long in* quantity.

4. A vowel has its *short, English sound*, when followed by a consonant in the same syllable; otherwise its *long sound*, without regard to quantity: *a* at the end of an accented syllable, has an indistinct sound, as in Columbia.

5. A single consonant or a mute and liquid between the vowels of the penult and final syllables, is joined to the latter; in other cases, the vowel of the accented syllable takes the consonant before and after it, except *u*, and the vowels *a, e* and *o*, before two vowels, the first of which is *e* or *i*; when it takes the former only.

6. Pronounce *es* final with the *e* protracted; *ch* like *k*; *ci, ti, si*, before a vow el, like *sh*; *æ, œ*, like *e*; *qu* like *kw*; *gu, su*, before a vowel in the same syllable, *like gwe, sw*.

CONTENTS AND ANALYSIS.

CHAPTER I.—INTRODUCTION.—Page 13.

1, BOTANY defined.—2, Its departments. Organography. 3, Vegetable Physiology. 4, Glossology. 5, Systematic Botany. 6, Relation to man—ultimate aim. 7, Its merits and claims. 8, Natural world—its divisions,—a, mutual relations. 9, Mineral defined. 10, Plant defined. 11, Animal defined,—a, the three kingdoms blend in one. 12, Vegetation universal,—a, effects of light upon it—and heat,—b, elevation above the sea—Peak of Teneriffe,—c, soil,—d, moisture,—e, extremes of heat—illustrations,—f, extremes of cold—illustrations,—g, light—illustrations. 13, Variety of the vegetable kingdom. 14, Causes which affect it,—a, plants adapted to localities. 15, Cultivation,—16, Cabbage, &c., for illustration. 17, Species dependent on cultivation,—a, conclusion.

CHAPTER II.—PLAN OF VEGETATION. ELEMENTARY ORGANS.—18.

18, Embryo. 19, Axis,—ascending—descending. 20, Bud,—its development, &c. 21, Axillary buds,—universal. 22, Bud a distinct individual,—a, illustration. 23, Branches,—a, plant compound,—b, reproductive. 24, Flower, origin of,—25, its nature and end,—a, illustration. 26, Decay,—a, a leaf the elementary organ. 27, Leaf consists of,—a, elementary tissues. 28, Chemical basis of the tissues—organic bases,—a, illustration. 29, Cellular tissue—parenchyma,—a, pith of elder,—b, c, cellular tissue how colored,—d, size of cells,—e, they become solid,—f, Raphides. 30, Woody tissue—its design,—a, illustration. 31, Glandular fibre—fossil coal. 32, Vasiform tissue,—a, articulated—continuous,—b, illustration. 33, Vascular tissue,—a, spiral vessels,—b, spiral thread,—c, its size,—d, situation of spiral vessels,—e, what they contain,—f, ducts,—g, closed—annular—reticulated—the office of these ducts. 34, Laticiferous tissue,—a, size, &c. 35, Epidermis—where it is not found. 36, Structure,—a, illustration. 37, Stomata. 38, Form. 39, Position,—a, size. 40, Surface. 41, Hairs—simple—branched,—a, position—downy—pubescent—hirsute—rough—tomentose—arachnoid—sericeous—velvety—ciliate. 42, Stings. 43, Prickles. 44, Glands—sessile—imbedded,—a, glandular hairs. 45, Receptacles of secretion.

CHAPTER III.—PRIMARY DIVISIONS OF THE VEGETABLE KINGDOM.—26.

46, Phænogamia—Cryptogamia,—47, their distinctions of tissue,—48, of cotyledons. 49, Further distinctions. 50, A species,—a, illustration—number of species known. 51, Varieties,—a, where they occur. 52, A genus,—a, illustration,—b, summary.

CHAPTER IV.—OF THE FLOWER.

§1. OF ITS PARTS AND THEIR ARRANGEMENT.—28.

53, Parts of the flower enumerated,—a, essential organs—perfect flower,—b, imperfect flower—sterile—fertile—neutral. 54, Perianth consists of—calyx—corolla—achlamydeous flowers. 55, Calyx defined—sepals. 56, Corolla defined—petals. 57, Stamens—definition of—office—andrœcium. 58, Pistils—office of—gynœcium. 59, Receptacle—order of the organs upon it. 60, Specimens. 61, A complete and regular flower,—a, theoretical number of the parts,—b, their theoretical position, — c, summary — correspondence between a flower and a leafy branch. 62, Apparent exceptions — examples of symmetrical flowers. 63, a, First cause of deviation, the development of one or more additional whorls — examples. — b, Second cause, the suppression of entire whorls — examples. — c, Third cause, the suppression of parts of whorls — examples. — d, Fourth cause, the union of parts of the same whorl — examples. — e, Fifth cause, the union of organs of different whorls — examples. — f, Sixth cause, unequal development of similar organs — examples. — g, Seventh cause, reconversion of organs — examples. — h, Eighth cause, development of axis — examples.

CHAPTER V. — OF THE FLOWER.

§ 2. STAMENS AND THE ARTIFICIAL CLASSES. — 34.

64, The stamens, basis of a classification — why. — 65, Definition — 66, parts — which essential. 67, Filament. 68, Anther — when sessile — connectile — analogy, — a, cells — dehiscence, — b, connectile, — c, stamen abortive. 69, Modes of attachment of anther to filament, 1st, innate; 2d, adnate; 3d, versatile; 4th, introrse — extrorse. 70, Pollen — forms, — a, structure — molecules. 71, Physiological structure — of the filament — connectile — anther — pollen. 72, Theoretical structure — proof from the transitions of stamens into pistils — examples. 73, Circumstances in which stamens vary — twenty-four Linnean Classes. 74, Number, — Etymology of the names of the Classes. Class 1st, 2nd, 3d, 4th, 5th, 6th, 7th, 8th, 9th, 10th, 11th. 2nd, Position, — 12th, 13th. 3d, Relative length, — 14th, 15th, 4th, Connection, — 16th, 17th, 18th, 19th, 20th. 5th, Absence, — 21st, 22d, 23d, 24th.

CHAPTER VI. — THE FLOWER.

§ 3. OF THE PISTIL AND THE ARTIFICIAL ORDERS. — 38.

75, The pistil, its position — structure. 76, Ovary — ovules. 77, Ovary simple — compound — carpels. 78, The style — number — connection. 79, Stigma — simple — compound. 80, Number of styles, orders founded upon. Order 1st, 2nd, 3rd, 4th, 5th, 6th, 7th, 8th, 9th, 10th, 11th, 12th. Note, — Orders of the class Didynamia — Tetradynamia — Orders of the 16th, 17th, 18th, 20th, 21st, and 22d classes. Orders of Syngenesia, Equalis — Superflua — Frustranea — Necessaria — Segregata. Orders of Polygamia, Monœcia — Diœcia. Orders of the 24th, class. 81, Ovules. 82, Placenta — structure — direction. 83, Physiological structure — of the ovary — style — stigma — without epidermis. 84, Theoretical structure explained, — sutures — ventral — dorsal, — a, illustration, — b, c, illustrations continued. 85, Propositions, — first — second — third — fourth. 86, These propositions when true. 87, Central placenta, — parietal placenta. 88, Free central placenta, — explanation. — a, Ovules proved to be analogous to buds. 89, Ovules enclosed — naked, — a, erect — ascending — pendulous — suspended. 90, Foramen — primine — secundine — nucleus — a, illustration. 91, Funiculus — chalaza.

§ 4. OF THE MUTUAL ACTION OF THE STAMENS AND PISTILS. — 44.

92, Their specific office — how accomplished, — a, illustrations, — tulip — Kalmia, &c. 93, Action of pollen upon the stigma — tubes. 94, Molecules — their destination.

CHAPTER VII. — THE FLOWER.

§ 5. OF THE CALYX. — 45.

95, Calyx — etymology — color. — 96, Sepals — monosepalous — polysepalous. — 97, Calyx — inferior — superior, — 98, caducous — deciduous — persistent, — 99, reduced — wanting — a, Pappus — pilose — plumose — setose — paleaceous.

§ 6. OF THE COROLLA. — 46.

100, The corolla — etymology. 101, Petals — polypetalous — monopetalous. 102, Claw — lamina, — 103, tube — limb, — 104, regular — irregular. 1, Corolla campanulate. 2, Infundibuliform. 3, Hypocratiriform. 4, Rotate. 5, Labiate — ringent — personate — galea. 105, Forms of polypetalous corollas, — 1, Cruciform. 2, Rosaceous. 3, Liliaceous. 4, Caryophyllaceous. 5, Papilionaceous — vexil'um — alæ — carinæ. 106, Physiological structure — colors.

§ 7. OF THE NECTARY AND DISK.—49.

107, Definition,—a, Nectary—labellum—spur,—b, Disk—hypogynous—epi gynous,—c, true character.

§ 8. OF ÆSTIVATION.—50.

108, Definition—vernation,—a, illustration. 1, Æstivation valvate,—2, Convolute,—3, Quincuncial,—4, Contorted,—5, Alternate,—6, Vexillary,—7, Induplicate,—8, Supervolute.

CHAPTER VIII—THE FRUIT.—51.

109, Its importance—design,—a, utility,—b, in respect to time—defined. 110 Analogues to ovary—changes,—a, examples.

§ 1. OF THE PERICARP.—52.

111, Fruit consists of,—a, naked seeds. 112, Pericarp defined—epicarp—sarcocarp,—a, illustration. 113, Physiology of growth,—114, of ripening,—a, acidification,—b, maturity. 115, Dehiscence—indehiscent—valves,—1, Dehiscence loculicidal,—2, Septicidal,—3, Septifragal,—4, Sutural,—5, Circumscissile. 116, Forms of pericarp.—1, Capsule,—a, cells—columella.—2, Silique.—3, Silicle.—4, Legume.—5, Follicle.—6, Drupe.—7, Nut.—8, Caryopsis.—9, Achenium.—10, Samara.—11, Pyxis.—12, Pome.—13, Pepo.—14, Berry—strawberry—black berry.—15, Strobile.

CHAPTER IX.—THE FRUIT.—57.

§ 2. OF THE SEED.

117, The seed defined,—a, its parts, 118, Integuments—testa—mesosperm—endopleura,—a, Testa, its substance—surface—form—appendages,—b, Coma distinguished. 119, Aril. 120, Hilum. 121, Seed orthotropous—anatropous. 122, Albumen—where most abundant—where wanting. 123, Embryo—124, its parts,—a, radicle,—b, plumule—its direction. 125, Cotyledon,—a, the number, &c. 126, Monocotyledons—endogens. 127, Dicotyledons—exogens distinguished. 128, Cotyledons many—none. 129, Embryos, number of.—Spores

§ 3. OF GERMINATION.—60.

130, The embryo, its importance,—a, germination defined,—131, the process explained. 132, The cotyledons. 133, Conditions of germination,—a, heat—b, water—c, oxygen—d, darkness. 134, Duration of the vitality of seeds. Note, Two examples of.

§ 4. OF DISSEMINATION.—61.

135, Remark—examples, Erigeron—a, Wings, hooks—b, Impatiens—Streams and oceans—Squirrels, birds.

CHAPTER X.—THE ROOT.—62.

136, Its definition—origin,—a, divisions. 137, Prone direction,—a, horizontal direction. 138, Number and extent. 139, How distinguished from stems. 140, Exceptions—adventitious buds—subterranean stems. 141, Collum—a, stationary. 142, Parts of the root,—a, caudex—b, fibres—c, spongioles. 143, Forms of roots. 144, Ramose—a, analogous to branches—illustration,—b, further illustration,—c, extent of roots. 145, Fusiform root—forked—tap root—premorse—napiform. 146, Fibrous,—a, fasciculated. 147, Tuberous,—a, palmate—b, granulated. 148, Use of fleshy roots. 149, Floating root. 150, Epiphytes—parasites.

PHYSIOLOGICAL STRUCTURE AND FUNCTIONS OF THE ROOT.—65

151, Internal structure. 152, Fibrils, structure—function. 153, Growth of root, 154, Its most obvious function—most important one—a, illustration. 155, Activity of absorption dependent on. 156, Part which absorbs, &c.—a, illustration,—b transplanting. 157, Force of absorption,—a, illustration. 158, Cause of absorption—not capillary attraction,—a, experiment in philosophy—exosmose—endosmose. 159, Requisite conditions,—a, how they exist in the root,—application 160, Use of absorption,—a, power of choice—illustration.

CHAPTER XI.—THE STEM, OR ASCENDING AXIS.—69.

161, Definition, — a, cause of its ascent. 162, Direction horizontal — erect — procumbent — ascending — subterranean. 163, Annual — perennial. 164, Distinction in regard to size, &c. — a, tree — b, shrub — c, herb. 165, Most distinctive property of stem. 166, Buds. 167, Leaf-bud. 168, The scaly envelopes. a, Scales not formed in hot climates, &c. — their design. 169, Bud terminal, developes a simple axis, — a, axillary, developes branches, — b, adventitious. 170, Branch. 171, Thorn, — a, its nature, &c. — b, distinguished from prickles. 172, Node — internode, — a, how formed — why the axis diminishes upwards. 173, Arrangement of branches, — 174, spiral, — a, modifications — circular, how caused. 175, Alternate — opposite — whorled. 176, Same of the branches. — a, Two classes of stems. 177, Subterranean — stemless plants — varieties. 178, Bulb, — a, tunicated — scaly, — b, how renewed, — c, bulblets. 179, Corm. 180, Tuber. 181, Rhizoma. 182, Creeper, — a, repent stems, their use. 183, Varieties of aerial stems. 184, Caulis. 185, Runner. 186, Scape, — a, culm. 187, Vine — a, tendril. 188, Twining stems — their direction. 189, Trunk — a, its dimensions. Note, Illustrations, b, its duration. Note, Illustrations. 190, Sucker. 191, Offset. 192, Stolon. 193, Plurality of trunks — a, Banyan — Mangrove.

§ 1. PHYSIOLOGICAL STRUCTURE OF THE EXOGENOUS STEM.—77.

194, Structure of herbaceous stems — 195, of the first year's growth of perennial stems, — a, basis of the distinction of Exogens and Endogens. 196, Exogens. 197, Endogens. 198, Parts of the exogenous structure. 199, Pith. 200, Medullary sheath. 201, Wood — number of layers. 202, Layer consists of, — a, arrangement of its parts. 203, Alburnum — duramen. 204, Medullary rays. 205, Bark, — its parts. — 206, its structure — cork — liber. 207, New layer, how formed, — a, outer layers, why shaggy — horizontal fibres, — b, qualities resident in, — c, its peculiar vessels.

§ 2. FUNCTIONS OF THE STEM.—80.

208, Conveyance of sap, — 209, through what portion — its course — elaboration — descent.

§ 3. OF THE ENDOGENOUS STRUCTURE.—81.

210, Its peculiarity. 211, Composition. 212, Each bundle consists of, — a, mode of increase.

CHAPTER XII.—THE LEAF.—82.

213, Its importance — character. 214, How distinguished. 215, Color — autumnal hues, — a, due to what — chlorophyll — changes, — b, color of flowers.

§ 1. VERNATION.—82.

216, Meaning of the term — leaf-bud how compacted, — a, illustration. 217, Forms of vernation, — 1, Equitant — 2, Obvolute — 3, Involute — 4, Revolute — 5, Convolute — 6, Plaited — 7, Circinate.

§ 2. ARRANGEMENT.—83.

218, In the bud — after the axis is developed, — 1, Scattered — 2, Alternate — 3, Opposite—4, Verticillate — 5, Fasciculate. 219, How these forms may be reconciled with the spiral, — a, illustration — b, Alternate explained, — c, Opposite or whorled explained. 220, Leaves radical — cauline — ramial.

§ 3. ORGANOGRAPHY.—84.

221, Nature of the leaf — lamina — sessile — petiolate. 222, Petiole — its form — 1, Compressed — 2 Winged — 3, Amplexicaul. 223, General form of the lamina — base — apex. 224, Leaf simple — compound. 225, Physiology. 226, Venation — its organs. 227, Midrib — nerves. 228, Veins — veinlets. 229, Modes of Venation. 1, Reticulate—2, Parallel — 3, Forked. 230, Varieties of reticulate venation, — 1, Feather-veined — 2, Radiate-veined — 3, Varieties of parallel venation.

§ 4. FORM OF FIGURE.—86.

231. Theory of — form dependent on venation. 232, Forms resulting from the

feather venation — 1, Orbicular — 2, Elliptic — 3, Oblong — 4, Ovate — 5, Lanceolate — 6, Obovate — 7, Spathulate — 8, Cordate — 9, Auriculate — 10, Hastate — 11, Sagittate — 12, Reniform — a, Forms dependent on the development of the tissue — 13, Runcinate — 14, Lyrate — 15, Pinnatifid — 16, Sinuate. 232, Forms resulting from radiate venation — 17, Palmate — 18, Digitate — 19, Pedate — 20, Lacinate — 21, Peltate — 22, Reniform, &c. 234, Forms of parallel-veined leaves — 23, Linear — 24, Oval, &c. — 25, Cordate — 26, Acerose.

§ 5. MARGIN. — 89.

235, How modified — 1, Entire — 2, Dentate — 3, Serrate — 4, Crenate — 5, Erose — 6, Undulate — 7, Spinous — 8, Incised — 9, Laciniate — 10, Crisped — 11, Repand.

§ 6. APEX. — 90.

236, Termination of leaf — 1, Acute — 2, Obtuse — 3, Acuminate — 4, Emarginate — 5, Retuse — 6, Mucronate.

§ 7. SURFACE — 90.

237, Terms descriptive of the epidermis on the leaf or elsewhere — 1, Glabrous — 2, Pubescent — 3, Rough — 4, Pilose — 5, Hoary — 6, Villose — 7, Woolly — 8, Tomentose — 9, Rugose — 10, Punctate.

§ 8. COMPOUND LEAVES. — 91.

238, Leaf becomes compound on what principle. 239, Leaflets — articulated. 240, Forms resulting from the feather-veined arrangement — 1, Pinnate — 2, Equally pinnate — unequally — cirrhose — 3, interruptedly — 4, Number of leaflets — trifoliate — single — 5, biplnnate — 6, tripinnate — 7, biternate — 8, triternate. 241, Forms resulting from radiate venation — 9, Quinate — 10, Septinate. 242, Leaf with regard to insertion — 1, Amplexicaul — 2, Perfoliate — 3, Decurrent — 4, Connate. 243, Combined terms, — a, the preposition sub.

§ 9. ASCIDIA, STIPULES, AND BRACTS. — 93.

244, Leaves of Teazel — Tillandsia — Arum. 245, Ascidia, — a, of the Sarracenia, how formed, &c. 246, Nepenthes. 247, Dischidia. 248, Dionæa. 249, Stipules, — varieties — positions. 250, Their nature. 251, Leaves stipulate — exstipulate — stipels. 252, Bracts, — 253, their nature. 254, Involucre — involucel, — 255, of the Compositæ. 256, Glume — awn — palæ — valves.

§ 10. DURATION. — 95.

257, Leaves temporary — 1, Fugacious — 2, Deciduous — 3, Persistent. 258, Fall of the leaf — previous changes. 259, Cause of defoliation.

§ 11. PHYSIOLOGICAL STRUCTURE. — 96.

260, Of the frame-work, — a, of the lamina. 261, Parenchyma disposed in two layers, — a, how covered. 262, Internal structure of the parenchyma. 263, Arrangement of the cells, — a, chlorophyll. 264, Stomata communicate with what — found on which surface. 265, Vessels of the latex — their course. 266, Leaf of Oleander — air cells.

§ 12. OF THE FUNCTIONS OF LEAVES. — 98.

267, Enumerated — result — latex. 268, Crude sap consists of. 269, Exhalation — 270, distinguished from evaporation — 271, takes place through the stomata — occurs only in the light — why — a, illustration. 272, Exhalation dependent on absorption — quantity — illustration. 273, Absorption, — a, illustration. — 274, by their lower surface — illustration. 275, Respiration — 276, consists in — 277, constant — the result — 278, illustration — 279, two periods of its greatest activity — a, in germination, — b, flowering — proportion of oxygen evolved. — Note, illustration. — 280, Life of the plant dependent on. 281, Digestion — the process. 282, Carbon — its sources, — a, illustration. — Plants blanched in the dark. 283, Fixation of carbon — relative amount absorbed and evolved. — Experiments of Dr. Daubeny. 284, Relation of animal to vegetable kingdom in regard to carbon — Reflections

CHAPTER XIII. — INFLORESCENCE. — 102.

255, Definition. 256, Position, — a, exceptions. 257, Peduncle — flower sessile 258, Peduncle simple — branched — Pedicel. 259, Scape. 290, Rachis. 291, Inflorescence solitary, — 292, centrifugal — centripetal, resulting from what. 293, Centripetal, — 294, centrifugal, — a, how indicated — all the flowers terminal, why. 295, Both modes combined — examples. 296, Varieties of centripetal inflorescence. 297, Spike — 298, Raceme — 299, Ament — 300, Spadix — 301, Corymb — 302, Umbel — 303, Head — a, of the Compositæ — compound flowers — 304, Panicle — 305, Thyrse — a, Compound umbel — Umbellet — Compound raceme, &c. 306. Varieties of centrifugal inflorescence, — 307, Cyme — a, its normal structure and development — b, inference, — 308, Fascicle — 309, Verticillaster. — a, Peduncle converted into a tendril.

CHAPTER XIV. — PRINCIPLES OF NUTRITION. — 106.

310, Four simple organic elements — their proportion. 311, Carbon. 312, Mineral ingredients — Agricultural Chemistry. 313, Sources of the simple elements. 314, Air. 315, Soil. 316, Water. 317, Ammonia. 318, Air plants, — three conditions requisite. 319, Irrigation — Draining. 320, Tillage — Amendments. 321, Fallow ground — Rotation of crops. 322, Light and Heat. 323, Digestion, &c. 324, Proper juice. 325, Products first developed. 326, Three general nutritive products — composition. 327, Sugar — Diastase. 328, Mutual transformations. 329, Secretions.

CHAPTER XV. — SYSTEMATIC BOTANY. — 111.

§ OF THE CLASSIFICATION OF PLANTS.

330, Systematic Botany defined. 331, Remarks on the extent of the field of botanic research. 332, Folly of studying individuals only. 333, Individuals grouped into species, — a, illustrations — clover — pine. 334, Species grouped into genera, — illustration. 335, Genera resolved into orders and classes. 336, Two methods of classifying the genera, — artificial — its basis, — natural — its basis. — 337, Comparative merits of the two — use of the artificial. 338, Value of the natural — obscurities now removed. 339, Remaining difficulties — artificial method how and why retained in this work. 340, Artificial arrangement consists of.

CHAPTER XVI. — § OF THE NATURAL SYSTEM. — 113.

341, Its aim, — 342, distinguished from the Artificial — what characters employed, — 343, advantages, — 344, yet to be fully consummated — some artificial characters yet necessary. 345, The first two grand divisions — Phænogamia — its characters, — Cryptogamia — characters, — 346, uncertainty of these characters, — approximation of groups. 347, Sporogens. 348, Subdivision of Phænogamia — Exogens — characters — Endogens — characters. 349, Classes, six — Exogens divided into two — Angiosperms — characters — Gymosperms — characters. 350, Endogens divided into two — Aglumaceous, characters — Glumaceous, characters. 351, Cryptogamia divided into two — Acrogens, characters — Thallogens, characters. 352, Affinities of the classes. 354, Sub-classes — Polypetalæ, characters — Monopetalæ, characters — Apetalæ, characters. 355, Orders — 356, how formed. 357, Alliances, groups, &c. 358. Extent of the orders. 359, Summary.

CHAPTER XVII. — § 1. NOMENCLATURE, &c. — 118.

360, Names of the orders Latin adjectives — derivation, — a, exceptions. 361, Etymology of generic names. 362, Of specific names, — 363, derivation — rules. § 2. BOTANIC ANALYSIS. 364, defined, — 365, proper state of plants for, — 366, importance of. 367, Process, — 368, with the learner. — Analytical tables. § 3. COLLECTING AND PRESERVING PLANTS. 369, Importance of, — a, hortus siccus. 370, Apparatus. 371, Directions for gathering, — 372, pressing, — 373, changing. 374, Arrangement of the specimens. 375, Genera arranged — how preserved. 376, Fruit, seed, and wood, how preserved.

BOTANY.

CHAPTER I.

INTRODUCTION.

1. BOTANY is the science which treats of the Vegetable Kingdom. It includes the knowledge of the habits, structure, and uses of plants, together with their nomenclature and classification.

2. Like its kindred sciences, it is resolved into distinct departments, according to the nature of the subjects to which it relates. That part which investigates the organic structure of vegetables, is called ORGANOGRAPHY, corresponding to Anatomy, in the science of Zoology.

3. That part of botany which relates to the phenomena of the vital functions of plants, is called VEGETABLE PHYSIOLOGY; including the consideration of their germination, growth, and reproduction. It has, therefore, a direct and practical bearing upon the labors of husbandry, in the propagation and culture of plants, both in the garden and in the field.

4. Another department, of essential importance, is GLOSS OLOGY, which relates to the explanation and application of botanical terms, whether nouns or adjectives, by which the organs of plants, with their numerous modifications, are designated.

5. A fourth department, called SYSTEMATIC BOTANY, arises from the consideration of plants, in relation to each other, their mutual affinities, and their endless diversities, whereby the 100,000 species, supposed to exist, may be arranged, classified, and designated, by distinctive characters and names.

6. Finally, in its extended sense, Botany comprehends, also, the knowledge of the relations of plants to the other departments of nature, particularly to mankind. The ultimate aim of its researches is, the development of the boundless resources of the vegetable kingdom for our sustenance, protection, and enjoyment; for the healing of our diseases, and the alleviation of our wants and woes.

7. This extensive department of Natural History, therefore, justly claims a large share of the attention of every individual, not only on account of the aid it affords to horticulture, to the employments of rural life, and to the healing art, but also for the intellectual and moral culture, which, among other kindred sciences, it is capable of imparting in an eminent degree.

a. No science more effectually combines pleasure with improvement, than Botany. It conducts the student into the fields and forests, amidst the verdure of spring, and the bloom of summer; — to the charming retreats of Nature, in her wild luxuriance, or where she patiently smiles under the improving hand of cultivation. It furnishes him with vigorous exercise, both of body and mind, which is no less salutary than agreeable, and its subjects of investigation are all such as are adapted to please the eye, refine the taste, and improve the heart.

8. The natural world, by distinctions sufficiently obvious, is divided into three great departments, commonly called the MINERAL, VEGETABLE, and ANIMAL KINGDOMS.

a. Vegetables, or plants, hold an intermediate position between animals and minerals: while they are wanting in both the intelligence and instinct of the former, they are endowed with a physical organization, and a living principle, whereby they are remarkably distinguished above the latter; they constitute the ultimate nourishment and support of the one, the vesture and ornament of the other.

9. A *mineral* is an inorganic mass of matter, that is, without distinction of parts or organs. A stone, for example, may be broken into any number of fragments, each of which will retain all the essential characters of the original body, so that each fragment will still be a stone.

10. A *plant* is an organized body, endowed with vitality but not with sensation, composed of distinct parts, each of which is essential to the completeness of its being. A *geranium* is com posed of organs, which may be separated or subdivided indefi-

nitely, but no one of the fragments, alone, will still be a complete plant.

11. Animals, like plants, are organized bodies, endowed with vitality, and composed of distinct parts, no one of which is complete in itself; but they are raised above either plants or minerals, by the power of perception.

a. These distinctions, long since suggested by the immortal Linnæus, are perfectly obvious and definite, in the higher grades of the animal and vegetable kingdoms; but, in descending the scale, we recognise a gradual and constant approach, in both, to inorganic matter, and consequently to each other; so that, in the lowest forms of life, all traces of organization disappear, and the three great kingdoms of nature, like three converging radii, meet, and blend in a common centre.

12. Vegetation, in some of its forms, appears to be *coextensive with the surface of the earth.* It springs up, not only from the sunny soil, moistened with rain and dew, but even from the naked rock, amidst the arid sands of the desert, in thermal and sulphurous springs, in arctic and alpine snows, and from the beds of seas and oceans.

a. Among the multitude of natural causes which affect the growth of vegetation, the action of the sun, through the light and heat which it imparts, is the most efficient. This is most powerful at the equator, and gradually diminishes in intensity, as we proceed from thence towards either pole. Vegetation, therefore, arrives at its highest degree of luxuriance at the equator, and within the tropics. In the temperate zones it is less remarkable for the beauty and variety of its flowers, and the deliciousness of its fruits, than in the torrid; yet it is believed to be no less adapted to promote the arts of civilised life, and the well-being of man in general. In still higher latitudes, plants become few, and of stinted growth, until finally, within the arctic circles, they apparently, but not absolutely, cease to vegetate.

b. Since climate is affected by elevation above the level of the sea, in the same manner as by increase of latitude, we find a similar diminution of vegetable activity, in ascending high mountains. Thus, the peak of Teneriffe, situated on a fertile island, within the tropics, is clothed, at different elevations, with plants peculiar to every latitude, in succession, from the torrid to the frigid zones, while the summit, being always covered with snow, is as barren as the region of the poles. So also the White Mountains, in New Hampshire, exhibit upon their summits a vegetation similar to that of Labrador, or even Greenland.

c. One of the first requisites for the growth of plants, is a soil, from which, by means of roots, they may derive their proper nutriment and support. But numerous species of lichens and mosses find their most congenial habitations upon the bare rock. *The coral island no sooner arises* to the surface, than it arrests the

2*

floating germs of vegetation, which soon clothe the rough rock with verdure of a humbler kind, and ultimately, by the growth and decay of successive generations, form a soil for the sustenance of the higher forms of vegetable life.

d. Another important requisite is moisture. But the arid sands of the great African desert are not absolutely destitute of vegetable life. Even there, certain species of Stapelia are said to flourish, and those dreary regions, where neither rain nor dew ever falls, are occasionally enlivened by spots of verdure, like islands in the ocean, composed of these and kindred plants.

e. Extremes of heat are not always fatal to vegetation. In one of the Geysers of Iceland, which was hot enough to boil an egg in four minutes, a species of Chara has been found, in a growing and fruitful state. A hot spring at the Island of Luzon, which raises the thermometer to 187°, has plants growing in it and on its borders. But the most extraordinary case of all, is one recorded by Sir J. Staunton. 'At the Island of Amsterdam a spring was found, the mud of which, far hotter than boiling water, gave birth to a species of liverwort.' Other similar instances are on record.

f. Nor are the extremes of cold fatal to every form of vegetation. The reindeer lichen, of Lapland, grows in vast quantities among almost perpetual snows And far in the arctic regions, the eternal snows are often reddened, for miles in extent, by a minute vegetable of the Algæ tribe, called red snow, of a structure the simplest that has yet been observed, consisting of a single round cell containing a fluid.

g. Light is also a highly important agent in vegetation; yet there are plants capable of flourishing in situations where it would seem that no ray of it ever entered. Mushrooms, and even plants of higher orders, have been found growing amidst the perpetual midnight of deep caverns and mines. Sea weeds of a bright green color have been drawn up from the bed of the ocean, from depths of more than 100 fathoms.

13. The vegetable kingdom is no less remarkable for its *rich and boundless variety*, than for its wide diffusion. Plants differ from each other in respect to form, size, color, habits, structure, and properties, to an unlimited degree, so that it would be difficult, indeed, to find two individuals, even of the same species, which should perfectly coincide in all these points.

a. Yet this variety is never abrupt, never capricious; but here, as in other departments of nature, uniform resemblances are so blended with it, as to lay an adequate foundation for Systematic Botany.

14. The same causes which affect the general increase of plants, exercise, also, an important influence in determining their *character*. Hence, every climate has not only its own *peculiar degree* of vegetable activity, but also its *peculiar species*.

a. Other causes, besides temperature, are efficient in determining the species of

any given locality, such as the qualities of the soil, the degree of moisture both of the earth and skies, the inclination of surface, rocks, shades, and winds, the combined action of which often becomes an exceedingly complicated matter. Now to each of these innumerable combinations of circumstances, the Creator has adapted the constitution of certain species of plants, so that each given locality may be expected to produce its own appropriate kinds. But since some species are also endowed with the power of accommodating themselves to a wide range of circumstances, these are found more extensively diffused, while others, without this power, are comparatively rare.

15. Vegetation is susceptible of important changes by cultivation. Many plants are improved, in every desirable quality, by accommodating themselves to the conditions of soils enriched and enlivened by art. Examples are seen in almost every cultivated species.

16. The cabbage, in its wild state, is a slender, branching herb, with no appearance of a head. The potatoe, in its native wilds of tropical America, is a rank, running vine, with scarcely a tuber upon its roots. All the rich and delicate varieties of the apple have sprung, by artificial means, from an austere forest-fruit. The numerous and splendid varieties of the Dahlia are the descendants of a coarse Mexican plant, with an ordinary yellow flower, of a single circle of colored leaves. The tulip and the geranium afford similar examples.

17. Changes, not only in the *qualities* of vegetation, are effected by culture, but also in the *species* themselves, through the substitution of the useful or the ornamental for the native products of the soil. Thus, in agricultural districts, almost the whole face of nature is transformed, by human skill and industry, from the wilderness to the fruitful field.

a. Hence it appears that there is scarcely a spot on earth which is not caused, by the quickening energy of the Creator, to teem with vegetable existence, in some of its numberless forms, while his goodness is conspicuous in rendering those tribes which are most subservient to the wants of man capable of the widest diffusion.

CHAPTER II.

PLAN OF VEGETATION.—ELEMENTARY ORGANS.

18. THE earliest and simplest state of the plant is an *embryo* contained in a seed. This consists essentially of two parts, the *radicle* and *plumule;* the former about to be developed into the root, the latter into the ascending plant with its appendages.

19. As soon as the process of germination commences, the radicle begins to extend itself downwards in the direction of the earth's centre, constantly avoiding the air and the light, forming the *descending axis,* or root. The plumule, taking the opposite direction, extends itself upwards, always seeking the light, and expanding itself, to the utmost extent of its power, to the influence of the atmosphere. This constitutes the *ascending axis,* or trunk, around which the leaves and their modifications are arranged.

20. At the commencement of its growth, the ascending axis is merely a *bud,* that is, a *growing point,* enveloped in rudimentary leaves, or scales, for its protection. As this growing point advances, the enveloping scales expand into leaves below, while new ones are constantly appearing, in succession, above. Thus the axis is always terminated by a bud.

21. By this process the axis is elongated, simply in one direction. But, besides this, there is also a bud (or buds), either visible, or in a rudimentary state, formed in the axil of each leaf.

a. These axillary buds are generally visible, either before or after the leaf has fallen. In some plants, however, they seldom appear; but their existence is inferred from the fact, that even in such cases, they are occasionally developed in extraordinary circumstances.

22. Each bud is a *distinct individual,* capable of an independent existence, in favorable circumstances, although severed from the parent stock.

a. The common practice of propagation by layers, offsets, engrafting, and *budding, is both a result and a proof of* this principle. A plant may be, and often is, in this manner, multiplied indefinitely, by the dissevered parts of itself, as well as by the seed.

23. But, remaining connected with the parent stock, axillary buds, a part or all of them, according to circumstances, are developed into *branches*, each of which may again generate buds and *branchlets* in the axils of its own leaves, in the same manner.

a. Thus, by the repetition of this simple process, the vegetable fabric is reared from the earth, a compound *being*, formed of as many united individuals as there are buds, and as many buds as there are branches and leaves, ever advancing in the direction of the growing points, by the deposition of matter derived from the cellular tissue, clothing itself with leaves as it advances, and enlarging the diameter of its axis by the deposition of matter elaborated by, and descending from, the leaves already developed, until it reaches the limits of the existence assigned it by its Creator.

b. But the plant, reared by this process alone, would consist only of those parts requisite to its own individual existence, without reference to the continuance of its species beyond its own dissolution. It would be simply an axis, expanded into branches and leaves. But the Divine command, which first caused the tribes of vegetation, in their diversified beauty, to spring from the earth, required that each plant should have its 'seed within itself,' for the perpetuation of its kind.

24. At certain periods of its vegetation, therefore, a change is observed to occur in the plant, in regard to the development of some of its buds. From the diminished or altered supply of sap, received from the vessels below, the growing point ceases to lengthen in the direction of the axis, but expands its leaves in crowded and concentric whorls; each successive whorl, proceeding from the outer to the inner, undergoing a gradual transformation from the original type (a leaf), according to the purpose it is destined to fulfil in the production of the seed. Thus, instead of a *leafy branch*, the ordinary progeny of a bud, a *flower* is the result.

25. A flower may, therefore, be considered as a transformed branch, having the leaves crowded together by the non-development of the axis, and moulded into more delicate structures, and tinged with more brilliant hues, not only to adorn and beautify the face of nature, but to fulfil the important office of reproduction.

a. In the common peony, for example, as the leaves approach the summit of the stem, they gradually lose their characteristic divisions, and, at length, jus-

below the flower, become simple *bracts*, still retaining every essential mark of a leaf. Next, by an easy gradation, they appear in the *sepals* of the calyx, the outer envelope of the flower, still essentially the same. Then, by a transition rather more abrupt, they pass into the delicate and highly colored *petals* of the corolla, retaining still the form and organisation of the *leaf*. To the petals next succeed those slender organs called *stamens*, known to be undeveloped leaves from the fact of their being often converted into petals. Lastly, those two central organs termed *pistils*, are each the result of the infolding of a leaf, the midrib and the united edges being yet discernible.

26. When the flower has accomplished its brief but important office in reproduction, its deciduous parts fall away, and the remaining energies of the plant are directed to the development of the germ into the perfect fruit. This being accomplished, the whole plant speedily perishes, if it be an annual, or, if not, it continues to put forth new branches, from other growing points, which, in their turn, are to be terminated by flowers and fruit the following year.

a. Such is a very brief outline of the plan of vegetation, or the process of nature in the germination, growth, fructification, and decay of plants. And it is impossible to contemplate it, without admiring that simplicity of design in the midst of the most diversified results which every where characterises the works of God. Every part of the vegetable fabric may be ultimately traced to one elementary organic form, of which the leaf is the type. The lamina, or blade, in various stages of transition, constitutes the several organs of fructification, while the united bases of all the leaves constitute the axis itself.

27. When we more minutely examine the internal organization of plants, we find their different parts, however various in appearance, all constructed of the same materials. The leaf, for example, consists of a *foot-stalk* prolonged into a *framework of veins*, a *fleshy substance* filling up the interstices, and a *cuticle*, or skin, covering the whole. Now this framework is composed of *woody fibre*, *aqueducts*, and *air-vessels*, all of which may be traced through the foot-stalk into the stem, where they equally exist, — this part of the leaf being only a prolongation of the stem. The fleshy substance is of the same nature with the pith of the stem, or the pulp of the fruit; and, finally, the cuticle corresponds exactly to the thin covering of the newly formed branches, of the various parts of the flower, and even of the roots.

a. These several kinds of structure, of which the various organs are composed, are called the *elementary tissues*. They are five in number; — *cellular tissue, woody tissue, vasiform tissue, vascular tissue,* and *laticiferous tissue.*

28. The *chemical basis* of the vegetable tissues is proved by

analysis to be oxygen, hydrogen, and carbon, with an occasional addition of nitrogen, the same simple elements as, by their varied combinations, constitute the air, water, and most animal substances. ·The *organic basis* is simple *membrane* and *fibre* Of one, or both, of these two forms, all the tissues are con structed.

a. If the fleshy portion of the leaf above mentioned, or the pulp of the fruit be closely examined, they will be found composed of numerous vesicles of extreme minuteness, adhering together. These vesicles, or bladders, consist of a delicate *membrane* enclosing a fluid, such as is seen on a large scale in the pulp of an orange. Now this membrane, composing the walls of the cells or vesicles, is one of the *elementary forms* of vegetable tissue. Again, if the stalk of a strawberry or geranium leaf be cut *around* but *not through*, and the two parts be thus pulled asunder for a short space, a number of glistening *fibres* will be seen running from one portion to the other. Under a microscope these appear to be spiral coils, partially straitened by being thus drawn out from the membranous tubes in which they were lying coiled up. Thus are we able to distinguish the elementary *membrane* and *fibre*, of which the various forms of vegetable tissue are composed.

29. CELLULAR TISSUE is so called, from its being composed of separate cells, or vesicles, adhering together. This kind of tissue is the most common, no plant being without it, and many being entirely composed of it. The form of the little cells which compose it, appears to be, at first globular or egg-shaped, but afterwards, being flattened at their sides, by their mutual pressure, they become cubical, as in the pith, or twelve-sided, the cross-section being six-sided; each cell assuming a form more or less regular, according to the degree of pressure exerted upon it by those adjacent. It is also called PARENCHYMA.

a. The cuttings of the pith of elder, or those of any kind of wood, will, under a microscope, exhibit irregular cells and partitions, resembling those of a honey comb. (Fig. 1, *a.*)

b. The vesicles of cellular tissue have no visible communications with each other, but transmit their fluids by invisible pores.

c. Cellular tissue is transparent and colorless in itself, but exhibits the brilliant hues of the corolla, or the rich green of the leaf, from the coloring matter contained within the cells.

d. The vesicles of this tissue are extremely variable in size. They are usually about $\frac{1}{300}$ of an inch in diameter, but are found of all sizes, from $\frac{1}{30}$ to $\frac{1}{3000}$ of an inch.

e. Although this tissue is usually soft and spongy, it sometimes acquires considerable *hardness by the deposition of solid* instead of fluid matter in the cells

This occurs in the prickles of the rose, the stones of the plum, peach, &c., and in the albumen of seeds.

f. In some plants, as in the Turkey rhubarb, &c., little bundles of crystals called *raphides* (from the Gr. ραφίδες, sewing needles,) are formed in the cells.

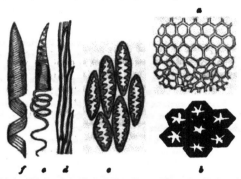

FIG. 1.— Forms of tissue; *a*, cutting of elder pith — cellular; *b*, cells from the gritty centre of the pear; *c*, from the stone of the plum — both strengthened by solid matter; *d*, woody fibre; *e*, spiral vessel with a single fibre partly drawn out; *f*, vessel with a quadruple fibre.

30. WOODY TISSUE, called also FIBRE, consists of slender, transparent, membranous tubes, tapering to a point each way, and adhering together by their sides, the end of one tube extending beyond that of another, so as to form continuous threads. It differs from cellular tissue, in the greater strength, and, at the same time, the greater tenuity, of its membrane. It seems designed for the transmission of fluid, as well as for giving firmness to those parts which need support. (Fig. 1, *d.*)

a. Tissue of this form constitutes the fibre of flax, hemp, &c., the ligneous substance of the stems and roots, the petioles, and veins of leaves, &c.

31. The most remarkable modification of the woody fibre, is that called *glandular.* It consists of little glandular points, arranged along the walls of the woody tubes. It occurs only in resinous wood, chiefly of the fir tribe (*Coniferæ*). It has frequently been detected by the microscope, in fragments of fossil coal, whence it is inferred that coal-beds originated from buried forests of the Coniferæ. *Witham on fossil vegetables, &c.*

32. VASIFORM TISSUE consists of large tubes, called *dotted* having numerous little pits, sunk in the thickness of its

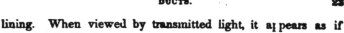

lining. When viewed by transmitted light, it appears as if riddled full of holes.

a. It is of two kinds; 1st, *articulated*, having its tubes interrupted by joints and partitions, as in the oak, vine, and in the monocotyledonous stems; 2d, *continuous*, without joints or partitions; often found in the roots of plants.

b. These are the largest vessels in the vegetable fabric; and their open mouths are particularly discernible in the cuttings of the oak, cane, &c. It is through these that the sap arises to the stem, and is conveyed to the leaves.

3ż. VASCULAR TISSUE consists essentially of *spiral vessels*, with their modifications.

a. The true *spiral vessel* much resembles the woody fibre in form, being a long, slender tube, tapering each way, but is thinner and weaker. Its peculiar mark is an elastic, spiral fibre, coiled up within it, from end to end.

b. The spiral thread is usually single, sometimes double, triple, &c. In the Chinese pitcher plant, it is quadruple. (Fig. 1, *f.*)

c. In size, spiral vessels are variable. Generally their diameter is about $\frac{1}{1000}$ of an inch; often not more than $\frac{1}{3000}$.

d. The situation of spiral vessels is in the medullary sheath, that is, just around the pith; also in every part which originates from it, such as the veins of leaves, petals, and other modifications of leaves, and especially in the petioles, from which it may be uncoiled, in the manner above described. (28, *a.*)

e. In their perfect state they contain air, which they transmit, in some way, from one to another.

f. Ducts are membranous tubes, with conical or rounded extremities, their sides being marked with transverse bars, rings, or coils, incapable of being unrolled without breaking.

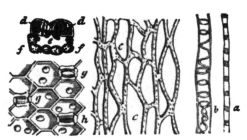

FIG 2 — Poras of tissue, &c.; *a*, annular ducts; *b*, spiral and annular at intervals; *c*, laticiferous tissue; *e*, stomata of iris — vertical section, *d*, *d*, green cells at the orifice; *f*, *f*, cells of the parenchyma, *e*, air-chamber; *g*, *g*, view of epidermis and stomata of yucca, *h*, stomata closed; small, luminous bodies in the cells.

g. In this modification of spiral vessels the tube is much lengthened, and the coil within it is either *closed*, that is, will not unroll, as in the ferns; or it is *annular*, that is, broken into distinct rings, as in the garden balsam; or it is *reticulated*, that is, branching, the branches crossing so as to form a net-work. The office of all these ducts is the same, — that of conveying fluid. It is only in the spiral vessel that we find air. (Fig. 2, *a, b.*)

34. LATICIFEROUS TISSUE is so called, from *latex*, the true nutritious sap, which it is destined to elaborate and convey. It consists of branched anastomosing (ανα, to and fro, στομα, a passage) tubes, lying chiefly in the bark, and the under side of leaves. (Fig. 2, *c.*)

a. These tubes are very irregular in form, direction, and position. They expand and contract at intervals, cross and recross the other tissues, and, proceeding from the inner parts, ramify upon the outer surface, and upon the hairs, forming meshes of inconceivable fineness. Their average diameter is about $\frac{1}{1800}$ of an inch. They are largest in plants which have a milky latex, or juice.

35. The EPIDERMIS, or skin, is a form of cellular tissue externally enveloping the plant. It is found upon every part exposed to the air, except the stigma of the flower, and the spongioles of the roots. *These* it does not cover, nor is it found upon those parts which habitually live under water. And, where the bark of the stem is rugged with seams and furrows, this organ is not distinguishable.

36. It consists of a tissue of flattened cells of various figures, filled with air. Usually there is but one layer of cells, but sometimes there are two or three, especially in tropical plants. The Oleander has four. Its office, in the economy of the plant, is, to check the evaporation of its moisture.

a. The delicate membrane, which may be easily stripped off from the leaf of the house-leek or the garden iris, is the epidermis. It is transparent, colorless, and, under the microscope, reveals its cellular structure.

37. The epidermis does not entirely exclude the tissues beneath it from the external air, but is perforated by certain apertures, called STOMATA (mouths), which open or close under the influence of the light. (Fig. 2.)

38. Stomata are usually of an oval form, bounded by a pair of kidney-shaped cells, containing a green matter. Sometimes they are round, and bounded by several cells. Many other *varieties of form* have been noticed

39. Stomata are always placed over, and communicate with, the *intercellular passages*, that is, the spaces between the cells of the tissue. They are never found on the midrib, or veins, of the leaf, or over any ligneous part of the structure. They are most abundant over the soft, green tissue of the leaves, young shoots, and the parts of the flower.

a. These organs are of a size so minute, that more than 100,000 of them have been counted within the space of a square inch. The largest known are about $\frac{1}{500}$ of an inch in length. Their function is intimately connected with respiration.

FIG. 3. — Hairs and glands; *a, c*, simple hairs; *b*, branched hair of the mullein; *d*, gland surmounted by a hair; *e*, gland at the top of a hair; *f*, prickles of the rose.

40. The surface of the epidermis is either *smooth*, or furnished with numerous processes, originating from itself, or from the cellular substance beneath it. These are of several classes, namely, *glands, hairs, prickles, stings,* &c.

41. HAIRS are minute expansions of the epidermis, consisting each of a single lengthened cell, or of a row of cells, placed end to end, containing air. They are *simple* or *branched.* (Fig. 3.)

a. Hairs are occasionally found upon the leaves, stem, and indeed upon any other part. In the cotton plant (Gossypium) they envelope the seed. They give various names to the surface, to which they are appended, according to their nature and appearance; thus it is said to be *downy*, or *pubescent*, when clothed with soft, short hairs; — *hirsute*, with longer hairs; — *rough*, with short, stiff hairs; — *tomentose*, when they are entangled and matted; *arachnoid*, when like cobwebs; — *sericeous*, when silky; — *velvety*, when they are short, soft, and dense; — *ciliate*, when long and fringed, like the eyelash.

42. STINGS are tubular and acute hairs, fixed upon minute glands *in the cuticle, which* secrete an acrid fluid. By the

slightest pressure this fluid is injected through the tube into the wound made by its point. Ex. nettle.

43. PRICKLES (Fig. 3) are also expansions of the epidermis, consisting of hardened cellular tissue (29, *e*). They are appended to the cuticle alone, and are stripped off with it. Unlike the *thorn* (171), they have no connection with the wood, nor do they disappear by cultivation. Ex. rose, bramble. (Rubus.)

44. GLANDS (Fig. 3) are minute bodies of cellular tissue, situated on various parts of the plant, generally serving to elaborate and discharge its peculiar secretions, which are oily, resinous, saccharine, acrid, &c.

a. They are either sessile, as in the cassia; or mounted upon a stipe, as in the passion flower; or imbedded in the leaf, causing it to appear *punctate*, as in the leaf of the lemon. Often the gland appears to be merely the expansion of a hair, either at its base or its summit. Such are called *glandular hairs.*

45. Analogous to glands, are those cavities formed in the cellular tissue, to serve as *receptacles of secretion.* Examples are seen in the rind of the orange and lemon, containing minute drops of a fragrant volatile oil. The turpentine of the fir balsam is stored up in large reservoirs of this kind.

CHAPTER III.

PRIMARY DIVISIONS OF THE VEGETABLE KINGDOM.

46. THE vegetable kingdom has long been considered by botanists under two great natural divisions, namely, PHÆNOGAMIA, or FLOWERING PLANTS, and CRYPTOGAMIA, or FLOWERLESS PLANTS.

47. Besides the obvious distinction made by the presence or the absence of the flower, these divisions are further distinguished by their structure. The Phænogamia abound with the *ligneous* and *vascular tissue,* while the Cryptogamia consist more generally of the *cellular.* Hence the former are also called *VASCULARES,* and the latter CELLULARE.

48. Again, the former are distinguished for producing seeds composed of determinate parts, as *cotyledons* (♯125) and embryo, while the latter produce certain minute bodies, called *spores*, having no such distinction of parts. Thus the Phænogamia are also called COTYLEDONOUS and the Cryptogamia ACOTYLEDO-NOUS plants.

49. Lastly, we find in the Phænogamia, a system of compound organs, such as root, stem, leaf, and flower, successively developed on a determinate plan (♯18 – 26), while in the Cryptogamia, a gradual departure from this plan commences, and they become simple expansions of cellular tissue, without symmetry or proportion.

a. In the following pages we shall first direct our attention exclusively to the compound organs of FLOWERING PLANTS; and since, in our descriptions of these organs, frequent references will be made to particular species and genera, for illustrations and examples, it seems proper to subjoin, in this place, a brief notice of these fundamental divisions also.

50. A SPECIES embraces all such individuals as may have originated from a common stock. Such individuals bear an essential resemblance to each other, as well as to their common parent, in all their parts.

a. Thus the white clover, (*Trifolium repens*) is a *species*, embracing thousands of contemporary individuals, scattered over our hills and plains, all of a common descent, and producing other individuals of their own kind from their seed. The innumerable multitudes of individual plants which clothe the earth, are, so far as known, comprehended in about 80,000 species.

51. To this law of resemblance in plants of a common origin, there are some apparent exceptions. Individuals from the same parent often bear flowers differing in color, or fruit differing in flavor, or leaves differing in form. Such differences are called VARIETIES. They are never permanent, but exhibit a constant tendency to revert to their original type

a. Varieties occur chiefly in cultivated species, as the apple, potatoe, tulip, Geranium, &c., occasioned by the different circumstances of soil, climate, and culture, to which they are subjected. But they continue distinct only until left to multiply spontaneously from seed, in their own proper soil.

52 A GENUS is an assemblage of species, with more points of agreement than of difference, and more closely resembling *each other than they resemble* any species of other groups.

3*

a. For example, the genus *Trifolium* includes the species *T. repens*, *T. pratense* &c., agreeing in structure and aspect so obviously, that the most hasty observer would notice their relationship. Also in the genus *Pinus*, no one would hesitate to include the white pine and the pitch pine (*P. strobus* and *P. rigida*), any more than he would fail to observe their differences.

b. Thus, the whole vegetable kingdom is, by the most obvious characters, distributed into *species*, and the species, by truly natural affinities, grouped into *genera*. These divisions constitute the basis of all the systems of classification in use, whether by natural or artificial methods.

₂*₂ To the admirer of nature, flowers are among the first subjects of attention, as mere objects of taste. They are conspicuous for their superior beauty, even in the vegetable kingdom, where all is beautiful. Yet, as objects of science, they merit a still higher regard, whether we consider the Creative skill displayed in their construction, or their important agency in the reproduction of the plant. But, to the practical botanist, an intimate knowledge of their organic structure is one of his *first* requisites, on account of the indispensable use of the floral organs in classification.

CHAPTER IV.

THE FLOWER.

§1. OF ITS PARTS, AND THEIR ARRANGEMENT.

53. A FLOWER may consist of the following members:—
1. The FLORAL ENVELOPES, called, collectively, the PERIANTH, (περι, around, ανθος, a flower); 2. The STAMENS; 3. The PISTILS; and, 4. The RECEPTACLE, or TORUS.

a. Of these, only the stamens and pistils are regarded in science as essential parts. These, together with the receptacle, are said to constitute a *perfect flower*, even when one or all other parts are wanting; because these two organs alone are sufficient for the perfection of the seed. In a popular sense, however, a perfect flower must possess *all* the organs above mentioned.

b. If the stamens or the pistils, either or both, be wanting, the flower is said to be *imperfect*. An imperfect flower is either *sterile*, having stamens only, or *fertile*, having pistils only, or *neutral*, having neither organs complete.

FIG. 4. — No. 1, Lily (Lilium Japonicum); 2, pink (Dianthus); 3, a stamen; 4, a pistil.

54. The FLORAL ENVELOPES, or PERIANTH, consist of one or more circles or *whorls* of leaves, surrounding the stamens. The outer of these whorls is called the *calyx*, and the other, if there be any, the *corolla*. The calyx may, therefore, exist without the corolla, but the corolla cannot exist without the calyx. If neither of them exist, the flower is said to be naked, or *achlamydeous* (α, privative, and χλαμυς, a cloak).

55. The CALYX (καλυξ, a cup), therefore, is the external envelope, the *cup*, of the flower, consisting of a whorl of leaves, with their edges distinct or united, usually green, but sometimes highly colored. The calyx-leaves are called SEPALS.

56. The COROLLA (Lat. *corolla*, diminutive of *corona*, crown) is the interior envelope of the flower, consisting of one or more circles of leaves, either distinct, or united by their edges, usually of some other color than green, and of a more delicate structure than the calyx. Its leaves are called PETALS

57. The STAMENS are those thread-like organs, situated just within the perianth and around the pistils. Their number varies from one to a hundred, but the most common number is five. Their office is, the fertilization of the seed. They are collectively called the *andrœcium* (ανδρες,* stamens, οικος, a house).

* The plural of ανηρ, a man, a term applied to the stamen, by Linnæus, in accordance with his favorite theory of the sexes of plants. The term γυνη, woman, is, on the same grounds, applied to the pistil

58. The PISTILS occupy the centre of the flower. They are sometimes numerous, but often only one. They are destined to bear the seed. Collectively, they are called *gynæcium* (γυνη, pistil, οικος, a house).

59. The RECEPTACLE is the *summit* of the flower-stalk, out of which the floral organs grow, and upon which they stand in concentric whorls, the *gynæcium* in the centre, the *andræcium* encircling it, the *corolla* next without, and the *calyx* embracing the whole

60 The principal parts of the flower are shown in the cuts (Figs. 4, 6, 7, &c.), or better by *specimens, with which, both here and throughout the work, the student should always be provided.*

The slender, thread-like organs seen at *a* (Fig. 4, No. 1), are the stamens, surrounding the pistil *b*; *c* is the perianth, consisting of two similar whorls, the outer one a calyx of three sepals, the inner a corolla of three petals, surrounding or enveloping the stamens and pistil; at *d* is the receptacle. At *a* (No. 2) is the inner envelope, the corolla; at *b* is the outer envelope, the calyx or cup, which seems to contain the rest of the flower like a *cup*; at *c*, below the calyx, are certain leafy appendages called *bracteoles* or bracts.

a. Let the pupil compare specimens of these and other flowers, whose parts are well developed, until he becomes familiar with the appearance of each organ, and can instantly apply its name.

61. A complete and regular flower, therefore, is made up of four sets of organs, arranged in concentric whorls. In regard to the *number* and *position* of the individual organs composing these whorls, it is important to observe,

a. First, that each set consists, *theoretically*, of the same number of organs, that is, if the sepals be 5, there should be 5 petals, 5 stamens, and 5 pistils; or, if 3 sepals, there should be 3 petals, 3 stamens, and 3 pistils, &c.

b. Secondly, the position of the organs in each set alternates with those of each adjacent set, that is, the sepals alternate with the petals, the petals with the stamens, and the stamens with the pistils.

c. Thus, in a word, the normal structure of the truly symmetrical flower, divested of all irregularities, consists of four concentric whorls of organs, the organs of each whorl being equal in number, and alternate in position with those of the other whorls (Fig. 5; 1). This structural arrangement, as will hereafter be seen, exactly coincides with that of the *leafy branch*, agreeably to the beautiful theory of the 'transformation of the leaves into the floral organs,' [*] to which allusion has already been made (§ 25). When the bud is developed into a branch, instead of a flower, the leaves are usually arranged in a simple spiral line. This spire may be broken up into equal circles or whorls, from causes to be hereafter explained. In either case the leaves of one spire, or circle, do not issue from the stem at points exactly over the *leaves* of the next circle below, but over the *intervals* between them.

62. This simple normal structure of the flower is, however, subject to many apparent exceptions, so that few, comparatively, are found perfectly conformable to it. Of these few the order Linaceæ affords good examples. In the flax (Linum) the flower is built upon the normal plan, consisting of 5 sepals, 5 petals, 5 stamens, and 5 pistils (each with 5 double carpels), all alternating with each other, according to the diagram (Fig. 5; 1).

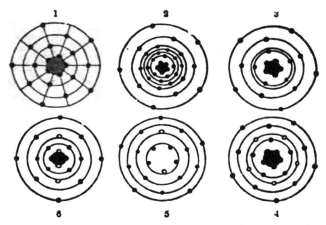

FIG. 5.— Plan of flowers; 1, of a regular and symmetrical flower, as the flax (Linum); 2. of the cherry, showing the four whorls of stamens; 3, of the primrose, showing the position of the suppressed row of stamens; 4, of the Samolus, showing the position of the 5 abortive stamens; 5, of a labiate flower, as the hemp-nettle (Galeopsis), where one stamen and one carpel is wanting; 6, of a cruciform flower, as mustard, where the stamens are in two whorls, two of those in the outer whorl and two carpels being suppressed.

63. If, with this adopted standard, we compare the numerous

---

[*] This theory was first suggested by Linnæus, the founder of Systematic Botany, and subsequently by Wolff and Goethe. After having been long unheeded by botanists, it has at length been revived by modern writers of the highest merit, and shown to be perfectly coincident with facts. 'The adoption of this theory, accordingly, has given a new aspect to botany, and rendered it one of the most philosophical and inductive of the natural sciences.' See Gray's Bot. Text-Book, Chap VIII, where this theory is clearly stated, and richly illustrated.

forms of floral structure which occur, we shall be able to trace out the features of the general plan, even among the widest deviations, and to learn the nature and causes of these deviations. Some of them are the following.

a. One or more additional whorls of the same organ may have been developed. For example, the flower of the Trillium, which, as in most liliaceous plants, is trimerous (τρεις, three, and μερος, part) in its parts, has 6 stamens, evidently in two whorls, and in the flower of the cherry (No. 2,) there are 20 stamens, which may be regarded as arranged in four whorls of fives. Other illustrations will occur to the student.

b. Some of the entire whorls may have been suppressed. For example, in the primrose there are 5 sepals, 5 petals, and 5 stamens, but the stamens are placed opposite the petals. This is to be attributed to the absence of an intermediate whorl of stamens, for in the Samolus, a plant of the same natural order, there is a circle of sterile filaments in the place of the absent stamens (Fig. 5; 3, 4).

c. Some of the parts of a whorl may have been suppressed. Such deficiencies are very common. In the sage, for example, and Monarda, three of the stamens are wanting, in place of which are two rudimentary filaments, and the third rudiment makes its appearance in some allied genera. In most of the Labiatæ but one stamen is wanting (Fig. 5; 5). In the carrot, caraway, and all the Umbelliferæ, the pistils are reduced from 5, the normal number, to 2.

d. The parts of the same whorl may have been united. Thus the sepals may be united at their edges in different degrees, as in the phlox, pink, &c. Or the petals may be thus united, as in the morning glory: or the stamens, as in the mallows tribe; or the pistils, which is extremely common. In short, scarcely a flower can be found in which some of these cohesions do not occur.

e. The organs of different whorls may have been conjoined, causing great disturbances in the symmetry of the flower. The calyx often, as in the currant, coheres with the whole surface of the ovarium (97), only becoming *free* at the summit, so that it seems to stand upon it. It is then said (but improperly) to be superior. Again, the stamens adhere to the petals in their lower part, so as to appear to grow out of them; they are then said (improperly) to be *inserted* into the corolla. In the Orchis tribe the stamens are consolidated with the pistil. The term *free* is used in opposition to these adhesions, just as the term *distinct* is used in opposition to the cohesion of the same organs with each other.

f. The organs of the same whorl may have been unequally developed. This is the case in the corollas of the pea and bean tribes, called papilionaceous (Lat. *papilio,* a butterfly), and in those of the mint tribe called labiate (Lat. *labium,* a lip).

g. Again, organs of one kind may have been reconverted into those of another kind, or into leaves. Such monstrosities are of frequent occurrence among cultivated plants, and may be regarded as proofs of the present doctrine of the floral structure. In all double flowers, as the rose, peony, tulip, &c., the stamens have been reconverted into petals. By still further changes, all parts of the flower tend towards a leafy character, rendering the resemblance of the flower to an undeveloped branch very obvious. Nay, in some cases, the whole flower-bud, after having given a slight indication of a floral character, is transformed into a leafy branch, showing that all parts of the flower are formed out of the same elements as the leaves.

h. Sometimes the flower-stalk is not effectually checked in its growth by the development of the flower, but is prolonged *through* it, and produces secondary flowers in the midst of the organs of the first. This is not unfrequent in the rose. Several instances of these malformations are exhibited below. (Fig. 6.)

k. This mode of studying the floral structure is deeply interesting and instructive, but our limits will not permit us to dwell upon it, nor is it necessary. The intelligent student will be able to extend the above illustrations by an examination of almost any flower, with reference to its deviations from the normal plan.

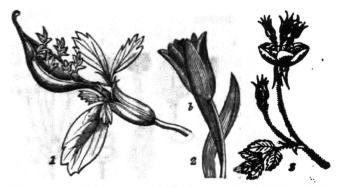

FIG. 6. — 1, From Lindley, — a flower of white clover, reverting to a leafy branch; 2, he drawn from a living specimen, — a tulip, b, a leaf arising from the peduncle, takes the position, form, and color (in part) of a sepal; 3, here drawn from a living specimen, — a rose (R. damascena) with the axis prolonged into secondary rose-buds.

⁂ In our detailed description of the flower, we shall commence with those organs which are deemed essential, their mysterious agency being indispensable to the perfection of the seed.

CHAPTER V.

THE FLOWER.

§2. OF THE STAMENS, AND THE ARTIFICIAL CLASSES.

64. THE stamens and pistils are situated within the floral
envelopes, and since one or both are always present, in every
species, at least, of the Phænogamous plants, they were seized
upon by Linnæus* as the basis of his beautiful arrangement,
called the Artificial System.

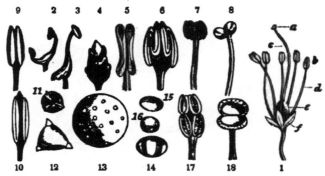

FIG. 7. – Forms of stamens, anthers, pollen, &c. 1. Stamens and pistil of a flower (Rho
dodendron Lapponicum), in their natural position ; *a*, stigma, *b*, anthers, *c*, sty e. *d*, filaments
e, ovary, *f*, calyx and receptacle ; 2, stamen of ginger ; 3, sage ; 4, Berberis ; 5, Vaccinian
amœnum, with the terminal pores ; 6, cucumber, with the sinuous lobes of the anther :
7, Polygonum ; 8, Lemna, anther bursting vertically ; 9, lily ; 10 Magnolia ; 1?, a four-celled
anther ; 18, anther of Alchemilla, bursting transversely.   Nos. 11, 12, 13, 14, 15, 16, various
magnified) forms of pollen-grains.

---

* Carl Von Linné, or Linnæus, the most eminent of naturalists, was the son of a clergy-
man, born in 1707, at Rhœshult, in the province of Smaland, Sweden. In his 24th year
while a member of the University of Upsal, he conceived the idea of that system of plants
which bears his name. In 1741 he became professor of medicine in the same University,
and in 1761, on account of his great literary attainments, was elevated to the rank of nobility.
He died in 1778. To him the natural sciences are under incalculable obligations, all of
which he classified and arranged anew. But the science of botany, especially, is indebted
to him for those discoveries and classifications, which have, more than any others, contri-
buted to its general diffusion. In his 'immortal work,' *Species Plantarum*, he enriched the
language of botany by a new nomenclature of species, and many new terms in the tech
nology of plants, for their more accurate description.

65. The STAMENS are those thread-like organs, seen in the midst of the flower, situated around the pistils and within the corolla, or the calyx, constituting the andrœcium.

66. The stamen (Fig. 4, No. 3) consists of three distinct parts; namely, the *filament, a;* the *anther, b;* and the *pollen, c.* The filament is sometimes wanting, the two latter are essential.

67. The FILAMENT (Lat. *filum,* a thread) is the stem, supporting the anther at or near its top, and is analogous to the stem of a leaf, or to the claw of a petal. When it is wanting, the anther, like a leaf or a petal in a similar case, is said to be *sessile.*

68. The ANTHER is generally situated at the summit of the filament, and is composed of two parallel lobes or cells, connected to each other and to the filament by the *connectile.* It is analogous to the blade of the leaf, each half blade being transformed into a lobe, and the midrib into the connectile.

a. Each cell of the anther usually opens by a longitudinal fissure, called the *dehiscence,* but sometimes, as in the potato, Pyrola, &c. by an aperture (pore) at the summit. In the Polygala, mallow, &c. the *two cells* are reduced to *one.*

b. The *connectile* is usually a mere prolongation of the filament terminating, not at the base, but at the summit of the anthers. In some cases it is prolonged above them, into a sort of appendage, as in the violet, silk-weed, &c.

c. The anther is sometimes wanting, and the filament in such cases cannot constitute a stamen, but is said to be *abortive,* or *sterile.*

69. In regard to the modes of attachment between the anther and the filament, we find the following variations; the anthers are said to be,

1. *Innate,* when they are attached to the filament by the base of the connectile.

2. *Adnate,* when they are attached to the filament by their back, so as to appear lateral; as in the Anemone, water-lily.

3. *Versatile,* when fixed by a single point to the connectile, from which they lightly swing; as in the grasses.

4. When the anthers are attached to the *inside* of the filament, or connectile, so that the line of dehiscence faces the pistils, they are called *introrse* (turned inward). But when they are attached to the outside of the connectile, so that the dehiscence faces the petals, they are called *extrorse* (turned outward). Examples of the former are seen in the violet; of the latter in the larkspur. These distinctions are of importance, as will hereafter be seen.

70. The POLLEN is, in appearance, a small, yellow dust, contained in the cells of the anther. When viewed with a microscope, it *appears to consist of grains* (granules) of various forms,

4

usually spherical, but in some plants cubical, in others triangular, in others still, polygonal, &c., always being of the same form
in the same species. (Fig. 7.)

a. Each grain of pollen has been ascertained to consist of a membranous sack
containing a fluid. In this fluid are suspended molecules of inconceivable
minuteness, possessed of a tremulous motion. When the membrane is exposed
to moisture, it swells and bursts, discharging its contents. (Fig. 12.)

71. *Physiological structure.* The filament consists of a bundle
of delicate ligneous tissue, with spiral vessels, surrounded by
cellular tissue, the same tissues which compose the stem of the
leaf (260). The same tissues have also been traced into the
connectile. The anther consists almost wholly of cellular tissue,
corresponding to the fleshy substance (parenchyma) of the leaf.
The pollen consists of disintegrated bladders of the same tissue.

72. *Theoretical structure.* Thus it is evident, as we have already seen, that
however much the stamen may differ in aspect from a leaf, they both have the
same original plan. This is further evident, from the gradual transition of stamens into petals, as seen in the water-lily or the double rose. In the former, the
process is so gradual that the outer whorls exactly resemble petals, except in having
the tops developed into yellow anthers, while in the rose we find organs in every
conceivable state of transition from stamens to petals. That the petals are modified leaves, will hereafter be more definitely shown (106).

FIG. 8.—Stamens of the water-lily gradually passing into petals.

73. The stamens vary in the different kinds of plants, in respect to their *number, position, relative length, connection,* and
presence. Upon these five different conditions of the stamens,
the TWENTY-FOUR ARTIFICIAL CLASSES of Linnæus are founded.

74. 1st. *Number.* The first eleven classes are founded upon
the *number* of the stamens—the stamens being also *free* (63,
c.), and of *equal length.* Their names are derived from the
Greek numerals combined with ανδρες (57, note), as follows:—

Class I, MONANDRIA (μονος, solitary,) includes all genera (52)
of plants with one stamen to each flower.

Class II, DIANDRIA (δις, twice), with two stamens to each flower.

III, TRIANDRIA (τρις, thrice), with three stamens.

IV, TETRANDRIA (τετρα, four times), with four stamens.

V, PENTANDRIA (πεντε, five), with five stamens.

VI, HEXANDRIA (ἑξ, six), with six stamens.

VII, HEPTANDRIA (ἑπτα, seven), with seven stamens.

VIII, OCTANDRIA (οκτω, eight), with eight stamens.

IX, ENNEANDRIA (εννεα, nine), with nine stamens.

X, DECANDRIA (δεκα, ten), with ten stamens.

XI, DODECANDRIA (δωδεκα, twelve), with twelve stamens.

2d. *Position.* The next two classes depend upon the position of the stamens, — the stamens being free and equal.

XII, ICOSANDRIA (εικοσι, twenty), includes those genera of plants which have twenty or more stamens to the flower, seated on the calyx (perigynous).

XIII, POLYANDRIA (πολυς, many), twenty or more stamens, seated on the receptacle (hypogynous).

3d. *Relative length.* The two following classes are founded upon the relative length of the stamens, together with their number.

XIV, DIDYNAMIA (δις, twice, δυω, two, νημα, a filament), includes plants with four stamens, of which two are long, and two are short.

XV, TETRADYNAMIA (τετρα, four times, δυω, νημα), with six stamens, of which four are long, and two are short.

4th. *Connection.* The five succeeding classes depend upon the connection of the stamens, in various ways.

XVI, MONADELPHIA (μονος, αδελφος, a brother), includes plants with the filaments united into one set or fraternity.

XVII, DIADELPHIA (δυω, αδελφος), into two sets or fraternities.

XVIII, POLYADELPHIA (πολυς, αδελφος), into many sets or fraternities.

XIX, SYNGENESIA, (συν, together, γενεσις, origin), stamens united by their anthers, into a tube.

XX, GYNANDRIA (γυνη, ¶ 57, note, ανηρ), stamens consolidated with the style.

POLYGAMIA (πολυς, many, γαμ
stamens and pistils are se
and united in others, eithe
three different plants.

XXIV. CRYPTOGAMIA (κρυπτος, conc
those genera of plants wher
tils are wanting, or at leas
called FLOWERLESS PLANTS.

a. Such are the twenty-four Linnean classes, in whic
table kingdom are included. Nothing could have beer
eleven. To distinguish them, we have only to count
classes are founded upon distinctions less simple, tho
understood. A good specimen flower of each class sho
ined, to illustrate the definitions, and fix them in the me
The following simple figures are emblematic of each
is required to apply the appropriate numbers and name

CHAPTER VI.

THE FLOWER.

§ 3 OF THE PISTIL, AND THE ARTIFICIAL ORDERS.

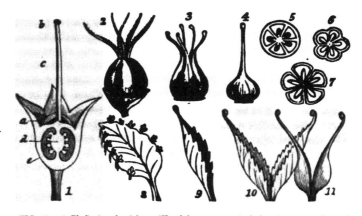

FIG. 10. — 1, Pistil of a whortleberry (Vaccinium amœnum); *b*, the stigma; *c*, style, *a*. the epigynous disk; *e*, perpendicular section of the ovary combined with the adherent (superior) calyx; *d*, the placenta with the ovules; 2, the gynœcium of a flower with 5 pistils, showing the carpels and styles distinct; 7, cross section of the same; 3, the carpels united and the styles distinct; 6, cross section of the same; 4, both carpels and styles united; 5, cross section of the same; 8, leaf of Bryophyllum, putting forth buds from its margin; 9, carpel of the garden cherry, reverting to the form of the leaf; 10, two such carpels; 11, two perfect carpels.

75. The pistil (or pistils) occupies the centre of the flower, at the termination of the axis. It consists of three parts, the *ovary*, or germ, *a*, (Fig. 4.) the *style*, *b*, and the *stigma. c.* The style is sometimes wanting, and the stigma then becomes *sessile* upon the ovary. (See also Figs. 10, 11.)

76. The OVARY (Lat. *ovarium*, a depository, from *ovum*, an egg) is the tumid and hollow part of the pistil, situated at its base, containing the *ovules*, or young seeds within its cavities. and destined to become the fruit.

77. The ovary is either simple or compound. When compound, it consists of two or more lobes or divisions, called CARPELS (*καρπος*, fruit), united together more or less closely

4*

Sometimes these divisions are very evident, being but slightly connected, while in other cases, all external marks of them disappear. When simple, it of course consists of a single carpel (Fig. 10.)

78. The STYLE is that prolonged columnar part of the ovary, or rather of each carpel, which bears the stigma at its top. The number of the styles, when they are not wanting, always equals the number of carpels: but when the carpels are closely united, the styles may be united also, into a single compound column, or they may even then remain distinct.

79. The STIGMA is the upper portion, or extremity, of the style, extremely various in form, but usually globular. Like the ovary and style, it is either simple or compound. When it is compound it consists of as many united lobes as there are carpels.

80. The number of distinct styles (or of stigmas, when the styles are wanting) constitutes the basis of the artificial orders, into which the first thirteen classes of Linnæus are subdivided. They are named from the Greek numerals prefixed to the termination *gynia*, (γυνη, 57, Note,) as follows.

Order 1. Monogynia, includes all the genera of plants in either of the first thirteen classes, with one style to the flower.

2. Digynia, with two styles to the flower.
3. Trigynia, with three styles.
4. Tetragynia, with four styles.
5. Pentagynia, with five styles.
6. Hexagynia, with six styles.
7. Heptagynia, with seven styles.
8. Octogynia, with eight styles.
9. Enneagynia, with nine styles.
10. Decagynia, with ten styles.
11. Dodecagynia, with eleven or twelve styles.
12. Polygynia, with more than twelve styles.[*]

---

[*] The orders of the remaining classes are founded upon characters not depend ing upon the pistil, and are as follows:—

The orders of class 14, Didynamia, are only two;
1. Gymnospermia, with seeds apparently naked.
2. Angiospermia, with seeds evidently in a seed-vessel, or pericarp.

81. The OVULES are certain little globular bodies, produced in the cells of the ovary, destined to become the seeds in the matured fruit. (Fig. 10; 1.)

82. The PLACENTA is that part of the ovary from which the ovules arise, and to which they are attached. It consists of a line, or fleshy ridge, placed in some angle of the cell. Its direction is always vertical, that is, parallel with the axis of growth (Fig. 10; 1, d.)

83. *Physiological structure.* The ovary and style are composed chiefly of one or more bundles of vascular tissue, imbedded in cellular tissue. The stigma consists of a loose cellular substance, called the conducting tissue, communicating with the placenta through the centre of the style. It is the only part of the ascending axis which is destitute of the *epidermis* (35).

84. *Theoretical structure.* The pistil, as before stated (25, a), is the modification of a leaf, or of a whorl of leaves, each leaf constituting a carpel. Each carpel has its own style and stigma, and is formed of a leaf folded together in such a way that the upper surface becomes the inner, and is turned towards the

---

The 15th class, Tetradynamia, is divided into two orders, which are distinguished by the form of the pod:—

1. Siliculosa, the fruit a silicle, or short pod.

2. Siliquosa, fruit a silique, or more or less elongated pod.

The orders of the 16th, 17th, 18th, 20th, 21st, and 22d classes are of the same name and character as the first 13 classes themselves, that is, they are founded upon the number of the stamens to the flower, thus:—

Order 1, Monandria, includes all Monadelphous plants, Diadelphous plants, &c. with one stamen to each flower.

2, Diandria, with two stamens to each flower, and so on.

The orders of the 19th class, Syngenesia, are five:—

Order 1. Equalis (equal), with the florets (flowers) of the head all perfect.

2. Superflua (superfluous), florets of the rays, or margin of the head pistillate, the rest perfect.

3. Frustranea (frustrated), florets of the margin neutral, the rest perfect.

4. Necessaria (necessary), florets of the margin pistillate and fertile, the rest staminate and sterile

5. Segregata (separated), each floret having its own proper calyx.

The orders of class 23d, Polygamia, are two, founded upon the same characters as the two preceding classes:—

1. Monœcia, where both separated and perfect flowers are found in the same individual.

2. Diœcia, where the different flowers occupy different individuals.

The orders of class 24th, Cryptogamia, are nine, the same as the natural orders of this grand division, is Filices, the ferns, Musci, the mosses, &c

_ compared with the pistil of the cherry, se
can be entertained that the two sides of the leaf corres,
ovary, the margins to the ventral suture, the midvein u
the lengthened summit of the leaf to the style and :
flower contains two such leaves, which always present tht
each other, as seen in the figure. This corresponds with
carpels, in which the ventral sutures of each are contiguot

c Many other plants, as the rose, Anemone, Ranuncul
transformations of the pistil, so that there can be no do
formed upon the same plan in all plants. *The ovary, ther
leaf; the style, the lengthened apex; and the stigma, a thickene
of the upper margin of the leaf.*

85. From this doctrine of the structure of tl
he student will be able and expected to dei
ropositions like the following.

a. First. A compound ovary consists of a whe
ives, their united edges all meeting in the
hering sides forming a kind of radiation from i
. Second. There must be as many cells as tl
. Third. The partitions between the cells, i
ments (*dissepio*, to separate,) must each t
t be vertical; they must be equal in numbe
alternate with the stigma, which is also doul
Again, the single carpel can have no true d
ver occur, it is regarded as an ---
-- (Fig. 11)

87. The placentæ are developed at each of the two edges of the carpellary leaf. If these edges be in their normal conditions, that is, united, there will be apparently but one placenta to the carpel, and that central. But if the edges be separate, there will necessarily be two placentæ to each carpel, the one to the right and the other to the left of the dorsal suture and style. They are then said to be parietal (*paries*, a wall).

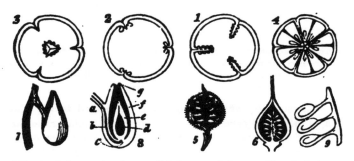

FIG. 11. — 1, Cross section of a one-celled, three-carpelled ovary with parietal placentæ, the dissepiments partially obliterated ; 2, dissepiments wholly obliterated ; 3, dissepiments obliterated, showing a free central placenta ; 4, a five-celled ovary with 5 false dissepiments, as in the flax · 5, vertical section of an ovary with parietal placentæ ; 6, with free central placentæ ; 7, an amphitropous ovule ; 8, vertical section of the same ; a, funiculus ; b, raphe ; c, chalaza ; d, nucleus ; e, secundine ; f, primine ; g, micropyle ; 9, anatropous ovules attached to the ovary.

88. But the placentæ are sometimes found in the *common centre* when there are *no dissepiments* (Fig. 11; 3, 6). This anomaly, which is called a *free central placenta*, is thus explained. The dissepiments were at first actually formed in the usual manner, but afterwards, by the rapid expansion of the shell, they were torn away and obliterated.

a. As the ovules are always developed by the placentæ, they, of course, grow out of the margins of the carpellary leaf, and are, therefore, understood to be analogous to buds. For, in the Bryophyllum, and some other plants, the true leaves do habitually develop buds at their margins (Fig. 10, 8), and in the mignu onette the ovules themselves have been seen transformed into leaves.

89. The ovules are almost always enclosed in the ovary. In the mignionette they are partially naked, and in the fir tribe, Coniferæ, entirely so, the carpellary leaf being open or wanting

a. The ovule is said to be *erect* when it grows from the base of the ovary *ascending*, when it grows from a little above the base; *pendulous*, when it hangs from the summit of the cavity, and *suspended*, when it hangs from a little below the summit.

90. In their early state, the ovules are quite soft, consisting of two sacks or integuments, containing a pulpy mass. and open only at their apex, where there is a passage left through both, called the *foramen*. The outer integument is called the *primine*, the other the *secundine*, and the central pulpy mass the *nucleus*. (Fig. 11; 8.)

a. The foramen may be detected even in the perfect seed, by soaking it in water, and then pressing out the fluid thus absorbed, which will be seen to issue from this little orifice. It has an important agency in the fertilization of the seed, which at this early period has no traces of the embryo (18).

91. The stalk by which the ovule is connected to the placenta, is called the *funiculus*, and its point of attachment to the nucleus of the ovule, the *chalaza*. Through these the ovule receives its nourishment from the placenta. (Fig. 11; 8, 9.)

§4. OF THE MUTUAL ACTION OF THE STAMENS AND PISTILS.

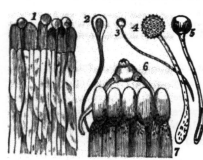

92. The specific use of the stamens and pistils is the fertilization of the seed (57, 58). This appears to be effected in the following manner. At the proper season, the anthers discharge the pollen contained in their cavities through their dehiscence or pores, into the air. Some of it thus falls upon the stigma.

FIG. 12. — 1, Section of the upper part of the style of the snap-dragon, the pollen tubes passing down between the cells; 2, 3, 4, 5, various forms of pollen, showing the tubes; 6, pollen of the Œnothera biennis, one of its tubes descending among the cells of the style.

a. The Author of nature makes special provision for the accomplishment of this function. Thus the anthers are generally placed above the stigma, the stamens being *longer* than the pistils when the flower is erect, as in the tulip, and *shorter*, when it droops, as in several species of the lily. In the mountain laurel (Kalmia), the anthers are confined in ten cavities in the corolla; at the proper season they are disengaged, and thrown forcibly against the stigma, by the elasticity of the filaments. In Monœcious and Diœcious plants, where the stamens are placed apart from the pistils in different flowers, the pollen is often conveyed to the pistil by insects in going from flower to flower in search of honey.

93. Soon after the pollen falls upon the stigma, the outer coat of each granule bursts (70, *a*) at one or more points, allowing the inner coat to pass through it in the form of a tube. This tube insinuates itself between the cells of the stigma, and passes down between the loose cells of the style, extending itself until it reaches the ovary, even when the style is of considerable length. When these tubes reach the ovary, they direct themselves towards the ovules in different parts, and enter the foramen, which at this time is turned towards the base of the style, and brought in contact with its conducting tissue (83).

94. As to the further action of the pollen grains, it is conjectured that the molecules which they contain (69, *a*) are conveyed by the tubes into each ovule, and that there developing themselves into new cells, and becoming fixed in their places, they constitute the embryo of the future plant. All that is certainly known, however, is, that the embryo first appears in the ovule shortly after the pollen tube enters it.

CHAPTER VII.

THE FLOWER.

§ 1. OF THE CALYX.

95. THE term *calyx* comes from the Greek, and signifies a *cup*. It is applied to the outer whorl of the floral envelopes, in reference to its common form and position. It is generally green, but is sometimes *colored*, that is, it is of some other color than green. It seems designed for the protection of the more delicate organs of the flower in æstivation (in the bud).

96 The divisions of the calyx are called *sepals*, which are

sometimes distinct, but generally cohere by their edges, to a greater or less extent, forming a cup as in the rose, or a tube as in the pink. The calyx is then said to be *monosepalous*, a term which must never be *literally* applied, since no true calyx can consist of merely a single sepal; when the sepals are not united in any degree, the calyx is said to be *polysepalous*.

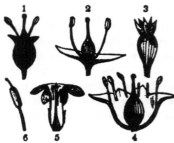

FIG. 13.—3, Ovary, with adherent (superior), persistent calyx; 1, vertical section of the same, showing the epigynous (Gr. *upon the pistil*) stamens; 2, calyx free (inferior), stamens hypogynous (Gr. *under the pistil*); 4, stamens on the calyx, that is, perigynous (Gr. *around the pistil*); 5, stamens on the corolla (perigynous); 6, stamen with the connectile continued beyond the anther.

97. If the calyx is *free*, that is, distinct from the ovary, as in the pink, it is said to be *inferior*, while the ovary is superior; but if the calyx be *adherent* to the sides of the ovary, so as to appear to grow out of its summit, as in the rose, it is said to be *superior*. (Fig. 13; 1, 3.)

98. In respect to duration, it is *caducous* when it falls off as soon as the flower is expanded, Ex. poppy; *deciduous*, when it falls off as the flower decays, Ex. water lily; and *persistent*, when it remains upon the germ after the corolla has fallen; Ex. rose, apple.

99. The calyx is sometimes reduced to a mere rim, and sometimes, when there is no corolla, the calyx is entirely wanting (54).

a. Again, the calyx is reduced to a whorl of mere hair-like processes, called *pappus*, or down. This kind of calyx is peculiar to the Compositæ, as the Asters, sunflower, &c., where the flowers are collected in heads so compact that the calyx has no room to develop itself in the usual manner. If the pappus consists of simple hairs, it is said to be *pilose;* if the hairs are feathery, *plumose;* if they are stiff, like bristles, *setose;* if dilated, so as to become chaffy, *paleaceous.*

§ 6. OF THE COROLLA.

100. *Corolla* is a Latin diminutive, signifying a chaplet

crown. It is fitly applied to that whorl of the floral envelopes situated between the calyx and the stamens, upon the delicate texture and hues of which chiefly depend the beauty of the flower.

101. The divisions of the corolla are called petals. Like the sepals of the calyx, they are either distinct, or united. by their adjacent edges to a greater or less extent, as in the morning glory. When they are distinct, the corolla is said to be *polypetalous;* otherwise, *monopetalous*, a term which is as greatly misapplied in this case as *monosepalous* is to the calyx, since no true corolla can consist simply of a single petal.

102. A petal consists of two parts; the *claw*, which is the narrow part at the base, answering to the stalk of a leaf, and the *lamina*, which is the expanded portion supported by the claw, and answers to the blade of the leaf. The claw is sometimes very long, as in the pink, and often is wanting, as in the rose.

103. When the petals are confluent into a monopetalous corolla, the united claws form that part of it which is called the *tube*, and the lamina constitute the upper, expanded portion of it, which is called the *limb* or border. Both of these parts are exhibited in the Phlox.

104. Monopetalous corollas are *regular* when all the parts correspond to each other in shape, size, and cohesion; and *irregular* when they do not. Both these kinds assume various forms (Fig. 14), which have received appropriate names, as follows:

1. *Campanulate* (bell-shaped), having the tube wide, and swelling abruptly at the base, as in the bell-flower (Campanula)

2. *Infundibuliform* (funnel-form), tubular at the base, but gradually enlarging towards the border. Ex. morning glory, tobacco.

3. *Hypocrateriform* (salver-form), the tube ending abruptly in a border spreading horizontally. Ex. Phlox.

4. *Rotate* (wheel-form), limb regular, or nearly so, spreading, with a very short or imperceptible tube. Ex. mullein.

5. *Labiate* (lipped). This corolla has its limb deeply cleft

into two irregular segments, called the upper and lower lip.   If the lips be widely separate, they are said to be *ringent* (*ringo*, to grin).   Ex. monkey-flower.   If the upper and lower sides are pressed together, *personate* (*persona*, a mask); Ex. snap dragon. If the upper lip is arched, it is termed the *helmet* or *galea*.   Ex Lamium.   This form of the corolla almost universally characterizes the plants of the large and important natural order Labiatæ.

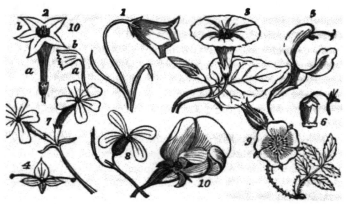

FIG. 14. — Forms of corollas; 1, Campanula rotundifolia; 2, tobacco; 3, Convolvulus; 4, Veronica; 5, sage; 6, Gaultheria procumbens; 7, Phlox; 8, cabbage; 9, rose; 10, Lathyrus.

105. Several forms of polypetalous corollas have also received appropriate names, and are described as follows.   The last only is irregular.

1. Cruciform (*crux*, a cross), consisting of four petals spreading at right angles to each other.   Plants with this corolla constitute the large natural order Cruciferæ, which corresponds to the 15th class in the artificial arrangement.   Of this kind is the mustard (Sinapis).

2. *Rosaceous*, like the rose.   A regular corolla, consisting of five or more petals, spreading horizontally, attached to the receptacle by very short claws.   Ex. rose, apple.

3. *Liliaceous*, like the lily.   The Perianth consists of six parts each gradually bending outwards in such a manner as to resemble the campanula'e.   Ex. lily, tulip (Fig. 4).

4. *Caryophyllaceous*, like the pink. This corolla consists of five petals, having long claws immersed in a tubular calyx. Ex. pink, cockle (Fig. 4).

5. *Papilionaceous*, butterfly-shaped. This corolla consists of five dissimilar petals, which have received names as follows; — the upper and largest is called the *banner* (*vexillum*); the two lateral ones beneath this, the *wings* (*alæ*); and the two lower ones cohering by their lower margins, the *keel* (*carina*). Examples, pea, bean, locust. Plants with this kind of corolla consti tute the greater part of the Leguminosæ, one of the most extensive and useful of the natural families.

106. PHYSIOLOGICAL STRUCTURE. The floral envelopes are found, in their physical organization, to agree with leaves, of which they are only modifications. They consist of thin expan- sions of cellular tissue, traversed by veins of delicate spiral vessels, all covered with an epidermis often having stomata. Their various colors are produced by secretions contained in the little bladders of the cellular tissue.

§ 7. OF THE NECTARY AND DISK.

107. These are terms which have been applied to certain anomalous forms of the floral organs, and are very variable in structure and position.

a. The NECTARY (*nectar*, honey) is properly an apparatus for the secretion of honey. In the violet, larkspur, columbine, &c., it consists of a prolongation of the petal into a *spur*. In the nasturtium it is a similar prolongation of the sepal In the passion flower, grass parnassus, gold-thread, &c., the nectaries are merely abortive stamens passing into petals. In the lady's slipper and other Orchida- ceous plants, the lower petal being inflated and larger than the rest of them, was called nectary by the Linnean school, but by modern writers the labellum, or lip.

b. The DISK is a term applied to certain little projections situated between the bases of the stamens and the pistils. Its more common form is that of a raised rim, either entire or variously lobed, surrounding the base of the ovary, that is, *hypogynous* (*ὑπό*, under, *γυνη*, the pistil), as in the peony, or it appears at the top of the ovary when the calyx is superior, and is then said to be *epigynous* (*ἐπί*, upon, *γυνη*), as in the Cornus.

c. The true character of the disk is little understood. It is supposed by Lindley to consist of stamens in a rudimentary state, as it is sometimes separated into a circle of glandular bodies, alternating with the true stamens

18. OF ÆSTIVATION.

108. ÆSTIVATION (*æstivus*, of summer) is a term used by botanists, to denote the relative arrangement of the several organs of the flower while yet undeveloped in the bud. It is the same to the flower-bud as VERNATION (*vernus*, of the spring) is to the leaf-bud.

a The different modes of æstivation may be best observed in sections of the bud, made by cutting it in a horizontal direction. The most common varieties are the following.

1. *Valvate;* applied to each other by the margins only; as the petals of the Umbelliferæ, the valves of a capsule, &c.

2. *Convolute;* when one is wholly rolled in another, as in the petals of the wall-flower.

3. *Quincuncial;* when the pieces are five in number, of which two are exterior, two interior, and the fifth covers the interior with one margin, and has its other margin covered by the exterior, as in Rosa.

4. *Contorted;* each piece being oblique in figure, and overlapping its neighbor by one margin, its other margin being, in like manner, overlapped by that which stands next it, as the corolla of Apocynum.

5. *Alternative;* when, the pieces being in two rows, the inner is covered by the outer in such a way that each of the exterior rows overlaps half of two of the interior, as in the Liliaceæ.

FIG. 15. — Æstivation of the corolla; 1, Hydrangea; 2, Cheiranthus; 3, Rose (single); 4, Oxalis; 5, Lilium; 6, Pisum; 7, Lysimachia; 8, Solanum; 9, calyx of the Rose. The last form, with 4 and 5, are also termed imbricate.

6. *Vexillary;* when one piece is much larger than the others, and is folded over them, they being arranged face to face, as in papilionaceous flowers.

7. *Induplicate;* having the margins bent abruptly inwards, and the external face of these edges applied to each other without any twisting; as in the flowers of some species of Clematis.

8. *Supervolute;* when one edge is rolled inwards, and is enveloped by the opposite edge rolled in an opposite direction; as the leaves of the apricot.

Of these forms of æstivation, the 4th, 5th, and 9th, are frequently designated by the general term *imbricate*, that is, edge overlapping edge.

CHAPTER VIII.

THE FRUIT.

109. THE *fruit* appears to be the ultimate object and aim of the whole vegetable organization; accordingly, when this is perfected, the process of vegetation ceases, the foliage withers, and the whole plant, if it be an annual, soon dies. But in the *fruit*, provision is made for the reproduction of the species, so that it is justly said to be 'the termination of the old individual, and the beginning of the new.'

a. The fruit is, therefore, the most important part of the plant. Although it does not, like the flower, serve to adorn the face of nature by the beauty of its form and color. yet, besides its own peculiar office of perpetuating vegetable life, it affords one of the principal means of subsistence to animals and to man.

b. The fructification, in respect to time, is subsequent to the flower, is always preceded by it, and, as has been sufficiently shown, is dependent upon it for its maturity and perfection. After having imbibed the pollen from the anthers, the pistil, or its ovary, continues to enlarge, and is finally matured in the form of the peculiar fruit of the plant. The *fruit* is, therefore, properly speaking, *the ovary brought to perfection.*

110. Such being the case, it follows that the fruit is constructed on the same general plan as the ovary, and its structure may be inferred with much accuracy, by the examination of the latter at the time of flowering. In many cases, however, the fruit undergoes such changes in the course of its growth from the ovary, as to disguise its real structure; so that an early examination would be even more safe in its results than a late one.

a. For example, the oak-acorn is a fruit with but one cell and one seed, although its ovary had three cells and six ovules. The change is produced by

5*

the non development of five of the ovules, while the sixth grows so rapidly as to obliterate the dissepiments and occupy the whole space. The same change also takes place in the hazle-nut. The ovary of the birch is two-celled and two-ovuled, but, by the suppression of one cell with its ovule, the fruit becomes one-celled.

§ 1. OF THE PERICARP.

111. The PRUIT consists of the *pericarp* and the *seed;* the former may be wanting, but the latter is essential.

a. Truly *naked* seeds are found in few plants, except the Coniferæ, where the pollen falls directly upon the ovules without the intervention of the pistil. The seeds of the sage and the borage, with their respective tribes, generally said to be *naked*, are not so in fact, for each seed being the product of an ovary with one ovule must necessarily be a one-seeded pericarp.

112. The PERICARP ($\pi\epsilon\varrho\iota$, around, $\kappa\alpha\varrho\pi o\varsigma$, fruit) is the covering or envelope of the seeds, of whatever nature it may be. It consists of three different parts. 1. The *epicarp* ($\epsilon\pi\iota$, upon) is the outer integument, or skin. 2. The *endocarp* ($\epsilon\nu\delta o\nu$, within) called also putamen or shell, is the inner coat, and the *sarcocarp* ($\sigma\alpha\varrho\xi$, flesh) is the intervening fleshy substance.

a. Thus, in the peach, the skin is the epicarp, the fleshy pulp the sarcocarp, and the shell of the stone the endocarp. In the apple or pear, the endocarp forms the glazed lining of the cells, the epicarp the epidermis, and the sarcocarp the intervening pulp.

113. The growth of the fruit depends upon the absorption of sap from the parts below. This fluid, finding no *growing axis* to be prolonged in the usual manner into a branch, is accumulated in the pistil and adjacent parts, is condensed by evaporation, and elaborated into cellular matter by the external surfaces, which still perform the functions of true leaves. Thus these parts become gradually distended into the form and dimensions of the fruit.

114. The process of ripening consists of certain chemical changes, effected by the combined action of heat, light, and air. In its earliest stages, the pericarp consists of a structure similar to that of leaves, being composed of cellular and ligneous tissue, with an epidermis and stomata (35, 37).

a. Secondly, the fleshy pulp, or sarcocarp, is developed, and becomes sour by absorbing from the air an excess of oxygen, which is the proper acidifying principle.

b. Lastly, when the fruit has attained its full growth, the pulp becomes gradually sweetened and softened, by the formation of sugar at the expense of the acids and of the ligneous matter, which before rendered it both sour and hard. These transitions are exemplified by the apple, plum, currant, &c., where the greater portion of nutritive matter is stored up in the pericarp; but in the fruit of the oak, chestnut, some of the grasses, &c., it is chiefly or entirely deposited in the seed.

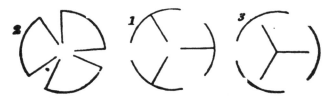

FIG. 16.—Modes of dehiscence; 1, Loculicidal; 2, Septicidal; 3, Septifragal. The straight lines represent the dissepiments.

115. *Dehiscence.* When the pericarp has arrived at maturity, it either remains permanently closed (indehiscent) as the acorn, or it separates into parts forming openings. These parts are called *valves*, and these openings, the *dehiscence.* Regular dehiscence is always vertical, and is called,

1. *Loculicidal* (*loculus*, a cell, *cædo*, to cut), when it takes place by the opening of the dorsal suture of each carpel directly into the cell. Ex. lily.

2. *Septicidal* (*septum*, a wall, and *cædo*), when it takes place through the dissepiments (which are doubled, § 85, *c*). Ex. mallows.

3. *Septifragal* (*septum*, and *frango*, to break), when the valves separate from the dissepiments, which remain still united in the axis. Ex. Convolvulus.

4. *Sutural* (*sutura*, a seam), when it takes place at one or both sutures, in a fruit with a simple carpel. Ex. pea.

5. An irregular dehiscence, called *circumscissile* (*circumscindo*, to cut around), occurs in the plantain, verbena, henbane, &c., where the top of the pericarp falls off like a lid. (Fig. 18; 16.)

116. The forms of the pericarp are exceedingly diversified, and have been studied by botanists with great attention. The following varieties are generally described in elementary works.

1. CAPSULE (a casket), is a term applied to those pericarps which are of a hard and woody texture, proceeding from a compound ovary, dehiscing at the side or top, by *valves*, or sometimes by pores only.

a. The capsule consists of only *one cell*, or is divided within by dissepiments (85, *c*) into many cells. The central pillar, or substance formed by the united placentæ is called the *columella*. To this the seeds are generally attached. The seed-vessels of the Lobelia, mullein, pink, poppy, bloodroot (Sanguinaria), are capsules.

FIG. 17.— Forms of fruit: 1, capsule of Rhododendron; 2, Nicotiana; 3, Colchicum; 4, Œnothera; 5, silique of Raphanus; 6, silicle of Capsella; 7, legume of the pea; 8, jointed legume (loment) of Desmodium; 9, follicle of Apocynum; 10, nut of oak; 11, drupe of Cerasus.

2. SILIQUE (a pod). This is a long, narrow pericarp of two valves, divided into two cells, by a false dissepiment formed by the extended placentæ. The seeds are attached to the edges of this dissepiment, alternating with its opposite sides. Ex. mustard, wallflower, and other Cruciferæ.

3. SILICLE (a little pod), differs from the silique, by being shorter, and more nearly oval. Ex. pepper-grass, shepherd's purse (Thlaspi). The silique and silicle are peculiar to plants with cruciform corollas.

4 LEGUME (also a pod), two-valved, one-celled, consists of a simple carpel, and thus differs essentially from the silique. It bears its seeds attached to the margin of each valve alternately, along the ventral suture only. Ex. pea, and all other plants of the great natural order Leguminosæ. The legume, therefore, accompanies the papilionaceous corolla.

5. FOLLICLE (a bag) is a pericarp with one valve and one

cell, opening by a sutural dehiscence on the inner side, and bearing seeds at the base, or along the suture. Ex. peony, col·umbine, silk-weed.

6. DRUPE (stone-fruit) is one-celled, one or two seeded, indehiscent, with a hard and bony *endocarp* (stone), and a moist and pulpy *epicarp* and *sarcocarp*. Ex. plum, cherry, peach. It also includes those fruits which have a fibro-fleshy, or even coriaceous epicarp, as the walnut, butternut, which kinds of fruit are called *drupaceous*.

7. The NUT is a hard, dry, indehiscent shell, proceeding from an ovary which is two or more celled, and two or more ovuled, but becoming by suppression one-celled, and one-ovuled (110, *a*). It differs from the Drupe, in wanting the soft, succulent covering. Instead of this it is seated in a kind of persistent involucre, called a cupule. Ex. chestnut, oak, beech, hazle.

8. CARYOPSIS (kernel). This is a thin, dry, one-celled pericarp, inseparable from the seed which it encloses. Ex. maize, wheat, Carex. When it is not inseparable from the seed, it is called a *utricle*, as in the pig-weed (Chenopodium).

9. An ACHENIUM is a small, dry, hard, one-celled pericarp, distinct from the seed which it contains. Ex. Borago, Ranunculus, Aster, and the Compositæ generally.

10. SAMARA (winged fruit). It consists of a dry, indehiscent, one-seeded pericarp, with a wing-like appendage. Ex. birch, maple.

11. A PYXIS (box) is a capsule which opens by a circumsessile dehiscence (115; 5), so as to appear like a little cup with a lid. Ex. plantain (Plantago), purslane (Portulaca).

12. POME (apple). This is a fleshy, indehiscent pericarp, formed of the permanent calyx, containing several cartilaginous carpels, or cells, which enclose the seeds. Ex. apple, pear, quince.

13. The PEPO (gourd) is an indehiscent, fleshy fruit, proceeding from a compound ovary, either one-celled, or entirely filled with pulp. Ex. cucumber, melon, pumpkin.

14. *Berry* (*Bacca*), a succulent, pulpy pericarp, holding the seeds loosely within, with no other covering than its own soft

mass. Ex. currant, whortleberry. The orange and lemon answer this definition, and are therefore berries.

FIG. 18.— Forms of fruit; 13, naked achenia of Fragaria on the surface of the enlarged,
fleshy receptacle; 14, drupaceous achenia of a Rubus on a fleshy, deciduous receptacle; 15,
samara of Acer; 16, pyxis of Hyoscyamus; 17, pome of Pyrus (pear); 18, berry of Ribes
(gooseberry); 19, section of the same enlarged; 20, strobile of Pinus; 21, cremocarp of the
Umbelliferæ, as Conium.

a. This definition cannot include the strawberry, which consists of an enlarged, fleshy receptacle, bearing numerous achenia upon its surface. Nor does
it include the blackberry, which, like the other species of the Rubus, is an aggregate fruit composed of united drupes. These fruits are called Etæria, by Mirbel.
(Fig. 18; 13, 14.)

15. STROBILE (cone). This is an aggregate fruit, consisting
of scale-like carpels spread open, with naked seeds on their
inner side, at base. Such is the fruit of the fir tribe, which is
on this account called Coniferæ

CHAPTER IX.

THE FRUIT.

§ 2. OF THE SEED.

117. THE seed is the ultimate product of vegetation, and contains the rudiments of a new plant, similar in all respects to the original.

a. The seed consists of three principal parts; — the INTEGUMENTS, the ALBUMEN, and the EMBRYO.

118. The INTEGUMENTS, or coverings, invest the seed immediately exterior to all its other parts. Although apparently single, they consist of several membranes, to each of which an appropriate name has been applied. The first, or outer membrane, is the TESTA; the second, the MESOSPERM; the third, the ENDOPLEURA, corresponding with the primine, &c. (90) of the ovule.

a. The testa is either papery (membranous), leathery (coriaceous), horny (crustaceous), bony, fleshy, or woody. Its surface is generally smooth, sometimes beautifully polished, as in the Indian shot (Canna), or columbine, and often highly colored, as in varieties of the bean, &c. It is sometimes expanded into wings, as in the Arabis, and sometimes into a tuft of hairs at one end, called *coma*, as in the silk-weed, or it is entirely enveloped in hairs, as in the cotton.

b. The *coma* must not be confounded with the *pappus* (99, *a*), which is a modification of the calyx, appended to the pericarp, and not to the seed, as in the *achenia* of the thistle, dandelion, and other Compositæ.

119. The *aril* is an expansion, proceeding from the summit of the funiculus, or seed-stalk (91), (or from the placenta when the funiculus is wanting) either partially or wholly investing the seed. A fine example is seen in that gashed covering of the nutmeg, called mace. In the celastrus it completely envelops the seed. In other seeds it is a mere scale. and often it is wanting.

120. The HILUM, or *scar*, is that point or mark left on the coats of the seed, by its separation from the funiculus (stalk). It is commonly called the *eye*, as in the bean, pea, maize, &c. (Fig. 11; 8, *a*.)

121. The hilum of the seed sometimes corresponds with the chalaza of the ovule. In this case the ovule, or seed, is said to be *orthotropous* (erect), Ex. candleberry (Myrica). More generally, however, the funiculus (91) extends beyond the hilum, passing under the integuments partly around the nucleus, before it is joined to it. The point of this final juncture is always the chalaza, and that part of the funiculus which then intervenes between the hilum and the chalaza is called the raphe. This form of the ovule, or seed, is called *anatropous* (inverted), and is exemplified in the apple. The raphe can, therefore, exist only in the anatropous seed, and serves to distinguish it. (See Fig. 11; No's 8 and 9.)

122. The ALBUMEN. Next within the integuments, there is a white substance called the albumen, consisting chiefly of starch. It constitutes the chief bulk of some seeds, as maize, wheat, rye, and serves to nourish the embryo in its nascent state. It abounds chiefly in those seeds which have but one cotyledon. It is wholesome and nutritious, even in poisonous plants. The albumen in some seeds is entirely wanting, particularly in the bean, pea, &c., the nutritious matter being all absorbed in the cotyledon.

123. The EMBRYO is an organized body, the rudiments of the young plant, situated within the integuments. To the growth of this all other parts of the seed are subservient. In some seeds the embryo is distinctly visible. Ex. bean, Convolvulus.

124. The embryo is divided into three parts; the *radicle*, the *plumule*, and *cotyledon*.

a. The *radicle* is the descending part of the embryo, destined to form the root (radix). In respect to position, it always points towards the foramen.

b. The PLUMULE is the ascending part of the embryo, or the rudiment of the ascending axis of the future plant. It is usually directed towards the chalaza.

125. The COTYLEDON is the bulky, porous, and farinaceous part of seeds, destined to form the first or seminal leaves of the young plant, as well as to afford nourishment to the plumule and radicle, before they can obtain it from the earth. In the bean, squash, cucumber, and most other plants, the cotyledons are conspicuous in rising above the ground.

a. The number of cotyledons is variable; and upon this circumstance is founded the most important and distinct division of the PHÆNOGAMIA, or FLOWERING PLANTS.

126. Monocotyledonous plants are those whose seeds have but one cotyledon,

or, if two are present, one is minute or abortive. Such plants are also called ENDOGENS (ενδον, inside, γινομαι, to originate or grow), because their stems increase by internal accretions (197). Such are the grasses, the palms, the Liliaceæ, &c., whose leaves are mostly constructed with parallel veins.

127. Dicotyledonous plants are such as bear seeds with two cotyledons These are also called EXOGENS (εξω, outside), because their stems increase by external accretions, including the bean tribe, the melon tribe, all our forest trees, &c. These are also distinguished at a glance, by the structure of their leaves which are reticulate-veined, that is, with veins dividing and uniting again, like network.

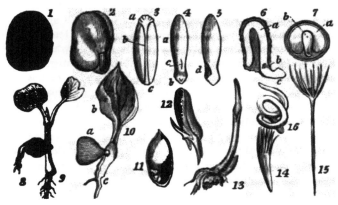

FIG. 19.— Structure of seeds and germination; 1, seed of a garden bean; 2, the same after germination is commenced and the skin thrown off; 3, seed of Triglochin (magnified); a, fungous chalaza, b, raphe, c, hilum; 4, embryo; a, cotyledon, b, radicle, c, fissure, beneath which lies the plumule; 5, vertical section of the same; d, the radicle seen beneath the fissure; 6, germinating seed of Alisma; a, cotyledon, b, plumule, c, radicle; 7, seed of Canna lutea, vertical section, a, albumen, b, embryo; 8, fruit of Mirabilis, showing the commencement of germination, the embryo protruding the radicle; 9, the same, having thrown off the pericarp and become a young plant; 10, germinating seed of Calla Æthiopica; a, seed, b, first leaf of plumule, c, radicle; 11, section of the fruit of a grass with the embryo at base; 12, the same after germination has commenced; 13, the germination completed, and the young plant formed; 14, embryo of Pinus, showing the numerous cotyledons; 15, the same after germination has commenced; 16, embryo of Cuscuta, having no cotyledon.

128. The pine and fir have seeds with from two to three cotyledons, while the dodder (Cuscuta) is almost the only example known of an embryo with no cotyledon.

129. A few plants, as the onion, orange, Coniferæ, &c., occasionally have two or even several embryos in a seed, while all the CRYPTOGAMIA, or flowerless plants, have no embryo at all, nor even seeds, but are reproduced from SPORES, (48) bodies analogous to the pollen grains of flowering plants.

§ 3. OF GERMINATION.

130. The embryo is the most important part of the seed. It is to the protection and nourishment of this alone, that all other parts of the seed, and even of the whole plant, are subservient, and if this be injured or destroyed, the ultimate object of the whole vegetable economy would seem to be defeated.

a. GERMINATION is a term denoting the first stages of vital action in the seed; the process is briefly described as follows:

131. When the seed is *planted* in a moist soil, at a moderate temperature, the integuments gradually absorb water, soften, and expand. The water is decomposed, its oxygen combines with the carbon of the starch which had been stored up in the tissues, carbonic acid is evolved, and the starch, at length converted into sugar for the nourishment of the embryo, which now begins to dilate and develope its parts. Soon the integuments burst, the radicle descends, seeking the damp and dark bosom of the earth, and the plumule arises, with expanding leaves, to the air and the light.* (See Fig. 19, explanations.)

132. As to the cotyledons, they either remain under ground at the centre of motion, as in all Monocotyledonous plants and in the oak, or, as in almost all Dicotyledonous plants, they arise above the surface with the ascending axis, become green, and perform the functions of digestion and respiration, like leaves, for the nourishment of the young plant.

133. The conditions requisite for the germination of the seed are heat, moisture, oxygen, and darkness.

a. Heat is a requisite condition of all vital actions, as well in the *sprouting* of a seed as in the *hatching* of an egg, and if it be not supplied from a source within,

---

* The phenomena of germination, in all its stages, may be observed in an interesting experiment. Let a few seeds, as of the flax or the pea, be enveloped in a lock of cotton, floating upon water in a bulb-glass. In a few days, the plumule ascends in its genial air, while the radicle shoots downwards in long silky fibres.

The ascent of the plumule in a direction contrary to gravity is a law in vegetation, as universal as the law of attraction in matter, and no less difficult to explain. From the two following experiments, it would seem to result both from the influence of the light and the law of gravitation. Professor Shultz planted some seeds of cabbage, mustard, and beans, in moss, and so arranged them that the only light which they could receive was from a mirror reflecting the solar rays upwards; they sent their stems downwards, and their roots upwards.

Mr. Knight placed vessels, containing earth with germinating seeds, upon the circumference of a large horizontal wheel, which was kept constantly and rapidly revolving for several days. The seeds grew, but instead of ascending perpendicularly, the axis of each plant inclined at an angle of 45°, or more, towards the centre of the wheel, in accordance with combined action of the centrifugal force of the wheel, and the attraction of the earth.

must be obtained from without. Different degrees of heat are required by different plants, but a temperature from 50° to 80° is most favorable to those of the temperate zones. Such is the genial warmth supplied by the sun.

b. Water is also requisite for softening the integuments, and for dissolving the dry nutriment stored up in the albumen, or the cotyledons. This is supplied in showers of rain and dew.

c. Oxygen is requisite, as seen above, for the conversion of starch into sugar, a process always depending upon the formation and evolution of carbonic acid, as well in the seed as in the laboratory of the chemist. This is supplied by the water and by the air.

d. And, finally, darkness is favorable, because it is through the influence of light, as will hereafter be shown, that plants absorb carbonic acid from the air, decompose it, retain the carbon itself, and give back the oxygen only. Light would therefore tend to increase the quantity of carbon, rather than diminish it Hence the seed should be *buried* in the soil.

134. The ripened seeds of most plants have the power of retaining their vitality for many years, if they are placed in circumstances which will neither cause them to germinate nor decay, such as a low or moderate temperature, with the absence of moisture. Thus the seeds of maize have been known to grow when 30 years old, rye 40 years, kidney beans 100 years, and the raspberry and beach plum after many centuries.*

§4. THE DISSEMINATION OF SEEDS

135. Is a subject highly curious and interesting; and when attentively considered, serves, like a thousand other cases in the works of Nature, to illustrate the wisdom and design of its great Author. By means of the coma, or pappus, already described, the seeds of the thistle, dandelion, and numerous other plants, are wafted by winds to considerable distances, across rivers, mountains, and even the ocean itself. The *Erigeron Canadense*, a weed now common on both sides the Atlantic, was supposed by Linnaeus to have been transported to Europe from Canada, of which country it is native.

a. Seeds are also furnished with wings for the same purpose. Others are provided with hooks, or beards, by which they lay hold of men or animals, and are thus scattered far and wide.

b. Some seeds, as the Impatiens, which are destitute of all such appendages, are thrown to some distance by the bursting of the elastic pericarp. Rivers, streams, and the currents of the ocean, are all means of transporting seeds from country to

---

* No instance of the longevity of seeds is more remarkable than that related by Dr. Lindley. 'I have before me,' says he, 'three plants of raspberries, raised from seeds which were taken from the stomach of a man whose skeleton was found 30 feet below the surface of the earth. He had been buried with some coins of the emperor Hadrian, and it is therefore probable that the seeds were 1600 or 1700 years old.'

Several years ago, in the State of Maine, about 40 miles from the sea, some men, in digging a well, threw up some sand from a remarkable layer, about 20 feet below the surface, and placed it by itself. A year or two afterwards several shrubs sprung up from this sand, grew, produced fruit, and proved to be the beach-plum.

country. Thus, the cocoa, and the cashew-nut, and the seeds of the mahogany, have been known to perform long voyages, without injury to their vitality. Squirrels, laying up their winter stores in the earth, birds, migrating from clime to clime, and from island to island, in like manner conspire to effect the same important end.

CHAPTER X.

THE ROOT.

136. THE ROOT is the basis of the plant, and the principal source of its nourishment. It originates with the *radicle* of the seed; the tendency of its growth is downwards, and it is generally immersed in the soil.

a. When the radicle has burst the integuments of the seed, and penetrated the soil, its body becomes divided into branches, or fibres; each of these is again divided and sub-divided into fibres, often exceedingly numerous and minute, ever extending and multiplying, until the vegetable has attained its full growth.

137. The prone direction of the root is accounted for by the extreme delicacy of the fibres, which renders them averse to the air and light, by their avidity for moisture, and by the effects of gravitation.

a. Although the primary direction of the roots is downward, they are not known to extend to any great depth. After having descended to a certain distance beneath the surface, they extend themselves horizontally, keeping at about a uniform depth, however great the irregularities of the surface.

138. The number and extent of the roots must always correspond to the demands of the vegetable, both for affording it nourishment, and for maintaining it in its erect position. It follows, therefore, that for every expanding leaf, or extending twig, there must be a corresponding increment of the roots and fibres beneath the soil.

139. Roots are generally distinguished from stems by their downward direction, by the presence of absorbing fibres, by the constantly irregular arrangement of their branches, and by the absence of buds, stomata, and pith.

140. To all these characteristics there are, however, exceptions. Thus, buds, in peculiar circumstances, are developed by the roots, sending up shoots, or suckers, around the parent stem. This does not happen in the natural or healthy state of the plant, but only when the life of the upper axis is partially or wholly myed, the roots remaining in full vigor, and elaborating more nourishment

than there is now demand for. Such buds are, therefore, merely *adventitious*. On this account it would seem that those *roots*, commonly so called, which do naturally and uniformly produce buds, are with propriety described by modern writers as *subterranean stems*; as the *root-stalk* of the sweet flag (Calamus), the *bulb* of the tulip, or the *tuber* of the potato.

141. The summit of the root, or that part which connects it to the ascending axis, is designated as the *collum*, or neck.

a. Strictly speaking, this is the only stationary part of the plant. Occupying the centre of motion between the ascending and descending axis, every enlargement that takes place upon its upper surface arises into the air, while all below it descends into the earth.

142. The parts of the root which require especial notice, are the *caudex*, *fibrils*, and *spongioles*.

a. The CAUDEX (stock) is the main body of the root.

b. The FIBRILS are the finer branches of the root, sent off from the caudex. These are the *true roots*.

c. The SPONGIOLES are the tender and delicate extremities of the fibrils; and, since the latter lengthen only by accretions made to these extremities, these are their *growing points*.

143. The form of the root is much diversified in different plants, but the principal varieties which have received distinctive names, are the following :—

144. *Ramose* (branching). This root consists of ramifications sent off from the main root, like the branches of a tree, but in no determinate order. Such are the roots of most trees and shrubs. (Fig. 20.)

a. There is a strong analogy between the roots of a tree and its branches. In many instances they may be made to perform, each the functions of the other; that is, the tree may be *inverted*, and the branches will become roots and the roots put forth leaves like the branches. The willow and the maple may be thus inverted without injuring their vitality.

b. A branch may often be made to put forth roots instead of leaves. If a branch (offset) of the willow or currant (Ribes) be inserted into the ground, either by the lower or the upper end, or by both at once, it will take root and flourish. Other trees, as the mulberry (Morus) may be multiplied by layers. A branch is bent and inserted into the ground by the apex. When it has taken root it is severed from the parent stock, and becomes a perfect tree.

c. The roots of a tree extend in all directions, and to considerable distances. This distance is at least equal to the extent of the branches, and often much greater. Those of the elm embrace an area of 300 feet diameter, of the poplar,

6*

400. Forest trees, being less exposed to the assaults of the wind, are much less firmly rooted than those in open situations.

145. *Fusiform* (spindle-shaped). It consists of a thick, fleshy caudex, tapering downwards, and also, for a short space, upwards. It sends off from the sides and extremity, thread-like fibrils, which are in fact its true roots, since they alone absorb nourishment from the ground. Ex. parsnip, radish.

FIG. 20. — Forms of the root; 7, branching roots of a tree; 8, root of Daucus; 9, Oxalis, 10, Orchis.

a. When the fusiform root divides into two principal branches, it is said to be *forked.* When it tapers from the *collum* downwards its whole length, it is called a *conical* or *tap* root. But its most remarkable variety is the

b. Premorse, in which the caudex terminates abruptly below, as if it had been *bitten off* (præmorsus). This is due to the fact that the lower extremity perishes after the first year. Ex. Viola pedata, and Scabiosa succisa.

c. The *napiform* (turnip-shaped) root is another variety of the fusiform, where the upper portion swells out, so that the diameter is greater than the length. Ex. turnip.

146. The *fibrous* root consists of numerous thread-like divisions, sent off directly from the base without any caudex. Such are the roots of most grasses, which multiply their fibres exceedingly in a light sandy soil.

a. A *fasciculated* root is a variety of the fibrous, with some of its fibres thickened, as in the crow-foot (Ranunculus), peony, Dahlia, &c.

117. A *tuberous* root consists of one or more fleshy knobs, or

tumors, situated at the base among the fibres. Ex. Orchis. This root must be distinguished from the *tuber*, which, like the potatoe, uniformly bears buds, and is now classed among stems.

a. A *palmate* (hand-shaped) root is a variety of the tuberous, where the knob is separated below into short, thickened processes, as in some species of the Orchis.

b. A *granulated* root consists of many small tubercular knobs, connected by fibres, as seen in the common wood sorrel. Some writers call this variety *moniliform* (*monile*, a necklace).

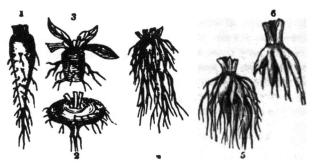

FIG. 21.—Forms of the root;—1, Raphanus; 2, Brassica rapa; 3, Scabiosa; 4, Pea; 5, Pæonia; 6, Dahlia.

148. All the above forms of fleshy roots appear to be reservoirs where the superabundant nutriment secreted by the plant, is accumulated and kept in store for the following year, or for the time of flowering.

a. To the varieties already mentioned, we may add several others, which are remarkably distinguished by their not being fixed in the soil.

149. The *floating* root is peculiar to plants which float loosely upon the surface of the water. Ex. Lemna, Callitriche. The latter, called water starwort, floats upon the surface only until flowering, after which it sinks to the bottom, fixes its roots in the mud, and there ripens its seeds.

150. *Aerial* roots are those which, instead of originating from portions of the plant beneath the surface of the ground, are produced from some portion in the open air. Of these roots, several varieties are remarkable. 1st, Those which are sent forth

from the joints of creeping or prostrate plants; as the ground-ivy, and the twin-flower (Linnæa). 2d, The roots of certain erect plants of the endogenous structure, originating from the stem high in air, descending and entering the soil. Of this class the screw-pine (Pandanus) is a remarkable example, whose aerial roots are often several feet in length before reaching the earth. Such roots, a few inches in length, are also seen in the common maize (Zea).

b. A third class of aerial roots is peculiar to the *epiphytes* (επι, upon, φυτον, a plant). These plants are fixed upon the trunk and branches of other species, and derive their nourishment chiefly from the air. Such are the long moss (Tillandsia), pendent from lofty trees, and many of the Orchidaceæ at the south. 4th, The roots of *parasites* are usually aerial. These are not only attached to other vegetables, but, penetrating their tissues, they derive nourishment from their juices. The Cuscuta and Mistletoe are examples.

PHYSICAL STRUCTURE AND FUNCTIONS OF THE ROOT

151 The internal structure of the root is similar to that of the stem (q. v.), except that there is often a greater proportion of cellular, fleshy matter, as in the beet. In Endogens the root is endogenous, in Exogens it is exogenous, but in the latter case it is always destitute of a pith.

152. The fibrils are in fact but subdivisions of the caudex, or main root. They consist of minute bundles of vasiform tissue (32), enclosed in a loose, cellular epidermis, except at the extremities (35), where the tissue is naked and becomes exceedingly loose and spongy. These (spongioles) have the property of powerfully absorbing water.

153. The growth of the root does not take place by the expansion of the parts already formed, but simply by the addition of new matter at the extremities, and by the formation of new layers upon the surface. This accounts for the facility with which it penetrates the crevices of the soil, and forces its way into the hardest earth.

154. The most obvious function of the root is the purely mechanical one of *fixing the plant in the earth*, and maintaining

ts posture. But its peculiar and most important function is
ABSORPTION. or drawing from the soil that food and moisture
which its growth absolutely requires.

a. Let any small growing plant be taken from the earth, and immersed by its
roots in a glass of water. If it be then exposed to the light of day, or especially
to the sun, the water will disappear from the glass more rapidly than could be
expected from evaporation alone. A plant of spearmint has thus been found to
absorb water at the rate of more than twice its own weight per day. The water
thus absorbed by the roots is mostly sent off again, or *exhaled* through the leaves
(a process called EXHALATION), only a small part of it, together with the salts
which it held in solution, being retained for the use of the plant.

155. The activity of *absorption* must, therefore, depend upon
the activity of *exhalation ;* and since the latter is dependent upon
the presence of light and heat, it follows that absorption will, in
general, be more active by day than by night.

156. The root does not absorb moisture by its whole surface,
indiscriminately, but only by the spongioles at the extremities
of the fibrils, where the pores are not obstructed by the epidermis. From the spongioles it is conducted by the vasiform tissue
of the fibril to the vessels of the main root, and immediately
carried up the stem, and distributed to all parts of the plant.

a. If a growing radish be placed in such a position that only the fibres at the
end may be immersed in water, the plant will continue to flourish. But if the
root be so bent that the fibrils shall be curved up to the leaves, and only the
curved body of the root be immersed, the plant will soon wither, but will soon be
again revived, if the fibres be relaxed and again submerged.

b. Hence, in transplanting trees, too much care cannot be taken to preserve,
uninjured, as many as possible of these tender, absorbing fibres.

157. The force with which plants absorb fluids by their roots
is very great, as is proved by experiment.

a. If the stem of a vine be cut off when the sap is ascending, and a bladder be
tied to the end of the standing part, it will in a few days become distended with
sap, even to bursting. Dr. Hales contrived to fix a mercurial gauge to a vine thus
severed, and found the upward pressure of the sap equal to 26 inches of mercury.
or 13 lbs. to the square inch.

158. The causes of the absorption of fluids, by the roots, have been the subject
of much inquiry. It has generally been said to be due to capillary attraction;
but, unfortunately for this theory, there are no capillary tubes in the vegetable
structure, but only closed cells, more or less elongated, through the membranous
walls of which the fluids must force their way. There is, however, a phenomenon

in Natural Philosophy, discovered by Dutrochet, which bears so strong a resemblance to absorption in Physiology, that late writers are generally agreed in explaining the latter by the former. It is, briefly, as follows:

a. Let the broad end of a tunnel-shaped glass be firmly covered with a piece of bladder, and the cavity within be filled with a solution of gum or sugar. If now the outer surface of the bladder be immersed in water, a passage of fluid will take place through the membrane into the glass, so that the volume of the solution will be much increased, while at the same time there will be a current in the opposite direction, the solution within passing into the water without, but in a much smaller quantity. If, on the other hand, the glass be filled with water and immersed in the solution, it will be partly emptied by this action. The principal current is termed ENDOSMOSE (flowing inwards), and the other EXOSMOSE (flowing outwards).

159. From the above experiment, and others of a similar nature, it is justly inferred, that the conditions requisite for the action of these two currents are, *two fluids of different densities, separated by a porous septum*, or partition. Wherever these conditions exist, the current exists also.

a. Now these conditions exist in the root. The spongiole is the porous septum; the water around it is one of the fluids, and the other is the fluid within, rendered dense by the admixture of the descending sap elaborated by the leaves. Now if the *absorption* be the *endosmose* resulting from these conditions, there must be the counter current, the *exosmose*, also. That this is actually the case, is proved by the fact that the peculiar products of the species may always be detected in the soil about the roots of the plant, and also, that a plant grown in water, always communicates some of its peculiar properties to the fluid in which it is immersed.

160. The use of absorption in the vegetable economy is not merely the introduction of so much water into the plant, but to obtain for its growth those mineral substances held in solution by the water, which constitute an important part of its food.

a. Now in accomplishing this object, the roots seem to be endowed with a certain power of *selection* or *choice*, which has not been satisfactorily explained. Thus, if wheat be grown in the same soil with the pea, the former will select the *silex* along with the water which it absorbs, for the construction of the more solid parts of its stem; while the latter will reject the silex, and appropriate to its use the *calcareous* matter which the water holds in solution.

b. The flowing of the sap from incisions, in early spring, depends upon *an excess of absorption over exhalation*. After the decay of the leaves in autumn, and the consequent cessation of exhalation, — the rootlets, being deep in the ground, below the influence of the frost, continue their action for a time, and an accumulation of sap in the vegetable takes place. Also, in early spring, before the leaves are developed, this action recommences, and the plant becomes gorged with sap, so that it will flow from incisions, as in the sugar-maple. But this flowing ceases as soon as the buds expand into leaves and flowers.

CHAPTER XI.

THE STEM, OR ASCENDING AXIS.

161. THAT part of the plant which originates with the plumule (124, *b*), and arises above the surface, expanding itself to the influence of the air and the light, is called the ASCENDING AXIS OR STEM.

a. The cause of its upward tendency is unknown (131, note), but is supposed to be in some way due to the principles of light and gravitation.

162. Although the first direction of the stem's growth is *vertical*, there are many plants in which it does not continue so, but extends in an oblique or horizontal direction, either just above the surface of the ground, or just beneath it. When the stem continues to arise in its original direction, it is said to be *erect*. When it grows horizontally upon the surface, it is said to be *procumbent, creeping, trailing*, &c. When it arises obliquely it is an *ascending* stem, and when it continues buried beneath the soil it is a *subterranean* stem.

a. The subterranean stem, and some varieties of the creeping, have usually been described as roots.

163. In regard to duration, the stem, like the root, is said to be *annual* when it lives but one season, afterwards dying, at least down to the root, and *perennial* when its existence is continued beyond one season, to an indefinite period of time.

164. In regard to the size and duration of the stem, plants are distinguished into trees, shrubs, and herbs.

a. A TREE is a plant with a perennial, woody stem, or trunk, which does not divide into branches for a certain distance above the ground. Ex. elm, palm.

b. A SHRUB is a plant of smaller dimensions than a tree, having a perennial, woody stem, which divides into branches at or near the ground, like the alder. A shrub of diminished size is termed an *undershrub*. Ex. whortleberry.

c. An HERB is a plant with an annual or perennial root, producing stems which, above the ground, are of annual duration only, and do not become woody. Ex. the grasses, mullein.

165. The most distinctive property of the stem is the forma tion and development of BUDS. At the commencement of its growth, the ascending axis is itself a bud.

166. BUDS are of two kinds, namely, the *leaf-bud*, containing the rudiments of a leafy branch, and the *flower-bud*, containing the same elements transformed into the organs of a flower, for the purposes of reproduction.

167. The leaf-bud consists of a minute, tender, *growing* point of cellular tissue, originating with the pith, surrounded and pro tected by a covering of imbricated scales and incipient leaves. (Fig. 22; 1.)

168. These scaly envelopes of the bud appear to be the rudimentary leaves of the preceding year, formed late in the season. arrested in their development by the frosts and scanty nutriment, and reduced to a scar and hardened state. If the bud of the maple or horse-chestnut (Æsculus) be examined, when swollen in spring, the student will notice a gradual transition from the outer *scales* to the evident *leaves* within.

a. It is an interesting illustration of designing Wisdom, that buds are furnished with scales only in wintry climates. In the torrid zone, or in hot-houses, where the temperature is equalized through the year, plants develope their buds into foliage immediately after their formation, without clothing them in scales. In annual plants, also, the buds are destitute of scales, not being destined to survive the winter. Hence it is evident that the transformation of autumnal leaves into scales, is a means ordained by the great Author of nature, to protect the young shoots, in their incipient stages, from cold and moisture,— an office which they effectually fulfil by their numerous downy folds, and their insoluble coat of resin. *

169. The original bud (plumule) of the embryo is at first developed into a *simple* stem, and being itself continually repro duced, is always borne at the termination of that stem; that is, *the axis is always terminated by a bud.*

a. Besides this, the axis produces a bud (21, a) in the axil of each leaf, that is, at the point just above the origin of the leaf-stalk. If these axillary buds remain inactive, the stem will still be *simple*, as in the mullein. In general, however,

---

* In many trees the scales of the buds are clothed with a thick down. In others, as in the horse-chestnut, balm of Gilead, and other species of poplar, the buds are covered with a viscid and aromatic resin, resembling a coat of varnish. A considerable quantity may be separated from a handful of such buds in boiling water.

some or all of them are developed, forming leafy divisions of the axis, which thus becomes *branched.*

b. Buds are said to be *adventitious* when they are neither terminal nor axillary. Such buds generally result from some unnatural condition of the plant, as maiming or disease, and may be formed in the internodes, or upon the roots (140), or from the trunk. or even from the leaves, as in the Bryophyllum.

170. A BRANCH, therefore, is a division of the axis, produced I y the development of an axillary bud.

171. A THORN, or spine, is a leafless, hardened, pointed, woody process, with which some plants are armed, as if for self-defence. Ex. Cratægus, locust.

a. The thorn appears to be an abortive growth of a bud, resulting from the imperfect development of the *growing point* only, while its leafy coverings perish. Some plants which naturally produce thorns become thornless by cultivation. In such cases the buds are enabled, by better tillage, to produce branches instead of thorns. Ex. apple, pear, gooseberry.

b. The thorn is distinguished from the prickle (43) by its woody structure, and its connection with the wood of the stem, while the prickle, as of the rose, consists of hardened cellular tissue, connected with the bark only.

172. That point in the stem where the leaf, with its axillary bud, is produced, is called the NODE, and the spaces between them the INTERNODES.

a. In the internodes the fibres of the stem are parallel, but at the nodes this order is interrupted in consequence of some of the *inner* fibres being sent off laterally into the leaf-stalk, occasioning, more or less, a jointed appearance. Hence, also, each internode contains fewer fibres, and is of a less diameter than those below it, so that the axis gradually diminishes upwards.

173. Since the branches arise from *axillary* buds, their arrangement upon the stem will depend upon that of the leaves, which, in all young plants, at least, are arranged with great symmetry and order.

174. It is a general law in the arrangement of the leaves and in deed of all other appendages, that they are disposed *spirally,* that is, in a line which winds around the axis like the threads of a screw.

a. But this arrangement is often so much disguised by disturbing causes that it can scarcely be recognized. The most common modification of it is the circular, which is readily explained. The spiral line is formed by the union of two motions, the circular and the longitudinal. The latter is produced in the growing plant by the advancement or lengthening of the axis. Now, if the latter be

7

interrupted from any cause, a circular arrangement is the consequence, — an arrangement so conspicuous in the organs of the flower (61, *a*, *b*, *c*), and in the leaves of the Stellatæ, and other plants.

175. When a *single* leaf arises at a node the arrangement is more obviously spiral, and is said to be *alternate*. When *two* arise at each node they are placed *opposite* to each other, and at right angles to the adjacent pairs. When *three or more* arise at each node they are disposed, of course, in a circle, and are said to be verticillate, or whorled.

176. In like manner, the arrangement of the branches, when divested of all disturbing causes, is found to be SPIRAL; that is, *alternate* in most plants, *opposite* in the ash, &c., or *verticillate* in the pine, &c.

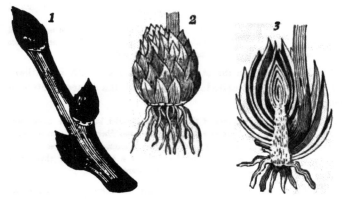

FIG. 22. — 1, Buds, terminal and lateral, with their scaly envelopes; 2, the scaly bulb of the lily, showing its analogy to the bud; 3, vertical section of the same.

a. The ascending axis is exceedingly various in form, size, position, and structure, *existing in every plant* under some one or other of its modifications. It has already been stated, that although its tendency is at first upwards, it does not always arise above the surface. Hence the primary division of this organ into *subterranean* and *aerial.*

177. The SUBTERRANEAN STEM was deemed a root by the earlier botanists, and those plants which possessed such stems only were called *acaulescent* or *stemless*, terms still in use, denoting merely the absence of *aerial* stems. The principal modifications are the bulb, corm, tuber, rhizoma, and creeper.

178. The **BULB** partakes of the nature of the bud. It consists of an oval mass of short, thickened scales, closely compacted in concentric circles and layers, emitting a stem from their midst, and roots from the base or *collum* (141).

a. Bulbs are said to be *tunicated* when they consist of concentric layers, each entire, and enclosing all within it, as in the Onion. But the more common variety is the *scaly* bulb, consisting of thickened concave scales, connected to gether at the base, as the lily, tulip.

b. The bulb is renewed annually, at the approach of winter, by the development of new bulbs in the axils of the scales, which increase at the expense of the old.

c. *Bulblets* are small, aerial bulbs, formed in the axils of the leaves upon the stem, which, when matured, fall to the ground, take root, and produce a perfect plant. The tiger-lily (Lilium bulbiferum) is an example, also several species of the onion. Such plants are termed bulbiferous.

FIG. 22.—Forms of the stem;—1, Allium; 2, Arum; 3, Solanum tuberosum; 4, San guinaria; 5, a spinous branch.

179. The **CORM** is the dilated, subterranean base of a stem, resembling the bulb in form and position, but differing in structure, being composed of a uniform and solid mass, without distinction of layers or scales. It has been improperly called a *solid bulb*. Ex. Arum, or Indian turnip.

180. The **TUBER** is an annual, thickened portion of a subterranean stem, provided with latent buds (called eyes), from which new plants arise the succeeding year. It is the development of buds, and the fact of its origin with the ascending axis, that places the tuber among stems instead of roots. The potatoe is an *example*.

181. The RHIZOMA, or rootstock, is a prostrate, thickened, rooting stem, either wholly or partially subterranean, often covered with scales, which are the rudiments of leaves, or marked with scars, which indicate the insertion of former leaves, and yearly producing both shoots and roots.   Such is the thickened, horizontal portion of the blood-root (Sanguinaria), sweet flag (Calamus), and the bramble (Rubus).

182. The CREEPER differs from the above only in size, consisting of slender branches, exceedingly tenacious of life, extending horizontally in all directions and to considerable distances beneath the surface, sending out roots and branches at intervals.   The witch-grass (Triticum repens) is an example.   Such plants are a sore evil to the garden.   They can have no better cultivation than to be torn and cut in pieces by the spade of the angry gardener, since they are thus multiplied as many times as there are fragments.

a. Repent stems of this kind are not, however, without their use.   They fre quently abound in loose, sandy soil, which they serve to bind down and secure against the inroads of water, and even of the sea itself.   Holland is said to owe its very existence to certain repent stems, by which its shores are apparently bound together.   Much of the surface of that country is well known to be even below the level of the sea.   To protect it from inundation, dikes of earth have been built, with immense labor, along the coast.   These dikes are overspread with a thick growth of such plants as the mat-grass, or Arundo arenaria, the Carex arenaria, and the Elymus arenarius, by the innumerable roots and creepers of which they are enabled to resist the washing of the waves

183. To AERIAL STEMS belong the following varieties;—caulis, runner, scape, vine, trunk, sucker, offset, and stolon.

184. CAULIS (stem) is the term commonly applied to the aërial stems of herbaceous plants, which are annual in duration, and destitute of woody tissue.   *Caulescent* and *acaulescent* are con venient terms, denoting, the former the presence, and the latter the absence of the caulis, or aerial stem.

185. RUNNER.   This is a prostrate, filiform stem, or shoot, extending itself along the surface of the ground, and throwing out roots and leaves at its extremity, which become a new plant, soon putting forth new runners in its turn.   Ex strawberry.

186. The SCAPE is a stem which springs from the summit of the root, or rootstock, and bears the inflorescence of the plant, but not its foliage.   Ex. Sarracenia, daffodil, several species of the Orchis, &c.   The foliage of such plants is usually *radical*, that is, springing from the root or subterranean stem

a. CULM (*culmus*) is a term by which the peculiar stems of the grasses, and similar plants are usually designated in descriptive botany. It seems, however, an unnecessary distinction.

187. VINE. This is a term denoting those stems which, being too weak to stand erect, creep along the ground, or any conven-ent support, and do not throw out roots like the runner. The vine sometimes supports itself on other plants, or objects, by means of *tendrils*, as the gourd, and most of its tribe (Cucurl ita-ceæ); the grape-vine, &c. Such plants are called *climbers.*

a. The *tendril* is a leafless, thread-like branch; or an appendage growing out of the petiole of the leaf; or it is the lengthened extremity of the midrib of the leaf. Its first growth is straight, and it remains so until it reaches some object, when it immediately winds and coils itself about it, and thus acquires a firm, though elastic hold. This beautiful appendage is finely exemplified in the Cucurbitaceæ and grape, above cited; also in many species of the pea tribe (Leguminosæ), where it is appended to the leaves.

188. The *twining* vine, or stem, having also a length greatly disproportionate to its diameter, supports itself on other plants or objects, by entwining *itself* around them, being destitute of tendrils. Thus the hop (Humulus) ascends into the air by foreign aid, and it is a curious fact that the direction of its windings is always the same, namely, with the sun, from right to left; nor can any artificial training cause it to reverse its course. This appears to be a general law among twining plants. Every individual plant of the same species revolves uniformly in one direction although opposite directions may characterize different species. Thus the Convolvulus revolves from left to right, against the sun.

189. TRUNK. This is the name given to the peculiar stems of trees. It is the central collum, or axis, which supports their branching tops, and withstands the assaults of the wind by means of the great firmness and strength of the woody or ligne-ous tissue in which it abounds.

a. The trunk often attains to great dimensions. The white pine (Pinus strobus) of the American forest, with a diameter of 6 or 7 feet, sometimes attains the height of 180, or even 200 feet, with a trunk straight, erect, and without a branch for more than two thirds its length. *

---

* At the first establishment of Dartmouth College, there was felled upon the college plain a tree of this species, measuring 210 feet in length. A Bombax of the South American forests, measured by Humboldt, was 120 feet in height, and 15 in diameter. The Dagon tree on the island of Teneriffe, is said to be 16 feet in diameter. Trees of the genus Adansonia, in Senegal and the Cape Verd Islands, have been found of more than 34 feet in diameter. The famous Chestnut tree on Mt. Etna, often mentioned by travellers, is 64 feet in diameter, and consequently near 200 feet circumference.

7*

b. In regard to duration, trees differ much, some attaining their growth in a few years and immediately decaying, while on the contrary, the ordinary age of trees is beyond the age of man, and some outlive many generations, as the oak, pine. *

190. The SUCKER is a branch proceeding from the stem, or root, beneath the surface, producing leaves, &c., and throwing out roots from its own base, becoming an independent plant. Ex. rose, raspberry.

191. An OFFSET is a short, lateral branch, terminated by a cluster of leaves, and capable of taking root when separated from the parent plant. Ex. house-leek (Sempervivum).

192. A STOLON is a branch which proceeds from an elevated part of the stem, and afterwards, descending to the earth, takes root, sends up new shoots, and finally becomes a new plant. It differs from the sucker, in originating above the ground and not below it.

FIG. 24. — Forms of the stem; 1, Fragaria; 2, Vitis; *b,* tendrils; 3, cirrhose leaf of Pisum, 4, Pyrola; 5, sucker.

193. A *plurality* of stems, or trunks, is observed in a few spe-cies of trees growing in tropical regions. The Banyan (*Ficus*

---

* It is recorded that a live oak, in Louisiana, lived 1000 years; a sycamore in Palestine, 1050 years; a pine in Asia Minor, 1800 years; a cedar on Mt. Lebanon, 2120 years, and the great chestnut on Mt. Etna, 2600 years. -It is also supposed that there are yet living, in the " garden of Gethsemane," some of the olives which witnessed our Saviour's passion; and at Terni, Italy, is an olive plantation supposed to have existed since the age of Pliny.

indica), and the black Mangrove (*Rhizophora mangle*) are mentioned as examples of this singular conformation.

a. The former originally arises with a single trunk. From the principal branches, when they have become so widely extended as to need additional support, long, leafless shoots are sent down. When these shoots reach the earth, they take root, and become new trunks, in all respects similar to the first. The branches thus supported still continue to advance, and other trunks to descend, until a single tree becomes a grove or forest. There is, in Hindostan, a tree of this kind, called the Banyan, which is said by travellers to stand upon more than 3000 trunks, and to cover an area of 7 acres. The Mangrove tree is a native of the West Indies. The new trunks of this tree are said to be formed from the seeds which germinate without becoming detached from the branches, sending down remarkably long, tapering radicles to the earth.

§ L. OF THE PHYSIOLOGICAL STRUCTURE OF THE EXOGENOUS STEM

194. The substance of *herbaceous* stems is soft and succulent, consisting almost wholly of cellular tissue, traversed longitudinally by some few bundles (strings) of woody fibre and vascular tissue, which diverge from the main stem into the leaves.

195. This is essentially the structure of the *first year's* growth of *perennial* plants also. Cellular tissue constitutes the framework of the yearly shoots of the oak, as well as of the annual pea, but in the former it becomes strengthened and consolidated by the deposition of ligneous fibre in subsequent years.

a. Plants differ in respect to the arrangement of these fibres and vessels, and in the mode of their increase; on this difference is based that first grand distinction of Phænogamous plants into Exogens and Endogens, to which allusion has already been made (126—7).

196. The division of EXOGENS (outside growers) includes all the trees and most of the herbaceous plants of temperate climates, and is so named because the additions to the diameter of the stem are made *externally* to the part already formed.

197. The division of ENDOGENS (inside growers), including the grasses, and most bulbous plants of temperate regions, and the palms, canes, &c. of the tropics, is named from the accretions of the stem being made *within* the portions already formed.

198. In the exogenous structure, the stem consists of the pith, wood, and bark.

199. The PITH (*medulla*) occupies the central part of the stem. It consists of a light, spongy mass of cellular tissue, is chiefly abundant in young plants, and appears to be serviceable only in the earlier stages of growth. It is then pervaded by fluids; but as the plant advances in age, it becomes dry, being filled with air only, and much diminished in volume.

FIG. 95.—Exogens,—oak, fir, &c.; Endogens, palm (American), Agave, &c.

200. Immediately around the pith is formed the MEDULLARY SHEATH, which is a thin, delicate membrane of vascular tissue (93), sending off a portion of its spiral vessels to the stalk and veins of each leaf. This, with the leaves, is the only part of exogenous stems which usually contains spiral vessels.

201. The WOOD is composed of concentric zones, or layers, pervaded and intersected by the medullary rays (204). The first, or inner layer, together with the pith and medullary sheath, *is the prod*uct of the first year. One new layer is formed each *successive* year, during the life of the plant; hence the whole

umber of layers, if counted at the base, will correctly indicate ae age of the tree.

202. Each woody layer is composed of ligneous fibre, vasiform tissue, and ducts (33, *f*). The first gives strength and solidity to the trunk, and determines the direction of the cleavage.

a. The *ducts* are always first formed and lie in the inner part next the centre, while the *fibres* are produced towards the end of the season, and are deposited in the outer parts of the zone. The former are distinguished by the large size of their open ends, while the woody fibres are more minute and compact. This circumstance renders the limits of each layer distinctly perceptible in a cross section of the stem.

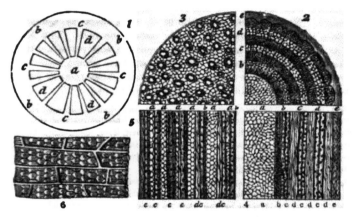

FIG. 26. — Sections of wood; 1, 2, 3, horizontal, 4, 5, vertical. 1, Exogenous stem of one year's growth; *a*, pith, *b*, bark, *c*, medullary rays, *d*, woody bundles of fibre and vessels; 2, stem of three years' growth, *a*, pith, *e*, bark, *b*, *c*, *d*, successive annual layers; 4, a, pith, b, spiral vessels of the medullary sheath, c, dotted ducts, d, woody fibre, e, bark ; 3, Endogenous stem, exhibiting the bundles of woody fibre, spiral vessels, and ducts, irregularly disposed in the cellular tissue ; 5, a, cellular tissue, b, spiral vessels on inner side of c, dotted ducts, d, woody fibre on the exterior side. 6, Laticiferous vessels of the bark.

203. The outer and more recent portion of the layers constitutes the ALBURNUM (*albus*, white), or sap-wood. This is usually of a softer structure and lighter color than the rest of the wood, and it is through the vessels of these layers alone that the sap ascends. The interior layers of the alburnum gradually harden by the deposition of solid secretions in their vessels, until they can no longer allow the passage of fluids through them. Thus the DURAMEN (*durus*, hard), or heartwood is formed, the texture of which is firm and durable. It is only the duramen which is useful in the arts.

204. The MEDULLARY RAYS are those fine lines which appear in a cross section of the stem, radiating from the pith to the bark, intersecting all the intervening layers. They consist of thin, firm plates of cellular tissue; being, like the pith, the remains of that tissue, which at the first constituted the whole of the stem.

a. These rays are quite conspicuous in vertical sections of the oak, or the maple, where they are sometimes called the *silver grain.*

205. The BARK is the external covering of the stem, consisting of several integuments, of which the outer is the *epidermis* (35), that next within the *cellular integument,* and the inner the *liber.*

206. The structure of the two outer integuments is chiefly cellular, and that of the inner, or liber, is both cellular and woody. The cellular integument is very thick in Quercus suber, and constitutes that useful substance. *cork.* The *liber* (Lat. the inner bark, hence a book, because it was manufactured into parchment) is usually thin, delicate, and strong, and has been often applied to useful purposes, as in those trees of Polynesia from which cloth, mats, and sails are made.

207. At the end of the spring a portion of the sap, now transformed into a viscid, glutinous matter called *cambium,* is deposited between the *liber* and the *wood,* becomes organized into cells, and forms a new layer upon each. Soon afterwards, the new layers are pervaded by woody tubes and fibres, which commence at the leaves and grow downwards. Thus the number of layers formed in the bark and wood will always be equal.

a. Since the growth of the bark takes place by *internal* accretions, it follows that the older layers must be carried outwards and continually expanded. Thus, although smooth and entire at first, they at length become shaggy and rough, with longitudinal furrows and ridges, and finally they are cast off, as in the hemlock, spruce, walnut, &c. Not unfrequently, however, the older layers are extended in horizontal grains, or fibres, encircling the stem, as in the white birch (Betula papyracea).

b. The peculiar virtues or qualities of the plant reside in the bark rather than in the wood; hence this is the part chiefly used for medicine, dyes, tannin, &c.

c. That vascular system which is peculiar to the bark, serving for the circulation of its fluids, is called the *laticiferous tissue* (34). It exists in the form of a complete network of vessels, through which the sap moves in all directions.

§ 2. FUNCTIONS OF THE STEM.

208. We have already stated (156) that the stem serves to convey the sap from the roots to the opposite extremities of the plant.

209 That portion of the stem which serves this important purpose is the alburnum (203). Through its ducts and fibres the sap is elevated to the *leaves*, with the vessels of which they communicate. Having been there elaborated by *exhalation* and *decomposition* into a certain nutritious fluid called *latex*, it descends by the laticiferous tissues of the liber. Of this descending sap a part is carried *inward* from the bark by the medullary rays, and thus diffused through the whole stem; the remainder descends to the roots, and is in the same manner diffused through their substance, both for their nourishment, and for the purpose of maintaining the conditions requisite for *endosmose* (159, *a*).

§ 2. OF THE ENDOGENOUS STRUCTURE.

210. In the *endogenous* stem there is no distinction of pith, wood, and bark, nor does a cross-section exhibit any concentric arrangement of annual layers. (Fig. 26; 3, 5.)

211. It is composed of the same tissues and vessels as that of the exogen, that is, of cellular tissue, woody fibre, spiral vessels, and ducts; the first existing equally in all parts of the stem, and the rest imbedded in it in the form of bundles.

212. Each bundle consists of one or more ducts, with spiral vessels adjoining their *inner* side next the centre of the stem, and woody fibres on their outer side, as in the exogen.

a. A new set of these bundles is formed annually, or oftener, proceeding from the leaves and passing downwards in the *central* parts of the stem, where the cellular tissue is most abundant and soft. After descending awhile in this manner, they turn outwards, and interlace themselves with those which were previously formed. Hence the lower and outer portions of the palms, and other endogens, become exceedingly dense and hard, even so as to resist the stroke of the axe.

b. The age of most endogenous trees, as the palms, would seem to be limited by this peculiarity of growth. The stem at length becomes incapable of further increase in diameter, and the lower portions of it so densely filled with the descending fibres as to become impervious to all succeeding ones, and the tree languishes and dies.

c Endogenous stems, both herbaceous and woody, are often hollow. with solid joints; as in the grasses and bamboo.

CHAPTER XII.

THE LEAF.

213. THE leaf constitutes the verdure of plants, and is by far the most conspicuous and beautiful object in the scenery of nature. It is also of the highest importance in the vegetable economy, being the organ of *digestion* and *respiration*.

214. The leaf is characterized by a thin and expanded form, presenting the largest possible surface to the action of the air and the light, which agents are indispensable to the life and increase of the plant.

215. The color of the leaf is almost universally green, which of all colors is the most agreeable to the eye; but its intensity varies by infinite shades, and is often finely contrasted with the more delicate tints of the flower. Towards maturity its verdure is changed, often to the most brilliant hues, as red, crimson, orange, yellow, giving our autumnal forest scenery a gaiety, variety, and splendor of coloring, which the wildest fancy could scarcely surpass.

a. The color of the leaf is due to minute globules, or grains, called *chlorophyll* (green leaf), adhering to the insides of the cells, just beneath the cuticle, and composed of carbon and hydrogen, with a small proportion of oxygen. Their change of color in autumn, is stated by Macaire to depend upon their oxydation. As the leaves in autumn absorb more oxygen by night than they evolve by day an excess is gradually added to the chlorophyll, which changes the green first to yellow, then to orange, red, and crimson successively, according to the quantity absorbed. The same effect may be produced by acids.

b. As flowers are modifications of leaves, it is probable that their various and splendid coloring is due to the same source, namely, the modifications of the chlorophyll by various degrees of oxydation, or by the presence of acids or alkalies in the cells.

§ 1. VERNATION.

216. A leaf-bud contains a collection of undeveloped leaves, folded together in such a manner as to occupy the least possible space. The particular manner in which the young leaves are folded in the bud varies in different species, and is called VERNATION.

a The vernation of the leaf is exhibited in a most interesting manner, by making, with a keen instrument, a cross-section of the bud in its swollen state, just before its expansion; or it may be well observed by removing the scales.

217. The forms of vernation are mostly similar to those of æstivation (108), and are expressed by similar terms. Some of the principal are the following:

FIG. 27. — Forms of vernation. The numbers agree with the corresponding paragraphs.

1. *Equitant*, overlapping each other in a parallel manner, without any involution, as in the leaves of the Iris.

2. *Obvolute*, one of the margins of each leaf interior and the other exterior to the margin of the leaf opposite. Ex. sage.

3. *Involute*, having the edges rolled inwards. Ex. apple, violet

4. *Revolute*, the margins rolled outwards or backwards. Ex. willow, rosemary
5. *Convolute*, the leaf wholly rolled up from one of its sides, as in the cherry.
6. *Plaited*, each leaf folded like a fan. Ex. vine, birch.
7. *Circinate*, when rolled downwards from the apex. Ex. sundew, fern.

§ 2. ARRANGEMENT.

218. In regard to their insertion upon the axis, the arrangement of the young leaves in the bud is nearly or quite circular, but by the development of the axis, this arrangement is modified in various ways, and the leaves are then said to be

1. *Scattered*, or irregular, as in the potatoe.
2 *Alternate*, one above the other, on opposite sides. Ex. pea
3. *Opposite*, two against each other at the same node (172). Ex. Hydrangia.
4. *Verticillate*, or whorled, more than two in a circle at each node. Ex. meadow lily.
5. *Fasciculate*, or tufted, in crowded whorls, or spires. Ex. Callitriche.

219 We have formerly shown how some of these modes of arrangement may be reconciled with the spiral (174, *a*), and we here add, that, in general, when the leaves are said to be scattered or alternate, they will be found, by the attentive observer, to be strictly, though perhaps irregularly, spiral;—always so in the annual shoot.

a Thus in the potato-vine, above cited, or in the house-leek, poplar, &c., if we commence at the lower leaf, and draw a line to the next above it, thence to the next and so on to the sixth leaf, we shall have gone just once around the stem,

8

describing one turn of an elongated spire, so that each *sixth* leaf only is   used exactly above the first.

b. In the *strictly* alternate arrangement, we shall have made one comp... turn on arriving at every third leaf. But this is rare. More commonly the third leaf is a little to the right or left of the perpendicular line on which the first is inserted, so that several turns must be made before we arrive at one which is exactly in that line.

c. The opposite, or whorled, arrangement may be referred to the non-develop ment of some of the internodes; but a better theory is that which supposes seve ral coördinate spires arising side by side: *two*, when the leaves are opposite, and *three*, or more, when they are whorled. For the leaves of the second pair, or whorl, are never placed exactly *above* those of the first, but above their *intervening spaces*, in accordance with the alternation of the petals with the sepals, &c. (61, *b*).

220. In regard to their position upon the plant, leaves are *radical*, when they grow out of the stem at or beneath the surface of the ground, so as to appear to grow from the roots; *cauline*, when they grow from the stem, and *ramial* (*ramus*, a branch) when from the branches.

§3. ORGANOGRAPHY.

221. A leaf may be regarded as an expansion of the two outer integuments of the bark (205) extended into a broad, thin surface by a woody framework, or skeleton, proceeding from the medullary sheath (200). This broadly expanded part is called the LAMINA, or BLADE of the leaf, and it is either *sessile*, that is, connected to the stem by its base, or it is *petiolate*, connected to the stem by a foot-stalk called the PETIOLE.

222. The petiole, therefore, where it exists, is the unexpanded part of the leaf, but like the claws of the petals (102), it is not an essential part, and is often wanting. Its form is rarely cylindric, but is usually flattened or channeled on the upper side. It is said to be

1. *Compressed*, when it is flattened in a vertical direction, so that it is agitated by the slightest breath of air, as in the aspen (Populus).

2. *Winged* (margined), when it is flattened or expanded later-ally into a border. Ex. orange.

3. *Amplexicaul* (sheathing), when it is dilated at the base into a margin which embraces or surrounds the stem, as in the Um-belliferæ.

223. The lamina is generally of a rounded oval outline, longer than wide, with equal *sides* but unequal *ends*. It is, however, subject to variety almost infinite in this respect. The end of the blade next the stem is the *base*, and that most remote, the *apex*.

224. A leaf is *simple* when its blade consists of a single piece, however cut, cleft, or divided; and *compound* when it consists of several distinct blades, supported by as many branches of a *compound* petiole.

225. The frame-work, or skeleton, of the lamina above mentioned, consists of the ramifying vessels of the petiole, while the lamina itself is, of course, *parenchyma* (29). These vessels are collectively called *veins*, from the analogy of their functions.

226. The manner in which the veins are divided and distributed is termed *venation*. The organs of venation, differing from each other only in size and position, may be termed the *midvein, veins, veinlets*, and *veinulets*. (The old terms *midrib* and *nerves*, being anatomically absurd, are here discarded.)

227. The *midvein* is the principal prolongation of the petiole, running directly through the lamina to the apex; as in the leaf of the birch. If there be several similar divisions of the petiole, radiating from the base of the leaf, they are appropriately termed the *veins;* and the leaf is said to be three-veined, five-veined, &c. Ex. maple.

228. The primary branches sent off from the midvein or the veins we may term the *veinlets;* and the secondary branches, or those sent off from the veinlets, are the *veinulets*.

229. There are three principal modes of venation which are, in general, characteristic of the three grand divisions of the vegetable kingdom.

1st. *Reticulate* or *net veined*, as in Exogens. The petiole is prolonged into the leaf in the form of the midvein, or several primary branches, dividing and subdividing into branchlets, which unite again, and by their frequent inosculations form a kind of network. Ex. maple, bean.

2nd. *Parallel-veined*, as in Endogens. In this kind of venation the veins are all parallel, whether proceeding from the base of the leaf to the apex, or sent off laterally from the midvein, and

are always connected by simple transverse veinlets. Ex. grass
lily.

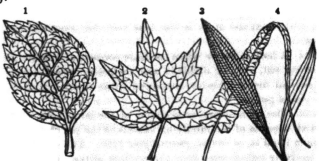

FIG. 96 — Forms of venation. 1, 2, Exogens; 3, Endogen; 4, acrogen.

3d. *Forked-veined,* **as** in the Cryptogamia, when the veins
divide and subdivide by forked divisions which do not unite
again. Ex. ferns.

230. Of the first kind of venation, the *reticulate,* there are two varieties which
deserve the most careful attention. The *feather-veined* and the *radiate-veined.*

1 The *feather-veined* leaf is that in which the venation consists of a midvein, giving off at intervals lateral veinlets and
branching veinulets. Ex. beech, chestnut.

2. In the *radiate-veined,* the venation consists of several
veins (¶ 227) of nearly equal size, radiating from the base
towards the circumference, each with its own system of veinlets
and veinulets. Ex. maple, crow-foot.

3 In parallel venation, the veins are either *straight,* as in the
linear leaf of the grasses, *curved,* as in the oval leaves of the
Orchis, or *transverse,* as in the Canna, Calla, &c.

§ 4. FORM OR FIGURE.

231. That infinite variety of beautiful and graceful forms for which the leaf is
distinguished, becomes intelligible to the student only when viewed in connection
with its venation. Since it is through the veins alone that nutriment is conveyed
for the development and extension of the parenchyma, it follows that there will
be the greatest extension of *outline* where the veins are largest and most numerous. Consequently, the form of the leaf will depend upon the direction of the
veins, and the vigor of their action, in developing the intervening tissue. For
this interesting theory we are indebted to Alphonse De Candolle.

a. In our description of individual forms, we shall select only the most remarkable, leaving others for explanation in the Glossary.

The most obvious arrangement is that which is founded upon the modes of the veining; but it should be premised that different forms of venation often give rise to the same outline.

232. Of FEATHER-VEINED leaves, the following forms depend on the length of the veinlets in relation to each other, and to the midrib. If the middle veinlets are longer than the rest, the leaf will be

1. *Orbicular* (roundish), as in Pyrola rotundifolia.

FIG. 29.—Figures of feather-veined leaves. The numbers refer to paragraphs. *a*, deltate leaf of Populus.

2. *Elliptical* (oval), as in Lespedeza prostrata; or

3. *Oblong* (narrow-oval). Ex. Arenaria lateriflora.

If the lower veinlets are longer than the rest, the leaf will be

4. *Ovate* (egg-shaped), as in the Mitchella repens, or

5. *Lanceolate* (lance-shaped), narrow, and tapering to each end. Ex. sweet-william.

When the veins are most developed towards the summit of the leaf, it becomes

6. *Obovate* (inversely egg-shaped), as in the walnut; or

7. *Spathulate* (shaped like a spathula), as in the daisy.

Again, if the lowest veinlets are longest, sending off veinulets backwards, the leaf will be

8. *Cordate* (heart-shaped), like the ovate form, with a hollow (sinus) at the base, as in the lilac.

9. *Auriculate*, having ear-shaped lobes at the base. Ex. sage.

10. *Hastate* (halbert-shaped), hollowed out at the base and sides. Ex. Bitter-sweet.

11. *Sagittate* (arrow-shaped), with pointed, descending lobes at base. Ex. Polygonum sagittatum; Sagittaria; &c.

12. *Reniform* (kidney-shaped), broad, rounded at the apex and hollowed at the base, as in the Asarum Canadense.

a. The following forms depend less upon the proportion of the veinlets than on the imperfect development of the tissue between them.

FIG. 30. — 13 — 16, figures of feather-veined leaves, the remainder of radiate-veined.

13. *Runcinate* (re-uncinate), having the margin extended at the veins into pointed segments, which curve backwards. Ex. Taraxacum.

14. *Lyrate* (lyre-shaped), with several deep, rounded sinuses, occasioned by deficiency of tissue between the lower veinlets ; water-cress (Sisymbrium).

15. *Pinnatifid* (feather-cleft), with deep sinuses between the veinlets, separating each margin of the leaf into oblong, parallel segments. Ex. Lepidium.

16. *Sinuate*, having deep, rounded openings between the veinlets, seen in the leaves of the white oak.

233. RADIATE-VEINED leaves assume many forms, depending upon the direction of the veins, and the quantity of the intervening tissue. Some of them are the following.

17. *Palmate* (palm-shaped), having five lobes, with as many veins (227) separated by deep divisions, so as to resemble the palm of the hand with the fingers. Ex. passion-flower.

18. *Digitate* (finger-shaped), having narrower and deeper segments than the palmate, as in the hemp.

19. *Pedate* (foot-shaped). The same as palmate, except that the two lateral lobes are themselves subdivided, as in the peony and passion-flower.

20. *Laciniate* (gashed), the veins and veinlets separate, as if the blade were cut and gashed with scissors. Ex. Ranunculus.

21. *Peltate* (shield-like), the veins radiating in all directions, and all connected by intervening tissue. This form is gener ally also orbicular, and appears to result from the union of the base-lobes. Ex. Podophyllum peltatum, Tropeolum, Brasenia.

22. *Reniform, broad-ovate, broad-cordate, &c.*, may also result from the radiate veining.

234. The form of PARALLEL-VEINED leaves is less diversified than that of the preceding classes, being

23. *Linear*, when the veins (and fibres) are straight, as in the grasses. This form may also occur in the feather-veined leaf by an equal development of all the veinlets as in Linaria vul garis, &c.

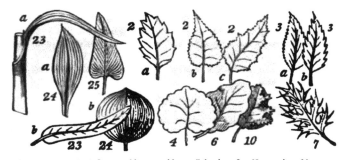

FIG. 31 — 23, 24, 25, figures of leaves with parallel veins; 2 — 10, margins of leaves.

24. *Oval, lanceolate, oblong*, or some kindred form, when the veins are curved, as in Carex, Cypripedium, Orchis, &c., or it may be

25. *Cordate*, when some of the lower veins are curved backwards and then upwards, as in Pontaderia, and even *sagittate*, when they are directed downwards at the base, as in the Sagit taria.

26 *Acerose* (needle-shaped), when there is little or no distinc
tion of lamina, petiole, or veins, as in the leaves of the pine.

235. The margin of the leaf is also modified chiefly by the
same causes which affect the form. It is said to be

1. *Entire*, when even-edged. This may result from the full development of the
tissue, or from a vein running parallel with the margin. Ex. lilac, lily.

2. *Dentate* (toothed), the tissue incomplete, having teeth with concave edges,
pointing outwards from the centre. Ex. hawkweed. If the teeth are very fine,
the margin is said to be *denticulate*. If the teeth are themselves toothed, it is
doubly dentate.

3. *Serrate*, having sharp teeth pointing forward like the teeth of a saw Ex
Rosa. If the serratures are very small, it is *serrulate*. If they are themselves
serrate, it is *doubly serrate*.

4. *Crenate*, notched with rounded or convex teeth, as in Glechoma. If such
notches are very small, it is *crenulate*.

5. *Erose* (gnawed), having the margin irregularly toothed, or jagged, as if
bitten by animals.

6. *Undulate* (wavy), the margin rising and falling like waves. Ex. Amaranthus.

7. *Spinous*, when the veins project far beyond the tissue in sharp spines, as in
the thistle. Such leaves are said to be *armed*, and the opposite corresponding
term is *unarmed*.

8. *Incised* (cut), margin divided by deep incisions.

9. *Laciniate* (torn), divided by deep and irregular gashes.

10. *Crisped*, margin much expanded and curled by a superabundance of tissue
as in the mallows.

11. *Repand*, having the margin slightly concave between the projecting veins.
Ex. Solanum nigrum.

236. In regard to the termination of a leaf at its apex, it is
said to be

1. *Acute*, when it ends with an acute angle.

2. *Obtuse*, when it ends with a segment of a circle.

3. *Acuminate*, ending with a long, tapering point.

4. *Emarginate*, having a small notch at the end.

5. *Retuse*, terminating with a round end, having the centre depressed.

6. *Mucronate*, abruptly terminated by a short, hard, bristly point, &c.

237. The following terms are employed in descriptive botany,
chiefly to denote the modifications of the surface (epidermis)

of the leaf. They are, however, equally applicable to the surface of any other organs. (41, a.)

1. *Glabrous, smooth;* denoting the absence of all hairs or bristles. Hydrangea
2. *Pubescent,* covered with soft hairs or down. Lonicera Xylosteum.
3. *Rough,* with hard, short, even points. Borago officinalis.
4. *Pilose,* with short, weak, thin hairs. Prunella vulgaris.
5. *Hoary,* white, with very short, dense hairs. Gnaphalium.
6. *Villose,* with long, thin hairs. Solidago altissima.
7. *Woolly,* with long, dense, matted hairs. Mullein.
8. *Tomentose,* with dense, short, and rather rigid hairs. Spirea tomentosa.
9. *Rugose,* the tissue between the reticulated veins convex. from its superabun
dance. Sage.
10. *Punctate,* dotted with pellucid glands (44, a). Hypericum punctatum.

§ 8. COMPOUND LEAVES.

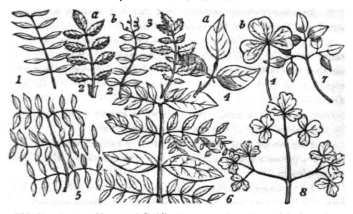

FIG. 32. — Compound leaves 4, Trifoliate leaves; a, pinnately, as of the bean; b, palmately, clover

238. When a *simple* leaf becomes a *compound* one, the division takes place upon the same principle as the separation of an
entire leaf into segments, lobes, and teeth, namely, from a deficiency of parenchyma; the number and arrangement of the
leaflets will therefore, in like manner, depend upon the mode of
veining.

239. The divisions of a compound leaf are called LEAFLETS,
and the same distinctions of outline, margin, &c., occur in them
as in simple leaves. In the truly compound leaf, each leaflet

(which is usually supported on a distinct stalk), is articulated (*articula*, a joint), with the main petiole, and separates from it in decay.

240. From the *feather veined* arrangement may result the following forms of compound leaves:

1. *Pinnate* (winged), where the petiole (midvein) bears a row of leaflets on each side, generally equal in number and opposite, as in the Acacia.

2. A pinnate leaf is said to be *equally pinnate* where the petiole is terminated by neither leaflet nor tendril, as the Cassia Marilandica, and *unequally pinnate* when it is terminated by an odd leaflet or by a tendril. Ex. rose, locust, pea. In the latter case the leaf is called *cirrhose*.

3. An *interruptedly pinnate* leaf has the leaflets alternately small and large, as in the potato, avens.

4. A pinnate leaf sometimes consists of as many as twenty cr thirty pairs of leaflets, as in the Astragalus. Sometimes the number of leaflets is but three, and the leaf becomes *ternate* or *trifoliate*, as in the ash; and, finally, it is sometimes, by the nondevelopment of the *pinnæ* (pairs) reduced to a single terminal leaflet, as in the lemon. Such a leaf is known to be compound by the articulation of the leaflet to the petiole.

5. A *bipinnate* leaf (twice pinnate), is formed when the leaflets of a pinnate leaf themselves become pinnate. Ex. Fumaria officinalis.

6. A *tripinnate* leaf (thrice pinnate), is formed when the leaflets of a bipinnate leaf become pinnate, Ex. Aralia spinosa. In the leaf of the honey-locust (Gleditschia), we sometimes find all these three degrees of division, namely, the pinnate, bipinnate, and tripinnate, curiously combined, illustrating the gradual transition of the simple to the most compound leaf.

7. A *biternate* leaf is formed when the leaflets of the ternate leaf become themselves ternate, as in Fumaria lutea.

8. A *triternate* leaf is formed when the leaflets of a biternate leaf become again ternate. Ex. Aquilegia.

241. The following forms of compound leaves may result from the division of a *radiate-veined* leaf; the *ternate, biternate,* &c., already mentioned:

9. *Quinate*, when there are five leaflets radiating from the same point of the petiole, as in Potentilla argentea.

10. *Septinate*, when there are seven leaves from the same point in the petiole, and so on.

242. With regard to insertion, the leaf is said to be

1. *Amplexicaul*, when its base surrounds or clasps the stem.

FIG. 33. — Modes of insertion.

2. *Perfoliate*, when the base lobes of an amplexicaul leaf are united together, so that the stem appears to *pass through* the leaf.

3. *Decurrent*, when the base lobes of the leaf grow to the stem below the point of insertion, so that the leaf seems to *run downwards* (Lat. *decurro*).

4. *Connate*, when the bases of two opposite leaves are united.

5. *Stellate*, verticillate, or whorled, when several leaves are arranged around the stem at the same node.

243. It is often found necessary, in the description of a plant, to combine two or more of the terms above mentioned, to express some intermediate figure or quality; thus *ovate-lanceolate*, signifying between ovate and lanceolate, &c.

a. The Latin preposition *sub* (under), prefixed to a descriptive term, denotes the quality which the term expresses, in a lower degree, as *subsessile*, nearly sessile, *subserrate*, somewhat serrate, &c.

§ 9. ASCIDIA, STIPULES, AND BRACTS.

244. In the teazel (Dipsacus) of our own fields, and in the Tillandsia, or wild pine of South America, there are hollows at the point of union between the leaf-stalk and the stem, capable of holding a considerable amount of water. The midrib and petiole of the leaves of the Arum, also, are channeled out in such a manner as to convey water to the axil.

245. But the most remarkable of all leaves are those which are hollowed out into the form of pitchers, called *ascidia*.

a. In the Sarracenia, a plant common in our own peat-bogs, these pitchers are evidently formed by the very deep channeling of the petiole, and the uniting together of the involute edges of its winged margin so as to form a complete vase, with a broad expansion at the top, which may be regarded as the true leaf

The *ascidia* thus formed are always full of water, in which insects are drowned being prevented from escaping by the deflexed hairs at the mouth.

246. The Nepenthes is a native of the East Indies. Its proper leaves are sessile and lanceolate. The midvein extends beyond the apex, like a tendril, to the length of six or eight inches. The extremity of this tendril is inflated into a hollow vessel similar to a pitcher, and usually contains about half a pint of pure water. It is furnished with a leafy lid, connected to it by a ligament which expands or contracts according to the state of the atmosphere, so that the cup is open in damp weather to receive moisture from the air, and closed in dry weather to prevent its evaporation.

FIG. 34.—Ascidia. 1, Sarracenia purpurea; 2, Nepenthes distillatoria; 3, Dischidia Raf-flesiar.a.

247. Another wonderful provision of this kind is observed in a plant growing in the forests of India, called Dischidia. It is a twining plant, ascending the tall trees to the distance of 100 feet from its roots, and destitute of leaves except near its top. These cannot, therefore, it would seem, derive much nourishment from the earth. The pitchers seem formed of a leaf with its edges rolled towards each other, and adherent, and its upper end, or mouth, is open to receive whatever moisture may descend into it, of which there is always a considerable quantity. But the greatest marvel in its structure is yet to be described. Several bundles of absorbent fibres, resembling roots (142, *b*), are sent out from the nearest parts of the stem and enter the pitchers and spread themselves through the cavity. The design of this apparatus scarcely needs be mentioned.

248. The leaf of Venus' fly-trap (Dionæa muscipula), native at the south, is also of a very curious construction. At the extremity of each leaf are two lobes bordered with spines. In the cavity between the lobes are several sharp points projecting upwards, and a gland which secretes a liquor attractive to insects. But when an unlucky fly, in search of food, alights upon it, the irritable lobes instantly close and impale him in their fatal embrace

249. STIPULES are certain leaf-like expansions situated on each side of the petiole, at its base. They are membranous, leathery, or spiny. They do not occur in every plant, but are pretty uniformly present in each plant of the same natural order Ex. pea, rose, Viola tricolor.

250. Stipules are generally supposed to be accessory leaves, although their

nature is certainly obscure. They are subject to the same laws of venation and form, perform the same functions, and are sometimes almost undistinguishable from the leaves themselves. They also (very rarely) develop buds in their axils.

a. When they grow from the stem itself, they may, therefore, be regarded a rudimentary *leaves*, but when from the base of the petiole, as is most common they are the undeveloped *leaflets* of a pinnate leaf, as in the rose.

251. When leaves are furnished with stipules they are said to be *stipulate,* and when without them they are *exstipulate* The stipules which are situated at the base of *leaflets* are called *stipels.*

FIG. 26.—Stipules, Bracts, &c. 1, *a,* stipule of grass; 2, *b,* of rose; 3, *c,* bract of Tilia 4, *d,* of a Campanula; 5, Stem, *a,* Involucre, *c,* involucel; 6, Cornus Canadensis, *a,* colored Involucre, *c,* flowers; 7, Arum, *a* spathe, *c,* spadix.

252. BRACTS, called also *floral leaves,* are leaf-like append-ages, intermediate between leaves and the floral organs. From leaves they are generally distinguished by their being placed near the flower, their smaller size, their difference in form, and often in color.

253. That bracts are of the same nature as leaves is perfectly evident, for so gradual is the transition between them that no absolute limits can be assigned. That they have a common origin with the sepals of the calyx also, is equally evident, — so imperceptibly do the latter pass into bracts; affording one of the strongest proofs of the doctrine of floral metamorphosis.

a. Bracts have received different names, according to their arrangement and situation. They constitute an

254. *Involucre,* when they are arranged in a whorl, and sur round several flowers. In the Phlox, and generally, it is green

but sometimes, as in the Cornus, it is colored like petals.   Situ
ated at the base of a compound umbel (305, *a*) it is called a
general involucre, at the base of a partial umbel a *partial involucre*, or *involucel*, both of which are seen in the Umbelliferæ.

255. In the Compositæ the involucre consists of imbricated bracts, often in
several whorls surrounding the base of the heads (compound flowers), as the
calyx surrounds a simple flower.

256. In the grasses, the bracts subsist under the common name of husk or
chaff, to which is attached the *awn* or *beard*. The bracts situated at the base of a
spikelet of flowers, are called the *glume*, corresponding to the involucre. Those
situated at the base of each separate flower are *paleæ*, answering to the calyx, or
corolla. The pieces, of which each glume or palea is composed (generally two)
are called *valves*.

§ 10. DURATION.

257. Leaves, although so universal an accompaniment of vegetation, are only
temporary appendages. They rapidly attain their growth, and in a great ma-
jority f cases flourish but a single season, at the end of which they perish,
although the plant on which they grew may continue to flourish for ages. To
mark their duration more accurately, leaves are said to be

   1. *Fugacious*, when they fall off early, before the end of summer.

   2. *Deciduous*, when they endure for a single season and fall in autumn.

   3. *Persistent*, or *evergreen*, when they remain through all seasons, retaining their
color until the new leaves of the following spring appear, so that the plant is
always verdant. In accordance with the last two distinctions, plants are said to
be DECIDUOUS, or EVERGREEN.

258. The *fall of the leaf* in temperate climates, occurs near the end of autumn,
and marks an important era in the year. The first symptoms of decay are seen
in the changes of color from green to various shades of gold and crimson. These
gorgeous hues, gradually fading, at length give place to a pale russet, the common color of the faded leaf.

259. *Defoliation*, or the separation of the leaf from the stem, is due to several
causes. During the latter part of the summer, the vessels become clogged by the
deposition of earthy and solid matter contained in the sap, until they can no longer admit the free circulation of the fluids through them. The whole structure
consequently loses its vitality, dries up, and withers, and is finally cast off at the
point of articulation, as a dead part is from the living body of an animal.

§ 11. PHYSIOLOGICAL STRUCTURE.

260. Since the frame-work of the leaf is merely a divergent
portion of the medullary sheath (200), it must consist essentially
of the same tissues, namely, spiral vessels accompanied bv
woody fibre, that is, *fibro-vascular* tissue.

a. The tissue of the lamina, in like manner, must essentially correspond with the outer integuments of the bark, of which it is but an extension. That peculiar form of *cellular tissue* of which it is composed is called *parenchyma.*

261. The parenchyma of the leaf exists in two layers, as might be inferred from the manner in which it is produced (221) In all those leaves which are ordinarily horizontal in position, one surface being upwards and the other downwards, these two layers are dissimilar in structure; but in those 'eaves where the lamina is vertical, as in the iris, they do not materially differ.

a. The whole structure is, of course, clothed with the epidermis.

262. The internal structure of the parenchyma is more complicated than would be at first supposed. A powerful microscope is necessary for its examination Let an exceedingly thin paring be taken from a vertical section of the lamina and submitted to the solar (or compound) microscope, in such a manner that the rays shall pass from section to section. Fig. 36 represents a magnified view of such a paring of the leaf of the lily, which may be regarded as characteristic of leaves in general.

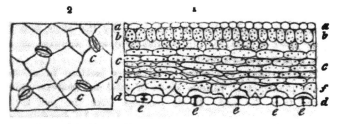

FIG. 36 — 1, Magnified section of a leaf of the lily ; 2, of the epidermis with stomata

263. The upper surface (*a, a*) is thus seen to consist of the flattened cells of the epidermis, arranged in a single layer. Just beneath this (*b, b*) is the more compact part of the parenchyma, consisting of a layer of oblong cells placed in such a position that their longer axis is perpendicular to the leaf's surface. Next below we meet with the parenchyma of the lower surface (*c, c*), composed of oblong cells arranged longitudinally, and so loosely compacted as to leave larger empty spaces between. Lastly, we find again the epidermis (*d, d*) of the under surface with stomata (*e, e*), opening into air-chambers.

a. Within all the vesicles of the parenchyma are seen adhering to the walls, the green globules (chlorophyll) which give color to the parenchyma,—dark green above, where it is more compact; but paler beneath, where the cells are more loose and separate.

264. The empty spaces between the cells, called intercellular, communicate with the external air by means of the stomata (37 — 39), which are generally found only in the *lower* surface. In those leaves, however, whose position is naturally vertical instead of horizontal, stomata are found equally on *both* surfaces. In other leaves, as in the Nymphæa, they are found upon the *upper* surface alone, the lower being in contact with the surface of the water.

265. The *vessels of the latex* (34) are distributed through the under layer of the parenchyma. These are prolongations of the ramified veins, which, having reached the edge of the leaf, double back upon themselves, pervade the lower surface, and are again collected into the petiole, through which they are finally returned into the bark.

266. A singular structure occurs in the Oleander of Barbary, and other plants of hot and arid regions. The epidermis on the upper surface is double and very compact, and there are few if any stomata on the lower surface, their places being supplied by cavities within its substance, opening outwards by a small aperture, and covered within by minute hairs. These peculiarities are adapted to the conditions of the air and soil in which the Oleander flourishes. The hairs absorb moisture from the air, which the cavities readily retain, while the double epidermis effectually restrains its evaporation.

§ 12. OF THE FUNCTIONS OF LEAVES

267. These are *exhalation, absorption, respiration,* and *digestion,* and the result of their combined action is the conversion of the *crude sap,* absorbed from the soil by the roots, into the *proper juice* or *latex,* for the nourishment and increase of the plant, with its various products.

268. The crude sap consists of water holding in solution minute quantities of various kinds of solid and gaseous matter derived from the soil. In its passage from the root to the leaves, its composition is somewhat modified by dissolving the previously formed secretions, which it meets with on its way.

269. EXHALATION is the process by which the superabundant water of the sap is given off to the atmosphere, so that the remaining sap is reduced, as it were, by *concentration,* and contains a greater proportion of solid matter. It is analogous to perspiration in animals.

270. It is to be distinguished from evaporation; the latter depending solely upon heat and the state of the air, and being, in plants, almost wholly restrained by the epidermis.

271. Exhalation appears to take place through the stomata

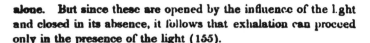

alone. But since these are opened by the influence of the l ght and closed in its absence, it follows that exhalation can proceed only in the presence of the light (155).

a. If a plate of glass be held near the *under* surface of an active leaf of the Hydrangea, in a still air, it will soon be covered with dew; but if the experiment be repeated by holding the glass over the *upper* surface, it will remain dry. Again, if the *light* be suddenly excluded from the plant in a state of active growth, it will immediately cease to transpire, whatever be the temperature; and if the stomata be then examined they will be found *closed.*

272. That exhalation and absorption by the roots are mutually dependent upon each other, has already been stated (155). The *quantity* of fluid discharged by the former may therefore be inferred from that of the latter. This has also been confirmed by experiment. A sunflower 3½ feet high, was ascertained by Hales to transpire from 20 to 30 oz. of water daily; a cabbage from 15 to 25 oz, &c. Experiments have also been made upon single leaves, recently plucked, with the petiole immersed in water. Thus a leaf of the sunflower, weighing 31 grains, absorbed and exhaled its own weight of water in 6 hours.

273. ABSORPTION is primarily the office of the roots (154), but in certain circumstances it is performed by the leaves also.

a. When the roots are imperfect, or wanting, or serve merely to fix the plant in its position, as in some aerial parasites, and in some of the Orchidaceæ, it is evident that the plant must derive its nourishment chiefly from the absorption performed by the leaves. Experiment also proves that the leaves of plants in general are capable of this function. Every one knows how plants, when parched and withered by drought, are revived by a shower which does not reach their roots, but only moistens their leaves.

274. The lower surface of the leaf appears to be chiefly instrumental in absorption. This is readily shown by experiment. Leaves with their lower surfaces in contact with the water, remain fresh much longer than others with their upper surfaces thus placed. Leaves of the white mulberry, with the upper surface only in contact with water, faded in six days, while others, reversed in position, lasted as many months.

275. RESPIRATION in plants is analogous to respiration, or *breathing,* in animals. In both it is equally constant and equally necessary. It is performed principally by the leaves, but is not confined to them, being partially performed by other parts also, even by the roots.

276. Respiration consists of the absorption of oxygen from the atmosphere, accompanied by the evolution of carbonic acid.

a. This process must not be confounded with another which occurs, of a contrary nature, treated of under the head of digestion.

277. Respiration appears to be going on constantly, by day and by night, during the life of the plant, even while it is actively engaged in the contravening process of the *fixation* of carbon. The result of it is, the removal of a certain superfluous portion of carbon, in a state of combination with oxygen,* from the nutritive substances of the plant, just as the same deleterious acid is removed from the blood of animals by breathing.

278. Let a few healthy plants be placed under a bell-glass containing air from which all the carbonic acid has been previously removed. After a few hours let the air be tested by shaking it with lime-water, and it will be found to contain carbonic acid, rendering the lime-water turbid. This effect will be produced, whether the bell-glass stand in the sunshine or in darkness, but the quantity of acid evolved will be found to be much greater in the darkness.

279. Respiration is carried on with peculiar activity during the two periods of *germination* and *flowering*.

a. In germination pure oxygen is absorbed, either from the air or water, or both, in the absence of light (133, d), and returned to the air combined with the superfluous carbon of the starch, which thus is converted into sugar for the nourishment of the young plant.

b. It is also equally active at the time of flowering, a large quantity of oxygen being converted into carbonic acid by the flower. By this process it seems that the starch previously contained in the disk (107), or receptacle (59), is changed into saccharine matter for the nutrition of the pollen and ovules (70, 81), the superfluous portion flowing off in the form of honey. And it has been ascertained that the quantity of oxygen evolved bears a direct proportion to the development of the disk.†

280. The life of the plant depends upon the continuance of respiration, for if it be surrounded by an atmosphere with too great a proportion of carbonic acid, or in a confined portion of air, which has become vitiated by its own action, and *excluded from the light*, its respiration is necessarily soon suspended, and it speedily perishes.‡

281. DIGESTION, in plants, consists properly of all these changes effected by the leaves in rendering the crude sap fit for the purposes of nutrition. But that process which is more par-

---

* Carbonic acid is composed of 6 parts (by weight) of carbon, combined with 16 parts of oxygen.

† Thus Saussure found that the flower of the Arum, while in bud, consumed 5 or 6 times its own volume of oxygen in 24 hours; during the expansion of the flower, 30 times, and during its withering, 5 times. When the floral envelopes were removed, he found that the quantity of oxygen consumed by the stamens and pistils in 24 hours, was, in one instance, 132 times their own bulk.

ticularly described under the head of digestion, consists in *the decomposition of carbonic acid by the green tissues of the leaves, under the stimulus of the light, the fixation of the solid carbon, and the evolution of pure oxygen.*

282. Carbon is one of the principal ingredients in the vegetable structure. The chief source from which plants obtain it is the atmosphere, which always contains it in the form of carbonic acid, evolved by combustion, by the respiration of animals, from the earth, &c.

a. 'Now if we place some fresh leaves in an inverted bell-glass, containing air charged with 7 or 8 per cent. of carbonic acid, and expose them to the direct light of the sun for a few hours, it will be found that a large proportion of the carbonic acid will have disappeared, and will be replaced by pure oxygen. But this change will not be effected in the dark, or by any degree of artificial light. Accordingly we find that plants which grow in the dark become *blanched* from the want of the proper supply of carbon, on which their green color depends.

283. We have before stated that this *fixation* of carbon in the substance of the plant, contravenes the process of respiration, in which carbon is *given off.* The former occurs only in the light of day, the latter by night as well as by day. But as to the *relative* amount of carbon thus absorbed by the former process, and evolved by the latter, there can be no reasonable doubt; for when we consider how large a portion of the tissues of every plant is solid carbon, and that too, derived chiefly from the atmosphere, it is evident that much more carbonic acid is, on the whole, consumed by vegetation than is evolved. In accordance with this are the results of the experiments of Dr. Daubeny, who has recently shown, that 'in fine weather, a plant, consisting chiefly of leaves and stems, if confined in a capacious vessel, and duly supplied with carbonic acid during sunshine, as fast as it removes it, will go on *adding to the proportion of oxygen present, as long as it continues healthy.'*

284. Thus are the two great kingdoms of nature rendered mutually subservient. each to the well-being, and even the existence, of the other. Animals require an atmosphere comparatively pure, although, by their respiration and decay, they are continually adding to the proportion of its deleterious gases. Plants, on the other hand, thrive by the decomposition of these gases and the restoration of pure oxygen to the air in their stead. It is impossible not to admire this beautiful arrangement of Providence, by which, as in a thousand other cases, the means and ends are rendered reciprocal, affording the highest proof of wisdom and design.

---

‡ Another view of respiration, different from the above, has been ably maintained; viz. that it is not a *vital action*, but only a necessary result of a *temporary suspension of* vital action. During the absence of the vivifying stimulus of the light, a part of the carbonic acid absorbed by day is lost, from the want of power to retain it, and a small quantity of oxygen is absorbed to recombine with some of the carbon recently set free. But as this theory does not account for the loss of carbonic acid by day as well as by night, and moreover supposes *imperfection* in the original design of the Creator, I have not yet seen fit to adopt it.

CHAPTER XIII.

INFLORESCENCE.

285. INFLORESCENCE is a term denoting the arrangement of the flowers upon a stem or branch.

286. In regard to *position* upon the stem, the inflorescence, like the leaf-bud, of which we have shown it to be a modification, is either *terminal* or *axillary*.

a. It is, however, in some plants, particularly in the potatoe tribe (Solanaceæ), situated *opposite* to a leaf. This irregularity is accounted for, if we suppose, with Lindley, that the flower-stalk, originating in the axil of the leaf next below, adheres to the internode (172) in its lower part, and does not separate from it until it is opposite the succeeding leaf.

287. The PEDUNCLE (flower-stalk) is that part of the stem on which the inflorescence is immediately supported. It bears no leaves, or, at most, only such as are reduced in size, and altered in form, called bracts (252). If the peduncle is wanting, the flower is said to be *sessile*.

288. The peduncle, like the stem of which it is a portion, may be either *simple* or *branched*. When it is simple it bears, of course, a single flower, but when it is divided into branches it bears several flowers, and its final divisions, each bearing a single flower, are called PEDICELS.

289. A SCAPE is a flower-stalk which springs from a subterranean stem, in such plants as are called stemless (177). Ex. Sarracenia, Taraxacum, Hyacinthus. Like the peduncle, of which it is a modification, it is leafless, or with bracts only, and may be either simple or branched.

290. The RACHIS (ραχις, the spine) is the *axis* of the inflorescence, or the main stem of a compound peduncle, along which the pedicels are arranged, as seen in the Plantago, currant, grape, and grasses.

291. The inflorescence is said to be *solitary* when it consists of a single terminal flower, as in Erythronium, or when but a single axillary flower is developed at the same node, as in Petunia, Convolvulus.

292. In regard to the evolution of the inflorescence, that is the mode of succession in the development of the flowers, bota nists have recently observed two important distinctions, namely, the *centripetal* and the *centrifugal*, the former resulting from axillary, and the other from terminal flowers.

293. In CENTRIPETAL inflorescence the evolution (blossoming) of the flowers commences with those of the *circumference* (or the base) and proceeds towards the *centre* (or the summit), as in the Umbelliferæ and the Cruciferæ.

a. The student will readily perceive that the *circumference* of a depressed (flat-tened, inflorescence corresponds to the *base* of a lengthened one; and also that the *centre* of the former answers to the *summit* of the latter. For when the axis, or rachis, is lengthened, it is the *centre* which it bears along with it at its apex, leav-ing the circumference at the base.

294. In CENTRIFUGAL inflorescence the blossoming com-mences with the terminal and central flower, and proceeds towards the lateral flowers, or those of the circumference. Ex. Hydrangea, elder, and the pink tribe

a. 'This mode of inflorescence is generally indicated by the presence of a soli-tary flower seated in the axils of the dichotomous (forked) branches.' All the flowers are considered terminal, because they do in fact (except the first which terminates the axis) terminate *lateral branches* successively produced at the node next below the primary flower. This is beautifully illustrated in Spergula.

FIG. 37. — Modes of inflorescence ; 1, centrifugal inflorescence (cyme) of Cerastium maximum ; 2, fascicle ; 3, centripetal inflorescence (corymb) ; 4, spike.

295. Sometimes we find these two modes of inflorescence combined in the same plant. In the Compositæ, as Dr. Gray remarks, the heads, which may be called the partial inflores-

1

cences, are centripetal, while the *general* inflorescence is centrifugal, that is, the central head is developed before the lateral ones. But in the Labiatæ the partial inflorescences (verticilasters, 309) are centrifugal, while the general inflorescence is centripetal.

296. Of centripetal inflorescence the principal varieties are, the spike, raceme, ament, spadix, corymb, umbel, head, panicle, and thyrse.

297. The SPIKE is an inflorescence consisting of several sessile flowers arranged along a common peduncle (rachis). Ex. Plantago, Verbascum.

298. The RACEME is the same as the spike, but having the flowers raised on pedicels, each being axillary to a bract, and blossoming in succession from the base upwards. The raceme may be either *erect*, as in Hyacinthus, Pyrola, or *pendulous*, as in the currant and black cherry.

299. The AMENT, or catkin, is a spike whose flowers are covered each with a scaly bract, instead of a calyx and corolla, and fall off together, all remaining still connected with the rachis. Ex. Salix, Betula.

300. The SPADIX is a spike with a fleshy rachis enveloped in a large bract, called spathe. Ex. Arum, Calla.

301. The CORYMB is the same as the raceme, having the lower pedicels so lengthened as to elevate all the flowers to nearly or quite the same level. Ex. wild thorn (Cratægus).

302. An UMBEL resembles the corymb, but the pedicels are of nearly equal length, and all arise from the same point in the common peduncle. Ex. Asclepias, Aralia hispida, onion.

303. A HEAD or CAPITULUM is similar to an umbel, but the flowers are sessile or nearly so upon the summit of the peduncle. Ex. button-bush, clover, globe-amaranth (Gomphrena).

a. But the more common kind of *capitulum* is that where the summit of the peduncle (*rachis*) is dilated into a broad disk (*receptacle*) bearing the sessile flowers upon its surface. This is the kind of inflorescence peculiar to the vast family of the Compositæ, and is equivalent to the *compound flowers* of the earlier botanists.

b. In the capitulum there is a general resemblance to the simple flower, the

•ays answering to petals, and the involucre (254) to the calyx. The flowers are called *florets*, those in the outer circle, *florets of the ray*, and those of the central portions, *florets of the disk.*

304. The PANICLE is a compound inflorescence, formed by an irregular branching of the pedicels of the raceme. Ex. oats, Poa, and many other grasses.

305. The THYRSE is the same as the panicle, having the lower branches rather shorter than those in the midst, and all of them very compact, as in the lilac (Syringa), horse-chestnut.

a. The umbel becomes *compound* when each pedicel becomes itself an umbel, as in most of the Umbelliferæ. In these cases the secondary umbels are called UMBELLETS, and sometimes *partial umbels.* See § 254.

By a similar decomposition, a raceme becomes a *compound raceme*, a corymb a *compound corymb*, &c.

FIG. 3? — Modes of inflorescence ; 1, raceme; 2, ament; 3, spadix ; 4, head ; 5, panicle 6, verticillaster; 7, thyrse.

306. Of the centrifugal inflorescence, the following varieties are described ; namely, cyme, fascicle, and verticillaster.

307. CYME. This inflorescence has the general aspect of the corymb, but is remarkably distinguished from it by its centrifugal evolution, and by its branches being repeatedly 2-forked and 3-forked, as exemplified in Hydrangea, Viburnum, chick weed.

a. The cyme is found only in plants with opposite leaves, and its normal structure and development are as follows ‘ The terminal flower, which is the first to be opened, is borne upon a peduncle of two or more nodes, which are, of course, transverse to each other (219, c). From one, or two, or all of these nodes, pairs of secondary, opposite peduncles arise, each of which, like the first, is binodal or multinodal, and terminated by a flower. Again, in the nodes of these secondary peduncles, may arise, in the same manner as before, pairs of tertiary peduncles, each to be terminated by a flower, and perhaps to bear still other peduncles, and so on.

b. Hence it is evident, that in each axil of the forked branches there should be a solitary flower. This, however, is often wanting. Irregularities may also be occasioned by the absence of other parts.

308. FASCICLE. This is a modification of the cyme, in which the flowers become crowded, and nearly sessile, as in sweet william, and other species of Dianthus.

309. VERTICILLASTER or VERTICIL, called also, though improperly, *whorl*, is a term denoting those reduced cymes which are peculiar to the Labiatæ, where two such cymes occupy the opposite axils of each pair of leaves.

a. Sometimes the peduncle, instead of producing flowers, is changed into a tendril, as in the vine.

CHAPTER XIV.

REVIEW OF THE PRINCIPLES OF NUTRITION.

310. It has already been shown, in the preceding chapters, that plants consist chiefly of *four simple organic elements;* viz. *carbon, oxygen, hydrogen,* and *nitrogen.* The first mentioned exists in a larger proportion, the last in a smaller, than either of the others. These four elements constitute about 94 *per cent* of all vegetable matter.

311. CARBON (essentially charcoal) enters so largely into the composition of plants, that it retains the exact form and texture of the wood after the other ingredients have been expelled by heat. On this element chiefly depends their solidity and strength. Its proportion is from 40 to 60 *per cent.* NITROGEN, although perhaps equally essential, is less abundant in the *tissues, and exists* largely only in certain important vegetable *products;* as *gluten,* legumine, albumen.

312. Besides these four universal elements, many other substances, earthy and mineral, are found in quantities greater or less in different species: thus forest trees and most other inland plants contain potassa; marine plants, soda, iodine, &c.; the grasses, silex and phosphate of lime; rhubarb and sorrel, oxalate of lime; the Leguminosæ, carbonate of lime.   Now all these ingredients, being found in plants, are inferred to be essential elements in the food which they require for healthy vegetation; and an inquiry into the sources from which they may be supplied, constitutes the chief object of *Agricultural Chemistry*.

313. It is evident that plants do not *create* a particle of matter, and therefore do not originate in themselves any of the ingredients which compose them; consequently they must obtain them from sources without.   These sources are obviously *air*, *earth*, and *water*.   Carbon is derived from the carbonic acid which the atmosphere contains, and from the decaying vegetable matter of the soil.   Oxygen is derived from the water, and from the carbonic acid of the atmosphere; hydrogen, from water and ammonia; and nitrogen, from ammonia alone, either drawn from the air or the soil.

314. The ATMOSPHERE contains about $\frac{1}{1000}$ part of carbonic acid, diffused throughout the whole extent; and, as this gas contains 27 *per cent.* of carbon, it may be demonstrated, that the whole atmosphere contains at least fourteen hundred billions of tons of solid carbon, derived from the sources mentioned in § 282, — an amount fully adequate to the vast and ceaseless drain made upon it by the vegetable kingdom.

315. SOIL consists of two classes of materials; viz. mineral and organic.   The former, called earths, consists of disintegrated and decomposed rocks, — all the various mineral substances which are found to enter into the composition of plants, as potassa, soda, silica, lime, &c., all of which are more or less soluble in water.   The organic materials consist of the remains of former tribes of plants and animals, mingled with the earths, which, having access to air, are decomposed, evolving carbonic acid and ammonia both to the air and the water.

316. WATER is composed of oxygen and hydrogen, in the proportion of 8 to 1 by weight.   Having pervaded the atmosphere

10

on the state of vapor and rain, and percolated through the soil, it holds in solution carbonic acid, ammonia, and many of the various minerals above mentioned.

317. AMMONIA consists of nitrogen and hydrogen, in the proportions of 14 to 3 by weight. It arises from decaying animal and vegetable matter, as above stated, and is also generated in the atmosphere, during storms, by the flashes of the electric fluid.

318 Thus it appears that the three compounds, water, carbonic acid, and ammonia, may yield to plants their four essential organic elements. And, since all of them are contained in the air, some plants are capable of subsisting on air alone ; but most species are dependent on water, earth, and air, and demand a copious supply. The external circumstances, therefore, first requisite to healthy vegetation are, —

1. Free access to an atmosphere which is often agitated by winds.

2. A proper supply of rain or river-water.

3. A soil possessing the peculiar minerals required by the species to be grown upon it, together with a certain proportion of vegetable mould.

319. The first of these is everywhere abundantly supplied by nature, and asks no aid from man. The second and third are often deficient, and are to be supplied by the labors of agriculture. By *irrigation*, streams of water are turned from their natural channels to add to the scanty moisture of fields parched with drought; while, by *drainage*, the inundated bog is converted into a luxuriant lawn.

320. The object of *tillage* is to pulverize and *lighten* the too compact soil, and thus expose every part to the oxygen of the air in order to hasten its decomposition. The object of *manuring* is mainly to increase the quantity of organic matter. By various *amendments*; as gypsum, lime, and pulverized charcoal, ammonia is powerfully attracted from the air, and yielded again to the water. Marl promotes the decomposition of the soil, and ashes add to the potassa which exists naturally in it being derived from the decomposition of the rocks which contain it, as *granite*, clay-slate, basalt, &c.

321. Soils are often improved by lying *fallow* for a season, thus allowing time to form by decomposition a fresh supply of that particular ingredient which had been exhausted by previous crops. On the same principle is explained the beneficial effects of a *rotation* of such crops as require different mineral substances in their composition.

322. But when all these materials have been supplied to the plant, still two other agents are requisite, without which the great work of vegetation will not go on. These life-giving principles are *light* and *heat*, both of which emanate in floods from the sun. Under their influence the *raw material* is received into the vessels of the plant, and assimilated to its own substance, — a process which can be fully comprehended only by HIM whose power is adequate to carry it on.

323. Under the influence of solar light, and a temperature above the freezing point, water is imbibed by the roots and raised into the tissues of the stem, dissolving, as it passes, small portions of gum or sugar previously deposited there. In this state it is crude sap. But passing on it enters the leaves, and is there subjected to the action of the *chlorophylle* (215. *a*), which chiefly constitutes the apparatus of digestion. Here it is concentrated by exhalation and evaporation, sending off quantities of pure water. Meanwhile the leaves are imbibing carbonic acid, decomposing it, retaining the carbon, and returning pure oxygen to the air.

324. Thus elaborated, the sap is now termed the PROPER JUICE, and consists of course of carbon and water, with a little nitrogen, and minute portions of the mineral substances mentioned above. From this juice are elaborated the *building material* of the vegetable fabric, and all its various products and secretions.

325. First, by the aid of light, *chlorophylle* is developed, clothing the plant in living green. Next *lignin* is produced, the peculiar principle of tissue, whether cellular, vascular, or woody, consisting of carbon with the exact elements of water, viz. oxygen and hydrogen.

326. Meanwhile gum, starch, and sugar, nutritive products common to all plants, are also developed from the *proper juice*,—

not all to be immediately employed in building up the tissues, but mostly to be stored away in reserve for future use. Such deposits are made in the root of the beet, tuber of the potato, and in the fruit of almost all plants. These three products, with lignin, are all composed of carbon with the elements of water, — gum and starch containing them in the same proportions.

327. Sugar is sometimes produced directly from the proper juice, as in the root of the beet, stalk of the maize and sugar-cane; but oftener, during germination, from the starch deposited in the seed. Its composition differs from that of starch, only in containing a larger proportion of the elements of water, or (what is the same thing) a smaller proportion of carbon. The trans-formation of starch into sugar appears to be dependent on the presence of a certain substance called "*diastase;* minute quan-tities of which exist in seeds, and about the *eyes* of the potato."

328. The similarity of these four *general products*, in chemical constitution, accounts for the facility with which they are con-verted into each other in the growing plant. Thus gum is converted into starch (in which state it is best adapted for pre-servation), and starch is converted into sugar (131). In flowering, sugar is rapidly consumed by the flower, — a portion of it being reconverted into starch, and deposited in the seed. Both gum and sugar appear to be converted into *lignin* during the growth of the tissues; and this substance, in the laboratory of the chemist, has been changed again into gum and sugar.

329. Among the numerous *secretions* of plants which our limits forbid us to consider, are the vegetable acids, containing more oxygen than exists in water; and the oils and resins, containing less than exists in water, or none at all These substances vary in the different species almos' to infinity, taking into their con-stitution, in addition to the four *organic elements*, minute portions of the mineral substances introduced by rain-water. Their peculiarities of flavor, odor, color, properties, &c. although so obvious to the senses, are occasioned by differences in constitu-tion often so slight as to elude the most delicate tests of the chemist.

CHAPTER XV.

SYSTEMATIC BOTANY.

§ 1. OF THE CLASSIFICATION OF PLANTS.

330. SYSTEMATIC BOTANY relates to the arrangement of plants into groups and families, according to their characters, for the purpose of facilitating the study of their names, affinities, habits, history, properties, and uses.

331. The student in botanical science is introduced into a boundless field of inquiry. The subjects of his research meet him at every step: they clothe the hill and the plain, the mountain and the valley. They spring up in the hedges and by the wayside; they border the streams and lakes, and sprinkle over its sur-face; they stand assembled in vast forests, and cover with verdure even the depths of the ocean; they are innumerable in multitude, infinite in variety. Yet the botanist proposes to acquaint himself with each individual of this vast king-dom, so that he shall be able readily to recognize its name, and all that is either interesting, instructive, or useful concerning it, whenever and wherever it is pre-sented to his view.

332. Now it is obvious, that if the student should attempt the accomplishment of this task by studying each *individual plant* in detail, whether with or without the aid of books, the longest life would scarcely be sufficient to make a begin-ning.

333. But such an attempt would be as unnecessary as fruitless. The Author of Nature has grouped these myriads of individuals into SPECIES (50). When he called them into existence in their specific forms, he endowed each with the power of *perpetuating its own kind and no other*, so that they have descended to us distinguished by the same differences of character and properties as at the begin-ning. When, therefore, the student has become acquainted with any one indi-vidual plant, he is also equally acquainted with *all others belonging to the same species.*

a. Thus a single stalk of *white clover* becomes a representative of all the mil-lions of its kind that grow on our hills and plains, and a single description of the *white pine* will answer, in all essential points, for every individual tree of that ancient and noble species, in all lands where it is found.

334. Again, the species themselves, although separated from each other by obvious differences, still are found to exhibit many constant affinities, whereby they are formed into larger groups, called GENERA (52). Thus the white clover and the red (Trifolium repens and T. pratense) are universally recognized as of different species, but of the same genus; and a single generic description of any

10*

one plant of the genus Trifolium will convey intelligence, to a certain extent concerning every other plant belonging to its 150 species.

335. Thus the whole vegetable kingdom is grouped into species, and the species themselves into genera. But natural affinities do not stop here. The genera are still too numerous for the purpose of clear and systematic study. The naturalist would therefore generalize still further, and reduce the genera to still fewer and larger tribes or groups. Accordingly he finds, on comparing the genera with each other, that they still possess some characters in common, although, perhaps, of a more general nature than those which distinguish them among each other. These general characters, therefore, serve to associate the genera into a systematic arrangement of Classes and Orders.

336. There are two independent and widely different methods of classifying the genera, which have generally been approved, namely, the Artificial System of Linnæus and the Natural System of Jussieu. The former has for its basis those characters which are derived from the organs of fructification, leaving all other natural affinities out of view. The latter, on the contrary, is founded upon all those natural affinities and resemblances of plants, by which Nature herself has distinguished them into groups and families.

337. In regard to the relative merit of these two systems, there is now no longer room for comparison. That of Linnæus is beautiful and ingenious, and, in the early stages of science, furnished the readiest means for determining the names of plants ; but this must be regarded as its principal use. Indeed, its author himself did not design it for any higher end, or claim for it any higher merit.

338. But, in acquiring a thorough and accurate knowledge of the vegetable kingdom, the Natural System is not only the best, but it is the *only* method which can be relied upon for this purpose. The obscurity and misconceptions which formerly embarrassed the science of the vegetable structure, so as to render this system unavailable, have now been so far removed by the labors of De Candolle and Lindley, in Europe, and of Drs. Torrey and Gray, of our own country, that it is brought generally within the scope of the ordinary mind, and shown to be founded in true philosophy. Accordingly, it is now generally adopted.

339. It was long supposed that no other arrangement than that of Linnæus could be practically useful in analysis, and consequently almost every local flora retains this arrangement even to this day. But the use of such an arrangement is objectionable, since it severs true affinities, confounds natural distinctions, often dispersing widely apart even the species of the same genus, thus preoccupying the young mind with views contrary to true science. In this work this system is wholly laid aside, and an attempt is made to prepare an analytical synopsis of the Flora, founded upon the Natural System, to some extent, indeed, artificial, yet exhibiting no *form* save that of the true system.

340. The *artificial arrangement* consists of *classes, orders, genera*, and *species.* The two latter are the same as in the natural system (50, 51), and the two higher divisions, classes and orders, have already been seen (74, 80) to be founded upon the number, situation, and connection of the stamens and pistils.

CHAPTER XVI.

OF THE NATURAL SYSTEM.

341. It is the aim of the Natural System to associate in the same divisions and groups, those plants which have the greatest - *general* resemblance to each other, not only in *aspect* and *structure*, but also in *properties*.

342. While the artificial arrangement employs only a *single* character in classification, the natural seizes upon *every* character in which plants agree or disagree with each other. Thus, those plants which correspond in the greatest number of points will be associated in the smaller and lower divisions, as species and genera, while those corresponding in fewer points will be assembled in divisions of higher rank.

343. By an acquaintance, therefore, with the characters of each of the families of the Natural System, we may at once determine to which of them any new plant belongs, what are its affinities with others, and what are its poisonous or useful properties.

344. Although the aim of this System is as above stated, yet the *full consummation* of it is still reserved for a future age. At present, though greatly advanced, we are still obliged to call in the aid of artificial characters, where Nature is as yet too profound for ordinary skill. Such aid is, for example, employed in the first subdivision of Angiosperms.

345. The first and highest division of the vegetable kingdom, namely, into the *Phænogamia* or *Flowering Plants*, and the *Cryptogamia* or *Flowerless Plants*, has already been noticed, and its distinctions explained, in Chapter III, and elsewhere. These grand divisions lie at the foundation of both the System of Linnæus and of Jussieu, and are truly founded in nature; for

The PHÆNOGAMIA
    1. Consist of a regular axis of growth with leafy appendages.
    2. They possess a woody and vascular structure.
    3 They develope flowers, and
    4. They produce seeds.　On the other hand
The CRYPTOGAMIA
    1. Are destitute of a regular axis and of true leaves.
    2. They possess a cellular structure only.
    3. They do not develope flowers, and
    4. They produce SPORES (129) instead of seeds.

346. These distinctive characters must not, however, be regarded as decisive in all cases; for the higher Cryptogamia, as the ferns, give indications both of a regular woody axis and of a vascular structure, while some of the lower Phænogamia can scarcely be said to produce flowers.　And, universally, so gradual are the transitions from family to family and tribe to tribe, that it is impossible to fix upon characters so definite as to completely circumscribe any one group, while at the same time, they exclude every member of surrounding and approximating groups.

347. There is a small and curious order of plants of comparatively recent discovery, native chiefly of the East Indies, which appear, from the most authentic accounts of them, to form the connecting link between the Flowering and Flowerless plants, combining a part of the characters of each, so that botanists are at a loss to which it belongs.　They possess a cellular structure, develope flowers immediately from the root, whence they are called Rhizanths (ῥίζη, a root, ανθοc, a flower); but their ovaries are said to be filled with spores instead of seeds, and hence they are also called Sporogens.　Ex. Rafflesia.

348. Again, the Phænogamia are very naturally resolved into two subdivisions, depending upon their manner of growth, called EXOGENS and ENDOGENS, whose distinctions are briefly as follows: —

EXOGENS,
    1. Growing by external accretions (196).
    2. Bearing leaves which have reticulated veins (229) and which fall off by an articulation.
    3. Seeds with two or more cotyledons (137) or *dicotyledonous,*

Endogens,
1. Growing by internal accretions (197).
2. Leaves parallel-veined (229) and decaying without falling off.
3. Seeds with one cotyledon (126) or *monocotyledonous.*

349. *Classes.* The groups above mentioned, comprising the whole vegetable kingdom, are again subdivided into six classes. The first two are formed from the subdivision Exogens, and are founded upon the presence or absence of the pericarp; namely

Class I. ANGIOSPERMÆ, (as the oak, rose,)
1. Ovules produced within an ovary, and
2. Fertilized by the action of the pollen through the stigma.
3. Becoming seeds enclosed in a pericarp.
4. Embryo with two opposite cotyledons.
Class II. GYMNOSPERMÆ, (as the pine, yew,)
1. Ovules produced naked beneath a scale-like carpel.
2. Fertilized by the direct action of the pollen without the stigma.
3. Becoming truly naked seeds, that is, destitute of a pericarp.
4. Embryo mostly with several whorled cotyledons.

350. The next two classes are formed from the subdivision Endogens, and are founded upon the presence and absence of glumes or husks; namely,

Class III. AGLUMACEÆ or AGLUMACEOUS ENDOGENS,
Plants of the endogenous structure with flowers constructed on the usual plan; perianth verticillate, of one or more whorls of petaloid organs, or wanting. Ex. lily, orchis, rush.
Class IV. GLUMACEÆ or GLUMACEOUS ENDOGENS,
Plants of the endogenous structure, the flowers invested in an imbricated perianth of glumes instead of a calyx; as the grasses, grains sedges.

351. The Cryptogamia are separated into two great classes, called Acrogens and Thallogens; the former including those tribes which make some approximation towards the Phænogamia, and the latter including the lowest tribes of the vegetable kingdom. As their names indicate, they are distinguished from each other by their manner of growth; thus,

Class V. ACROGENS (growing from *acpoc,* the summit or point) have a regular stem, or axis, which grows by the extension of the point, or apex only, without increasing at all in diameter, generally furnished with leaves, and composed of cellular tissue and ducts Ex. ferns, mosses, club-mosses, and the Equisetaceæ.

Class VI. THALLOGENS, consisting merely of cellular tissue, with a tendency to grow into a flat expansion called *thallus*, but having no distinction of root, stem, leaves, or flowers. Ex. Lichens, seaweeds, liverworts, fungi.

352. *Affinities of the Six Classes.* These may be represented to the sight by the following arrangement·

Angiosperms.

Gymnosperms.   Aglumaceæ.

Acrogens.   Glumaceæ.

Thallogens.

Angiosperms stand in the highest rank, as they justly merit, by their superior organization. These are nearly allied to Gymnosperms by their mode of growth; and, on the other hand, to Aglumaceæ by their mode of flowering. Gymnosperms are intimately connected with Acrogens through Equisetaceæ of the latter, which stands intermediate; and the Aglumaceæ approach the Glumaceæ, almost indefinitely, through the Junceæ (rushes). Between the Acrogens and Thallogens a close relationship is established through the Musci (mosses), while the *sporogens* form the connecting link between the Endogens and the lowest tribes of vegetation, as the Fungi. Thus, from the highest rank we descend to the lowest, through Gymnosperms and Acrogens on the one hand, and through Aglumaceæ and Glumaceæ on the other, forming a *circle* of affinities.

353. The mutual relations of the six classes with the higher divisions, are presented in the following synopsis:

VEGETABLE KINGDOM;	PHÆNOGAMIA;	EXOGENS;	Class I. ANGIOSPERMS.
			Class II. GYMNOSPERMS.
		ENDOGENS;	Class III. AGLUMACEOUS.
			Class IV. GLUMACEOUS.
	CRYPTOGAMIA; ··········		Class V. ACROGENS.
			Class VI. THALLOGENS.

354. SUB-CLASSES. The classes are next to be broken up into smaller divisions. In effecting this object most writers have employed artificial methods, since no natural one, founded upon clear and comprehensive distinctions, has yet been devised. Thus Angiosperms, which class is by far the largest of the six, is divided into three *sub-classes*. POLYPETALÆ, or POLYPETALOUS EXOGENS, flowers with distinct petals; MONOPETALÆ, or MONOPETALOUS EXOGENS, flowers with united petals; APETALÆ, or APETALOUS EXOGENS, flower with no floral envelopes, or with a calyx only.

355. ORDERS, or FAMILIES, are the most important of all the natural associations. On the accuracy and distinctness of the

characters of *these*, botanists have bestowed the highest degree of attention, and the student's progress will depend chiefly upon his acquaintance with them.

356. Orders are formed by associating together those genera which are the most nearly allied *to each other*, or to some *one genus* previously assumed as the *type*. Therefore, as the species form genera, so genera form orders.

357. In systematic works, the orders are also associated on natural principles into alliances, groups, &c., which are intermediate between these and the sub-classes, and are designated numerically, thus, group 1st, group 2d, &c., or by names derived from a leading order.

358. In regard to their extent, the orders differ very widely, some consisting of a single genus, as Sarraceniaceæ, while others comprehend hundreds of genera, as Compositæ. For convenience' sake the larger orders are broken up into sub-orders, or tribes.

359. The Natural System, with its classes and subordinate divisions, may be exhibited in one view;

The VEGETABLE KINGDOM is separated

    1st, into Grand Divisions and Subdivisions.
    2nd, " Classes.
    3d, " Sub-classes, Alliances, and Groups.
    4th, " Orders and Sub-orders.
    5th, " Genera and Sub-genera.
    6th, " Species and Varieties, and
    7th, " Individuals.

CHAPTER XVII.

§1. NOMENCLATURE.

360. THE names of the *Orders* are Latin adjectives, (feminine, plural, to agree with *plantæ*, plants, understood,) usually derived from the name of the most prominent, or leading genus, in each, by changing or prolonging the termination into *aceæ*, as *Rosaceæ*, the rose tribe, *Papaveraceæ*, the poppy tribe, from Rosa and Papaver.

a. Earlier names, however, derived from some leading character in the Order, and with various terminations, are still retained. Thus, *Compositæ*, with compound flowers; *Labiatæ*, with labiate flowers.

361. *Generic* names are Latin substantives, arbitrarily formed, often from some medicinal virtue, either supposed or real, or from some obvious character of the genus; sometimes from the native country of the plants, or from the name of some distinguished botanist, or patron of botany, to whom the genus is thus said to be dedicated. Also the ancient classic names, either Latin or Greek, are often retained. Examples of all these modes of construction will be hereafter seen.

362. *Specific* names are Latin adjectives, singular number, and agreeing in gender with the name of the genus to which they belong. They are mostly founded upon some distinctive character of the species; as *Gerardia glauca*, glaucous-stemmed Gerardia; *G. purpurea*, purple-flowered Gerardia; *G. tenuifolia*, slender-leaved Gerardia. Frequently the species is named after some other genus, which, in some respect, it resembles; as *Gerardia quercifolia*, oak-leaved Gerardia. *G. delphinifolia*, larkspur-leaved Gerardia.

363. Species, like genera, are also sometimes named in commemoration of distinguished persons. The rules given by Lindley, for the construction of such names, are, 1st, If the person is the discoverer, the specific name is a substantive in the genitive case, singular number; as, *Lobelia Kalmii*, Kalm's Lobelia; *Pinus Fraseri*, Fraser's pine. 2d, If the name is merely conferred in honor of the person to whom it is dedicated, it is an adjective ending in *nus, na, num;* as *Erica Linnæana*, Linnæus's heath; *Rosa Lawrenciana*, Miss Lawrence's rose. In these cases, and in all others where the specific name is derived from proper names, or where it is substantive, as it often is, it should begin with a capital letter.

§2. BOTANICAL ANALYSIS.

364. The application of the rules of Systematic Botany to the natural plant, in order to ascertain its affinities, place, name, &c. is called *botanical analysis*.

365. In order to be in a proper state for this kind of examination plants should be in full blossom, and fresh, that is, not with-

ered or decayed. A good lens is requisite for the examination of the minute parts of the structure, or of the flower.

366. The analysis of plants is a constant object of pursuit with the practical botanist. Without this exercise, the study of authors will be of little avail. A more accurate and useful knowledge of a plant can be acquired in a few minutes, by a careful examination of the living specimen, or even of the dried, than by committing to memory the most elaborate descriptions found in books. During the flowering months, the learner will often in his walks meet with plants in blossom, with which he is yet unacquainted. And he who is duly interested in his pursuit, will by no means fail to seize and analyse each specimen while the short hour of its bloom may last, and to store his memory with the knowledge of its names, habits, and uses. Thus, in a few seasons, or even in one, he will have grown familiar with nearly, or quite, every species of plants in his vicinity.

367. Let us now suppose the pupil in possession of a specimen of an unknown plant in full blossom. In order to study it by the aid of authors, a point immediately requisite is its name. Now, having learned by examination the organic and physiological structure of the flower, leaves, stem, &c., the experienced botanist, who has at his command the characters of all the Natural Families, will at once determine to which of them the plant belongs.

368. But this is not to be expected of the pupil who is supposed to be yet, in a measure, unacquainted with the characters of the orders. He must be guided to the place which his specimen holds in the classification, by a longer course of inquiry and comparison. For the assistance of the learner, therefore, and for the convenience of all, we are happy to be able to add a full series of ANALYTICAL TABLES, which, with proper use, will seldom fail of conducting them almost immediately, to the object of their research. See the directions.

§ 3. OF COLLECTING AND PRESERVING PLANTS

369. The student in botanical science should give an early and persevering attention to the collection and preservation of specimens of as many species of plants as he can procure. The advantages to be derived from such collections, either in refreshing the memory by reviewing them, or in instituting a more thorough examination at one's leisure, are such as will afford an abundant compensation for all the labor requisite in preparing them.

a. Such a collection of dried specimens of plants is called an HERBARIUM, or by the more significant title, hortus siccus (dry garden).

370. The apparatus requisite for the accompaniment of this object is, 1st, a close tin box, 20 inches in length, and of a portable form; 2d, a portable press, consisting of two boards of light material 12 by 18 inches, opening and shutting by hinges, like the cover of a book, and secured by springs (even a large book is a good substitute); 3d, a quantity of smooth, bibulous paper, of large size (a dozen or more quires of printing paper); 4th, eight or ten boards of the same size as the paper; 5th, a small screw-press, or several lead weights of various sizes, from 15 to 30 pounds each.

11

371. In gathering plants for this purpose, or *specimens*, as they are called, the smaller and herbaceous plants should be taken up with a portion of the roots, while from larger plants there should be selected a shoot, with complete representations of the leaves and flowers. They may be preserved for several days, without withering, in the tin box, or they may at once be laid between several thicknesses of the paper, and enclosed in the portable press. It is always desirable that they be gathered in a dry day; if not, they should be freed from dampness before being committed to the paper and press.

372. In drying the specimens, great care is required, that they may preserve well their natural appearance, form, and color. It is generally recommended that they be carefully spread out, as nearly in their natural position as possible, between 8 or 10 thicknesses of paper, and then submitted to pressure between the boards. The degree of pressure should never be such as to crush their parts, and may be easily regulated by the screw, or by the number and size of the weights used. Cotton batting may be used to equalize the pressure.

373. As often as once a day they should be taken from the press, transferred to fresh and dry paper, and returned, until they are thoroughly dried, when they are ready to be transferred to the cabinet. The true secret of preserving specimens *with all their colors* is to extract the moisture from them by pressure in an abundance of dry, bibulous paper *as soon as possible*.

374. The next object with the collector is the arrangement of his specimens. For this purpose, each one is first to be fastened to a sheet of firm white paper, about 10 inches by 18, either by glue or with loops of paper of the same kind, or they may be stitched to the paper with a fine needle. The latter mode, if done skilfully, is preferable. Then let all those specimens which belong to the same genus be collected together and placed within a folded sheet of colored paper, with the name of the genus and each species written on the outside. Each sheet should also be labelled with the names of the plant, the locality, time of gathering, habits, &c.

375. The genera are next to be collected together into orders, each order being wrapped or folded in a still larger sheet, of a different color from that which enfolds the genera, having the name of the order, with a catalogue of its genera on the outside. Thus arranged, the orders are to be laid away upon the shelves of a cabinet, or packed in a chest. To protect the plants from the attacks of insects, pieces of camphor gum are to be placed among them, or a piece of sponge saturated with the oil of turpentine. To save them from decay, they should be kept dry, and well ventilated.

376. Fruits and seeds which are too large to be pressed with the plants, and also truncheons of wood, are to be preserved separately, in a cabinet

INDEX AND GLOSSARY.

———

.•. The figures refer to paragraphs.

A ; (α privative) in composition signifies without.

Abortion ; an imperfect development of any organ. Abbreviations, 123.

Absorption, 157, 158, 272, 273, 274.

Acaulescent, 194.

Accessory ; something added to the usual parts.

Accretion ; the growing of one thing to another.

Accumbent ; lying upon. In the Cruciferæ it denotes the radicle lying upon the edges of the cotyledons.

Acerose, 234, 26.

Achenium, 116, 9.

Achlamydeous, 54.

Acicular ; needle-shaped.

Acine ; a separate grain or carpel of a collective fruit.

Acotyledonous, 48.

Acrogens, 351.

Aculeate ; armed with prickles.

Acuminate, 236, 3.

Acute, 236, 1.

Adherent, 97.

Adnate ; growing to or upon, 69, 2.

Æstivation, 108.

Aggregate ; assembled closely together.

Agramaceous, 360.

Ala, 105, 5.

Alburnum, 203.

Albumen, 122.

Alternative, 108, 5.

Alveolate ; with partitions like a honey-comb.

Ament, 299.

Amplexicaul, 222, 3.

Anastomosing ; the uniting of vessels ; inosculating.

Anatropous, 121.

Anciptal ; two-edged

Androecium, 57, 65.

Androgynous ; with both stamens and pistils.

Angiosperms, 349.

Anthelmintic ; expelling or killing worms

Animal, definition of, 11.

Antiseptic ; efficacious against putrefaction.

Anther, 68.

Apetalæ, 354. Apetalous, without petals.

Appressed ; pressed closely upon something else.

Apterous ; without wings (or margins).

Aquatics ; growing in or belonging to the water.

Arachnoid ; 41, a.

Arboreous ; tree-like.

Arborescent ; belonging to a tree.

Areolæ ; having the surface divided into little spaces, or areas.

Aridity ; dryness.

Aril, 119.

Aristate ; bearded, as in the glumes of barley.

Armed, 235, 7.

Aroma ; the spicy quality of a thing.

Articulation ; a joint ; the place where one thing is joined to another.

Artificial Classes, 73.

Artificial Orders, 80.

Ascidia, 245.

Ascending ; arising obliquely, assurgent.

Assurgent ; arising in an oblique direction.

Attenuate ; rendered slender or thin.

Auriculate, 232, 9.

Awn, 256.

Axil (arm-pit); the angle between the petiole and branch, on the upper side

Axillary ; growing out of the axila.

Axis, ascending, 19.

Axis, descending, 19.

Baccate; berry-like, covered with pulp.
Banner, 105, 5.
Bark, 205.
Beak; a hard, short point, like the beak of a bird.
Bearded; with long awns or hairs.
Berry, 116, 14.
Bicuspidate; with two points.
Bidentate; with two teeth.
Biennial; of two years' duration.
Bifid; two-cleft.
Bifoliate; with two leaves.
Bilabiate; two-lipped.
Bifurcate; two-forked.
Binate; growing two together.
Bipinnate, 240, 5.
Bipinnatifid; twice pinnatifid.
Bisaccate; with two tumors or sacks.
Biternate, 240, 7.
Bivalved; two-valved.
Botany defined, 1.
Brachiate; with opposite spreading branches (arms).
Bracteate; having bracts.
Bracteolæ; little bracts.
Bracts, 252.
Branchlets; small branches.
Branch, 170.
Bristles; rigid hairs.
Bud, 20, 22. 165—169.
Bulb, 178.
Bulbiferous, 178, e.
Bulblets, 178, c.
Bulbous; having bulbs.

Caducous, 98.
Cæspitose; turfy, growing in tufts.
Calycine; of a calyx.
Calyculated; having bracteoles resembling an external or additional calyx.
Calyptra; (an extinguisher) applied to the cover of the theca of some mosses.
Calyx, 55, 95.
Cambium, 207.
Campanulate, 104, 1.
Campylotropous; denotes that the ovule is curved upon itself.
Canaliculate; channelled, or furrowed.
Canescent; hoary, approaching to white.
Capillary; very slender, hair-like.
Capitate; growing in a head.
Capsule, 116, 1.
Carina, 105, 5. Carinate, keel-shaped.
Caryopsis; a small, 1-celled, indehiscent pericarp, adhering to the seed which it encloses, as in the grasses. 116, 8.
Carpels, 77.
Carpophore; the axis of the fruit in the Umbelliferæ.
Cartilaginous; gristly.
Caryophyllaceous, 105, 4.
Cathartic; purgative.
Catkin, 299.
Caudate; with a tail-like appendage.
Caudex, 142, a.

Caulescent, 184.
Cauline, 220.
Caulis, 184.
Cellular; composed of cells.
Cellular tissue, 29.
Cellulares, 47.
Cernuous; nodding.
Chaffy; with chaff like processes.
Chalaza, 91.
Chemical basis of vegetable tissue, 28.
Chlorophyll, 215, a.
Chromulæ; green coloring-matter or particles.
Ciliæ; hairs like those of the eyelash.
Ciliate, 41, a.
Circinate, 217, 7.
Circumscissile, 115, 5.
Cirrhose, 240, 2.
Clavate; club-shaped.
Claw, 102.
Climbers, 187.
Cochleate; resembling the shell of a snail.
Cohering; connected.
Collum, 141.
Columella, 116, a.
Colored; not green.
Columnar; formed like columns.
Column; the consolidated stamens and pistils of Orchidaceæ.
Coma, 118, a.
Commissure; the inner face of the carpels of Umbelliferæ.
Compound leaves, 238.
Comose; a kind of inflorescence, having a tuft of sessile bracts on the top of it.
Compound leaves; consisting of several leaflets.
Compressed, 222, 1.
Concave; hollow.
Concentric; points or lines at equal distance from a common centre.
Concrete; hardened, or formed into one mass.
Confluent; running into one another.
Conjugate; joined in pairs.
Connate; joined together at the base, 242, 4.
Connectile, 68, b.
Connivent; converging.
Conoid; like a cone.
Contorted; 108, 4, twisted.
Convolute, 108, 2.
Convex; rising spherically.
Coral Islands, 12, c.
Cordate, 234, 25.
Coriaceous; leathery, thick, and tough.
Corm, 179.
Cornute; horned.
Corolla, 56, 100.
Corona (a crown); the expanded cup like disk of the Narcissus, &c.
Corymb, 301.
Corymbose; arranged like a corymb
Costate; ribbed.

Cotyledon, 125.
Cotyledonous plants, 48.
Creeper, 182.
Crenate, 235, 4.
Crenulate, 235, 4.
Crisped, 235, 10.
Cruciform, 105, 1.
Cryptogamia, 345.
Cucullate; hooded, cowled.
Culm, 156, a.
Cultivation, effects of, 15.
Cuneate; wedge-shaped.
Cupule; the cup, or involucre, of the acorn, and of all amentaceous plants.
Cuspidate; like the point of a spear. A leaf is cuspidate when suddenly contracted to a point.
Cuticle; the epidermis; scarf-skin.
Cyathiform; cup-shaped; concave.
Cylindraceous; like a cylinder in form.
Cyme, 307. Cymose, like a cyme.
Cyanic, of the blue series; i. e. white, red, blue any color save yellow or ochroleous.
Decandrous; with 10 stamens.
Deciduous, 98.
Declinate; turned towards one side.
Decompound; more than once compounded, as bipinnate, &c.
Decumbent; lying down, or leaning on the ground.
Decurrent, 242, 3.
Decussate; crossing each other at right angles.
Deflexed; bent downwards.
Defoliation, 259.
Dehiscence, 68, a, 115.
Deltoid; shaped like the Greek letter Δ.
Dentate, 235, 2.
Denticulate, 235, 2.
Depressed; pressed inward or flattened from above.
Diandrous; with two stamens.
Diadelphous; having the stamens united in 2 sets.
Diaphanous; transparent.
Dichotomous; branching by two equal divisions; forked.
Diclinous; (stamens and pistils) in separate flowers.
Dicotyledonous plants, 127.
Didymous; two united.
Didynamous; having two long stamens and two short ones in the same flower.
Diffuse; wide-spread, scattered.
Digestion, 281.
Digitate, 233, 18.
Digynous; with two pistils.
Dioecious; bearing staminate flowers on one individual, and pistillate on another.
Discoid; in the Compositæ, when the flowers are all tubular in the same head.
Disk, 107, b; also, the centre of a head in the Compositæ.
Dissected; cut into 2 parts.

Dissepiment; the partitions by which the cells of the pericarp are separated.
Dissemination of seeds, 135.
Distichous; leaves or flowers in two opposite rows.
Distinct, 63, e.
Divaricate; spreading in a straggling manner.
Dodecandrous; having twelve stamens
Dorsal, 54 (on the back).
Drupe, 116, b.
Ducts, 33, f.
Duramen, 203.

Echinate; beset with prickles.
Elementary organs, 29, &c.
Elliptical, 232, 2.
Elongated; exceeding the common length
Emarginate, 236, 4.
Embryo, 123, 124, 130.
Emollient; softening.
Endocarp, 112.
Endogenous structure 210, 211.
Endogens, 126, 197, 348.
Endopleura, 118.
Endosmose, 156, a.
Endostome; inner mouth or perforation.
Ensiform; sword-shaped, two-edged.
Entire, 235, 1.
Epicarp, 112.
Epidermis, 35.
Epigynous, 107, b.
Epiphytes, 150, b.
Equitant, 217, 1.
Erose, 235, 5.
Esculent; eatable.
Etiolated; blanched or whitened.
Exhalation, 269, 271.
Exogenous structure, 198, 199, &c.
Exogens, 127, 196, 348.
Exosmose, 156, a.
Exotic; foreign; not native.
Exserted; projecting or extending out of the flower or sheath.
Exsiccated; dried up.
Exstipulate, 251.
Extrorse, 68, 4.

Fæcula; the nutritious part of wheat and other fruits.
Falcate; sickle-shaped; linear and curved
Farinaceous; mealy.
Fascicle, 306.
Fasciculated, 146, a.
Fastigiate; having a flat or level top.
Favose; deeply pitted.
Feather-veined, 230, 1.
Febrifuge; efficacious against fever.
Fecundation; the act of making fruitful
Ferruginous; iron-colored; rusty.
Fibrils, 142, b, 152.
Fibro-vascular tissue, 260.
Fibrous, 146.
Filament, 67.
Filiform; shaped like a thread.

Fimbriate; fringed.

Fistular or fistulous; tubular.

Flabelliform; fan-shaped.

Flexuous; bent in an undulating manner.

Floating root, 149.

Floral envelopes, or perianth, 54.

Floral leaves, 232.

Florets, 303, b.

Flosculous; consisting of many tubular monopetalous flowers, or florets.

Flower, origin of, 24.
   "   consists of, 53.
   "   physiological structure of, 106.
   "   normal structure of, 61, c.

Flower-bud, 166.

Foliaceous; having the form of leaves.

Follicle, 116, 5.

Foot-stalks; the stalks of either flowers or leaves.

Foramen, 90.

Fork-veined, 229, 3.

Free, 97.

Free central placenta, 68.

Fringed; having a border like a fringe.

Frond; the leaves of the ferns, palms, &c. have been generally so called.

Fruit, 109, 110.
   "   growth of, 113.
   "   ripening of, 114.
   "   consists of, 111.

Frutescent; shrubby.

Fugacious, 257, 1.

Fungous; of the substance of the Fungi.

Funiculus, 91.

Furcate; forked.

Fusiform, 145.

Galea; (104, 5) the arched upper lip of a labiate flower.

Geminate; doubled.

Genus, 52.

Germ; the old name of the ovary.

Germination, 130 — 133.

Gibbous; swelled out, protuberant.

Glabrous, 237, 1.

Glands, 44.

Glandular fibre or tissue, 31.

Glaucous; sea green; pale bluish green with a powder or bloom.

Globose; round or spherical.

Glossology, 4.

Glumaceæ, 350.

Glume, 256.

Granular; 147, b, formed of grains or covered with grains.

Gregarious; herding together.

Grooved; furrowed or channelled.

Groups, 357.

Gymnosperms, 349.

Gynandrous; having the stamens and styles combined in one body.

Gynœcium, 58.

Hairs, 41.

Hastate 232, 10.

Habit; the general aspect or external features of a plant, by which it is known at sight.

Head, 303.

Helmet or Galea, 104, 5.

Herb, 164, c.

Herbarium, 369, 370.

Heterogamous; flowers not all perfect some being neutral or pistillate.

Hexandrous; having six stamens.

Hilum, 120.

Hirsute, 41, a.

Hispid; rough, with stiff hairs.

Hoary, 237, b.

Homogamous; flowers all tubular, similar and perfect, as in some of the Compositæ.

Homogeneous; having a uniform nature or composition.

Hooded; curved or hollowed at the end into the form of a hood.

Hot springs, 12, a.

Hyaline; crystalline, transparent.

Hybrid; partaking of the nature of two species.

Hypocrateriform, 104, 3.

Hypogynous, 107, b.

Imbricate; placed one over another, like the tiles upon a roof, 106, 8.

Incised, 235, 8.

Incrassated; becoming thicker by degrees.

Indehiscent, 115.

Indigenous; native of.

Induplicate, 108, 7.

Incumbent; lying against or across. In the Cruciferæ it denotes that the radicle is applied to the back of one of the cotyledons.

Indusium; the membrane that encloses the theca of ferns.

Inferior, 97.

Inflated; tumid and hollow, blown up like a bladder.

Inflexed; bending inward.

Inflorescence, 265, &c.
   "   centripetal, 2&c.
   "   centrifugal, 294.

Infundibuliform, 104, 2.

Innate, 68, 1.

Inserted into; growing out of.

Integument, 118.

Intercellular passages, 39.

Internode, 172.

Introrse, 69, 4.

Involucel, 254.

Involucre, 254.

Involute, 217, 3.

Irregular; unequal in size or figure.

Keel, 105, 5.

Kidney-shaped, 232, 12 (reniform).

Labellum, 107 a.

Labiate, 104, a.
Laciniate, 233, 20.
Lactescent; milky or juicy.
Lamina, 102.
Lanate; woolly.
Lanceolate, 232, 5.
Lateral; relating to the side.
Latex, 265.
Laticiferous tissue, 34, 267, c.
Leaf consists of, 27.
  " form of, 231.
  " color of, 215.
  " margin of, 235.
  " surface of, 237.
  " functions of, 267.
  " duration of, 257
Leaf-bud, 167.
Leaflets, 239.
Leaves, arrangement of, 218.
Legume, 116, 4.
Leguminous; having legumes.
Lenticular; lens-shaped.
Liber, 205, 206.
Ligneous; woody.
Ligula, or ligule; the membrane at the top of the sheath of grasses, &c.
Ligulate; strap-shaped.
Liliaceous, 105, 3.
Limb, 103.
Linear, 234, 23.
Linnean Classes, 73, 74.
Linnean Orders, 80.
Loculicidal, 115, 1.
Loment; a jointed legume.
Lunate; crescent-shaped.
Lyrate, 232, 14.

Marescent; withering on the plant.
Marginal; on the margin.
Medulla; pith.
Medullary rays, 204.
Medullary sheath, 200.
Membranous, or membranaceous; with the texture of membrane.
Mericarp; half-fruit.
Mesosperm, 118.
Midrib, 226. Midvein, 226.
Mineral defined, 9.
Monadelphous; stamens all united.
Monandrous; with one stamen.
Moniliform, 147, b.
Monocotyledonous plants, 126.
Monœcious; stamens and pistils apart, in separate flowers on the same plant.
Monopetalæ, 354.
Monopetalous, 101.
Monosepalous, 96.
Mucronate, 236, 6.
Multifid; many-cleft.
Muricate; with hard short points.

Naked ovules, or seeds, 111, a.
Napiform, 145, a.
Narcotic; producing sleep or torpor.
Natural System, 341.

Nectariferous; producing honey.
Nectary, 107, a.
Nerves, 227.
Net-veined, 229.
Nodding; in a drooping position.
Node, 172.
Normal; regular, according to rule.
Normal structure of plants, 61.
  " causes of deviation from, 63.
Nucleus, 90.
Nut, 116, 7. Nutrition, principles of, 310.

Ob, in composition implies inversion, as obovate, inversely-ovate, &c.
Oblong, 232, 3.
Obovate, 232, 6.
Obvolute; 217, 2.
Obsolete; indistinct, as if worn out.
Obtuse; blunt. Ochroleucous, yellowish
Octandrous; with eight stamens. [white.
Octogynous; with eight styles.
Officinal: used in or belonging to the shops.
Offset, 191.
Oleaginous, oily.
Operculum; the lid to a pyxis, &c.
Opposite, 218, 3.
Orbicular, 232, 1.
Orders, 355.
  " names of, 360.
Ordinal; relating to the Orders.
Organic bases, 28.
Organography, 2.
Orthotropous, 121.
Oval, 234, 24.
Ovary, 76, 77.
Ovate; egg-shaped (surface), as a leaf
Ovoid; egg-formed (solid), as a fruit.
Ovules, 81, 89.

Paleaceous, 99, a.
Palea, 256.
Palmate, 147, a.
Panduriform; fiddle-shaped, rounded at the ends, narrow in the middle.
Panicle, 304.
Papilionaceous, 105, 5.
Papillose; producing small glandular excrescences.
Pappus, 99, a.
Parasitic; growing upon or nourished by another.
Parallel-veined, 229, 2.
Parenchyma, 29, 261.
Parietal placenta.
Pectinate; comb-like, with long, narrow segments.
Pedate; when the palmate leaf has the two lateral lobes cut into two or more segments.
Pedicel, 268.
Pedicellate; furnished with a pedicel.
Peduncle, 267, 268.
Pellucid; transparent.
Peltate, 233, 21.

Pendulous; drooping, hanging down.
Pentagonal; with 5 sides and 5 angles.
Pentandrous; with 5 stamens.
Pepo, 116, 13.
Perennial; enduring three years or more.
Perfoliate, 242, 2.
Perianth, 54.
Pericarp, 112.
Perigynous; inserted into the calyx.
Peristome; the rim or border surrounding the orifice of the theca of a moss.
Permanent; same as persistent.
Persistent, 98.
Personate, 104, 5.
Petal, 101.
Petaloid; resembling petals.
Petiolate, 221.
Petiole, 221, 222.
Phænogamia, 46, 345.
Pilose, 99, a.
Pinnæ; (wings) the segments of a pinnate leaf.
Pinnate, 240, 1.
Pinnatifid, 232, 15.
Pistil, 56, 75.
    " physiological structure of, 83.
    " theoretical structure of, 64.
Pistillate; bearing pistils.
Pith, 199.
Placenta, 87.
Plaited, 217, 6.
Plant defined, 10.
Plicate; folded like a fan.
Plumose, 99, a.
Plumule, 124, b.
Pod; legumes, siliques, &c.
Pollen, 70.
Polyandrous; with many stamens.
Polyadelphous; stamens united in several sets.
Polygamous; having staminate or pistillate and perfect flowers on the same tree.
Polygynous; with many pistils.
Polypetalæ, 354.
Polypetalous, 101.
Polysepalous, 99.
Polyspermous; many-seeded.
Pome, 116, 12.
Pores; apertures of perspiration in the cuticle.
Premorse, 145, b.
Prickles, 43.
Primine, 90.
Prismatic; formed like a prism, with 3 or more angles.
Procumbent; trailing on the ground.
Proliferous; forming young plants about the roots.
Prostrate; trailing flat on the ground.
Pubescent, 41, a.
Pulp; the soft, juicy, cellular substance found in berries and other fruits.
Pulverulent; powdery.
Punctate, 237, 10.

Pungent; stinging or pricking.
Putamen, 112.
Pyriform; pear-shaped.
Pyxis, 116, 11.

Quinate, 241, 9.
Quincuncial, 108, 3.

Raceme, 208.
Racemose; resembling a raceme.
Rachis, 290.
Radiate; when the outer flowers of an inflorescence are largest, or furnished with rays.
Radiate-veined, 230, 2.
Radical, 220.
Radicle, 124, a.
Ramial, 220.
Ramose, 144.
Raphe, 121.
Raphides, 29, f.
Receptacle, 56.
Recurved; bent or curved backwards.
Reflexed; curved backwards and downwards.
Reniform, 232, 12.
Repand, 235, 11.
Respiration, 275 — 280.
Resupinate; inverted.
Reticulate, 229, 1.
Retrorse; bent backwards.
Retuse, 236, 5.
Revolute, 217, 4.
Rhizoma, 191.
Rhomboid; oval and angular in the middle.
Rib [costa]; ridge caused by projecting veins, &c.
Ringent, 104, 5.
Root, 136 — 160.
    " growth of, 153.
    " forms of, 143, &c.
    " use of, 154.
    " physiological structure of, 151.
Rosaceous, 105, 2.
Rostrate; with a beak.
Rosulate; arranged in a radiant manner like the petals of a double rose.
Rotate, 104, 4.
Rugose, 237, 9.
Runcinate, 232, 13.
Runner, 186.

Saccate; with a bag or sack.
Sagittate, 232, 11.
Samara, 116, 10.
Sap, 268.
Sapwood, 203.
Sarcocarp, 112.
Scabrous; rough.
Scale; the bracts of the Compositæ.
Scape, 186, 260.
Scarious; dry, colorless, membranaceous.
Scorpoid; when racemes are revolute before expansion, as Drosera, &c.

Scattered, 218, 1.
Secund; turned to one side.
Secundine, 99.
Scrobiculate; pitted or furrowed.
Seed, 117, &c.
  " vitality of, 134.
Segments; parts or divisions.
Seminal; of the seed.
Sepals, 96.
Septicidal, 115, 2.
Septifragal, 115, 3.
Septinate, 241, 10.
Septum; a partition.
Sericeous, 41, a.
Serrate, 235, 3.
Serrulate, 235, 3.
Sessile, 221.
Setaceous, or setose; bristly.
Setose, 99, a.
Sheath; the lower part of the leaf or leaf-
  stalk which surrounds the stem.
Shrub, 164, b. Signs, 126.
Silicle, 116, 3.
Silique, 116, 2.
Sinuate, 232, 16.
Sinus; the recesses formed by the lobes
  of leaves, &c.
Soporific; inducing sleep.
Sori; the patches of fructification on the
  back of the fronds of ferns.
Spadix, 300.
Spathe; the sheath surrounding a spadix
  or a single flower.
Spathulate; obovate, with the lower end
  much narrowed and tapering.
Species, 50.
Specific names, 362.
Spermoderm; skin of a seed.
Spike, 297.
Spines, 171.
Spinous, 235, 7.
Spiral vessels, 33, a.
Spongioles, 142, c.
Spores, 129.
Sporogens, 347.
Sporules or spores, 129.
Spur, 107, a.
Stamens, 57, 65, 73.
  " consist of, 66.
  " and pistils, use of, 92.
Staminate; with stamens only, barren.
Standard; same as vexillum or banner.
Stellate, 242, 5.
Stem, 161.
  " functions of, 208, 209.
Sterile; barren, unfruitful.
Sternutatory; exciting to sneezing.
Stigma, 79.
Stings, 42.
Stipe; the stalk of a pod, of a fungus, &c.
Stipela, 251.
Stipitate; borne on a stipe.
Stipules, 249.
Stipulate, 251.
Stolon, 193.

Stoloniferous; bearing stolons.
Stomata, 37, 38, 39.
Straight-veined; where the principal
  veins pass direct to the margin.
Striæ; small streaks, channels or furrows.
Striate; with striæ, slightly furrowed, &c.
Strigose; clothed with short, stiff, and
  appressed hairs.
Strobile, 116, 15.
Style, 79.
Stylopodium; a kind of disk which is
  epigynous and confluent with the style.
Sub; in composition, it denotes a lower
  degree of the quality, as sub-sessile,
  nearly sessile, &c.
Submersed; under water.
Subulate; awl-shaped.
Succulent; thick, juicy, and fleshy.
Suffrutescent; somewhat shrubby.
Suffruticose; same as the last.
Sulcate; furrowed or grooved.
Superior, 97.
Suture, 84.
Symmetrical, 61, c.
Syncarpous; when the fruit consists of
  united carpels.
Syngenesious; when the anthers are uni
  ted into a tube, as in Compositæ.
Systematic botany, 330.

Tap root, 145, a.
Tendril, 187, a.
Terete; rounded or cylindric.
Terminal; borne at the summit.
Ternate, 240, 4.
Testa, 118.
Tetradynamous; with 2 short and 4 long
  stamens.
Tetragynous; with 4 pistils.
Tetrandrous; with 4 stamens.
Thallogens, 351.
Thallus; that part of Lichens which
  bears the fructification.
Theca; the vessels which contain the
  sporules of the Cryptogamia.
Thorn, 171.
Throat; the orifice of the tube of the
  corolla.
Thyrse, 305.
Tomentose, 41, a.
Toothed; dentate.
Torose; uneven or undulating on the sur-
  face.
Torus; receptacle, 59.
Trailing; creeping or lying on the ground.
Transverse; cross-wise.
Tree, 164, a.
Triandrous; with 3 stamens.
Tricuspidate; having three points.
Tridentate; three-toothed.
Trifid; three-cleft.
Tripinnate, 239, 6.
Triternate, 239, 8.
Truncate; blunt, as if cut square off.
Trunk, 159.

Tube, 103.
Tuber, 180.
Tuberiferous; bearing tubers.
Tuberous, 147.
Tubular, 103.
Tunicated, 178, a.
Turbinate; shaped like a top.
Turgid; swollen.

Umbel, 302, a.
Umbellet, 305, a.
Umbilicate; depressed in the centre.
Unarmed, 235, 7.
Uncinate; hooked at the end.
Undulate; wavy.
Unguis; the claw, as of a petal.
Unilateral; one-sided.
Utricle, 116, 8.

Valvate, 108, 1.
Valves, 115.
Varieties, 51.
Vascular tissue, 33.
Vasculares, 47.
Vasiform tissue, 32.

Vegetable physiology, 3.
Vegetable kingdom, variety of, 13.
Vegetation, its diffusion, 14, a.
Veins, 298.
Veinlets, 298.   Veinulets, 298.
Velvety; clothed with a dense, soft pub
    escence.
Venation, 296.
Ventral, 84.
Vernation, 216.
Verticillaster, 309.
Verticillate, 218, 4.
Vesicular; bladdery.
Vexillary, 108, 6.
Vexillum, 105, 5.
Villose; villous; clothed with long hairs.
Vine, 187.
Viscid; clammy, sticky.
Vittæ; receptacles of secretion in the
    seed of Umbelliferæ.

Whorled, 175.
Winged, 222, 2.
Wood, 201, 202.
Woody tissue, or fibre, 30.

ABBREVIATIONS AND SIGNS,

OFTEN USED IN DESCRIPTIVE BOTANY.

ach. achenia.
æst. æstivation.
alter. alternate.
amplex. amplexicaul.
anth. anther.
axill. axillary.
cal. calyx.
caps. capsule.
cor. corolla.
decid. deciduous.
diam. diameter.
ellip. elliptical.
emarg. emarginate.
epig. epigynous.
f. or *ft.* feet.

fil. filaments.
fl. flower; *fls.* flowers.
fr. fruit.
hd. head; *hds.* heads.
hyp. hypogynous.
imbr. imbricate.
inf. inferior.
invol. involucre.
irreg. irregular.
leg. legume.
lf. leaf; *lvs.* leaves.
lfts. leaflets.
lom. loment.
opp. opposite.
ova. ovary.

ped. peduncle.
pet. petals.
perig. perigynous.
perig. perigynium.
recep. receptacle.
reg. regular.
rhiz. rhizoma.
rt. root.
sds. seeds.
seg. segments.
sep. sepals.
st. stem.
sta. stamens.
stig. stigmas.
sty. styles.

The names of the *months*, and of *states* and *countries*, are often abbreviated, and always in the same manner as in other works; thus, Apr. April; Jn. June; Mass. Massachusetts; N. Y. New York; Ia. or Ind., Indiana, &c.

The following Signs are also in general use ·

① An annual plant.
① A biennial plant.
♃ A perennial herb.
♄ A plant with a woody stem.
♂ A staminate flower or plant.
♀ A pistillate flower or plant.

☿ A perfect flower, or a plant bearing perfect flowers.
☿ Monœcious, or a plant with staminate and pistillate flowers.
♀ ♂ Diœcious; staminate and pistillate flowers on separate plants.
♀ ☿ ♂ Polygamous; the same species with staminate, pistillate, and perfect flowers.
● Wanting or none.
∞ Indefinite, or numerous.
●— Cotyledons accumbent.
● | Cotyledons incumbent. } Used only in the Cruciferæ.
0>> Cotyledons conduplicate. }
§ A naturalized plant.
† A plant cultivated for ornament.
‡ A plant cultivated for use. This, with the two last, are placed at the end of a description. In other situations they have their usual signification as marks of division or reference. In measure of length, or other dimensions, the following signs are adopted in this work : —

    f (without the period) A foot.
    ′ (a single accent) An inch.
    ″ (a double accent) A line (one twelfth of ′).

! The note of exclamation, now common in botanical works, is used in contrariety to the note of interrogation (?). It denotes, in general, *certainty from personal observation.* Affixed to a locality, it denotes that the writer has examined specimens either in or from that place. Affixed to the name of an individual, it denotes that the writer has examined specimens supplied by him. In this work the note of affirmation is used only where the fact stated or implied is somewhat *new,* or might otherwise have been regarded as doubtful.

Authors' names, when of more than one syllable, are usually abbreviated by writing the first syllable and the next following or last consonant. The following are nearly all the names thus abridged in this work : —

Adans. Adanson.	Eagel. Engelman.	Nutt. Nuttall.
Agh. Agardh.	Forsk. Forskahl.	Pers. Persoon.
Ait. Aiton.	Frœl. Frœlich.	Pall. Pallas.
Arn. Arnott.	Gært. Gærtner.	Pav. Pavon.
Bart. Barton.	Ging. Gingins.	Poir. Poiret.
Benth. Bentham.	Gmel. Gmelin.	Ph. Pursh.
Berl. Berlandier.	Gron. Gronovius.	R. Br. Robert Brown.
Bernh. Bernhardi.	Hedw. Hedwig.	Raf. Rafinesque.
Brongn. Brongniart.	Hoffm. Hoffmann.	Rich. Richard.
Bigl. or Bw. Bigelow.	Hook. Hooker.	Schw. Schwenitz.
Boehm. Boehmer.	Juss. Jussieu.	Scop. Scopoli.
Bong. Bongard.	Lam. Lamark.	Ser. Seringe.
Br. Brown.	Lec. Le Conte.	Schk. Schkuhr.
Cass. Cassini.	Lindl. Lindley.	Sm. Smith.
Cav. Cavanilles.	Linn. * Linnæus.	Spr. Sprengel.
Darl. Darlington.	Lk. Link.	Sw. Swartz.
DC. De Candolle.	Lehm. Lehmann.	T. & G. Torrey & Gray.
Desf. Desfontaines.	Mart. Martius.	Torr. Torrey.
Desv. Desvaux.	Mentz. Mentzel.	Tourn. Tournefort.
Dew. Dewey.	Michx. Michaux.	Traut. Trautvetter.
Duh. Duhamel.	Mill. Miller.	Willd. Willdenow.
Endl. Endlicher.	Mirb. Mirbel.	Walt. Walter.
Ehrh. Ehrhart.	Mœnch. Mœnchansen.	
Ell. Elliot.	Muhl. Muhlenberg.	

ANALYTICAL TABLES.

The object of scientific tables is usually twofold. First, philosophical ; — to exhibit in one condensed view the affinities and differences of the several subjects to which they relate, by bringing them into immediate comparison and contrast. Second, practical ; — to aid the student in his researches by affording him an abridged method of analysis. The analytical tables which accompany this flora may subserve both these purposes, but they are designed chiefly for the latter ; viz.

---

* In this flora, wherever no authority is added to the generic or specific name, *Linn.* is to be understood.

as an expeditious method of botanical analysis. They are the result of much labor and investigation, since each character employed required a previous examination of all the species included under it. The process of analysis by these tables consists of a simple series of *dilemmas* or *alternatives*; the decision being, in almost all cases, to be made merely between *two opposite* or *obviously distinct* characters. These decisions or *dilemmas* being, moreover, few in number, conduct the student with absolute precision (provided the tables be free from error, and the specimen a good one) to the order or genus to which his plant belongs, by once or a few times reading across the page. The advantages thus afforded will be duly appreciated, at least by those who have hitherto been subjected to the drudgery of reading through whole pages of dry generic descriptions, and that too, often, without arriving at any satisfactory conclusion.

In regard to the generic characters employed in the tables, it will be observed that they are drawn from *leaves, fruit, flower*, or any portion of the plant which suited our convenience, — our only inquiry being after those which appear to us the most *obvious* and *constant*. It should be remarked, however, that in many instances these characters are not strictly applicable to *all* the known species belonging to those particular orders or genera, but only to those which are described in this work; that is, found in the United States, north of lat. 36½°. In our choice of terms we have always, of necessity, studied the utmost brevity of expression, but have used none but such as are explained in the glossary or in the body of the work.

Although the manner in which these tables are to be used will in general be obvious at a glance, yet it may not be unprofitable to attend to the following directions and illustrations. We will suppose the student to be in possession of an unknown plant which he wishes to study by the aid of the Flora, in other words to analyze. To this end, he first determines to which of the six great classes of the natural system it belongs, — either by his previous knowledge of their characters, which should be thoroughly understood, or by an appeal to the first synoptical view, page 131. Thus he inquires, —

1. Is the plant a flowering or flowerless one? *Ans.* Flowering. It belongs therefore to Phænogamia. Turn next to the 2d couplet.

2. Are the leaves net-veined, &c. or parallel-veined, &c.? *Ans.* Net-veined, and the flowers are not *completely* 3-parted; that is, the petals, sepals and stamens may be in 3's, but the *pistil* is single. The plant belongs, therefore, to Exogens. Turn next to the 3d couplet.

3. Stigmas present, &c. or stigmas 0, &c.? *Ans.* Stigmas present. The plant, therefore, belongs to Class I. Angiosperms. Next ascertain to which of the subclasses it belongs, by consulting the 6th triplet.

6. Corolla with distinct petals, — united petals, — or wanting? *Ans.* Distinct. The plant will therefore be found among the Polypetalous Exogens. Now turn to the 7th couplet, and inquire,

7. Is the plant an herb, or a shrub, &c.? *Ans.* A shrub. Turn then to couplet 48th.

48. Leaves opposite, or leaves alternate? *Ans.* Alternate. Turn to 56.

56. Oligandrous, or polyandrous? *Ans.* Oligandrous. 63.

63. Ovary free, &c. or adherent, &c.? *Ans.* Free. 64.

64. Corolla irregular, &c. or regular? *Ans.* Regular. 65.

65. Shrubs climbing, or erect? *Ans.* Erect. 66.

66. Stamens — how many, and how situated? *Ans.* 6, opposite to the petals. The plant therefore belongs to Order VI. Berberidaceae. The pupil now turns to that Order, and inquires in the conspectus, —

Is the plant an herb or a shrub? *Ans.* A shrub, and hence belongs to Genus I. Berberis. Turn finally to that genus, and study the species.

Again, suppose that, by a similar process, we had traced an unknown plant to the Natural Order Rosaceae. We should then turn to that Order (XLVIII.), and inquire,

Ovaries free, or adherent? *Ans.* Free.

Naked or enclosed, &c.? *Ans.* Naked.

Are they 3—50 in number or 1 only? *Ans.* 3—50.

Are they in fruit follicles, a compound berry, or achenia? *Ans.* Achenia.

On a dry receptacle or juicy? *Ans.* Dry.

Caudate with the persistent style or not? *Ans.* Not caudate.

Leaves simple or compound. *Ans.* Compound.

Is the plant caulescent or acaulescent? *Ans.* Acaulescent, and the genus is Waldsteinia. Turn lastly to that genus, and learn the species. Further illustrations are perhaps unnecessary.

A SYNOPSIS OF THE NATURAL SYSTEM;

BEING

A KEY TO THE CLASSES AND ORDERS, FOR THE READY ANAL
YSIS OF ANY PLANT EMBRACED IN THIS FLORA.

§ 1. Classes and Subclasses.

: Flower ng plants, or PHÆNOGAMIA. . . . 2
: Flowerless plants, or CRYPTOGAMIA. . . . 5
1 Leaves net-veined. Flowers never *completely* 3-parted. EXOGENS. . . . 3
2 Leaves parallel- (rarely net) veined. Flowers 3-parted. ENDOGENS. . . . 4
   3 Stigmas present. Sds. enclos d in se d-vessels. . . . 6. ANGIOSPERMS
   3 St g.0. Sds. n k. (Pine. Fir. &c.) CONIFERÆ. CXXXI. GYMNOSPERMS. I
4 Fls. with no glumes. Perianth whorled or wanting. . . . 127. AGLUMACEÆ. II
4 Fls. with green g umes. no perianth and 1-seeded fr. . . . 147. GLUMACEÆ. IV
   5 Stems and leaves distinguishable. . . . 149.   -   -   -   - ACROGENS. V
   5 Stems and leaves confounded together. . . . 151.   -   - THALLOGENS. V.
6 Corolla with distinct petals. . . . 7 - - - - - **Polypetalous Exogens.**
6 Corolla with united petals. . . . 69 - - - - **Monopetalous Exogens.**
6 Corolla (and often the calyx also) wanting. . . . 97 - - **Apetalous Exogens.** :

§ 2. Orders of the Polypetalous Exogens.

7 Herbs (§ 164). . . . 8
7 Shrubs, trees or under-shrubs. . . . 48
8 Leaves alternate or all radical. . . . 9
8 Leaves cauline, opposite, at least the lower ones. . . . 35
   9 Leaves furnished with stipules. . . . 32
   9 Leaves destitute of stipules. . . . 10
10 Polyandrous—stamens 17—200, indefinite. . . . 11
10 Oligandrous—stamens few and definite. . . . 16
   11 Stamens hypogynous—situated on the receptacle. . . . 12
   11 Stamens perigynous—situated on the calyx or corolla. . . . 15
12 Sepals 3—5. Leaves centrally peltate, in water. . . . 13
12 Sepals 3—9. Leaves neither peltate nor tubular. . . . 14
12 Sepals 5, persistent. Leaves tubular, pitcher-form. - - SARRACENIACEÆ. X
12 Sepals 2, deciduous. Juice usually colored. - - - PAPAVERACEÆ. XI
   13 Petals 3 or 4, in one row. Lvs. oval, floating. - - CABOMBACEÆ. VII
   13 Petals many, in several rows. Lvs. round, erect. - NELUMBIACEÆ. VIII
14 Pistils (or pistil) simple and distinct, few or many. - - RANUNCULACEÆ. I
14 Pistil compound, large 12—30-celled. (In water.) - - - NYMPHÆACEÆ. IX
   15 Petals 5. Styles several, distinct. - - - ROSACEÆ. XLVIII
   15 Petals 5. Styles united into one. - - - LOASACEÆ. LVI
   15 Petals numerous, in several rows. Styles united. - - CACTACEÆ. LXI
16 Flowers very irregular, one-sided or two-sided. . . . 17
16 Flower regular or nearly so. . . . 19
   17 Filaments 6—8, united below into 2 sets. . . . 18
   17 Filaments 5, united only at top. - - - - BALSAMINACEÆ. XXVIII
   17 Filaments 8, distinct. Leaves simple, peltate. - TROPÆOLACEÆ. XXIX
   17 Filaments 8, distinct. Lvs. biternate, with tendrils. SAPINDACEÆ. XLIV
15 Leaves much dissected and divided. - - - - - - FUMARIACEÆ. XII
18 Leaves simple, entire. - - - - - - - POLYGALACEÆ. XVI
   19 Ovary superior—free from the calyx or nearly so. . . . 20
   19 Ovary inferior—wholly adherent to the calyx. . . . 30
20 Sepals 2. Fleshy herbs. - - - - - PORTULACACEÆ. XXIV
20 Sepals 3 or more. Herbs leafy, green. . . . 21
20 Sepals 3 or more. Herbs leafless and not green. . . . 87 - - (LXXVIII.
   21 Stamens hypogynous—situated on the receptacle. . . . 22
   21 Stamens perigynous—situated on the calyx or corolla. . . . 28
22 Sepals, petals and stamens symmetrical. . . . 23
22 Sepals, petals and stamens unsymmetrical. Fruit a pod. . . 26

23 Leaves simple. . . . 24
23 Leaves compound. . . . 25
24 Petals persistent. Ovary 1-celled. Leaves radical. - - DROSERACEÆ. XVIII.
24 Petals fugacious. Ovary 5—10-celled. Leaves cauline. - LINACEÆ. XXVI.
24 Petals deciduous. Ovary 5-celled. Leaves cauline. . . . .87 - - (LXXVIII.)
    25 Juice acrid. Sepals valvate in the bud. - - - LIMNANTHACEÆ. XXX.
    25 Juice acid. Sepals imbricate in the bud. - - - - OXALIDACEÆ. XXXI.
    25 Juice bitter. Leaves pinnate, dotted. - - - - - RUTACEÆ. XXXIV
26 Flowers cruciform, regular. Stamens tetradynamous. - - CRUCIFERÆ. XIII.
26 Flowers rather irregular. Stamens not tetradynamous. . . . 27
    27 Calyx closed in the bud. Pod closed until ripe. CAPPARIDACEÆ. XIV.
    27 Calyx open in the bud. Pod open before ripe. - - - RESEDACEÆ. XV.
28 Stamens 5, opposite to the five petals. . . . 78 - - - - - - (LXXXV.)
28 Stamens alternate with the petals when of the same number. . . . 29
    29 Styles 3—20, as many as the sepals. - - - - CRASSULACEÆ. LXIV.
    29 Styles 2, fewer than the sepals. - - - - - - SAXIFRAGACEÆ. LXV.
    29 Styles 4, fewer than the sepals. Parnassia. - - DROSERACEÆ. XVIII.
30 Flowers 5-parted, in simple or compound umbels. . . . 31
30 Flowers 1, 2, 3 or 4-parted, not in umbels. - - - - - ONAGRACEÆ. LV.
    31 Styles 2, forming a 2-partible, dry fruit. - - - UMBELLIFERÆ. LXVII.
    31 Styles 3 or 4 (rarely 2), forming a berry or drupe, - ARALIACEÆ. LXVIII.
32 Flowers both regular and perfect. . . . 33
32 Flowers either irregular or monœcious. . . . 34
    33 Stamens ∞, united into a column with the 5 styles. MALVACEÆ. XXXVII.
    33 Stamens 10, united only at base, free from the styles. OXALIDACEÆ. XXXI.
    33 Stamens many or few, distinct, perigynous. - - - ROSACEÆ. XLVIII.
34 Ovary free, 3-celled, forming a capsule. - - - - - - VIOLACEÆ. XVII.
34 Ovary free, 1-celled, forming a legume. - - - - - LEGUMINOSÆ. XLVII.
34 Ovary adherent, 3-celled. Flowers monœcious. - - BEGONIACEÆ. LVIII.
    35 Leaves furnished with (either large or small) stipules. . . . 36
    35 Leaves destitute of stipules. . . . 37
36 Petals 5, twisted in the bud, larger than the sepals. - GERANIACEÆ. XXVII.
36 Petals 5, not twisted in the bud, very small. - - ILLECEBRACEÆ. XXII.
36 Petals 2 or 3, not twisted. - - - - - - - - - ELATINACEÆ. XXV.
    37 Flowers very irregular. - - - - - - - - - POLYGALACEÆ. XVI.
    37 Flowers regular, or but slightly unequal. . . . 38
38 Ovary or ovaries superior—free from the calyx. . . . 39
38 Ovary wholly adherent to the calyx tube. . . . 40
38 Ovary adhering to the calyx tube by the angles only. - MELASTOMACEÆ. LI.
    39 Ovaries many, distinct, simple, caudate. - - - - RANUNCULACEÆ. L
    39 Ovary 1, simple. Leaves 2 only in Podophyllum. - BERBERIDACEÆ. VI.
    39 Ovary compound. . . . 41
40 Involucre 4-leaved, white, subtending the small cyme. - CORNACEÆ. LXIX.
40 Involucre none. Leaves numerous, simple. - - - - ONAGRACEÆ. LV.
40 Involucre none. Leaves 3 only, compound. Panax. ARALIACEÆ. LXVIII.
    41 Sepals 2, fewer than the petals. - - - - - PORTULACACEÆ. XXIV
    41 Sepals 3—5. Style and stigma 1. . . . 42
    41 Sepals 3—5. Styles and stigmas several. . . . 43
42 Sepals equal, combined into a tube. - - - - - - - LYTHRACEÆ. LII
42 Sepals unequal, nearly distinct. - - - - - - - - CISTACEÆ. XIX.
    43 Stamens hypogynous—on the receptacle. . . . 44
    43 Stamens perigynous—on the calyx. . . . 45
44 Stem tumid at the nodes. Leaves not punctate. - CARYOPHYLLACEÆ. XXIII.
44 Stem often ancipital. Lvs. with pellucid and black dots. HYPERICACEÆ. XX.
    45 Stamens 20 or more, indefinite. Exotic. - - - MESEMBRYACEÆ. LXII.
    45 Stamens fewer than 20, definite. . . . 46
46 Pistils (follicles) distinct, as many as the sepals. - - CRASSULACEÆ. LXIV.
46 Pistils 2—5, partly or completely united. . . . 47
    47 Styles 3—5. Embryo coiled. - - - - - CARYOPHYLLACEÆ. XXIII.
    47 Styles 2. Embryo straight. - - - - - - SAXIFRAGACEÆ. LXV.
48 Leaves opposite. . . . 49
48 Leaves alternate. . . . 58
    49 Flowers irregular. - - - - - - - - - HIPPOCASTANACEÆ. XLIII.
    49 Flowers regular. . . . 50

50 Stamens 4,—as many as the sepals and petals. . . . 57
50 Stamens 5,—as many as the sepals and petals. . . . 56
50 Stamens 6—100, more than the sepals and petals. . . . 51
  51 Ovary free (or half-free and 4-carpelled). . . . 52
  51 Ovary adherent to the calyx tube. . . . 56
52 Stamens perigynous. Stigmas 2 or 4. . . . 54
52 Stamens hypogynous. Stigmas 1 or 3. . . . 53
  53 Stigmas distinct. Leaves not punctate. - - - - - CISTACEÆ. XIX
  53 Stamens polyadelphous. Leaves punctate. - - - HYPERICACEÆ. XX
54 Leaves palmate-veined (or compound). Fruit a samara. ACERACEÆ. XLII
54 Leaves feather-veined, simple. Fruit a capsule. - - SAXIFRAGACEÆ. LXV
  55 Stamens opposite to the petals. Vines with tendrils. - - VITACEÆ. XI
  55 Stamens alternate with the sepals. Tendrils none. CELASTRACEÆ. XLV
56 Sepals, petals and stamens ∞. - - - - - - - CALYCANTHACEÆ. XLIX
56 Sepals and petals 4, stamens 8. Fuchsia. - - - - - ONAGRACEÆ. LV
56 Sepals and petals 5, stamens ∞. - - - - 4 - - - MYRTACEÆ. L.
  57 Flowers in cymes. Large shrubs or trees. - - - CORNACEÆ. LXIX.
  57 Flowers in spikes or fascicles. Small parasites. LORANTHACEÆ. LXX.
58 Oligandrous—stamens few and definite. . . . 63
58 Polyandrous—stamens 20 or more. . . . 59
  59 Sepals 5 (rarely more), as many as the petals. . . . 61
  59 Sepals 3, petals 6—9. . . . 60
60 Petals imbricated in the bud. Stipules membranaceous. - MAGNOLIACEÆ. II.
60 Petals valvate in the bud. Stipules none. - - - - - ANONIACEÆ. IV.
  61 Filaments united into a tube. - - - - - MALVACEÆ. XXXVII.
  61 Filaments distinct, perigynous. - - - - - - ROSACEÆ. XLVIII
  61 Filaments distinct, hypogynous. . . . 62
62 Leaves with stipules, dotless, cordate. Flowers small. TILIACEÆ. XXXVIII
62 Leaves without stipules, dotless. Flowers large. TERNSTRŒMIACEÆ. XXXVI.
62 Leaves without stipules, pellucid-punctate. - - - AURANTIACEÆ. XXXV
  63 Ovary free from the calyx—superior. . . . 64
  63 Ovary adherent to the calyx—inferior. (Flowers symmetrical). . . . 68
64 Corolla more or less irregular. Fruit a pod (legume). LEGUMINOSÆ. XLVII.
64 Corolla regular. Fruit not leguminous. . . . 65
  65 Climbing without tendrils. Stamens 12—18. - - MENISPERMACEÆ. V
  65 Climbing without tendrils. Stamens 5. - - - - CELASTRACEÆ. XLV.
  65 Climbing with tendrils. Stamens 5, with a crown. PASSIFLORACEÆ. LVII.
  65 Erect shrubs or trees. . . . 66
66 Stamens 4 or 5, opposite the petals. - - - - - - RHAMNACEÆ. XLVI.
66 Stamens 6, opposite the petals. - - - - - - - BERBERIDACEÆ. VI.
66 Stamens 2—10, alternate with the petals if the same in number. . . . 67
  67 Lvs. pinnate, punctate. Ov. separate or 2-celled. ZANTHOXYLACEÆ. XXXII.
  67 Lvs. (mostly) pinnate, dotless. Ov 1-celled, with 3 styles. ANACARD. XXXIII.
  67 Leaves simple. Seeds 4 or 5. Nemopanthes. . . . 81 - - - (LXXIX).
  67 Leaves simple. Seeds 8—12. - - - - - - § ESCALLIONEÆ. LXV.
68 Flowers (in late autumn) 4-parted: petals linear. - HAMAMELACEÆ. LXVI.
68 Flowers (in June) 4-parted: petals lanceolate. - - - CORNACEÆ. LXIX.
68 Flowers 5-parted. Styles 2. - - - - - - - - - GROSSULACEÆ. LX.
68 Flowers 5-parted. Styles 5. - - - - - - - - - ARALIACEÆ. LXVIII.

§ 3. Orders of the Monopetalous Exogens.

  69 Stamens as many as the lobes of the corolla. . . . 70
  69 Stamens 6—12,—more numerous than the lobes of the corolla. . . . 87
  69 Stamens 2—4,—fewer than the lobes of the corolla. . . . 88
70 Flowers in dense heads (compound) surrounded by an involucre. . . . 71
70 Flowers separate, or not furnished with an involucre. . . . 72
  71 Stamens 4, distinct. - - - - - - - - - - - DIPSACEÆ. LXXIV.
  71 Stamens 5, united by the anthers. - - - - - COMPOSITÆ. LXXV.
72 Calyx superior—adherent to the ovary. . . . 73
72 Calyx inferior—free from the ovary. . . . 76
  73 Stamens cohering by the anthers. . . . 74
  73 Stamens distinct. . . . 75
74 Flowers regular. Vines with tendrils. - - - - - CUCURBITACEÆ. LIX
74 Flowers irregular. Tendrils none. - - - - - - LOBELIACEÆ. LXXVI

75 Leaves alternate. Flowers 5-parted. - - CAMPANULACEÆ. LXXVII.
75 Leaves opposite. Flowers 5-parted. - - - - CAPRIFOLIACEÆ LXXI.
75 Leaves opposite. Flowers 4-parted. - - - - RUBIACEÆ. LXXII.
76 Plants with a milky juice. Ovaries 2. follicular. . . . 86
76 Plants with a watery juice. Ovaries not follicular. . . 77
   77 Stamens opposite to the lobes of the corolla 78
   77 Stamens alternate with the lobes of the corolla. . . . 79
78 Ovary with 5 styles and but one seed. - - - PLUMBAGINACEÆ. LXXXV.
78 Ovary with 1 style and many seeds. - - - - - - PRIMULACEÆ. LXXXII
   79 Shrubs, trees or under-shrubs. . . . 80
   79 Herbs. Leaves opposite or all radical. . . . 82
   79 Herbs. Leaves alternate, cauline. 84
   79 Herbs (vines) leafless. - - - - - - - § CUSCUTINEÆ. XCVIII.
80 Stamens hypogynous or slightly cohering to the base of the corolla. . . . 81
80 Stamens inserted on the corolla tube inside. - - - SOLANACEÆ. XCIX.
80 Stamens inserted on the summit of the corolla tube. DIAPENSIACEÆ. XCVII.
   81 Ovary forming a dry, many-seeded capsule. - ERICACEÆ. LXXVIII.
   81 Ovary forming a fleshy, 4—6-seeded drupe. - AQUIFOLIACEÆ. LXXIX
82 Leaves with stipules. Corolla bud valvate. - - - RUBIACEÆ. LXXII.
82 Leaves without stipules. . . . 83
   83 Ovary 1-celled, opening by a lid. Lvs. radical. PLANTAGINACEÆ. LXXXIV.
   83 Ovary 1-celled, opening by 2 valves. Leaves cauline. GENTIANACEÆ. C.
   83 Ovary 3-celled, opening by 3 valves. Lvs. cauline. POLEMONIACEÆ. XCVI.
84 Ovary 4-parted, separating into 4 achenia. - - - BORRAGINACEÆ. XCIV.
84 Ovary compound, not separating, 1-celled. - - HYDROPHYLLACEÆ. XCV.
84 Ovary compound, not separating, 2- or 3-celled. . . . 85
   85 Cor. bud twisted-imbricate. Embryo less than albumen. POLEMON. XCVI.
   85 Cor. bud twisted-plicate. Embryo larger than albumen. CONVOLV. XCVIII.
   85 Corolla bud imbricated, not twisted. Verbascum. SCROPHULARIACEÆ. XCI.
   85 Corolla bud plicate or induplicate-valvate. - - - SOLANACEÆ. XCIX.
86 Flowers (in umbels) with a 5-lobed corona. - - ASCLEPIADACEÆ. CII.
86 Flowers (in cymes, &c.) with no corona. - - - - - APOCYNACEÆ. CI.
   87 Stamens distinct. Style 1. - - - - - - - ERICACEÆ. LXXVIII.
   87 Stamens distinct. Styles 4. - - - - - - - EBENACEÆ. LXXX.
   87 Stamens united, all into one set. - - - - - STYRACACEÆ. LXXXI.
   87 Stamens united into 2 equal sets (3 & 3). . . . 18 - - - - - (XII)
   87 Stamens united into 2 unequal sets (9 & 1). . . . 34 - - - - (XLVII.)
88 Herbs. Calyx adherent to the ovary. . . 89
88 Herbs. Calyx free from the ovary. . . . 90
88 Shrubs or trees. Calyx free from the ovary. . . . 94
   89 Stamens 4. Linnæa. - - - - - - - - - CAPRIFOLIACEÆ. LXXI.
   89 Stamens 3. - - - - - - - - - - - VALERIANACEÆ. LXXIII.
90 Plants brown, leafless. - - - - - - - OROBANCHACEÆ. LXXXVII.
90 Plants verdant, leafy. . . . 91
   91 Leaves all radical and the corolla spurred. - LENTIBULACEÆ. LXXXVI.
   91 Leaves cauline, or if radical the corolla spurless. . 92
92 Ovary deeply 4-lobed, forming 4 achenia. - - - - - LABIATÆ. XCIII.
92 Ovary entire, but splitting into 1—4 little nuts. - - - VERBENACEÆ. XCII.
92 Ovary entire, capsular, 2-carpeled. . . . 93
   93 Corolla bud valvate. Capsule (falsely) 4-celled. PEDALIACEÆ. LXXXIX.
   93 Corolla bud convolute. Capsule 2-celled. - - - ACANTHACEÆ. XC.
   93 Corolla bud imbricate. Capsule 2-celled. - - SCROPHULARIACEÆ. XCI.
94 Flowers regular, diandrous. . . . 96
94 Flowers irregular. . . . 95
   95 Shrub climbing or tree diandrous. - - - BIGNONIACEÆ. LXXXVIII.
   95 Tree tetradynamous. Paulownia. - - - SCROPHULARIACEÆ. XCI.
96 Corolla imbricated and twisted in the bud. - - - - - JASMINACEÆ. CIII.
96 Corolla valvate in the bud. - - - - - - - - OLEACEÆ. CIV.

§ 4. Orders of the Apetalous Exogens.

97 Herbs with alternate leaves, or leafless. . . . 98
97 Herbs with opposite or verticillate leaves. . . . 106
97 Shrubs or trees. Leaves alternate. . . . 115
97 Shrubs or trees. Leaves opposite. . . . 125
98 Stipules sheathing the stem. Lvs. simple, entire. - - POLYGONACEÆ. CX.
98 Stipules not sheathing, serrate. Leaves pinnate. . . . 15 - - - (XLVIII.)
98 Stipules 0, or if any, not sheathing or serrate. . . . 99
   99 Flowers with a regular calyx or an involucre. . . . 100

99 Flowers with no calyx or involucre.... 105
100 Calyx adherent to the ovary. Stamens 6—12. - - ARISTOLOCHIACEÆ. CV.
100 Calyx adherent to the ovary. Stamens 5. - - - - SANTALACEÆ. CXIII.
100 Calyx free from the ovary.... 101.
 101 Pistils entirely distinct. Stam ns 00.... 14 - - - - - - - - - (I)
 101 Pistils united into a compound ovary.... 102
102 Ovary 1-celled with 1 seed..... 103
102 Ovary 3-celled with 3 seeds. - - - - - - - EUPHORBIACEÆ. CXXI.
102 Ovary 5-celled with 00 seeds.... 29 - - - - - - (LXIV.)
102 Ovary 6—10-celled with 6—10 see ls. - - - - - PHYTOLACCACIÆ. CXI.
 103 Pistil 1. Embryo straight. - - - - - - - URTICACEÆ. CXXX.
 103 Pistils 2—5. Embryo coiled.... 104.
104 Calyx and imbricated bracts dry and scarious. - - AMARANTACEÆ. CVIII.
104 Calyx (and bracts also if any) green. - - - - CHENOPODIACEÆ. CVI.
 105 Flowers on a spadix with a spatha.... 130 - - - - (CXXXIV.)
 105 No spadix or spathe. Stamens 6 or 7. - - - SAURACEÆ. CXVII.
 105 No spadix or spathe. Stamens 2 or 3. - - - PODOSTEMIACEÆ. CXX.
105 Calyx adherent to the ovary.... 30 - - - - - - - (LV.)
105 Calyx (or involucre) free..... 107
106 Calyx 0. Involucre 0. Aquatic. - - - - - CALLITRICHACEÆ. CXIX.
 107 Leaves verticillate. - - - - - - - - CERATOPHYLLACEÆ. CXVIII.
 107 Leaves opposite.... 108
108 Herbs with a milky juice. Fruit 3-seeded. - - - EUPHORBIACEÆ. CXXI.
108 Herbs with a watery juice.... 109
 109 Stipules present. Leaves lobed or compound. § CANNABINEÆ CXXX.
 109 Stipules present. Leaves simple, entire.... 36 - - - - (XXI.)
 109 Stipules none.... 110
110 Stamens 00, several times more than the sepals.... 14 - - - - - (I.)
110 Stamens 8—10, twice more than the sepals.... 111
110 Stamens as many as the sepals or fewer.... 112
 111 Calyx tubular, enclosing the 1-seeded utricle. - SCLERANTHACEÆ. CVII.
 111 Calyx spreading; capsule 00-seeded..... 30 - - - - - - (LV.)
112 Calyx large, colored, funnel-form limb entire. - - NYCTAGINACEÆ. CIX.
112 Calyx small or minute, 3—5-lobed.... 113
 113 Calyx and imbricated bracts dry and scarious. - AMARANTACEÆ. CVIII.
 113 Calyx not scarious nor bracted.... 114
114 Stamens alternate with the sepals. Flowers perfect.... 78 - (LXXXII.)
114 Stamens opposite to the sepals. Flowers perfect.... 44 - - (XXIII.)
114 Stamens opposite to the sepals. Flowers dioecious. - URTICACEÆ. CXXX.
 115 Flowers not in aments, with a calyx and mostly perfect.... 116
 115 Flowers imperfect, the sterile only in aments.... 120
 115 Flowers imperfect, both the sterile and fertile in aments.... 121
116 Stamens alternate with the sepals, of the same number..... 117
116 Stamens opposite to the sepals, or more numerous.... 118
 117 Leaves serrate.... 66 - - - - - - - - - - (XLVI.)
 117 Leaves entire, covered with whitish scurf. - - - ELÆAGNACEÆ. CXV.
 117 Leaves entire, smooth, evergreen. - - - - - EMPETRACEÆ. CXXII.
118 Sepals 3, with 6 stamens. Avine. - - - - - ARISTOLOCHIACEÆ. CV.
118 Sepals 4, with 8 stamens. Erect shrubs. - - - - THYMELACEÆ. CXIV.
118 Sepals 6, with 9 stamens in 3 rows. - - - - - - LAURACEÆ. CXII.
119 Sepals 5—9, with 5—9 stamens in one row.... 119
 119 Leaves pinnately compound, punctate..... 67 - - - - - (XXXII.)
 119 Leaves simple. Calyx adherent to the ovary. - SANTALACEÆ. CXIII.
 119 Leaves simple. Calyx free from the ovary. - - ULMACEÆ. CXVI.
120 Leaves simple. Nut or nuts in a cup or burr. - - CUPULIFERÆ. CXXIV.
120 Leaves pinnate. Nut naked. - - - - - - JUGLANDACEÆ. CXXIII.
 121 Plants with a milky juice. Fruit fleshy. - - - § MOREÆ. CXXX
 121 Plants with a watery juice. Fruit dry.... 122
122 Aments globular, pendulous..... 123
122 Aments cylindrical, or oval.... 124
 123 Aments racemed. Nutlet 2-celled, several seeded. BALSAMIFLUÆ CXXVIII.
 123 Aments solitary. Nutlets 1-celled, 1-seeded. - PLATANACEÆ. CXXIX.
124 Ovary 1-celled, 1-ovuled, 1-seeded in fruit. - - - MYRICACEÆ. CXXVI.
124 Ovary 2-celled, 2-ovuled, 1-seeded in fruit. - - - BETULACEÆ. CXXV.
124 Ovary many-ovuled, many-seeded in fruit. - - - SALICACEÆ. CXXVII.
 125 Calyx 3- or 4-parted.... 126
 125 Calyx 5-parted: fruit a double-samara.... 54 - - - - - (XLII.)
126 Stamens 2. Fruit a single samara.... 96 - - - - - - - (CIV.)

126 Stamens 4. Shrub evergreen, in soil. *Buxus.*   - EUPHORBIACEÆ. CXXI.
126 Stamens 4. P r site on trees, evergreen. . . . 57   - - - - - - - (LXX.)
126 Stamens 8. Shrubs with scurfy leaves. - - - - - ELEAGNACEÆ. CXV.

§ 5. Orders of the Aglumaceous Endogens.

127 Shrubs (climbing). . . . 140
127 Herbs. . . . 128
128 Perianth complete, of 6 parts (or 4, white), never on a spadix. . . . 132
129 Perianth complete, but obscured in a small roundish head. . . . 146
128 Perianth none or incomplete, mostly on a spadix. . . . 129
129 Plants terrestrial (or on a thick scape). . . . 130
129 Plants growing in water. . . . 131
130 Spadix with a spathe, or on a scape. - - - - - ARACEÆ. CXXXIV.
130 Spadix 1 with no spathe and on a leafy stem. - - TYPHACEÆ. CXXXVI.
131 Root floating in water, plant on the surface. - - LEMNACEÆ. CXXXV.
131 Root fixed in the mud. Plant submersed. - NAIADACEÆ. CXXXVII.
132 Perianth adhering to the ovary (in the perfect flowers). . . . 133
132 Perianth free from the ovary. . . . 138
133 Fls. monoecious or dioecious. Aquatic. HYDROCHARIDACEÆ. CXXXIX.
133 Flowers perfect. Plants terrestrial. . . . 134
134 Flowers irregular. . . . 135
134 Flowers regular. . . . 136
135 Stamens 1 or 2, adhering to the pistil (gynandrous). ORCHIDACEÆ. CXL.
135 Stamens 1, free from the pistil. *Indian shot.* - - CANNACEÆ. CXLIV.
135 Stamens 3, anthers extrorse. - - - - - - IRIDACEÆ. CXLIX.
136 Stamens 3. . . . 137
136 Stamens 6. - - - - - - - - - - AMARYLLIDACEÆ. CXLVII.
137 Anthers introrse. Perianth woolly outside. HÆMODORACEÆ. CXLVIII.
137 Anthers extrorse. Perianth smooth outside. - - IRIDACEÆ. CXLIX.
138 Petals and sepals similarly colored. . . . 139
138 Petals and sepals dissimilarly colored. . . . 145
139 Styles and stigmas 3, distinct. . . . 141
139 Styles or sessile stigmas united into 1. . . . 142
140 Flowers in spikes or panicles. Capsule 3-cornered. - - DIOSCOREACEÆ. CL.
140 Flowers in umbels. Berry globose.   } - - - SMILACEÆ. CLI.
141 Leaves net-veined, petiolate. Fruit a berry. }
141 Leaves parallel-veined. Capsule 00-seeded. - - MELANTHACEÆ. CLV.
141 Lvs. rush-like. Fruit splitting into 1 or 2-seeded parts. ALISM. CXXXVIII.
142 Perianth colored, juicy and withering. . . . 143
142 Perianth dry, green ; or if colored, scarious. - - - - JUNCACEÆ. CLVI.
143 Flowers regular, hexand.ous. . . . 144
143 Flowers irregular or triandrous. Aquatic. - PONTIDERIACEÆ. CLIV.
144 Perianth woolly or scurfy outside, tubular. - - HÆMODORACEÆ. CXLVIII.
144 Perianth smooth outside, mostly 6-parted. - - - - LILIACEÆ. CLIII.
145 Styles or stigmas 3. Leaves net-veined. - - - TRILLIACEÆ. CLII.
145 Style and stigma 1. Leaves parallel-veined. - COMMELINACEÆ. CLVII.
145 Styles many. Leaves somewhat net-veined. ALISMACEÆ. CXXXVIII.
146 Petals conspicuous, yellow. - - - - - - - - XYRIDACEÆ. CLVIII.
146 Petals inconspicuous, white. - - - - - - - ERIOCAULONACEÆ. CLIX.

§ 6. Orders of the Glumaceous Endogens.

147 Stems mostly solid. Sheaths of the leaves entire. - CYPERACEÆ. CLX.
147 Stems hollow. Sheaths split to the joints. - - - GRAMINEÆ. CLXI.

§ 7. Orders of the Cryptogamia.

148 Plants consisting of woody and vascular tissue. . . . 149
148 Plants consisting of cellular tissue only. . . . 150
149 Fruit terminal, cone-like. Leaves sheath-like. - EQUISETACEÆ. CLXII.
149 Fruit axillary or in spikes. Leaves 1-veined. - LYCOPODIACEÆ. CLXIII.
149 Fruit borne on the veiny, often contracted leaves. - - FILICES. CLXIV.
149 Fruit radical or nearly so, of two kinds. - - - - MARSILEACEÆ. CLXV.
150 Leaves veinless, distinct from each other and from the axis. MUSCI.
150 Leaves veiny, mostly confluent into one expansion. HEPATICÆ.
151 Plants with no distinct axis of growth. . . . 152
151 Plants having a distinct axis with whorled branches. CHARACEÆ. CLXVI.
152 Aërial, dry, crustaceous, on trees, rocks. &c. LICHENS.
152 Aërial, succulent, often ephemeral, never green. FUNGI.
152 Aquatic, consisting of simple vesicles or lobed fronds. ALGÆ.

FIRST GRAND DIVISION,

PHÆNOGAMIA, OR FLOWERING PLANTS.

—

*Plants consisting of a regular axis of growth with leafy appendages,
composed of a cellular, vascular and ligneous structure;
developing flowers and producing seeds.*

———

SUBDIVISION FIRST.
EXOGENS, OR DICOTYLEDONOUS PLANTS.

STEM composed of distinct bark and pith, with an intervening layer
of woody fibre and vessels. GROWTH by annual, concentric, external zones or layers. LEAVES mostly with reticulated veins, and falling off by an articulation. SEPALS and PETALS in 5s and 4s much
oftener than in 3s. EMBRYO with 2 opposite, or several whorled
cotyledons.

———

CLASS I. ANGIOSPERMS.

OVULES produced within an OVARY and fertilized by the pollen
through the medium of the PISTIL, becoming SEEDS enclosed in a
PERICARP. EMBRYO with two opposite cotyledons.

·

SUBCLASS I. POLYPETALÆ.

Floral envelops usually consisting of both calyx and corolla, the
latter composed of distinct petals.

———

ORDER I. RANUNCULACEÆ.—CROWFOOTS.

Herbs with an acrid, colorless juice.
Leaves mostly alternate and much divided, with half-clasping petioles.
Calyx.—Sepals mostly 5, sometimes 3, 4 or 6, mostly deciduous, and imbricated in æstivation.
Corolla.—Petals 3—15, hypogynous, sometimes irregular or 0.
Stamens ∞, distinct, hypogynous. Anthers, adnate or innate.
Ovaries ∞, rarely solitary or few, distinct, seated on the torus.
Fruit either dry achenia, or baccate, or follicular.
Embryo minute, at the base of horny or fleshy albumen.

Genera 61, species about 1000 (Lindley), mostly natives of cold, damp climates. Europe is supposed
to contain one-fifth of the species, North America one-seventh, India one-twenty-fifth, South America
one-seventeenth. Africa very few, and New Holland but 19.

Properties.—Almost all the genera contain an acrid juice highly prejudicial to animal life, but easily
decomposed and deprived of its activity by a heat of 212 deg. They also lose their poisonous qualities in
drying. This order is rich in ornamental cultivated plants.

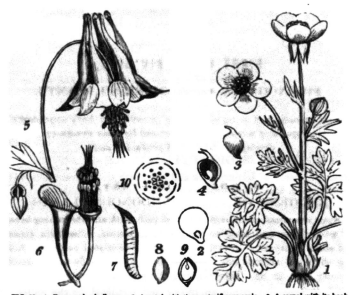

FIG. 39.—1. Ranunculus bulbosus. 2. A petal with the nectariferous scale. 3. A carpel with its beak. 4. Vertical section, showing the erect ovule. 5. Aquilegia Canadensis. 6. Torus, with the stamens and pistils, and a petal attached. 7. Follicle. 8. Seed. 9. Its vertical section, showing the funiculus and minute embryo. 10. Cross section of the flower, showing the arrangement of the 5 ovaries, the 20 stamens in 4 series, &c.

Conspectus of the Genera.

				Pet. 1-lipped.	Trollius.	7
			Flowers { caulina.	Pet. 2-lipped.	Helleborus.	8
			solitary, { radical.		Coptis.	9
				white.	Cimicifuga.	15
		Pet. small.	Flowers racemose,	dark purple.	Zanthorhiza.	19
		Petals 0.	Calyx white.		Isopyrum.	10
	Sepals		Calyx bright yellow.		Caltha.	6
	equal .			plane 3-lobed.	Nigella.	22
		Petals conspicuous,	all spurred behind.	Aquilegia.	11
folliculær.	Sepals	colored, . .	upper one large, vaulted.		Aconitum.	13
	unequal, . .	green and foliaceous.	upper one spurred behind.	Delphinium.	12
			Petals plane.		Pæonia.	21
				palmate.	Trautvetteria.	16
			Leaves alternate,	3—2-ternate.	Thalictrum.	17
	Petals 0, or	Invol. 0. . .	Leaves opposite. Mostly climbers.	Clematis.	1
	inconspicu.	Involucre calyx-like, near the flower. .	Hepatica.	3
		Involucre leaf-like, remote from flower. .	Anemone.	2
			with no nectary, scarlet. .	Adonis.	4
			with nectariferous, tubular claws.	Myosurus.	18
	acheniate.	Petals conspicuous, . .	with a nectariferous scale at base.	Ranunculus.	5
		solitary, many-seeded. Flowers racemose. .	Actæa.	14
Carpels	baccate,	numerous, 1—2-seeded. Flowers solitary. .	Hydrastis.	20

1. CLEMĀTIS.

Gr. κληρα, a tendril; climbing by tendrils, or twining petioles instead.

Calyx 4-(rarely 5, 6 or 8-) sepaled, colored, pubescent; corolla 0, or smaller than the calyx; filaments ∞, shorter than the sepals; ovaries 4—20; styles longer than the stamens; achenia caudate with the long, plumose, permanent style.—♃ *Mostly climbing. Leaves mostly compound and opposite.*

§ *Sepals 4. Petals several, minute.* ATRAGENE. DC.

1. C. VERTICILLÁRIS. DC. (Atragene Americana, *Sims.*) *Whorl-leaved Virgin's Bower.*

Climbing; *lrs.* in 4s., verticillate, ternate; *lfts.* cordate, nearly entire; *ped.* 1-flowered; *sepals* very large, acute.—A handsome climber in highland woods, Vt., (*Dr. Phelps*) to N. Car. W. to the Rocky Mts. Stem ascending trees 15f by means of its twisting petioles. At each node is a whorl of four 3-foliate leaves, and 2 large purple flowers. Leaflets acute, 1—2' by ½—1'. Sepals thick, 15" by 5". Filaments about 24, outer ones (petals?) dilated, spatulate tipped with imperfect anthers. May, Jn.

§§ *Petals 0.* CLEMATIS *proper.*

2. C. VIRGINIÀNA. *Virgin's Bower.*

St. climbing; *lvs.* ternate; *lfts.*, ovate, cordate, acuminate, lobed and cut dentate; *fls.* often ♀ ♂, paniculate.—A common, hardy climber in hedges and thickets, Can. to Ga. and the Miss. Stem 8—15 f. in length, supporting itself on fences and brushwood by means of the long petioles. Leaflets 2—3' by 1½—2', with mucronate teeth. Sepals 4, white, oval-oblong, obtuse. Stamens 28—36. Panicles large, axillary, dichotomous. Fruit furnished with long, plumose tails (caudæ), appearing in large, downy tufts. Aug. †

3. C. VIORNA. *Leather Flower.*

St. climbing; *lrs.* pinnately divided; *lfts.* ovate-lanceolate, acute at each end, entire or 3-lobed; *fls.* solitary, campanulate; *sep.* thick and leathery, acuminate.—In woods, Penn. to Ill. (*Jenny*) and Ga. Stem 10—15f in length, cylindrical, striate. Leaves opposite, decompound, consisting of 9—12 leaflets, Flowers axillary, purple, large, nodding. Peduncle 3—6' long, with a pair of small, simple, entire leaves near the middle. Fruit with long, plumose tails, Jn. Jl. †

4. C. OCHROLEUCA. Ait. (C. sericea. *Michx.*) *Erect Clematis.*

St. herbaceous, erect, simple, silky-pubescent; *lrs.* undivided, ovate, entire, silky beneath; *fls.* pedunculate, terminal, solitary, inclined to one side; *cal* silky outside.—Mts. and river banks, N. Y. to Ga. An erect species, 12—18' high. Leaves subsessile, 2—4' long, two-thirds as wide, with prominent veins, upper surface becoming glabrous. Flowers yellowish white (ochroleucous), campanulate in form. May, Jn.

5. C. CRISPA. *Crisp-flowered Clematis.*

St. climbing; *lrs.* pinnate and ternate; *lfts.* ovate-lanceolate, very acute, 3-lobed or entire; *fls.* solitary; *sep.* acuminate, revolute, thick, with undulate and crisped margins.—Va. to Flor. Stem striate, 6—8f long. Flowers a third smaller than in C. Viorna, pale-purple, campanulate. Sepals spreading or revolute at the end. Peduncles axillary, shorter than the leaves. Achenia with naked (not plumose) caudæ. Jl. †

6. C. FLAMMULA. *Sweet Virgin's Bower.*—*Lvs.* pinnate; *lfts.* smooth, entire, orbicular-oval, oblong or linear, acute.—From France. A fine climber for arbors, &c., very ornamental and sweet-scented. Flowers white. Jl.—Oct. †

7. C. FLORIDA. *Large-flowered Virgin's Bower.* *Lvs.* 2-foliate and decompound; *segments* ovate, acute, entire; *sep.* acuminate and glabrous; *involucre* 0.—From Japan. Vine 12f long, with large, white and yellow flowers. Jn. Sept. †

8. C. VITICELLA. *Lvs.* 3-foliate and decompound, lobes or segments entire *sep.* obovate.—From Spain. This, as well as the preceding species, is often double-flowered. Vine 20f long. Flowers purple. †

Observation.—All the species are ornamental, and of easy culture. They require only a common soil and are propagated by layers, cuttings, or from the seed.

2. ANEMONE.

Gr. ανεμος, wind; most of the species grow in elevated or windy places.

Involucre remote from the flower. of 3 divided leaves; calyx regular, of 5—15 colored sepals; corolla 0; stamens ∞, much shorter than the sepals; ovaries ∞, free, collected into a roundish or oval

head; achenia ∞, mucronate.—♃ *Lvs. radical. Scapes with leaf
like involucres.*

1. A. NEMORÒSA (and quinquefolia. *Linn.*). *Wood Anemone.*
Lvs. ternate; *lfts.* undivided, or with the middle one 3-cleft, and lateral ones
2-parted, incisely dentate; *invol.* similar to the leaves, petiolate; *st.* 1-flowered.
—A common and interesting little plant, found in old woods, hedges, and sometimes in open fields. Root creeping. Stem 6—9' high, erect. The involucre
consists of 3 petiolate leaves, placed in a whorl near the top of the stem, its
bracts cut-toothed and lobed, the lateral segments cleft, sometimes quite to the
base, so as to render the leaf quinate. At the top of the stem is a single
white flower, purplish outside. Apr. May.

2. A. CYLINDRICA. Gray.
Whole plant pubescent; *lvs.* ternate, *lateral lfts.* 2-parted to the base, *middle
one* deeply 3-cleft, segments all linear, cuneate below, cut-dentate and lobed at
apex; *lvs. of invol.* petiolate; *ped.* 2—6, rarely 1, all naked; *sep.* 5; *ach.* woolly,
in a long, cylindric head.—Dry, hilly places, Mass. W. to Ia. Not common.
Scape about 2f high. Leaves about 2—3' wide, and similar in their divisions
to those of *Ranunculus acris.* Naked flower-stalks 8—10' long, umbellate, but
little diverging. Flowers large, solitary. Petals pale yellow, obovate, obtuse.
Heads of fruit 1½' long. May, Jn.

3. A. VIRGINIÀNA. *Virginian Anemone.*
Lvs. ternate; *lfts.* subpetiolate, ovate-lanceolate, cut-dentate, acuminate,
lateral ones 2-lobed, middle ones 3-lobed; *invol.* foliaceous, petiolate; *fr.* in
oblong heads.—A tall species in dry woods and hilly pastures. Can. to Car.
Scape erect, 2—3 f. high, round, hairy, dividing above into about 3 long,
parallel peduncles, middle one naked, lateral ones each with an involucel of 2
bracts. Leaves 2—3' by 3—4', on radical petioles 6—10' long, petioles of the
bracts much shorter. Flowers solitary, yellowish-green. Fruit woolly, in
heads 1' long. July.
ß. alba. Oakes. *Fls.* larger; *sep.* white.—Ledges, Vt. *Dr. Robbins.*

4. A. HUDSONIÀNA. Rich. (A. multifida, *DC.* and 1st edit.)
Hairy; *lvs.* 3—5-parted to the base, segments cuneate, laciniately dentate,
scape 1, 2 or 3-flowered; *invol.* and *involucels* similar, 2-leaved, on short petioles;
sep. 5—8, oval, obtuse.—On rocky ledges, shores of Onion River, Colchester
and Burlington, Vt. *Dr. Robbins.* Watertown, N. Y. *Dr. Crewe.* Rare
Scape 6—10' high, simple, or dividing below the middle. Leaves mostly in 5
segments distinct to the base, about 1' diam., each segment ½' long, in 3 linear
lobes, petioles 1—2' long. Flower small, white, varying to purple. Heads of
fruit oval or globose. Jn.

5. A. PENNSYLVÀNICA. (A. aconitifolia. *Michx.* A. dichotoma. *Linn.*)
St. dichotomous; *lvs.* 3—5-parted and incisely dentate; *invol.* and *involucels*
leaf-like, sessile, 3-parted, the lobes lanceolate, acute, incisely serrate; *sep.* 5;
fr. in globose heads.—Shores and rocky places, Penn. N. to Arctic Am. Rare.
Scape 15—20' high, dividing above the middle into about 3 shortish peduncles,
the middle one naked, the other 2 each with a 2-leaved involucel, the involucre at their base. Flowers white, 12—14'' diam. Carpels hairy, compressed,
as long as the curved style. Jn. Jl.

6. A. PÀTENS. (A. Ludoviciana. *Nutt.*)
Silky-villous; *lvs.* 3-parted or divided, segments cuneiform, 3-cleft and
incised, lobes lance-linear; *invol.* subulately dissected; *sep.* 5—6, erect.— Dry
hills, Ill.! W. to Rocky Mts. Stem 6—10' high. Leaves smoothish above, segments 1—3' long, 1—2'' wide. The dissected involucre concave or cup-shaped.
Sepals 1' long, silky outside, pale dull purple. Tail of the carpels near 2' long.

7. A. CORONÀRIA. *Poppy-leaved Anemone.*—*Lvs.* ternate, with multifid segments and linear, mucronate lobes; *sep.* 6, oval, close.—From Levant. A
hardy, flowering plant, with large, single or double variegated flowers. May. †

8. A. HORTENSIS. *Star Anemone.*—*Lvs.* 3-parted, with crenate, cut-dentate
lobes; *invol.* sessile of oblong, entire or cut leaflets; *sep.* 10—12, oblong.—

From Italy. A fine garden species, with double and semi-double varieties of red, white and blue flowers. May. †

Observation.—Many other foreign species are ornamental, and perhaps rarely cultivated. They all prefer a fresh, loamy soil.

3. HEPATICA. Dill.

Gr. ἥπατος, of the liver; from the fancied resemblance of the leaf.

Involucre of 3 entire, ovate, obtuse bracts, resembling a calyx, situated a little below the flower; calyx of 5—9 petaloid sepals, disposed in 2 or 3 rows; corolla 0; achenia awnless.

H. TRILŎBA. Chaix. (Anemone Hepatica. *Linn.*) *Liverwort.*
Lvs. trilobate, the lobes entire; *scape* 1-flowered, hairy.—Woods, Can. to Car. This little plant is one of the earliest harbingers of spring, often putting forth its neat and elegant flowers in the neighborhood of some lingering snow-bank. The root consists of numerous and strong fibres. Leaves all radical, on long, hairy petioles, smooth, evergreen, coriaceous, divided into 3 lobes, which suggests all its names. Flowers on scapes 3—4' long, solitary, numerous, generally blue, but frequently in varieties of white and flesh-color. In cultivation they become double. In respect to the form of the leaves there are two varieties :—
α. *obtusa*, lobes obtuse, rounded.—Prefers the south side of hills.
β. *acuta*, lobes acute.—Prefers the north side of hills.

4. ADONIS.

Feigned to have sprung from the blood of Adonis, when wounded by the boar.

Sepals 5, appressed; petals 5—15, with naked claws; achenia in a spike, ovate and pointed with the hardened, persistent style.

A. AUTUMNÀLIS. *Pheasant's Eye.*
St. branching; *fls.* 5—8-petaled; *carpels* crowned with a very short style, and collected into an ovate or subcylindric head; *pet.* longer than the calyx.—A fine, hardy annual, from Europe, naturalized in some parts of N. Y. Stem thick. Leaves pinnately parted, with numerous linear segments. Flowers crimson, 1¼' diam. Seeds to be sown in autumn, in a light soil. †§

5. RANUNCŬLUS.

Lat. rana, a frog; from the aquatic habitat of some species.

Calyx of 5 ovate sepals; corolla of 5 roundish, shining petals, each with a nectariferous scale or pore at the base inside; filaments ∞. much shorter than the petals; achenia ∞, crowded in a roundish or oblong head.—*Herbs mostly* ♃, *with yellow flowers.*

*Leaves all undivided.

1. R. FLAMMŬLA. *Small Spearwort.*
St. declinate; *lvs.* smooth, linear-lanceolate or ovate-lanceolate, lower ones petiolate; *ped.* opposite the leaves.—An aquatic herb, growing in ditches and swamps, Can. to N. Car. W. to Ill.! Root fibrous. Stem 6—18' long, more or less decumbent, succulent. Leaves 3—6' in length, ½—1' wide, entire or with a few teeth, thickened at the acute summit. Flowers solitary, of a golden yellow, on peduncles ½ as long as the leaves. It abounds in a very acrid juice. Jn.—Aug.

2. R. REPTANS. *Creeping Crowfoot.*
Very small, smooth; *st.* creeping, geniculate, rooting; *nodes* 1-flowered; *lvs.* subulate, smooth, entire, remote.—A slender species, creeping on river banks and other wet places, Hanover, N. H., (*Mr. T. Rickard,*) W. to Oregon. Stem 6—10' long, round, rooting at the joints. Leaves fleshy, 6—12" in length, mostly very narrow, and acute at each end. Flowers on axillary peduncles. Sepals spreading, obtuse. Petals obovate, yellow, fading to white. Nectary covered by a scale. Achenia very smooth. Jl.
β *ovalis.* Bw. *Lvs.* oval and lanceolate; *pet.* 5—10.

γ. *filiformis.* DC. (R. filiformis. *Michx.*) *St.* filiform, very long, with linear leaves and small flowers.

3. R. PUSILLUS. Poir. β. *muticus.* T. & G. *Puny Crowfoot.*
Erect; *lvs.* all petiolate, lower ones ovate, upper ones linear lanceolate; *pet* mostly but 3, scarcely longer than the calyx; *carpels* ovate, pointless, smooth, in small globose heads.—In wet grounds, N. Y. and Penn. Stems slender, weak, 6—12' high, dichotomously branched. Lower leaves subcordate, ½—1' long, ⅓ as wide, petioles 1—3" long, upper ones 1—1½" long, ¼ as wide, with minute, remote teeth. Flowers very small, yellow, on long peduncles. May.

4. R. CYMBALARIA. Ph. *Sea Crowfoot.*
Very small, smooth; *st.* filiform, creeping, rooting at the joints; *lvs.* reniform-cordate, crenate-dentate; *ped.* solitary. mostly 2-flowered; *pet.* spatulate; *ach.* oblong.—In salt marshes on the sea-coast, N. J. to Arctic Am. and at Salina, N. Y. Stem round, sending out runners from the joints. Leaves radical, ½—1' diam., on long petioles. Scapes 2—6' high. each with 2 or 3 small, bright yellow flowers, and as many obtuse bracts. Nectary naked. Jn.

* * *Leaves divided.*

5. R. ABORTIVUS. *Round-leaved Crowfoot.*
Smooth; *radical lvs.* roundish, cordate at base, crenate, petiolate; *cauline lvs.* ternate or pedate, angular, with linear segments, *upper ones* sessile; *cal.* a little longer than petals, reflexed.—A very pretty species in woods, Can. to Ark. remarkable for the dissimilarity of the root and stem leaves. Stem 8—16' high, nearly naked. Root leaves 8—18" diam., quite regularly margined with crenate divisions, and on petioles 2—5' long. Lower stem leaves pedate, with a pentangular outline; upper in 3 deep segments. Flowers small, yellow. Fruit in globose heads. May. Jn.

6. R. SCELERATUS. Ph. *Celery Crowfoot. See also Addenda, p. 638.*
Smooth; *lower lvs.* 3-parted, segments 3-lobed, crenately subincised; *stem lvs.* 3-parted, segments crenately incised, *upper ones* simple, lanceolate, entire; *carpels* in an oblong head.—Grows in wet places, Can. to Car. Stem rather thick, hollow, much branched, 1—1½' high. Lower petioles 3—5' long, with rather large, palmately 3—5-parted leaves. Floral leaves or bracts mostly simple, lanceolate and entire. Flowers numerous, small, yellow. Calyx deflexed. This is one of the most acrid of the tribe, and will raise blisters upon the skin. Jn. Jl.

7. R. RECURVATUS. *Wood Crowfoot.*
St. erect, and with the petioles, covered with spreading hairs; *lvs.* 3-parted. hairy, segments oval, unequally incised, the lateral ones 2-lobed; *cal.* recurved; *pet.* linear-lanceolate; *ach.* uncinate.—About 1 f. high, in damp woods, Lab. to Ga., pale green, branching above. Leaves 1½—2' long, 2—3½' wide, on petioles 3—6' long. Upper leaves subsessile and 3-parted quite to the base. Flowers small, with inconspicuous, pale yellow petals. Carpels ovate, tipped with minute, hooked beaks. May.—Jl.

8. R. ACRIS. *Butter-cups. Crowfoot. Yellow Weed.*
St. erect, many flowered; *lvs.* more or less pubescent, deeply trifid, the segments laciniate, *upper ones* with linear segments; *ped.* round; *cal.* hairy spreading; *carpels* roundish, smooth, compressed; *beak* short, recurved.—This is the most common species from Penn. to Hudson's Bay, in meadows and pastures, rapidly and extensively spreading. Stem 2f high, round, hollow, mostly hairy. Leaves 1½—3' diam., upper ones in 3 linear segments. Flowers large, golden yellow. Jn.—Sept.
β. *Fl's.* double, the *pet.* excessively multiplied.—Gardens.

9. R. BULBOSUS. *Bulbous Crowfoot.* (Fig. 39.)
Hairy; *st.* erect, bulbous at the base; *radical lvs.* ternate, *lfts.* petiolate, incisely dentate, each about 3-cleft; *ped.* furrowed; *cal.* reflexed.—This is another acrid species, very common in pastures, mow-lands, &c. Root fleshy. Stem leafy, furrowed, 6—18' high, hollow, thickened at the base into a sort of bulb, and dividing above into upright peduncles, with golden-yellow flowers. It is well distinguished from R. acris by its reflexed sepals, and its furrowed

peduncles. The lobes of the root leaves are also rounded rather than acute at apex. May, Jn. ⚥

10. R. fascicularis. Muhl. *Early Crowfoot.*

St. erect, branched; *lvs.* pubescent, ternate, the middle segments deeply 3-cleft, lateral ones remotely 3-lobed; *cal.* villous, spreading, shorter than the petals.—Rocky woods and hills, Penn. to Wiscon. N. to Can. Root a fascicle of fleshy fibres. Radical leaves on petioles 3—8' long, so divided as to appear almost pinnate; upper leaves 3-parted, nearly sessile. Flowers large. Petals yellow, cuneate-obovate, with a scale at base as broad as the transparent claw. Apr. May.

11. R. Pennsylvanicus. (R. hispidus. *Ph.*) *Bristly Crowfoot.*

St. erect, and with the petioles covered with stiff, spreading hairs; *lvs.* villous, ternate, *lfts.* subpetiolate, deeply 3-lobed, incisely serrate; *cal.* reflexed, rather longer than the roundish petals.; *carpels* tipped with a short, straight style.—A very hairy species, in wet grounds, Can. and U. S. Stem 1½—3f high. Leaves 2—3' diam.; leaflets strongly veined and with spreading segments. Flowers numerous, small, bright yellow. Fruit in dense oblong or cylindrical heads. Jn.—Aug.

12. R. repens. (R. intermedius. *Eaton.* R. Clintonii. *Beck.*)

St. branching from the base, prostrate, radicating at the joints; *lvs.* trifoliate, *lfts.* petiolate, cuneiform, 3-lobed, cut dentate; *ped.* furrowed; *cal.* spreading; *carpels* with a broad, not recurved point.—In moist or shady places, Can. and U. S. Stems 1—3 or 4f long, generally nairy at base, the early flowering branches erect. Petioles hairy, long. Leaves hairy on the veins, dark green. Flowers middle size, bright yellow. Petals often emarginate. May—Jl.

β. *linearilobus.* DC. *St.* very long, floriferous; *lobes of lvs.* very narrow.

γ. *Marilandicus.* T. & G. *St.* and *petioles* densely hirsute with soft hairs; *fts.* distinctly petiolulate.

13. R. Purshii. Richardson. *Floating Crowfoot.*

Floating; *st.* long; *submerged lvs.* cleft into numerous capillary segments, *emersed ones* reniform, 3—5-parted, the lobes variously divided; *sep.* reflexed, naIf as long as the petals; *carpels* smooth, with a short, straight, ensiform style; *hds.* globose.—Ponds, sluggish streams, and muddy places, Can., U. S. Stem 1—2f or more in length, fistulous. Leaves pentangular in outline, ½—1½ diam., those below most finely divided; petioles ½—2' in length. Flowers bright yellow, emerging on forked, striate peduncles. May, Jn.

β. (R. fluviatilis. *Bw.* R. lacustris. *Beck.*) *Lvs.* all capillaceous-multifid; *fls.* as large as in R. acris.

14. R. aquatilis. β. *capillaceus. River Crowfoot.*

St. floating; *submerged lvs.* filiformly dissected; *pet.* obovate, larger than the calyx, white; *carpels* transversely rugose.—Ponds and sluggish streams, Arctic Am. to S. Car. W. to Rocky Mts. The whole plant is submerged except the flowers, and perhaps a few of the upper leaves. Stem 1—2f or more in length, slender, weak, round, smooth, jointed. Leaves divided dichotomously into numerous, hair-like segments, in outline roundish and ½—1' diam. Peduncles thick, 1—1½' long. Flowers smaller than in R. acris. Petals rather narrow, white, except the yellow claws. Jl. Aug.

Observation.—Several of the above mentioned species are double-flowered in cultivation, as Nos. 8, 9 and 12. Of foreign species, R. Asiaticus, the garden Ranunculus, with large double flowers varying to every hue, and R. aconitifolius, with white double flowers, are sometimes, but not generally, found in our gardens.

6. CALTHA.

Gr. καλαθος, a goblet; the yellow calyx may well be compared to a golden cup.

Calyx colored, of 5 orbicular sepals, resembling petals; corolla 0: stamens ∞, shorter than the sepals; follicles 5—10, oblong, compressed, erect, many-seeded.—♃ *Aquatic and very glabrous.*

C. palustris. *Marsh Marigold. Cowslip.*

St. erect; *lvs.* cordate, suborbicular, crenate.—Wet meadows, Can. to Car. W. to Oregon. Root large, branching. Stem about 1½ high, hollow, round,

13

dichotomous. Lower leaves 2—4' wide, on long semicylindric petioles, upper ones sessile, all of a dark shining green, veiny and smooth. Flowers of a golden yellow in all their parts, 1½' diam., few and pedunculate. Outer row of filaments clavate, twice longer than the inner. The young leaves are in great request in spring, for greens. May.

β. *integerrima*. (C. integerrima. *Ph.*) *Les.* entire; *sep.* obovate, obtuse.
γ. *plena*, with double flowers. Cultivated in gardens.

7. TROLLIUS.

Germ. trol or trollen, globular; alluding to the form of the flowers.

Sepals 5—10—15, roundish ovate, colored, deciduous; petals 5—25, small, linear, tubular at base; stamens ∞, much shorter than the sepals; follicles ∞, subcylindric, sessile, many-seeded.—♃ *Smooth, with palmate leaves.*

1. T. LAXUS. Salisb. (T. Americanus *Muhl.*) *American Globe Flower.*
Sep. 5, oblong, spreading; *pet.* 15—25, shorter than the stamens.—In swamps, Can. to Penn. Not common. About 1f high. Leaves deeply cleft into 5 segments, which are lobed and cut-dentate. Sepals yellow, resembling petals, 4—5" long. Petals very small, orange-colored. Follicles about 10, crowned with the persistent styles. This is the only American species. Jn.

2. E. EUROPÆUS. *European Globe Flower.*—Erect, branched, leafy; *lvs.* deeply cleft or divided, segments cuneate at base, acute, incisely lobed and toothed; *fls.* solitary, erect, large, globular; *ped.* long, naked; *sep.* closely converging; *pet.* equaling the stamens. Native of Europe. Stem 2—3f high. Flowers of a rich yellow. A very ornamental plant, of easy culture from seeds or roots. May, Jn.†

3. T. ASIATICUS. *Asiatic Globe Flower.*—Erect; *lvs.* deeply divided into 5 broad segments; segments laciniately lobed and toothed; *fls.* terminal, solitary, pedunculate; *sep.* spreading; *pet.* longer than the stamens.—Native of Asia. Plant about 2f high, with ample foliage and large, deep orange-color ed flowers—yellow in some of its varieties. Jn.†

8. HELLEBORUS.—Adans.

Ελειν, to cause death; βορα, food; the poisonous qualities are well known.

Sepals 5, mostly greenish, persistent; petals 8—10, very short, tubular, 2-lipped; stamens ∞; stigmas 3—10, orbicular; follicles cohering at base, many-seeded.—♃ *Lvs. coriaceous, divided. Fls. large, nodding.*

H. VIRIDIS.—*Green Hellebore.*
Glabrous; *radical lvs.* pedately divided, segments lanceolate, acute, serrate; *cauline lvs.* few, palmately parted, nearly sessile; *peds.* often in pairs; *sep.* roundish ovate, acute, green.—A European plant, ♃ on Long Island. Stem 2—3 f. high, thick. Apr.†

9. COPTIS.—Salisb.

Gr. κεπτω, to cut; from the numerous divisions of the leaves.

Sepals 5—6, oblong, concave, colored, deciduous; petals 5—6 small, cucullate, obconic; stamens 20—25; follicles 5—10, stipitate, rostrate, diverging in a stellate manner, 4—6-seeded.—*Low herbs with radical leaves, and a long, slender, perennial, creeping rhizoma.*

C. TRIPOLIA.—*Goldthread.*
Lvs. 3-foliate; *scape* 1-flowered; *pet.* much smaller than the sepals.—Penn N. to Arctic Am. Stem subterranean, extensively creeping, golden yellow, very bitter and tonic. Leaves all radical, leaflets sessile, 4—8" long, crenate-mucronate, smooth, coriaceous, common petiole 1—2' long. Peduncles 3—4 high, with a single, minute bract above the middle, bearing a single white starlike flower. The 5 or 6 yellow petals are barely distinguishable by their color among the white stamens. May. Medicinal

10. ISOPŸRUM.

Sepals 5, petaloid, deciduous; petals 5. small, tubular, sometimes 0; stamens 10—40; ovaries 3—20; follicles subsessile, acuminate with the style. 2-several-seeded.—*Delicate herbs, with leaves 2—3-ternate, segments 2—3-lobed. Fls. pedunculate, axillary and terminal, white.*

I. **BITERNÀTUM.** Torr. and Gray. (Enemion. *Raf.*)
Low, erect, glabrous; *petioles* auricled at base; *lrs.* membranaceous; *pet.* 0: *carpels* 3—6, broadly ovate, divaricate, sessile, strongly veined, 2-seeded; *sds.* obovate, compressed, smooth and shining.—2l Damp shades, Western States. Root fibrous. Stems several, 4—10' high. Leaves mostly biternate, petiolules longer than the petioles, segments cuneate-obovate, 4—6" long. Flowers on slender peduncles 1—2' long. May.

11. AQUILEGIA.

Lat. *aquila*, the eagle: the spurred petals resemble the talons of a bird of prey.

Sepals 5, equal, ovate, colored, spreading, caducous; petals 5, tubular. dilated at the mouth, the outer margin erect, the inner attached to the torus, extending behind into a long, spurred nectary; stamens 30—40, the inner ones longer and sterile; styles 5; follicles 5, many seeded.—2l *Fls. nodding.*

1. **A. CANADENSIS.** *Wild Columbine.* (Fig. 39.)
Glabrous; *divisions of the leaves* 3-parted, rather obtuse, incisely dentate; *sep.* rather acute, longer than the corolla; *spurs* straight, longer than the limb; *sta.* and *sty.* exserted.—This beautiful plant grows wild in most of the States, in dry soils, generally on the sunny side of rocks. It is cultivated with the greatest ease, and is much more delicate in foliage and in the hues of its flowers, than the common blue Columbine. Stem branching, a foot high, with ternate, lobed leaves. Flowers terminal, scarlet without and yellow within, pendulous, much embellished by the numerous descending, yellow stamens and styles. Fruit erect. May.

2. **A. VULGÀRIS.** *Common Columbine.*—Spurs incurved; *sts.* leafy, many-flowered; *lrs.* nearly smooth, glaucous, biternate; *sty.* a little longer than the stamens.—From Europe. Stem 1—2f high, with a profusion of handsome, smooth foliage, and large purple flowers. Leaflets bifid and trifid, with rounded lobes. In cultivation the flowers become double by the multiplication of the hollow, spurred petals. They also vary in color through all shades from purple to white. Jn.†

12. DELPHINIUM.

Gr. δελφιν, a dolphin; from the fancied resemblance of the flower.

Sepals 5, colored, the upper one spurred; petals very irregular, the two upper ones terminating behind in a tubular, nectariferous spur. enclosed in the spur of the calyx; styles 1—5; follicles 1—5. —*Showy herbs, with leaves much divided. Fls. blue, red or purple, never yellow.*

1. **D. CONSOLÌDA.** *Branching Larkspur.*
St. suberect, smooth, with spreading branches; *fls.* few, loosely racemed, *ped.* longer than the bracts; *sty.* 1; *carpel* solitary, smooth.— The common larkspur of the gardens, sparingly naturalized, fields and roadsides. Leaves in numerous linear divisions. Jn. Jl. It has numerous varieties of double and semi-double flowers of various colors.§ †

2. **D. EXALTÀTUM.** *American Larkspur.*
Petioles not dilated at base; *lrs.* flat, 3-cleft below the middle, segments cuneiform, 3-cleft at the end, acuminate, the lateral ones often 2-lobed; *rac.* straight; *spur longer than the calyx.*—Native of the Middle States, rarely of the *Northern.* Stem 3—4f high, straight, erect. Flowers of a brilliant purplish

blue. It is deservedly esteemed in the flower-garden, and is of the easiest culture. Jl. Aug.†

3. D. TRICÖRNE. Michx. *Three-fruited Larkspur.*

Petioles slightly dilated at base; *lvs.* 5-parted, divisions 3—5-cleft, lobes linear, acutish, *st.* shorter than the sepals, lower ones 2-cleft, densely bearded inside; *spur* ascending, straight, as long as the calyx; *carpels* 3, spreading in fruit.—Uplands, Penn. to Mo. and Ark. Plant 6—18' high, nearly smooth, Leaves roundish in outline, on long petioles. Flowers 6—8, light blue, in a rather loose panicle.

4. D. AZUREUM. Michx. *Azure Larkspur.*

Pubescent or nearly smooth; *st.* erect; *lvs.* 3—5-parted, many-cleft, with linear lobes; *petioles* some dilated at base; *rac.* strict; *pet.* shorter than sepals, lower one densely bearded, 2-cleft; *spur* ascending.—Native in Wis. and Ark. A very variable species, cultivated in gardens. Stem 2—4 f. high. Flowers azure-colored.†

5. D. GRANDIFLÖRUM. *Large Blue Larkspur.*—*Lvs.* palmate, many-parted, lobes linear, distant; *pedicels* longer than bract; *pet.* shorter than calyx.—A superb perennial species, from Siberia. Flowers double or single, in racemes, of brilliant dark blue, with a tinge of purple.†

6. D. ELÄTUM. *Bee Larkspur.*—*Lvs.* downy, 5-lobed, lobes cuneate at base, trifid, cut-dentate; *spur* inflexed.—Native of Siberia. Stem 5 or 6f high. Flowers blue, and when viewed at a little distance, resembling the *bee* in form.†

Observation.—A few other species may perhaps be found in gardens. All are showy plants, of the easiest culture.

13. ACONITUM.

Gr. ακονιτος, without dust; because the plants grow on dry rocks.

Sepals 5, irregular, colored, upper one vaulted; petals 5, the 3 lower minute, the 2 upper on long claws, concealed beneath the upper sepal, recurved and nectariferous at the apex; styles 3—5; follicles 3—5.—24 *Lvs. digitate or palmate. Fls. in terminal spikes.*

1. A. UNCINÄTUM. *American Wolf's-Bane.*

St. flexuose; *panicle* rather loose, with divergent branches; *lvs.* palmate, 3—5-parted, with rhomboidal-lanceolate, cut-dentate divisions; *galea* (upper sepal) exactly conical, rostrate; *spur* inclined, somewhat spiral; *ova.* villous.— A cultivated, poisonous plant, also native N. Y. to Ga. Stem 2f high. Leaves coriaceous, dark green, 4—5' wide. Flowers large, purple, 3 or 4 near the summit of each branch. Jl. Aug.

2. A. NAPELLUS. (A. delphinifolia. DC.) *Monk's-Hood.*—*St.* straight, erect; *lvs.* deeply 5-cleft, cut into linear segments, furrowed above; *upper sep.* arched at the back, *lateral ones* hairy inside; *ova.* smooth.—A poisonous plant, cultivated among flowers. It is a tall, rank perennial, making quite a consequential appearance. Stem 4f high, with a long spicate inflorescence at its termination. Flowers dark blue, surmounted by the vaulted upper sepal, as if hooded in a monk's cowl. Aug.—There are varieties with flowers white, rose-colored, &c.

14. ACTÆA.

Gr. ακτη, the elder; which plant these herbs resemble in foliage.

Calyx inferior, of 4 roundish, deciduous sepals; corolla of 4—8 spatulate, unguiculate petals; filaments about 30, dilated above; anthers 2-lobed, introrse; stigmas sessile; ovary ovoid; berry globose, with a lateral furrow, 1-celled; seeds many, smooth, compressed.—24 *with ternately divided lvs. Fls. white.*

1. A. RUBRA. Bigelow. (A. Americana. Ph.) *Red Bane-berry.*

Lvs. twice and thrice ternate; *rac.* hemispherical; *pet.* acute; *pedicels of the fruit* slender; *berries* red, ovoid-oblong.—Not uncommon in rocky woods, Penn. to Lab. W. to Rocky Mts. Stem 1½—2f high, dividing into 2 branches, one of which usually bears leaves only, the other leaves and a cluster of flowers.

Leaves 2 or 3-ternate, with ovate-lanceolate leaflets, variously lobed and cut. Petioles 4—7 long, smooth, and slightly glaucous, like the whole plant. Flowers 20—40, in a short dense raceme. Berries bright red, on slender pedicels. May.

2. A. ALBA. Bw. (A. Americana. *β*. alba. *Ph.*) *White Bane-berry.* *Lvs.* twice and thrice ternate; *rac.* oblong; *pet.* truncate; *pedicels of the fruit* thicker than the peduncles; *berries* white.—Grows in rocky woods, common, Can. to Ga., much like the last in foliage. Plant 1½—2 f. high, bearing 2 compound leaves and a cluster of flowers. Leaflets 1—2′ long, ½ as wide, acuminate. Raceme 1—3′ long, 1½′ thick, the pedicels ½′ long, at length purple, and about as thick as the purple peduncles,—characters which, as well as the milk-white fruit, readily distinguish this species from the last. May.

15. CIMICIFUGA.

Lat. *cimex*, a bug, *fugo*, to drive away : alluding to its offensive odor.

Sepals 4—5; petals 3—8, sometimes wanting; stamens ∞, anthers introrse; follicles 1—8, oblong, many-seeded.—♃ *Lvs. ternately divided. Fls. white, in long slender racemes.*

1. C. RACEMOSA. Ell. (Actæa. *Linn.* Macrotys. *Raf.*) *Black Snake-root.* *Lvs.* ternately decompound; *lfts.* ovate-oblong, incisely serrate; *rac.* very long; *pet.* 2, forked, slender; *sty.* 1; *capsule* follicular, dry, dehiscent, ovate.—A tall, leafy plant, with the aspect of an Actæa, found in upland woods. Stem 4—8 f. high, with long, panicled racemes of white sepaled and monogynous flowers. Petals 4—6, small. Stamens about 100 to each flower, giving the raceme the appearance of a long and slender plume. Flowers very fetid. Jn. Jl.

2. C. AMERICANA. Michx. (C. podocarpa. *Ell.* Actæa podocarpa. *DC.*) Glabrous; *lvs.* triternate, segments ovate, terminal one cuneiform at base, 3-parted or 3-cleft and incised; *pet.* concave, sessile, 2-lobed, nectariferous at base; *ova.* 2—5, stiped, obovate and pod-shaped in fruit; *sds.* flat, scaly.—Woods, Penn. to N. Car. Stem 3—6f high. Leaflets 2—4′ long, with coarse, unequal, mucronate serratures. Flowers smaller than in C. racemosa, in a long panicle of racemes. Follicles abruptly beaked, 6—8-seeded.

16. TRAUTVETTERIA. Fisch. and Meyer.

Named in honor of Trautvetter, a German botanist.

Sepals 4—5; petals 0; stamens ∞. petaloid; anthers introrse; carpels 15—20, membranaceous and indehiscent, 3-carinate, 1-seeded, tipped with the short, hooked style.—♃ *Lvs. palmately lobed.*

T. PALMATA. Fisch. and Meyer. (Cimicifuga. *Hook.*) *St.* slender, terete, smooth, branched above; *lvs.* few, rugose and reticulate veined, palmately 5—9-lobed. upper ones sessile, lower on long petioles, lobes lanceolate, acute, incisely serrate; *fls.* cymose.—Prairies, la. S. to Tenn. Plant 2—5f high. Radical leaves 4—6′ wide, 3—5′ long. the petioles twice as long. Stem leaves 2—4, remote. Flowers many. Sepals orbicular, concave, caducous, white. Stamens conspicuous, white. Jl. Aug.

17. THALICTRUM.

Said to be from *θαλλω*, to be green.

Calyx colored, of 4—5 roundish, concave, deciduous sepals; corolla 0; filaments ∞, compressed, dilated upwards, longer than the calyx; ovaries numerous (4—15), with sessile stigmas; achenia awnless, ovoid.—♃ *Lvs. ternately divided. Fls. often ♀ ♂.*

1. T. DIOICUM. *Early Meadow Rue.* Very smooth; *lvs.* decompound; *lfts.* roundish, with obtuse lobes; *filaments* aliform; *fls.* ♀ ♂. Herb 1—2f high, meadows and woods, British Am. to Car. Stem striate, jointed. Leaflets paler beneath, with 5—7 rounded lobes or teeth. Flowers in long-stalked panicles. Sepals 5, obtuse, purplish. The

13°

barren flowers with numerous slender filaments and yellow anthers, the fertile ones smaller, with shorter stamens. Fruit oval, striate. May.

2. T. CORNÙTI. (T. Corynellum. *DC.*) *Meadow Rue.*

Lfts. obtusely 3-lobed, paler underneath; *fs.* ♀ ♂; *filaments* clavate; *fr.* sessile, striate.—A handsome herbaceous plant, common in meadows. Stem 3– 4f high smooth, hollow, jointed, furrowed. Leaves resembling those of the columbine (Aquilegia), green above, smooth, several times compounded. Leaflets 1—2' long, ¾ as wide. Petioles sheathing at base. Panicles large and diffuse. The barren flowers have numerous club-shaped stamens, with oblong yellow anthers. Fertile flowers smaller and less crowded. Jn. Jl.

3. T. ANEMONÒIDES. Michx. (Anemone thalictroides. *Linn.*) *Rue Anemone.*

Floral lvs. petiolate, simple, whorled, resembling an involucrum; *radical lvs.* biternate; *fls.* umbeled.—Woods and pastures, Northern, Middle, and Western States. The root of this little herbaceous plant consists of several oblong tubercles. Stem erect, 6—8' high, slender, bearing several white flowers at top in a sort of umbel. Leaves ½—1' long, ⅓ as wide, cordate at base, 3-lobed, on petioles ½—1½' long; radical common petioles 2—4' long. Apr. May.

18. MYOSÙRUS. Dill.

Gr. μυς, μυος, mouse, ουρα, tail; alluding to the long spike of caryels.

Sepals 5, produced downwards at base below their insertion; petals 5, with slender, tubular claws; stamens 5—20; achenia very closely spicate on the elongated torus.—① *Lvs. linear, entire, radical. Scapes 1-flowered.*

M. MINÌMUS. (M. Shortii. *Raf.*) *Mouse-tail.*

Prairies and bottoms, Ill., *Mead!* to La. and Oreg., *Nuttall.* A diminutive plant, remarkable for its little terete spikelet of fruit, which is often an inch long. Leaves 1—3' long, 1—2'' wide. Scape a little taller, with a single minute pale-yellow flower at top. Apr.

19. ZANTHORHÌZA.

Gr. ξανθος, yellow, ριζα, root.

Sepals 5; petals 5, of 2 roundish lobes, raised on a pedicel; stamens 5—10; ovaries 5—10, beaked with the styles, 2—3-ovuled; follicles mostly 1-seeded, seed suspended.—*Suffruticose; st. and bark yellow and bitter. Lvs. pinnately divided. Rac. axillary, compound. Fls. small, dark purple, often* ♀ ♂ ♂.

Z. APIIFOLIA. L'Her. (Z. simplicissima. *Michx.*)

River banks, Penn. to Ga. Root thick. Stem short, woody, leafy above. Leaves glabrous, about 8' long, including the long petioles. Leaflets 5, 2– 3' long, sessile, incisely lobed and dentate. Racemes many-flowered, appearing with the leaves. Follicles spreading, 1½'' long. March, April.

20. HYDRASTIS.

Gr. ὑδωρ, water; the plant grows in watery places.

Sepals 3, ovate, petaloid, equal; corolla 0; stamens ∞, a little shorter than the sepals; baccate fruit composed of numerous, aggregate, 1-seeded acinea.—♃ *with 2 lvs. and 1 flower.*

H. CANADENSIS. *Turmeric-root.*

The only species. It grows in bog meadows, Can. to Car. and Ky.! Rare. Root of a deep yellow color internally. Stem 6—9' high, becoming purplish, hairy above. Leaves 2 only, alternate, on the upper part of the stem, petiolate, emarginate at base, palmate, with 3–5 lobes. Peduncle terminal, solitary, 1-flowered. Sepals reddish white, of short duration. Fruit red, juicy, resembling the raspberry. Seeds nearly black. May, Jn.

21. PÆONIA

The physician Pæon, according to mythology, first used this plant in medicine, and cured Pluto with it.

Sepals 5, unequal, leafy, persistent , petals 5 ; stamens ∞ (mostly changed to petals by cultivation); ovaries 2—5 ; style 0 ; stigmas double, persistent ; follicles many-seeded.—♃ *Rt. fasciculate. Lvs biternate. Fls. large, terminal, solitary.*

1. P. OFFICINALIS. *Common Pæony.—St.* erect, herbaceous ; *lower lvs.* bipinnately divided ; *lfts.* ovate-lanceolate, variously incised ; *fr.* downy, nearly straight.—The splendid pæony has long been cultivated in every part of Europe and in this country. This species is said to be native of Switzerland. It is a hardy perennial, requiring very little pains for its cultivation. Among its varieties the *double red* is the most common. The *white* is truly beautiful. The *flesh-colored* and the *pink* are also favorites. May, Jn.

2. P. ALBIFLORA. *White-flowered Pæony.—Lfts.* elliptic-lanceolate, acute, entire, smooth ; follicles recurved, smooth.—Native of Tartary. Whole plant dark, shining-green and smooth. Flowers smaller than the last, but truly elegant and fragrant. Petals white. Calyx brown, with 3 green, sessile bracts at base. Nine or ten varieties with flowers single and double, white, rose-colored, &c., are now mentioned in the catalogues of American gardeners.

3. P. ANOMÁLA. *Jagged-leaved Siberian Pæony.—Lfts.* with many lanceolate segments, smooth ; *follicles* depressed, smooth ; *cal.* bracteolate.—From Siberia. Distinguished by the long, narrow segments of the leaflets. Flowers concave, rose-colored. Follicles usually 5.

4. P. MOUTAN. *Chinese Tree Pæony.—St.* shrubby, ♃ ; *lfts.* oblong-ovate, glaucous and somewhat hairy beneath, *terminal one* 3-lobed ; *ova.* 5, distinct surrounded by the very large disk.—From China. The woody stem branches into a bush 3—4f high. Leaves large, on long stalks. Flowers very large, always double in cultivation, fragrant and truly splendid. This plant is remarkable for producing the largest form of disk in the vegetable kingdom.

5. P. PAPAVERACEA. *Chinese Poppy-flowered Pæony.—St.* shrubby, ♃ ; *lfts.* oblong-ovate, glaucous and slightly hairy beneath, *terminal one* 3-lobed ; *ova.* about 5, closely united into a globose head.—From China. Resembles the last in foliage, but is remarkably distinguished from all the other species by its united carpels. Flowers white, with a purple centre, often single in cultivation. Other species and varieties are cultivated, rarely in this country, amounting to about 150 in all.

22. NIGELLA.

Lat. Niger, black ; the color of the seeds, which are used in cookery.

Calyx of 5 sepals, colored ; corolla of 5 3-cleft petals ; styles 5 ; capsules 5, follicular, convex.—① *European herbs. Lvs. in many linear and subulate segments.*

1. N. DAMASCENA. *Fennel Flower.—Fls.* in a leafy involucre ; *anth.* obtuse : *carpels* 5, smooth, 2-celled, united as far as the ends into an ovoid-globose capsule.—Native of S. Europe. A hardy annual of the gardens, to which have been applied the gentle names of " ragged lady," " devil in a bush," &c. Leaves twice and thrice pinnatifid, as finely cut as those of the Fennel. Flowers terminal, solitary, encompassed and over-topped by a circle of leaves divided like the rest. They are often double, white or pale-blue. Jn.—Sept.

2. N. SATIVA. *Nutmeg Flower.—St.* hairy, erect ; *fls.* naked ; *anth.* obtuse : *capsules* muricate.—From Egypt. Rather smaller than the last. Jn.—Sept.

ORDER II. MAGNOLIACEÆ—MAGNOLIADS.

*Trees or shrubs.
Lvs. alternate, coriaceous, simple, entire or lobed, never serrate.
Stip. membranaceous, either convolute in the leaf-bud, or placed face to face.
Fls. solitary, large and showy, mostly odorous and perfect.*

Cal.—Sepals 3—6, deciduous, colored like the petals.
Cor.—Petals 6—12, hypogynous, in several rows, imbricate in æstivation.
Sta. indefinite, hypogynous, distinct, with short filaments, and adnate anthers.
Ova. several, in many rows upon an elongated torus.
Fr. follicular or baccate, 1—2-seeded
Sds. attached to the inner suture of the carpels, from which (in Magnolia) they are suspended by a long delicate funiculus.

An order consisting of 11 genera and 65 species, including some of the most splendid and majestic forest trees. The southern and western states seem to be the region of the most of them. China, Japan, and the Indies contain a few

Properties.—The bark of the species mentioned below contains an intensely bitter principle, which is tonic and stimulating, and the corollas are aromatic beyond almost all other flowers.

Genera.

Carpels dehiscent by the dorsal suture, seeds pendulous. . : : : : :   *Magnolia.*   1
Carpels indehiscent, seeds enclosed, not pendulous.   : : : : :   *Liriodendron* 2

1. MAGNOLIA.

In honor of Pierre Magnol, a French botanist, author of ' Botanicum Montpelliense,' &c.

Sepals 5, often 0 or petaloid; petals 6—12, caducous; carpels 2-valved, 1—2-seeded, imbricated into a cone; seeds baccate, sub-cordate, and suspended, when mature, by a long funiculus.—*A superb genus, consisting mostly of large trees with luxuriant foliage, and large, fragrant flowers.*

1. M. GLAUCA. *White Bay.*
Lvs. oval, glaucous beneath; *pet.* obovate, tapering to the base.—This species is native in N. Eng., particularly at Gloucester, Mass., thence to La. and Mo. The tree is about 25f in height, remarkable only for the beauty of its foliage and flowers. The leaves are smooth, entire, of a regular, elliptical form, remarkably pale beneath. Flowers terminal, white, solitary, of 3 sepals and several concave petals, appearing in July.

2. M. ACUMINÀTA. *Cucumber Tree.*
Lvs. oval, acuminate, pubescent beneath; *pet.* obovate, obtusish.—Grows near the Falls of Niagara, but is more abundant in the Southern States. It is a noble forest tree. Trunk perfectly straight, 4—5f diam., 60—80f high. bearing an ample and regular summit. Leaves very acuminate. Flowers 5 -6' diam., bluish, sometimes yellowish-white, numerous, and finely contrasted with the rich dark foliage. Cones of fruit about 3' long, cylindric, bearing some resemblance to a small cucumber. May.

3. M. UMBRELLA. Lam. (M. Tripetala. *Linn.*) *Umbrella Tree.*
Lvs. deciduous, cuneate-lanceolate, silky when young; *sep.* 3, reflexed; *pet.* 9, narrow-lanceolate, acute.—A small tree, 20—30f high; common in the Middle and Southern States, extending north to southern N. Y. Branches irregular. Leaves 16—20' by 6—8', often appearing whorled at the ends of the branches in the form of an *umbrella.* Flowers terminal, white, 7—8' diam. Fruit conical, 4—5' long, of a fine rose-color when ripe. The wood is soft and porous, and of little use in the arts. May, June.

4. M. GRANDIFLÒRA,
Native of the Southern States, is the noblest species of the genus. Its great height (80 f.), its shining, dark-green leaves, its fragrant, white flowers a foot in diameter, form a combination of rare magnificence.†

2. LIRIODENDRON.

Gr. λειριον, a lily ; δενδρον, a tree.

Sepals 3, caducous; petals 6; carpels imbricated in a cone, 1—2-seeded; seeds attenuated at apex into a scale.—*Trees, with large and fragrant flowers.*

L. TULIPIFÈRA. *Tulip Tree. White Wood. Poplar.*
A fine tree, one of the most remarkable of the American forests. Can. to La., especially abundant in the Western States. It is ordinarily about 80f high, with a diam. of 2 or 3f, but along the Ohio and Mississippi rivers it grows much larger. Near Bloomington, Ia., I measured a tree of this species which had been recently felled. Its circumference, 4 feet from the ground.

was 23f; 30 feet from the ground its diameter was 5f; the whole height
125f. The trunk is perfectly straight and cylindric. At top it divides rather
abruptly into coarse, crooked, rather unsightly branches. Leaves dark green,
smooth, truncate at the end, with two lateral lobes, 3—5′ in length and breadth,
on long petioles. In May and June it puts forth numerous large and brilliant
flowers, greenish-yellow, orange within, solitary, 4—6′ diam. The wood is
extensively used as a substitute for pine.

Order IV. ANONACEÆ.—Anonads.

Trees or shrubs.
Lvs. alternate, simple, entire, without stipules.
Fls. usually green or brown, axillary, large, shorter than the leaves.
Cal.—Sepals 3—4, persistent, often united at base.
Cor.—Petals 6, in two rows, hypogynous, æstivation valvate.
Sta. indefinite, densely covered. *Fil.* short. *Anth.* adnate, extrorse.
Ova. numerous, closely packed. *Sty.* short or 0. *Stig.* simple.
Fr. dry or succulent, 1—many-seeded, distinct or aggregated. *Sds.* anatropous.

Genera 52, species 320, chiefly native within the tropics of both hemispheres. Four species are
found in the United States, all of the following genus. Plants generally aromatic in all their parts.

UVARIA.

Lat. *uva*, grape; from the resemblance of the fruit of some species.

Sepals 3, united at base; petals 6, in 2 rows; carpels oblong,
baccate, often torulose, pulpy within; seeds several.—*Aromatic shrubs
or trees.*

U. **triloba.** Torr. and Gr. (Anona. *Linn.*) *Pawpaw.*
Lvs. obovate-oblong, acuminate; *pet.* dark-purple, exterior orbicular, 3 or
4 times as long as the sepals.—A small and beautiful tree, 15—20f high, on
banks of streams, Middle, Southern and Western States. Branches and leaves
nearly glabrous, the latter 8—12′ by 3—4′, very smooth and entire, tapering to
very short petioles. Fruit about 1′ thick and 3′ long, ovoid-oblong, about
8-seeded, yellowish, fragrant, eatable, ripe in October. Flowers in March,
Apr.

Order V. MENISPERMACEÆ.—Menispermads.

Shrubs twining or climbing, with alternate, entire leaves.
Fls. small, in panicles or racemes, usually diœcious.
Cal.—Sepals 3—6, in a double series, 3—4 in each, imbricated in æstivation, hypog., deciduous.
Cor.—Petals 1—8, hypogynous, usually as many as the sepals, rarely 0.   [many.
Sta. distinct or monadelphous, equal in number to the petals and opposite to them, or 3 or 4 times as
Anth. innate and consisting of 4 globose lobes.
Ova. usually solitary, sometimes 3—4. *Fr.* a drupe, globose-reniform.

Genera 31, species 175, most of them natives of tropical Asia and America. The only northern genus
is Menispermum.

Properties.—A few plants of this order contain a bitter principle in their roots. A foreign species
of *Menispermum* yields the columbo of the shops, which is a valuabl. tonic; another genus, *Anamirta
Cocculus,* of India, furnishes the *Indian cockle,* so intoxicating to fishes.

MENISPERMUM.

Gr. μην, the moon; σπερμα, seed · from the crescent form of the seed

Flowers ♀♂; sepals 4—8, in a double row; petals 4—7, minute,
retuse; ♂ Stamens 12—20. ♀ Ovaries and styles 2—4; drupes
1-seeded; seeds lunate and compressed.

M. **Canadense.**—*Moon-seed.*
St. climbing; *lvs.* roundish, cordate, angular, peltate, the petiole inserted
near the base; *rac.* compound; *pet.* 6—7, small.—In woods and hedges near
streams, Can. to Car. W. to the Miss. Stems round, striate, 8—12f long.
Leaves 4—5′ diam., generally 5-angled, smooth, pale beneath, on petioles 3—5′
long. Flowers in axillary clusters, small, yellow. Drupes about 4″ diam.,
black, resembling grapes. The root is perennial, and in medicine has the properties of a tonic. Jl.
: *β. lobatum,* has the leaves lobed.

Order VI. BERBERIDACEÆ.—Berberids.

Herbs or shrubs, with alternate, usually exstipulate, simple or compound leaves
F's. solitary, racemose or panicled, perfect.
Cal.—Sepals 3—4—6 imbricate in 2 rows, often reinforced by petaloid scales.
Cor. hypogynous. P. 4, 1—6 times as many as the sepals and opposite to them.
Sta. as many or twice as many as the petals, and opposite to them.
Anth. generally opening by recurved valves, extrorse.
Ova. 1-celled, solitary, simple. Sty. often lateral. Stig. often lateral or peltate.
Fr. berried or capsular.
Sds. one or few, attached to the bottom of the cell, or many, attached to lateral placentæ.

Genera 12, species 100, inhabiting the temperate zones. Some genera, as the Podophyllum and Jeff.r sonia, possess cathartic properties. Others as the Berberis, contain in their fruits malic and oxalic acid.

Conspectus of the Genera.

Leaves not peltate. { Petals 2, flowers on a scape.			Jeffersonia. 3
{ Petals 6, with a scale at base.			Leontice. 4
Herbs perennial. { Leaves peltate ; stamens 6o.			Podophyllum. 2
Shrubs, with yellow flowers and irritable filaments.			Berberis. 1

1. BERBĒRIS.

Calyx of 6, obovate, spreading, colored sepals, with the three outer ones smaller ; corolla of 6 suborbicular petals, with 2 glands at the base of each ; filaments 6, flattened ; anthers 2 separate lobes on opposite edges of the connectile ; style 0 ; berry oblong, 1-celled ; seeds 2 or 3.—*Fine hardy shrubs.*

B. **vulgāris.** *Berberry Bush.*
Spines 3-forked ; *lrs.* simple, serratures terminated by soft bristles ; *ra.* pendulous, many-flowered ; *pet.* entire.—A well known bushy, ornamental shrub, in hard, gravelly soils, Northern States. Grows 3—8f high. Leaves 1½—2′ long, ⅔ as wide, round-obtuse at apex, tapering at base into the petiole, and remarkably distinguished by their bristly serratures. Flowers yellow, a dozen or more in each hanging cluster. Stamens irritable, springing violently against the stigma when touched. Berries scarlet, very acid, forming an agreeable jelly when boiled with sugar. The bark of the root dyes yellow. Jn. ∮?

2. PODOPHYLLUM.

Gr. ποvς, ποδος, a foot ; φυλλον, a leaf ; alluding to the long, firm petioles.

Sepals 3, oval, obtuse, concave, caducous ; petals 6—9, obovate, concave ; stamens 9—18, with linear anthers ; berry large, ovoid, 1-celled, crowned with the solitary stigma.—♃ *Low, rather poisonous herbs. Lvs.* 2. *Fl.* solitary.

P. **peltātum.** *May Apple. Wild Mandrake.*
In woods and fields, common in Middle and Western States, rare in N. Eng. Height about 1f. It is among our more curious and interesting plants. Stem round, sheathed at base, dividing into 2 round petioles, between which is the flower. Leaves oftener cordate than peltate, in 5—7 lobes, each lobe 6′ long from the insertion of the petiole, 2-lobed and dentate at apex. Flowers pedunculate, drooping, white, about 2′ diam. Petals curiously netted with veins. Fruit ovoid-oblong, large, yellowish, with the flavor of the strawberry. The root is cathartic. May.

3. JEFFERSONIA. Bart.

In honor of President Jefferson, a patron of science.

Sepals 4, colored, deciduous ; petals 8, spreading, incurved ; stamens 8, with linear anthers ; stigma poltate ; capsule obovata, stipitate, opening by a circumscissile dehiscence.—*Scape simple,* 1-*flowered. Lvs.* 2-*parted or binate*

J. **diphylla.** Barton.
A singular plant. 8—14′ high, Middle and Western States. Rhizoma horizontal. Each petiole bears at the top a pair of binate leaves, which are placed base to base, and broader than long, ending in an obtuse point, glaucous

beneath. Scape as long as the petioles. Flowers large, regular, white. The capsule opens only half round, and has therefore a persistent lid. Apr.—The plant has in Ohio the reputation of a stimulant and antispasmodic, and is then significantly termed *rheumatism root*.

4. LEONTICE.

Gr. λεων, a lion; the leaf is likened to a lion's foot-track.

Calyx free from the ovary, of 3—6 green sepals; corolla of 6 petals, each bearing a scale attached to the claw at base within; stamens 6; cells of the anther dehiscent at edge; pericarp membranaceous (caducous), 2—4-seeded; seeds erect, globose.

L. THALICTROÏDES. (Caulophyllum. *Michx.*) *Pappoose Root.*
Smooth; lvs. biternate and triternate; lfts. oval, petiolate, unequally lobed, the terminal one equally 3-lobed.—A smooth, handsome plant, in woods, Can. to Ky. Plant glaucous, purple when young. Stem 1—2½f high, round, dividing above into 2 parts, one of which is a 3-ternate leaf-stalk, the other bears a 2-ternate leaf and a racemose panicle of greenish flowers. Leaflets paler beneath, 2—3' long, lobed like those of the Thalictrum or Aquilegia. Seeds 2 (mostly 1 by abortion), naked after having burst the caducous, thin pericarp resembling berries on thick stipes. May.

ORDER VII. CABOMBACEÆ.—WATERSHIELDS.

Herbs aquatic, with floating, entire, centrally peltate leaves.
Fls. axillary, solitary, small. Sep. 3—4, colored inside.
Cor.—Petals 3—4, alternate with the sepals.
Sta. hypogynous, either 6, or more than 17. Anth. adnate.
Ova. 2 or more. Stig. simple.
Fr. indehiscent, tipped with the hardened style.
Sds. globular, pendulous. Embryo minute, 2-lobed, external to an abundant, fleshy albumen.
Genera 2, species 3. American water-plants, extending from Cayenne, S. America, to N. England.
Property—Slightly astringent.

BRASENIA. Schreb.

Calyx of 3—4 sepals, colored within, persistent; corolla of 3—4 petals; stamens 18—36; ovaries 6—18; carpels oblong, 2-(or by abortion 1-)seeded.—2/ *Aquatic. The stem, peduncles, and under surface of the leaves are covered with a viscid jelly.*

B. PELTATA. Pursh. (Hydropeltis purpurea. *Mx.*) *Water Target.*
It inhabits muddy shores and pools, often in company with the water-lily Can. to Ga. and Ark. Leaves peltate, elliptical, entire, 2—3' by 1—1½', with the long, flexible petioles inserted exactly in the centre, floating on the surface of the water, smooth and shining above. Flowers arising to the surface, on long, slender, axillary peduncles. Petals purple, about 3" long. July.

ORDER VIII.—NELUMBIACEÆ.—WATER-BEANS.

Herbs aquatic, with peltate, fleshy, radical lvs. Rhizoma prostrate.
Fls. large, solitary, on long, erect scapes. Sep. 4—5.
Cor.—Petals 00, in many rows, arising from without the disk.
Sta. 00, in several rows; filaments petaloid; anth. adnate, introrse.
Ova. 00, separate, each with a simple style and stigma.
Fr.—Nuts generally 1-seeded, half sunk in hollows of the very large torus.
Sds. destitute of albumen, and with a highly developed embryo.
This order comprises but a single genus with 2 species, two of which inhabit the still waters of tropical regions, and the other, of the U. S. The nuts are eatable, and indeed all the other parts of the plant.

NELUMBIUM. Juss.

Characters of the genus the same as those of the order.

N. LUTEUM.
Lvs. peltate, orbicular, entire; anth. with a linear appendage.—A magnificent flowering plant, peculiar to the stagnant waters of the south and west.

but occasionally met with in Ct. and N. Y. Rhizoma creeping in mud in depths of water from 2 or 3 to 6 f. From this arise the simple scapes and petioles to the surface. Leaves 10—18' diam., the petioles inserted on one side of the centre. Flowers several times larger than those of Nymphæa odorata, and without fragrance. Petals concave, of a brilliant white, becoming yellow towards the base. The nuts imbedded in the torus are about the size of acorns, and are used for food by the Indians. June.

ORDER IX. NYMPHÆACEÆ.—WATER LILIES.

Herbs aquatic, with peltate or cordate leaves from a prostrate rhiz oma.
Fls. large, showy, often sweet-scented.
Cal — ¿ Sepals and petals numerous, imbricated, gradually passing into each other. *Sep.* persistent.
Cor. — ? *Pet.* inserted upon the disk which surrounds the pistil.
Sta. numerous. in several rows upon the disk. *Fil.* petaloid. *Anth.* adnate, introrse.
Ova. many-celled, many-seeded, surrounded by a fleshy disk.
Sds. attached to the spongy placenta, and enveloped in a gelatinous arli.

Genera 5, species 20, inhabiting the northern hemisphere. Their general aspect is that of an endogen, but they have two foliaceous cotyledons. The stems of Nymphæa contain a powerful astringent principle, which is removed by repeated washing in water, after which they are tasteless and may be used for food.

Genera.

Flowers white or rose-color, ...   Nymphæa. 1
Flowers yellow, ...   Nuphar. 2

1. NYMPHÆA.

The Greek Nymph or Naiad of the waters.

Sepals 4—5; petals 00, inserted on the torus at its base; stamens gradually transformed into petals; stigma surrounded with rays; pericarp many-celled, many-seeded.—♃ *Aquatic.*

N. ODORÀTA. *Water Lily.*
Lvs. orbicular, cordate, entire, with veins prominent beneath; *cal.* 4-sepaled, equaling the petals; *stig.* 15—20-rayed.—One of the loveliest of flowers, possessing beauty, delicacy and fragrance in the highest degree. Ponds and sluggish streams, N. Am. E. of R. Mts. Rhizoma thick, in mud where the water is of 3—8 or 10f in depth, sending up leaves and flowers to the surface. Leaves 5—6' diam., dark shining green above, cleft at the base quite to the insertion of the long petiole. Sepals colored within. Petals lanceolate, 1½—2' long, of the most delicate texture, white, tinged with purple. Filaments yellow, dilated gradually from the inner to the outer series so as to pass insensibly into petals. (§ 72.) July.
β. rosea. Ph. Petals stained with purple. *Mass.*

2. NUPHAR. Smith.

Sepals 5 or 6, oblong, concave, colored within; corolla of numerous small petals furrowed externally. and inserted with the numerous, truncated, linear stamens on the torus; stigma discoid, with prominent rays; pericarp many-celled, many-seeded.—♃ *Aquatic.*

1. N. ADVÈNA. Ait. (Nymphæa Adv. *Mx.*) *Yellow Pond Lily.*
Lvs. oval, rounded at apex, with rounded, diverging lobes at base; *sep.* 6; *pet.* 00; *stig.* 12—15-rayed, margin crenate.—Very common in sluggish streams and muddy lakes, Can. to Ga. W. to Oreg. A well looking and very curious plant, but from its filthy habits it has been called, with some justice, the *frog lily.* The rhizoma is large, creeping extensively. Leaves large, dark green, shining above, and, when floating, pale and slimy beneath. Petioles half round. Flowers rather large and globular in form, erect, on a thick, rigid stalk. Three outer sepals yellow inside, and the three inner entirely yellow, as well as the petals and stamens. Jn. Jl.
β. tomentosa. T. & G. (N. tomentosa. *Nutt.*) *Lvs.* canescently tomentose beneath

2. KALMIANA. Ait. (Nymphæa Kalmiana. *Michx.* Nuphar lutei.
β. Kalmiana. T. & G.) *Kalm's Pond Lily.*
"*Floating lvs.* oblong, cordate, lobes approximate; *submersed lvs.* membranaceous, reniform-cordate, the lobes divaricate, margin waved, apex retuse;"
stig. 8—12-rayed, somewhat crenate.—A smaller species, with small yellow
flowers, growing in similar situations with the last, N. States.—Dr. Robbins,
from whose MSS. the above is quoted, thinks it wholly distinct from N. lutea,
Smith, or any other species. Petiole subterete; upper leaves 2—3' long 1½ 2½'
wide, lower leaves 3—4' diam. Jl.

ORDER X. SARRACENIACEÆ.—WATER PITCHERS.

Herbs aquatic, perennial in bogs, with fibrous roots.
Lvs. radical, with a hollow, urn-shaped petiole and lamina articulated at summit.
Fls. large, solitary, or several on scapes.
Cal.—Sepals 5, persistent, with a 3-leaved involucel at base. *Æst.* imbricate.
Cor.—Petals 5, unguiculate, hypogynous, concave.
Sta. 60, hypogynous. *Anth.* oblong, adnate, introrse.
Ova. 5-celled, placentæ central. *Sty.* single. *Stig.* dilated, peltate, 5-angled.
Fr. capsular, 5-celled, 5-valved, crowned with the broad persistent stigma.
Sds. 60, minute.

An order consisting of only 2 genera, (one inhabiting the bogs of N. America, the other in Guiana,)
and 7 species.

SARRACENIA. Tourn.

In memory of Dr. Sarrasen of Quebec, the discoverer of the genus.

Calyx of 5 sepals, with 3 small bracts at base; petals 5, deciduous;
stigma very large, peltate, persistent, covering the ovary and stamens; capsule 5-celled, 5-valved, many-seeded.

S. PURPUREA. *Side-saddle Flower.*

Lvs. (ascidia) radical, decumbent, inflated, contracted at the mouth, winged
on the inner side, ending in a broad-cordate, erect lamina.—One of the most
curious of plants. Grows in wet meadows and about mud lakes, Lab. to Flor.
Leaves 6—9' long, rosulate, ever-green, composed of a hollow, pitcher-form
petiole (?) swelling in the middle, with a wing-like appendage extending the
whole length inside, from ½—1' wide, and extended on the outside of the mouth
into a lamina (?), covered above with reversed hairs. Their capacity when
of ordinary size is about a wine-glass, and they are generally full of water
with drowned insects. Scape 14—30' high, terete, smooth, supporting a single
large, purple, nodding flower, almost as curious in structure as the leaves. Jn.
β. heterophylla. Torr. (S. heterophylla. *Eaton.*)—Scape rather shorter;
sep. yellowish-green; *pet.* yellow.—Northampton, Ms. *Mr. R. M. Wright;*
Leaves scarcely different.

ORDER XI. PAPAVERACEÆ.—POPPYWORTS.

Plants herbaceous, generally with a colored juice.
Lvs. alternate, simple or divided, without stipules.
Fls. solitary, on long peduncles, never blue.
Cal.—Sepals 2, rarely 3, deciduous, imbricated in æstivation.
Cor.—Petals 4, rarely 5 or 6, hypogynous.
Sta. often 60, but some multiple of 4, rarely polyadelphous. *Anth.* innate.
Ova. solitary. *Sty.* short or 0. *Stig.* 2, or if more, stellate upon the flat apex of ovary.
Fr. either pod-shaped, with 2 parietal placentæ, or capsular with several.
Sds. 60, minute, Embryo minute, at the base of oily albumen.

An order consisting of 18 genera and 130 species, more than two-thirds of which are natives of Europe.
The order is characterised by active narcotic properties, principally resident in the turbid juices. The
seeds are commonly rich in fixed oil. Several of the species are highly ornamental in cultivation.

Conspectus of the Genera.

	{ Leaves armed with prickly teeth. { Stigmas concave.			*Argemone.* 3
	{ { Stigmas convex.			*Meconopsis.* 2
{ yellow.	{ Leaves unarmed, entirely green, cauline.			*Chelidonium.* 4
{ orange-red.	Leaves radical, reniform. Capsule terete.			*Sanguinaria.* 1
{ white.	Leaves unarmed, cauline. Capsule globose.			*Papaver.* 5
Petals { colorless.	Leaves multifid with linear segments. Capsule terete.			*Eschscholtzia.* 6

1. SANGUINARIA.

Lat. *sanguis,* blood; all parts abound in a red juice.

Sepals 2, caducous; petals 8, in 2 series, those of the outer series

14

longer; stamens 00; stigma 1—2-lobed, sessile; capsule pod-like, oblong, 1-celled, 2-valved, acute at each end, many-seeded.—♃ *Juice orange red.*

S. CANADENSIS. *Blood-root.*

An interesting flower of woods, groves, &c., appearing in early spring Rhizoma fleshy, tuberous, and when broken or bruised exudes a blood-colored fluid, as also does every other part of the plant. From each bud of the root-stalk there springs a single large, glaucous leaf, and a scape about 6' high, with a single flower. Whole plant glabrous. Leaf kidney-shaped, with roundish lobes separated by rounded sinuses. Flower of a quadrangular outline, white, scentless, and of short duration. The juice is emetic and purgative. Apr. May.

2. CHELIDONIUM.

[Its departure.
Gr. Χελιδων, the swallow; being supposed to flower with the arrival of that bird, and to perish with

Sepals 2, suborbicular; petals 4, suborbicular, contracted at base; stamens 24—32, shorter than the petals; stigma 1, small, sessile, bifid; capsule silique-form, linear, 2-valved, 1-celled; seeds crested.— ♃ *with yellow juice.*

C. MAJUS. *Celandine.*

Lvs. pinnate; *lfts.* lobed, segments rounded; *fls.* in umbels.—A pale green, fleshy herb found under fences, by road-sides, &c., arising 1—2f high. Leaves smooth, glaucous, spreading, consisting of 2—4 pairs of leaflets with an odd one. Leaflets 1½—2½' long, ½ as broad, irregularly dentate and lobed, the partial stalks winged at base. Umbels thin, axillary, pedunculate. Petals elliptical, entire, yellow, and very fugacious, like every other part of the flower. The abundant bright yellow juice is used to cure itch and destroy warts. May.—Oct. ⚥

3. ARGEMONE.

Gr. αργεμα, a disease of the eye, which this plant was supposed to cure.

Sepals 3, roundish, acuminate, caducous; petals 6, roundish, larger than the sepals; stamens 00, as short as the calyx; stigma sessile, capitate, 6-lobed; capsule obovoid, opening at the top by valves.— ① *Herbs with yellow juice.*

A. MEXICANA. *Horn Poppy.*

Lvs. repand-sinuate or pinnatifid, with spiny teeth; *fl.* solitary, erect, axillary; *cal.* prickly; *caps.* prickly, 6-valved.—A weed-like plant, native at the south and west, ⚥ at the north. Stem 2—3f high, branching, armed with prickly spines. Leaves 5—7' or 8' long, sessile, spinose on the margin and veins beneath. Flowers axillary and terminal, on short peduncles, about 2' diam., yellow. The juice becomes in air a fine gamboge-yellow, and is esteemed for jaundice, cutaneous eruptions, sore eyes, fluxes, &c. July.⚥

β. *Fls.* ochroleucous.—γ. *Fls.* larger, white.

4. MECONOPSIS. Viguier.

Gr. μηκων, a poppy; οψις, resemblance.

Sepals 2; petals 4; stamens 00; style distinct; stigmas 4—6, radiating, convex, free; capsule obovate, 1-celled, opening by 4 valves at apex.—♃ *Herbs with a yellow juice.*

M. DIPHYLLA. DC. (Chelidonium. *Michx.* Stylophorum. *Nutt.*)

Lvs. pinnately divided, glaucous beneath, segments 5—7, ovate-oblong sinuate, *cauline* 2, opposite, petiolate; *ped.* aggregated, terminal; *caps.* 4-valved, echinate-setose.—Woods, Western States! Plant 12—18' high. Leaves large 8' by 6', on petioles about the same length; terminal segments somewhat confluent. Peduncle about 3' long. Petals deep yellow. May.

5. PAPAVER.

Celtic, *papa*, *pap*; a superific food for children, composed of poppy-seeds, &c.

Sepals 2, caducous; petals 4; stamens ∞; capsule 1-celled, opening by pores under the broad, persistent stigma.—*Exotic herbs, mostly ⊕, with white juice abounding in opium.*

1. P. SOMNIFERUM. *Opium Poppy.*
Glabrous and glaucous; *lvs.* clasping, incised and dentate; *sep.* glabrous; *cap.* globose.—A plant with large, brilliantly white flowers, double in cultivation. Stem 1½—3f high. Leaves 4—8′ by 2—3′, with rather obtuse dentures. Every part, but especially the capsule, abounds with a white juice powerfully narcotic, and which when hardened in the sun, forms the *opium* of the shops. For this drug, it is extensively cultivated in Europe and southern Asia. Jn. Jl. †§

2. P. DUBIUM. *Dubious Poppy.*
Canlescent; *st.* hispid with spreading hairs; *lvs.* pinnately parted, segments incised; *ped.* clothed with appressed hairs; *sep.* hairy; *caps.* obovoid-oblong, glabrous.—Sparingly naturalized in cultivated grounds, Penn. Stem about 2f high. Flowers light red or scarlet. Jn. Jl. §

3. P. RHŒAS. *Common Red Poppy.*—St. many-flowered, hairy; *lvs.* incisely pinnatifid; *capsules* smooth, nearly globose.—Distinguished from the last species chiefly by its more finely divided leaves and its globular capsule. About 2f high. Flowers very large and showy, of a deep scarlet red. Varieties are produced with various shades of red and particolored flowers, more or less double. Jn. Jl. †

4. P. ORIENTALE. *Oriental Poppy.*—St. 1-flowered, rough; *lvs.* scabrous, pinnate, serrate; *capsules* smooth. — Native of Levant. Stem 3f high. Flowers very large, and of a rich scarlet color, too brilliant to be looked upon in the sun. Jn. †

6. ESCHSCHOLTZIA.

Named for Eschscholtz, a German botanist, well known for his researches in California.

Sepals 2, cohering by their edge, caducous; petals 4; stamens ∞, adhering to the claws of the petals; stigmas 4—7, sessile, 2—3 of them abortive; capsule pod-shaped, cylindric, 10-striate, many-seeded.—⊙ *Leaves pinnatifid, glaucous. The juice, which is colorless, exhales the odor of hydrochloric acid.*

1. E. DOUGLASII. Hook. (Chryseis Californica, of *Lindl.* and of 1st *edit.*)—St. branching, leafy; *torus* obconic; *cal.* ovoid, with a very short, abrupt acumination; *pet.* bright yellow, with an orange spot at base.—A very showy annual, common in our gardens. Native of California, Oregon, &c. The foliage is smooth, abundant and rich, dividing in a twice or thrice pinnatifid manner into linear segments. Flowers 2′ broad. †

2. E. CALIFORNICA. Hook. (Chryseis crocea, *Lindl.* and of 1st *edit.*)—St. branching, leafy; *torus* funnel-form, with a much dilated limb; *cal.* obconic, with a long acumination; *fls.* orange-yellow.—From California. Leaves and color of flowers as in the preceding, except the latter are more of a reddish-orange hue. †

ORDER XII. FUMARIACEÆ.—FUMEWORTS.

Plants herbaceous, with brittle stems and a watery juice.
Lvs. mostly alternate, multifid, often furnished with tendrils.
Fls. irregular, purple, white or yellow. Sep. 2, deciduous.
Cor.—Petals 4, hypocynous, parallel, one or both of the outer saccate, 2 inner cohering at apex.
Sta. 6, diadelphous; *fil.* dilated; *anth.* adnate, extrorse. 2 outer 1-celled, middle 2-celled.
Ova. superior, 1-celled; *sty.* filiform; *stig.* with one or more recurva.
Fr. either an indehiscent nut 1—2-seeded, or a pod-shaped capsule many-seeded.
Sds. shining, arilled. *Albumen* fleshy.

Genera 16, species 110,—some of them beautiful and delicate, inhabiting thickets in the temperate regions of the northern hemisphere. They possess no remarkable action upon the animal economy.

Conspectus of the Genera.

		{ Fruit a pod-shaped capsule . . : *Corydalis* 3
{ only 1 of the outer, acc. or spurred.	{ Fruit a subglobose nut . . . : *Fumaria* 4
{ distinct, { 2 outer equally saccate or spurred . . . . . . : *Dielytra*. 1
Petals { united, base bigibbous, apex 4-lobed. Climbing herbs. . . . . : *Adlumia*. 2

1. DIELYTRA. Borkhausen.

Gr. δις, double; ελυτρον, wing-case; in allusion to the two spurs.

Sepals 2, small; petals 4, the 2 outer equally spurred or gibbous at the base; stamens united in 2 sets of 3 in each; pod 2-valved, many-seeded.—♃.

1. D. CUCULLARIA. DC. (Corydalis Cucullaria. *Pers.*) *Dutchman's Breeches.*

Rt. bulbiferous; *rac.* 4—10-flowered, secund; *spurs* divergent, elongated, acute, straight.—Woods, Can. to Ky. A smooth, handsome plant. Rhizoma bearing triangular, small, pale-red bulbs under ground. Leaves radical, multifid, somewhat triternate, smooth, with oblong-linear segments, the petioles rather shorter than the scape. Scape slender, 6—10′ high. Flowers scentless, nodding, whitish, at summit yellow. Pedicels short, axillary to a bract, an ¹ with 2 minute bracteoles near the flower. Spurs about as long as the corolla. April, May.

2. D. CANADENSIS. DC. (D. eximia. *Beck.* Corydalis Canadensis. *Goldie.*) *Squirrel Corn. Dutchman's Breeches.*

St. subterranean, tuberiferous; *tubers* globose; *rac.* simple, secund, 4—6-flowered; *spurs* short, rounded, obtuse, slightly incurved.—A smooth, pretty plant, common in rocky woods, Can. to Ky. The rhizoma bears a number of roundish tubers about the size of peas, and of a bright-yellow color. Leaves radical, subglaucous, biternate, the leaflets deeply pinnatifid, segments linear-oblong, obtuse, 6—8″ long. Scape 6—8′ high, bearing a few odd-looking flowers. Corolla white, tipped with yellow, 5″ long. Calyx minute. Stamens 3 on each lip. May, Jn.

3. D. EXIMIA. DC. (Corydalis formosa. *Ph.*) *Choice Dielytra.*

Rhizoma scaly-bulbiferous; *lvs.* numerous; *rac.* compound, the branches cymose; *spurs* very short, obtuse, incurved; *stigma* 2-horned at apex.—A fine species, on rocks, &c., found by *Dr. Sartwell*, in Yates Co., N. Y. (S. to N. Car.) Leaves radical, 10—15′ high, somewhat triternate, with incisely pinnatifid segments, but quite variable. Scape 8—12′ high, with several (4—8) cymes, each with 6—10 purplish, nodding flowers. Corolla 8—10″ long, broad at base. Bracts purplish, at base of pedicels. Jn.—Sept. †

2. ADLUMIA. Raf.

Named for John Adlum, Washington, D. C., a cultivator of the vine.

Sepals 2, minute; petals 4, united into a fungous, monopetalous corolla, persistent, bigibbous at base, 4-lobed at apex; stamens united in 2 equal sets; pod 2-valved, many-seeded.—② *Climbing.*

A CIRRHOSA. Raf. (Fumaria fungosa. *Willd.* Corydalis. *Pers.*) *Mountain Fringe.*

A delicate climbing vine, native of rocky hills, Can. to N. Car. Stem striate, many feet in length. Leaves decompound, divided in a pinnate manner, ultimate divisions 3-lobed, smooth, their foot-stalks serving for tendrils. Flowers very numerous, in axillary, pendulous, cymose clusters, pale-pink. Calyx minute. Corolla slightly cordate at base, of 4 petals united into a spongy mass, cylindric, compressed, tapering upwards, 2-lipped. Fine for arbors. Jn.—Aug.†

3. CORYDALIS. DC.

Greek name of the Fumitory, from which genus this was taken.

Sepals 2, small; petals 4, one of which is spurred at the base;

stamens 6, diadelphous; filaments united into 2 equal sets by their broad bases which sheath the ovary; pod 2-valved, compressed, many-seeded.—*Lvs. cauline. Pedicels racemose, bractless.*

1. C. GLAUCA. Ph. (Fumaria glauca. *Curtis.*) *Glaucous Corydalis.*
St. erect, branched; lvs. glaucous, bipinnate, segments cuneate-obovate, 3-lobed; pods linear, as long as the pedicels.—②. A smooth, delicate plant, in mountainous woods, Can. to N. Car., covered with a glaucous bloom. Root fusiform. Stem 1—4f high. Leaflets nearly 1' long and ¼' wide, cut into 3 obtuse lobes. Flowers terminal, on the subpaniculate branches. Calyx of 2, ovate, acuminate sepals, between which, placed crosswise, is balanced the cylindrical, ringent corolla, beautifully colored with alternating shades of red and yellow. May—Aug.

2. C. AUREA. Willd. (Fumaria aurea. *Muhl.*) *Golden Corydalis.*
St. branching, diffuse; lvs. glaucous, bipinnate, lobed, the lobes oblong-linear, acute; bracts linear-lanceolate, dentate, 3 times as long as the peduncle; rac. secund, opposite the leaves and terminal; pod terete, torulose.—① In rocky shades, Can. to Mo. S. to Ga. Stem 8—12' high, with finely divided leaves. Flowers bright yellow, about half as long as the torulose pods which succeed them. May—Aug.

4. FUMARIA.

Lat. *fumus*, smoke; from its disagreeable smell.

Sepals 2, caducous; petals 4, unequal, one of them spurred at the base, filaments in 2 sets each with 3 anthers; nut ovoid or globose, 1-seeded and valveless.—*Lvs. cauline, finely dissected.*

F. OFFICINALIS. *Fumitory.*
St. suberect, branched, and spreading; lvs. bipinnate; lfts. lanceolate, cut into linear segments; rac. loose; sep. ovate-lanceolate, acute, about as long as the globose, retuse nut.—A small, handsome plant, in sandy fields and about gardens, introduced from Europe. Stem 10—15' high, smooth as well as the leaves. Leaflets cut into segments dilated upwards. Flowers small, rose-colored, nodding, the pedicels becoming erect in fruit, and twice as long as the bracts. July, Aug. §

ORDER XIII. CRUCIFERÆ.—CRUCIFERS.

Plants herbaceous, very rarely suffruticose, with alternate leaves.
Fls. yellow or white, rarely purple, without bracts, generally in racemes. [a cross.
Cal.—Sepals 4, deciduous.
Cor. of 4 regular petals, their claws inserted into the receptacle, and their limbs spreading in the form of
Sta. 6, 2 of them upon opposite sides, shorter than the other 4. [sepiment.—*stig.* two.
Ova. composed of two united carpels, with two parietal placentæ united by a membranous false dis-
Pr a silique or silicle usually 2-celled.
Sds. attached in a single row to each side of the placenta; —albumen 0.
Embryo, with the two cotyledons variously folded on the radicle.

Genera 172, species 1800.—This is a very natural order, larger than any of the preceding. The greater part of the species are found in the temperate zones. About 100 are peculiar to this continent.

Properties.—The Cruciferæ as a class are of much importance to man. They furnish several alimentary articles which are very nutricious, as the turnip, cabbage or cauliflower; several others are used as condiments; as mustard, radish, cochlearia, &c. They all possess a peculiar acrid, volatile principle, diapnoic and through every part, often accompanied by an etherial oil abounding in sulphur. They are also remarkable for containing more nitrogen than other vegetables, for which reason ammonia is generally evolved in their putrefaction. In medicine they are eminently stimulant and antiscorbutic. None are really poisonous, although very acrid. The root of Isatis tinctoria affords a blue coloring matter.

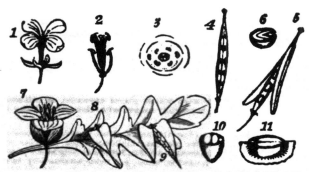

FIG. 10.—1. A flower of Sinapis nigra. 2. The stamens (4 long and 2 short) and pistil. 3. Plan of the flower,—stamens in 2 rows, outer row half wanting. 4. A silique.—5, partly open, showing the septum with seeds attached. 6. Cross section of a seed, cotyledons conduplicate (0 >>). 7. Flower (enlarged) and leaf of Capsella. 8. A silicle.—9, open, showing the narrow septum with seeds. 10. Cross section of a seed, cotyledons incumbent (0 ‖). 11. Section of a winged seed of Arabis Canadensis, cotyledons accumbent 0 ‖.

Conspectus of the Genera.

* Ornamental exotics not culinary.

Fruit siliculose ;	{ Petals equal.	{ Silicle. { Some of the stamens toothed.	*Alyssum.*	8
		{ 2-celled. { Stamens all toothless.	*Lunaria.*	9
		{ Silicle 1-celled, 1-seeded.	*Isatis.*	11
	{ Petals unequal, the 2 outer ones larger.		*Iberis.*	10
Fruit siliquose ;	{ Seeds flat(o=).	{ Stigma capitate.	*Cheiranthus.*	21
		{ Stigmas cornute.	*Matthiola.*	22
	{ Seeds ovoid (oꟷ). Calyx furrowed at base.		*Hesperis.*	19

** Plants native or naturalized, and culinary exotics.

Flowers cranic.	Fruit siliculose.	Silicle jointless.	{ ovate and compressed,	{ emarginate,	{ 4—60-seeded.	*Thlaspi.*	1
					{ 2-seeded.	*Lepidium.*	3
				{ entire at the apex.		*Draba.*	4
			{ ovoid or globose.	{ Leaves cauline.		*Cochlearia.*	5
				{ Leaves radical.		*Subularia.*	6
			{ triangular, cuneate at base.			*Capsella.*	2
		Silicle with 2 joints, upper joint ovate or ensiform.				*Cakile.*	25
	Fruit siliquose.	Valves veinless.	{ Seeds in a single row.	{ Pods linear. { Seeds not bordered, { Seeds bordered.	{ o=. { o‖.	*Cardamine.* { *Sisymbrium.*	16 { 18
				{ Pods lanceolate. Leaves few.		*Cheiranthus.*	21
			{ Seeds in a double row,	{ wingless.		*Dentaria.*	17
				{ winged on the margin.		*Nasturtium.*	12
		Valves with one central vein, flat, linear.				*Turritis.*	14
		Valves 0. Silique indehiscent, transversely celled.				*Arabis.*	15
						Raphanus.	23
Flowers yellow.	Fruit siliquose.	Calyx erect or closed.	{ Seeds in a single row,	{ globose. { oblong (0=). { oblong (0 ‖).	{ Pods dehiscent. { Pods indehiscent. { Leaves lyrate. { Leaves undivided.	*Brassica.* { *Raphanus.* { *Barbarea.* { *Erysimum.*	24 { 26 { 13 { 20
			{ Seeds in a double row.			*Turritis.*	14
		Calyx spreading.	{ Seeds ovoid or oblong (0 ‖).			*Sisymbrium.*	18
			{ Seeds globose (0 >>).			*Sinapis.*	23
	Fruit siliculose, . .	{ oblong or somewhat terete.				*Nasturtium.*	12
		{ obovoid or subglobose.				*Camelina.*	7

SECTION I. SILICULOSÆ. (§ 80, note.)

1. THLASPI. Dill.

Gr. θλαω, to compress; on account of the compressed or flattened silicles.

Calyx equal at base ; petals equal ; silicle short, flat, emarginate at the apex, many-seeded ; valves carinate, often winged on the back ; cotyledons accumbent (0=).—*Lvs. undivided. Fls. white.*

1. T. ARVENSE. *Penny Cress.*

Lvs. oblong, coarsely dentate, smooth, *silicle* roundish-obovate, shorter than *the pedicel; stig. subsessile.*—In cultivated, stony fields, Can. and Northern States. *The whole plant smooth, 8—18′ high, branched. Leaves 1—2′ long, ⅓ as wide, the cauline slightly arrow-shaped with small obtuse auricles, wavy and toothed* at margin. Flowers small, in terminal racemes. Silicles large, flat, with dilated wings. The plant has a disagreeable flavor of garlic. June. §

2. T. ALLIACEUM.

Lvs. oblong, obtuse, somewhat dentate, upper ones sagittate-amplexicaul, with acute auricles; *silicles* ovate-ventricose; *stig.* subsessile.—In cultivated fields, Western States, not common. Stems 6—10' high. Lower leaves petiolate. Flowers smaller than in T. arvense, in terminal racemes. This also savors of garlic. May—Jl. §

3. T. TUBEROSUM. Nutt.

Rt. tuberiferous and fibrous; *st.* pubescent, simple, short; *lvs.* rhomboid-ovate, obscurely dentate, smooth and sessile, radical ones petiolate; *silicle* suborbicular.—♃ Penn. Stem not more than 4—5' high. Flowers rather large, rose-colored. Apr. May.

2. CAPSELLA. Vent.

Diminutive from *capsa*, a chest or box; alluding to the fruit.

Calyx equal at base; silicles triangular-cuneiform, obcordate, compressed laterally; valves carinate, not winged on the back; septum sublinear; style short; seeds ∞; oblong, small, 0∥.— ① *Fls. white. A troublesome weed.*

C. BURSA-PASTORIS. Mœnch. (Thlaspi Bursa-pastoris. *Linn.*) *Shepherd's purse.*

Found everywhere, in fields, pastures, and roadsides. Stem 6—8—12' high, nearly smooth in the upper part, hirsute below, striate, branching. Root-leaves rosulate, 2—5—8' long, ¼ as wide, cut-lobed, on margined petioles; segments about 13. These leaves are sometimes wanting, (when the weed is crowded,) or only dentate. Stem-leaves much smaller, very narrow, with two small, acute auricles at base, half clasping the stem. Flowers small, in racemes, which are finally 3—12' long. Silicle smooth, triangular, emarginate at the end, and tipped with the style. April—Sept. §

3. LEPIDIUM. R. Br.

Gr. λεπις, a scale; from the resemblance of the silicle.

Sepals ovate; petals ovate, entire; silicles oval-orbicular, emarginate; septum very narrow, crossing the greater diameter; valves carinate, dehiscent; cells 1-seeded, 0∥ or 0—.—*Fls. white.*

1. L. VIRGINICUM. *Wild Pepper-grass.*

Lvs. linear-lanceolate, incisely serrate, smooth; *st.* paniculately branched above; *sta.* 2—4; silicles orbicular, emarginate; seeds 0—.—① In dry fields and road-sides, U. S. Stem rigid, round, smooth, 1f high. Leaves 1—2' by 1—3", acute, tapering at base into a petiole, upper ones sessile, lower pinnatifidly cut. Flowers and silicles very numerous, in a panicle of racemes. *Fls.* very small, mostly diandrous; silicles lens-shaped, 1½" diam. with a notch at the end. Taste pungent, like that of the garden pepper-grass. Jn.—Oct.

2. L. CAMPESTRE, R. Br. (Thlaspi campestris. *Linn.*) *Yellow Seed.*

Cauline lvs. sagittate, denticulate; *silicles* ovate, winged, emarginate, scaly, punctate.—♀. In waste places and dry fields, especially among flax. Stem strictly erect, round, minutely downy, 6—10' high, branching. Leaves 1' long, ½ as wide, acute, with 2 lobes at base, upper one clasping the stem, all minutely velvety. Flowers small. Silicles 1½" long, numerous, in long racemes. Jn. Jl. §

3. L. RUDERALE.

Lvs. cauline, incised, those of the branches linear, entire; *fls.* apetalous, and with but 2 stamens; *silicles* broadly oval or suborbicular, emarginate, wingless; *cotyl.* 0∥.—Dry fields, Mich., Ia., Mo. Stem 10—15' high. Racemes many. Flowers remarkable for wanting the petals, which are always present in our other species.

4. L. SATIVUM. *Peppergrass.*—*Lvs.* variously divided and cut; branches without spines; *silicles* orbicular, winged.—① Native of the East. Stems 1—2f

high, very branching. Silicles 2—3″ broad, very numerous. A well known garden salad. July. ‡♀

4. DRABA.

Gr. ἀραβις, acrid, biting; from the taste of the plant.

Calyx equal at base; petals equal; filaments without teeth; silicle oval-oblong, entire, the valves flat or convex; cells 2, many-seeded; seeds not margined.

1. D. VERNA. (Eriophila vulgaris. *DC.*) *Whitlow Grass.*
Scape naked; *lvs.* oblong, acute, subserrate, hairy; *pet.* bifid; *stig.* sessile; *silicle* oval, flat, shorter than the pedicel.—① A little early-flowering plant in grassy fields, Can. to Va. Leaves all radical, lanceolate, ½—1½′ long, ½ as wide, with a few teeth towards the end. Scape a few inches high, with a raceme of 5—15 small, white flowers. Calyx spreading. Petals cleft half way down. Silicles about a line wide and 3—4″ long, with deciduous valves. Apr. May.

2. D. ARABISANS. Michx. (Arabis. *Ph.*)
St. leafy, somewhat branched and pubescent; *lvs.* lanceolate, acutely dentate; *silicle* oblong-lanceolate, smooth, longer than the pedicel; *sty.* very short.—Lake shores, among rocks, Vt., N. Y., Mich. Stems several from the same root, 6—8′ high. Radical leaves 1′ or more in length, attenuate at base, with a few slender, spreading teeth; cauline leaves somewhat clasping. Flowers white, in a short raceme. Silicle elongated (1½′ long), acuminate, contorted, and might be called a silique. May.

3. D. CAROLINIÀNA. Walt. (D. hispidula. *Michx.*)
St. leafy at base, hispid, naked and smooth at the top; *lvs.* ovate-roundish, entire, hispid; *silicles* linear, smooth, longer than the pedicels, corymbose.—② Sandy fields, Conn., *Dr. Robbins*, R. I., *Mr. G. Hunt*, S. to Ga. Stem 1—3′ high, very hairy. Leaves clustered on the lower part of the stem, very hairy. Petals white, twice as long as the sepals. Silicle ½′ long, lance-linear, many-seeded. Stigma subsessile. Apr. Jn.

4. D. RAMOSISSIMA. Desv. (Alyssum dentatum. *Nutt.*)
Minutely pubescent; *sts.* numerous; *lvs.* linear-lanceolate, with remote and slender teeth, upper ones entire; *rac.* corymbosely paniculate; *silicle* lanceolate, about the length of the pedicel, and tipped with the style ½ as long.—♃ On rocks, Harper's Ferry, Va., west to Ky. Stems slender, 4—10′ long, with tufted leaves at top. Leaves about 1′ long, with 1 or 2 teeth on each side. Flowers white. Silicles 3—5″ in length, ascending. Apr. May.

5. D. NEMORÀLIS. Ehrh.
St. pubescent, branched; *lvs.* oval, hirsute, cauline lanceolate, toothed; *pet.* emarginate; *silicles* oblong-elliptical, the length of the pedicels; *sds.* nearly 30.—♃ Mich. Mo. Plant slender, 8—10′ high. Stem with few branches. Leaves mostly radical. Racemes much elongated in fruit, with very long pedicels. Flowers minute, yellowish white. May.

6. D. CUNEIFOLIA. Nutt.
Hirsute-pubescent; *st.* branching and leafy below; *lvs.* sparingly toothed, *radical* spatulate-oblong, *cauline* few, oblong, ovate, somewhat attenuate at base; *rac.* rather elongated in fruit; *silicles* oblong-lanceolate, minutely hispid, twice as long as the pedicels; *pet.* emarginate. *T. & G.*—♃ Grassy places about St. Louis, &c., *Nuttall.* Plant 3—8′ high. Flowers much larger than in the preceding. Petals white, nearly thrice longer than the sepals. Silicles about ½′ long and 30-seeded. March, Apr.

7. D. BRACHYCARPA. Nutt.
Minutely pubescent; *radical lvs.* roundish-ovate, petiolate; *cauline* oblong or linear, slightly dentate or entire; *rac.* many flowered, straight, elongated in fruit; *pet.* obovate, entire; *silicle* oval, glabrous, about as long as the pedicels 10—12 seeded.—♃ Grassy places near St. Louis. Stem much branched and leafy. Silicles 2—3″ long, March, Apr.

5. COCHLEARIA. Tour.

Lat. cochlear, a spoon; referring to the concave leaves.

Calyx equal at base, spreading; petals entire; stamens without teeth; silicle sessile, oblong or ovoid-globose, with ventricose valves; seeds many, not margined; 0—.—*Fls. white.*

1. C. ARMORACIA. *Horse Radish.*—*Radical lvs.* oblong, crenate; *cauline* long, lanceolate, dentate or incised, sessile; *silicle* elliptic.—♃ A common garden herb, native of Europe. Root fleshy, large, white, very acrid. Stem 2—3f high, angular, smooth, branching. Radical leaves near a foot long, ⅓ as wide, on long, channeled petioles. Lower stem-leaves often cut in a pinnatifid manner, upper toothed or entire. Flowers small, in corymbose racemes. The root is a well known condiment for roast beef and other viands. Jn.

β. aquatica. (C. aquatica. *Eaton* and *1st edit.*) *Lvs.* all pinnatifid, the lower ones doubly and finely so. Wet places, often submerged.§

2. C. OFFICINALIS. *Scurvy Grass.*—*Radical lvs.* cordate, petiolate, *cauline* ovate, angular or dentate; *silicles* oval-globose, half as long as the pedicel.— ♃ Native of Europe and of Arctic Am. Stem 8—12′ high. Root leaves 4—18′ long, ⅓ as wide. Flowers racemed. Occasionally cultivated for its powerful antiscorbutic properties. Jn.

6. SUBULARIA.

Named in reference to the linear-subulate leaves.

Silicle oval, valves turgid, cells many-seeded; stigma sessile; cotyledons linear, curved.—① *Aquatic, acaulescent herbs.*

8. AQUATICA. *Awlwort.*

A small plant growing on the muddy shores of ponds in Maine, *Nutt.*, and near the White Mts., *Pickering.* Leaves all radical, entire, subulate, an inch in length. Scape 2—3′ high, racemose, with a few minute, white flowers, on slender pedicels only 2″ in length. Jl.

7. CAMELINA. Crantz.

Gr. χαμαι, dwarf; λινον, flax.

Calyx equal at base; petals entire; silicle obovate or subglobose, with ventricose valves and many-seeded cells; styles filiform, persistent; seeds oblong, striate, not margined, 0 ‖.

C. SATIVA. Crantz. (Myagrum. *Linn.*) *Gold-of-pleasure. False Flax.* *Lvs.* lanceolate, sagittate at base, subentire; *silicle* obovate-pyriform, margined, tipped with the pointed style.—① In cultivated fields. Stem 1½—2¼ f. high, straight, erect, branching. Leaves roughish, 1—2′ long, clasping the stem with their acute, arrow-shaped lobes. Flowers small, yellow, in paniculated racemes. Silicles 2—3″ long, on pedicels 2—3 times as long.—Said to be cultivated in Germany for the oil which is expressed from the seeds. Jn.§

8. ALYSSUM.

Gr. α. privative; λυσσα, rage; supposed by the ancients to allay anger.

Calyx equal at base; petals entire; some of the stamens with teeth; silicle orbicular or oval, with valves flat or convex in the centre; seeds 1—4 in each cell.

1. A. SAXATILE. *Rock Alyssum. Madwort.*—*St.* suffruticose at base, subcorymbose; *lvs.* lanceolate, entire, downy; *silicle* ovate-orbicular, 2-seeded; *sds.* margined.—An early-flowering garden perennial, native of Candia. Stem 1f high, with numerous yellow flowers in close corymbose bunches. Apr. May.†

2. A. MARITIMUM. Lam. *Sweet Alyssum.*—*St.* suffruticose and procumbent at base; *lvs.* linear-lanceolate, acute, somewhat hoary; *pods* oval, smooth.— ♃ Sweet-scented garden plant, with fine leaves and small white flowers. Stem a foot in length. Flowers from Jn. to Oct.—All the species of Alyssum are of easy culture in common loamy soils. †

9. LUNARIA.

Lat. luna, the moon; from the broad, round silicles.

Sepals somewhat bisaccate at base; petals nearly entire; stamens without teeth; silicle pedicellate, elliptical or lanceolate, with flat valves; funiculus adhering to the dissepiment.

1. L. REDIVIVA. *Perennial Satin Flower* or *Honesty.*—St. erect, branching; lvs. ovate, cordate, petiolate, mucronately serrate; *silicles* lanceolate, narrowed at each end.—♃ From Germany. Stem 2—3f high. Flowers light purple. Jn. †

2. L. BIENNIS. DC. *Honesty.*—St. erect; lvs. with obtuse teeth; *silicles* oval, obtuse at both ends.—♁ These are large, hairy plants, native of Germany. Stems 3—4f high. Leaves cordate. Flowers lilac-colored. The broad, round, silvery silicles are the most remarkable feature of the plants. May, Jn. †

10. IBERIS.

Most of the species are native of *Iberia*, now Spain.

The 2 outside petals larger than the 2 inner; silicles compressed truncate, emarginate, the cells 1-seeded.—*None of the species are N. American.*

1. L. UMBELLATA. *Purple Candy-tuft.*—Herbaceous, smooth; lvs. linear-lanceolate, acuminate, lower ones serrate, upper ones entire; *silicles* umbellate, acutely 2-lobed.—This and the following species are very popular garden annuals, very pretty in borders, and of very easy culture. I. umbellata is from S. Europe. Stem 1f high. Flowers purple, terminal, in simple umbels, and like the rest of the genus remarkable for having the 2 outer petals larger than the 2 inner ones. Jn. Jl. †

2. I. AMARA. *Bitter Candy-tuft.*—Herbaceous; lvs. lanceolate, acute, somewhat toothed; fls. corymbed, becoming racemed; *silicles* obcordate, narrowly emarginate.—① Native of England. Stem 1f high. Flowers white. Jn. Jl.†

3. I. PINNATA. *Winged-leaved Candy-tuft.*—Herbaceous, smooth; lvs. pinnatifid; rac. corymbose, but little elongated after flowering.—① From S. Europe. Plant 1f high. Flowers white. Jn.—Aug. †

4. I. SAXATILIS. *Rock Candy-tuft.*—Shrubby; lvs. linear, entire, somewhat fleshy, rather acute, smooth or ciliate; fls. in corymbs.—① From S. Europe. Nearly 1f high. Flowers white. Apr.—Jn.†

Obs.—Twenty-four species of the Iberis have been described, others of which are equally ornamental with those above mentioned.

11. ISATIS.

Gr. ισαζω, to make equal; supposed to remove roughness from the skin.

Silicle elliptical, flat, 1-celled (dissepiment obliterated), 1-seeded, with carinate, navicular valves, which are scarcely dehiscent.—*None of the species are N. American.*

1. TINCTORIA. *Woad.*—Silicles cuneate, acuminate at base, somewhat spatulate at the end, very obtuse, 3 times as long as broad.—① The Woad is native of England. It is occasionally cultivated for the sake of its leaves, which yield a dye that may be substituted for indigo. The plant grows about 4 f. high, with large leaves clasping the stem with their broad bases. Flowers yellow, large, in terminal racemes. May—Jl. ‡

SECTION 2. **SILIQUOSÆ.** (§ 80, *note*.)

12. NASTURTIUM. R. Br.

Lat. nasus tortus; from the effect of these acrimonious plants upon the nose.

Sepals equal at base, spreading; silique subterete, mostly curved upwards, sometimes short so as to resemble a silicle; valves veinless; seeds in a double row, 0═.—-*Aquatic herbs.*

1. N. OFFICINALE. R. Br. (Sisymbrium Nast. *Linn.*) *English Water Cress.*
Les. pinnate; *lfts.* ovate, subcordate, repand; *pet.* white, longer than the calyx.—♃ Brooks and ponds. Stems decumbent, 1f long, thick, with axillary branches. Leaves of 3—7 leaflets; leaflets broad, often cordate, rather acute, obtusely toothed, terminal one largest. Flowers corymbed. Siliques less than 1' long. Jn.—It is beginning to be cultivated in the vicinity of our cities as a salad. §‡

2. N. AMPHIBIUM. R. Br. (Sisymbrium. *Linn.*) *Amphibious Water Cress.*
Lrs. oblong-lanceolate, pinnatifid or serrate; *rt.* fibrous; *pet.* longer than the calyx; *silique* elliptical, acute at base, tipped with the mucronate style.—♃ Banks of the Mohawk, *Dr. Robbins.* Rare. Stem 1—2f high, furrowed. Leaves variable, immersed ones pinnatifid or pectinate, upper ones serrate. Flowers yellow, minute, in a long, dense raceme. Silique half as long as the spreading or reflexed peduncle, pointed with the short style. Jn. Jl.

3. N. PALUSTRE. DC. *Marsh Water Cress.*
Lrs. pinnately lobed, amplexicaul, lobes confluent, dentate, smooth; *rt.* fusiform; *pet.* as long as the sepals; *silique* spreading, turgid, obtuse at each end.—♃ In wet places. Stem 1—2f high, erect, branched above. Leaves 2—3' long, all more or less pinnatifid, smooth, except a few ciliæ at base. Flowers numerous, minute, yellow. Silique 3—4" long, on pedicels of equal length. Jn.—Aug.

4. N. HISPIDUM. DC. (Sisymbrium. *Poiret.*) *Hispid Water Cress.*
St. villous; *lrs.* somewhat villous, runcinate-pinnatifid, lobes rather obtusely dentate; *siliques* (rather silicles) ovate, tumid, pointed with the style, scarcely more than half as long as the pedicels; *pet.* scarcely as long as the calyx.—♃ Banks of streams, Walpole, N. H., Conn. to Penn. Stem angular, branched, 1—3f high, with many paniculate racemes above. Leaves 3—6' long. Flowers minute, yellow. Silicles 1" long, on pedicels 2—3" long and somewhat spreading.

5. N. NATANS. DC. *β. Americanum.* Gray. *Floating Water Cress.*
Emersed lrs. serrate, oblong-linear, undivided, immersed ones doubly pinnatifid, with capillary segments; *pet.* twice as long as the calyx; *siliques* obovate, twice as long as the style.—♃ In water, Can. and U. S. Stem long, submerged. Flowers white, middle size. Jl.

6. N. SYLVESTRE. (Sisymbrium vulgare. *Pers.*) *Creeping Water Cress.*
Lrs. pinnately divided, segments lanceolate, incisely serrate; *pet.* longer than the calyx; *siliques* oblong, torulose; *sty.* very short.—Banks of the Delaware near Philadelphia. *Nuttall.* §

13. BARBAREA. R. Br.

In honor of St. Barbara, who discovered (what no one has since perceived) its medicinal virtues.

Sepals erect, subequal at base; silique columnar, 2—4-cornered; valves concave-carinate; seeds in a single series; 0=.--*Lrs. lyrately pinnatifid. Fls. yellow.*

B. VULGARIS. R. Br. (Erysimum Barbarea. *Linn.*) *Winter Cress.*
Lower lrs. lyrate, the terminal lobe roundish, *upper ones* obovate, pinnatifid at base, crenate or repand-dentate; *siliques* obscurely 4-cornered. —♃ In old fields, also brook-sides, Northern States, W. to Oregon, common. Whole plant glabrous. Stem furrowed. 1—2f high, branching above. Leaves 1—3—4' long, dark green, shining, on clasping petioles, the terminal lobe 1—1½' diam., upper ones sessile, all with obtuse teeth. Flowers on pedicels ½' long, in terminal racemes. Siliques slender, ½' long, curved upwards. May, Jn.

14. TURRITIS. Dillon.

Lat. *turris*, turreted; from the pyramidal form of the plant.

Sepals erect, converging : petals erect ; silique long. linear, 2-edged; valves plane : seeds in a double series, 0=.—*Fls. cyanic.*

1. T. GLABRA. *Smooth Tower Mustard.*

St. erect; *radical lvs.* petiolate, dentate, with ramose hairs, *cauline ones* broad-lanceolate, sagittate, half-clasping, glaucous, smooth; *siliques* erect.— Shores of Lake Superior, W. to the Rocky Mts. Naturalized about New Haven. *Eaton.* Stem round, simple, 1½f high. Leaves 1—2′ long. Siliques 2—3′ long, very narrow. Flowers pale sulphur-yellow. May.

β ? T. & G. *Lvs.* all linear-lanceolate and glabrous, *radical ones* remotely repand-denticulate, *cauline* entire.—Watertown, N. Y., on rocks. *Torrey & Gray.*

2. T. BRACHYCARPA. TORR. & Gray.

Glabrous and glaucous; *radical lvs.* spatulate, dentate, *cauline ones* linear-anceolate, sagittate and subamplexicaul; *siliques* short, linear-oblong; *pedicels* pendulous in flower, spreading in fruit.—② Lake shores, Mich. Stem 1—2f high, often purplish, as well as the foliage. Flowers rather large, pale purple. Siliques 1′ long, spreading.

15. ARABIS.

Said to derive its name from Arabia, its native country.

Sepals erect; petals unguiculate, entire; silique linear, compressed; valves 1-veined in the middle; seeds in a single row in each cell.— *Fls. white.*

1. A. CANADENSIS. (A. falcata. *Minx.*) *Sickle Pod.*

Cauline lvs. sessile, oblong-lanceolate, narrow at base, pubescent; *pedicels* pubescent, reflexed in the fruit; *silique* subfalcate, veined, pendulous; *sds.* winged.—♃ On rocky hills, Can. to Ga. W. to Ark. A plant remarkable for its long, drooping pods, which resemble a sickle blade, or rather a curved sword blade. Stem 2—3f high, slender, round, smooth. Leaves 1—3′ long, ⅓ as wide; the lowest early marescent, middle and upper ones sessile or clasping, with narrow bases, remotely denticulate. Flowers small, white. Pods slender, flattened, nearly 3′ long. Jn.

2. A. LYRATA. (Sisymbrium arabidoides. *Darl.*)

St. and *upper lvs.* smooth and glaucous; *radical lvs.* lyrately pinnatifid, often pilose; *st.* branched at base; *pedicels* spreading; *siliques* erect.—② On rocky hills, Can. to Va. Stems often many, united at base, 6—12′ high. Root-leaves numerous, rosulate, 1—3′ long, ⅓ as wide, petiolate, lower stem-leaves pinnatifid or sinuate-dentate, upper ones sublinear and subentire. Flowers middle size. Siliques when mature 1½—2′ long, less than 1″ wide. Apr.-May.

3. A. LÆVIGATA. DC. (Turritis lævigata. *Muhl.*)

Smooth and glaucous; *radical lvs.* obovate and oblong, tapering to a petiole, dentate, *stem lvs.* linear-lanceolate, amplexicaul, obtuse, *upper ones* entire; *pedicels* about as long as the calyx, erect; *siliques* very long, linear, at length spreading and pendulous; *sds.* winged.—♃ In rocky woods and low grounds, Can. to Ark. Stem 1—2f high, round, smooth, simple or branched above. Root-leaves often purplish, ⅓—1½′ long, ⅓ as wide, with acute teeth. Stem-leaves 2—5′ long, ⅓ as wide, upper ones entire. Flowers in long racemes. Siliques 2—3′ long, scarcely 1″ wide. May.

4. A. HIRSUTA. Scop. (Turritis. *Linn.*)

Erect, branching; *lvs.* mostly dentate, hirsute, *radical ones* oblong-ovate tapering to a petiole, *cauline ones* oval or lanceolate, sagittate; *siliques* straight, erect.—① Found in low, rocky grounds, Can. to Va. W. to Oregon. Stems 2 or more from the same root, round, hairy at base, near a foot high, dividing into very slender and parallel branches. Leaves scarcely dentate, sessile, with heart-shaped or arrow-shaped bases, upper ones acute. Flowers greenish-white. Siliques straight, 1—2′ long. Jn.

5. A. HETEROPHYLLA. Nutt.

Nearly smooth; *radical lvs.* spatulate, toothed, *upper ones* linear, sessile, entire; *silique* long and spreading; *pet.* linear-oblong, exceeding the calyx.— Near Paris, Me., and the White Mts., N. H. Radical leaves somewhat pilose with simple hairs, upper ones linear, about 2′ long, and 1—2″ wide. Siliques about 3′ long. *Nuttall.*

6. A. DENTÀTA. TORR. & Gray.

Plant somewhat scabrous; *radical lrs.* obovate, petiolate, unequally and sharply dentate; *cauline ones* oblong, amplexicaul; *pet.* minute, spatulate, as long as the sepals; *stig.* subsessile; *silique* short.— ① River banks, Western States! Stem slender, ascending, 1—2f high. Petals white, with a purplish tinge. Siliques 1' in length. Apr.

16. CARDAMINE.

Gr. καρδια, heart, δαμαω, to strengthen; from its stomachic properties.

Calyx a little spreading; silique linear, with flat, veinless valves, narrower than the dissepiment, and often opening elastically; stigma entire: seeds not margined, with a slender funiculus, 0—.—*Mostly* 24. *Fls white.*

1. C. HIRSÙTA. (C. Pennsylvanica. *Muhl.*) *Pennsylvanian Cardamine.*

Lrs. pinnate or lyrately pinnatifid; *lfts.* entire, or sparingly repand-denticulate, those of the radical leaves oval-oblong, of the cauline linear-oblong, the terminal one longest, about 3-lobed; *pet.* oblong-cuneate; *siliques* erect, with a very short style.—① or 24 A variable plant common in wet places throughout the U. S. Stem 8—16' high, mostly smooth. Leaflets 3—5 pairs, 4- 12" long, smoothish. Fls. small. Siliques about 1' long, 12—18 seeded. Jn.

2. C. VIRGINICA. (C. hirsuta. β. *Hook.*) *Virginian Cardamine.*

Lrs. lyrately pinnate; *lfts.* with a single tooth on one or both sides; *pet.* ne rly twice as long as the calyx; *rac.* strictly erect; *stig.* sessile; *silique* long, is curved, erect.—② A small and delicate species, much resembling the last, b.. probably distinct. Found on dry hill-sides, Vt. Ct. to Ky. and Mo. Stem 4 - 8' high, slender, leafy. Leaflets 2—4 pairs with a trilobate odd one, oval, 1- 2" in length, those of the upper leaves 3—5" long, but very narrow. Petals small. Siliques filiform, 1' long. Jn.

3. C. PRATENSIS. *Field Cardamine.*

St. erect or decumbent, simple; *lrs.* pinnately 7—15-foliate; *lfts.* petiolate, subentire, lower ones suborbicular, upper linear-lanceolate; *sty.* distinct.— 24 Swamps, N. Y. to Arctic Am. Whole plant smooth. Stem round, striate, 10—16' high. Leaves few, 1½—2' long including the petiole. Leaflets of the root-leaves 1—3" diam., of the cauline 3—6" by ½". Flowers large, few, in a terminal raceme. Petals white or rose-color. Siliques nearly 1' in length, erect. Apr. May.

4. C. ROTUNDIFOLIA. Michx. (C. rhomboidea. *DC.* Arabis. *Pers. Nutt.*)

Glabrous or somewhat hairy; *lrs.* entire or repand-toothed, *radical ones* orbicular-ovate, on long petioles, *cauline* oval or oblong-lanceolate, petiolate below, sessile above, dentate.—24 Another variable species with rather large, white or reddish flowers. Stems 6—12' high, angular or striate, mostly erect. Leaves of root 10—18" diam., on petioles 2—4' long. Racemes about 3' long, 12—20-flowered. Petals 2—4 times as long as the calyx. Siliques spreading, 6—12" long. Apr. May.

α. T. & G. *Rt.* mostly tuberiferous; *st.* erect; *lower stem lrs.* rhomboid-oval; *pet.* large.—Wet meadows, Conn. Vt. *Dr. Robbins.*

β. T. & G. *Rt.* mostly fibrous; *st.* decumbent, branching; *lrs.* all petiolate; *pet.* smaller, purplish.—Shaded springs and rivulets, N. Y.

4. C. BELLIDIFOLIA. (C. rotundifolia. *Br.* not *Michx.*)

Lrs. smooth, *radical ones* orbicular-ovate, nearly entire, petiolate; *cauline ones* entire or 3-lobed; *siliques* erect.—A minute species on the summits of the White Mts. *Abel Storrs!* &c., also Arc. Am. to Calif. Stem 1½—3' high. Leaves mostly radical, broadly oval or ovate, ¼' long. on petioles as long as the stems. Fascicles corymbose, each of 3 or 4 white flowers. Petals oval, obtuse, about twice as long as the calyx. Jl.

17. DENTARIA.

Lat. dens, a tooth; from the tooth like projections of the rhizoma.

Sepals converging; silique lanceolate, with flat, veinless, revolute valves, opening elastically; placentæ not winged; seeds in a single row, ovate, not bordered; funiculus slender, 0=.—*Rhizoma* ♃. *Lvs. divided, often but 2 or 3. Fls. white or purplish.*

1. D. DIPHYLLA. *Pepper Root.*

St. 2-leaved; *lfts.* ternate, subovate, unequally and incisely dentate; *rhiz.* dentate.—In woods and wet meadows, Can. to Car. and to the Miss. Stem about 1f high, round, smooth, with 2, nearly opposite, ternate leaves above the middle. Leaflets on very short stalks, the lateral ones oblique, all with rounded, mucronate, unequal teeth. Flowers racemed, large, white; the petals much larger than the calyx. The rootstock is long and large in proportion to the plant, beset with teeth, with a pungent, aromatic taste. May.

2. D. LACINIATA. Muhl. (D. concatenata. *Michx.*)

Rhiz. moniliform; *cauline lrs.* 3, 3-parted, the divisions lanceolate or linear-oblong, incisely toothed or pinnatifid, *lateral ones* lobed.—In woods, Can. and U. S. The rootstock consists of several tubers of a pungent taste. Stem 1f high, smooth, simple. Leaves usually in a whorl about half-way up, the segments with very irregular, mucronate teeth, rarely subentire, lateral ones cut nearly to the base, rendering the leaf almost quinate. Root-leaves sometimes 0. Flowers racemed, purplish. Apr. May.

3. D. MAXIMA. Nutt.

St. tall; *lrs.* alternate, 5—7, remote, the margin a little roughened; *lfts.* somewhat oval, incisely and acutely dentate, lateral ones lobed.—Western N. Y. and Penn. Tubers of the rhizoma concatenate. Stem often nearly 2f high. Flowers pale purple.

4. D. HETEROPHYLLA. Nutt.

Rhiz. moniliform, with oblong tubers; *radical lrs.* on long petioles, deeply and obtusely lobed, lobes crenate-dentate with abruptly mucronate teeth, *cauline lrs.* 2, rarely 3, alternate, petiolate, ternately divided, segments linear-lanceolate, entire or rarely toothed, rough-edged.—Woods, Penn. to Ky. Stem 8—12' high. Cauline leaflets 1—2' long, 2—3" wide. Corymb with about 9 pale purple flowers. Jn.

18. HESPÉRIS.

Gr. ἐσπερις, evening; when the flower is most fragrant.

Calyx closed, furrowed at base, shorter than the claws of the petals; petals bent obliquely, linear or obovate; silique 4-sided, 2-edged or subterete; seeds not margined; stigmas forked, with the apices converging.

1. H. MATRONÁLIS. *Rocket.*

St. simple, erect; *lrs.* lanceolate, ovate, denticulate; *pet.* emarginate, mucronate; *pedicels* as long as the calyx.—A fine garden perennial, said to be found native about Lake Huron. Stem 3—4f high. Flowers purple, often double, and white in *β. hortensis.* †

2. H. APRICA. *Siberian Rocket.*—*St.* erect, simple, pubescent; *lrs.* oblong, obtuse, entire, ciliate-hispid; *pedicels* as long as the calyx.—♃ From Siberia. Stem a foot high. Flowers purple. May. Jn. †

19. SISYMBRIUM. Allioni.

Calyx mostly spreading, equal at base; petals unguiculate, entire; silique subterete; valves concave; style very short; seeds in a single series, ovoid; cotyledons 0‖, sometimes oblique.

1. S. OFFICINÀLE. Scop. (Erysimum. *Linn.*) *Hedge Mustard.*

Lrs. runcinate; *rac.* slender, virgate; *siliques* subulate, erect, closely appressed to the rachis.—① A common and troublesome weed, in fields, road-sides, rubbish, &c., Can. and U. S. Stem 1—3f high, round, more or less hairy, with spreading branches. Lower leaves 3—8' by 1—3', the lower segments *placed at right angles* to the midvein, or pointing backwards, the terminal segment largest. Upper leaves in 3 lanceolate segments plated at right angles. Flowers small, yellow, terminating the raceme, which becomes 1—2f long and environed by the appressed, sessile pods. Jn.—Sept. Medicinal. ♦

2. S. THALIÀNA. Gay. (Arabis. *Linn.*) *Thalian Hedge Mustard.*
Lrs. subdentate and pilose, *radical ones* numerous and petiolate, oblong. *cauline ones* lanceolate ; *cal.* much shorter than the pedicels ; *siliques* ascending, twice longer than the pedicels.—① Rocks and sandy fields, Vt. to Ga. W. to Ky. Stem 4—12′ high, erect, with slender, erect branches, striate, pilose, often purple at base. Root leaves rosulate, 1—2′ long ; cauline denticulate, ciliate, sessile, 6—12″ by 1—3″. Pedicels spreading, 3—5″ long. Flowers small, white. Siliques slender, straight, 7—10″ long. Styles scarcely any. May.

3. S. TERES. T. & G. (Cardamine. *Michx.*)
St. erect, branched ; *lrs.* all somewhat lyrately pinnatifid ; *siliques* short, linear, acuminate, on very short peduncles ; 0 !.—① Shores of Lake Champlain, Vt. Plant about 8′ high, slightly scabrous with very short hairs. Siliques erect, terete, 4″ in length, beaked with the short, slender style. Seeds 00.

4. S. CANESCENS. Nutt.
Lrs. bipinnately divided, canescent, lobes oblong or lanceolate, subdentate or obtuse ; *pet.* about equaling the calyx ; *siliques* oblong-linear, shorter than the pedicels.—① Arctic Sea to Flor. Plant 1—2f high, often nearly smooth. Leaves about 3′ long, sessile, segments 5—7 pairs, finely divided. Fls. very small. Siliques often erect, on spreading pedicels. Variable.

20. ERYSÌMUM.

Gr. ερυω, to cure ; from its salutary medicinal properties.

Calyx closed ; siliques columnar, 4-sided ; stigma capitate ; seeds in a single series ; cotyledons oblong, 0‖.

1. E. CHEIRANTHÖIDES.
Pubescence minute, appressed, branched ; *lrs.* lanceolate, denticulate or entire ; *silique* erect, spreading, twice longer than the pedicels ; *stig.* small, nearly sessile.—① By streams and in wet grounds, U. S. and Can., not common. Stem erect, 1—2f high, often branched, and, with the leaves, scabrous. Leaves acute at each end, 1—2′ long, ⅓ as wide. Flowers small, yellow, in long racemes. Siliques ½′ to near 1′ in length, linear, and somewhat spreading. Jl.

2. E. ARKANSÀNUM. Nutt. *Yellow Phlox.* *False Wall-Flower.*
Scabrous, with an appressed pubescence ; *st.* simple ; *lrs.* linear-lanceolate, remotely dentate, sessile, lower ones runcinate-toothed ; *inflorescence* racemose, corymbed at summit ; *siliques* long, 4-angled, suberect ; *stig.* capitate.— ② A fine plant with large, showy flowers, resembling the wall-flower. Banks of Scioto, *Sullivant.* Arkansas, *Nuttall.* Bluffs of the Wabash ! *Wood.* Ill. *Mead!* Stem 1—3f high, slender. Leaves 2—3′ by 3—6″. Sepals straw-color. Petals large, bright orange-yellow. Siliques 3′ long. Jn. Jl.

21. CHEIRANTHUS.

Arabic kheyry, the name of a certain plant, and Gr. ανϑος, flower.

Calyx closed, 2 of the sepals gibbous at base ; petals dilated ; silique terete or compressed ; stigma 2-lobed or capitate ; seeds flat, in a single series, often margined, 0=.

1. C. HESPERIDOÏDES. T. & G. (Hesperis pinnatifida. *Michx.*)
Glabrous ; *lower lrs.* lyrate-pinnatifid, *upper* lanceolate, attenuate at base, unequally and sharply serrate-dentate, acuminate ; *pedicels* as long as the calyx ; *pet.* obovate-spatulate, obtuse ; *silique* terete ; *stig.* capitate ; *sds.* margined.— ⁊ Penn. to Ill.! S. to Ark. Stem slender, furrowed, 2—3f high. Leaves thin, 3—5′ long, ⅓ as wide, those of the stem scarcely petiolate. Racemes terminal and axillary. Calyx shorter than the claws of the violet-colored petals. Siliques torulose, 15—20″ long ; seeds oblong, plano-convex, with a narrow border. May, Jn.

2. C. Cheiri.—*Wall-Flower.*—*St.* somewhat shrubby and decumbent at base; *lvs.* entire or slightly dentate, lanceolate, acute, smooth; *branches* angular; *pet.* obovate; *siliques* erect, acuminate.—♃ From S. Europe. A popular garden flower, admired for its agreeable odor and its handsome corymbose clusters of orange or yellow flowers. Plant 1—2f high. Jn.

22. MATTHIÖLA. R. Br.

In honor of P. A. Matthioli, physician to Ferdinand of Austria, and botanic author.

Calyx closed, 2 of the sepals gibbous at base; petals dilated; siliques terete; stigmas connivent, thickened or cornute at the back.-- *Herbaceous or shrubby, oriental plants, clothed with a hoary, stellate pubescence.*

1. M. annuus. R. Br. (Cheiranthus. *Linn.*) *Ten-weeks Stock.*—*St.* herbaceous, erect, branched; *lvs.* hoary-canescent, lanceolate, obtuse, subdentate; *silique* subcylindrical, without glands.—① A fine garden flower from S. Europe. Stem 2f high, and, with the leaves, covered with a soft, stellate pubescence. Flowers variegated. Jn.†

2. M. incānus. R. Br. (Cheiranthus. *Linn.*) *Purple July Flower.*—*St.* shrubby at base, erect, branched; *lvs.* lanceolate, entire, hoary-canescent; *siliques* subcylindrical, truncate and compressed at apex, without glands.—② One of the most popular flowers of the genus, native of England, &c. Stem 2f high. Flowers purple.—Several varieties are enumerated, as the Double-flowered, Brompton Stock, and Brompton Queen. Jn.†

3. M. fenestrālis. R. Br. (Cheiranthus. *Linn.*) *Window July Flower.*—*St.* suffruticose, erect, simple; *lvs.* crowded, recurved, undulate, downy; *siliques* downy, without glands, broadest at base.—From S. Europe. Plant 1f high. Flowers numerous, large, purple. Jl. Aug. †

4. M. Græcus. R. Br. (Cheiranthus. *Linn.*) *Grecian Stock.*—*St.* herbaceous, erect, branched; *lvs.* lanceolate, glabrous; *siliques* somewhat compressed, without glands.—② From Greece. Plant about 1f high, distinguished from the remainder of the genus by its smooth foliage. Flowers white, appearing all summer.†

23. SINĀPIS.

Sepals equal at base, spreading; petals ovate, with straight claws, siliques subterete; valves veined; style short and subulate, or ensiform; seeds in a single series, subglobose, 0 >>.—*Fls. always yellow.*

1. S. nigra. *Black Mustard.*
Lower lvs. lyrate, *upper* linear-lanceolate, entire, smooth; *silique* smooth, somewhat 4-angled, appressed to the rachis of the raceme.—① In cultivated grounds and waste places. Stem 3—6f high, round, smooth, striate, branching. Leaves all petiolate, lower ones variously lobed and dentate, upper ones pendulous and entire. Sepals and petals sulphur-yellow. Pods very numerous, nearly 1' long, beaked with the 4-sided styles. Seeds 00, small, globose, nearly black, well known as a condiment. Jn. Jl. ‡§

2. S. arvensis. *Field Mustard.*
St. and *lvs.* hairy; *silique* smooth, many-angled, torose, about 3 times longer than the slender, ancipital style.—① Naturalized in N. Y., *T. & G.*, and in Vt., *Dr. Robbins.* Lower leaves large, sublyrate-pinnatifid, upper ones oblong-ovate, all repand-toothed. Silique somewhat spreading, 1½' long. Seeds large and black. Jn.—Aug.§

3. S. alba. *White Mustard.*—*Lvs.* lyrate, smoothish; *siliques* hispid, torose, shorter than the ensiform beak; *sds.* large, pale yellow.—① Native of Europe. Stem 2—5f high, thinly hirsute. Leaves all lyrately pinnate, dentate, petiolate. *Siliques* spreading, about 4-seeded. The seeds are used for about the same purposes as those of S. nigra, much esteemed in medicine. Jn. Jl. ‡

24. BRASSICA.

Celtic, *bresic*, the cabbage.

Sepals equal at base, (mostly) erect; petals obovate; filaments without teeth; silique subcompressed, valves concave, with a central vein; style short, subterete, obtuse; seeds globose, in a single (often double) row; 0>>.—*Fls. yellow.*

1. B. CAMPESTRIS. *Cole.*

Lvs. somewhat fleshy and glaucous, the *lower* lyrate-dentate, subciliate, *upper ones* cordate-amplexicaul, acuminate.—① Native of Sweden, naturalized in cultivated fields and waste places.   Stem 1½—3f high, round, smooth above, with a few scattered, reversed hairs below.   Lower leaves 3—7' long, ½ as wide, the terminal lobe greatly exceeding the lateral ones; upper smaller, entire, with rounded, clasping lobes at base, tapering to an obtuse point.   Racemes 1—2f long.   Sepals erect, spreading.   Corolla yellow, 4—5" diam.   Siliques 1½' long, with the style ½'.   Seeds small, dark brown.   Jn. Jl.✦
   β. *Rutabaga.* (*Swedish Turnip.*)—*Rt.* tumid, napiform, subglobose, yellowish.—Cultivated like the common turnip; but after a thorough experiment it is conceded by farmers to be inferior in value to that root, although it grows to an enormous size. ‡

2. B. RAPA.—*Radical lvs.* lyrate, rough, not glaucous, *cauline ones* incised, *upper* entire, smooth. ‡
   β. *depressa.* (*Common Turnip.*)—*Rt.* depressed-globose or napiform, contracted below into a slender radicle.—② Long cultivated for the table, &c., in gardens and fields.   Stem 2—4f high, and, with the leaves, deep green.   Upper leaves amplexicaul.   Pods 1' long.   Seeds small, reddish-brown.   Jn. ‡

3. B. OLERACEA. (*Cabbage.*)—*Lvs.* very smooth and glaucous, fleshy, repand-toothed or lobed.—② Native of Europe, where it grows on rocky shores and cliffs, with no appearance of a head, forming a surprising contrast with the cultivated varieties.   The excellence of the cabbage as a pot-herb needs no encomium. ‡
   β. *bullata.* (*Savoy Cabbage.*)—*Lvs.* curled, subcapitate when young, finally expanding.
   γ. *botrytis-cauliflora.* (*Cauliflower.*)—*St.* low; *hds.* thick, compact, terminal; *fls.* abortive, on short, fleshy peduncles. ‡
   δ. *botrytis-asparagoides.* (*Broccoli.*)—*St.* taller; *hds.* subramose; *branches* fleshy at the summit, consisting of clusters of abortive flower-buds. ‡
   ε. *capitata.* (*Head Cabbage. York Cabbage.*)—*St.* short; *lvs.* concave, packed in a dense head before flowering; *rac.* paniculate. ‡

Section 3.—LOMENTACEÆ.

25. CAKILE.

Silicle 2-jointed, the upper part ovate or ensiform; seed in the upper cell erect, in the lower pendulous, sometimes abortive.—① *Maritime herbs.*

C. MARITIMA. Scop. (Bunias edentula. *Br.*) *Sea Rocket.*
   Upper joint of the silicle ensiform or ovate-ensiform.—Native of the seacoast! and of the lake shores of N. Y.   A smooth, succulent plant, branching and procumbent, 6—12' long.   Leaves sinuate-dentate, oblong-ensiform, caducous.   Flowers on short, fleshy peduncles, in terminal spikes or racemes, corymbosely arranged.   Petals purple, obtuse at end.   Silicle smooth, roundish, lower joint clavate-obovate upper with one elevated line on each side. Jl. Aug.

26. RAPHANUS.

Gr. ρα, quickly, φαιρω, to appear from its rapid growth

Calyx erect; petals obovate, unguiculate; siliques terete, torose, not opening by valves, transversely jointed or divided into cells; seeds large, subglobose, in a single series, 0 >>.

1. R. RAPHANISTRUM. *Wild Radish.*
Lvs. lyrate; *silique* terete, jointed, smooth, becoming in maturity 1-celled, longer than the style.—① Naturalized in cultivated fields and by road-sides, but rare. Stem glaucous, branching, 1—2f high, bristly. Leaves rough, dentate, petiolate or sessile. Calyx bristly. Pods yellow, blanching as they decay. Jn. Jl. §

2. R. SATIVA. *Garden Radish.—Lower lvs.* lyrate, petiolate; *silique* torose, terete, acuminate, scarcely longer than the pedicels.—① A well-known salad root, from China. Stem 2—4f high, very branching. Lower leaves 6—10' long. Flowers white or tinged with purple. Pods 1—2' long, thick and fleshy. The principal varieties are the *turnip radish,* root subglobose; *common radish,* root oblong, terete; *black Spanish radish,* root black outside. Jn. Aug. ‡

ORDER XIV. CAPPARIDACEÆ.—CAPPARIDS.

Herbs, shrubs or even trees, destitute of true stipules.
Lvs. alternate, petiolate, either undivided or palmately divided.
Fls. solitary or racemose. *Sep.* 4.
Cor.—Petals 4, cruciate, unguiculate, hypogynous, more or less unequal.
Sta. 4—12, or some multiple of four, almost perigynous.
Torus small, often elongated, bearing a single gland.
Ova. often stipitate, of 2 united carpels. *Sty.* united into one. *Stig.* discoid.
Fr. either pod-shaped, and dehiscent, or fleshy and indehiscent. *Placentæ* usually 2.
Sds. many, reniform. *Albumen* 0. *Embryo* curved. *Cotyl.* foliaceous.

Genera 28, species 340,—chiefly tropical plants. They are more acrid in their properties than the Cruciferæ, but otherwise much resemble them. One species of Polanisia is used as a vermifuge.

Conspectus of the Genera.

Torus minute { Stamens 6.	*Cleome.*	2
Torus minute { Stamens 8—32.	*Polanisia.*	3
Torus linear and elongated like a stipe. Stamens 6.	*Gynandropsis.*	1.

1. GYNANDROPSIS. DC.

Gynandria, a Linnean class, οψις, appearance.

Sepals distinct, spreading; petals 4; stamens 6, the filaments adnate below to the linear, elongated torus its whole length; pod linear-oblong, raised on a long stipe, which rises from the top of the torus.—① *Lvs. digitate. Fls. racemed.*

G. PENTAPHYLLA. DC. (Cleome. *Linn.*)
Middle lvs. petiolate, 5-foliate, *floral* and *lower ones* 3-foliate; *lfts.* oovate, entire or denticulate.—In cultivated grounds, Penn., &c. Stem simple, 2—3f high. Flowers of a very singular structure. Pedicels about 1' long, slender. Calyx small. . Petals white, ⅓ as long as their filiform claws. Stamens 1' long, spreading, apparently arising from the midst of the long styloid torus. Pod 2' long.§

2. CLEOME.

Sepals sometimes united at base; petals 4; torus minute or roundish; stamens 6—4; pod subsessile or stipitate.—*Herbs or shrubs. Lvs. simple or digitate. Fls. racemed or solitary.*

C. PUNGENS. *Spiderwort.*—Glandular-pubescent; *st.* simple, and with the petioles, aculeate; *lvs.* 5—9-foliate, on long petioles; *lfts.* elliptic-lanceolate, acute at each end, obscurely denticulate; bracts simple; *fls.* racemed; *sep.* distinct; *pet.* on filiform claws; *sta.* 6, twice longer than the petals.—A common garden plant, with curious purple flowers. Stem 3—4f high. Jl. Aug. †

3. POLANISIA. Raf.

Sepals distinct, spreading; petals 4, unequal; stamens 8—32; filaments filiform or dilated at the summit, torus minute; pods linear.—① *Strong-scented herbs.*

P. GRAVEÖLENS. Raf. (Cleome dodecandria. *Michx.*)
Viscid-pubescent; *lvs.* ternate; *lfts.* elliptic-oblong; *fls.* axillary, solitary; *sta.* 8—12; *capsule* oblong, lanceolate, attenuate at base.—A strong-scented plant, found on gravely shores! Vt. to Ark. Stem 1f high, branching, striate. Leaflets 1—1½' long, ⅓ as wide, nearly entire and sessile; common petiole 1' long. Flowers in terminal racemes. Petals yellowish-white, narrowed below into long claws. Filaments slender, exserted. Pods 2' long, glandular-pubescent, siliquose, viscid like every other part of the plant. Jl.

ORDER XV. RESEDACEÆ.—MIGNIONETTES.

Herbs with alternate, entire or pinnate leaves. *Stip.* minute, gland-like.
Fls. in racemes or spikes, small and often fragrant.
Cal.—Sepals somewhat united at base, unequal, green.
Cor.—Petals lacerated, unequal.
Sta. 4—40, inserted on the disk. *Torus* hypogynous, one-sided, glandular.
Ova. sessile, 3-lobed, 1-celled, many-seeded. *Placentæ* 3, parietal.
Fr. a capsule, 1-celled, opening between the stigmas before maturity.

Genera 6, species 41, inhabiting the countries around the Mediterranean Sea, having no very remarkable properties. Reseda Luteola contains a yellow coloring matter, and other species are very fragrant.

RESÉDA.

Lat. reseda, to calm; the plants are said to relieve pain.

Sepals many, petals of an equal number, each bearing one or more stamens; torus large, fleshy, bearing the ovary, with several stamens and styles.

1. R. LUTEÖLA. *Dyer's Weed.*
Lvs. lanceolate, entire, with a tooth on each side at base; *cal.* 4-cleft.—①
Nearly naturalized in Western N. Y. Stem about 2f high. The flowers are without petals, arranged in a long spike, which, as Linnæus observes, follows the course of the sun, inclining east, south and west by day, and north by night.—It affords a useful yellow dye, also the paint called Dutch-pink. §

2. R. ODORÁTA. *Mignionette.—Lvs.* entire, 3-lobed; *sep.* shorter than the petals.—A well known and universal favorite of the garden, native of Egypt. The flowers are highly fragrant and no boquet should be considered complete without them. The variety *frutescens* is by a peculiar training raised to the height of 2 feet with the form of a tree. The species *phytcuma*, native of Palestine, has a calyx longer than the petals.

ORDER XVI POLYGALACEÆ.—MILKWORTS.

Plants herbaceous or shrubby, sometimes twining.
Lvs. alternate, or rarely opposite, mostly simple, always without stipules.
Fls. perfect, unsymmetrical. *Pedicels* with 3 bracts.
Cal.—Sepals 5, very irregular, 3 exterior, 2 interior (wings) larger and petaloid.
Cor.—Petals 3, hypogynous, the anterior (keel) larger than the others.   [the claws of the petals
Sta.—6—8. *Fil.* combined in a tube which is split on the upper side, and coherent to some extent with
Ova. superior, compressed, 2-celled, one cell often abortive. *Sty.* curved and often cucullate.
Fr. loculicidal or indehiscent. *Sds.* pendulous.

Genera 19, species 485, very equally distributed, each division of the globe having two or three genera peculiar to it. The properties of the Polygalaceæ have not been well determined. Some of the genera possess a bitter matter and a milky juice which is emetic, expectorant and diuretic. Polygala is the only northern genus.

POLYGÁLA. Tourn.

Gr. πολυ, much, γαλα, milk, said to favor the lacteal secretions of animals.

Sepals 5, persistent, 2 of them wing-shaped and petaloid; petals 3, cohering by their claws to the filaments, lower one carinate; capsule obcordate, 2-celled, 2-valved, 2-seeded; seeds carunculate.—*The N. American species herbaceous. Lower petal (keel) mostly tipped with a crest.*

* *Spikes ovate, globose or oblong, dense, obtuse.*

1. P. SANGUINEA. (P. purpurea. *Nutt.*) *Caducous Polygala.*
St. branching at top; *lvs.* linear, alternate; *fls.* beardless, in alternate, ob-

.ong spikes; *calycine wings* obovate.—① An erect plant, 6—12' high, found in meadows and wet grounds, Mass. to La., and known at once by its short, reddish, cylindric spike of flowers. Stem angular, with fastigiate branches, each ending in a smaller spike than that of the main stem, but rising above it in height. Flowers purple, caducous. Jl.—Oct.

2. P. Nuttalli. T. & G. (P. sanguinea. *Nutt.*) *Nuttall's Polygala.*
St. erect, somewhat fastigiate; *lvs.* linear; *spikes* rather loose, ovoid-globose; *calycine wings* elliptic-obovate, attenuate at base, twice longer than the fruit; *crest* minute.—① Martha's Vineyard, *Oakes.* R. I. *Olney!* to La. Stem 6—10' high. Leaves 6—8'' by 1—2'', acute. Spikes 5—10'' long, 4—6'' diam. Wings of the calyx rose-red. Seeds black. Aug.

3. P. CRUCIATA. *Cross-leaved Polygala.*
St. erect, somewhat fastigiate, winged at the angles; *lvs.* verticillate in 4s, linear-oblong, punctate, spikes ovate, dense, obtuse, sessile or nearly so; *crest* minute.—① In sphagnous swamps and other low grounds. Stem 3—12' high, very slender, smooth, slightly winged at the 4 angles. Leaves 2—10'' or more long, 1—2'' wide (upper ones the largest), obtuse, tapering to the base, with small, resinous dots. Spikes capitate, about the size of the last. Wings of calyx greenish-purple, much dilated at apex. Aug.

4. P. LUTEA. *Yellow Polygala.*
St. simple or branching; *root lvs.* spatulate, obtuse, attenuate at base, *cauline ones* lanceolate, acute; *rac.* ovate, obtuse, dense; *fls.* pedicellate; *wings* ovate, mucronate; *keel* with a minute crest.—② Sandy plains, N. J. to Flor. Stem 8—12' high, generally with a few long spreading branches. Flowers bright yellow, longer than the bracts. Style dilated in the middle and with a stipitate gland. Jn.—Oct.

5. P. INCARNATA. *Flesh-colored Milkwort.*
Glaucous; *st.* erect, slender, mostly simple; *lvs.* few, scattered, linear-subulate; *spike* oblong, terminal; *wings* lanceolate, cuspidate; *claws of the petals* united into a long, cleft tube.—① Dry soils N. J. to Flor. W. to Ark. Stem 1—2' high. Leaves 4—6'' long, remote. Spikes 1—1½' long. Flowers] ale rose-color or flesh-color. The slender corolla tube nearly twice as long as .he wings, the keel with a conspicuous crest. Jn. Jl.

* * *Spikes elongated or racemose.*

6. P. VERTICILLATA. *Whorl-leaved Polygala.*
St. branched, erect; *lvs.* linear, verticillate; *spikes* linear, stalked; *fls.* alternate, crested; *calycine wings* roundish.—① Found on dry hills, U. S. and Can. Stem very slender, square, 6—8' high. Leaves in whorls of 5 (r 6, 4—10'' long, 1'' wide, alternate on the branches. Flowers small, greenish-white, in very slender racemes 5—10'' long, which are higher upon the bran :hes than upon the main stem. Jl.—Oct.

7. P. AMBIGUA. Nutt. *Dubious Polygala.*
St. erect, with virgate branches; *lvs.* linear, *lower ones* verticillate, *upper* alternate; *spikes* dense, on long peduncles; *calycine wings* roundish.—① Dry fields and woods, Mass. to Va. Stem 9—15' high, angular, smooth, much branched. Leaves sessile, tapering to the base, 4—10'' by 1''. Racemes spicate, acute, about 1' long, 20—30-flowered, on peduncles 1½—2½' long. Flowers small, greenish-white, tinged with purple. Jl.—Nearly allied to P. verticillata.

8. P. SENEGA. *Seneca Snake-root.*
St. erect, smooth, simple, leafy; *lvs.* alternate, lanceolate, tapering at each end; *fls.* slightly crested, in a terminal, spike-form, slender raceme.—♃ Woods, Western States, rare in Eastern. Root ligneous, branched, contorted, about ¼' thick, ash-colored. Stems 8—14' high, several from the same root. Leaves 1—3' long, ⅓ as wide, numerous, scattered. Flowers white, in a filiform spike 1—3' long. Sepals obtuse, larger than the petals. The root has a sweetish, nauseous taste, soon becoming pungent and hot. Jl.—A valuable stimulating expectorant.

9. P. POLYGAMA. Walt. (P. rubella. *Willd.*) *Bitter Polygala.*
Sts. simple, numerous; *lvs.* linear, oblong, mucronate, alternate below;

vec. terminal and lateral; *fls.* sessile, those of the stem winged, those of the root apterous.—♃ Fields and pastures, Can. to Flor. and La. Stems crowded, many from the same root, angular, smooth. Leaves smooth, lower obovate, upper linear-lanceolate, obtuse, sessile. Flowers, crested, purple, smaller than the last. Wings of the calyx obtuse. Anthers 8, in 2 equal parcels. Bracts small, subulate, caducous. Terminal racemes with perfect flowers, radical racemes prostrate or subterraneous, wingless and nearly apetalous. Jn. Jl.— Bitter and tonic.

*** *Flowers large, few.*

10. P. PAUCIFOLIA. *Fringed Polygala.*
St. simple, erect, naked below; *lvs.* ovate, acute, smooth; *terminal fls.* large, crested, *radical ones* apterous.—♃ A small, handsome plant, with a few rather large purple flowers. Woods and swamps, Brit. Am. to Ga. Stems 3—4' high, with its acute leaves mostly near the top, 2—4 flowers above them. Calyx of 5 leaves, the upper one gibbous at base. Corolla mostly purple, with a purplish crest on its middle lobe. The radical flowers are either close to the ground or subterraneous, smaller, greenish, wanting the wings of the calyx. May.

ORDER XVII. VIOLACEÆ.—VIOLETS.

Plants herbaceous or shrubby.
Lvs. simple, alternate, sometimes opposite, stipulate, involute in vernation.
Cal.—Sepals 5, persistent, slightly united, elongated at base, the 2 lateral interior.
Cor.—Petals 5, commonly unequal, the inferior usually spurred at base.
Sta. 5, usually inserted on the hypogynous disk. *Fil.* dilated, prolonged beyond the *anth.*
Ova. of 3 united carpels, with 3 parietal placentæ. *Sty.* 1, declinate. *Stig.* cucullate.
Fr. a 3-valved capsule. *Sds.* many, with a crustaceous testa and distinct chalaza.

Genera 11, species 300, mostly inhabitants of the Northern Temperate Zone. The roots of almost all the Violaceæ possess emetic properties, and some are valued in medicine. The ipecac of the shops is partly the product of certain Brazilian species of Ionidium. Several species of the violet are cultivated for the beauty of their flowers. Of the 4 genera found in North America, only 2 are found in the Northern States.

Genera.

Sepals unequal, more or less auricled at base. . . . . : *Viola.* 1
Sepals nearly equal, not auricled at base. . . . . : . : *Solea.* 2

1. VIÓLA.

Sepals 5, oblong, acute, equal, auricular at base; petals 5, irregular, the upper one (lower by resupination) broadest, spurred at base, the 2 lateral equal, opposite; stamens approximate; anthers connate, the lobes diverging at base; capsule 1-celled, 3-valved, seeds attached to the valves.—♃ *Low herbaceous plants, acaulescent or caulescent. Peduncles angular, solitary, 1-flowered, recurved at the summit so as to bear the flowers in a resupinate position.*

* *Acaulescent. Flowers blue.*

1. V. SELKIRKII. Goldie. *Selkirk's Violet.*
Lvs. cordate, crenately serrate, minutely hirsute above, smooth beneath; the sinus deep and nearly closed; *stig.* triangular, margined, distinctly beaked; *spur* nearly as long as the lamina, thick, very obtuse.—Grows on woody hills and mountains, Mass., N. Y., Can. A small, stemless violet, with small pale blue flowers conspicuously spurred. The radical, heart-shaped leaves are rather numerous and longer than the peduncles. The lateral petals bearded, and with the upper one striate with deep blue.

2. V. CUCULLATA. Ait. (V. affinis. *Le Conte.*) *Hood-leaved Violet.*
Smooth, sometimes more or less pubescent; *lvs.* cordate, cucullate at base, crenate; *stip.* linear; *inferior* and *lateral petals* bearded.—This is one of the more common kinds of violet, found in low, grassy woods, from Arctic Am. to Flor. Leaves on long petioles, heart-shaped, remarkably rolled at the base into a hooded form. The late leaves are crenate-reniform. Flowers light blue or purple, with scapes somewhat 4-sided, longer than the leaves. Petals twisted,

veiny, entire, white at the base, the lateral and upper ones marked with a few blue striæ. Very variable in respect to pubescence. May.

β. sororia. T. & G. (V. sororia. *Willd.*) Nearly smooth; *lvs.* exactly cordate; *fls.* small.

γ. reniformis. Pubescent; *lvs.* broadly reniform.

δ. alba. T. & G. Nearly smooth; *fls.* white.—R. L. *Olney!*

3. V. **sagittata.** Ait. *Arrow-leaved Violet.*

Lvs. oblong-lanceolate, sagittate-cordate, subacute, often incisely dentate at base, serrate-crenate, smooth or slightly pubescent; *ped.* longer than the leaves; *lower* and *lateral pet.* densely bearded.—On dry hills, Can. to Flor. W. to Ark. Leaves varying from oblong-sagittate to triangular-hastate, on margined petioles, acute or not. Scapes 3–5' long. Sepals lanceolate, acute. Petals entire, veiny, purplish-blue, white at base. Stigma rostrate, margined. Apr.—Jn.

4. V. **ovata.** Nutt. *Ovate-leaved Violet.*

Lvs. ovate, crenate, ciliate, abruptly decurrent on the short petiole, pubescent; *lateral pet.* bearded; *stig.* a little rostrate.—On dry hills, N. J. Leaves many, mostly hairy on both sides, sometimes nearly smooth, ⅓ as wide as long, acute or not, upper ones often laciniate-dentate at base. Sepals ciliate, oblong-ovate, deeply emarginate behind. Petals entire, veiny, pale-purple, obovate, the lateral ones with dense white beard. Spur broad. Apr. May.

5. V. **palmata.** *Palmated Violet.*

Pubescent; *lvs.* cordate, lobed in a hastate or palmate manner, the lobes crenate and toothed, the middle one much the largest; *lateral pet.* bearded.—In upland pastures, Can to Ark. Stem 3—6' high. Root-stock scaly. Petioles hairy. The early leaves are ovate, entire, the later and perfect are often purple beneath, variously lobed and cleft, the middle lobe always the largest and longest, with 2 or 3 each side. Peduncle sub-4-angled, 3—6' long. Stipules lanceolate. Petals purple, entire, veiny, white at the base, upper ones smaller, lateral ones densely bearded, and marked with blue striæ. May.

6. V. **pedata.** *Pedate Violet.*

Nearly glabrous; *rt.* premorse; *lvs.* pedate, 5—9-parted, segments linear-lanceolate, mostly entire; *stig.* large, obliquely truncate; *beak* obscure.—Dry woods and pastures, Can. to Ill. and to Flor. Rhizoma fleshy, ending abruptly as if cut or bitten off. Leaves thick, divided into about 7 obtuse, narrow segments. Petioles with long, ciliate stipules at base. Peduncles sub-4-angled, much longer than the leaves. Petals pale blue, white at base, all of them beardless and entire. Apr. May.

7. V. **delphinifolia.** Nutt. *Larkspur-leaved Violet.*

Nearly glabrous; *lvs.* pedate, 7—9-parted, with linear 2—3-cleft segments; *stig.* thick, distinctly beaked; 2 *upper petals* pubescent, 3 *lower* emarginate; *spur.* saccate, short.—♃ Prairies and bottoms, Ill.! and Mo. Root thick. Leaves often finely divided with many dissected segments. Stipules acuminate, subentire. Peduncles a little longer than the leaves. Flowers rather smaller than in the last, of a rich blue. Mar. Ap.

8. V. **palustris.** *Mountain Violet.*

Lvs. reniform-cordate; *stip.* broadly ovate, acuminate; *stig.* margined; sepals ovate, obtuse; *caps.* oblong-triangular; *sds.* ovate, dark green.—Summits of the White Mts. About 3' high, pubescent. Leaves crenate, 1' by ¾'. Flowers small, pale blue, on peduncles longer than the leaves and bibracteate near the middle. Rhizoma creeping, scaly. Jn.

9. V. **odorata.** *Sweet* or *English Violet.*—Stolons creeping; *lvs.* cordate, crenate, nearly smooth; *sep.* obtuse; *lateral pet.* with a hairy line.—Native of England. It is well characterized by its long, trailing, leafy runners. The leaves are truly heart-shaped. Stipules lanceolate, toothed. Peduncles longer than the leaves, bracted. Flowers small, fragrant.—Several garden varieties are known, distinguished by the form and color of the flowers; viz. the purple, white, and blue flowered, the double white, double purple, and double blue flowered, and the Neapolitan with pale blue flowers. Apr. May. †

** *Acaulescent. Flowers white.*

10. V. **BLANDA.** Wild.   (V. clandestina. *Ph.* V. amœna.   *Le Conte.*) *Bland or Sweet-scented Violet.*

Lvs. cordate, slightly pubescent ; *petiole* pubescent ; *fls.* white.—Found in meadows, Can. to Penn. The rhizoma is slender and creeping. Leaves close to the earth, nearly round, cordate or ovate, and sometimes with a rounded sinus so as to appear reniform. Petioles half round. Peduncles sub-4-sided, longer than the leaves. Petals white, greenish at base, upper and lateral ones marked with a few blue lines, generally beardless. Fls. small, fragrant. May.

11. V. **LANCEOLATA.** *Lance-leaved Violet.*

Lvs. smooth, lanceolate, narrowed at base into the petiole, obtusish, sub-crenate. Found in wet meadows, Can. to Tex. Rhizoma creeping. Leaves very narrow, and, with the stalk, 3—5' long. Petioles half round. Peduncles sub-4-sided. Petals white, greenish at base, upper and lateral ones marked with blue lines, generally beardless. Flowers small. May.

12. V. **PRIMULÆFOLIA.** *Primrose Violet.*

Lvs. lance-ovate, abruptly decurrent at base; *bracts* lance-linear ; *pet.* acute, nearly equal, beardless.—Found in damp soils, Mass. to Ky. Rhizoma creeping. Leaves sometimes subcordate, rather obtuse, mostly smooth, longer than their stalks. Petals obovate, acute flat, marked with purple lines at base, generally beardless, as long as the bracts. Flowers small, white, on sub-4-sided stalks. May in N. Eng.

3. acuta. T. & G. (V. acuta. *Bw.*)—Smooth; *lvs.* ovate; *pet.* acute, lateral ones nearly beardless. Mass.

*** *Acaulescent. Flowers yellow.*

13. V. **ROTUNDIFOLIA.** Michx. *Round-leaved Violet.*

Lvs. orbicular-ovate, cordate, slightly serrate, nearly smooth, with the sinus closed; *petiole* pubescent; *cal.* obtuse. –A small yellow violet, found in woods, N. Eng. to Tenn. Leaves nearly round, with a deep, narrow sinus at base, obscurely and remotely serrated. Veins and petioles pubescent. Peduncles as long as the claws, sub-4-sided, bracted in the middle. Petals yellow, marked at base with brown lines. Flowers small.

**** *Caulescent.*

14. V. **CANADENSIS.** *Canadian Violet.*

Smooth; *lvs.* cordate, acuminate, serrate; *ped.* shorter than the leaves; *stip.* short, entire.—A large species, found in woods, British Am. to Car., often a foot in height. Stem subsimple, terete, with lance-ovate, membranaceous stipules. Leaves alternate, the lower on very long petioles, acute or obtuse. Peduncles sub-4-sided, terminal, with minute bracts. Flowers large, nearly regular. Petals white or light blue, yellowish at base, the upper ones purple without and marked with blue lines, lateral ones bearded. Flowering all summer.

15. V. **PUBESCENS.** Ait. *Common Yellow Violet.*

Villous-pubescent ; *st.* erect, naked below; *lvs.* broad-cordate, toothed ; *stip.* ovate, subdentate.—A large yellow violet, found in dry, stony woods, Can. to Ga. and Mo. Root fibrous. Stem simple, more or less pubescent, somewhat triangular and fleshy, bearing a few leaves at the top, leafless below. Leaves broad-ovate, cordate, or deltoid ; obscurely dentate, obtuse, on short stalks. Stipules large, ovate, wavy. Flower-stalks rather shorter than leaves, downy, axillary, solitary, with 2 subulate bracts. Petals yellow, lateral ones bearded, and with the upper one marked with a few brown lines. The plant varies in pubescence, sometimes even glabrous. Height very variable, 5—20'. May– Jn.

β. eriocarpa. Nutt. (V. eriocarpa. *Schw.*) *Capsule* densely villose.

γ. scabriuscula. T. & G. (V. scabriuscula. *Schw.*) *St.* decumbent, branching from the root, and with the smaller leaves somewhat scabrous.

16 V. **HASTĀTA.** Michx.

Smooth, simple, erect, leafy above; *lvs.* deltoid-lanceolate, hastate or broadly ovate-acuminate, dentate; *stip.* ovate, minute, ciliate, dentate; *lower pet.* dilated, obscurely 3-lobed, *lateral ones* slightly bearded; *sep.* lanceolate, with a very short spur.—Pine woods, Penn. to Flor. Stem 6—10' high. Peduncles shorter than the leaves. Flowers yellow. May.

17. V. MUHLENBERGII. Torr. *Muhlenberg's Violet.*

St. weak, assurgent; *lvs.* reniform-cordate, upper ones crenate, rather acuminate; *stip.* lanceolate, serrate, ciliate.—A spreading, slender species, in swamps, &c., U. S., N. to Lab. Stems branched below, 6—8' long, with large stipules cut into fringe-like serratures. Leaves 6 -10" diam., younger ones involute at base. Petioles longer than the leaves, and shorter than the axillary peduncles. Bracts linear, alternate, on the upper part of the stalk. Petals entire, pael purple, the lateral ones bearded. Spur porrected, very obtuse. Stigma rostrate. May.

18. V. ROSTRATA. *Beaked Violet.*

Smooth; *st.* terete, diffuse, erect; *lvs.* cordate, roundish, serrate, upper ones acute; *stip.* lanceolate, deeply fringed; *pet.* beardless; *spur* longer than the corolla.—A common violet in moist woods, Can. to Ky., well characterized by its long, straight, linear, obtuse nectary, which renders the large flowers similar to those of the larkspur. Stem 6—8' high, branching below. Petioles much longer than the leaves. Stipules almost pinnatifid. Peduncles slender, very long, axillary. Flowers pale blue. May.

19. V. STRIATA. Ait. (V. ochroleuca. *Schw.*) *Striped Violet.*

Smooth; *st.* branching, nearly erect; *lvs.* roundish-ovate, cordate, the upper ones somewhat acuminate, crenate-serrate; *stip.* large, ciliate-dentate, oblong-lanceolate; *spur* somewhat porrected.—Wet grounds, U. S. and Can. Stem 6—12' high, half round. Leaves 1—1½' wide, on petioles 1—2' long. Stipules conspicuous, laciniate. Peduncles axillary, often much longer than the leaves. Corolla large, yellowish-white or ochroleucous, lateral petals densely bearded, lower one striate with dark purple. Stigma tubular. Jn.

20. V. ARVENSIS. Ell. (V. tenella. *Muhl.* V. bicolor. *Ph.* V. tricolor. β. arvensis. *DC.*)

St. 3-angled, erect; *lvs.* orbicular-spatulate, smooth, subdentate, upper ones ovate-spatulate; *stip.* foliaceous, pinnatifid, very large; *pet.* longer than the calyx, bluish-white.—① A rare species, though widely dispersed from N. Y. South to Ga., and W. to Mo., on dry hills. Stem pubescent on the angles, 2—4' high. Leaves 3—5" diam., shorter than the petioles, with about 5 obtuse teeth or angles; cauline ones more narrow, sometimes entire. Stipular segments linear-oblong, as long as the leaves. Peduncles 4-angled, twice longer than the leaves. Petals twice longer (scarcely longer T. & G.) than sepals, lateral ones bearded, lowest with 5 striæ. May.

21. V. TRICOLOR. *Tricolored Violet. Pansey. Heart's-ease.*—*St.* angular, diffuse; *lvs.* oblong-ovate, lower ones ovate-cordate, deeply crenate; *stip.* runcinately pinnatifid or lyrate, the terminal segment as large as the leaves; *spur* short, thick.—Gardens, where its pretty flowers are earliest in spring, and latest in autumn. Flowers variable in size, the 2 upper (lower) petals purple, the 2 lateral white and with the lower, striate, all yellow at base. †

22. V. GRANDIFLORA. *Great Purple Violet.*—*St.* 3-cornered, simple, procumbent; *lvs.* ovate-oblong, crenate, shorter than the peduncles; *stip.* lyrate-pinnatifid; *fls.* large.—Native of Switzerland. A large and beautiful species, with dark purple flowers, 1—2' diam. Whole plant smooth, 6—12' long. Stipules ½—1' long. Flowering all seasons but winter. †

2. SOLEA. Gingins.

Sepals nearly equal, not auriculate; petals unequal, the lowest 2 lobed and gibbous at base, the rest emarginate; stamens cohering, the lowest 2 bearing a gland above the middle; capsule surrounded at base by the concave torus; seeds 6—8, very large.—24 *Lvs. cauline, alternate.*

S. CONCOLOR. Gingins. (Viola concolor. *Ph.*) *Green Violet.*

A strictly erect plant, in woods, Western N. Y. to Me., S. to Car. Stem

1—2f high, simple, and, with the leaves, somewhat hairy. Leaves 4—6' by 1½—2¼', lanceolate, acuminate, subentire, tapering to short petioles. Peduncles very short, 1—5-flowered, axillary. Flowers small, greenish white. Calyx about as long as the corolla. Lower petal twice larger than the others. Capsule nearly 1' in length. Apr. May.

ORDER XVIII. DROSERACEÆ.—SUNDEWS.

Plants herbaceous, delicate, often covered with glands.
Lvs. alternate, with stipulary fringes, circinate in vernation.
Fld. when young, circinate. Sep. 5, persistent, equal, imbricate in æstivation.
Cor.—Petals 5, hypogynous, marcescent.
Sta. distinct, marcescent usually equal in number to the petals.
Ova. single. Sty. 3—5, either wholly distinct or slightly united, bifid or branched.
Fr. a capsule, 1—3-celled, usually many-seeded. Sds. sometimes ariled.

Genera 7, species 80, scattered over the whole globe, wherever marshes are found. Their leaves are usually furnished with glandular hairs, and are entire, alternate or crowded. Attached to this order is the genus Parnassia, regarded by some as forming a separate order. It is variously located by different botanists. We follow Torrey and Gray, after De Candolle, in placing it here. Some peculiarity exists in the arrangement and structure of the stamens in this genus, which will be mentioned farther on.
No remarkable properties have been discovered belonging to plants of this order.

Conspectus of the Genera.

Stamens { hypogynous, all perfect and { 5 in number. Styles 3—5.	. . . . .	Drosera. 1
{ 10—15 in number. Style 1.	. . . . .	Dionæa. 2
{ perigynous, inner row 5 perfect ones, outer row 5 groups of imperfect ones.	.	Parnassia. 3

1. DROSERA.

Gr. δροσος, dew; from the dew-like secretion.

Sepals 5, united at base, persistent; petals 5; stamens 5, with adnate anthers; styles 3—5, each 2-parted; capsule 3—5-valved, 1-celled, many-seeded.—*Small aquatic herbs.*

1. D. ROTUNDIFOLIA. *Round-leaved Sundew.*
Lvs. radical, nearly round, depressed; *petioles* hairy; *scapes* erect, bearing a simple raceme.—♃ This curious little plant is not uncommon in bogs and muddy shores of ponds and rivers. It is at once distinguished by the reddish glandular hairs with which the leaves are beset, and which are usually tipped with a small drop of a clammy fluid, appearing like dew glistening in the sun. Leaves small, lying flat on the ground, narrowed into the elongated petiole. Scape 5—8' high, at first coiled inward. Flowers arranged on one side, very small, white. Aug.

2. LONGIFOLIA. *Long-leaved Sundew.*
Lvs. radical, spatulate and obovate, tapering at base into a long, smooth petiole; *scape* bearing a simple raceme.—♃ A more slender and delicate species, in similar situations with the last. Leaves slender, ascending, cuneiform, oblong, crenate, beset with numerous hairs tipped with dew-like drops,—length including the petiole 1—3'. Scape ascending at base, bearing a cluster of small, yellowish-white flowers, and arising 3—8'. Jn.—Aug.

3. D. FILIFORMIS. *Thread-leaved Sundew.*
Lvs. filiform, very long; *scape* nearly simple, longer than the leaves, many-flowered; *pet.* obovate, erosely denticulate, longer than the glandular calyx; *sty.* 2-parted to the base.—♃ Grows in wet sandy places, much larger than the preceding species. The leaves are destitute of a lamina, are suberect, nearly as long as the scape, beset with glandular hairs except near the base. Scape about a foot high, with large, purple flowers. Aug. Sept.

4. D. LINEARIS. Goldie. *Linear-leaved Sundew.*
Lvs. linear, obtuse; *petioles* elongated, naked, erect; *scapes* few-flowered, about the length of the leaves; *cal.* glabrous, much shorter than the oval capsule; *sds.* oval, shining, smooth.—① Borders of lakes, Can., Mich. to the Rocky Mts. *Hooker. T. & G.* Scape 3—6' high, with about 3 small flowers. Leaves about 2" wide, clothed with glandular hairs which are wanting on the petiole. Jl. Aug.

16

2. DIONÆA. Ellis.

Dionæa is one of the names of Venus.

Sepals 5, ovate, oblong, spreading; petals 5, obovate, with pellucid veins; stamens 10—15; style 1; stigmas 5, connivent, many-cleft; capsules indehiscent, breaking irregularly, 1-celled, many-seeded.— ♃ glabrous. *Lvs. radical, sensitive, closing convulsively when touched. Scape umbellate.*

D. MUSCIPULA. Ell. *Venus' Fly-trap.*—Native of the Southern States. Sometimes cultivated in a pot of bog earth placed in a pan of water. Leaves rosulate, lamina roundish, spinulose on the margins and upper surface, instantly closing upon insects and other objects which light upon it. (See Part I. § 248.) Scape 6—12' high, with an umbel of 8—10 white flowers. Apr. May. †

3. PARNASSIA. Tourn.

Named for Mount Parnassus, the abode of the Muses, Graces, &c.

Sepals 5, united at base, persistent; petals 5, persistent, nearly perigynous; stamens perigynous, in 2 series, the outer indefinite in number, united in 5 groups, sterile, the inner 5 perfect; capsule 1-celled, 4-valved; seeds very numerous, with a winged testa.—♃ *herbs with radical lvs. and 1-flowered scapes.*

1. P. CAROLINIÀNA. *Grass of Parnassus.*

Sterile filaments in 5 clusters, 3 in each, distinct to near the base, surmounted with little spherical heads; *pet.* much exceeding the calyx, marked with green veins; *lvs.* radical or sessile on the scape, broad-oval, with no sinus at the base.—An exceedingly elegant and interesting plant, growing in wet meadows and borders of streams, U. S. to Can. Root fibrous. Leaves about 7-veined, broad-oval or ovate, smooth, leathery, radical ones long-stalked, cauline ones sessile, clasping, a few inches above the root. Scapes about 1f high, with a handsome regular flower about 1' diam. Jl. Aug.

2. P. PALUSTRIS.

Lvs. all cordate, the cauline one (if any) sessile; *scales* (bundles of sterile stamens) smooth, with numerous slender, pellucid setæ.—Bogs and lake shores, Mich. to Lab. and W. to the Rocky Mts. Scapes about 6' high, naked or with a single clasping leaf near the base. Flowers white. Sepals oblong-lanceolate. Petals marked with 3—5 green or purple veins. Each scale is distinguished by 10—15 whitish hair-like bristles.

ORDER XIX. CISTACEÆ.—ROCK ROSES.

Plants herbaceous or shrubby. Branches often viscid.
Lvs. entire, opposite or alternate, usually feather-veined.
Fls. white, yellow, or red, very fugacious, in one-sided racemes.
Cal.—Sepals 5, unequal, the 2 inner with a twisted æstivation.
Cor.—Petals 5, hypogynous, crumpled in æstivation.
Sta. indefinite, hypogynous, distinct. *Anth.* innate.
Ova. distinct, or many-celled. *Sty.* single. *Stig.* simple. [seeding from the middle of the valves.
Fr. capsular, either 1-celled with parietal placentæ, or imperfectly 3—5-celled, with dissepiments pro

Genera 7, species 185, found most abundant in the north of Africa or south of Europe. They possess no interest on account of their properties.

Conspectus of the Genera.

Petals 5 { large and showy, or wanting.	. . . .	Helianthemum. 3
{ minute. Delicate shrubs.	. . . .	Hudsonia. 1
Petals 3, { near-lanceolate.	. . . .	Lechea. 1

1. LECHÈA.

In memory of John Leche, a Swedish botanist.

Sepals 5, the 2 outer minute; petals 3, lanceolate, small; stamens 3—12; stigmas 3, scarcely distinct; capsule 3-celled, 3-valved; placentæ nearly as broad as the valves, roundish, each 1—2-seeded — ♃ *Suffruticose, branching plants. Stipules 0.*

1. L. MAJOR. Michx. (L. villosa. *Ell.* L. minor. *Linn.*) *Larger Pinweed.*
Erect, hairy; *branches* villous, radical ones prostrate; *cauline lvs.* elliptical, mucronate, those of the radical branches roundish, minute; *fls.* small, numerous, in fasciculate racemes, somewhat 1-sided.—In dry woods, U. S. and Can Stem 1—2f high, rigid, brittle, hairy, purple, paniculately branched. Leaves of the stem about 1' long, alternate, opposite, or even verticillate on the prostrate branches, crowded. Flowers brownish-purple, inconspicuous. Capsule roundish, about the size of a large pin-head. Jl. Aug.

2. L. MINOR. Lam. *Smaller Pinweed.*
Erect, smoothish, branched; *lvs.* linear-lanceolate, acute; *panicle* leafy, its branches elongated; *fls.* in nearly simple racemes; *caps.* rather large.—Grows in dry, sandy grounds, U. S. and Can. Stem 8—12' high, often decumbent at base. Stem leaves, 6—10'' by 2—3'', alternate, sparingly ciliate and revolute at the margin, those of the long slender branches minute. Flowers nearly twice as large as in L. major. Petals brownish purple, cohering at apex. Capsule also rather larger than in L. major. Jn.—Sept.

3. L. THYMIFOLIA. Ph. *Thyme-leaved Pinweed.*
Frutescent; *sts.* decumbent at base, hoary with appressed hairs, very branching and leafy; *root lvs.* on the short radical branches, imbricate, elliptical, very small; *cauline lvs.* linear or oblanceolate, often whorled. Sea-coasts, Mass.! to N. J. Stem about 1f high, rigid and very bushy. Upper leaves about 1' long, erect and crowded. Flowers in terminal and axillary clusters, on very short pedicels. Petals brown. Capsules globose. Jl.—Sep.

2. HELIANTHEMUM.

Gr. ἥλιος, the sun, ἄνθος, a flower.

Sepals 5, the 2 outer smaller; petals 5, or rarely 3, sometimes abortive; stamens 00; stigmas 3, scarcely distinct; capsule triangular, 3-valved, opening at top; seeds angular.—*Fls. yellow.*

1. H. CANADENSE. Michx. (Cistus Canad. *Willd.*) *Frost Plant. Rock Rose.*
St. ascending; *branches* erect, pubescent; *lvs.* alternate, without stipules, lanceolate, acute, hairy; *petaliferous fls.* few, large, terminal, *apetalous ones* lateral, solitary or racemose.—In dry fields and woods, Can. to Flor. Stem about 1f high, at length shrubby at base. Leaves 8—12'' long, ⅓ as wide, entire, subsessile. Flowers with large bright yellow petals, in a terminal corymb. The axillary flowers later, very small, with very small petals, or apetalous. Stamens declinate. Capsule smooth, shining, those of the apet. fls. not larger than a pin's head. Seeds few, brown. Jn.—Sep.

2. H. CORYMBOSUM. Michx. (Heteromeris cymosa. *Spach.*)
St. branching, canescent, erect; *lvs.* lance-oblong, canescently tomentose beneath; *fls.* in crowded, fastigiate cymes, the primary ones on elongated, filiform pedicels, and with *petals* twice longer than the calyx; *sep.* villous-canescent, outer ones linear, obtuse, inner ovate, acute.—Sterile sands, N. J. to Ga. Plant somewhat shrubby, very tomentose when young, at length diffusely branched, about 1f high. Primary flowers about 1' diam. Secondary ones small, apetalous. Jn.—Aug.

3. HUDSONIA.

In honor of Wm. Hudson, author of "Flora Anglica."

Sepals 3, united at base, subtended by 2 minute ones at base; petals 5; Stamens 9—30; style filiform, straight; capsule 1-celled, 3-valved, many-seeded.—*Low shrubs with very numerous branches, and minute exstipulate lvs.*

1. H. TOMENTOSA. Nutt. *Downy Hudsonia.*
Hoary-tomentose; *lvs.* ovate, imbricate, acute, shorter than the intervals of the stem; *fls.* subsessile; *pet.* obtuse.—Shores of the ocean and lakes, N. J to N. H.! and Wisc., &c. Plant consisting of numerous slender, ascending

stems from the same root, and a multitude of tufted branches, all covered with whitish down. Leaves about 1—2″ in length, closely appressed to the stem. Flower small, yellow, on pedicels not longer than the leaves. May.

2. H. ERICOÏDES. *Heath-like Hudsonia.*

Hoary-pubescent; *lvs.* acerose-subulate; *ped.* longer than the leaves, filiform, hairy; *sep.* acutish.—A very delicate shrub, found in pine barrens, Mass. to Va. Stem ½f high, erect, with numerous short, compound, procumbent branches. Leaves needle-like, scattered, 2—4″ long. Flowers yellow, shorter than the peduncles. Capsule oblong, pubescent. May.

ORDER XX. HYPERICACEÆ.—St. John's-worts.

Herbs, shrubs or trees, with a resinous juice, and often with angular branches.
Lvs. opposite, entire, mostly punctate with pellucid dots, and black glands. *Stip.* 0.
Fls. perfect, mostly yellow, with cymose inflorescence.
Cal.—Sepals 4—5, distinct or cohering, persistent, unequal, dotted.
Cor.—Petals 4—5, hypogynous, æstivation twisted, veins oblique, dotted.
Sta. hypogynous, indefinite, in 3 or more parcels, *Anthers* versatile.
Ova. single, superior. *Style* slender. *Stigma* simple.
Fr. a capsule or berry, many-celled. *Seeds* indefinite, minute.

Genera 19, species 376, very generally distributed, presenting a very great variety in habit, and flourishing in all kinds of localities. The juice of many species is considered purgative and febrifugal.

Conspectus of the Genera.

Petals and sepals
{ 5. Hypogynous glands { 2. . . . . *Hypericum.* 2.
                { 0. . . . . *Elodea.* 3.
{ 4. Hypogynous glands 0. . . . . *Ascyrum.* 1

1. ASCYRUM.

Gr. a, privative, σκυρος, roughness; i. e., a smooth plant.

Sepals 4, the 2 outer usually larger; petals 4; filaments slightly united at base into several parcels; styles 2—4, mostly distinct; capsule 1-celled.—*Plants suffruticose. Lvs. punctate with black dots. Fls. yellow,* 1—3, *terminal on each branch. Pedicels bibracteolate.*

1. A. CRUX-ANDREÆ. (A. multicaule, *Michx.*) *St. Peter's-wort.*

St. much branched at base; *branches* suberect, ancipital above; *lvs.* obovate or linear-oblong, obtuse; *inner sep.* minute, roundish; *pet.* linear-oblong; *sty.* 1—2.—Sandy woods, N. J. to La. Stem about 1f high, thickly clothed with leaves which are ½—1½′ long, of very variable width. Flowers pale-yellow, on very short pedicels, with 2 bracteoles close to the calyx. Petals exceeding the sepals and stamens. July.

2. A. STANS. Michx. (A. hypericoides. *Linn.*)

St. straight, erect, ancipital or winged, branched above; *lvs.* oblong, obtuse, sessile; *outer sep.* cordate, orbicular, longer than the 2 lanceolate, interior ones; *sty.* 3.—Swamps in pine barrens, N. J. to La. Stem 1—2f high. Leaves 1—1½′ long, ⅓ as wide, somewhat glaucous. Flowers usually 3 together, much larger than in the preceding. Yellow. Jl. Aug.

2. HYPERICUM.

Sepals 5, connected at base, subequal, leaf-like; petals 5, oblique; Stamens ∞ (sometimes few) united at base into 3—5 parcels, with no glands between them; styles 3—5, distinct or united at base, persistent.—*Herbaceous or shrubby plants. Lvs. punctate, with pellucid dots, opposite, entire. Fls. solitary, or in cymose panicles, yellow.*

* *Stamens* 20—100, *polyadelphous. Herbs.*

1. H. PYRAMIDATUM. Ait. (H. ascyroides. *Willd.*) *Giant Hypericum.*

St. branching, somewhat quadrangular; *lvs.* sessile, oblong-ovate, acute, smooth; *sty.* as long as the stamens.—⚦ A large flowering species, found on dry hills, also on river banks, Ohio and Penn. to Car. Stem 3—5f high, scarcely angular, smooth, rigid, herbaceous. Branches corymbose and erect, or late cal, axillary, opposite. Leaves acute, not acuminate, those of the stem 2¼—5' long, ½ as wide, of the branches about half these dimensions. Flowers 1½' diam. Petals obovate, ½—⅓' wide. Stamens capillary, 100 or more. Capsules 1' long, ovoid-conical, tipped with the 5 styles. Seeds 00. Jl. Aug.

2. H. PERFORÀTUM. *Common St. John's-wort.*
St. 2-edged, branched; *lvs.* elliptical, with pellucid dots; *sep.* lanceolate, half as long as the petals.—2¼ A hardy plant, prevailing in pastures and dry soils, Can. and U. S., much to the annoyance of farmers. Stem 1—2f high, brachiate, erect, round, with 2 opposite, elevated lines extending between the nodes. Leaves 6—10" long, ⅓ as wide, ramial ones much smaller, all obtuse, the dots as well as veins best seen by transmitted light. Flowers numerous, deep yellow, in terminal panicles. Petals and sepals bordered with fine dark colored glands. Jn. Jl. §

3. H. CORYMBÒSUM. Muhl. (H. punctatum. *Beck.*) *Spotted St. John's.*
St. erect, round, smooth, branching; *lrs.* clasping, oblong-oval, obtuse, covered with black dots; *cymes* terminal, brachiate, dense-flowered, corymbose · *sep.* ovate, acute.—2¼ In wet meadows and damp woods, N. Eng. to Ark. Stem 1½—2f high. Leaves 1—2' long, nearly ½ as wide, with pellucid punctures besides the black dots. Flowers small, numerous, pale-yellow, petals nearly 3 times as long as the sepals, with oblong black dots. Stigmas orange-red, on distinct styles. Jn. Jl.

4. H. ANGULÒSUM. *Michx.*
St. simple below, corymbosely branched above, sharply 4-angled; *lrs.* somewhat ovate, closely sessile, scarcely punctate; *cymes* leafless, compound; *fls.* alternate, solitary on the ultimate branches; *sep.* lanceolate, acute, half as long as the petals.—2¼ Cedar swamps, N. J. to Flor. Stem nearly 2f high. Leaves nearly 1' long, ⅓ as wide, rather distant. Petals obovate, brownish-red, with a single lateral tooth near the apex. Jn.—Aug.

5. H. ELLIPTÌCUM. *Hook.* (H sphærocarpon. *Bart.*)
St. quadrangular, simple, cymose at summit; *lrs.* elliptical, obtuse, somewhat clasping, pellucid-punctate; *cyme* pedunculate; *sep.* unequal; *sty.* united at base.—2¼ Low grounds, Uxbridge, Ms., *Rickard!* R. I. *Olney,* to N. Y. and Penn. Stem 8—16' high, slender, colored at base. Leaves 8—13" by 2—4", somewhat erect, about as long as the internodes. Cymes of about a dozen flowers, generally 1—2' above the highest pair of leaves. Central flowers subsessile. Petals acutish, orange-yellow, 2—3" long; sepals shorter. Stigmas minute. July.

* * *Stamens 20—100, polyadelphous. Shrubs.*

6. H. KALMIÀNUM. *Kalm's St. John's-wort.*
St. corymbosely branched; *branches* somewhat 4-sided, two of the angles slightly winged; *lrs.* linear-lanceolate, very numerous, obtuse, attenuate at base; *cymes* 3—7-flowered, fastigiate; *sep.* half as long as petals.—Rocks below Niagara Falls! &c. A shrubby species a foot or more in height. Leaves an inch in length, slightly revolute on the margin, 1-veined, minutely and thickly punctate, sessile. Branches slender and delicate. Flowers yellow. Stamens very numerous. Aug.

7. H. PROLIFÌCUM.
Branching; *branches* ancipital, smooth; *lrs.* oblong-lanceolate, obtuse, narrowed at base, crenulately waved at edge; *cymes* compound, leafy; *sep.* unequal, leafy, ovate, cuspidate; *pet.* obovate, a little larger than sepals; *sty.* at length distinct.—A highly ornamental shrub, 2—4f high, prairies and creek shores, Mid. and West. States! Leaves 2—2½' long, 4—6'' wide. Flowers ½' diam., orange-yellow, in an elongated inflorescence. Stamens 00. Jl. Aug. †
β. T. & G. *Lrs.* much smaller; *capsule* attenuate at summit.

8. H. ADPRESSUM. Barton.
St. shrubby at base, 2-winged above; *lrs.* oblong-linear, sessile, with pellucid punctures; *fls.* 15—20, in a leafless cyme; *sep.* unequal, half as long as the oblong-obovate petals; *sty.* united; *caps.* 3-celled.—Swamps, R. I. ! Pa.

16*

to Aik. About 2f high. Leaves 1—2' by 2—4", often somewhat lance-shaped.
Flowers about 6" diam., with very numerous stamens. Aug Sept.

9. H. AUREUM. Bertram. (H. amœnum. *Ph.*) *Golden Hypericum.*
Branches spreading, ancipital; *lvs.* oblong, obtuse, attenuate at base,
glaucous beneath; *fls.* few together, subsessile; *pet.* coriaceous, reflexed; *sty.* 3,
connate, persistent on the ovoid-conic capsule.—A beautiful shrub, 2—4f high,
native of S. Car. and Ga. Flowers large, orange-yellow. Stamens 100 or
more. Capsule red. Jn.—Aug.

10. H. NUDIFLORUM. Michx.
St. shrubby at base; *branches* winged; *lvs.* ovate-oblong, sessile, obtuse;
cymes leafless, pedunculate; *central fls.* shortly pedicellate; *pet.* obovate, longer
than the linear sepals; *sty.* united.—Wet grounds, Penn. to La. Plant 1—2f
high, with numerous 4-sided branches. Leaves thin, about 2' long, with minute
reddish dots. Flowers small and rather loose in the cyme. Aug. †

* * * *Stamens* 5—20, *distinct.*

11. H. MUTILUM. (H. quinquenervium. *Walt.*) *Small St. John's-wort.*
St. erect, usually much branched, often subsimple, quadrangular; *lvs.*
obtuse, ovate-oblong, clasping, 5-veined, minutely punctate; *cymes* leafy; *pet.*
shorter than the sepals; *sta* 6—12.—① Damp, sandy soils, Can. to Ga. W. to
Ia ! Stem 3—6—9' high. Leaves closely sessile, apparently connate, 4—8" by
2—5", outer veins obscure. Flowers minute. Jl. Aug.

12. H. CANADENSE. *Canadian St. John's-wort.*
St. quadrangular, branched; *lvs.* linear, attenuated to the base, with pellucid and also with black dots, rather obtuse; *sep.* lanceolate, acute, longer
than the petals; *sta.* 5—10.—① Wet, sandy soils, Can. to Ga. Stem 8—15'
high, slightly 4-winged. Lower branches opposite, upper pair forked. Leaves
about 1' by 1—2", sometimes linear-lanceolate, radical ones obovate, short.
Flowers small, orange-colored. Ovary longer than the styles. Capsule red,
longer than the sepals. Jl. Aug.

13. H. SAROTHRA. Michx. (Sarothra gentianoides. *Linn.*)
St. and *branches* filiform, quadrangular; *lvs.* very minute, subulate; *fls.*
sessile.—Dry fields and roadsides, U. S. and Can. Stem 4—8' high, branched
above into numerous, very slender, upright, parallel branches, apparently leafless, from the minuteness of the leaves. Flowers very small, yellow, succeeded
by a conical, brown capsule which is twice the length of the sep. Jl. Aug.

14. H. DRUMMONDII. Torr. & Gray. (Sarothra. *Grev. & Hook.*)
Branches alternate, square above; *lvs.* linear, very narrow, acute, longer
than the internodes; *fls.* pedicellate; *sta.* 10—20; *sep.* lanceolate, shorter than
the petals, but longer than the ovate capsule.—Near St. Louis, &c. Plant
more robust than the last, nearly 1f high, very branching. Leaves ¼' long.
Flowers about 4" diam.

3. ELODEA. Adans.
Gr. ἑλώδης, marshy ; from the habitat of the plants.

Sepals 5, equal, somewhat united at base; petals 5, deciduous,
equilateral; stamens triadelphous, the parcels alternating with 3 hypogynous glands; styles 3, distinct; capsule 3-celled.—♃ *Herbs with
pellucid-punctate leaves, the axils leafless.*

1. E. VIRGINICA. Nutt. (E. campanulata. *Ph.* Hypericum. *Linn.*)
St. erect, somewhat compressed, branching; *lvs.* oblong, amplexicaul;
sta. united below the middle, with 3 in each set.—Swamps and ditches, U. S.
and Can. Whole plant usually of a purplish hue, 9—20' high. Leaves 1½—2½'
long, ⅓ as wide, upper ones lanceolate, lower oblong-ovate, all very obtuse
glaucous beneath. Cymes terminal and axillary. Flowers 5" diam., orange-
yellow. Petals about twice longer than the calyx. Glands ovoid, orange-
colored. Capsules ovoid-oblong, acutish. Jl. Sept.

2. E. PETIOLATA. Pursh. (Hypericum. *Walt.*)
Lvs. oblong, narrowed at base into a petiole; *fls.* mostly in 3s, axillary,

nearly sessile; *fl.* united above the middle; *capsules* oblong, much longer than the sepals.—Swamps, N. J. to Ky. Stem about 2f high. Leaves 1—3' long, rounded-obtuse. Flowers smaller than in the last, of a dull orange-color, Aug. Sept.

Order XXII. ILLECEBRACEÆ.—Knotworts.

Plants herbaceous or suffruticose, branching.
Lvs. sessile, entire. *Stipules* and *bracts* scarious. *Fls.* minute.
Cal.—Sepals 5, distinct or coherent at base, persistent.
Cor.—Petals minute, inserted between the sepals, often wanting.
Sta. equal in number to the sepals (sometimes less or more) inserted into the perigynous disk.
Ova. superior, 1-celled. *Sty.* 2—5, either partially or wholly combined.
Fr. a utricle, *sd. solitary,* attached to base of cell; or a many-seeded capsule.
    Genera 24, species 100, found mostly around the Mediterranean. Seven of the genera have been found in N. America. A slight astringency is their only known property.

Conspectus of the Genera.

1 { or sterile filaments none.	. . . .	*Anychia.* 1
	minute, resembling sterile filaments.	*Paronychia.* 2
Petals { conspicuous, white or rose-colored.	*Spergula.* 3

1. ANYCHIA. Michx.

Gr. ονυξ, the finger nail; a supposed remedy for the maladies of that organ.

Calyx of 5, ovate-oblong, connivent sepals, callous, subsaccate at the apex; corolla 0; filaments 2—5, distinct; stigma subcapitate; utricle enclosed in the sepals.—① *Small herbs, with dichotomous branches, Lrs. stipulate.*

A. dichotóma. Michx. (Queria Canadensis. *Linn.*) *Forked Chickweed.*
St. at length much branched, erect; *lrs.* lanceolate, cauline ones opposite, ramial ones alternate; *fls.* about as long as the stipules, terminal ones subfasciculate.—Dry woods and hills, Can. and N. Eng. to Ark. Stem 4—10. high, round, slender, pubescent above, with dichotomous, filiform branches. Leaves 2—9" by ⅓—2", acute or obtuse, with ovate-acuminate, scarious stipules at base situated at each fork of the stem. Flowers axillary, solitary, or in terminal clusters of 3 or more, very small, white. Jn.—Aug.
β. capillacea. Torr.—Smooth; *branches* capillary; *lrs.* oblong, obtuse, cuneiform at base. Ia.! Ill.!

2. PARONYCHIA. Tourn.

Etymology similar to the foregoing.

Sepals united at base, acuminate cuspidate at apex, the lining membrane colored and cucullate or saccate at summit; petals (sterile fil?) very narrow and scale-like; stamens 5; styles more or less united; stigmas 2; utricle 1-seeded included in the calyx.

1. P. Jamesii. Torr. & Gray.
Cæspitose, much branched; *lvs.* linear-subulate, scabrous; *fls.* few, in small, dense, dichotomous cymes, the central ones sessile; *pet.* (or setæ) alternate with the fertile filaments; *sep.* linear, with a minute cusp.—Prairies, Mason Co., Ill. *Mead.* R. Mts. *James. Nutt.* Stems about ½f long. Flowers small.

2. P. dichotóma. Nutt. (Achyranthes. *Linn.*)
Cæspitose, densely branching; *lrs.* acerose-mucronate, glabrous, 2-grooved each side; *cymes* compound, diffuse, without central *fls.*; *setæ* much shorter than the stamens.—Rocks, Harper's Ferry. Stems 6—12' high. Leaves 1' by ⅓", crowded. Style bifid ⅓ its length. Jl.—Nov.

3. SPERGÚLA.

Lat. *spergo,* to scatter; from the dispersion of the seeds.

Sepals 5, nearly distinct; petals 5, entire; stamens 5—10; styles 3—5; capsules superior, ovate, 3—5-valved, many-seeded.—① *Herbs with flowers in loose cymes. Lrs. stipulate.*

1. S. ARVENSIS. *Corn Spurry.*

Lvs. verticillate, linear-subulate; *sta.* 10; *sty.* 5; *ped.* reflexed in fruit; *sds.* reniform, angular, rough.—A common weed in cultivated grounds, Can. to Ga. Root small. Stem round, branched, with swelling joints, beset with copious whorled leaves, somewhat downy and viscid. Two minute stipules under each whorl. Cyme forked, the terminal (central) peduncles bending down as the fruit ripens. Petals white, longer than the calyx, capsule twice as long. Seeds many, with a membranous margin. May—Aug. §

2. S. RUBRA. T. & G. (Arenaria rubra. *Linn.*)

St. decumbent, much branched; *lvs.* linear, slightly mucronate; *stipules* ovate, membranous, cleft; *sep.* lanceolate, with scarious margins; *pet.* red or rose-color; *sty.* 3; *sds.* compressed, angular, roughish.—A common and variable species, found in sandy fields! Can. to Flor. &c. Stems a few inches in length, slender, smooth, spreading on the ground, with small, narrow leaves and dry, sheathing stipules. Flowers small, on hairy stalks. May—Oct.

ORDER XXIII. CARYOPHYLLACEÆ.—CLOVEWORTS.

Herbs, with the stems swelling at the nodes.
Lvs. opposite, entire, destitute of stipules. Fls. regular.
Cal.—Sepals 4—5, distinct, or cohering in a tube, persistent.
Cor.—Petals 4—5, (sometimes none) either unguiculate and inserted upon the pedicel of the ovary or without claws and inserted on the outside of a fleshy disk.
Sts. twice as many as the petals, rarely equal or few; *anthers* introrse.
Ova. often stipitate. *Styles* stigmatose the whole length of their inner surface.
Fr. a 1-celled capsule or imperfectly 2—5-celled, opening at the apex by twice as many teeth as there are stigmas. Seeds numerous.

Genera 53, species 1055. Eleven of the genera are North American, the remainder are found in the temperate and frigid climates of the Eastern Continent. Properties unimportant. The order is noticeable chiefly for the beauty of a few of the cultivated species.

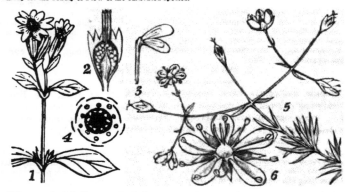

FIG. 41.—1. Lychnis diurna. 2. Vertical section, exposing the 5 styles, its placenta and seeds of the 1-celled capsule. 3. A petal, with its long claw, its bifid lamina, and its 2-toothed crown. 4. Cross section of the flower, showing the arrangement of its parts. 5. Arenaria stricta, showing the spreading cyme. 6. A flower enlarged—calyx not tubular.

Conspectus of the Genera.

			{ Sepals partly united.				Adenarium.	4
		{ entire. { Sepals entirely distinct.					Arenaria.	3
	{ Styles 3. { Petals 5, { bifid.						Stellaria.	1
not	{	{ Petals 0.					Molluge.	6
tubular.	{ Styles 4.	Petals 4—5 or 0, entire.					Sagina.	5
	{ Styles 5.	Petals 5, bifid.					Cerastium.	2
		{ Calyx calyculate with 2—4 scales at base.					Dianthus.	10
		{ Calyx without scales at the base.					Saponaria.	9
	{ Styles 2.						Silene.	7
Calyx { tubular	{ Styles 3.						Lychnis.	8
	{ Styles 5.							

TRIBE 1.—ALSINEÆ.

Sepals distinct or nearly so. Petals without claws inserted on the outside of the disk. Stamens inserted on the margin of the disk.

1. STELLARIA.

Lat. *stella*, a star :—from the stellate or star like flowers.

Sepals 5, connected at base ; petals 5, 2-parted ; stamens 10, rarely fewer; styles 3, sometimes 4 ; capsule superior, 1-celled, 3-valved, many-seeded.—*Small grass-like herbs, in moist, shady places. Fls. in forked cymes.*

1. S. MEDIA. Smith. (Alsine. *Linn.*) *Chickweed.*
Lvs. ovate; *st.* procumbent, with an alternate, lateral, hairy line ; *sta.* 3—5 or 10.—A common weed in almost every situation N. of Mex., flowering from the beginning of spring to the end of autumn. Stems prostrate, branched, brittle, round, jointed, leafy, and remarkably distinguished by the hairy ridge extending from joint to joint, in an alternate manner. Flowers small, white. The seeds are eaten by poultry and the birds. ♀

2. S. LONGIFOLIA. Muhl. (S. graminea. *Bw.*)
Lvs. linear, entire; *cyme* terminal, spreading, with lanceolate, scarious bracts; *cal.* 3-veined, about equal to the petals.—U. S., N. to Arc. Circ. The stems are of considerable length, very slender and brittle, supported on other plants and bushes. Leaves alternate at base. Flowers in a divaricate, naked cyme, very elegant, white, appearing in 10 segments like the other species. Three acute, green veins singularly distinguish the sepals. Jn. Jl.

3. S. PUBERA. Michx.
St. decumbent, pubescent in one lateral or two opposite lines; *lvs.* oblong-oval, acute, sessile, somewhat ciliate; *fs.* on short, filiform, recurved pedicels.—♃ In rocky places, Penn. and Ky. to Ga. Stem 6—12′ long, often diffusely spreading. Leaves 1—2½′ by 4—10″, with minute, scattered hairs. Flowers ½′ diam., axillary and terminal, large, with 10 stamens and 3 styles. Apr.—Jn.

4. S. LONGIPES. Goldie. (S. palustris. *Rich.* Micropetalon. *Pers.*)
Smooth and shining; *st.* more or less decumbent, with ascending branches; *lvs.* linear-lanceolate, broadest at base, acute; *peduncles* and *pedicels* filiform, cymose, with ovate, membranous bracts at base; *sep.* with membranous margins, obscurely 3-veined, scarcely shorter than the petals.—♃ Lake-shores, N. Y.! and Mich. Petals white, 2-parted. Flowers in loose cymes, the terminal peduncle, or the middle one, the longest. Jn.—Aug.

5. S. BOREALIS. Bigelow. (S. lanceolata. *Turr.* Micropetalon. *Pers.*)
St. weak, smooth; *lvs.* veinless, broad-lanceolate, acute; *ped.* at length axillary, elongated, 1-flowered; *pet.* 2-parted (sometimes wanting), about equal to the veinless sepals.—① Wet places. N. H., N. Y., N. to Artic Am. A spreading flaccid plant. Stem 6—12—15′ long, with diffuse cymes both terminal and axillary. Leaves 8—15″ long, 1-veined. Petals when present white, small, at length about as long as the lanceolate, acute sepals. Capsules longer than the calyx. Jn. Jl.

6. S. AQUATICA. Pollich. (S. borealis. *Darl.*)
Nearly glabrous; *st.* slender, decumbent; *lvs.* oblong, acute, with manifest veinlets; *sep.* lanceolate, very acute, 3-veined, rather longer than the bifid petals; *caps.* ovoid, about equaling the calyx; *sty.* 3.—♃ Swampy springs, Penn. *Dr. Darlington.* Md. *Dr. Robbins.* Also Rocky Mts. A very slender plant, 6—12′ long, with inconspicuous flowers. Leaves 6″ by 2—3″. May.

2. CERASTIUM. *Linn.*

Gr. *κερας*, a horn ; from the resemblance of the capsules of some of the species.

Calyx of 5. ovate, acute sepals ; corolla of 5 bifid petals ; stamens

10, sometimes 5 or 4, the alternate ones shorter; styles 5; capsule superior, cylindrical or roundish, 10-toothed; seeds numerous.

* *Petals scarcely longer than the calyx.*

1. C. VULGÁTUM. *Mouse-ear Chickweed.*

Hairy, pale green, cæspitose; *lvs.* attenuated at the base, ovate, or obo-vate-obtuse; *fls.* in subcapitate clusters; *sep.* when young, longer than the pedi-cels.—① Fields and waste grounds, Can. and U. S., flowering all summer. Stems 6—12′ long, ascending, mostly forked. Leaves 5—8″ by 3—5″, mostly very obtuse, lower ones tapering to the base. Flowers in dense, terminal clus-ers, the terminal (central) one solitary, always the oldest. Sepals mostly green, a little shorter than the corolla. Petals white, appearing in 10 segments.

2. C. VISCÓSUM. (and C. semidecandrum. *Linn.*) *Sticky Chickweed.*

Hairy, viscid, spreading; *lvs.* oblong-lanceolate, rather acute; *fls.* in bose cymes; *sep.* scarious and white on the margin and apex, shorter than the pedicels.—♃ Fields and waste grounds, U. S. and Can. Plant more deeply green than the last. Stems many, assurgent, dichotomously-cymose. Leaves 5—9′ long, ⅓—½ as wide, radical ones subspatulate. Flowers white, in diffuse cymes. Petals hardly as long as the sepals, obovate, bifid. Jn.—Aug

β. *semidecandrum.* T. & G. Stamens 5.—Mass. to Ia.!

** *Petals much longer than the calyx.*

3. C. ARVENSE. (C. tenuifolium. *Ph.*) *Field Chickweed*

Pubescent, somewhat cæspitose; *lvs.* linear-lanceolate, acute, often longer than the internodes; *cyme* on a long, terminal peduncle, few flowered; *pet.* more than twice longer than the calyx; *cap.* scarcely exceeding the sepals.— Rocky hills. Stems 4—10′ high, decumbent at base. Leaves 9—15″ long, 1—2″ wide. Flowers white, rather large. Capsule usually a little longer than the calyx. May—Aug.

4. C. OBLONGIFOLIUM. TORR. (C. villosum. *Muhl.*)

Villose, viscid above; *st.* erect or declined; *lvs.* oblong-lanceolate, most-ly obtuse, and shorter than the internodes; *fls.* numerous, in a spreading cyme; *pet.* twice as long as the sepals; *cap.* about twice as long as the calyx.—♃ Rocky places. Stems 6—10′ high, thick. Leaves 9—12″ by 3—5″, tapering from base to an acute or obtuse apex. Flowers larger than either of the fore-going, white, in two or three-forked cymes. Apr.—Jn.

5. C. NUTANS. Raf.

Viscid and pubescent; *st.* weak, striate-sulcate, erect; *lvs.* lanceolate, narrow, shorter than the internodes; *fls.* many, diffusely cymose, on long, filiform, nodding pedicels; *pet.* nearly twice as long as the calyx.—① Low grounds, Vt. to Ill.! and La. Pale green and very clammy. Stems 8—15′ high, branched from the base. Leaves 1—2′ long, ½ as wide. Flowers white. Cap-sules a little curved, nearly thrice longer than the calyx. May.

3. ARENARIA.

Lat. *arena,* sand ; in which most species grow.

Sepals 5, spreading; petals 5, entire; stamens 10, rarely fewer; styles 3; capsule 3-valved, 1-celled, many-seeded.—*Fls. terminal. Sty. rarely 2 or 4.*

1. A. SQARRÓSA. Michx.

Cæspitose; *st.* few-flowered; *lower lvs.* squarrose-imbricate, crowded, *upper ones* few, all subulate, channeled, smooth; *pet.* obovate, three times longer than the obtuse, veinless sepals.—♃ In sandy barrens, N. Y. *Robbins,* to Ga. Stems 6—10′ high, pubescent, much divided at base into simple branches. Leaves about ½′ long, obtuse, sessile. Flowers white, in small terminal cymes. Sepals green. Capsules obtuse. Apr.—Sept.

2. A. STRICTA. Michx. *Straight Sandwort.*

Glabrous, diffuse; *st.* branched from the base; *lvs.* subulate, linear, erect; *pet. much longer than* the calyx; *sep.* ovate-lanceolate, acute, 3-veined; *cymes* few-flowered, with spreading branches.—♃ Sterile grounds Arc. Am. to Car. Stem 8—10' high. Leaves 5—8" long, very narrow and acute, rigid, sessile, 1-veined, much fasicled in the axils. Petals obovate-oblong, twice as long as the sepals, white. May, Jn.

3. A. GREENLANDICA. Spreng. (A. Glabra. *Bw.*) *Greenland Sandwort.*
Glabrous; *sts.* numerous, low, filiform, suberect; *lvs.* linear-subulate, flat, spreading; *pedicels* 1-flowered, elongated, divaricate; *sep.* veinless, ovate, obtuse, membrane-margined, much shorter than the petals.—♃ Summits of high mountains, N. H! N. Y., N. to Greenland. It grows in tufted masses, consisting of exceedingly numerous stems about 3' high, and sprinkled over with large (8" diam.) white flowers with yellow stamens. Aug.

4. A. SERPYLLIFOLIA. *Thyme-leaved Sandwort.*
St. dichotomous, spreading; *lvs.* ovate, acute, subciliate; *cal.* acute, substriate; *pet.* shorter than the calyx; *caps.* ovate, 6-toothed.—① By roadsides, and in sandy fields, Ms. to Ga. Stems numerous, downy, with reflexed hairs, a few inches in length. Leaves 2—3" long, ⅛ as wide. Flowers on axillary and terminal peduncles. Petals white, oval, mostly much shorter than the 3—5-veined, acuminate, hairy sepals. Jn.

5. A. LATERIFLORA. *Side-flowering Sandwort.*
Erect., slightly pubescent; *lvs.* oval, obtuse; *ped.* lateral, 2—3-flowered. —♃ A slender, upright species, found in damp, shady grounds, N. States, and Brit. Am. Stem 6—10' high, nearly simple. Leaves elliptical, rounded at each end, 6—10" long, ⅓ as wide, on very short petioles. Peduncles terminal and lateral, 2—3' long, dividing into 2 or more filiform pedicels, one of them with 2 bracteoles in the middle. Flowers 4" diam., white. Petals more than twice as long as sepals. Jn.

3. ADENARIUM. Raf.

Sepals 5, united at base; petals 5, unguiculate, entire; stamens 10, inserted into a glandular disk; styles 3—5; capsule 3—5-valved, many-seeded.—① *Herbs of the sea-coast, with fleshy leaves.*

A. PEPLÖIDES. DC. (Arenaria. *Linn.* Honckenya. *Ehrh.* and *1st. edit.*) *Sea Chickweed.*—Very fleshy; *st.* creeping, with erect, subsimple branches; *lvs.* ovate, obtuse. veinless, exceeding the petals.—Abundant on the Atlantic coast! N. J. to Lab. Upright stems a foot high. Leaves 5—7—10" long, ⅓ as wide, abruptly pointed, clasping at base, shorter than the internodes. Flowers small, white, axillary, on short pedicels. Jl.

5. SAGINA.
Lat. *sagina*, any kind of food or nourishment,

Sepals 4—5, united at base; petals entire, 4 or 5, or 0; stamens 4—10; styles 4—5; capsule 4—5-valved, many-seeded.—*Fls. solitary.*

1. S. PROCUMBENS. *Creeping Pearlwort.*
St. procumbent; glabrous; *pet.* very short; *sta., sep.* and *pet.* 4 or 5.—②
A small weed, with slender, creeping stems 3 or 4' long, found in damp places, R. I.! N. Y. to S. Car., W. to Oregon. Leaves very small, linear, mucronate-pointed, connate or opposite. Flowers white and green, axillary, on peduncles longer than the leaves. Jn.

2. S. DECUMBENS. T. & G. (Spergula saginoides. *Linn.*) *Pearlwort.*
St. decumbent, ascending, mostly glabrous; *lvs.* linear-subulate, very acute; *ped.* much longer than the leaves; *pet.* and *sep.* 5; *sta.* 10.—① Sandy fields, U. S. and Can. Stem 2—3' long. Flowers axillary and terminal. Petals white, hardly as large as the sepals. Jl. Apparently a variety of S. procumbens. ⸹?

3. S. APETALA.
Erect and pubescent; *lvs.* linear-subulate; *ped.* elongated, ascending in fruit; *sep.* and *sta.* 4; *pet.* very minute or 0.—① Sandy fields, N. J., Penn

Stems numerous, filiform, 2—4' high. Sepals acute, shorter than the capsule
May Jn.

6. MOLLŪGO.

Calyx of 5 sepals, inferior, united at base, colored inside; corolla
0; stamens 5, sometimes 3 or 10; filaments setaceous, shorter than,
and opposite to the sepals; anthers simple; capsule 3-celled, 3-valved,
many-seeded; seeds reniform.—*Lvs. at length apparently verticillate,
each whorl consisting of* 1 *or* 2 *large, substipulate leaves, with several
axillary, smaller ones.*

M.   VERTICILLATA.   *Carpet-weed.*
   Lvs cuneiform, acute ; *st.* depressed, branched; *pedicels* 1-flowered, subumbellate; *sta.* mostly but 3.—① A small, prostrate plant, in dry places throughout N. Am. Stems slender, jointed, branched, lying flat upon the ground. At
every joint stands a whorl of wedge-shaped or spatulate leaves of unequal size,
usually five in number, and a few flowers, each on a solitary stalk which is
very slender and shorter than the petioles. Flowers small, white. Jl.—Sep.

TRIBE 2.—SILENEÆ.

*Sepals united into a cylindrical tube. Petals clawed, inserted with the stamens
upon the stipe of the ovary.*

7. SILĒNE.

Silenus was a drunken divinity of the Greeks, covered with slaver, as these plants are with a viscid secretion.

Calyx tubular, swelling, without scales at base, 5-toothed ; petals 5,
unguiculate, often crowned with scales at the mouth, 2-cleft; stamens
10; styles 3; capsule 3-celled, many-seeded.

   * *Calyx vesicular, inflated ; petals scarcely crowned.*

1. S. ACAULIS.   *Stemless Campion.*
   Low and densely cæspitose; *lrs.* linear, ciliate at base; *ped.* solitary,
short, 1-flowered; *cal.* campanulate, slightly inflated; *pet.* obcordate, crowned.
—♃ A little turfy plant, 1—3' high, on the White Mts., N. H., and throughout
Arctic Am. Stems scarcely any. Leaves numerous, ¼' long. Flowers purple.

2. S. STELLATA.   Ait.   (Cucubalus stellatus. *Linn.*) *Stellate Campion.*
   Erect, pubescent; *lrs.* in whorls of 4s, oval-lanceolate, acuminate; *cal.*
loose and inflated; *pet.* fimbriate.—♃ An elegant plant, woods and prairies,
Can. to Car., W. to Ill.! and Ark. Stem 2—3f high, paniculately cymose. Leaves
2—3' long, ⅓ as wide, tapering to a long point, sessile. Calyx pale-green, with
more deeply colored veins. Petals white, lacerately fringed, claws webbed
at base. Jl.

3. S. NIVEA.   DC.   (Cucubalus niveus. *Nutt.*) *Snowy Campion.*
   Minutely puberulent, erect, simple or dichotomous above ; *lrs.* oblong-
lanceolate, acuminate; *fls.* few, terminal ; *cal.* inflated, with short and obtuse
teeth; *pet.* 2-cleft, with a small bifid crown ; *caps.* stiped.—♃ in moist places,
Penn., Ohio, near Cincinnati, (*Clark!*) Ill. Stem slender, leafy. 1½—3f high,
generally forked near the top. Leaves 2—3' by ⅓—⅓', tapering to a very slender
point, floral ones lance-ovate. Flowers 1—3. Calyx reticulated. Petals white.

4. S. INFLATA.   Smith.   (Cucubalus Behen. *Linn.*) *Bladder Campion.*
   Glabrous and glaucous; *lrs.* ovate-lanceolate; *fls.* in cymose panicles,
drooping; *cal.* ovoid-globular, reticulated with veins.—♃ in pastures about
fences, Charlestown, Ms.! &c. Stem erect, about 2f high. Leaves 1½—3' long,
⅓ as wide, rather acuminate. Petals white, cleft half-way down. Calyx remarkably inflated, and reticulated with pale purple veins. Jl.—The young
shoots and leaves may be used as a substitute for asparagus.

   ** *Calyx not inflated. Petals crowned.*

5. S. ANTIRRHINA.   *Snap-dragon Catch-fly.*
   Nearly smooth; *st.* erect; *lrs.* lanceolate, acute, sub-ciliate; *ped.* trifid,

3-flowered; *pet. emarginate; cal.* ovate.- ♃ Road-sides and dry soils, Can. and U. S. Stem slender, branching, with opposite leaves, about a foot in height. Leaves about 2' long, the upper ones very narrow, all sessile and scabrous on the margin. A few of the upper internodes are viscidly pubescent above their middle. Flowers small, red, in loose, erect cymes. Jl.

6. S. NOCTURNA. *Nocturnal Catch-fly.*
St. branching, hairy below; *lvs.* pubescent, with long cilia at base, *lower ones* spatulate, *upper* lance-linear; *fls.* appressed to the stem, in a dense one-sided spike; *cal.* cylindrical, almost glabrous, reticulated between the veins; *pet.* narrow, 2-parted.—① Near New Haven, Ct., *Robbins.* to Penn. Va. Flowers white, greenish beneath. Jl. § †

7. S. NOCTIFLORA. *Night-flowering Catch-fly.*
Viscid-pubescent; *st.* erect, branching; *lower lvs.* spatulate, *upper* linear; *ca.* cylindrical, ventricose, the alternate veins veinleted; *teeth* subulate, very long; *pet.* 2-parted.—From Europe, introduced into our cultivated grounds! Flowers rather large, white, expanding only in the evening, and in cloudy weather. § †

8. S. PENNSYLVANICA. Michx. *Pennsylvanian Catch-fly.*
Viscid-pubescent; *sts.* numerous; *lvs. from the root* spatulate or cuneate, *of the stem* lanceolate; *cyme* few-flowered; *pet.* slightly emarginate, sub-crenate. —♃ Dry, sandy soils, N. Eng.! to Ky. and Ga. Stem decumbent at base, nearly 1f high, with long, lanceolate leaves, and terminal, upright bunches of flowers. Calyx long, tubular, very glutinous and hairy. Petals wedge-shaped, red or purplish. Jn.

9. S. VIRGINICA. *Virginian Catch-fly.*
Viscid-pubescent; *st.* procumbent or erect, branching; *fls.* large, cymose; *cal.* large, clavate; *pet.* bifid, broad, crowned.—♃ Gardens and fields, Penn. to to Ga. Stem 1—2f high, often procumbent at base. Leaves oblong, a little rough at the margin. Cymes dichotomous. Stamens and pistils exserted. Petals red, large. Jn. †

10. S. REGIA. Sims. *Splendid Catch-fly.*
Scabrous, somewhat viscid; *st.* rigid, erect; *lvs.* ovate-lanceolate; *cyme* paniculate; *pet.* oblanceolate, entire, erose at the end; *sta.* and *stig.* exserted.— ♃ A large species, beautiful in cultivation, native Ohio, *Sullivant!* to La. Stems 3—4f high. Leaves 2—3', by 8—15''. Flowers very large, numerous. Calyx tubular, 10-striate, 1' long. Petals bright-scarlet, crowned. Jn. Jl. †

11. S. ARMERIA. *Garden Catch-fly.*
Very smooth, glaucous; *st.* branching, glutinous below each node; *lvs.* ovate-lanceolate; *fls.* in corymbose cymes; *pet.* obcordate, crowned; *cal.* clavate, 10-striate.—① Introduced from Europe. A popular garden flower. Stem 1—1½f high, many-flowered. Leaves 1½—2½' long, ½ as wide; internodes elongated. Calyx ½' long, a little enlarged above. Petals purple, laminæ half as long as calyx. Jl.—Sept. § †

8. LYCHNIS.

Gr. λυχνος, a lamp . some cottony species having been used as lamp-wicks.

Calyx tubular, 5-toothed, ovoid or cylindrical; scales 0; petals 5, anguiculate, limb slightly cleft; stamens 10; pistils 5; capsule 1-celled, or 5-celled at the base, with a 5-toothed dehiscence.—*Corolla sometimes crowned.*

1. L. GITHAGO. Lam. (Agrostemma Githago. *Linn.*) *Corn Cockle.*
Hairy; *st.* dichotomous; *ped.* elongated; *lvs.* linear; *cal.* longer than the corolla; *pet.* entire, without the corona.—① A well known handsome weed, growing in fields of wheat, or other grains and of a pale green color. Stem 2—3f high. Leaves 3—5' by ⅛—¼', fringed with long hairs. Flowers few, large, of a dull purple, on long, naked stalks. Seeds roundish, angular, purplish black. Jl. §

2. L. CHALCEDONICA. *Scarlet Lychnis or Sweet William.*—Smoothish; *fls.* fasciculate; *cal.* cylindric, clavate, ribbed; *pet.* 2-lobed.—24 A fine garden-flower, native of Russia. Stem 1—2f high, with dark-green, ova e-lanceolate, acuminate leaves, and large, terminal, convex, dense fascicles of deep-scarlet flowers. It has varieties with *white flowers*, and also with *double.* Jn. Jl. †

3. L. FLOSCUCULI. *Ragged Robbin.*—Smoothish; *st.* ascending, dichotomous at summit; *fls.* fascicled; *cal.* campanulate, 10-ribbed; *pet.* in 4 deep, linear segments.—24 Native of Europe. Stem 1—2f high, rough-angled, viscid above. Leaves lanceolate, smooth. Flowers pink, very beautiful, with a brown, angular, smooth calyx. Capsule roundish, 1-celled. Jl.—Sept. †

4. L. CORONATA. *Chinese Lychnis.*—Smooth; *fls.* terminal and axillary, 1—3; *cal.* rounded, clavate, ribbed; *pet.* laciniate.—Native of China. Stem 1—2f high. Petals of lively red, remarkable for their large size. There are varieties with *double red*, and *double white* flowers. †

5. L. DIURNA.—*St.* dichotomous-paniculate; *fls.* ♂ ♀; *pet.* half-bifid, lobes narrow, diverging; *caps.* ovoid-globose.—Native of Britain, almost naturalized! Stems about 2f high, pubescent. Leaves 1—3' long, elliptic-ovate, acute Flowers light-purple, middle size. Jl.—Sep. †

6. L. CORONARIA. DC. (Agrostemma coronaria. Linn.) *Mullein Pink. Rose Campion.*—Villose; *st.* dichotomous; *ped.* long, 1-flowered; *cal.* campanulate, veined.—24 Native of Italy. Whole plant covered with dense wool. Stem 2f high. Flowers purple, large. Varieties are *white-flowered, red-double-flowered*, &c. †

Obs.—Other species rarely found in collections are *L. fulgens* with scarlet flowers; *L. secaria*, with pink flowers; *L. alpina*, low, with pink flowers, &c.

9. SAPONARIA.

Lat. *sapo*, soap ; the mucilaginous juice is said to make soap.

Calyx tubular, 5-toothed, without scales; petals 5, ungui. alate ; stamens 10 ; styles 2 ; capsule oblong, 1-celled. *Petals often crowned.*

1. S. OFFICINALIS. *Common Soap-wort.*
Lvs. lanceolate, inclining to elliptical; *fls.* in paniculate fascicles; *cal.* cylindrical; *crown of the petals* linear.—24 By roadsides, New Eng. to Ga. A hardy, smooth, succulent plant, with handsome, pink-like flowers. Stem 1—2f high. Leaves 2—3' long, ½ or more as wide, very acute. Flowers many, flesh colored, often double. The plant has a bitter taste, with a saponaceous juice Jl. Aug. §

2. S. VACCARIA. *Fly-trap.*
Lvs. ovate-lanceolate, sessile; *fls.* in paniculate cymes; *cal.* pyramidal, 5-angled, smooth; *bracts* membranaceous, acut .—① Gardens and cultivated grounds. Whole plant smooth, a foot or more high. Leaves broadest at base, 1—2' long, ½ as wide, tapering to an acute apex. Flowers on long stalks, pale-red. Capsule 4-toothed. Seeds globose, black. Jl. Aug. § †

10. DIANTHUS.

Gr. Διος ανθος, the flower of Jove, alluding to its preëminent beauty and fragrance.

Calyx cylindrical, tubular, striate, with 2 or more pairs of opposite, imbricated scales at base; petals 5, with long claws, limb unequally notched ; stamens 10 ; styles 2, tapering, with tapering, revolute stigmas ; capsule cylindric, 1-celled.

1. D. ARMERIA. *Wild Pink.*
Lvs. linear-subulate, hairy; *fls.* aggregate, fascicled; *scales of the calyx* lanceolate, subulate, as long as the tube.—① Our only native species of the pink, found in fields and pine woods, Mass. to N. J.! Stem erect, 1—2f high, branching. Leaves erect, 1—2' long, 1—3'' wide at the clasping base, tapering to a subulate point. Flowers inodorous, in dense fascicles of 3 or more.

Calyx and its scales ½' long. Petals small, pink-colored, sprinkled with white, crenate. Aug.

2. D. BARBÁTUS. *Sweet William* or *Bunch Pink.*—Lvs. lanceolate; *fls.* aggregate, fascicled; *scales of the calyx* ovate-subulate, as long as the tube.—♃ An ornamental flower still valued as in the times of old Gerarde, "for its beauty to deck up the bosoms of the beautiful, and garlands and crowns for pleasure." Stems 1½f high, thick. Leaves 3—5' by ½—1', narrowed to the clasping base. Flowers in fastigiate cymes, red or whitish, often greatly variegated. May.—Jl. †

3. D. CHINÉNSIS. *China Pink.*—*St.* branched; *lvs.* linear-lanceolate; *fl.* solitary; *scales* linear, leafy, spreading, as long as the tube.—⚥ Native of China. An elegant species, well characterized by its leafy, spreading scales, and its large, toothed or crenate, red petals. The foliage, like the other species, is evergreen, being as abundant and vivid in winter as in summer. †

4. D. PLUMÁRIUS. *Single Pink. Pheasant's-eye.*—Glaucous; *st.* 2—3-flowered; *fls.* solitary, *calyx teeth* obtuse; *scales* ovate, very acute; *lvs.* linear; rough at the edge; *pet.* many-cleft, hairy at the throat.—♃ Native of Europe. From this species probably originated those beautiful pinks called pheasant's eye, of which there are enumerated in Scotland no less than 300 varieties. Flowers white and purple. Jn.—Aug. †

5. D. CARYOPHYLLUS. *Carnation. Bizarres, Picotees, Flakes, &c.*—Lvs. linear-subulate, channeled, glaucous; *fls.* solitary; *scales* very short, ovate; *pet.* very broad, beardless, crenate.—Stem 2—3f high, branched. Flowers white and crimson; petals crenate. This species is supposed to be the parent of all the splendid varieties of the carnation. Over 400 sorts are now enumerated by florists, distinguished mostly by some peculiarity in color, which is crimson, white, red, purple, scarlet, yellow, and arranged in every possible order of stripes, dots, flakes, and angles.

6. D. SUPÉRBUS. *Superb Pink.*—Lvs. linear-subulate; *fls.* fastigiate; *scales* short, ovate, mucronate; *pet.* pinnate.—♃ A singularly beautiful pink, native of Europe. Stem 2f high, branching, with many flowers. Petals white, gashed in a pinnate manner beyond the middle, and hairy at the mouth. Jl.—Sept.

Obs.—Other species of this admirable genus are occasionally cultivated, but the varieties of Nos. 4 and 5 are by far the most common. The "Monthly Pink," common in house cultivation, with bright green, channeled, linear leaves, short, caespitose stems, pink red, double flowers, appears to be a variety of D. Carthusianorum.

ORDER XXIV. PORTULACACEÆ.—PURSLANES.

Herbs succulent or fleshy, with entire leaves and no stipules.
Cal.—Sepals 2, united at base.
Cor.—Petals 5, sometimes more or less, imbricated in aestivation.
Sta. variable in number. Filaments distinct. Anthers versatile or introrse.
Ova. superior 1-celled. Sty. several, stigmatose along the inner surface.
Fr. a pyxis, dehiscing by a lid or capsule, loculicidal, with as many valves as stigmas.

Genera 12, species 194, inhabiting dry places in every quarter of the world. They possess no remarkable properties.

Conspectus of the Genera.

{ Capsule 3-valved.			Ta'inum. **3**
Stamens { 9—20. { Pyxis dehiscing transversely.			Portulaca. **1**
{ 5, opposite the petals.			Claytonia. **2**

1. PORTULÁCA. Tourn.

Sepals 2, the upper portion deciduous; petals 5 (4—6), equal; stamens 8—20; styles 3—6-cleft or parted; pyxis subglobose, dehiscing near the middle, many-seeded.—*Low, herbaceous, fleshy. Fls. expanding only in sunshine.*

1. P. OLERÁCEA. *Purslane.*
Lvs. cuneate; *fls.* sessile.—① A prostrate, fleshy weed, more common in our gardens than desirable. Stem thick and succulent, much branched, and spreading, smooth. Leaves fleshy, sessile, rounded at the end. Flowers yellow. The herbage of the plant is of a reddish-green color. Sometimes used as a pot-herb. Jn.—Aug. §

2. P. PILOSA, β. *Scarlet-flowered Purslane.*—*Sts.* ascending, much branched; *branches* suberect, enlarged upwards; *lvs.* linear, obtuse, the axils villose with long, woolly hairs; *fls.* terminal, sessile, 1 or few together, surrounded by an irregular circle of leaves and dense tufts of wool; *pet.* obovate; *sta.* about 15.— A very delicate plant, with purple stems, and large, bright purple flowers.— *P. australis*, with broader leaves and scarlet fls. is also popular in house cultivation. The species are mostly natives of S. Africa. †

2. CLAYTONIA.

In memory of John Clayton, a botanist of Virginia.

Sepals 2, ovate or roundish; petals 5, emarginate or obtuse; stamens 5, inserted on the claws of-the petals; stigmas 3-cleft; capsule 3-valved, 2—5-seeded.—*Small, fleshy, delicate, early-flowering plants.*

1. C. CAROLINIÀNA. Michx. *Spring Beauty.* ·
Lvs. ovate-lanceolate; *sep.* and *pet.* obtuse; *rt.* tuberous.—♃ A delicate little plant, flowering in April, common in woods and rocky hills, Can. to N. Car. W. to the Miss. Root a compressed, brown tubercle, buried at a depth in the ground, equal to the height of the plant. Root-leaves very few, if any, spatulate. Stem weak, 2—3' high, with a pair of opposite leaves half-way up, which are 1—2' by ½—¼', entire, tapering at base into the petiole. Flowers in a terminal cluster, white, with a slight tinge of red, and beautifully penciled with purple lines. Apr. May.

2. C. VIRGINÌCA. *Virginian Spring Beauty.*
Lvs. linear, or lance-linear; *sep.* rather acute; *pet.* obovate, mostly emarginate or retuse; *ped.* slender, nodding.—♃ In low, moist grounds, Mid. and S. States. W. to Mo., *Everett!* rare in N. Eng. Tubercle or cormus as .arge as a hazelnut, deep in the ground. Stem 6—10' long, weak. with a pair of opposite, very narrow leaves 3—5' long. Flowers 5—10, rose-colored, with deeper colored veins, in a terminal cluster. Sepals acute or obtuse. Petals often elliptical, subacute. Apr. May.

3. TALINUM. Adans.

Sepals 2, ovate, concave, deciduous; petals 5, sessile; stamens 10 —20, inserted with the petals into the torus; style trifid; capsule subglobose, 3-valved, many-seeded.

1. T. TERETIFOLIUM.
St. simple or branched, short and thick; *lvs.* terete, subulate, crowded at the summit of the stem, on short branches; *ped.* elongated; *fls.* in a dichotomous cyme; *pet.* purple.—♃ An interesting little plant, on rocks, Penn. *Dr. Darlington!* to Ark. Rhizoma or perennial stem firm and fleshy, with fibrous roots. Branches 1—3' long. Leaves 1—2' long, incurved, fleshy. Bracts ovate-lanceolate, minute. Peduncles 5—8' high. Flowers small, ephemeral. Stamens about 20. Jn.—Aug.

2. T. PATENS.—*Spreading-flowered Talinum.*—*St.* erect or decumbent at base slender; *lvs.* ovate, flat, fleshy; *panicle* terminal, with spreading, dichotomous peduncles.—♃ Native in S. America. A handsome plant, sometimes cultivated. Stem 1—2f high, round, purple, terminating in a naked, spreading panicle of small purple flowers. Leaves 2—3' long, tapering to the base. Aug.—Oct

ORDER XXV. ELATINACEÆ.—WATER PEPPERS.

Herbs small, annual, with opposite leaves and membranaceous stipules. *Fls.* minute, axillary.
Cal.—Sepals 2—5, distinct or slightly coherent at base, persistent.
Cor.—Petals hypogynous, as many as the sepals.
Sta. equal in number to, or twice as many as the petals. *Anth.* introrse.
Ova. 2—4-celled. *Stigmas* 2—5, capitate; placentæ in the axis.
Fr. capsular. *Seeds* numerous.

*Genera 4, species 22, fo*und in every part of the globe, growing in marshes. The following is the only northern genus :

ELATINE.

Gr. ελατη, fir; from the resemblance of the slender leaves of some species.

Stigmas sessile, minute.

E. AMERICÀNA. Arn. (Crypta minima. *Nutt.* Peplys Americana. *Ph.*)
Mud Purslane.—*St.* diffuse, procumbent, striate, rooting from the joints, with assurgent branches; *lvs.* cuneate-oval or obovate, obtuse, entire; *sty.* 0; *cp., pet., sta.* and *stig.* 2—3, as well as the cells and valves of the capsule; *stig.* very minute.—A small mud plant, on the borders of ponds and rivers! U. S. Flowers axillary, sessile, solitary. Corolla minute, closed. Jl.—Sep.

ORDER XXVI. LINACEÆ—FLAXWORTS.

Plants herbaceous or suffrutescent.
Lvs. entire, sessile, alternate, sometimes nearly opposite, without stipules.
Fls. terminal, usually in corymbs or panicles, regular and symmetrical.
Cal.—Sepals 3, 4 or 5, distinct, or more or less united; æstivation strongly imbricated.
Cor.—Petals equal in number to sepals, hypogynous, unguiculate; æstivation twisted.
Sta. 3, 4 or 5, united at base into a hypogynous ring, which is often toothed, opposite the petals.
Ova. of as many cells as sepals or styles. *Stig.* capitate.
Sds. solitary in each cell, compressed, suspended. *A.* bumen 0.

Genera 3, *species* 90. A very important order in the arts. The Linum has a very tenacious fibre in its bark, which is wrought into thread and cloth, forming the *linen* of commerce. Some species are cathartic, and yield from their seeds a fine mucilage. Only one genus need be mentioned here, viz.:

LINUM.

Celtic *llin,* a thread; hence λινον, Eng. *linen,* flax.

Sepals, petals, stamens and styles 5, the latter rarely 3; capsules 5-celled; cells nearly divided by a false dissepiment. (Fig. 11., No. 4.)

1. L. RIGIDUM. *Stiff-leaved Flax.*
St. angular, branching; *lvs.* alternate, rigid, linear, acute; *fls.* panicled; *sep.* ovate-lanceolate, acuminate, and with the bracts, glandularly fimbriate-serrate on the margins; *caps.* globose, shorter than the calyx.—① Near New Haven, Conn., *Robbins!* R. I., found by the *Prov. Bot. Assoc.* Stem 10—16' high, erect, with many suberect branches above. Leaves 4—7" by ⅓—1", scabrous on the margin. Sepals 3-veined. Flowers 6—8" diam., sulphur-yellow.

2. L. VIRGINIÀNUM. *Virginian Flax.*
St. branching above, erect; *lvs.* alternate. linear-lanceolate, those of the root oblong, upper ones acute; *panicles* corymbose, terminal, with the flowers racemose on the branches; *sep.* broad-ovate, mucronate; *caps.* depressed, scarcely longer than the calyx.—① Woods, hills, &c., U. S. and Can. Stem about 2f high, slender, leafy, terete, glabrous. Leaves 6—10" by 1—2", with one distinct vein. Flowers 4—6" diam., yellow, on short pedicels. Sepals 1-veined. Jl.

β.? diffusum. Wood.—*St.* angular, diffusely branched; *branches* and *lanceolate lvs.* spreading; *fls.* very small (scarcely 2" diam.)—Wet prairies, Ia.! Quite different in habit and may prove a new species.

3. L. USITATISSIMUM. *Common Flax.*
St. branching above; *lvs.* alternate, linear-lanceolate, acute; panicle corymbose; *sep.* ovate, acute, 3-veined at the base, membranaceous on the margin; *pet.* crenate.—① Introduced and somewhat naturalized in fields. Stem 1—2f high, with 3-veined leaves, and many large, handsome, blue flowers. Jn. Jl.— This important plant has been cultivated from remote antiquity, (see Gen. xli. 42) for the strong fibres of the bark, which are manufactured into *linen.* The seeds yield *linseed oil,* so extensively used in mixing paint, printers' ink, &c. They are also medicinal. §‡

4. L. PERENNE. *Perennial Flax.*—Glabrous, with virgate branches; *lvs.* linear, acute, scattered; *fls.* supra-axillary and terminal; *sepals* oval, margins membranaceous, shorter than the globose capsule; *petals* retuse, blue, 3 or 4 times the length of the sepals.—24 Native West of the Miss. (perhaps not within the
17*

limits of this Flora), also of Europe and Asia. Not uncommon in gardens. Flowers large, blue. †

Order XXVII. GERANIACEÆ.—Gerania.

Stems herbaceous or suffrutescent, tumid and separable at the nodes.
Lvs. opposite, (at least the lower ones,) mostly stipulate, petiolate, palmately veined.
Fis.—Peduncles terminal or opposite the leaves, sometimes axillary.
Cal.—Sepals 5, persistent, veined, one sometimes saccate or spurred at base.
Cor.—Petals 5, hypogynous or perigynous, unguiculate; æstivation twisted.
Sta. usually monadelphous, hypogynous, twice or thrice as many as the petals.
Ova. 1 of 3 united carpels, 2-ovuled, alternate with sepals, upon an elongated axis, from which they separate.
Fr.— ? rate in fruit, curving upwards on the persistent style.

Genera 4, species 608. The Cape of Good Hope is the favorite habitation of some of the most important genera. Most species of the beautiful Pelargonia are native of that region alone.

Conspectus of the Genera.

Stamens 10, { all perfect.	. . . .	Geranium. 1
{ 5 perfect, with 5 shorter and imperfect.	. . . .	Erodium. 2
{ 7 perfect, corolla irregular.	. . . .	Pelargonium 8

1. GERANIUM.

Gr. γερανος, a crane; the beaked fruit resembles a crane's bill.

Sepals and petals 5, regular; stamens 10, all perfect, the 5 alternate ones longer, and each with a nectariferous gland at its base; fruit rostrate, at length separating into 5 long-styled, 1-seeded carpels; styles smooth inside, at length recurved from the base upwards and adhering by the point to the summit of the axis.—*Herbaceous, rarely shrubby at base. Peduncles 1, 2 or 3-flowered.*

1. G. MACULÀTUM. *Spotted Geranium.*
St. erect, angular, dichotomous, retrorsely pubescent; lvs. 3—5-parted, lobes cuneiform and entire at base, incisely serrate above, radical ones on long petioles, upper ones opposite, on short petioles; pet. entire; sep. mucronate-awned.—Woods, &c., U. S. and Can., but rare in N. Eng. A fine species, worthy a place among the parlor "geraniums." Stem 1—2′ high. Leaves 2—3′ diam., cleft ¾ way down, 2 at each fork. Flowers mostly in pairs, on unequal pedicels, often somewhat umbeled on the ends of the long peduncles. Root powerfully astringent. Apr.—Jn.

2. G. ROBERTIÀNUM. *Herb Robert.*
St. diffuse, hairy; lvs. 3—5-parted to the base, the segments pinnatifid, and the pinnæ incisely toothed; sep. mucronate-awned, half the length of the entire petals.—♃ Smaller and less interesting than the preceding, in dry, rocky places, Can. to Va. and Ky. It has a reddish stem, with long, diffuse, weak branches. Leaves on long petioles, somewhat hairy, outline 1½—3′ diam, with pinnatifid segments. Flowers small, pale purple. Capsules small, rugose, keeled. Seeds smooth. The plant has a strong disagreeable smell. May.—Sept.

3. G. PUSILLUM. *Weak Crane's-bill.*
St. procumbent; lvs. reniform or roundish, deeply 5—7-parted, lobes 3-cleft, linear; sep. hairy, acuminate, about as long as the emarginate petals.—① A delicate, spreading species, growing in waste grounds, pastures, &c., L. I. and Western N. Y. Torr. Stem weak, if long, branching, covered with short, deflected hairs. Leaves opposite, divided almost to the base into 5 or 7 lobes, these again variously cut. Peduncles axillary, forked, bearing 2 purplish-red flowers in Jn. and Jl.

4. G. CAROLINIÀNUM. *Carolinian Crane's-bill.*
St. diffusely branched; lvs. deeply 5-parted, lobes incisely toothed; ped. rather short and clustered on the ends of the branches; sep. mucronate-awned, as long as the emarginate petals.—① Fields and hills throughout Can. and U. S. Stems pubescent, diffuse, 8—15′ long, swelling at the joints. Leaves ¾ -1½ diam., hairy. Flowers small, rose-colored, in pairs, and somewhat fasciculate. Seeds minutely reticulated, reddish brown, 1 in each hairy, beaked carpel. Jl.—Perhaps too near the following species.

5. G. DISSECTUM. Willd. *Wood Crane's-bill.*

St. diffuse, pubescent; *lrs.* deeply 5-parted, lobes 3-cleft, incisely dentate; *ped.* dichotomous; *pedicels* hairy; *sep.* mucronately awned, scarcely as long as the emarginate petals; *beak* hairy; *carp.* rugose.—① rocky places, N. Sts.! A small spreading plant, 8—12' long. Leaves pentagonal in outline, 1½—2' diam., divisions and their segments oblong-linear, submucronate. Peduncles 6—10' long, with 4 bracts at the fork. Pedicels 6—10' long. Sepals 3-veined. Petals purplish, deeply notched, a little longer than the sepals. Jn. Jl.

6. G. SANGUINEUM. *Bloody Geranium.*—*S'.* erect, diffusely branched; *ped.* longer than the petioles; *lrs.* opposite, 5-parted, orbicular in outline, lobes trifid, with linear segments; *carpels* bristly at top.—A beautiful species native of Europe, deemed worthy of culture by many a florist. Grows about a foot high. Leaves orbicular, deeply divided into 5 or 7, 3-fid lobes. Flowers large, round, of a deep red or blood-color. †

2. ERODIUM. L'Her.

Gr. spadix, a heron; from the resemblance of the beaked fruit to the heron's bill.

Calyx 5-leaved; petals 5; scales 5, alternate with the filaments and nectariferous glands at the base of the stamens; filaments 10, the 5 alternate ones abortive; fruit rostrate, of 5 aggregate capsules, each tipped with the long, spiral style, bearded inside.

1. E. MOSCHATUM. L'Heritier. (Geranium moschatum. *Linn.*) *Musk Geranium.*—*St.* procumbent; *lrs.* pinnated with stalked, ovate, unequally serrated segments; *ped.* downy, glandular; *pet.* equaling the calyx.—① Native of England. Sometimes cultivated for the strong, musky scent of its herbage. A foot high. Leaves large. Flowers small, purple. May—Jl.

2. E. CICONIUM. L'Her. (G. ciconium. *Linn.*) *Heron's-bill Geranium.*—*St.* ascending; *lrs.* pinnate; *lfts.* pinnatifid, toothed; *ped.* many-flowered; *pet.* oblong, obtuse.—① From S. Europe. Stem about 1f high. Flowers purple.

3. PELARGONIUM. L'Her.

Gr. πελαργος, a stork; from the resemblance of the beaked fruit to the stork's bill.

Sepals 5, the upper one ending in a nectariferous tube extending down the peduncle with which it is connected; *pet.* 5, irregular, longer than the sepals; filaments 10, 3 of them sterile.—*A large genus of shrubby or herbaceous plants, embracing more than 300 species and innumerable varieties, nearly all natives of the Cape of Good Hope. Lower lvs. (in plants raised from the seed) opposite, upper ones alternate.*

* *Stem scarcely any. Root tuberous.*

1. P. FLAVUM. *Carrot-leaved Geranium.*—*St.* very simple; *lrs.* decompound, laciniate, hairy, segments linear; *umbel* many-flowered.—Flowers brownish-yellow. From the Cape of Good Hope, as well as the other species.

2. P. TRISTE. *Mourning Geranium.*—*Lrs.* hairy, pinnate; *lfts.* bipinnatifid, divisions linear, acute. A foot high. Flowers dark green, in simple umbels.

** *Stem elongated, herbaceous or suffruticose.*

3. P. ODORATISSIMUM. *Nutmeg-scented Geranium.*—*St.* short, fleshy; *lvs.* roundish, cordate, very soft · *branches* herbaceous, long, diffuse.—Valued chiefly for the powerful, aromatic smell of the leaves, the flowers being small, whitish.

4. P. ALCHEMILLOIDES. *Lady's-mantle Geranium.*—*St.* villous; *lrs.* cordate, villous, 5-lobed, palmate; *ped.* few-flowered; *stip.* sessile.—Stem 6' high, diffuse, very hairy, with deflexed bristles. Flowers pink-colored.

5 P. TRICOLOR. *Three-colored Geranium.*—*S'.* suffruticose, erect; *lrs.* lanceolate, villous, cut-dentate, trifid; *upper pet.* glandular at base.—Stem 1½f high. This species is distinguished for its beautifully variegated flowers. Petals roundish and nearly uniform in shape, but very different in color; the 3 lower

ones are white, slightly veined, the 2 upper of a rich purple, almost black at base.

6. P. CORIANDRIFOLIUM. *Coriander-leaved Geranium.*—St. herbaceous, biennial, somewhat downy; lvs. bipinnate, smooth, lobes linear, subpinnatifid.—Stem diffuse, 1f high. Distinguished by the finely divided leaves and large flowers. The 2 upper petals much the largest, obovate, veined with purple, the 3 lower, of which the middle one is often wanting, are narrow and of a pure white.

*** *Leaves neither divided nor angular; stem fruticose.*

7. P. GLAUCUM. *Glaucous-leaved Geranium.*—Very smooth and glaucous; lvs. lanceolate. entire, acuminate; ped. 1—2-flowered.—Stems 3f high, shrubby and branched. The plant is remarkably distinguished by its leaves. Peduncles axillary, with 1 or 2 elegant flowers. Petals obovate, of a delicate blush-color, with red veins.

8. P. BETULINUM. *Birch-leaved Geranium.*—Lvs. ovate, unequally serrate, smoothish; stip. ovate-lanceolate; ped. 2—4-flowered.—Stem shrubby, 3f high. The plant is well named for its leaves. Flowers pale-pink, with deep red veins.

9. P. ACETOSUM. *Sorrel-leaved Geranium.*—Lvs. very smooth, obovate, crenate, somewhat fleshy; ped. few-flowered; pet. linear.—Stem shrubby, 3f high. Named for the acid flavor of the leaves. Flowers pink.

**** *Leaves either angular or palmately lobed; stem fruticose.*

10. P. ZONÀLE. *Horse-shoe Geranium.*—Lvs. cordate-orbicular, obsoletely lobed, toothed, marked with a concentric zone.—Stem thick, shrubby, 2—3f high. One of the most popular of all the species. Leaves always marked with a dark concentric stripe of various shades. The flowers are of a bright scarlet, umbeled, on long peduncles. It has many varieties, of which the most remarkable is—

β. *marginale;* silver-edged, the leaves of which are bordered with white.

11. P. INQUINANS. *Scarlet Geranium.*—Lvs. round-reniform, scarcely divided, crenate, viscid; umbels many-flowered; pet. obovate, cuneate.—Justly admired for the vivid scarlet of its numerous flowers. The name alludes to the reddish, clammy moisture which stains the fingers in handling the soft, downy branches.

12. P. PELTÀTUM. *Ivy-leaved Geranium.*—Lvs. 5-lobed, entire, fleshy, smooth, more or less peltate; umbels few-flowered.—Stem climbing, several feet in length. Whole plant very smooth. A beautiful species, with umbels of very handsome purplish flowers.

13. P. TETRAGÓNUM. *Square-stalked Geranium.*—Branches 4-cornered, fleshy; lvs. cordate, bluntly lobed, somewhat toothed; pet. 4, the upper ones pale-pink, with crimson veins, the 2 lower small, white.—Leaves small, rounded, notched, with scattered hairs.

14. P. GRANDIFLÒRUM. *Large-flowered Geranium.*—Smooth, glaucous; lvs 5-lobed, palmated, cordate at base, the lobes dentate towards the end; pet. three times as long as the calyx.—Distinguished for the size and beauty of the flowers, which are white, the 2 upper ones elegantly veined, and tinged with red, larger than the rest.

15. P. GRAVEÒLENS. *Rose-scented Geranium.*—Lvs. palmately 7-lobed, lobes oblong, bluntly toothed, revolute, and very rough at the edge; umbels many-flowered, capitate.—Nectary about half as long as calyx. Leaves very fragrant. Flowers purple.

16. P. RADÙLA. *Rasp-leaved Geranium.*—Lvs. palmate, rough; lobes narrow, pinnatifid, revolute at edge, with linear segments; umbels few-flowered; nectary nearly as long as the calyx.—Distinguished for its large rough leaves deeply divided into linear segments, and with a mint-like fragrance. Flowers purple.

17. P. QUERCIFOLIUM. *Oak-leaved Geranium.*—Lvs. cordate, pinnatifid, with rounded recesses, lobes obtusely crenate branches and petioles hispid.—Leaves rough, often spotted. Flowers purplish.

Obs.—The above are among the more distinguished and popular species of this vast and favorite genus Innumerable varieties produced from seeds and propagated by cuttings are equally common and often of superior beauty. No genus seems to be regarded with such universal favor for green-house plants as this. The species and their multitudes of hybrid creations, produced by modern ingenuity, are cultivated with assiduous attention by nearly every family which makes the least pretensions to taste throughout the civilized world.

ORDER XXVIII. BALSAMINACEÆ—JEWEL WEEDS.

Herbs annual, with succulent stems and a watery juice.
Lvs. simple, without stipules. *Fls.* very irregular and unsymmetrical.
Cal.—Sepals 5, deciduous, the 2 upper connate, the lowest spurred or gibbous.
Cor.—Petals 5, hypogynous, united by pairs, or rarely 5, distinct.
Sta. 5, hypogynous. *Filaments* subulate. *Anthers* 2-celled.
Ova. 5-celled, compound. *Stigmas* sessile.
Fr. capsular, 5-celled, bursting elastically by 5 valves. *Sds.* several in each cell. *Embryo* straight.

Genera 2, species 110. With regard to its properties and uses, this order is of no importance, but some of its species are highly ornamental.

1. IMPATIENS.

Impatient, with respect to the irritable capsules.

Sepals colored, apparently but 4, the 2 upper being united, the lowest gibbous and spurred; petals apparently 2, each of the lower being united to the 2 lateral ones; anthers cohering at the apex; capsule often 1-celled by the obliteration of the dissepiments, 5-valved, bursting elastically.—*Stems smooth, succulent, tender, sub-pellucid, with tumid joints.*

1. I. PALLIDA. Nutt. (I. noli-tangere. *Michx.*) *Touch-me-not.*
Lvs. oblong-ovate, coarsely and obtusely serrate, teeth mucronate; *ped.* 2–4-flowered, elongated; *lower gibbous sep.* dilated-conical, broader than long, with a very short, recurved spur; *fls.* pale yellow, sparingly maculate.—① Wet, shady places, U. S.! and Can. Stem 2–4f high, branched. Leaves 2–5' long, ⅓ as wide, with large, obtuse teeth, each tipped with a very short mucro. Flowers large, mostly in pairs. Two outer sepals pale green, callous-pointed, the rest pale yellow, the lower produced into a conic nectary, ending in a spur ⅓' long. Capsules oblong-cylindric, 1' long, bursting at the slightest touch when mature, and scattering the seed. Aug.

2. I. FULVA. Nutt. (I. noli-tangere. *β. Michx.*) *Jewel-weed.*
Lvs. rhombic-ovate, obtusish, coarsely and obtusely serrate, teeth mucronate; *ped.* 2–4-flowered, short; *lower gibbous sep.* acutely conical, longer than broad, with an elongated, recurved spur; *fls.* deep orange, maculate with many brown spots.—① In wet, shady grounds, Can. to Ga., more common than the last. somewhat glaucous.—Stem 1½–3f high. Leaves 1–3' long, ⅓ as wide, having like the last, a few filiform teeth at the base. Flowers about 1' in length, the recurved spur of the lower sepal ⅓' long. Capsule as in the last. Aug.

3. I. BALSAMINA. *Garden Balsamine.*—*Lvs.* lanceolate, serrate, upper ones alternate; *ped.* clustered; *spur* shorter than the flower.—① From the E. Indies. I. is one of the most beautiful of garden annuals, forming a showy pyramid of finely variegated, carnation-like flowers. The prevailing colors of the petals are red and white, but the former varies in every possible shade of crimson, scarlet, purple, pink and flesh-color. The flowers are often double.

ORDER XXIX. TROPÆOLACEÆ—TROPHYWORTS.

Plants herbaceous, smooth, climbing or twining, with a pungent, watery juice.
Lvs. peltate or palmate. *Fls.* irregular.
Cal.—Sepals 5, colored, united, the lower one spurred.
Cor.—Petals 5, the three lower ones stalked, the 2 upper inserted on the calyx.
Sta. 5, distinct, unequal.
Ova. of 3 united carpels. *Style* 1. *Stigmas* 3. *Sds.* large. Albumen 0.
Fr. separating into 3 indehiscent, 1-seeded nuts.

Genera 2, species 40, natives of S. America. They possess the same antiscorbutic properties as the Crucifera. The fruit of the following species is pickled and used as a substitute for capers.

TROPÆOLUM.

Lat. *tropæum*, a trophy; the leaf resembles a shield, the flower an empty helmet.

Character essentially the same as of the order.

T. MAJUS. *Nasturtion. Indian Cress.*—Lvs. peltate, roundish, repand on the margin, with the long petiole inserted a little one side of the centre; pet. obtuse, the two upper distant from the 3 lower, which are fimbriate at base, and contracted into long claws.—① Native of Peru. Stem at length climbing by means of its long petioles several feet. Leaves a fine example of the peltate form, about 2' diam. Flowers large and showy, orange-colored, with blotches of deeper shade. They are eaten for salad. June—Oct.

ORDER XXX. LIMNANTHACEÆ.

Herbs annual, with an acrid, watery juice. *Lvs.* alternate, pinnatifid.
Stipules 0. *Flowers* regular.
Cal.—Sepals 3—5, united at base, persistent, valvate in æstivation.
Cor.—Petals 3—5, marcescent, inserted upon an hypogynous disk.
Sta. twice as many as petals and inserted with them. *Filaments* opposite the sepals, with a small process outside the base.
Ova. of 3—5 distinct carpels. *Sty.* united. *Stig.* simple.
Fr. 3—5 achenia, rather fleshy. *Seeds* solitary.

Genera 2, species 3, mostly natives of the temperate parts of North America. They have no very remarkable properties. Floerkea is the only northern genus.

FLŒRKEA. Willd.

Named in honor of Floerke, a German botanist.

Sepals 3, longer than the 3 petals; stamens 6; ovaries 3, tuberculate, style 2-cleft.—① *small, aquatic, with pinnately divided leaves.*

F. PROSERPINACÖIDES. Lindl. (F. uliginosa. *Muhl.*) *False Mermaid.*
Grows in marshes and on river and lake shores, Vt. to Penn. W. to Mo. Stems decumbent, less than a foot in length, weak and slender. Leaves alternate, upper ones, or those above the water, pinnately 5-parted, lower or submersed ones mostly 3-parted, all on slender petioles 1—3' in length. Flowers axillary, pedunculate. Petals white, small, about half as long as the sepals. Achenia large, 2 or 1, roundish.

ORDER XXXI. OXALIDACEÆ.—WOOD SORRELS.

St. low, herbaceous, with an acid juice, and alternate, compound leaves.
Stip. rarely present. *Fls.* regular and symmetrical.
Cal.—Sepals 5, persistent, equal, sometimes slightly cohering at the base.
Cor.—Petals 5, hypogynous, equal, unguiculate, deciduous, twisted in æstivation.
Sta. 10, hypogynous, more or less monadelphous, those opposite the petals longest.
Ova.—Carpels 5, united, opposite the petals.
Fr. capsular, usually membranous, 5-lobed and 5-celled.

Genera 6, species 235, inhabiting hot and temperate regions. The stem and leaves generally contain free oxalic acid. The order is represented in the Northern States by the following genus only.

OXALIS

Gr. ὀξύς, sour; from the acid taste of most species.

Sepals 5, distinct or united at base; petals much longer than the calyx; styles 5, capitate; capsule oblong or subglobose; carpels 5, 1—several-seeded.—*Mostly ♃ with trifoliate leaves.*

1. O. ACETOCELLA. *Common Wood Sorrel.*
Acaulescent; *scape* longer than the leaves, 1-flowered; *lfts.* broad-obcordate, with rounded lobes; *sty.* as long as the inner stamens; *rt.* dentate, scaly.—Woods and shady places, Can. and Northern States. Leaves palmately 3-foliate, on long, weak stalks, purplish beneath. Peduncles longer than the leaves, each with a nodding scentless flower whose petals are white, yellowish at the base, delicately veined with purple. The whole plant has an agreeable, acid taste. Jn.

2. O VIOLACEA.   *Violet Wood Sorrel.*

Acaulescent, smooth; *scape* umbelliferous; *pedicels* subpubescent; *fls.* nodding; *tips of the cal.* fleshy; *sty.* shorter than the outer stamens.—An elegant species, in rocky woods, &c., throughout the U. S. Bulb scaly. Scape nearly twice taller than the leaves, 5—8' high. Leaves palmately 3-foliate, sometimes none; leaflets nearly twice as wide as long, with a very shallow sinus at the very broad apex. Umbel of 3—9 drooping flowers. Petals large, violet-colored, striate. May.

3. O. STRICTA.   *Yellow Wood Sorrel.*

Caulescent; *st.* branching, erect; *ped.* umbelliferous, longer than petioles; *sty.* as long as the inner stamens.—① Fields, U. S. and Can. The plant varies in height, from 3—8' or more, according to the soil. Stem leafy, round, smooth, succulent. Leaves palmately 3-foliate, numerous, scattered on long stalks. Umbels on long, axillary stalks, about the length of the petioles. Flowers small, yellow, appearing all summer. Capsules sparingly hirsute, with spreading hairs.

4. O. CORNICULATA.   *Ladies' Wood Sorrel.*

Caulescent; *st.* creeping, radicating, diffusely branching; *lfls.* pubescent; *ped.* 2 or more-flowered, shorter than the petioles; *pet.* cuneiform, erose at the apex; *sty.* long as the inner stamens.—Grows in cultivated grounds, U. S. and Can. Resembles the last, but "is undoubtedly distinct." *Robbins.* Stems leafy, prostrate, a foot or more in length. Sepals pubescent, half as long as the emarginate, yellow petals. Capsules densely and closely pubescent. May, and after.

ORDER XXXII. ZANTHOXYLACEÆ.

Trees or shrubs, without stipules.
Lvs. alternate or opposite, pinnate, rarely simple, with pellucid dots.
Fs. regular, polygamous, gray, green, or pink. *Sep.* 3—6, small, cohering at the base.
Cor.—Petals longer than the sepals, of the same number or 0.
Sta. alternate with petals, of the same number, seldom twice as many; in the pistillate flowers either wanting or imperfect. *Anthers* introrse.
Ova. usually of the same number as sepals, stipitate, distinct or united.
Fr baccate, membranaceous or drupaceous, or 2-valved capsules.

  Genera 20, species 110, chiefly of tropical America, only 2 genera being native in the United States.
  Properties.—Bitter, aromatic and stimulant; properties residing chiefly in the bark.

Conspectus of the Genera.

{ trees, with 21—41-foliate leaves.	*Ailanthus.*	3
Unarmed { shrubs, with 3-foliate leaves.	*Ptelea.*	2
Prickly shrubs.	*Zanthoxylum.*	1

1. ZANTHOXYLUM.

Gr. ξανθος, yellow, ξυλον, wood; from the color of the wood.

☿ Calyx inferior, 5-parted; corolla 0; stamens 3—6; pistils 3—5: carpels 3—5, 1-seeded; ♀ like the ☿ but wanting the stamens; ♂ like the ☿ but wanting the pistils.—*Leaves pinnately 3—5-foliate.*

Z. AMERICÀNUM. Miller. (Z. fraxineum. *Willd.*) *Prickly Ash.*

Prickly; *lfls.* ovate, subentire, sessile, equal at the base; *umbels* axillary.—A shrub 10 or 12f high, found in woods in most parts of the U. S. The branches are armed with strong, conical, brown prickles with a broad base. Leaflets about 5 pairs with an odd one, smooth above, downy beneath; common petioles with or without prickles. Flowers in small dense umbels, axillary, greenish, appearing before the leaves. The perfect and the staminate ones grow upon the same tree, and the pistillate upon a separate tree. The bark is bitter, aromatic and stimulant, used for rheumatism and to alleviate the tooth-ache. Apr. May.

2. PTELEA.

♀ ☿ ♂ Sepals 3—6, mostly 4, much shorter than the spreading petals; ♂ stamens longer than the petals and alternate with them, very short and imperfect in ♀; ovary of 2 united carpels; styles unit:d. short or 0; stigmas 2; fruit 2-celled, 2-seeded samaræ, with a broad, orbicular margin.—*Shrubs with* 3—5-*foliate leaves. Fls. cymose.*

P. TRIFOLIÁTA. *Shrubby Trefoil.*
　Lvs. 3-foliate, *lfts.* sessile, ovate, short-acuminate, lateral ones inequilateral, terminal ones cuneate at base; *cymes* corymbose; *sta.* mostly 4; *sty.* short.—An ornamental shrub, 6—8f high, Western States! rare in Western N. Y. Leaflets 3—4½' by 1½—1½', the peduncles rather longer. Flowers white, odorous, nearly ¼' diam. Samara nearly 1' diam.

3. AILANTHUS.

　♀ ♀ ♂ Sepals 5, more or less united at base; petals 5 : ♀ stamens 2—3; ovaries 3—5; styles lateral; fruit a 1-celled, 1-seeded samara with oblong margins; ♂ stamens 10; ♀ ovaries, styles and samaræ as in ♀.—*Oriental. Trees and shrubs with pinnate leaves. Fls. in panicles.*

　A. GLANDULÓSA. Desf. *Tree of Heaven.*—*Lvs.* glabrous, unequally pinnate; *lfts.* ovate or oblong-lanceolate, acuminate, shortly petiolulate, with one or two obtuse, glandular teeth each side at base, *terminal one* long-petiolate.—A tree of large dimensions, and with extremely rich and luxuriant foliage, native of China and Japan. Trunk straight, with a smooth, brown bark. Leaves 3—5f in length, with 10—20 pairs of leaflets and an odd one. Flowers in terminal panicles, greenish, perfecting seed in our climate.—The tree is of extremely rapid growth, and is becoming common in our streets and shrubberies. †

ORDER XXXIII. ANACARDIACEÆ.—SUMACHS.

Trees or shrubs, with a resinous, gummy, caustic, or even milky juice.
Lvs. alternate, simple or ternate or unequally pinnate, without pellucid dots.
Fls. terminal or axillary, with bracts, commonly dioecious.
Cal.—Sepals 3—5, united at base, persistent.
Cor.—Petals same number as sepals, sometimes 6, imbricate in æstivation.
Sta. as many as petals, alternate with them, distinct, on the base of the calyx.
Ova. 1-celled, free. Ovule one. Styles 3 or 4. Stigmas 3.
Fr. a berry or drupe, usually the latter and 1-seeded.

　Genera 41, species 95, chiefly natives of tropical regions, represented in the United States by the genus Rhus only.

　Properties.—These plants abound in a resinous juice, which is often poisonous, but is used as an indelible ink in marking linen, and as an ingredient in varnish. Even the exhalations from some of the species are deemed poisonous. The Cashew nut is the product of a small tree of both Indies. When fresh the kernel is full of a milky juice, and has a most delicious taste, but the coats are filled with a caustic oil which blisters the skin, and kills warts.

RHUS.

Said to be from *ρεω*, to flow; because used in hæmorrhage.

　Calyx of 3 sepals united at base; petals and stamens 5; styles 3, stigmas capitate; fruit a small, 1-seeded, subglobose, dry drupe. —*Small trees or shrubs. Leaves alternate, mostly compound. Flowers often by abortion* ♀ ♂ *or* ♀ ♀ ♂.

　　* *Leaves pinnate.*

　1. R. GLABRA. *Smooth Sumach.*
　Lvs. and *branches* glabrous; *lfts.* 6—15 pairs, lanceolate, acuminate, acutely serrate, whitish beneath; *fr.* red, with crimson hairs.—Thickets and waste grounds, U. S. and Can. Shrub 6—15f high, consisting of many straggling branches, smooth, except its fruit. Leaflets about 3' long, ⅓ as wide, sessile except sometimes the terminal odd one. Flowers in terminal, thyrsoid, dense panicles, greenish-red, ♀ ♂. Fertile ovaries clothed with grayish down, which in the fruit becomes crimson, and contains malic acid (bi-malate of lime, *Prof. Rogers*), extremely sour to the taste. Jn. Jl.—The bark of this and other species may be used in tanning. The drupes dye red. Lands long neglected are sometimes overrun by this shrub.

2. R. TYPHINA. *Stag-horn Sumach.*
Branches and *petioles* densely villous; *lfts.* 6—15 pairs, oblong-lanceolate, acuminate, acutely serrate, pubescent beneath; *fr.* red, with crimson hairs. — A larger shrub than the former, attaining the height of 20f, in rocky or low barren places, U. S. and Can. Stem with straggling, thick branches. Leaves at length 2—3f long. Leaflets 2—4' long, ⅓ as wide, sessile, except the terminal odd one. Flowers in terminal, thyrsoid, dense panicles, yellowish-green, often ♀ ♂ or ♀ ♀ ♂. Drupes compressed, compact, the crimson down very acid. Jn.—The wood is aromatic, of a sulphur-yellow, and used in dyeing.

β. laciniata.—*Lfts.* very irregularly coherent and incised; *panicles* partly transformed into gashed leaves. Hanover, N. H. *Rickard.*

3. R. COPALLINA. *Mountain Sumac.*
Branches and *petioles* pubescent; *lfts.* 4—10 pairs, oval-lanceolate, mostly entire, unequal at base, common *petiole* winged; *fls.* in dense panicles; *drupes* red, hairy.—A smaller shrub, not half the height of the last, in dry, rocky places, U. S. and Can. Common petiole about 6' long, expanding into a leafy margin, between each pair of leaflets. Leaflets 1—3' long, near ⅓ as wide, dark green and shining on the upper surface. Panicles of flowers terminal, sessile, thyrsoid, ♀ ♂, greenish Drupes acid. Jl.
β. leaflets coarsely and unequally serrate. N. Y. *Barratt.*

4. R. VENENATA. DC. (R. vernix. *Linn.*) *Poison Sumac. Dog-wood.*
Very glabrous; *lfts.* 3—6 pairs, oval, abruptly acuminate, very entire; *panicles* loose, pedunculate; *drupes* greenish-yellow, smooth.—A shrub or small tree of fine appearance, 10—15f high, in swamps, U. S. and Can. Trunk several inches diam., with spreading branches above. Petioles wingless, red, 6—10' long. Leaflets about 3' long, nearly ⅓ as wide, sessile, except the odd one. Panicles axillary, ♀ ♂, those of the barren tree more diffuse. Flowers very small, green. Drupes as large as peas. Jn. The whole plant is very poisonous to the taste or touch, and even taints the air to some distance around with its pernicious effluvium.

* * *Leaves ternate.*

5. R. TOXICODENDRON. *Poison Oak. Poison Ivy.*
Erect or decumbent; *lvs.* pubescent; *lfts.* broadly oval, acuminate, entire or sinuate-dentate; *fls.* in racemose, axillary, subsessile panicles; *drupes* smooth, roundish.—Can. and U. S. A small shrub, 1—3f high, nearly smooth in all its parts. Leaflets 2—6' long, ⅓ as wide, petiolate, the common petiole 4—5' long. Flowers small, ♀ ♂. Drupes pale brown. Poisonous, but less so than the last.

β. radicans. Torr. (R. radicans. *Linn.* and of *1st edit.*) *Poison Ivy. St* climbing 3—20 or 50f! by myriads of radicating tendrils.—It seems now generally conceded that this is but a variety. Certainly, if so, it is a very remarkable one. In damp, shady places. Poisonous.

6. R. AROMATICA. Ait. *Sweet Sumac.*
Lfts. sessile, incisely crenate, pubescent beneath, lateral ones ovate, terminal one rhomboid; *fls.* in close aments, preceding the leaves; *drupe* globose, villous.—A small, aromatic shrub, 2—6f high, in hedges and thickets, Can. and U. S. Leaflets 1 2' long, ⅓ as wide, sessile, the common petiole an inch or two in length. Flowers yellowish, with a 5-lobed, glandular disk. Drupes red, acid. May.

* * * *Leaves simple.*

7. R. COTINUS. *Venetian Sumac.*—*Lvs.* obovate, entire; *fls.* mostly abortive; *pedicels* finally elongated and clothed with long hairs.—A small shrub, 6f high, native in Ark. according to *Nuttall*, remarkable chiefly for the very singular and ornamental appearance of its long, diffuse, feathery fruit-stalks, showing in the distance as if the plant were enveloped in a cloud of smoke. Flowers small, in terminal, compound panicles. Leaves smooth, entire, much rounded at the end. In Italy the plant is used for tanning. †

18

ORDER XXXIV. RUTACEÆ.—RUEWORTS.

Herbs, or generally shrubs and trees, with punctate lvs. and no stipules.
Fls. perfect. Sep. 4—5. Pet. 4—5, rarely 0.
Sta. as many, or twice or thrice as many as petals, inserted on the outside of a cup-like disk.
Ova. 3—5-lobed, 3—5-celled; styles united or distinct only at base.
Fr. usually separating into its component, few-seeded carpels.

Genera 47, species 400, usually inhabiting the warmer parts of the temperate zone on the Eastern continent, and the equatorial parts of S. America. They are characterized by a powerful odor and intense bitterness, often febrifugal and anthelmintic. Dictamnus abounds in a volatile oil, diffusing an inflammable gas.

Genera.

Sepals permanent.   Petals equal . . . . Ruta. 1
Sepals deciduous.   Petals unequal . . . . Dictamnus. 2

1. RUTA.

Calyx of 4—5 sepals united at base; petals 4—5, concave, obovate, distinct; torus surrounded by 10 nectariferous pores; stamens 10; capsule lobed.—♃ *Herbaceous or shrubby, mostly European.*

R. GRAVĒŌLENS.   *Common Rue.*—Suffruticose, nearly glabrous; lvs. 2 and 3-pinnately divided, segments oblong, obtuse, terminal ones obovate-cuneate, all entire or irregularly cleft; fls. terminal, corymbose; pet. entire.—Native of S. Europe.   Stem branched, 3—4f high.   Leaflets 6—10″ by 2—4″, conspicuously dotted.   Corolla yellow, 6″ diam.   Jn.—Sept. ‡

2. DICTAMNUS.

Calyx of 5 deciduous sepals; petals 5, unguiculate, unequal; filaments declinate, with glandular dots; capsules 5, slightly united.—♃ *Herbs, native of Germany.*

D. ALBUS.   Willd.   (and D. Fraxinella. *Link.*) *Fraxinella.*—St. simple; lvs. pinnate, the rachis more or less winged; fls. in a large, terminal, erect panicle.—In gardens.   Stems 1—2f high.   Flowers showy, white, varying to rose-color and purple.   The whole plant emits a lemon-scented, aromatic, volatile oil, which is so abundant in hot weather as to render the air around it inflammable. †

β. *rubra.*   Flowers purple; rachis of the leaves winged. †

ORDER XXXV. AURANTIACEÆ.—ORANGES.

Trees or shrubs, glabrous, abounding in little transparent receptacles of volatile oil.
Lvs. alternate, articulated with the petiole which is frequently winged.
Cal.—Sepals 3—5, united into a short, urceolate or campanulate cup.
Cor.—Petals 3—5.
Sta. as many as the petals, or some multiple of their number, in a single row, hypogynous.
Ova. compounded of several united carpels.   Style 1.
Fr.—A berry (orange), many-celled, pulpy, covered with a thick rind.
Sds. attached to the inner angle of each carpel.   Albumen 0.

Genera 20, species 85, nearly all natives of tropical Asia, and are naturalized throughout all tropical regions, and cultivated in all civilized countries for their beauty and fragrance, both of flowers and fruit.

Properties.—These fruits contain free *citric* and *malic* acid, and their pulp is grateful to the taste. The rind contains an aromatic, volatile oil, which is tonic and stomachic.   The rind of the lime yields the *oil of Bergamot*, and the flower of the orange the *oil of Neroli.*

CITRUS.

Gr. κιτρια, the citron; the fruit of one of the species.

Sepals and petals in 5s; anthers 20, or some other and higher multiple of 5, versatile, the connectile articulated to the filament; filaments dilated at base, polyadelphous; berry 9—18-celled.—*A noble genus of trees and shrubs, all tropical, combining in its species, beauty of form, with shining, ever-green foliage, odoriferous flowers, and fragrant and delicious fruit.*

1. C. LIMŌNUM.   *Lemon Tree.*—Petioles somewhat winged, articulated with the lamina (which is thus shown to be the terminal or odd leaflet of a reduced

compound leaf); *lft.* oblong, acute, dentate; *sta.* 35; *fr.* oblong-spheroid, with a thin rind and very acid pulp.—A tree about 15f in height, which, when laden with its golden fruit, suspended among its dark green leaves, makes a most beautiful appearance. It is a native of tropical regions, and is easily cultivated in our climate if protected during winter. †

2. C. LIMETTA. *Lime Tree.—Petioles* not winged; *leaf (leaflet)* ovate-orbicular, serrate; *sta.* 30; *fr.* globose, with a sweet pulp, and a protuberance at top. This like most other species, is native of Asia. Height about 8f, with a crooked trunk, diffuse branches with prickles. Berry 1½′ diam., of a greenish-yellow, shining surface. †

3. C. AURANTIUM. *Sweet Orange Tree.—Petiole* winged; *leaf (leaflet)* oblong, acute, crenulate; *sta.* 20; *fr.* globose, with a thin rind and sweet pulp.—A middle-sized evergreen tree, with a greenish-brown bark. When filled with its large, round, golden fruit (sometimes to the number of 20,000, *Lindley*), it is one of the most beautiful objects in nature. It is easily cultivated in the green house. †

4. C. MEDICA. *Citron Tree.—Petioles* not winged; *leaf (leaflet)* oblong, acute; *sta.* 40; *fr.* oblong-spheroid, rugose, with an acid pulp.—Commonly about 8f high. Fruit 6′ in length, very fragrant. †

5. C. DECUMANA. *Shaddock Tree.—Petioles* winged; *leaf (leaflet)* obtuse, emarginate; *fr.* very large, with a thick rind.—A tree 15f in height. Wings of the petioles as broad as the leaves. Fruit grows to the diameter of 7—8′, weighs 14 pounds, and is of a yellowish-green color. †

Obs.—In a splendid work entitled "The Natural History of Oranges," written in French by Risso of Nice in 1818, there are described 169 varieties, and 105 of them figured. They are arranged as Sweet Oranges, of which there are described 42 varieties : Bitter and Sour Oranges. 32; Bergamots. 5; Limes 8; Shaddocks. 6; Lumes, 12; Lemons, 46; Citrons, 17. The most successful methods of cultivation are by cuttings.

ORDER XXXVI. TERNSTRŒMIACEÆ.—TEAWORTS.

Trees or shrubs, with alternate, coriaceous, exstipulate leaves.
Fls. axillary or terminal, white, rarely red or pink.
Cal.—Sepals 5 or 7, concave, coriaceous, deciduous, the inner often the largest.
Cor.—Petals 5, 6 or 9, not equal in number to the sepals.
Sta. ∞, hypogynous. *Filaments* distinct, or united into one or more sets.
Ova. superior, with several cells. *Styles* 3—7, more or less combined.
Fr. 2—7-celled, capsular. *Sds.* large, few, attached to the axis.

Genera 33, species 130. Beautiful flowering plants. 60 or 70 of them natives of S. America, 4 of N America. the remainder of China and E. Indies. Their properties are in general little known. The tea, so extensively used as a beverage in the civilized world is the leaf of 2 or 3 species of Thea. It contains a peculiar extractive matter and a stimulating, essential oil, which becomes narcotic in some hot climates.

Genera.

Sepals {5, equal. Trees.   *Gordonia.* 1
       {5—7, unequal, the inner ones largest. (Shrubs.)   *Camellia.* 2

1. GORDONIA.

In honor of James Gordon, a distinguished nurseryman of London.

Sepals 5, roundish, strongly imbricated; petals 5; styles united into one; capsule woody, 5-celled; cells 2-seeded; seeds winged.— *Trees with large, white flowers.*

G. PUBESCENS. L'Her. (Franklinia Americana. *Marsh.*) *Franklinia.—Lvs.* serrate, deciduous, oblong-cuneiform, shining above, canescent beneath; *sep.* and *pet.* silky outside.—A tree 30—50f high in Ga. and Flor., or an ornamental shrub in cultivation at the north, admired for its large, white flowers, with yellow stamens and rich fragrance. May—Aug.

2. CAMELLIA.

In honor of G. J. Kamel, a Jesuit, author of some botanical works.

Sepals imbricated, the inner ones larger; petals sometimes adhering at base. filaments ∞, shorter than the corolla, united at base; styles united; stigmas 3—5, acute.—*Ornamental shrubs, native of China and Japan.*

C. Japonica. *The Plant. Japan Rose.*—Lvs. ovate, acuminate, acutely serrate, glabrous and shining on both sides, coriaceous and firm, on short petioles; *fls.* terminal and mostly solitary; *pet.* obovate, of a firm texture; *sta.* about 50, mostly changed to petals in cultivation; *stig.* unequally 5-cleft. A lofty tree in Japan, its native country, a splendid flowering shrub with us, of difficult cultivation, requiring protection in our climate. Flowers varying from white to red, resembling the rose but wanting its fragrance. Over 300 varieties are enumerated.

Order XXXVII. MALVACEÆ.—Mallows.

Herbs, shrubs or trees, with alternate, stipulate, divided leaves. Hairs stellate or none.
Fls. axillary, showy, regular, often with an involucel at the base.
Cal.—Sepals generally 5, more or less united at base, valvate in æstivation.
Cor.—Petals equal in number to the sepals, hypogynous.
Sta. indefinite, monadelphous. *Anthers* 1-celled, bursting transversely.
Ova. of several carpels arranged in one or more rows around a common axis.
Sty. as many as the carpels, either united or distinct.
Fr. capsular or baccate; carpels one or more-seeded, united or distinct.
Sds. sometimes (as in Gossypium) hairy. *Embryo* curved.

Genera 37, species 1000. A somewhat important class of plants, forming about one-fiftieth of all the flowering plants of tropical valleys.—But few are natives of the temperate, and none of the frigid zone. In the Northern States they are all herbs. The most important product of the order is cotton.
Properties.—Generally abounding in mucilage, and destitute of any deleterious qualities.

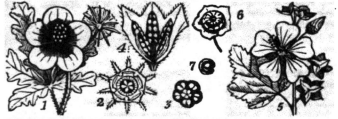

FIG 41.—1. Hibiscus Trionum. 2. Cross section of the flower, showing the arrangement of its parts. 3. Cross section of the 5-celled capsule. 4. Capsule open by its 5 valves. 5. Malva sylvestris. 6. Its fruit consisting of 10 carpels arranged in a circle. 7. Section of one of the carpels showing the curved embryo.

Conspectus of the Genera.

of 3 oblong bracteoles which {	are nearly distinct.	Malva.	1
	are united.	Lavatera.	3
of 3 cordate bracteoles. {	Capsule 3—5-celled.	Gossypium.	2
	Carpels 60, distinct.	Malope.	6
	united.	Althæa.	4
of 6—15 bracteoles which are {	nearly distinct.	Hibiscus.	5
Involucel {	Cells 3—6-seeded.	Abutilon.	7
wanting. Carpels united. {	Cells 1-seeded.	Sida.	8

1. MALVA.

Gr. μαλαχη, soft; on account of the soft mucilaginous properties.

Calyx 5-cleft, the involucel mostly 3-leaved; carpels ∞, 1-celled, 1 seeded, indehiscent, arranged circularly.

1. M. rotundifolia. *Low Mallow.*
St. prostrate; lvs. roundish, cordate, obtusely 5-lobed; *ped.* in fruit reflexed; *cor.* twice as long as the calyx.—♃ Common in cultivated grounds. Root fusiform. Stems numerous, a foot or more long. Leaves of a fine, delicate texture, somewhat reniform, crenate, with 5—7 shallow lobes, and on long, hairy stalks. Peduncles axillary, aggregate. Petals pale pink, deeply notched. Fruit depressed-globose, composed of the numerous carpels arranged circularly. The child sportively calls them *cheeses*, a name which their form very naturally suggests. *Jn.—Oct.* §

2. M. sylvestris. *High Mallow.* (Fig. 41, 5.)
St. erect; lvs. 5—7-lobed, lobes rather acute; *ped.* and petioles hairy.—

♃ Native of England. A popular garden flower of the easiest culture, often springing up spontaneously in fields and roadsides, Mid. and W. States! Height 3f. Flowers reddish purple, with veins of a darker hue. The whole plant, especially the root, abounds in mucilage. Jn.—Oct. ♀ †

3. M. HOUGHTONII. TORR. & Gray. *Houghton's Malva.*

St. erect, hirsute; *lvs.* strigose, ovate, truncate at the base, lower ones cordate, all undivided, coarsely crenate; *panicle* terminal, diffuse, many-flowered; *pet.* purple; *carpels* 10—15.—Prairies and bottoms, Ill. Mead! &c. A handsome but rather rough species, 2—3f high. Root fusiform. Leaves 2—3' by 1—2', on long, hairy petioles, thick. Flowers nearly as large (1½' diam.) as those of M. sylvestris. Jl. Aug.

4. M. MAURITIANA. *Ivy-leaved Mallow.*—St. erect; lvs. 5-lobed, obtuse; *petioles* and *pedicels* smoothish, or downy on the upper side.—① From S. Europe. A tall species, 4—6f high. Stem smooth. Flowers purple, with deeper colored veins. †

5. M. MOSCHATA. *Musk Mallow.*—St. erect; *radical lvs.* reniform, incised, *cauline ones* many-parted, the segments linear; *ped.* and *cal.* hairy.—Native of Britain. Stems 2f high, branched. Flowers large and handsome, rose-colored. The whole herb gives out a musk-like odor in favorable weather. Jl.

6. M. CRISPA. *Curled or Crisped-leaved Mallow.*—St. erect; lvs. angular-lobed, dentate, crisped, smooth; *fls.* axillary, sessile.—① A tall, straight, simple, erect plant from Syria. Gardens, almost naturalized. Stem 5—6f high. Leaves large, roundish, margins abundantly crisped and curled. Flowers white, not conspicuous. Jn.—Aug. †

2. GOSSYPIUM.

A word said to be from the Arabic. gos, a silky substance

Calyx obtusely 5-toothed, surrounded by an involucel of 3 cordate **leaves**, deeply and incisely toothed; capsule 3—5-celled; seeds involved in cotton.—*Fls. yellow.*

1. G. HERBACEUM. *Common Cotton Plant.*—Lvs. 5-lobed, with a single gland below, lobes mucronate; *cotton* white.—① This is the species commonly cultivated in the Southern States. It is an herbaceous plant, about 5f high. The flowers like those of all the other species are yellow. Leaves cut half way down into 3 large and 2 small, lateral, rounded, pointed lobes. Gland on the midvein at its back, half an inch from the base. Jl. †

2. G. BARBADENSE. *Sea Island Cotton Plant.*—Lvs. 5-lobed with 3 glands beneath, upper ones 3-lobed; *seeds* black; *cotton* white.—② Native and cultivated in the W. Indies. A larger plant than the foregoing. Sown in Sept. and Oct. An acre yields an average product of 270 pounds of this cotton.— These plants are ornamental in cultivation. †

3. LAVATÉRA.

Named in honor of the two Lavaters, physicians of Zurich.

Calyx surrounded at base with a 3-cleft involucel; carpels ∞, i celled, 1-seeded, indehiscent, arranged circularly.

1. L. ARBOREA. *Tree Mallow.*—Lvs. 7-angled, downy, plicate; *pedicels* axillary, 1-flowered, clustered, much shorter than the petiole.—② A splendid plant for borders or shrubberies, from Europe. Height about 6f. Flowers purple. Sept. Oct. †

2. L. THURINGIACA. *Gay Mallow.*—Lvs. somewhat downy; lower ones angular, upper ones 3-lobed, the middle lobe largest.—♃ From Germany. Height 4f. Flowers light blue. Sept.

4. ALTHÆA.

Gr. αλθω, to cure; the mucilaginous root is highly esteemed in medicine.

Calyx surrounded at base by a 6—9-cleft involucel; carpels ∞, 1-**seeded, indehiscent,** arranged circularly around the axis.

18*

1. A. OFFICINÀLIS. *Marsh Mallow.*

Lvs. soft-downy on both sides, cordate-ovate, dentate, somewhat 3-lobed, all entire; *ped.* much shorter than the leaves, axillary, many-flowered.—♃ A European plant, naturalized on the borders of our salt marshes. Stem 3f high, erect, firm, covered with thick, woolly down, with alternate, velvet-like leaves. Flowers large, axillary and terminal, pale purple. The root, as well as the other parts of the plant, abounds in mucilage, and in medicine is often used as an emollient to promote suppuration. Sept. ‡

2. A. ROSEA. Cav. (Alcea rosea. *Linn.*) *Hollyhock.*—*St.* erect, hairy; *lvs.* cordate, 5—7-angled, rugose; *fls.* axillary, sessile.—① Native of China ? A tall plant, very commonly cultivated in gardens. Numerous varieties have been noticed, with single, double, and semi-double flowers, of various shades of coloring, as white, rose-colored, flesh-colored, dark red, and even a purplish black, purple, yellow, straw-color, &c. †

3. A. FICIFOLIA. Cav. (Alcea ficifolia. *Linn.*) *Fig-leaved Hollyhock.*—*St.* erect, hairy; *lvs.* palmate, 7-lobed beyond the middle, lobes oblong, obtuse, angularly toothed.—Native of Levant. Stem tall as the above. Flowers orange colored. †

5. HIBISCUS.

Calyx 5-cleft, surrounded by a many-leaved involucel; stigmas 5; capsule 5-celled; cells several-seeded.

1. H. MOSCHEÙTOS. T. & G. (H. Moscheutos and palustris. *Linn.*) *Marsh Hibiscus.*

Herbaceous, simple, erect; *lvs.* ovate, obtusely dentate, hoary-tomentose beneath; *ped.* long, axillary, or connected with the petiole.—♃ A tall, showy plant, in brackish marshes by the sea or near salt springs, and on wet prairies, U. S. and Can. Stem round, downy, 4—6f high. Leaves 4—6' by 3—4', often with two lateral lobes. Flowers larger than those of the hollyhock, rose-colored, purple in the centre. Peduncles usually distinct from the petiole, often some of them united with it, and jointed above the middle. Styles 1' longer than the stamens. Aug.

β. (H. incanus, *Wendl* ?) *Fls.* larger; *pet.* (4—5' long) of a light sulphuryellow with a purple base. Marshes, Indiana !

2. H. VIRGINÌCUS. *Virginian Hibiscus.*

Lvs. acuminate, cordate-ovate, serrate-dentate, upper and lower ones undivided, middle ones 3-lobed; *ped.* axillary, and in terminal racemes; *fls.* nodding; *pistils* declinate.—♃ Marshes near the sea, L. I. to Ga. The whole plant scabrous-tomentose, about 3f high. Leaves 2—2½' by 1½', some of them somewhat 3-lobed. Flowers 2—3' diam., red or rose-color. Capsule hispid, acuteangled. Aug.

3. H. MILITÀRIS. Cav. *Halbert-leaved Hibiscus.*

Glabrous; *lvs.* hastately 3-lobed, lobes acuminate, serrate; *cor.* tubular-campanulate; *caps.* smooth, ovoid-acuminate.—Middle and Western States. Stem 3—4f high. Leaves cordate at base, 4 or 5' long, rendered somewhat hastate by a small lobe each side at base. Petals flesh-color, with a purplish base, 2—3' long. Peduncles with the joint above the middle. Jl. Aug.

4. H. MANÌHOT. *Hand-leaved Hibiscus.*

Not prickly; *lvs.* palmately divided into 5—7 linear, acuminate, coarsely dentate lobes; *ped.* and *involucel* hispid; *bracts of the involucel* 5—7, ovate or lanceolate, acutish, persistent, entire; *cal.* split on one side; *capsule* densely hirsute, acuminate.—♃ Western States. A beautiful herb, 4—5f high. Leaves cordate, lobes 6—10' long, ½—1½' wide, separated to near the base, about as long as the petioles. Teeth largest near the summit. The flowers are of an exceedingly rich sulphur-yellow; purple in the centre. Petals 3—4' long. Jl. Aug.

5. H. COCCÍNEUS. Walt. (H. speciosus. *Ait.* and 1*st. edit.*) *Scarlet Hibiscus.*— Very smooth; *lvs.* palmate, 5-parted; *lobes* lanceolate, acuminate, remotely serrate above; *cor.* expanding; *cap.* smooth, ovoid.—♃ A splendid flower, native of damp soils, in Georgia, &c., and is raised from seeds in our gardens

Root perennial. Stem herbaceous, 5—9f high. Segments of the leaves 6' long, very acuminate. Flowers of a bright carmine red. Petals slender at the base, 4—5' long. Column still longer, slender and terete. Jl.—Oct. †

6. H. GRANDIFLÒRUS. Michx. *Great-flowering Hibiscus.*—Lvs. cordate, 3-lobed, coriaceous, tomentose, hoary beneath; cor. expanding; caps. tomentose, truncated.—♃ Southern States. Stems 5—7f high. Leaves and flowers very large, the latter, when expanded, nearly a foot in diameter. Petals flesh-colored, red at the base. Jl.—Oct. †

7. H. SYRÌACUS. *Syrian Hibiscus.*—Lvs. cuneiform, ovate, 3-lobed, dentate; pedicels scarcely longer than the petiole; involucel about 8-leaved.—A beautiful, hardy, free-flowering shrub, from Syria, 5—10f high. Flowers purple. There are varieties with white, red and striped flowers, both single and double. †

8. H. TRIÒNUM. *Flower of an Hour.*—Lvs. dentate, lower undivided, upper 3-parted, lobes lanceolate, middle one very long; cal. inflated, membranaceous, veined.—① From Italy. An exceedingly beautiful flower, branching, 1—2f high. Flowers large, numerous, but soon withering. Petals of a rich chlorine yellow, the base of a deep brown. † (Fig. 41, 1.)

9. H. ESCULÈNTUS. *Edible Hibiscus or Okra.*—Lvs. cordate, 5-lobed, obtuse, dentate; petiole longer than the flower; involucel about 5-leaved, caducous.—Native of W. Indies. Plant herbaceous, 2—3f high, nearly glabrous. Petiole with a hairy line on the upper side, nearly 1f in length. Lamina 8—10' broad. The flowers 1—2' long, on a short peduncle. Petals greenish-yellow. The large, mucilaginous pods are used for pickles, or served up with butter.

6. MALÖPE.

Calyx surrounded by a 3-leaved involucel; carpels irregularly aggregated, 1-seeded.

M. MALACÖÌDES.

"*Lvs.* ovate, crenate; *stip.* oblong-lineai *ped.* axiliary, 1-flowered.—① Penn. *Muhl.* Stem 1—1½f high, sparingly branched, clothed with white hairs above. Leaves hairy on the veins beneath, nearly glabrous above. Petioles 1' long. Bracteoles setaceous. Carpels hispid, in a depressed, globular head. Petals yellow." *Torrey & Gray* suppose it may prove a species of Malva.

7. ABUTÌLON. Dill.

Calyx 5-cleft, without an involucel, often angular; ovaries 5, many-seeded; styles many-cleft; capsule of 5 or more carpels, arranged circularly, each 1-celled, 1—3-seeded.

A. AVICENNÆ. (Sida Abutilon. *Linn.*) *Indian Mallow.*
Lvs. roundish-cordate, acuminate, dentate, velvety-tomentose; *ped.* shorter than the petiole, solitary; *carpels* about 15, 3-seeded, inflated, truncate, 2-beaked. —① Native in both Indies and naturalized in most of the states, inhabiting waste places, &c. Stem branched, 3—4f high. Leaves 4—6' diam., deeply cordate at base, abruptly acuminate at apex, very soft and velvety at surface. Flowers yellow, near 1' broad. Jl. §

8. SIDA.

Calyx 5-cleft, without an involucel. ovary 5—many-celled; capsule of 5 or more 1-seeded carpels; radicle superior.

1. S. SPINÒSA.
St. rigid, branched, minutely pubescent; *lvs.* ovate-lanceolate, serrate, with a spinose tubercle at the base of the petiole; *stip.* setaceous; *fls.* axillary; *carpels* birostrate.—① Sandy fields and roadsides, Middle, Southern and Western States! Plant bushy. 8—16' high. Leaves 9—15" long, ½ as wide, mostly obtuse at each end. Petals yellow, obovate, of short duration. Jl. Aug.

2. S. NAPÆA. Cav. (Napæa lævis. *Linn.*)
St. slender, glabrous; *lvs.* palmately 5-lobed, nearly glabrous, lobes oblong-linear, acuminate, coarsely toothed; *ped.* many-flowered; *carpels* 10, near

moa vine. By cultivation it sports into endless varieties, differing in the form, color, size, and flavor of the fruit, and in respect to the hardiness of its constitution. In N. England its cultivation is chiefly confined to the garden and as a dessert fruit; but there are extensive vineyards in the Middle and Western States, for the production of wine. The vine is propagated by cuttings. Varieties without end may be raised from the seed, which will bear fruit the fourth or fifth year. A vineyard, it is said, will continue to produce fruit for 200 years.

2. AMPELOPSIS. Michx.

Gr. αμπελος, a vine, οψις, appearance; from its resemblance.

Calyx entire; petals 5, distinct, spreading; ovary 2-celled, cells 2-ovuled; style very short; berry 2-celled, cells 1—2-seeded.

A. QUINQUEFOLIA.
Lvs. quinate, digitate; lfts. oblong, acuminate, petiolate, dentate, smooth.
—A vigorous climber, found wild in woods and thickets. It has long been cultivated as a covering for walls, and is best known by the name of *Woodbine.*
By means of its radicating tendrils, it supports itself firmly upon trees, ascending to the height of 50f. In the same manner it ascends and overspreads walls and buildings. The large, quinate leaves constitute a luxuriant foliage of dark, glossy green. Flowers inconspicuous, greenish, in dichotomous clusters. Berries dark blue, smaller than peas, acid. Jl.

ORDER XLII. ACERACEÆ.—MAPLES.

Trees or shrubs with opposite, usually simple and palmate-veined leaves.
Stipules 0. Fls. often polygamous, in axillary corymbs or racemes.
Cal.—Sepals 5, rarely 4—9, more or less united, colored, imbricate in æstivation.
Cor.—Petals 5, rarely 4—9, hypogynous; sometimes 0.
Sta. hypogynous, 3—12, usually 8. Anthers introrse or versatile.
Ova. 2-lobed, compounded of 2 united carpels.
Fr. a double samara with opposite wings, thickened at the lower edges.
   Genera 3, species 60. The sap of several species of the Maple yields sugar by evaporation.

Genera.

Flowers mostly polygamous. Leaves simple. . . . . : Acer. 1
Flowers diœcious. Leaves compound, pinnate. . : : : : . : ; : : Narundo. 2

1. ACER. Mœnch.

Lat. acer, sharp, vigorous; the wood was anciently manufactured into weapons of war.

Calyx 5-cleft; corolla 5-petaled or 0; stamens 8; styles 2; samaræ 2, winged, united at base, by abortion 1-seeded.—Lvs. simple.

§ *Flowers corymbose, &c. Trees.*

1. A. RUBRUM. *Red Maple. Swamp Maple.*
Lvs. palmately 5-lobed, cordate at base, unequally and incisely toothed, the sinuses acute, glaucous beneath; fls. aggregate, about 5 together, on rather long pedicels; ova. smooth.—The red maple is a common tenant of low woods and swamps throughout the Atlantic States. It is a tree somewhat above the middle size. The trunk is covered with a smooth bark, marked with large, white spots, becoming dark with age. In spring, the appearance of the tree is remarkable for the deep crimson flowers with which it is thickly clothed. Each bud produces a fascicle of about 5 flowers. Stamens much exserted. The fertile flowers are succeeded by a red fruit, furnished with a pair of wings resembling those of some insect. The wood is hard and compact, and is much used in cabinet work, particularly that well-known and handsome variety called *curled maple.* Mar. Apr.

2. A. DASYCARPUM. Ehrh. (A. eriocarpum. *Mx.*) *White Maple.*
Lvs. palmately 5-lobed, truncated at base, unequally and incisly toothed, with obtuse sinuses, white and smooth beneath; fls. in crowded, simple umbels, with short pedicels and downy ovaries.—This species much resembles the last, but its leaves are larger, and the winged fruit is also larger than that of the red maple or of any of the following species. It is a tall tree, 50f. in height, not uncommon in the N. England forests. The flowers are of a yellowish green

color, as also the fruit. The wood is white, softer and less esteemed than that of other species. The sap yields sugar in smaller proportion than the sugar maple.

3. A. saccharinum. *Sugar Maple. Rock Maple.*

Lvs. palmately 5-lobed, subcordate at base, acuminate, remotely toothed, with rounded and shallow sinuses, glaucous beneath; *fls.* pedunculate, pendulous.—This fine tree is found throughout U. S., but most abundant in the primitive soils of N. England, constituting the greater part of some of its forests. It is a tree of lofty proportions, 70f in height, with a trunk 3f diam. The bark is of a light-gray color, rough and scaly. The branches become numerous and finely ramified in open situations, and in summer are clothed with a foliage of uncommon luxuriance and beauty, on which account it is more extensively cultivated as a shade tree than any other, not even excepting the majestic and favorite elm. Maple sugar, perhaps the most delicious of all sweets, is mostly the product of this species. An ordinary tree will yield 5—10 pounds in a season. The wood is very strong and compact, and makes the best of fuel. It is sometimes curled like the red maple, but oftener presents that beautiful arrangement of fibre, called *bird's-eye maple*, which is highly esteemed in cabinet-work. The flowers are exceedingly abundant, and, suspended on long, thread-like pedicels, are most delicately beautiful. Apr.

4. A. nigrum. Mich. f. *Black Maple. Sugar Tree.*

Lvs. palmately 5-lobed, cordate, with the sinus closed, lobes divaricate, sinuate-dentate, paler beneath, with the veins beneath and petioles pubescent; *fls.* corymbose, on long, slender pedicels; *fr.* glabrous, turgid at base, the wings diverging.—A large tree, in mountainous situations, Vt. to Ia. ! Resembles the last, but is distinct. *Robbins., Tully.* Trunk 30—50f high, with a shaggy bark. Leaves 3—5' diam., dark-green above, the 2 inferior lobes much smaller. Flowers pendulous, on long peduncles, yellowish. Fruit with wings 1' in length, pale-yellow, and more diverging than in A. saccharinum. The sap, like the last mentioned tree, yields sugar abundantly. Apr.

§ § *Flowers in racemes. Mostly shrubs.*

5. A. pennsylvanicum. (A. striatum. Lam.) *Striped Maple. Whistle-wood.*

Lvs. with 3 acuminate lobes, rounded at base, sharply denticulate, smooth; *rac.* simple, pendulous.—A small tree or shrub 10—15f high, Can. to Ga., and Ky., but most abundant in our northern woods. The bark is smooth, and beautifully striped length-wise with green and black. Flowers large, yellowish-green, succeeded by long clusters of fruit, with pale-green wings. The smaller branches are straight and smooth, easily separated from the bark in spring, and are often manufactured by the boys into certain wind instruments. Hence it is called whistle-wood. In Europe it is prized in ornamental gardening. May.

6. A. spicatum. Lam. *Mountain Maple Bush.*

Lvs. about 5-lobed, acute, dentate, pubescent beneath; *rac.* erect, compound.—A shrub of smaller stature than the last, found in mountain or hilly woods throughout the country. The bark is a light gray. Leaves small, rough, divided into 3 or 5 lobes, which are somewhat pointed, with large, sharp teeth, and more or less cordate at base. Flowers greenish, numerous and minute, in cylindric, oblong, close, branched clusters, becoming pendulous with the winged fruit. Jn.

7. A. pseudo-platanus. *Sycamore.—Lvs.* cordate, 5-lobed, glabrous and glaucous beneath, segments or lobes acute, unequally dentate; *fls.* in .ong, pendulous racemes; *samara* glabrous.—Native of Northern Europe. An ornamental tree, 40—50f high, with very large, dark green leaves. A beautiful variety with striped leaves is also cultivated. Apr. May. †

2. NEGUNDO. Mœnch.

Flowers ♀ ♂; corolla 0; ♀ flowers racemed, ♂ fascicled ; calyx, stamens and fruit as in the last genus.—*Leaves compound, pinnately 3—5-foliate.*

N. ACERÖIDES. Mœnch. (Acer Negundo. *Linn.*) *Ash-leaved Maple. Box Elder.*

Lvs. ternate and 5-pinnate; *lfts.* ovate, acuminate, remotely and une-qually dentate; ♀ racemes long and pendulous, *barren fls.* corymbose; *fr.* ob-long, with large wings dilated upwards.—A handsome tree, 20—30f in height, with irregular, spreading branches, growing in woods. The trunk is a foot or more in diameter, and when young, covered with a smooth, yellowish-green bark. Leaflets serrated above the middle, petiolate, the terminal one largest, all slightly pubescent. Wings of the samara approximate, broadest towards the end. Apr.

ORDER XLIII. HIPPOCASTANACEÆ.—BUCKEYES.

Trees or shrubs. Leaves opposite, rarely alternate, compound, without stipules.
Fls. showy, with the pedicels articulated.
Cal. campanulate, of 5 united sepals.
Cor.—Petals 5, (one of them sometimes abortive,) unequal, hypogynous.
Sta. 5—8, distinct, unequal, inserted upon a disk with the petals.
Ova. roundish, 3-cornered, 3-celled, crowned with a single, filiform, conical style.
Fr. roundish, coriaceous, with 1—3 large, roundish, smooth seeds.

Genera 3, native of N. America and Northern India. The species are generally ornamental trees, with astringent properties residing in the bark. The seeds contain much starch, and are nutritive but bitter. Only the following genus is found in the Northern States, and even this is not indigenous in N. Eng.

ÆSCÚLUS.

Calyx campanulate or tubular, 5-lobed; corolla irregular, 4—5-petaled; stamens, ovary and fruit, as expressed in the order.—*Trees, with palmately 5—7-foliate leaves. Flowers in thyrse-like panicles.*

1. Æ. GLABRA. Willd. (Pavia pallida. *Spach.* P. Ohiensis. *Michx.*) *Ohio Buckeye.*
Lfts. 5, oval or oblong, acuminate, serrate or serrulate; *fls.* in lax, thyr-soid panicles; *cor.* 4-petaled, spreading, with the claws as long as the calyx; *sta.* longer than the corolla; *fr.* echinate.—A small, ill-scented tree, along the banks of the Ohio and its tributaries. Leaflets 3—6′ long, ½ as wide, subsessile, or abruptly contracted at base to short stalks. Flowers yellowish-white, small, slightly irregular. Fruit about ¾′ diam.

2. Æ. FLAVA. Ait. (Pavia flava. *DC.*) *Big Buckeye. Sweet Buckeye.*
Lfts. 5—7, oblong-ovate or elliptic-ovate, acuminate, serrulate, pubescent beneath; *fls.* in thyrsoid, pubescent panicles, about 6 on each division of the peduncle; *cal.* campanulate, not half the length of the corolla; *pet.* very unequal, connivent, longer than the stamens; *fr.* unarmed.—A large tree, 30—70f high, common in the Western and Southern States. Leaflets 4—7′, by 1—3′. Flowers pale yellow. Fruit globose, uneven on the surface, but not prickly, 2—2½′ diam, with 1 or 2 large brown seeds. Apr. May.

3. Æ. PAVIA. (Pavia rubra. *Lam.*) *Small Buckeye.*—*Lfts.* 5, oblong-lan-ceolate, cuneate at base, abruptly and shortly acuminate, finely serrate; *fls.* very irregular, in a lax, thyrsoid raceme, *pet.* 4, erect, as long as the stamens.—A beautiful shrub, 6—10f high, native of the Southern States. Flowers large, red, glabrous. Apr. May. †

4. Æ. PARVIFLÓRA, Walt. (Æ. machrostachya. *Michx.*) native at the South, a beautiful shrub, with numerous small, white flowers, in a long, slender, thyr-soid raceme, is rarely cultivated.

5. Æ. HIPPOCASTÁNUM. *Horse Chestnut.*—*Lvs.* digitate, of 7 obovate leaflets; *pet.* 5; spreading; *fr.* prickly.—A noble tree, justly admired for its majestic proportions, and for the beauty of its foliage and flowers. It is a native of the north of Asia, but is now known throughout Europe and in this country, and is a frequent ornament of courts and avenues. It is of rapid growth, and attains *the height of 40 or 50f.* In June it puts forth numerous pyramidal racemes or *thyrses of flowers,* of pink and white, finely contrasting with the dark green of *its massy foliage.* The leaves are digitate, with 7 obovate, acute, serrate leaf-lets. The fruit is large, mahogany-colored, and eaten only by deer.

ORDER XLIV. SAPINDACEÆ.—SOAPWORTS

Trees, shrubs or herbs, the latter furnished with tendrils.
Lvs. alternate, usually compound and without stipules.
Fls. small, usually polygamous. Sep. 4—5, distinct, imbricated in æstivation.
(or—Petals as many as the sepals, sometimes 1 less, (or rarely wanting,) inserted outside the hypogynous disk which lies at the bottom of the calyx.
Sta. 4 or 10; fil. distinct; anth. into ∞
Ova. of 3 united carpels; sty. partly or completely united.
F. a 3-celled capsule or samara, or often fleshy and indehiscent.
Sds. 1—3 in each cell, usually arilled, without albumen.

CARDIOSPERMUM.

Gr. κάρδια, heart, σπέρμα, seed ; the globose seeds marked with a large, cordate hilum.

Sepals 4, the 2 outer smallest ; petals 4, each with an emarginate scale above the base ; the 2 lower remote from the stamens, their scales crested ; glands of the disk 2, opposite the lower petals ; stamens 8, unequal ; style trifid ; capsule membranous, inflated.—*Climbing herbs with biternate leaves. Lower pair of pedicels changed to tendrils.*

C. HALICACABUM.   *Heart-seed.   Balloon-vine.*
Plant nearly glabrous ; *leaflets* ovate-lanceolate, incisely lobed and dentate ; *fruit* pyriform-globose, large, bladder-like.—Native on the Missouri and its branches. *Torr. & Gr.* Naturalized in the W. States. *Mead.* A curious vine, 4—6f in length, with remarkably large, inflated, membranous capsules. Jl. †

ORDER XLV. CELASTRACEÆ.—STAFF-TREES.

Shrubs, or rarely trees, with opposite or alternate leaves. Fls. not always perfect.
Cal.—Sepals 4—5, united at base, imbricated.   which surrounds the ovary.
Cor.—Petals as many as sepals, inserted by a broad base under the margin of the flat, expanded disk
Sta. as many as the petals and alternate with them, inserted on the margin of the disk.
Ova. superior, immersed in and adhering to the disk.
Fr. a capsule or berry. Seeds either with or without an arillus.

Genera 37, species 174, chiefly native of the temperate zone of both hemispheres. They possess acrid and bitter properties, sometimes emetic and stimulant.

Genera.

Shrubs with leaves	opposite,	{ compound (ternate).	.	.	.	.	.	.	.	*Staphylea.* 1
		{ simple.	.	.	.	.	.	.	.	*Euonymus.* 2
	alternate,	simple.	.	.	.	.	.	.	.	*Celastrus.* 3

TRIBE 1. STAPHYLEÆ.

Leaves pinnate, opposite.   Seeds not ariled.   Cotyledons thick.

1. STAPHYLEA.

A Greek word, meaning a cluster of grapes ; from the form of the fructification.

Fls. ☿ ; calyx of 5, colored, persistent sepals ; petals and stamens 5 ; styles 3 ; capsules 2—3, membranous and inflated.

S. TRIFOLIA.   *Bladder-nut.*
Lvs. ternate ; *rac.* pendulous ; *pet.* ciliate below ; *fr.* ovate.—A handsome shrub, 6—8f high, in moist woods and thickets. Can. to Car. and Tenn. Leaflets oval-acuminate, serrate, pale beneath, with scattered hairs. Flowers white, in a short, drooping raceme. The most remarkable feature of the plant is its large, inflated capsules, which are 3-sided, 3-parted at top, 3-celled, containing several hard, small nuts or seeds, with a bony, smooth and polished testa. May

TRIBE 2. EUONYMEÆ.

Leaves simple.   Seeds usually ariled.   Cotyledons leafy.

2. CELASTRUS.

Flowers sometimes polygamous ; calyx flat, of 5 united sepals ; corolla spreading, of 5 sessile petals : capsule subglobose, or 3-angled, 3-celled ; seeds with an arillus, 1—2 in each cell.—*Climbing shrubs, with deciduous leaves, and minute, deciduous stipules.*

C. SCANDENS.   *Staff-tree.*
Unarmed; st. woody, twining ; *lvs.* oblong, acuminate, serrate ; *rac.*
19

minal; *fls.* diœcious.—A climbing shrub in woods and thickets, the stems wining about other trees or each other, ascending to a great height. Leaves alternate, stipulate, petiolate, smooth. Flowers in small racemes, greenish-white Seeds covered with a scarlet aril, and contained in a 3-valved capsule, continuing upon the stem through the winter. Jn.

3. EUONÝMUS.

Calyx flat, of 5, (sometimes 4 or 6) united sepals; corolla flat, inserted on the outer margin of a glandular disk; stamens 5, with short filaments; capsule colored, 5-angled, 5-celled, 5-valved; seeds ariled.—*Shrubs, erect or trailing, with opposite leaves.*

1. E. ATROPURPUREUS. Jacq. *Spindle Tree. Burning Bush.*

Branches smooth; *lvs.* elliptic-lanceolate, acuminate, finely serrate, puberulent beneath; *ped.* compressed, many-flowered; *fls.* usually pentamerous.—A smooth shrub, 4—10f high, in shady woods, U. S. E. of the Miss. Leaves 2—5′ long, ⅓ as wide, mostly acute at base, on petioles ⅓—1′ long. Peduncles opposite, slender, 1—2½′ long, each with a cyme of 3—6 flowers. Corolla dark-purple, about 2½″ diam. Capsule crimson, smooth. Seeds covered in a bright red aril. Jn.

2. E. AMERICÁNUS. *Burning Bush.*

Branches smooth, 4-angled; *lvs.* oval and elliptic-lanceolate, subentire at margin, acuminate, acute or obtuse at apex, smooth; *ped.* round, about 3-flowered; *fls.* mostly pentamerous.—Shrub of smaller size than the preceding, with small leaves, in moist woods, U. S. and Can. Leaves ⅓—2′ long, ⅓ as wide, coriaceous. Peduncles longer than the leaves, 2, 3, or 4-flowered. Flowers a little larger than in No. 1, yellow and pink, the parts in 3s, 4s or 5s. Capsule dark red, warty. Seeds with a bright red aril. Jn.

3. E. EUROPÆUS.—*Lvs.* oblong-lanceolate, serrate, glabrous; *ped.* compressed, 3-flowered; *fls.* usually tetrandrous.—Native of Europe. A handsome shrub, 4—12f high, sometimes found in shrubberies, although certainly not superior in elegance to E. Americanus. May—Jl.

Order XLVI. RHAMNACEÆ.—BUCKTHORNS.

Shrubs or trees, often spiny. *Leaves* simple, alternate. *Stipules* minute or 0.
Fls. small, axillary or terminal, greenish, sometimes diœcious.
Cal.—Sepals 4 or 5, united at base, valvate in æstivation.
Cor.—Petals 4 or 5, distinct, cucullate or convolute, inserted into the orifice of the calyx, sometimes 0.
Sta. opposite the petals, 4 or 5.
Ova. superior, or half superior, with an erect ovule in each cell.
Fr. a capsule, drupe or berry.

   Genera 42, species 280, distributed throughout all countries, except those in the frigid zones. Many are native of the U. States. Ceanothus is peculiar to N. America.

   Properties.—The berries of many species of Rhamnus are violent purgatives. The Zizyphus Jujuba, yields the well-known *jujube paste* of the shops. The leaves of Ceanothus have been used as a substitute for tea.

Genera.

Calyx free from the ovary; petals plane; flowers minute.   :   :   :   :   :   :   Rhamnus. 1
Calyx adherent to the ovary at base; petals unguiculate.   :   :   :   :   :   Ceanothus. 2

1. RHAMNUS.

Calyx urceolate, 4—5-cleft; petals 4—5, emarginate, inserted upon the calyx; ovary free, 2—4-celled; styles 2—4, more or less united; fruit drupaceous. 3—4-seeded.—*Small trees or shrubs. Lvs. mostly alternate. Fls. minute.*

1. R. CATHARTICUS. *Buckthorn.*

Shrub erect, with thorny branches; *lvs.* ovate, doubly serrate; *fls.* tetrandrous, ♀ ♀ ♂ and ♀ ♂, fasicled; *fr.* subglobose, 4-seeded.—A shrub, 10—15f high, in mountains and woods, Mass. and N. Y., rare. Leaves nearly smooth, 1--2′ long, ⅓ as wide, in crowded clusters at the ends of the branchlets. Flowers small, numerous, green. Sepals reflexed, petals entire. Fruit black, globose, and with the inner bark, powerfully cathartic. This shrub is sometimes used for hedges. ‡

2. R. ALNIFOLIUS. L'Her. (R. franguloides. *Michx.*) *Alder-leaved Buckthorn.*

Shrub erect, with unarmed branches; *lvs.* oval, acuminate, serrate, pubescent on the veins beneath; *ped.* aggregate, 1-flowered; *fls.* mostly pentandrous; *cal.* acute; *sty.* 3, united, very short; *fr.* turbinate, black.—A shrub 3—4f high, common in rough pastures and hills, Penn. to Can. Leaves 1—3' long, ½ as wide, acute at base. Flowers mostly apetalous. Berries about as large as currants, black, 3-seeded. May, Jn.

2. CEANOTHUS.

Calyx tubular. campanulate. 5-cleft. separating transversely after flowering; petals 5. saccate-arched, with long claws; stamens mostly exserted; style mostly 3-cleft; capsule obtusely triangular. 3-celled, 3-seeded. surrounded at base by the persistent tube of the calyx.—*Shrubby and thornless.*

1. C. AMERICANUS. *Jersey Tea. Red-root.*

Lvs. oblong-ovate, serrate, 3-veined; *panicles* axillary, elongated.—A small shrub, with a profusion of white blossoms, found in woods and groves, U. S. Very abundant on the barrens at the West. Stems 2—4f high, slender, with reddish, round, smooth branches. Leaves thrice as long as broad, very downy, with soft hairs beneath. Flowers minute, white, in crowded panicles from the axils of the upper leaves. Stamens enclosed in the curiously vaulted corolla. The root, which is large and red, is sometimes used for coloring. The leaves have been used as a substitute for tea. Jn.

2. C. OVÀLIS. Bw. *Oval-leaved Crunothus.*

Lvs. oval-lanceolate, with glandular serratures, 3-veined, veins pubescent reneath; *thyrse* corymbose, abbreviated.—Burlington, Vt., *Robbins*, W. to Mich. Shrub 2—3f high. Leaves smooth and shining, 1—3' long, ¼—½ as wide, mostly acute at each end, crenately serrate, the serratures tipped with black, glandular points. Thyrse short, almost hemispherical, 1½' diam. Flowers white, larger than those of the last. May

ORDER XLVII. LEGUMINOSÆ.—LEGUMINOUS PLANTS.

Herbs shrubs or trees. Lvs alternate, usually compound, margins entire.
Stipules 3- at the tumid base of the petiole. *Stipels* commonly 2.
Cal —sepals generally 5, more or less united, often unequal.
Cor —Petals 5. either papilionaceous or regular, perigynous.
Sta. diadelphous, monadelphous or distinct. *Anthers* versatile.
Ova superior, single and simple. *Style* and *stigma* simple.
Fr. a legume, either continuous (1-celled), or (a *loment*) jointed into 1-seeded cells.
Sds. solitary or several, destitute of albumen.

The genera and species of this vast order were estimated by Mr. Bentham, in 1845, as follows:

	Suborder 1. Papilionaceæ,	280 genera,	6800 species.
"	2. Cæsalpineæ,	80 "	700 "
"	3. Mimoseæ,	20 "	1000 "
	Total,	467 "	6500 "

Geography.—The Leguminosæ are distributed throughout all lands, with the exception of a few unimportant islands. from the equator to either of the frigid zones. Of its 6500 species now known, about 800 are natives of the United States and Territories.

Properties —No family of the vegetable kingdom possesses a higher claim to the attention of the naturalist than the Leguminosæ, whether we regard them as objects of ornament or utility. Of the former we might mention the splendid varieties of Cassia, with their purple flowers, the Acacias, with their airy foliage and silky stamens, the pride of India, Colutea and Cæsalpina, with a host of others, which, like the sweet pea, are redolent with perfume. Of the latter, the beans, peas, lentils, clover and lucerne, are too well known to require particular commendation. Among timber trees the Rosewood (a Brazilian species of Mimosa), the Laburnum, whose wood is durable and of an olive-green color, and the locust (Robinia) of our own country, are pre-eminent.

The following are a few of the important officinal products of this order. In medicine: *liquorice* is the product of the root of Glycyrrhiza glabra of S Europe. The purgative *senna* consists of the leaves of Cassia senna, C. acutifolia. C Æthiopica and other species of Egypt and Arabia. C. Marylandica is also a cathartic, but more mild than the former. The sweet pulp *tamarind*, is the product of a large and beautiful tree (Tamarindus Indica) of the E. and W. Indies. Resins and balsams: *Gum senegal* is yielded by Acacia Verek of the river Senegal. *Gum Arabic*, by several species of Acacia of Central Africa. *Gum Tragacanth*, by Astragalus verus, &c. of Persia. *Balsam Copaira* is the product of several species of Copaifera, natives of Brazil and W. India. *Balsam Tolu* of Myrospermum toluiferum of Peru, and *balsam Peru* of M. peruiferum of the same country. Dyes &c: *Indigo*, the most valuable of all, (but a violent poison,) is the product of several southern species of Indigofera, as 1 and of the W. Indies, and 1. argentea of Egypt. *Brazil-wood*, from Cæsalpina Brasiliensis. *Log-wood* from Hæmatoxylon Campeachianum, of Campeachy, and *Red-sandal-wood* from Pterocarpus santalinus of Egypt, &c., &c.

FIG. 43.—1. Lathyrus odoratus. 2. The stamens, the upper one free, and with the style, turned upwards. 3. The legume. 4. A seed, showing the embryo with the two large cotyledons. 5. Hedysarum boreale, a leaf and jointed legumes (loments). 6. One of the joints open, showing the seed. 7. Trifolium pratense. 8. The legume and part of the calyx. 9. Section of the seed. 10. A flower enlarged.

Conspectus of the Genera.

* Corolla papilionaceous,

				turgid,	1-celled.	Phaca.
					sub-2-celled.	Astragalus.
					Stem erect.	Tephrosia.
			∞-seeded,	compr's'd.	St. twining.	Apios.
		Legumes	1-seeded...	Stamens 5.		Petalostemon.
				Stamens 10.		Dalea.
	Herbs.		Fls. racemed.	Sta. 5 & 5.		Æschynomene.
		Loment 2—10-jnt.	Flowers umbellate.	Sta. 9 & 1.		Hedysarum.
						Coronilla.
				Erect.	Fls. yellow.	Colutea.
			unarmed.		Fls. blue.	Amorpha.
	unequally.	Shrubs and trees,	armed with stipular spines.	Twining. Bracts colored.		Wistaria.
						Robinia.
Leaves					Seeds oval.	Vicia.
pinnate		cirrhose.	Style villous	outside.	Sds. globose.	Pisum.
	abruptly,		Style glabrous.	inside, next the free sta.		Lathyrus.
		not cirrhose.	Stem erect.			Ervum.
						Vicia.
			1-seeded, indehiscent.	Calyx bibracteolate.		Lespedeza.
			2-seeded. None of the flowers apetalous.	Calyx naked.		Psoralea.
			2-seeded in the apetalous flowers.			Melilotus.
						Amphicarpæa.
					Keel and sta. circinate.	Phaseolus.
		Legumes	∞-seeded.	Cal. 5-toothed.	Keel long-clawed.	Clitoria.
				Calyx 4-toothed.		Galactia.
	pinnately.	Loment	2-jointed and one-seeded.			Stylosanthes.
			several-jointed, several-seeded.			Desmodium.
Leaves			Stamens diadelph. (9 & 1.)	Leg. falcate or spiral.		Medicago.
3-foliate	palmately.	Herbs.	Stamens 10, distinct, equal.	Leg. included in cal.		Trifolium.
		Trees.				Baptisia.
		Herbs.				Laburnum.
Leaves simple.		Shrubs.				Crotalaria.
Leaves palmately 5—15-foliate.		Trees.				Genista.
						Cercis.
						Lupinus.

** Corolla not papilionaceous.

Herbs.	Corolla regular.	Unarmed and glabrous.	.	.	.	.	.	.	*Darlingtonia.* 46
	Corolla irregular.	Armed with uncinate spines.	.	.	.	.	.	.	*Schrankia.* 29
Shrubs.	Corolla regular.	.	.	.	.	.	.	.	*Cassia.* 29
	unarmed.	.	.	.	.	.	.	.	*Mimosa.* 28
Trees	armed with triple spines.	.	.	.	.	.	.	.	*Gymnocladus.* 27
		.	.	.	.	.	.	.	*Gleditschia.* F

Suborder 1. PAPILIONACEÆ.

Petals papilionaceous, imbricate in æstivation, the upper one external. Stamens mostly 10 and diadelphous.

1. LATHYRUS.

Calyx campanulate, the 2 upper sepals shortest; stamens 10, diadelphous (9 and 1); style flat, dilated above, ascending, bent at a right angle with the ovary, pubescent or villous along the inside next the free stamen; legume oblong, several-seeded.—*Herbaceous, mostly climbing. Lvs abruptly pinnate, of 1—several pairs of leaflets. Petioles produced into tendrils. Peds. axillary.*

1. L. VENOSUS. Muhl.
St. 4-cornered, naked; *stips.* semi-sagittate, lanceolate, very small; *ped.* 8—16-flowered, shorter than the leaves; *lfts.* 5—7 pairs, somewhat alternate obtusish, mucronate.—♃ In shady grounds, Can. and U. S. Stem erect, 2—3f high, mostly smooth. Leaflets 1½—2' long. Peduncles axillary, many-flowered, about the length of the leaves. Corolla purple. Legumes flat and narrow. Jn. Jl.

2. L. OCHROLEUCUS. Hook. (L. glaucifolius. *Beck.*)
St. slender; *ped.* 7—10-flowered, shorter than the leaves; *upper segments of the calyx* truncate, angular; *lfts.* about 3 pairs, broadly ovate; *stip.* semi-cordate.—♃ A small, delicate species, very rare, in shady places and on river banks, N. J. to Wisc.! N. to the Arctic circle. Stem 2—3f long, leaning or climbing on other plants. Leaflets 1—1½' long, ½ as wide, larger than the stipules. Peduncles axillary, shorter than the leaves. Corolla yellowish-white (ochroleucous.) Jn. Jl.

3. L. PALUSTRIS. *Marsh Lathyrus.*
St. winged; *stip.* semi-sagittate, large, ovate, mucronate; *lfts.* in 2 pairs, oblong-ovate, mucronate; *ped.* 3—5-flowered, larger than the leaves.—♃ A slender climber, found in wet meadows and thickets, N. Eng. to Or. Stem slender, square, broadly winged at the angles, supported by the tendrils. Leaves pinnate-cirrhose, leaflets broad or narrow-ovate. Flowers drooping, rather large, variegated with blue and purple. Jn. Jl.

4. L. MARITIMUS. Bw. (Pisum maritimum. *Ph.*) *Beach Pea.*
St. quadrangular, compressed; *petioles* flat above; *stip.* sagittate; *lfts.* numerous, subalternate, ovate; *ped.* many-flowered.—A pale green, creeping plant, resembling the common pea, found on sandy shores, N. Y. to Lab., W. to Or. Stem rigid, 1—2f in length. Stipules connate. Leaves ending in a branching tendril, the lower pairs of leaflets largest. Flowers large, blue. Pod hairy. May—July.

5. L. MYRTIFOLIUS. Muhl.
St. quadrangular, winged, weak and flexuous; *stip.* semi-sagittate, ovate-lanceolate, acuminate; *lfts.* 2 pairs, oblong-lanceolate, acute, mucronate, veinless; *ped.* longer than the leaves, 4—5-flowered.—♃ A little climber, on river banks, Can. to Md. *Robbins.* Stem about 3f long. Leaflets 1—2' long, ½ as wide. Flowers pale purple. Jl. Aug.

6. L. LATIFOLIUS. *Everlasting Pea.*—*Ped.* many-flowered; *lfls.* 2, lanceolate; *joints* membranous, winged.—♃ A very showy plant for gardens and arbors, native of England. Stem 6f long, climbing, winged between the joints. Flowers large, *pink, clustered* on a peduncle 6—10' in length. Jl. Aug.

7 L. ODORATUS. *Sweet Pea.*—*Ped* 2-flowered; *lfls.* 2, ovate-oblong; *leg.*
19*

nirsute.—① A well known garden flower, native of Sicily. The flowers appear in June, are large, variegated with red and white. Very fragrant.

8. L. **sativus.** *Chick Pea.*—*Peduncles* 1-flowered; *lfts.* 2—4; *leg.* ovate, compressed, with 2 winged margins at the back.—① Native of S. Europe, where it has been sometimes cultivated for food; but it proves to be a slow poison both to man and beast, producing ultimately entire helplessness, by rendering the limbs rigid, but without pain.

2. VICIA.

Celtic *gwig*, whence Gr. βικιον, Lat. *vicia*, Fr. *vesce*, and Eng. *vetch*.

Calyx tubular, with the 3 inferior segments straight and longer than the 2 above; vexillum emarginate; stamens 10, diadelphous (9 and 1); style filiform, bent at right angles with the ovary, villous beneath the stigma on the outside (next the keel); legume oblong, several-seeded.—*Herbaceous, mostly climbing. Leaves abruptly pinnate, with several pairs of leaflets and a branching tendril. Peduncles axillary.*

1. V. **Americana.** *Muhl. American Vetch.*
Smooth; *ped.* 4—8-flowered, shorter than the leaves; *stip.* semi-sagittate, deeply dentate; *lfts.* 10—14, elliptic-lanceolate, obtuse, mucronate, veined, somewhat alternate; *legumes* oblong-linear, compressed, reticulated.—N. Y. W. to the R. Mts. Stems slender, 1—3f long. Leaflets 1' by 5″, subsessile. Flowers blue or purple. Lower calyx teeth broad-lanceolate, much longer than the 2 upper. Style very hairy at the summit. May.

2. V. **Caroliniana.** *Walt. Carolinian Vetch.*
Ped. many-flowered; *fls.* distant; *teeth of the calyx* shorter than the tube, the two upper very short; *sty.* hairy at the summit; *lfts.* 8—12, linear-oblong, smoothish; *leg.* not reticulated, oblong.—Woods and river banks. A slender climber, 4—6f long. Leaflets about 8″ by 2—3″. Flowers pale-blue, the banner tipped with deep purple. May.

3. V. **cracca.** *Tufted Vetch.*
Fls. in imbricated spikes; *lfts.* lanceolate, pubescent; *stip.* semi-sagittate, linear-subulate, entire.—A slender climber, 2—3f long, about fences, hedges, thickets, &c., lat. 39° to Can. Stem square, downy. Leaves of many pairs of downy, mucronate leaflets, with a branched tendril at the end of the principal stalk. Leaflets 6—8″ by 2—3″, petiolulate. Flowers blue and purple, in a long, dense, one-sided raceme. July.

4. V. **tetrasperma.** Loisel. (V. pusilla. *Muhl.* Ervum. *Linn.*) *Slender Vetch.*
Ped. about 2-flowered; *calyx teeth* lanceolate, shorter than the tube; *leg.* smooth, 4-seeded; *lfts.* 4—6, small, linear; *stip.* lanceolate, semi-sagittate.—① Slender and delicate plants, banks of streams, &c., Can. to Penn. Stems almost filiform, 1—2f long. Leaflets 5—10″ by 1″, acute or obtuse. Flowers very small, bluish-white, on filiform peduncles. Legumes 4—6″ long, 4, sometimes 5-seeded. Jl.

5. V. **sativa.** *Common Vetch. Tares.*
Fls. solitary or in pairs, subsessile; *lfts.* 10—12, oblong-obovate, often linear, retuse, mucronate; *stip.* semi-sagittate, subdentate, dotted; *leg.* erect, roundish, reticulated, smooth.—① A slender, climbing plant, found in cultivated fields, introduced from Europe. Stem decumbent or climbing, 2—3f long. Leaflets 8—12″ by 1—4″, lower ones near the base of the petiole. Flowers pale purple, half as long as the leaves. Legumes 1—2′ long. Jn. ☽

6. V. **Faba.** Willd. (Faba vulgaris. *Mœnch.*) *Coffee Bean. Windsor Bean, &c.*—*St.* rigidly erect, with axillary, many-flowered racemes; *lfts.* 2—4, oval, entire, mucronate or acute; *tendrils* obsolete; *stip.* semi-sagittate, dentate at base.—Native of Egypt. This species is frequently found in gardens, but not so much admired as formerly for the table. Stem simple, 1—2f high. Flowers

white. with a large black spot on each of the alæ. Legume torulose. Seeds very large, with the large hilum at one end. (See Fig. 19, 1, 2.) †

3. ERVUM.

Calyx deeply 5-cleft. the segments acute, linear, and nearly equal, about the length of the corolla ; stigma capitate, smooth ; style fili-form ; legume oblong, 2—4-seeded.—① *Lvs. abruptly pinnate, of many leaflets and a terminal tendril.*

E. HIRSUTUM. *Hairy or Creeping Vetch.*
Lf's. linear, truncate, mucronate ; *stip.* semi-sagittate, narrow ; *ped.* 3—6-flowered, shorter than the leaves ; *leg.* hirsute, 2-seeded.—A creeping weed in cultivated fields, N. Y. to S. Car. Stem very slender, 1—3f long. Leaflets 3—20, 4—8" long, hardly 1" wide, broadest above. Peduncles axillary, 3—6-flowered. Calyx segments rather shorter than the bluish-white corolla. Legumes short, with roundish. compressed brown seeds. Jn. § †

4. PISUM.

Celtic. *wn,* Lat. *pisum,* Eng. *pea,* Fr. *pois.*

Calyx segments leafy, the upper 2 shortest ; vexillum large, re-flexed : stamens 10, diadelphous (9 and 1) ; style compressed, cari-nate. villous on the upper side ; legume oblong, tumid, many-seeded ; seeds globose, with an orbicular hilum.—*Herbaceous, climbing. Lvs. abruptly pinnate, ending with branching tendrils.*

P. SATIVUM. *Common Garden Pea.*—*Lfls.* ovate, entire, usually 4 ; *stip.* ovate, semi-cordate at base, crenate ; *ped.* several-flowered.—① One of the most valuable of leguminous plants, smooth and glaucous. Stem 2—5f long, nearly simple, climbing by tendrils Leaflets 2—3' long, ¼ as wide, obtuse, mucronate. Stipules rather larger than the leaflets. Flowers 2 or more, on ax-illary peduncles, large, white. This plant has been cultivated from time im-memorial, so that its native country is unknown. There are many varieties. Jn. ‡

5. PHASEOLUS.

Lat. *phaseolus,* a little boat ; which the pods may be said to resemble.

Calyx sub-bilabiate, upper lip 2-toothed, lower 3-toothed ; keel with the stamens and style spirally twisted ; legume compressed and fal-cate, or cylindric, many-seeded ; seeds compressed, reniform.—*Her-baceous, twining or trailing. Lvs. pinnately trifoliate. Lfls. stipellate*

1. P. DIVERSIFOLIUS. Pers.
St. prostrate, diffuse, scabrous with recurved hairs ; *lfls.* angular, 2—3-lobed or entire ; *ped.* longer than the leaf, few-flowered ; *lower tooth of the cal.* longer than the tube ; *leg.* pubescent, broadly linear, cylindric.—① A creeping or climbing plant, 3—5f long, on sandy shores and prairies, Can. and U. S. Leaflets 1—2' long, ¼ as wide, with scattered hairs beneath, often variously and very obtusely lobed. Peduncles 2—8-flowered, 3—6' long. Corolla pur-plish. Legumes become black when ripe, 5—7-seeded. Aug.—Oct.

2. P. HELVOLUS. (and P. vexillatus. *Linn.*)
St. slender, twining ; *lfls.* between oblong-ovate and linear, entire ; *ped.* slender, several times longer than the leaves. few-flowered ; *leg.* straight, cylin-dric. 8—10-seeded.— ♃ Sandy fields, N. Y. to Flor. and La. Stem 3—5f long. Leaflets 1—2' by ¼—1'. Peduncles 4—8' long, 4—7-flowered. Calyx with 2 bracts at base. Corolla purplish, vexillum large, roundish. Legume 2—3' long, very narrow, subfalcate. Aug. Sept.

3. P. PERENNIS. Walt. *Wild Bean Vine.*
Twining, pubescent ; *rac.* paniculate, mostly in pairs, axillary ; *lfts.* ovate. acuminate, 3-veined ; *leg.* pendulous, falcate. broad-mucronate.- ♃ A slender, twining vine, in dry woods, Can. and U. S., common. Stem 4—7f long, somewhat branching. Leaflets 1½—3½' long, ⅔ equal width ; termina

e, lateral ones unequally enlarged at base outside, under
racemes 1—3 together, 6—12' long, loose, often unfruitful.
violet. Legume about 2' long, ½' wide, with compressed,
e seeds. July, Aug.

RMUS. Torr. & Gray.
trorsely hirsute; *lfts.* linear-oblong, not lobed, as long as the
reticulated on both surfaces; *stip.* subulate; *ped.* much lon-
; *hds.* few-flowered; *leg.* very hirsute, about 5-seeded. *T.*
Ill., *Mead.* Also Ark. and La. Stem 2—4f long, prostrate.
-5". Pods about 1' long, ½ as wide. Aug.

Pole Bean. Kidney Bean. String Bean.—St. twining;
; *rac.* solitary, shorter than the leaves; *pedicels* in pairs; *cal.*
ts at base; *leg.* pendulous, long-mucronate; *seed* reniform,
ghtly colored.—① Native of E. Indies. Universally culti-
t only for the mature fruit, but for the young pods, which
rite dish called *string beans.* Stem 5—8f long, twining
owers mostly white. July.

s. *Scarlet Pole Bean.—St.* twining; *lfts.* ovate-acute; *rac.*
he leaves; *pedicels* opposite; *cal.* longer than the 2 appressed
. pendulous; *seeds* reniform.—① Native of S. America.
wining against the sun. Flowers scarlet, numerous and
it not so generally admired as the last. July.

Lima Bean.—St. twining; *lfts.* ovate, deltoid, acute; *rac.*
es; *ped.* in pairs; *cal.* longer than its 2 bracts at base; *leg.*
somewhat lunate; *seeds* large, much compressed, purplish-
E. Indies. Stem 6—8f long. Flowers small, whitish.
ultivated. July.

arf Kidney Bean. Bush Bean. White Field Bean.—St.
ing, erect; *lfts.* broad-ovate, acute; *cal.* shorter than its 2
pendulous, compressed, rugose.—① Native of India. Stem
white. Seeds white, small, but there are many varieties

1. W. FRUTESCENS. DC. (W. speciosa. *Nutt.* Glycine frutescens. *Linn.*) St. pubescent when young, at length glabrous; *lfts.* 9—13, ovate or elliptic-lanceolate, acute, sub-pubescent; *wings* with 2 auricles at base; *ova.* glabrous.—An ornamental, vigorous vine, in rich alluvion, Southern and Western States. Stems several yards long, climbing over bushes, &c. Leaflets 1—2' by ⅓—1. Flowers nearly as large as those of the sweet pea, numerous, in racemes 3—6 or 8' long, sheathed in very conspicuous bracts. Seeds spotted. Apr. May. †

2. W. CONSEQUANA. Benth. *Chinese Wistaria.—Lfts.* 9—13, ovate-lanceolate, silky-pubescent; *rac.* terminal, nodding, loosely many-flowered.—A splendid flowering vine from China. Stem of rapid growth, 12' or more in length. Flowers in long, pendulous clusters. May. Jn. †

8, GALACTIA.

Gr. γαλα, milk: alluding to the juice of some of the species.

Calyx bibracteolate, 4-cleft, the segments of nearly equal length, upper one broadest; pet. oblong: vexillum broadest and incumbent; keel petals slightly cohering at top.—*Herbs prostrate or twining, sometimes shrubby. Lvs. pinnately trifoliate. Rac. axillary.*

1. G. GLABELLA. Michx. St. mostly prostrate, nearly glabrous; *lfts.* elliptic-oblong, emarginate at each end, sub-coriaceous, shining above, a little hairy beneath; *rac.* pedunculate, about the length of the leaves; *fls.* pedicellate.—In arid soils, N. J. to Flor. Stem 2—4f long. Leaflets 10—20'' by 5—10'', varying in form from elliptic through oblong to ovate. Flowers rather large, reddish-purple, greenish externally. Aug. Sept.

2. G. MOLLIS. Michx. St. mostly twining, softly pubescent; *lfts.* oval, obtuse, nearly smooth above, softly villose and whitish beneath; *rac.* longer than the leaves, pedunculate, fasciculate; *fls.* on very short pedicels; *leg.* villose.—Dry soils, Md. to Ga. Stem several feet long. Leaflets about 1' long, ½ as wide. Flowers about half as large as in the last. Aug.

9. CLITORIA.

Calyx bibracteolate, tubular, 5-toothed, segments acuminate; vexillum large, spreading. roundish, emarginate; keel smaller than the wings, acute, on long claws; legume linear-oblong, torulose, many seeded.—♃ *Mostly twining. Lvs. pinnately 3—5-foliate. Fls. very large, solitary or several together.*

C. MARIANA. Glabrous; *st.* suberect or twining, suffruticose; *lfts.* 3, oblong-ovate or lanceolate, obtuse, lateral ones petiolulate; *ped.* short, 1—3-flowered; *bracteoles* and *bracts* very short; *leg.* torulose, 3—4-seeded.—Dry soils, N. J.! to Flor. Stem 1—3f long, round, slender, branched. Leaflets rather remote, about 1' by ½'. Corolla pale blue, 2—2½' in length, calyx ½', bracteoles 2''. Jl. Aug.

10. AMPHICARPÆA. Ell.

Gr. αμφι, around, καρπος, fruit: in reference to the ovary sheathed at base.

Calyx tubular, campanulate, 4-toothed (or 5-toothed, the upper 2 united) with nearly equal segments; petals oblong; vexillum with the sides appressed; stigma capitate: ovary on a sheathed stipe; legume flat, 2—4-seeded.—① *Slender, twining. Lvs. pinnately trifoliate. The upper fls. complete, but usually barren, the lower apetalous and fruitful.*

A. MONOICA. Nutt. (Glycine monoica. *Linn.*) *Pea Vine.* St. hairy; *lfts.* ovate, acute, smooth · *rac.* of the stem with pendulous

flowers; *radical ped.* with apetalous, fertile flwers.—A
woods and thickets, Can. and U. S. Stem twining, rough
length. Leaflets very thin, 1—3' long, ⅓ as wide, lateral
. Racemes axillary, few-flowered. Flowers pale purple.
noothish, with 3—4 dark purple seeds. Radical legumes
with one large, compressed, brown seed. Jl.—Sept.

11. ROBINIA.

In memory of John Robin, herbalist to Louis XIV.

mpanulate, 5-cleft, the 2 upper segments more or
exillum large; alæ obtuse; stamens diadelphous
rded inside; legume compressed, elongated, many-
*d shrubs with stipular spines. Lvs. unequally pin-
, in axillary racemes.*

CACIA. *Locust Tree.*
ned with stipular prickles; *lfts.* ovate and oblong-ovate;
oth, as well as the legumes.—Native in Penn. and the more
ern States, and abundantly naturalized in N. Eng. In the
and lightness of its timber, and the beauty of its foliage
:eeded by few trees of the American forest. West of the
imes attains the height of 80f with a diameter of 3 or 4.
dom exceeds half these dimensions. The pinnate leaves
ametry of form, each composed of 8—12 pairs of leaflets,
d. These are oval, thin, nearly sessile, and very smooth.
us, pendulous clusters, diffusing an agreeable fragrance.
ith 5 or 6 small brown seeds. When young, the tree is
hich disappear in its maturity. May.

nt. *Clammy Locust.*—*Stipular spines* very short; *branch-*
glandular-viscid; *lfts.* ovate; *rac.* crowded.—This beauti-

T. Virginiàna. Pers. (Galega. Linn.) *Goa.'s R.u. Cal-gut.*
Erect, villous; *lfs.* numerous, oblong, acuminate; *rac.* terminal, subsessile; *leg.* falcate, villous.—♃ Plant 1—2f high, with beautiful white and purple flowers, found in dry sandy soils, Can., Ia., Ill., S. to Flor. Stem simple. Leaflets 15—27, 10—13" by 2—3", mucronate, straight-veined, odd one oblong-obcordate, petiolules 1" long. Stipules subulate, ½' long, deciduous. Flowers large, in a dense, terminal raceme. Calyx very villous. Banner white, keels rose-colored, wings red. Jl.

14. PSORALEA.

Gr. ψωραλεος, leprous or scaly; alluding to the glandular dots.

Calyx 5-cleft, campanulate, segments acuminate, lower one longest; stamens diadelphous, rarely somewhat monadelphous; legume as long as the calyx, 1-seeded, indehiscent.—♃ *or* ♄ *Often glandular. Lvs. various. Stip. cohering with the base of the petiole. Fls. cyanic.*

1. P. FLORIBUNDA. Nutt.
Canescent, much branched, destitute of glands; *lvs.* palmately 3—5-foliate; *lfts.* oblong-obovate, varying to linear; *stip.* setaceous; *rac.* slender, 40—50-flowered, twice longer than the leaves; *pedicels* as long as the flowers and longer than the small, ovate, acuminate bracts; *vex.* roundish; *leg.* smooth.—Alluvial soil, Ill. *Mead.!* and Ark. W. to the Rocky Mts. Stem 2—4f high, the branches spreading. Leaflets 1—2' by 2—4", common petiole ½—1' long. Flowers bluish purple, nearly as large (3" long) as in the two following. Jn.

2. P. ESCULENTA. Ph.
Hirsute, erect, branching; *lvs.* palmately 5-foliate, *lfts.* lanceolate; *spikes* axillary, dense; *cal. seg.* lanceolate, a little shorter than the corolla; *leg.* ensiform, beaked; *rt.* thick and fusiform.
β. (P. ESCULENTA. *Nutt.*) Nearly acaulescent; *lfts.* oblong-obovate.—Mo. near the lead mines. Stem a few inches high. Leaflets 1—2' long, nearly half as wide. Flowers pale blue. The root is about 1' diam., rather insipid, but is eaten by the Indians, either raw or boiled. Jn. Jl.

3. P. EGLANDULÒSA. Ell. (P. melilotoides. *Michx.*)
St. much branched; *lfts.* oblong-lanceolate, finely dotted with glands; *spikes* oblong; *bracts* broadly-ovate, acuminate, and with the calyx hairy; *leg.* roundish, transversely wrinkled.—Dry soils, Ia.! to Ark. Slender, 2f high, spreading. Leaflets 2—2½' long, ⅓ as wide, obtuse, longer than the petioles. Flowers blue. Pods 2" diam. Jn. Jl.

4. P. ONOBRỲCHIS. Nutt.
Pubescent; *lfts.* ovate, acuminate; *rac.* elongated; *cal.* much shorter than corolla, teeth small, obtuse, equal; *leg.* ovate, transversely wrinkled.—Low grounds and thickets Western States! Stem rigidly erect, nearly simple, 3—5f high. Leaflets 2—4' long, nearly ⅓ as wide. Flowers small, pedicellate, blue. Pods exceeding the calyx, rostrate. Jn. Jl.

15. AMORPHA.

Gr. a, privative, μορφη, form; alluding to the deficiencies of the corolla.

Calyx subcampanulate, 5-cleft; vexillum concave, unguiculate, erect; wings and keel 0; stamens exserted; legume oblong, somewhat curved at the point, scabrous with glandular points, 1—2 seeded. —*Shrubs or half-shrubby American plants. Lvs. unequally pinnate, punctate. Fls. bluish white, in virgate racemes.*

1. A. FRUTICÒSA.
Pubescent or nearly glabrous, shrubby or arborescent; *lfts.* 9—13, oval, petiolulate, very obtuse, the lower pair remote from the stem; *cal. teeth* obtuse, short, lower one acuminate and rather the longest; *leg.* 2-seeded.—A shrub or small tree, 6—16f high, Wis. *Lapham!* to La. and Flor., W. to Rocky Mts. Leaves 3—5' long, leaflets about 1' by ½', rather remote from each other and

from the stem, petiolules scarcely 2'' long. Spicate racemes terminal, solitary or fascicled, 3—4' long. Vexillum purple, emarginate. Jn.

2. A. CANESCENS. Nutt. *Lead Plant.*
Suffruticose and canescently villose ; *lfts.* small, numerous, and crowded, ovate-elliptical, subsessile, mucronate ; *spikes* aggregate ; *fls.* subsessile ; *cal. teeth* equal, ovate, acute ; *vex.* bright blue ; *leg.* 1-seeded.—A beautiful species, 2—4f high, in dry, sandy soils ! Wis. to La. and Rocky Mts., and is supposed to prefer localities of lead ore. Leaves 2—3' long. Leaflets coriaceous, 16—24 pairs, obtuse at base, 4—6'' by 1—2'. Spikes 2—3' long. Jl. Aug.

16. DALEA.

In honor of Thomas Dale, an English botanist of the last century.

Calyx subequally cleft or toothed ; petals unguiculate, claws of the wings and keel adnate to the staminate tube half way up ; vexillum free, the limb cordate ; stamens 10, united into a cleft tube ; ovary 2-ovuled ; legume enclosed in the calyx, indehiscent, 1-seeded.—*Mostly herbaceous and glandular-punctate. Lvs. odd-pinnate. Stipels 0, stipules minute, setaceous. Spikes mostly dense.*

D. ALOPECUROIDES. Willd. (D. Linnæi. *Michx.* Petalostemon. *Ph.*)
Glabrous and much branched ; *lfts.* 8—14 pairs, linear-oval, obtuse or retuse, mucronate, punctate beneath ; *spikes* pedunculate, oblong-cylindric, terminal, silky ; *bracts* about equaling the acuminate segments of the calyx.—① Prairies and bottoms, Ill. ! Mo., Car. Plant about 2f high, bushy and leafy and pale green. Leaflets not more than 4'' by 1'', sessile, and nearly in mutual contact. Spikes 1—2' long. Vexillum white, wings and keel pale violet.

17. PETALOSTEMON. Michx.

Alluding to the union of the petals and stamens.

Calyx 5-toothed, nearly equal ; petals 5, on filiform claws, 4 of them nearly equal, alternate with the stamens and united with the staminate tube ; stamens 5, monadelphous ; tube cleft ; legume 1-seeded, indehiscent, included in the calyx.—4 *Lvs. unequally pinnate, ex-stipellate. Fls. in dense, pedunculate, terminal spikes or heads.*

1. P. CANDIDUM. Michx. (Dalea. *Willd.*)
Glabrous, erect ; *lfts.* 7—9, all sessile, linear-lanceolate, mucronate, glandular beneath ; *spikes* on long peduncles ; *bracts* setaceous ; *vex.* broadly cordate, the other petals ovate.—A fine-looking plant, in dry prairies Southern and Western States ! Stem 2—4f high, sparingly branched, slender. Leaflets 9—18'' by 3—5'', terminal one largest. Flowers small, white, crowded in dense spikes which are 1—3' long. Jl.

2. P. VIOLACEUM. Michx. (Dalea. *Willd.*)
Minutely pubescent, erect ; *lfts.* 5, linear, glandular beneath ; *spikes* pedunculate, oblong or subglobose ; *vex.* cordate, the other petals oblong, obtuse at base.—A beautiful plant, of similar habits with the last. Stem slender, striate, subsimple, 1½—2f high. Leaflets about 1' by 1'', all sessile. Spikes 1—8, very dense, ½— 1½' long. Petals of a bright violet purple. Jl. Aug.

18. TRIFOLIUM. Tourn.

Gr. τριφυλλον, (three-leaved) ; Lat. *trifolium* ; Fr. *trefle* ; Eng. *trefoil.*

Calyx tubular or campanulate, 5-toothed, persistent ; petals more or less united at the base, withering ; vexillum reflexed ; alæ oblong, shorter than the vexillum ; carina shorter than the alæ ; stamens 10, diadelphous (9 & 1) ; legume short, membranous, mostly indehiscent, covered by and scarcely longer than the calyx, 2—4-seeded ; seeds roundish.—*Herbs. Lvs. palmately trifoliate ; lfts. with straight, scarcely reticulated veins. Flowers in dense heads or spikes.*

* *Heads not involucrate. Flowers pedicellate, deflexed when old.*

1. T. REPENS. *Creeping* or *White Clover* or *Trefoil.* *Shamrock.*
St. creeping, diffuse; *lfts.* obcordate, denticulate; *stip.* narrow, scarious, *hds.* subumbellate, on very long, axillary peduncles; *leg.* about 4-seeded; *cal. teeth* shorter than the tube.—♃ In all soils, mountainous, meadow or rocky, throughout N. Am. Stems several from the same root, extending 6—12', rooting at the joints. Peduncle angular, much longer than the leaves. Flowers white. May—Sep.—Highly valued for pasturage.

2. T. REFLEXUM. *Buffalo Clover.*
Pubescent; ascending or procumbent; *lfts.* obovate or oblong-obovate, serrulate, some of them emarginate; *stip.* leafy, semi-cordate; *hds.* many-flowered; *leg.* about 4-seeded—♃ ? Prairies and meadows, Western! and Southern States. Stem 8—16' high. Leaflets subsessile, 7—8'' by 4—5''; petioles 1—2' long. Heads large and handsome. Peduncles 1—3' long. Vexillum rosered. Apr.—Jn.

3. T. STOLONIFERUM. Muhl. *Running Buffalo Clover.*
Glabrous, creeping; branches axillary, ascending, short; *lfts.* broadly obcordate, denticulate; *stip.* leafy, ovate-lanceolate, acuminate; *fls.* loose, umbellate-capitate; *leg.* about 2-seeded.—♃ Fields and woods, Western States! Stems 6—12' long, several together. Branches 3—4' high, generally with one head, which is 1' diam. Leaflets 6—10'' by 5—9''. Flowers white, erect, but in fruit all reflexed. May, Jn.

* * *Heads not involucrate. Flowers never deflexed nor yellow.*

4. T. ARVENSE. *Hare's-foot Trefoil.*
Hds. cylindrical, very hairy; *calyx teeth* setaceous, longer than the corolla; *lfts.* narrow-obovate.—① A low plant, abundant in dry, sandy fields. Stems much branched, round, hairy, 6—12' high. Leaves hairy, on short petioles, of 3 narrow leaflets, ½—1' long. Stipules ovate-lanceolate, acute, often red. Heads of white or pale red flowers, spiked, ½—1½' long, very soft and downy, the slender, equal calyx teeth being densely fringed with fine, silky, reddish hairs, and projecting far beyond the corolla. Jl. Aug. Common in N. Eng.

5. T. PRATENSE. *Common Red Clover.* (Fig. 43, 7.)
Spikes dense; *sts.* ascending; *cors.* unequal; *lower teeth* of the calyx longer than the four others, which are equal; *lfts.* oval, entire.—② This is the common red clover, so extensively cultivated in grass lands, with herds-grass (*Phleum pratense*) and other grasses, and often alone. Stems several from the same root, hairy. Leaves ternate, the leaflets ovate, lighter colored in the centre, entire and nearly smooth. Stipules ovate, mucronate. Flowers red, in short, ovate spikes or heads, sweet-scented. Corollas monopetalous. Flowers all summer. §

6. T. MEDIUM. *Zig-Zag Clover.*
St. suberect, branching, flexuous, nearly glabrous; *lfts.* oblong or elliptical, subentire; *stip.* lanceolate, acuminate; *hds.* of *fls.* ovoid-globose, pedunculate; *cal. teeth* setaceous, hairy.—♃ In meadows, Danvers, Mass. *Oakes.* Heads of flowers larger than in T. pratense. Corollas deep purple. Leaves of a uniform green. §

7. T. INCARNATUM. *Flesh-colored Clover.*—St. erect, flexuous; *lfts.* ovate-orbicular, obtuse or obcordate, sessile, crenate, villous; *spikes* dense, oblong, obtuse, leafless; *cal. teeth* setaceous, villous.—① A fine species from Italy, occasionally cultivated as a border flower, and has been proposed (*Dr. Dewey,* Rep. Herb. Pl. Mass.) for cultivation as a valuable plant for hay.

* * * *Heads not involucrate. Flowers never deflexed, yellow.*

8. T. PROCUMBENS. *Yellow Clover* or *Trefoil.*
St. procumbent or ascending; *lfts.* obovate-cuneate, or obovate-orbicular, obtuse or retuse, denticulate, terminal one petiolulate; *stip.* ovate-lanceolate, acuminate, much shorter than the petioles; *hds.* small. subglobose, on shortish peduncles; *cor.* yellow; *sty.* 3 or 4 times shorter than the 1-seeded legumes.—① In dry soils, N. H.! to Va. Stems many from the same root, slender, more or

one often subcordate, lateral ones unequally enlarged at base outside, under surface scabrous. Racemes 1—3 together, 6—12' long, loose, often unfruitful. Corolla purple and violet. Legume about 2' long, ¼' wide, with compressed, reniform, dark purple seeds. July, Aug.

4. P. LEIOSPERMUS. Torr. & Gray.
St. slender, retrorsely hirsute ; *lfts.* linear-oblong, not lobed, as long as the petiole, hirsute and reticulated on both surfaces; *stip.* subulate ; *ped.* much longer than the leaves; *hds.* few-flowered ; *leg.* very hirsute, about 5-seeded. *T. & G. abr.*—Prairies, Ill., Mead. Also Ark. and La. Stem 2—4f long, prostrate. Leaflets 1—2' by 3—5''. Pods about 1' long, ¼ as wide. Aug.

5. P. VULGARIS. *Pole Bean. Kidney Bean. String Bean.*—*St.* twining; *lfts.* ovate acuminate ; *rac.* solitary, shorter than the leaves; *pedicels* in pairs ; *cal.* as short as its 2 bracts at base ; *leg.* pendulous, long-mucronate ; *seed* reniform, variously, often brightly colored.—① Native of E. Indies. Universally cultivated in gardens, not only for the mature fruit, but for the young pods, which constitute that favorite dish called *string beans.* Stem 5—8f long, twining against the sun. Flowers mostly white. July.

6. P. MULTIFLORUS. *Scarlet Pole Bean.*—*St.* twining; *lfts.* ovate-acute ; *rac.* solitary, as long as the leaves ; *pedicels* opposite ; *cal.* longer than the 2 appressed bracts at base; *leg.* pendulous; *seeds* reniform.—① Native of S. America. Stem 6—10f long, twining against the sun. Flowers scarlet, numerous and very brilliant. Fruit not so generally admired as the last. July.

7. P. LUNATUS. *Lima Bean.*—*St.* twining; *lfts.* ovate, deltoid, acute; *rac.* shorter than the leaves; *ped.* in pairs; *cal.* longer than its 2 bracts at base; *leg.* scimetar-shaped, or somewhat lunate; *seeds* large, much compressed, purplish-white.—Native of E. Indies. Stem 6—8f long. Flowers small, whitish. Much valued and cultivated. July.

8. P. NANUS. *Dwarf Kidney Bean. Bush Bean. White Field Bean.*—*St.* smooth, very branching, erect ; *lfts.* broad-ovate, acute; *cal.* shorter than its 2 oracts at base; *leg.* pendulous, compressed, rugose.—① Native of India. Stem 1f high. Flowers white. Seeds white, small, but there are many varieties. Much cultivated. June.

6. APIOS.

Gr. name for the wild pear, which the root resembles in form.

Calyx campanulate, obscurely bilabiate, the upper lip of 2 very short, rounded teeth, the 2 lateral teeth nearly obsolete, the lower one acute and elongated ; keel falcate, pushing back the broad, plicate vexillum at top ; ovary sheathed at base.—2♃ *Twining, smooth Root bearing edible tubers. Leaves pinnately 5—7-foliate.*

A. TUBEROSA. Ph. (Glycine Apios. *Linn.*) *Ground Nut.*
St. twining ; *lvs.* pinnate, of 7 ovate-lanceolate leaflets; *rac.* shorter than the leaves.—Thickets and shady woods, Can. and U. S., twining about other plants. Stem round, 2—4f in length. Leaves rather numerous, each consisting of 3 (rarely 2) pairs of leaflets and an odd, terminal one. These are ovate, narrow, more or less pointed, smooth, on short pedicels. Racemes axillary, solitary, 1—3' long, crowded. Flowers dark purple. To the root are appended oval, fleshy tubers, which are very nutritious, and would perhaps be cultivated had we not the potato. Jl., Aug.

7. WISTARIA. Nutt.

In memory of Caspar Wistar, M. D., President of Am. Phil. Soc.

Calyx bilabiate, upper lip emarginate, the lower one 3 sub-equal teeth ; vexillum with 2 callosities ascending the claw and separating above ; wings and keel falcate, the former adhering at top ; legume torulose ; seeds many, reniform —*Twining, shrubby plants, with pinnate leaves. Rac. large, with large, colored bracts. Fls.* lilac-colored

1. W. FRUTESCENS. DC. (W. speciosa. *Nutt.* Glycine frutescens. *Linn.*) *St.* pubescent when young, at length glabrous; *lfts.* 9—13, ovate or elliptic-lanceolate, acute, sub-pubescent; *wings* with 2 auricles at base; *ova.* glabrous.—An ornamental, vigorous vine, in rich alluvion, Southern and Western States. Stems several yards long, climbing over bushes, &c. Leaflets 1—2' by ½—1. Flowers nearly as large as those of the sweet pea, numerous, in racemes 3—6 or 8' long, sheathed in very conspicuous bracts. Seeds spotted. Apr. May. †

2. W. CONSEQUANA. Benth. *Chinese Wistaria.*—*Lfts.* 9—13, ovate-lanceolate, silky-pubescent; *rac.* terminal, nodding, loosely many-flowered.—A splendid flowering vine from China. Stem of rapid growth, 12f or more in length. Flowers in long, pendulous clusters. May. Jn. †

8, GALACTIA.

Gr. γαλα, milk: alluding to the juice of some of the species.

Calyx bibracteolate, 4-cleft, the segments of nearly equal length, upper one broadest; pet. oblong: vexillum broadest and incumbent; keel petals slightly cohering at top.—*Herbs prostrate or twining, sometimes shrubby. Lvs. pinnately trifoliate. Rac. axillary.*

1. G. GLABELLA. Michx. *St.* mostly prostrate, nearly glabrous; *lfts.* elliptic-oblong, emarginate at each end, sub-coriaceous, shining above, a little hairy beneath; *rac.* pedunculate, about the length of the leaves; *fls.* pedicellate.—In arid soils, N. J. to Flor. Stem 2—4f long. Leaflets 10—20' by 5—10'', varying in form from elliptic through oblong to ovate. Flowers rather large, reddish-purple, greenish externally. Aug. Sept.

2. G. MOLLIS. Michx. *St.* mostly twining, softly pubescent; *lfts.* oval, obtuse, nearly smooth above, softly villose and whitish beneath; *rac.* longer than the leaves, pedunculate, fasciculate; *fls.* on very short pedicels; *leg.* villose.—Dry soils, Md. to Ga. Stem several feet long. Leaflets about 1' long, ⅔ as wide. Flowers about half as large as in the last. Aug.

9. CLITORIA.

Calyx bibracteolate, tubular, 5-toothed, segments acuminate; vexillum large, spreading. roundish, emarginate; keel smaller than the wings, acute, on long claws; legume linear-oblong, torulose, many seeded.—♃ *Mostly twining. Lvs. pinnately 3—5-foliate. Fls. very large, solitary or several together.*

C. MARIANA. Glabrous; *st.* suberect or twining, suffruticose; *lfts.* 3, oblong-ovate or lanceolate, obtuse, lateral ones petiolulate; *ped.* short, 1—3-flowered; *bracteoles* and *bracts* very short; *leg.* torulose, 3—4-seeded.—Dry soils, N. J.! to Flor. Stem 1—3f long, round, slender, branched. Leaflets rather remote, about 1' by ⅔' Corolla pale blue, 2—2½' in length, calyx ½', bracteoles 2''. Jl. Aug.

10. AMPHICARPÆA. Ell.

Gr. αμφι, around, καρπος, fruit: in reference to the ovary sheathed at base.

Calyx tubular, campanulate, 4-toothed (or 5-toothed, the upper 2 united) with nearly equal segments; petals oblong; vexillum with the sides appressed; stigma capitate; ovary on a sheathed stipe; legume flat, 2—4-seeded.—① *Slender, twining. Lvs. pinnately trifoliate. The upper fls. complete, but usually barren, the lower apetalous and fruitful.*

A. MONOICA. Nutt. (Glycine monoica. *Linn.*) Pea Vine. *St.* hairy; *lfts.* ovate, acute, smooth · rac. of the stem with penunlous peta'iferous. barren flowers; *radical ped.* with apetalous, fertile flowers.— A very slender vine, in woods and thickets, Can. and U. S. Stem twining, rough backwards, 4—8f in length. Leaflets very thin, 1—3' long, ⅓ as wide, lateral ones oblique at base. Racemes axillary, few-flowered. Flowers pale purple. Cauline legumes smoothish, with 3—4 dark purple seeds. Radical legumes often subterraneous, with one large, compressed, brown seed. Jl.—Sept.

11. ROBINIA.

In memory of John Robin, herbalist to Louis XIV.

Calyx short, campanulate, 5-cleft, the 2 upper segments more or less coherent; vexillum large; alæ obtuse; stamens diadelphous (9 & 1); style bearded inside; legume compressed, elongated, many-seeded.—*Trees and shrubs with stipular spines. Lvs. unequally pinnate. Fls. showy, in axillary racemes.*

.. R. PSEUDACACIA. *Locust Tree.*
Branches armed with stipular prickles; *lfts.* ovate and oblong-ovate; *rac.* pendulous, smooth, as well as the legumes.—Native in Penn. and the more Southern and Western States, and abundantly naturalized in N. Eng. In the durability, hardness and lightness of its timber, and the beauty of its foliage and flowers, it is exceeded by few trees of the American forest. West of the Alleghanies it sometimes attains the height of 80f with a diameter of 3 or 4. In N. England it seldom exceeds half these dimensions. The pinnate leaves have a beautiful symmetry of form, each composed of 8—12 pairs of leaflets, with one at the end. These are oval, thin, nearly sessile, and very smooth. Flowers in numerous, pendulous clusters, diffusing an agreeable fragrance. Pod narrow, flat, with 5 or 6 small brown seeds. When young, the tree is armed with thorns which disappear in its maturity. May.

2. R. VISCOSA. Vent. *Clammy Locust.*—*Stipular spines* very short; *branchlets, petioles* and *leg.* glandular-viscid; *lfts.* ovate; *rac.* crowded.—This beautiful tree is from the South, where it attains the height of 40f. The flowers numerous, rose-colored, in erect, axillary clusters, with the thick, dark green foliage, render this tree one of the most brilliant ornaments of the park or the garden. Apr.

3. R. HISPIDA. *Rose Acacia.*—*Stipular spines* almost wanting; *shrub* mostly hispid; *rac.* loose, suberect.—A beautiful shrub, native of the Southern States. It is cultivated in our gardens for the sake of its numerous, large, red flowers. Height 3—5f. May.
β. *rosea* has its branches nearly smooth.

12. COLUTEA.

Calyx 5-toothed; vexillum with 2 callosities, expanded, larger than the obtuse carina; stigma lateral, under the hooked summit of the style, which is longitudinally bearded on the back side; legume inflated, scarious.—*Shrubs with unequally pinnate leaves.*

C. ARBORESCENS. *Bladder Senna.*—*Lfts.* elliptical, retuse; *vex.* shortly gibbous behind.—A hardy, free-flowering shrub, native of Italy, &c., growing almost alone on the summits of Mt. Vesuvius. Stems 8—12f high. Leaflets about 9. Flowers large, yellow, with a broadly expanded banner. In medicine the leaves are used instead of senna. Jn.—Aug. †

13. TEPHROSIA.

Gr. τεφρος, ash-colored; in allusion to the color of the foliage.

Calyx with 5, nearly equal, subulate teeth; bracteoles 0; vexillum *large, orbicular;* keel obtuse, cohering with the wings; stamens diadelphous *(in the following species)* or monadelphous; legume linear, much compressed, many-seeded.—*Herbs and shrubs, with unequally pinnate leaves.*

T. Virginiana. Pers. (Galega. Linn.) Goa.'s Rue. Cat-gut.

Erect, villous; lf.s. numerous, oblong, acuminate; rac. terminal, subsessile; leg. falcate, villous.—⁴ Plant 1—2f high, with beautiful white and purple flowers, found in dry sandy soils, Can., Ia., Ill., S. to Flor. Stem simple. Leaflets 15—27, 10—13" by 2—3', mucronate, straight-veined, odd one oblong-obcordate, petiolules 1" long. Stipules subulate, ½' long, deciduous. Flowers large, in a dense, terminal raceme. Calyx very villous. Banner white, keels rose-colored, wings red. Jl.

14. PSORALEA.

Gr. ψωραλεος, leprous or scaly: alluding to the glandular dots.

Calyx 5-cleft, campanulate, segments acuminate, lower one longest; stamens diadelphous, rarely somewhat monadelphous; legume as long as the calyx, 1-seeded, indehiscent.—⁴ or ♄ *Often glandular. Lvs. various. Stip. cohering with the base of the petiole. Fls. cyanic.*

1. P. floribunda. Nutt.

Canescent, much branched, destitute of glands; lvs. palmately 3—5-foliate; lfts. oblong-obovate, varying to linear; stip. setaceous; rac. slender, 40—50-flowered, twice longer than the leaves; pedicels as long as the flowers and longer than the small, ovate, acuminate bracts; vex. roundish; leg. smooth. —Alluvial soil, Ill. Mead.! and Ark. W. to the Rocky Mts. Stem 2—4f high, the branches spreading. Leaflets 1—2' by 2—4", common petiole ½—1' long. Flowers bluish purple, nearly as large (3" long) as in the two following. Jn.

2. P. esculenta. Ph.

Hirsute, erect, branching; lvs. palmately 5-foliate, lfts. lanceolate; spikes axillary, dense; cal. seg. lanceolate, a little shorter than the corolla; leg. ensiform, beaked; rt. thick and fusiform.

β. (P. esculenta. Nutt.) Nearly acaulescent; lfts. oblong-obovate.—Mo. near the lead mines. Stem a few inches high. Leaflets 1—2' long, nearly half as wide. Flowers pale blue. The root is about 1' diam., rather insipid, but is eaten by the Indians, either raw or boiled. Jn. Jl.

3. P. eglandulosa. Ell. (P. melilotoides. Michx.)

St. much branched; lfts. oblong-lanceolate, finely dotted with glands; spikes oblong; bracts broadly-ovate, acuminate, and with the calyx hairy; leg. roundish, transversely wrinkled.—Dry soils, Ia.! to Ark. Slender, 2f high, spreading. Leaflets 2—2½' long, ½ as wide, obtuse, longer than the petioles. Flowers blue. Pods 2" diam. Jn. Jl.

4. P. Onobrychis. Nutt.

Pubescent; lfts. ovate, acuminate; rac. elongated; cal. much shorter than corolla, teeth small, obtuse, equal; leg. ovate, transversely wrinkled.— Low grounds and thickets Western States! Stem rigidly erect, nearly simple, 3—5f high. Leaflets 2—4' long, nearly ½ as wide. Flowers small, pedicellate, blue. Pods exceeding the calyx, rostrate. Jn. Jl.

15. AMORPHA.

Gr. α, privative, μορφη, form; alluding to the deficiencies of the corolla.

Calyx subcampanulate, 5-cleft; vexillum concave, unguiculate, erect; wings and keel 0; stamens exserted; legume oblong, somewhat curved at the point, scabrous with glandular points, 1—2 seeded. —*Shrubs or half-shrubby American plants. Lvs. unequally pinnate, punctate. Fls. bluish white, in virgate racemes.*

1. A. fruticosa.

Pubescent or nearly glabrous, shrubby or arborescent; lfts. 9—13, oval, petiolulate, very obtuse, the lower pair remote from the stem; cal. teeth obtuse, short, lower one acuminate and rather the longest; leg. 2-seeded.—A shrub or small tree, 6—16f high, Wis. Lapham! to La. and Flor., W. to Rocky Mts. Leaves 3—5' long, leaflets about 1' by ½', rather remote from each other and

cies than either of the preceding, found in woods, U. S. and Can. Stem branching, erect, 4—5f high. Leaflets 3' long, widest at base, smooth, entire, pointed. Stipules of the leaves ovate, long-acuminate, of the leaflets awl-shaped. Flowers large, purple, with conspicuous bracts. Pods in about six joints, long, pendulous, rough. Aug.

8. D. MARILANDICUM. Boott. (D. obtusum. *DC.* H. Marilandicum. *Linn.*)
Plant erect, branching, hairy; *lfts.* ovate, obtuse, subcordate at base; *stip.* subulate; *panicle* terminal; *joints of the loment* roundish, reticulate, hispid —Woods, N. States to Flor. Stem 2—3f high. Leaflets ½—1' long, ½ as wide. Flowers violet-purple, small. Loment 1—3-jointed. Aug.

9. D. CILIARE. DC. (H. ciliare. *Willd.*) *Fringed Desmodium.*
Plant erect, slender, subpubescent; *lrs.* crowded, on short, hairy petioles; *lfts.* small, ovate, short-stalked, pubescent beneath, ciliate on the margin; *stip.* filiform, caducous; *panicle* terminal, the lower branches much longer; *joints of the loment* 2 or 3, half-orbicular, hispid, reticulate.—Woods, N. Eng. to La. Height 2f. Flowers purple. Aug.

10. D. RIGIDUM. DC. (H. rigidum. *Ell.*)
Erect, branching, rough-pubescent; *lfts.* ovate-oblong, obtuse, terminal one the longest; *petiole* short, hairy; *stip.* acuminate, ciliate, caducous; *rac.* paniculate, very long; *leg.* with 2—3 semi-oval or semi-obovate joints.—Hills and woods, Mass. to La. Stem 2—3f high, often with numerous, long, erect, rigid branches. Leaflets 1—3' long, ½ as wide, rather coriaceous, reticulately veined. Flowers violet-purple. Aug.

11. D. PANICULATUM. DC. (H. paniculatum. *Linn.*)
Plant erect, smooth; *lfts.* thin, oblong-lanceolate; *stip.* subulate; *panicle* terminal, with long and slender pedicels; *bracts* lanceolate; *joints of the loment* rhomboidal.—A handsome species, near 3f in height, found in woods, U. S. and Can. Stem slender, striate. Leaves of 3, smooth, narrow leaflets, broadest at the base, tapering to an obtuse point, about 3' in length, with subulate, deciduous bracts. Pods 4—5 jointed, large. Flowers purple, numerous. Jl. Aug.

12. D. ROTUNDIFOLIUM. DC. (H. rotundifolium. *Linn.*)
St. prostrate, hairy; *lfts.* suborbicular, hairy on both sides; *bracts* broadly ovate, acuminate; *rac.* few-flowered; *joints of the loment* subrhomboidal.—A hairy, prostrate plant, 2—3f in length, found in rocky woods throughout the U. S. Leaves of 3 roundish leaflets, pale beneath ½ —2' diam., on hairy stalks. Stipules cordate, reflexed, hairy. Flowers purp.e, in axillary and terminal racemes. Pods about 6-jointed. Aug.

13. D. HUMIFUSUM. Beck. (H. humifusum. *Muhl.*) *Prostrate Desmodium.*
St. procumbent, striate, nearly smooth; *lfts.* oval, sub-pubescent; *stip.* persistent; *rac.* axillary and terminal; *leg.* of 2—4 obtusely 4-angled joints.— Woods, Waltham, Mass. *Bigelow,* Penn. *Muhl.* A species much resembling the last, but the whole plant is much smoother, with smaller and narrower bracts. Stem 2—3f long. Leaflets oval or ovate, subacute. Aug.

14. D. VIRIDIFLORUM. Beck. (Hedys. virid. *Linn.*)
St. erect, densely pubescent and scabrous above; *lfts.* ovate, mostly obtuse, scabrous above, softly villous beneath; *stip.* ovate-lanceolate, acuminate, caducous; *panicle* very long, leafless; *cal.* very hairy, upper lip bifid; *leg.* of 3—4 triangular joints.—Alluvial soils, N. Y. to Flor. and La. Stem 3—4f high, rigid, branched. Leaflets 2—3' long. Corolla violet, turning green in withering. Legume 1—2' long.

15. D. LÆVIGATUM? DC. (H. lævigatum. *Nutt.*)
Glabrous; *st.* simple, erect; *lvs.* on long petioles; *lfts.* ovate or oblong-ovate, rather obtuse; *stip.* subulate, minute and deciduous; *panicle* terminal, nearly simple; *fls.* in pairs, on elongated pedicels; *bracts* ovate, very small; *upper lip of calyx* emarginate, segments of the lower lip lanceolate, lowest one acuminate, half as long as the corolla.—Woods, N. J. *Nutt.* Harper's Ferry! The smoothest of our Desmodia, 2—3f high. Leaflets rather coriaceous, 1—2½' long, ½—1½' wide. Pedicels 5—8" long. Flowers purple. Sept.—My specimens, as well as those of Nuttall, are without fruit.

16. D. SESSILIFOLIUM. Torr. & Gray. (H. sessilifolium. *Torr.*)
St. erect, tomentose-pubescent; *lvs.* sessile; *lfls.* linear or linear-oblong, obtuse at each end, scabrous above, softly tomentose beneath; *stip.* subulate; *panicle of spicate rac.* very long; *bracts* minute; *leg.* small, hispid, of 2—3 semiorbicular joints.—Woods, Western States and Texas. Stem 2—3f high. Leaflets about 2' by ½'. Flowers small, numerous and crowded. Aug.

17. D. STRICTUM. DC. (H. strictum. *Pursh.*)
Erect, slender, nearly glabrous and simple; *lvs.* petiolate; *lfls.* linear, elongated, coriaceous and reticulately veined, mucronate; *stip.* subulate; *panicles* slender, few-flowered; *leg.* hispid, incurved, of 1—3 lunately triangular joints with a filiform isthmus.—Pine barrens, N. J. to Flor. and La. Stem about 3f high. Leaflets 2—3' by 2—3", longer than the petioles. Flowers small, purple, on very slender pedicels. Aug.

28. LESPEDÊZA. Michx.

In honor of Lespedez, governor of Florida, who protected Michaux in his travels there.

Calyx 5-parted, bibracteolate, segments nearly equal; keel of the corolla very obtuse, on slender claws; legume (loment) lenticular, compressed, small, unarmed, indehiscent, 1-seeded.—*Genus taken from Hedysarum.* 4 *Lvs. palmately trifoliate, reticulate-veined.*

§ *Flowers all complete and fertile, in dense spikes. Corolla ochroleucous or white, with a purple spot on the vexillum, scarcely longer than the calyx.*

1. L. CAPITÁTA. Mx. (L. frutescens. *Ell.* Hedysarum frutescens. *Willd.*) *Bush Clover.*—*Lfts.* elliptical, obtuse, silky-pubescent; *stip.* subulate; *fascicles of fls.* ovate, subcapitate, shorter than the leaves, axillary; *loments* hairy, shorter than the villous calyx.—An erect, hairy, half-shrubby plant, in dry soils, Can. to Car. Stem nearly simple, villous, 2—4f high. Leaves numerous, on short petioles, consisting of 3 coriaceous leaflets. Leaflets 1—1½' by 3—6", nearly smooth above, covered with silky pubescence beneath. Aug. Sept.
β. *angustifolia.* Ph. (L. angustifolia. *Ell.*)—*Lfts.* linear, smooth above.

2. L. HIRTA. Ell. (Hedysarum hirtum. *Linn.*)
Villous and pubescent; *lfts.* roundish-elliptic; *rac.* capitate, axillary, oblong, longer than the leaves; *cor.* and *loment* about as long as the calyx.—Plant 2—4f high, found in dry woods, Can. and U. S., erect, branching and very hairy. Leaves less numerous than in the last, on very short stalks, consisting of 3 oval leaflets hairy beneath. Peduncle hairy, becoming longer than the raceme. Flowers reddish-white, crowded. Aug. Sept.

§§ *Flowers of two kinds, complete and apetalous, the latter chiefly bearing the fruit. Corolla violet or purple, much longer than the calyx.* LESPEDEZARIA. T. & G.

3. L. PROCUMBENS. Michx. (Hedysarum repens. *Willd.*)
S. procumbent, villous; *lf's.* oval, upper surface smooth; *rac.* short, on very long, setaceous peduncles; *loments* roundish, pubescent.—Dry woods and sandy fields, Mass. to La. Plant pubescent in all its parts. Stems several from the same root, slender, 2—3f long. Leaves consisting of 3 oblong or roundish leaflets, on hairy stalks. Flowers purple, in short, raceme-like heads, axillary, the lower ones apetalous, and on short, the upper on very long, thread-like peduncles. Aug.

4. L. REPENS. Torr. & Gray. (H. repens. *Linn.*) *Creeping Lespedeza.*
St. prostrate, diffuse, nearly smooth; *lfls.* oval or obovate-elliptical, smooth above, on very short petioles; *ped.* axillary, filiform, simple, few-flowered, lower ones bearing apetalous flowers; *leg.* suborbicular, subpubescent.—Dry soils, Can., *Hooker*, N. J. and Southern States! Probably it will yet be found in N. Y. Stems very slender, numerous. Leaflets 5—9" by 3—5", obtuse. Peduncles 2—3' long. Aug. Sept

5. L. VIOLACEA. Pers. (H. violaceum. *Linn.*) *Violet Lespedeza.*
Erect or diffuse, branching; *lfls.* elliptic or oval-oblong, obtuse or emar-

iling the petiole, more or less pubescent beneath; rac. axillary, er ones with apetalous flowers; fls. in pairs; leg. ovate, smooth-than the calyx.—Dry woods, Can. and U. S. Root creeping as clustered, slender, 8—14' long. Apetalous flowers few, the ldom producing fruit. Leaflets 6—12" by 4—9". Petioles ollas small, violet, pedicellate. Legume rhomboidal. Jl. Aug. .. divergens. Ph.) Ped. filiform, divergent, much longer than unfruitful; leg. reticulate.—Leaflets ovate.

LIFLORA. Michx. (L. violacea, β. T. & G.) ranching, puberulent; lfts. small, oblong-oval, obtuse, mucro the petioles; fls. glomerate, on peduncles much shorter than it the base apetalous and fertile; lower segment of the calyx in ers much longer than the others; leg. orbicular-ovate, reticu-ch longer than the calyx.—Woods, Can.! to Flor., Ohio! and slender, 1—2f high, with numerous, crowded, small leaves. —6 or 8" by 1—9". Flowers numerous, mostly apetalous. diam. Aug. Sept.

JULATA. Pers. (L. violacea, γ. T. & G.) rigid, simple, glabrous; petioles nearly erect; lfts. sublinear, beneath, strongly reticulated and mucronate; fls. fasciculas peduncles; segments of the calyx of nearly equal length; leg. id, acute.—N. J.! to Ill.! and La. Stem 2f or more high, slen-ded. Leaflets 10—18' by 1½—3", a little broadest in the mid-end, upper ones smaller. Flowers all complete in some speci-is in others. Corolla violet. Legume 1½" diam. Aug.

us. Nutt. ched, tomentose-pubescent; lfts. oval or roundish, longer than axillary, many-flowered, equaling or exceeding the leaves in fls. few; leg. hairy, ovate, acuminate, longer than the subulate r soils, N. Y. to La. A variable plant, 2—3f high. Leaves

31. L U P Í N U S. Tourn.

Lat. lupus, a wolf; because it overruns the field and devours its fertility. (Doubtful)

Calyx deeply bilabiate, upper lip 2-cleft, lower entire or 3-toothed; wings united towards the summit; keel acuminate; stamens monadelphous, the filamentous sheath entire; anthers alternately oblong and globose; legume coriaceous and torulose.—*Herbs. Lrs. palmately* 5—15-*foliate.*

1. L. PERENNIS. *Common Lupine.*

Rt. creeping, perennial; *lfts.* 7—9, oblanceolate, mucronate; *fls.* alternate; *cal.* without appendages, upper lip emarginate, lower entire.—♃ Grows wild abundantly in sandy woods and hills, Lake Champlain to Wis. *Lapham!* S. to Ga. It is a beautiful plant, much cultivated in gardens. It is often called sun-dial, from the circumstance of its leaves turning to face the sun from morning till night. Stem erect, soft, smoothish, a foot high. Leaves soft, downy, on long stalks. Lfts. 1½—2′ by 4—6″, lanceolate, broadest above the middle. Flowers blue, varying to white, in a terminal spike or raceme. May, June.

2. L. POLYPHYLLUS. Lindl. *Many-leaved Lupine.*—Tall; *lfts.* 11—15, lanceolate, sericeous beneath; *fls.* alternate, in a very long raceme; *pedicels* longer than the lanceolate, deciduous bracts; *cal.* ebracteolate, both lips subentire; *lg.* densely hairy.—♃ A splendid ornament of the garden, from Oregon. Stem 3—5′ high. Racemes a foot or more long. Flowers scattered (subverticillate in *β. grandifolius*, Lindl.), white, purple or yellow in different varieties. †

3. L. NOOTKATENSIS. Donn. *Nootka Sound Lupine.*—St. villous, with long, spreading hairs; *lfts.* oblong-lanceolate, mucronate, attenuate at base, sericeous beneath; *cal.* very hairy, both lips nearly entire; *bracts* linear, hairy, longer than the calyx.—A handsome species, from the N. W. Coast, 2—3f high, in gardens. Leaflets about 7. Flowers purple. †

4. L. ARBOREUS. *Tree Lupine.*—Fruticose; *fls.* in whorls; *cal.* appendaged, lips acute, entire.—A handsome exotic shrub, 6f high, with large yellow flowers. †

Obs.—Several annual species are occasionally sown in gardens, as *L. a lius*, with white flowers; *L. pilosus*, with rose-colored flowers; *L. luteus*, with yellow flowers, and *L. h rsutus*, with blue flowers, and an appendaged calyx.

32. L A B U R N U M. Benth.

Calyx campanulate, bilabiate: upper lip 2, lower 3-toothed; vexillum ovate, erect, as long as the straight wings; filaments diadelphous (9 & 1); legume continuous, tapering to the base, several-seeded.—*Oriental thornless shrubs or trees. Lrs. palmately trifoliate. Fls. mostly yellow.*

1. L. VULGARE. (Cytisus Laburnum. *Linn.*) *Golden Chain.*—Arborescent; *lf's.* oblong-ovate, acute at base, acuminate; *rac.* simple, elongated, pendulous; *leg.* hirsute.—A small, ornamental tree, 15f high, from Switzerland. Flowers numerous, large, in racemes 1f long. †

2. L. ALPINUM. (Cytisus alpinus. *Linn.*) *Scotch Laburnum.*—Arborescent; *lfts.* oblong-ovate, rounded at base; *rac.* long, simple, pendulous; *leg.* glabrous. ′—A beautiful tree, 30f high, native of various alpine regions of Europe. Like the former, it develops numerous, brilliant yellow flowers, in long, drooping clusters.—There are varieties with ochroleucous, white, and even purple flowers. †

33. B A P T I S I A. Vent.

Gr. βαπτω, to dye; a use to which some species are applied.

Calyx 4—5-cleft half-way, persistent; petals of about equal length, somewhat united; vexillum orbicular, emarginate; stamens 10, distinct, deciduous: legume inflated, stipitate, many (or by abortion few)-seeded.—♃ *Lrs. palmately 3-foliate, or simple.*

1. B. TINCTORIA. R. Br. (Sophora. *Linn.* Podalyria. *Lam.*) *Wild Indigo*
Glabrous, branching; *lrs.* palmately 3-foliate, subsessile; *lfts.* roundish-obovate, acute at base, very obtuse at apex; *stip.* setaceous, caducous; *rac.*

loose, terminal; *leg.* subglobose.—A plant with bluish-green foliage, frequent in dry soils, Can. and U. S. Stem very bushy, about 2f high. Leaflets about 7″ by 4—6″, emarginate, petiole 1—2″ long. Flowers 6—12 or more in each raceme. Petals 6″ long, yellow. Legume about as large as a pea, on a long stipe, mostly 1-seeded. Jl.—Sept.

2. B. LEUCOPHAEA. Nutt. *Ochroleucous Baptisia.*
Villous; *petioles* almost 0; *lfts.* oblanceolate, varying to obovate; *stip.* and *bracts* large, triangular-ovate, persistent; *rac.* secund, with numerous flowers drooping on long pedicels; *leg.* ovoid or roundish, inflated.—Dry, rich soil, Southern! and Western States! Stem 2—3f high, smoothish when old. Leaflets 2—3′ by ⅓—2′, stipules more than half as large. Raceme 40—60-flowered. Pedisels 1—2′ long. Corollas very large, ochroleucous. Apr.

3. B. LEUCANTHA. Torr. & Gray. (B. alba. *Hook.*) *White-flowered Bapt.*
Glabrous and glaucous; *lrs.* on short petioles; *lfts.* cuneiform-obovate, obtuse; *rac.* long, erect; *bracts* caducous; *leg.* inflated, stipitate.—Very conspicuous in prairies, &c., Mich. Ia.! to Ark. Stem thick, 2—3f high, branches about 3, towards the summit. Racemes terminal, of large, white flowers, 6—24′ long, showy. Leaflets 1—2 long, ⅓ as wide, turning bluish-black in drying. Jn. Jl.

4. B. AUSTRALIS. R. Br. (B. cœrulea. *Nutt.*) *Blue-flowered Baptisia.*
Glabrous; *petioles* short; *lfts.* obovate, or somewhat oblong, obtuse; *stip.* lanceolate, rather longer than the petioles, distinct at base; *rac.* long, erect; *bracts* caducous; *ped.cels* rather shorter than the calyx; *leg.* oblong-oval, stipe long as the calyx.—Alluvial soils, Ohio river, *Clark!* Harper's Ferry! to Ga. and La. Stem 2—3f high, branched. Petioles 1—6″ long. Leaflets 1½—3′ by ⅓—1′, sometimes acute. Stipules ½—1′ long. Flowers indigo-blue, large. Pod about 2′ long. Jn.—Aug.

34. CERCIS.
Gr. κερκις, a weaver's shuttle; from the form of the legumes.

Calyx broadly campanulate, 5-toothed; petals scarcely papilionaceous, all distinct; wings longer than the vexillum and smaller than the keel petals; stamens 10, distinct: legume compressed, with the seed-bearing suture winged; seeds obovate.—*Trees with simple, cordate leaves and rose-colored flowers.*

C. CANADENSIS. *Judas Tree. Red-bud.*
Lvs. broadly ovate-cordate, acuminate, villous on the veins beneath.—A handsome tree, 20—30f high, Mid. and W. States. The wood is finely veined with black and green, and receives a fine polish. Leaves 3—4′ by 4—5′, entire, smooth, 7-veined, on petioles 1—2′ long. The flowers appear in advance of the leaves, usually in abundance, in small, lateral clusters. Corolla bright purple. May.—The young twigs will dye wool a nankeen color. The old author Gerarde in compliance with the popular notion of his time, says " This is the tree whereon Judas did hang himself, and not on the elder tree, as it is said."

SUBORDER 2.—CÆSALPINÆ.
Corolla not papilionaceous, irregular. Stamens 10 or fewer, all distinct.
35. CASSIA.
From the Hebrew word Katzioth.

Sepals 5, scarcely united at base, nearly equal; petals 5, unequal, but not papilionaceous; stamens 10, distinct; 3 upper anthers often sterile, 3 lower ones beaked; legume many-seeded.—*Trees, shrubs or herbs. Lvs. simply, abruptly pinnate.*

1. C. MARILANDICA. *American Senna.*
Plant smooth; *lfts.* 6—9 pairs, oblong-lanceolate, mucronate, an obovoid gland near the base of the common petiole; *fls.* in axillary racemes and terminal panicles.—2⅟ This beautiful plant is frequently met with in alluvial soils, (*U. S.*) growing in close masses, 3—6f high. Stem round, striate, often with

scattered hairs. Petioles channeled above, and distinguished by the pedicelled gland near the base. Leaflets 1—2' by 4—9''. Racemes in the upper axils, forming a leafy panicle. Petals bright-yellow, 3 erect and 2 declined. In medicine it is a mild cathartic. Aug.

2. C. Chamæcrista. *Sensitive Pea. Dwarf Cassia.*

St. erect or decumbent; *lfts.* 8—12 pairs, oblong-linear, obtuse, mucronate; *gland* on the petiole subsessile; *fascicles of flowers* supra-axillary, subsessile; *anthers* 10, all fertile.—① An elegant plant, in dry soil, Mass. Mid. W. and S. States. Stem ½—2f high, round, pubescent. Leaflets crowded, 4—8'' by 1—2½'', smooth, subsessile. Flowers large, 2, 3 or 4 in each fascicle. Bracts lance-subulate, as are also the stipules, persistent. Petals bright yellow, the 2 upper ones with a purple spot. Aug.—The leaves possess considerable irritability.

3. C. nictitans. *Wild Sensitive Plant.*

St. erect or procumbent; *lfts.* 6—15 pairs, oblong-linear, obtuse, mucronate, sessile; *gland on the petiole* slightly pedicellate; *fls.* small, 2 or 3 in each supra-axillary, subsessile fascicle; *sta.* 5, subequal.—In dry sandy soils, Mass. to La. Stem about 1f long, slender, a little branching. Leaflets crowded, 4—6'' by 1—2'', common petiole 1—2' long, with the gland a line or two below the lowest pair of leaflets. Flowers very small, pale yellow, on short pedicels. Jl. —The leaves are quite sensitive, closing by night and when touched.

36. GYMNOCLÁDUS. Lam.

Gr. γυμνος, naked, κλαδος, a shoot; for its coarse, naked shoots in winter.

Flowers ♀ ♂. ♂ Calyx tubular, 5-cleft, equal; petals 5, inserted into the summit of the tube; stamens 10, distinct. ♀ Calyx and corolla as above; style 1: legumes 1-celled, oblong, very large, pulpy within.—*A slender, unarmed tree, with unequally bipinnate lvs. Lfts. ovate, acuminate.*

G. Canadensis. Lam. *Coffee Tree.*

Grows in Western N. Y., Ohio, Ia.! &c., on the borders of lakes and rivers. Height 50f, with a trunk 15' diam., straight and simple to the height of 25f, covered with a rough, scaly bark, and supporting a rather small, but regular head. The compound leaves are 2—3f long, and 15—20' wide, being doubly compounded of a great number of dull green leaflets. Single leaflets often occupy the place of some of the pinnæ. Flowers large and white, succeeded by large, curving pods containing several hard, gray seeds. The wood is reddish, fine-grained and strong, and is valuable in architecture, and cabinet-work. May—Jl.

37. GLEDITSCHIA.

In honor of John G. Gleditsch, a botanical writer, Leipzig, about 1750.

Flowers ♀ ♀ ♂. Sepals equal, 3—5, united at base; petals 3—5; stamens 3—5, distinct, opposite the sepals, sometimes by abortion fewer or 0; style short; legume continuous, compressed, often intercepted between the seeds by a quantity of sweet pulp.—*Trees, with supra-axillary, branched spines. Lvs. abruptly pinnate and bipinnate often in the same specimen.*

G. triacanthus. *Honey Locust.*

Branches armed with stout, triple spines; *lfts.* alternate, oblong-lanceolate, obtuse; *leg.* linear-oblong, compressed, intervals filled with sweet pulp.— This fine tree, native from Penn. to Mo. and La., is becoming common in cultivation. In favorable circumstances it attains the height of 70f, undivided half its length, with a diameter of 3—4f. The thorns with which its branches are armed in a most formidable manner, are 2—3' long, ligneous, often having 2 secondary ones branching from the sides. Foliage light and elegant. Leaflets about 15, 1—1½' long, ¼ as wide, 1, 2 or 3 of them frequently transformed, either partly or wholly, into smaller leaflets (§ 240, 6). Flowers small, white, succeeded by flat, *crooked*, hanging pods 12—18' long, of a dull red. Seeds flat, hard, brown, imbedded in a fleshy substance, at first sweet but becomes sour. Jn.

Suborder 3.—MIMOSEÆ.

Sepals and petals valvate in æstivation, subregular. Stamens 5—200
Embryo straight.—Leaves abruptly pinnate or bipinnate.

38. MIMOSA.

Gr. μιμος, a buffoon; the leaves seems sporting with the hand that touches them.

Flowers ♀ ☿ ♂. ♀ Calyx 5-toothed; corolla 0, or 5-toothed; sta.
4—15; legume separated into 1-seeded joints; ♂ like the perfect, but
without ovaries or fruit.—♃ *Hbs. and shrubs, natives of tropical Amer. &c*

M. PUDÍCA. *Sensitive Plant.*—*St.* prickly, more or less hispid; *lvs.* digitate-
pinnate; *pinnæ* 4, of many (20 or more) pairs of linear leaflets.—Native of Brazil. Stem shrubby, about a foot high. Leaflets about 3″ long, very numerous
Flowers small, capitate.—It is occasionally cultivated for the interest excited
by its spontaneous motions,—the leaves bending, folding, and apparently shrink-
ing away from the touch of the hand.

39. SCHRANKIA. Willd.
In honor of Francis de Paula Schrank, a German botanist.

Flowers ☿ ♂; calyx minute, 5-toothed; petals united into a funnel-
shaped, 5 cleft corolla; stamens 8—10, distinct or monadelphous; legume
echinate, dry, 1-celled, 4-valved, many-seeded;—♃ *Prickly herbs. St.
procumbent. Lvs. sensitive, bipinnate. Fls. in spherical heads, purplish.*

S. UNCINÀTA. Willd. (Mimosa horridula. *Michx.*) *Sensitive Brier.*
St. angled, grooved; *pinnæ* 6—8 pairs; *lfts.* numerous, minute, elliptical,
reticulated beneath; *hds.* solitary, on peduncles shorter than the leaves; *leg.*
very densely clothed with prickles.—Dry soils, Clark Co., Mo. *Mead*, and
Southern States. Stem 2—4f long, and with the petioles and peduncles armed
with short, sharp prickles turned downwards. Leaflets about 2″ by ¼″. Peduncles 2—3′ long, heads ½—¾′ diam. May—Jl.

40. DARLINGTONIA. DC.
In honor of Hon. Wm. Darlington, of Penn., author of Flora Cestrica, &c.

Flowers ☿; calyx campanulate, 5-toothed; petals 5, distinct; sta
mens 5, distinct; style filiform; stigma minute, funnel-shaped;
legume lanceolate, dry, 2-valved, 4—6-seeded.—♃ *Unarmed and glabrous herbs. Lvs. abruptly bipinnate; lfts. very numerous. Fls. white, in
axillary, pedunculate heads.*

D. BRACHYLÒBA. DC. (Desmanthus. *Benth.*)
ε. *Illinoensis.* T. & G. (Mimosa Illinoensis. *Michx.*) *Pinnæ* 6—11
pairs, with a gland between the lowest pair only; *stems* numerous, diffuse; *leg.*
slightly falcate.—Prairies and bottoms, Ill. to La. Stems 2—3f high, simple,
striate. Leaves 2—4′ long. Leaflets linear-oblong, subfalcate, obtuse at each
end, 2¼″ by ¼″. Legumes crowded, ¾′ long. Jn. Aug.—This genus is reduced
by *Bentham* to Desmanthus, *Willd.*, but there are numerous genera based on
less important distinctions than this; *e. g.* Vicia and Lathyrus.

Order XLVIII. ROSACEÆ.

Trees, shrubs or herbs. Leaves alternate.
Stipules usually large or conspicuous, sometimes none.
Fls. regular, commonly showy, rarely diœcious.
Cal.—Sepals 5, rarely fewer, united, often reinforced by as many bracts.
Cor.—Petals 5, regular, rarely wanting, inserted on the disk which lines the orifice of the calyx.
Sta. ∞, usually numerous, arising from the calyx, distinct.
Ova. superior, 1 or several, distinct, 1-celled, often coherent to the sides of the calyx and each other.
Styles distinct or united. *Fruit* a drupe, pome, achenia or follicle.

This order, as here constituted, consists of three suborders, which by Lindley are regarded as separate
orders; viz. *Amygdaleæ*, *Pomeæ* and *Rosaceæ* proper, to which is added *Chrysobalaneæ*, not represented
in this flora. The genera and species in each suborder are estimated by Lindley as follows:

	genera		species	
Chrysobalaneæ,	11	"	50	"
Amygdaleæ,	5	"	110	"
Pomeæ,	16	"	200	"
Rosaceæ proper,	50	"	6??	"
Total,	82	"	8??	"

A large proportion of these are natives of temperate climates north o the equator.

Properties.—A highly important order, whether we regard its delicious fruit, its medicinal products, or the beauty of its flowers. None of its species (excepting those of the Almond tribe) are unwholesome. An astringent principle characterizes the family, residing chiefly in the bark and the root. The roots of the blackberry have been used in medicine as an astringent; those of Gillenia, as an emetic; Agrim... art, as a vermifuge. The petals of Rosa damascena yield the well known fragrant oil, called attar of rose. The Almond, Peach, &c., abound in prussic acid, a deadly poison, residing chiefly in the kernels.—Of the Rosaceæ, as ornamental flowering shrubs, it is scarcely necessary to speak.

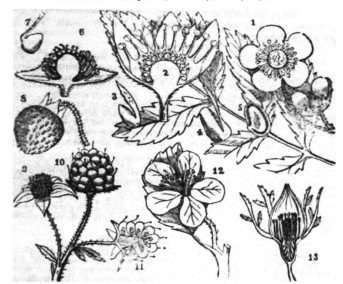

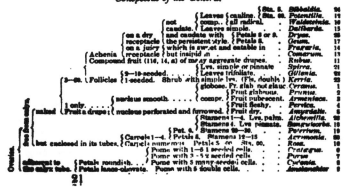

FIG. 44.—1. Potentilla argula, flower and leaf. 2. Vertical section of a flower with the petals removed showing the perigynous disk, stamens, ovaries, &c. 3. Enlarged ovary and style. 4. Mature ovary. 5. Section showing the seed and funiculus. 6. Vertical section of a flower of Fragaria, showing the perigynous stamens, the ovaries, &c. 7. Enlarged carpel. 8. Fruit, consisting of the enlarged receptacle with the achenia external. 9. Perigynous stamens of Rubus Idæus. 10. Fruit, the fleshy carpels aggregated. 11. Section of the fruit. 12. Flower of the apple tree. 13. Vertical section of a rose, showing the distinct carpels in the calyx tube.

Conspectus of the Genera.

SUBORDER I.—AMYGDALEÆ.

Ovary solitary. Fruit a drupe. Seeds mostly solitary. Calyx deciduous

1. CERÁSUS. Juss.

Name from *Cerasus*, a town in Pontus, whence originated the garden cherry.

Calyx 5-cleft, regular, deciduous; petals much spreading; stamens 15—30; drupe globose, succulent, very smooth, destitute of a glaucous bloom; nucleus subglobose, smooth.—*Trees or shrubs. Lvs. conduplicate in æstivation.*

* Flowers in racemes.

1. C. SEROTINA. DC. (C. Virginiana. *Michx.* Prunus. *Ehrh.*) *Black or Wild Cherry.*—*Lvs.* deciduous, oval-oblong, acuminate, unequally serrate, smooth, shining above; *petioles* with 2—4 glands; *rac.* spreading, elongated.—A large forest tree, throughout the U. S. Trunk 50—80f high, of uniform size and undivided to the height of 20—30f, 2—4f diam. Bark black and rough. Leaves 3—5' long, ½ as wide, with 1—2 pairs of reddish glands at base. In May and June it puts forth numerous cylindric clusters of white flowers. Fruit nearly black when mature, bitterish, yet pleasant to the taste, and is greedily devoured by birds.—The wood, extensively used in cabinet-work, is compact, fine-grained, and receives a high polish. The bark has a strong, bitter taste, and has been used in medicine as a tonic.

2. C. VIRGINIÀNA. DC. (C. serotina. *Hook.* Prunus. *Linn.*) *Choke Cherry.* *Lvs.* smooth, sharply serrate, oval, deciduous, the lower serratures glandular, veins bearded on each side towards the base; *petiole* with 2 glands; *rac.* lax, short, spreading; *pet.* orbicular.—A small tree or shrub, 5—20f high, in woods and hedges. Bark grayish. Leaves 2—3' long, ½ as wide, with a short, abrupt acumination, and spreading, subulate serratures. Flowers white, appearing in May. The fruit (cherries) is abundant, of a dark red color, very astringent to the taste, yet on the whole agreeable.

** Flowers subumbellate or solitary.

3. C. PENNSYLVANICA. Ait. (Prunus borealis. *Ph.*) *Wild Red Cherry.* *Lvs.* oblong-ovate, acuminate, finely serrate, membranous, smooth; *umbels* corymbose, with elongated pedicels; *drupe* small, ovoid-subglobose.—A small tree, common in woods and thickets in the Northern States. The trunk rarely exceeds 25f in height, with a diameter of 6—8'. Bark smooth, reddish-brown. Leaves 2—5' long, ½ as wide, the fine teeth mostly glandular, apex tapering to a long acumination. Flowers white, on long (2½') slender pedicels collected into a sort of umbel. Fruit red, very acid.—This tree is of rapid growth, and quickly succeeds a forest-clearing if neglected. May.

4. C. PUMÍLA. Michx. (Prunus depressa. *Ph.*) *Sand Cherry.* *Lvs.* lanceolate, oval or obovate, acute, subserrate, smooth, paler beneath; *umbels* few-flowered, sessile; *drupe* ovoid.—A small, trailing shrub, in gravelly soils, Can. and U. S. Branches ascending, 1—2f high. Leaves 2—3' long, ½ as wide, very acute at each end. Flowers white, 3, 4 or 5 in each umbel, the pedicels smooth, 1' in length. Fruit small, dark red, acid but agreeable to the taste. May.

5. C. AVIUM. Mœnch. (Prunus. *Linn.*) *Duke Cherry. Ox-heart. English Cherry. Bigareau, &c.*—*Branches* erect or ascending; *lvs.* oblong-obovate, recumirate, hairy beneath; *umbels* sessile, with rather long pedicels; *drupe* ovoid-globose, subcordate at base.—Cultivated in gardens, fields, &c., common. Trunk 20—50f in height, with an oblong or pyramidal head. Leaves 3—6' long, ½ as wide, on petioles 1—2' long, often with 2 glands. Flowers expanding with the leaves, white. Drupes various shades of red, firm but juicy. May.—The fruit is well known and appreciated. About 75 varieties are published in American catalogues. ‡

6. C. VULGÀRIS. Mill. (Prunus Cerasus. *Linn.*) *Sour Cherry. Large Red Cherry. Morello, &c.*—*Branches* spreading; *lvs.* ovate-lanceolate, acute at apex, narrowed at base, nearly smooth; *umbels* subsessile, with short pedicels; *drupe*

globose.—A smaller tree than the preceding, much cultivated. Trunk 15—20f high, with a roundish, compact head. Branches slender. Leaves 2—3' long, ⅔ as wide, unequally serrate, on petioles ⅓ as long, with 2 glands. Flowers white, expanding sooner than the leaves, 2 or 3 from each bud, on pedicels ⅓' long. Fruit large, various shades of red, acid or subacid. Apr.—In Prince's Catalogue, 1844, these two species are transposed (perhaps by mistake). About 125 varieties are there published, of which 50 belong to the present species. ‡

2. PRUNUS. Tourn.

Calyx 5-cleft, regular, deciduous; petals much spreading; stamens 15—30; ovary 2-ovuled; drupe ovate, fleshy, smooth, generally cov ered with a glaucous bloom; nucleus compressed, smooth.—*Small trees or shrubs. Lvs. convolute in vernation.*

1. P. AMERICÀNA. Marsh. (Cerasus nigra. *Loisel.*) *Red Plum. Yellow Plum.*
Somewhat thorny; *lvs.* oblong-oval and obovate, abruptly and strongly acuminate, doubly serrate; *drupes* roundish-oval, reddish-orange, with a thick, coriaceous skin.—Hedges and low woods, U. S. and Can., often cultivated for its sweet, pleasant fruit, which is about the size of the damson. Shrub 10—15f high. Leaves 2—3' long, ⅔ as wide, petioles ⅓—½' long, mostly with 2 glands at the summit. Flowers preceding the leaves, 3—4 in each of the numerous umbels, white. Drupes nearly destitute of bloom, ripe in Aug. Flowers in May. ‡

2. P. INSITITIA. *Wild Bullace Tree.*
Lvs. ovate-lanceolate or oblanceolate, tapering to the petiole, acute, serrate, pubescent-villous beneath; *branches* somewhat spiny; *fls.* naked, generally in pairs; *cal.* segments entire, obtuse; *pet.* obovate; *fruit* globular.—A European shrub or small tree, 15—20f high, naturalized " on the banks of Charles River, in Cambridge, road-sides at Cohasset, and other places in the vicinity of Boston." *Emerson*, Rep. trees and shrubs of Mass. The leaves and flowers are from separate, but adjacent buds, the former 1—1½' long, with short petioles. Petals white. Fruit black, covered with a yellowish bloom. §

3. P. MARÍTIMA. Wang. (P. littoralis. *Bw.*) *Beach Plum*
Lvs. oval or obovate, slightly acuminate, sharply serrate; *petioles* with 2 glands; *umbels* few-flowered; *pedicels* short, pubescent; *fr.* nearly round.—A small shrub, abundant on the sea-beach, particularly on Plum Island! at the mouth of Merrimac river. Very branching. Leaves 1—3' long, downy-canescent beneath when young, becoming at length nearly smooth. Flowers white, 2—5 in each of the numerous umbels. Fruit globular, eatable, red or purple, little inferior in size to the common garden plum, ripe in Aug., Sept. Fl. in May.

4. P. SPINÒSA. *Black Thorn. Sloe.*—Branches thorny; *fls.* solitary; *cal.* campanulate, lobes obtuse, longer than the tube; *lrs.* pubescent beneath, obovate-elliptical, varying to ovate, sharply and doubly dentate; *drupe* globose.—Hedgerows and cultivated grounds, Penn. *Pursh.* A thorny shrub, 12—15f high, native of Europe. §

5. P. CHICÀSA. Michx. (Cerasus. *DC.*) *Chickasaw Plum.*—Branches spinose; *lrs.* oblong-lanceolate or oblanceolate, glandular-serrulate, acute, nearly smooth; *umbels* 2—3-flowered, pedicels short, smooth; *drupe* globose.—A fine fruit-shrub, native of Arkansas, &c., often cultivated. Height 8—12f, with a bushy head. Leaves 1—2' long, ⅓ as wide, petioles about ⅓' long. Flowers small, white, expanding with the leaves, in Apr. Fruit red, or yellowish-red, tender and succulent, ripe in July. There are several varieties. ⨍

6. P. DOMÉSTICA. *Common Garden Plum. Damson Pl.*—Branches unarmed; *lrs.* oval or ovate-lanceolate, acute; *pedicels* nearly solitary; *drupe* globose, oval, ovoid and obovoid.—This long cultivated tree or shrub is said to be a native of Italy. It rarely exceeds 15f in height. Leaves quite variable in form, 1—3' long, ⅔ as wide, sometimes obtuse, on petioles about 1' in length. Flowers white, generally but one from a bud, expanding while the leaves are but half grown, in Apr. and May. Fruit black, varying through many colors to white, covered with a rich glaucous bloom, ripe in Aug. About 150 varieties are published in the catalogues of American gardeners. ‡

3. ARMENĬĀCA. Tourn.
Named from Armenia, its native Country.

Calyx 5-cleft, deciduous; petals 5; drupe succulent, pubescent; nucleus compressed, smooth, margins sulcate, one obtuse and the other acute.—*Small trees. Lvs. convolute in æstivation.*

1. A. vulgàris. Lam. (Prunus Armeniaca. *Willd.*) *Common Apricot.—Lvs.* broadly ovate, acuminate, subcordate at base, denticulate; *stip.* palmate; *fls.* sessile, subsolitary, preceding the leaves; *drupe* somewhat compressed, subglobose, large.—Occasionally cultivated in gardens, &c. Tree 10—15f high. Leaves 2—3' long, ⅔ as wide, smooth, petioles nearly 2' long, with several glands. Flowers white, Apr. Fruit purplish-yellow, &c., 1—2' diam., ripe Jl. Aug. There are about 20 varieties. †

2. A. dasycarpa. DC. (Prunus. *Ehrh.*) *Black Apricot.—Lvs.* ovate, acuminate, doubly serrate; *petioles* with 1 or 2 glands; *fls.* pedicellate; *drupe* subglobose.—This species is from Siberia. The tree or shrub is about the size of the last, hardy and thrifty. Leaves smooth above, pubescent on the veins beneath, 2—3' long, ⅔ as wide, on petioles near 1' long. Flowers white, preceding the leaves, distinctly pedicellate. Fruit dark purple when mature, in July. *Fls.* Apr. ‡ Neither species is yet common.

4. PERSĬCA. Tourn.
Named from Persia, its native country.

Calyx 5-cleft, tubular, deciduous; petals 5; drupe fleshy, tomentose or smooth; nucleus somewhat compressed, ovate, acute, rugosely furrowed and perforated on the surface.—*Small trees. Lvs. conduplicate in æstivation.*

1. P. vulgàris. Mill. (Amygdalus Persica. *Willd.*) *Common Peach.—Lvs.* lanceolate, serrate, with all the serratures acute; *fls.* solitary, subsessile, preceding the leaves; *drupe* tomentose.—Tree or shrub, 8—15f high. Leaves 3—5 long, ⅔ as wide, smooth, petioles short, with 1 or 2 glands. Flowers rose-color with the odor of Prussic acid. Fruit large, 1—2½' diam., yellowish, tinged with purple, densely tomentose.—About 200 varieties of this delicious fruit are now named and described in the catalogues of American nurserymen. In order to attain its proper flavor in the Northern States, the peach requires protection in the spring months. The double-flowered peach is a highly ornamental variety, blossoming in May. ‡

2. P. lævis. (Amygdalus Persica. *Willd.*) *Nectarine.—Lvs.* lanceolate, serrate, the serratures all acute; *fls.* solitary, subsessile, appearing before the leaves; *drupe* glabrous.—Closely resembles the peach tree in form, foliage and flowers. The fruit is 1—3' diam., smooth, yellow, purple, red, &c. Of its numerous (about 25)varieties, about a fourth are *clingstones,*—flesh adhering to the stone, and the remainder *freestones* or *clearstones,*—flesh free, or separating from the stone.‡

5. AMYGDĂLUS. Willd.

Calyx 5-cleft, campanulate, deciduous; petals 5; drupe not fleshy, compressed; nucleus perforate and furrowed, ovate, compressed, one edge acute, the other broad-obtuse.—*Trees or shrubs. Lvs. conduplicate in æstivation.*

1. A. commūnis. Willd. *Almond.—Lvs.* lanceolate, serrate, with the lower serratures glandular; *fls.* sessile, in pairs, appearing before the leaves.—From Barbary. Scarcely cultivated in this country for the fruit, which we receive mostly from S. Europe. A double-flowered variety is highly ornamental in shrubberies. †

2. A. nana. *Dwarf single-flowering Almond.—Lvs.* ovate, attenuate at base, simply and finely serrate; *fls.* subsessile, appearing before the leaves.—A very ornamental shrub, from Russia. Height about 3f, branching. Leaves 3—6 long, ⅔ as wide, smooth, acuminate at each end. Flowers numerous. Petals *oblong obtuse,* rose-colored, often double. May, Jn. †

3. A. PUMILA. *Dwarf double-flowering Almond.*—Lvs. lanceolate, doubly serrate; *fls.* pedicellate.—Native of China. A low shrub, highly ornamental, common in cultivation. Stems 2—3f high, branching. Leaves 3—5′ by ½—1′, acute at each end, smooth. Flowers very numerous, clothing the whole shrub in their roseate hue, while the leaves are yet small. May, Jn. †

SUBORDER 2—POMEÆ.

Ovaries 2—5 (rarely 1,) cohering with the sides of the persistent calyx and with each other. Fruit a pome.

6. CRATÆGUS.

Gr. *σφατος*, strength; on account of the firmness of the wood.

Calyx urceolate, limb 5-cleft; petals 5; stamens ∞; ovaries 1—5, with as many styles; pome fleshy, containing 1—5 bony, 1-seeded carpels, and crowned at the summit by the persistent calyx and disk.— *Trees or shrubs, armed w th thorns. Lvs. simple, often lobed. Bracts subulate, deciduous. Fls. corymbose.*

1. C. COCCINEA. (C. Crus-galli. *Bw.* C. glandulosa. *Willd.*) *Crimson-fruited Thorn. White Thorn.*—Lvs. broadly ovate, acutely serrate and sub (9)-lobed, thin and smooth, subacuminate, abrupt at base; *petioles* long, slender, and (with the calyx) smooth and subglandular; *sty.* 3—5.—A thorny shrub or small tree, 10—20f high, in thickets by streams, &c., Can. and U. S. Branches crooked and spreading, branchlets and thorns whitish. Thorns stout, rigid, sharp, a little recurved, about 1½′ long. Leaves 1½—2½′ long, ⅔ as wide, lobed, or (rather) coarsely, doubly acuminate-serrate. Petioles very slender, ½ as long as the lamina. Flowers white, in paniculate, lateral corymbs of about 12. Fruit 3—5″ diam., bright purple, eatable in Sept. *Fls.* May.

2. C. CRUS-GALLI. (Mespilus. *Lam. &c.*) *Cock-spur Thorn.* Lvs. obovate-cuneiform or oblanceolate, subsessile, serrate, coriaceous, shining above; *spines* very long; *corymbs* glabrous; *sep.* lanceolate, subserrate; *sty.* 1 (2 or 3).—Hedges and thickets, Can. and U. S., rare. Shrub 10—20f high, much branched. Thorns 2—3′ long, straight, sharp, and rather slender. Leaves 1—2½′ long, ½—⅔ as wide, tapering and entire at base, mostly obtuse at apex; petioles 1—5″ long. Flowers white, fragrant, in corymbs of about 15, on very short, lateral branchlets. Fruit pyriform, dull red, 2—3″ diam., persistent during winter, unless eaten by birds. Jn.

β. *pyracanthifolia.* Ait.—Lvs. oblong-lanceolate, petioles ½′ long.

3. C. PUNCTATA. Jacq. (C. latifolia. *DC.* Mespilus. *Spach.*) *Thorn.* Lvs. cuneiform-obovate, doubly and often incisely serrate, entire at base and narrowed to a petiole, veins straight and prominent, pubescent beneath; *corymbs* and *cal.* villose-pubescent; *sty.* 3 (1 or 2); *fr.* globose, punctate.—Borders of woods, U. S. and Can. Tree 12—25f high. Branches wide-spreading, crooked, covered with cinerous bark. Thorns stout, sharp, 1—2′ long, sometimes wanting. Leaves 1½—2½′ long, ⅔ as wide, acute or short acuminate; petioles ½—1′ long. Flowers white, in somewhat leafy, compound corymbs of 8—15. Fruit 5—8″ diam., red or yellowish, eatable in Sept. *Fl.* May, Jn.

4. C. TOMENTOSA. (C. pyrifolia. *Ait.* C. lobata. *Bosc.* C. flava. *Hook.*) *Black Thorn.*—Lvs. oval, or elliptic-ovate, narrowed at base into a margined petiole, subplicate, incisely and doubly serrate, smoothish above, tomentose beneath; *corymbs* large, tomentose when young; *sty.* 3—5; *fr.* pyriform.— Thickets and hedges, Can. S. to Ky. and Car.—A large shrub, 12—15f high, armed with sharp thorns 1—2′ long. Leaves 3—5′ long, ½—⅔ as wide, acute at apex; margined petiole ½—1′ long. Fls. large, fragrant, white, in a leafy corymb of 8—12. Fruit 4—6″ diam., orange-red, eatable but rather insipid. May, Jn.

β. (Torr. & Gray.) Lvs. strongly plicate, nearly smooth, smaller.

5. C. OXYCANTHA. *Hawthorn. English Thorn.* Lvs. obovate or broad ovate, obtuse, 3—7-lobed, serrate, smoothish, shining above; *stip._* large, incisely dentate; *corymbs* glabrous; *sty.* 1—3; *fr.* ovoid,

21*

small.—Hedges, &c., sparingly naturalized. Shrub very branching, 8—18f high. Thorns slender, very sharp, axillary, ½' long. Leaves 1½—2' long, nearly as wide. lower ones deeply lobed; petioles ¼—1' long, with 2 leafy stipules at base. Flowers white. Fruit 2—3" diam., purple.—Used for hedges (extensively in Europe). There are several varieties. § ‡

6. C. PARVIFLÒRA. Ait. (C. tomentosa. *Michx.* Mespilus laciniata. *Walt.*)

Thorns slender; *lvs.* coriaceous, pubescent, cuneate-obovate, subsessile, incisely serrate; *fls.* subsolitary; *cal.* with the *pedicels* and *branchlets* villous-tomentose; *sep.* laciniate, foliaceous; *sty.* 5; *fr.* large, roundish-obovoid, with 5 bony, 1-seeded nuts.—Sandy woods, N. J. and Southern States. A much branched shrub, 4—7f high. Leaves 1—2' by ½—1', the upper surface shining and nearly glabrous when old. Fruit greenish-yellow, near ½' diam., and eatable when ripe. Apr. May.

7. C. CORDÀTA. Ait. (C. populifolia. *Walt.*) *Washington Thorn.*

Thorns glabrous and glandless; *lvs.* cordate-ovate, somewhat deltoid, acuminate, incisely lobed and serrate, with long and slender petioles; *sep.* short; *sty.* 5; *fr.* small, globose-depressed.—Banks of streams, Va. to Ga., cultivated in the Middle States for hedge-rows. Shrub 15—20f high, the branches with very sharp and slender thorns 2—3' long. Leaves often deeply 3—5-lobed, about 2' by 1½'. Pomes ¼' diam., numerous, red. Jn. § ‡

7. PYRUS.

Celtic *peren;* Anglo-Saxon *peri;* Fr. *poire:* Lat. *pyrus;* Eng. *pear.*

Calyx urceolate, limb 5-cleft; petals 5, roundish; styles 5 (2 or 3), often united at base; pome closed, 2—5-carpeled, fleshy or baccate; carpels cartilaginous, 2-seeded.—*Trees or shrubs. Lvs. simple or pinnate. Fls. white or rose-colored, in cymose corymbs.*

§ *Leaves simple. Cyme simple. Styles united at base.*

1. P. CORONÀRIA. (Malus. *Mill.*) *Crab Apple. Sweet-scented Crab-tree.*

Lvs. broad-ovate, rounded at base, incisely serrate, often sublobate, smoothish, on very slender petioles; *pet.* unguiculate; *sty.* united and wooly at the base; *fr.* as well as the *fls.* very fragrant, corymbose.—Borders of woods, Mid. West. and South. States. A small tree, 10—20f high, with spreading branches. Leaves 2—3' long, ¾ as wide, resembling those of Cratægus coccinea; petioles ½—1' long. Flowers very large, rose-colored, in loose corymbs of 5—10. Fruit as large (1—1½' diam.) as a small apple, yellowish, hard and sour, but esteemed for preserves. May. ‡

2. P. ANGUSTIFOLIA. Ait. (Malus. *Michx.*)

Glabrous; *lvs.* lance-oblong, acute at base, slightly dentate-serrate, shining above; *sty.* distinct; *fr.* small.—Penn. and S. States. A tree 15—20f high, resembling the last, but with smaller leaves and fruit. Apr. May.

3. P. MÀLUS. *Common Apple Tree.*—*Leaves* ovate, or oblong-ovate, serrate, acute or short-acuminate, pubescent above, tomentose beneath, petiolate; *corymbs* subumbellate: *pedicels* and *calyx* villose-tomentose; *pet.* with short claws: *sty.* 5, united and villose at base; *pome* globose.—Native in Europe and almost naturalized here. Tree 20—25f high (in thickets 25—40). Branches rigid, crooked, spreading. Bark rough and blackish. Leaves 2—3' long, ⅔ as wide, petioles ½—1' long. Flowers expanding with the leaves, fragrant, large, clothing the tree in their light roseate hue, making ample amends for its roughness and deformity.—The Romans had 22 varieties (*Pliny*) but the number is now greatly increased. Probably nearly 1000 varieties are cultivated in the U. S. ‡

4. P. COMMÙNIS. *Pear Tree.*—*Leaves.* ovate-lanceolate, subserrate, glabrous above, pubescent beneath, acute or acuminate; *corymbs* racemose; *cal.* and *pedicels* pubescent; *sty.* 5, distinct and villose at base; *pome* pyriform.—Tree usually taller than the apple, 20—35f high. Bark rough, blackish. Branches ascending. Leaves 2—3½' long, ⅔ as wide; petioles 1—2' long. Flowers white, small.—Native in Europe, where, in its wild state, the fruit is small and impalatable. The Romans cultivated 36 varieties (*Pliny*), but, like the apple, varieties without end are now raised from the seed of this delicious fruit. ‡

§ § *Leaves simple. Cymes compound. Styles united at base.*

5. P. ARBUTIFOLIA. Linn. f. (Mespilus. *Linn.* Aronia. *Pers.*) *Choke Berry.*
Lvs oblong-obovate or oval-lanceolate, obtuse or acute, crenate-serrulate, smooth above, tomentose beneath when young, attenuate at base into a short petiole; *ped.* and *cal.* when young, tomentose; *fr.* pyriform or subglobose, dark red.—Low, moist woodlands, U. S. and Can. A shrub 5—8f high. Leaves 1 —2' long, ½ as wide, often subacuminate, subcoriaceous, serratures small, with a glandular, incurved point; petioles 2—4" long. Flowers white, in compound, terminal corymbs of 12 or more. Fruit astringent, as large as a currant. May Jn. †
β. **melanocarpa**. Hook. (P. melanocarpa. *Willd.*)—Lvs., *cal.* and *ped.* glabrous or nearly so; *fr.* blackish-purple.—Swamps. Height 2—4f.

§ § § *Leaves pinnate. Cymes compound. Styles distinct.*

6. P. AMERICANA. DC. (Sorbus Americana. *Ph.*) *Mountain Ash.*
Lfts. oblong-lanceolate, acuminate, mucronately serrate, smooth, subsessile; cymes compound, with numerous flowers; *pome* small, globose; *sty.* 3—5. —A small tree in mountain woods, N. Eng. and Mid. States. Trunk 15—20f high, covered with a reddish-brown bark. Leaves 8—12' long, composed of 9 —15 leaflets. Leaflets 2—3½' by ½—1', subopposite, often acute, on petioles 1" in length. Flowers small, white, in terminal cymes, of 50—100 or more. Fruit scarlet, 2—3" diam., beautiful. May. †
β. *microcarpa*. T. & G. (P. microcarpa. *DC.* Sorbus microcarpa. *Ph.*)— *Fr.* smaller.

7. P. AUCUPARIA. *English Mountain Ash.*—*Lfts.* as in P. Americana, except that they are always smooth on both sides, and, with the serratures, less acute at apex; *fls.* corymbose; *fr.* globose.—Native of Europe. A tree 20—40f high, often cultivated as well as the last species, for its ornamental clusters of scarlet berries. It is a tree of larger size and rougher bark than the last, but is hardly to be distinguished by the foliage, flowers or fruit. †

8. CYDONIA. Tourn.

Named for *Cydonia*, a town in Crete, from whence it was brought.

Calyx urceolate, limb 5-cleft; petals 5; styles 5; pome 5-carpeled; carpels cartilaginous, many-seeded; seeds covered with mucilaginous pulp.—*Trees or shrubs. Lvs. simple. Fls. mostly solitary.*

C. VULGARIS. Pers. (Pyrus Cydonia. *Willd.*) *Quince.*—Lvs. oblong-ovate, obtuse at base, acute at apex, very entire, smooth above, tomentose beneath; *ped.* solitary, and, with the *cal.*, woolly; *pome* tomentose, obovoid.—Shrub 8—12f (rarely 20f) high, with crooked, straggling branches. Leaves about as large as those of the pear tree. Flowers white, with a tinge of purple, large, terminal. Fruit large, lengthened at base, clothed with a soft down, yellow when ripe, highly esteemed for jellies and preserves.—The plant is reared from layers.

10. AMELANCHIER. Medic.

Calyx 5-cleft; petals 5, oblong-obovate or oblanceolate; stamens short; styles 5, somewhat united at base; pome 3—5-celled; cells partially divided, 2-seeded.—*Small trees or shrubs Lvs. simple, serrate. Fls. racemose, white.*

A. CANADENSIS. Torr. & Gray. (Mespilus. *Linn.* Aronia. *Pers.* Pyrus Botryapium. *Linn. f.* Mespilus arborea. *Michx.*) *Shad Berry. June Berry. Wild Service Berry.*—Lvs. oval or oblong-ovate, often cordate at base, acuminate or cuspidate or mucronate, sharply serrate, smooth; *rac.* loose, elongated; *seg. of the cal.* triangular-lanceolate, nearly as long as the tube; *pet.* linear-oblong or oblanceolate; *fr.* purplish, globose.—A small tree or shrub, found in woods, U. S. and British Am., rarely exceeding 35f in height. Leaves alternate, 2—3' long, downy-tomentose when young, at length very smooth on both sides, very acute and finely serrate. Flowers large, white, in terminal racemes, appearing in April and May, rendering the tree quite conspicuous in the yet naked forest. Fruit pleasant to the taste, ripening in June.

β. oblongifolia. T. & G. (A. ovalis. *Hook.*)—Shrubby; lvs. (blcng-oval, mucronate, and with small, sharp serratures; *rac.* and *fls.* smaller; *pet.* oblong-obovate, thrice longer than the calyx.

γ. rotundifolia. T. & G. (Pyrus ovalis. *Willd.*)—Lvs. broad-oval; *pet.* linear-oblong.—Shrub 10—20f high.

δ. alnifolia. T. & G. (Aronia alnifolia. *Nutt.*)—Shrubby or arborescent; lvs. orbicular-oval, rounded or retuse at each end, serrate only near the apex; *pet.* linear-oblong; *sta.* very short.

Suborder III.—R O S A C E Æ   P R O P E R.

Ovaries solitary or several, distinct; fruit achenia or follicular.

10. R O S A.

Celtic *rhos*, red; Gr. *podor*; Lat. *rosa;* Eng. *rose.*

Calyx tube urceolate, fleshy, contracted at the orifice, limb 5-cleft, the segments somewhat imbricated in æstivation, and mostly with a leafy appendage; petals 5, (greatly multiplied by culture); achenia 90, bony, hispid, included in and attached to the inside of the fleshy tube of the calyx.—*Shrubby and prickly. Leaves unequally pinnate. Stipules mostly adnate to the petiole.*

* *Native species.*

1. R. CAROLINA. (R. Caroliniana. *Bw.*) *Carolina Rose. Swamp Rose.*
   St. glabrous, with uncinate, stipular prickles; *lfts.* 5—9, oblong-lanceolate or elliptical, acute, sharply serrate, glaucous beneath, not shining above, *petioles* hairy or subaculeate; *fls.* corymbose; *fr.* depressed-globose, and with the *peduncles* hispid.—A prickly (not hispid) shrub, in swamps and damp woods, Can. and U. S., 4—8f high, erect and bushy, with reddish branches. Prickles mostly 2 at the base of the stipules. Leaflets 1—2' long, ½ as wide, rather variable in form. Flowers in a sort of leafy corymb of 3—7. Petals obcordate, large, varying between red and white. Fruit dark red. Jn. Jl.

2. R. LUCIDA. Ehrh. (R. Caroliniana. *Mx.* not *Bw.*) *Shining or Wild Rose.*
   St. armed with scattered, setaceous prickles, those of the stipules straight; *lfts.* 5—9, elliptical, imbricate, simply serrate, smooth and shining above; petioles glabrous or subhispid; *fls.* generally in pairs (1—3); *fr.* depressed-globose, and with the peduncles, glandular hispid.—Shrub 1—3f high, in dry woods or thickets throughout the U. S., slender, with greenish branches. Leaflets 1—1½' long, ½ as wide, acute or obtuse, odd one petiolate, the others sessile. Sepals often appendiculate, as long as the large, obcordate, pale red petals. Fruit small, red. Jn. Jl.

   β. T. & G. (R. parviflora. *Ehrh.*)—Lfts. ova , mostly very obtuse, paler beneath; *petioles* smooth or pubescent.

3. R. NITIDA. Willd. *Shining or Wild Rose.*
   St. low, densely armed with straight, slender, reddish prickles; *lfts.* 5—9, narrow-lanceolate, smooth and shining, sharply serrate; *stip.* narrow, often reaching to the lower leaflets; *fls.* solitary; *cal.* hispid; *fr.* globose.—In swamps. N. Eng. States. Stems 1—2f high, reddish from its dense armor of prickles. Leaflets 1—1½' long, ½ as wide, subsessile, odd one petiolulate. Stipules 5—8'' long, adnate to the petiole, each side. Flowers with red, obcordate petals. Fruit scarlet. Jn.

4. R. BLANDA. Ait. (R. gemella. *Linn.*) *Bland Rose.*
   Taller; *st.* armed with scattered, straight, deciduous prickles; *lfts.* 5—7, oblong, obtuse, serrate, smooth, but not shining above, paler and pubescent on the veins beneath, petiole unarmed; *stip.* dilated; *fls.* mostly in pairs (1—3); *fr.* globose, smooth, as well as the short peduncles.—Shrub found on dr .-, sunny hills, Northern and Middle States. Stems 2—3f high, with reddish bark. Flowers rather large. Sepals entire, shorter than the reddish, emarginate petals. Bracts large, downy. Jn.

5. R. SETIGERA. Michx. (R. rubifolia. R. Br.) *Michigan or Prairie Rose.*
   Branches elongated, ascending, glabrous; *spines* few, strong, stipular; *lfts.*

large, 3—5, ovate; *stip.* narrow, acuminate; *fls.* corymbose; *cal.* glandular, segments subentire; *sty.* united; *fr.* globose.—This splendid species is a native of Michigan, and other States W! and S. About 20 varieties are enumerated in cultivation. They are hardy, of rapid growth, and capable of being trained 12—20f. Flowers in very large clusters, changeable in hue, nearly scentless, and of short duration.

＊＊ *Naturalized species.* ＊

6. R. RUBIGINÖSA. (R. suaveolens. *Ph.*) *Eglantine. Sweet Brier.*

St. glabrous, armed with very strong, recurved prickles; *lfts.* 5—7, broad-oval, with ferruginous glands beneath; *fls.* mostly solitary; *fr.* ovoid, oval or obovoid; *ped.* glandular-hispid.—A stout, prickly shrub, 4—10f high, naturalized in fields and road-sides, throughout the U. S. The older stems are bushy, much branched, 1' diam., the younger shoots nearly simple, declined at top. Leaflets ¼—1' long, ⅔ as wide, unequally and sharply serrate, acute, bright green above, rusty beneath, and when rubbed, very fragrant. Flowers light red, 1—2' diam., fragrant. Fruit orange-red. Jn.—Of this beautiful species there are about 25 cultivated varieties, single and double.

7. R. CINNAMOMEA. *Cinnamon Rose.*

St. tall, with ascending branches; *spines of the younger stems* numerous, scattered, *of the branches* few, larger, stipular; *lfts.* 5—7, oval-oblong, rugose, cinereous-pubescent beneath; *stip.* undulate; *sep.* entire, as long as the petals; *fr.* smooth, globose.—Native of Oregon. Stem 5—12f high, with reddish bark. Flowers mostly double, purple.

＊＊＊ *Exotic species.* † *Prickles straight, mostly acerose.*

8. R. GALLICA. *Common French Rose.*—*St.* and *petioles* armed with numerous, fine, scattered prickles; *lfts.* mostly 5, elliptical or broad-oval, thick; *fls.* erect; *sel.* 5 or more, large, spreading; *sep.* ovate; *fr.* ovoid, and with the *ped.*, hispid. —The common red rose of gardens, from which have originated not less than 300 varieties, known in cultivation, and registered in catalogues, as the *velvet, carmine, carnation, &c.* Many of them are beautifully variegated, as the *tricolor* and *picotée.* The dried petals are used in medicine, and from them are extracted tinctures for cookery. Jn. Jl.

9. R. PIMPINELLIFOLIA. Ser. (R. spino-issima. *Linn.*) *Scotch* or *Burnet Rose.*—*St.* densely armed with straight, acerose prickles: *lfts.* 5—9, roundish. obtuse, smooth, simply serrate; *fls.* small, usually roseate, but changing in the numerous varieties to white, red or yellow.—Native of Scotland and other parts of Europe. These shrubs are but 2—3f high, with small, delicate leaflets. Flowers numerous, globular, very fine. May, Jn.

10. R. EGLANTERIA. Ser. (R. lutea. *Mill.*) *Yellow Rose. Austrian Eglantine* —*St.* with a cinerous bark, branches red, both armed with straight, slender, scattered prickles; *lfts.* 5—7, small, broad-oval or obovate, smooth, shining above, sharply serrate; *cal.* nearly naked and entire; *prt.* large, broad-obcordate.—From Germany. Shrub about 3f high, bushy. Flowers numerous, of a golden-yellow, very fugacious, of less agreeable fragrance than the leaves. There are many varieties, both single and double, variegated with red. Jn.

11. R. ALPINA. *Alpine or Boursault Rose.*— *Younger shoots* echinate with numerous weak prickles, *older ones* smooth, rarely armed with strong prickles *lfts.* 5—11, ovate or obovate, sharply and often doubly serrate; *stip.* narrow, apex diverging; *ped.* deflexed after flowering, and with the calyx hispid o smooth; *sep.* entire, spreading; *fr.* ovoid, pendulous, crowned with the conni vent calyx.—Hardy, vigorous, climbing. with pink, red or crimson flowers.

＊＊＊ *Exotic species.* †† *Prickles falcate, strong.*

12. R. DAMASCÈNA. *Damask Rose.*—*St.* branching and bushy, armed with unequal spines, mostly stipular, cauline ones broad, falcate or hooked; *lfts.* large, broadly elliptical, downy-canescent; *sep.* reflexed; *fr.* ovoid, elongated.—Native of the Levant. Shrub 3—4f high. Flowers rather numerous, of a delicate, pale roseate hue, usually with very numerous petals, and a delicious fragrance. Among its numerous varieties is the common *monthly*, low, blooming at all seasons.

13. R. CANINA. *Dog Rose.—Prickles* remote, strong, compressed, falcate; *lfts.* 5—9, with acute, incurved, and often double serratures; *stip.* rather broad, serrulate; *ped.* and *cal.* smooth or hispid; *sep.* after flowering, deflexed and deciduous; *fr.* ovoid, red.—Native of Europe. Shrub 4—8f high.

β. *Burboniana.* Ser.—*Lfts.* ovate, subcordate, simply dentate; *fls.* purple, double and semi-double; *pet.* concave; *sep.* entire.—A splendid class of roses, of which more than 100 varieties are cultivated. They are hardy, with ample and glossy foliage.—18 other varieties are described by Seringe in DC.

14. R. CENTIFOLIA. *Hundred-leaved* or *Provens Rose.—Prickles* nearly straight, scarcely dilated at base; *lfts.* 5—7, ovate, glandular-ciliate on the margin, subpilose beneath; *flower-bud* short-ovoid; *sep.* spreading (not deflexed) in flower; *fr.* ovoid; *cal.* and *ped.* glandular-hispid, viscid and fragrant.—From S. Europe. Shrub 2—4f high, very prickly. Flowers usually of a pink color, but varying in hue, form and size, &c., through a hundred known varieties.

15. R. MOSCHATA. *Musk Rose.—Shoots* ascending and climbing; *prickles* cautine, slender, recurved; *lfts.* 5—7, lanceolate, acuminate, smoothish, discolored; *stip.* very narrow, acute; *fls.* often very numerous; *ped.* and *cal.* subhispid; *sep.* subpinnatifid, elongated and appendiculate; *fr.* ovoid, red.—Native of ———. Stems trailing or climbing 10—12f. Flowers peculiarly fragrant, rather large, white, produced in panicles.

16. R. ALBA. *White Garden Rose.*—Slightly glaucous; *prickles* slender, recurved, sometimes wanting; *lfts.* roundish-ovate, shortly acuminate; *petioles* and *veins* subtomentose, glandular; *sep.* pinnatifid; *pet.* spreading; *fr.* ovoid, nearly smooth.—From Germany. Shrub 5—8f high. Flowers large, corymbose, sweet-scented, generally pure white, but often, in its numerous varieties, tinged with the most delicate blush.

17. R. MULTIFLORA. *Many-flowered* or *Japan Rose.—Branches, ped.* and *cal.* tomentose; *shoots* very long; *prickles* slender, scattered; *lfts.* 5—7, ovate-lanceolate, soft and slightly rugose; *stip.* pectinate; *fls.* corymbose, often numerous; *flower bud* ovoid-globose; *sep.* short; *sty.* exserted, scarcely cohering in an elongated, pilose column; *pet.* white, varying through roseate to purple.—Japan. Shrub with luxuriant shoots, easily trained to the height of 15—20f.

18. R. INDICA. *Chinese Monthly* or *Bengal Rose.*—Erect or climbing, purplish; *prickles* strong, remote; *lfts.* 3—5, ovate, acuminate, coriaceous, shining, smooth, serrulate, discolored; *stip.* very narrow; *fls.* solitary or paniculate; *ped.* often thickened, and, with the *cal.* smooth, or rugose-hispid; *sta.* inflexed; *fr.* turbinate?—Splendid varieties, blooming from Apr. to Nov. Flowers of every hue from pure white to crimson.

β. *Lawrenciana.* (R. Lawrenciana. Lindl. R. Ind. t. acuminata. Ser.) *Miss Lawrence's Rose.—St.* and *branches* aculeate, bristly and subglabrous; *lfts.* ovate, purplish beneath; *pet.* obovate-acuminate.—A class of varieties with very small flowers, pink to deep purple.

19. R. BRACTEATA. *Macartney Rose.—Branches* erect, tomentose; *prickles* recurved, often double; *lfts.* 5—9, obovate, subserrate, coriaceous, smooth and shining; *stip.* fimbriate-setaceous; *fls.* solitary, terminal; *ped.* and *cal.* tomentose; *fr.* globose, large, orange.—Varieties with cream-colored, white, to scarlet flowers.

20. R. SEMPERVIRENS. *Evergreen Rose.—St.* climbing; *prickles* subequal; *lfts.* persistent, 5—7, coriaceous; *fls.* subsolitary or corymbose; *sep.* subentire, elongated; *sty.* coherent into an elongated column; *fr.* ovoid or subglobose, yellow, and with the *ped.* glandular hispid.—Allied to the following, but its leaves are coriaceous and evergreen, persistent until January.

21. R. ARVENSIS. *Ayrshire Rose.—Shoots* very long and flexile; *prickles* unequal, falcate; *lfts.* 5—7, smooth or with scattered hairs, and glaucous beneath, deciduous; *fls.* solitary or corymbose; *sep.* subentire, short; *sty.* cohering in a long, glabrous column; *fr.* ovoid-globose, smoothish.—England. The shoots grow 15—20f in a season and are very hardy. Flowers white to blush, crimson and purple.

* * *Exotic species.* ††† *Unarmed.*

22. R. BANKSIÆ. *Banks' Rose.*—Smooth; *lfts.* lanceolate, crowded, 3—5,

scarcely serrate; *sip.* deciduous; *fs.* umbellate; *fr.* globular, nearly black.—
From China. Thornless shrubs, with small, cup-shaped flowers. Not hardy.

(ibt.—This beautiful genus includes, according to Seringe, 146 species; but the varieties produced by cultivation amount to near 1600.

11 RUBUS.
Celtic *rub*, red : the color of the fruit of some species.

Calyx spreading. 5-parted; petals 5, deciduous; stamens ∞, inserted into the border of the disk; ovaries many, with 2 ovules, one of them abortive; achenia pulpy, drupaceous. aggregated into a compound berry; radicle superior.— ♃ *Half shrubby plants. Stems usually ♂, and armed with prickles. Inflorescence imperfectly centrifugal. Fr. esculent.*

§ *Fruit inseparable from the juicy, deciduous receptacle.* BLACKBERRIES.

1. R. VILLÒSUS. Ait. *High Blackberry.* .
Pubescent, viscid and prickly; *st.* angular; *lfts.* 3—5, ovate, acuminate, serrate, hairy both sides; *petioles* prickly; *cal.* acuminate, shorter than the petals; *rac.* loose, leafless, about 20-flowered.—A well known, thorny shrub, Can. and U. S. Stems tall and slender, branching, recurved at top, 3—6f high. Leaflets 2½—4' by 1½—2½', terminal one on a long petiolule, the others on short ones or none. Pedicels slender, 1' long. Petals white, obovate or oblong, obtuse. Fruit consisting of about 20 roundish, shining, black, fleshy carpels, closely collected into an ovate or oblong head, subacid, well-flavored, ripe in Aug. and Sept.

β. frondosus. Torr. (R. frondosus. *Bw.*)—*Lfts.* incisely serrate; *rac.* with a few simple leaves or leafy bracts at base; *fls.* about 10 in each cluster, the terminal one opening first, as in all the species, the lowest next, and the highest but one last. Fruit more acid and with fewer carpels.

2. R. HISPÌDUS. (R. sempervirens. *Bw.*) *Bristly Blackberry.*
St. slender, reclining or prostrate, hispid with retrorse bristles; *lvs.* 3-foliate, rarely quinate, smooth and green both sides; *lfts.* coarsely serrate, obovate, mostly obtuse, subcoriaceous; *ped.* corymbose, many-flowered, with filiform pedicels and short bracts; *fls.* and *fr.* small.—In damp woods, Can. to Car. Stems slender, trailing several feet, with suberect branches 8—12' high. Leaflets 1—8' long, ½ as wide, nearly sessile, persistent through the winter, on a (1—3') long, common petiole. Flowers white. Fruit dusky-purple, sour. May, Jn.

β. setosus. T. & G. (R. setusus. *Bw.*)—*Lfts.* oblanceolate, rather narrow, 1½—2½' long, tapering, and (like the variety *a*) entire at base, sharply serrate above. Fruit red.

3. R. CANADENSIS. (R. trivialis. *Ph.*) *Low Blackberry. Dewberry.*
St. procumbent or trailing, subaculeate; *lvs.* 3-foliate, rarely quinate; *lfts.* elliptical or rhomboid-oval, acute, thin, unequally cut-serrate; *pedicels* solitary elongated, somewhat corymbed; *fr.* large, black.—Common in dry, stony fields Can. to Va., trailing several yards upon the ground. Leaflets light green and membranaceous, nearly sessile, 1—1½' long, ½ as wide, common petioles 1—2' long, pubescent or a little prickly. Flowers large, on slender pedicels. Petals obovate, white, twice as long as the calyx. Fruit ½—1' diam., very sweet and juicy, in July and Aug. Fl. May.

4. R. CUNEIFOLIUS. Ph. *Wedge-leaved Blackberry.*
St. erect, shrubby, armed with recurved prickles; *lvs.* 3-foliate, and with the young branches and petioles pubescent beneath; *lfts.* cuneate-obovate. entire at base, dentate above, subplicate, tomentose beneath; *rac.* loose, few-flowered.—A low shrub, 2—3f high, in sandy woods, Long Island, *Torrey,* to Flor. Petioles often prickly. Leaflets rarely 5. 1—2' long, ½ as wide, obtuse, or with a short acumination. Petals white or roseate, 3 times as long as the calyx. Fruit black, juicy, well-flavored, ripe in Jl. Aug. Fl. May. Jn.

§ § *Fruit concave beneath, separating from the dry, conical, persistent receptacle.* RASPBERRIES.

* *Leaves simple.*

5. R. ODORÀTUS. *Rose-flowering Raspberry.* *Mulberry.*
St. erect or reclining, unarmed, glandular-pilose; *lvs.* palmately 3—5-lobed.

unequally serrate; *fls.* large, in terminal corymbs; *pet.* orbicular, purple.—A fine flowering shrub, 3—5f high, in upland woods, U. S. and Brit. Am., common. Leaves 4—8' long, nearly as wide, cordate at base, lobes acuminate, petioles 2—3' long, and, with the branches, calyx and peduncles, clothed with viscid hairs. Flowers nearly 2' diam., not very unlike a rose, save the (100—200) stamens are whitish. Fruit broad and thin, bright red, sweet, ripe in Aug. *Fl.* Jn. Jl.†

6. R. CHAMÆMÓRUS. *Dwarf Mulberry. Cloudberry.*
Herbaceous; *st.* decumbent at base, erect, unarmed, 1-flowered; *lvs.* cordate-reniform, rugose, with 5 rounded lobes, serrate; *sep.* obtuse; *pet.* obovate, white.—An alpine species with us, found by *Dr. Robbins* on the White Mts. and by *Mr. Oakes* in Me. Flowers large. Fruit large, yellow or amber color, sweet and juicy, ripe in Aug. *Fl.* May, Jn.

7. R. NUTKÁNUS. Moçino. *Nootka Sound Rubus.*
St. shrubby, somewhat pilose, with glandular hairs above; *lvs.* broad, 5-lobed, unequally and coarsely serrate; *ped.* few-flowered; *sep.* long-acuminate, shorter than the very large, round-oval, white petals.—A fine species, Mich., Wis. to Oreg., &c., with very large, showy, white flowers. It has received some notice in cultivation, and a few other species of this section also. †

* * *Leaves 3—7-foliate.*

8. R. IDÆUS. *Garden Raspberry.*
Hispid or armed with recurved prickles; *lvs.* pinnately 3 or 5-foliate; *lfts.* broad-ovate or rhomboidal, acuminate, unequally and incisely serrate, hoary-tomentose beneath, sessile, odd one petiolulate; *fls.* in paniculate corymbs; *pet.* entire, shorter than the hoary-tomentose, acuminate calyx.—Many varieties of this plant are cultivated for the delicious fruit. Stems shrubby, 3—5f high. Leaflets smoothish above, 2—4' long, ⅔ as wide. Flowers white, in lax, terminal clusters. Fruit red, amber color or white.—Plants essentially agreeing with the above described were found at Cambridge, Vt., in woods, also at Colebrook, Ct., by *Dr Robbins.* ‡

9. R. STRIGÓSUS. Michx. (R. Idæus. *Nutt.*) *Wild Red Raspberry.*
Plant shrubby, strongly hispid; *lvs.* pinnately 3 or 5-foliate; *lfts.* oblong-ovate or oval, obtuse at base, coarsely and unequally serrate, canescent-tomentose beneath, odd one often subcordate at base, lateral ones sessile; *cor.* cup-shaped, about the length of the calyx.—In hedges and neglected fields, Can. and N. States, very abundant. Stem without prickles, covered with strong bristles instead. Leaflets 1½—2½' long, ⅔—⅘ as wide, terminal one distinctly petiolulate. Flowers white. Fr. hemispherical, light red, and of a peculiar rich flavor, in Jn.—Aug. *Fl.* May.

10. R. OCCIDENTÀLIS. *Black Raspberry. Thimble-berry.*
Plant shrubby, glaucous, armed with recurved prickles; *lvs.* pinnately 3-foliate; *lfts.* ovate, acuminate, sublobate or doubly serrate, hoary-tomentose beneath, lateral ones sessile; *fls.* axillary and terminal, *fr.* black.—A tall, slender bramble, 4—8f high, in thickets, rocky fields, &c., Can. and U. S. Plant not hispid. Leaflets 2—3' long, ½—⅔ as wide, nearly white beneath, odd one distinctly petiolulate, common petiole terete, long. Flowers white, lower ones solitary, upper corymbose. Fruit roundish, glaucous, of a lively, agreeable taste, ripe in July. *Fl.* May. ‡

11. R. TRIFLÓRUS. Rich. (R. saxatilis. *Dw.*) *Three-flowered Raspberry.*
St. shrubby, unarmed, declined; *branches* herbaceous, green; *lvs.* 3 or 5-foliate; *lfts.* nearly smooth, thin, rhombic-ovate, acute, unequally cut-dentate, odd one petiolulate; *stip.* ovate, entire; *ped.* terminal, 1—3-flowered; *pet.* erect, oblong-obovate.— Moist woods and shady hills, Penn. to Brit. Am. Stems flexuous, smooth, reddish. Petioles very slender, 1—2' long. Leaflets 1—2' by ½—1', lateral ones sessile, oblique or unequally 2-lobed. Petals white, rather longer than the triangular-lanceolate, reflexed sepals. Fruit consisting of a few large, dark-red grains, acid, ripe in Aug. *Fl.* May.

12. R. ROSÆFOLIUS. *Rose-leaved Rubus or Bridal Rose.*—Erect, branching, armed with nearly straight prickles; *lvs.* pinnately 3—7-foliate; *lfts.* ovate-lanceolate, subplicate, doubly serrate, smooth beneath, velvety above; *stip.*

minute, subulate; *sep.* spreading, long-acuminate, shorter than the narrow-obovate, emarginate petals; *sty.* 00.—A delicate house-plant, with snow-white double flowers. Native of Mauritius.

12. POTENTILLA.

Lat. *potentia,* power; in allusion to its supposed potency in medicine.

Calyx concave, deeply 4—5-cleft, with an equal number of alternate, exterior segments or bracteoles; petals 4—5. obcordate; stamens 00; filaments slender; ovaries collected into a head on a small, dry receptacle; styles deciduous; achenia 00.—*Herbaceous or shrubby. Lvs. pinnately or palmately compound. Fls. solitary or cymose, mostly yellow.*

* *Leaves palmately trifoliate.*

1. P. NORVEGICA. *Norwegian Potentilla or Cinquefoil.*

Hirsute; *st.* erect, dichotomous above; *lfts.* 3, elliptical or obovate, dentate-serrate, petiolulate; *cymes* leafy; *cal.* exceeding the emarginate petals.—Old fields and thickets, Arc. Am. to Car. Stem 1—4f high, covered with silky hairs, terete, at length forked near the top. Cauline petioles shorter than the leaves. Leaflets ¼—1½' by ¼—¾', (lower and radical ones very small,) often incised. Stipules large, ovate, subentire. Flowers many, crowded, with pale yellow petals shorter than the lanceolate, acute, hairy sepals. Jl.—Sept.

β.? hirsuta. T. & G. (P. hirsuta. *Michx.*)—*Hairs* loose, silky; *st.* slender, erect, sub-simple; *lower and middle lvs.* equal, long-petiolate; *lfts.* roundish-obovate, sessile, incisely dentate; *fls.* few, petals rather conspicuous, nearly as long as the calyx.—Dry fields. With reluctance I adopt the views of Torrey & Gray in regard to this plant.

2. P. TRIDENTATA. Ait. *Trident or Mountain Potentilla.*

Smooth; *st.* ascending, woody and creeping at base; *lfts.* 3, obovate-cuneat., evergreen, entire, with 3 large teeth at the apex; *cymes* nearly naked; *pet.* twice longer than the calyx.—On the White Mts.! and other alpine summits in the N. States. Flowering stems 6—12' high, round, often with minute, appressed hairs. Petioles mostly longer than the leaves. Leaflets sessile, 9—18" by 4—6", coriaceous, smooth. Flowers with white, obovate petals. Carpels and achenia with scattered hairs. Jn. Jl.

3. P. MINIMA. Haller.

St. pubescent, ascending, mostly 1-flowered; *lvs.* trifoliate; *lfts.* obovate, obtuse, incisely serrate, with 5—9 teeth above; *pet.* longer than the sepals.—Alpine regions of the White Mts. Stems numerous and leafy, 1—3' high. Leaflets with the margins and veins beneath hairy. Flowers small. Petals obcordate. Bracteoles oval-obtuse, narrowed at the base.

* * *Leaves palmately 3 or 5-foliate.*

4. P. CANADENSIS. (P. sarmentosa. *Willd.*) *Common Cinquefoil.*

Villose pubescent; *st.* sarmentose, procumbent and ascending; *lvs.* palmately 5-foliate, the leaflets obovate, silky beneath, cut-dentate towards the apex, entire and attenuate towards the base; *stip.* hairy, deeply 2 or 3-cleft, or entire; *pedicels* axillary, solitary; *bracteoles* of the calyx longer than the segments, and nearly as long as the petals.—Common in fields and thickets, U. S. and Can. Stems more or less procumbent at base, from a few inches to a foot or more in length. Flowers yellow, on long pedicels. Calyx segments lanceolate or linear. Apr.—Aug.

β. pumila. T. & G. (P. pumila. *Ph.*)—Very small and delicate, flowering in Apr. and May.—I cannot perceive any difference between this and the above, except its diminutive size and early flowering. In dry, sandy soils. Stems about 3' high.

γ. simplex. T. & G. (P. simplex. *Michx.*)—Plant less hirsute; *st.* simple, erect or ascending at base; *lfts.* oval-cuneiform; flowering in June—Aug.—In richer soils. Stems 8—14' high. Leaflets about 1' long, ½ as wide.

5. P. ARGENTEA. *Silvery Cinquefoil.*

St. ascending, tomentose, branched above; *lfts.* oblong-cuneiform, with a

22

.rge, incised teeth, smooth above, silvery-canescent beneath, sessile; *fls.*
:ymose corymb; *pet.* longer than the obtusish sepals.—A pretty species, on
r rocky hills, Can. and N. States, remarkable for the silvery whiteness of
lower surface of the leaves. Stems 6—10' long, at length with slender
.ches. Leaflets 5—9" by 1—2', with 2 or 3 slender, spreading teeth each
:; upper ones linear, entire. Flowers small. Calyx canescent. Petals yel-
l. Jn.—Sept.

⁕⁕⁕ *Leaves pinnate.*

6. P. FRUTICÒSA. (P. floribunda. *Ph.*) *Shrubby Cinquefoil.*
St. fruticose, very branching, hirsute, erect; *lfts.* 5—7, linear-oblong, all
essile, margin entire and revolute; *pet.* large, much longer than the calyx.—A
.ow, bushy shrub, in meadows and rocky hills. Northern States and Brit. Am.
Stems 1—2f high, with a reddish bark; petioles shorter than the leaves. Leaflets ½—1½' (mostly 1') by 2—3" wide, acute, crowded, pubescent. Stipules
nearly as long as the petioles. Flowers ½—1½' diam., yellow, in terminal clusters. Jn.—Aug.

7. P. ANSERÌNA. *Silver-weed. Goose-grass.*
St. slender, creeping, prostrate, rooting; *lvs.* interruptedly pinnate; *lfts.*
many pairs, oblong, deeply serrate, canescent beneath; *ped.* solitary, 1-flowered,
very long.—A fine species on wet shores and meadows, N. Eng. to Arctic Am.
Stems subterraneous, sending out reddish stolons 1—2f long. Petioles mostly
radical, 6—10' long. Leaflets 1—1½' by 3—6", sessile, with several minute
pairs interposed. Peduncles as long as the leaves. Fls. yellow, 1'diam. Jn.—Sept.

8. P. ARGÙTA. Ph. (P. confertiflora. *Hitchcock.* Boottia sylvestris. *Bw.*)
False Avens. White-flowered Potentilla.—*St.* erect; *radical lvs.* on long
petioles, 7—9-foliate, *cauline* few, 3—7-foliate; *lfts.* broadly ovate, cut-serrate;
fls. in dense, terminal cymes.—Along streams, &c., Can. and N. States, W. to
the Rocky Mts. Stems 2—3f high, stout, terete, striate, and with nearly the
whole plant very pubescent. Radical leaves 1f or more long. Leaflets 1—2'
.ong, ⅔ as wide, sessile, odd one petiolulate. Fls. about 8" diam. Petals roundish, yellowish white, longer than the sepals. Disk glandular, 5-lobed. May, Jn.

9. P. PARADOXA. Nutt. (P. supina. *Mr.*)
Decumbent at base, pubescent; *lvs.* pinnate; *lfts.* 7—9, obovate-oblong,
incised, the upper ones confluent; *stip.* ovate; *ped.* solitary, recurved in fruit;
pet. obovate, about equaling the sepals; *ach.* 2-lobed, the lower portion chiefly
composed of starch-like albumen.—River banks, Ohio to Oregon. *Nuttall* in
T. & G. Fl. p. 437.

13. COMÀRUM.
Gr. κομαρος, the strawberry tree, which this plant resembles.

Calyx flat, deeply 5-cleft, with bracteoles alternating with the segments; petals 5, very small; stamens numerous, inserted into the
disk; achenia smooth, crowded upon the enlarged, ovate, spongy, persistent receptacle.—4 *Lrs. pinnate.*

C. PALUSTRE. *Marsh Cinquefoil.*
In spagnous swamps, N. States! Wisc.! to the Arctic Circle. Stems creeping at base, 1—2f high, nearly smooth, branching. Leaflets 3, 5 and 7, crowded,
1½—2½' long, ⅓ as wide, oblong-lanceolate, hoary beneath, obtuse, sharply serrate, subsessile; petiole longer than the scarious, woolly, adnate stipules at
base. Flowers large. Calyx segments several times larger than the petals.
Petals about 3" long, ovate-lanceolate, and, with the stamens, styles and upper
surface of the sepals, dark purple. Fruit permanent. Jn.

14. FRAGARIA.
Lat. fragrans, fragrant; on account of its perfumed fruit.

Calyx concave, deeply 5-cleft, with an equal number of alternate,
exterior segments or bracteoles; petals 5, obcordate; stamens ∞;
achenia smooth, affixed to a large, pulpy, deciduous receptacle.—4
‛—iferous. Lvs. trifoliate. Fls. on a scape, white.

1. F. Virginiana. Ehrh. (F. Canadensis. *Michx.*) *Scarlet* or *Wild Strawberry.*—Pubescent ; *cal. of the fruit* erect-spreading ; *ach.* imbedded in pits in the globose receptacle ; *ped.* commonly shorter than the leaves.—Fields and woods, U. S. and Brit. Am.　Stolons slender, terete, reddish, often 1f or more long, rooting at the ends.　Petioles radical, 2—6' long, with spreading hairs. Leaflets 3, oval, ob use, coarsely dentate, subsessile, 1—2½' long, ¾ as wide, lateral ones oblique.　Scape less hairy than the petioles, cymose at top.　Flowers in Apr. and May.　Fruit in Jn. Jl., highly fragrant and delicious when ripened in the sun.

2. F. vesca. *Alpine, Wood,* or *English Strawberry.*
　Pubescent ; *calyx of the fruit* much spreading or reflexed ; *ach.* superficial on the conical or hemispherical receptacle which is without pits ; *ped.* usually longer than the leaves.—Fields and woods, Northern States, &c.　Stolons often creeping several feet.　Leaves pubescent, and flowers as in F. Virginiana.— Numerous varieties are cultivated in gardens, where the fruit is sometimes an ounce or more in weight.　Fl. Apr. May.　Fr. Jn. Jl.

3. F. Chilensis. Ehrh.　*Chili Strawberry.*—*Lfts.* villose-silky beneath, rugose, coriaceous, broadly obovate, obtuse, serrate ; *ped.* and *cal.* silky ; *pet.* large, spreading.—From Oregon and California.　Not generally cultivated.

Obs.—Other species with varieties are sometimes found in gardens ; as *F. elatior,* the hautbois S., with tall, thin leaves, tall and strong scapes, and fruit greenish white tinged with purple ; *F. grandiflora,* the pine-apple S. (made a synonym of F. Chilensis by DC.) with firm, crenate leaves, large flowers and large, globose fruit, varying from whitish to purple.

15. DALIBARDA.
Named by Linnæus, in honor of *Dalibard,* a French botanist.

Calyx inferior, deeply 5—6-parted, spreading. 3 of the segments larger ; petals 5 ; stamens numerous ; styles 5—8, long. deciduous ; fruit achenia, dry or somewhat drupaceous.—♃ *Low herbs.　St. creeping.　Lvs. undivided.　Scapes 1—2-flowered.*

D. repens. *False Violet.*
　Diffuse, pubescent, bearing creeping shoots ; *lvs.* simple, roundish-cordate, crenate ; *stip.* linear-setaceous ꝑ *cal.* spreading in flower, erect in fruit.—In low woods, Penn. to Can.　Creeping stems 1 or 2' to 10 or 12' in length.　Leaves 1—2' diam., rounded at apex, cordate at base, villose-pubescent, on petioles 1, 2 or 3' long.　Scapes 1-flowered, abou' as long as the petioles.　Petals white, obovate, longer than the sepals.　Jn.

16. WALDSTEINIA. Willd.
Named by Willdenow, in honor of *Franz de Waldstein,* a German botanist.

Calyx 5-cleft, with 5 alternate. sometimes minute and deciduous bracteoles ; petals 5 or more, sessile, deciduous ; stamens 00, inserted into the calyx ; styles 2—6 ; achenia few, dry, on a dry receptacle. —♃ *Acaulescent herbs, with lobed or divided radical lvs., and yellow fls.*

W. fragarioides. Traut.　(Dalibarda. *Michx.* Comaropsis. *DC.*) *Dry Strawberry.*—*Lvs.* trifoliate ; *lfts.* broad-cuneiform, incisely dentate-crenate, ciliate ; *scapes* bracteate, many-flowered ; *cal. tube* obconic.—A handsome plant, in hilly woods, Can. to Ga., bearing some resemblance to the strawberry.　Rhizoma thick, scaly, blackish.　Petioles 3—6' long, slightly pubescent.　Leaflets 1—3' diam., nearly sessile, dark shining green above, apex rounded and cut into lobes and teeth.　Scape about as high as the leaves, divided at top, bearing 2—6 flowers ½' diam.　Petals varying from 5—10!　Jn.

17. GEUM.
*Gr.*γευω, to taste well ; in allusion to the taste of the roots.

Calyx 5-cleft, with 5 alternate segments or bracteoles, smaller and exterior ; petals 5 ; stamens 00 ; achenia 00, aggregated on a dry receptacle, and caudate with the persistent, mostly jointed, geniculate and bearded style.—♃

les articulated and geniculate, upper joint deciduous.

.E.   *Water Avens.   Purple avens.*

 st. subsimple; *radical lvs.* lyrate; *stip.* ovate, acute; *fls.* nod-
g as the erect calyx segments; *upper joint of the persistent style*
e plant, with drooping, purple flowers, conspicuous among
meadows, Northern and Mid. States. Rhizoma woody, creep-
' high, paniculate at top. Root leaves interruptedly pinnate,
e, 4—6' long, terminal leaflet large, roundish, lobed and cre
em leaves 1—3, 3-foliate or lobed, subsessile. Flowers sub-
 purplish-brown. Petals broad-obcordate, clawed, purplish-
Jn.—The root is aromatic and astringent.

TUM. Ait.   *Yellow Avens.*
adical lvs. interruptedly pinnate; *cauline* 3—5-foliate; *lfts.* obo-
obed and toothed; *stip.* large and erect; *bracteoles* linear, shorter
pet. roundish, longer than the calyx; *sty.* smooth, upper joint
oist or dry, N. States and Brit. Am. Stem hispid at base, 2—
nous, and with spreading hairs at summit. Root leaves 5—8'
 lyrate, the terminal leaflet largest, obovate and lobed. Flow-
ther large, yellow. Receptacle densely pubescent. Jl. Aug.

INIÁNUM. (G. album. *Gmel.*)   *White Avens.*
 radical lvs. pinnate, ternate, or even rarely simple; *cauline*
bed, all unequally and incisely dentate, nearly smooth or softly
ect; *pet.* not exceeding the calyx; *sty.* glabrous; *recep.* densely
 and thickets, Can. and U. S. Stem simple or branched,
.   Leaves very variable in form, lower ones often 3-foliate,
) appendaged petioles. Stipules mostly incised. Upper leaves
ssile. Flowers rather small, white. Peduncles in fruit long
Jl.

OPHYLLUM. Willd.   *Large-leaved Yellow Avens.*
tical lvs. interruptedly lyrate-pinnate, the terminal leaflet much
lish-cordate; *cauline* with minute, lateral leaflets, and a large,

18. SANGUISORBA.
Lat. sanguis, sorbere, q. d. to absorb blood; the plant is esteemed a vulnerary.

Calyx tube 4-sided, 2 or 3-bracted at base; limb 4-parted; petals 0; stamens 4, opposite the calyx segments; filaments dilated upwards; style 1, filiform; achenium dry, included in the calyx.—*Herbs with unequally pinnate leaves.*

S. Canadensis. *Burnet Saxifrage.*
Glabrous; *lfts.* oblong, cordate, obtuse, serrate; *spikes* dense, cylindric, very long; *sta.* much longer than the calyx.—2↳ In wet meadows, U. S. and Brit. Am., and cultivated in gardens. Stem 2—3f high, smooth, striate, sparingly branched. Stipules leafy, serrate. Leaflets 2—4' long, ⅓—½ as wide, petiolate, mostly stipellate. Spikes 3—6' long, terminating the long, naked branches. Bracteoles 3. Calyx greenish white, resembling a corolla. Aug.

19. POTERIUM.
Literally a drinking vessel, and hence a beverage; from the use of the plant.

Fls. ♂. Calyx tube contracted at the mouth, 3-bracteate, limb 4-parted; petals 0; stamens 20—30; ovaries 2; style filiform; achenia dry, included in the calyx.—*Herbs with unequally pinnate leaves.*

P. sanguisorba. *Burnet.*
Herbaceous; *st.* unarmed, angular, and with the leaves, smooth; *lfts.* 7—11, ovate or roundish, deeply serrate; *spikes or hds.* subglobose, the lower flowers staminate.—2↳ Occasionally cultivated as a salad, but is now less valued in medicine than formerly. It is said by Hooker to be native about Lake Huron.

20. AGRIMONIA.
Gr. αγρος, a field, ροχος, alone; a name of dignity for its medicinal qualities.

Calyx tube turbinate, contracted at the throat, armed with hooked bristles above, limb 5-cleft; petals 5; stamens 12—15; ovaries 2; styles terminal; achenia included in the indurated tube of the calyx.
—2↳ *Lvs. pinnately divided. Fls. yellow, in long, slender racemes.*

1. A. Eupatoria. *Agrimony.*
Hirsute; *lvs.* interruptedly pinnate, upper ones 3-foliate; *lfts.* ovate, oval or oval-lanceolate, coarsely dentate; *stip.* large, dentate; *pet.* twice longer than the calyx.—Road sides, borders of fields, Can. and U. S., common. Stem 1—3f high, branching, leafy. Leaflets 3, 5, 7, with small ones interposed, nearly smooth beneath, 1½—3' long, ½ as wide, sessile, terminal one with a petiolule 1—3' long. Racemes 6—12' long, spicate. Flowers yellow, about 4" diam., in very short pedicels. Calyx tube curiously fluted with 10 ribs, and surmounted with reddish, hooked bristles. Jl.

β. *hirsuta.* Torr.—Smaller and more hairy.
γ. *parviflora.* Hook. (A. parviflora. *DC.*)—Less hairy; *fls.* smaller, on longer pedicels.

2. A. parviflora. Ait. (A. suaveolens. *Ph.*)
St. and *petioles* hirsute; *lvs.* interruptedly pinnate; *lfts.* numerous, crowded, pubescent beneath, linear-lanceolate, equally and incisely serrate; *stip.* acutely incised; *rac.* spicate-virgate; *fls.* small; *pet.* longer than calyx; *fr.* hispid.—Woods and dry meadows, Penn.! to S. Car. W. to Ia. and Tenn. Stem 3—4f high, the hairs spreading, brownish and glandular. Leaflets 2—3' by ⅓—½', with smaller ones intermixed. Petals yellow. The plant has an agreeable balsamic odor. Aug.

21. SPIRÆA.
Gr. σπειρα, a cord or wreath; the flowers are, or may be used in garlands.

Calyx 5-cleft, persistent; petals 5, roundish; stamens 10—50, exserted; carpels distinct, 3—12, follicular, 1-celled, 1—2-valved, 1—10-seeded; styles terminal.—2↳ *Unarmed shrubs or herbs. Branches and lvs. alternate. Fls. white or rose-color, never yellow.*

* *Leaves without stipules.*

ŌSA. *Hardhack.*
tomentose; *lvs.* simple, ovate-lanceolate, smoothish above,
rac. short, dense, aggregated in a dense, slender, terminal
.—A small shrub, very common in pastures and low grounds,
stem very hard, brittle, consequently troublesome to the scythe
Leaves 1½—2′ long, ½ as wide, dark green above, rusty white
mentum beneath, crowded, and on short petioles. Flowers
ous, with conspicuous stamens, light purple, forming a slen-
ster of some beauty. The persistent fruit in winter furnishes
bird. Jl. Aug.

OLIA. (S. alba. *Bw.*) *Queen of the Meadow. Meadow-sweet.*
ous; *lvs.* oblong, obovate or lanceolate, sharply serrate; *rac.*
less dense, terminal panicle; *carpels* 5.—A small shrub in
, U. S. and Brit. Am. Stems 3—4f high, slender, purplish,
moot⁾, 1½—3′ long, ¼—½ as wide, acute at each end, petiolate,
leaves in the axils. Flowers white, often tinged with red,
with conspicuous stamens, in a more or less spreading pani-

us. *Goat's Beard.*
 lvs. membranaceous, tripinnate; *lfts.* oblong-lanceolate, acu-
al ones ovate-lanceolate, doubly and sharply serrate; *fls.* ♀ ♂,
arpels 3—5, very smooth.—On the Catskill Mts., N. Y. to Ga.
Stem 4—6f high, branching. Flowers very small, white,
der racemes, forming a large, compound panicle. Jn. Jl.
ong, virgate racemes. Georgetown, D. C. *Robbins.*

ŌSA. Raf. (S. chamædrifolia. *Ph.*) *Corymbose Spiræa.*
r oblong-oval, incisely and unequally serrate near the apex,
ute tomentose beneath; *corymbs* large, terminal, pedunculate,
und, dense, often leafy; *sty.* and *carpels* 3—5.—Mountains,
Co., **Va.** *Dr. Robbins*, to Ky. S. to Flor. Stem slightly pu-

quisite beauty, in meadows and prairies, Mich. Ia. ! to Car. Stem 4—8f high. Flowers numerous and exceedingly delicate. Jn. Jl. †

9. S. FILIPENDULA. *Pride of the Meadow.*—Herbaceous, smooth; *lvs.* interruptedly pinnate; *lfts.* pinnatifidly serrate, 9—21, with many minute ones interposed; *stip.* large, semicordate, serrate; *corymb* on a long, terminal peduncle.—A very delicate herb, often cultivated. Stems 1—3f high. Leaves 3—6' long, leaflets 1—2' long, linear, the serratures tipped with short bristles. Flowers white, 4 or 5″ diam. Petals oblong-obovate. Jn.

10. S. ULMARIA. *Double Meadow-sweet.*—Herbaceous; *lvs.* 3—7-foliate, with minute leaflets interposed; *lateral lfts.* ovate-lanceolate; *terminal one* much larger, palmately 5—7-lobed, all doubly serrate, and whitish-tomentose beneath; *stip.* reniform, serrate; *panicle* corymbose, long-pedunculate.—In gardens, where the numerous white flowers are mostly double. Jl. †—Other species of this beautiful genus are sometimes cultivated.

22. GILLENIA. Mœnch.

Gr. γελαω, to laugh; on account of its exhilarating qualities.

Calyx tubular-campanulate, contracted at the orifice, 5-cleft; petals 5, linear-lanceolate, very long. unequal; stamens 10—15, very short; carpels 5, connate at base; styles terminal, follicles 2-valved. 2—4-seeded.—♃ *Herbs with trifoliate, doubly serrate leaves.*

1. G. TRIFOLIATA. Mœuch. (Spiræa. *Linn.*) *Indian Physic.*

Lfts. ovate-oblong, acuminate; *stip.* linear-setaceous, entire; *fls.* on long pedicels, in pedunculate, corymbose panicles.—In woods, western N. Y. to Ga. A handsome shrub, 2—3f high, slender and nearly smooth. Lower leaves petiolate, leaflets 2—4' long, ¼ as wide, pubescent beneath, subsessile. Flowers axillary and terminal. Petals rose-color or nearly white, 8″ by 2″. Seeds brown, bitter. Jn. Jl.—Root said to be emetic, cathartic or tonic, according to the dose.

2. G. STIPULACEA. Nutt. (Spirea. *Ph.*) *Bowman's Root.*

Lfts. lanceolate, deeply incised; *radical lvs.* pinnatifid; *stip.* leafy, ovate, doubly incised, clasping; *fls.* large, in loose panicles.—Western N. Y. to Ala. Readily distinguished from the former by the large, clasping stipules. Flowers fewer, rose-colored. Jn.—Properties of the root like the former.

23. KERRIA. DC.

In honor of Wm. Ker, a botanical collector, who sent plants from China.

Calyx of 5 acuminate, nearly distinct sepals; corolla of 5 orbicular petals; ovaries 5—8, smooth, globose; ovules solitary; styles filiform; achenia globose.—*A slender shrub, native of Japan. Lvs. simple, ovate, acuminate, doubly serrate, without stipules. Fls. terminal on the branches, solitary or few together, orange-yellow.*

K. JAPONICA. DC. (Corchorus Japonica. *Willd.*) *Japan Globe Flower.*—Common in gardens, &c. Stems numerous, 5—8f high, with a smooth bark. Leaves minutely pubescent, 2—3' by 1—1½', with a very sharp, slender point. Petioles 3—5″ long, ` Flowers double in cultivation, and abortive, globose, near 1 diam. †

24. SIBBALDIA PROCUMBENS. Linn.—Mts. of Vt. *Pursh.* Also Can. to Greenland.

25. DRYAS INTEGRIFOLIA. Vahl.—White Mts., N. H. *Pursh.* Also N. to Greenland.

26. ALCHEMILLA ALPINA. Linn.—White Mts., N. H., Green Mts., Vt., and Greenland, according to *Pursh.* These three species, whose leading characters are indicated in the "Conspectus of the Genera," have never, to my knowledge, been attributed to N. Eng. by any botanist except on the authority of Pursh, which in this case, Drs. Torrey & Gray (p. 432) think to be "extremely doubtful."

ORDER XLIX. CALYCANTHACEÆ.—CALYCANTHS.

Shrubs with square stems exhibiting 4 axes of growth surrounding the central one.
Lvs. opposite, entire, simple, without stipules. *Fls.* solitary, axillary.
Cal.— { Sepals numerous, colored, in several rows, confounded with the petals, all united below into a
Cor.— { fleshy tube or cup.
Sta. ∞, inserted into the fleshy rim of the calyx; inner row sterile.
Ova. indefinite, inserted on the disk which lines the calyx tube.
Fr.—Achenia hard, enclosed in the calyx tube as in genus Rosa.

The order consists of but 2 genera, Calycanthus, American, and Chimonanthus of Japan. The species
are 6. The flowers are highly aromatic, and the same quality resides in the bark.

CALYCANTHUS.

Gr. καλυξ, calyx, ανθος, a flower; from the character.

Lobes of the calyx imbricated in many rows, lanceolate, somewhat coriaceous and fleshy, colored; stamens unequal, about 12 outer ones fertile; anthers extrorse. *The bark and leaves exhale the odor of camphor. Fls. of a lurid purple.*

C. FLORÍDUS. *Carolina Allspice. Sweet-scented Shrub.*

Lvs. oval, mostly acute or acuminate, tomentose beneath; *branches* spreading; *flowers* nearly sessile.—Fertile soils along streams, Va. and all the S. States. Not uncommon in gardens farther north. The species of Willdenow and Elliot are all referred to this of Linnæus, by Torrey and Gray, as follows:

β. (C. lævigatus *Willd.*)—*Lvs.* oblong or. ovate-lanceolate, acuminate or gradually acute, glabrous; *branches* erect. †

γ. (C. glaucus *Willd.*)—*Lvs.* oblong or ovate-lanceolate, acuminate, glaucous and glabrous beneath; *branches* spreading. † ·

δ. (C. inodorus. *Ell.*)—*Lvs.* lanceolate, scabrous and shining above, smooth below; *branches* spreading.

ORDER L. MYRTACEÆ.—MYRTLEBLOOMS.

Trees and *shrubs*, without stipules.
Lvs. opposite entire, punctate, usually with a vein running close to the margin.
Cal. adherent below to the compound ovary, the limb 4–6-cleft, valvate.
Cor.—Petals as many as the segments of the calyx.
Sta. indefinite. *Anthers* introrse. *Style* and *stigma* simple.
Fr. with many seeds.

A fine order, of 46 genera and 1300 species, native of warm or torrid countries, especially of S. America
and the E. Indies.

Properties.—A fragrant or pungent volatile oil, residing chiefly in the pellucid dotting of the leaves pervades the order. The Caryophyllus aromaticus, native of Arabia, a tree about 20 feet in height, yields the
clove (*clou* Fr. a nail) which is the *dried flower. Cajeput oil* is distilled from the leaves of the Melaleu-
ca Cajeputi, native of the E. Indies A kind of *gum kino* is obtained from Eucalyptus resinifera, also a
native of India The root of the Pomegranate yields an extract which is an excellent vermifuge.—All the
genera are exotic with us.—Many of them are highly ornamental in culture.

Genera.

Fruit 2–3-celled Leaves evergreen, with a marginal vein. . . . . **Myrtus.** 1
Fruit many-celled. Leaves deciduous, without the marginal vein. . . . . . **Punica.** 2

1. MYRTUS. Tourn.

Gr. μυρον, perfume.

Calyx 5-cleft; petals 5; berry 2–3-celled; radicle and cotyledons distinct.—*Shrubs with evergreen leaves.*

M. COMMÚNIS. *Common Myrtle.—Lvs.* oblong-ovate, with a marginal vein; *fls.* solitary; *invol.* 2-leaved.—This popular shrub is a native of S. Europe. In our climate it is reared only in houses and conservatories. Among the ancients it was a great favorite for its elegance of form, and its fragrant, evergreen leaves. It was sacred to Venus. The brows of bloodless victors were adorned with myrtle wreaths, and at Athens it was an emblem of civic arthority. Leaves about 1' by ⅓'. Flowers white. Jl. Aug.†

2. PUNICA.

Lat. Punica, Carthagenian, or. of Carthage, where it first grew.

Calyx 5-cleft; petals 5; berry many-celled, many-seeded; seeds *incoste;* placenta parietal.—*Deciduous trees and shrubs.*

P. GRANÀTUM. *Pomegrana'e.*—Arborescent; *lvs.* lanceolate, w..h no marginal vein.—A thorny bush when wild, from S. Europe, where it is sometimes used for hedges like the hawthorn. Leaves lanceolate, entire. smooth, 2—3' by 5—10'', obtuse. The flowers are scarlet, large, and make a fine appearance. The fruit is large, highly ornamental and of a fine flavor. Much care is requisite for its cultivation. It requires a rich loam, a sunny situation, protected by glass. In this way double flowers of great beauty may be produced. †

P. NANA. *Dwarf Pomegranate.*—Shrubby; *lvs.* linear-lanceolate, acute.—Native of the W. Indies, where it is used as a hedge plant. Shrub 4—6f high with smaller purple flowers, often double. †

ORDER LI. MELASTOMACEÆ.—MELASTOMES.

Trees, shrubs or herbs with square branches, and usually exstipulate.
Lvs. opposite, entire and undivided, without dots and with several veins.
Cal.—Sepals 4—6, united, persistent, the tube urceolate, cohering with only the angles of the ovary.
Cor—Petals as many as the segments of the calyx, twisted in æstivation.
Sta. twice as many as the petals, sometimes of the same number, the filaments inflexed in æstivation.
Anth. before flowering contained in the cavity between the cal. and the sides of ova.
Fr. capsular or baccate.

Genera 118, species 1200. The order is represented in the United States by a single genus, the remainder being natives chiefly of India and tropical America. No plant of this order is poisonous. All are slightly astringent.

RHEXIA.

Gr. ῥῆξις, a rupture; some of the species are good vulneraries.

Calyx 4-cleft, swelling at the base: petals 4; stamens 8, 1-celled; style declined; capsule 4-celled, nearly free from the investing calyx tube; placentæ prominent; seeds numerous.—♃ *Lvs. opposite, exstipulate,* 3-*veined.*

1. R. VIRGÍNICA. *Meadow Beauty. Deer Grass.*

St. with 4-winged angles; *lvs.* sessile, oval-lanceolate, ciliate-serrulate, and with the stem clothed with scattered hairs; *cal.* hispid.—Grows in wet grounds, Mass. to Ill.! and La. Stem 1f or more high, often 3-forked above. Leaves with 3 (rarely 5 or 7) prominent veins, 1—3' long, about ½ as wide, acute. Flowers large, in corymbose cymes. Petals bright purple, obovate, hispid beneath, caducous. Anthers long and prominent, crooked, golden yellow above with a purple line beneath. Style somewhat longer than the stamens, a little declined. Jl. Aug.

2. R. MARIÁNA. *Maryland Deer Grass.*

St. nearly terete, covered with bristly hairs; *lvs.* lanceolate, acute, attenuate at base into a very short petiole and, with the calyx, clothed with scattered hairs.—In sandy bogs, N. J. to Flor. The whole plant is hispid, even the petals externally. Stem 1—2f high, slender, and generally without branches. Leaves often narrowly oblong, serrate-ciliate. Petals large, obovate, purple. Jn.—Sept.

ORDER LII. LYTHRACEÆ.—LOOSESTRIFES.

Herbs, rarely *shrubs,* frequently with 4-cornered branches.
Lvs. opposite, rarely alternate. entire, with neither stipules nor glands.
Cal. tubular, the limb 4—7 lobed, sometimes with as many intermediate teeth.
Cor.—Petals inserted into the calyx between the lobes, very deciduous, or 0.
Sta. equal in number to the petals, or 2—4 times as many, inserted into the calyx.
Ova. superior, enclosed in the calyx-tube. 2—4-celled *Sty.* united into one.
Fr.—Capsule membranous, enveloped in the calyx, usually by abortion 1-celled.
Sds. small, ∞, attached to a central placenta. *Albumen* 0.

Genera 35, species 300. Some of the species are found in temperate climes, but most of them are tropical. Lythrum salicaria, nati e of Europe, N. Holland and U. S., is used for tanning where it abounds. All the species are astringent.

Conspectus of the Genera.

		{ 8 horns. Petals 8.	.	.	.	.	.	Hypobrichia. 5
	{ with 4 teeth and { 4 short horns.		.	.	.	.	.	Ammannia. 1
{ campanulate, { with 5 teeth and 5 long horns.			.	.	.	.	.	Decodon. 3
{ cylindrical, with minute, intermediate horns.			.	.	.	.	.	Lythrum. 2
Calyx { ventricose, gibbous at base, intermediate horns ∞.			.	.	.	.	.	Cuphea. 4

1. AMMANNIA.

Named in honor of John Ammann, native of Siberia, prof. of bot. St. Petersburg.

Calyx campanulate, 4—5-toothed or lobed, generally with as many horn-like processes alternating with the lobes; petals 4 or 5, often 0; stamens as many, rarely twice as many as the calyx lobes; capsule 2—4-celled, many-seeded.—① *In wet places. Sts. square and lvs. opposite, entire. Fls. axillary.*

1. A. HUMILIS. Michx. (A. ramosior. *Linn.*) *Low Ammannia.*
St. branched from the base, ascending; *lvs.* linear-oblong or lanceolate, obtuse, tapering at base into a short petiole; *fls.* solitary, closely sessile, all the parts in 4s; *sty.* very short.—An obscure and humble plant in wet places, Ct. to Ga. W. to Oregon. Stems square, procumbent at base, 6—10′ high. Flowers minute, one in the axil of each leaf, with 4 purplish, caducous petals. Calyx with 4 short, horn-like processes, alternating with the 4 short lobes of the limb. Aug. Sept.

β. (T. & G. A. ramosior. *Michx.*)—*Lvs.* subsessile, cordate-sagittate at base; *fls.* about 3 in each of the lower axils, solitary above.—In N. J., where, it is said by T. & G., to grow with and pass into the other variety.

2. A. LATIFOLIA. (A. ramosior. *Linn.*)
St. erect, branching; *lvs.* linear-lanceolate, dilated and auriculated at the sessile base; *fls.* crowded and apparently verticillate, upper subsolitary and pedunculate; *cal.* 4-angled, 4-horned; *sep. pet. sta. and cells of capsule* 4.—Wet prairies, Western States. Stem 1—2f high. Leaves 2—3′ by 2—5″. Flowers purple. Jl.—Sept.

2. LYTHRUM.

Gr. λυθρον, black blood; referring to the color of the flower.

Calyx cylindrical, striate, limb 4—6-toothed, with as many intermediate, minute processes; petals 4—6, equal; stamens as many, or twice as many as the petals, inserted into the calyx; style filiform; capsule 2-celled, many-seeded.—*Mostly* ♃, *with entire leaves.*

1. L. HYSSOPIFOLIA. (L. hyssopifolium. *Bw.* and 1*st edit.*) *Grass-poly.*
Glabrous, erect, branching; *lvs.* alternate or opposite, linear or oblong-lanceolate, obtuse; *fls.* solitary, axillary, subsessile; *pet.* and *sta.* 5 or 6.—① A slender, weed-like plant, found in low grounds, dried beds of ponds, &c., Mass. and N. Y., near the coast, rare. Plant 6—10′ high, with spreading, square branches. Leaves sessile, acute at base, pale green, each with a single small flower sessile in its axil. Petals pale purple. Calyx obscurely striate, with short lobes. Jl.

2. L. ALATUM. Ph. *Wing-stem Lythrum.*
Glabrous, erect, branched; *st.* winged below; *lvs.* lance-ovate, sessile, broadest at base, alternate and opposite; *fls.* axillary, solitary.—Damp grounds, Southern and Western States, common! Stem 1—2f high, strict, the wings narrow. Leaves 1—2′ long, ¼ as wide. Calyx tube 12-striate, 12-toothed, alternate teeth cornute. Corolla purple, wavy, 6-petaled. Stamens 6, included. Jn. Jl.

3. L. LINEARE. *Linear-leaved Lythrum.*
St. slender, somewhat 4-angled, branched above; *lvs.* linear, mostly opposite and obtuse; *fls.* nearly sessile; *pet.* and *sta.* 6.—Swamps, near the coast, N. J. to Flor. Stem 2—4f high, the angle sometimes slightly winged. Leaves 1—2′ by 2—4″, rather fleshy. Flowers small, nearly white.

4. L. SALICARIA. *Loose-strife.*
More or less pubescent; *lvs.* lanceolate, cordate at base; *fls.* nearly sessile, in a long, somewhat verticillate, interrupted spike; *pet.* 6 or 7; *sta.* twice as many as the petals.—♃ An ornamental plant, native in wet meadows, Can. and N. Eng., rare. Stem 2—5f high, branching. Leaves 3—6′ long, ¼ as wide, gradually acuminate, entire, on a short petiole, opposite, or in verticils of 3, upper ones reduced to sessile bracts. Flowers large, numerous and showy. Petals purple Jl. Aug. †

5 L. VIRGATUM. *Austrian Lythrum.*—*St.* erect, branched, virg: · · *lvs.* lanceolate, acute at each end, floral ones small ; *fls.* about 3 in each a: d t of the virgate raceme ; *sta.* 12.—A fine species for the garden, native of Au stia. Stem 3—4f high. Flowers purple. Jn.—Sept. †

3. DECODON. Gmel.

Gr. δεκα, ten, οδους, a tooth ; from the horn-like teeth of the calyx.

Calyx short, broadly campanulate, with 5 erect teeth, and 5 elon gated, spreading, horn-like processes ; stamens 10, alternate ones very long ; style filiform ; capsule globose, included, many-seeded.—⁊ *Lvs opposite or verticillate, entire. Fls. axillary, purple.*

D. VERTICILLATUM. Ell. (Lythrum. *Linn.* Nesæa. *Kunth.*)
Swamps throughout the U. S. and Can. Stem woody at base, often prostrate, and rooting at the summit, 3—8f in length, or erect and 2—3f high, 4—6 angled. Leaves opposite or in whorls of 3, lanceolate, on short petioles, acute at base, 3—5' long, gradually acuminate and acute at apex. Fl vers in axillary, subsessile umbels of 3 or more, apparently whorled, const tuting a long, leafy, terminal and showy panicle. Petals 5 or 6, large, and of a fine purple. Jl. Aug.

α. pubescens.—*St.* and *lvs.* beneath pubescent. R. Island.
β. lævigatum.—Glabrous and bright green. Most common in N. Eng.

4. CUPHEA.

Gr. κυφος, curved or gibbous ; in reference to the capsule.

Calyx tubular, ventricose, with 6 erect teeth, and often as many intermediate processes ; petals 6—7, unequal ; stamens 11—14, rarely 6—7, unequal ; style filiform ; capsule membranaceous, 1—2-celled, few-seeded.—*Herbaceous or suffruticose. Lvs. opposite, entire. Fls. axillary and terminal.*

C. VISCOSISSIMA. Jacq. (Lythrum petiolatum. *Linn.*)
Herbaceous, viscid-pubescent ; *lvs.* ovate-lanceolate, petiolate, scabrous ; *fls.* on short peduncles ; *cal.* gibbous at base on the upper side, 12-veined, 6-toothed, very viscid.—① Wet grounds. Pittsfield, Mass., *Hitchcock,* Cambridge, N. Y., *Stevenson,* to Ga. and Ark. Stem 9—18' high, with alternate branches. Leaves somewhat repand, 1—2' long, ⅓ as wide, on petioles ⅓ as long. Flowers solitary, one in each axil, irregular. Calyx often purple, ventricose. Petals violet, obovate. Stamens included. Capsule bursting length-wise before the seeds are ripe. Aug.

5. HYPOBRICHIA. Curtis.

Calyx 4-lobed, without accessory teeth ; petals 0 ; stamens 2—4 ; ovary 2-celled ; stigma 2-lobed, subsessile ; capsule globose, bursting irregularly, many-seeded.—*A submersed aquatic herb. Lvs. opposite, crowded, linear. Fls. axillary, sessile, minute.*

H. NUTTALLII. Curt. (Peplis diandra. *Nutt.*)
A little inhabitant of ponds and sluggish streams, Ill., *Mead, Buckley,* to Mo. and La. Its habit is similar to a Callitriche. Stem mostly submersed, 10 20' long. Leaves 10—15" by 1—2", very numerous. Jn.—Aug.

ORDER LV. ONAGRACEÆ.—ONAGRADS.

Plants herbaceous, sometimes shrubby, with alternate or opposite leaves.
Fls. axillary, or in terminal spikes or racemes.
Cal.—Sepals 4, (3—6) united below into a tube, the lobes valvate in æstivation.
Cor.—{ Petals 4, (2—6) inserted with the 4 or 5 (1—2—3—6) stamens into the throat of the calyx. *Pollen*
Sta.—{ triangular, often cohering by threads.
Ova. coherent with the tube of the calyx ; placenta in the axis.
Fr. baccate or capsular, 2—4 celled, many-seeded. *Albumen* none.

Genera 36, species 320, particularly abundant throughout America, more rare in the Old World. They posse is no remarkable properties. Many genera are ornamental, and one, the well known Fuchsia, is so to a high degree.

To this order is appended the suborder *Haloragea*, consisting of aquatic herbs of a low grade, the flowers being imperfect or reduced to solitary parts or organs.

FIG. 45.—1. Flower of Œnothera fruticosa. 2. Plan of the flower. 3. Section of the 4-celled capsule of Œ. biennis. 5. Hippuris vulgaris; 6, its flower, with 1 stamen, 1 ovary, 1 style. 4. Vertical section of its 1-seeded fruit. 7. Circæa Lutetiana. 8. The flower enlarged. 9. Plan of the flower. 10. Vertical section of the 2-celled and 2-seeded fruit.

Conspectus of the Genera.

		Pet. scarcely clawed;	Seeds comose.	*Epilobium.* 1
			Seeds naked.	*Œnothera.* 2
			Claws unarmed.	*Gaura.* 5
	Herbs. Pet. clawed;	Claws with 2 teeth.		*Clarkia.* 3
Sta. 8;	Beautiful green-house shrubs.			*Fuchsia.* 4
Fls. perf. Stamens 4; styles united into 1.				*Ludwigia.* 6
by 4s; Flowers monœcious; aquatic; leaves multifid.				*Myriophyllum.* 9
by 3s; flowers apetalous; aquatic; leaves pectinate				*Proserpinaca.* 8
by 2s; flowers complete and regular; leaves dentate.				*Circæa.* 7
Parts of fl. arrang'd by 1s; flowers apetalous; aquatic; leaves verticillate.				*Hippuris.* 10

TRIBE 1. ONAGREÆ.

Flowers perfect, the parts arranged in 4s (rarely 3s); pollen connected by threads.

1. EPILOBIUM.

Gr. επι, upon. λοβον, a pod, ιον, a violet; i. e. a violet growing upon a pod.

Calyx tube not prolonged beyond the ovary, limb deeply 4-cleft, 4-parted and deciduous; petals 4; stamens 8, anthers fixed near the middle; stigma often with 4 spreading lobes; ovary and capsule linear, 4-cornered, 4-celled, 4-valved; seeds 00, comose, with a tuft of long hairs.—24.

1. E. ANGUSTIFOLIUM. (E. spicatum. *Lam.*) *Willow Herb. Rose-bay.*

St. simple, erect; *lvs.* scattered, lanceolate, subentire, with a marginal vein; *rac.* long, terminal, spicate; *pet.* unguiculate; *sta.* and *sty.* declined; *stig.* with 4 linear, revolute lobes.—In newly cleared lands, low waste grounds, Penn. to Arctic Am. Stem 4—6f high, often branched above. Leaves sessile, smooth, 2—5' long, ⅓ as wide, acuminate, with pellucid veins. Flowers numerous and showy, all the parts colored, petals deep lilac-purple, ovary and sepals (5—6" long) pale glaucous-purple. Jl. Aug.

β. *canescens.*—Flowers of a pure white in all their parts; ovaries silvery-canescent.—Danville, Vt. *Miss M. L. Tinole!*

2. E. COLORATUM. Muhl. *Colored Epilobum.*

St. subterete, puberulent, erect, very branching; *lvs.* mostly opposite, lanceolate, dent-serrulate, acute, subpetiolate, smooth, often with reddish veins; *pet.* small, 2-cleft at apex; *cal.* campanulate; *sty.* included; *stig.* clavate; *ovules* in a single row.—Ditches and wet, shady grounds, British Am. to Ga. W. to Oregon. Stem 1—3f high, becoming very much branched. Leaves 2—4' long, ⅓ as wide, with minute, white dots, upper ones alternate and sessile, lower on short petioles. Flowers numerous, axillary. Pedicels 1—2" in length, ovaries 4—6", capsules 20", very slender. Petals rose-color, twice longer than the sepals. Jl.—Sept.—Scarcely distinct from the next.

3. E. PALUSTRE. *Marsh Epilobium.*

St. terete, branching, somewhat hirsute; *lvs.* sessile, lanceolate, subdenticulate, smooth, attenuate at base, rather acute, lower ones opposite; *pet.* small, obcordate, twice longer than calyx; *sty.* included; *stig.* clavate; *caps.* pubescent.—In swamps and marshes, Penn. to Artic Am., W. to Oregon. Stem 1—2f high, very branching. Leaves mostly alternate, 1—3' long, ⅛ as wide, entire, or with a few minute teeth. Flowers numerous, axillary. Petals rose-color. Capsules 1—2' long, on short pedicels. Aug.

β. albiflorum. Lehm. (E. lineare. *Muhl.*) *St.* slender, at first simple, branched at top; *lvs.* linear, entire, margin revolute; *capsules* canescent.

4. E. MOLLE. Torr. (E. strictum. *Muhl.*) *Soft Epilobium.*

Plant clothed with a dense, soft, velvet-like pubescence; *st.* terete, straight, erect, branching above; *lvs.* opposite (alternate above), crowded, sessile, mostly entire and oblong-linear, obtusish; *pet.* deeply emarginate, twice longer than the calyx; *stig.* large, turbinate; *caps.* elongated, subsessile.—① Swamps, Mass. to N. J., rare. Stem 1—2f high. Leaves numerous, 8—15" by 1—4". Flowers rose-color. Capsules 3' long. Sept.

5. E. ALPINUM. *Alpine Epilobium.*

St. creeping at base, usually with 2 pubescent lines, few-flowered; *lvs.* opposite, oblong-ovate, subentire, obtuse, sessile or subpetiolate, smooth; *stig.* undivided; *caps.* mostly pedicellate.—Mountains, Northern States to Artic Am. Stem 6—12' high. Leaves often slightly petiolate and denticulate, lower obtuse, middle acute, and upper acuminate. Flowers smaller than in E. molle, reddish-white.

β. natans. Hornem.—*St.* large, nodding at the summit; *lvs.* oblong, denticulate.

2. ŒNOTHÈRA.

Gr. οινος, wine, θηρα, to hunt; the root is said to cause a thirst for wine.

Calyx tube prolonged beyond the ovary, deciduous, segments 4, reflexed; petals 4, equal, obcordate or obovate, inserted into the top of the tube; stamens 8; capsule 4-celled, 4-valved; stigma 4-lobed; seeds many, naked.—*Herbs with alternate leaves.*

1. Œ. BIENNIS. *Common Evening Primrose. Scabish.* (Fig. 45.)

St. erect, hirsute; *lvs.* ovate-lanceolate, repand-denticulate; *fls.* sessile, in a terminal, leafy spike; *calyx tube* 2 or 3 times longer than the ovary; *sta.* shorter than petals; *caps.* oblong, obtusely 4-angled.—① and ② Common in fields and waste places, U. S. and British Am. Stem mostly simple, 2—5f high, with whitish, scattered hairs. Leaves 3—6' by ¼—1½', roughly pubescent, slightly toothed, sessile on the stem, radical ones tapering into a petiole. Flowers numerous, opening by night and continuing but a single day. Petals large, roundish, obcordate. Seeds very numerous, 2 rows in each cell. Jn.—Aug.

β. muricata. (Œ. muricata. *Ph.*) *St.* muricate or strigosely hirsute, red; *pet.* scarcely longer than the stamens. Stem 1—2f high.

γ. grandiflora. (Œ. grandiflora. *Ait.*) *Pet.* much longer than the stamens, rather deeply obcordate. Stem branched. †

2. Œ. FRUTICOSA. *Perennial Evening Primrose.*

St. pubescent or hirsute; *lvs.* oblong-lanceolate, repand-denticulate; *rac.* leafy, or naked below, corymbed; *caps.* oblong-clavate, 4-angled, pedicellate —♃ In sterile soils, Mass., Ct., N. Y. to Flor. and Western States. Stem hard, rigid, (not shrubby) branched, purple, 1—3f high. Leaves variable in pubescence, form and size., 1—3' by 3—9", sessile, minutely punctate. Flowers few or many, 1½' diam. in a terminal, bracteate, mostly pedunculate raceme. Calyx tube longer than the ovary. Petals broad-obcordate, yellow. Jn.—Aug.

β. ambigua. *Lvs.* membranaceous; *pet.* longer than broad.

3. Œ. PUMILA (& Œ. pusilla? *Michx.*) *Dwarf Evening Primrose.*

Low, pubescent; *st.* ascending; *lvs.* lanceolate, entire, obtuse, attenuate at base; *spike* loose, leafy, naked below; *calyx tube* shorter than the subsessile, oblong-clavate, angular ovary.—♀ A small, half-erect plant, common in grass lands, Can. to S. Car. Stem 6—10' long, round, slender, simple. Leaves 1—1½' by 2—3", radical ones spatulate, petiolate. Flowers yellow, 6' diam., opening in succession 1 or 2 at a time. Jn.—Aug.

ANTHA. Michx. *Golden Evening Primrose.*
, slender; *fls.* small, crowded, spicate; *calyx tube* equal in
, longer than the segments; *pet.* broadly obovate, emarginate,
mens; *caps.* smooth, pedicellate, clavate, the alternate angles
② Western N. Y. to Mich. Stem 12—18' long, purple.
obtuse, attenuate at base, denticulate, radical ones spatulate,
orange-yellow. Jn. Jl.

us. Michx.
ften decumbent at base, much branched; *lvs.* linear-lance-
rhat denticulate; *fls.* large, in terminal corymbs; *calyx tube*
vary; *pet.* longer than the stamens; *caps.* canescent, with
gles, tapering at base.—♃ Montauk Point, L. I. *Torrey,* to
e. Stem 1—2f high. Flowers yellow, 1' diam. Capsules
o a slender pedicel. May—Jl.

'A.
, diffusely branched or subsimple, assurgent; *lvs.* pubescent,
te-dentate or incised; *fls.* axillary, solitary, sessile; *cal.* vil-
r than the ovary; *caps.* prismatic.—① Fields, N. J. to La.
Leaves often pinnatifid. Flowers about ¼' diam., terminal,

. (Œ. minima. *Ph.*) Low, simple, 1-flowered; *lvs.* nearly
:ns, N. J., &c.
SA. Nutt.—Minutely pubescent, mostly erect and branched;
muate at base, lower ones petiolate; *fls.* large, in a long,
tube longer than the ovary; *caps.* obovoid-clavate, pedicel-
and Tex. Stem 2—3f high. Flowers white or rose-colored,

A. Nutt.
th; *st.* erect and virgately branched; *lvs.* linear-lanceolate,
otely denticulate, somewhat petiolate; *fls.* in a long raceme;

4. FUCHSIA.

In honor of Leonard Fuchs, an excellent German botanist of the 15th century.

Calyx tubular-infundibuliform, colored, deciduous: limb 4-lobed; petals 4, in the throat of the calyx, alternate with its segments; disk glandular, 8-furrowed; baccate capsule oblong, obtuse, 4-sided.—*Mostly shrubby. S. American plants of great beauty.*

1 F. COCCINEA. Ait. (F. Magellanica. *Lam.*) *Ladies' Ear-drop.*—*Branches* smooth; *lrs.* opposite and in verticils of 3s, ovate, acute, denticulate, or short petioles; *fls.* axillary, nodding; *sep.* oblong, acute; *pet.* convolute, half as long as calyx.—Native of Chili A very delicate and beautiful green-house shrub. 1—6f high. Flowers on long, filiform pedicels. Calyx scarlet, much longer than the included, violet-purple petals. Stamens crimson, much exserted. Berry purple.

2. F. GRACILIS. Lindl.—*St.* suffruticose, often simple; *lrs.* opposite, ovate, petiolate, slightly acuminate, glandular-dentate; *fls.* opposite, solitary, pendulous, longer than the leaves; *pet.* nearly as long as the sepals and much broader.—Chili. A beautiful parlor plant, quite common. Stem 2—3f high, thick. Flowers larger, but less elegant than in the former, with a red calyx and crimson corolla. †

5. GAURA.

Gr. γαυρος, superb; a term characteristic of the flowers.

Calyx tube much prolonged above the ovary, cylindric: limb 4-cleft; petals 4, unguiculate, somewhat unequal, inserted into the tube: stamens 8, declinate, alternate ones a little shorter: ovary oblong, 4-celled, one only proving fruitful: nut usually by abortion 1-celled, 1—4-seeded.—*Herbaceous or shrubby. Lrs. alternate. Fls. white and red, rarely trimerous.*

1. G. BIENNIS. *Biennial Gaura.*
St. branched, pubescent; *lrs.* lanceolate, remotely dentate; *spike* crowded; *calyx tub* as long as the segments; *pet.* rather declinate and shorter than sepals; *fr.* subsessile, 8-ribbed, alternate ribs minute.—A beautiful biennial, on the dry banks of streams, Can. to Ga. rare. Stem 3—5f high. Leaves sessile, pale green, acute at each end. Flowers numerous, sessile. Calyx reddish. Corolla at first rose-color, changing to deep red. Stigma 4-lobed. Fruit rarely maturing more than 1 seed. Aug.

2. G. FILIPES. Spach. *Slender-stalked Gaura.*
St. pubescent, paniculate and naked above; *lvs.* linear-oblong, repand-dentate, lower ones almost pinnatifid; *branches* of the panicle very slender, naked, with tufted leaves at their base; calyx canescent, longer than the petals.—Dry grounds, S. and W. States! Stem rigid, 3—5f high, very leafy just below the panicle. Leaves 1—3' long, 2—6" wide, tapering at each end. Petals oblong-spatulate, rose-color, or white. July, Aug.

6. LUDWIGIA.

In honor of C. D. Ludwig, prof of botany at Leipsic, about 1750.

Calyx tube not prolonged beyond the ovary, limb 4-lobed, mostly persistent: petals 4, equal, obcordate, often minute or 0: stamens 4, opposite the sepals: style short: capsule short often perforated at top, 4-celled, 4-valved, many-seeded, and crowned with the persistent calyx lobes.—*⁂ Herbs, in wet grounds. Lrs. entire.*

1. L. ALTERNIFOLIA *Seed-box. Bastard Loosestrife.*
Erect, branched, nearly or quite smooth; *lrs.* alternate, lanceolate, sessile, pale beneath; *ped.* axillary, solitary, 1-flowered, 2-bracted above the middle; *pet.* scarcely as large as the spreading, acuminate sepals; *caps.* large, with 4 winged angles, crowned with the colored calyx.—Shady swamps. Stem 1—3f

high, round, with a strong bark and several branches. Leaves with marginal veins, 2—3′ long, ½—1′ wide. Capsule convex at apex, the angles conspicuously winged. Sepals large, purplish. Petals large, yellow, showy. Jl. Aug.

2. L. HIRTELLA. Raf. (L. hirsuta. *Ph.*)

Hairy, erect, sparingly branched ; *lvs.* alternate, ovate-oblong, sessile, obtuse ; *fl.* axillary, solitary, pedicellate, with two bracteoles below it ; *sep.* nearly as long as the petals; *cap.* subglobose, 4 angled and winged.—Moist soils, N. J. to Flor. Stem 1—3f high. Leaves numerous, hairy both sides, ½—1½′ by 2—8″. Flowers yellow, about ½′ diam. Calyx spreading, and, with the capsule, villous. Jn.—Sept.

3. L. LINEÀRIS. Walt. (Isnardia. *DC.*)

Glabrous, slender, with angular branches ; *lvs.* lance-linear, acute at each end ; *fls.* axillary, solitary, sessile ; *pet.* obovate-oblong, slightly longer than the sepals, but much shorter than the elongated, obovoid-clavate, 4-sided capsules.—Swamps, N. J. and S. States. An erect, smooth plant, 1—2f high, often sending out runners at the base with obovate leaves. Fls. sometimes apetalous. Jl.—Sept.

4. L. SPHÆROCARPA. Ell. (Isnardia. *DC.*)

Erect, smooth or nearly so ; *lvs.* alternate, lanceolate, acute, attenuate at base; *fls.* axillary, subsolitary, on very short pedicels ; *pet.* minute or wanting, as well as the bracteoles; *sep.* as long as the small, subglobose capsule.—In water, S. to Ga. partly submerged, or in very wet grounds, near Boston, Ms. Stem 2—3f high, branching, angular. Margin of the leaves rough, sometimes remotely and obscurely denticulate. Fls. greenish, inconspicuous. Jl.—Sept.

5. L. POLYCARPA. Short & Peter.

Glabrous, erect, much branched and often stoloniferous; *lvs.* lance-linear, gradually acute at each end ; *fls.* apetalous, axillary, solitary, with two subulate bracteoles at base ; *caps.* 4-angled, truncated above, tapering below, crowned with the 4-lobed stylopodium.—Swamps, Western States ! Stem 1—3f high. Leaves 2—3′ by 2—4″, ten times longer than the flowers. Aug.—Oct.

6. L. PALUSTRIS. Ell. (Isnardia. *Linn.*)   *Water Purslain.*

Prostrate and creeping, smooth and slightly succulent; *lvs.* opposite, ovate, acute, tapering at base into a petiole ; *fls.* sessile, axillary, solitary ; *pet.* 0, or very small.—In U. S. and Can., creeping in muddy places, or floating in water. Stem round, reddish, 10—18′ long, often sparingly branched. Leaves, including their slender petioles, about 1½′ by ½′, ovate-spatulate. Calyx lobes and style very short. Petals when present, flesh-color. Capsule 1—2″ long, abrupt at each end, with 4 green angles. Jn.—Sept.

TRIBE 2. **CIRCÆEÆ.**

Flower regular, all its parts in 2s.

7. CIRCÆA.

Circæa was supposed to have used these plants in her enchantments.

Calyx slightly produced above the ovary, deciduous, limb 2-parted ; petals 2, obcordate ; stamens 2, opposite the sepals ; capsule obovoid, uncinate-hispid or pubescent, 2-celled, 2-seeded ; styles united. ♃ *Lvs.* opposite.

1. C. LUTETIÀNA. *Larger Enchanter's-Nightshade.* (Fig. 45.)

St. erect, pubescent above; *lvs.* ovate, subcordate, acuminate, slightly repand-dentate, opaque, longer than the petioles; *bracts* none ; *fr.* reflexed, hispid, uncinate.—Damp shades and thickets, Can. to Car. W. to Ill.! Stem 1—2f high, sparingly branched, tumid at the nodes. Leaves dark green, smooth or slightly pubescent, 2—4′ long, ½ as wide, petiole 8—15″ long. Flowers small, rose-color, in long, terminal and axillary racemes. Fruit obcordate, with conspicuous hooks. Jn. Jl.

2. C. ALPINA. *Alpine Enchanter's-Nightshade.*

Smooth ; *st.* ascending at base, weak; *lvs.* broad-cordate, membranaceous, dentate, as long as the petioles; *bracts* setaceous; *capsule* pubescent.—A small.

delicate plant, common in wet, rocky woodlands in mountainous districts, N. Eng., Brit. Am. W. to Or. Stem diaphanous, juicy, 5—10' high. Leaves 1— 3' long, ½ as wide, acute or acuminate, with small, remote teeth, pale green and shining. Flowers white, rarely reddish, minute, in terminal racemes. Jl.

SUBORDER,—HALORAGEÆ.

Plants small, aquatic. Flowers minute, axillary, sessile. Calyx entire, or 2—4-lobed. Petals 3—4, often 0. Stamens 1—8, inserted with the petals into the summit of the calyx. Ovary inferior, 1—4-celled. Fruit dry, indehiscent, 1—4-celled. Seed pendulous, 1 in each cell.

8. PROSERPINÁCA.

Lat. *Proserpina*, a Roman goddess, from some fancied resemblance.

Calyx tube adherent to the ovary, 3-sided, limb 3-parted; petals 0; stamens 3; stigmas 3; fruit 3-angled, 3-celled, bony, crown'd with the permanent calyx.— ♃ *Aquatic. Lrs. alternate.*

1. P. PALUSTRIS. *Spear-leaved Mermaid-weed.*
Lrs. linear-lanceolate, sharply serrate above the water, those below (if any) pinnatifid.—Ditches, swamps and ponds, often partly submersed, N. Eng. to Ark. Root creeping. Stems ascending at base, 6—20' high, striate, round ▸ ish. Leaves 10—15' by 2—3'', acute at each end, lower ones on short petioles, and if growing in water, pinnatifid with linear segments. Flowers greenish, sessile, 1—3 together in the axils of the upper leaves, succeeded by a very hard, triangular nut. Jn. Jl.

2. P. PECTINACEA. Lam. (P. palustris, β. *Mx.*) *Cut-leaved Mermaid-weed.*
Lrs. all pectinate, with linear-subulate segments; *fr.* obtusely 3-angled ·—Sandy swamps, in Ms.! (rare) S. to Flor. Stems 5—10' high, ascending a base from long, creeping roots. Leaves all finely and regularly divided into very narrow segments. Styles 0. Stigmas attenuate above. Fruit rather smaller (less than 1'' diam.) than in P. palustris, rugose when mature. Jl. Aug

9. MYRIOPHYLLUM. Vaill.

Gr. *μυριος*, a myriad, *φυλλον*, a leaf; from the numerous divisions of the leaf.

Flowers ♂, or frequently ⚥; calyx 4-toothed in the ⚥ and ♀, 4 parted in the ♂; petals 4, often inconspicuous or 0; stamens 4—8 stigmas 4, pubescent, sessile: fruit of 4 nut-like carpels cohering by their inner angles —♃ *Submersed, aquatic herbs. Submersed lrs. parted into capillary segments. Upper fls. usually ♂, middle ones ⚥, lower ♀.*

1. M. SPICÁTUM.—*Spiked Water-Milfoil.*
Lrs. in verticils of 3s, all pinnately parted into capillary segments; *fls.* in terminal, nearly naked spikes; *floral lrs.* or *bracts* ovate, entire, shorter than the flowers; *lowest ones* subserrate and larger; *pet.* broadly ovate; *sta.* 8; *carpels* smooth.—N. Eng. to Ark., in deep water, the flowers only rising above the surface. Stems slender, branched, very long. Leaves composed of innumerable, hair-like segments, always submerged. Flowers greenish, sessile. Jl. Aug.

2. M. VERTICILLÁTUM. *Water-Milfoil.*
Lrs. in verticils of 3s, lower ones pinnately parted into opposite, capillary or setaceous segments; *fls.* in terminal, leafy spikes; *floral lrs.* pectinate-pinnatifid, much longer than the flowers; *pet.* oblong-obovate; *sta.* 8; *carpels* smooth.—In stagnant water, Can. to Flor. W. to Oregon. Stem long, less slender than in the last, only the upper part emerging. Flowers small, green, axillary, with conspicuous floral leaves. Sepals acute. Anthers oblong. Jl. Aug.

3. M. HETEROPHYLLUM. *Michx.* (Potamogeton verticillatum. *Walt.*)
Lrs. in verticils of 5s, the lower ones pinnately parted into capillary lobes; *spikes* terminal, nearly naked; *floral lrs.* ovate-lanceolate, serrate, longer than the flowers, crowded; *pet.* oblong; *sta.* 4—6; *carpels* scabrous, with 2 slight ridges on the back.—In sluggish water, Can. to Flor. and Tex., rare. Stem

23*

thick, branching. Leaves very various, lowest floral ones pectinately divided. Petals somewhat persistent. Sepals minute. Bracteoles serrulate. Jn.—Sept.

4. M. AMBIGUUM. Nutt. (M. natans. *DC.*) *Water Milfoil.*
Lvs. alternate, submersed ones pinnate, with capillary segments, middle ones pectinate, upper linear, petiolate, toothed or entire; fls. mostly ♀; pet. oblong, somewhat persistent; sta. 4; carpels smooth, not ridged on the back.—In ponds and ditches, Penn. to Mass.! Stems floating, upper end emerged with the minute flowers, and linear floral leaves. But in other situations it varies as follows:

β. *limosum.* Nutt. (M. procumbens. *Bw.*)—*St.* procumbent and rooting; leaves all linear, rigid, often entire.—Muddy places! where it is a small, creeping and branching plant.

γ. *capillaceum.* Torr.—Lvs. all immersed and capillary.—Ponds!

5. M. TENELLUM. Bw.
Erect and almost leafless; floral lvs. or bracts alternate, minute, entire, obtuse; fls. ♂; pet. linear; sta. 4; carpels smooth, not ridged.—About the edges of ponds and rivers, Providence, R. I. Olney! northern part of N. Y. to Newfoundland. Rhizoma prostrate, creeping, sending up several stems or scapes, which are simple, and 4—12′ high. Flowers small, purplish-white, sessile, alternate, a little shorter than the bracts, the upper ones ♂. Jl.

6. M. SCABRATUM. Michx.
Lvs. pinnatifid, in whorls of 4s and 5s; fls. verticillate, axillary; upper fls. ♂, with 4 stamens; lower ones ♀; fr. 8-angled, the ridges tuberculate.—Plymouth, Mass. Oakes. Block Island, Robbins, S. and W. States. Stem 6—12′ high. Segments of the leaves linear-capillary.

10. HIPPURIS.

Gr. ιππος, a horse, ουρα, tail.

Calyx with a minute, entire limb, crowning the ovary; corolla 0; stamen 1, inserted on the margin of the calyx; anther 2-lobed, compressed; style 1, longer than the stamen, stigmatic the whole length, in a groove of the anther; seed 1.—♃ *Aquatic herbs. St. simple. Lvs. verticillate, entire. Fls. axillary, minute.*

H. VULGARIS. *Mare's-tail.* (Fig. 45.)
Lvs. in verticils of 8—12, linear, acute, smooth, entire; fls. solitary, often ♀ ♀ ♂.—In the borders of ponds and lakes. Penn. to Arctic Am., very rare. Rhizoma with long, verticillate fibres. Stem erect, jointed, 1—2′ high. The flowers are the simplest in structure of all that are called perfect, consisting merely of 1 stamen, 1 pistil, 1 seed in a 1-celled ovary, and with neither calyx, lobes or corolla. May, Jn.

ORDER LVI. LOASACEÆ.—LOASADS.

Plants herbaceous, hispid, with pungent hairs secreting an acrid juice.
Lvs. opposite or alternate, usually more or less divided. Stipules 0.
Fds. axillary, 1-flowered. Sep. united, 5, persistent, equal.
Cor.—Petals 5 or 10, cucullate, inserted into the recesses of the calyx.
Sts. ∞, inserted with the petals, distinct or adhering in several sets.
Ova. adherent to the calyx more or less, 1-celled, with 3—5 parietal placentæ. Sty. 1.
Sds. many or few, anatropous.
Genera 15, species 70, natives of America.

MENTZELIA.

Named by Linn. in honor of C. Mentzel, physician to the Elector of Brandenburg.

Calyx tubular, limb 5-parted; petals 5—10, flat, spreading; stamens indefinite, 30—200; ovary inferior; styles 3, filiform, connate, and often spirally twisted; stigmas simple, minute; capsule 1-celled, many-seeded.—*Branching herbs. Lvs. alternate.*

1. M. LINDLEYI. Torr. & Gray. (Bartonia aurea. *Lindley.*) *Golden Bartonia.*—Hispid; lvs. ovate-lanceolate, pinnatifid, lobes often dentate; fls. solitary

or nearly so, terminal; *pet.* broadly obovate, very abruptly acuminate; *filaments* filiform, and, with the seeds, numerous.—① Native of California. Stems decumbent, branching, 1—3f in length, with golden yellow flowers 2—3 inches in diameter, the beauty of which is greatly heightened by the innumerable thread-like, yellow stamens.

2. M. OLIGOSPERMA. Nutt.

Very rough with barbed hairs; *st.* dichotomous; *lvs.* ovate-lanceolate, tapering to very short petioles, lobed or incisely dentate; *pet.* entire, cuspidate, expanding in sunshine; *sta.* 20 or more, shorter than the petals; *caps.* 3—5-celled.—♃ Dry or rocky places, Pike Co., Ill., *Mead*, and Mo. to Tex. Root tuberous. Stems 1f high, divaricately branched. Leaves 10—15″ by 6—8″, upper ovate. Flowers solitary, of a deep, golden yellow, 8—10″ diam., very fugacious. Capsule cylindric, very small. May—Jl.

ORDER LVII. PASSIFLORACEÆ.—PASSIONWORTS.

Plants herbaceous or shrubby, usually climbing. *Lvs.* alternate, often glandular. *Stip.* foliaceous. *Fls.* axillary or terminal, often with a 3-leaved involucre. *Cal.*—Sepals 4—5, united below into a tube, the sides and throat of which are lined with a ring of filamentous processes, which appear to be metamorphosed petals. *Cor.*—Petals 5, arising from the throat of the calyx outside the crown. *Sta.* 5 monadelphous, surrounding the stipe of the ovary. *Ova.* superior, on a long stipe, 1-celled. Styles 3. *Fr.* stalked, within the calyx, many-seeded.

Genera 12, species 216, natives of tropical America, but cultivated in many other countries as ornamental flowers. The fruit of the Granadilla (Passiflora multiformis) is eaten in the W. Indies, and highly valued as a dessert, but the root is poisonous.

PASSIFLÖRA.

[the Savior's passion.
Lat. *flos passionis*; the several parts of the flower were superstitiously compared to the instruments of

Calyx colored, deeply 5-parted, the throat with a complex, filamentous crown; petals 5, sometimes 0; stamens 5, connate with the stipe of the ovary; anthers large; stigmas 3, large, clavate, capitate; fruit a pulpy berry.— *Climbing herbs or shrubs.*

1. P. CŒRULEA. *Common Passion-flower.*—Shrubby; *lrs.* palmately and deeply 5-parted; segments linear-oblong, entire, lateral ones often 2-lobed; *pet.* glandular, with a 3-bracteolate involucre near the flower; *bracteoles* entire; *fil. of the crown* shorter than the corolla.—Native of Brazil, where it grows to the thickness of a man's arm and the height of 30f. Flowers large and beautiful, blue externally, white and purple within, continuing but one day. Fruit ovoid, yellow. Admired in cultivation.

2. P. INCARNÄTA. *Flesh-colored Passion-flower.*

Lvs. deeply 3-lobed, lobes oblong, acute, serrate; *petioles* with 2 glands near the summit; *bracteoles of the involucre* 3, obovate, glandular; *crown* triple.—Native from Va. to Flor. Stem climbing 20—30f. Flowers large and showy. Petals white. Two outer rows of filaments long, purple, with a whitish band, the inner row of short rays, flesh-colored. Berry pale yellow, of the size of an apple, eatable. May—July.

3. P. LUTEA. *Yellow Passion-flower.*

Lvs. glabrous, cordate, 3-lobed, obtuse; *petioles* without glands; *ped.* mostly in pairs; *pet.* narrower and much longer than the sepals.—A slender climber, 5—10f long, in woods and thickets, Ohio and S. States. Leaves yellowish-green, nearly as broad as long. Flowers small, greenish-yellow. Corona in 3 rows, the inner row a membranous disk with a fringed border. Fruit dark-purple. May—Jl.

ORDER LVIII. BEGONIACEÆ.—BEGONIADS.

Herbaceous plants, or succulent undershrubs, with an acid juice. *Lvs.* alternate, toothed, rarely entire, oblique at the base. *Stipules* large, scarious. *Fls.* pink-colored, in cymes, monœcious or diœcious. *Cal.* adherent, colored. [pistillate. *Sep.* in the staminate 4, in the pistillate 3 or 4. *Pet.* smaller than sepals, 2 in the staminate, 2 or 4 in the *Sta.* (stam. fls.) indefinite, distinct or combined. *Anth.* collected into a head, 2-celled

Obs. (rint. fls.) winged, 3-celled, with 3 large placentæ meeting in the axis. Stig. 3, 3-lobed, somewhat Fr. capsular, winged, 3-celled, many-seeded. Sds. minute, without albumen.   [spiral.

Genera 2, species 156, common in the West Indies, S. America and East Indies—none N. American. The roots are astringent and slightly bitter.

DIPLOCLINIUM. Lindl.

Gr. διπλοος, double, κλινη, couch; alluding to the double placentæ.

Fls. ♂·—♂ Sepals orbicular, colored like the petals, but larger; pet. oblong, acute; sta. combined in a column; anth. in a globose head. ♀ Sepals 3, lanceolate, larger than the 2 petals; stig. lobes distinct, spiral, erect; caps. wings unequal; placentæ double, or two in each cell.—*Evergreen, succulent undershrubs.*

D. EVANSIANUM. Lindl. (Begonia discolor. *Willd.* and 1st *edit.*)—Glabrous; st. branched, tumid and colored at the joints, succulent; lvs. large, slightly angular, mucronate-serrate, cordate-ovate, very unequal at base, petiolate, with weak, scattered prickles, and straight, red veins, the under surface deeply reddened; fls. pink-colored in all their parts except the golden yellow anthers and stigmas; ♀ larger than the ♂ and on peduncles twice as long.—From China. †

ORDER LIX. CUCURBITACEÆ.—CUCURBITS.

Herbs succulent, creeping or climbing by tendrils
Lvs. alternate, palmately-veined, rough.   Fls. monœcious or polygamous, never blue.
Cal. 5-toothed.   [reticulated veins.
Cor.—Petals 5, united with each other and cohering to the calyx, very cellular, strongly marked with
Sta. 5, distinct, more generally cohering in 3 sets.   Anth. very long and wavy or twisted.
Ova. inferior, 1-celled, with 3 parietal placentæ often filling the cells.
Fr. a pepo or membranous.   Seeds flat, with no albumen, often ariled.

Genera 60, species 270, natives of tropical regions, only a few being found in the temperate zones of Europe and America. A highly important order of plants, affording some of the most delicious and nutritive of fruits. A bitter, laxative principle pervades the group, which is so concentrated in a few as to render them actively medicinal. The officinal *colocynth* is prepared from the pulp of Cucumis Colocynthis, a powerful drastic poison.

Conspectus of the Genera.

		{ 1-seeded.				*Sicyos.*	1
	{ Fruit membranaceous, echinate,	{ 4-seeded.				*Echinocystis.*	2
	{	{ Seeds thin at edge.				*Lagenaria.*	5
{ white.	{ Fruit a pepo with a ligneous, smooth rind.					*Cucumis.*	6
		{ Indehiscent. { Seeds thick at edge.				*Cucurbita.*	7
	{ Fruit a pepo, { dehiscing elastically on one side.					*Momordica.*	4
Flowers { yellow.	{ Fruit a small, oval, many-seeded berry.					*Melothria.*	3

1. SICYOS.

Gr. σικυος, the ancient name of the cucumber.

Flowers ♂·—♂ Calyx 5-toothed; corolla rotate, 5-petaled; stamens 5, monadelphous or at length triadelphous; anthers contorted. ♀ Calyx 5-toothed, campanulate; petals 5, united at base into a campanulate corolla; styles 3, united at base; fruit ovate, membranaceous, hispid or echinate, with 1 large, compressed seed.—⊕ *Climbing herbs, with compound tendrils. Sterile and fertile fls. in the same axils.*

S. ANGULATUS. *Single-seed Cucumber.*

• St. branching, hairy; lvs. roundish, cordate with an obtuse sinus, 5-angled or 5-lobed, lobes acuminate, denticulate; ♀ much smaller than the ♂.—Can. and U. S. A weak, climbing vine, with long, spiral, branching tendrils. Leaves 3–4' broad, alternate, on long stalks. Flowers whitish, marked with green lines, the barren ones in long-pedunculate racemes. Fruit ½' long, ovate, spinous, 8–10 together in a crowded cluster, each with one large seed. Jl.

2. ECHINOCYSTIS. Torr. & Gray.

Gr. εχινος, sea urchin, κυστις, bladder; alluding to the spiny, inflated fruit.

Flowers monœcious. *Sterile fl.*—Calyx of 6 filiform-sutulate segments, shorter than the corolla; petals 6, united at base into a rotate-campanulate corolla; stamens 3, diadelphous. *Fertile fl.*—Cal. and cor. as above; abortive fil. 3, distinct, minute; style very short; stig. 2, large; fruit roundish, inflated, echinate, 4-seeded.—① *A climbing herb with branched tendrils.*

E. LOBÁTA. T. & G. (Sicyos. *Michx.* Momordica echinata. *Muhl.*) A smoothish, running vine, in rich river soils, Can. to Penn. and Mo. Stem deeply furrowed, with long, 3-parted tendrils placed nearly opposite the long petioles. Leaves membranaceous, palmately 5-lobed, cordate at base, lobes acuminate, denticulate. Flowers small, white, the barren ones very numerous, in axillary racemes often 1f long; fertile ones solitary or several, situated at the base of the raceme. Fruit 1—2′ in length, setose-echinate, at length dry and membranaceous, with 4 large seeds. Jl.--Sept.

3. MELOTHRIA.

Gr. μηλον, a melon, ϑριος, a certain food.

Flowers ♀ ♀ ♂ or ♂. Calyx infundibuliform-campanulate, limb in 5 subulate segments; petals 5, united into a campanulate corolla. ♂ Stamens 5, triadelphous. ♀ Stigmas 3; fruit a berry, ovoid, small, many-seeded.—*Tendrils simple.*

M. PENDÚLA.
Lvs. roundish-cordate, 5-lobed or angled, slightly hispid; *fls.* axillary, the sterile in small racemes, the fertile solitary, on long peduncles.—N. Y. to Ga and La. A slender vine, climbing over other vegetables. Leaves small (1—2′ diam.) Flowers small, yellowish. Style short, surrounded by a cup-shaped disk. Fruit small, oval. Jl.

4. MOMORDICA.

Lat. *mordeo, momordi,* to chew; from the chewed appearance of the seeds.

Flowers ♂. ♂ Calyx 5-cleft; petals 5, united at base; stamens 5, triadelphous. ♀ Calyx and corolla as in the ♂; style 3-cleft; pepo fleshy, bursting elastically; seeds compressed, with a fleshy arillus.

M. BALSAMINA. *Common Balsam Apple.—Lvs.* palmately 5-lobed, dentate, naked, shining; *ped.* solitary, filiform, 1-flowered, with an orbicular-cordate, dentate bract above the middle; *fr.* roundish-ovoid, angular, tuberculate, bursting elastically on one side.—From E. Indies. Occasionally cultivated for the balsamic and vulnerary fruit. Stem slender, climbing by simple tendrils. Flowers pale-yellow. Fruit orange-color, as large as a goose-egg. Aug.

5. CUCUMIS.

Said to be from the Celtic *cucc,* a hollow vessel.

Flowers ♂ or ♀. Calyx tubular-campanulate, with subulate segments; corolla deeply 5-parted. ♂ Stamens 5, triadelphous. ♀ Style short; stigmas 3, thick, 2-lobed; pepo fleshy, indehiscent; seeds ovate, flat, acute and not margined at the edge.—*Creeping, or climbing by tendrils. Fls. axillary, solitary, yellow.*

1. C. SATÍVUS. *Cucumber.—St.* prostrate, rough; *tendrils* simple; *lvs.* subcordate, palmately 5-angled or lobed, lobes subentire, acute, terminal one longest; *fr.* oblong, obtusely prismatic, prickly, on a short peduncle. -① Native of Tartary and India, whence it was first brought to England in 1573. It is now universally cultivated for the table, either fresh or pickled. Gathered and eaten before maturity. Jn.—Sept.—Many varieties.

2. C. MELO. *Musk Melon.—St.* prostrate, rough; *tendrils* simple; *lvs.* subcordate, roundish, obtuse, palmately 5-angled, lobes rounded, obtuse, obscurely denticulate; *fls.* ♀ ♀ ♂, the ♀ on short peduncles; *fr.* oval or subglobose, longitudinally torulose.—① Native of Asia, whence it was first brought to England in 1570. Generally cultivated for the juicy, yellowish, delicately flavored flesh of the mature fruit. Jn. Jl.—Varieties numerous.

3. C. Anguria. *Prickly Cucumber.*—*St.* prostrate, slender, hispid; *tendrils* simple; *lvs.* palmately and deeply sinuate-lobed, cordate at base; *fr.* oval-ovoid or subglobose, echinate.—① Native of Jamaica. Cultivated for the green fruit, which is about the size of a hen's egg and used for pickles. Jl. Aug.

4. C. Citrullus. Ser. (Cucurbita. *Linn.*) *Water Melon.*—*St.* prostrate, slender, hairy; *tendrils* branching; *lvs.* palmately 5-lobed, very glaucous beneath, lobes mostly sinuate-pinnatifid, all the segments obtuse; *fls.* solitary, on hairy peduncles, bracted at base; *fr.* elliptical, smooth, discolored.—① Native of Africa and India. Generally cultivated for its large and delicious fruit. Jn.—Aug.

5. C. Colocynthis. *Colocynth.*—*St.* prostrate, subhispid; *lvs.* cordate-ovate, cleft into many obtuse lobes, hairy-canescent beneath; *tendrils* short; *fls.* axillary, pedunculate; ♀ with a globose, hispid calyx tube and campanulate limb, with small petals; *fr.* globose, yellow when ripe, about as large as an orange, and intolerably bitter. The extract is the colocynth of the shops, poisonous, but medicinal.—From Japan.

6. LAGENARIA. Ser.

Gr. λαγηνος, a flagon or bottle; from the form of the fruit.

Flowers ♂. Calyx campanulate, 5-toothed; petals 5, obovate. ♂ Stamens 5, triadelphous; anthers very long, contorted. ♀ Stigmas 3, thick, 2-lobed, subsessile; pepo ligneous, 1-celled; seeds ariled, obcordate, compressed, margin tumid.—*Mostly climbing by tendrils.*

L. vulgaris. Ser. *Calabash. Bottle Gourd.*—Softly pubescent; *st.* climbing by branching tendrils; *lvs.* roundish-cordate, abruptly acuminate, denticulate, with 2 glands beneath at base; *fls.* axillary, solitary, pedunculate; *fr.* clavate, ventricose, at length smooth.—① Native within the tropics, often cultivated—the hard, woody rind of the fruit being used as ladles, bottles, &c. Flowers white Jl. Aug.

7. CUCURBITA.

A Latin word, signifying a vessel; from the form of the fruit of some species.

Flowers ♂. Corolla campanulate; petals united and coherent with the calyx. ♂ Calyx 5-toothed; stamens 5, triadelphous, anthers syngenecious, straight, parallel. ♀ Calyx 5-toothed, upper part deciduous after flowering; stigmas 3, thick, 2-lobed; pepo fleshy or ligneous, 3—5-celled; seeds thickened at margin, obovate, compressed, smooth.—*Fls. mostly yellow.*

1. C. Pepo. *Pumpkin.*—Hispid and scabrous; *st.* procumbent; *tendrils* branched; *lvs.* (very large) cordate, palmately 5-lobed or angled, denticulate; *fls.* axillary, ♂ long-pedunculate; *fr.* very large, roundish or oblong, smooth, furrowed and torulose.—① Native of the Levant. Long cultivated as a useful kitchen vegetable or for cattle. Flowers large, yellow. Fruit sometimes 3f diam., yellow when mature, yielding sugar abundantly. Jl.

2. C. Melopepo. *Flat Squash.*—Hairy; *st.* procumbent, with branched tendrils; *lvs.* cordate, palmately somewhat 5-lobed, denticulate; *fls.* pedunculate; *fr.* depressed-orbicular, the margin mostly torulose or tumid, smooth or warty.—Native country unknown. Cultivated for its fruit, a well known kitchen vegetable. There are many varieties in respect to the fruit.

3. C. verrucosa. *Warted Squash. Club Squash. Crook-neck Squash, &c.*—Hairy, procumbent; *lvs.* cordate, palmately and deeply 5-lobed, denticulate, terminal lobe narrowed at base; *fls.* pedunculate, large; *fr.* roundish elliptic, or clavate, often elongated and incurved at base.—① Mentioned by Nuttall as long cultivated by the Indians west of the Mississippi. Common in our gardens, with numerous well known varieties of the fruit. Jl.

4. C. ovifera. *Egg Squash.*—*Lvs.* cordate, angular, 5-lobed, denticulate, pubescent; *cal.* obovate, with a short neck, limb deciduous after flowering; *fr.* obovate, striped with lines lengthwise.—Native of Astrakan. Herbage and flowers similar to those of C. pepo, but less scabrous.

Order LX. GROSSULACEÆ.—Currants.

*Shr. either unarmed or spiny. Lvs. alternate, lobed, plaited in vernation.
Fls. in ax. dusky racemes, with bracts at their base.
Cal.—4—5-sepal., 4—5-cleft, regular, colored, marescent, imbricate in æstivation.
Cor.—Petals inserted in the throat of the calyx, small, distinct, as many as sepals.
Sta. as many as petals and alternate with them, very short ; anthers introrse.
Ova. 1-celled, with 2 parietal placentæ ; ovules numerous ; styles 2.
Fr. a 1-celled berry (the cell filled with pulp) crowned with the remains of the flower.
Sds. semitransp., the embryo minute, radicle next the micropyle.*

Genus 1, species 52. The gooseberries and currants are natives of the N. temperate zone of both continents, but unknown in the tropics or S. hemisphere, except S. America.

Properties.—The berries contain a sweet, mucilaginous pulp, together with malic or citric acid. They are always wholesome and usually esculent.

RIBES.

Character the same as that of the Order.

* *Stems unarmed.* CURRANTS.

1. R. FLORIDUM. L'Her. *Wild Black Currant.*

Lvs. subcordate, 3—5-lobed, sprinkled on both sides with yellowish, resinous dots ; *rac.* many-flowered, pendulous, pubescent ; *cal.* cylindrical ; *bracts* linear longer than the pedicels ; *fr.* obovoid, smooth, black.—A handsome shrub in woods and hedges, Can. to Ky. common, 3—4f high. Leaves 1—2' long, the width something more, lobes acute, spreading, 3, sometimes with 2 small additional ones ; dots just visible to the naked eye. Petioles 1—2' long. Flowers rather bell-shaped, greenish yellow. Fruit insipid. May, Jn.

2. R. PROSTRATUM. L'Her. (R. rigens. *Michx.*) *Mnn'ain Currant.*

St. reclined ; *lvs.* smooth, deeply cordate, 5—7-lobed, doubly serrate, reticulate-rugose ; *rac.* erect, lax, many-flowered ; *cal.* rotate ; *berries* globose, glandula.-hispid, red.—A small shrub, on mountains and rocky hills, Penn. to Can., ill-scented, and with ill-flavored berries—sometimes called *Skunk Currant.* Prostrate stems, with erect, straight branches. Leaves about as large as in No. 1, lobes acute. Petioles elongated. Racemes about 8-flowered, becoming erect in fruit. Bracts very short. Flowers marked with purple. Berries rather large. May.

3. R. RUBRUM. *Common Red Currant.*

Lvs. obtusely 3—5-lobed, smooth above, pubescent beneath, subcordate at base, margin mucronately serrate ; *rac.* nearly smooth, pendulous ; *cal.* short, rotate ; *bracts* much shorter than the pedicels ; *fr.* globose, glabrous, red.— Woods, St. Johnsbury, Vt. *Mr. Carey,* Wisconsin, *Lapham!* N. to the Arctic ocean. Cultivated universally in gardens.

β. (*White Currant.*) *Fr.* light amber-colored, larger and sweeter.

4. R. NIGRUM. *Black Currant.*—*Lvs.* 3—5-lobed, punctate beneath, dentate-serrate, longer than their petioles ; *rac.* lax, hairy, somewhat nodding ; *cal.* campanulate ; *bracts* nearly equaling the pedicels ; *fr.* roundish-ovoid, nearly black.—Native of Europe, &c. Cultivated and esteemed for its medicinal jelly. Flowers yellowish.—This species much resembles R. floridum.

5. R. AUREUM. Ph. *Missouri, or Golden Currant.*

Plant smooth ; *lvs.* 3-lobed, lobes divaricate, entire or with a few large teeth ; *petioles* longer than the leaves ; *bracts* linear, as long as the pedicels ; *rac.* lax, many-flowered ; *cal.* tubular, longer than the pedicels, segments oblong, obtuse ; *pet.* linear ; *fr.* smooth, oblong or globose, yellow, finally brown. —Mo. W. to Oregon. A beautiful shrub, 6—10f high, common in cultivation. Flowers numerous, yellow, very fragrant. Apr. May.

** *Spinescent or prickly.* GOOSEBERRIES.

6. R. CYNOSBATI. *Prickly Gooseberry.*

St. prickly or not ; *subaxillary spines* about in pairs ; *lvs.* cordate, 3—5-lobed, softly pubescent, lobes incisely dentate ; *rac.* nodding 2—3-flowered ; *calyx tube* ovate-cylindric, longer than the segments ; *pet.* obovate, shorter than the calyx segments ; *berries* prickly.—A handsome shrub, Northern and Western States, about 4f high, in hedges and thickets, mostly without prickles, but armed with 1—3 sharp spines just below the axil of each leaf. Leaves 1½—2½' diam. Petioles downy. Flowers greenish-white. Fruit mostly covered with long prickles, brownish-purple, eatable. May, Jn.

7. R. ROTUNDIFOLIUM. Michx. (R. triflorum. *Willd.*) *Wild Gooseberry.*
St. without prickles; *subaxillary spines* mostly solitary, short; *lvs.* roundish, smooth, 3—5-lobed, incisely dentate; *ped.* smooth, 1—3-flowered; *cal.* cylindrical, smooth; *pet.* spatulate, unguiculate; *sta.* exserted, smooth, much longer than the petals; *sty.* hairy, exserted, deeply 2—3-cleft; *berries* smooth.—In woods, N. H. to N. Car. and Mo. Shrub 3—4f high. Stems with a whitish bark. Leaves 1—2' diam. mostly truncate at base, shining above. Petioles ciliate, 1—3' long. Petals white. Fruit purple, delicious, resembling the garden gooseberry. May.

8. R. LACUSTRE. Poir. *Swamp Gooseberry.*
St. covered with prickles; *subaxillary spines* several; *lvs.* deeply 3—5-lobed, cordate at base, lobes deeply incised; *rac.* 5—8-flowered, pilose; *cal.* rotate; *berries* small, hispid.—In swamps, Northern States, and British Am. Shrub 3—4f high. Stems reddish from the numerous prickles, which differ from the spines only in size. Leaves shining above, 1½—2½' diam. Petioles ciliate, hispid, longer than the leaves. Flowers green. Fruit covered with long prickles, dark-purple, disagreeable. May.—The older stems are unarmed, save with a few spines.

9. R. HIRTELLUM. Michx. (R. triflorum. *Dw.* R. saxosum. *Hook.*)
St. unarmed, rarely prickly; *subaxillary spines* short, solitary, or nearly so; *lvs.* roundish, cordate, 3—5-lobed, toothed, pubescent beneath; *ped.* short, 1—2-flowered; *calyx tube* smooth, campanulate: segments twice longer than the petals; *sta.* longer than either; *sty.* hairy, 2-cleft; *fr.* smooth.—In rocky woods, N. H. and Mass. to Wisconsin, N. to Hudson's Bay. Leaves 9—18" diam., generally cleft half way to the middle. Flowers nodding, greenish. Fruit purple. May, Jn.

10. R. UVA-CRISPA. (R. Grossularia. *Willd.* and 1st *edit.*) *English or Garden Gooseberry.*—*St.* prickly; *lvs.* roundish, 3—5-lobed, hairy beneath, on short, hairy petioles; *ped.* hairy, 1-flowered; *cal.* campanulate; *sty.* and *ova.* hairy; *fr.* smooth or hairy, globose.—Native in England, and long cultivated until there are several hundred varieties, with red, white, green and amber fruit, often weighing an ounce or more each. Apr.

ORDER LXI. CACTACEÆ.—INDIAN FIGS.

St. succulent and shrubby, usually angular or 2-edged.
Lvs. almost always wanting, when present, fleshy, smooth and entire.
Fls. sessile, usually showy and of short duration.
Cal.—{ Sepals and petals numerous, often indefinite and confounded with each other, the sepals from
Cor.—{ the surface, and the petals from the summit of ovary.
Sta. indefinite. *Fil.* long and filiform. *Anth.* ovate, versatile.
Ova. inferior, fleshy, 1-celled, with parietal placentæ.
Sty. single, filiform, with several anthers in a star-like cluster.
Fr. succulent, 1-celled, many-seeded.
Sds. without albumen, with thick, foliaceous cotyledons, or often with scarcely any.

Genera 16, species about 900, all peculiarly American, no one having ever been found in any other quarter of the globe. They are chiefly confined within the tropics, only two or three species having been found beyond them. The prickly Pear (Opuntia vulgaris) is the only species found native as far north as New York.

Conspectus of the Genera.

Flowers { tubular-campanulate, rose-colored, &c. { Axis cylindric. . . . . **Cereus.** 2
{ { Axis globose. . . . . **Melocactus.** 3
{ somewhat rotate, yellow. . . . . **Opuntia.** 1

1. OPUNTIA. Tourn.

Opuntiana was a country near Phocis, where this was said to be naturalized.

Sepals and petals numerous, adnate to the ovary, not produced into a tube above it; stamens 00, shorter than the petals; style with numerous, thick, erect stigmas; berry umbilicate at apex, tuberculate; cotyledons semiterete.—*Shrubby plants, with articulated branches, the joints usually broad and flattened, with fascicles of prickles regularly arranged upon the surface*

O. vulgaris. Mill. (Cactus opuntia. *Linn.*) *Prickly Pear.*

Prostrate, creeping; *joints* ovate; *prickles* numerous in each fascicle, often with several subulate spines; *fls.* yellow.—A curious, fleshy plant, native in rocky and sandy places, Mass. to Flor. W. to Ia.! It is often cultivated. The singular form resembles a series of thick, fleshy leaves, 4—6' long, 1 as wide, growing from the tip or sides of each other, and armed with orange-colored spines. The flowers come forth from the edge of the joints, large, bright-yellow, and succeeded by a smooth, crimson, eatable fruit. †

2. CEREUS. DC.

Sepals very numerous, imbricated, adnate to the base of the ovary, and united into a long tube above it, the outer shorter, the inner petaloïd: stamens 00, coherent with the tube; style filiform, with many stigmas; berry scaly, with the remains of the sepals; cotyledons none?—*Fleshy shrubs, with woody, cylindrical, grooved axes, armed with clusters of spines. Fls. from the clusters of spines.*

1. C. PHYLLANTHUS. DC. (Cactus. *Linn.*) *Spleenwort.*—*Branches* ensiform, compressed, serrate; *fls.* with the terete, slender tube much longer than the limb of the petals.—From S. America. The articulations of the stem are 2f or more long, 2' wide, weak, bordered with large, obtuse serratures, and traversed lengthwise by a central, cylindrical, woody axis. Flowers white, 9—12' long, expanding by night, fragrant. †

2. C. PHYLLANTHÖIDES. DC. (Cactus. *Linn.*)—*Branches* ensiform, compressed, obovate, with spreading, rounded teeth; *fls.* arising from the lateral crenatures of the branches; *tube* shorter than the limb of the petals.—From Mexico. A splendid flowerer, with leaf-like, fleshy joints, each 6—10' long, 1—2' wide. Flowers rose-colored, 4' in length, expanding by day. †

3. C. TRUNCATUS. DC. (Cactus. *Linn.*)—Branching; *joints* short-compressed, serrate, truncate at the summit; *fls.* arising from the summit of the joints; *sty.* longer than the stamens or reflexed petals.—From Brazil. A very distinct species, a foot or more high. Joints 2—3' long, 1—1½' wide, leaf like. Flowers 2—3' long, pink-colored. †

4. C. GRANDIFLÔRUS. DC. (Cactus. *Linn.*)—Creeping, rooting; *st.* with about 5 angles; *fls.* terminal and lateral, very large, nocturnal; *pet.* spreading, shorter than the linear-lanceolate sepals.—From the W. Indies. Stems cylindric or prismatic, branching, the angles not very prominent. Flowers expanding by night, and enduring but a few hours, 8—12' diam. Sepals brown without, yellow within; petals white.—A magnificent flower, but of difficult culture. †

5. C. FLAGELLIFORMIS. DC. (Cactus. *Linn.*) *Snake Cactus.*—*St.* creeping, with about 10 angles, hispid; *fls.* lateral, diurnal; *tube* slender, longer than the limb of the petals.—From S. America. Stem about the size of the little finger, cylindric, indistinctly articulated, 2—5f long. Flowers of a lively pink color, smaller than those of the last, and continuing in bloom several days. †

Obs.—Many other species of this curious genus are occasionally reared in the parlor or the green-house, —so many that to notice them individually would transcend our limits.

3. MELOCACTUS.

Compounded of *melon* and *cactus*; from its form.

Calyx tube adherent to the ovary, lobes 5—6, petaloid; petals as many as sepals, united with them into a long cylindric tube; stamens and style filiform; stigma 5-rayed; berry smooth, crowned with the withered calyx and corolla.—*Suffruticose, fleshy, leafless. Spadix simple, crowning the globular, deeply furrowed axis. Flowers terminal.*

M. COMMÛNIS. *Turk's Cap. Melon Thistle.*—*Axis* ovate-subglobose, dark green, 12—18-angled; *ribs* straight; *spines* fasciculate, subequal.—Native of the Caribbean Islands. This remarkable plant appears like a large, green melon, with deep furrows and prominent ribs, and is full of juice. It is surmounted with a spadix (cephalium), which is cylindric, tuberculate, densely tomentose, bearing the red flowers at summit. †

ORDER LXII. MESEMBRYACEÆ.—FICOIDS.

Plants fleshy, of singular and various forms, yet often beautiful.
Lvs. mostly opposite, thick and oddly shaped.   [tion.
Fls. solitary, axillary and terminal, remarkable for their profusion, numerous, brilliant, and of long dura-
Cal.—Sepals varying from 4 to 8, but usually 5, somewhat connected at base.
Cor.—Petals indefinite, colored, in many rows.
Sta. indefinite, distinct, arising from the calyx.
Ova. inferior or nearly superior, many-celled. Stigmas numerous.
Caps. many-celled, opening in a stellate manner at the apex.
Sds. more commonly indefinite, attached to the inner angle of the cells.

    Genera 5, species 375, chiefly natives of the arid, sandy plains of the Cape of Good Hope. The species are much cultivated for ornament.

MESEMBRYANTHEMUM.

Gr. μεσημβρια, mid-day, ανθος; flowers expanding at midday.

Character essentially the same as that of the order.

1. M. CRYSTALLINUM. *Ice-plant.*—*Rt.* biennial; *lvs.* large, ovate, acute, wavy, frosted, 3-veined beneath.—A popular house plant, from Greece. It has a creeping stem, 1f or more in length, and, with the leaves, is covered over with frost-like, warty protuberances, giving the plant a very singular aspect. Flowers white, appearing all summer. †

2. M. CORDIFOLIUM. *Heart-leaved Ice-plant.*—Procumbent, spreading; *lvs.* petiolate, opposite, cordate-ovate; *cal.* 4-cleft, 2-horned.—♃ An interesting plant in *house* cultivation, from Cape Good Hope. The whole plant fleshy and succulent like others of its kind. Flowers pink-colored. Calyx thick, green, the horns opposite. Capsule translucent, marked at summit with cruciform lines. †

ORDER LXIV. CRASSULACEÆ.—HOUSELEEKS.

Plants herbaceous or shrubby, succulent. Lvs. entire or pinnatifid. Stip. o.
Fls. sessile, usually in cymes.
Cal.—Sepals 3—20, more or less united at base, persistent.
Cor.—Petals as many as the sepals, distinct, rarely cohering.
Sta. as many as the petals and alternate with them, or twice as many.
Ova. as many as the petals and opposite them. Fil. distinct. Anth. 2-celled, bursting lengthwise.
Fr.—Follicles as many as the ovaries, each opening by the ventral suture, many-seeded.

    Genera 22, species 456, chiefly natives of the warmer regions of the globe, particularly the Cape of Good Hope. About 20 are found in North America. They grow in the thinnest and dryest soil, on naked rocks, sandy deserts, &c. They have no peculiar property except a slight acridity. Many are highly ornamental.

Conspectus of the Genera.

In 4s : { Stamens 4.				Tillæa. 1
{ Stamens 5.				Bryophyllum. 5
In 5s : { stamens 10 : { Carpels distinct.				Sedum. 2
{ Carpels united.				Penthorum. 4
Floral organs arranged { In 12s.				Sempervivum. 3

1. TILLÆA. Michx.

In memory of Mich. Ang. Tilli, an Italian botanist; died 1740.

Calyx of 3 or 4 sepals united at base; petals 3 or 4, equal; stamens 3 or 4; caps. 3 or 4, distinct, follicular, opening by the inner surface, 2 or many-seeded.—① *Very minute, aquatic herbs. Lvs. opposite.*

T. SIMPLEX. Nutt. (T. ascendens. *Eaton.*) *Pigmy-weed.*
    St. ascending or erect, rooting at the lower joints; *lvs.* connate at base, linear-oblong, fleshy; *fls.* axillary, solitary, subsessile, their parts in 4s; *pet.* oval or oblong; *carpels* 8—10-seeded.—Near East Rock, New Haven, Ct. (*Dr. Robbins*), and Philadelphia, on muddy banks, rare. Stem 1—3′ high. Leaves 2—3″ long. Flowers as large as a pin's head. Petals oval, flat, acute, twice as long as the oval, minute calyx, longer than the stamens and fruit, and of a greenish-white color. Jl.—Sept.

2. SEDUM.

Lat. *sedere*, to sit; the plants, growing on bare rocks, look as if sitting there.

Sepals 4—5, united at base; petals 4—5, distinct; stamens 8—

10; carpels 4—5, distinct, many-seeded, with an entire scale at the base of each.—*Mostly herbaceous. Inflorescence cymose. Fls. mostly pentamerous.*

1. S. TELEPHIÖIDES. Michx.
Lvs. broadly lanceolate, attenuate at base, subdentate, smooth; *cymes* dense, corymbose; *sta.* 10, the petals, sepals and carpels in 5s.—Found on rocks, lake and river shores, N. Y., N. J., Harper's Ferry, Va.! &c. Stem a foot high. Leaves 1—2′ long, ⅓ as wide. Flowers numerous, purple, in a terminal, branching cyme. Jn.—Aug.—Like the other species, very tenacious of life. My specimens, gathered several months since at Harper's Ferry, are still growing in the dry papers.

2. S. TERNÀTUM. Michx. *Stone-crop.*
Lvs. ternately verticillate, obovate, flat, smooth, entire, the upper ones scattered, sessile, lanceolate; *cyme* in about 3 spikes; *fls.* secund, the central one with 10 stamens, the rest with only 8.—♃ In Can. West, Penn. the Southern and Western States, *Plummer!* Cultivated in N. Eng. Stems 3—8′ long, branching and decumbent at base, assurgent above. Cyme with the 3 branches spreading and recurved, the white flowers loosely arranged on their upper side. Jl. Aug. †

3. S. TELEPHIUM. *Common Orpine. Live-forever.—Rt.* tuberous, fleshy, white; *st.* 1 or 2f high, erect; *lvs.* flattish, ovate, obtuse, serrate, scattered; *cyme* corymbose, leafy.—♃ From Europe. Cultivated and nearly naturalized. Stems simple, leafy, round, smooth, purplish. Leaves sessile, fleshy. Flowers white and purple, in dense, terminal, leafy tufts. Aug. †

4. S. ANACAMPSÈROS. *Evergreen Stone-crop.—Rt.* fibrous; *st.* decumbent; *lvs.* cuneiform, attenuate at base; *cymes* corymbose, leafy.—♃ Native of Europe, growing there in crevices of rocks. Stems reddish and decumbent at base, erect and glaucous above. Lvs. fleshy, bluish green. Fls. purple. Jl. †

5. S. ACRE. *English Moss. Wall Pepper.—*Procumbent, spreading, branching from the base; *lvs.* very small, somewhat ovate, fleshy, crowded, alternate, closely sessile, obtuse, nearly erect; *cyme* few-flowered, trifid, leafy.—From Great Britain. In cultivation it spreads rapidly on walls, borders of flower-beds, &c. densely covering the surface. Flowers yellow. The whole plant abounds in an acrid, biting juice. †

3. SEMPERVÌVUM.

Lat. *semper vivere,* to live forever; in allusion to their tenacity of life.

Sepals 6—20, slightly cohering at base; petals as many as sepals, acuminate; stamens twice as many as petals; hypogynous scales lacerated; carpels as many as the petals.—♃ *Herbaceous plants or shrubs, propagated by axillary offsets. Lvs. thick, fleshy.*

1. S. TECTÒRUM. *House-leek.—Lvs.* fringed; *offsets* spreading.—A well known plant of the gardens, with thick, fleshy, mucilaginous leaves. It sends out runners with offsets, rarely flowering. It is so succulent and hardy that it will grow on dry walls, and on the roofs of houses (tectorum). It is sometimes placed in the borders of flower-beds.

2. S. ARBOREUM. *Tree House-leek.—St.* arborescent, smooth, branched; *lvs.* cuneiform, smoothish, bordered with soft, spreading ciliæ.—A curious and ornamental evergreen, from the Levant. Stem very thick and fleshy, branching into a tree-like form, 8—10f high (1—3f in pots). Fls. yellow, rarely appearing.

4. BRYOPHYLLUM. Salisb.

Gr. βρυω, to grow, φυλλον, a leaf: i. e. germinating from a leaf.

Calyx inflated, 4-cleft scarcely to the middle; corolla monopetalous, the tube long and cylindrical, 4-sided and obtuse at base; limb in 4, triangular, acute lobes; seeds many.—*An evergreen, fleshy,*

t, native of the E. Indies. Lvs. opposite, unequally them sometimes simple. Fls. greenish-purple.

Salisb.—Not uncommon in house cultivation, requiring but well-drained pot of rich loam. Stem thick, green, about 2f -5-foliate, with thick, oval, crenate leaflets. Flowers in a nicle, pendulous, remarkable for the large, inflated calyx, and exserted corollas.—This plant is distinguished in vegetable Fig. 10, 1, and § 88, a.

5. PENTHORUM.

ænts, five; on account of the 5-parted, angular capsule.

epals united at base; petals 5 or 0; capsules of 5 5-angled, 5-celled and 5-beaked; seeds 00, minute.— *cculent) herbs. Lvs. alternate. Fls. yellowish, cymose.*

Virginia Stone-crop.

and angular above; *lvs.* nearly sessile, lanceolate, acute at ally serrate; *fls.* in unilateral, cymose racemes.—A hardy uty, in moist situations, Can. and U. S. Stem 10—16' high, branches. Leaves 2—3' by ¼—1', membranaceous, smooth, nally serrate. Racemes several, recurved at first, at length e flowers arranged on their upper side, constituting a corym- le yellowish-green cyme. Pet. generally wanting. Jl—Sept.

LXV. SAXIFRAGACEÆ.—SAXIFRAGES.

alternate or opposite, sometimes stipulate.
ering more or less, persistent.
the sepals, inserted between the lobes of the calyx.
elled, opening longitudinally.
2-carpels, cohering at base, distinct and divergent above.
—2-celled, many-seeded.
, native of temperate and frigid climes in both continents. As a tribe their ræats species are among our most ornamental, cultivated plants.

2. PENNSYLVANICA. *Tall Saxifrage.*

Lvs. radical, oblong-lanceolate, rather acute, tapering at base, denticulate; scape nearly leafless; branches alternate, with close cymes forming a diffuse panicle; fls. pedicellate; pet. linear-lanceolate, but little longer than the calyx. —Larger than the foregoing, common in wet meadows, Me. to Ohio. Leaves fleshy, pale green, 5—8' by 1 —2', on a broad petiole. Scape 2—3f high, gross, hollow, hairy and viscid, branched into a large, oblong panicle of yellowish green flowers of no beauty. May.

3. S. AIZOÖN. Jacq.

Lvs. mostly radical, rosulate, spatulate, obtuse, with cartilaginous, white teeth, and a marginal row of impressed dots; fls. corymbose-paniculate; cal. (and ped. glandular-viscid) tube hemispherical, as long as the 5-toothed limb; pet. obovate; sty. divergent, longer than the calyx.—Southern shores of Lake Sup. (*Pitcher* in T. & G. I. p. 566) on shady, moist rocks. Stem 5—10' high. Fls. white. Jl.

4. S. AIZÖIDES.

Cæspitose, leafy; lvs. alternate, linear-oblong, more or less ciliate, slightly mucronate, thick, flat, mostly persistent; flowering stems annual; fls. paniculate, sometimes solitary; sep. ovate, slightly coherent with the ovary; pet. oblong, longer than the sepals; stigmas depressed; caps. rather thick, as long as the styles.—In the clefts of rocks, Willoughby Mt., Westmore, Vt. 500f above W. Lake, *Wood*, N. to the Arctic sea. Barren stems short, with densely crowded leaves; flowering ones ascending, 2—4' long, with scattered leaves. Leaves 4—6' long, about 2'' wide. Pedicels bracteate. Flowers yellow, dotted.

5. S. OPPOSITIFOLIA. *Opposite-leaved Saxifrage.*

Lvs. opposite, rather crowded, obovate, carinate, ciliate, obtuse, punctate, persistent; fls. solitary; cal. free from the ovary; pet. large, obovate, 5-veined, longer than the stamens.—In the same locality as the above, *Wood*. Stems purplish, very branching and diffuse. Leaves bluish-green, 1—2'' in length, narrowed and clasping at base. Flowering stems annual, 1—3' long. Flowers light purple, large and showy.

Obs.—I discovered this and the foregoing species in the above locality, in Aug. 1845, when they had passed flowering.

6. S. RIVULARIS.

St. weak, ascending, 3—5-flowered; radical lvs. petiolate, reniform, crenately lobed; cauline lanceolate, subentire; calyx lobes broad-ovate, nearly as long as the ovate petals, but much shorter than the thick, short-beaked capsules. —White Mts., N H., *Oakes*, N. to Arc. Am. A very small species, with white, bracteate flowers. Stems about 2' high, annual, with alternate leaves.

2. SULLIVANTIA. Torr.

In honor of Wm. S. Sullivant, author of Musci Alleghanenses, &c.

Calyx campanulate, coherent with the base of the ovary; segments ovate, acute; petals oval-spatulate, unguiculate, inserted on the summit of the calyx tube, and twice as long as its lobes; stamens 5, inserted with the petals, shorter than the calyx; capsule 2-beaked, 2-celled; seeds 00, ascending; testa wing-like, not conformed to the nucleus.—4 Lvs. mostly radical, palmate-veined. Fls in a loose panicle, small, white.

S. OHIÖNIS. Torr.

A diffuse, weak-stemmed plant, first discovered in Highland Co., Ohio. by him whose name it bears. Stem annual, very slender, 8—16' long, ascending, glandular. Radical leaves roundish, cordate, lobed and toothed, 1—2' diam., on long petioles. Cauline leaves mostly very small, bract-like, cuneate at base, 3—5-toothed at summit. May, Jn.

3. HEUCHÉRA.

In honor of Prof. Heucher, botanic author, Wittenberg, Germany.

Calyx 5-cleft, coherent with the ovary below, segments obtuse; co-

24*

f 5 small, entire petals, inserted with the 5 stamens
f the calyx; capsule 1-celled, 2-beaked, many-seeded.
al, long-petioled.

ICANA. *Alum-root.*
scent; *lvs.* roundish, cordate, somewhat 7-lobed, lobes short
nate-dentate, teeth mucronate; *panicle* elongated, loose; *pedi-*
d short, obtuse; *pet.* spatulate, about as long as the calyx;
l.—A neat plant, rare in the southern parts of N. Eng. and
the W.! and S. Leaves all radical, 2—3½' diam., on peti-
h Scape 2—4f high, paniculate, nearly ⅓ this length. Pe-
ered. Calyx campanulate, more conspicuous than the purplish-
y, Jn.—Root astringent, hence the common name.

ENS. Ph. (H. grandiflora. *Raf.*)
, minutely-pubescent above, and with the long petioles, gla-
glabrous, orbicular-cordate, 7—9-lobed, lobes rounded, and
cronate, ciliate teeth; *ped.* cymose, dichotomous, joints flexu-
ulate; *fls.* large; *pet.* longer than the included stamens; *sty.*
enn., Md! Va. Scape 1—2f high. Leaves 3—5' diam. the
n a few scattered hairs. Flowers 5—6" long, purple. May, Jn

RDSONI. R.-Br.
d) and petioles hairy and rough; *lvs.* orbicular-cordate, with
7-lobed, lobes obtuse, incisely crenate, ciliate; panicle rather
omewhat oblique; *pet.* ciliolate, somewhat unequal, about the
ls; *sta.* a little exserted; *sty.* included.—Prairies and bottoms,
Can. Scape 1—2f high. Leaves glabrous above, veins b-
wers 6—7" long. May.

4, MITELLA. Tourn.

A Lat. diminutive from *mitra*, a mitre. See *Tiarella.*

campanulate; petals 5, pectinately pinnatifid, insert-

stolons creeping.—Common in rocky woods Can. to Penn., and generally associated with *Mitella diphylla*, which plant, in its general aspect, it much resembles. The scape arises from a creeping root-stock about 10' high, often bearing a leaf. Leaves 2—3' long, ¼ as wide, hairy, and on hairy petioles 4—6 long. Racemes 1—2½' long; fls. wholly white, with minute bracts. May, Jn.

6. CHRYSOSPLENIUM. Tourn.

Gr. χρυσος, gold, σπλην, the spleen; on account of the medicinal qualities.

Calyx adnate to the ovary, 4—5-lobed, more or less colored inside; corolla 0; stamens 8—10, superior, short; styles 2; capsule obcordate, compressed, 1-celled, 2-valved, many-seeded.—*Small, aquatic herbs.*

C. AMERICANUM. Schw. (C. oppositifolium. *Michx.*) *Water-carpet.*
Lvs. opposite, roundish, slightly crenate, tapering to the petiole.—A small plant, in springs and streams, spreading upon the muddy surface. Stem square, 3—6 inches long, divided in a dichotomous manner at top. Leaves opposite, ½' in length, smooth. Calyx 4-cleft, greenish-yellow, with purple lines. Corolla 0. Stamens 8, very short, with orange-colored anthers, which are the only conspicuous part of the flower. The terminal flower is sometimes decandrous. Apr. May.

SUBORDER 2.—ESCALLONIEÆ.

7. ITEA.

Gr. name for the willow; from a resemblance of foliage.

Calyx small, with 5, subulate segments; petals 5, lance-linear, inflexed at the apex, inserted on the calyx; stamens 5, inserted into the calyx; styles united; capsule 2-celled, 2-furrowed, 8—12-seeded. —*A shrub with alternate, simple leaves, and a simple, spicate, terminal raceme of white flowers.*

I. VIRGINICA.
Margins of swamps and sluggish streams, N. J. and Penn. to Flor. Shrub about 6f high. Leaves 1½—3' long, oval-acuminate, serrulate, on short petioles. Capsule oblong, acuminate with the style, its two carpels separating in maturity. May, Jn.

SUBORDER. 3.—HYDRANGEÆ.

Petals valvate. Capsules 2-celled. Leaves opposite, exstipulate. SHRUBS

8. HYDRANGEA.

Gr. ὑδωρ, water. αγγιον, a vessel; because the cultivated species require so copious a supply of water

Marginal flowers commonly sterile, with a broad, rotate, 4—5-cleft, colored calyx, and with neither petals, stamens nor styles. *Fertile fl.* Calyx tube hemispherical, adherent to the ovary, limb 4—5-toothed, persistent; petals ovate, sessile; stamens twice as many as the petals; capsule 2-beaked, opening by a foramen between the beaks; seeds numerous.—*Shrubs with opposite leaves. Fls. cymose, generally radiant.*

1. H. ARBORESCENS. (H. vulgaris. *Michx.*) *Common Hydrangea.*
Lvs. ovate, obtuse or cordate at base, acuminate, serrate-dentate, nearly smooth; fls. in fastigiate cymes.—An elegant shrub, native in the Middle and Western States! cultivated in the Northern, attaining the height of 5 or 6f on its native shady banks. Fertile flowers small, white, becoming roseate, very numerous. The cultivated varieties have either the marginal flowers radiate, or all sterile and radiate. †

2. H. QUERCIFOLIA. Bartram. *Oak-leaved Hydrangea.*—Lvs. deeply sinuate-lobed, dentate, tomentose beneath; cymes paniculate, radiant, the *sterile flowers* very large and numerous.—A beautiful shrub, native of Flor., not uncommon

ht 4—5f. Leaves very large. Sterile flowers with roundish
becoming reddish, very showy. †

ι. *Changeable Hydrangea.*—*Lvs.* elliptical, narrowed at each
e, strongly veined, smooth ; *cymes* radiant ; *fl.* mostly sterile.
of China, where it has long been cultivated. Stems 1—3f
ge. Barren flowers very numerous and showy, at first green,
ily through straw-color, sulphur-yellow, white, purple, and
t flowers are central and much smaller. It thrives in large
with loam, abundantly watered. The flowers endure sever-

BORDER 4.—**P H I L A D E L P H E Æ.**

) in æstivation. Capsule 3—4-celled, loculicidal. Shrubs.

9. PHILADELPHUS.

Name from Philadelphus, king of Egypt.

arted, half-superior, persistent ; corolla 4—5-petaled ;
amens 20—40, shorter than the petals ; capsule 4-
with loculicidal dehiscence ; seeds many, arilled.—
ring shrubs. *Lvs, opposite, exstipulate.*

avs. Willd. (P. inodorus. *Michx.*) *Large-flowered Syringa.*—
inate, denticulate, 3-veined, axils of the veins hairy ; *fig. 4,*
ided.—A very showy shrub, 6f high, native at the South, cul-
ries. Branches smooth, long and slender. Flowers large, in
of 2 or 3, white, nearly inodorous. Calyx divisions conspicu-
and much longer than the tube. Jn.—The upper leaves are
uite narrow. †

us. *False Syringa.*—*Lvs.* ovate, subdentate, smooth ; *fy.* dis-
3. Europe. A handsome shrub, often cultivated in our shrub-

desolations of autumn and winter, this alone puts forth its yellow blossoms. The small branches were formerly used for "divining rods," to indicate the presence of the precious metals and of deep springs of water, and there are even at this day, persons who deem a denial of these virtues to the witch hazel, as offence little short of heresy.

ORDER LXVII. UMBELLIFERÆ.—UMBELLIFERS.

St. herbaceous, hollow, furrowed.

Lea. usually divided, simple or compound, with sheathing petioles.

Flo. arranged in umbels, mostly white, often yellow, pink, blue, or greenish.

Cal. adhering to the ovary, entire or 5-toothed.

Cor.—Petals 5, usually inflected at the point, imbricate in æstivation.

Sta. 5, alternate with the petals and inserted with them upon the disk.

Ova. inferior, 2-celled, surmounted by the fleshy disk which bears the stamens and petals.

Sty. 2, distinct, or united at their thickened bases. *Stig.* simple.

Fr. dry, consisting of 2 coherent carpels, separating from each other by their faces (commissure) into two halves (merocarps).

Carpophore.—the slender, simple, or forked axis by which the carpels are borne, cohering to it by the faces of the commissure.

Ribs.—A definite number of ridges traversing the carpels, the larger ones (primary) alternating with the smaller (secondary).

Vittæ.—Little linear receptacles of colored volatile oil, imbedded in the substance of the pericarp, just beneath the intervals of the ribs and the commissure.

Genera 287, species 1500.—This is a vast and well defined natural order, native of damp places, waysides, groves, &c., in the cool parts of the world. Very few are found in tropical countries except upon the mountains.

Properties aromatic, stimulant and carminative, depending upon a volatile oil residing in the vittæ of the fruit, in the roots, &c. The herbage is frequently pervaded by an acrid, narcotic principle, rendering it very poisonous. Of this nature is the Conium maculatum (hemlock), Cicuta virosa, Æthusa Cynapium (fools parsley), besides many others which have, at least, a suspicious character. But the fruit is never poisonous, and is usually stimulant and aromatic, as caraway, anise, dill, coriander, &c. Even the roots and herbage of other species are wholesome and nutritive, as the carrot parsnep, sweet cisely, celery and Archangelica. The gum-resin, assafœtida, exudes from incisions in the Ferula of Persia; the gum galbanum is the product of Galbanum officinale, an Indian species.

The genera of the Umbelliferæ are numerous, and not easily distinguished. The characters by which De Candolle has more successfully than any other author divided this order into tribes and genera, are chiefly founded upon the number and development of the ribs, the presence or absence of the vittæ, and the form of the albumen, particularly at the commissure. These parts, therefore, minute as they are, will require the special attention of the student.

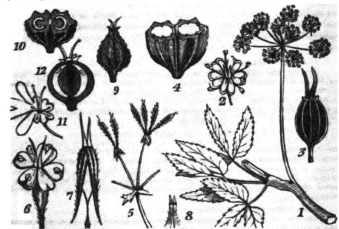

FIG. 46.—1. Zizia aurea, with its compound, naked umbel, &c. 2. A flower enlarged. 3. The fruit, with its filiform ribs and two persistent styles. 4. Cross section, showing the two carpels with the vittæ and flat commissure. 5. Umbel of Osmorhiza longistylis in fruit. 6. A flower enlarged. 7. The fruit with the mericarps separating from the base and supported by the bifid carpophore. 8. Summit of the fruit of O. brevistylis. 9. Fruit of Conium maculatum, with the undulate-crenulate ribs. 10. Cross section, showing the grooved commissure and involute albumen. 11. Radiant flower of Coriandrum. 12. Vertical section of the globose fruit, showing the minute embryo.

Conspectus of the Genera.

• Plants native or naturalized.

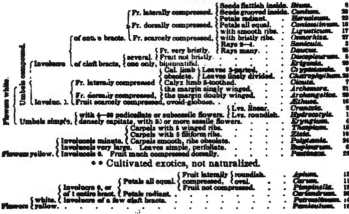

SUBORDER 1.—**ORTHOSPERMÆ.**

The inner surface of the seeds and albumen flat or nearly so.

1. HYDROCOTYLE.

Gr. ὕδωρ, water, κοτυλη, a vessel; the concave leaf often holds water.

Calyx obsolete; petals equal, ovate, spreading, entire, the point not inflected; styles shorter than stamens; fruit laterally flattened, the commissure narrow; carpels 3-ribbed, without vittæ.—*Herbaceous, creeping, usually aquatic plants. Umbels simple. Involucre few-leaved.*

1. H. AMERICÀNA. *Pennywort.*

Smooth and shining; *st.* filiform, procumbent; *lvs.* reniform-orbicular, slightly lobed, crenate; *umbels* sessile, 3—5-flowered; *fr.* orbicular.—2/ A small, delicate plant, growing close to the moist earth beneath the shade of other vegetables, Can. to S. Car. Stems branching, 2—6' long. Leaves thin, 1—2' diam., on petioles 2—3' long. Flowers greenish-white, small, nearly sessile, in simple, capitate, sessile, axillary umbels. Jn.—Aug.

2. H. INTERRUPTA. Muhl. (H. vulgaris. *Michx.*)

Smooth; *lvs.* peltate, orbicular, crenate; *umbels* capitate, proliferous, subsessile, about 5-flowered; *fr.* acute at base.—2/ In wet places, New Bedford, Mass. *T. A. Greene,* rare. Root and stem creeping. Leaves almost centrally peltate, thin, 8—10'' diam. Petioles 2—3' long. Peduncles longer than the petioles. Flowers subsessile, in close umbels which become whorls in interrupted spikes by other umbels being successively produced on the extending peduncle. Jn.

3. H. UMBELLÀTA. *Umbellate Pennywort.*

Smooth; *lvs.* peltate, orbicular, crenate, emarginate at base, on long petioles; scapes about as long as the petioles; *umbels* simple, often proliferous; *fls.* pedicellate.—2/ In ponds and bogs, Mass. ! to La., rare. Stems creeping, often submersed, several inches long. Leaves 8—12'' diam., notched at base so as to appear reniform. Petioles a little eccentric, and with the scapes slender, floating or erect, and 4—6' long. Umbels 20—30-flowered, the upper pedicel often prolonged and umbellate. May—Jl.

4. H. RANUNCULÔIDES. Linn. f. (H. cymbalarifolia. *Muh.*)

Glabrous; *les.* roundish-reniform, 3—5-lobed, crenate; *petioles* much longer than the peduncles; *umbels* 5—10-flowered, capitate; *fr.* roundish, smooth.—In water, Penn. to Ga. Stems weak, 1—2f long. Leaves 1—2' diam., the middle lobe smaller than the others. Petioles 2—3' long. Peduncle about 1' long. Jl. Aug.

2. CRANTZIA. Nutt.

In honor of Prof. Crantz, author of a monograph on the Umbelliferæ.

Calyx tube subglobose, margin obsolete; petals obtuse; fruit subglobose, the commissure excavated, with 2 vittæ; carpels unequal, 5-ribbed, with a vitta in each interval.—*Small, creeping herbs with linear or filiform, entire leaves. Umbels simple, involucrate.*

C. LINEÀTA. Nutt. (Hydrocotyle. *Michx.*)

Les. cuneate-linear, sessile, obtuse at apex, and with transverse veins, shorter than the peduncles.—Muddy banks of rivers, Mass.! to La. Stems several inches long, creeping and rooting in the mud. Leaves 1—2' by 1—2'', often linear and appearing like petioles without laminæ. Umbels 4—6-flowered. Peduncles ½ longer than the leaves. Involucre 4—6-leaved. Fruit with red vittæ. May—Jl.

3. SANICÜLA. Tourn.

Lat. *sanare*, to cure; on account of the reputed virtues as a vulnerary.

Flowers ♀ ☿ ♂; calyx tube echinate, segments acute, leafy; petals obovate, erect, with a long, inflected point; fruit subglobose, armed with hooked prickles; carpels without ribs; vittæ numerous.—♃ *Umbel nearly simple. Rays few, with many-flowered, capitate umbellets. Involucre of few, often cleft leaflets, involucel of several, entire.*

8. MARILANDICA. *Sanicle.*

Les. 5-parted, digitate, mostly radical; *lfts.* or segments, oblong, incisely serrate; *sterile fls.* pedicellate, *fertile sessile*; *calyx segments* entire.—In low woods, thickets, U. S., and Can., common. Stem 1—2f high, dichotomously branched above, smooth, furrowed. Radical leaves on petioles 6—12' long, 3-parted to the base, with the lateral segments deeply 2-parted. Segments 2—4' long, ½ as wide, irregularly and mucronately toothed. Cauline leaves few, nearly sessile. Involucres 6-leaved, serrate. Umbels often proliferous. Umbellets capitate. Flowers mostly barren, white, sometimes yellowish. Fruit densely clothed with hooked bristles. Jn.

4. ERYNGIUM. Tourn.

Gr. ερυγειν, to belch; a supposed remedy for flatulence.

Flowers sessile, collected in dense heads; calyx lobes somewhat leafy; petals connivent, oblong, emarginate with a long inflexed point; styles filiform; fruit scaly or tuberculate, obovate, terete, without vittæ or scales.—*Herbaceous or suffruticose. Fls. blue or white bracteate; lower bracts involucrate, the others smaller and paleaceous.*

1. E. AQUÀTICUM. *Button Snake-root.*

Les. broadly linear, parallel-veined, ciliate with remote soft spines; *bracts* tipped with spines, those of the involucels entire, shorter than the ovate-globose heads.—Low grounds on prairies, Ia.! Ill., &c. A remarkable plant, appearing like one of the Endogenæ. Very glaucous. Stem simple, 1—5f high. Leaves often 1—2f long, ½—1½' wide. Heads pedunculate, ½—1' diam. Flowers white, inconspicuous. Jl. Aug.

2. E. VIRGINIÀNUM. Lam. (E. aquaticum. *Michx.*)

Les. linear-lanceolate, uncinately serrate, tapering to both ends; *invol.* of 7—8 linear leaflets, longer than the heads, 3-cleft or spinose-dentate; *scales* tricuspidate.—♃ Marshes, N. J. to Ohio, *Prof. Lock!* and La. Stem hollow, 3—4f high, branched above. Leaves 6—10' by 5—10'', upper ones much small-

er. Heads numerous, less than 1' diam. Flowers pale blue or nearly white.
Jl. Aug.

5. DISCOPLEURA. DC.

Gr. δισκος, the disk ; πλευρα, a rib; i. e. the disk and ribs (of the fruit) united.

Calyx teeth subulate, persistent ; petals ovate, entire, with a minute, inflexed point ; fruit ovate, often didymous ; carpels 5-ribbed the 3 dorsal ribs filiform, subacute, prominent, the 2 lateral united, with a thick accessory margin ; intervals with single vittæ, seeds subterete.—① *Lvs. much dissected. Umbels compound. Bracts of the involucre cleft. Fls. white.*

D. CAPILLACEA. DC. (Ammi. *Spreng.*) *Bishop-weed.*
Erect or procumbent; *umbels* 3—10-rayed; *lfts of the invol.* 3—5, mostly 3-cleft; *fr.* ovate.—In swamps near the coast, Mass. ! to Ga. Stem much branched, 1—2f high. Leaves very smooth, ternately dissected, with subulate, spreading segments. Umbels axillary, pedunculate, spreading. Involucre leaflets about 3, with setaceous segments. Involucels filiform, longer than the umbellets. Jl.—Nov.

6. BUPLEURUM. Tourn.

Gr. βους, an ox, πλευρον, a rib; from the ribbed (veined) leaves of some of the species.

Calyx margin obsolete ; petals somewhat orbicular, entire, with a broad, closely inflexed point ; fruit laterally compressed ; carpels 5-ribbed, lateral ones marginal ; seed teretely convex; flattish on the face.—*Herbaceous or shrubby. Lvs. mostly simple. Invol. various. Fls. yellow.*

B. ROTUNDIFOLIUM. *Modesty. Thorough-wax.*
Lvs. roundish-ovate, entire, perfoliate; *invol.* 0; *involucels* of 5, ovate, mucronate bracts; *fr.* with very slender ribs, intervals smooth, mostly without vittæ.—In cultivated grounds and fields, N. Y. and Penn. and Ia. ! rare. Stem 1f or more high, branching. Leaves 1—3' long, ½ as wide, rounded at base, acute at apex, very smooth. Umbels 5—9-rayed. Involucels longer than the umbellets. Fruit crowned with the wax-like shining base of the styles (stylopodium). Jl. Aug.

7. CICUTA.

A Latin name used by Virgil (Ec. 2d and 5th) but of unknown application.

Calyx margin of 5, broad segments ; petals obcordate, the points inflected ; fruit subglobose, didymous ; carpels with 5 flattish, equal ribs, 2 of them marginal ; intervals filled with single vittæ, commissure with 2 vittæ ; carpophore 2-parted ; seeds terete.—♃ *Aquatic poisonous herbs. Leaves compound. Stems hollow. Umbels perfect. Invol. few-leaved or 0. Involucels many-leaved. Fls. white.*

1. C. MACULATA. *Water Hemlock. Spotted Cowbane.*
St. streaked with purple; *lower lvs.* triternate and quinate; *upper* biternate ; segments lanceolate, mucronately serrate; *umbels* terminal and axillary.
—Common in wet meadows, U. S. and Can. Stem 3—6f high, smooth, striate, jointed, hollow, glaucous, branched above. Petioles dilated at base into long abrupt, clasping stipules. Leaflets or segments 1—3' long, ½—¾' wide, finely serrate, the veins mostly running to the notches, rarely to the points ! Umbels rather numerous, naked, 2—4' broad. Involucels of 5—6 short, narrow, acute bracts. Fruit 1½" diam., 10-ribbed, crowned with the permanent calyx and styles. Jl. Aug.—The thick, fleshy root is a dangerous poison, but sometimes used in medicine.

2. C. BULBIFERA. *Bulbiferous Cicuta. Narrow-leaved Hemlock.*
Axils of the branches bulbiferous; *lvs.* biternately divided ; *lfts.* linear, with *remote, divergent* teeth; *umbels* terminal and axillary.—In wet meadows, Penn.

to Can. Stem 3—4f high, round, striate, hollow, green, branching. Leaves various, those of the stem generally biternate, of the branches ternate. Leaflets or segments 2—4' long, 1—4" wide, linear or lance-linear, smooth, with slender teeth. Bulblets often numerous, opposite, and within the axils of the bracteate petioles. Umbels terminal. Involucre 0. Umbellets of close, small, white flowers, and slight involucels. Aug.

8. SIUM.

Celtic *siv*, water: that is, a genus of aquatic plants.

Calyx margin 5-toothed or obsolete ; petals obcordate, with an inflexed point ; fruit nearly oval , carpels with 5 obtusish ribs, and several vittæ in each interval : carpophore 2 parted.— ♃ *Aquatic. Les. pinnately divided. Umbels perfect, with partial and general many-leaved involucra. Fls. white.*

1. S. LATIFOLIUM. *Water Parsnep.*
   St. angular, sulcate; *lfts.* oblong-lanceolate, acutely serrate, acuminate ; *cal.* teeth elongated.—A tall plant in swamps and ditches, N. J. to Ia.! and Can. Stem about 3 high, smooth, hollow, with 7 deep-furrowed and prominent angles. Leaflets or segments 4—6' long, 1—2' broad, equally serrate, in about 4 pairs, with an odd one, those submerged, if any, pinnatifid. Petioles embracing the stem. Umbels large, with many-flowered rays. Flowers small, white. Jl. Aug.

2. S. LINEARE. *Michx.* (S. latifolium, β. *lineare. Bw.*)
   St. angular, sulcate; *lfts.* 9—11, linear and lance-linear, finely serrate, acute ; *cal* teeth acute.—More common than the last, in swamps, N. J. to Ia.! and Can. Stem 2—4f high, smooth, with 7 prominent angles. Leaflets 2—4' long, 2—4" (rarely 10") wide, the odd and lower ones petiolulate, middle pairs sessile. Umbels 1½—2½' broad. Involucre of 5 or 6 linear bracts, ½ as long as the 15—21 rays. Umbellets with numerous, small, white flowers. Fruit roundish, crowned with the broad, yellowish stylopodium. Jl. Aug.

9. CRYPTOTÆNIA. DC.

Gr. κρυπτω, to conceal, ταινια, a wreath or border; from the obsolete border of calyx.

Margin of the calyx obsolete : petals with an inflexed point ; fruit linear-oblong or ovate-oblong : : carpels with 5 obtuse ribs ; carpophore free, 2-parted ; vittæ very narrow, twice as many as the ribs. — ♃ *Lrs. 3-parted, lobed and toothed. Umbels compound, with very unequal rays. Invol. 0. Involucels few-leaved. Fls. white.*

C. CANADENSIS. DC. (Sison Canadense. *Linn.*) *Hone-wort.*
   Lrs. smooth ; *lfts.* or segments rhomboid-ovate, distinct, entire or 2—3-lobed, doubly serrate, lateral ones oblique at base; *umbels* numerous, irregular, axillary and terminal.—Common in moist woods. Stem erect, 1—2f high. Lower petioles 2—6' long, clasping. Leaflets 3, 2—3' long, 1—2' wide, petiolulate. Umbels paniculate, of 3—5 very unequal rays. Umbellets of 4—6 unequal pedicels and minute involucels. Flowers small, white. Fruit near 3" long, oblong-elliptic. Jl.

10. ZIZIA. Koch.

Calyx margin obsolete or minutely toothed : petals carinate, apex acuminate, inflexed ; fruit roundish or oval, didymous ; carpels 5-ribbed, lateral ribs marginal ; intervals with 1—3 vittæ, commissure with 2—4 ; carpophore 2-parted : seeds plano-convex.— ♃ *Lvs. divided. Umbels perfect. Invol 0. Involucels few-leared. Fls. yellow.*

1. Z. AUREA. Koch. (Smyrnium. *Linn.* Thaspium. *Nutt.?*) *Golden Alexanders.*
   Lrs. biternate ; *lfts.* oval-lanceolate, serrate; *umbellets* with short rays.—Hills and meadows, U. S. and Can. Stems 1—2f high, branching above, rather slender, erect, hollow, angular-furrowed, smooth as well as every other part of

the plant, and furnished with few leaves. The lower leaves are on long peti-cles, the leaflets with coarse serratures, and sometimes quinate. The umbels are about 2 inches broad, of 10—15 rays, the umbellets ¼ inch broad, dense. Flowers numerous, orange-yellow. Fruit oval, brown, with prominent ribs. Root black, tufted. June.

2. Z. INTEGERRIMA. DC. (Smyrnium. *Linn.*) *Golden Alexanders.*
Lvs. biternate; *lfts.* oblique, oval, entire, smooth and glaucous.—Rocky woods, &c., N. Y. to Ohio and La., rare. Stem 1—2f high, branching above. Radical leaves often triternate, cauline biternate, all petiolate. Segments 1—1½' long, ½ as wide, mucronate, lateral oblique at base, odd one often 2—3-lobed. Umbels terminal, loose, on a long peduncle. Rays unequal, slender, spread-ing, 1—3' long, with minute involucels. Fruit roundish, compressed laterally May, Jn.

11. CARUM
From *Caria*, the native country of the plant, according to Pliny.

Calyx margin obsolete; petals obovate, emarginate, the point in flexed; styles dilated at base, spreading; fruit oval, compressed lat-erally; carpels 5-ribbed, lateral ribs marginal; intervals with single vittæ, commissure with 2.—*Herbs with dissected leaves. Umbels per-fect. Involucra various. Fls. white.*

C. CARVI. *Caraway.—Lvs.* somewhat bipinnatifid, with numerous linear segments; *invol.* 1-leaved or 0; involucels 0.—Native of Europe, &c. Stem about 2f high, branched, smooth, striate. Lower leaves large, on long petioles, with tumid, clasping sheaths. Umbels on long peduncles, involucrate bract, when present, linear-lanceolate. Jn.—Cultivated for its fine aromatic fruit, so well known in domestic economy. ‡

12. PIMPINELLA.

Calyx limb obsolete; petals obcordate, a little unequal; disk 0; flowers perfect or diclinous; styles capillary, as long as fruit; fruit ovate, ribbed, with convex intervals.—*European herbs, mostly 4, with pinnately, many-parted leaves, and white flowers. Umbels compound. Invol. 0.*

P. ANISUM. *Anise.—Radical lrs.* incisely trifid; *cauline* ones multifid, with narrow-linear segments, all glabrous and shining; *umbels* large, many-rayed.—Native of Egypt. The aromatic and carminative properties of the fruit are well known.

13. APIUM.
Celtic *apon*, water; the plants grow in watery situations.

Calyx margin obsolete; petals roundish, with a small, inflexed point; fruit roundish, laterally compressed; carpels 5-ribbed, the lateral ribs marginal; intervals with single vittæ; carpophore undi-vided —*European herbs. Umbels perfect, naked.*

A. GRAVEÓLENS. *Celery.—Lower lrs.* pinnately dissected, on very long peti-oles, segments broad-cuneate, incised; *upper lrs.* 3-parted, segments cuneate, lobed and incisely dentate at apex.—② Native of Britain. Stem 2—3f high, branching, furrowed. Radical petioles thick, juicy, 1f in length. Umbels with unequal, spreading rays. Flowers white.—The stems when *blanched* by being buried, are sweet, crisp and spicy in flavor, and used as salad. Jn.—Aug. ‡

14. PETROSELINUM. Hoffm.
(Gr. πετρα, σελινον, stone-parsley; from its native habitat.

Calyx margin obsolete; petals roundish, with a narrow, inflexed *point;* fruit ovate, compressed laterally; carpels 5-ribbed; intervals with single vittæ, commissure with 2; carpophore 2-parted.—*European herbs. Umbels perfect. Invol. few-leaved. Involucel many-leaved.*

P. sativum. Hoffm. (Apium Petroselinum. *Willd.*) *Parsley.*—Lvs. decompound, segments of the lower ones cuneate-ovate, terminal ones trifid, all incised, cauline segments lance-linear, subentire; *involucels* of 3—5 subulate bracts.—⊙ From Sardinia and Greece. Stem 2—4f high, branched. Leaves smooth and shining, with numerous, narrow segments. Petals white. Jn.—Cultivation has produced several varieties. Esteemed as a pot-herb, for soups, &c.‡

15. THASPIUM. Nutt.

From the Isle of *Thaspia*, which gave name to the ancient allied genus Thapsia.

Calyx margin 5-toothed; petals elliptic, with a long, inflexed point; fruit elliptical, not compressed laterally; carpels convex, with 5 winged ribs; intervals with single vittæ, commissure with 2.—♃ *Umbels without an involucre. Involucels 3-leaved, lateral.*

1. T. cordatum. Nutt. (Smyrnium cordatum. *Mx.* Zizia cordatum. *Dc.*) *Radical lvs.* simple, cordate, crenate, *cauline ones* ternate, stalked; *segments* acute, serrate; *umbels* terminal.—Shady hills and barrens, U. S. and Can. Stem erect, slightly branched, smooth, 2—3f high. Root leaves on long stalks, roundish-heart-shaped, the rest ternate, becoming only 3-parted above, all light green. Umbels dense with yellow flowers. Fruit black, oval, with 3 prominent, paler, winged ridges on each side. May, Jn.

β. atropurpureum. (Thapsia trifoliata. *Linn.*) *Fls.* dark purple.—N. J., Penn.

3. T. barbinode. Nutt. (Ligusticum barbinode *Michx.*) *St.* pubescent at the nodes; *lower lvs.* triternately divided, *upper* biternately, segments cuneate-ovate, acute or acuminate, unequally and incisely serrate, entire towards the base; *umbels* terminal and opposite the leaves; *fr.* elliptical, the ribs alternately broader.—River banks, Can. and U. S. Stem 2—3f high, angular and grooved, branching above. Leaves smooth, upper ones subopposite; segments 1—2′ by ¼—1¼′. Rays about 2′ long, each about 20-flowered. Petals deep yellow. Jn.

16. ÆTHUSA.

Gr. αιθω, to burn; on account of its poisonous acridity.

Calyx margin obsolete; petals obcordate, with an inflexed point; fruit globose-ovate; carpels with 5 acutely carinated ribs; lateral ones marginal, broader; intervals acutely angled, with single vittæ, commissure with 2.—⊙ *Poisonous herbs. Invol.* 0. *Involucels* 1-sided. *Fls. white.*

Æ. cynapium. *Fool's Parsley.* *Lvs.* bi- or tripinnately divided, segments cuneate, obtuse; *involucels* 3-leaved, pendulous, longer than the partial umbels.—In waste grounds, Ms., rare. Stem about 2f high, green, striate. Leaves with numerous, narrow, wedge-shaped segments, uniform, dark green, flat. Leaflets of the involucels linear, long, deflected, and situated on the outside. Jl. Aug.—The plant somewhat resembles *parsley*, but is distinctly marked by the involucels, and by its disagreeable odor. It is said to be poisonous.

17. LIGUSTICUM.

One species was said to be native of *Liguria* ; hence the name.

Calyx teeth minute or obsolete; petals obovate, emarginate, with an inflexed point; fruit nearly terete, or slightly compressed laterally; carpels 5-ribbed, with numerous vittæ.—♃ *Lvs. ternately divided. Involucels many-leaved. Fls. white.*

1. L. Scoticum. *Sea Lovage.* *Stem lvs.* biternate, the *upper ones* ternate; *lateral lfts.* oblique, the *terminal one* rhomboid; *bracts of the involucres* numerous, linear.—Sea coast. Root thick,

tapering. Stem a foot high, nearly simple, striate, smooth. Leaves petiolate Leaflets 1—2½' long, dark green, smooth and shining, ent're at base, serrate above. Fruit 4—5" long. Jl.

2. L. ACTÆFOLIUM. Michx. (Thaspium. *Nutt.*)

Lvs. triternate, with ovate, dent-serrate leaflets; *umbels* numerous, paniculate; *invol.* and *involucels* of about 3, short, subulate leaves.—Banks of the St. Lawrence. *Michx.* Topsfield and Scituate, Mass. *Oakes. Russel.* S. States, rare. Plant 3—6f high. Leaflets 2—3' long, lateral ones trapeziform. Umbels on long, verticillate peduncles, terminal one abortive.

18. CONIOSELINUM. Fisch.
Name compounded of *Conium* and *Selinum.*

Calyx teeth obsolete; petals obovate, with an inflected point; fruit compressed on the back; carpels with 5 winged ribs, lateral ones marginal and much the broadest; intervals with 1—3 vittæ, commissure with 4—8.—② *Smooth. St. hollow. Lvs. on very large, inflated petioles. Invol. various. Involucels 5—7-leaved.*

C. ? CANADENSE. Torr. & Gray. (Selinum. *Michx.* Cnidium. *Spr.*)

Lvs. ternately divided, divisions bipinnate, with oblong-linear lobes; *invol.* 0, or 2—3-leaved; *fr.* oblong-oval; *vittæ* solitary in the dorsal intervals, 2—3 in the lateral.—In wet woods, Maine to Wisconsin! but not common. Stem 3—5f high. Leaves much compounded, the ultimate segments pinnatifid with linear-oblong lobes. Umbels compound. Petals white, spreading. Styles slender, diverging. Fruit about 2" long. Aug. Sept.

19. FŒNICULUM. Adans.
Lat. diminutive of *fœnum,* hay; from the resemblance of its odor.

Calyx margin obsolete; petals revolute, with a broad, retuse apex; fruit elliptic-oblong, laterally subcompressed; carpels with 5 obtuse ribs, marginal ones a little broader; intervals with single vittæ, commissure with 2.— *Umbels perfect, with no inrol. or involucels.*

F. VULGARE. Gaert. (Anethum. Willd.) *Fennel.—Lrs.* biternately dissected, segments linear-subulate, elongated; *rays of the umbel* numerous, unequal, spreading; *carpels* turgid, ovate-oblong.—Native of England, &c. Cultivated in gardens. Stem 3—5f high, terete, branched. Leaves large and smooth, finely cleft into numerous, very narrow segments. Flowers yellow. Jl.—The seeds are warmly aromatic. ‡

20. ARCHANGELICA. Hoffm.
So named for its preeminence in size and virtues among the Umbelliferæ.

Calyx teeth short; petals equal, entire, lanceolate, acuminate, with the point inflexed; fruit dorsally compressed, with 3 carinate, thick ribs upon each carpel, and 2 marginal ones dilated into membranaceous wings; vittæ very numerous.—24 *Umbels perfect. Involucels many-leaved.*

1. A. ATROPURPUREA. Hoffm. (Angelica triquinata. *Mx.*) *Angelica.*

St. dark purple, furrowed; *petioles* 3-parted, the divisions quinate; *lfts.* incisely toothed, odd leaflet of the terminal divisions rhomboidal, sessile, the others decursive.—Among the largest of the umbelliferæ, well known for its aromatic properties, common in fields and meadows, Northern and Western States. Stem 4—6f high, 1—2½' in thickness, smooth, hollow, glaucous. Petioles large, inflated, channeled on the upper side, with inflated stipules at base. Leaflets cut-serrate, the terminal one sometimes 3-lobed, the lateral ones of the upper division decurrent. Umbels 3, terminal, spherical, 6—8' diam. without the involucre; umbellets on angular stalks and with involucels of subulate bracts longer than the rays. Flowers greenish white.

2. A. HIRSUTA. Torr. & Gray. (Angelica. *Muhl.*)

St. striate, the summit with the umbels tomentose-hirsute; *lvs.* bipinnate.

ly divided, the divisions quinate, segments oblong, acutish, the upper pair connate but not decurrent at base.—Dry w ods-, N. Y. to Car. Stem simple, erect, straight, 3—5f high. Leaves on petioles 6—10 long. Leaflets 1—2¼' long, ¼ as wide, mostly ovate-oblong, often tapering at base. Umbels 3 or 4. on long, velvety peduncles, 2—4' broad. Rays unequal, spreading, densely tomentose. Involucre 0. Involucels of 4—6 bracts, about as long as the rays. Jl. Aug.

3. A. PEREGRINA. Nutt.

St. striate, pubescent at summit; *lvs.* ternately divided, the divisions quinate, segments incisely serrate; *umbel* with many slender rays; *invol.* 0; *involucels* of many leaflets, as long as the umbellets.—Sea coast, Me. and Mass *Pickering.* Marginal ribs of the fruit thick and obtuse.

4. A. OFFICINALIS. Hoffm. (Angelica, *Linn.*) *Garden Angelica.*

St. smooth, round, striate; *lvs.* bipinnately divided into lobate, subcordate, acutely serrate segments, the terminal one 3-lobed; *sheaths* large and saccate. Said to be native in Labrador, &c. Cultivated in gardens occasionally for the sake of the stalks, which are to be blanched and eaten as celery. ‡

21. ARCHEMÓRA. DC.
A fanciful name from *Archemorus*, who, according to mythology, died by swallowing a bee.

Calyx limb 5-toothed; petals obcordate with an inflexed point; fruit oval, lenticular. compressed on the back; carpels with 5 ribs, marginal ones broadly winged; intervals with single large vittæ, commissure with 4—6; seeds flat.—⁴ *Invol.* 0. or *few-leaved. Involucels many-leaved.*

A. RIGIDA. DC. (Œnanthe. *Nutt.*) *Water Dropwort. Cow-bane.*

St. rigid, striate, smooth; *lvs.* pinnately divided, smooth, *lfts.* 3—11, oblong-lanceolate or ovate, entire or remotely toothed, sessile; *umbels* spreading, smooth.—Swamps, Mich. to Flor. and La. Stem 2—4f high, slender, terete. Leaflets 2—4' by 3—9", varying in outline in the same plant. Umbels 2—3, of many slender rays. Petals white. Fruit with subequal greenish ribs, and large, purple vittæ filling the intervals. Commissure white. Sept.—Said to be poisonous.

ß. (Œnanthe ambigua. *Nutt.*) *Lfts.* long-linear, mostly entire.

22. PASTINÀCA. Tourn.
Lat. *pastus.* food or repast; from the nutritive properties of the root.

Calyx limb 5-toothed; petals broad-lanceolate. with a long inflexed point; fruit much compressed. oval, with a broad margin; carpels with 5 nearly obsolete ribs; intervals with single vittæ; carpophore 2-parted; seeds flat.—⅔ *Rt. fusiform. Invol.* mostly 0; *involucels* 0 or *few-leaved. Fls. yellow.*

P. SATIVA. *Common Parsnep. Wild Parsnep.*

Lvs. pinnate, downy beneath; *lfts.* oblong, incisely toothed, the *upper one* 3-lobed.—⅔ The parsnep is said to have been introduced, but it grows wild abundantly in fields, by fences, &c. The root is fusiform, large, sweet-flavored, esculent, as every one knows, in its cultivated state, but in its wild state becomes hard, acrid and poisonous, and much dwindled in size. Stem 3f high, erect. furrowed, smooth, branching. Umbels large, terminal. Flowers yellow, small. Fruit large, flat. The abundance of saccharine matter in the cultivated root, renders it wholesome and nutritious. Jl.

23. HERACLÈUM.
Named after the hero Hercules; it being a rank, robust plant.

Calyx limb of 5 small. acute teeth : petals obcordate, with the point inflexed, often radiant in the exterior flowers, and apparently deeply 2-cleft; fruit compressed, flat, with a broad, flat margin. and 3 obtuse, dorsal ribs to each carpel; intervals with single vittæ; seeds

25*

fist.—*Stout herbs, with large umbels. Invol. deciduous. Involucels many-leared.*

H. LANÀTUM.
Lvs. ternate, petiolate, tomentose beneath; *lfts.* petioled, round-cordate, lobed; *fr.* orbicular.—Penn. to Lab. W. to Oreg. A large, coarse-looking, umbelliferous plant, growing about moist, cultivated grounds. Stem about 4f high, thick, furrowed, branching, and covered with spreading hairs. Leaves very large, on channeled stalks. Leaflets woolly underneath, irregularly cut-lobed and serrated. At the top of the stem and branches are its huge umbels, often a foot broad, with spreading rays, and long-pointed, lanceolate involucels. Involucre of lanceolate, deciduous leaflets. Petals deeply heart-shaped, white. Jn.

24. POLYTÆNIA. DC.

Calyx limb 5-toothed; petals with a long inflexed point; fruit oval, glabrous, lenticularly compressed on the back, with a thickened, corky margin; ribs obscure or obsolete; commissure with 4—6 vittæ; seeds plano-convex.—*A smooth herb, with bipinnately divided leaves. Invol.* 0. *Involucel of setaceous bracts.*

P. NUTTALLII. DC.
Prairies and barrens, Western States! &c. Stem furrowed, scabrous or nearly smooth. Lower leaves on long petioles, segments incisely toothed, upper ones 3-cleft, lobes entire or with lateral teeth. Umbels terminal and opposite the leaves, about 2' broad. Fruit large, (3" long) tumid and smooth, with a thick, corky pericarp, and the flavor of turpentine. May.

25. D A U C U S. Tourn.
Δαυκος, the ancient Greek name of the *carrot.*

Calyx limb 5-toothed, petals emarginate with an inflected point; the 2 outer often largest and deeply 2-cleft; fruit oblong; carpels with 5 primary, bristly ribs, and 4 secondary, the latter more prominent, winged, and divided each into a single row of prickles, and having single vittæ beneath; carpophore entire, free.—② *Invol. pinnatifid. Involucels of entire or 3-cleft bracts. Central fl. abortive.*

D. CARÒTA. *Carrot.*
St. hispid; *petioles* veined beneath; *lvs.* tripinnate or pinnatifid, the segments linear, acute; *umbels* dense, concave.—The word *kar* in Celtic signifies *red*, hence carrot. Naturalized in fields and by roadsides, abundant in the Mid. States. Root fusiform. Stem 2—3f high, branching. Leaves numerous, divided in a thrice pinnatifid manner, pale green. Umbels large and very compact, with white flowers blooming all the summer. Cultivation has produced several varieties. Jl.—Sept. ◊ ‡

SUBORDER 2'.—**C A M P Y L O S P E R M Æ.**
The inner surface of the seed deeply furrowed, or with involute margins.

26. CHÆROPHYLLUM.

Calyx limb obsolete; petals obovate, emarginate, point inflexed; fruit laterally compressed: carpels with 5 obtuse, equal ribs; intervals with 2 vittæ, commissure deeply sulcate.—*Lvs.* bi- or tricrnate, segments incisely cleft or toothed. *Invol.* 0, or few-leaved. *Involucel many-leaved.*

C. PROCUMBENS. Lam. (Scandix procumbens. *Linn.*)
Decumbent or assurgent, nearly glabrous; *segments of the lvs.* pinnatifid, with oblong, obtuse lobes; *umbels* diffuse, few-flowered, often simple; *invol.* 0; *fr.* linear-oblong.—① or ② Moist woods, Ohio, *Clark!* Ky. *Short,* to N. J. *Stems* 1—2f long, pubescent when young, diffuse, slender. Segments of the

leaves about 4″ by 1″. Umbels quite irregular, often with leaves in the place of the involucre. Rays 1—4, 1—4-flowered, about 2′ long. Petals white. Apr. May.

27. OSMORHIZA. Raf.

Gr. οσμη, perfume, ριζα, root; from the aniseto, aromatic root.

Calyx margin obsolete; petals oblong, nearly entire, the cuspidate point inflexed; styles conical at base; fruit linear, very long, clavate, attenuate at base; carpels with 5 equal, acute, bristly ribs; intervals without vittæ; commissure with a deep, bristly channel.—*Lrs. biternately divided, with the umbels opposite. Invol. few-leaved; involucels 4—7-leaved   Fls. white.*

1. O. LONGISTÝLIS. DC. (Uraspermum. Claytoni. *Nutt.*) *Sweet Cicely.*
  Sty. filiform, nearly as long as the ovary; *fr.* clavate.—A leafy plant, very common in woods, Can. to Va., 1—3f high, with inconspicuous umbels of white flowers. Root branching, fleshy, of an agreeable, spicy flavor. Stem erect, branching above, nearly smooth. Root leaves on long, slender stalks, the upper stem leaves sessile, both decompound, the ultimate divisions often pinnate; leaflets irregularly divided by clefts and sinuses into lobes and teeth, the lobes broadly ovate, slightly pubescent. Involucres of linear bracts longer than the rays. Fruit blackish, an inch in length, much more acute at the base than at the summit, crowned with the persistent styles. May, Jn.

2. O. BREVISTÝLIS. DC. (U. hirsutum. *Bw.*) *Short-styled Cicely.*
  Sty. conical, scarcely as long as the breadth of the ovary; *fr.* somewhat tapering at the summit.—Common in woods, Can. to Penn. W. to Oreg. The general aspect of this species is very similar to that of the preceding, but the root is destitute of the anise-like flavor of that species, being disagreeable to the taste. The plant is more hairy, and with more deeply cleft divisions in the leaves. Involucre deciduous. Umbels with long, diverging rays, of which but few prove fertile. The fruit is similar to the last, but crowned with convergent, not with spreading styles. May, Jn.

28. CONIUM.

Gr. κωνειον, hemlock, from κωνος, a top; because it causes dizziness.

Calyx margin obsolete: petals obcordate, with an acute, inflected point; fruit ovate, laterally compressed: carpels with 5, acute, equal, undulate-crenulate ribs, lateral ones marginal; intervals without vittæ; seeds with a deep, narrow groove on the face.—*Poisonous herbs. Lvs. decompound. Invol. and involucels 3—5-leaved, the latter unilateral. Fls. white.*

C. MACULÀTUM. *Poison Hemlock.*
  St. spotted; *lvs.* tripinnate; *lfts.* lanceolate, pinnatifid; *fr.* smooth.—Grows in waste grounds, way-sides. A well known poisonous plant. Stem much branched, about 4f high, very smooth, round, hollow, with purplish spots. The lower leaves are very large, several times pinnate, bright green, on long, sheathing foot-stalks. Umbels terminal, the involucre of 6—8 lanceolate bracts, the involucels with the inner half wanting. Flowers small, white. Fruit with undulate or wrinkled ribs. The plant is a powerful narcotic, exhaling a disagreeable odor when bruised. Used in medicine. Jl. Aug. ♂

SUBORDER 3.—CŒLOSPERMÆ.

Seeds incurved at base and apex.

29. ERIGENIA. Nutt.

Gr. ηριγενεια, daughter of the early spring; for its early flowering.

Calyx limb obsolete: petals flat, entire; fruit contracted at the commissure; carpels 3-ribbed, ovate-reniform.—*Rt. tuberous. Radi-*

cal leaf triternately decompound. Involucrate lvs. solitary, biternately compound. Involucels of 3—6 entire, linear-spatulate bracts.

E. BULBÒSA. Nutt. (Sison. *Michx.* Hydrocotyle composita. *Ph.*)

A small, early-flowering herb, along the shady banks of streams, Western N Y. (*Torr. & Gray.*) W. to Ohio! and Mo. Plant 4—6' high, with 2—4 leaves, the lower one radical, numerously divided, the divisions incisely cleft into narrow segments; the upper ones bract-like, similarly divided, each subtending a 3-rayed umbel of white flowers. March, Apl.

30. CORIANDRUM.

Gr. κορις, a bug; on account of the smell of the leaves.

Calyx with 5 conspicuous teeth; petals obcordate, inflexed at the point, outer ones radiate, bifid; fruit globose; carpels cohering, with 5 depressed, primary ribs, and 4 secondary, more prominent ones; seeds concave on the face.—① *Smooth. Invol.* 0 *or* 1*-leaved. Invo lucels* 3*-leaved, unilateral.*

C. SATIVUM. *Coriander.—Lvs.* bipinnate, lower ones with broad-cuneate leaflets, upper with linear ones; *carpels* hemispherical.—Native of Europe, &c. This well known plant is cultivated chiefly for the seeds, which are used as a spice, as a nucleus for sugar-plums, &c. Stem 2f high. Leaves numerously divided, strong-scented. Umbels with only the partial involucra. Flowers white. Jl. ‡

ORDER LXVIII. ARALIACEÆ.—ARALIADS.

Trees, shrubs or herbs, with the habit of umbellifers.
Cal. superior, entire or toothed.
Cor —Petals 5—10 deciduous, rarely 0, valvate in æstivation.
Sta. equal in number to the petals, and alternate with them. *Anth.* introrse.
Ova. crowned with a disk, 2 or many-celled. Ovules solitary. Styles as many as cells.
Fr. baccate or drupaceous, of several one-seeded cells.

Genera 21, species 160, nearly allied to Umbelliferæ, from which they are distinguished chiefly by the several-celled ovary and fleshy fruit. They are natives of northern temperate climes of both hemispheres.—Several species are well known in medicine. &c., as ginseng, spikenard, sarsaparilla, &c. The latter is sometimes substituted for the sarsaparilla of the shops.

Conspectus of the Genera.

		Flowers perfect.	Aralia.	1
compound.	Flowers polygamous.		Panax.	2
Leaves	simple, angular and lobed.		Hedera.	3

1. ARALIA.

Calyx tube adherent to the ovary, limb short, 5-toothed or entire; petals 5, spreading, apex not inflexed; stamens 5—10; styles 5, spreading; berry crowned with the remains of the calyx and styles, mostly 5-celled and 5-seeded.—*Lvs. compound. Fls. in simple, solitary or racemose panicles.*

1. A. NUDICAULIS. *Wild Sarsaparilla.*

Nearly stemless; *lf.* solitary, decompound; *scape* naked, shorter than the leaf; *umbels few.*—♃ A well known plant, found in woods, most abundant in rich and rocky soils, Can. to Car. and Tenn. It has a large, fleshy root, from which arise a leaf-stalk and a scape, but no proper stem. The former is long, supporting a single, large, compound leaf, which is either 3-ternate or 3-quinate. Leaflets oval and obovate, acuminate, finely serrate. The scape is about a foot high, bearing 3 simple umbels of greenish flowers. Jn. Jl.

2. A. RACEMÒSA. *Pettymorrel. Spikenard.*

St. herbaceous, smooth; *lvs.* decompound; *ped.* axillary, branching, umbelled.—♃ In rocky woods, Can. to the S. States. Stem 3—4f high, dark green or reddish, arising from a thick, aromatic root. The leaf-stalks divide into 3 partitions, each of which bears 3 or 5 large, ovate, serrate leaflets. Umbels numerous, arranged in branching racemes from the axils of the leaves or branches. The root is pleasant to the taste, and highly esteemed as an ingredient in small beer, &c. July.

3. A. hispida. *Wild Elder. Bristly Aralia.*

St. shrubby at base, hispid; *lvs.* bipinnate; *lfts.* ovate, cut-serrate; *umbels* on long peduncles.—♃ Common in fields, about stumps and stone-heaps, N. Eng. to Va. Stem 1—2f high, the lower part woody and thickly beset with sharp, stiff bristles, the upper part branching herbaceous. Leaflets many, ending in a long point, ovate, smooth. Umbels many, simple, globose, axillary and terminal, followed by bunches of dark-colored, nauseous berries. The plant exhales an unpleasant odor. Jl. Aug.

A. spinosa. *Angelica Tree.*

Arborescent; *st.* and *petioles* prickly; *lvs.* bipinnate; *lfts.* ovate, acuminate, sessile, glaucous beneath; *umbels* numerous, forming a very large panicle; *fls.* small, few-leaved.—Damp woods, Penn. to Flor. and La. Shrub 8—12f high, with the leaves all crowded near the summit. Flowers white. Aug.—Emetic and cathartic. ‡

2. PANAX.

Gr. παν, all, ακος, a remedy; i. e. a panacea, or universal remedy.

Diœciously polygamous. ♀ Calyx adnate to the ovary, limb short, obsoletely 5-toothed; petals 5; stamens 5, alternate with the petals; styles 2—3: fruit baccate, 2—3-celled; cells 1-seeded. ♂ Calyx limb nearly entire; petals and stamens 5.—*Herbs or shrubs. Lvs. 3 (in the herbaceous species), palmately compound. Fls. in a solitary, simple umbel.*

1. P. trifolium. *Ground Nut. Dwarf Ginseng.*

Rt. globose, tuberous; *lvs.* 3, verticillate, 3—5-foliate; *lfts.* wedge-lanceolate, serrate, subsessile; *sty.* 3; *berries* 3-seeded.—Common in low woods, Can. to S. States. The globular root is deep in the ground, and nearly ¼' diam., connected with the stem by a short, screw-like ligament. The stem arises 3—6' above the surface, smooth, slender, simple. At the summit is a whorl of 3 compound leaves, with a central peduncle terminating in a little umbel of pure white flowers. Leaflets generally 3, nearly or quite smooth. Barren and fertile flowers on different plants, the latter without stamens, succeeded by green berries, the former with a single, abortive style. May.

2. P. quinquefolium. *Ginseng.*

Rt. fusiform; *lvs.* 3, verticillate, 5-foliate; *lfts.* oval, acuminate, serrate, petiolate; *ped. of the umbel* rather shorter than the common petioles.—Not uncommon in rocky or mountainous woods, Can. to the mountains of the Southern States. Root whitish, thick and fleshy. Stem round, smooth, 1f high, with a terminal whorl of 3 compound leaves, and a central peduncle bearing a simple umbel. Leaves on round and smooth foot-stalks, consisting of 5, rarely 3 or 7 obovate leaflets. The flowers are small, yellowish, on short pedicels. The barren ones borne on separate plants have larger petals and an entire calyx. The fertile ones are succeeded by berries of a bright scarlet color. The root is in little estimation as a drug with us, but it enters into the composition of almost every medicine used by the Chinese and Tartars. Jn. Jl.

3. HEDERA.

Celtic hedra, a cord: from the vine-like habit.

Calyx 5-toothed; petals 5, dilated at the base, berry 5-seeded, surrounded by the permanent calyx.—*European shrubby plants, climbing or erect, with simple, evergreen leaves and green flowers.*

H. helix. *English Ivy.*—*St.* and *branches* long and flexible, attached to the earth or trees or walls by numerous radicating fibres; *lvs.* dark green, smooth, with white veins, petiolate, lower ones 5-lobed, upper ovate; *fls.* in numerous umbels, forming a corymb; *berry* black, with a mealy pulp.—Native of Britain. There are several varieties in gardens. †

ORDER LXIX. CORNACEÆ.—Cornels.

Trees and *shrubs* seldom *herbs*, without stipules.
Lvs. opposite (alternate in one species), with pinnate veinlets. Hairs fixed by the centre.
Cal.—Sepals adherent to the ovary, the limb minute, 4 or 5-toothed or lobed.
Cor.—Petals 4 or 5, distinct, alternate with the teeth of the calyx.
Sta. of the same number as petals and alternate with them.
Ova. 1 or 2-celled. *Fruit* a baccate drupe, crowned with the calyx.

Genera 9, species 43. They are natives throughout the temperate zone of both continents. The order is distinguished for its bitter and astringent bark. That of Cornus florida is an excellent tonic, similar in its action to the Peruvian bark. Cornus is the only N. American genus.

CORNUS.

Lat. *cornu*, a horn; from the hardness of the wood of some species.

Calyx 4-toothed, segments small; petals 4, oblong, sessile; stamens 4; style 1; drupe baccate, with a 2 or 3-celled nut.—*Trees, shrubs or perennial herbs. Lvs. (mostly opposite), entire. Fls. in cymes, often involucrate. Floral envelops valvate in æstivation.*

* *Flowers cymose. Involucre 0. Shrubs.*

1. C. STOLONIFÈRA. Michx. (C. alba. *Wang.*) *White-berried Cornel or Dog-wood.*—*St.* often stoloniferous; *branches* spreading, smooth; *shoots* virgate; *lvs.* broad-ovate, acute, pubescent, hoary beneath; *cymes* naked, flat; *berries* white.—A small tree, N. and W. States and Can., 8—10f in height, with smooth, slender, spreading branches, which are commonly red, especially in winter. It often sends out from its base prostrate and rooting stems, with erect shoots. Leaves distinctly veined, minutely pubescent, and whitish tomentose beneath, petiolate and pointed. Flowers in terminal cymes, white, followed by bluish-white drupes. According to Dr. Bigelow, it sometimes blossoms twice a year. May, Jn.

2. C. SERICEA. *Red Osier.*
Branches spreading; *branchlets* woolly; *lvs.* ovate, rounded at base, acuminate, ferruginous, pubescent beneath; *cymes* depressed, woolly; *drupes* bright blue.—U. S. and Can. A variety has leaves tapering at base. A shrub about 8f high, with opposite, dusky, purple branches, and dark-red shoots. Leaves 2—4' long, ⅓ as wide, varying from ovate and oval to lanceolate, nearly smooth above, with rather prominent veins; petioles ½—1' long. Flowers yellowish-white appearing in June.

3. C. CIRCINÀTA. *Round-leaved Cornel or Dog-wood.*
Branches verrucose; *lvs.* orbicular or very broadly oval, white tomentose beneath; *cymes* spreading, depressed; *drupes* light-blue.—A shrub some 6f high, Can. to Md., W. to Ia. Stem greyish, upright, with opposite, cylindrical, green, spotted or warty branches. Leaves large, about as broad as long, opposite, acuminate, covered with a white, thick down on the under side. Flowers white. Berries hollowed at base, soft, crowned with the remains of the style. Jn.

4. C. PANICULÀTA. *White or Panicled Cornel.*
Branches erect, smooth; *lvs.* ovate-lanceolate, acuminate, acute at base, scabrous above, hoary beneath; *cymes* paniculate; *drupes* white.—A handsome shrub, 10f high, common in low woodlands and thickets, N. and W. States and Can. It has numerous and very branching stems, covered with a greyish bark, the shoots chestnut-colored. Leaves small, (1—2' long, ½—1' wide). Petioles 1—4" long. Flowers small, white in all their parts, in many small, conical cymes, succeeded by small drupes.

5. C. ALTERNIFOLIA.
Lvs. alternate, oval, acute, hoary beneath; *branches* alternate, verrucose; *drupes* purple, globose.—A small tree, N. and W. States and Can., about twice the height of the last, in moist woods. The branches are smooth, even, spreading from the upper part of the stem, and forming a depressed summit. Bark greenish, marked with warty streaks. Leaves irregularly scattered along the *branches, oval*-lanceolate, acute, entire, veined, whitish underneath, on rather *long stalks. Flowers* pale buff-color, in a loose cyme. Jn.

** *Flowers umbellate.   Involucre 4-leaved, petaloid.*

6. C. FLORIDA. *Flowering Dog-wood.*

Arboreus; *lvs.* opposite, ovate, acuminate, entire; *fls.* small, in a close, cymose umbel or head, surrounded by a very large, 4-leaved, obcordate involucre.—A tree from 20—30f in height, very ornamental when in flower. Woods, U. S. and Can. The wood is very hard and compact, covered with a rough bark, which is extremely bitter, and used in medicine as a tonic. The leaves, which at flowering-time are but partially expanded, are acutely ovate, nearly smooth, veiny, pale underneath. The true flowers are inconspicuous, greenish-yellow, but the involucre is very large and showy, of veiny, white, obovate leaves, ending in a callous point, which is turned up or down so abruptly as to give an emarginate appearance to the leaf. Drupe red. May.

7. C. CANADENSIS. *Low Cornel or Dog-wood.*

Herbaceous, low; *upper lvs.* whorled, veiny, on short petioles; *st.* simple. —A small, handsome plant, common in woods, nearly throughout N. Am. N. of lat. 39°, remarkable for its large, white involucre. Rhizoma creeping, woody. The flowering stems erect, 4—8' high, bearing 2 small stipules in the middle, and a whorl of 6 leaves at the top, two of which are larger, placed a little lower and opposite. An umbellate cyme of flowers arises from the centre of the whorl, and with its large, showy involucre of 4 white leaves, might easily be taken for a single flower. They are succeeded by a bunch of red berries. The barren stems support a whorl of 4 equal leaves. May, Jn.

ORDER LXX. LORANTHACEÆ.—LORANTHS.

St. parasitical, half-shrubby, dichotomous.
Lvs. evergreen, opposite, fleshy, without stipules.
Fls. dioecious and small, whitish or greenish-yellow, sometimes perfect and brilliant.
Cal. adnate to the ovary in perfect flower, limb 3—5-cleft or obsolete.
Cor. of 3—4 or 5 petals, cohering in a tube, sometimes distinct, inserted into the epigynous disk.
Sta. as many as the petals and opposite to them, or to the sepals when the pet. are 5.
Ova. 1-celled, with a single suspended ovule. *Sty.* simple or 0.
Fr. baccate, with one anatropous seed.

Genera 22, species 412, about equally distributed throughout the tropical regions of Asia and America. They possess the remarkable property of rooting firmly on other plants and living upon their juices. They are slightly astringent.

VISCUM. Tourn.

♂ or ♀ ♂.—♂ Calyx with 4 (3—5) triangular, erect segments, valvate in æstivation; anthers as many as the sepals, and inserted on them; corolla 0. ♀ Limb of the calyx obsolete; petals 4, fleshy, epigynous; stamens 0; stigma sessile; berry fleshy, 1-seeded.—*Lvs. very rarely alternate or scale-like.*

V. FLAVESCENS. Ph. (V. album. *Walt.* V. verticillatum. *Nutt.*) *Misseltoe.* *Branches* opposite, sometimes verticillate, terete; *lvs.* cuneate-obovate, 3-veined, obtuse; *spikes* axillary, solitary, about as long as the leaves; *berries* white, semi-transparent.—A yellowish green, succulent parasite inserted on the branches of aged trees, N. J. W. to Ia. and the Southern States! Stems 1—1½f high, rather thick, much branched. Leaves 9—16'' by 4—9'', smooth and entire, on short petioles. Flowers small, sterile ones mostly 3-parted. Berry with a viscous pulp. Apr. May.

SUBCLASS II. MONOPETALÆ.

Floral envelops consisting of both calyx and corolla, the latter composed of petals more or less united (monopetalous).

ORDER LXXI. CAPRIFOLIACEÆ.—HONEYSUCKLES.

Shrubs rarely herbs, often twining, with opposite leaves and no stipules.
Fls. cymose and often fragrant.
Cal. adherent to the ovary (superior), the limb 5- (rarely 4-) cleft or toothed.
Cor. tubular or rotate, regular or irregular. [The tube.
Sta. as many, or one less than as many as the lobes of the corolla, alternate with them and inserted on
Ova. 2- (rarely 4 or 5-) celled. *Style* 1. *Stig.* 1—4.
Fr. baccate, fleshy or dry, crowned with the persistent calyx lobes. *Seeds* pendulous.

Genera 14, species 225, chiefly natives of the northern temperate regions, and occasionally found in the alpine parts of the tropical zone.

Properties.—The fever-root (Triosteum perfoliatum) is a mild cathartic, and in large doses emetic; the dried and roasted berries are sometimes substituted for coffee. The leaves and bark of the elder are both emetic and cathartic; the flowers are sudorific, and the berries laxative. The beauty and fragrance of the honeysuckles in cultivation are well known to every one.
The order consists of two distinct tribes; Loniceræ and Sambuceæ.

Conspectus of the Genera.

		{ few-seeded.	.	.	.	.	.	Lonicera.	1
	{ Berry 1—3-celled, { many-seeded.	.	.	.	.	.	.	Diervilla.	2
{ Shrubs.	{ Berry 4-celled, 2-seeded.	.	.	.	.	.	.	Symphoricarpus.	4
	{ Stamens 4. Trailing, evergreen.	.	.	.	.	.	Linnæa.	5
{ tubular. { Herbs.	{ Stamens 5. Stem erect, simple.	.	.	.	.	.	Triosteum.	3
	{ simple leaves.	.	.	.	.	.	Viburnum.	7
Corolla { rotate, regular. Shrubs with { pinnate leaves.	.	.	.	.	.	.	Sambucus.	8

TRIBE 1. LONICEREÆ.

Corolla tubular, the limb often irregular. Style filiform.

1. LONICERA.

In honor of Adam Lonicer, a physician of Frankfort, in the 16th century.

Calyx 5-toothed, tube subglobose; corolla infundibuliform or campanulate, limb 5-cleft, often labiate; stamens 5, exserted; ovary 2—3 celled; berry few-seeded; stigma capitate.—*A genus of climbing or erect shrubs, with opposite and often connate leaves.*

§ *Stems climbing. Flowers sessile, verticillate.* CAPRIFOLIUM.

1. L. HIRSÙTA. Eaton. (C. pubescens. *Goldie.*) *Hairy Honeysuckle.*
Lvs. hairy above, soft-villose beneath, veiny, broad-oval, abruptly acuminate, the upper pair connate-perfoliate; *fls.* in verticillate spikes; *cor.* ringent; *fil.* bearded.—A shrubby climber, rather rare, in woods, N. Eng. to Mich. and Can., twining about trees to the height of 15—20f. The whole plant is more or less hairy. Leaves pale green, not shining, the edges and the upper side ciliate with scattered hairs. The flowers are large, numerous, greenish-yellow, in whorled, axillary and terminal clusters. Limb of corolla spreading. Style and stamens exsert. Jn.

2. L. PARVIFLÒRA. Lam. (C. parviflorum. *Ph.*) *Small-flowered Honeysuckle.*
Lvs. smooth, shining above, glaucous beneath, oblong, all sessile or connate, the upper pair perfoliate; *fls.* in heads of several approximate whorls; *cor.* ringent; *tube* short, gibbous at base; *fil.* bearded.—A small, smooth, shrubby climber, in rocky woods, Can. and U. S. Stem 8—10f long. Leaves wavy and revolute on the margin, very glaucous on the under side. Flowers rather small. Corolla 1' in length, yellow, tinged with dull red, gibbous at the base, the short limb in curled segments. Stamens and style exserted. Berries orange-colored. May, Jn.

β. ? *Sullivantii. Lvs.* pubescent beneath, all except the upper pair distinct, the lower ones petiolate.—Ohio, *W. S. Sullivant!* S. Car. *Miss Carpenter*; Perhaps distinct.

3. L. FLAVA. Sims. (C. Fraseri. *Ph.*) *Yellow Honeysuckle.*

Lvs. ovate, glaucous beneath, with a cartilaginous margin, upper pair connate-perfoliate; *spikes* terminal, of close whorls; *cor.* smooth, tube slender, gibbous at base, limb somewhat ringent; *fil.* smooth.—A beautiful shrub, scarcely twining, mountains, N. Y. to Ga. W. to Wisconsin. Often cultivated. Leaves deciduous, obtuse, abruptly contracted at base, except the upper perfoliate pair. Flowers in heads of about 10, fragrant. Corolla an inch or more in length, the tube much longer than the limb, bright yellow. Upper lip much broader than the lower, in 4 segments. Jn. Jl. †

4. L. GRATA. Ait. (C. gratum. *Ph.*) *Evergreen Honeysuckle.*

Lvs. evergreen, obovate, smooth, glaucous beneath, the upper pair connate-perfoliate; *fls.* in sessile, terminal and axillary whorls; *cor.* ringent, tube long, slender, not gibbous at base.—A beautiful climbing species, damp woodlands, N. Y., Penn. and Western States. Leaves opposite or in 3s, margin revolute. Flowers large and very fragrant, 5 or 6 in each whorl. Corolla pale yellow within, becoming reddish without. Stamens exserted. Berries red. The leaves are very obtuse, ending in a short, abrupt point. Jn.

5. L. SEMPERVIRENS. Ait. (C. sempervirens. *Michx.*) *Trumpet Honeysuckle.*

Lvs. oblong, evergreen, the upper ones connate-perfoliate; *fls.* in nearly naked spikes of distant whorls; *cor.* trumpet-shaped, nearly regular, ventricose above.—In moist groves and borders of swamps, N. Y. to Flor. and La. Common in cultivation, where few flowers are found more beautiful, although they are deficient in fragrance. Stem woody, twining with the sun. Leaves ovate or elliptical, of a dark, perennial green above. Corolla trumpet-shaped, nearly 2' long, dilated at the mouth, with 5 short, nearly regular segments, of a fine scarlet without and yellow within. May—Aug. †

**6. L. PERICLYMENUM. Tourn. (C. Periclymenum. *Linn.*) *Woodbine Honeysuckle.*—*Lvs.* deciduous, all distinct, elliptical, on short petioles; *fls.* in ovate, imbricate, terminal heads; *cor.* ringent.—A woody climber, native of Europe, cultivated and nearly naturalized. Flowers yellow and red, fragrant, succeeded by red berries. May—Jl. †

β. quercifolia. (*Oak-leaved Honeysuckle.*) *Lvs.* sinuate-lobed.

**7. L. CAPRIFOLIUM. (Caprifolium Italicum. *R. & S.*) *Italian Honeysuckle.*—*Lvs.* deciduous, the upper pair perfoliate-connate; *fls.* in a terminal verticil; *cor.* ringent.—Native of Europe. Greatly admired in cultivation for its beauty and fragrance. Flowers of various hues, red, yellow and white. Jn.—Aug. †

§ § *Stem erect. Flowers pedunculate, geminate.* XYLOSTEUM.

8. L. CILIATA. Muhl. (Xylosteum ciliatum. *Ph.*) *Fly Honeysuckle.*

Lvs. ovate, subcordate, ciliate; *corolla limb* with short and subequal lobes; *tube* saccate at base; *sty.* exserted; *berries* distinct.—A branching, erect shrub, 3—4f high, found in woods, Me. to Ohio and Can. Leaves thin, oblong-ovate, often cordate at the base, somewhat ciliate on the margin, and villose beneath when young. Flowers pale straw-yellow, in pairs at the top of the peduncle, with an obtuse spur turned outwards at the base. Berries ovoid, red, in pairs but not connate, 3—5-seeded. June.

9. L. OBLONGIFOLIA. Hook. (X. oblongifolium. *Goldie.*)

Lvs. oblong or oval, velvety-pubescent beneath; *corolla limb* deeply bilabiate; *tube* gibbous at base; *ped.* long, filiform, erect; *berries* connate or united into one, globose, purple, bi-umbilicate.—A shrub 3—4f high, in swamps, Can. and N. Y. Leaves almost sessile, 1—2' long, peduncles of equal length. Corolla hairy, greenish-yellow outside, purplish inside, the lower lip nearly entire, the upper one 4-lobed, erect. Berries marked with the remains of the two calyces. Jn.

**10. L. COERULEA. (X. villosum. *Mx.* X. Solonis. *Eat.*) *Blue-fruited Honeysuckle.*—*Lvs.* oval-oblong, ciliate, obtuse, villous both sides, at length smoothish; *ped.* short, reflexed in fruit; *bracts* longer than the ovaries; *berries* connate or united into one, deep blue.—A low shrub, in rocky woods, Mass. and N. Y., N. to Hudson's Bay. Stem 2f high, with small leaves and pairs of

26

small, yellow flowers, which are longer than their peduncles. Leaves ovate, oval, obovate and oblong, ending abruptly. May, June.

11. L. TARTARICA. *Tartarian Honeysuckle.*—*Stems* erect, much branched; *lvs.* ovate, cordate, obtuse, smooth, shining and dark green above, paler beneath, entire, on short petioles; *peduncles* axillary, solitary, 2-flowered; *segments of the corolla* oblong, obtuse, equal.—An elegant and much admired shrub, from Russia. Grows from 4 to 10f high. Leaves 1—2′ by ½—1½′, coriaceous. Flowers small, pale purple, varying to pure white, fragrant. Apr.—Jn. †

2. DIERVILLA. Tourn.

In honor of Dierville, a French surgeon, discoverer of the original species.

Calyx tube oblong, limb 5-cleft; corolla twice as long, funnel shaped; limb 5-cleft and nearly regular; stamens 5; capsular fruit 2-celled (apparently 4-celled from the projecting placentæ), many-seeded.—*Shrubs, with opposite, serrate, deciduous leaves.*

D. TRIFIDA. Mœnch. (D. Tournefortii. *Michx.* D. Canadensis. *Muhl.*) *Bush Honeysuckle.*—*Lvs.* ovate, acuminate, on short petioles; *ped.* axillary and terminal, 1—3 flowered; *caps.* attenuate above.—A low shrub, not uncommon in hedges and thickets, Can. to Car. Stem about 2f high, branching. Leaves 2—4′ by 1—1½′, finely serrate, ending in a long, narrow point. Ovaries slender, 4—5″ long, about half the length of the greenish yellow corolla. Stamens and style much exserted. Stigma capitate. Jn.

3. TRIOSTEUM.

Gr. τρεις, three; οστεον, a bone; from the three bony seeds.

Calyx tube ovoid, limb 5-parted, segments linear, nearly as long as the corolla; corolla tubular, gibbous at base, limb 5-lobed, subequal; stamens 5, included; stigma capitate, lobed; fruit drupaceous, crowned with the calyx, 3-celled, 3-seeded; seeds ribbed, bony.—♃ *Herbaceous, rarely suffruticose.*

T. PERFOLIÀTUM. *Fever-wort.*
Lvs. oval-acuminate, connate; *fls.* axillary, verticillate or clustered.—A coarse, unattractive plant, growing in rocky woods. Stem simple, stout, erect, round, hollow, 3—4f high, covered with soft, clammy hairs. Leaves 6′ by 3′, entire, abruptly contracted at base, yet always connate, nearly smooth above, pubescent beneath. Flowers sessile, in clusters of 5 or 6. Corolla dull purple, viscid-pubescent, the limb in 5 rounded lobes. Fruit a rather dry drupe, somewhat 3-sided, crowned with the long, leafy, spreading calyx segments, orange-colored when mature, containing 3 bony nuts or seeds. June.—The root is large and fleshy, and in much repute in medicine, having many of the properties of Ipecacuanha.

4. SYMPHORICARPUS.

Gr. συν, together; φερω, to bear; καρπος, fruit; bearing fruit in close clusters.

Calyx tube globose, limb 4—5-toothed; corolla funnel-shaped or bell-shaped, the limb in 4—5 subequal lobes; stamens 4—5, inserted on the corolla; stigma capitate; berry globose, 4-celled, 2-seeded (2 opposite cells abortive).—*Small shrubs, with entire lvs. and small fls.*

1. S. RACEMÒSUS. Michx. (Symphoria. *Pers.*) *Snow-berry.*
Fls. in terminal, loose, interrupted, often leafy racemes; *cor.* campanulate, densely bearded within; *sty.* and *sta.* included.—A smooth, handsome shrub 2—3f high, common in cultivation, and native in Western N. Y., Canada, &c. Leaves oval or oblong, the margin often wavy, nearly or quite smooth, paler beneath, on short petioles. Corolla rose-color, the throat filled with hairs. Berries large, round or ovoid, of a snowy white, and very ornamental when mature. July, Aug.

2. S. OCCIDENTÀLIS. R. Br. *Wolf-berry.*
Lvs. ovate, obtusish; *spikes* dense, axillary and terminal, subsessile, and

ding; *cor.* somewhat funnel-form, densely bearded inside; *sta.* and bearded style exserted.—Woods, Mich. to Wis. *Lapham !* and Can.   Shrub 2—4f high. Leaves 1—3' by ½—2'; pubescent or nearly glabrous, paler beneath. Corolla rather larger and more expanded than in the last, purplish-white. Berries white.   July.

3. S. vulgaris. Michx. (Lonicera symphoricarpus. *Linn.* Symphoria glomerata. *Nutt.*)—*Lvs.* roundish-oval; *spikes* axillary, subsessile, capitate and crowded; *cor.* campanulate, lobes nearly glabrous; *sta.* and *bearded style* included.—River banks, Penn. to Mo. and S. States.   Shrub 2—3f high. Branches purplish and often pubescent.   Leaves 1—2' by ½—1½', somewhat pubescent.   Corolla greenish-red.   Berries purple.

5. LINNÆA. Gron.

In honor of Carl Von Linne, the most profound of naturalists, ancient or modern.

Calyx tube ovate, limb 5-parted, deciduous ; bracteoles at base 2 ; corolla campanulate, limb subequal, 5-lobed ; stamens 4, 2 longer than the other 2 ; berry dry, 3-celled, indehiscent, 1-seeded (2 cells abortive).—*A trailing, evergreen herb, widely disseminated throughout the northern temperate zone.*

L. borealis. Gron. *Twin-flower.*
The only species, native of moist, shady, rocky soils, generally in evergreen woods, from lat. 39° to the Arc. Sea.   It has long, creeping, filiform, brownish stems, rooting and branching their whole length, and covering the ground in large patches.   Leaves small, opposite, petiolate, roundish, with obtuse lobes or teeth, and scattered hairs.   Peduncles filiform, slightly hairy, about 3' high (the only erect part of the plant), the lower part leafy, the upper furnished with a pair of minute, linear, opposite bracts, and terminating with 2 pedicellate, nodding flowers.   The corolla is rose-colored and very fragrant.   Jn.

TRIBE 2. SAMBUCEÆ.

Corolla regular, rotate.   Stigmas 3—5, nearly sessile.

6. SAMBŪCUS.

Lat. *sambuca*, a musical instrument, said to have been made of the elder.

Calyx small, 5-parted ; corolla 5-cleft, segments obtuse ; stamens 5 ; stigma obtuse, small, sessile ; berry globose, pulpy, 3-seeded.— *Shrubs or perennial herbs, with pinnate, or bipinnate lvs.   Fls. in cymes.*

1. S. Canadensis. *Common Elder.*
St. shrubby; *cymes* 5-parted; *lvs.* nearly bipinnate; *lfts.* oblong-oval, acuminate, smooth.—A common, well known shrub, 6—10f high, in thickets and waste grounds, U. S. and Can.   Stem filled with a light and porous pith, especially when young.   Leaflets in 3 or 4 pairs with an odd one, serrate, the lower ones often binate or trifoliate.   Petioles smooth.   Flowers numerous, in very large (2f broad in la !) level-topped cymes, white, with a heavy odor.   Berries dark purple.   May—Jl.

2. S. pubens. Michx. *Panicled Elder.*
S. shrubby; *cymes* paniculate or pyramidal; *lfts.* oval-lanceolate, acuminate, in 2 or 3 pairs, with an odd one, and, with the petiole, pubescent beneath. —A common shrub, in hilly pastures and woods, Hudson's Bay to Car., growing about 6f high, often more or less.   Leaves simply and unequally pinnate. Leaflets sharply serrate, very pubescent when young.   Flowers in a close, ovoid thyrsus or panicle.   Corolla white.   Berries scarlet, small.   Jn.
β. *leucocarpa.   Berries* white.—Catskill Mountains.   T. & G.

7. VIBURNUM.

Lat. *viere*, to tie ; for the pliancy of the twigs of some of the species.

Calyx small, 5-toothed, persistent ; corolla limb 5-lobed, segments obtuse ; stamens 5, equal, longer than the corolla ; stigmas mostly sessile ; drupe 1-seeded.—*Shrubs or small trees.* *Lvs. simple, petiolate.* *Fls. cymose, sometimes radiant.*

* *Cymes radiant, the marginal flowers much larger than the others and neutral.*

1. V. LANTANÖIDES. *H. bble-bush. Wayfaring Tree.*

Lis orbicular-cordate, abruptly acuminate, unequally serrate; *pet.* and *veins* covered with a ferruginous down ; *cymes* sessile ; *fr.* ovate.—A shrub, very ornamental when in flower. It is rather common in the rocky woods of N. Eng. and N. Y., which it adorns in early spring with its large cymes of brilliant white flowers. Height about 5f. Branches long and crooked, often trailing and rooting. Leaves very large, covered with a rusty pubescence when young, at length becoming green, the dust and down remaining only upon the stalk and veins. The radiant sterile flowers of the cyme are near 1' diam., from a greenish color becoming white, flat, with 5 rounded lobes. Inner flowers much smaller, fertile. May.

2. V. OPÙLUS. *B. Americana.* Ait. T. & G. (V. Oxycoccus. *Ph.*) *High Cranberry.*—Smooth ; *lvs.* 3-lobed, 3-veined, broader than long, rounded at base, lobes divaricate, acuminate, crenately toothed ; *petioles* glandular ; *cymes* pedunculate.—A handsome shrub, 8—12f high, in woods and borders of fields, N. States and Brit. Am. Stems several from the same root, branched above. Leaves with large, remote, blunt teeth, the stalks with 2 or more glands at base channeled above. Cymes bordered with a circle of large, white, barren flowers, like the preceding species. Fruit resembles the common cranberry in flavor, and is sometimes substituted for it. It is red, very acid, ripens late, remaining upon the bush after the leaves have fallen. June.

β. ruseum. *Guelder Rose. Snow-ball.*—*Lvs.* rather acute at base, longer than broad, lobes acuminate, with acuminate teeth ; *petioles* glandular ; *fls.* all neutral, in globose cymes.—Native of Europe. This variety is the popular shrub so generally admired and cultivated as a companion of the Lilac, Snowberry, Philadelphus, &c. Its dense, spherical cymes are wholly made up of barren flowers.

* * *Cymes not radiant.* *Flowers all similar and fertile.* *Leaves lobed or incised.*

4. V. ACERIFOLIUM. *Maple-leaved Viburnum.* *Dockmackie.*

Lvs. subcordate, acuminate, 3-veined, 3-lobed, acutely serrate ; *pet.* without glands ; *cymes* on long peduncles.—A shrub, 4—6f high, with yellowish green bark, growing in woods, Can. and U. S. Leaves broad, rounded and sometimes cordate at base, divided into 3 acuminate lobes with sharp serratures, a form not very unlike that of the maple leaf, the under surface, as well as the younger branches a little downy. Branches straight, slender, very flexible, ending with a pair of leaves and a long-stemmed, cymose umbel of white flowers. Fruit oval, compressed. Stamens much exserted. June.

5. V. PAUCIFLÒRUM. Pylaie. *Few-flowered Viburnum.*

Nearly smooth in all its parts ; *lvs.* roundish, slightly 3-lobed or incised at summit, mostly 5-veined from the base ; *cymes* small and pedunculate, terminating the very short lateral branches ; *fl.* much shorter than the corolla.—A small shrub, with white flowers, Mansfield Mt., Vt. *Macrae,* White Mts., N. H. *Robbins,* N. to Newfoundland.

6. V. LENTÀGO. *Sweet Viburnum.*

Lvs. ovate, acuminate, acutely and finely uncinate-serrate ; *petiole* with undulate margins.—A common, tree-like shrub, in rocky woods, Can. to Ga. and Ky. Height 10—15f. Leaves smooth, conspicuously acuminate, about 3' long and half as wide, their petioles with a curled or wavy, dilated border on each side. Flowers white, in broad, spreading cymes, succeeded by well-flavored, sweetish berries of a glaucous black. Jn.

7. V. NUDUM. *Naked-stalked Viburnum.* *Withe Rod.*

Smooth ; *lvs.* oval-oblong, revolute at the edge, subcrenulate ; *pet.* naked ; *cymes* pedunculate.—A shrub or small tree, 10—15f high, in swamps, U. S. *Leaves* elliptical, punctate, coriaceous, the margin more or less rolled, nearly entire, smooth as well as every other part, and when full grown, 3 or 4 inches ong. Cymes large, on peduncles an inch or two in length, with caducous bracts. Flowers numerous, white. Berries dark blue, covered with a glaucous bloom, sweetish when ripe. June.

7. *cassinoides*. (V. pyrifolium. *Lam.*) Lvs. ovate, oval or often rhomboidal, acuminate, acute, obtuse or even emarginate on the same twig; *margin* finely serrate; *fr.* oblong-ovoid.

8. V. PRUNIFOLIUM. *Black Haw. Sloe.*

Lvs. smooth, roundish-obovate, acutely serrate, with uncinate teeth; *petioles* margined with straight, narrow wings.—In woods and thickets, N. Y. to Ga. A shrub or small tree, 10—20f high, the branches spreading, some of them often stinted and naked, giving the plant an unthrifty aspect. Leaves about 2' long and nearly as wide, on short petioles, slightly margined. Cymes rather large, terminal, sessile. Flowers white, succeeded by oval, blackish berries which are sweet and eatable. June.

9. V. DENTÀTUM. *Arrow-wood.*

Nearly smooth; *lvs.* roundish-ovate, dentate-serrate, subplicate, on long stalks; *cymes* pedunculate.—A shrub, 8—12f high, not uncommon in damp woods and thickets, Can. to Ga. It is called arrow-wood from the long, straight, slender branches or young shoots. Leaves roundish, 2—3' diam., the upper pair oval, the veins beneath prominent, parallel and pubescent in their axils. Flowers white, succeeded by small, roundish, dark blue berries. June.

10. V. PUBESCENS. *Downy Viburnum.*

Lvs. ovate, acuminate, dentate-serrate, subplicate, villous beneath and somewhat hairy above, on short stalks; *stipules* 2, subulate; *cymes* pedunculate; *fr.* oblong.—In dry, rocky woods and thickets, Can. to Car. rare. A shrub, about 6f high. Leaves about 2' long, each with a pair of short, hairy, subulate appendages (stipular?) at the base of the very short petiole. Cymes small, few-flowered. Flowers rather larger than those of the foregoing species, white. Jn.

11. V. TINUS. *Laurestine.*—Lrs. ovate, entire, their veins with hairy tufts beneath.—An exceedingly beautiful evergreen shrub, from Europe. Height 4—5f. Leaves acute, veiny, dark shining green above, paler beneath. Flowers white, tinged with red, very showy. Degrees of pubescence variable.

ORDER LXXII. RUBIACEÆ.—MADDERWORTS.

Trees, shrubs, and *herbs.* Lvs. opposite, sometimes verticillate, entire.
Stip. between the petioles, sometimes resembling the leaves.
Cal.—Tube more or less adherent (superior or inferior), limb 4—5-cleft.
Cor. regular, inserted upon the calyx tube, and of the same number of divisions.
Sta. inserted upon the tube of corolla, equal in number and alternate with its segments.
Ova. 2 (rarely more) celled. *Style* simple or partly divided.
Fr. various. *Seeds* one, few or many in each cell.

Genera 299. Species 2972. It is generally divided into two Suborders, viz. *Stellate* and *Cinchoneæ,* to which a third, *Loraniæ* (which has no representatives at the North) is appended by Torrey and Gray. The species of the first Suborder, Stellatæ, are common in the northern parts of both continents; the two other Suborders chiefly prevail in warm or torrid regions.

Properties.—A very important family, furnishing many useful products. The madder, one of the most important of dyes, is furnished by the root of Rubia tinctoria. A similar coloring matter is possessed by several species of Galium. Among the *Cinchoneæ* we find Cinchona and Cephaelis furnishing two of the most valuable of all medicines. Peruvian bark, a powerful febrifuge, well known and appreciated everywhere, is the product of several species of the former, viz. Cinchona micrantha, C. condaminea, C. lanccolata, C. magnifolia &c., all natives of Peru. Their febrifugal properties depend upon the presence of two alkalies Cinchonia and Quinia, both combined with Kinic acid. Ipecacuanha, the prince of all emetics, is the product of the root of Cephaelis ipecacuanha, a little shrubby plant with creeping roots, in the damp forests of Brazil. Several other species of Cinchoneæ afford substitutes for the true ipecac.

Coffee is the hard albumen of the seeds of Coffea Arabica, a tree of moderate size, with a light brown trunk and a conical shaped head. Leaves shining, light green. Flowers white, fragrant. The berries black when ripe. Coffee is said to have been used in Ethiopia from time immemorial. In Paris and London it seems not to have been in general use earlier than the year 1700, but since that time, enough has been drank in Europe and America to float the British navy.

Conspectus of the Genera.

Leaves (and intermedular leaf like stipules) in verticils.		*Galium.*	1
	Ovary with 2 corollas.	*Mitchella.*	2
	Capsule 2-celled, many seeded	*Hedyotis.*	3
	Ovary Carpels 2, 1-seeded, 1 indehiscent.	*Spermacoce.*	4
Leaves Herbs. Sample Carpels 2, 1-seeded, both indehiscent.		*Diodia.*	5
adherent. opposite. Shrubs, with flowers in globose heads.		*Cephalanthus.*	6
Ovary partly free from the calyx. Leaves opposite.		*Spigelia.*	7

SUBORDER 1.—**S T E L L A T Æ.**

Calyx wholly adherent (superior) to the ovary which is two-celled, two-seeded. Leaves verticillate. Herbs.

1. GALIUM.

Gr. γαλα, milk; the flowers of one species (G. verum) are used in coagulating milk.

Calyx minute, 4-toothed; corolla rotate, 4-cleft; stamens 4, short; styles 2; carpels 2, united, 1-seeded, indehiscent.—*Herbs, with slender, 4-angled stems. Lvs. verticillate.*

* *Fruit smooth.*

1. G. ASPRELLUM. Michx. *Rough Cleavers or Clivers.*

St. diffuse, very branching, rough backwards; *lvs.* in 6s, 5s or 4s, lanceolate, acuminate or cuspidate, margin and midvein retrorsely aculeate; *ped.* short, in 2s or 3s.—♃ Common in thickets and low grounds, Can. and Northern States. Stem weak, 2—5f long, leaning on other plants, and closely adhering to them by its minute, retrorse prickles. Leaves 5—8″ by 2—3″. Flowers white, small and numerous. Fruit minute, smooth, often slightly hispid when young. Jl.

2. G. TRIFIDUM. *Dyers' Cleavers. Goose-grass.*

St. decumbent, very branching, roughish with retrorse prickles; *lvs.* in 5s and 4s, linear-oblong or oblanceolate, obtuse, rough-edged; *parts of the flower* mostly in 3s.—♃ In low, wet grounds, Can. and U. S. It is one of the smallest of the species. Leaves 3—6″ by 1—2″, often cuneate at base. Peduncles mostly in 3s, and axillary. Flowers small, white. Jl.

β. tinctorium. Torr. (G. tinctorium. *Linn.*)—*St.* nearly smooth; *lvs. of the stem* in 6s, *of the branches* in 4s; *ped.* 2—3-flowered; *parts of the flower* in 4s.—A somewhat less slender variety than the first. The root is said to dye a permanent red.

γ. latifolium. Torr. (G. obtusum. *Bw.*)—*Lvs.* in 4s, oblanceolate, obtuse; *ped.* 3-flowered; *parts of the flower* in 4s.

3. G. VERUM. *Yellow Bedstraw.*

Erect; *lvs.* in 8s, grooved, entire, rough, linear; *fls.* densely paniculate.—♃ Found in dry, open grounds, in the vicinity of Boston, probably introduced. *Bigelow.* Root long, fibrous. Stem slender, erect, 1—2f high, with short, opposite, leafy, unequal branches. Leaves deflexed, linear, with rolled edges. Flowers numerous, small, yellow, in small, dense, terminal panicles. Jn.— The roots dye red. The flowers are used in England to curdle milk. §

4. G. CONCINNUM. Torr. & Gray.

St. decumbent, diffusely branched, retrorsely scabrous on the angles; *lvs.* in 6s, linear, glabrous, 1-veined, scabrous upwards on the margins; *ped.* filiform, twice or thrice trichotomous, with short pedicels; lobes of the corolla acute. Dry woods and hills, Mich., Ky. *T. & G.* Ia.! Stems very slender, 10—15′ high. Leaves in numerous whorls, 5—8″ by 1″, slightly broader in the middle. Flowers minute and numerous, white. Jn.

** *Fruit hispid.*

5. G. APARINE. *Common Cleavers.*

St. weak, procumbent, retrorsely prickly; *lvs.* in 8s, 7s or 6s, linear-oblanceolate, mucronate, rough on the midvein and margin; *ped.* axillary, 1—2-flowered.—① In wet thickets, Can. and Northern States to Ia. *Plummer!* Stems several feet long, leaning on other plants, and closely adhering by their hooked prickles to everything in their way. Leaves 12—20″ by 2—3″. Flowers numerous, small, white. Fruit rather large, armed with hooked prickles. Jn.— The root will dye red. The herbage is valued as a domestic remedy. § ?

6. G. TRIFLORUM. Michx. *Tri-flowering Galium.*

St. weak, often procumbent, smoothish, shining; *lvs.* in 5s and 6s, lanceolate, acuminate-cuspidate, 1-veined, scarcely ciliate on the margin; *ped.* elongated, axillary, 3- (rarely 2) flowered at the extremity; *fls.* pedicellate; *fr.*

aispid with hooked hairs.—24 Grows in moist thickets and woods, Can. and U. S. Stem 1—3f long, slightly branched. Leaves 1—2' long, ⅓ as broad, often obovate. Flowers greenish-white, small. Fruit whitish with its turbinate clothing. Jl.

7. G. BOREALE. (G. septentrionale. *Bw.*) *Northern Galium.*

St. erect, smooth; lvs. in 4s, linear-lanceolate, rather acute, 3-veined, smooth; fls. in a terminal, pyramidal panicle.—24 Grows in rocky, shady places, Northern States and Brit. Am. Stems 1f or more high, several together, branched above. Leaves 12—20'' by 2—9'', tapering to an obtusish point. Flowers numerous, small, white, in a thyrse-like panicle at top of the stem. Fruit small. Jl.

8. G. PILÓSUM. Ait. (G. puncticulosum. *Michx.*) *Hairy Galium.*

St. ascending, hirsute on the angles; lvs. in 4s., oval, indistinctly veined, hirsute both sides and punctate with pellucid dots; ped. several times forked, each division 2—3-flowered; fls. pedicellate.—24 A rare species, found in dry woods and sterile soils, Mass.! to Ia.! and Tex. Stem 1—2f high, acutely 4-angled, mostly with few, short, spreading branches, sometimes much branched. Leaves 9—12'' by 4—8'', obtusish, very hairy as well as the stem and fruit. Flowers purplish. Jn.

9. G. CIRCÆZANS. Michx. *Circæa-like Galium.*

St. erect or ascending, smooth; lvs. in 4s, oval or ovate-lanceolate, 3-veined, smoothish, ciliate on the margins and veins; ped. divaricate, few-flowered; fr. subsessile, nodding.—24 Grows in woods, U. S. and Can. Stem about 1f in height, with a few short branches near the top, or simple. Leaves 1—2' by 4—8''. Flowers on very short, reflexed pedicels, scattered along the (usually 2) branches of the dichotomous peduncle. Fruit covered with little hooks as in Circæa. Jl.—The leaves have a sweet taste like liquorice.

β.? *lanceolatum.* Torr. (G. Torreyi. *Bw.*) Very smooth; lvs. lanceolate; fr. sessile.—A fine variety ? with larger leaves (2' or more in length). Flowers purple.

γ.? *montanum.* T. & G. (G. Littelli. *Oakes.*) Dwarf; lvs. obovate.—White Mts. *Oakes.*

SUBORDER 2.—CINCHONEÆ.

Calyx adherent to the ovary. Leaves opposite (rarely verticillate). Stipules between the petioles, often united with them into a sheath.

2. MITCHELLA.

In honor of Dr. John Mitchell, an English resident in Virginia.

Flowers 2 on each double ovary; calyx 4-parted; corolla funnel-shaped, hairy within; stamens 4, short, inserted on the corolla; stigmas 4; berry composed of the 2 united ovaries.—*Evergreen herbs, smooth and creeping, with opposite leaves.*

M. REPENS. *Partridge Berry.*

St. creeping; lvs. roundish-ovate, petiolate.—A little prostrate plant found in woods throughout the U. S. and Can. Stem furnished with flat, coriaceous, dark green leaves, and producing small, bright red berries, remarkably distinguished by their double structure, and remaining on the plant through the winter. The corollas are white or tinged with red, very fragrant. Fruit well flavored but dry and full of stony seeds. Jn.

3. HEDYÓTIS.

Gr. ἡδύς, sweet, (ους) ορος, the ear; said to cure deafness.

Calyx tube ovate, limb 4-parted; corolla 4-lobed; stamens 4, inserted on the corolla; stigma 2-lobed; capsule 2-celled, many-seeded. —*Herbs, rarely shrubs. Lvs. opposite. Stip. connate with the petiols.*

§ *Corolla hypocrateriform, with a long tube, limb glabrous.* *Peduncles* 1-*flowered.* Houstonia. Linn.

1. H. cœrulea. Hook. (H ıus. cœrulea. *Linn.*) *Dwarf Pink. Innocence.* *Radical lvs.* ovate-spatulate, petiolate; *sts.* erect, numerous, dichotomous; *ped.* filiform, 1—2-flowered.—An elegant little plant, found in moist grounds, fields and road-sides, Can. and U. S. Its blossoms appear early, and are usually found in patches of considerable extent, covering the surface of the ground with a *cerulean* hue. The cauline leaves are small, opposite, lance-ovate. Stems very slender, forked, 3—5′ high, each branch bearing a flower. Corolla pale blue, yellowish at the centre. May—Aug.

2. H. minima. T. & G. (Houstonia. *Beck.*) Glabrous, simple or dichotomously branching; *lvs.* linear-spatulate, much attenuated to the base; *ped.* axillary and terminal, often longer than the leaves; *sds.* 10—15 in each cell, oval, smooth, concave on the face.--Prairies, &c., Mo. ! Tenn.! to La. Very small and delicate, 1—3′ high. Leaves about 5″ by 1″. Flowers rose-color. Mar.—May.

§ § *Corolla infundibuliform, often hairy inside.* *Flowers in terminal* *racemes.* Amphiotis. DC.

3. H. ciliolàta. Torr. (Hous. Canadensis. *Muhl.*) *Clustered Dwarf Pink* *Radical lvs.* ovate, obtuse, narrow at the base, ciliate on the margin; cauline ones ovate-spatulate, sessile; *corymbs* terminal, pedicellate; *ped.* trichotomous; *divisions of the calyx* lance-linear.—Banks of lakes and rivers, Ontario! Niagara! W. to Ohio. A little plant, stouter than the last. Root-leaves numerous, stem-leaves few. Stems smooth, 4-angled, branched above, and bearing a corymbose cluster of numerous pale purple flowers. Calyx half-adherent, its lobes about half as long as the tube of the corolla. May—July.

4. H. longifolia. Hook. (Houstonia longif. *Gaert.*) *Long-leaved Dwarf* *Pink.*—*Radical lvs.* oval-elliptic, narrowed to each end; *cauline* linear or lance-linear, 1-veined; *fls.* in small, paniculate cymes.—♃ Dry hills, N. and Mid. States! to Ark. and Flor. Much more slender than the next, Stems erect, 5—12′ high, 4-angled, smooth or ciliate on the angles. Leaves 9—15″ by 2—3″, cauline sessile, rather acute at each end, all smooth. Flowers 2 or 3 together, on very short pedicels, pale-purple, with deeper-colored striæ in the throat. Jn. Jl.

β.? (H. tenuifolia. *Nutt.*) *St.* very branching; *lvs.* very narrow; *ped.* filiform; *fls.* smaller.

5. H. purpùrea. Torr. & Gray. (Houstonia. *Linn.*) *St.* ascending, clustered, branching, 4-angled; *lvs.* ovate-lanceolate, 3—5-veined, closely sessile; *cymes* 3—7-flowered, often clustered; *calyx segments* lance-linear, longer than the capsule.—Mid. and W. States! in woods and on river banks. A very delicate flowerer, about 1′ high. Leaves 1—2′ long, ½ as wide. Corolla (purple, *T. & G.*) white, scarcely tinged with purple. May—Jl.

§ § § *Corolla rotate, tube very short.* *Seeds 50—60 in each cell.* *Flowers* *mostly glomerate in the axils of the leaves.* Elatinella. *Torr. & Gray.*

6. H. glomerata.. *Creeping Green-head.* *St.* assurgent, branching; *lvs.* oblong-lanceolate, pubescent, narrowed at the base into a short petiole or sessile; *fls.* glomerate in the axils and terminal. A plant varying in size from 1—2′ to as many feet, found in swamps, &c., N. Y. to La. Leaves ½′ in length, apparently connate from the stipules adhering to each side of the petiole. Stipules 2-cleft into narrow subulate divisions. Calyx in 4 deep, leafy divisions which are much longer than the white, rotate corolla. Stamens scarcely exserted. Style very short. Capsule opening crosswise. Jn.—Sept.

4 SPERMACÒCE.

Gr. σπερμα, seed, ακωη, a point; alluding to the pointed seeds.

Calyx tube ovoid, limb 2—4-parted; corolla tubular, limb spreading, 4-lobed, stamens 4; stigma 2-cleft; fruit dry, 2-celled, crowned with the calyx; seeds 2, peltate, furrowed on the face.—*Mostly herbaceous and tropical. Fls. small, axillary, sessile, whorled*

S. GLABRA. Michx.

Glabrous, procumbent at base; *lvs.* opposite, lanceolate, entire; *whorls* many-flowered; *cal.* 4-toothed (rarely 5); *cor.* funnel-form, short, hairy in the throat; *anthers* included in the tube; *stig.* subsessile.—River banks, Western States! Stem 1—2' long, teretc, with 4 prominent lines, branched. Leaves 2—3' by ⅓—1', tapering to each end. Flowers white, 8—20 in a whorl, subtended by the subulate bracts of the stipules. Jl. Aug.—Resembles some of the Labiatæ.

5. DIODIA.

Gr. δις, twice. οδους, tooth; alluding to the two calyx teeth crowning the ovary.

Calyx, corolla, stamens, style and fruit, as in Spermacoce, except that the (2 or 3) 1-seeded, separable carpels are both indehiscent; seeds oval, peltate.—*American, chiefly tropical herbs, with the habit of Spermacoce in all respects save the indehiscent carpels.*

1. D. VIRGINIÀNA. (Spermacoce. A. Rich.)

Procumbent, nearly glabrous or hirsute; *lrs.* lanceolate-linear, sessile, entire; *bristles of the stipules* longer than the sheaths; *fls.* solitary, opposite; *corolla tube* thrice longer than the calyx; *sta.* exserted; *sty.* deeply 2-cleft, the lobes filiform.—♃ Damp places, Ill. to Ga.! and La. Stem 1—2f long, somewhat 4-sided. Leaves 1—2' by 2—4'', 1-veined, often with smaller ones fascicled in the axils. Corolla 5'' long, hairy inside. May—Sept.

2. D. TERES. Walt. (Spermacoce diodina. Michx.)

Procumbent or ascending, hairy or scabrous; *lrs.* linear-lanceolate, sessile, rough-edged, acute, much longer than the sheaths or fruit; *fls.* solitary or several in each axil; *cor.* funnel-form, with a wide tube, twice longer than the calyx; *fr.* somewhat hairy and 4-sided.—Sandy fields, N. J. to Ill. Mead! and Southern States. Stems rather rigid, much branched, 5—18' long, brownish. Leaves about 1' by 2''. Corolla reddish-white, shorter than the reddish-brown bristles.

6. CEPHALANTHUS.

G. κεφαλη, a head. ανθος, a flower: flowers growing in dense heads.

Calyx limb 4-toothed; corolla tubular, slender, 4-cleft; stamens 4; style much exserted.—*Shrubs with opposite leaves and short stipules. Fls. in globose heads, without an invol.*

C. OCCIDENTÀLIS. *Button Bush.*

Lrs. opposite, and in 3s, oval, acuminate, entire, smooth; *hds.* pedunculate.—A handsome shrub, frequenting the margins of ponds, rivers and brooks, U. S. and Can. It is readily distinguished by its spherical heads of flowers which are near 1' diam., resembling the globular inflorescence of the sycamore (Platanus occidentalis). Height about 6f. Leaves spreading, entire, 3—5' by 2—3'. The flowers are tubular. with long, projecting styles, and are inserted on all sides of the round receptacle. July.

SUBORDER 3.—SPIGELIEÆ.

Calyx persistent, almost entirely free from the ovary. Leaves opposite, with intermediate stipules.

7. SPIGELIA.

In honor of Adrian Spigelius, Prof. of Anat. and Surg. at Padua, 1578—1625.

Calyx 5-parted. segm linear-subulate; cor. narrowly funnel-form, limb 5-cleft, equal; stam. 5; anth. convergent; caps. didymous, 2-celled, few-seeded.—*Herbaceous or suffrutescent. Lvs. opposite. Stipules small, interpetiolar. Fls. sessile, in terminal spikes.*

S. MARYLANDICA. *Pink-root. Worm-grass.*

Erect, simple, nearly glabrous; *st.* square; *lvs.* sessile, ovate-lanceolate, acute or acuminate, margin and veins scabrous-pilose; spikes 3—8-flowered; *cor.* tube 4 times longer than the calyx; *anth.* exserted; *lobes* of the *cor.* lanceolate; *caps.* glabrous, shorter than the calyx.—♃ In woods, Penn. to Flor. W. to Ill. *Mead*, and Tenn. *Miss Carpenter!* An elegant dark green herb, a foot high. Leaves 3—4' by 1½—2½', entire, often ovate-acuminate, the stipules scarcely perceptible. Flowers 1½—2' long, somewhat club-shaped, scarlet without, yellow within. Style exserted. June.—A celebrated anthelmintic.

ORDER LXXIII. VALERIANACEÆ.—VALERIANS.

Herbs, with opposite leaves and no stipules.
Cal. adherent, the limb either membranous or resembling a pappus.
Cor. tubular or funnel-form, 4—5-lobed, sometimes spurred at base.
Sta. distinct, inserted into the corolla tube, alternate with, and generally fewer than its lobes.
Ova. inferior, with one perfect cell, and two abortive ones.
Sds. Solitary, pendulous, in a dry, indehiscent pericarp.

Genera 12, species 185, widely diffused in temperate climates. The true *valerian* of the shops, used in hysteria, epilepsy, &c., is a product of Valeriana officinalis. The roots of several other species possess a heavy odor, and are tonic, antispasmodic, febrifugal, &c. The *spikenard* (John xii. 3, &c.) of old, valued as a perfume and a stimulant, is from the root of Nardostachys Jatamansi.

Genera.

Limb of the calyx at length a plumose pappus, deciduous, ... *Valeriana.* 1
Limb of the calyx toothed and persistent, or obsolete. . . . . *Fedia.* 2

1. VALERIĀNA.

Dedicated to king *Valerius*, a patron and friend of botanists. *Linn.*

Calyx at first very small, at length forming a plumose pappus; corolla funnel-form, regular, 5-cleft; stamens 3; fruit 1-celled, 1 seeded.—♃ *Lvs. opposite, mostly pinnately divided. Fls. in close cymes.*

1. V. SYLVATICA. β. *uliginosa. Wild Valerian.*

St. erect, striate, simple; *radical lvs.* ovate or subspatulate, undivided • *cauline ones* pinnately divided, segments ovate-lanceolate, entire or subserrate, the terminal one often dentate; *lobes of the stig.* minute, 2 or 3; *fr.* ovate, compressed, smooth.—Stem 1—3f high. Swamps, Vt.! to Mich., very rare. Plant nearly smooth. Leaves ciliate with scattered hairs; those of the root petioled, sometimes auriculate at base, those of the stem with 4—8 lateral segments and a large terminal one. Flowers numerous, rose-colored, appearing in July.

2. V. PAUCIFLŌRA. Michx.

Glabrous, erect or decumbent, often stoloniferous at base; *radical lvs.* ovate, cordate, slightly acuminate, on long petioles, crenate-serrate; *cauline* pinnately 3—7-parted; *lfts.* ovate, terminal one much the largest; *cymules* few-flowered, corymbose; *corolla tube* long and slender.—Ohio! to Va. and Tenn. Stem mostly simple, 1—2f high. Leaves of the succors mostly undivided, 1—1½' by ½—1½', petioles 1—4' long. Flowers pale purple, ½' in length. Jn. Jl.

3. V. CILIĀTA. Torr. & Gray.

Simple, smooth and somewhat fleshy; *lvs.* lance-linear, some of them pinnately cleft into 3—7 lance-linear, acute segments, margins densely and minutely ciliate, mostly attenuated to the base; *cauline ones* few, with linear segments; *panicle* compound; *fr.* compressed, 4-ribbed, crowned with the late calyx limb of 10 or 12 plumose setæ.—Low grounds, Can., Wis.! Ohio! Root yellowish, fusiform. Stem 1—3f high. Root-leaves many, 3—6' long, segments 2—4" wide. Flowers white, in a close panicle, which is greatly expanded in fruit. June.

2. FEDIA. Adans.

Perhaps from *fædus*; on account of the strong odor of some of the species.

Calyx limb 3—6-toothed and persistent, or obsolete; corolla tubular, 5-lobed, regular; stamens 2 or 3; fruit 2 or 3-celled, 1-seeded. —① *Lvs. opposite, sessile. One or two cells of the fruit abortive.*

1. F. FAGOPÌRUM. Torr. & Gray. (Valerianella radiata. *Mench.*) *Wild Corn-salad or Lamb Lettuce.—St.* dichotomous, nearly smooth; *lvs.* oblong-spatulate, subentire; *fr.* 3-sided, obscurely 2—3-toothed at the summit. -Western N. Y. to Ohio! Stem 8—18' in height. Bracts lanceolate, acute. Fruit resembling that of buck-wheat (Polygonum Fagopyrum) in form, containing one large seed and two empty cells. Flowers white. June.

2. F. RADIÀTA. Michx.
Lvs. entire, or toothed towards the base, obtuse; *fls.* white; *fr.* ovoid, pubescent, somewhat 4-angled, obscurely 1-toothed at apex; *empty cells* not divergent, but with a groove between them; *fertile cell* flattish, broader than the other 2.—Low grounds, Mich.! Ohio! to La. Stem 6—12' high, dichotomous like the other species, smooth. Leaves oblong, more or less tapering to the base, 1—2' by 2—4''. Fruit less than 1'' long, at length nearly smooth. May.

3. F. OLITÒRIA. Vahl. *Pawnee Lettuce.*
Fr. compound, oblique, at length broader than long, not toothed at apex; fertile cell larger than both the others; empty cells united, but with a groove between; *lvs.* spatulate-obtuse, radical ones petiolate; *fls.* pale blue.—Naturalized in some portions of the U. S. Stem smooth, 8—12' high, dichotomous. Leaves mostly entire. Flowers in dense cymules. Fruit 1' diam. June. ‡

4. F. UMBILICÀTA. W. S. Sullivant.
Fruit subglobose, inflated, apex 1-toothed, the anterior face deeply umbilicate, sterile cells several times larger than the fertile one; *bracts* subspatulate-linear, not ciliate.—Columbus, Ohio, *Sullivant!* Plant smooth in all its parts, 1—2f high, many times dichotomous. Leaves oblong, obtuse, clasping, dilated and coarsely dentate at base, 1½—3' by 3—10''. Flowers in numerous cymules, corymbosely arranged. Fruit nearly 1'' diam., with 1 rib at the back produced into a tooth at apex, and a conspicuous depression in front.

ORDER LXXIV. DIPSACEÆ.—TEASELWORTS.

Herbs or low shrubs, with whorled or opposite leaves.
Fls. collected upon a common receptacle and surrounded by a many-leaved involucre.
Cal. adherent, often pappus-like, surrounded by a scarious involucel.
Cor. tubular, somewhat irregular, the limb 4—5-parted.
Sta. 4, alternate with the lobes of the corolla, often unequal. *Anthers* distinct.
Ova. inferior, one-celled, one-ovuled. *Style* one, simple.
Fr. dry, indehiscent, with a single suspended seed.

Genera 6, species 150. The order is nearly allied to the Compositæ. The species are all natives of the temperate regions of the Eastern continent, none of them American. Their properties are unimportant. One of the species below is useful in dressing cloth.

1. DIPSACUS.

Gr. διψαω, to thirst; alluding to the water held in the axils of the leaves.

Flowers in heads; involucre many-leaved; involucel 4-sided; calyx superior: corolla tubular, 4-cleft; fruit 1-seeded, crowned with the calyx.—② *Plants large, hairy or prickly. Lvs. opposite, connate (sometime distinct) at base.*

1. D. SYLVESTRIS. Mill. *Wild Teasel.*
Lvs. connate, sinuate or jagged; *hds.* cylindrical; *bracts of the involucre* longer than the head of flowers, slender and pungent, bent inwards.—A tall, naturalized, European plant, growing in hedges and by road-sides, Mass. to Ia.! Stem about 4f high, angled and prickly, with the opposite, lance-shaped leaves united around it. Flowers bluish, in a large oval or cylindrical head whose bracts or scales are not hooked as in the next species, but straight. July. §

2. D. FULLÒNUM. *Fullers' Teasel.—Lvs.* connate entire or serrate *hd.* cylindrical; *bracts* hooked; *invol.* spreading.—A cultivated, European plant. Root fleshy, tapering. Stem erect, furrowed, prickly, hollow, about 5f high. Leaves two at each node, united at their bases around the stem in such a way as to hold a quantity of water. Flowers whitish, in large, oval or ovoid heads. Cultivated for the use of the clothiers (*fullonum*), who employ the heads with their hard, hooked scales to raise the nap upon woollen cloths. For this purpose they are fixed around the circumference of a revolving drum. Flowers in July. ‡

2. SCABIOSA.

Lat. *scabies*, leprosy; plants said to be useful in cutaneous diseases.

Flowers in heads; involucre many-leaved; involucel nearly cylindrical, with 8 little excavations; calyx limb consisting of 5 setæ, sometimes partially abortive.—♃ *Large, mostly European herbs with opposite leaves.*

1. S. succisa. *Devil's-bit.*—*Rt.* premorse; *stem lvs.* remotely toothed; *hds. of fls.* nearly globose; *cor.* in 4 equal segments.—In gardens, though rarely cultivated. The stem is about 1f high. Corolla violet. †

2. S. atropurpurea. *Mourning Bride.*—*Lvs.* pinnatifid and incised; *hds of fls.* radiant; *receptacle* cylindric; *outer crown of the seed* short, lobed and crenate.—A beautiful species, 2—4f high, with dense heads of dark purple flowers. †

Order LXXV. COMPOSITÆ.—Asterworts.

Plants herbaceous or shrubby.
Lvs. alternate or opposite, without stipules. simple though often much divided.
Fls. collected into a dense head (capitulum), upon a common receptacle, surrounded by an involucre o. many bracts (scales).
Cal. closely adherent to the ovary, the limb wanting, or membranaceous and divided into bristles, hairs &c., called *pappus.*
Cor. superior, consisting of 5 united petals, either ligulate or tubular.
Sta. 5, alternate with the lobes of the corolla. *Anth.* cohering into a cylinder.
Ova. inferior, 1-celled, 1-ovuled. *Style* 2-cleft, the inner margins of the branches occupied by the stigmas.
Fr. an achenia, dry, indehiscent, 1-seeded, crowned with the pappus.

This is the most extensive and most natural of all the orders of the vegetable kingdom, always distinguished at sight by the capitate flowers and the united anthers. It comprehends 1005 genera (at present known, 1846), and about 8000 species; being nearly one-ninth of all the species of flowering plants. The general inflorescence is centrifugal, that is, the central or terminal heads are first developed, while the inflorescence of the heads is centripetal, the outer flowers first expanding. In color the flowers are various; sometimes those of the disk and ray are of different colors, again they are all of the same, but in the former case the disk florets are almost always yellow.

This immense order is diffused throughout all countries of the globe, but in very different proportions. According to Humboldt, they constitute about one-seventh of the Phænogamous Flora of Germany, one-eighth of France, one-fifteenth, of Lapland, one sixth, of North America (north of Mexico), and one-half, of Tropical America. In New Holland they are in the proportion of about one-sixteenth, according to Brown, while in the island of Sicily they are one-half. The Liguliflora are said to be most abundant in cold regions, and the Tubuliflora in hot regions. The Labiatiflora are almost exclusively confined to South America. In the northern parts of the world the Composite are universally herbaceous, but towards the tropics they gradually become frutescent and even trees. In Chili they are generally shrubs, and on the island of St. Helena they are trees.

Properties, &c.—The Composite furnish comparatively few useful products. A bitter principle pervades the whole, which, when combined with resin and astringent mucilage, becomes *tonic* and *febrifugal*, as in the camomile, colt's-foot, thoroughwort, goldenrod, &c. Some are *anthelmintic* from the prevalence of the resinous principle, as tansey, Artemisia, Vernonia. Others are aromatic and extremely bitter, as wormwood and all the species of Artemisia. Other species are very acrid, as mayweed. The Jerusalem artichoke (Helianthus tuberosus), the vegetable oyster (Tragopogon), the true artichoke (Cynara), lettuce, dandelion and a few others, are the only species useful for food. The order abounds in ornamental plants of the highest interest to the florist, and of easy culture. Among these are the splendid Dahlias and Chinese Chrysanthema, with the numerous progeny of Aster, Helianthus, Xeranthemum, Coreopsis and multitudes of others, constituting the richest ornaments of the autumnal flower garden.

The inflorescence of the Composite is peculiar, and its real nature often complex and obscure. The following definitions of terms are given with reference to this order only, and if understood, will remove many difficulties that lie in the student's way in the investigation of this subject.

Capitulum or head (compound flower of the earlier botanists); a collection of flowers (florets) on a common receptacle (rachis), as in Aster, Helianthus, &c.
Involucre (calyx by analogy) is the lower and outer envelope of the head.
Scales; the modified leaves or bracts composing the involucre.
Monophyllous involucre; where the scales are united by their edges.
Polyphyllous involucre; where the scales are distinct.
Simple involucre; where the scales are equal and arranged in a single row.   [short ons.
Calyculate involucre; where a single row of scales is surrounded at base by an outer row of very
Imbricated involucre; where the scales are in several rows, the outer ones becoming gradually shorter.
The *Receptacle* or *rachis* is the dilated extremity of the peduncle, inclosed by the involucre, and upon which the flowers stand. It is
Columnar, flat, conical or *depressed,* according to its form;
Paleaceous or *chaffy,* where the flowers are subtended by chaffy scales which are analogous to bracts;
Alveolate, where it presents the appearance of a honey-comb, each flower having been surrounded by a membranous rim or involucel;
Areolate, where the alveoli are reduced to a mere line;
Fimbrillate, where the alveoli are split into teeth or bristles;
Naked, when smoothish, being destitute of chaff, alveoli, bristles, &c.

The *flowers* are moreover said to be
Of the disk, where they stand in or near the centre of the head;
Of the ray, where they stand in the outer circle or circumference of the head;
Ligulate (strap-shaped), when the limb is split on one side and spread open in the form of a strap.
Tubular, when they are monopetalous with a regular limb. The *heads* are termed
Homogamous, where they consist wholly of perfect flowers;
Heterogamous, where the flowers of the disk are perfect or staminate, while those of the ray or margin are pistillate or neutral;
Radiant, where the flowers are all ligulate, as in the dandelion;
Radiate, where those of the ray or margin only are ligulate, the rest being tubular, as in Aster;
Discoid, where all the flowers are tubular, as in the *Boneset*:
Monœcious, where the same head has both staminate and pistillate flowers;
Heterocephalous, where the same individual plant has some of its heads wholly of staminate, and others wholly of pistillate flowers.

D œcious, where the same species has some of its individuals with staminate heads only, and others with pistillate heads only. The anthers are usually *appendiculate*, that is, prolonged at the summit into a membranous appendage. The achenia are termed
Rostrate, when they are prolonged at the summit into a slender neck supporting the pappus, as in the dandelion:
Compressed, when they are flattened parallel with the *diameter* of the head;
Obcompressed, when flattened parallel with the *circumference* of the head.

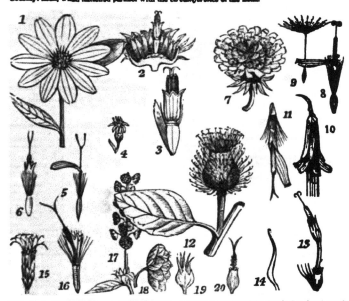

FIG. 67.—1. Helianthus strumosus—head radiate. 2. Vertical section of the head, showing the scales of the involucre, and a single disk-flower remaining upon the convex receptacle. 3. A perfect disk-flower magnified, showing the achenium, the 2 awns of the pappus, the 5-toothed tubular corolla, the 5 stamens raised around the branched style, and the chaff-scale at base. 4. Head (radiate) of Solidago cæsia. 5. A pistillate, ligulate flower of the ray. 6. A perfect disk fl. 7. A (radiant) head of Taraxacum Dens-leonis. 8. A perfect, ligulate fl. 9. Achenium, with its long beak and plumose pappus. 10. A (radiant) head of Nabalus altissimus. 11. A flower. 12. Lappa major, head discoid. 13. A flower. 14. One of the hooked scales. 15. A (discoid) head of Eupatorium purpureum. 16. A flower. 17. Ambrosia artemisiæfolia. 18. Staminate head enlarged. 19. Pistillate involucre enlarged. 20. The fertile flower.

Conspectus of the Genera.

		Corollas cyanic.	Leaves alternate.	1
	discoid.		Leaves opposite or verticillate.	2
		Corollas yellow.		3
		Rays yellow. .	Leaves alternate.	4
			Leaves opposite or all radical.	5
Heads	radiate.		Leaves opposite or all radical.	6
	radiant.	Rays cyanic. .	Leaves alternate.	7
	radiant.			8

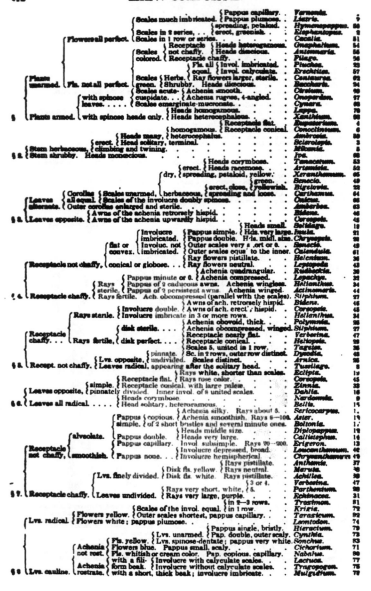

		Scales much imbricated. { Pappus capillary.	Vernonia. . 9
		{ Pappus plumose.	Liatris. . 7
		Scales in 2 series, . . . { spreading, petaloid.	Hymenopappus. 58
		Scales in 1 row or series. . { erect, greenish.	Elephantopus. 2
	Flowers all perfect.		Cacalia. 81
		{ Receptacle { Heads heterogamous.	Gnaphalium. 54
		Scales { not chaffy. { Heads diœcious.	Antennaria. 55
		colored. { Receptacle chaffy.	Filago. 56
		{ Fls. all { Invol. imbricated.	Pluchea. 23
		equal. { Invol. calyculate.	Erechtites. 57
		Scales { Herbs. { Ray flowers larger, sterile.	Centaurea. 92
Plants	Fls. not all perfect.	green. { Shrubby. Heads diœcious.	Baccharis. 24
unarmed.		Scales acute- { Achenia smooth.	Cirsium. 95
	with spinose	cuspidate. . . { Achenia rugose, 4-angled.	Onopordon. 97
	leaves.	Scales emarginate-mucronate. .	Cynara. 98
		{ Heads homogamous. .	Lappa. 70
§	Plants armed.	with spinose heads only. { Heads heterogamous. .	Xanthium. 25
			Eupatorium. 4
		{ Receptacle flat.	Conoclinium. 6
		{ homogamous. { Receptacle conical.	Ambrosia. 80
		Heads many, { heterocephalus. .	Sclerolepis. 3
	erect. { Head solitary, terminal. .	Milleria. 5
§§.	Stem herbaceous, { climbing and twining. .	Mikania. 5
	Stem shrubby. Heads monœcious. .	Iva. 88
		{ Heads corymbose.	Tanacetum. 53
		erect. { Heads racemose.	Artemisia. 52
		dry, { spreading, petaloid, yellow.	Xeranthemum. 95
		{ green.	Senecio. 49
	Corollas { Scales unarmed, herbaceous, { erect, close, yellowish.	Bigelovia. 22
Leaves	all equal. { Scales of the involucre doubly spinose.	Carthamus. 64
alternate.	Outer corollas enlarged and sterile. .	Oxlinus. 86
		Amberica. 63
		{ Awns of the achenia retrorsely hispid.	Bidens. 46
§§.	Leaves opposite. { Awns of the achenia upwardly hispid.	Coreopsis. 45
		{ Heads small.	Solidago. 19
	Involucre { Pappus simple. { Hds. very large.	Inula. 21
	imbricated. { Pappus double. H'ls. midl. size.	Chrysopsis. 20
	flat or { Involuc. not { Outer scales very s ort or 0.	Senecio. 49
	convex. { imbricated. { Outer scales equ' to the inner.	Calendula. 61
		{ Ray flowers pistillate. .	Helenium. 36
	Receptacle not chaffy, { conical or globose. . . { Ray flowers neutral.	Leptopoda. 43
		{ Achenia quadrangular. .	Rudbeckia. 35
		{ Pappus minute or 0. { Achenia compressed.	Lepachys. 36
		Rays { Pappus of 2 caducous awns. Achenia wingless.	Helianthus. 34
		sterile. { Pappus of 2 persistent awns. Achenia winged.	Actinomeris. 44
§§.	Receptacle chaffy. { Rays fertile. Ach. obcompressed (parallel with the scales).	Silphium. 37
		{ Awns of ach. retrorsely hispid.	Bidens. 46
		{ Involucre double. { Awns of ach. erect, hispid.	Coreopsis. 45
	Rays sterile. { Involucre imbricate in 3 or more rows.	Helianthus. 34
		{ Achenia obovoid, thick. .	Polymnia. 26
		disk sterile. . . . { Achenia obcompressed, winged.	Silphium. 27
	Receptacle	{ Receptacle nearly flat. .	Verbesina. 47
	chaffy. . . { Rays fertile, { disk perfect. . { Receptacle conical. .	Heliopsis. 48
		{ Scales 5, united in 1 row. .	Tagetes. 35
		Lvs. opposite, { pinnate. { Sc. in 2 rows, outer row distinct.	Dysodia. 48
§§.	Recept. not chaffy. { Leaves radical, appearing after the solitary head.	Arnica. 26
			Tussilago. 8
		{ Rays white, shorter than scales.	Eclipta. 19
		{ Receptacle flat. { Rays rose color.	Coreopsis. 45
		simple. { Receptacle conical, with large palea.	Zinnia. 33
	Leaves opposite, { pinnately divided. Inner invol. of 8 united scales.	Dahlia. 16
		Heads corymbose. .	Nardosmia. 9
§§.	Leaves all radical . . . . { Head solitary, heterogamous.	Bellis. 14
		{ Achenia silky. Rays about 5.	Sericocarpus. 1.
		Pappus { copious. { Achenia smoothish. Rays 8—100.	Aster. 10
		simple. { of 2 short bristles and several minute ones.	Boltonia. 1?
		alveolate. { Heads middle size. .	Diplopappus. 10
		Pappus double. { Heads very large.	Callistephus. 14
	Receptacle	Pappus capillary. Invol. subsimple. Rays 20—200.	Erigeron. 11
	not chaffy, { smoothish. { Involucre depressed, broad.	Leucanthemum. 46
		Pappus none. . . { Involucre hemispherical.	Chrysanthemum 49
		{ Rays pistillate.	Anthemis. 37
		{ Disk fls. yellow. { Rays neutral.	Maruta. 36
	{ Lvs. finely divided. { Disk fls. white. Rays pistillate.	Achillea. 38
		{ 3 or 4.	Verbesina. 47
§§.	Receptacle chaffy. { Leaves undivided. { Rays very short, white, { 5.	Parthenium. 28
		{ Rays very large, purple.	Echinacea. 31
		{ in 2—3 rows.	Trastmon. 51
		{ Scales of the invol. equal, in 1 row.	Krigia. 72
	Lvs. radical. { Flowers yellow. { Outer scales shortest, pappus capillary.	Taraxacum. 99
		Flowers white; pappus plumose.	Leontodon. 74
		{ Pappus single, bristly.	Hieracium. 79
		{ Lvs. unarmed. { Pap. double, outer scaly.	Cynthia. 73
		Fls. yellow. { Lvs. spinose-dentate; pappus very white.	Sonchus. 83
	Achenia { Flowers blue. Pappus small, scaly.	Cichorium. 71
	not rost. { Fls. whitish or cream color. Pap. copious, capillary.	Nabalus. 80
	Achenia { with a fili- { Involucre with calyculate scales.	Lactuca. 77
	form beak. { Involucre without calyculate scales.	Tragopogon. 75
§§.	Lvs. cauline. { rostrate. { with a short, thick beak; involucre imbricate.	Mulgedium. 78

Suborder 1.—TUBULIFLORÆ.

Corolla of the perfect or disk flowers tubular, regular, the limb 5-cleft or lobed.

Tribe 1. VERNONIACEÆ.

Heads discoid, homogamous. Branches of the style subulate, hispid throughout.

1. VERNONIA. Schreb.

Named for Wm. Vernon, an English botanist who traveled in America in search of plants.

Flowers all tubular; involucre semicylindric, of ovate, imbricated scales; receptacle naked; pappus double. the exterior chaffy; the interior capillary.—♃ *Herbs or shrubs. Lvs. mostly alternate.*

1. V. NOVEBORACENSIS. Willd. *New York Vernonia. Iron-weed.*
Lvs. numerous, lanceolate, serrulate, rough; *cyme* fastigiate; *scales of involucre* filiform at the ends.—A tall, showy plant with numerous large, dark purple flowers, found in meadows and other moist situations, U. S. Stem branching at top, reddish, 3—6f high. Leaves crowded, paler beneath, radical ones often lobed. Cymes terminal, flat-topped, compound. Scales and corollas deep purple, the former ending in long, thread-like appendages. Sept.
β. *præalta.* Less. (V. præalta. *Willd.*) *St.* and *lvs.* beneath pubescent; *scales* : early destitute of the filiform appendages.—Rather taller than the preceding.

2. V. FASCICULATA. Michx. *Iron-weed.*
St. tall, striate or grooved, tomentose; *lvs.* narrow-lanceolate, tapering to each end, serrulate, lower ones petiolate; *hds.* numerous, in a somewhat fastigiate cyme; *invol.* ovoid-campanulate; *scales* appressed, mucronulate or obtuse.—Woods and prairies Western States, very common! A coarse, purplish-green weed 3—10f high. Leaves 4—8' by 1—2', smooth above. Cymes compact, or loose. Heads large, or small. Corollas showy, dark purple, twice longer than the involucre. Jl. Aug.
β. Taller and more branching, with smaller heads.—Woods, Ia. !

2. ELEPHANTOPUS.

Gr. ελεφας, elephant, πους, foot; alluding to the form of the leaves in some species.

Heads 3—5-flowered, glomerate; flowers all equal; involucre compressed, the scales about 8, oblong, dry, in 2 series; corolla palmate-ligulate, 5-cleft, segments acuminate; achenia ribbed, hairy; pappus chaffy-setaceous.—♃ *Erect, with alternate, subsessile leaves. Corolla violet purple.*

E. CAROLINIANUS. Willd.
St. branched, leafy, hairy; *lvs.* scabrous and somewhat hairy, ovate or oval-oblong, obtuse, crenate-serrate, lower ones on petioles, upper ones subsessile; *hds.* terminal and subterminal.—Dry soils, Penn. Ohio! to Flor. and La. Stem 20—30' high, flexuous, the branches divaricate. Lower stem leaves 5—7' by 3—5', upper about 2' by 1½', the highest oblong, smaller, subtending the small heads in the form of an involucre. Jl.—Sept.

Tribe 2. EUPATORIACEÆ.

Heads discoid or radiate. Branches of the style much elongated, obtuse, minutely pubescent towards the summit outside. Anthers not cordate. Leaves mostly opposite.

Section 1. Heads discoid, homogamous.

3. SCLEROLEPIS. Cass.

Gr. σκληρος, hard, λεπις, a scale.

Head many-flowered; scales of the involucre equal, linear, in 2 series; receptacle naked; corolla 5-toothed, enlarged at the throat; branches of the style much exserted; achenia 5-angled, crowned with a cup-shaped pappus of 5, obtuse, horny scales.—♃ *Aquatic, glabrous, simple, with* 1—3 *terminal heads.* Lvs. verticillate. Fls. purple.

S. VERTICILLÀTA. Cass. (Sparganophorus. Mickx.)
In shallow water, N. J. to Flor. Stem decumbent at base, 1—2f high. Leaves in numerous whorls of about 6, linear-setaceous, entire, 1' in length. Head commonly solitary, at top of the stem. Jl. Sept.

4. EUPATORIUM.
Dedicated to Eupater, king of Pontus, who first used the plant in medicine.

Flowers all tubular; involucre imbricate, oblong; style much exserted, deeply cleft; anthers included; receptacle naked, flat; pappus simple, scabrous; achenia 5-angled.—♃ *Herbs, with* opposite *or* verticillate *leaves.* Hds. corymbose. Fls. of the cyanic series, that is, white, blue, red, &c. never yellow.

* Leaves verticillate. Flowers purple.†

1. E. FISTULÒSUM. Barratt. (E. purpureum. Willd. in part. E. maculatum. Linn. in part. E. purpureum. γ. angustifolium. T. & G.) Trumpet-weed.—St. fistulous, glabrous, glaucous-purple, striate or fluted; lvs. in about 12 whorls of 6s, largest in the middle of the stem, rather finely glandular-serrate; midvein and veinlets livid purple; corymb globose, with whorled peduncles.—Thickets, U. S. and Can., very abundant in the Western States! Height 6—10f, hollow its whole length. Leaves, including the 1' petiole, 8' by 2'. Corymb often 1f diam. Flowers purple. The glaucous hue and suffused redness of this majestic plant are most conspicuous in flowering-time. It does not appear to possess the acrid properties of E. maculatum. July—Sept.

2. E. MACULÀTUM. (E. purpureum, β Darl.) Spotted Eupatorium.
St. solid, striate, hispid or pubescent, greenish and purple, with numerous glands and purple lines; the glands on the stem and leaves give out an acrid effluvium in flowering-time; lvs. tripli-veined, 3—5 in a whorl.—Low grounds, U. S. and Can. Stem 4—6f high. Leaves petiolate, 6—7' by 3—4', strongly serrate. Flowers purple. July—Sept.

β. urticæfolium. Barratt. Height 4—5f; st. solid, slender; lvs. thin, much longer than the usual form of E. maculatum.

3. E. PURPÙREUM. Linn. not of DC. Willd. nor Ph. (E. verticillatum. Willd.)
St. solid, glabrous, green, sometimes purplish, with a purple band at the joints about 1' broad; lvs. feather-veined, in whorls of 3, 4 and 5, smooth above, with a soft pubescence beneath along the midvein and veinlets, coarsely serrate.—Dry woods or meadows, common, U. S. and Can. Stem 6f high. Leaves thin and soft, 9—12' (including the 1' petiole) by 3—4'. Corymb lax, pale purple, varying to whitish. Aug. Sept.

β. album. Barratt. (E. trifoliatum. Darl.) Fls. dull white; lvs. 5 in a whorl, large and distant.—It occurs also with 4 leaves in a whorl—a tall variety, upper leaves subfalcate; also with 3 leaves in a whorl—tall and slender.

4. E. TERNIFOLIUM. Ell. (in part.)
St. solid, somewhat hispid and glandular, greenish, with purple dots and lines; lvs. mostly 3 in a whorl, the upper and last whorls smooth and finely serrate.

β. vesiculosum. Barratt. St. striate, purplish, solid, 2—3f high.—Abundant in meadows and pastures. A handsome variety, with a profusion of purple flowers in a large, spreading corymb. The leaves present a vesicular appearance in a remarkable degree on their upper surface.

** Leaves opposite. Heads 3—5-flowered.

5. E. HYSSOPIFOLIUM. Narrow-leaved Eupatorium.
Lvs. opposite (the upper ones alternate), often verticillate, linear-lanceo-

late, tripli-veined, punctate, lower ones subserrate, upper ones entire.—A more delicate species, smooth in all its parts, or minutely pubescent, in dry fields, Mass.! to La. Stem about 2f high, branching, with numerous narrow leaves, which are mostly opposite, and a spreading corymb at the summit. Heads 5-flowered. Outer scales shortest, the others shorter than the purplish flowers. Aug. Sept.

6. E. LEUCOLÉPIS. T. & G. (E. glaucescens. β. leucolepis. DC. E. linearifolium. Mx.)—St. mostly simple; lvs. lanceolate or linear, obtuse, closely sessile, serrate, lower ones obscurely tripli-veined; corymb fastigiate, canescent; hds. 5-flowered; scales 8—10, scarious at the summit, as long as the flowers.—Sandy fields, N. J. to La. Stem 2—3f high. Leaves 1½—2½ by ½—1', glaucous-green both sides, divaricate with the stem, upper ones linear and entire. Corolla dilated at mouth, with short, obtuse lobes, white. Aug.—Oct.

7. E. ALTISSIMUM. (Kuhnia glutinosa. DC.) Goldenrod Eupatorium. St. pubescent-tomentose, tall, corymbose at the summit; lvs. lanceolate, acutely serrate above the middle, pubescence tapering to each end, subsessile, conspicuously 3-veined; hds. 5-flowered; scales 8—12, obtuse, pubescent.— Woods and sandy soils, Penn. and Western States, Plummer! Stem round, striate, 3—7f high. Leaves 3—4' by ½—1'. much resembling those of Solidago Canadensis; small ones often fascicled in the axils. Corymb compound, consisting of many simple, subcapitate ones. Corollas whitish, nearly twice as long as the scales. Sept. Oct.

8. E. ALBUM. (E. glandulosum. Michx.) White-flowered Eupatorium. St. pubescent; lvs. ovate-lanceolate, sessile, scabrous or pubescent, acute, obscurely 3-veined; corymb fastigiate; hds. 5-flowered; scales 8—14, lance-linear, setaceously acuminate, scarious on the margin, and much longer than the flowers; ach. glandular.—Sandy fields, Penn. to La. Stem about 2f high, numerously divided above. Leaves 2—3' by ½—1', upper ones entire and alternate. Involucre concealing the flowers, and with them copiously sprinkled with resinous dots, whitish. Aug.—Oct.

9. E. TEUCRIFOLIUM. Willd. (E. verbenæfolium. Mx. E. pubescens. Pers.) Hairy Eupatorium.—Lvs. opposite, sessile, distinct, ovate, rough, veiny, the lower ones doubly serrate, the upper ones subserrate or entire; st. paniculate, pubescent, with fastigiate, corymbose branches above.—Mass.! to La. Plant hairy, 2—3f high, with a somewhat panicled corymb of white flowers. The upper leaves are often entire. Involucre 5-flowered, with twice as many scales in two rows. Closely allied to the following, but is much more rough. Aug.

10. E. SESSILIFOLIUM. Sessile-leaved Eupatorium. Lvs. opposite, sessile, distinct, amplexicaul, ovate-lanceolate, rounded at the base, very smooth, serrate; st. smooth.—Plant 2—4f high, in rocky woods, Mass. to Ia.! and Ga. Stem slender, erect, branching at top into a corymb with white flowers. Leaves large, tapering regularly from the somewhat truncate base to a long point, with small serratures, paler beneath. Flower-stalks downy. Heads 5-flowered, with twice as many scales in two rows. Sept.

11. E. ROTUNDIFOLIUM. Willd. Hoarhound. Lvs. opposite, sessile, distinct, roundish-ovate, subcordate at base, 3-veined and reinleted, coarsely serrate, scabrous above. pubescent beneath; hds. about 5-flowered, inner scales acuminate, as long as the flowers.—A slender species, in dry fields, N. J. and S. States. Stem 2—3f high, roughish. Leaves 1—2½' long, ½ as wide, mostly obtuse. Heads fastigiate-corymbose. Involucre very pubescent, outer scales shorter than the inner. Flowers white. Pappus longer than corolla. Styles much exserted. Aug. Sept.

12. E. PUBESCENS. Muhl. (E. ovatum. Bw.) Hairy Eupatorium. St. hirsute; lvs. opposite, sessile, ovate, acute, obtusely dentate, rough, pubescent; corymb fastigiate; invol. about 8-flowered.—A large, rough plant, 3—4f high, growing in dry grounds, N. H.! to Penn. Distinguished by its opposite, broadly ovate leaves, and its strong pubescence. Involucre of about 12 pubescent scales, the outer much the shortest. Aug

27*

*** *Leaves opposite. Heads 8—20-flowered.*

13. E. PERFOLIÀTUM. *Thoroughwort. Boneset.*

Lvs. connate-perfoliate, very pubescent.—A common, well known plant, on low grounds, meadows, U. S. and Can. Abundant. Stem 1—5f high, round, rough and hairy. Each pair of leaves are so united at the base as to constitute a single lamina, centrally perforated by the stem, and placed at right angles to it; they are rough, rugose, serrate, tapering to a long point, and both combined, are 8—14' in length. Heads about 12-flowered, clustered in large, terminal corymbs. Corollas white. Aug.—The plant is bitter, and is used in medicine as a tonic.

14. E. RESINÒSUM. Torr.

St. minutely tomentose; *lvs.* linear-lanceolate, closely sessile, tapering to a long acumination, divaricate with the stem, slightly viscidly glandular both sides; *corymb* fastigiate, compound; *hds.* 10—15-flowered; *scales* obtuse, hoary-tomentose.—Wet, sandy soils, N. J., Penn. Stem 2—3f high, growing in tufts. Leaves 3—6' by 3—6''. Aug. Sept.—This singular species appears to be nearly confined to the pine barrens of N. J., where it was first found by *Dr. Torrey.*

15. E. AGERATÖIDES. *Nettle-leaved Eupatorium.*

St. smooth, somewhat branched; *lvs.* on long petioles, subcordate, ovate, acuminate, dentate, 3-veined, nearly smooth; *corymbs* compound; *invol.* simple, smooth.—Rocky hills and woods, Can. and U. S. Stem round, 2—4f high, and with the whole plant nearly smooth. Leaves large, 3—6' long, 2—4' broad at base, coarsely toothed, petioles 1—2' long. Heads numerous, in small clusters, constituting a compound corymb. Involucre scales mostly in a row, containing 12 or more flowers of a pure white. Aug. Sept.

16. E. AROMATÍCUM. *Aromatic Eupatorium.*

St. rough, pubescent, corymbose at summit; *lvs.* petiolate, opposite, subcordate, lance-ovate, acute, 3-veined, obtusely serrate, smoothish; *invol.* simple, pubescent.—A handsome species, in low woods, Mass. to La. Whole plant slightly pubescent, about 2f high. Leaves 2—4' long, ½ as wide, on petioles less than an inch long. Heads of the flowers large, 10—15-flowered, white and aromatic, in small corymbs. Scales about equal. Aug. Sept.

17. E. SERÒTINUM. Michx.

St. pube· t, diffusely branched; *lvs.* petiolate, lance-ovate, acute, sharply serrate, tri] ·ined, nearly glabrous; *corymbs* compound; *hds.* 12—15-flowered; *scales* . ·—12, scarious-edged, very pubescent. Ill. *Mead,* to Ga. Stem 4—6f high, somewhat paniculate above. Leaves 4—6' by ½—1½', upper ones nearly entire, and somewhat scattered, lower ones opposite, with large, irregular serratures. Sept. Oct.

5. MIKANIA. Willd.

In honor of Professor Mikan, of Prague.

Flowers all tubular; involucre 4—6-leaved, 4—6-flowered; receptacle naked; pappus capillary, simple, scabrous; anthers partly exserted; achenia angled.—*Mostly climbing herbs. Lvs. opposite.*

M. SCANDENS. Willd. *Climbing Boneset.*

St. smooth; *lvs.* cordate, repand-toothed, acuminate, the lobes divaricate, rather unequal; *hds.* in pedunculate, axillary corymbs.—A climbing plant of wet thickets, Mass.! to Ga., rather rare. Every part smooth. Leaves 2—3' by 1—2', on petioles 1—2' long, apex tapering to a long point. Branches short, nearly naked, each bearing a small corymb of whitish, or pink-colored flowers. Aug. Sept.

6. CONOCLINIUM. DC.

Gr. κωνος, cone, κλινη, bed or receptacle.

Heads many-flowered; receptacle conical. Character otherwise as in Eupatorium.—♃ *Herbaceous or suffruticose. Lvs. opposite, petiolate, serrate. Fls. blue or purple, in crowded corymbs.*

C. cælestinum. DC. (Cœlestina cœrulea. *Spreng.* Eupatorium cælestinum. *Linn.*)—Herbaceous, nearly glabrous, much branched; *lvs.* deltoid-ovate, truncate or subcordate at base, tapering to an obtusish apex, crenate-serrate, veiny; *petioles* slender, about half as long as the lamina; *corymbs* numerous, subumbellate; *scales* numerous, setaceous.—Hedges, thickets, roadsides, &c., Penn. and S. and W. States! Stem 1—2½f high, terete, with opposite branches. Leaves 1—2½' long, ⅔ as wide. Flowers 20—50 in a head, of a light or sky-blue, reddish in fading. Aug. Sept.

7. LIÁTRIS.

Gr. λι, an emphatic prefix, ετρως, invulnerable; used as a vulnerary.

Flowers all tubular; involucre oblong, imbricate; receptacle naked; pappus plumose, copious; achenia obconic, 10-striate; styles much exserted.—♃ *herbs or shrubs. Root tuberous. St. simple. Lvs. alternate. Fls. cyanic.*

* *Heads 16—60-flowered.*

1. L. squarròsa. Willd. *Blazing Star.*
Smooth or scabrous-pubescent; *lvs.* linear, lower ones attenuated at base, *rac.* flexuous, leafy; *hds.* few, sessile or nearly so; *invol.* ovate-cylindric; *scales* large, squarrose-spreading, outer larger, leafy, inner mucronate-acuminate, scarcely colored; *fls.* numerous, *pappus* plumose.—A splendid plant, native (in N. Y. according to *Prof. Eaton*) Penn. to Flor. and W. States! Stem 2 —3f high, thickly beset with long, linear leaves. Heads 5—20, with brilliant purple flowers. Aug. †

2. L. cylindracea. Michx.
St. low, slender and very leafy, smooth or somewhat hirsute; *lvs.* rigid, linear, mostly 1-veined; *hds.* few, sessile or pedicellate, cylindrical, 15—20-flowered; *scales* short, close, rounded or obtuse and abruptly mucronate at apex.—Prairies and barrens, Mich. to Mo. Stem 6—18' high. Leaves 2—5' by 2—4". Heads 1' long, rarely solitary, sometimes 10 or 12, mostly about 5. Flowers bright purple.

3. L. scariòsa. *Gay Feather.*
Scabrous-pubescent; *lvs.* lanceolate, lower on long petioles, upper linear and much smaller; *hds.* remotely racemed; *invol.* globose-hemispherical; *scales* obovate, very obtuse, purplish; *fls.* numerous; *pappus* scabrous.—A beautiful plant, 4—5f high, in woods and sandy fields, Mass. (*Rickard!*) to Ill.! and La. Stem rather stout, whitish above. Leaves numerous, entire, lower 3—9' long, upper 1—3' by 1—3", rough-edged. Heads 5—20, 1' diam., in a long raceme, each 20—40-flowered. Corolla purple. Aug. †

* * *Heads 5—15-flowered.*

4. L. graminifolia. Willd. Torr. & Gray. *Grass-leaved Liatris.*
Glabrous or with scattered hairs; *st.* slender and simple; *lvs.* linear, 1-veined; *hds.* 7—12-flowered, spikes or racemes sometimes paniculate below; *involucre* acute at base; *scales* many, obtuse, appressed, outer row shorter; *ach.* hairy.—N. J. to Ala.
β.? *dubia.* (L. pilosa. β. dubia. *Ph.* L. dubia. *Bart.*) *Inflorescence* sometimes compound below, or partly paniculate.—Pine barrens, N. J. Stem 2—3f high. Leaves 3—6' by 2—4". Heads rather small. Sept. Oct.

5. L. spicàta. Willd. *Slender-spiked Liatris.*
Lvs. lance-linear, smoothish, punctate, ciliate, lower ones narrowed at base; *hds.* in a long, terminal spike, nearly sessile; *lfts. of the invol.* oblong, obtuse; *fls.* about 8; *pappus* scabrous-plumose.—Native from N. J. and Mich.? to Flor. and La. Abundant in prairies. A beautiful species, often cultivated. Stem 2—5f high. Heads numerous, with bright purple flowers. Aug. †
β. *resinosa.* T. & G. (L. resinosa. *Nutt.*) *Plant* smaller; *hds.* about 5-flowered.

6. L. pycnostàchya. Michx. *Thick-spiked Liatris.*
Simple, more or less hirsute, very leafy; *lvs.* rigid, ascending, straight,

ower ones long, lanceolate, veined, obtuse, upper short, narrow-linear; *spike* dense and thick, long and bracted below; *hds.* numerous, cylindrical, sessile, 5-flowered; *scales* appressed, with acute, scarious and colored squarrose tips.— Prairies, Ill.!, to Tex. A stout species, distinguished from L. spicata chiefly by its acute, squarrose scales and few-flowered heads. Stem 3—5f high. Spikes cylindrical, 10—20′ long.

β. T. & G. (L. brachystachya. *Nutt.*) *St.* and *invol.* nearly glabrous.

Section 2. Heads radiate.

8. TUSSILAGO.

Altered from the Lat. *tussis*, cough ; considered a good expectorant.

Heads many-flowered ; flowers of the ray ♀, those of the disk ♂; involucre simple ; receptacle naked ; pappus capillary.—♃ *Lvs. radical. Fls. yellow, with very narrow rays.*

T. Farfàra. *Colt's-foot.*
A low plant, in wet places, brook sides, N. and Mid. States, and is a certain indication of a clayey soil. Scape scaly, about 5′ high, simple, appearing with its single, terminal, many-rayed, yellow head, in March and April, long before a leaf is to be seen. Leaves arising after the flowers are withered, 5—8′ by 3—6′, cordate, angular, dentate, dark green above, covered with a cotton-like down beneath, and on downy petioles. §?

9. NARDOSMIA. Cass.

Gr. νάρδος, spikenard, οσμη, smell; from the fragrance of the flowers.

Heads many-flowered, somewhat ♀ ♂; flowers of the ray ♀, of the disk ♀, but abortive in the sterile plant ; involucre simple ; receptacle flat, naked ; pappus capillary.—♃ *Lvs. radical. Fls. cyanic. The ray flowers of the sterile heads are in a single row ; of the fertile heads in several, but very narrow.*

N. palmàta. Hook. (Tussilago. *Ait.*)
Scape with a fastigiate thyrse or corymb; *lvs.* roundish-cordate, 5—7-lobed, tomentose beneath, the lobes coarsely dentate.—In swamps, Fairhaven, Vt., *Robbins.* Sunderland, Mass., *Hitchcock.* W. to R. Mts. Very rare. A coarse, acaulescent plant, with large, deeply and palmately-lobed leaves, and a stout scape covered with leaf-scales and 1—2f high. The heads are fragrant, numerous, with obscure rays, those of the barren plants almost inconspicuous. May.

Tribe 3. ASTEROIDEÆ.

Heads radiate, rarely discoid. Branches of the style more or less flattened and linear, equally pubescent above outside. Leaves mostly alternate.

Section 1. Heads radiate. Rays cyanic.

10. ASTER.

Gr. αστηρ, a star; from the radiated flowers.

Involucre oblong, imbricate ; scales loose, often with green tips, the outer spreading ; disk flowers tubular, ♀ ; ray flowers ♀, in one row, generally few (6—100), ligulate, oblong, 3-toothed at apex, finally revolute ; receptacle flat, alveolate ; pappus simple, capillary, scabrous ; achenium usually compressed.—*A large genus of ♃ herbs, very abundant in the U. S., flowering in late summer and autumn. Lvs. alternate. Disk fls. yellow, changing to purple, ray flowers blue, purple or white, never yellow.*

§ *Scales imbricate, with appressed, greenish tips. Rays 6—15. Lower leaves cordate, petiolate. Heads corymbose.* BIOTIA. DC.

1. A. corymbòsus. Ait. (Eurybia corymbosa. *Cass.*) *Corymbed Aster. St. corymbose-fastigiate, smooth; branches hairy; lvs. ovate, acutely ser*

rate, acuminate, the lower ones cordate, petiolate; *petioles* naked; *invol.* oblong, imbricate with closely appressed, obtuse scales.—Common in dry woods, N. and Mid. States. Stem 2f high, smooth, often reddish, more or less flexuous Leaves large, mostly smooth, lower ones cordate-acuminate, with sharp serratures, middle ones ovate, upper ones becoming lanceolate. Flowers in a broad, flat-topped corymb, large, very open, with about 6 long, narrow, white rays. Aug

2. **A. MACROPHYLLUS.** Willd. (Eurybia macrophylla. *Cass.*) *Large-leaved Aster.*—*St.* branched, diffuse; *lvs.* ovate, petiolate, serrate, rough, upper ones ovate-lanceolate, sessile, lower ones cordate, petiolate; *petioles* somewhat winged; *invol.* cylindric, closely imbricate with oblong, acute scales.—Distinguished for its very large root leaves which are 6—10′ by 3—5′. Grows in woods, N. States and Can. Stem furrowed, 1—2f high. Leaves nearly smooth. Rays about 13, white or pale blue. Sept.

§ § *Scales imbricated, with spreading, green tips. Rays 12—30. Pappus bristles rigid, some of them thickened upwards. Heads large, corymbose. Lower leaves never cordate, cauline sessile, rigid.* CALLIASTRUM. T. & G.

3. **A. RADULA.** Ait. *Rasp-leaved Aster.*
St. erect, simple below, angular; *lvs.* lanceolate, acuminate, narrowed towards the base, sessile, serrate, rugose and rough; *invol.* imbricate, scales appressed, with small, spreading green tips.—Moist groves and hedges, Me. to Penn.! Not common. Height 1—3f. Distinguished for its stiff, narrow, sharply serrate leaves which abundantly clothe the straight, smooth stem. Branches nearly naked, undivided, each having a single large head, rarely more. Rays numerous, short, white or purplish. The lower leaves are sometimes ovate-lanceolate. Aug. Sept.

4. **A. SPECTABILIS.** Ait. *Showy Aster.*
St. erect; *lvs.* somewhat scabrous, oblong-lanceolate, sessile, entire, lower ones serrate in the middle; *branches* corymbose; *hds.* hemispherical, with numerous, squarrose-spreading, ciliate scales.—A low Aster of pine barrens, Mass.! to Ky. Stem straight, 1—2f high, branching above into a nearly simple corymb of 10—15 heads, which are large and showy, with many long, blue rays. Sept.—Nov.

5. **A. GRACILIS.** Nutt. *Slender Aster.*
St. minutely-pubescent, corymbose at summit; *lvs.* oblong-lanceolate, incisely and remotely serrulate, narrowed to the sub-clasping base; *corymb* loose, spreading; *scales* linear-oblong, whitish, with green, spreading tips; *rays* about 12.—Pine barrens, N. J. Stems clustered, 12—14′ high, purplish, leafy, slender. Leaves 1—2½′ long, glabrous, opaque, lower ones somewhat spatulate. Corymb simple or compound. Rays pale violet, about as long as the involucre. Sept.

§ § § *Scales green, or with green tips. Rays 00. Pappus bristles soft, none of them thickened upwards. Achenia compressed.* ASTER *proper.*

* *Lower leaves cordate, petiolate. Heads paniculate.*

6. A. CORDIFOLIUS. *Heart-leaved Aster.*
St. paniculate, smoothish; *lower lvs.* cordate, hairy beneath, sharply serrate, acuminate, petiolate; *petioles* winged; *invol.* closely imbricate, the scales with short, green tips.—Common in rocky woods, N. and W. States. Stem smooth below, more or less pubescent above, a little flexuous, striate, 2f high, with a handsome panicle of racemes at top of numerous, rather small flowers. Rays 10—15, pale blue varying to white. Lower leaves large, cordate, with a deep sinus at base, the serratures very acute, the summit ending in a long, acute point, slightly rough above, hairy and paler beneath. Petioles more or less winged, hairy. Above, the leaves are gradually reduced to small or minute bracts. Sept.

7. A. SAGITTIFOLIUS. *Arrow-leaved Aster.*
St. with racemose branches above, smooth; *lvs.* oblong-lanceolate, acuminate, sessile, serrate in the middle, radical ones ovate, oblong, cordate-sagittate,

serrate, petiolate; *invol. scales* loose, lanceolate.—Low woods, N. and W. States
and Can. Stem 2—4f high, dividing into many ascending, rigid branches, with
numerous and crowded heads, forming a compound panicle of racemes. Heads
small, each with about 12 rays, which are white or with various shades of
blue. Leaves becoming smaller above, lanceolate and even linear. Sept.

8. **A. UNDULÀTUS.** *Wave-leaf Aster.*
St. paniculate, hispid; *branches* secund, leafy, 1-flowered; *lvs.* oblong-
cordate, amplexicaul, very entire, hairy, somewhat undulate or crenate-serrate,
lower ones ovate, cordate, subserrate, with winged petioles.—Native of dry
woods, U. S. Plant rough, about 2f high, with slender branches. Lower
leaves on long winged petioles, cordate, acuminate, upper ones becoming narrow-ovate and clasping. Flowers pale blue, solitary, forming a loose panicle
of somewhat one-sided racemes. Aug. Sept.

9. **A. AZURÈUS.** Lindl. (A. Oolentangiensis. *Riddell.*)
Scabrous; *st.* and *racemose paniculate branches* rigid; *lvs.* lance-ovate, cordate, slightly serrate, on hairy petioles, middle and upper ones lanceolate and
linear, acute at each end, sessile, entire, highest subulate; *hds.* broadly obconic;
scales oblong-linear, acute, appressed.—Woods and prairies, Western States.
Stem about 2f high. Leaves of several forms between the lowest cordate to
the small, subulate, numerous floral ones of the slender branches. Racemes
rather remote, panicled, with middle-sized heads. Rays blue.

10. **A. SHORTII.** Hook. *Short's Aster.*
Slender and nearly glabrous, simple or somewhat branched above; *lvs.*
lance-ovate, cordate, petiolate, long-acuminate, subentire, upper ones sessile
and obtuse at base; *hds.* middle-size, racemose or racemose-paniculate, rather
numerous; *invol.* broad-campanulate; *scales* scarious, close, green-tipped, shorter
than the disk flowers.—A distinct and beautiful species, on rocky banks of
streams, Ohio! to Ark. Stem a little flexuous, 2—4f high. Lower leaves about
5' by 1½', the others successively diminished upwards to the flowers where they
are minute. Rays violet blue.

* * *Lower leaves never cordate. Cauline leaves clasping and cordate or
auriculate at base.*

11. **A. PATENS.** (A. amplexicaulis. *Willd.*) *Spreading Aster.*
St. simple, paniculate above, pubescent; *lvs.* lanceolate, cordate, clasping
the stem, acuminate, scabrous on the margin, pubescent; *panicle* loose, few-
flowered; *scales* imbricate, lanceolate, lax, the points herbaceous.—Grows in
moist grounds, Northern States. Stem 2—3f high, slender, branching above
into a loose, terminal panicle. Leaves large, (3—6' long) on the stem, becoming small and bracteate on the branches. Heads solitary on the ends of the
leafy branchlets, large, with 20—30 violet-colored rays. Aug.—Nov.

12. **A. LÆVIS.** (A. mutabilis. *Linn.* A. amplexicaulis. *Muhl.*) *Smooth Aste-.*
Very smooth; *st.* angular; *branches* simple, 1-flowered; *lvs.* subamplexicaul, remote, oblong, entire, shining, radical ones subserrate; *invol.* closely
imbricate, the scales broadly-linear, rigid, thickened and herbaceous at the
apex.—A very smooth and beautiful species, 2—3f high, growing in low grounds.
Stem polished, green, often somewhat glaucous. Leaves rather fleshy, broadest
at base, the lower ones tapering to a winged petiole. Flowers large and showy,
with numerous rays of a fine blue, becoming purple. Sept.—Nov.
β. *levigatus.* (A. lævigatus. *Willd.*) *Lvs.* long, linear-lanceolate.
γ. *cyaneus.* (A. cyaneus. *Ph.*) *St.* and *lvs.* conspicuously glaucous.—These
are beautiful varieties, especially the latter, which is perhaps the most beautiful of all the asters.

13. **A. CONCINNUS.** Willd., not of Nees. *Elegant Aster.*
St. simple, paniculate at the summit, pubescent; *lvs.* lanceolate and lance-
linear, narrowed and clasping at the base, remotely serrate, upper ones entire;
invol. closely imbricate, scales green at the tip.—Woods, Northern States!
A slender species, 1—2f high. Branches of the panicle rather short and remote. Leaves 3—5' long, acuminate, varying from ½—1' in width, smooth ex-

eept the mid-vein beneath; branch leaves few and much smaller. Heads middle-size, with 10—15 bluish purple rays. Sept.—Nov.

14. A. PUNICEUS. *Red-stalked Aster.*

St. hispid, paniculate; *lvs.* amplexicaul and more or less auriculate at base, lanceolate, serrate, roughish above; *invol.* loose, longer than the disk, the scales linear-lanceolate, long and revolute, nearly equal and 2-rowed.—A large, handsome aster, common in swamps and ditches, sometimes in dry soils, N. States and Can. Stem 4—6f high, generally red, (at least on the south side). furrowed, hispid. Lower leaves with remote serratures, rough-edged and rough on the upper surface, all acuminate and narrowed at base. Flowers large and showy. Rays 50—80, long and narrow, pale purple. Aug.—Oct.

15. A. PRENANTHÖIDES. Muhl. *Prenanthes-like Aster.*

St. hairy or pubescent above, corymbose-paniculate; *lvs.* oval-lanceolate, serrate, acuminate, attenuate at base into a long winged petiole which is auriculate at the insertion; *invol.* imbricated with several rows of linear, green-tipped, spreading scales. Grows in low woods, N. Y. to Ky. Stem 2—3f high, with a terminal, corymbose panicle of large heads on short peduncles. Rays showy, pale blue.—Leaves remarkable for the long, winged petiole, which is dilated at its base into rounded, auriculate segments. Branch leaves smaller, nearly entire. Sept.—Nov.

16. A. AMETHYSTINUS. Nutt. *Amethystine Aster.*

Hirsute; *st.* racemose-paniculate; *lvs.* linear-lanceolate, entire, rough, acute, with somewhat auriculate appendages at the clasping base; *invol.* of equal scales.—Eastern Mass., *Nuttall,* &c. Heads small, with azure rays. Aug.—Oct.

17. A. NOVÆ ANGLIÆ. *New England Aster.*

Fls. terminal, crowded, somewhat fastigiate; *st.* hispid, paniculate; *lvs.* linear-lanceolate, amplexicaul, auriculate at base; *scales of involucre* equal, lax, linear-lanceolate, rather longer than the disk.—A large and beautiful aster, in fields, meadows and shades, more common in the W. States! than in N. Eng.! Stem 4—6f high, straight, erect, viscidly hairy, colored. Leaves very numerous, narrow, entire, with 2 auricular appendages at base. Flowers large, in a kind of loose, paniculate corymb. Ray-flowers deep purple, numerous (75—100). Sept. †

*** *Leaves neither cordate nor auriculate, the margin serrate.*

† *Scales not spreading.*

18. A. TRADESCANTI. *Tradescant's Aster.*

Branches virgate, paniculate; *lvs.* lanceolate, remotely serrate, sessile smooth; *invol.* closely imbricate; *st.* round, slender, smooth.—A fine species with numerous leaves, growing in fields, Mass. to La. Stem rigid, brownish about 3f high, with numerous slender, racemose branches. Lower stem-leaves narrowly lanceolate, 4' long, gradually reduced in size upwards. Heads small, numerous, with pale purple rays. Aug.—Oct.

β. fragilis. T. & G. (A. fragilis. *Willd.*) Cauline leaves serrulate or entire, short; heads much scattered on the branches.

19. A. MISER. Ait. T. & G. (A. miser, divergens, diffusus and pendulus, *Ait.*) *Starved Aster.*—*St.* racemose-paniculate, hairy or pubescent; *lvs.* sessile, lanceolate, sharply serrate in the middle; *invol.* imbricated with acute scales; *rays* short.—A very variable species common in old fields, hedges, U. S. and Can. In height it varies from 6 to 30', and in luxuriance proportionately to the moisture or fertility of the soil. The stem is very branching or nearly simple, bearing a large, compound, racemose panicle or a few simple racemes Leaves varying from narrow-lanceolate to broad-oval, 1—5' in length. Heads usually numerous, small, with small white or purple rays.

β. diffusus. Branches spreading, diffuse; *lvs.* elliptical-lanceolate, more or less narrowly so, midvein hairy beneath; *hds.* often sessile, forming short, crowded spikes or long virgate ones.

γ. hirsuticaulis. T. & G. (A. hirsuticaulis. *Lindl.*) *St.* hirsute; *lvs.* long and narrow, midvein hirsute; *lds.* racemose or spicate, upper in short, dense bunches; *scales* linear.

20. A. **SIMPLEX.** Willd. (A. salicifolius. *Darl.*)   *Willow-leaved Aster.*
    Glabrous; *st.* racemose-paniculate above; *lvs.* lanceolate, acuminate, entire, the margins scabrous, lower ones serrate; *scales* loosely imbricated, linear-subulate.—Another variable species in low grounds, U. S. and Can. Stem 1—5f high, somewhat corymbose. Leaves 2—4′ by 5—10″, very smooth both sides, tapering to a slender point; those of the branches and branchlets proportionately smaller. Heads rather few, middle size, on the short branchlets. Sept.
    β. altior. Branches hirsute or pubescent; *lds.* above the middle size, with blue rays.—Stem 4—6f high. ·
    γ. humilior. Branches pubescent, with short, crowded spikes of small heads; *rays* pale blue.—Stem 1—2f high.
    δ. recurvatus. Diffuse, with long, spreading or recurved branches; *lds.* loosely racemed; *rays* bluish-white.—Western!

21. A. **TENUIFOLIUS.**   *Narrow-leaved Aster.*
    St. smooth, erect, paniculate-branching, with 1-flowered branchlets; *lvs.* linear-lanceolate, tapering at each end, long-acuminate, entire, with roughish margins, the lower ones often serrate in the middle; *invol. scales* lax, acute.—Grows in moist fields, Can. to Va. Stem leaves 2—4′ long, those of the branches and branchlets proportionately smaller. Heads small, with numerous (20—30), long, pale purple rays. Sept.

22. A. **GREENEI.** Torr. & Gray.   *Greene's Aster.*
    St. glabrous, racemosely branched; *lvs.* glabrous, subclasping, remotely appressed-serrulate, scabrous above, lower narrow-lanceolate, upper short, numerous, ovate-lanceolate; *lds.* rather small, on short, bracted peduncles.—Near Boston. *Dr. B. D. Greene, Dr. Pickering.* Cauline leaves 3—5′ long, ramial leaves much smaller.

23. A. **NOVI-BELGII.**   *New York Aster.*
    Glabrous; *st.* terete, stout, often glaucous; *lvs.* rather rigid, lanceolate, acute, the lower subserrate and subclasping; *lds.* racemose or corymbose; *scales* rather loosely imbricated, lanceolate, subequal, with acute green tips; *rays* numerous.—A smooth, handsome Aster in Western and Southern States, not common. Stem 2—4f high, with few, straight, somewhat corymbose branches. Leaves 4—6′ long, tapering to each end, rough-edged, upper ones much smaller. Heads rather large. Rays pale blue, expanding 9—12″. Aug.—Oct.

       † † *Scales spreading or squarrose at tip.*

24. A. **LAXUS.**   *Loose-stalked Aster.*
    St. loosely corymbose-panicled above; *lvs.* linear-lanceolate, acuminate, rough-edged, lower ones subserrate, those of the stem subreflexed, of the branches much spreading; *invol.* imbricate, scales lanceolate, acute, reflexed at the apex. Fields, Mass., N. Y. Stem 2—3f high, with small, bluish flowers. Sept. Oct.

25. A. **LAXIFOLIUS.** Nees.   *Loose-leaved Aster.*
    St. scabrous; *rac.* compound; *branches* racemose at the summit or slightly compound; *lvs.* linear-lanceolate, elongated, mucronately serrulate, attenuate at each end, clasping at base, scabrous above; *scales* squarrose.
    γ. lætiflorus. T. & G. *St.* slender, with long, filiform, spreading branches; *lvs.* rather rigid and very scabrous; *rac.* loose, the pedicels nearly leafless.—Ohio and Wis. Described by *Drs. Torrey & Gray* as a very graceful plant of considerable size, with very long, narrow leaves, and numerous, long, showy, pale purple rays.

26. A. **LONGIFOLIUS.** Lam. (A. lævigatus. *Ph.*)   *Long-leaved Aster.*
    Glabrous; *st.* very branching, branches many-flowered; *lvs.* subamplexicaul, linear-lanceolate, entire, lower ones serrate, smooth; *invol. scales* lax, lanceolate, nearly equaling the disk.—Fields and thickets, N. Y. to Car. Stem 3f high. Leaves pale below, shining above, smooth both sides, the lower ones 4—6′ long. Heads numerous, showy, with 25—30 light blue rays. Nov.

* * * * *Leaves neither cordate nor auriculate, the margin entire or subentire.*

† *Scales erect.*

27. A. sericeus. Vent. (A. argenteus. *Michx.*) *Silk-leaved Aster.*

Sts. slender, clustered, glabrous below, silky-pubescent and branched above; *lvs.* clothed on both sides with a dense, appressed, silky-canescent pubescence, lance-oblong, entire, acute and mucronate, sessile; *hds.* large, mostly solitary, terminal on the short, leafy branchlets; *scales* lanceolate, silky-canescent like the leaves, spreading at tip.—A singularly elegant Aster, with shining, silvery foliage, prairies! and river banks! Wis. and Iowa, to Miss. Stem 1—2f high. Lower leaves 2—3′ by ⅓—1½′, the upper much smaller. Rays deep violet-blue. Aug.—Oct. †

28. A. concolor. *One-colored Aster.*

St. subsimple, erect, pubescent; *lvs.* lance-oblong, entire, mucronate, grayish, with a minute, silky pubescence both sides, upper ones cuspidate-acuminate; *rac.* terminal, virgate, simple or somewhat compound, elongated; *scales* lanceolate, silky, acute, appressed.—Pine barrens, N. J. to Flor. A slender and virgate plant, 1—3f high, sometimes branched below. Root often tuberous. Leaves 1½′ by ⅓′, reduced in size upwards. Heads in a long raceme, with blue rays and a rust-colored pappus.

29. A. turbinellus. Lindl.

Smooth or slightly scabrous; *branches* and *branchlets* very slender; *lvs.* lanceolate, tapering to each end, acute, slightly clasping, entire, those of the branches linear, and of the branchlets subulate; *invol.* turbinate, acute at base, as long as the disk flowers; *scales* imbricated in many rows, linear, obtuse with short green tips.—Woods and river bottoms, Ill. Mead, Mo., &c. to La. Stems 2f high, with the branches numerous and somewhat corymbose. Lower leaves 3—5′ by ⅓—1½′, the others gradually reduced upwards to the scales of the obconic or top-shaped involucre. Heads middle-size, with blue rays and brownish pappus. Sept.

30. A. dumosus. *Bushy Aster.*

Nearly smooth; *branches* racemose-panicled; *lvs.* numerous, smooth, linear, sessile, entire or subserrate, those of the branches very short; *invol.* cylindrical, closely imbricate.—About 2f high, in dry shades and borders of woods U. S. Stem much branched, smooth or slightly pubescent, with long, linear leaves, those of the branches smaller and becoming very minute.—Heads middle sized, scattered, solitary, with about 24 purplish white rays. Quite variable. Sept.

β. *foliosus.* (A. foliosus. *Ait.*) *St.* racemose-compound; *lvs.* acute, often serrulate; *scales* narrower, subacute.

γ. *strictior.* (A. fragilis. *Lindl.*) Somewhat paniculate; *branch leaves* rather numerous and appressed.

31. A. carneus. *Flesh-colored Aster.*

Smooth; *st.* dividing into many straight, racemose, leafy branches; *lvs.* uniform, linear-lanceolate, acuminate, entire, the lower ones tapering to a sessile base, the upper amplexicaul; *scales* acute, much shorter than the disk.—A handsome bushy Aster by fences, &c. (Claremont!) N. H., W. to Ia. Rare. Stem about 2f long, often purple. Stem leaves 3—5′ by ⅓—⅓′, branch leaves much smaller. Heads numerous, middle-size, somewhat secund, each with 20—30 pale purple, narrow rays. Sept. Oct.

32. A. graminifolius. Ph. *Grass-leaved Aster.*

Subpubescent; *st.* slender, branched above; *lower lvs.* very numerous, narrow-linear; *ped.* slender, 1-flowered; *scales* linear-subulate, loose, scarcely imbricated.—N. H. Eddy. High cliffs. Willoughby Lake, Vt.! Branches simple leafy, naked at the end, 1-flowered, somewhat corymbose. Rays 15—25, much longer than the disk, purple or rose-colored.

†† *Outer scales spreading or squarrose.*

33. A. ericoides. *Heath-like Aster.*

Nearly or quite smooth; *branches* virgate, spreading, paniculate; *lvs.* linear or linear-lanceolate, very smooth, those of the branches subulate and approximate, short, of the stem long, of the root oblong-spatulate; *invol.* somewhat squarrose.—Grows in rocky fields, in most of the States. Stem 1—3f high, with numerous brittle branches and branchlets forming a thick bush and terminated each by a single pale purple flower. Leaves rather numerous, the cauline ones 3' in length. Heads small, about 20-rayed. Sept.

34. A. MULTIFLŌRUS. *Many-flowered Aster.*

Hairy or pubescent; *st.* diffusely branched; *lvs.* linear, entire, sessile, pubescent, margins subciliate; *invol.* imbricate, squarrose, linear or spatulate, with oblong, ciliate scales.—A very branching, diffuse species, with very numerous, small flowers crowded on the racemose branches, each with about 12 white rays. Stem variously pubescent, 1—2f high. Leaves 1—2' long, obtuse, very narrow. Rocks and dry fields, U. S. Variable. Sept.

35. A. PRÆALTUS. Poir. (A. salicifolius. *Ait.*) *Willow-leaved Aster.*

St. corymbosely-paniculate, with hairy lines above; *lvs.* lanceolate, closely sessile or subamplexicaul, smooth and shining above, with a rough margin, subserrate or entire, acute, the lower ones narrowed towards the base; *invol.* loosely imbricated with acute, green-tipped, linear scales.—Common in moist woods and by streams (N. H. to Wis. *Lapham!*), varying from 2 to 3f in height. The stem is slender, often flexuous, green or often purple, dividing above into flowering branches, arranged in a sort of corymbose panicle of large and showy blue flowers. Aug.—Oct.

36. A. ELŌDES. Torr. & Gray. *Swamp Aster.*

Glabrous and very smooth; *branches* corymbose-paniculate; *lvs.* linear-lanceolate, entire, shining, thick, upper ones somewhat clasping; *invol.* closely imbricated in several rows of linear, green-pointed, spreading scales.—In swamps, Mass. to Va. Stem 1—2f high, with very smooth foliage and large, showy, blue flowers. Aug. Sept.

37. A. OBLONGIFOLIUS. Nutt. (and A. graveolens. *Nutt.*) *Oblong-leaved Aster.*—*St.* rigid, diffusely branched, hairy; *branches* spreading, with loose and irregular branchlets; *lvs.* oblong-lanceolate, acute, mucronate, partly clasping, entire, rough-edged, abruptly reduced on the branches and branchlets; *hds.* solitary, terminal on the slender branchlets; *involucre scales* nearly equal, green, spreading.—Prairies, &c. Western States! Plant 1—2f high, often glandular-viscid. Cauline leaves 12—20'' by 3—5''; those of the branches 6'' by 2'', of the branchlets 3'' by ¼'', indistinguishable from the scales. Rays purple. Pappus brownish. Sept. Oct.

§ § § § *Scales imbricated, scarious on the margins, destitute of green tips.*

38. A. ACUMINĀTUS. Michx. *Acuminate Wood Aster.*

St. simple, flexuous, angular, branching into a corymbose panicle above; *lvs.* broad-lanceolate, narrowed and entire at the base, serrate and acuminate; *invol. scales* lax, linear.—Mountains and woods, Can., N. Eng., N. Y. Stem a foot high, rough, downy. Leaves large, unequally and remotely serrate above, and ending in a long, acuminate point. Panicle corymbose, terminal, few-flowered, nearly or quite naked. The leaves are mostly situated just below the corymb, sometimes scattered. Heads rather large, with about 15 long, white rays. Aug.

39. A. NEMORĀLIS. Ait. (A. lædifolius. *Ph.* A. uniflorus. *Mx.*) *Wood Aster.*— *Branches* corymbed or 0; *ped.* 1-flowered, nearly naked, filiform; *lvs.* linear-lanceolate, acute at each end, veinless, revolute-margined, roughish; *scales of the involucre* very acute, loose, shorter than the disk; *rays* about 20.— A handsome plant, in swampy woods, N. H. *Storrs!* Mass. *Robbins!* to N. J. Rather rare. Stem slender, 10—20' high. Leaves numerous, 10—18'' by 2—4'', rarely subdentate. Heads large, few, often but one, terminating the simple axis. Rays large, white or pale purple. Sept. Oct.

40. A. PTARMICŌIDES. T. & G. (Heliastrum album. *DC.* Chrysopsis alba. *Nutt.*)—*St.* corymbose-fastigiate above; *lvs.* linear-lanceolate, acute, rough-margined, entire, lower ones dentate, attenuated into a short petiole; *rays* short.—A very distinct Aster, low and leafy, found in rocky soils, by streams and lakes, Vt. *Robbins!*, to Mo. Rare. Stems clustered, simple, each bearing a spreading panicle of heads which are below the middle size and furnished with snow-white rays. July—Sept.

41. A. FLEXUOSUS. Nutt. (A. sparsiflorus. *Ph.*) *Few-flowered Aster.*

St. branching, slender, flexuous, very smooth; *lvs.* long and succulent, the lower ones sublanceolate-linear, upper ones subulate; *branches* leafy, 1-flowered; *invol. scales* lanceolate, acuminate, appressed; *rays* numerous, shorter than the involucre. Grows in salt marshes, Mass. to Flor. The whole plant very smooth, 1f high, with large, purple flowers; disk yellow. Aug.—Oct.

42. A. LINIFOLIUS. (A. subulatus. *Michx.*) *Sea Aster.*

St. paniculate, much branched from the base; *lvs.* long, linear, very acute, the uppermost subulate; *invol.* cylindric with subulate scales; *radical hds.* minute.—An annual species, found in salt marshes, Mass. to Car. Stem 12—18′ high, very smooth, thick, reddish. Leaves smooth, sessile. The plant is very branching, with numerous short-rayed, small, purple flowers. Aug.

11. SERICOCARPUS. Nees.

Gr. σηριχος, silken, χαρπος, fruit; from the character of the genus.

Heads few-flowered; ray-flowers 4—6, ♀; disk-fls. 6—10, ♀; involucre oblong, imbricated; scales appressed, with green, spreading tips; receptacle alveolate; achenium obconic, very silky; pappus simple.—♃ *Herbs with alternate leaves and close corymbs. Rays white.*

1. S. SOLIDAGINEUS. Nees. (Aster solidaginoides. *Michx.*)

Smooth; *lvs.* linear-lanceolate, obtuse, entire, sessile, obsoletely 3-veined, rough on the margin; *corymb* fastigiate; *hds.* aggregate. subsessile, 5-rayed; *scales* obtuse, white, with green tips.—In woods, Can. to La. Not common. Stem slender, simple, about 2f high. Leaves smooth, pale green, 1—2′ by 3-5″. Heads rather small, in a level-topped corymb. Involucre oblong. Scales imbricate, appressed, with conspicuous green tips. Rays long, white. Jl. Aug.

2. S. CONYZOIDES. Nees. (Aster. *Willd.* Conyza asteroides. *Linn.*)

St. somewhat pubescent, simple, corymbose at top; *lvs.* oval-lanceolate, smooth beneath, slightly 3-veined, narrowed at base, acute, the upper ones sessile, nearly entire, the lower narrowed into the petiole, serrate; *invol.* cylindrical, the scales oval, obtuse, appressed, slightly reflexed at summit; *rays* 5, short.—Common in woods and thickets, Mass. to Flor. Stems somewhat 5-angled, 1—2f high. Leaves somewhat fleshy. Ray short, but longer than the disk. white. July, Aug.

12. DIPLOPAPPUS. Cass.

Gr. διπλοος, double, παππος, pappus; from the character.

Heads many-flowered: ray-fls. about 12, ♀; disk-fls. ♀; involucre imbricate; receptacle flat, subalveolate; pappus double, the exterior very short, interior copious, capillary; achenium compressed.—♃ *Lvs. entire, alternate. Rays cyanic. Disk yellow.*

1. D. LINARIIFOLIUS. Hook. (Aster linariifolius. *Linn.*)

St. straight, roughish; *branches* 1-flowered, fastigiate; *scales of invol.* imbricate, carinate, as long as the disk; *lvs.* linear, entire, 1-veined, mucronate, carinate, rough, rigid, those of the branches recurved.—A handsome species, in dry woods, along streams, U. S. and Can., rather rare. Stems subsimple, purplish, about a foot high, decumbent at base. Leaves numerous, rigidly upright or recurved, obtuse, with a small, mucronate point, pale beneath, shining above. Branchlets near the top, leafy, each with one rather large and showy, violet-colored head. Aug. Sept.

2. D. UMBELLATUS. Hook. (A. amygdalinus. *Michx.* A. umbellatus. *Ait.*)

St. smooth, straight, simple; *corymb* fastigiate; *lvs.* long, lanceolate, smooth, attenuate-acuminate at each end, rough on the margin; *invol. scales* obtusely lanceolate.—A tall, handsome plant, growing in low grounds, river banks and fields, N. Eng. to La. Stem 3—4f high (in dry fields but 1—2), purplish, channeled, simple, smooth, branching above into a large, level-topped, compound corymb of showy flowers. Leaves narrow, entire, 4—6' in length. those of the branchlets smaller. Rays about 12, white. Disk yellow. Aug. Sept.

β. *amygdalinus*. *St*. roughish above, green ; *branches of the corymb* divaricate *lvs*. broader.—Quite different in aspect from variety *a*. Common.

3. D. CORNIFOLIUS. Less.. (Aster cornifolius. *Muhl.*)

St. smooth below, scabrous and slightly paniculate above, few-flowered; *lvs*. elliptical, acuminate, entire, tapering to the base, with scattered hairs, rough-edged ; *invol. scales* imbricate, shorter than the disk. Grows in woods, N. and Mid. States. Whole plant nearly smooth, erect, 1—2f high. Leaves acute at the base, paler beneath, on very short stalks. Flowers few, large; outer scales very short. Rays about 10, white. July, Aug.

13. ERIGÈRON.

Gr. ηρ, the spring. γερων, an old man ; because it is hoary early in the season.

Heads many-flowered, subhemispherical ; ray-flowers ♀ very numerous (40—200), narrow, linear ; flowers of the disk ☿; receptacle flat, naked ; involucre nearly in 1 row ; pappus generally simple.— *Herbs with alternate leaves. Rays cyanic.*

§ *Rays longer than the involucre. Mostly* ♃.

1. E. BELLIDIFOLIUM. Muhl. (E. pulchellum. *Mx.*) *Robin's Plantain.*

Hirsute ; *radical lvs.* obovate, obtuse, subserrate ; *stem lvs.* remote, mostly entire, lance-oblong, acute, clasping; *hds.* 3—7, in a close, terminal corymb; *rays* nearly twice longer than the involucre, linear-spatulate.—Dry fields and thickets, U. S. and Can. Stem erect, simple, sometimes stoloniferous, 1—2f high. Leaves 2—3' by 6—9'', mostly broadest above the middle. Rays 60—100, bluish (rarely reddish)-purple. This is our earliest species, flowering in May and June. Resembles the following.

2. E. PHILADELPHICUM. (E. purpureum. *Ait.*) *Narrow-rayed Robin's Pl.*

Pubescent or hirsute ; *lvs.* thin, lower spatulate, crenate-dentate, upper oblong-oblanceolate, narrowed to the clasping (sometimes cordate-auriculate) base, subserrate ; *hds.* few, on long, slender peduncles ; *rays* very numerous, filiform, more than twice longer than the involucre.—Woods and pastures throughout N. Am. Stem slender, 1—3f high. Leaves 2—4' by 6—9'', lower much attenuated at base, upper acute. Rays 150—200! reddish-purple or flesh-colored, nearly as slender as hairs. Jn.—Aug.

β.? *Ricardi. Cauline lrs.* cordate-ovate. Meriden, N. H. *Rickard!*

γ. *St.* stout, with coarsely serrate leaves.

3. E. HETEROPHYLLUM. Muhl. (E. annuum. *Pers.*) *Common Fleabane White-weed.*—*St.* hispid with scattered hairs, branching ; *lvs.* hirsute, coarsely serrate, the lowest ovate, contracted at base into a winged petiole, stem leaves ovate-lanceolate, sessile, acute, the highest lanceolate; *rays* very numerous and narrow.—A common weed, in fields and waste grounds, Can. to Penn. and Ky. Stem thick, 2—4f high, striate, terminating in a large, diffuse, corymbose panicle of large heads. Rays white or purplish, 100 or more, short. June.—Aug.

4. E. STRIGÒSUM. *Fleabane. White-weed. Daisy.*

Hairy and strigose ; *lrs.* lanceolate, tapering to each end, entire or with a few large teeth in the middle, lower ones 3-veined and petiolate ; *panicle* corymbose ; *pappus* double.—A rough weed in grassy fields, Can. and U. S. Stem about 2f high, slender, furrowed, with close, short, stiff hairs, and bearing a large, loose corymb. Leaves also with close-pressed bristles, sessile. Rays very narrow, white. June—Oct.

β. (E. integerrifolium. *Bw.*) *St.* simple, smooth ; *lvs.* entire, pubescent ; *fls.* corymbed. Rays 100—150.

§§ *Rays shorter than the involucre. Plants* ① *or* ②.

5. E. DIVARICÀTUM. Michx.

Decumbent and diffusely branched, hirsute ; *lvs.* linear and subulate ; *hds.*

very small, loosely corymbose; *rays* minute.—Dry soil, Western States ! S. to La. Plant of a greyish or bluish aspect, 3—6' high, but at length spreading 1—2f. Leaves 4—12'' by ⅓—1''. Rays purplish. June—Aug.

6. E. CANADENSE. *Canadian or Common Fleabane.*

Invol. oblong; *rays* numerous, (40—50), crowded, minute; *pappus* simple; *st.* hairy, paniculate; *lvs.* lanceolate, lower ones subserrate.—A very common annual plant of no beauty, growing by roadsides and in fields, throughout N. Am. Stem ⅓—9f! high, branching, hairy and furrowed. Leaves very narrow, with rough edges. Flowers white, very numerous, small, of mean appearance, irregularly racemose upon the branches, and constituting a large, oblong panicle. The plant varies greatly in size, according to the soil. Aug.—Nov.

14. CALLISTÉPHUS. Cass.

Gr. καλλος, beautiful. στεφος, a crown; characteristic of the pappus.

Ray-flowers ♀, numerous; disk-flowers ☿; involucre hemispherical; receptacle subconvex; pappus double, each in 1 series, outer series short, chaffy-setaceous, with the setæ united into a crown; inner series of long, filiform, scabrous, deciduous setæ.—① *Exotica. Lvs. alternate.*

C. CHINENSIS. Ness. (Aster Chinensis. *Linn.*) *China Aster.*—*St.* hispid; *branches* divergent, 1-flowered; *lvs.* ovate, coarsely dentate, petiolate, cauline ones sessile, cuneate at base.—Said to be originally from China. Stem about 18' high, with long branches, each terminated by a single, large head. Rays dark purple. Disk yellow. July—Sept.—Cultivation has produced many beautiful and even splendid varieties, double and semi-double, with white, blue, red, flaked and mottled rays. †

15. BELLIS.

Lat. bellus, pretty; a term quite appropriate to the genus.

Heads many-flowered; rays ♀; disk ☿; involucre hemispherical, of equal scales; receptacle subalveolate, conical; pappus 0.—*Low herbs, either ① and caulescent or* ♃ *and acaulescent. Hds. solitary.*

B. PERENNIS. *Garden Daisy.*—*Root* creeping; *scape* naked, single-flowered; *lvs.* obovate, crenate.—♃. Native of England and other parts of Europe, nearly naturalized in some parts of N. England in cultivated grounds. Scape 3 or 4' high, with a single white flower which is single, double or quilled in the different varieties. Blossoms in the spring and summer months.

16. DAHLIA.

In honor of Andrew Dahl, a Swedish botanist, pupil of Linnæus.

Involucre double, the outer series of many distinct scales, the inner of 8 scales united at base; receptacle chaffy; pappus 0.—♃ *Splendid Mexican herbs. Lvs. pinnate, opposite.*

• 1. D. VARIABILIS. Desf. (D. superflua. *Ait.*)—*St.* green; *rachis of the lvs.* winged; *lfts.* ovate, acuminate, serrate, puberulent or nearly smooth; *outer invol.* reflexed; *ray fls.* ♀, sterile or fertile.—These superb and fashionable plants are natives of sandy meadows in Mexico. They have coarse and roughish leaves, resembling those of the common *elder*, but the flowers are large and beautiful, sporting into innumerable varieties, single and double, of every conceivable shade of scarlet, crimson, purple, red, rarely yellow, blooming from July until arrested by frost.

2. D. COCCINEA. Cav. (D. frustranea. *Ait.*)—*St.* frosty, or hoary, hollow; *lvs.* with the *rachis* naked; *lfts.* roughish beneath; *outer invol.* spreading; *rays* neuter. -Stems about 4f high. Foliage rather glaucous. Rays scarlet, saffron-color or yellow, never purple or white.—The Dahlias are generally cultivated by the divisions of the tuberous roots, which, as soon as the frost blackens the tops, are to be taken up and preserved through the winter in a dry place, free from frost.

17. BOLTONIA.

In honor of J. B. Bolton, author of "Ferns of Great Britain," &c., 1788.

Heads many-flowered; ray flowers ♀, in a single series, those of the disk tubular, ♀; scales in 2 series, appressed, with membranous margins; receptacle conic, punctate; achenia flat, 2 or 3-winged; pappus of minute setæ, 2 (—4) of them usually lengthened into awns. —♃ *Glabrous, branching herbs. Lvs. lanceolate, entire, sessile. Hds. loosely corymbose. Rays purplish-white.*

1. B. ʀʟᴀsᴛɪғᴏʟɪᴀ. L'Her.

Lvs. lanceolate and oblanceolate, acute, tapering to the narrow base, lower ones sometimes serrate; *hds.* on short peduncles, in a somewhat contracted corymb; *branches* leafy; *ach.* obcordate, conspicuously winged, pubescent, with 2 awns nearly its own length.—Prairies and banks of streams, Ill. *Jenney!* Penn. to N. Car. This plant resembles an Erigeron, but is very smooth, 3—6f high. Stem leaves 2—4' by ⅓—1'; branch leaves of the same form but smaller. Rays about 30, expanding 9''. Jl. Aug.

2. B. ᴀsᴛᴇʀᴏ̈ɪᴅᴇs. L'Her. (B. diffusa. *Ell.?* Chrysanthemum Carolinianum. *Wall.*)—*Lvs.* linear-lanceolate, obtuse or acute, all entire, narrowed to the base, those of the branches subulate, minute; *hds.* on long peduncles, in a diffuse and loosely paniculate corymb; *branches* and *branchlets* very slender and nearly naked; *ach.* ovate or somewhat obcordate, smooth, 2-awned.— Prairies, &c. Ia.! Ill. to Ga. and La. A very smooth plant, between an Aster and an Erigeron, with a diffusely branched summit, 3—7f high. Leaves 2—5' or 6' by ⅓—1', reduced upwards to setaceous bracts 1—2'' in length. Heads terminating the filiform branchlets. Rays expanding 7''. Aug. Sept.

Section 2. Heads radiate. Rays yellow.

18. SOLIDĀGO.

Lat. *solidari*, to 'nite; from the vulnerary qualities of the plants.

Flowers of the ray about 5, ♀, remote, of the disk ♀; involucre oblong, imbricate, with appressed scales; receptacle punctate, narrow; pappus simple, capillary, scabrous.—♃ *Herbs, very abundant in the U. S. Stem erect, branching near the top. Lvs. alternate. Hds. small, with 1—15 (very rarely 0) small rays. Fls yellow (one species whitish), expanding in the autumnal months.*

§ 1. *Stems much branched, corymbose. Leaves all linear, entire, sessile.*

1. S. ʟᴀɴᴄᴇᴏʟᴀᴛᴀ. Ait. *Grass-leaved Goldenrod.*

St. angular, hairy, much branched; *lvs.* linear-lanceolate, entire, 3-veined, rough-margined, slightly hispid on the veins beneath; *corymbs* terminal, fastigiate.—In woods and meadows, Can. and U. S. Distinguished from most other species by its flat-topped corymb. Stem 2—4f high, with numerous, very long and narrow leaves, which are distinctly 3-veined and acutely pointed, smaller ones often fascicled in the axils. Flowers in terminal, crowded clusters. Involucre ovate. The whole plant is fragrant. Sept.

2. S. ᴛᴇɴᴜɪғᴏʟɪᴀ. Ph.

St. angular, smooth, with many fastigiate branches; *lvs.* linear, spreading, obscurely 3-veined, scabrous on the margin, the axils leafy; *corymb* terminal, consisting of clustered heads; *rays* about 10, scarcely as long as the disk. —Meadows near the sea-coast, Mass.! to La. Also Wis. *Dr. Lapham!* A very slender species, distinguished from S. lanceolata by the extreme narrowness of the leaves and the thinner, more open corymb, which is often reduced to a few heads. The leaves bear tufts of smaller ones in their axils, and are punctate with resinous dots. Aug—Oct.

§ 2 *Stem simple, corymbose above. Lower leaves lanceolate, petiolate.*

3. S. ʀɪɢɪᴅᴀ. *Hard-leaved Goldenrod.*

St. stout, rough and hairy; *lvs.* ovate-oblong, rough with minute hairs those of the upper part of the stem very entire, lower ones serrate; *flowering branches* paniculate, with close, short racemes; *rays* elongated; *involucre scales* obtuse.—A tall species, in dry fields and rocky woods, Ct. to Mo. and Tex. Abundant in the western prairies! Stem 3—5f high, round, striate, with rigid leaves, of which the radical ones are sometimes near a foot long. Heads larger than in any other species described in this Flora. Rays 7—9, about 3' by 1", deep yellow. Aug. Sept.

4. S. Ohiensis. Riddell.

Glabrous; *lower lvs.* lanceolate, obtuse, entire or serrulate above, tapering to long petioles, *upper* oblong-lanceolate, abruptly acute, sessile, entire; *hds.* numerous, 15—20-flowered, rather large, in a dense, fastigiate corymb.—Mead ows and prairies, western N. Y. to Ia.! A perfectly smooth species, 2—3f high Stem simple, reddish, leafy. Leaves of a firm texture, the radical 6—8' by 1—1½', on petioles of equal length, middle cauline, about 2' by 5". Heads about 6-rayed. Sept. Oct.

5. S. Riddellii. Frank. (S. Mexicana. *β. Hook.*) *Riddell's Solidago.*

Stout and nearly glabrous, corymbosely branched; *radical lvs.* very long, lance-linear, entire, acute, on long, margined, carinate petioles, *cauline lvs.* clasping at base, arcuate, carinate, narrow, acute, entire; *branches* leafy; *hds.* 20—24-flowered, densely clustered in a compound, fastigiate corymb.—Wet prairies Ohio! Wis. to Mo., not uncommon A well marked species, 15—30' high. Radical leaves 12—18' long, almost grass-like, cauline 3—6' by ¼', with a strong mid-vein, and generally much recurved. Rays small, 6—9. Sept.

§ 3. *Heads in glomerate, axillary clusters.*

6. S. squarrosa. Muhl. *Ragged Goldenrod.*

St. stout, simple, erect, thickly pubescent above; *lvs.* smooth, lower ones very broad, oval-spatulate, serrate, acute, upper ones lanceolate-elliptic, highest, entire; *rac.* glomerate, rigid and pubescent; *scales* squarrose with spreading green tips; *hds.* many-flowered; *rays* 10—12, elongated.—A handsome species, found on rocky hills, Can. to Penn. Stem 3—5f high. Heads very large, forming a long terminal spike of short, dense, axillary fascicles or racemes. Sept.

7. S. cæsia. Ait. (S. axillaris. *Ph.*) *Blue-stemmed Goldenrod.*

St. erect, round, smooth and glaucous, often flexuous; *lvs.* smooth, linear-lanceolate, lower ones serrate; *rac.* axillary, erect.—A very elegant species, in thickets and dry woods, Can. and U. S. Stem 1—3f high, of a bluish-purple color, terete and slender, somewhat flexuous, simple or branched. Leaves 2—5' long, ending in a long point, sessile, glaucous beneath. Racemes axillary, numerous, short. Flowers of a deep, rich yellow. Rays 5—7, once and a half the length of the involucre. Aug.

β. flexicaulis. (S. flexicaulis. *Ph.* not of *Linn.*) *St.* flexuous, angular; *lvs.* ovate-lanceolate, longer than the subcapitate racemes.—Leaves about 2' by ¼'. Rays pale yellow.

8. S. latifolia. Muhl. (S. macrophylla. *Bw.* S. flexicaulis. *β. Ph.*)

St. somewhat flexuous, angular, smooth; *lvs.* broadly ovate, acuminate at each end, deeply serrate, smooth; *petioles* marginal; *rac.* axillary and terminal.—A singular and very distinct species, common in dry woods and by rocky streams, U. S. and Can. Stem slender, not always perfectly smooth about 2f high. Leaves 3—5' by 2—4', with acute, often long-acuminate serratures. Clusters very short, axillary, the stem ending with a long terminal one. Heads few. Sept.

9. S. bicolor. (Aster bicolor. *Nees.*) *Two-colored Goldenrod.*

Hairy; *st.* simple; *lvs.* elliptical entire, acute at each end, lower ones serrate, short-stalked; *rac.* short, dense, leafy, erect; *invol. scales* obtuse.—In woods and dry hills, Can., N., Mid. & W. States. A species remarkably distinguished among the solidagos by having white rays. Stem generally simple, 2f high, a little hairy. Leaves hairy on both sides, mostly entire, gradually reduced in size upwards. Flowers in numerous close, short, axillary clusters, forming a long, terminal, interrupted spike. Rays about 8, very short, yellow-ish-white, obscure. July. Aug.

β. *concolor.* T. & G. (S. hirsuta. *Nutt.*) Flowers all yellow.—Penn.

§ 4. *Heads in erect, terminal, simple or compound racemes, not secund.*

10. S. puberula. Nutt.

Plant puberulent; *st.* simple, terete; *lvs.* lanceolate, entire, attenuated at each end, radical ones subserrate; *rac.* spicate, axillary, erect and condensed, *ped.* pubescent; *invol. scales* linear-lanceolate, acute; *rays* about 10, elongated.—Found in low woods, Maine, Ms. Stem straight, purplish, 2—3f high, terminating in a long, thyrsoid spike of dense, appressed racemes. Leaves very minutely pubescent both sides, the lowest on dense, winged stalks. Heads rather large, bright yellow. Aug. Oct.

11. S. stricta. Ait. *Upright Goldenrod.*

Smooth; *st.* strict, erect, simple; *cauline lvs.* lanceolate, very entire, rough-edged, *radical ones* serrate, very long; *rac.* paniculate, erect; *ped.* smooth.—In wet woods, Northern States. Stem (and every other part) very smooth, about 2f high. Leaves 2—4—8' by ½—½—1', lower attenuated at base into a long, winged petiole. Panicle terminal, close, composed of short, dense, appressed racemes. Heads 12—18-flowered. Aug.

12. S. speciosa. Nutt. *Showy Goldenrod.*

St. smooth, simple; *lvs.* lanceolate, entire and scabrous on the margin, thick, the radical and lower ones subserrate, very broad; *rac.* erect, numerous, forming a terminal, thyrsoid panicle; *pedicels* shorter than the involucre, pubescent; *rays* large, 6—8.—Woods, Mass.! to Ohio and Ga. A very tall, showy species, sometimes 6f high. Stem stout, often purple, furrowed. Leaves ample, some of them 6' by 3'. Heads exceedingly numerous, with conspicuous rays of a rich yellow, in a large, showy, pyramidal panicle. Aug. Oct.

β. *angustata.* T. & G. (S. erecta. *DC.*) Panicle slender, spicate.—N. J.

13. S. thyrsoidea. Meyer. (S. virgaurea. *Bw.*) *Thyrsoid Goldenrod.*

St. simple, flexuous, very smooth, pubescent above; *lvs.* smooth, ovate, sharply serrate, acute, the lower ones on long petioles, the upper ones subsessile, lanceolate; *rac.* mostly simple, short; *hds.* large, with conspicuous rays.—A very fine goldenrod, in woods on the sides of the White Mts., and at Franconia Notch, N. H.! Also "on the sides of Killington Peak and Mansfield Mt., Vt." *Robbins.* It is remarkable for the long, slender stalks of the lower, ovate leaves, and for the large, deep yellow heads which exceed in size those of most other species. Stem 1—3f high, racemes axillary and terminal, usually in a thyrse-like panicle. Aug.

14. S. virgaurea. *European Goldenrod.*

St. flexuous, furrowed, pubescent at top; *stem lvs.* lanceolate, serrate, *lower ones* oval, attenuated at both ends; *rac.* erect, ray elongated, flowers large.—This is the only species common to the two continents. One of its numerous varieties is seen scattered here and there on the lower summits of the White Mts.! scarcely on Mt. Washington peak. The flowers are very few, often one only, but larger than those of most other species, and of a rich golden-yellow. Stem often purple, 2—3' high, simple, with axillary and terminal flowers. Aug.

15. S. humilis. Ph. *Low or Humble Aster.*

Glabrous; *st.* simple, erect; *radical lvs.* oblanceolate, petiolate, obtuse and crenate-serrate at apex, the *cauline* lanceolate, acute, the *upper* linear, entire; *rac.* simple or paniculate; *scales* oblong; *rays* short.—On limestone rocks, at Winooski Falls, Colchester, also on the Winooski and Onion rivers, Vt. *Robbins.* Stem 6—12' high, somewhat glutinous. Raceme slender, strict. Leaves of the stem about 2' by 3—4", serrulate. Heads middle size, 6—8-rayed. Aug. Sept.

β. Taller; *hds.* more numerous, in short, glomerate clusters, forming a dense, slender, interrupted raceme. Near the Willey House! White Mts.

§ 5. *Heads in secund racemes. Leaves evidently tripli-veined.*

16 S. nemoralis. *Field Aster.*

St. subtomentose; *cauline lvs.* oblanceolate, sessile, hispid, nearly entire, commonly with tufts of smaller ones in their axils, *radical ones* subcuneiform, serrate; *rac.* paniculate, secund; *rays* 5—7.—A common, starved-looking species with a greyish, dusty aspect, bearing a dense panicle of deep yellow flowers. Height 1—2f. In dry, sterile fields and by roadsides, U. S. and Can Heads small, but with conspicuous rays. Panicle composed of many short racemes, inclining to one side, or often of a single, terminal, recurved one Often the stem divides into branches, each bearing a panicle. Sept.

17. S. Canadensis. *Canadian Goldenrod.*
St. downy; *lvs.* lanceolate, serrate, 3-veined, rough; *rac.* paniculate, secund, recurved; *rays* short.—In old fields, hedges, U. S. and Brit. Am. Common. From 18' to 5f high. Stem furrowed, terminated by a copious panicle which inclines to one side. Leaves sessile, 3' long, sometimes nearly entire, and perhaps a little downy. Heads almost innumerable, very small, with very obscure, yellow rays. Aug.—Oct.
β. *procera.* T. & G. (S. procera. *Ait.*) *St.* villous; *lvs.* rough, villous beneath; *hds.* larger and with larger rays. In low grounds, 4—7f high. Leaves distinctly 3-veined.

18. S. serotina. Willd. *Smooth Goldenrod.*
St. round, striate, smooth; *lvs.* linear-lanceolate, acuminate, serrate; *rac.* secund, recurved, paniculate; *ped.* pubescent; *hds.* small, 15—20-flowered.—A smooth species, in meadows and thickets, U. S. and Can. Stem 3—6f high, very smooth, often glaucous or purple. Leaves 3—5—7' long, about ¼ as wide, smooth, margin scabrous, slightly toothed, upper ones entire. Flowers numerous, forming a more or less compact panicle inclined at summit. Rays about 8, small. Sept.—Variable.

19. S. gigantea. Ait. *Gigantic Goldenrod.*
St. smooth, striate; *lvs.* lanceolate, serrate, margin rough, scabrous above and on the margin and on the veins beneath; *rac.* paniculate; *branches* pubescent; *ped.* and *pedicels* hairy.—A large, showy species, in low, open grounds, U. S. and Can. Stem green, sometimes purplish, 4—7f high, often much branched above. Leaves 2—4—7' long, about ¼ as wide, acuminate at each end, often with divergent teeth. Heads about as large as in the last. Panicle often diffuse, on spreading, leafy branches. Aug.—Oct.

20. S. ciliaris. Willd.
St. angular, smooth; *lvs.* elliptic-lanceolate, subserrate and scabrous on the margin, smooth both sides, sub-3-veined; *rac.* paniculate; *pedicels* (elongated) smooth; *bracts* (often) ciliate; *rays* short.—In the public lands about Ft. Niagara, N. Y. In these specimens the whole plant is smooth except the margins of the leaves. Stem 2—3f high, striate. Leaves subcoriaceous, radical ones petiolate. Racemes thin, spreading. Aug. Sept.

21. S. Missouriensis. Nutt.
Glabrous, low, simple, slender; *lvs.* lance-linear, tapering to each end, very acute and rough-edged, lower ones with acute, slender serratures, radical oblanceolate, petiolate; *rac.* small, in a dense, pyramidal or somewhat corymbose panicle; *hds.* small, 12—15-flowered.—A delicate species, 1—2f high, in dry prairies, Ill. and Mo.! Leaves smooth and shining, lower 3—4' by 3—5'', the others gradually reduced upwards to minute bracts. Rays about 8. Jl. Aug.

§ 6. *Heads in secund racemes. Leaves feather-veined, all entire.*

22. S. sempervirens. (S. lævigata. *Ait.*) *Evergreen Goldenrod.*
St. smooth; *lvs.* lanceolate, somewhat succulent, smooth, entire and scabrous on the margin, closely sessile; *rac.* paniculate; *pedicels* scabrous-pubescent; *rays* elongated.—Marshes along the coast, and river banks, within the influence of the water. Stem 3—6f high, purplish, somewhat glaucous, with numerous long and narrow leaves. Heads large. Rays about 8, long and narrow. Sept.

23. S. odora. Ait. *Sweet-scented Goldenrod.*
St. round, pubescent, slender; *lvs.* linear-lanceolate, acute, abrupt, and sessile at base, very entire, smooth, punctate with pellucid dots, rough-edged, *rac.* paniculate.—In dry, fertile woodlands and sunny hills, U. S. and Can

Stem 2—3f high, yellowish-green. Leaves 1½—3' by 3—5", with a strong, yellowish midvein, but no veinlets. Panicle inclined. Racemes 2—3' long spreading, each generally with a leaf at base, and a simple row of small heads on the upper side. Jl.—Sept.—The only species of solidago which has properties generally considered either agreeable or useful. The leaves are aromatic and yield by distillation a fragrant volatile oil from the pellucid reservoirs. They are a good substitute for tea, and have been exported to China.

§ 7. *Heads in secund racemes. Leaves feather-veined, the lower serrate.*

24. S. PATÜLA. Muhl. *Spreading Goldenrod.*

St. smooth, angular-striate; *lvs.* elliptic, acute, serrate, very scabrous above, smooth beneath, lower ones oblong-spatulate; *rac.* paniculate, spreading; *pedicels* pubescent.—In wet places, Can. N. and W. States, not common. Stem 2—4f high, virgate, often purple, strongly angled, with leafy branches at top. Stem leaves 1—2' long, ½ as wide, radical ones 2 or 3 times larger, all perfectly smooth beneath, although quite rough backwards above. Racemes short, on the ends of the spreading branches, with large heads. Sept.

25. S. NEGLECTA. Torr. & Gray. *Neglected Solidago.*

St. smooth; *lvs.* rather thick, smooth, varying from ovate-lanceolate to narrow-lanceolate, tapering to both ends, feather-veined, entire, the lower and radical ones serrate; *rac.* secund, dense, somewhat spreading, on elongated, slender, suberect branches, which are somewhat leafy at base; *ach.* smooth.—Grows in swamps, N. H.! to Ia., rather rare. Stems 2—4f high, straight, round, dividing at top into several nearly erect branches, forming an elongated panicle. The leaves are sometimes nearly tripli-veined, often very scabrous on the margin. Racemes short. Heads middle-size, 10—20-flowered. Scales obtuse. Aug. Sept.

26. S. ARGÜTA. Ait. *Sharp-notched Goldenrod.*

St. erect, straight, smooth; *lvs.* smooth, acutely and unequally serrate, with diverging teeth, cauline ones elliptical, sessile, highest ones entire and small, radical ones oblong-ovate, attenuate at base into winged petioles; *rac.* paniculate, secund, dense; *hds.* middle size; *rays* about 10.—In meadows and woods, U. S. (from lat. 38°) N. to the Arc. Circle. A smooth plant 2—3f high, with a large, dense, corymbose panicle of very numerous heads. Racemes recurved, a finger's length, the compound pedicels roughish, bracted. Aug. Sept.

β. *juncea.* (S. juncea. *Ait.*) Lrs. lanceolate, lower ones serrate, upper ones entire; *st.* brownish, striate; *rays* twice as long as the involucre; *panicle* less dense.

27. S. MUHLENBERGII. Torr. & Gray. (S. arguta. *Muhl.*)

St. furrowed, glabrous; *lrs.* smooth both sides, strongly and sharply serrate, the radical ones ovate, petiolate, cauline ones elliptical-lanceolate, acuminate at each end; *rac.* secund, short, remote, axillary, spreading; *pedicels* pubescent; *hds.* 15—20-flowered; *scales* linear, obtuse.—In damp woods and thickets, N. H.! to Penn. Stem 2—3f high, generally simple, bearing a long, open panicle. Leaves large, notched with very acute or acuminate teeth, feather-veined. Heads middle size, with 6—8 rather large rays. Aug.—Oct.

28. S. ALTISSIMA. (S. rugosa. *Willd.*) *Tall Goldenrod.*

St. erect, hairy; *lrs.* lanceolate, lower ones deeply serrate, rough and wrinkled.—A very variable species, the tall, rough varieties of which are common about the borders of fields, in hedges, U. S. and Brit. Am. Stem rough with hairs, erect, 3—5f high, much branched at top. Leaves variously toothed or serrate, numerous both upon the stem and branches. Branches widely spreading, each terminating in a recurved panicle with the flowers turning upwards. Scarcely two of the plants look alike. The branches are very widely spread, or but little diverging, with few and scattered heads, or with numerous heads; the leaves are equally or unequally serrate, hairy or woolly. Aug.—Oct.

29. S. LINOIDES. Solander.

Smooth; *st.* slender, simple; *lrs.* lanceolate, finely serrate and scabrous on the margin, radical ones petiolate, upper entire; *hds.* small, in short, secund, at length spreading racemes; *scales* oblong-linear, obtuse, appressed; *rays* 1—4, short.—A small species, near Boston, *Greene* in N. Am. Fl., ii. 216. Stem 12—20' high. Leaves 1—5' by 3—6'' wide. Panicle small, usually turned to one side. Sept. Oct.

30. S. ULMIFOLIA. Muhl. *Elm-leaved Solidago.*

St. glabrous, with hairy branches; *lrs.* thin, elliptic-ovate, serrate, acuminate, sessile, tapering to the base, smooth above, villous beneath; *rac.* paniculate, recurved-spreading; *ped.* villous; *rays* 3—5, short.—In woods and low grounds, Northern and Western States! A very distinct species, more resembling the *elm* in its slender, arched branches than in its leaves. Stem striate, about 3f high, rarely with scattered hairs. Radical leaves tapering to winged petioles, and hairy both sides, with coarse and unequal serratures, upper ones entire, middle ones about 3' by 1½'. Rays deep yellow. Aug. Sept.

31. S. ELLIPTICA. Ait.

St. erect, glabrous, leafy; *lrs.* elliptical, acute at each end, obscurely serrate, glabrous, upper ones sessile, entire; *racemes* short, recurved, paniculate; *hds.* middle size, about 7-rayed; *scales* narrow, acute.—Salt marshes, R. I. *Olney!* Near New York, *T. & G.* Stem 3—5f high, bearing a close, somewhat leafy, pyramidal panicle. Leaves 2—4' by ½—1½', rough-edged, the serratures appressed and rather remote. Rays oblong, rather large, pale yellow. Oct.

19. ECLIPTA.

Heads many-flowered; ray fls. ♀ numerous narrow; disk ♀ tubular, mostly 4-toothed; scales 10—12, in 2 rows, leafy, lance-ovate; receptacle flat; chaff bristly; achenia somewhat angular or 2-edged; pappus 0.—① *Herbs strigose with rigid hairs, erect or procumbent. Lvs. opposite, axillary and terminal, solitary. Fls. white.*

E. ERECTA. (E. procumbens. *Michx.*)

St. often decumbent; *lrs.* lanceolate or lance-oblong, tapering to each end, subserrate; *ped.* much longer than the heads; *scales or leaves of the involucre* acuminate.—Damp soils, Md. to Ill. *Mead!* S. to Flor. Stem often rooting at the lower joints, 1—3f long, with an elastic, thread-like fibre. Leaves 8—14'' by 2—5'', rough, obscurely tripli-veined. Heads small, with minute flowers and short rays. The juice turns black, and is said to dye wool black. Jn.—Sep.

β. *brachypoda.* T. & G. (E. brachypoda. *Michx.*) *Ped.* scarcely longer than the heads.

20. CHRYSOPSIS. Nutt.

Gr. χρυσος, gold, οψις, appearance; for the showy, yellow flowers.

Heads many-flowered; ray-flowers ♀, disk-flowers ♀; involucre imbricate; receptacle subalveolate, flat; pappus double, the exterior short, interior copious, capillary; achenium hairy, compressed.—♃ *Hairy herbs, with alternate and entire leaves.*

1. C. FALCATA. Ell. (Inula falcata. *Ph.*)

Woolly and villous; *lrs.* sessile, linear, very acute, subfalcate, spreading, veins pilose on both sides; *hds.* in axillary corymbs; *invol.* pilose.—A low, leafy plant, in dry, sandy soil, near the sea, Mass. to N. J. Stem thick, leafy, about 8' high. Heads small, bright yellow, in crowded, axillary corymbs. Rays 3-toothed at the apex. Sept. Oct.

2. C. MARIANA. Nutt. (Inula Mariana. *Linn.*)

Hairy; *lrs.* oblong-lanceolate, serrate, the upper ones sessile, acute, the lower ones spatulate and generally obtuse; *corymb* simple; *invol.* viscidly pubescent.—Sandy barrens, N. J., Md.! to Flor., common. The stem and leaves are clothed with scattered, long, silky hairs. Plant about 2f high. Lower leaves taper at base into petioles. The corymb of flowers is terminal, nearly or quite simple. Heads large, 16—20 rayed, yellow, on viscid glandular peduncles. Aug. Oct.

3. C. VILLOSA. Nutt. (Amellus. *Ph.* Diplopappus. *Hook.*)

Erect, leafy, villous-pubescent and strigose; *lrs.* entire, sessile, ciliate to

low, lower ones oblong-spatulate, upper ones oblong-linear or lanceolate; *hds.*
large, solitary and terminal, somewhat fastigiately corymbose; *scales* linear-
subulate.—Prairies, Ill. to Oreg.   Stem 1—2f high   Leaves 1—2′ by 3—5″,
whitish and rough.   Rays about 25, oblong-linear, entire, golden yellow. JL Sept.

21. INÜLA.

Heads many-flowered ; involucre imbricate ; ray-flowers numerous,
♀, disk flowers ☿ ; receptacle naked ; pappus simple, scabrous ; anthers with 2 bristles at base.—♃ *Coarse European herbs, with alternate leaves and yellow flowers.*

1. HELENIUM.   *Elecampane.*
Lvs. amplexicaul, ovate, rugose, downy beneath; *invol. scales* ovate.—A
large herbaceous, coarse-looking plant, common by road-sides, N. Eng. to
Ill.   Stem 4—6f high, furrowed, branching and downy above.   Radical leaves
very large (1—3f by 6—12′), serrate, those of the stem clasping.   Flowers large,
solitary, terminal, of a bright yellow.   Rays linear, with 2 or 3 teeth at the
end.   The medicinal virtues of the plant have long been esteemed.   These are
tonic and expectorant.   Flowers in July and Aug.

Section 3. Heads discoid.

22. BIGELOVIA. DC.
In honor of Dr. Jacob Bigelow, the well-known author of " Florula Bostoniensis," &c.

Heads 3—4-flowered, the flowers all tubular, ☿; involucre cylindrical, as long as the flowers ; scales rigid, linear, closely imbricated ;
receptacle pointed by a scale-like cusp ; achenia obconic, hirsute ;
pappus bristles in one series.—♃ *Glabrous, slender.   Lvs. alternate,
entire.   Hds. fastigiately corymbose, with yellow fls. and colored scales.*

B. VIRGATA. DC. (Chrysocoma virgata. *Nutt.*)
Smooth in all its parts ; *st.* virgately branched from near the base ; *branches* corymbose-fastigiate above ; *lvs.* narrowly linear, 1-veined, the cauline linear-spatulate ; *scales* glutinous.—Swamps, N. J., &c.   A plant resembling Solidago tenuifolia in aspect, 1—2f high.   Leaves 2—3′ by 1—2″, rather firm and
somewhat remote.   Fls. bright yellow, the scales also yellowish.   Aug.—Oct.

23. PLUCHEA. DC.

Heads many-flowered, those of the margin ♀, of the centre ☿ but
sterile ; involucre imbricated ; receptacle flat, naked ; style undivided ; pappus capillary, simple.—*Strong-scented herbs, with alternate
leaves and corymbs of purple flowers.*

1. P. CAMPHORATA. DC. (Conyza camph. *Muhl.*   C. Marilandica.
Michx.)   *Lvs.* ovate-lanceolate, somewhat pubescent, acute, serrate, ser
ratun s mucronate ; *fls.* in crowded corymbs.—♃ A fleshy, strong-scented plant,
native of salt marshes, Mass. to Flor.   Stem a foot high, thick, downy, with alternate leaves and axillary branches.   Flowers light purple.   Aug.

2. P. FŒTIDA. DC. (Baccharis. *Linn.*   Conyza camphorata. *Ph.*)
Erect, nearly glabrous, very leafy ; *lvs.* broadly lanceolate, acute or acuminate at each end, petiolate, feather-veined, obtusely subserrate ; *hds.* numerous, in paniculate corymbs ; *scales* ovate-lanceolate, acute.—A strong-scented
plant, in open, hilly grounds, Western States.   Stem 1—2f high, subsimple.
Leaves 4—7′ by 1½—3′, sprinkled with minute dots ; petioles ½—1′ long.   Heads
numerous.   Aug.—Oct.

24. BACCHÄRIS.
From Bacchus, wine ; from its fragrance resembling that of wine.

Heads many-flowered, ♀ ♂ : involucre imbricate, cylindric or ovate,
*with subcoriaceous, ovate scales ; sterile flowers with the stamens ex

serted; receptacle naked; pappus capillary.—*Shrubby plants, with alternate leaves and white flowers.*

B. **halimifolia.** *Groundsel Tree.*
Shrubby; *lvs.* obovate, incisely dentate above, the highest ones lanceolate; *panicle* compound, leafy; *fascicles* pedunculate.—This is almost the only arborescent plant of this order found in the Northern States. It is 6—12f high, growing on sea-coast and river alluvion. Every part is covered with white dust. The fertile heads growing upon separate plants are in large, loose, terminal panicles, and furnished with very long, slender pappus. Corollas white. Sept. The beauty of this shrub entitles it to cultivation.

Tribe 4. SENECIONIDEÆ.

Heads radiate or discoid. Branches of the style linear, hairy or hispid at the apex, which is either truncated or produced into a conical or elongated appendage. Leaves opposite or alternate.

Section 1. Heads radiate.

25. ARNICA

Involucre of equal. lanceolate scales, 1 or 2 rowed: ray-flowers ♀, disk ♀; receptacle flat. with scattered hairs; pappus single, rigid and serrulate.—♃ *St.* simple. *Lvs. opposite. Fls. yellow.*

A. **mollis.**
St. pubescent, erect; *lvs.* pubescent. becoming nearly glabrous, thin, veiny, dentate, ovate-lanceolate and oblong; radical ones stalked, cauline sessile; *hds.* few; *invol.* hairy, with acuminate scales; *ach.* hairy.—An alpine plant found in ravines on the White Mts., and also, according to *Torrey & Gray*, on the Mts. in Essex Co., N. Y. Stem 1—2f high, with several pairs of sessile leaves, and 1—5 yellow heads of middle size. Leaves 2—5 inches in length, the upper ones broad at the base, the lower tapering to a winged petiole, often acute but not acuminate.

26. POLYMNIA.
The name of one of the ancient Muses; why applied to this plant is not obvious.

Involucre double, outer of 4 or 5 large, leafy scales, inner of 10 leaflets, concave; ray-flowers pistillate. few; disk sterile; receptacle chaffy; pappus none.—♃ *Clammy herbs. Lvs. opposite. Fls. yellow.*

1. P. **canadensis.** *Leaf-cup.*
Viscid-villous; *lvs.* denticulate, petiolate, acuminate, lower pinnatifid, upper 3-lobed or entire.—A coarse, broad-leaved, hairy-viscid plant, 3—5f high, Niagara Falls! Stem with opposite leaves and spreading branches. Flowers light-yellow, the rays short, surrounded by the concave leaflets of the double calyx in such a manner as to form a sort of cup, hence called leaf-cup. Leaves feather-veined, 3—8' long, and nearly as wide, lobes deeply divided and acuminate. Heads ⅓' diam. June.

2. P. **uvedalia.** *Yellow Leaf-cup.*
Lvs. opposite, 3-lobed, acute, decurrent into the petiole, lobes sinuate-angled; *rays* elongated.—In highland woods. Stem 3—6f high. Lower leaves very large. Flowers large, yellow, the rays much longer than the involucre. July.—Neither of these plants has been found in N. Eng., and they are rare in N. Y., but not uncommon in the Western States!

27. SILPHIUM.

Heads many-flowered; ray-flowers numerous, in 2 or 3 rows, fertile, outer row ligulate; disk flowers sterile: involucre campanulate, scales in several series, leafy and spreading at summit; receptacle small, flat, chaffy: achenia broad flat, obcompressed, crowned with a

2-toothed pappus.—♃ *Stout, coarse, resinous herbs.* *Hds. large. Fls yellow.*

1. S. LACINIÀTUM. *Rosin-weed. Polar Plant.*
Very rough, with white, hispid hairs; *lvs.* alterna e, pinnately parted, .ower petiolate, segments sinuate-lobed or entire; *hds.* spicate, distant; *scales of involucre* ovate, appendaged and squarrose at apex.—Western States ! to Tex., producing columns of smoke in the burning prairies by its copious resin. Stem 3—10f high. Lower leaves 1—2f long, much divided, resembling those of some thistles. Heads 4—8, very large, with large, yellow rays. Jl.—Sept.

2. S. TEREBINTHINACEUM. *Prairie Burdock.*
St. and *ped.* glabrous; *lvs.* mostly radical, ovate and ovate-oblong, cordate, dentate-serrate, obtuse, scabrous, on long petioles; *hds.* few, paniculate; *scales* roundish and oval, glabrous.—Prairies, Western ! and Southern States. Plant exuding resin. Stem 4—8f high, nearly naked and simple. Leaves erect, scabrous, rigid, 1—2f long, 7—16′ wide. Involucre globose. Rays 15—25, 1′ long. Achenia narrowly 2-winged. July—Sept.
β. *pinnatifidum.* T. & G. (S. pinnatifidum. *Ell.*) *Lvs.* more or less deeply lobed or pinnatifid.

3. S. TRIFOLIÀTUM. *Ternate-leaved Silphium.*
St. glabrous and often glaucous, terete or 6-sided; *cauline lvs.* lanceolate, acute, scabrous above, smooth below, remotely dentate, on very short petioles, verticillate in 3s or 4s; *upper ones* opposite; *hds.* loosely cymose, on rather long peduncles; *scales* broadly ovate, rather obtuse, smooth; *ach.* oval, with 2 short teeth.—Dry woods and prairies, Ohio, *Sullivant !* and Southern States. Stem 4—6f high.—Leaves 4—6′ by 1—2″. Rays 12—16, expanding about 2½′.

4. S. INTEGRIFOLIUM. Michx.
Scabrous; *st.* quadrangular, striate, simple; *lvs.* opposite, sessile, ovate-lanceolate, entire or slightly dentate; *hds.* in a close corymb; *scales* squarrose; *ach.* roundish, broadly winged, with 2 long teeth.—Western States ! S. to Ga. Stem very rigid, 3—7f high. Leaves rigid, broad and clasping at base, 3—6′ long, ½ as wide, rather variable in form. Heads middle-size. Rays 12—20, 1′ in length. Achenia twice as large as in the preceding species. July, Aug.
β. *ternatum.* Wood. *St.* 6-sided; *lvs.* ternately verticillate.—Prairies ! with the common form ; apparently connecting this with S. trifoliatum, from which it is nevertheless quite distinct in habit.

5. S. PERFOLIÀTUM (and S. connatum. *Linn.*) *Cup-plant.*
St. square; *lvs.* large, thin, opposite, connate-perfoliate, ovate, coarsely toothed, narrowed towards the base; *hds.* in a trichotomous cyme, the central on a long peduncle; *scales* ovate, obtuse, squarrose; *ach.* broadly obovate, winged, emarginate.—Along streams, &c., Mich. ! to Tenn. A coarse, unattractive plant, quite distinct, although variable. Stem 4—7f high. Leaves 8—14′ by 4—7′, the upper pairs forming a cup with their connate bases. Heads large, with 15—25 rays.

28. PARTHENIUM.
Gr. παρθενος, a virgin; from its medicinal efficacy.

Heads many-flowered ; ray-flowers 5, somewhat ligulate, fertile ; disk-flowers tubular, sterile : involucre hemispherical ; scales in 2 series, outer ovate, inner orbicular ; receptacle conical, chaffy ; achenia 5, compressed, cohering with 2 contiguous paleæ.—*American herbs with alternate leaves.*

P. INTEGRIFOLIUM.
St. pubescent, striate, erect; *lvs.* hispid-scabrous, lance-ovate, coarsely dentate-crenate, coriaceous, lower petiolate, upper sessile; *hds.* many, tomentose, corymbed.—♃ Dry soils, Middle and Western States ! Stem rigid, 3—5f nigh. Radical petioles 1f long. Leaves 4—12′ long, ½ as wide. Heads white, with 5 very short, cucullate, white rays. July—Sept.

29. HELIOPSIS.

Gr. ἥλιος, the sun, ὄψις, appearance; flowers radiant like the sun.

Involucre imbricate, with ovate, subequal scales; rays linear, large, ♀; disk ☿; receptacle chaffy, conical, the paleæ lanceolate; achenia 4-sided; pappus 0.—♃ *Lvs. opposite. Hds. large. Fls. yellow.*

H. LÆVIS. Pers. (Helianthus. *Linn.*) *Ox-eye.*

St. smooth; *lvs.* ovate-oblong, coarsely serrate, petiolate, 3-veined, smooth beneath, upper ones usually lanceolate, lower ones more or less truncate at base.—A large, symmetrical plant, in hedges and thickets, U. S. Stem angular, striate, di- or trichotomously branched above, 3—5f high. Leaves 2—6′ by 1—4′, acute, distinctly 3-veined. Branches thickened at the summit, each terminating with a large, solitary, yellow head. Rays lanceolate, broad at base and obtuse at summit. June, Jl.

β. gracilis. T. & G. (H. gracilis. *Nutt.*) Small and slender; *lvs.* scabrous, ovate-lanceolate, acute at base.—2f high.

γ. scabra. T. & G. (H. scabra. *Hook.*) *St.* and *lvs.* scabrous and yellowish-green; *lvs.* somewhat deltoid, distinctly truncate at base.—6f high. Common in Ia.!

30. RUDBECKIA.

Dedicated to the celebrated Olaus Rudbeck, prof. of Botany at Upsal, Sweden.

Involucre scales nearly equal, leafy, in a double row, 6 in each; ray-flowers neutral; disk perfect; receptacle conic, with unarmed paleæ or chaff; pappus 0, or a 4-toothed margin.—♃ *Lvs. alternate. Hds. large. Rays yellow.*

* *Disk pale green or purplish.*

1. R. LACINIÀTA.

Glabrous; *lower lvs.* pinnate, segments 3-lobed, *upper ones* ovate; pappus crenate.—In the edges of swamps and ditches, Can. and U. S. A tall, showy plant, resembling Helianthus, from which, however, it is readily distinguished by its conical disk. Stem round, branching, 6—8f high. Leaves alternate, ample, rough, upper ones generally ovate, the rest variously divided, toothed or cut, petiolate. Flowers large, terminal. Rays 1—2′ long, oblanceolate, bright yellow, spreading or drooping. Aug.

2. R. SUBTOMENTÒSA. Ph.

St. branching, tomentose-pubescent; *lvs.* petiolate, hispid-scabrous above, softly subtomentose beneath, serrate, the lower deeply 3-lobed or 3-parted, upper undivided, ovate, acuminate; *hds.* corymbose; *scales* numerous, spreading; *disk* purplish-brown; *rays* large, spreading.—A coarse, rough species, 3—4f high, prairies, &c., Western and Southern States. Stem angular, marked with brown lines. Leaves 3—5′ long, on petioles 1—2¼″ long. Rays deep or orange yellow, 10—15, about 1′ long. July, Aug.

* * *Disk dark purple.*

3. R. TRILÒBA.

Hirsute; *branches* panicled, spreading; *lower cauline lvs.* mostly 3-lobed, coarsely serrate, acuminate; *upper* ovate-lanceolate, somewhat clasping, serrate or entire; *radical ones* ovate or oval, obtuse, crenate-dentate or incisely lobed, petiolate; *hds.* rather small, disk dark purple, ovoid; *rays* about 8, broad-oval, rather longer than the linear, reflexed scales.—Fields, Middle and Western States. A handsome species, 2—4f high, very branching. Leaves 2—4′ long 3-veined. Rays deep yellow, 6—10″ long, ⅓ as wide. Chaff cuspidate-awned at the summit. Aug. Sept.

4. R. HIRTA. *Rough Cone-flower.*

Very hirsute or hispid; *st.* simple or somewhat branched; *ped.* naked; *lvs.* ovate-spatulate, 3-veined, petiolate, denticulate, the upper ones sessile, ovate-lanceolate; *invol. scales* numerous, narrow, imbricated in 3 rows; *rays* spreading.—A showy plant, in dry soils, Mass. *Rickard!* Western N. Y.! to La. and Ia.! Stems subsimple or branching from the base, covered with prickly prominences, each branch leafless towards the summit and bearing a large head with 12—15 bright yellow rays. These are an inch long, and surround a broadly conical disk of dark purple or dark brown chaff and flowers. July Sept.†

5. R. FULGIDA. Ait. (R. chrysomela. *Michx.* R. spatulata. *Ph.*)
St. hirsute, with rigid hairs; *branches* slender, naked above; *lvs.* strigose pubescent, remotely denticulate, radical petiolate, ovate, cauline lance-oblong, tapering to the sessile, subclasping base; *scales* oblong, spreading, as long as the spreading rays· *chaff* linear-oblong, obtuse.—Mountains, Penn. to Ohio! and Ga. Stem 1—⅓f high. Rays 12—14, scarcely longer than the leafy involucre, deep orange-yellow. July, Oct. †

6. R. SPECIÒSA. Wender.
St. hispidly hirsute; *branches* slender, elongated, naked above; *lvs.* scabrous-pubescent, strongly dentate, radical ones broadly ovate, 5-veined, on long petioles, cauline ovate and lanceolate, 3-veined, upper sessile; *scales* lance-linear, much shorter than the spreading rays; *chaff* linear-oblong, acute.—Borders of woods, Ill. *Jenney!* Ohio! to Penn. A large and very showy species, 2—4f high. Leaves rather thin, radical 4—5′ by 3—4′, the teeth mucronulate; petioles 6—10′ long. Rays about 18, oblong,linear, bright yellow. Aug. —Oct. †

31. ECHINACEA. Mœnch.
Gr, εχινος, the hedgehog ; from the character of the palea.

Involucre scales imbricated in 2 rows; ray-flowers neutral; disk-flowers ♀; receptacle conic, with rigid, mucronate paleæ; pappus 0. —♃ *Lvs. alternate. Rays purple, pendulous.*

1. E. PURPUREA. Mœnch. (Rudbeckia *Linn.*) *Purple Cone-flower or Comb-flower.*—Very rough; *lower lvs.* broad-ovate, attenuate at base, remotely toothed; *cauline ones* lanceolate-ovate, acuminate, nearly entire; *rays* very long, deflexed, bifid.—Thickets and barrens, Western! and Southern States. A tall, handsome plant, often cultivated. Stem 4f high, branched, sulcate. Leaves 4—8′ long, ⅓ as wide, rough, with short, stiff bristles, 3-veined. Heads large, solitary, on long peduncles. Disk thickly beset with the stiff, pointed, brown chaff. Rays about 15, 2—3′ long, pendulous. July—Sept.—Root black, pungent, medicinal. †
β. Rays nearly white.—Ill. *Dr. Mead in T. & G. Fl. Am.*

2. E. ANGUSTIFOLIA. DC. (Rudbeckia. pallida. *Nutt.*)
St. hispid, subsimple, slender, naked above; *lvs.* entire, hispidly pubescent, 3-veined, lower ones lanceolate, petiolate, upper lance-linear, sessile; *scales* in about 2 rows, short; *rays* 12—20, slender, drooping.—Prairies and marshes, Ill. Mo. to Tex. Plant of a more slender habit than the last, 2—3f high. Leaves 3—6′ by 3—6″. Petioles 0—8″ long. Heads on long, naked peduncles. Rays 1—2′ long, purple, varying to white. Disk brown. May—July.

32. LEPACHYS.

Involucre in one series of linear scales; ray flowers few, neutral; disk perfect; receptacle columnar, chaffy; chaff obtuse and bearded at apex; pappus 0; fertile achenia compressed, 1—2-winged.—♃ *Lvs. alternate, pinnately divided. Hds. of flowers yellow, with long, drooping rays.*

L. PINNATA. Torr. & Gray. (Rudbeckia. *Mx.* Obeliscaria. *Cass.*)
Scabrous; *lvs.* all pinnate, the divisions 3—7, some of the lower ones 2-parted, the rest undivided; *rays* elongated.—In dry soil, Western N. Y., Western! and Southern States. Stem 2—4f high, slender, furrowed and hispid. Heads very showy. Rays yellow, about 2′ in length, pendulous, the disk ovate, purple.

33. ZINNIA.
Dedicated to John Godfrey Zinn, a German botanist, 1587.

Involucre scales oval, margined, imbricate; rays 5, persistent, entire, ♀; disk flowers ☿; receptacle chaffy, conical; pappus of the disk of 2 erect awns.—① *Native at the South, &c.	Lvs. opposite, entire.*

1. Z. ELÉGANS.—*Hds.* on long peduncles; *lvs.* cordate, ovate, sessile-amplexicaul; *st.* hairy; *chaff* serrated.—Native of Mexico. Jl. Aug.—Several varieties are known in cultivation; viz. *a. violacea,* with violet-colored flowers; *β. alba,* flowers white; *γ. purpurascens,* flowers purple; *δ. coccinea,* flowers of a brilliant scarlet. †

2. Z. MULTIFLÓRA.—*Hds.* on long peduncles; *lvs.* ovate-lanceolate, on short petioles.—Native of the Southern States.—Z. pauciflorum, with bright yellow flowers, is also found in cultivation. †

34. HELIANTHUS.

Gr. ἥλιος, the sun, ἄνθος, flower; from the resemblance of the flowers.

Heads many-flowered; ray-fls. neutral, disk ☿; scales of the involucre imbricated in several series; receptacle flat or convex, the chaff persistent, embracing the achenia; pappus of 2 chaffy awns, deciduous; achenia compressed or 4-sided, not margined.—*Herbs mostly ♃, rough. Lvs. opposite, the upper often alternate, mostly tripli-veined. Rays yellow, disk yellow or purple.*

§ *Disk dark purple.*

1. H. ANNUUS. *Common Sunflower.—Lvs.* all cordate, 3-veined, only the lowest opposite; *ped.* thick; *fls.* nodding.—This well-known annual is from S. America. It grows in any soil, but its magnitude is increased by the fertility of it in direct proportion, until it reaches the height of 10 and even 20 feet. The common height may be stated at 7 feet. The enormous size of the flowers with their broad rays of brilliant yellow are too well known to require description. An edible oil has been expressed from the seeds. Jl.—A splendid variety occurs with the flowers *all* radiate.

2. H. ANGUSTIFOLIUS. *Narrow-leaved Sunflower.*

Erect, slender, glabrous or hispidly hirsute; *lvs.* sessile, linear, tapering to a long point, 1-veined, rigid, opposite, the upper often alternate, margin subdenticulate, often revolute; *hds.* pedunculate, few; *scales* lance-linear, the long point spreading; *chaff* linear, 3-toothed.—Sandy or rocky places, N. J.! Ky. and Southern States, common. Stem 2—3f high, subsimple. Leaves 2—5′ by 3—6″, broadest at the abrupt base. Rays 12—18, expanding about 2′. Disk flowers brown at the summit.

3. H. RIGIDUS. Desf. (H. scaberrimus. *Ell.* H. crassifolius. *Nutt.*) *Stiff-leaved Sunflower.—St.* erect, rigid, simple or with few branches, scabrous or smoothish, nearly naked above; *lvs.* lanceolate, tapering to each end, petiolate, mostly opposite, tripli-veined, serrulate or entire, rigid, scabrous both sides; *hds.* few; *scales* ovate, acute, regularly imbricate, shorter than the disk chaff obtusish; *pappus* 2-squamæ.—Prairies, &c., Wis. *Lapham,* to Mo., La., &c. Plant 2—4f high. Leaves 3—6′ by ¼—1′, very rough with papillose hairs, but less so than in H. divaricatus. Rays 12—20, expanding 2—3′, light yellow.

§ § *Disk yellow.* * *Leaves opposite.*

4. H. LÆTIFLÓRUS. Pers. *Splendid Sunflower.*

St. rough and branched above; *lvs.* oval-lanceolate, acuminate, serrate, tripli-veined, very rough on both sides, on short petioles, upper ones often alternate; *scales* ovate-lanceolate, ciliate, appressed, a little shorter than the disk; chaff entire or 3-toothed; *rays* 12—20—Barrens, &c., Ia. Ohio. *Torr. & Grey.* A rough, but showy plant, 3—4f high. Leaves thick, 5—8′ by 1¼—2½′. Rays nearly 2′ in length. Disk yellow. Aug.—Oct. †

5. H. OCCIDENTÁLIS. Riddell. *Western Sunflower.*

St. slender, simple, nearly naked above; *lvs.* opposite, oval, scabrous, obscurely serrate, contracted at base into long, hairy petioles, upper ones small and few, entire; *hds.* pedunculate; *scales* lance-oval, appressed.—Sand prairies,

29*

Western States! Stem 3—5f high, scape-like, slender. Leaves 3—5' by 1—2', upper ones 1—2' long. Heads few, middle size. Rays 12—15, light yellow. Jl.—Sept.

6. H. CINEREUS. Torr. & Gray. *Sullivant's Sunflower.*

" Rough with a cinerous pubescence; *lvs.* ovate-oblong, rather acute, appressed to the stem, serrulate, sessile, the lower narrowed to a winged petiole; *ped.* naked, slender; *scales* lanceolate, canescent; *ach.* villous at summit.

β. *Sullivantii.* Larger and more branched; *st.* scabrous-hirsute; *lvs.* obscurely serrate, acute, the uppermost entire, often alternate.—Near Columbus, Ohio. *Sullivant.* Stem 2—3f high, virgate, sometimes a little branched, bearing a few heads nearly as large as those of H. mollis." *Fl. Am.* II. *p.* 324.

7. H. MOLLIS. Lam. (H. canescens. *Michx.*) *Soft-leaved Sunflower.*

St. villous; *lvs.* ovate, acuminate, sessile, cordate and clasping, entire or subserrate, tomentose-canescent, opposite, upper ones sometimes alternate; *scales of involucre* lanceolate, villose-canescent; *chaff* entire, acute and canescent above.—Prairies and barrens, Ohio! Ia.! Mo., common. A hoary and villose species, 2—4f high, nearly simple. Leaves 3—5' long, ¼—⅓ as wide. Rays 15—25, ½—1' by ¼' wide. Jl.—Sept.

8. H. DORONICOÏDES. Lam. *Leopard's-bane Sunflower.*

St. branched, rough or hirsute above; *lvs.* opposite, petiolate, or the upper alternate and subsessile, ovate and ovate-lanceolate, acuminate, tripli-veined, serrate, scabrous above, smooth or pubescent beneath; *scales* lance-linear, ciliate, a little longer than the disk; rays 12—15.—A large species, common in the Western! and Southern States. Stem 4—7f high, with irregular, alternate branches. Leaves 3—10' by 1—3'; petioles ½—1' long. Rays very showy, 15—20'' by 4—6''. Jl.—Sept.

β. *pubescens.* (H. pubescens. Hook.) *Lvs.* tomentose beneath, subsessile.

9. H. STRUMÒSUS. *Downy Sunflower.*

St. smooth below, scabrous above; *lvs.* ovate-lanceolate, acute, serrate, scabrous above, smooth or tomentose-canescent beneath; *hds.* few, about 10, rayed; *scales* ciliate, equaling the disk, squarrose-spreading at tip.—⚈ Grows in swamps, &c., Can. and U. S. Stem 3—5f high, erect, branching above. Leaves petiolate, with an acute point and close serratures, the lower surface varying in the degree of pubescence. Rays bright yellow, an inch or more in length. Scales hairy. Jl.

10. H. DECAPETÀLUS. *Ten-rayed Sunflower.*

Lvs. opposite, ovate, acuminate, remotely serrate, 3-veined, scabrous above, smooth or nearly so beneath; *invol. scales* lanceolate-linear, subciliate, spreading, nearly equal; *rays* 10—12, pale yellow.—Copses, along streams, Can. to Car. and Ky. Stem 3—4f high, purplish. Involucre varying in all degrees of leafiness between the present form and the variety following. Aug.†

β. *frondosus.* Hook. (H. frondosus. *Linn.*) *Outer scales* larger and leaf-like.

11. H. TRACHELIIFOLIUS. Willd.

St. strict, branching, scabrous or with scattered hairs; *lvs.* opposite, those of the branches mostly alternate, thin, scabrous both sides, tripli-veined, appressed, serrate, acuminate, petiolate, lower ones ovate, middle lance-ovate, upper lance-linear; *scales* lance-linear, attenuate-acuminate, longer than the disk, loosely spreading, ciliate; *chaff* slightly 3-toothed; rays 12—15.—A tall, handsome species, not uncommon in thickets, &c., Ohio! and Ia.! Stem purplish, 3—6f high. Leaves 3—6' by ½—3½'; petioles ½—1½' long. Heads middle size, at top of the slender, suberect branches. Rays expanding 2—3'.

12. H. HIRSÙTUS. Raf. (H. diversifolius. *Ell.*)

S'. simple or dichotomous above, scabrous, hirsute; *lvs.* opposite, petiolate, subserrate, 3-veined, ovate-lanceolate, obtuse at base, acuminate, very scabrous above, hirsute beneath; *scales* ovate-lanceolate, acuminate, hirsute, as long *as* the disk; rays 11—15.—Dry soils, Western and Southern States. Stem 4—7f high, with irregular, alternate branches. Leaves 3—10' by 1—3', petioles ½—1' long. Rays very showy, 15—18' by 4—6''. Jl.—Sept.

β. *pubescens.* (H. pubescens. *Hook.*) *Lvs.* tomentose beneath, subsessile.

13. H. DIVARICÀTUS. *Spreading Sunflower.*
St. smooth, branching or simple ; lvs. nearly opposite, sessile, ovate-lanceolate, 3-veined, scabrous above, smooth beneath ; *panicle* trichotomous, slender. few-flowered.—Not uncommon in rocky woods, brooksides, U. S. and Bri.. Am Stem 5f high, glaucous. Leaves rather abrupt at base, tapering to a long acute point, with obtuse serratures. Flowers large, although small for the genus, few, yellow and very showy. The panicle is either 2 or 3-forked. This plant is much improved by cultivation. Aug. Sept.
β. *zaberrimus.* St. subsimple ; lvs. thick, exceedingly rough and rigid Barrens, Ia. !
γ. *Lvs.* ternately verticillate. Otherwise as in β. Barrens, Ia. !

§ § *Disk yellow.* * * *Leaves alternate.*

14. H. GIGANTEUS. (H. altissimus. *Linn.*)
Lvs. alternate (the lowest opposite), lanceolate, acuminate, serrate, scabrous, obscurely 3-veined, tapering at base into short, ciliate, winged petioles *scales of the involucre* lanceolate-linear, ciliate ; *pappus* of 2 short, slightly fringed scales.—Can. to Car. and Ky., in low grounds and thickets. Stem 4—8f high, purplish, branching above into a corymbose panicle of large, yellow flowers. Leaves 2—5' by ⅓—1', opposite or alternate in various degrees. Rays 12—20. Variable.
"β. *ambiguus.* T. & G. *Lvs.* nearly all opposite, sessile and rounded at base —L. l." *Torr. & Gray.*

15. H. TOMENTÒSUS. Michx. *Downy-leaved Sunflower.*
St. stout, pubescent, branched above ; lvs. thin, large, acuminate, obscurely serrate, scabrous above, tomentose or nearly glabrous beneath, lower ones ovate, petiolate, upper ovate-lanceolate, subsessile ; hds. long-pedunculate ; scales lance-linear, long-acuminate, villous, squarrose ; chaff 3-toothed, hirsute at summit.—Dry soil, Ill. to Ga. A large species, 4—8f high, with ample leaves and flowers. Leaves 6—12' by 2—6', some of them tripli-veined. Ray elliptical lanceolate, 18" by 5". Aug.—Oct. †

16. H. GROSSE-SERRATUS. Martens. *Notch-leaved Sunflower.*
St. smooth and glaucous ; lvs. mostly alternate, lanceolate or lance-ovate, long-acuminate, sharply serrate, scabrous above, hoary and softly pubescent beneath, abruptly contracted into naked petioles; scales lance-subulate, loosely imbricated, sparingly ciliate, as long as the disk.—Ohio, *Sullivant,* &c., Ia. *Plummer,* Ill., *Mead,* to La. Allied to H. giganteus. Stem 4—6f high. Leaves 6—9' by 1—2' broadest near the base, lower ones rather coarsely serrate. Rays 15—20, expanding near 3'. Aug. Sept.

17. H. TUBERÒSUS. *Jerusalem Artichoke.—Lvs.* 3-veined, rough, lower ones opposite, cordate-ovate, upper ovate, acuminate, alternate ; *petioles* ciliate at base.—♃ Native of Brazil. The plant has been cultivated for the sake of its tuberous roots, which are used as a substitute for potatoes. It is naturalized in borders of fields, hedges, &c. Sept. § ‡

35. TAGÈTES.

Named for Tages, a Tuscan divinity, son of Genius and grandson of Jupiter.

Heads heterogamous ; involucre simple, tubular, of 5 united scales ; ray-flowers 5, persistent : receptacle naked ; pappus of 5 erect awns —① *Herbs of tropical America. Lvs. pinnately divided.*

1. T. PATÙLA. *French Marigold.—St.* erect, with spreading branches; *segments of the leaves* linear-lanceolate ; *ped.* elongated, subcylindric, one-flowered; *invol.* smooth. Plant about 2f high. Flowers about 2f high. Flowers yellow.

2. T. ERECTA. *African Marigold.—Segments of the leaves* lanceolate, ciliate-serrate; *ped.* 1-flowered, ventricose and thickened at the summit ; *invol.* angular.—The heads are twice larger than in T. patula, and on shorter peduncles. —These are well known and popular garden flowers with several varieties.

36. HELENIUM.

Named for the celebrated Helen, who is said to have availed herself of its cosmetic properties.

Involucre double, the outer of leafy, narrow scales, the inner chaffy; rays pistillate; pappus of several 5-awned, chaffy leaves; receptacle globose. naked in the disk, and chaffy in the ray only; ray-flowers half 3-cleft; seed villose.—*Lvs. alternate, decurrent. Rays yellow.*

H. ᴀᴜᴛᴜᴍɴᴀʟᴇ. *American Sneeze-wort.*
Lvs. lanceolate-serrate, smooth or slightly pubescent, decurrent; *fls.* loosely corymbose.—♃ In damp places. Stem 2—3f high, branching, strongly winged by the decurrent leaves. Leaves tapering to each end or elliptic-lanceolate, more or less deeply serrate. Flowers large, numerous, terminal, with drooping rays, each ending in 3 obtuse teeth, and longer than the large, globose disk. The plant is very bitter. Aug.
β. canaliculatum. T. & G. (H. canaliculatum. *Lam.*) *Rays* concave, canaliculate or 3-furrowed.

37. ANTHÊMIS.

Involucre hemispherical, with nearly equal scales; rays numerous, pistillate; receptacle chaffy, convex or conic; achenia crowned with a slight border.—*European herbs, with much divided leaves.*

1. A. ᴀʀᴠᴇɴsɪs. *Corn Chamomile.*
Sᵗ. erect, hairy; *lvs.* bipinnatifid, hairy and canescent, segments linear-lanceolate; *ach.* crowned with a narrow margin; *chaff of the receptacle* lanceolate, cuspidate, longer than the flowers.—② Grows in dry, cultivated fields. A pilose, inodorous plant, somewhat naturalized in the Northern States. Stems diffusely branching, 8—15' high. Heads large, solitary on the leafless, downy summits of the branches. Disk yellow, rays white. July. ⚥

2. A. ɴᴏʙɪʟɪs. *Chamomile.*—*St.* prostrate, branching from the base, woolly; *lvs.* decompound-pinnatifid, segments linear, subulate; *chaff* scarious, lanceolate, scarcely as long as the flowers.—♃ Native of Britain and other parts of Europe. Grows wild occasionally in fields, and is cultivated in gardens. The strong and agreeable scent of the chamomile is well known, also its tonic and anodyne qualities, which chiefly reside in the flowers. July—Sept.

38. MARÛTA. Less.

Involucre hemispherical, imbricated; rays neutral; disk perfect; eceptacle conical, chaffy (at least at the summit); pappus 0; anchenia smooth.—*European herbs, with alternate, much divided leaves.*

M. ᴄᴏᴛᴜʟᴀ. DC. (Anthemis. *Linn.*) *May-weed.*
Sᵗ. erect, nearly smooth; *lvs.* bipinnatifid, segments linear-subulate; *chaff* ristly, shorter than the flowers.—① Naturalized in all waste places, in hard, ry soils, especially by roadsides, in patches of great extent, presenting almost a uniform whitish surface when in blossom. Stem branching, diffuse, a foot high, with alternate leaves divided and subdived into a multitude of segments. Flowers solitary, on terminal, striated stalks. The plant is ill-scented. Linnæus says it is grateful to toads, drives away fleas, and is annoying to flies. June—Sept. ⚥

39. ACHILLÊA.

Named after Achilles, a disciple of Chiron, who first used the plant.

Involucre ovoid. of unequal. imbricated scales; rays 5--10, short, pistillate; receptacle flat, chaffy; achenia without a pappus.— ♃ *European herbs, with much divided, alternate leaves.*

1. A. ᴍɪʟʟᴇғᴏʟɪᴜᴍ. *Millfoil. Yarrow.*
Lrs. bipinnatifid, with linear, dentate, mucronate segments; *invol.* and *st.* furrowed.—The yarrow abounds in fields, pastures, &c N. Eng. to Oregon and *Arc. Am.* It is called also millfoil, from its leaves being cut and parted into so

numerous divisions and subdivisions. Stem a foot high, branching at top into a dense, flat-topped corymb of white or rose-colored flowers. It has an agreeable, pungent taste and smell. June—Sept.

2. A. PTARMICA. *Sneezewort.*
Lvs. linear, acuminate, equally and sharply serrate, smooth.—Found in moist grounds and shady places, Can. and N. Y. *Pursh.* Mass. *Nicholls.* Plant about 15' high, branching at top into a diffuse corymb of white flowers. The leaves are remarkably distinct from the yarrow. The dried powder of the leaves used as snuff provokes sneezing. A variety with double flowers occurs, which is quite ornamental in pots. Aug. †

40. LEUCANTHÊMUM.

Gr. λευκος, white, ανθος, flower; the heads have large, white, conspicuous rays.

Involucre broad, depressed, imbricated ; rays pistillate, numerous ; receptacle flat, naked ; achenia striate ; pappus 0.—*Herbs, with alternate leaves.*

L. VULGÀRE. Lam. (Chrysanthemum Leucanthemum. *Linn.*) *Whiteweed. Ox-eye Daisy.*—*Lvs.* amplexicaul, lanceolate, serrate, cut-pinnatifid at base ; *st.* erect, branching.—♃ The common white-weed is an annoyance to farmers, rapidly overspreading pastures and neglected fields, U. S. to Arc. Am. Stems 2f high, simple or with one or two long branches, furrowed. Leaves comparatively few and small, obtuse, the lower ones petiolate, with deep and irregular teeth ; upper ones small, subulate, those of the middle sessile, clasping, deeply cut at base, with remote teeth above. Heads large, terminal, solitary. Disk yellow. Rays numerous, white. July.—Sept.

41. PYRETHRUM.

An ancient name of a certain plant, supposed to be of Anthemis.

Involucre hemispherical, imbricate ; scales with membranous margins ; receptacle naked ; pappus a membranous margin crowning the achenia.—*European herbs, chiefly perennial, with alternate leaves.*

P. PARTHENIUM. *Feverfew.*—*Lvs.* petiolate, flat, tripinnate, the segments ovate, cut ; *ped.* branching, corymbose ; *st.* erect ; *invol.* hemispherical, pubescent.— Several varieties of the Feverfew are cultivated, and are in great favor with many florists, on account of their fine pyramidal form, surmounted with a corymb of pure white, double flowers, which retain their beauty for several weeks.

42. CHRYSANTHÊMUM.

Gr. χρυσος, gold, ανθος ; many species bearing golden colored flowers.

Involucre imbricate, hemispherical, the scales with membranous margins ; receptacle naked ; pappus 0.—*Ornamental plants from China and other eastern countries.* *Lvs. alternate, lobed.*

1. C. CORONARIUM.—*St.* branched ; *lvs.* bipinnatifid, broader at the summit, acute.—① Native of S. Europe and N. Africa. The variety with double flowers is frequently cultivated as a hardy annual. Stem about 3f high, striate, smooth, erect, with alternate, clasping leaves. Flowers large, terminal, solitary. Aug.

2. C. CARINÀTUM. Willd. (C. tricolor. *Andr.*) *Three-colored Chrysanthemum.* *Lvs.* bipinnate, fleshy, smooth ; *invol. scales* carinate.— ① Native of Barbary. Heads large and beautiful ; disk purple, rays white with a yellow base. A variety has flowers entirely yellow. Jl.—Oct.

3. C SINENSE. Sabine. (Pyrethrum Sinense *DC.*) *Chinese Chrysanthemum.* *Lvs.* coriaceous, stalked, sinuate-pinnatifid, dentate, glaucous ; *rays* very *long.*—A native of China, where it has been long cultivated and highly esteemed for its beauty. A great number of varieties have been produced, with double, semidouble and quilled flowers of every possible shade of color. It is of very easy culture in any common soil. The plants are propagated by divisions, by

suckers and by cuttings. Although they grow in any soil, yet it is better to give them a rich loam, and water them with liquid manure.

43. LEPTOPÖDA. Nutt.

Gr. λεπτος, *slender,* πους, *foot; alluding to the elongated peduncle.*

Heads many-flowered; rays neutral, cuneate, 3—4-cleft; disk ☿; scales spreading, numerous, attenuate; receptacle conical; chaff 0; pappus of 8—10 fringed squamæ.—♃ *North Am·rican herbs, with the habit of Helenium.*

L. BRACHYPÖDA. T. & G. (Helenium quadridentatum. *Hook.*) *False Helenium.—Sl.* leafy, corymbose at summit; *lvs.* decurrent, lanceolate, subentire, the lower toothed, obtuse; *hds.* on short peduncles; *scales* lance-linear; about half as long as the 8—12 drooping rays; *disk* brownish-purple.—A plant separated from Helenium only on account of its sterile rays. It grows in damp soil, from the southern counties of Ill. to Tex. and Southern States. Stem about 2f high. Heads several or numerous. Rays broadest at summit, rathe deeply and irregularly toothed, 7—9″ by 4—5″

Section 2. Heads discoid and radiate in the same genus.

44. ACTINOMÊRIS. Nutt.

Gr. ακτιν, *a ray,* μηρις, *a part; partially radiate.*

Heads many-flowered, ray flowers 4—14, rarely 0; involucre scales foliaceous, subequal, in 1—3 series; receptacle conical or convex, chaffy; achenia compressed, flat, obovate, 2-awned.—♃ *Plants tall. St. winged with the decurrent leaves. Hds. corymbose, yellow.*

1. A. HELIANTHÖIDES. Nutt. (Verbesina. *Michx.*)
Sl. hirsute, winged except near the base; *lvs.* alternate, ovate-lanceolate, decurrent, acuminate, serrate, hirsute and scabrous; *corymb* contracted; rays 6—14, long and narrow; *scales* erect.—Rather common in barrens and prairies, Western States! It is a rough plant, with the aspect of a Helianthus. Stem conspicuously winged with the decurrent leaves, 2—4f high. Leaves 2—4′ by 6—14″, grayish. Rays often irregular, 1′ long. Jn. Jl.

2. A. SQUARRÖSA. Nutt. (Coreopsis alternifolia. *Linn.*)
St. tall, winged, branching above, somewhat pubescent; *lvs.* alternate, often opposite, oblong-lanceolate, elongated, tapering to each acute or acuminate end, scabrous, decurrent on the petiole and stem; *hds.* small; *scales* spreading or reflexed; *rays* 4—8; *receptacle* very small.—Dry, alluvial soils western N. Y., and Western States! common. It is a tall, unsightly weed, 5—10f high. Leaves 6—14′ by 1—3′, sharply serrate, especially the lower. Rays ½′ long. Aug.—Oct.

45. COREOPSIS.

Gr. κορις, *a bug,* οψις, *appearance; from the concavo-convex, 2-horned achenia.*

Involucre double, each 6—10-leaved; receptacle chaffy; achenia compressed, emarginate, each commonly with a 2-toothed pappus.— *Lvs. mostly opposite. Rays rarely wanting*

§ Rays 0.

1. C. DISCOIDEA. Torr. & Gray. *Rayless Tick-seed.*
Glabrous, much branched, erect; *lvs.* ternate, long-petiolate; *lfts.* ovate-lanceolate, strongly dentate, petiolulate; *hds.* loosely paniculate, on slender peduncles; *outer involucre* 3—5 linear-spatulate, leaf-like bracts, *inner* of many linear, appressed scales; *ach.* linear-oblong, tapering below, twice longer than the 2 erect awns which are hispid upwards!—Ohio, *Sullivant!* to La. Stem and branches purplish. Terminal leaflets 3—5′ by ½—1½′, lateral much smaller. Heads small (½′ diam.), about 30-flowered. Jl.—Sept.

§ § *Rays and disk yellow.*

2. C. TRICHOSPERMA Michx. *Tick-seed Sunflower.*

St. glabrous, square, dichotomous; *lvs.* pinnately 5—7-parted, briefly petiolate, segm. lanceolate, incised or serrate; *scales* of the outer invol. ciliate, linear, long as the inner; *rays* entire, large; *ach.* narrow-cuneate, 2-awned.— ① in wet grounds, N. Y., Mass. to Car. A smooth, branching plant, 1—2f high, with a panicle of large, showy, yellow heads. Branches and leaves mostly opposite. Leaves thin, the upper 3-cleft, subsessile. Leaflets narrow, tapering to a long point, with a few unequal, remote serratures. Achenia ¼' long, crowned with stout, hispid awns. Jl. Aug.

3. C. ARISTÖSA. Michx.

Sparingly pubescent; *lvs.* pinnately 5—9-parted, segments lance-linear, incisely serrate or pinnatifid; *hds.* small, with conspicuous rays; *outer invol.* of 10—12 linear, green scales, about as long as the inner, villous at base; *awns* about as long as the achenium.—Low woods, Western States! Stem obtusely 4-angled, 2—3f high. Leaves thin, 4—6' long, petioles ½—1'. Rays 8, orange-yellow, expanding 1½'.

β. (C. involucrata. *Nutt.?*) *Scales of the outer involucre* about 13, a third longer than those of the inner.—Ia.!

4. C. TRIPTÈRIS. (Chrysostemma. *Less.*)

Glabrous; *st.* simple, tall, corymbose at summit; *lvs.* coriaceous, opposite, petiolate, 3—5-divided, divisions linear-lanceolate, entire, acute; *hds.* small, on short peduncles.—A tall, smooth, elegant species, in dry soils, Southern and Western States, common in Ia.! Stem 4—8f high, slender, terete. Divisions of the leaves 3—5' by ⅓—1¼'. Rays spreading, ½' long. Outer scales linear, obtuse, spreading, as many as the inner. Jl.—Oct.

5. C. VERTICILLÀTA. *Whorl-leaved Coreopsis.*

Glabrous, branched; *lrs.* 3-divided, closely sessile, divisions pinnately parted, segments linear, obtuse; *rays* acute or (in cultivation) obtuse and 2 or 3-toothed; *ach.* obovate, slightly 2-toothed.—24 Moist places, Md. and Western States! Stem 1—3f high. Leaflets apparently verticillate in 6s. Heads with bright yellow rays, near 1' long. Outer scales oblong-linear, obtuse, united at base. June—Aug.

6. C. PALMÀTA. Nutt. (Calliopsis. *Spreng.*)

Nearly smooth; *st.* branched, angled and striate, very leafy to the summit; *lrs.* sessile, deeply 3-cleft, rigid, lobes linear, acutish, entire or again cleft; *rays* obovate-oblong; *ach.* linear-elliptic, incurved.—Dry prairies, W. States! Stem 1—2f high, sometimes much branched. Leaves 1—2½' long, some of them undivided, lobes 2—4" wide. Heads 1 or several, with yellow rays. Outer scales linear-oblong, obtuse. Jn. Jl.

7. C. LANCEOLÀTA. *Lance-leaved Coreopsis.*

St. ascending, often branched below; *lower lvs.* oblanceolate, petiolate, the *upper* lanceolate, sessile, all entire, with scabrous margins; *hds.* solitary, on very long, naked peduncles; *rays* 4—5-toothed at apex; *ach.* suborbicular, with 2 small teeth.—24 Native of the Southern States! Heads showy. Rays about 8, 1' by ½'. Jn.—Aug. †

§ § § *Rays or disk purple.*

8. C. DRUMMONDII. T. & G. (Calliopsis. *Don.*) *Drummond's Coreopsis.*— Pubescent; *lvs.* pinnately divided, sometimes simple, segments (or leaves) oval, entire; *scales* lanceolate-acuminate; *rays* unequally 5-toothed, twice longer than the involucre; *ach.* obovate, incurved, scarcely toothed.—① From Texas. Stems 10—20' high. Rays large, yellow, with a purple spot at base. †

β. *atrosanguinea*, a garden variety, with dark orange flowers.

9. C. TINCTORIA. *Dyer's Coreopsis.*—*Lvs.* attenuate, *radical ones* subbipinnate; *lfts.* oval, entire, smooth; *cauline* subpinnate, *lfls.* linear; *rays* two-colored; *ach.* naked.—A handsome border annual, native of the Upper Missouri. Stem 1—3f high, with light, smooth foliage. Heads with yellow rays, beautifully colored with purple at their base. Flowering all summer. †

10. C. rosea. Nutt. (Calliopsis. *Spreng.*) *Rose-red Tickseed.*
St. branched; *lvs.* opposite, 1-veined, linear, entire; *ped.* short; *outer scales* very short; *rays* obscurely tridentate.—♃ A very delicate species, in wet grounds, Mass. *Dr. Robbins!* to N. J. Stem slender, erect, 8—16′ high. Leaves 1—2′ long, scarcely 1″ wide, clothing the stem. Heads few, small. Rays rose-color, varying to white. Disk light yellow. J′. Aug.

46. BIDENS.
Lat. *bidens*, two-toothed; the achenia have 2 (or more) barbed teeth.

Involucre nearly equal, double, scaly or leafy at the base; rays few, neutral; disk perfect; receptacle chaffy, flat; pappus of 2—4 awns, rough backwards; achenia quadrangular.—*Lvs. opposite. Rays often wanting.*

* *Rays inconspicuous or* 0.

1. B. frondòsa. *Leafy Burr-Marigold.*
Fls. discoid; *outer invol.* 6 times as long as the flower, its leaflets ciliate at base; *lower lvs.* pinnate, *upper ones* ternate, lanceolate, serrate.—① A common weed, in moist, cultivated fields throughout Can. and U. S., often called beggar ticks, from the 2-horned achenia which adhere to every one who passes by it. Stem 2f high, sending out many spreading branches. Lower leaves in 3s or 5s. Flowers in clusters at the end of the branches, without rays, yellow, surrounded by a large and leafy involucre. Aug. Sept.

2. B. connàta. Willd. (B. tripartita. *Bw.*) *Trifid Burr-Marigold.*
Smooth; *lvs.* lanceolate, serrate, slightly connate at base, lower ones mostly trifid; *outer scales* longer than the head, leafy; *ach.* with 3 awns.—① In swamps and ditches, Can., N. Eng. to Mo. Stem 1—3f high, smooth and 4-furrowed, with opposite branches. Leaves opposite, smooth, serrate, the lower ones often divided into 3 segments, the rest generally entire, lanceolate, sharply serrate and somewhat connate. Flowers terminal, solitary, without rays, consisting only of the tubular, yellow florets, surrounded by a leafy involucre. Aug.

3. B. cernua. *Water Burr-Marigold.*
Fls. subradiate, cernuous; *outer invol.* as long as the flower; *lvs.* lanceolate, subconnate, dentate.—In swamps and ditches, Can. to Penn. Stem 1—2f high, purplish, branched, round at base, striate above, branches opposite, leaves opposite, somewhat connated at base. Flowers yellowish-green, finally drooping, generally with yellow rays about 8 in number. Aug.—Oct.

4. B. bipinnàta. *Spanish Needles.*
Smooth; *lvs.* bipinnate; *lfls.* lanceolate, pinnatifid; *hds.* subradiate; *outer invol.* the length of the inner.—① Grows in waste places, N. Y. *Sartwell*, Conn. *Robbins.* Stem 2—4f high, branching, smooth. Leaves bipinnately dissected, nearly smooth. Heads of flowers on long peduncles, each with 3—4 (or 0) obscure, obovate, yellow rays. Jl.—Sept.

** *Rays conspicuous.*

5. B. chrysanthemòides. *Radiate Burr-Marigold.*
Fls. radiate; *rays* 3 times as long as the nearly equal involucre; *lvs.* oblong, attenuate at each end, connate at base, dentate.—① A low plant, with large, yellow-rayed flowers, in muddy places, Can. and U. S. Stem 6—20′ high, round and smooth. Leaves smooth, with few remote teeth, narrow, opposite, with narrow, connate bases. Flowers commonly erect, rays about 8, large, spreading. Scarcely distinct from *B. cernua.* Sept. Oct.

6. B. Beckii. Torr. *Beck's Burr-Marigold.*
St. subsimple; *submersed lvs.* capillaceous-multifid; *emersed ones* lanceolate, connate, acutely serrate or laciniate; *fls.* radiate; *rays.* longer than the involucre.—♃ In water, N. Y. *Sartwell*, &c. Vt. *Chandler*, N. to Can. Stem 2—3f long, simple or with minute, slender branches above. Lower leaves dissected as in *Ranunculus aquatilis*; upper 1—2′ long, ½ as wide, deeply serrate. Head solitary, terminal, yellow. July. Aug.

47. VERBESINA.

Heads few or many-flowered ; rays ♀, few or 0 ; disk ☿ ; scales in 2 or more series, imbricated, erect ; chaff concave or embracing the flowers ; achenia compressed laterally, 2-awned.—♃ *American plants, sometimes shrubby. Lvs. often decurrent, serrate or lobed. Hds. solitary or corymbose.*

1. V. Sɪᴇɢᴇsʙᴇᴄᴋɪᴀ. Michx. (Coreopsis alata. *Ph.* Actinomeris alata. *Nutt.*)—*St.* 4-winged ; *lvs.* opposite, ovate or lance-ovate, serrate, acuminate, tripli-veined, tapering to a winged petiole ; *hds.* radiate, in trichotomous cymes ; *rays* 1—5 ; *ach.* wingless.—Roadsides and dry fields, Western and Southern States, common. Stem 4—6f high. Leaves 5—8' by 3—4', thin. Heads about 25-flowered, with yellow corollas and yellow, lanceolate rays, the latter about ½' long. Aug. Sept.

2. V. Vɪʀɢɪɴɪᴄᴀ. *Virginian Crown-beard.*
St. narrowly winged, pubescent above ; *lvs.* alternate, lanceolate or lance-ovate, subserrate, scabrous, acute or acuminate, tapering to the sessile base ; lower ones decurrent ; *corymbs* compound, dense ; *rays* (oval) and *disk fls.* white ; *ach.* winged.—Dry woods, Penn. to La. Stem (3—5f high) and leaves beneath often more or less tomentose. Heads about 20-flowered, the 3 or 4 rays scarcely ½' long. Aug. Sept.

48. DYSODIA. Cav.

Heads many-flowered ; rays ♀ : disk ♂ ; involucre of a single series of partially united scales, usually calyculate ; achenia elongated, 4-angled, compressed ; pappus scales chaffy, in one series, fimbriately and palmately cleft into bristles.—① *Lvs. mostly opposite and pinnately parted or toothed. Hds. paniculate or corymbose. Fls. yellow.*

D. ᴄʜʀʏsᴀɴᴛʜᴇᴍöɪᴅᴇs. Lagasca. (Tagetes pappusa. *Vent.* Bœbera chr. *Willd.*)—*St.* glabrous, much branched ; *lvs.* pinnately parted, lobes linear, toothed ; *hds.* terminal ; *scales* campanulate, united at base ; bracts at base 7—9, linear ; *pappus bristles* slender, as long as the involucre.—Prairies, &c., Ill., Mo. to La. An ill-scented plant, above 1f high, resembling a Tagetes. Flowers bright yellow.

49. SENECIO
Lat. *senex,* an old man ; the word is synonymous with Erigeron.

Involucre of many equal leaflets or invested with scales at base, the scales withered at the points ; receptacle not chaffy ; pappus simple, capillary and copious.—*A vast genus, embracing* 600 *species of herbs and shrubs. Lvs. alternate. Fls. mostly yellow, exceeding the involucre.*

* Heads discoid.

1. S. ᴠᴜʟɢÂʀɪs. *Common Groundsel.*
St. paniculate, erect, angular ; *lvs.* sinuate-pinnatifid, dentate, amplexicaul.—A common weed growing about houses, in waste grounds, rubbish, &c. Introduced from Europe. Stem 18' high, leafy, branching, generally smooth. Leaves alternate, thin, bright green, the radical ones stalked. Flowers without rays, terminal, scattered, yellow, appearing all summer. ⚥

* Heads radiate.

2. S. ᴀᴜʀᴇᴜs. *Golden Senecio.*
Radical lvs. ovate, cordate, crenate-serrate, petiolate, *cauline ones* pinnatifid, dentate, terminal segments lanceolate ; *ped.* subumbellate, thick ; *rays* 8—12. —♃ A handsome and very variable plant, in meadows, woods, &c. (U. S. and Brit. Am.) with golden-yellow flowers. Stem smoothish, striate, erect, 1—2f high, simple or branched above, terminating in a kind of umbellate, simple or compound corymb. Lower stem leaves lyrate, upper ones few and slender

Peduncles more or less thickened upwards. Scales linear, acute, purplish at apex. Rays 4—5″ long, spreading. May—Aug.

β. Dalsamitæ. (S. Balsamitæ. *Muhl.*) *St.* villous at base; *lvs.* few, small and distant, pubescent, radical ones oblong-lanceolate; *ped.* villous at base. Rocky hills and pastures.

γ. gracilis. (S. gracilis. *Ph.*) *Radical lvs.* orbicular, on long petioles, *cauline* few, linear-oblong, incisely dentate; *ped.* short, pilose, with small, few-rayed heads.—A slender state of the species, on rocky shores.

δ. obovatus. (S. obovatus. *Willd.*) *Radical lvs.* obovate; *ped.* elongated.—Meadows, &c.

ε. lanceolatus. Oakes. *Radical lvs.* lanceolate, acute, *cauline* lanceolate, pinnatifid at base.—Shady swamps, Vt. *Robbins.*

3. S. PSEUDO-ELEGANS. DC. (S. elegans. *Linn.*) *Purple Jacobœa.—Lvs.* equal, pinnatifid, pilose-viscid, spreading; *ped.* somewhat scaly; *invol.* calyculate with leafy scales; *scales* mostly withered at the tips.—Native of the Cape of Good Hope. A beautiful plant in cultivation. Flowers of the disk yellow, of the rays of a most brilliant purple. A variety has double flowers, with colors equally brilliant. Another variety has white flowers. Jn.—Aug. †

Section 3. Heads discoid.

50. HYMENOPAPPUS. L'Her.
Gr. ὑμεν, a membrane, παππος, pappus; from the character.

Heads many-flowered; flowers all ☿, tubular; scales 6—12, in 2 series, oval, obtuse, membranaceous, colored; receptacle small, naked; anthers exserted; achenia broad at the summit, attenuate to the base; pappus of many short, obtuse, membranaceous scales in one series.—☉ or ♃ *North American, villose herbs. St. grooved and angled. Lvs. alternate, pinnately divided.*

H. SCABIOSÆUS. L'Her.
Hoary-villose, or nearly glabrous; *lvs.* pinnately or bipinnately parted, segments linear or oblong, entire or sparingly toothed; *hds.* collected in simple corymbs; *scales of the invol.* obovate, 7—11, white, greenish at base, undulate on the margin, longer than the disk; *cor.* deeply lobed; *ach.* pubescent.—Ill. *Mead !* and Southern States. Stem 1—2f high, whitish with soft cotton when young, at length purplish and glabrous. Segments 1—1½′ by 1—2″, rather acute. Heads whitish, about 21-flowered. (Apr. May. *T. & G.* Aug. *Mead*)

51. CACALIA.
An ancient Gr. name of an uncertain plant.

Flowers all ☿; involucre cylindric, oblong, often calyculate with small scales at the base; receptacle not chaffy; pappus capillary, scabrous.—*Mostly* ♃. *Smooth. Lvs. alternate. Hds. of fls. corymbed. mostly cyanic.*

1. C. SUAVEÖLENS. *Wild Caraway.*
Glabrous; *st.* striate-angular; *lvs.* petiolate, hastate-sagittate, serrate, smooth, green on both sides; *fls.* corymbed, erect; *invol.* many-flowered.—♃ Western N. Y. to Conn., *Robbins !* to Ga. ! and Ill. Stems 4—5f high, striate, leafy. Radical leaves on long stalks, pointed; cauline ones on winged stalks. Flowers whitish, in a terminal, compound corymb. Scales and peduncles smooth, with setaceous bracts beneath the involucre, and beneath the divisions of the peduncles. Aug.—Resembles a Nabalus.

2. C. ATRIPLICIFOLIA. *Orache-leaved Caraway.*
St. herbaceous; *lvs.* petiolate, smooth, glaucous beneath, radical ones cordate, lentate, cauline ones rhomboid, sub-bidentate on each side; *fls.* corymbed, erect; *invol.* 5-flowered.—N. Y. to Ga. and Ill. ! Stem 3—5f high, round, leafy, subramose. Leaves alternate, the lower ones as large as the hand, with large, unequal teeth. Heads of flowers small, ovoid-cylindric, whitish, loosely corymbose at the top of the branches. Jl.—Sept.

3. C. **reniformis**. Muhl.
St. sulcate-angled; *lrs.* palmately veined, nearly smooth, green both sides, petiolate, lower ones reniform, upper flabelliform; *corymb* compound, fastigiate; *hds.* 5-flowered.—Woods Ia.! Ill., Penn., S. to Car. Stem 3—6f high, nearly simple, glabrous. Leaves 3—12′ by 5—18′, repand-dentate, lower petioles very long. Scales of involucre 5, obtuse, whitish. July.

4. C. **tuberósa**. Nutt.
St. angular-sulcate; *lrs.* oval or ovate, strongly 5—7-veined, obtuse or subacute, entire or repand-denticulate, not glaucous, lower ones tapering into long petioles, upper ones on short petioles; *hds.* in compound corymbs.—Marshes, Western States! Stem 2—5f high, branched above. Leaves rather thick, 3—7′ long, ½ as wide, veined like those of the plantain. Hea:!s oblong, 5-leaved and 5-flowered, white. May—Jl.

5. C. **coccinea** Curt. (Emilia sagittata. *DC.*) *Scarlet Cacalia. Tassel Flower.—Radical lrs.* ovate-spatulate, *cauline* amplexicaul, crenate; *invol.* ovate-cylindric, scales linear, at length reflexed; *ach.* ciliate; *pappus* in several rows.—A pretty garden flower, native of the E. Indies, &c. Stem 1f or more high. Flowers bright scarlet. Jn.—Sept. A bed or patch sown thickly makes a fine appearance.

52. ARTEMISIA.

Probably from *Artemis*, one of the names of the goddess Diana.

Involucre ovoid, imbricate, with dry, connivent scales; receptacle naked or subvillous; disk flowers numerous, ♀, tubular, ray flowers few, often without stamens, and with a subulate corolla or 0; achenia with a small disk; pappus 0.—*Bitter herbs. Lvs. alternate. Cor. yellow.*

§ 1. *Receptacle naked. Disk flowers sterile.*

1. A. **Dracúncŭlus**. *Taragon.*—*Lrs.* smooth, lanceolate, acuminate at each end; *hds.* subglobose, pedunculate, erect.—A culinary herb, native of S. Europe. Stem herbaceous, 2—3f high. Jl. Aug.—It is of the easiest culture, and is used for pickles, salad, and for seasoning soup. ‡

2. A. **dracunculoïdes**. Ph. (A. cernua. *Nutt.*)
Erect, much branched, whitish pubescent when young; *lower lvs.* 3-cleft, *upper* entire, narrowly linear, attenuated at both ends; *hds.* globose, small, nodling, pedicellate, in paniculate racemes; *scales* with scarious margins.—St. Louis, Mo. to the Saskatchawan. Stem shrubby, 6—8f high, with numerous slender branches. Leaves 1—4′ by 1—3″, radical trifid or sometimes 2 or 3 times trifid.

3. A. **boreális**. Pallas. (A. spithamæa. *Ph.*)
Cæspitose, silky-villose or smoothish; *st.* simple (6—10′ high); *lower lvs.* petiolate, linear-lanceolate, entire towards the base, ternately, pinnately or bipinnately parted above, with linear lobes, *upper lvs.* linear, 3—5-cleft or entire; *hds.* hemispherical, spicate or racemose-paniculate.—Keweena Point. Lake Superior. *Dr. Houghton* in T. & G. Fl. ii. 417.

4. A. **Canadensis**. Michx. *Sea Wormwood.*
St. erect or decumbent; *lrs.* pinnatifid with linear segments; *fls.* subglobose, sessile, in crowded panicles resembling spikes.—⁊ Shores of the great lakes. Plum Island, *Bigelow.* Near Amherst College, *Hitchcock.* Willoughby Mt., Vt., *Wood.* Stem 2—4f high, much branched, sulcate, brownish, mostly erect. Leaves all much divided into linear-setaceous segments. Heads numerous, small, forming a large panicle of racemes. Scales with a membranous margin. Aug.

5. A. **caudàta**. Michx.
St. herbaceous, simple, densely and pyramidally paniculate; *radical* and *lower cauline lvs.* subbipinnate, pubescent, *upper ones* subpinnate, segments subsetaceous, alternate; *hds.* ovoid-globose, pedicellate, erect.—On the sea coast, N. H. to R. I. Stem 3—5f high, strict. Leaves in many attenuated and some-

what fleshy segments. Heads small. Outer scales ovate, inner scarious, elliptical. Aug. Sept.

§ 2. *Receptacle naked. Flowers all fertile.*

6. A. ABROTÀNUM. *Sntherncood.*—*St.* erect; *lower lvs.* bipinnate; *upper ones* capillary, pinnate; *invol.* downy, hemispherical.—A well known shrubby plant in gardens, about 3f high. Leaves alternate, much divided into very narrow, linear segments. Flowers numerous, nodding, yellow. Native of S. Europe. ‡

7. A LUDOVICIÀNA. Nutt.
Herbaceous, canescently tomentose; *lvs.* lanceolate, lower incisely and remotely serrate or subpinnatifid, upper entire; *hds.* ovoid, subsessile, arranged in a simple, slender, leafy panicle.—Lake and river shores, Mich. to Mo. W. to Oreg. Stem 2—5f high, simple or branched. Leaves quite variable in size, and also in pubescence, sometimes nearly smooth. Heads small and crowded. Aug. Sept.

8. A. VULGÀRIS. *Common Artemisia or Mugwort.*
Lvs. tomentose beneath, cauline ones pinnatifid, segments lanceolate, acute, subdentate, floral ones entire, linear-lanceolate; *hds.* erect, ovoid, subsessile; *invol.* tomentose.—♃ Introduced from Europe and naturalized in fields, roadsides, banks of streams, &c., Vt. N. H. Stem 2—3f high, branching into a panicle of spicate racemes. Leaves very variable, but never nearly so attenuated as in A. Canadensis. Heads few, purplish.

9. A. BIENNIS. Willd.
Erect, herbaceous, smooth; *lvs.* bipinnately parted, upper ones pinnatifid, all with linear, acute and mostly incised lobes; *hds.* sessile, arranged in a close, narrow, leafy panicle of short spikes.—Ohio to Mo. and the Saskatchawan. *T. & G.* Aug.—Oct.

§ 3. *Receptacle villous or hairy. Flowers all fertile.*

10. A. ABSINTHIUM. *Common Wormwood.*
Lvs. multifid, clothed with short, silky down, segments lanceolate; *hds.* hemispherical, drooping; *receptacle* hairy.—Naturalized in the mountainous districts of New England, growing among rubbish, rocks and by roadsides. Stems angular, branched, with erect racemes of nodding, yellow flowers. The whole plant is proverbially bitter, and of powerful medicinal qualities as a tonic, stomachic, &c. § ‡

11. A. PONTICA. *Roman Wormwood.*—*Lvs.* tomentose beneath, cauline ones bipinnate, leaflets linear; *hds.* roundish, stalked, nodding.—Common in gardens, where it arises 3 or 4f, with simple branches and racemes of yellow flowers. Head with 24 flowers, those of the ray about 6. From Austria. †

53. TANACÈTUM.
Said to be a corruption of *a θανατος,* deathless; for the durable flowers.

Involucre hemispherical, imbricate, the scales all minute; receptacle convex, naked; pappus a slight, membranous border; achenia with a large, epigynous disk.—*Lvs. alternate, much dissected. Fls. yellow.*

T. VULGÀRE. *Tansey.*—*Lvs.* pinnately divided, segments oblong-lanceolate, pinnatifid and incisely serrate; *hds.* fastigiate-corymbose.—♃ Native of Europe, and naturalized in old fields and roadsides. Stems clustered, 2—3f high, branched above into a handsome corymb of yellow flowers. Aug.—The whole plant has a strong and aromatic smell and a very bitter taste. The seeds are anthelmintic. A variety called *double tansey* occurs, with dense and crisped leaves. §

54. GNAPHALIUM.
Gr. γναφαλον, cotton or wool; from the soft, cottony surface of the herbage.

Heads discoid, heterogamous; involucre imbricate with scarious, colored scales; marginal flowers subulate, pistillate, in several rows;

central flowers ♀; receptacle flat, naked; pappus simple, scabrous, capillary.—*Lvs. alternate, entire.*

1. G. POLYCEPHALUM. Michx. *Fragrant Life-everlasting.*
Lvs. linear-lanceolate, acute, scabrous above, whitish tomentose beneath, as well as the paniculate stem; *hds.* capitate, corymbose; *scales of the involucre* ovate-lanceolate, acute.—① Common in fields, &c., Can. and U. S. It is distinguishable by its strong, agreeable odor, and its brownish color. Stem 1—2f high, whitish, with a cottony down, much branched. Leaves sessile, cottony beneath. Flowers in crowded clusters at the ends of the branches. Involucre with whitish scales and yellow flowers. Aug.

2. G. DECURRENS. Ives. *Decurrent-leaved Life-everlasting.*
St. erect, stout, much branched, viscidly pubescent; *lvs.* linear-lanceolate, very acute, *decurrent*, naked above, white and woolly beneath; *hs.* in dense, roundish, terminal clusters.—♃ A stout species, covered with a dense, hoary pubescence. It grows in hilly pastures, &c., N. H., Vt. to N. J. Stem 2f high, with scattered leaves and spreading branches. Leaves on the upper side green, scabrous and viscid. Scales whitish, with yellow corollas. Aug.

3. G. PURPUREUM. (G. Americanum. *Willd.*) *Purple Life-everlasting.*
St. erect, simple or branched from the base, tomentose; *lvs.* linear-spatulate or obovate-spatulate, downy-canescent beneath, green above; *hds.* sessile, crowded, terminal and axillary.—Grows in sandy fields and pastures, N. H. to Ia. and La. Stem 8—12' high, sending out shoots at the base. Heads with purplish scales and yellow corollas. June.

4. G. ULIGINÓSUM. *Mud Life-everlasting.*
St. diffusely branched, woolly; *lvs.* linear-lanceolate; *hds.* in terminal, crowded, leafy clusters; *involucre scales* one-colored, inner acute; *ach.* smooth. —① A small, spreading plant clothed with whitish down, common in sandy places where water occasionally stands, N. Mid. and W. States. Stem 4—6' high. Leaves numerous, acute, narrowed at the base. Scales of the involucre oblong, obtuse, yellowish. Aug.

55. ANTENNARIA. Br.

Name in allusion to the bristles of the pappus, which resemble *antennæ.*

Heads diœcious; involucre of imbricate, colored scales; pistillate corollas filiform; receptacle subconvex, alveolate; pappus simple, bristly.—♃ *Tomentose. Lvs. alternate. entire. Hds. corymbose, with white scales.*

1. A. MARGARITACEA. Br. (Gnaphalium. *Linn.*) *Common Life-everlasting.*
St. erect, simple, corymbosely branched above; *lvs.* linear-lanceolate, acute, 3-veined, sessile, and beneath the stem woolly; *corymbs* fastigiate; *invol. scales* elliptic, obtuse, opaque, white.—♃ Named for its dry, imperishable, pearl-white flower-scales. In fields and pastures, U. S. and Brit. Am. Stem 1—2f high, and with its numerous, scattered leaves, clothed with white and cotton-like down. Heads of flowers numerous, hemispherical, much adorned by the fadeless, imbricated scales of which the outer are downy at the base. Flowers yellow. July.—The plant is slightly fragrant.

2. A. PLANTAGINIFOLIA. Br. (Gnaphalium. *Linn.*) *Mouse-ear Life-everlasting.*—*Stolons* procumbent; *st.* simple; *radical lvs.* ovate, mucronate, 3-veined, silky-canescent; *stem lvs.* small, lanceolate; *invol. scales* ovate, obtuse. —♃ Borders of woods, &c., U. S. and Brit. Am., common. Whole plant whitish with down. Stem 5—8' high, often branched at base into several from the same root. Root-leaves much larger than those of the stem, rounded at apex, tapering to a petiole. Stem-leaves few, bract-like. Heads clustered, terminal, purplish-white. May—Aug.
β. *dioica.* (A. dioica. *Br.*) *Radical lvs.* much smaller, spatulate; *stem lvs.* linear-lanceolate; *inner scales* linear, acute.—Abundant in old pastures. Apr. July. A variety (too?) strongly marked.

56. FILÂGO. Tourn.

Apparently from the Lat. *Ala,* on account of the cottony fibres or hairs.

Heads heterogamous; involucre of a few villous scales; marginal flowers ♀; receptacle columnar, naked at the apex, chaffy at base; achenia terete; central ones with a hairy pappus.—*Downy-canescent herbs. Lvs. alternate, entire.*

F. GERMANICA. (Gnaphalium. *Linn.*) *German Cud-weed.*
St. dichotomous or proliferously branched above; *lvs.* linear-lanceolate, acute, crowded, erect; *hds.* few-flowered, in dense, capitate clusters, terminal and lateral; *scales* cuspidate, passing insensibly into the paleæ of the receptacle, each with a pistillate flower in the axil.—① A European plant, sparingly naturalized in fields and roadsides, Ms. Stem 6—10′ high. Scales straw-color, with a green line outside. July—Oct.

57. ERECHTÍTES. Raf.

Gr. ερεχθω, to trouble; the species are troublesome weeds.

Flowers all tubular, those of the margin pistillate, of the disk perfect; involucre cylindrical, simple, slightly calyculate; receptacle naked; pappus of numerous fine, capillary bristles.—① *Lvs. simple, alternate. Fls. corymbose, whitish.*

E. HIERACIFOLIUS. Raf. (Senicio hieracifolius. *Linn.*) *Fire-weed.*
St. paniculate, virgate; *lvs.* oblong, amplexicaul, acute, unequally and deeply toothed with acute indentures; *invol.* smooth; *ach.* hairy.—① A well known, rank weed, growing in fields, (Can. and U. S.) particularly and abundantly in such as have been newly cleared and burnt over, and hence it is called fire-weed. Stem thick and fleshy, branching, 3f high, roughish. Leaves of a light green, large, irregularly cut into many deep and acute teeth. Flowers terminal, crowded, destitute of rays, white. Involucre large and tumid at base. Aug. Sept.

58. XANTHIUM.

Heads heterocephalous. *Sterile.*—Involucre imbricate; anthers approximate, but distinct; receptacle chaffy. *Fertile.*—Involucre 2-leaved, clothed with hooked prickles, 1 or 2-beaked, 2-flowered; stamens 0.—① *Coarse weeds with alternate leaves.*

1. X. STRUMARIUM. *Clotweed.*
St. unarmed, branching; *lvs.* cordate, lobed, 3-veined, unequally serrate, rough; *fr.* elliptical, armed with uncinate, stiff thorns, and ending with 2, spreading, straight horns.—A coarse, rough plant, in old fields, &c., N. Eng. and Mid. States. Stem branched, bristly, spotted, 2—3f high. Leaves large, on long stalks, rigid. Sterile flowers few together, terminal, globular, green. Fertile, in sessile, axillary tufts. Fruit a hard, 2-celled burr, near an inch long, covered with stiff, hooked prickles, which, like those of the common burdock, serve to disperse the seeds. Aug. ♂?

2. X. ECHINÁTUM. Murray. (X. macrocarpon. *DC.*)
St. rough and strigose, spotted; *lvs.* scabrous, obscurely lobed, obtuse, broad and subcordate at base, repand-toothed; *fr.* oval, densely armed with rigid, uncinate bristles; *horns* incurved.—A very coarse plant in marshes near the seacoast, Mass. to Car. Stem thick, 1—2f high. Fruit very large, hairy. Aug.—Oct.

3. X. SPINÓSUM. *Prickly Clotweed.*
St. branched; *spines* at base of the leaves triple, slender; *lvs.* ovate-lanceolate, cuneate at base, petiolate, 3-lobed or dentate or entire, under surface and veins above whitish; twice longer than the spines; *invol.* oblong, with slender, uncinate spines.—Roadsides and fields, Mass. to Penn.! and Ga. Plant about 1f high, very conspicuously armed with straw-colored spines ½—1′ long. Heads sessile, sterile in the upper, fertile in the lower axils. Sept—Nov

59. AMBROSIA.

Gr. αμβροσια, food of the gods ; a term strangely applied.

Heads heterocephalous. *Sterile.*—Involucre of several united scales, hemispherical, many-flowered ; anthers approximate, but distinct ; receptacle naked. *Fertile.*—Involucre 1-leaved, entire or 5-toothed, 1-flowered, corolla 0 ; styles 2 ; stamens 0.—*Herbaceous plants with mostly opposite lvs. and unsightly flowers.*

1. A. TRIFIDA. *Tall Ambrosia.*

Hairy, rough ; *lvs.* three-lobed, serrate, the lobes oval-lanceolate, acuminate ; *fr.* with 6 lines below the summit.—① A very tall, herbaceous plant, not very common in N. Eng., found in hedges and low grounds in the valley of Connecticut river. In the W. States it is abundant! Stem 5—10f high, erect, branching, furrowed. Leaves opposite, in 3 large, deep lobes with long points and close serratures. Flowers mean and obscure, in long, leafless spikes, axillary and terminal. Aug.

β. *integrifolia.* T. & G. (A. integrifolia. *Muhl.*) *Lvs.* ovate, acuminate, serrate, bristly on both sides, ciliate at base, often some of them 3-lobed ; *rac.* terminal, single or ternate.

2. A. ARTEMISIÆFOLIA. (A. elatior. *Ph.*) *Hog-weed.*

Lvs. twice-pinnatifid, nearly smooth ; *petioles* ciliate ; *rac.* terminal, panicled ; *st.* virgate.—① A common and troublesome weed of the gardens, fields, &c. (Can. to Ga.) far more worthy of its English than its Latin name. Stem 2—3f high, branching, pubescent when young. Leaves with segments acute and parallel. Barren flowers, small, green, in terminal racemes, the fertile ones sessile about the axils of the upper leaves. Aug. Sept.

3. CORONOPIFOLIA. Torr. & Gray.

Canescent-strigose, branched ; *lvs.* crowded, rigid, the lower opposite, bipinnatifid, upper lanceolate, sessile, pinnatifid ; *sterile hds.* spicate, *fertile* clustered at the base of the sterile spikes, in the axils of the upper leaves ; *fr.* hairy. —① Prairies, Wis. to Tex. Stem 1—5f high, at length very branching and leafy. Aug. Sept.

4. A. BIDENTATA. Michx.

Very hirsute ; *st.* branching ; *branches* simple ; *lvs.* crowded, mostly alternate, closely sessile or partly clasping, undivided, oblong, with a single tooth or short lobe on each side near the base ; *sterile hds.* densely spicate, *fertile* axillary ; *fr.* 4-angled, acutely pointed, the 4 ribs produced in 4 short spines. ① Prairies, Ill. to La. Stem 1—3f high, with numerous leaves and very dense, terminal spikes. Jl.—Sept.

60. IVA.

A name of barbarous origin.

Heads discoid ; involucre 3-leaved ; marginal flowers 5, fertile, the others sterile ; receptacle hairy ; achenia obconic, obtuse ; pappus 0. —*Herbs or shrubs. Lower lvs. opposite.*

1. FRUTESCENS. *High-water Shrub. Marsh Elder.*

St. shrubby ; *lvs.* lanceolate, punctate, deeply serrate, rough.—♃ In the borders of salt marshes, Mass. to La., common. Stem thick, 3—8f high, with numerous opposite branches. Leaves numerous, 3-veined, upper ones entire. Flowers green, small, drooping, in close, leafy clusters. Aug.

TRIBE 5. CYNAREÆ.

Heads ovoid, discoid, rarely radiate, homogamous (rarely diœcious), or heterogamous with the marginal flowers in a single series. Style in the perfect flowers often tumid near the summit.

61. CALENDULA.

Lat. *calends,* the first day of the month ; some species blossom monthly.

Heads radiate ; involucre of many equal leaves, in about 2 series ;

receptacle naked ; achenia of the disk membranaceous ; pappus 0.—
An oriental genus of annual herbs.   Lvs. alternate.

C. OFFICINALIS. *Pet Marigold.*—Viscid-pubescent ; *st.* erect, branched ; *lvs.*
oblong, acute, mucronate, sessile, subdentate and scabrous-ciliate on the margin ; *hds.* terminal, solitary ; *ach.* carinate, muricate, incurved.—A common
and handsome garden plant, from S. Europe. It has *double, lemon-colored,* and
other varieties. Flowers large and brilliant, generally orange-colored. Jn.—Sept.†

62. CENTAUREA.

The centaur Chiron, it is said, cured with these his foot wounded by Hercules.

Heads discoid ; involucre imbricate ; ray flowers larger than the
rest, sterile, often wanting ; receptacle bristly ; pappus of filiform,
scabrous bristles in several series.—*A genus of oriental herbs, with alternate leaves.*

1. C. NIGRA. *Black Knapweed.*
St. erect, branched, pubescent above; *lower lvs.* angular-lyrate, *upper* lan-
ceolate, dentate ; *scales of the involucre* ovate, with an erect, capillary fringe.—
♃ A troublesome weed in meadows and pastures, Mass. Introduced from Eu-
rope. Stem about 2f high, simple or oftener divided into elongated branches.
Heads few, large, terminal, solitary. Scales dark brown. Flowers purple.
Jl. Aug. §

2. C. CYANUS. *Blue-bottle. Bachelor's-button.*
St. erect, branching, downy; *lvs.* linear, entire, downy, the lowest subdentate ; *invol. scales* serrate.—① Introduced from Europe, cultivated and sparingly
naturalized in old fields. It is a hardy annual, justly popular for its handsome
flowers, which are very variable in color. Heads ovoid, solitary on the ends
of the branches. Jl.—Sept. §

3 C. AMERICANA. Nutt. (C. Nuttallii. *Spreng.*)
St. erect, sulcate, sparingly branched; *lower lvs.* oblong-ovate, repand-dentate, *upper ones* lanceolate, acute, *all* sessile and glabrous ; *hds.* few or solitary,
very large ; *ped.* thickened at summit; *invol.* depressed-globose, scales with a
pectinate-pinnate, reflexed appendage.—① Native in Ark. and La., naturalized
in Ill. *Mead.* Cultivated in gardens. Stem 2—4f high, with very showy, pale-
purple heads. §†

63. AMBERBOA. DC.

Heads discoid ; involucre imbricate ; ray-flowers wanting or larger
than the rest, sterile ; pappus of oblong or obovate paleæ, attenuated
to the base, all similar, rarely small or 0.—*Eastern herbs, with alternate leaves.*

1. A. MOSCHATA. Willd. (Centaurea. *Linn.*) *Sweet Sultan.*—*Lvs.* lyrate-
dentate ; *invol.* subglobose, smooth ; *scales* ovate ; *ray-flowers* scarcely enlarged,
not exceeding the disk ; pappus 0.—A handsome border annual from Persia
Flowers purple. A variety has white flowers. July—Oct.

2. A. ODORATA. *a. amboracea.* DC. (Centaurea suaveolens. *Willd.*) *Yellow
Sweet Sultan.*—*Lower lvs.* broadly subspatulate, dentate, *upper* lyrate at base ;
hds. globose ; *ray-fls.* enlarged upwards, longer than the disk ; *pappus* chaffy, a
little shorter than the fruit.—From Levant. Leaves scarcely pinnatifid. Flowers yellow.
β. *glauca.* (Centaurea glauca. *Willd.*) *Lvs.* often deeply pinnatifid; *flowers*
purple.

64. CARTHAMUS.

Arabic *qorthom,* to paint; from its coloring property.

Heads discoid ; involucre imbricated, outer bracts foliaceous,
flowers all tubular and ⚥, filaments smooth ; pappus 0 ; receptacle
with setaceous paleæ ; achenia 4-angled.—*Oriental herbs.*

C. TINCTORIUS. *Common Saffron.*—*St.* smooth; *lrs.* ovate-lanceolate, sessile, spinose-denticulate.—① Native of Egypt, but long cultivated in other lands on account of its orange-colored flowers. Stem branching, striate, 1—2f high. Leaves subamplexicaul, smooth and shining. Heads large, terminal, with numerous long and slender flowers. The latter are useful in coloring, and as a nursery medicine. July.

65. XERANTHEMUM.

Gr. ξηρος, dry. ανθος; on account of its dry, imperishable flowers.

Heads discoid; involucre hemispherical, with radiant, colored, opaque, scarious scales; receptacle paleaceous; pappus paleo-setaceous.—① *Native of S. Europe.*

X. ANNUUM. Willd. *Eternal Flower.*—*St.* erect, branched; *lvs.* oblong-lanceolate, obtusish, alternate, entire; *hds.* large, terminal, solitary; *scales of the involucre* obtuse, scarious, inner ones of the ray spreading, lanceolate, obtuse.— A singularly beautiful plant, half hardy, of easy culture. Stem 2—3f high. The radiant involucre scales are of a rich purple, but there are varieties with red, white, blue and yellow rays. The splendid flowers retain their beauty through the winter.

66. CNICUS. Vaill.

Gr. κνιζω, to prick; well applied to these herbs.

Heads discoid; involucre ventricose, imbricate with doubly spinous scales; ray-flowers sterile; receptacle very hairy; pappus in 3 series, the outer 10-toothed, the 2 inner each 10-bristled.—*Oriental herbs.*

C. BENEDICTUS. *Blessed Thistle.*—*Lrs.* somewhat decurrent; dentate and spiny; *invol.* doubly spinous, woolly, bracteate.—① Native of Persia, Tauria and Greece. About 2f high, with yellow flowers. Sparingly naturalized. June.—It was formerly in great estimation in medicine, but is now considered worthless. ‡

67. ONOPERDON. Vaill.

Gr. ονος, and περδω, the application to the present noble genus is not obvious.

Heads discoid, homogamous; involucre ventricose, imbricate with spreading, spinous scales; receptacle deeply alveolate; pappus copious, capillary, scabrous; achenia 4-angled.—*Large, branching herbs, with decurrent leaves.*

O. ACANTHIUM. *Cotton Thistle.*
Invol. scales spreading, subulate; *lrs.* ovate-oblong, decurrent, sinuate, spinous, woolly on both sides.—② This fine looking thistle occurs naturalized in waste grounds, and is about 3f in height. The whole plant has a white, cottony appearance. Stem winged by the decurrent leaves, which are unusually large. Involucre round, cottony, spinous. Flowers purple. July, Aug.

68. CYNARA.

Gr. κυων, a dog; the stiff, hard spines of the invol. resemble a dog's teeth.

Heads discoid, homogamous; involucre dilated, imbricate, scales fleshy, emarginate, pointed; receptacle setaceous; pappus plumose; achenia not beaked.—*Natives of the Old World.*

1. C. SCOLYMUS. *Garden Artichoke.*
Lvs. subspinose, pinnate and undivided; *invol. scales* ovate.—⚄ Native of S. Europe, naturalized in gardens and cultivated grounds. A well known garden esculent. The parts used are the receptacle, the lower part of the involucre and the upper portion of the stalk. It is cultivated from suckers placed in rows 3 feet apart. Aug. Sept. §‡

2. C. CARDUNCULUS. *Cardoon.*—*Lvs.* spiny, all pinnatifid; *invol. scales* ovate.—⚄ Native of Canada. Flowers purple. This plant is blanched by

having earth heaped up around it, and then the petioles become tender, crisp eatable, like celery. Aug. Sept. ‡

69. CIRSIUM. Tourn.

Gr. κιρσος, a swelling of the veins, which this plant was supposed to heal.

Heads discoid, homogamous ; involucre subglobose, of many rows of spinose-pointed, imbricated scales ; receptacle bristly ; style scarcely divided ; pappus copious, plumose ; achenia compressed, smooth.— *Herbs with alternate leaves, generally armed with spinose prickles. Fls. cyanic.*

1. C. ARVENSE. Scop. (Cnicus arvensis. *Ph.*) *Canada Thistle.*
Lrs. sessile, pinnatifid, spinous ; *st.* panicled ; *invol.* round or ovate, with minute spines, scales close-pressed, ovate-lanceolate.—♃ A very common thistle in fields, roadsides and waste places, N. Eng. to Ohio. It is one of the severest pests of the farmer, requiring his constant vigilance to extirpate it from his fields. In England it is called *cursed thistle.* Root creeping, very long and exceedingly tenacious of life. Stem 3f high, with a branching panicle at top. Leaves alternate, thickly beset with thorns. Heads rather small, purple, the involucre nearly thornless, and is the only part of the plant that can be safely handled. July. ◊

2. C. DISCŎLOR. Spreng. (Cnicus discolor. *Muhl.*) *Tall Thistle.*
Lrs. sessile, pinnatifid, rough-haired, downy beneath, the segments 2-lobed, divaricate, spinose ; *invol.* globose, the scales ovate, appressed, with spreading spines at the tips.—♂ A slender thistle, 3—5f high, much branched, and leafy at the summit. Found in thickets, N. Eng. to Ill. Heads terminating the branches, an inch in diameter, with reddish-purple corollas. July, Aug.

3. C. MUTICUM. Michx. (Cnicus glutinosus. *Bw.*) *Glutinous Thistle.*
Lrs. pinnatifid with divaricate segments ; *invol.* ovate, with unarmed, villous-arachnoid, glutinous scales.—♂ A fine looking thistle found in damp soils, Can. and U. S. Stem branching, 3—7f high. Leaves armed with spines at each angle. Heads ½' diam., with deep purple corollas, the scales webbed and glutinous on the back. Aug. Sept.

4. C. HORRIDŬLUM. Michx. (Cnicus horridulus. *Linn.*) *Yellow Thistle.*
Lrs. sessile, pinnatifid, acutely cut, spinose ; *hds.* invested with an external involucre of about 20 very spinose bracts ; *scales* unarmed.—♂ Found in meadows and hills, N. Eng. to Flor. The stem is 1—3f high, invested with wool. Leaves somewhat clasping, woolly and hairy, armed with stiff spines. Heads large, with yellowish-white corollas surrounded by a whorl of lanceolate or linear leaflets tipped with stiff thorns, the scales webbed. Aug.

5. C. PUMĪLUM. Spreng. (Cnicus odoratus. *Muhl.*) *Pasture Thistle.*
Hairy ; *lrs.* green on both sides, clasping, oblong-lanceolate, pinnatifid, the segments irregularly lobed, ciliate, spinose ; *invol.* round-ovate, spinose, naked.—♂ A common, low, turgid thistle in roadsides, pastures, N. Eng. and Mid. States. Stem 1—2f high, stout, striate, with 1—3 very large heads of fragrant, purple flowers. Aug.

6. C. LANCEOLĀTUM. Scop. (Cnicus. *Ph.* Carduus. *Linn.*) *Common Thistle.*
Lvs. decurrent, pinnatifid, hispid, the segments divaricate and spinose ; *invol.* ovate, villous ; *scales* lanceolate, spreading.—♂ Common in borders of fields, roadsides, N. Eng. and Mid. States, always distinguished by the decurrent leaves. Stem 3—4f high, winged by the decurrent leaves which are white and woolly beneath, armed with formidable spines at all points. Fls. numerous, large, purple. Involucre scales webbed, each ending in a spine. July—Sept.

7. C. ALTISSĬMUM. Spreng. (Carduus. *Linn.* Cnicus. *Willd.*)
Tall, branched, villose-pubescent, leafy ; *lrs.* whitish beneath, spinose-ciliate, sessile, lanceolate-oblong, sinuate-dentate, lower ones pinnatifid, petiolate : *lobes* or *teeth* spinescent ; *invol.* ovoid-oblong ; *scales* close-compressed, ovate-lanceolate, outer ones with a spreading spine at apex.—Fields and barrens, Penn. and Western States ! common. Stem 3—8f high. Leaves 6—8' by 1—6'.

Heads about 1' diam., with linear-lanceolate bracts at base. Flowers purple or purplish-white. Aug.

8. C. VIRGINIÁNUM. Michx. (Carduus. *Linn.* Cnicus. *Ph.*)

Slender and mostly simple; *lvs.* sessile, lanceolate, margin revolute, entire or repand-dentate, teeth spinescent, or sometimes remotely sinuate-lobed or pinnatifid, upper surface glabrous, under surface tomentose-canescent; *hds.* small; *invol.* subglobose; *scales* tipped with a short, spreading prickle.—Woods, Ohio, *T. & G.*, and Southern States. Plant about the size of the Canada thistle, clothed with an arachnoid pubescence, with few or many heads (sometimes but one) which are about ½' diam. Flowers purple. Apr.—Sept.

70. LAPPA. Tourn.

Lat. *lappa*, a burr, from Gr. λαβειν, to lay hold of; a characteristic term.

Heads discoid, homogamous; involucre globose, the scales imbricated and hooked at the extremity; receptacle bristly; pappus bristly, scabrous, caducous.—② *Coarse, European herbs. Lvs. alternate, large.*

L. MAJOR. Gaert. (Arctium Lappa. *Linn.*) *Burdock.*

Lvs. cordate, unarmed, petioled.—Common in waste and cultivated grounds, fields, N. Eng. Mid. and W. States. Each plant is a large, conical, ill-scented and coarse looking mass of vegetation, surmounted by a branching, irregular panicle of ovoid heads with tubular corollas of an exceedingly delicate pink color. The leaves are very large, with wavy edges. This plant is an instance of design in the dissemination of seeds, such as cannot be mistaken. The scales of the involucre all end in a minute, firm hook, which seizes hold of everything that passes by. Thus men and animals are made the unwilling agents of scattering widely the seeds of this unsightly plant. July, Aug. ⚥

β. Leaves pinnatifid.—Penn. *Dr. Darlington.*

SUBORDER 2.—LIGULIFLORÆ.

Flowers all perfect, ligulate, in a radiatiform or radiant head.

TRIBE 6. CICHORACEÆ.

Branches of the style uniformly pubescent. Plants with a milky juice. Leaves alternate.

71. CICHORIUM. Tourn.

The Egyptian name *chikouryeh*, whence Gr. κιχωρη, and Eng. *succory*.

Involucre double, the outer of 5 leafy scales, the inner of about 8 linear ones; receptacle chaffy; pappus scaly; achenia not rostrate, obscurely 5-sided.—*Oriental herbs with bright blue fls., about 20 in a head.*

1. C. INTYBUS. *Succory.—Fls.* in pairs, axillary, sessile; *lower lvs.* runcinate. —⚥ A European plant 2—3f high, with large, showy, sky-blue flowers, naturalized in grass fields, by roadsides, and becoming quite common in many localities. Stem round, with few long branches, rough. The upper leaves become cordate acuminate, sessile, inconspicuous, only the radical ones runcinate. The flowers are 1—2' diam., and placed rather remote on the long naked-ish branches. Corollas flat, 5-toothed. The root is used in France as a substitute for coffee. July—Sept. ⚥

2. C. ENDIVIA. *Endive.—Ped.* axillary, in pairs, one of them elongated and 1 headed, the other very short, about 4-headed; *hds.* capitate.- A hardy annual, from the E. Indies, esteemed and cultivated for salad. The French physicians have recently found it a remedy for jaundice. ⚥

72. KRIGIA. Schreb.

Dedicated to Dr. Daniel Krieg, a German botanist.

Involucre many-leaved, nearly simple, equal; receptacle naked; pappus double, or consisting of 5 broad, membranous scales surrounding 5—8 bristles several times as long as the 5-angled achenia.—*Small acaulescent herbs. Heads solitary, with 20—30 yellow flowers.*

K. Virginica. Willd. (Hyoseris. *Mx.* Cynthia. *Beck.*) *Dwarf Dandelion*
Scape 1-flowered; *lvs.* lanceolate, lyrate, smooth; *invol.* smooth.—① This
little plant is found on sandy hills and by roadsides, Can. to La.   Scapes 1—
several, smooth, slender, 1—8' high. After flowering it becomes longer than
the leaves. The primary leaves are roundish, entire; the rest irregularly lyrate.
Scales of the involucre 10—15, linear-lanceolate, arranged in a somewhat simple series. Corollas yellow. Ach. turbinate, scabrous, reddish-brown. May—Jl.

73. CYNTHIA. Don.
One of the names of Diana; its application to this genus is not obvious.

Involucre nearly simple, of equal, narrow scales; receptacle flat,
alveolate; pappus double, the outer minute, scaly; inner copious,
capillary; achenia short.—♃ *Lvs. alternate or all radical. Hds. with*
15—20 *yellow flowers.*

C. Virginica. Don. (Tragopogon. *Linn.* Krigia amplexicaulis. *Nutt.*)
St. mostly simple, scape-like; *radical lrs.* sublyrate or pinnatifid, on short,
winged petioles; *cauline ones* lanceolate, amplexicaul, entire.—In barrens and
dry soils, Western N. Y. to Ill.! &c. The plant is smooth and glaucous. Stem
1—2f high, often dichotomously divided, with 1—2 clasping leaves at the forks.
Radical leaves 3—5' long, sometimes nearly entire. Heads terminal on the
bracteate and subumbellate peduncles, with deep yellow flowers. Scales united
at base in a somewhat double series. May—July.

74. LEONTODON.
Gr. λεων, a lion, οδους, a tooth: in reference to the deeply toothed leaves.

Involucre imbricate, the outer scales very short; receptacle naked; pappus plumose, persistent on the somewhat rostrate achenia.—
Acaulescent herbs, with white fls. many in a head.

L. autumnalis. (Apargia. *Willd.*) *Autumnal Hawkweed.*
Scape branching; *ped.* scaly; *lvs.* lanceolate, dentate-pinnatifid, smoothish.
—A European plant, naturalized and common in the eastern parts of N. England, growing in grass-lands and by roadsides. The flower resembles those of
the dandelion (Taraxacum). Root large, striate, hol-
low, decumbent at base, 6—18' high, with a few branches and scattered scales.
Leaves all radical, spreading, 6' long, with deep, round sinuses, and covered
with remote hairs. Heads 1' diam. yellow, appearing from July to Nov. ⚤

75. TRAGOPOGON.
Gr. τραγος, a goat, πωγων, beard; in allusion to the tawny, showy pappus.

Involucre simple, of many leaves; receptacle naked; pappus plumose, achenia longitudinally striate, contracted into a long, filiform
beak.—② *European herbs, with long, linear, grass-like lvs.*

T. porrifolius. *Salsify. Vegetable Oyster.—Invol.* much longer than the corolla; *lvs.* long, linear, undivided, straight; *ped.* thickened upwards. Stem 3—4f
high. Flowers terminal, solitary, large, bluish-purple. This exotic is cultivated
in gardens for the root, which is long, tapering and nutritious. When properly
prepared it has a mild, sweetish taste, which has been compared to that of the
oyster. ‡

76. CATANANCHE.
Gr. κατα, αναγκη, from necessity: it *must necessarily* be admired (?).

Involucre imbricated, scarious; receptacle paleaceous; pappus paleaceous, 5-leaved; paleæ awned.—① *Oriental herbs, with alternate,*
lanceolate lvs.

C. cœrulea - -*Lvs.* linear and lanceolate, villous, somewhat bipinnatifid at
base; *lower scales of the involucre* ovate, mucronate.—From S. Europe. A handsome annual, 2—3f high. Heads solitary, on long peduncles with blue, spreading, *ligulate* corollas toothed at apex. Jl.—Sept. †

77. LACTÚCA.

Lat. *lac*, milk; from the milky juice in which all the species abound.

1. L. ELONGÀTA. *Wild Lettuce. Trumpet Milkweed.*

Lvs. smooth and pale beneath, lower ones amplexicaul, runcinate, upper lanceolate, entire, sessile; *hds.* racemose-paniculate.—A common, rank plant, growing in hedges, thickets, &c., where the soil is rich and damp. Stem hollow, stout, 3—6f high, often purple, bearing a leafless, elongated, sometimes corymbose-spreading panicle of numerous heads of flowers. Leaves very variable, the lower 6—12′ long, commonly deeply runcinate, often narrow-lanceolate, with a few narrow-lanceolate divisions. Corollas yellow. Achenia oblong, compressed, about the length of the beak. July, Aug.

β. integrifolia (L. integrifolia. *Bw.*) *Lvs.* nearly all undivided, lanceolate, sessile, the lowest often sagittate at base.

γ. sanguinarea (L. sanguinarea. *Bw.*) *Lvs.* runcinate, amplexicaul, mostly pubescent, glaucous beneath; *fls.* purple.—Stem 2—3f high, often purple (but this character is not peculiar to this variety).

2. L. SATÌVA. *Garden Lettuce.*—*St.* corymbose; *lvs.* suborbicular, the cauline ones cordate. The varieties of this exotic are every where well known and cultivated for salad. It is annual, with very smooth, yellowish-green foliage, which in one variety (*capitata*) is so abundant as to form heads like the cabbage. Heads numerous, small, with yellowish corollas. The milky juice contains opium, and if this salad be eaten too freely, unpleasant narcotic effects are the consequence. ‡

78. MULGEDIUM. Cass.

Lat. *mulgeo* to milk; in allusion to the lactescent qualities of the plants.

Involucre somewhat double. the outer series of scales short and imbricated; receptacle naked, faveolate; pappus copious, soft, capillary, crowning the short-beaked achenia.—*Lvs. mostly spinulose. Hds. with many yellow or cyanic flowers.*

1. M. ACUMINÀTUM. DC. (Sonchus acuminatus. *Willd.*)

Radical lvs. subruncinate; *cauline ones* ovate, acuminate, petiolate, dentate; *hds.* loosely paniculate, on somewhat scaly peduncles.—In hedges and thickets, N. Y. to Ia.! and S. States. A smooth plant, 3—6f high, with the stem often purplish. Leaves 3—6′ long, the lower ones often deltoid-hastate or truncate at the base, sinuate-denticulate, narrowed at base into a winged petiole. Heads small. Peduncles with a few scale-like bracteoles. Scales dark purple, with blue corollas. Pappus white, on the short-beaked, ovate-acuminate achenia. Aug. Sept.

2. M. LEUCOPHÆUM. DC. (Sonchus floridanus. *Ait.* Agathyrsus leucophæus. *Don.*)—*Lvs.* numerous, lyrate-runcinate, coarsely dentate; *hds.* paniculate, on squamose-bracteate peduncles.—Moist thickets, N. and W. States. A tall, leafy plant, nearly smooth. Stem 4—10f high. Leaves 5—12′ long, irregularly divided in a runcinate or pinnatifid manner, the segments repand-toothed, the radical ones on long stalks, the upper ones sessile, often undivided. Heads small, with pale blue or yellowish corollas, a tawny-white pappus, and arranged in a long, slender panicle. Aug. Sept.

3. M. FLORIDÀNUM. DC. (Agathyrsus. *Beck.* Sonchus. *Linn.*)

Glabrous; *st.* erect, paniculate above. purple or glaucous; *cauline lvs.* runcinately pinnate-parted, segments few, sinuate-dentate, upper ones triangular, acute or acuminate; *panicle* loose, erect, compound.—Western! and Southern States, hedges and waste grounds. A handsome plant with a terminal panicle of blue flowers. Stem 3—5f high. Leaves 4—8′ long, variable in form. Heads small. Rays expanding 9″. Jl.—Sept.

79. HIERACIUM. Tourn.

Gr. *ἱεραξ*, a hawk; supposed to strengthen the vision of birds of prey.

Involucre more or less imbricated, ovoid, many-flowered; receptacle subalveolate-fimbrillate; scales very unequal; pappus of scabrous fragile, copious, 1-rowed bristles.—4 *Lvs. alternate, entire or toothed.*

§ *Involucre imbricated.*

1. H. CANADENSE. Michx. (H. Kalmii. *Spreng.*) *Canadian Hawkweed.*
St. erect, subvillose, leafy, many-flowered; *lvs.* sessile, lanceolate or oblong-ovate, acute, divaricately and acutely dentate, the upper ones somewhat amplexicaul, with an obtuse base; *panicles* axillary and terminal, corymbose, downy.—In open dry or rocky woods. N. Eng., N. Y., Can. Stem stout, 1—2f high, more or less pubescent, the peduncles downy but not glandular. Leaves somewhat pubescent or hairy. Heads large and showy, with yellow flowers. Involucre sometimes with a few glandular hairs. Aug.

§ § *Involucre calyculate.*

2. H. VENÒSUM. *Veiny-leaved Hawkweed.*
Scape or *st.* naked or with a single leaf, smooth, paniculate; *lvs.* obovate, somewhat acute, entire, a little hairy above, nearly glabrous beneath, ciliate on the margin, veins colored; *invol.* glabrous, about 20-flowered; *ach.* linear.—In woods, &c. N. Eng. to W. States. Stem 1—2f high, dark brown, slender. Panicle diffuse, several times dichotomous, corymbose. Heads rather small, on slender pedicels, with bright yellow flowers. Jl. Aug.

3. H. GRONOVII. *Gronovius' Hawkweed.*
St. leafy, hirsute, paniculate; *invol.* and *pedicels* glandular-pilose; *radical lvs.* obovate or oblanceolate, entire, strigose, the midvein beneath very villous; *upper ones* oblong, closely sessile.—A hairy plant, found on dry hills, Can. and U. S. Stem about 2f high, furnished with a few leaves below, naked above and bearing a narrow, elongated panicle. Lower leaves tapering into a long stalk. Flowers yellow, on glandular, slender pedicels. Achenia tapering upwards to a slender point, but scarcely rostrate. Aug. Sept.

4. H. SCABRUM. Michx. (H. Marianum. *Willd.*) *Rough Hawkweed.*
St. leafy, scabrous and hispid; *lvs.* elliptic-obovate, scabrous and hirsute lower ones slightly dentate; *ped.* thick, and with the *invol.* densely glandular hispid; *hds.* 40—50-flowered.—A rough plant, on dry hills, borders of woods Can. to Car. and Ky. Stem 1—3f high, round, striate, rather stout. Lowe leaves petiolate, upper sessile, subacute, often purplish as well as the stem Heads large, with yellow flowers. Achenia obtuse at apex, bright red, with *a* tawny pappus. Aug.

5. H. PANICULÀTUM. *Slender Hawkweed.*
St. slender, leafy, paniculate, whitish tomentose below; *lvs.* lanceolate, glabrous, membranaceous, acute; *panicle* diffuse; *ped.* very slender; *hds.* 10—20-flowered.—A smooth, slender plant, in damp woods, Can. to Ga. Stem 1—3f high, several times dichotomous. Leaves thin, 2—4' long. Heads small, numerous, with yellow flowers. Pedicels long and filiform, forming a very diffuse panicle. Aug.

6. H. LONGIPÌLUM. Torr. (H. barbatum. *Nutt.*)
Plant densely pilose with long, straight, ascending, bristly hairs; *st.* strict, simple, smoothish and nearly leafless above; *lvs.* crowded near the base of the stem, oblong-lanceolate, attenuated to the base, entire; *hds.* glandular-tomentose or hispid, 20—30-flowered, in a small, terminal panicle.—Barrens and prairies, Western States. Plant 1—2f high, remarkable for the long, brownish, straight hairs with which the lower part is thickly clothed; otherwise it more nearly resembles the last. July—Sept.

7. H. AURANTIÀCUM.—*St.* leafy, hispid; *fls.* densely corymbose; *lvs.* oblong, somewhat acute, pilose, hispid.—Native of Scotland. Flowers numerous, large, orange-colored. One of the few species worthy of cultivation. †

80. NABÀLUS. Cass.

" Nomen omnino sensus expers, forte mutandum." *De Candolle*

Involucre cylindric, of many linear scales in one row, calyculate with a few short, appressed scales at base; receptacle naked pappus copious, capillary, brownish, 2-rowed persistent; achenia not beaked, smooth, striate.—*Erect herbs, with a thick, tuberous, bitter root. Hds 5—18-flowered, not yellow, although often straw-colored.*

§ *Heads pendulous, glabrous. Leaves very variable in the same species.*

1. N. ALBUS. Hook. (Harpalyce. Don. Prenanthes. *Linn.*) *Lion's-foot. White Lettuce.—St.* smooth and somewhat glaucous, corymbose-panicu-.ate above; *radical lvs.* angular-hastate, often more or less deeply lobed; *stem lvs.* roundish-ovate, dentate, petioled, the lobes or leaves obtuse; *hds.* pendulous; *invol.* of 8 scales, 9—12-flowered.—A conspicuous and not inelegant plant, in moist woods and shades, N. Eng. to Iowa, and Can. to Car. Stem stout, 2—4f high, purplish, often deeply so in spots. The leaves are very variable, the lowest 3—5-lobed or only hastate, the uppermost lan eolate, and between these the intermediate forms, hastate and ovate, all irregularly toothed. Scales purplish. Corollas whitish. Pappus brown. Some of the varieties have the reputation of curing the rattlesnake's bite. Aug.

β. *serpentaria.* (Prenanthes serpentaria. *P.*) *Radical lvs.* palmate-sinuate, those of the stem on long petioles, with the middle segment 3-parted; *upper lvs.* lanceolate.

2. N. ALTISSIMUS. Hook. (Harpalyce. *Don.* Prenanthes. *Linn.*) *Tall Nabalus.—St.* smooth, slender, straight, paniculate above; *lvs.* more or less deeply 3—5-cleft, all petiolate, angular, denticulate and rough-edged, the lobes acuminate; *hds.* pendulous; *invol.* of 5 scales and about 5-flowered.—A tall species, with cylindric, yellowish, nodding flowers, found in woods, Newfoundland to N. Eng. and Ky. Stem 3—5f high, bearing a narrow and elongated panicle. Heads in short, axillary and terminal racemes. Aug.

β. *ovatus.* Riddell. *Cauline lrs.* nearly all ovate, on slender petioles.

γ. *cordatus.* (Prenanthes cordata. *Willd.*) *L's.* cordate, on slender petioles.

δ. *deltoidea.* (Prenanthes deltoidea. *Ell.*) *Lrs.* deltoid, acuminate, acutely denticulate.

ε. *dissectus.* T. & G. *Lrs.* mostly 3-parted or divided, segments entire or deeply cleft into 2 or 3 narrow lobes.

3. N. FRASĚRI. DC. (P. rubicaulis. *Ph.*) *Fraser's Nabalus. St.* smooth, corymbosely paniculate above; *lrs.* subscabrous, mostly deltoid, often pinnately lobed, on winged petioles, the upper ones lanceolate, subsessile; *invol.* of about 8 scales, 8—12-flowered; *pappus* straw-colored.—♃ In dry, hard soils, Conn. and Mid. States (rare) to Flor. Stem 2—4f high, with paniculate branches. The leaves are as variable as in our other species, sometimes all being lanceolate, with only irregular indentures instead of lobes. Heads drooping, with purplish scales and cream-colored corollas. It is most effectually distinguished from N. albus by the more lively color of the pappus. Aug.

4. N. NANUS. DC. (P. alba. β. nana. *Bw.*) *St.* simple, low, smooth; *lrs.* on slender petioles, the lowest variously lobed or parted, the others successively deltoid-hastate, ovate and lanceolate *hds.* in small, axillary and terminal clusters, forming a slender, racemose panicle; *invol.* greenish-purple, of about 8 scales and 10—12 flowers; *pappus* dingy white.—This form of Nabalus is common on the White Mts., N. H.! where we find it with the same sportive character of foliage as appears in other species. Stem 5—10' high. Heads with whitish flowers. Aug.

5. N. BOOTTII. DC. *Boott's Nabalus. St.* simple, dwarf; *lower lrs.* subcordate or hastate-cordate, obtuse, the *middle ones* oblong, *the upper* lanceolate, mostly entire; *hds.* slightly nodding; *invol.* 10—18-flowered, of 10—15 obtuse, proper scales calyculate at base with lax, linear scales half their length; *pappus* straw-color.—White Mts., N. H., Whiteface Mt., Essex Co. N. Y. *Macrae.* N. Am. Fl., II., 482. Stem 5—8' high, bearing the heads in a subsimple raceme. Flowers whitish and odorous.

6. N. VIRGÀTUS. DC. (Prenanthes. *Mx.* Sonchus. *Desf.* Harpalyce. *Beck.*) Glabrous and glaucous, slender and simple; *lower lrs.* sinuate-pinnatifid, petiolate, *middle ones* toothed sessile, *upper* entire, partly clasping, gradually

inate, subulate bracts; *hds.* clustered, in a long, compound,
t secund raceme; *invol.* with about 8 scales and 10 flowers;
ored.—A remarkably slender, wand-like species, in sandy
r. Stem 2—4f high, racemose half its length. Leaves gradu-
simplified from the base upward, as in most of the species.

ads nodding or erect, hairy. Leaves undivided.

MOSUS. Hook. (Prenanthes. *Mx.* Harpalyce. *Don.*)
imple, slender; *lvs.* all undivided, lower oval-lanceolate, **sharp**-
etiolate, upper ovate-lanceolate, subclasping, entire; *hds.* in
, arranged in a long, interruptedly spicate panicle; *invol.* of
9—12 flowers; *pappus* straw-color.—N. J., N. W. States and
f high. Flowers pale red-purple.
vs. deeply and irregularly pinnatifid.

DINEUS. DC. (Prenanthes crepidinea. *Michx.*)
rous; *st.* tall, stout, corymbosely paniculate; *lvs.* large, irreg-
tioles winged, lower ones oblong-ovate, somewhat hastate o
ong-lanceolate; *hds.* nodding, in small pedunculate and pani
ol. hairy, of 11—14 scales, with 25—35 flowers; *pappus* tawny
kets, Western States! One of the largest species. Stem 5—8
–12′ by 2¼—7′, obtuse or acute. Heads large but not nume
scales and yellowish flowers. Aug.—Oct.

. T. & G. (N. Illinoensis. *DC.* Prenanthes asper. *Michx*
lli. *Poir.*)—*St.* strict, simple, scabrous; *lvs.* simple, scabrous-
e, lower ones oblong-oval, on margined petioles, upper lance-
-linear, subentire, sessile; *hds.* erect, in small fascicles, in a
l, compound raceme; *invol.* strongly hirsute, of 7—10 scales
lowers; *pappus* straw-color.—Dry prairies and barrens, West-
inner! common! Stem 2—4f high, nearly smooth. Leaves
ent or glabrous. Raceme 1—2f long. Fls. ochroleucous. Sept

perfected seeds and seed-down, the airy, globular form of which is very conspicuous among the tall grass. The leaves in spring furnish an excellent potherb. April—Nov. §

83. SONCHUS.

Gr. σογφος, hollow or soft; in allusion to the tender, feeble stem.

Involucre imbricate, of numerous unequal scales, at length tumid at the base; receptacle naked; pappus of simple, copious, white-silky hairs in many series; achenia not rostrate.—*Lvs. mostly spinulose. Heads with many yellow flowers.*

1. S. OLERACEUS. *Common Sow Thistle.*—Lrs. sagittate-amplexicaul, runcinate, subspinulose, dentate; *ped.* downy; *invol.* at length smooth.—① A sordid looking plant, native of Europe, naturalized in waste grounds, among rubbish, &c. The whole plant has a glaucous hue. Stem angular, hollow, fragile, 2—3f in height. Leaves apparently clasping, with large, retreating lobes at base, wavy and serrated in a runcinate manner, the teeth ending in weak spines. Involucres dilated at base, with yellow corollas. Sept. §

2. S. ASPER. Vill. (S. spinulosus. *Bw.*) *Rough Sow Thistle.*
St. glandular-hispid above; *lrs.* cordate-amplexical, oblong-lanceolate, undulate, spinulose, dentate; *ped.* subumbellate.—Found in similar situations with the former, but less common, U. S. Stem 1—2f high, smooth except at the summits of the branches, where it is covered with stiff hairs, each supporting a little gland at top. Leaves with numerous short, spiny teeth, wavy or slightly runcinate, the upper ones clasping so as to appear perfoliate. Scales with few, scattered hairs. Aug. Sept.

3. S. ARVENSIS. *Corn Sow Thistle.*
Rt. creeping, perennial; *t.* glabrous, erect; *lrs.* runcinate-pinnatifid, spinulose-dentate, cordate-clasping at base, with short and obtuse auricles; panicle umbellate-corymbose; *ped.* and *invol.* hispid; *ach.* somewhat 4-angled, the ribs transversely rugulose.—2 Waste grounds, naturalized, Eastern Mass. and Southern N. Y., rare. Stem angular, about 2f high. Heads large, with deep yellow flowers. §

ORDER LXXVI. LOBELIACEÆ.—LOBELIADS.

Herbs or shrubs with a milky juice. *Lvs.* alternate, without stipules.
Fls. axillary and terminal.
Cal. superior, the limb 5 lobed or entire.
Cor.—Limb irregular, 5-lobed, the tube inserted into the calyx.
Sta. 5, inserted with the corolla and alternate with its lobes.
Anth. coherent into a tube. *Pollen* oval.
Ova. adherent to the calyx tube. *Style* simple. *Stig.* surrounded with a fringe.
Fr. a capsule, 2 or 3-(rarely 1-)celled, many-seeded.

Most abundant in countries near the tropics, as W. Indies, Brazil and the Sandwich Islands, but they are found also throughout the temperate zones.

Properties.—All the species are poisonous, being pervaded by an acrid, narcotic juice. The common *Indian tobacco* (Lobelia inflata) is an exceedingly active medicine, emetic, sudorific and expectorant. It should be used, however, with great caution, since "less than a teaspoonful of the seeds or the powdered leaves would destroy life in a few hours." *Dr. Gray.* The other species produce similar effects, but in a less degree.

Genera.

Corolla tube cleft on the upper side to near the base, limb subbilabiate. . . . . . *Lobelia.* 1
Corolla tube short, entire, limb bilabiate. . . . . *Clintonia.* 2

1. LOBELIA.

In honor of Matthias de Lobel, physician and botanist to James I. Died 1616.

Corolla tubular, irregular, cleft nearly to the base on the upper side; stamens with the anthers united above into a curved tube; stigma 2-lobed; capsule opening at the summit; seeds minute.—*Herbaceous plants, with the fls. axillary and solitary, or in terminal, bracted racemes*

1. L. CARDINÀLIS. *Cardinal Flower.*
St. erect, simple, *lvs.* ovate-lanceolate, finely serrate, acute or §

rminal, bracted, secund raceme; *sta.* longer than the corolla.
f superior beauty, frequent in meadows and along streams,
o Ill.! Stem 2—4f high, often quite glabrous as well as the
ves 2—4' by 8—15", usually denticulate. Flowers on short
unerous, in a superb, nodding raceme. Bracts linear-subu-
than the flowers. Corolla deep scarlet, near 2' in length.

glabrous; *lvs.* entire.—Potsdam, N. Y.!
he segments rather narrower.—Mass.!

A. *Indian Tobacco.*
anched, erect; *lvs.* ovate-lanceolate, sessile, serrate, pilose;
In fields and woods, Can. and U. S. Root fibrous. Stem
angular, simple, becoming branched in proportion to the
growth, 10—15' high. Leaves elliptical, sessile, hairy and
n leafy spikes, axillary, peduncled. Corolla small, pale blue,
urgid capsule in the calyx. July—Sept.—This plant is ren-
he Thomsonian physicians, in whose practice it appears to
ately used. Its specific action, as above stated, is that of a
n small doses it is powerfully expectorant. To its salivating
ly owing the driveling of horses in autumn.

ANNA. *Dortmann's or Water Lobelia.*
rged, linear, entire, fleshy, 2-celled, obtuse; *scape* simple,
in a terminal raceme, remote, pedicellate, nodding.—2 A
rowing in ponds, N. States to Ga., the flowers only rising
Stem erect, hollow, nearly leafless, long, bearing above the
of 3 or 4 remote, pedicellate flowers. Leaves mostly radical,
submerged, having 2 longitudinal grooves. Flowers droop-
uly.

LITICA. *Blue Cardinal Flower.*
aple; *lvs.* oblong-lanceolate, acute or acuminate, unequally
hirsute; *rac.* leafy; *cal.* hispidly ciliate, with the sinuses re-

Maine, *Miss Towle!* to Niagara! Stem 6—12' high, commonly simple. Leaves mostly linear, sessile, an inch long and 1—2'' wide, upper ones entire, lower with minute teeth. Flowers remote, alternate, on axillary pedicels which are but little shorter than the leaf-like bracts. Corolla pale-blue, the 3 lower segments obovate. Aug.

8. L. LEPTOSTACHYS. DC. *Slender-spiked Lobelia.*

Glabrous; *st.* erect, virgate, simple; *lvs.* oblong-lanceolate, minutely denticulate, rather acute, sessile; *fls.* subsessile, small, in a long, slender spike; *cal. segments* lanceolate-acuminate, longer than the tube of the corolla; *bracts* lance-linear, denticulate, much longer than the pedicels.—Prairies, Western States! common. Stem 1—2f high. Leaves 1—2' by by 4—8''. Raceme 6—12 in length, the bracts and sepals rather conspicuous. Flowers light blue. Much resembles L. spicata. July.

9. L. NUTTALLII. DC. (L. gracilis. *Nutt.*) *Nuttall's Lobelia.*

Glabrous; *st.* erect, very slender, almost filiform, subsimple; *lvs.* few and remote, subentire, radical linear-spatulate, cauline linear, rather acute; *fls.* few, remote; *pedicels* twice longer than the corolla or the 2 subulate bracts at base; *cal. segments* shorter than the tube of the corolla.—An exceedingly slender plant, around sandy swamps, N. J.! 1—2f high., often branched. Leaves 6—12'' by 1—1¼''. Pedicels 3—10'' long, blue as well as the 2 brs. July, Aug.

2. CLINTONIA. Douglass.

Calyx 5-sepaled, subequal; corolla bilabiate, lower lip cuneate, 3-lobed, upper erect, 2-parted; stamens incurved, united into a tube; capsule silique-form, dry, chartaceous, 1-celled, many-seeded, dehiscent by 3 strap-shaped valves.—① *Procumbent herbs with small leaves and axillary, solitary flowers.*

C. ELÉGANS. Doug.—Glabrous, sparingly branched; *st.* slender, angular; *lvs.* sessile, ovate, 3-veined; *ova.* sessile, long-acuminate, triangular, contorted, much longer than the leaves; *cor.* blue, with a white spot in the middle of the lower lip.—Native of the Rocky Mts., &c. A beautiful annual, with flowers of the most brilliant blue. †

ORDER LXXVII. CAMPANULACEÆ.—BELLWORTS.

Herbs with a milky juice, alternate leaves and without stipules.
Fls. mostly blue, showy. *Cal.* superior, generally 5-cleft, persistent.
Cor. regular, campanulate, generally 5-cleft, withering, valvate in æstivation.
Sta. inserted with the corolla upon the calyx, equal in number to, and alternate with, its lobes.
Anth. Distinct, 2-celled. *Pollen* spherical.
Ova. adherent to the calyx, 2 or more-celled. *Style* covered with collecting hairs.
Fr.—Capsule crowned with the remains of the calyx, loculicidal. *Seeds* many.

Genera 28, species 500, chiefly abounding in the northern temperate zone and in South Africa. Of its 289 species, according to Alphonse DeCandolle, only 19 inhabit the torrid zone. The campanulaceæ are interesting chiefly for their beauty, being destitute of any important known properties.

Genera.

Calyx tube short. Corolla campanulate or subrotate. . . . . *Campanula.* 1
Calyx tube long, prismatic. Corolla rotate. . . . . *Specularia.* 2

1. CAMPANULA.

Lat. *campanula,* a little bell; from the form of the flowers.

Calyx mostly 5-cleft; corolla campanulate. or subrotate, 5-lobed, closed at base by the broad valve-like bases of the 5 stamens; stigma 3—5-cleft; capsule 3—5-celled, opening by lateral pores.—*Mostly* 4 *Fls. generally in racemes, sometimes spicate, or few and axillary*

1. C. ROTUNDIFOLIA. *Rock Bell-flower. Hair Bell.*

St. weak, slender; *radical lvs.* ovate- or reniform-cordate; *cauline ones* linear, entire; *fls.* few, nodding.—An exceedingly delicate plant, with blue, bell-shaped flowers. On damp rocks and rocky streams, N. States and Brit. Am. Stem a foot or more high, smooth. The root leaves generally decay on the opening of the flowers, so that a specimen with these (7—10'' by 4—7'') is rare

minal, bracted, secund raceme; *sta*. longer than the corolla.
superior beauty, frequent in meadows and along streams,
Ill.! Stem 2—4f high, often quite glabrous as well as the
ves 2—4' by 8—15", usually denticulate. Flowers on short
merous, in a superb, nodding raceme. Bracts linear-subu-
than the flowers. Corolla deep scarlet, near 2' in length.

glabrous; *lvs.* entire.—Potsdam, N. Y.!
e segments rather narrower.—Mass.!

. *Indian Tobacco.*
nched, erect; *lvs.* ovate-lanceolate, sessile, serrate, pilose;
In fields and woods, Can. and U. S. Root fibrous. Stem
angular, simple, becoming branched in proportion to the
growth, 10—15' high. Leaves elliptical, sessile, hairy and
leafy spikes, axillary, peduncled. Corolla small, pale blue,
urgid capsule in the calyx. July—Sept.—This plant is ren-
he Thomsonian physicians, in whose practice it appears to
ately used. Its specific action, as above stated, is that of a
small doses it is powerfully expectorant. To its salivating
ly owing the driveling of horses in autumn.

ANNA. *Dortmann's or Water Lobelia.*
ged, linear, entire, fleshy, 2-celled, obtuse; *scape* simple,
in a terminal raceme, remote, pedicellate, nodding.—♃ A
rowing in ponds, N. States to Ga., the flowers only rising
Stem erect, hollow, nearly leafless, long, bearing above the
of 3 or 4 remote, pedicellate flowers. Leaves mostly radical,
submerged, having 2 longitudinal grooves. Flowers droop-
ly.

LITICA. *Blue Cardinal Flower.*
ple; *lvs.* oblong-lanceolate, acute or acuminate, unequally

Maine, *Miss Thirle!* to Niagara! Stem 6—12' high, commonly simple. Leaves mostly linear, sessile, an inch long and 1—2" wide, upper ones entire, lower with minute teeth. Flowers remote, alternate, on axillary pedicels which are but little shorter than the leaf-like bracts. Corolla pale-blue, the 3 lower segments obovate. Aug.

8. L. LEPTOSTACHYS. DC. *Slender-spiked Lobelia.*

Glabrous; *st.* erect, virgate, simple; *lvs.* oblong-lanceolate, minutely denticulate, rather acute, sessile; *fls.* subsessile, small, in a long, slender spike; *cal. segments* lanceolate-acuminate, longer than the tube of the corolla; *bracts* lance-linear, denticulate, much longer than the pedicels.—Prairies, Western States! common. Stem 1—2f high. Leaves 1—2' by by 4—8". Raceme 6—12 in length, the bracts and sepals rather conspicuous. Flowers light blue. Much resembles L. spicata. July.

9. L. NUTTALLII. DC. (L. gracilis. *Nutt.*) *Nuttall's Lobelia.*

Glabrous; *st.* erect, very slender, almost filiform, subsimple; *lvs.* few and remote, subentire, radical linear-spatulate, cauline linear, rather acute; *fls.* few, remote; *pedicels* twice longer than the corolla or the 2 subulate bracts at base; *cal. segments* shorter than the tube of the corolla.—An exceedingly slender plant, around sandy swamps, N. J.! 1—2f high., often branched. Leaves 6—12" by 1—1½". Pedicels 3—10" long, blue as well as the 2. ↄ ↄs. July, Aug.

2. CLINTONIA. Douglass.

Calyx 5-sepaled, subequal; corolla bilabiate, lower lip cuneate, 3-lobed, upper erect, 2-parted; stamens incurved, united into a tube; capsule silique-form, dry, chartaceous, 1-celled, many-seeded, dehiscent by 3 strap-shaped valves.—① *Procumbent herbs with small leaves and axillary, solitary flowers.*

C. ELEGANS. Doug.—Glabrous, sparingly branched; *st.* slender, angular; *lvs.* sessile, ovate, 3-veined; *ora.* sessile, long-acuminate, triangular, contorted, much longer than the leaves; *cor.* blue, with a white spot in the middle of the lower lip.—Native of the Rocky Mts., &c. A beautiful annual, with flowers of the most brilliant blue. †

ORDER LXXVII. CAMPANULACEÆ.—BELLWORTS.

Herbs with a milky juice, alternate leaves and without stipules.
Fls. mostly blue, showy. *Cal.* superior, generally 5-cleft, persistent.
Cor. regular, campanulate, generally 5-cleft, withering, valvate in æstivation.
Sta. inserted with the corolla upon the calyx, equal in number to, and alternate with, its lobes.
Anth. Distinct, 2-celled. *Pollen* spherical.
Ova. adherent to the calyx, 2 or more-celled. *Style* covered with collecting hairs.
Fr.—Capsule crowned with the remains of the calyx, loculicidal. *Seeds* many.

Genera 28, species 500, chiefly abounding in the northern temperate zone and in South Africa. Of its 300 species, according to Alphonse DeCandolle, only 19 inhabit the torrid zone. The campanulaceæ are interesting chiefly for their beauty, being destitute of any important known properties.

Genera.

Calyx tube short. Corolla campanulate or subrotate. . . . . *Campanula.* 1
Calyx tube long, prismatic. Corolla rotate. . . . . *Specularia.* 2

1. CAMPANULA.

Lat. *campanula,* a little bell; from the form of the flowers.

Calyx mostly 5-cleft; corolla campanulate, or subrotate, 5-lobed, closed at base by the broad valve-like bases of the 5 stamens; stigma 3—5-cleft; capsule 3—5-celled, opening by lateral pores.—*Mostly* ♃ *Fls. generally in racemes, sometimes spicate, or few and axillary*

1. C. ROTUNDIFOLIA. *Rock Bell-flower. Hair Bell.*

St. weak, slender; *radical lvs.* ovate- or reniform-cordate; *caul.* linear, entire; *fls.* few, nodding.—An exceedingly delicate plant, with ↄ shaped flowers. On damp rocks and rocky streams, N. States and Stem a foot or more high, smooth. The root leaves generally opening of the flowers, so that a specimen with these (7—10" by ↄ

reduced to the minate, subulate bracts; *hds.* clustered, in a long, compound, virgate, somewhat secund raceme; *invol.* with about 8 scales and 10 flowers; *pappus* straw-colored.—A remarkably slender, wand-like species, in sandy soils, N. J. to Flor. Stem 2—4f high, racemose half its length. Leaves gradually reduced and simplified from the base upward, as in most of the species. Sept. Oct.

§ § *Heads nodding or erect, hairy.   Leaves undivided.*

7. N. RACEMÒSUS. Hook. (Prenanthes. *Mx.* Harpalyce. *Don.*)

Glabrous, simple; slender; *lvs.* all undivided, lower oval-lanceolate, sharp-y denticulate, petiolate, upper ovate-lanceolate, subclasping, entire; *hds.* in nodding fascicles, arranged in a long, interruptedly spicate panicle; *invol.* of 8—9 scales, with 9—12 flowers; *pappus* straw-color.—N. J., N. W. States and Can. Stem 2—4f high. Flowers pale red-purple.

6. T. & G. *Lvs.* deeply and irregularly pinnatifid.

8. N. CREPIDINEUS. DC. (Prenanthes crepidinea. *Michx.*)

Nearly glabrous; *st.* tall, stout, corymbosely paniculate; *lvs.* large, irregularly toothed, *petioles* winged, lower ones oblong-ovate, somewhat hastate o-deltoid, upper oblong-lanceolate; *hds.* nodding, in small pedunculate and panicled clusters; *invol.* hairy, of 11—14 scales, with 25—35 flowers; *pappus* tawny—Fields and thickets, Western States! One of the largest species. Stem 5—8f high. Leaves 4—12' by 2½—7', obtuse or acute. Heads large but not numerous, with brown scales and yellowish flowers. Aug.—Oct.

9. N. ASPER. T. & G. (N. Illinoensis. *DC.* Prenanthes asper. *Michx.* Choudrilla Ill. *Poir.*)—*St.* strict, simple, scabrous; *lrs.* simple, scabrous, pubescent, dentate, lower ones oblong-oval, on margined petioles, upper lance-oblong and lance-linear, subentire, sessile; *hds.* erect, in small fascicles, in a slender, elongated, compound raceme; *invol.* strongly hirsute, of 7—10 scales and with 11—14 flowers; *pappus* straw-color.—Dry prairies and barrens, Western States, *Dr. Skinner!* common! Stem 2—4f high, nearly smooth. Leaves 3—5' long, pubescent or glabrous. Raceme 1—2f long. Fls. ochroleucous. Sept.

81. TROXÌMON. Nutt.

Gr. τροξιμος, eatable; applied to this genus with little propriety.

Heads many-flowered; involucre campanulate, scales loosely imbricate, lance-ovate, membranaceous, in 2—3 rows; achenia oblong-linear, compressed, glabrous, not rostrate; pappus setaceous, copious, white.—♃ *Lvs. all radical.   Scape bearing a single, large, showy head with yellow flowers.*

T. CUSPIDÀTUM. Ph. (T. marginatum. *Nutt.*)

Rt. fusiform; *lvs.* linear-lanceolate, acuminate, margins tomentose, often undulate; *scales* acuminate-cuspidate, erect, smooth, in 2 series, the outer nearly equal to the inner.—Prairies, Wis., *Lapham*, Ill., *Mead*, W. to the Rocky Mts., *Nuttall.* Apr.—Jn.

82. TARAXÀCUM. Desf.

Gr. ταραχτικος, cathartic: on account of its once celebrated medicinal properties.

Involucre double, the outer of small scales much shorter than the inner, appressed row; receptacle naked; achenia produced into a long beak crowned with the copious, white, capillary pappus.—*Acaulescent herbs, with runcinate leaves.*

T. DENS-LEÒNIS. Less. (Leontodon Taraxacum. *Linn.*) *Dandelion.*

Outer scales of the involucre reflexed; *lrs.* runcinate, smooth, dentate.— ♃ In all open situations, blossoming at all seasons except winter. Leaves all radical, and examples of that peculiar form termed runcinate, that is, re-uncinate, the teeth or claws inclining backwards towards the base of the leaf rather than the summit. Scape hollow, round, bearing a single yellow head. After the flower is closed and decayed, the scape rises higher and bears a head of perfected seeds and seed-down, the airy, globular form of which is very conspicuous among the tall grass. The leaves in spring furnish an excellent pot-herb. April—Nov. ⚲

83. SONCHUS.

Gr. σομφος, hollow or soft; in allusion to the tender, feeble stem.

Involucre imbricate, of numerous unequal scales, at length tumid at the base; receptacle naked; pappus of simple, copious, white-silky hairs in many series; achenia not rostrate.—*Lvs. mostly spinulose. Heads with many yellow flowers.*

1. S. OLERACEUS. *Common Sow Thistle.*—Lvs. sagittate-amplexicaul, runcinate, subspinulose, dentate; *ped.* downy; *invol.* at length smooth.—① A sordid looking plant, native of Europe, naturalized in waste grounds, among rubbish, &c. The whole plant has a glaucous hue. Stem angular, hollow, fragile, 2—3f in height. Leaves apparently clasping, with large, retreating lobes at base, wavy and serrated in a runcinate manner, the teeth ending in weak spines. Involucres dilated at base, with yellow corollas. Sept. ⚲

2. S. ASPER. Vill. (S. spinulosus. *Bw.*) *Rough Sow Thistle.*
St. glandular-hispid above; *lvs.* cordate-amplexical, oblong-lanceolate, undulate, spinulose, dentate; *ped.* subumbellate.—Found in similar situations with the former, but less common, U. S. Stem 1—2f high, smooth except at the summits of the branches, where it is covered with stiff hairs, each supporting a little gland at top. Leaves with numerous short, spiny teeth, wavy or slightly runcinate, the upper ones clasping so as to appear perfoliate. Scales with few, scattered hairs. Aug. Sept.

3. S. ARVENSIS. *Corn Sow Thistle.*
Rt. creeping, perennial; *st.* glabrous, erect; *lvs.* runcinate-pinnatifid, spinulose-dentate, cordate-clasping at base, with short and obtuse auricles; panicle umbellate-corymbose; *ped.* and *invol.* hispid; *ach.* somewhat 4-angled, the ribs transversely rugulose.—♃ Waste grounds, naturalized, Eastern Mass. and Southern N. Y., rare. Stem angular, about 2f high. Heads large, with deep yellow flowers. ⚲

ORDER LXXVI. LOBELIACEÆ.—LOBELIADS.

Herbs or shrubs with a milky juice. *Lvs.* alternate, without stipules.
Fls. axillary and terminal.
Cal. superior, the limb 5 lobed or entire.
Cor.—Limb irregular, 5 lobed the tube inserted into the calyx.
Sta. 5, inserted with the corolla and alternate with its lobes.
Anth. coherent into a tube　*Pollen* oval.
Ova. adherent to the calyx tube　*Style* simple.　*Stig.* surrounded with a fringe.
Fr. a capsule, 2 or 3-(rarely 1-)celled, many-seeded.

Most abundant in countries near the tropics, as W. Indies, Brazil and the Sandwich Islands, but they are found also throughout the temperate zones.

Properties.—All the species are poisonous, being pervaded by an acrid, narcotic juice. The common *Indian tobacco* (Lobelia inflata) is an exceedingly active medicine, emetic, sudorific and expectorant. It should be used, however, with great caution, since "less than a teaspoonful of the seeds or the powdered leaves would destroy life in a few hours." *Dr. Gray.* The other species produce similar effects, but in a less degree.

Genera.

Corolla tube cleft on the upper side to near the base, limb subbilabiate. . . . . . *Lobelia.* 1
Corolla tube short, entire, limb bilabiate. . . . . *Clintonia.* 2

1. LOBELIA.

In honor of Matthias de Lobel, physician and botanist to James I. Died 1616.

Corolla tubular, irregular, cleft nearly to the base on the upper side; stamens with the anthers united above into a curved tube; stigma 2-lobed; capsule opening at the summit; seeds minute.—*Herbaceous plants, with the fls. axillary and solitary, or in terminal, bracted racemes*

1. L. CARDINALIS. *Cardinal Flower.*
St. erect, simple, *lvs.* ovate-lanceolate, finely serrate, acute or acuminate,

31*

sessile; *fls.* in a terminal, bracted, secund raceme; *sta.* longer than the corolla. —A tall species of superior beauty, frequent in meadows and along streams, Can. to Car., W. to Ill.! Stem 2—4f high, often quite glabrous as well as the whole plant. Leaves 2—4′ by 8—15″, usually denticulate. Flowers on short pedicels, few or numerous, in a superb, nodding raceme. Bracts linear-subulate, much shorter than the flowers. Corolla deep scarlet, near 2′ in length. Jl. Aug. †

    β. Whole plant glabrous; *lvs.* entire.—Potsdam, N. Y.!
    γ. Cor. white, the segments rather narrower.—Mass.!

    2. L. INFLĀTA. *Indian Tobacco.*
    St. hairy, branched, erect; *lvs.* ovate-lanceolate, sessile, serrate, pilose; *caps.* inflated.—⚇ In fields and woods, Can. and U. S. Root fibrous. Stem erect, very rough, angular, simple, becoming branched in proportion to the luxuriance of its growth, 10—15′ high. Leaves elliptical, sessile, hairy and veiny. Flowers in leafy spikes, axillary, peduncled. Corolla small, pale blue, leaving an oval, turgid capsule in the calyx. July—Sept.—This plant is rendered famous by the Thomsonian physicians, in whose practice it appears to be too indiscriminately used. Its specific action, as above stated, is that of a violent emetic. In small doses it is powerfully expectorant. To its salivating property is probably owing the driveling of horses in autumn.

    3. L. DORTMANNA. *Dortmann's or Water Lobelia.*
    Lvs. submerged, linear, entire, fleshy, 2-celled, obtuse; *scape* simple, nearly naked; *fls.* in a terminal raceme, remote, pedicellate, nodding.—⚇ A curious aquatic, growing in ponds, N. States to Ga., the flowers only rising above the water. Stem erect, hollow, nearly leafless, long, bearing above the surface a raceme of 3 or 4 remote, pedicellate flowers. Leaves mostly radical, spreading, obtuse, submerged, having 2 longitudinal grooves. Flowers drooping, pale blue. July.

    4. L. SYLPHILITĪCA. *Blue Cardinal Flower.*
    St. erect, simple; *lvs.* oblong-lanceolate, acute or acuminate, unequally serrate, somewhat hirsute; *rac.* leafy; *cal.* hispidly ciliate, with the sinuses reflexed.—♃ A fine, showy plant, but inferior in beauty to L. cardinalis, growing in wet meadows and along streams, U. States and Can., more common in the Western States. Stem erect, 2—4f high, simple, angular, with short hairs. Leaves lanceolate, broader at base, acute at each end, somewhat erosely dentate, pilose. Flowers large, on short peduncles, each solitary in the axil of an ovate-lanceolate bract. Corolla bright blue or purplish. Capsule half superior. July.

    5. L. PUBERŪLA. Michx. *Downy Lobelia.*
    Pubescent; *st.* erect, simple; *lvs.* ovate-oblong or elliptical, obtuse, sessile, repand-denticulate; *rac.* spicate, secund; *cal.* ciliate, the segments longer than the tube of the corolla.—♃ Native of mountains, &c., N. Y. to Ga. Stem 12— 30′ high, scarcely furrowed. Leaves covered with a short, downy or silky pubescence, 1—2′ inches in length and half as wide, the lower ones broadest towards the end. Flowers large, on very short pedicels, each solitary in the axil of an ovate-lanceolate bract, forming a somewhat one-sided raceme, leafy below. Calyx hairy at base. Corolla of a bright purplish-blue. July.

    6. L. SPICĀTA. Lam. (L. Claytoniana. *Mx.* L. pallida. *Muhl.*) *Clayton's Lobelia.*—Puberulent; *st.* erect, simple; *lvs.* oblong, sessile, mostly obtuse, obscurely denticulate, radical ones spatulate; *fls.* (small) in a long slender raceme; *pedicels* as long as the flowers or entire, subulate bracts; *scp.* subulate, as long as the tube of the corolla.—♃ Fields and prairies, Can. and U. S. Stem 1½—2f high, somewhat grooved, few-leaved, ending in a long, wand-like raceme. Flowers numerous, crowded, each axillary to a short, inconspicuous bract. Corolla pale blue, the palate bidentate. Aug.

    7. L. KALMIA. *Kalm's Lobelia.*
    Smooth; *st.* simple, slender, erect; *radical lvs.* spatulate, *stem lvs.* linear-lanceolate, obtuse, remotely denticulate; *rac.* lax, few-flowered, leafy; *fls.* pedicelled.—A small and delicate species, inhabiting the rocky banks of streams

Maine, *Miss Towle!* to Niagara! Stem 6—12' high, commonly simple. Leaves mostly linear, sessile, an inch long and 1—2'' wide, upper ones entire, lower with minute teeth. Flowers remote, alternate, on axillary pedicels which are but little shorter than the leaf-like bracts. Corolla pale-blue, the 3 lower segments obovate. Aug.

8. L. LEPTOSTACHYS. DC. *Slender-spiked Lobelia.*

Glabrous; *st.* erect, virgate, simple; *lvs.* oblong-lanceolate, minutely denticulate, rather acute, sessile; *fls.* subsessile, small, in a long, slender spike; *cal. segments* lanceolate-acuminate, longer than the tube of the corolla; *bracts* lance-linear, denticulate, much longer than the pedicels.—Prairies, Western States! common. Stem 1—2f high. Leaves 1—2' by by 4—8''. Raceme 6—12 in length, the bracts and sepals rather conspicuous. Flowers light blue. Much resembles L. spicata. July.

9. L. NUTTALLII. DC. (L. gracilis. *Nutt.*) *Nuttall's Lobelia.*

Glabrous; *st.* erect, very slender, almost filiform, subsimple; *lvs.* few and remote, subentire, radical linear-spatulate, cauline linear, rather acute; *fls.* few, remote; *pedicels* twice longer than the corolla or the 2 subulate bracts at base; *cal. segments* shorter than the tube of the corolla.—An exceedingly slender plant, around sandy swamps, N. J.! 1—2f high., often branched. Leaves 6—12'' by 1—1½''. Pedicels 3—10'' long, blue as well as the 2 - i lvs. July, Aug.

2. CLINTONIA. Douglass.

Calyx 5-sepaled, subequal; corolla bilabiate, lower lip cuneate, 3-lobed, upper erect, 2-parted; stamens incurved, united into a tube; capsule silique-form, dry, chartaceous, 1-celled, many-seeded, dehiscent by 3 strap-shaped valves.—① *Procumbent herbs with small leaves and axillary, solitary flowers.*

C. ELEGANS. Doug.—Glabrous, sparingly branched; *st.* slender, angular; *lvs.* sessile, ovate, 3-veined; *ova.* sessile, long-acuminate, triangular, contorted, much longer than the leaves; *cor.* blue, with a white spot in the middle of the lower lip.—Native of the Rocky Mts., &c. A beautiful annual, with flowers of the most brilliant blue. †

ORDER LXXVII. CAMPANULACEÆ.—BELLWORTS.

Herbs with a milky juice, alternate leaves and without stipules.
Fls. mostly blue, showy. *Cal.* superior, generally 5-cleft, persistent.
Cor. regular, campanulate, generally 5-cleft, withering, valvate in æstivation.
Sta. inserted with the corolla upon the calyx, equal in number to, and alternate with, its lobes.
Anth. Distinct. 2-celled. *Pollen* spherical.
Ova. adherent to the calyx. 2 or more-celled. *Style* covered with collecting hairs.
Fr.—Capsule crowned with the remains of the calyx, loculicidal. *Seeds* many.

Genera 28, species 500, chiefly abounding in the northern temperate zone and in South Africa. Of its 300 species, according to Alphonse DeCandolle, only 19 inhabit the torrid zone. The campanulaceæ are interesting chiefly for their beauty, being destitute of any important known properties.

Genera.

Calyx tube short. Corolla campanulate or subrotate. . . . . *Campanula.* 1
Calyx tube long, prismatic. Corolla rotate. . . . . *Specularia.* 2

1. CAMPANULA.

Lat. *campanula*, a little bell; from the form of the flowers.

Calyx mostly 5-cleft: corolla campanulate, or subrotate, 5-lobed, closed at base by the broad valve-like bases of the 5 stamens; stigma 3—5-cleft; capsule 3—5-celled, opening by lateral pores.—*Mostly* ♃ *Fls. generally in racemes, sometimes spicate, or few and axillary*

1. C. ROTUNDIFOLIA. *Rock Bell-flower. Hair Bell.*

St. weak, slender; *radical lvs.* ovate- or reniform-cordate; *cauline ones* linear, entire; *fls.* few, nodding.—An exceedingly delicate plant, with blue, bell-shaped flowers. On damp rocks and rocky streams, N. States and Brit. Am. Stem a foot or more high, smooth. The root leaves generally decay on the opening of the flowers, so that a specimen with these (7—10'' by 4—7'') is rather rare. Cauline leaves smooth, linear, 9' long and scarcely a line in width. Flowers terminal, in a loose panicle, drooping. Root creeping, perennial. Jn. Jl.

2. C. aparinöides. Ph. (C. erinoides. *Michx.*) *Prickly Bell-flower.*

St. flaccid, slender, branching above, triangular, the angles inversely aculeate; *lvs.* linear-lanceolate; *fls.* terminal.—A slender annual, found in wet meadows, Can. and Wisc.! to Ga. Stem 12—18' high, its 3 angles rough backwards, by means of which it supports itself upright among the grass. Leaves smooth on the upper surface, denticulate, the margin and veins rough backwards. Flowers small, white, on thread-like, flexuous peduncles at the top of the stem. June—Aug.

3. C. Americāna. (C. acuminata. *Michx.*) *American Bell-flower.*

St. erect; *lvs.* ovate-lanceolate, acuminate, uncinately serrate, the lower ones often cordate; *petioles* ciliate; *fls.* axillary, sessile; *sty.* exsert.—A tall, erect, ornamental species in fields, hills, &c., in Western N. Y.! and Penn. to Ill.! common. Also cultivated in gardens. Stem 2—3f high, nearly smooth. Leaves ending in a long point, smooth, with fine teeth. Flowers blue, flat, on short stalks or sessile, numerous, solitary or several in each upper axil, forming a terminal, leafy raceme. Corolla spreading. Aug. †

4. C. Illinoensis. Fresen. in DC.

St. angular, with spreading branches; *lvs.* ovate-lanceolate, long-acuminate, sharply serrate, reflexed, upper ones hairy; *fls.* sessile, 1—3 together in the upper axils; *cal. segments* subulate, serrate at base, spreading; *cor.* rotate; *caps.* prismatic-clavate.—Prairies of Illinois. Stem 3—5f high. Segments of the corolla hairy outside near apex. Capsule opening by 3 pores.

5. C. glomerāta. *Clustered Bell-flower.*—*St.* angular, simple, smooth; *lvs.* scabrous, oblong-lanceolate, cordate-sessile, lower petiolate; *fls.* glomerate, in a dense head; *cal. lobes* acuminate, half as long as the funnel-shaped corolla.—A European species, cultivated in gardens, naturalized at Danvers, Vt., *Oakes.* It is a handsome plant, about 2f high, with numerous bell-shaped flowers of an intense violet-blue, varying to pale purple. In cultivation it has many varieties. § †

6. C. medium.—*St.* simple, erect, hispid; *lvs.* lanceolate, obtusely serrate, sessile, 3-veined at base; *fls.* erect.—② An ornamental border flower, from Germany, and of the easiest culture. Root biennial. Stem several feet in height, undivided, rough with bristly hairs. Flowers very large, the base broad, limb reflexed, of a deep blue. Several varieties occur with double or single flowers, of blue, red, purple and white corollas. June—Sept. †

7. C. persicifolia. *Peach-leaved Bell-flower.*—*St.* angular, erect; *lvs.* rigid, obscurely crenate-serrate, radical oblong-obovate, cauline lance-linear; *fls.* large, broadly campanulate.—A beautiful species, native of Europe, with very large, blue (varying to white) flowers. †

8. C. planiflöra. DC. (C. nitida. *Ait.*)—Very glabrous; *st.* simple; *lvs.* sessile, coriaceous, shining, radical crowded, ovate or obovate, obtuse, crenulate, cauline linear-lanceolate, acute, subentire; *fl.* in a spicate raceme; *cal. lobes* ovate, acute, ½ as long as the campanulate-rotate corolla.—Native about Hudson's Bay, *Pursh.* A smooth species, with numerous blue flowers. †

9. C. lanuginosa, with ovate, crenate, rugose and somewhat woolly leaves is sometimes cultivated, and also a few other species.

2. SPECULARIA. Heist.

Lat. *speculum,* a mirror: alluding to the fig. or of S. speculum.

Calyx 5-lobed, tube elongated; corolla rotate, 5-lobed; stamens 5, distinct, half as long as the corolla, filaments hairy, shorter than the anthers: style included, hairy; stigmas 3: capsule prismatic, 3-celled, dehiscing in the upper part.—① *Fls. axillary and terminal, sessile, erect.*

1. S. perfoliāta. Lam. (Campanula amplexicaulis. *Mx.* and *of 1st edit.*)

St. simple, rarely branched, erect; *lvs.* cordate, crenate, amplexicaul; *fls.* sessile, aggregate, axillary.—Plant somewhat hairy, a foot high, found in fields and roadsides, Can., N. Eng.! to Ga. and Ill.! The strict, upright stem, is furnished with distant, short, alternate, heart-reniform, veiny, stem-clasping leaves, containg 1—4 crowded flowers in the concavity of their upper surface. Flowers axillary and terminal, the upper clusters larger. Corolla blue or purple, with spreading segments, calyx segments acute, lanceolate. June, July.

2. S. SPECÜLUM. *Venus' Looking-glass.*—*St.* diffuse, very branching; *lvs.* oblong-crenate; *fls.* solitary; *scales* at the base of the corolla sometimes wanting. —A pretty border flower named from the form of the blue corolla, which resembles a little, round, concave mirror (speculum). Aug.

ORDER LXXVIII. ERICACEÆ.—HEATHWORTS.

Plants shrubby or suffruticose, sometimes herbaceous.
Lvs. simple, alternate or opposite, mostly evergreen, entire or toothed, without stipules.
Inflorescence various. *Cal.* inferior or superior, 5- (seldom 4—6-) leaved or cleft, rarely entire.
Cor. regular or somewhat irregular, 4—5- (rarely 6-) cleft, the petals rarely almost distinct.
Sta. Generally distinct and inserted with the corolla [appendaged.
Anth. as many or twice as many as the lobes of the corolla, 2-celled, generally opening by pores, often *Embryo* straight, lying in the axis of, or in the end of fleshy albumen.

Genera 66, species 1066, diffused throughout all countries, but comparatively rare in the torrid zone. The true Ericaceæ (Heaths) are chiefly natives of the Cape of Good Hope, there being none in Asia, New Holland, and but one or two in America. The Tribe Vacciueæ are chiefly natives of N. America.

Properties.—The Ericaceæ are, in general, astringent and diuretic. Some of them yield a stimulating and aromatic resinous matter. The Bearberry, (Arctostaphylos Uva-ursi) is a well-known remedy in nephritic complaints. An infusion of the leaves is astringent, demulcent and diuretic. Similar properties are also possessed by the Pipsissiwa (Chimaphila umbellata). The species of Rhododendron and Kalmia are pervaded by a narcotic principle, rendering them (particularly their leaves) often actively poisonous. The honey collected from their flowers by bees appears to have been so to some of the soldiers in the retreat of the immortal ten thousand (Xenophon's Anabasis). The berries of the Vacciueæ (Whortleberries, Blueberries and Cranberries,) and of Gaultheria procumbens (spicy Wintergreen) are esculent and wholesome.

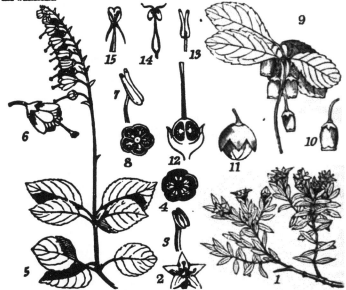

FIG. 68.—1. Azalea procumbens. 2. A flower enlarged. 3. A stamen much enlarged, showing the longitudinal dehiscence of each of the cells. 4. Cross section of a 5-celled capsule of Rhododendron, showing the inflexed margins of the valves. 5. Pyrola secunda. 6. A flower enlarged. 7. A stamen enlarged, showing the terminal tubes and pores. 8. Cross section of a 5-celled, many-seeded capsule. 9. Gaultheria procumbens. 10. A flower enlarged. 11. A berry. 12. Vertical section of the ovary, showing the five, fleshy calyx. 13. Anther of Vaccinium Vitis Idæa. 14. Stamen of Arctostaphylos Uva-ursi. 15. Awned stamen of a Vaccinium.

Conspectus of the Genera.

Cor. urceolate. Erect undershrubs,	*Vaccinium.*	1
segments reflexed.	*Oxycoccus.*	2
Ovary adherent to the calyx tube. Cor. deeply 4-cleft, segments spreading.	*Chiogenes.*	3
Berry the matured, fleshy calyx. Seeds ∞.	*Gaultheria.*	7
Drupe the matured ovary, 5-seeded.	*Arctostaphylos.*	4
opening betw. cells (septicidal. § 115, 1.2).	*Menziesia.*	6
Corolla ovoid.	Sta. 10.	*Andromeda.*	5
Capsule op'ng into the cells (loculicidal). Sta. 8.	*Erica.*	16
anthers free. Prostrate undershrub.	*Epigæa.*	9
Corolla salver-form, holding the anthers in 10 pits.	*Kalmia.*	10
Petals united.	Auth. opening by clefts.	*Azalea.*	11
Corolla funnel or bell-form. Anth. opening by pores.	*Rhododendron.*	13
Lvs. deciduous, serrate.	*Clethra.*	8
Caps. 3-celled. Lvs. evergreen, entire.	*Leiophyllum.*	14
Ovary free.	regular. Capsule 5-celled.	*Ledum.*	16
Petals subdistinct, very irregular.	*Rhodora.*	12
Flowers racemed.	*Pyrola.*	17
Flowers solitary.	*Moneses.*	18
Flowers corymbed.	*Chimaphila.*	19
Flower solitary.	*Monotropa.*	20
Petals distinct. Flowers racemed.	*Hypopitys.*	21
destitute of leaves or verdure. Petals united. Flowers racemed.	*Pterospora.*	22

(Shrubs verdant, erect or prostrate — Herbs — verdant, leaves mostly all radical)

SUBORDER 1. V A C C I N E Æ.

Ovary adherent to the tube of the calyx, becoming a berry or drupe-like fruit. Shrubs with scattered leaves.

1. VACCINIUM.

Calyx superior, 5-toothed; corolla urceolate, campanulate or cylindric, limb 4—5-cleft, reflexed; stamens twice as many as the lobes of the corolla, generally included; anthers with 2 awns on the back, or awnless; style erect, longer than the stamens; berry invested with the calyx, 4 or 5 (rarely 10)-celled, cells many-seeded.—*Shrubs or undershrubs with scattered lvs. Fls. solitary or racemose, white or reddish. Fr. generally eatable.*

§ *Flowers racemose.   Corolla urceolate, ovoid or oblong-cylindric.*

1. V. RESINÔSUM. Ait. *Black Whortleberry or Huckleberry.*

Branches cinereous-brown, villose when young; *lvs.* oblong-ovate or oblong-lanceolate, rather obtuse, entire, petiolate, with resinous dots beneath; *rac.* lateral, secund; *pedicels* short, subbracteolate; *cor.* ovoid-conic, at length subcampanulate, 5-angled; *berries* black.—This common shrub of our woods and pastures is about 2f high, very branching. Leaves 1—2′ long, ½ as wide, rarely acute, shining beneath with resinous patches and spots. Petioles 1″ in length. Flowers in lateral, dense, corymbose clusters, small, drooping. Corollas contracted at the mouth, greenish or yellowish-purple, longer than the stamens but shorter than the style. Berries black, globose, sweet and eatable, ripe in August. May.

β. *Lvs.* and berries covered with a glaucous bloom.

γ. *Lvs.* larger; *pedicels* longer than the corolla.

2. V. CORYMBÔSUM. (V. fuscatum. *Ait.*) *Blue Bilberry. High Whortleberry.*—*Flowering branches* nearly leafless; *lvs.* oblong-oval, acute at each end, mucronate, subentire, pubescent when young; *rac.* short, sessile; *cor.* ovoid-cylindrical.—A tall shrub, 4—8f nigh, growing in shady swamps and by mud ponds. Branches few, the young ones green or purplish. Leaves smooth on both sides except a slight pubescence on the veins beneath, tipped with a glandular point, formed by the prolonged midvein. Flowers numerous, nodding, generally appearing in advance of the leaves. Pedicels shorter than the corollas, with colored scales or bracts at base. Corolla large for the genus, purplish-white, slightly contracted at the mouth. Stamens included. Style often exserted. Berries large, black, often with a tinge of purple, subacid. Jn.

β. ? (V. dismorphum. *Michx.*) *Fls.* and *fr.* much smaller; *cal.* very obtuse; *sty.* exserted; *berries* black.—Grows with the other; frequent!

3 V. **virgàtum**. Muhl. (V. Pennsylvanicum. *Darl. Deck.* and 1*st* edit. *in part.*)—*Blue Whortleberry.*—*Branches* angular, green, *lvs.* oblong or elliptic-lanceolate, sessile, mucronate, often serrulate, smooth and shining on both sides; *rac.* numerous, dense-flowered, subterminal, sessile, mostly naked; *cor.* ovoid. — Hilly woods and thickets, N. Eng.! to Va. W. to Wis. *Lapham!* Stem 1—3f high, bushy. Leaves pale green, 12—18'' by 7—10'', often slightly pubescent when young, thin, at length very smooth. Flowers in numerous, small racemes, on the upper, nakedish branchlets; pedicels shorter (1—3'') than the corolla. Corolla yellowish and reddish-white, longer than the stamens, but equaling or shorter than the styles. Berries bluish-black, sweet. May, Jn.

4. V. **Pennsylvanicum** Lam. (V. tenellum. *Ait.*) *Common Low Blueberry.*—*Branches* green, with 2 pubescent lines; *lvs.* subsessile, crowded, elliptic-oblong, acute at each end, minutely serrulate, thin, glabrous and shining, with the veins beneath puberulent; *fls.* in short, bracteate, dense, subterminal racemes; *cor.* ovoid-cylindrical.—Thickets and pastures in hard soils, Can. to Penn., common in N. Eng. A low under-shrub, 6—12' high, growing in dense patches. Leaves 8—12'' by 4—6''. Flowers reddish-white, 3'' long. Bracts mostly colored. Berries large, blue, sweet and nutritious. May.

β. *Lvs.* dark green; *berries* black and shining, destitute of bloom.—With variety *a.*

5. V. **ligustrinum**. Michx. *Privet Whortleberry.*

Branches angular, slender, strict, erect; *lvs.* subsessile, erect, thick, lanceolate, veiny, pubescent, mucronate, serrulate; *fascicles* sessile; *pedicels* very short, glomerate; *cor.* ovoid-oblong.—Mountains, Penn. to Va. *Pursh,* who remarks that the leaves are very variable, the corolla reddish-purple, and the berries black. May, June.

6. V. **buxifolium**. Salisb.

St. low; *lvs.* obovate, crenate-dentate, smooth; *rac.* axillary and terminal, dense, subsessile; *cor.* orbicular-ovoid; *fl.* glandular; *stig.* capitate.—Near Winchester, Va. Stem 6—10' high. Corolla white, with purple lines. Calyx bracteate.

§ § *Flowers solitary. Corolla urceolate.*

7. V. **uliginòsum**. *Mountain Blueberry.*

Procumbent; *lvs.* obovate, very obtuse, entire, smooth, glaucous and veiny beneath; *fls.* mostly solitary, axillary; *cor.* ovoid-globose, 4-cleft; *sta.* 4; *anth* awned at the base.—A low, alpine shrub, White Mts. Stems with numerous, rigid branches. Leaves 3'' by 2'', broadest near the apex, scarcely petiolate, crowded near the ends of the branches, and of a bluish-green. Flowers half as long as the leaves, subsessile, sometimes 2 together. Berries oblong, deep-blue, crowned with the style. June, July. (Apr. May. *Beck.*)

8. V. **cæspitòsum**. Michx. *Turfy Vaccinium.*

Dwarf, cæspitose; *lvs.* obovate, attenuate at the base, thin, serrate, reticulate with veins, shining; *ped.* subsolitary, 1-flowered; *cal.* very short; *cor.* oblong, suburceolate; *pores of the anthers* long-tubular.—White Mts., N. H. *Oakes,* N. to Hudson's Bay. Stem a few inches high. Flowers numerous, nodding, on short pedicels. Anthers with 2 long awns at the back. Berries large, glabrous, blue, eatable.

§ § § *Corolla campanulate. Leaves deciduous.*

9. V. **stamineum**. (V. stamineum and album. *Ph.*) *Deerberry.*

Young branches pubescent; *lvs.* oval-lanceolate, acute, glaucous beneath; *pedicels* solitary, axillary, nodding; *cor.* campanulate-spreading, segments acute, oblong; *anth.* exserted, 2-awned near the base.—Dry woods, Can. to Flor. Shrub 2—3f high, very branching. Leaves 1—2' long, ½—⅔ as wide, broadest in the upper half, mostly rounded at base and on very short petioles; those on the slender, flowering branches very much smaller. Flowers on long, slender pedicels, arranged in loose, leafy racemes. Corolla white, spreading, stamens conspicuously exserte!, but shorter than the style. Berries large, greenish-white, bitter. May, June.

β. (V. elevatum. *Banks.*) *Lvs.* pale, pubescent beneath; *fls.* smaller. N. I.

10. **V. frondosum.** Willd. (V. glaucum. *Mx.*) *Blue Tangles. High Blueberry.*—*Lvs.* oblong-obovate, obtuse, entire, glaucous beneath, covered with minute, resinous dots; *rac.* loose, bracteate; *pedicels* filiform, bracteate near the middle; *cor.* ovoid-campanulate, including the stamens.—Grows in open woods, N. Eng. to Car. A shrub 3—5f high, with round, smooth and slender branches. Leaves twice as long as wide, tapering to each end but broadest in the upper half, the margin slightly revolute. Racemes lateral, few-flowered Pedicels 5—10″ in length. Flowers small, nearly globose, reddish-white, succeeded by large, globose, blue and sweet berries, covered with a glaucous bloom when mature. May, June.

11. **V. Canadense.** Rich. *Canadian Blueberry.*
Branches reddish-green, pubescent, leafy; *lvs.* subsessile, elliptic-lanceolate or oblong, acute at each end, villose beneath, tomentose on the veins above, entire; *rac.* fasciculate, sessile, subterminal; *corymb* campanulate; *cal. lobes* acute.—A shrub 8—12′ high, not uncommon in rocky fields and thickets, N. H.! Me.! to Hudson's Bay and to the Rocky Mts. Leaves 8—12″ by 3—5″. Flowers about 3″ long. Style and stamens included. Berries blue and sweet, similar to those of V. tenellum. May.

12. **V. dumosum.** Andrews. (V. frondosum. *Michx.* V. hirtellum. *Bw.*)
Branchlets, lvs. and *pedicels* sprinkled with minute bristles and resinous dots; *lvs.* obovate-oblong, subsessile, subcoriaceous, obtuse, mucronate, entire or ciliate-serrulate; *rac.* bracted with small, floral leaves; *pedicels* bracteolate in the middle; *cor.* cylindric-campanulate, including the stamens and style.—Swamps and thickets, Uxbridge, Mass. *Robbins!* S. to Flor. A small shrub 1½ high, with leafy racemes. Leaves about 16″ by 7″, cuneate at base, shining but minutely hispid above. Flowers white or purplish, each from the axil of a small, roundish-ovate leaf. Berries black, insipid, large (shining, *Don.*, hairy, *Bw*).

2. OXYCOCCUS. Pers.
Gr. οξυς, acid. κοκκος, berry.

Calyx superior, 4-cleft; corolla 4-parted, with sub-linear, revolute segments; stamens 8, convergent; anthers tubular, 2-parted, opening by oblique pores; berry globose, many-seeded.—*Slender, prostrate shrubs, with alternate, coriaceous lvs. and eatable fruit.*

1. **O. palustris.** Pers. (O. vulgaris. *Ph.* and 1*st. edit.* Vaccinium Oxycoccus. *Linn.*) *Common Cranberry.*—*St.* filiform, prostrate; *lvs.* ovate, entire, revolute on the margin; *pedicels* terminal, 1-flowered; *segments of the corolla* ovate.—A prostrate under shrub, found in alpine bogs, Can. and N. States. Stems creeping extensively, smooth, purple, with erect branches. Leaves somewhat remote, 2—3′ long, and half as wide, smooth and shining above, paler beneath. Flowers several together on the summits of the branches. Pedicels red, an inch in length, with 2 nearly opposite bracts in the middle. Corollas light pink, the 4 segments abruptly reflexed. Stamens purple. Fruit smaller than in the next species, crimson, ripe in Oct. Flowers in June.

2. **O. macrocarpus.** Pers. (V. macrocarpon. *Ait.*) *Larger Cranberry.*
St. creeping, filiform; *lvs.* oblong, scarcely revolute, obtuse, glaucous beneath; *pedicels* axillary, elongated, 1-flowered; *segments of the corolla* linear-lanceolate.—A prostrate, shrubby plant, in sphagnous swamps and meadows. Stems 8—15′ in length, brown, with ascending branches. Leaves numerous, 4—6″ by 2—3″, rounded at each end, on very short petioles, smooth both sides, subentire. Flowers flesh-colored, pedicels 5—15″ long, solitary in the axils of the upper leaves, the 4 segments generally abruptly reflexed. Berry large, bright scarlet, ripe in Oct. Flowers in June.

3. CHIOGENES. Salisb.
Gr. χιων, snow. γενος, offspring; in allusion to its evergreen habit

Calyx 4-cleft, persistent; cor. broadly campanulate, limb deeply 4-cleft; stam. 8, included, anth. fixed by the base, the 2 cells awnless on the back, bicuspidate at apex, opening longitudinally; ovary adherent, except at the summit, 4-celled; fruit white, 4-celled, many-seeded.—*A prostrate, evergreen undershrub, with alternate leaves. Fls. solitary, axillary.*

C. HISPIDULA. Gray. (Vaccinium hispidulum. *Linn.* Gaultheria hisp. *Muh.* and 1*st. edit.* Phalerocarpus serpyllifolia. *Don.* Glyciphylla hisp. *Raf.* &c., &c.) *Mountain Boxberry.*—A delicate woody creeper, in old shady woods, mountains, N. Eng. to Newfoundland, W. to the R. Mts. Stems ligneous, slender, creeping extensively, with numerous branches, and clothed with short, appressed, reddish hairs. Leaves numerous, alternate, roundish-oval, 4—6″ by 3—4″, abruptly acute, dark evergreen above, paler beneath. Corolla white, its parts in 4s. The leaves and white berries have an agreeable spicy flavor like those of *Gaultheria procumbens.* May, June.

SUBORDER 2.—ERICINEÆ.

Ovary free from the calyx. Testa conformed to the nucleus of the seed. Mostly shrubs. Leaves often evergreen.

4. ARCTOSTAPHÝLOS. Adans.

Gr. αρκτος, a bear. σταφυλος, a cluster of grapes; that is, bear-berry.

Calyx 5-parted, persistent; corolla ovoid, diaphanous at the base, limb with 5 small, recurved segments; drupe with a 5-celled putamen, the cells 1-seeded.—*Trailing shrubs, with alternate leaves.*

1. A. UVA-URSI. Spreng. (Arbutus Uva-ursi. *Linn.*)
St. procumbent; *lvs.* entire, obovate, smooth, alternate, on short petioles, evergreen, coriaceous, shining above, paler beneath; *fls.* in short, terminal, drooping clusters; *drupe* globular, about as large as a currant, deep red, nearly insipid, the nucleus consists of 5 bony seeds firmly united together.—A shrub growing on mountains, in the N. States and British America. Stem prostrate except the younger branches, which arise 3—8′. The leaves are about an inch in length, 2—3″ wide, often spatulate in form; medicinally they are astringent, and much valued in nephritic complaints.

2. A. ALPINA. Spreng. (Arbutus alpina. *Linn.*) *Alpine Bear-berry.*
Procumbent; *lvs.* thin, deciduous, obovate, acute, serrate, ciliate when young; *fls.* in short, terminal racemes; *bracteoles* ovate, broad, ciliate, about equaling the pedicel.—On the alpine regions of the White Mts., *Robbins.* Flowers white. Berries black.

5. ANDROMĒDA.

Named for Andromeda of ancient fable.

Calyx minute, 5-parted, persistent; corolla ovoid-cylindric; the limb 5-cleft, reflexed; stamens 8—10; capsule 5-celled, 5-valved, the dissepiments produced from the middle of the valves.—*Shrubs prostrate or erect. Lvs. mostly alternate.*

§ 1. *Cal. naked. Caps. valves bifid. Minute evergreen shrubs.* CASSIOPE

1. A. HYPNOIDES. (Cassiope. *G. Don.*) *Moss-like Cassiope.*
St. filiform, spreading; *lvs.* evergreen, subulate, smooth, crowded; *ped.* solitary, terminal; *cor.* globose, campanulate.—One of the smallest and most delicate of shrubs, a tree in miniature, resembling some of the mosses, found on the alpine summits of the White Mts.! Stems woody, much branched at base, 2—3′ high. Leaves minute, evergreen, spirally arranged, and so closely as to conceal the stems. Flowers small, but large in proportion, nodding; peduncles colored, smooth, round, an inch long in fruit. Calyx purple. Corolla light red, twice as long as the calyx, lobes erect. Stamens included. June.

§ 2. *Cal. bracteate at base. Capsule valves double. Lvs. evergreen, entire.* CASSANDRA.

2. A. CALYCULÀTA. (Cassandra. *G. Don.*) *Bracted Cassandra.*
Erect; *lvs.* oval-oblong, obtuse, obsoletely serrulate, subrevolute, ferruginous beneath; *rac.* terminal, leafy, subsecund.—An evergreen shrub, 2—4f high, flowering early, in wet situations, Can. and most of the U. S. The leaves are coriaceous, shining, dotted, about an inch long and half as wide, those of the racemes not half as large. Flowers numerous, 20—30 in each raceme, white, each from the axil of a small leaf. Calyx double, the outer of 2 bracts, the inner of 5 acute sepals. April, May.

§ 3. *Cal. naked. Anth. 2-awned. Lvs. evergreen, revolute.* EUANDROMEDA

3. A. POLIFÒLIA. *Marsh Andromeda. Wild Rosemary.*
Erect; *lvs.* entire, linear-lanceolate, coriaceous, revolute on the margins, glaucous beneath; *fls.* subglobose, in a dense, terminal *corymb.*—A beautiful evergreen shrub, 1—2f high, growing by the side of ponds and in swamps, N. Eng. to Wisc. *Lapham!* N. to Arc. Am. Leaves very smooth, 2—3' long and less than ¼' wide, on very short petioles, dark green and smooth above, bluish white beneath. Flowers in pendulous clusters. Calyx white, tipped with red Corolla rose-colored. June.

§ 4. *Cal. naked. Anthers 4-awned. Leaves mostly deciduous.* ZENOBIA.

4. A. RACEMÒSA. (Zenobia. *G. Don.*) *Clustered Zenobia.*
Lvs. oval-lanceolate, acute, glabrous, serrulate; *rac.* terminal, secund elongated, sometimes branched; *cal.* acute; *cor.* cylindric; *anth.* 4-awned at the summit.—A shrub 4—6f high, growing in wet woods, Can. to Flor. W. to Ky. It is remarkable for its naked racemes, 2—4' in length, consisting of about a dozen flowers, which are arranged in a single row, with much regularity. Leaves 1—2' in length, ⅓ as wide, minutely notched. Pedicels short, with two ovate-acuminate bracts at the base of the colored calyx. Corolla white, 4 or 5 times as long as the calyx. Anthers 2-cleft, about half as long as the corolla. Jn. Jl.

§ 5. *Cal. naked. Anthers awnless. Caps. valves simple. Leaves (mostly) deciduous.* LEUCOTHOË.

5. A. MARIÀNA. (Leucothoë. *G. Don.*) *Maryland Leucothoë.*
Glabrous; *lvs.* oval, subacute at each end, flat, entire, subcoriaceous, paler beneath; *flowering branches* leafless; *pedicels* fasciculate; *calyx lobes* linear, foliaceous; *cor.* ovate-cylindric; *sta.* 10; *fil.* villous.—Woods and dry, sandy soils, N. J. to Flor., common. A beautiful shrub, 2—3f high, with very smooth, deciduous foliage, and large, white or pale red flowers. Capsule depressed-globose. Seeds angular. June, July.

§ 6 *Corolla subglobose. Capsule with 5 supernumerary valves.* LYONIA.

6. A. LIGUSTRINA. Muhl. (Lyonia paniculata. *Nutt.*) *Panicled Lyonia.*
Pubescent; *lvs.* obovate-lanceolate, acuminate, finely serrulate; *fls.* somewhat paniculate, in terminal, leafless racemes; *anth.* awnless.—A deciduous shrub, 4—8f high, in swamps, &c., Middle and Southern States. Leaves abruptly acuminate, paler beneath, 2—3' long and nearly half as wide, on short petioles. Flowers small, nearly globose, white, in dense panicles, succeeded by globular capsules. June.

§ 7. *Sepals acuminate. Capsule pyramidal, pentangular. Leaves acid.* OXYDENDRON.

7. A. ARBÒREA. (Oxydendron. *DC.* Lyonia. *Don.*) *Sorrel Tree.*
Arborescent; *branches* terete; *lvs.* petiolate, oblong, acuminate, serrate; *panicles* terminal, consisting of numerous spicate racemes; *fls.* pedicellate, secund, spreading, at length reflexed; *cor.* ovate-oblong, pubescent externally.—Ohio, Penn., along the Alleghany Mts. to Flor. A fine tree, 40—50f high, trunk 10—15' diam. Bark thick and deeply furrowed. Leaves 4—5' by 1½—2', villous when young, at length smooth, with a distinctly acid taste. Flowers white. Capsule pyramidal, 5-sided. June, July. †

6. MENZIÉSA. Smith.

In honor of Menzies, companion of Vancouver in his voyage round the world.

Calyx deeply 5-cleft; corolla ovoid 4—5 cleft; stamens 8—10, inserted into the receptacle; capsule 4—5-celled, the dissepiments made by the introflexed margins of the valves; seeds many.—*Low, heath-like, shrubby plants, with evergreen leaves.*

1. M. TAXIFOLIA. Robbins. (M. cœrulea. *Swartz.* Phyllodoce tax. *Salisb.* Andromeda tax. *Pall.* Andromeda cœrulea. *Linn.*) *Mountain Heath.—St.* prostrate at base; *lvs.* linear, obtuse, with minute, cartilaginous teeth; *ped.* terminal, aggregate, one-flowered; *fls.* campanulate, decandrous; *cal.* acute.—A small shrub, a few inches high, found on the summit of the White Mts. It resembles a Heath in its flowers and some of the fir tribe in its leaves and stems. Stem decumbent at base, with crowded, scattered leaves above, which are 5—7″ in length. Flowers drooping, purple, at the top of the highest branch, on colored peduncles. Calyx in 5 segments, purplish. Corolla of 5 segments, emarginate, rather longer than the stamens. July.

2. M. GLOBULÁRIS. Salisb.
Branches and *pedicels* with scattered hairs; *lvs.* oval-lanceolate, ciliate above and on the veins beneath, apex tipped with a gland; *cal.* 4-cleft; *cor.* globose; *sta.* 8; *caps.* 4-celled, 4-valved.—Mountains Penn. to Car. Abundant near Winchester, Va. *Pursh.* Shrub 4f high. Flowers yellowish-brown, nodding and mostly solitary on each terminal pedicel. June.

7. GAULTHERIA. Kalm.

Named for one Gaulthier, a French physician at Quebec.

Calyx 5-cleft with 2 bracts at the base; corolla ovoid-tubular, limb with 5 small, revolute lobes; filaments 10, hirsute; capsule 5-celled, invested by the calyx which becomes a berry.—*Suffruticose. mostly American plants. Lvs. alternate, evergreen. Pedicels bibracteolate.*

G. PROCUMBENS. *Box-berry. Checker-berry. Wintergreen.*
St. with the procumbent branches erect or ascending; *lvs.* obovate, mucronate, denticulate, crowded at the top of the stem; *fls.* few, drooping, terminal.—A little shrubby plant, well known for its spicy leaves and its well-flavored scarlet berries. Common in woods and pastures, Can. to Penn. and Ky. The branches ascend 3′ from the prostrate stem, or rhizoma, which is usually concealed. Leaves thick, shining, acute at each end, with remote and very obscure teeth. Corolla white, contracted at the mouth. Filaments white towards the corolla. Fruit well flavored, consisting of the capsule surrounded by the enlarged calyx, which becomes of a bright scarlet color. June—Sept.

8. CLETHRA. Gaert.

Gr. name of the alder, which these plants somewhat resemble.

Calyx 5-parted, persistent; petals 5; stamens 10, exserted; style persistent; stigma 3-cleft; capsule 3-celled, 3-valved, enclosed by the calyx.—*Shrubs and trees. Lvs. alternate, petiolate. Fls. white, racemose.*

C. ALNIFOLIA. *Sweet-pepper Bush.*
Lrs. cuneiform-obovate, acute, acuminately serrate, green on both sides, smooth or slightly pubescent beneath; *fls.* in terminal, elongated, simple or branched racemes; *bracts* subulate.—A deciduous shrub, 4—8f high, growing in swamps. Leaves 2—3′ long, ¼ as broad above, with a long, wedge-shaped base, tapering into a short petiole. Racemes 3—5′ long. Peduncles and calyx hoary-pubescent, the former 3″ in length, and in the axil of a bract about as long. Corolla white, spreading, about equaling the stamens and styles. Jl. Aug.

9. EPIGÆA.

Gr. επι, upon, and γη, the earth; from its prostrate habit.

Calyx large, 5-parted, with 3 bracts at base; corolla hypocrateriform, tube villous within, limb 5-parted, spreading; stamens 10; anthers dehiscent by 2 longitudinal openings; capsule 5-celled. 5-valved.—*Suffruticose trailing. Lvs. evergreen.*

E. REPENS. *Trailing Arbutus. May Flower.*

Lvs. cordate-ovate, entire; *cor. tube* cylindrical.—♃ Woods, Newfoundland to Ky. and Penn. This little shrubby plant grows flat upon the ground, 10—15′ in length, covered with a hairy pubescence in all its parts. Leaves alternate, 2—2¼′ by 1¼′, roundish at the end and abruptly tipped with a very short point. Flowers very fragrant, white or tinged with various shades of red, in small clusters on short stalks. Calyx green, supported by 3 large bracts at base. Tube of the corolla hairy within, longer than the calyx, the border in 5, rounded, spreading segments. Apr. May.

10. KALMIA.

Named by Linnæus in honor of Peter Kalm, prof. at Abo, Finland.

Calyx 5-parted; corolla with 10 prominences beneath and 10 corresponding cavities within, including the 10 anthers; border 5-lobed; filam. elastic; capsule 5-celled, many-seeded.—*Beautiful shrubs, natives of N. America. Lvs. entire, evergreen, coriaceous. Fls. in racemcss corymbs, white and red.*

1. K. LATIFOLIA. *Mountain Laurel. Calico Bush.*

Lvs. alternate and ternate, oval-lanceolate, acute at each end, smooth and green on both sides; *corymbs* terminal, viscidly pubescent.—One of our most beautiful shrubs, sometimes attaining the height of a small tree. It is found in all the Atlantic States from Maine to Georgia, and W. to Ohio and Ky. in woods. The wood is usually very crooked, fine-grained and compact. The leaves are 2—3′ long, smooth and shining, acute at each end and entire. Flowers in splendid corymbs, white or variously tinged with red, abundant. The corolla has a short tube with a spreading limb 9—10″ diam. and a 5-lobed margin. Leaves narcotic, and poisonous to some animals. May, June.

2. K. ANGUSTIFOLIA. *Narrow-leaved Laurel. Sheep-poison.*

Lvs. ternate and opposite, elliptic-lanceolate, obtuse at each end, smooth; *corymbs* lateral; *bracts* linear-lanceolate.—A beautiful little shrub, smaller than the foregoing, 2—4f in height, in marshes and by ponds, Can. to Car. W. to Ky. The leaves are acutely and narrowly elliptic, with rounded ends, entire, smooth, 1—2′ long and ½ as wide, on short petioles. The flowers are of a deep purple, growing in small, axillary fascicles and apparently whorled among the leaves, in structure resembling those of the last species, but about half as large. Bracts minute, about 3 at the base of each pedicel. This is also said to be poisonous to cattle. June.

3. K. GLAUCA. *Glaucous Kalmia. Swamp Laurel.*

Branches ancipitous; *lvs.* opposite, subsessile, lanceolate, polished, glaucous beneath, revolute at the margin; *corymbs* terminal, the peduncles and bracts smooth.—A delicate shrub, 2f high, found in swamps, &c., Penn., Ky., N. Eng. N. to Arc. Am. Stem slender, the branches rendered distinctly 2-edged by an elevated ridge extending from the base of each opposite leaf to the next node below. Leaves smooth and shining, white underneath, about an inch in length. Flowers 8—10 in each corymb. Corolla about ½′ diam., pale purple. Calyx red, as is also the very slender peduncle. At the base of each peduncle is a pair of concave, obtuse bracts. June.

β. *rosmarinifolia. Leaves* linear, more revolute, green beneath.

11. AZALEA.

Calyx 5-parted, lobes equal; cor. subcampanulate, 5-parted, regular; sta. equal. erect, shorter than the corolla; anth. dehiscing laterally from the apex; ovary roundish; sty. straight, included; capsule 2—3-celled, 2—3-valved, many-seeded.--*A little branching, pro* cumbent shrub, with opposite, petiolate, evergreen, entire lvs. Pedicels terminal, solitary, 1-flowered. Cor. rose color.

　　A. PROCUMBENS. (Loiseleuria. Desv. Rhododendron. 1st edit.)
　　An exceedingly delicate shrub, native on the alpine summits of the White Mts., N. H.! Stems 3—6′ long, very branching and leafy. Leaves elliptical, thick, shining, not more than 3″ by 1″, margin strongly revolute. Flowers glabrous, on very short, purple pedicels, in the midst of the leaves. Jn. Jl.

12. RHODÓRA.

Gr. ῥόδον, a rose; the shrub bears only flowers at flowering time.

Calyx 5-toothed, persistent; cor. adnate to the calyx, deeply divided into 3 segments, upper one much the broadest, 2—3-lobed at the apex, in æstivation enfolding the 2 lower, entire segments; sta. 10, declinate; fil. unequal; anth. opening by 2 pores; caps. 5-celled, 5-valved; cells many-seeded; dissepiments formed by the introflexed margins of the valves.—*A shrub with deciduous, alternate leaves, and pale purple flowers.*

　　R. CANADENSIS. (Rhododendron Rhodora. Don.)
　　A handsome, flowering shrub, in bogs, mountain or plain, Can. to Penn., frequent. Stems 2—3f high, clothed with a smooth brown bark, each dividing at top into several erect, flowering branches. Each branch, while yet naked of foliage, bears a terminal cluster of 3—5 sessile flowers. Corolla 1′ long, about equaling the deflected stamens and style. Leaves obovate-oblong, downy-canescent beneath. Apr. May.

13. RHODODENDRON.

Gr. ῥόδον, a rose. δένδρον, a tree.

Calyx deeply 5-parted, persistent; cor. infundibuliform or campanulate, regular or irregular, 5-lobed; sta. 5—10, mostly declinate and exserted; anth. opening by 2 terminal pores; capsule 5-celled, 5-valved, opening at the summit; dissepiments introflexed from the margin of the valves.—*Shrubs with alternate, entire, evergreen or deciduous leaves. Fls. mostly in terminal, corymbose clusters. Cor. variously shaded from blue through purple to white.*

§ 1. *Corolla infundibuliform, tube long, cylindric, more or less viscid; limb unequal, spreading; stamens 5 or 6. Lvs. deciduous.* AZALEA.

　　1. R. NUDIFLORUM. Torr. (Azalea nudiflora. Linn.) *Swamp Pink.*
　　Lvs. oblong-lanceolate and oblanceolate; fls. rather naked, slightly viscid; tube of the corolla longer than the lobes; sta. much exserted.—A beautiful and fragrant flowering shrub, 4—6f high, rather frequent in the forests and thickets of the Northern States as well as the Southern. Stems crooked, much branched. Leaves 2—3′ by 1½′, margins ciliate, upper surface with minute, scattered hairs, lower paler and pubescent, with the midvein hispid. Flowers appearing before the leaves are fully grown, in rather naked umbels. Pedicels 6—8″ long Calyx minute, with rounded, ciliate segments. Corolla tube 8—10″ in length hairy, and, with the spreading, unequal limb variously shaded from pale pink to purple. Stamens purple, declinate, twice as long as the corolla. Style nearly 3 times as long. It varies in the number of stamens, color of corolla, &c Apr. May. †

　　2. R. VISCÓSUM. Torr. (Azalea viscosa. Linn.) *Clammy Swamp Pink.*
　　Lvs. obovate and oblong-lanceolate; fls. accompanied with leaves, very viscid; tube of the corolla twice as long as the lobes; sta. a little exserted.—Less frequent than the last, in rocky woods, Can. to Ga., W. to Ky. Shrub 4—6f high, much branched above, the branches hispid. Leaves 1—2′ long and about half as wide, smoothish, hispid, ciliate on the petiole, midvein and margin.

Flowers fragrant, in terminal umbels, on hairy pedicels, not appearing until the leaves are fully grown. Calyx minute and ciliate. Cor-lla white, tube an inch in length, clothed with glandular hairs, and very clammy, limb unequal. Stamens and style declined, the latter longest. May, June. †

ß. *glaucum*. Ph. (Azalea glauca. *Lam.*) *Lvs.* glabrous, glaucous beneath.

3. R. NITIDUM. Torr. (Azalea nitida. *Linn.*) *Shining Swamp Pink.*

Lvs. oblanceolate, coriaceous, smooth both sides, shining above, margins revolute; *fls.* accompanied with leaves, viscid; *tube* much longer than the segments; *sta.* exserted.—In mountain swamps, N. Y. *Storrs!* to Va. Shrub 3—6f high, with nearly smooth branches. Leaves dark green above, 1—2' in length, ½ as wide, roundish and submucronate above, tapering at the base into very short petioles, midvein hispid beneath. Flowers large, pale pink, fragrant. Tube an inch or more long, glandular-hairy. Pedicels ½' long. Calyx segments obsolete. Style 2' long. Jn. Jl.

4. R. CALENDULACEUM. Torr. (Azalea calendulacea. *Mx.*) *Flame Azalea.*

Branchlets subvillose; *lrs.* oblong, attenuated to the base, mucronate, pubescent both sides, ciliate on the margin; *corymbs* nearly leafless; *cal. teeth* oblong; *tube of the cor.* hirsute, not viscid, shorter than the lobes.—A splendid flowering shrub, in mountains and woods, Penn. to Ohio, *Sullivant!* and Ga. Stems 3—6f high. The leaves at flowering time are about 3' by 1½' or smaller. Flowers large and numerous, the corolla 2¼' in length, limb expanding nearly 2', usually yellow and bright crimson, showing at distance like flame. May.—Cultivation has produced numerous varieties, of every shade from golden-yellow to dark crimson, single and double. †

5. R. ARBORESCENS. Torr. (Azalea arborescens. *Ph.*) *Tree Azalea.*

Branches smooth; *lvs.* obovate, obtuse, both sides glabrous, glaucous beneath, margin ciliate, veins nearly glabrous; *corymbs* leafy; *cal. lobes* oblong, acute; *cor. tube* not viscid, longer than the lobes; *sta.* and *sty.* exsert.—Rivulets near the Blue Mts., Penn., *Pursh.* Shrubs 10—20f high. Flowers large, rose color, scales of the flower-buds large, yellowish-brown, with a fringed, white border. May—Jl.

6. R. HISPIDUM. Torr. (A. hispida. *Ph.*) *Hispid Azalea.*

Branches hispid; *lvs.* lanceolate, acuminate at each end, hispid above, glabrous beneath, glaucous both sides, ciliate, midvein beneath hispid; *corymbs* leafy; *pedicels* glandular-pilose; *cal. teeth* oblong, obtuse; *cor.* very viscid, tube scarcely longer than the lobes; *sta.* and *sty.* exsert.—Shrub of a bluish appearance, very upright, 10—15f high, mountains and lake shores, N. Y. and Penn. *Pursh.* Flowers white, bordered with red, tube reddish-white. Stamens often 10. Jl. Aug.

7. R. FLAVUM. Don. (Azalea pontica. *Linn.*) *Yellow Azalea.*—*Lvs.* oblong-obovate; *rac.* corymbose, with deciduous bracts; *cor.* viscid-pilose without, not leafy; *sta.* exsert, declinate.—Native of the Caucasian Mts. A splendid shrub, with golden-yellow flowers, varying to orange, copper-color or ochroleucous. †

§ 2. *Corolla campanulate, tube short, smooth, lobes rounded-obtuse : spreading. Leaves coriaceous, evergreen.*

8. R. MAXIMUM. *American Rose Bay.*

Lvs. obovate-oblong, smooth, coriaceous, discolored beneath, subrevolute on the margin; *corymbs* terminal; *cor.* somewhat campanulate, unequal; *pet.* roundish.—A splendid flowering shrub, not uncommon in N. England, but most abundant in the Middle States. The stems are crooked, 6—15f or more in height. The young leaves are downy, becoming very smooth when full grown and 4—7' long, entire, thick and leathery, permanent, remaining on the stem 2 or 3 years. Corymbs 15—20-flowered, in the midst of the evergreen leaves. At first each cluster appears in the form of a large compound bud enveloped in numerous bracts. These bracts are near an inch long, abruptly acuminate. Corolla pink or rose-colored, sometimes dotted with yellow, 1½—2' diam. The wood is hard and fine grained. July, Aug.

ß. *album*. Ph. (R. Purshii. *Don.*) *Cor.* white, segments oblong-obtuse.

9. R. Lapponicum. Wahl. (Azalea Lapponica. *Linn.*) *Lapland Rhododendron.*—Dwarf; *lrs.* elliptical, roughened with excavated punctures; *fls.* in terminal, leafy clusters, campanulate, limb spreading, 5-lobed; *sta.* 5, exserted.—An erect shrub 8—10′ high, native of the White Mts.! Branches numerous, with a rough bark. Leaves about 5″ by 2½″, acute, with an obtuse angle, revolute, ferruginous beneath, all fasciculated at the summits of the branches. Flowers 7—9″ diam. Peduncle bracted at base. Calyx pubescent. Corolla deep purple, regular, lobes roundish, as broad as the leaves. Style very long, ascending. June, July.

10. R. arboreum. Smith.—*St.* arborescent; *lrs.* lanceolate, glabrous, with shining white spots beneath; *fls.* densely corymbose; *caps.* pubescent, 8—10-celled.—A most beautiful tree or shrub, from the Himmaleh Mts. Flowers purple, red, white, cinnamon-color, &c. †

11. R. Ponticum.—*Lrs.* oblong-lanceolate, attenuated to each end, smooth and scarcely paler beneath; *corymbs* short, terminal; *cor.* campanulate-rotate; *cal. lobes* subacute, very short.—From Asia Minor. Flowers large, often 2′ diam., purple—but in cultivation very variable. †

12. R. Indicum. Sweet. (Azalea Indica. *Linn.*)—*Branchlets, petioles, veins* and *sepals* strigose, but not glandular; *lrs.* cuneate-lanceolate, ciliate, acuminate at each end; *fls.* terminal, 1—3 together, on short pedicels.—From Java. Flowers scarlet, purple, crimson, flame-color, &c., in cultivation very brilliant. †

14. LEIOPHYLLUM. Pers.

Gr. λειος, smooth, φυλλον, leaf.

Calyx 5-parted, equaling the length of the capsule, pet. 5, ovate-oblong; sta. 10, exserted; fil. subulate; cells of anthers dehiscing by a lateral cleft; ovary globose; sty. filiform; caps 5-celled, 5-valved, many-seeded.—*Small, smooth shrubs, with erect branches. Lvs. alternate, entire, oval, coriaceous. Corymbs terminal. Fls. white.*

L. buxifolium. Ell.

St. erect; *lrs.* oval or obovate, subsessile; *caps.* glabrous.—Pine barrens, N. J. to Car. Shrub 8—12′ high, much branched. Leaves 4—5″ by 2 or 3″, very smooth and shining, margin strongly revolute. Flowers numerous and small. May, June.

15. LEDUM.

Calyx minute, 4-toothed; corolla 5-petaled, spreading; stamens 5—10, exserted; anthers opening by 2 terminal pores; capsule 5-celled, 5-valved, opening at the base.—*Shrubs. Lvs. alternate, evergreen, entire, ferruginous-tomentose beneath, coriaceous. Fls. in terminal corymbs, white.*

L. palustre (and L. latifolium. Ait.) *Labrador Tea.*

Lvs. elliptic-oblong or oblong-linear; *sta.* 5—10, more or less exserted.—Mountain bogs, Penn. to Lab and Greenland, White Mts.! Not uncommon. A shrub 2—3f high, readily known by its leaves, which are smooth above, clothed beneath with a dense, ferruginous down, and strongly revolute or replicate at the margin. The petioles and the younger twigs are also covered with down. Leaves 1—2′ long, nearly ⅓ as wide. Corymbs terminal, of about a dozen flowers. Petals 5, white. Pedicels nearly as long as the leaves Stamens 5—10, as long as the petals. Style somewhat declined. July.

β. *angustifolium. Lrs.* narrower, almost linear; *sta.* mostly 10.

16. ERICA.

Gr. ερικω, to break: in allusion to the brittleness of the branches and stems.

Calyx 4-cleft; cor. tubular, globose, ovoid, urceolate, campanulate or hypocrateriform, limb short, 4-lobed; stam. 8; style filiform; caps. 4, rarely 8-celled, 4-valved, loculicidal; seeds 2—∞ in each cell.

affixed to the axillary placentæ, usually conformed to the smoothish or shining testa.—*European, or chiefly South African shrubs, branching, mostly brittle. Lvs. linear, acerose, margin revolute, verticillate, rarely alternate. Fls. axillary, solitary, verticillate, or terminal, corymbose or capitate, mostly nodding. Cor. of the cyanic series, from purple through red to white, very rarely orange or yellow.*

Obs.—Of this vast and beautiful genus, 429 species are described by Mr. Bentham in the Prodromus of DC., Part vii., pp., 613—693. All these species have been cultivated in Europe, and many in this country, but their successful culture is attended with more care than that of most other plants, and they have never as yet received general attention. To describe so few species as the limits of this work would permit, where so many are rarely and none generally met with, would be of little satisfaction to the student.

SUBORDER 3.—PYROLEÆ.

Ovary free from the calyx. Petals nearly distinct. Fruit a capsule. Mostly herbaceous.

17. PYRÓLA. Salisb.

Lat. diminutive from *Pyrus*; as the leaves (of P. elliptica) resemble those of the pear-tree.

Calyx 5-parted; pet. 5, equal; sta. 10; anth. large, pendulous, fixed by the apex, 2-horned at base, opening by 2 pores at top; sty. thick; stig. 5-rayed, 5-tubercled at apex; caps. 5-celled, 5-valved, opening at the angles, many-seeded.—*Low, scarcely suffruticose, evergreen herbs. Lvs. radical or nearly so, entire. Scape mostly racemose.*

§ 1. *Stamens ascending. Style declinate, longer than the petals.*

1. P. ROTUNDIFOLIA. *Round-leaved Pyrola.*
Lvs. orbicular-ovate, entire or crenulate, shorter than the dilated petiole *scape* 3-angled; *segments of the cal.* lanceolate, acute; *stig.* clavate, obscurely 5-toothed.—Common in woods, Can. to Car. W. to Wisc. Leaves all radical, round or inclining to ovate, nearly 2′ in diameter, smooth and shining, with conspicuous, reticulate veins. Petioles margined, as long as, and sometimes much longer than, the leaf. Scape 6—12′ high, bracteate at base and in the middle. Flowers drooping, large, fragrant, white, in an oblong, terminal raceme.

2. P. ASARIFOLIA. Michx. *Asarum-leaved Pyrola.*
Lvs. reniform-orbicular, coriaceous, entire or crenulate, shorter than the dilated petiole; *scape* angular, furrowed; *rac.* lax, many-flowered; *segments of the cal.* ovate, acute, appressed; *stig.* clavate, with the disk elongated and 5-lobed.—In old woods, Can. and N. States. Leaves all radical, 1½—1½′ diam., smooth and shining, conspicuously cordate at base, longer than, but not twice as long as, the margined petioles. Scape 5—10′ high, purplish, bracteate at base and near the middle, racemose one half its length. Flowers nodding, remote, large, deeply tinged with purple in all their parts. Style of about the same length and curvature as pedicel. June.

3. P. CHLORANTHA. Swartz. *Green-flowered Pyrola.*
Lvs. orbicular, crenulate, half as long as the narrow petiole; *rac.* few-flowered: *segments of the cal.* very short, obtuse; *pet.* oblong; *pores of the anth.* tubular; *stig.* clavate, with the disk elongated, and 5-lobed.—In woods, Can. and N. States, common. Leaves smaller than in either of the preceding species, often perfectly orbicular, but more frequently inclining to ovate, ½—1′ diam, smooth, shining, coriaceous. Petioles 1—2′ long. Scapes erect, angular, 8—12′ high, bearing a long, open raceme. Flowers nodding, large, remote, pedicels ½′ long, each in the axil of a very short bract. Petals greenish-white. Anther tubes conspicuous. June, July.

4. P. ELLIPTICA. Nutt. *Pear-leaved Wintergreen.*
Lvs. elliptical, membranaceous, obscurely dentate, longer than the petioles; *scape* mostly naked; *cal.* small, with ovate, obtuse segments; *pores of the anth.* short, tubular.—In woods, Can. and N. States to Wisc. Leaves 1—2′ long, more than half as wide, mostly acute and subentire, thin, smooth and light

green. Scape 5—9' high, slender, seldom bracteate, bearing short racemes. Flowers nodding, very fragrant. Pedicels longer than the bracts, but only half as long as the declinate, recurved style. Petals white. July. (*See Appendix.*)

§ 2. *Stamens erect. Style straight.*

5. P. secunda. *One-sided Pyrola.*

Lvs. ovate, acute, subserrate, longer than the petiole ; *rac.* secund.—In dry woods, Can. and N. States. Stem 2—3' high, bearing one or two fascicles of leaves near the summit. Leaves broadly ovate, acute at each end, with appressed, pointed serratures. Petioles 1' long. Peduncles scape-like, 5—7 high, bearing a 1-sided cluster of 10—15 greenish-white flowers. Petals oblong, shorter than the style. June, July.

6. P. minor. *Smaller Pyrola.*

Lvs. roundish-ovate, coriaceous, repand-crenulate ; *petiole* dilated at base, shorter than the lamina ; *rac.* subspicate ; *bracts* equaling or exceeding the pedicel ; *cal. lobes* short, subacute ; *sty.* included ; *stig.* 5-lobed.—White Mts., N. H., Mich. and Brit. Am. Scape angular. Leaves mucronulate at apex. Corolla globose, white, slightly tinged with purple.

18. MONÊSES. Salisb.

Calyx 5-parted ; cor. 5-parted, rotate ; sta. 10, regular, 2-spurred at base, at length inverted. opening by 2 pores at apex ; sty. rigid ; stig. peltate, radiately 5-cleft or lobed ; caps. 5-valved, 5-celled, many-seeded.—♃ *Low, simple, smooth. Lvs. at top of the stem roundish, crenulate, petiolate, veiny. Peduncle terminal, one-flowered, longer than the stamens. Fls. white.*

M. grandiflòra. Salisb. (Pyrola uniflora. *Linn.* and 1*st edit.*) Woods, among mosses, &c., Keene, N. H., *Bigelow.* Dexter, Jeff. Co., N. Y., *Vasey:* Brit. Am. Root creeping. Stem ascending, very short. Leaves 7—9" diam. Scape or peduncle about 3' high, slender, with a bract near the middle. Flower 9" diam. June.

19. CHIMAPHILA.

Gr. χειμα, winter. φιλεω, to love ; equivalent to the English name, *Wintergreen.*

Calyx 5-parted ; pet. 5, spreading ; sta. 10 ; fil. dilated in the middle ; anth. as in Pyrola ; sty. short, thick ; caps. 5-celled, opening from the summit ; seeds ∞.—*Small, suffruticose, evergreen plants, with the habit of Pyrola. Lvs. cauline, serrate, evergreen, opposite or irregularly verticillate. Fls. terminal.*

1. C. umbellàta. Nutt. (Pyrola. *Linn.*) *Prince's Pine. Pipsissiwa.*

Lvs. cuneate-lanceolate, serrate, in 4s—6s ; *fls.* corymbose ; *bracts* linear subulate ; *sty.* immersed in the ovary.—♃ In dry woods, flowering in July A common and beautiful evergreen, N. Eng. to Ohio ? and Can. Leaves in 2 or more irregular whorls, 2—3' long, ⅓ as wide, remotely and distinctly serrate on short petioles, coriaceous, shining, of a uniform dark green color. Pedun cle terminal, erect, 3—4' long, bearing 4—7 light purple flowers on nodding pedicles 8" long. Both this and the following species are tonic and diuretic *Bw.* July.

2. C. maculàta. Pursh. (Pyrola. *Linn.*) *Spotted Wintergreen.*

Lvs. lanceolate, acuminate, rounded at base, remotely serrate, discolored opposite or in 3s ; *ped.* corymbose, 2—3-flowered ; *fil.* woolly.—Can. to Car. Ohio. in sandy woods. Habits much like the last, but it is readily distinguished by its variegated leaves. Stem 3—4' high. Leaves 1—2' long, ⅓ as wide marked with a whitish line along the midvein and veinlets. Flowers purplish white, on nodding pedicels. June, July.

SUBORDER 4.—**MONOTROPEÆ.**

Ovary free from the calyx. Leafless herbs, destitute of verdure.

20. MONOTRŎPA.

Gr. μονος, one, τρεπω, to turn; term inapplicable, as the genus is now modified.

Calyx represented by 1—3 bracts; pet. 5, erect, persistent, gibbus at base; sta. 10; fil. persistent, alternating with 10 reflexed appadages of the torus; stig. orbicular, naked; caps. 5-celled.—*Para st. c herbs. St. or scape 1-flowered, scentless.*

M. UNIFLÒRA. *Indian Pipe. Bird's-nest.*
St. short; *scales* approximate; *fl.* nodding; *fr.* erect.—Common in woods, C. to Car. W. to Ill. A small, succulent plant, about 6' high, yellowish-white in all its parts. Stem furnished with sessile, lanceolate, semi-transparent leaves, or bracts, and bearing a large, terminal, solitary flower. Common in woods, near the base of trees, on whose roots it is said to be parasitic. Jn.

21. HYPOPÏTYS. Dill.

Gr. ὑπω, under, πιτυς, a pine tree; its place of growth.   ,

Sepals 4—5, colored; pet. as many as sepals, a little longer and of the same color, erect, deciduous, gibbous at base; sta. 8—10; fil. subulate, persistent; anth. 2-celled, small; stig. discoid, umbilicate; caps. 4—5-celled, 4—5-valved, many-seeded.—*Parasitic herbs, of a tawny white. Root scaly. St. simple. Fls. racemed, lateral ones tetramerous, terminal one pentamerous.*

1. H. MULTIFLÒRA. Scop. (H. Europæa. *Don.* Monotropa. *Linn.*)
Pet., sta. and *sty.* hirsute; *caps.* oval-oblong.
β. *Americana.* DC. (H. Europæa. *Nutt.*) *Plant* smaller, yellowish-brown.—In pine woods, Can., Penn., Car., UC. Is not this rather a variety of the following? It seems to be lost to recent botanists.

2. H. LANUGINÒSA. (Monotropa. *Michx.* and 1*st edit.*) *Pine Sap.*
Plant clothed with a velvet-like pubescence; *pedicels* much longer than the flower; *caps.* subglobose.—Woods, N. Y.! Can to Car. W. to Wisc.! The whole plant is of a tawny white, similar to the last. The root is a tangled mass of fibres. Scape 6—10' high, with many concave scales, covered with down. Flowers 7—12, in a terminal raceme, yellowish, drooping at first, becoming erect. Pedicels 1—2" long, bracts and flowers 3 times as long. Only the terminal flower is generally decandrous; the lateral ones have 8 stamens and 4 petals. Woods. Aug.

22. PTEROSPŎRA.

Gr. πτερος, a wing, σπορα, a seed; alluding to the winged seeds.

Calyx 5-parted; corolla roundish-ovoid, the limb 5-toothed and reflexed; stamens 10; anthers peltate, 2-celled, 2-awned; capsule 5-celled, 5-valved; seeds very numerous, minute, winged at the apex.—*♃ Plant leafless, brownish-red. Fls. racemed.*

P. ANDROMEDEA. Nutt. (Monotropa procera. *Ea.*) *Albany Beech-drops.*
In various parts of N. Y.! and Vt., rare. First discovered by Dr. D. S. C. H. Smith, near Niagara Falls, 1816. Scape 12—30' high, dark purple, clothed with short, viscid wool. Raceme 6—12' long, with 50 or more nodding flowers. Pedicels irregularly scattered, 6—8" long, axillary to long, linear bracts. Corolla shorter than the pedicels, somewhat campanulate, open at the throat, white, tipped with red at the summit. July.

ORDER LXXIX. AQUIFOLIACEÆ.—HOLLYWORTS.

Shrubs or trees, with evergreen, alternate or opposite, simple, coriaceous, exstipulate leaves.
Fls. small, white or greenish, axillary, solitary or clustered, sometimes diœcious.
Cal.—Sepals 4—6, imbricate in æstivation.
Cor. regular, 4—6-cleft or parted, hypogynous, imbricate in æstivation.
Sta. inserted into the tube of the corolla and alternate with its segments. *Anth.* adnate.
Ova. free from the calyx, 2—6-celled, with a solitary, suspended ovule in each cell.
Fr. drupaceous, with 2—6 stones or nucules. *Albumen* large, fleshy.

Genera 11, species 110, natives of America and S. Africa, only one, Ilex (the Holly), being found in Europe.

Properties.—The bark and leaves of Prinos verticillatus (black alder) are eminently astringent and tonic, as well as those of the holly. The berries are emetic and purgative. The leaves of Prinos glaber, and Ilex Paraguensis are used for tea.

Conspectus of the Genera.

Leaves { unarmed. { Petals united, mostly hexamerous.	. . . . . .	Prinos. 3
{ Petals distinct, mostly pentamerous.	. . . . .	Nemopanthus. 2
{ spinose, evergreen, coriaceous.	. . . .	Ilex. 1

1. ILEX.

The ancient Lat. name of the Holm Oak, the derivation uncertain.

Calyx 4—5-toothed, persistent; corolla subrotate, 4—5-parted; stamens 4—5; stigmas 4—5, subsessile, united or distinct; berry 4—5-seeded.—*Shrubs and trees. Lvs. alternate and spinose-dentate. Fls. often* ♂ ♀ *by abortion.*

1. OPÁCA.

Lvs. evergreen, oval, acute at end, with strong, spinous teeth, coriaceous, smooth and shining; *fascicles of fls.* lax, peduncles compound; *cal.* teeth acute; *fr.* ovate; *fls.* small, greenish-white.—A tree of middle size, quite generally diffused throughout the U. S. from Mass.! to La. It is chiefly interesting for its foliage, which is of an exceedingly rich, shining, perennial green. The flowers appear in June, in scattered clusters at the base of the older branches, and the fertile ones are succeeded by red berries which remain until late in autumn. The wood is fine grained and compact, useful in turnery, &c.

2. NEMOPANTHUS. Raf.

Gr. νημα, thread. πους, foot-stalk, ανθος; that is, a flower on a filiform peduncle.

Calyx minute; petals 5, distinct, linear, oblong; stamens 5; ovary hemispherical; stigmas 3—4, sessile; fruit a 3—4-celled, subglobose berry.—*Shrub, with alternate, entire, deciduous leaves. Fls. mostly diœcio-polygamous by abortion.*

N. CANADENSIS. Raf. (Ilex. *Michx.) Canadian Holly.*

Lvs. deciduous, oval, very entire, smooth, mucronate-pointed; *ped.* nearly solitary, very long; *fr.* somewhat 4-sided.—A shrub, 4—6f high, with smooth branches, growing in damp or rocky woods, Can., N. Eng.! to Mich. Leaves oval or ovate-oblong, about 2' long, on petioles ⅓ as long. The flowers, growing on long, slender, axillary peduncles which are seldom divided, are small, greenish-white. Segments of the corolla acute, long as the stamens. Ovary of the barren flowers pointed, of the fertile with a 4-lobed stigma. Berries dry, red. May, June.

3. PRINOS.

Gr. πριω, to saw; alluding to the serrated leaves.

Flowers often ♂♀ or ♂♀♀; calyx mostly 6-cleft; cor. 6-parted; sta. 4—6; berry roundish, much longer than the calyx; seeds bony, convex on one side, angular on the other.—*Shrubs. Lvs. alternate. Peduncels axillary, 1-flowered.*

1. P. VERTICILLÁTUS. (P. Gronovii. *Michx.) Winter Berry. Black Alder.*

Lvs. deciduous, oval, serrate, acuminate, pubescent beneath; *fls.* axillary, the fertile ones aggregate, the barren subumbellate.—This shrub is found in moist woods or swamps, Can. and most of the States, usually growing about

8f' high. Leaves narrowed at base into a short petiole, uncinately serrate, with prominent, pubescent veins beneath. Flowers white, dioecious, small, in imperfect umbels or heads, sometimes monoecious. Berries scarlet, in little bunches (apparently verticillate), roundish, 6-celled and 6-seeded, permanent. Jl.

2. P. AMBIGUUS. Michx. *Dubious Winter Berry.*

Lrs. deciduous, oval, entire, acuminate at both ends; *parts of the fls.* in 4s, the sterile ones crowded, the fertile solitary.—A shrub or small tree, 8—15f high, in wet grounds, Mid. States. Bark whitish, smooth. Leaves elliptic-oval, mucronate-pointed, petiolate, subpubescent beneath, 1—2' long and half as wide. Flowers polygamous, 4—5-cleft, the fertile ones on long peduncles. June.

3. P. LÆVIGĀTUS. Pursh.

Lrs. deciduous, lanceolate, appressed-serrulate, glabrous on both sides, shining above, minutely pubescent on the veins beneath; *fls.* hexamerous, ♀ axillary, subsessile, ♂ scattered, pedicellate.—In swamps and marshes, Northern and Western States! S. to N. J. Shrub 6—9f high, with grayish and warty branches. Leaves 2—3' by 8—12'', acute at each end; petioles 6—10'' long. Flowers mostly solitary, the sterile on pedicels near 1' long, the fertile pedicels scarcely ⅓ as long. Berries large, red. June.

4. P. LANCEOLĀTUS. Pursh.

Lrs. lanceolate, acute at each end, finely and remotely serrulate, glabrous both sides; ♂ *flowers* aggregated, triandrous, ♀ generally in pairs, pedunculate, 6—numerous; *berries* small, scarlet.—Barrens and marshes, Western (*Riddell*) and Southern States! Shrub 6—8f high.—I am wholly unacquainted with this species as a western plant, but have specimens collected in N. J.? by *Dr. Robbins.* The leaves are paler beneath, 2—3' (including the petiole 3—6'') by ⅓—1½', veins beneath pubescent, ferruginous. Pedicels of the barren flowers are ¼' long, of the fertile ¼' long. June.

5. P. GLABER. *Ink Berry. Evergreen Prinos.*

Lrs. evergreen, coriaceous, cuneate-lanceolate, glabrous, shining, serrate at the end.—A beautiful shrub 3—4f high, found in swamps, Mass.! R. I.! to N. Y. and Car. Leaves very smooth, leathery, shining, 1—1½' by 5—7'', broadest above the middle. Pedicels subsolitary, 1—3-flowered. Flowers white, mostly 6-parted. Berries roundish, black and shining. June, July.

ORDER LXXX. EBENACEÆ.—EBONADS.

Trees or *shrubs* without milky juice and with a heavy wood.
Lvs. alternate, exstipulate, coriaceous, entire. *Inflorescence* axillary.
Fls by abortion dioecious, seldom perfect. *Cal.* free, 3—6-cleft, divisions nearly equal, persistent.
Cor. regular, 3—6-cleft, often pubescent imbricate in æstivation.
Sta. twice or four times as many as the lobes of the corolla.
Ova. with 3 or more cells. *Style* with as many divisions.
Fr. a fleshy, oval or globose berry.

Genera 9, species 160, mostly natives of the Indies and the tropics, one only being found as far north as New York.

Properties.—Diospyrus is remarkable for the hardness and dark color of the wood. Ebony is the wood of D Ebenus, Ebenaster, and other species, natives of Africa. The fruit of the species below is eatable when fully ripe, although extremely bitter and astringent before maturity. The bark is eminently febrifugal and astringent.

DIOSPŸROS.

Gr. Διος πυρος, the fruit of Jove; the fruit, although excellent, hardly merits the name.

Fls. ♂♀. Cal. 4—6-lobed; cor. tubular or campanulate, 4—6-cleft, convolute in æstivation. ♂ sta. 8—50. mostly 16; fil. shorter than the anthers; ova. abortive; sty. 0. ♀ sta. mostly 8, without anthers; sty. 2—4-cleft; berry ovoid or globose, 4—12, mostly 8-celled. cells 1-seeded.—*A large genus of shrubs or trees, mostly tropical.*

D. VIRGINIĀNA. *Persimmon Tree.*

Lrs. elliptic, abruptly acuminate, entire, smooth, petiole, veins and margin puberulent; *rac.* axillary, 3—1-flowered, pedicels shorter than the flowers; *crl.* 4-parted; *s'a.* 8.—In woods, lat. 42° to La., frequent. A shrub or small tree at the North, a tree of large dimensions at the South Leaves 3—5' long, entire, glaucous beneath. Flowers obscure, pale greenish-yellow, the fertile ones succeeded by a round, orange-red fruit as large as the garden plum, and containing 6—8 stony seeds. They are rendered sweet and palatable by the frost, although very austere when green. The bark is tonic and astringent. Jn.

Order LXXXI. STYRACACEÆ.

Trees or shrubs with alternate, simple leaves destitute of stipules.
Fls. or racemes solitary, axillary, bracteate.
Cal. 5, rarely 4-lobed, imbricated in æstivation.
Cor. 5 rarely 4 or 6-lobed, imbricated in æstivation.
Sta. definite or ∞, unequal in length, usually cohering. Anth. innate, 2-celled.
Ova. adherent, 2—5-celled, the partitions sometimes hardly reaching the centre.
Fr. drupaceous, generally with but one fertile cell. Sds. 5—1.

Genera 6, species 115, sparingly distributed through the tropical and subtropical regions of both continents, only a few in colder latitudes. Storax and benzoin, two fragrant gum resins, regarded as stimulant and expectorant, are the products of two species of Styrax, viz. of S. officinale, a Syrian tree, and S. benzoin, native of Malay and the adjacent islands.

HALESIA. Ellis.
In honor of the learned and venerable Stephen Hales, D.D., F.R.S., 1730.

Calyx obconic, briefly 4-lobed ; cor. inserted into the calyx, campanulate, with a narrow base, 4-cleft or 4-parted ; sta. 8—12, connate into a tube below ; sty. filiform, pubescent ; fruit dry, 4-winged, wings equal or alternately smaller ; seeds 1—3.—*N. American shrubs.*

1. H. TETRAPTÉRA. *Four-winged Snowdrop Tree.*
Lvs. elliptic-acuminate, serrulate; *fascicles* 3-flowered, lateral, leafless, from the wood of the preceding year; *cal.* subentire; *sta.* 12; *fr.* with 4 equal wings.—Native of S. Car. to Flor., *Miss E. Carpenter!* Branches leafy at the summit. Leaves thin, 2—5' by 1½—2', obtuse or acute at base. Flowers pendulous, white, about 10'' long. †

2 H. DIPTÉRA. *Two-winged Snowdrop Tree.*
Lvs. oblong-ovate, obtuse, acuminate at each end, serrulate, softly pubescent beneath; *fascicles* 2—3-flowered, lateral ; *pedicels* and *cal.* pubescent; *sta.* 8; *fr.* with the alternate wings half as large or obsolete.—Native in Car. and Ga. Leaves somewhat larger than in the last, with rather smaller flowers. Corolla white. †

Order LXXXII. PRIMULACEÆ.—
PRIMWORTS.

Plants herbaceous, annual or perennial, sometimes suffruticose.
Lvs. usually radical, otherwise mostly opposite Stipules 0.
Fls. on scapes and in umbels or variously arranged in the axils of the leaves.
Cal. 5 (rarely 4)-cleft, inferior, regular, persistent.
Cor. 5 (rarely 4)-cleft, regular.
Sta. inserted on the tube of the corolla, as many as its lobes and opposite to them.
Ova. 1-celled, with a free, central placenta. Style and Stigma simple.
Fr.—Capsule many-seeded, the fleshy placenta attached only to the base of the cell.

Genera 28, species 215, common in the northern temperate regions, growing in swamps, groves, by rivulets and often among the snow of cloud-capped mountains. Many are beautiful, and highly prized in culture. Properties unimportant.

FIG. 49.—1. Primula Mistassinica. 2. A flower laid open showing the 5 stamens inserted on the tube of the corolla. 4. Plan of the flower. 6. Ovary and calyx. 5. Vertical section of the ovary, showing the free central placenta.

Conspectus of the Genera.

	7-cleft. Leaves in one whorl.	.	*Trientalis.* 4
	5-cleft. Flowers paniculate.	.	*Samolus.* 10
Corolla white.	4-cleft. Plants 2 inches high.	.	*Centunculus.* 9
	Stamens 6. Racemes axillary.	.	*Naumburgia.* 6
Corolla yellow.	Stamens 5.	.	*Lysimachia.* 7
Corolla scarlet.	Plant prostrate. Fls. solitary, axillary.		*Anagallis.* 8
undivided. Corolla wanting, calyx colored, white.	.	.	*Glaux.* 5
cauline. pectinate-pinnatifid, submersed in water.	.	.	*Hottonia.* 1
	Corolla segments spreading.	.	*Primula.* 2
Leaves all radical. Scape umbellate.	Corolla segments reflexed.	.	*Dodecatheon.* 3

1. HOTTONIA.

In honor of Peter Hotten, professor in the University of Leyden, died 1709.

Calyx 5-parted; cor. salver-form, with a short tube and a flat, 5-lobed limb; sta. inserted on the tube of the corolla, included; stig. globose; caps. globose-acuminate.—♃ *Fleshy, aquatic herbs, with pectinate-pinnatifid, submersed, radical lvs.*

H. INFLÀTA. Ell. (H. palustris. *Ph.* not *Linn.*)   *Water Feather.*
Scape articulate, the internodes and lower parts inflated; *fls.* verticillate, pedunculate.—A curious aquatic plant, in swamps and stagnant waters, Ms., R. I. and Ct., N. Y. to Flor. Stem immersed, round, thick, spongy, with a whorl of long and beautifully pectinate leaves at or near the surface of the water. Peduncles or scapes several (6—10) together, arising in a sort of umbel from the top of the stem, 8—10′ long, inflated between the joints, forming the most remarkable feature of the plant. Flowers small, white, in numerous verticils, generally 4 in each. Pedicels ½′ long. June.

2. PRIMÜLA.

Lat. *primus.* first; because its blossoms appear earliest in spring.

Corolla salver-form, with an open orifice; capsule opening with a 10-cleft dehiscence; stamens 5, not exserted; stigma globose.—*Herbs (mostly European) with radical lvs. Fls. in an involucrate umbel on a scape, showy.*

1. P. MISTASSINÏCA. Michx. (Fig. 49.)
Lvs. spatulate, dentate or crenate, obtuse or acute, attenuate at base; *invol.* 1—8-flowered; *bracts* 3 times shorter than the pedicels, linear-subulate; *cal.* much shorter than the tube of the corolla; *cor.* salver-form lobes obcordate.—Shores of Seneca Lake, N. Y., *Dr. Sartwell!* Cliffs, Willoughby lake, Vt.! throughout Brit. Am. A very delicate plant, about 3′ high. Leaves about 5, 5—8″ by 3—4″, almost petiolate. Flowers 5″ diam., white. Pedicels 7″ in length.

2. P. FARINÒSA. *β. Americana.* Torr. *Bird's-eye Primrose.*
Lvs. narrow, veiny, elliptic-lanceolate, obtuse, denticulate at apex, attenuate at base, under surface covered with a yellowish-white, farinaceous dust *invol.* farinaceous, 3—20-flowered, shorter than the pedicels; *bracts* long-acumi nate; *cal. segments* lanceolate, acute; *cor.* salver-form, lobes obcordate, bifid c obtuse.—Shores of Lakes Huron and Superior, *Nutt.*, *Houghton*, N. to lat. 66°. Scape 6—12′ high. Flowers purple or flesh-color.

3. P. AURICÜLA. *Auricled Primrose* or *Auricula.*—*Lvs.* obovate, entire or serrate, fleshy; *scape* many-flowered, central, as long as the leaves; *invol.* of short leaves; *cal.* powdery.—♃ Native of the Alps. A well known favorite of the florist. The cultivated varieties are innumerable, and many of them of exquisite beauty and fragrance. May. †

4. P. ELATIOR. Jacq. *Oxlip Primrose.*—*Lvs.* toothed, rugose, hairy on each side; *umbel* many-flowered, with the outer flowers nodding; *cor.* flat.—♃ Native of Britain. Flowers yellow, scentless, in a simple umbel elevated upon a scape a foot high. Apr. May. †

5. P. OFFICINÀLIS. Jacq. (P. veris. *Cam.*) *Cowslip Primrose.*—*Lvs.* toothed. rugose, hairy beneath; *umbels* many-flowered, flowers all nodding; *cal.* angular, *cor.* concave.—♃ Native of Britain. Flowers yellow. The plant smells strongly

of anise. Leaves are used as a potherb, and are recommended for feeding silkworms. Its varieties may be increased by raising from the seed. June. †

6. P. PURPUREA. Royl. *Purple Primrose.*—Lrs. lanceolate, obtuse, very smooth, covered beneath with yellowish farina, margin undulate, revolute; *scape* thick, glabrous, longer than the leaves; *invol.* 00-flowered, as long as the pedicels, farinaceous beneath; *cor. segments* obovate, obtuse, not emarginate.—Native of the Mountains of Napaul, Asia. Flowers dark purple. †

7. P. CALYCINA. Duby. *Double-cupped Primrose.*—Lrs. lanceolate, thin, smooth, entire, acute, surrounded with a white margin; *invol.* 3—5-flowered, as long as the pedicels; *cal. tube* ventricose; *cor. lobes* obcordate, emarginate.—Native of Mts. in Austria. Flowers purple, very beautiful. †

8. P. GRANDIFLORA. Lam. (P. vulgaris. *Huds.*) *Common Primrose.*—Lrs. obovate, oblong, rugose, villous beneath, toothed; *umbel* radical; *fl. stalks* as long as the leaves; *cor.* flat.—♃ Native of Europe. An interesting garden plant, esteemed for its early flowering, and for its being prolific in variation. In its wild state its flowers are yellow and single, but by cultivation they become double, and in the numerous varieties, red, pink, white, orange, purple, &c., and the umbels, in numerous instances, are on a scape. The roots and leaves smell of anise seed, and when dried and powered, are used as a snuff, and also as an emetic. The number of varieties is vast, and is readily increased by cultivation from seed. April. †

3. DODECATHEON.

Gr. δοδεκα, twelve, Σεος, god; alluding to its curious flowers which are about 12?

Calyx 5-parted, reflexed; cor. tube very short, limb rotate, 5-parted, segments reflexed: sta. 5, inserted into the throat of the corolla; fil. very short; anth. large, acute, connivent at apex; style exserted; caps. oblong-ovoid, 5-valved, many-seeded.—♃ *with radical, oblong lvs., an erect, simple scape, and a terminal umbel of nodding flowers.*

1. D. MEADIA. *American Cowslip, or Mead's Cowslip.*
Lrs. oval or oblong, obtuse, attenuate at base into a marginal petiole, glabrous, entire or repandly dentate; *scape* 9—20-flowered; *bracts of the invol.* ovate, inner ones lanceolate; *sep.* lanceolate, acute, entire; *fil.* united into a tube much shorter than the subulate anthers.—A singular, elegant herb, on prairies, dry or rocky soils, Penn. to Ind. *Dr. Skinner!* Ill., Wisc. and throughout the Western States. Whole plant very smooth. Leaves all radical, 7—10′ by 1½—2½′, on the margin usually undulate or repand-toothed. Scape 1—2f high. Involucre much shorter than the pedicels which are very slender, 1—2′ in length. Corolla white or purplish, abruptly reflexed. Anthers 5″ long, yellow, purple at base. May, Jn. † I have specimens gathered in Tenn. by *Miss E. Carpenter*, with leaves perfectly even and entire.

2. D. INTEGRIFOLIUM. Michx.
Lrs. ovate or lanceolate, subspatulate, subentire, obtuse, petiolate; *scape* few-flowered; *fls.* suberect; *bracts of the invol.* lanceolate or linear acute; *cal. segments* lanceolate, acute, entire; *fil.* connate into an elongated tube, nearly as long as the anthers.—A much smaller plant than the preceding. *Nuttall.*—I have specimens of this species? collected in Ill. by *Rev. E. Jenney.* Scape 6—10′ high, thick. Leaves 3—5′ long, on naked petioles. Flowers 5—11, somewhat crowded, on shorter pedicels. Corolla bluish-white.—It may be only a variety of D. Meadia.

4. TRIENTALIS.

Lat. triens, the third part of a foot (4′); alluding to the height of the plant.

Calyx and cor. 7-(6—8-) parted, spreading; sta. 7 (6—8); fr. capsular, somewhat fleshy, many-seed.—*St. low, simple. Lvs. subverticillate. Ped. 1-flowered.*

T. AMERICANA. Ph. (T. Europæa. *Michx.*) *Chickweed Wintergreen.*
St. erect, simple, leafless at base; *lvs.* glomerate, few, narrow-lanceolate,

serrulate, acuminate; *sep.* linear, acuminate. This little plant is common in the rocky woods of N. Eng., N. Y., and Brit. Am. Stem 3—6' high, with an irregular whorl of 4—8 lanceolate, smooth and shining leaves at the top. In the midst of these are 1—4 white, star-like flowers, borne on simple, filiform pedicels. The leaves are mostly acuminate at each end, 3' long and 1' wide. Segments of the corolla longer than the acute calyx leaves. Seeds attached to a central, spongy placenta. May, Jn.

5. GLAUX.

Gr. γλαυκος, bluish or glaucous; from the hue of the plant.

Calyx campanulate, 5-lobed, colored; corolla 0; stamens 5; capsule roundish, surrounded by the calyx, 5-valved, 5-seeded.—♃ *Maritime, branching, glabrous.*

     G. MARITIMA. *B'ack Saltwort.*

A small, smooth, fleshy plant, found occasionally in the salt marshes on our seacoast, Can. to N. J. Root perennial. Stem more or less procumbent at base, 4—6' high, smooth, branching, and very leafy. Leaves 1' in length, roundish-ovate, obtuse, entire, nearly or quite sessile, smooth, fleshy and darkly glaucous. Flowers small, sessile, axillary, solitary. Calyx white, tinged with red. July.

6. NAUMBURGIA. Mœnch.

Calyx and cor. deeply 5—6-parted; pet. linear-lanceolate, spreading. separated by minute, intervening teeth; sta. 5—6, inserted into the base of the corolla, exserted; anth. cordate; caps. globose, 5-valved; seeds few, on a globose placenta.—♃ *with opposite lvs.* Fls. *small, in dense, thyrsoid racemes.*

     N. THYRSIFLORA. M rch. (Lysimachia. *Linn.* and 1st *edit.* L. capitata. *Ph.*)—An erect, smooth herb, about 2f high, in swamps, Mass., Vt., N. Y.! W. to Ohio! N. to Arc. Am. Leaves many pairs, sessile, lanceolate, acute, entire, punctate, somewhat canescent beneath, 2—3' by ½—1'. Racemes somewhat capitate, on filiform, axillary peduncles. Flowers yellow. Stamens much exserted, united into a tube at base. Ju.

7. LYSIMACHIA.

Calyx 5-parted, rotate or campanulate, tube very short; sta. 5, inserted into the corolla at base; fil. often somewhat connate, or with intervening sterile ones; caps. globose, 5—10-valved, opening at the apex; seeds 00.—*Herbs mostly* ♃, *with opposite or verticillate, entire lvs.*

     1. L. STRICTA. Ait. (L. racemosa. *Mx.* Viscum terrestre. *Linn.*)

Simple or branched, erect; *lvs.* opposite or ternate, lanceolate or lance-linear, glabrous, punctate, acute, sessile; *fls.* verticillate, in a long, lax, term'. nal raceme; *pet.* lanceolate, spreading.—♃ In low, wet grounds, Can., N. Eng. to Va and Ohio. Common. Plant smooth, 1—2f high, bearing at top a regular, cylindric or conical raceme, 6—8' long. Peduncles an inch in length, quite spreading, each with a subulate bract at base. Stamens 2 long and 3 short, united at base. Flowers yellow, spotted with purple. Capsules 5-seeded. After flowering it throws out bulblets from the axils of the leaves, which will produce new plants the following spring. July.

     2. L. CILIATA. *Fringe-stalk Loosestrife. Heart-leaved Loosestrife.*

Subsimple, erect; *lvs.* opposite, rarely quaternate, ovate, subcordate or ovate-lanceolate, petioles ciliate upper side; *fls.* nodding, mostly opposite; *sta.* distinct, with 5 abortive filaments.—♃ In gravely soils and near streams, U. S. and Can. Root creeping. Stem somewhat 4-sided, 2—3f high, simple or with a few opposite branches. Leaves large, pointed, somewhat cordate at base, on petioles fringed with cilea, the upper ones apparently quaternate. Flowers large, yellow, axillary. Stamens inserted into a ring, nearly equal, with 5 alternate and intermedia e, rudimentary filaments or teeth. July.

3. L. **hybrida** Michx. *Hybrid Loosestrife.*

Smooth, erect, branched above; *lvs.* narrowly lanceolate, acute at each end, opposite, petioles ciliate, short; *fls.* nodding; *ped.* axillary; *sta.* united in a very short tube at base, with intermediate processes.—♃ Grassy meadows and prairies, Can., N. H.! to Car., W. to Ia.! and Ill. *Mead.* Its stem and flowers resemble those of L. ciliata, from which species it is chiefly distinguished by its narrower, lanceolate, never cordate leaves, and its more numerous and leafy branches, each of which bears a whorl of 4 leaves and 4 flowers at the ex 1. Stamens with intermediate processes. Jl.

4. L. **quadrifolia.** *Four-leaved Loosestrife.*

Simple, erect; *lvs.* verticillate, in 4s, rarely in 5s or 3s, sessile, lanceolate, acuminate, punctate; *ped.* axillary, 1-flowered, in 4s (3s or 5s); *pet.* oval, obtuse.—♃ In low grounds, river banks, Can. to Car. and Ky. Stem 18' high, somewhat hairy, simple, with many whorls of 4—5 leaves, each bearing a flower-stalk in its axil. Corolla yellow, with purple lines. Stamens unequal, united at base into a short tube. Anth. purple. Jn.

5. L. **heterophylla.** Michx. *Various-leaved Loosestrife.*

Erect, subsimple; *lvs.* opposite, linear-lanceolate, acute, entire, margin not reflexed, ciliate at base and on the short petiole, lower ones shorter, oval, obovate or even roundish; distinctly petiolate; *fls.* on long peduncles, opposite or apparently quaternate at top of the stem; *sta.* subequal, with intervening teeth.—Hills and woods, Ohio, *Clark!* Ill. *Jenney!* to Ga. Stem 12—18' high, branched from the base if at all. Leaves about 2½' by 5", lower 1' by ½', all paler beneath. Flowers very similar to those of L. ciliata. Jn. Jl.

6. L. **longifolia.** Pursh. (L. revoluta. *Nutt.*) *Prairie Moneywort.*

St. slender, 4-angled, flexuous, branched above; *lvs.* opposite, linear-oblong, sessile, margin revolute; *fls.* opposite or mostly quaternate and terminal on the stem and branches; *sep.* lance-linear, acuminate; *pet.* longer than the calyx, roundish-ovate, erose-dentate, abruptly acuminate.—Common in low prairies, W. States! The large yellow flowers are very conspicuous among the grasses. Stems 12—20' high, purple. Leaves 2—3' by 2—3", coriaceous, deep green. Flowers numerous, 9" diam., of a brilliant yellow. Anth. large. Jl.

8. ANAGALLIS.

Gr. αναγελαω, to laugh; it is said to be medicinally efficacious in expeding hypochondria.

Calyx 5-parted; cor. rotate, deeply 5-parted, longer than the calyx, tube 0; sta. 5, hirsute; anth. introrse; caps. globose, membranaceous, circumscissile.—*Herbs with square stems and (mostly) opposite lvs. Ped. axillary, solitary.*

A. **arvensis.** *Scarlet Pimpernel. Poor Man's Weather-glass.*

Procumbent, branched; *lvs.* broad-ovate, opposite or ternate, sessile; *ped.* longer than the leaves; *sep.* linear-lanceolate, about equaling the petals; *pet.* crenate-glandular.—A beautiful trailing plant, in fields, roadsides, &c., U. S. (except the colder parts of N. Eng.) and in almost all other countries. Stem 6—20' long, with elongated branches, or simple. Leaves 6—8" by 4—6". Flowers opposite, small but beautiful, with scarlet petals, opening at 8 o'clock, A. M., and closing at 2 P. M., in damp weather not open at all. Jn.—Aug.

9. CENTUNCŬLUS.

Calyx 5-parted; cor. urceolate-rotate, 4-cleft, shorter than the calyx; sta. 4, beardless, united at base; caps. globose, circumscissile; seeds very minute.—ⅅ *Very diminutive, with alternate lvs. Fls. axillary, solitary, subsessile.*

C. **minimus.** (C. lanceolatus. *Michx.*) *Bastard Pimpernel.*

Erect, branched; *lvs.* subsessile, ovate or lance-ovate, obtusish, entire, alternate, lower opposite; *sep.* linear-subulate, equaling the capsule.—Wet places, Ill. *Mead!* and Southern States. Scarcely more than 1—2 high (4—6", *Riddell*). Leaves about 2" by 1". Flowers reddish? Jl.

33*

10. SAMÓLUS.

Celtic *san*, salutary, *mos*, a pig ; a specific for the diseases of swine, says Pliny.

Calyx partly adherent, 5-cleft ; cor. hypocrateriform, 5-cleft ; sta. 5, alternating with 5 scales (sterile filaments) ; caps. dehiscent at top by 5 valves, many-seeded.—*Herbs with alternate leaves.   Fls. corymbous: or racemose.*

1. S. VALERANDI.   *Water Pimpernel.*

St. subsimple ; *lvs.* ovate, obtuse, radical ones on long petioles, upper sessile ; *fls.* racemed ; *pedicels* with a minute, obtuse bract near the middle ; *pet.* twice longer than the sepals.—♃ In wet, gravely places, throughout the world, frequent but not abundant.   Stem 1f high, round, with alternate, entire, broadly lanceolate leaves an inch in length, and tapering to short petioles.   Racemes terminal, long.   Flowers small, white, their pedicels having a small bract near the middle.   Corolla twice the length of the calyx, the tube broad, the limb bearing the barren filaments between its deep-cleft, obtuse lobes, and the true stamens opposite the latter and alternate with the former.   July—Sept.

2. S. FLORIBUNDUS. Kunth.   *Many-flowered Samolus.*

St. branched above ; *lvs.* obtuse, entire, radical obovate-spatulate, petiolate, round-obtuse, cauline oblong, obtuse, lower petiolate, upper subsessile ; *fls.* paniculate-corymbose ; *pedicels* filiform, with a minute, acute bract near the middle ; *pet.* scarcely longer than the sepals.—R. 1. *Olney!* to La., Ohio, *Clark* I and Ia.!   Stem 10—15' high.   Lower leaves 14" by 6", upper 8" by 5".   Flowers numerous, white, twice smaller than in the last.   Pedicels 6—8" in length. July—Sept.

ORDER LXXXIV. PLANTAGINACEÆ.—RIBWORTS.

Plants herbaceous, usually acaulescent.   *Lvs.* mostly rosulate.   *Fls.* in spikes.
Cal. 4-cleft, persistent.
Cor. membranaceous, the limb 4-parted, persistent.
Sta. 4, inserted into the tube of the corolla, alternate with its segments.   *Anth.* versatile.
Ova. 2 celled, sessile   *Style* single.
Fr.—Pyxis membranaceous, cells 1—2 or several-seeded.

Genera 3, species 130, scattered throughout all countries of the globe.   Properties unimportant

PLANTÁGO.

Calyx 4 (rarely 3)-parted ; corolla marescent, with a 4-cleft, reflected border ; stamens mostly exserted and very long ; capsule ovoid, 2-celled, circumscissile.—*Acaulescent herbs.   Lvs. all radical Fls. spicate.*

* *Leaves dilated.*

1. P. CORDÁTA. Lam.   *Heart-leaved Plantain.*

Lvs. cordate-ovate, broad, smooth, somewhat toothed ; *spikes* very long, flowers somewhat imbricate, lower ones scattered, with ovate, obtuse bracts.— ♃ Can. to Tenn. and N. J.   Well marked by its broadly heart-shaped leaves and its elongated spikes which are 6—8' long, and on scapes twice as high. Leaves 6' long, more or less cordate at base.   Corolla white, with obovate segments.   Pyxis a third longer than the calyx, with 2 seeds in each cell.   Jn. Jl.

2. P. MAJOR.   *Common Plantain or Ribwort.*

Lvs. ovate, smoothish, somewhat toothed, with rather long footstalks ; *scape* round, *fls.* imbricated ; *sds.* numerous.— ♃ This species is a native of Japan. Europe and America, is very common, always at the door and by the wayside.   The leaves are reputed a good external application for wounds, &c. The seeds are eaten by sparrows and other small birds.   The root consists of long fibres.   Leaves broad, flat, with about 7 veins, each containing a strong fibre, which may be pulled out.   Scape 1—3f high, with a very long (5—20'), cylindric spike.   Fls. white, inconspicuous, appearing in succession all summer.

3. P. LANCEOLÁTA.   *Lance-leaved Plantain.*

Lvs. lanceolate, tapering at each end ; *spikes* ovate. naked. *scape* angular

2↓ Can. to Ga. Common in pastures and grass-lands. Easily known by its longer leaves tapering at the base into a broad stalk, and with from 3 to 5 strong ribs; by its shorter spike (1—2' long), with dark colored calyxes and whitish, projecting stamens, and its slender, upright stalk (8—15' long) with prominent angles. Flowering from May to Oct. It is freely eaten by cattle.

4. P. MEDIA. *Hoary Plantain.*

Lvs. ovate, pubescent; *spike* short, cylindric; *scape* round; *sds.* solitary in each cell of the pyxis.—2↓ Grows in pastures and roadsides, N. Y. and Penn., flowering all summer. It has broad, flat leaves 2' long, covered with a hoary down, and with short footstalks. Spikes shorter than those of *P. major*, being 1—3' long and about half a foot high. Flowers white, with pink filaments and yellow anthers.

5. P. VIRGINICA. *Virginian or Lesser Plantain.*

Lvs. obovate-lanceolate, hoary-pubescent, subdenticulate; *scape* angular; *spike* cylindric, pubescent, with flowers somewhat remote.—A biennial species, on sandy or stony hills in the southern parts of N. England and N. Y. to La., much smaller than the preceding. The whole plant is covered with a soft, gray pubescence. Scape 4—8' high, very hairy Leaves 2—3' long, narrowed at base into the petiole, obtuse at the end. Corolla yellowish, with very acute segments including the stamens. Jl.

6. P. CUCULLATA. Lam. (P. maxima. *Jacq.*) *Hood-leaved Plantain.*

Lvs. ovate, slightly denticulate, 9-veined, cucullate at base, contracted into a long petiole; *scape* terete; *spike* cylindrical, short, dense-flowered; *sta.* exserted.—In wet, rocky situations, Me., *Pursh.* Scape 1—3f high, with a spike 2—3' in length. Leaves large, conspicuously rolled in or hooded at base.—A doubtful native of this country.

7. P. GLABRA. Nutt. *Smooth Plantain.*

Lvs. glabrous, ovate, denticulate; *scape* slender, somewhat compressed, nearly as long as the leaves; *fls.* scattered; *bracts* ovate, acuminate.—In arid places, Ohio, *Frank.*, Mo., *Nuttall.*

8. P. GNAPHALIŌIDES. Nutt. (P. lagopus. *Ph.* not of *Linn.*)

Whole plant clothed with a long silky wool; *lvs.* linear-lanceolate, entire, very acute; *spike* long, cylindric and dense-flowered; *sta.* included; *caps.* 2-celled, 2-seeded.—Ark.

β. Nutt. (P. aristata. *Michx.*) *Bracts* very long and spinulose.—Prairies and roadsides, Ill., *Mead.*

* * *Leaves linear.*

9. P. MARITIMA. (P. pauciflora. *Ph.*)

Lvs. linear, channeled, nearly entire, woolly at base; *spike* cylindrical, close; *scape* round.—Grows in salt marshes along the coast, Me. to N. J. It has a large, perennial root sending up a scape varying in height from 3' to a foot, and numerous, very fleshy, dark green, linear leaves deeply grooved on the inside, and 6—10' long. Spike slender, of numerous, sub-imbricate, whitish flowers. Aug.

10. P. PUSILLA. Nutt.

Minutely pubescent; *lvs.* linear-subulate, flat, entire, somewhat fleshy; *scape* terete, slender, longer than the leaves; *spike* interrupted, subcylindrical, loose-flowered below; *bracts* ovate, acute, as long as the calyx.—① A diminutive species, low grounds, Penn. Leaves crowded, about 1' long the scape 2—3', spicate ⅓ its length.

ORDER LXXXV. PLUMBAGINACEÆ.—LEADWORTS.

Plants herbaceous or suffruticose, variable in appearance
Lvs. undivided, alternate, or sometimes all radical and the flowers on a scape.
Cal. tubular, 5 toothed, plaited, persistent.
Cor. regular, hypocrateriform, of 5 petals united at base or sometimes almost distinct.
Sta. 5, hypogynous and opposite the petals or inserted on their claws.
Ova. 1-celled, free from the calyx. *Styles* 5 (seldom 3 or 4).
Pr. an utricle, or dehiscent by valves. Seed inverted.

Genera 6, species 160, mostly seaside or salt marsh plants, found in all latitudes.

Properties.—The root of Statice Limonium is one of the best and most powerful of all astringents. The species of Plumbago are acrid and escharotic, so much so, that the roots of P. Europæum are used by Lindley to be employed in Europe by beggars, to raise blisters on the face, in order to excite compassion.

Genera.

Inflorescence an involucrate head.	*Armeria.* 1
Inflorescence a panicle of racemes.	*Statice.* 2

1. ARMERIA. Willd.

Flowers collected in a dense head; invol. 3—many-leaved; cal. tubular-campanulate, 5-angled, with 5 shallow lobes, carious and plaited; petals 5, almost distinct; sta. 5, inserted on the base of the petals; styles 5. distinct; fr. indehiscent, invested with the calyx.— ♃ *Lvs. radical, mostly linear. Scape simple, appendaged above.*

A. VULGÀRIS. Willd. (Statice Armeria. *Linn.*) *Thrift.*

Scape terete, smooth, *lvs.* linear, flat, obtuse; *outer bracts of the invol.* ovate, acute, shorter than the sheathing appendage at their base.—A neat and elegant plant, native near the sea-coast, Brit. Am. *Hook.* N. Eng.? Middle and Southern States. Often cultivated. Leaves 3—4′ by 2—3″, numerous, crowded. Scape about 1f high, bearing a singular sheath at top; formed, according to Lindley, by the adherent bases of the involucral leaves. Involucre about 3-leaved. Flowers showy, rose-colored. Jn.—Aug. †

2. STATICE.

Gr. σταίζω, to stop; for, used medicinally, it stops the diarrhœa, says Pliny.

Flowers scattered in a paniculate or spicate inflorescence, otherwise essentially the same as in Armeria.—♃ *Lvs. radical or cauline, dilated, mostly entire. Invol. 0.*

1. S. LIMÒNICM. (S. Carolinianum. *Wa't.*) *Marsh Rosemary.*

Scape terete, paniculate; *lvs.* all radical, ovate-lanceolate, undulate, smooth, obtuse, mucronate below the tip.—Salt marshes, R. I. *Olney!* to Md.! and Car. Scape about a foot high, with several lanceolate, clasping bracts, and supporting at top a broad, branching panicle composed of close, secund spikes of sessile, blue flowers. Petals obovate, unguiculate, bearing the stamens on their claws. Leaves narrow, lanceolate, broader in the upper half, smooth, veinless, on long petioles. The root is large, ligneous, strongly astringent, much valued in medicine. Aug.—Oct.

ORDER LXXXVI. LENTIBULACEÆ.—BUTTERWORTS.

Plants herbaceous, aquatic. *Lvs* radical, undivided, or compound, root-like and bearing vesicles of air. *Scapes* generally simple, naked or with bracts, with few or many showy flowers. *Cal.* inferior, of 2 or 5 sepals united or distinct at base. *Cor.* irregular, bilabiate, personate, spurred. *Sta* 2, included within the corolla and inserted on its upper lip. *Anth.* 1-celled. *Ova.* 1-celled, with a free, central placenta. *Style* 1. *Stigma* cleft. *Fr.*—Capsule many-seeded. *Sds.* minute; embryo none

Genera 4, species 175, natives of swamps, pools and rivulets, diffused throughout nearly all countries. Properties unimportant.

Genera.

Calyx 4—6-cleft, capsule 5-celled.	*Pinguicula* 1.
Calyx 2-parted, subequal. Capsule 1-celled.	*Utricularia* 2.

1. PINGUICULA.

Lat. pinguis, fat; from the greasy appearance of the leaves.

Calyx bilabiate, upper lip trifid, lower bifid; cor. bilabiate or rarely subregular, upper lip bifid or 2-parted, lower trifid or 3-parted, spurred at base beneath; sta. 2, very short; stig. sessile, 2-lobed; caps. erect; seeds ∞.—♃ *In wet places. Lvs. radical, rosulate, entire. Scapes 1-flowered, nodding.*

P. VULGÀRIS. *Butterwort.*

Lvs. ovate or elliptic, obtuse, unctious-puberulent above; *scape* and *cal. subpubescent; cor. lips* very unequal, lobes obtuse; *spur* cylindrical, shorter than the corolla.—Wet rocks and thin, damp soils, N. Y. (near Rochester, *Dewey, Beck.*) N. to Arctic Am. *Hooker.* Scape 6—8' high, with solitary, nodding flowers. Leaves all springing from the root, fleshy, spatulate or ovate, with a tapering base, fleshy and unctious to the touch. Corolla with a purple tube, lined with soft hairs. Flowering early in April and May.

2. UTRICULARIA.

Lat. *utricula,* a little bottle ; alluding to the air-vessels appended to the roots.

Calyx 2-parted, subequal ; corolla irregularly bilabiate, personate, spurred ; sta. 2 ; stig. bilabiate ; caps. globular, 1-celled.—*Herbs aquatic, loosely floating or fixed in the mud. Lvs. radical, multifid or linear and entire, mostly furnished with little inflated vesicles. Scape erect*

§ *Floating. Leaves capillaceous, multifid. Roots few or 0. Branches producing turions at apex.*

1. U. INFLÀTA. Walt. (U. ceratophylla. *Mx.*) *Whorled Bladderwort.*
Upper lvs. in a whorl of 5 or 6 at the surface of the water ; *petiole* and *midrein* inflated, *lower lvs.* capillaceous, dissected, submerged ; *scape* 4—5-flowered.—24 In ponds, Mass. to Car., W. to Ohio. The proper stem (rhizoma?) is very long, branching, suspended in the water by a single, irregular whorl of 5 or 6 floating, inflated leaves which are oblong, cleft, and pinnatifid at the end. Flowers 4—5 together upon a scape 8' in length, pedunculated, with sheathing bracts. Spur nearly as long as the corolla, appressed to the lower lip, striate, emarginate. Cor. yellow, the upper lip broad-ovate, entire, lower 3 lobed. Aug.

2. U. VULGÀRIS. (U. macrorhiza. *Le Conte.*) *Common Bladderwort.*
Lvs. all submersed, capillaceous, multifid, fibrillose or setaceous ; *vesicles* numerous, small ; *st.* or *rhizoma* very long, floating ; *scape* simple, 5—11-flowered ! *spur* conical, obtuse, shorter than the corolla.—24 In stagnant pools, U. S. and Can. Floating stems several feet long, very branching. Leaves very numerous, 1' in length. Utricles furnished with a fringed, valvate aperture, usually inflated. Scape 5—10' high, stout, arising out of the water. Flowers alternate, showy, yellow, 5—6" long, lower lip larger, with a projecting palate, striped with brown. Jn. Jl.

3. U. INTERMEDIA. Hayne.
Lvs. all submersed, in 2 rows, dichotomously many-parted, without vesicles, roundish in outline, segments ciliolate-denticulate ; *leafless branches* with lateral vesicles and terminal turions ; *fls.* few ; *lips* entire ; *spur* conical, acute, appressed to the lower lip.—24 Pools, Mass. *Robbins !* R. I. *Pror. Frank. Soc.* Floating stems branched, a foot long. Leaves about 3" long, numerous. Turions much larger than the vesicles, green, scaly, producing new plants like bulbs Flowers 4—5" long, sulphur-yellow.

4. U. STRIÀTA. Le Conte.
Lvs. numerously subdivided, submerged, with vesicles ; *scape* 2—6-flowered, with a few scales ; *flowers* large, yellow, upper lip broad, divided into 3 lobes, the middle lobe striate with red, lower lip crenate, sides reflexed, having dark spots upon the palate ; *spur* slender, obtuse, with a notch at the end, pressed against the lower lip of the corolla and nearly as long.—① Native of swamps, Mass. to Flor. Root submerged, slightly attached to the mud. Leaves (radicles?) few, capillary, appendaged with few air vessels. Scape a foot high, generally with 2 flowers. June.

5. U. GIBBA. Linn. (and *Le Conte.*)
Minute, floating, leafless ? with few utricles and turions ; *scape* about 2 (4—7, *Le C.*)-flowered, naked, segments of the yellow corolla roundish, upper lip emarginate, lower subtrilobate, middle lobe crenate, subrevolute ; *spur* gibbous in the middle.—24 In pools, R. I., *Olney*, Mass., N. Y. to Car. Submersed stems dichotomous. Scapes 2—3' high, generally with but 2 small, yellow flowers. Spur swelling outward in the middle. Jl.

6. U. FORNICÀTA. Le Conte. (U. minor. *Ph.*)
Sts. numerous, fibrillose-branched, floating, utriculate ; *scape* naked, 1—2-flowered ; *upper lip of the corolla* 3-lobed, central lobe incurved over the palate,

lateral lobes appressed, *lower lip* entire; *spur* incurved, conical, obtuse, appressed to the lower lip.—Stagnant waters, N. Y. to Ohio, *Clark !* S. to Ga. The plant appears to be leafless, unless the hair-like lateral branches be considered leaves. Utricles very small. Scapes 3—5' high, filiform, dividing near the top, with a very minute bracteole at the fork. Flowers yellow.

7. U. PURPUREA. Walt. *Purple-flowered Bladderwort.*

St. long, floating, branched; *lvs.* submersed, fibrillose, verticillate, pinnately dissected, segments capillary, utriculate; *scape* assurgent, 2—3-flowered *upper lip* roundish-truncate, *lower lip* larger, its lateral lobes cucullate, smaller than the central; *spur* conical, flattened, appressed to and shorter than the upper lip.—① Pools, Mass.! to Flor. W. to Wis. Readily known by the large, bright purple flowers. Stem 1—3f long. Leaves about 1½' long. Utricles small. Scape 3—5' high. Corolla 8'' broad, the spur 3'', greenish. Aug.

§ § *Roots fixed in the mud or earth. Leaves simple, linear, with few utricles.*

8. U. RESUPINATA. Green. (U. Greenii. *Oakes.*)

Sts. creeping, fibrillose, rooting; *lvs.* linear-capillary, erect, undivided and entire; *scapes* numerous, simple, 1-flowered, with a minute clasping bract near the top; *spur* obtuse, cylindric, ascending, shorter than the elongated tube of the corolla.—Muddy shores of ponds, Tewksbury, (*Green,*) Plymouth and Uxbridge, Mass., *Robbins !* Leaves generally numerous, 6—15'' high, the bract 1' below the flower. Corolla light purple, 4'' long, lips roundish, entire, remote from the spur. Jl.

9. U. SUBULATA. (U. setacea. *Michx.*)

St. fibrillose, creeping, rooting, utriculate; *lvs.* few and minute, among the fibrillose roots, entire, linear, petiolate, glandular-obtuse, sometimes 0? *scapes* few, filiform, 1—5-flowered; *bracts* ovate, clasping; *pedicels* 4—5 times longer than the ovate, obtuse, veined sepals; *cor. upper lip* ovate, entire, *lower* 3-lobed; *spur* acute, appressed to and nearly equaling the lower lip.—A minute species, in swamps and ditches, Can., Mass.! to Flor. and La. Scapes 2—4' high. Leaves 2—3'' by ¼''. Flowers yellow, 3—4'' broad. Jn.

10. U. CORNUTA. Michx. *Horned Bladderwort.*

Scape rooting, rigidly erect, scaly, with about 2 sessile flowers; *spur* subulate, acute, longer than the corolla.—♃ Native of damp, boggy soils, but never floating, growing abundant in its localities, Can. to Mass. and Mich. Scape 2—12' high, leafless, but furnished with a few small, distant, pale, ovate scales, and bearing at the top 2—3 large, yellow flowers issuing from between bracts. The calyx consists of 2, ovate, colored leaves. Lower lip of the corolla much larger than the upper, broad, deflexed, emarginate; spur rigid, acute twice the length of the calyx. Jl. Aug.

11. U. PERSONATA. Le Conte. *Snap-dragon Bladderwort.*

St. strict, simple; *roots* few; *lvs.* 0; *fls.* 2—10; *pedicels* shorter than the acute lobes of the calyx; *cor. upper lip* oblong, obovate, emarginate; *palate* very large; *spur* linear-subulate, subacute, dependent, as long as the rest of the corolla.—Bogs, N. Eng. to Flor. Resembles the last, but distinguishable by its more numerous and smaller flowers, acute sepals, emarginate upper lip, and the more slender and acute spur of the yellow corolla. Stem or scape 10—18' high.

ORDER LXXXVII. OROBANCHACEÆ.—BROOMRAPES.

Plants herbaceous, leafless, growing parasitically upon the roots of other plants.
St. furnished with scales and bearing solitary or spicate flowers.
Cal. 4—5-toothed. inferior, persistent. *Cor* irregular, persistent, imbricate in æstivation.
Sta. 4, didynamous. *Anth.* 2-celled, cells distinct, parallel, often bearded at base.
Ova. 1-celled, free from the calyx, with 2 parietal placentæ. *Style* 1. *Stigma* 2-lobed.
Fr.—Capsule enclosed within the withered corolla, 1 celled, 2-valved, each valve bearing one simple or 2-lds. very numerous and minute. [lobed placenta in the middle.

Genera 12, species 136, mostly natives of the northern temperate zone. Properties astringent and bitter.

OROBANCHE.

Gr. οροβος, a vetch, αγχειν, to choke ; being supposed hurtful to the vetch and other plants.

Calyx 2—5-cleft, the segments often unequal ; corolla ringent, limb

4—5-lobed; ovary seated in a fleshy disk; capsule ovoid, acute, 1-celled, 2-valved, many-seeded.—♃ *Stems mostly simple.*

1. O. UNIFLÒRA. (O. biflora. *Nutt.*) *One-flowered Broomrape.*

Scapes in pairs, naked, each 1-flowered.—A small, leafless plant, with the general aspect of a Monotropa, found in woods and thickets, Can. and U. S. Root short, thick, smooth, scaly, surmounted by a stem not exceeding ½' in length. This divides at its top, generally into 2, scape-like, erect, round, simple, naked peduncles 4—5' high, downy, purplish-white, with a nodding flower at the top, of the same hue. A dozen or more such flower stalks are often found clustered together. June, July.

2. O. AMERICÀNA. *American Broomrape.*

St. simple, covered with oval-lanceolate, imbricated scales; *spike* smooth, terminal; *corolla* recurved; *sta.* exsert.—Woods, Can. to Ga. and La. Stem very thick, 4—6' high, very smooth, brownish-yellow, leafless, closely imbricated with pale, polished, oval scales. The spikes are dense. Bracts pale and smooth like the scales of the stem. Corolla tubular, bent downwards, the upper lip vaulted, yellow. Calyx irregularly divided into jagged segments, with 9 bracts at base. July.

2. EPIPHÈGUS. Nutt.

Gr. ἐπι, upon, φηγος, the beech; being supposed parasitical on the roots of that tree.

Monœciously polygamous; calyx abbreviated, 5-toothed. ♂ Corolla ringent, compressed, 4-cleft, lower lip flat. ♀ Corolla minute, 4-toothed, deciduous; capsule truncate, oblique, 1-celled, 2-valved, opening only on one side.—♃ *with virgate, simple branches which are floriferous their whole length.*

E. VIRGINIÀNUS. Eaton. (E. Americanus. *Nutt* and *1st edit.*) *Beech-drops.*

St. branched, leafless; *fls.* remote, alternate; *cor.* 4-toothed, deciduous.- This is said to be a parasitic plant growing from the roots of beech trees. Woods, Can. to Car. and Ky. Root a scaly ball covered with stiff, short and brittle radicles. Stem a foot high, with slender and irregular branches given off the whole length of it. Instead of leaves it has only a few small, ovate scales; one at the base of each branch. Flowers alternately scattered on each branch, the upper ones barren, with recurved corollas, brownish-white, with darker stripes above. Fertile ones smaller, deciduous. The whole plant is of a dull red color. Aug. Sept.

ORDER LXXXVIII. BIGNONIACEÆ.—BIGNONIADS.

Trees, shrubs, or rarely *herbs,* often climbing or twining.
Lvs. opposite, either simple or compound, without stipules.
Fls. terminal. *Cal.* divided or entire, sometimes spathaceous.
Cor.—Tube broad, with an irregular, 5-lobed or bilabiate limb.
Sta. 5, 1 or 3 sterile, often didynamous. *Anthers* 2-celled.
Ova. 2-celled, seated in a fleshy disk. *Style* 1. *Stigma* of 2 plates.
Fr.—Capsule coriaceous, 1—2-celled, 2-valved, many-seeded.
Sds. generally winged, destitute of albumen.

Genera 44, species 450, mostly North American. Others are diffused in all countries, particularly within the tropics. Several of the Brazilian species of Bignonia afford a valuable timber. But this order is best known for the beauty of its flowers.

Genera.

Calyx 5-toothed. Stamens 4, fertile. Climbing shrubs	Tecoma. 1
Calyx 2-parted. Stamens 2 fertile, with 2 or three abortive. Trees.	Catalpa. 2

1. TECÒMA. Juss.

Calyx campanulate, 5-toothed; cor. tube short, throat dilated, limb 5-lobed, subbilabiated or equal; sta. 4, didynamous, with the rudiment of a fifth; anth. cells 2, diverging; caps. 2-celled, 2-valved; seeds winged.—*Trees or shrubs, often climbing. Lvs. opposite, digitate or unequally pinnate.*

1. T. RADICANS. Juss. (Bignonia radicans. *Linn.*) *Trumpet Flower.*
Climbing by radicating tendrils; *lvs.* unequally pinnate; *lfts.* 4—5 pairs,
ovate, acuminate, dentate-serrate, puberulent beneath along the veins; *corymbi*
terminal; *cor. tube* thrice longer than the calyx; *sta.* included.—A splendid
climber in woods and thickets, along rivers. Penn. to Flor. W. to Ill.! Stem
90—80f! in length, ascending trees. Leaves 10—15' long, leaflets 2—3' by
1—2'. Flowers 2½' long, of a bright scarlet red. Fruit about 3' long, curved.—
One variety has yellow-scarlet flowers, another bright scarlet. Jn.—Aug. †

2. T. GRANDIFLÔRA. Delaun. (Bignonia Chinensis.) *Chinese Trumpet-flow-
er.*—Climbing, glabrous; *lvs.* unequally pinnate, *lfts.* 3—5 pairs, ovate-acuminate, dentate-serrate; *panicle* terminal; *pedicels* nodding, biglandular; *cor. tube*
scarcely longer than the 5-cleft calyx.—Native of China and Japan. Flowers
of a rich scarlet, shorter and broader than in T. radicans. †

2. CATALPA. Scop.

Calyx 2-parted; corolla campanulate, 4—5-cleft, the tube inflated;
stamens 2 fertile, 2 or 3 sterile; stigma 2-lipped; capsule 2-celled,
long, cylindric.—*Trees. Lvs. opposite or ternate-verticillate, simple, petiolate. Panicles terminal.*

C. BIGNONIÔIDES. Walt. 1788. (C. cordifolia. *Jaum.*) *Catalpa.*
Lvs. membranaceous, ovate-cordate, pubescent beneath, acuminate, subentire; *branches of the panicle* di-trichotomous; *cal. lips* mucronate.—A fine,
wide-spreading tree, native in the Southern States, but cultivated in many
places at the North, for ornament and shade. In favorable circumstances, it
attains the height of 50f, with a diameter of nearly 2f. It exhibits a wide-
spreading top, with comparatively few branches. Its leaves are beautifully
heart-shaped, and smooth, resembling those of the lilac, but much larger. In
color the bark is a light, shining gray. In May it puts forth blossoms in great
profusion. Their form is campanulate, color white, with yellow and violet
spots. Capsule cylindric, pendent, a foot in length; seed winged.

ORDER LXXXIX. PEDALIACEÆ.—PEDALIADS.

Herbs mostly strong-scented and glandular-hirsute. *Stipules* 0.
Lvs. opposite or alternate, undivided, angular or lobed. *Fls.* axillary, solitary, large.
Cal. 5-cleft, nearly equal.
Cor. hypogynous, irregular, tube ventricose, limb 3—5-lobed, mostly bilabiate.
Sta. 4 (with the rudiment of a 5'th), didynamous.
Ova. 1—2-celled, of 2 carpels. *Style* 1. *Stigma* divided.
Fr. drupaceous or capsular, often 2—4-horned, sometimes with 4—8 spurious cells formed by the divergent
lobes of the placenta cohering with the walls of the pericarp.
Sds. few, large, wingless.

Genera 12, species 27, natives of tropical America. &c. Some of them have been introduced into the
United States.

Genera.

Corolla 5-lobed. Leaves suborbicular.	*Martynia.* 1
Corolla 3-lobed. Leaves ovate-lanceolate.	*Sesamum.* 2

1. MARTYNIA.

In honor of John Martyn, botanical author and professor, Cambridge, Eng. 1768.

Calyx 5-cleft, 2—3-bracteolate at base; cor. campanulate, tube
gibbous at base, limb 5-lobed, unequal; sta. 5, one rudimentary and
sterile, 4 didynamous; caps. coriaceous, ligneous, 4-celled, 2-valved, each
valve terminating in a long, hooked beak.—① *chiefly southern, branching, viscid-pilose. Lvs. opposite, petiolate, subcordate, roundish.*

M. PROBOSCIDEA. Glox. (M. alternifolia. *Lam.*) *Unicorn Plant.*
Branches mostly decumbent; *lvs.* cordate, entire, suborbicular, villous, upper ones alternate; *fls.* on long, axillary peduncles; *beaks* much longer than the
capsule.—Native along rivers, Penn. to La. Stem 1—2f long. Leaves paler
beneath. Corolla pale, dull yellow, very large, the limb nearly as broad as the
leaves, spotted with brownish-purple. Sta. bright yellow, exserted. Aug. Sept.

Other ornamental species are *M. diandra*, flowers pink, spotted with purple;
and *M. lutea*, flowers deep yellow.

2 SESÁMUM.

Calyx 5-parted; corolla campanulate, 3-cleft, the lower lobe the longest; stigma lanceolate; capsule 2-celled, the cells divided by the inflexed edges of the valves.—① *of India.*

S. INDICUM. DC. *Oily-seed.—Lvs.* lanceolate-ovate, lower ones 3-lobed, upper ones undivided, serrate.—Native of E. India. Stem erect, about 18′ high. Leaves alternate, entire. Flowers axillary, subsessile. Corolla pale purple. The seeds yield an excellent oil which will keep several years without injury. It is used in cookery for all the purposes of sweet oil. Five pounds of the seed yield about one pound of oil. The leaves are emollient.

ORDER XC. ACANTHACEÆ. ACANTHADS.

Herbs or shrubs with opposite, simple, undivided, exstipulate leaves.
Fls. opposite or alternate, spicate, 3-bracteate, showy.
Cal. of 5 sepals united more or less, persistent. *Cor.* 5-lobed, subequal or bilabiate.
Sta. 4, didynamous, or more usually only 2 fertile with 2 rudimentary.
Ova. 2-celled, with the placenta parietal, adhering in the axis. *Styles* united.
Pr.—Capsule 2-celled, cells 2 or many-seeded.
Sds. roundish, supported by hooked, ascending processes of the placenta, without albumen.

Genera 105, species 750, chiefly tropical, only a few species ever extending into the United States. They are mostly mere weeds, but many are highly ornamental. Properties mucilaginous and slightly bitter, but of little importance to man.

Genera.

Stamens 2. Flowers in pedunculate, axillary clusters. . . . . *Dicliptera.* 1
Stamens 4, didynamous. Flowers sessile, axillary. . . . . *Ruellia.* 2

1. DICLIPTÈRA. Juss.

Gr. δις, double, καλυπτηρ, a cover; alluding to the 2 remarkable valves.

Calyx 5-parted, often 2—3-bracteolate; cor. bilabiate, upper lip emarginate, lower 3-cleft; fil. 2, each with a double anther; stig. 1; caps. attenuated below, half 2-celled, with 2 elastic valves; dissepiment growing from the centre of each valve; seeds 4, lenticular.— *Herbaceous or shrubby.*

D. AMERICANA. Wood. (Justicia pedunculosa. *Michx.* Dianthera Americana. *Linn.*)—*St.* simple; *lvs.* linear-lanceolate, acute at each end; *spikes* capitate, dense, on long, opposite or alternate, axillary peduncles.—24 On sluggish streams, Can. to Ga., W. to Ohio! Ia. *Plummer!* Ill. *Mead.* Stem 1—3′ high, grooved. Leaves opposite, 2—5′ by ½—1′, wavy, glabrous, contracted to a short petiole. Peduncles about as long as the leaves. Flowers pale purple, small. Valves of the capsule recurved elastically when mature, lower half attenuate and abortive; processes of the placentæ ascending, supporting the lens-shaped seeds in their grooves. Jl. Aug.

2. RUELLIA.

In honor of John Ruelle, physician to Francis I., and botanic author.

Calyx 5-parted, generally bibracteate at base; cor. subcampanulate with a slender tube and a 5-lobed limb; sta. 4, didynamous, approximating by pairs; caps. attenuated to each end, bursting by elastic, tooth-like valves; seeds few.—24 or ♄. *Fls. axillary and terminal.*

1. R. STREPENS. (R. hirsuta. *Ell.* R. ciliosa. *Willd.?* R. hybrida. *Ph.?* R. oblongifolia. *Mx.?*)—Hirsute, erect, branched; *lrs.* ovate, oval or oblong, acutish, margin entire, undulate, lower ones briefly petiolate, upper sessile; *fls.* subsessile, axillary, 2—3 together; *cal. segments* linear-subulate, hispid, half as long as the slender tube of the corolla.—In dry barrens and prairies, Ohio! Ind.! Ill.! to Penn. and Southern States. Stem 8—24′ high, often simple, clothed with white, bristly hairs. Leaves 1½—2½′ long, ½—¾ as wide. Calyx segments 8—10″ long. Corolla caducous, limb light bluish-purple, 1—1½′ broad, funnel-shaped, with a long and slender tube. July.—Varies to subglabrous.

3. Corolla smaller (limb 1' broad); plant smoothish.

Obs. There is much confusion in this genus. All my specimens, collected by myself and others in Ohio, Indiana, Illinois, Missouri, Tennessee and Alabama, are plainly referable to R. strepens alone.

ORDER XCI. SCROPHULARIACEÆ.—FIGWORTS.

Herbs, undershrubs, or rarely *shrubs,* scentless or fœtid, rarely aromatic.
Lvs. opposite, verticillate or alternate *Fls.* axillary or racemose, rarely spicate.
Cal.—Sepals 4 or 5, unequal, more or less united at base, inferior, persistent.
Cor. bilabiate, personate or otherwise irregular, the lobes imbricate in æstivation.
Sta. 4, didynamous, rarely with the rudiment of the 5th; sometimes 2 only, the 3 others either rudimentary.
Ova. free, 2-celled, many-seeded. *Style* simple. *Stigma* 2-lobed. (tary or wholly wanting.
Fr.—Capsule 2-celled, 2-valved, with central placenta.
Sds. indefinite, albuminous. *Embryo* straight.

Genera 176, species 1814, found in every part of the world, from the equator to the regions of perpetual frost. They constitute about 1-36 of the Phænogamia of N. America. *Lindley.*

Properties.—Generally acrid, bitter and deleterious plants. The most remarkable official species of the tribe is the foxglove (*Digitalis*), which exercises a wonderful control over the action of the heart, in regulating its pulsations. It is also employed in cases of dropsy, hemorrhage, &c. Taken in excess it speedily causes death. The Veronica Virginica, (Culvers Physic) and Linaria vulgaris (toad-flax) are purgative and emetic. Numerous species are cultivated for ornament. Nearly all of them turn black in drying.

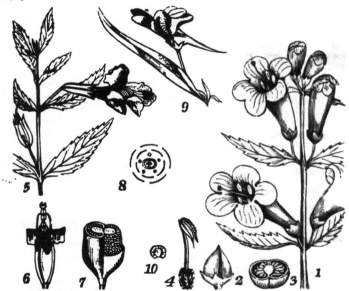

FIG. 50.—1. Dasystoma pubescens. 2. Mature fruit. 3. Cross section of the 2-celled capsule. 4. A stamen, enlarged. 5. Mimulus ringens. 6. Calyx with the corolla partly removed, showing the didynamous stamens in pairs, with the stigma above the highest pair. 7. Sections of the 2-celled, many-seeded capsule. 8. Plan of the flower, showing the position of the 5th rudimentary filament. 9. Linaria vulgaris, leaf and personate-bilabiate, spurred flower. 10. A winged seed.

Conspectus of the Genera.

Flowers diandrous.	Corolla deeply and variously lobed and colored.		*Schizanthus.* 1
	Corolla 4-cleft. Flowers spicate Peduncle scape-like.		*Synthris.* 17
		spurred at base.	*Linaria.* 3
	Corolla personate-bilabiate.	saccate at base.	*Antirrhinum.* 4
Herbs with alternate leaves.		Corolla yellow	*Pedicularis.* 12
		Bracts green.	Corolla purple. *Schwalbea.* 21
	Corolla ringent-bilabiate.	Bracts lobed and colored.	*Castilleja.* 22
Fls. didynamous.	Corolla large, tubular-campanulate, subequal.		*Digitalis.* 12
	Corolla equally 5-cleft. Minute mud plants.		*Limosella.* 16
Flowers pentandrous.	Corolla rotate, nearly regular.		*Verbascum.* 2

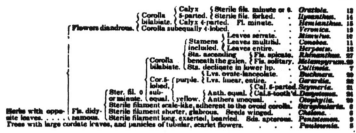

TRIBE 1. SALPIGLOSSEÆ.

1. SCHIZANTHUS. Ruiz & Pavon.

Gr. σχιζα, to cut, ανθος, a flower; in allusion to the numerous division of the showy corolla.

Corolla irregular, the upper lip 5-cleft, external in æstivation, lower much smaller, 3-parted; filaments 4, 2 of them sterile; capsule 2-celled.—① *from Chili. Lvs. pinnatifid, alternate. Cymes supra-axillary.*

S. PINNÁTUS. Ruiz & Pavon.—*Lvs.* 1—2-pinnately cleft; *cor. tube* shorter than the calyx, middle segment of the posterior lip, 2-lobed, cucullate, lateral segment falcate-spatulate, middle segment of the anterior lip emarginate, lateral 4-lobed; *sta.* exserted.—Plant 1—2f high, with delicate and beautiful flowers in clusters opposite the leaves. Calyx and peduncles viscid-pubescent. Corolla purple and yellow, with a dark spot in the midst. †

TRIBE 2. VERBASCEÆ.

2. VERBASCUM.

Lat. barba, beard; a name significant of the beard with which the plant is covered.

Corolla rotate, 5-lobed, unequal; stamens 5, declinate, all perfect; capsule ovoid-globose, 2-valved.—① *rarely* ♃ *or suffruticose. Lvs. alternate. Fls. in spikes or paniculate racemes.*

1. V. THAPSUS. *Common Mullein.*
Lvs. decurrent, densely tomentose on both sides; *rac.* spiked, dense; three of the *stamens* downy, two of them smooth.—The tall, dense, club-shaped spikes of the common mullein are very conspicuous in every slovenly field and by all roadsides, U. S. and Can. Stem erect, 3—5f high, woolly, its angles winged by the decurrent base of the leaves, generally simple, occasionally with one or two branches above. Leaves very rough with dense wool on both sides. Flowers rotate, of a golden yellow, nearly sessile. Notwithstanding its frequency, he mullein is generally supposed to have been introduced. Jn.—Aug.

2. V. BLATTARIA. *Moth Mullein.*
Lvs. clasping, oblong, smooth, serrate; *ped.* 1-flowered, solitary.—① Grows in waste grounds, roadsides, N. Eng. to Ia.! not common. Stem 3f high, branching above, bearing a terminal, leafy raceme 2—4' long. Lower leaves oblong, obovate; upper ones cordate-ovate, all coarsely and doubly serrate. Flowers on pedicels near an inch in length. Corolla yellow or white, marked with brown at the back. Stamens unequal, purplish, the filaments all hairy. Jn. Jl.

3. V. LYCHNITIS.
Whitish subtomentose; *st.* and *paniculate branches* angular; *lvs.* green above, crenate, lower petioled, narrowed to the base, upper ones sessile; *panicle* pyramidal, *fascicles* loosely many-flowered; *cal.* small, with lance-subulate segments; *fil.* with white wool.—Banks of the Delaware, near Philadelphia, *Pursh.* Ky. *McMurt.* Leaves very canescent beneath. Flowers pale yellow

4. V. PHŒNICEUM. (V. ferrugineum. *Andr.*)—*Sl.* pubescent or glabrous; *lvs.* smooth above, radical petiolate, ovate or oblong, serrate or entire, cauline few; *rac.* glandular-pilose, simple or subramose; *pedicels* solitary, remote, many times longer than the calyx; *fl.* with purple wool.—Native of Europe and Asia. Quite variable both in the form of the leaves and the color of the flowers, the latter being violet, red or copper color. †

TRIBE 3. **ANTIRRHINEÆ.**

Corolla tubular, often saccate or calcarate (spurred). Capsule dehiscing by pores. Inflorescence centripetal. Leaves (at least the lower) opposite or verticillate (upper often alternate). *Benth.*

3. LINARIA. Juss.

Lat. *linum*, flax ; from the resemblance of the leaves of some of the species.

Calyx 5-parted; corolla personate, upper lip bifid, reflexed; lower lip 3-cleft; throat closed by the prominent palate; tube inflated, with a spur behind; capsule 2-celled, bursting at the summit.—*Herbs Lower lvs. generally opposite, upper alternate. Fls. solitary, axillary often forming terminal, leafy racemes.*

1. L. VULGÀRIS. Mill. (Antirrhinum Linaria. *Linn.*) *Toad Flax. Snap-dragon.*—*Lvs.* linear-lanceolate, crowded; *spikes* terminal; *fls.* dense, imbricate; *cal.* smooth, shorter than the spur.—♃ A very showy plant common by roadsides, N. Eng. to Ky. Stems erect, smoothish, 1—2f high, very leafy and with numerous, short, leafy branches. Flowers large and numerous, crowded in a long, terminal spike. Corolla of a curious and grotesque form, furnished with a long tail or spur, the mouth closed by a prominent palate from the under lip. By lateral pressure it opens, closing with a spring when the pressure is removed. Color a brilliant yellow except the palate which is of a rich orange. July, Aug.

2. L. CANADENSIS. Dumont. (Antirrhinum Canadense. *Linn.*) *Canadian Snap-dragon.*—*Lvs.* scattered, erect, linear, obtuse; *fls.* racemed; *st.* simple; *scions* procumbent.—A smaller, annual species in roadsides, fields, Can. to Car. and Ky. Stem very slender, nearly simple, curving upwards from the decumbent base, about a foot high, smooth, furnished with small, remote leaves. A few leafy, prostrate or ascending shoots are given off from the base of the stem. Flowers small, blue, in a loose raceme at the end of the stems. Throat closed by the light blue palate. Spur filiform, as long as the corolla. Jn.—Sept.

3. A. ELATÌNE.

Procumbent, hairy; *lvs.* alternate, hastate, entire; *ped.* solitary, very long. —Fields, Can. to Car. A small, slender species. Stem creeping, 1—2f in length. Leaves 6—8″ by 3—4″, with a conspicuous auricle each side at base. Corolla yellow, the upper lip bright purple beneath, on long stalks. Calyx hairy, as well as the whole plant. Jn.—Sept.

4. L. TRIORNITHOPHŎRUM. Willd. (Antir. triorn. *Linn.*) *Three-bird Snap-dragon.* —Erect, spreading, smooth and glaucous; *lvs.* all verticillate in 3s or 4s, broad-lanceolate, acute; *fls.* interruptedly racemose, generally verticillate, on long pedicels.—♃ Native from Portugal to Austria. A showy plant 3—4f high, remarkable for the form and hue of the corolla which resembles three little birds seated in the spur. †

5. L. BIPARTÌTA. Willd. (Antirrhinum bipartita. *Vent.*)—Glabrous, erect; *lvs.* linear; *pedicels* much longer than the calyx; *sep.* lance-linear, acute, membranaceous at the margin; *cor. upper lip* deeply 2-parted; *spur* slender, arcuate. —A beautiful plant from Barbary. Corollas 8—10″ long, violet-blue, palate orange. †

4. ANTIRRHINUM.

Gr. αντι, like, *ριν*, a nose ; from the resemblance of the flowers to the snout of some animal.

Calyx 5-sepaled; cor. gibbous (not spurred) at base, the upper lip bifid, reflexed, lower trifid, closed by the prominent palate; caps. valveless, dehiscent by 3 pores.—*European herbs with the lower lvs opposite, the upper alternate. Inflorescence as in Linaria.*

1. A. MAJUS.　*Great Snap-Dragon.—Lvs.* lanceolate, opposite; *fls.* racemed; *sep.* glandular-hairy, lanceolate, a·ute.- An elegant and popular garden flower, native of England. Grows 1 or 2· .igh. Flowers large, pink-colored, the lower lip white and the mouth yellow, with a gibbous prominence at base beneath. There are varieties with scarlet, scarlet and white, and double flowers. †

2. A. ORONTIUM. *β.* grandiflorum. Chav.—Glabrous or hairy above, spreading; *lvs.* oblong-lanceolate; *fls.* remote, subsessile, upper ones subracemose; *cal. segments* equaling the corolla, and ovoid and very oblique capsule.—Native of Europe, Asia, and North America? *Bentham.* A showy garden plant, 1—2' high.　Corolla 6'' long, rose-color or white, with purple spots and veins. †

Tribe 4. CHELONEÆ.

Corolla tubular, not saccate or spurred. Capsule 2—4-valved. Calyx segments or lobes imbricate in æst. Inflorescence compound (general centripetal, partial centrifugal). *Benth.*

5. PAULOWNIA. Siebold.

Calyx deeply 5-cleft, fleshy; cor. tube long, declinato, enlarged above, limb oblique, with rounded segments; sta. 4, arched downwards, with no rudiment of a 5th; caps. ligneous, acuminate, valves septiferous in the middle; seeds 00, winged.—Tree, native of Japan.

P. IMPERIALIS. Sieb.　(Bignonia tomentosa. *Thunb.*)—A splendid tree with the habit of Catalpa, recently introduced in cultivation in this country! Branches crooked, nearly horizontal.　Leaves 7—12' by 4—9', opposite, petiolate, broad cordate-ovate, entire or somewhat trilobate, villous-canescent both sides, smoothish above when full grown. Panicles large, terminal, many-flowered.　Corolla 1½—2' long, between violet and rose-color, striped and spotted within. †

6. SCROPHULARIA.

So named from the resemblance of the roots to scrofulous tumors.

Calyx in 5 acute segments; corolla subglobose, limb contracted, sub-bilabiate, lip with an internal, intermediate scale (sterile filament); capsule 2-celled; valves with 2 inflated margins.—*Herbs or suffruticose, often fœtid. Lvs. opposite. Cymes in simple or compound terminal, thyrsoid panicles.*

S. NODOSA (and S. Marilandica. *Linn.* S. lanceolata. *Pursh.*)　*Figwort.* Glabrous; *st.* angled; *lvs.* ovate, ovate-oblong, or the upper lanceolate, acute, serrate or subincised, base broadly cordate or rounded or acutish; *thyrse* oblong, leafless or scarcely leafy at base; *cymes* pedunculate, loosely many-flowered; *cal. segments* broadly ovate, obtuse, slightly margined; *sterile anth.* broadly orbicular. *Benthan.*—24 In woods and hedges, Can. U. S. and Cal.; also in Europe.　Rare in N. Eng. Stem square, 4—6' high, with paniculate, opposite branches above.　Leaves 3—7' long, smooth, thin, often long-acuminate.　Flowers ovoid, 3—4'' long. Limb very small, sublabiate, having a green scale (sterile filament) adnate to the upper side. July—Oct.—The plant is quite variable in the form of the upper leaves and in the development of the panicle; but having observed it in numerous localities in the Middle and Western States, I cheerfully concur in the present view of Mr. Bentham.

7. COLLINSIA. Nutt.

Named by Mr. Nuttall in honor of Z. Collins, Esq., of Philadelphia.

Calyx 5-cleft; corolla bilabiate, orifice closed, upper lip bifid, lower trifid, with the middle segment carinately saccate and closed over the

34*

declinate style and stamens; capsule ovoid or globose, with 2 membranaceous, bifid valves; seeds large, concavo-convex.—① *with oppo-site or verticillate lvs., axillary and terminal inflorescence.*

C. VERNA. Nutt. *Vernal Collinsia. Tall Pink.*
Minutely puberulent; *lowest lvs.* ovate or oblong, petiolate, middle and upper sessile, ovate-lanceolate, cordate-amplexicaul, dentate, floral ones lance-linear, entire; verticillasters 2—6-flowered; *pedicels* many times longer than the flowers.—Banks of streams, shaded or open, N. Y. near Utica, *Gray,* to Ohio, *Locke!* Ia., *Plummer!* A tender herb, 8—18' high, branched from the base. Leaves 1—2' by ½—1', dilated at base. Pedicels 1—1½' long. Corolla 5" long, variegated with blue and white.

8. CHELONE.

Gr. χελωνη, a tortoise; from a fancied resemblance of the flower to the head of that animal.

Calyx deeply 5-parted, with three bracts at base; corolla inflated, bilabiate, the fifth filament abortive, smooth above, shorter than the rest; anthers woolly; caps. valves entire; seeds broadly membranaceous, winged.—♃ *with opposite lvs., distinguished from Pentstemon chiefly by the seeds.*

C. GLABRA. *Snake-head. Salt-rheum Weed.*
Smooth; *lvs.* opposite, oblong-lanceolate, acuminate, serrate; *fls.* densely spiked.—A plant of brooks and wet places (Can. and U. S.), with flowers shaped much like the head of a snake, the mouth open and tongue extended. Stem mostly simple, 2f high, erect. Leaves opposite, of a dark and shining green above, with irregular serratures, and sessile or nearly so. Flowers large, in a short, terminal, dense spike. Corolla white, often tinged with red, inflated, contracted at the mouth, with short, gaping lips. Filaments hairy. Style long, exsert, bending downwards. Aug. Sept.
β. *purpurea.* (C. purpurea. *Mill?*) *Lvs.* distinctly petiolate, acuminate; *cor.* rose-purple.—This variety prevails in the Western States! It is larger in its leaves and flowers. Petioles ½—1' long. Flowers very beautiful.

9. PENTSTEMON.

Gr. πεντε, στημον, five stamens (4 perfect and 1 abortive); from the character of the flower.

Calyx deeply 5-cleft; corolla ventricose, bilabiate; the fifth filament sterile, bearded, longer than the rest; anthers smooth; seeds ∞, angular, not margined.—♃ rarely ♄, of *N. America, branching, paniculate. Lvs. opposite. Fls. showy, red, violet, blue or white.*

1. P. PUBESCENS. Soland. (Chelone Pentstemon. *Linn.*) *Beard-tongue.*
Hirsute or glabrous; *radical lvs.* ovate or oblong, petiolate, *cauline* lanceolate-oblong or lance-ovate, serrulate, sessile; *panicle* loose; *cor. tube* dilated upwards, upper lip shortest; *sterile sta.* longitudinally bearded.—River banks, bluffs, hills and barrens, Western N. Y! to Ohio! Ia. and Ill. A handsome plant, 1—2f high. Stem round, smooth below, supporting a loose, oppositely branched panicle of bluish-purple flowers. Corolla 1' in length, the barren filament broadest at end. June.
α. *Lvs.* narrow and thinly pubescent.
β. (P. lævigatus. *Soland.*) *Lvs.* dilated and subamplexicaul, glabrous.

2. P. DIGITALIS. Nutt. (Chelone digitalis. *Sweet.*) *Fox-glove Pentstemon.*
Very glabrous or rarely puberulent; *radical lvs.* petiolate, oval-elliptic or oblong, *cauline* lanceolate, dilated and amplexicaul at base, serrate or rarely entire; *panicle* loose; *ped.* erect, spreading; *cor. tube* campanulate-dilated upwards, upper lip scarcely shorter than the lower; *sterile sta.* longitudinally bearded.—Rich soils, Ohio, Ia! to Tenn., *Miss Carpenter!* Large and splendid, 3f high. Leaves 6½' by 2', broadest at base and tapering to a long point. Flowers numerous. Corolla 15" long, bluish-purple, varying to white. Jn. Jl.—I a n strongly inclined to regard this also as a luxuriant variety of P. pubescens.

3. P. GRACILIS. Nutt. (Chelone gracilis. *Spreng.*) *Slender Beard-tongue.*
Glabrous; *radical lvs.* petiolate, elliptic-oblong or lance-oblong, *cauline*
linear-lanceolate, amplexicaul, entire or remotely serrulate; *panicle* pubescent,
slender; *ped.* erect; *cal. segments* ovate-lanceolate, acuminate; *cor. tube* long
and narrow, scarcely dilated upwards, upper lip shorter than the lower; *sterile*
sta. longitudinally bearded.—River bottoms, near Chicago, Ill., *Mead*, also Mo.
and C. W. Plant simple, glaucous, 2f high. Peduncles 3—7-flowered. Flowers nodding, 9—10" long, pale blue. June.

4. P. GRANDIFLORUS. Fraser. (Chelone grandiflora. *Spr.*) *Great-flowered*
Beard-tongue.—Erect, glabrous and glaucous; *radical lvs.* petiolate, obovate-oblong, *cauline* broadly ovate, the highest orbicular, amplexicaul, not connate; *panicle* long, slender and racemose, interrupted; *ped.* short, solitary or
fasciculate, rigid; *cor.* broadly campanulate; *sterile fil.* dilated and puberulent
at apex.—Ill. near Prairie du Chien, *Riddell*, Mo., *Nuttall.* Stem 3f high.
Flowers 1—3 together in the upper axils. Corolla 15" long, much dilated at
the mouth, variously shaded with blue and purple.

5. P. SPECIOSUS. Doug. *Showy Pentstemon.*—Erect, glabrous, glaucous; *radical lvs.* petiolate, oblong-spatulate, *cauline* sessile, lanceolate; *panicle* elongated,
slender, virgate, secund; *cal. segments* ovate-oblong, acuminate, margin membranaceous; *cor. tube* enlarged upwards; *sterile fil.* filiform, glabrous.—Oregon.
Height 3—4f. Flowers 1½' long, blue. †

6. P. CAMPANULATUS. Willd. (Chelone campanulata, augustifolia, rosea,
atropurpurea, *of authors.*)—Glabrous; *lvs.* acutely serrate, lance-linear or lance-
ovate, long-acuminate, often dilated at base; *panicle* long, loose and secund;
cor. tube ventricose above, lobes subequal; *sterile fil.* bearded.—Mexico. A very
variable species, 2—3f high, with large flowers varying from light purple to
dark red or purple. †

7. P. BARBATUS. Nutt. *Scarlet Pentstemon.*—Glabrous and glaucous; *lvs.* entire, lower oblong, upper lance-linear; *panicle* long and loose; *cor. tube* long,
scarcely dilated upwards; *lower lip* and *sterile fil.* densely bearded.—Mexico.
Height 2—3f. Corolla scarlet, 13" long. †

Obs.—P. gentianoides, with the panicle long, leafy at base, flowers violet, scarlet, &c., and a few other
species are rarely found in gardens.

TRIBE 5. GRATIOLEÆ.

Corolla tubular, not saccate or spurred. Capsule bivalved, rarely indehiscent.
Calyx lobes or segm. imbricate in æst. Inflor. centripetal, uniform. *Benth.*

10. MIMULUS.

Gr. μιμω, an ape; from the resemblance of the ringent or grinning corolla.

Calyx prismatic, 5-toothed: corolla ringent, the upper lip reflected
at the sides; palate of the lower lip prominent; capsule 2-cel.ed,
many-seeded; stigma thick, bifid.—*Herbs prostrate or erect, with square
stems and opposite lvs. Ped. axillary, solitary, 1-flowered.*

1. M. RINGENS. *Monkey Flower.*
Lvs. sessile, smooth, lanceolate, acuminate; *ped.* axillary, longer than the
flowers.—2f A common inhabitant of ditches and mud soils, Can. and U. S.,
with large, blue, ringent flowers Stem erect, square, smooth, about 2f high.
Leaves sessile, opposite, serrate, acute, lanceolate. Peduncles about as long as
he leaves, square, curved upwards, axillary and opposite. Calyx tubular, 5-
angled and 5-toothed. Corolla pale blue, yellow within. Jl. Aug.

2. M. ALATUS. *Wing-stem Monkey Flower.*
Lvs. petiolate, smooth, ovate, acuminate; *ped.* axillary, shorter than the
flowers; *st.* winged at the 4 corners.—2f In N. Y. to Ia., *Plummer!* and S.
States. This, like the last species, inhabits ditches and other wet places, and
grows to nearly the same height. The square stem, erect, smooth and winged
at the 4 angles, affords an adequate distinction. Leaves stalked, ovate. Flowers ringent, on short stalks, light purple. Calyx teeth rounded, mucronate. Aug.

3. M. LUTEUS. (M. rivularis, lyratus, variegatus and guttatus *of authors.*)

Yellow Monkey Flower.—Ascending or erect; *lvs.* orbicular-ovate or oblong, lower long-petiolate, sublyrate, upper sessile or clasping, many-veined; *ped.* .onger than the leaves; *cal. tube* ovoid, upper tooth largest; *cor. tube* broad, twice longer than the calyx.—California and Chili. Flowers yellow, often spotted with rose or purple. †

4. **M. CARDINÀLIS.** Doug. *Cardinal Monkey Flower.*—Erect, branched, villous; *lvs* ovate, erose-dentate, narrowed and amplexicaul at base, many-veined; *ped.* longer than the leaves; *cal. tube* large, inflated; *cor. lobes* reflexed. -California. Stem loosely branched, 2—3f high. Corolla scarlet, the tube hardly longer than the calyx, limb large and brilliant. †

11. CONOBEA. Aublet.

Calyx 5-parted, equal; upper lip of the corolla 2-lobed, lower lip 3-parted; fertile sta. 4; anth. approximating by pairs, cells parallel; caps. globose, ovoid, valves breaking away from the placentiferous dissepiment; seeds 00, ovoid.—*American, branching herbs, with opposite lvs. Ped. axillary, solitary or in pairs, 1-flowered, 2 bracteoles near apex.*

C. **MULTIFIDA.** Benth. (Capraria. *Michx.* Leucospora. *Nutt.*)
Low, diffusely branched, puberulent; *lrs.* petiolate, pinnately dissected, segments linear or cuneate, lobed or entire, obtuse; *cor. lobes* entire; *caps.* ovoid, valves at length 2-parted.—①? Along the banks of the Ohio, (*Clark!*) and other western rivers, common. A plant 4—6' high, with finely divided leaves, and of a grayish aspect. Leaves 1' long, in 5 or 7 segments, the petiole as long as the flowers. Corolla greenish, hardly exceeding the calyx. Capsule 1½" long. TL

12. HERPESTIS. Gaert.

Gr. ἑρπηστης, a creeper; from the prostrate habit of the planta-

Calyx 5-parted, unequal; cor. subbilabiate, upper lip emarginate or 2-lobed, lower 3-lobed; sta. 4. didynamous, parallel; caps. 2-furrowed, 2-celled, valves parallel with the dissepiment, the margins inflexed; seeds 00, small.—*Obscure weeds with opposite lvs. Ped. 1-flowered, axillary or subracemose, often with 2 bracteoles near the calyx.*

1. H. **ROTUNDIFOLIA.** Ph. (Monneria rotundifolia. *Michx.*)
St. mostly glabrous, creeping; *lrs.* orbicular-obovate, entire, glabrous, many-veined; *pedicels* ebracteate, 1—3-together, 2 or 3 times longer than the calyx; *lower cal. seg.* ovate; *cor.* ¼ longer than the calyx.—A prostrate mud plant, in ponds, Ill. *Mead!* to La. Stem 1f in length. Leav.s 6—12" diam., about 9-veined, sessile. Peduncles thick, half as long as the leaves. Calyx 2—3" in length. Flowers blue.

2. H. **MICRANTHA.** Pursh.
Glabrous, prostrate; *lrs.* oval, entire, sessile or clasping, obscurely many-veined; *pedicels* ebracteate, nearly as long as the leaves; *cal. lwer seg.* cordate; *cor.* scarcely longer than the calyx.—Borders of pools and rivers, N. J. Penn. to Va. A minute weed, 2—4' in length, with minute flowers. Leaves about 3" by 2", 5—7-veined. Flowers less than 2" long, blue? Aug.

3. H. **AMPLEXICAULIS.** Ph. (Monneria amplexicaulis. *Michx.*)
St. floating, woolly; *lrs.* amplexicaul, ovate, obtuse, entire, many-veined, glabrous above; *ped.* solitary, shorter than the calyx; *cal. lower seg.* cordate; *cor.* ¼ longer than the calyx; *hypogynous disk* long. 10-toothed at apex.—Swamps and ditches, N. J. to La. A few inches in length, with leaves 6—8" long. Flowers nearly 5" long. Style dilated at the end. Aug.

4. H. **MONNERIA.** Humboldt. (H. cuneifolia. *Ph.* Monneria cuneifolia. *Michx.*)—Glabrous, fleshy, prostrate; *lrs.* cuneate-obovate, obscurely crenate or entire; *pedicels* as long as the leaves, with 2 bracteoles near the calyx; *lower cal. seg.* ovate.—An obscure weed, on inundated banks, Penn. to Car. *Beck,* and La. *Eaton.* Leaves 6—8" long, obscurely veined, subclasping. Flowers very small, pale purple. Aug.

13. GRATIOLA.

Lat. *gratia.* favor; alluding to its medicinal virtues.

Calyx 5-parted, subequal ; cor. upper lip entire or slightly bifid, lower trifid, the palate not prominent ; sta. 2, fertile, mostly with 3 sterile filaments ; caps. 2-celled, 4-valved, valves inflexed at margin. —*Herbs with opposite lvs. Ped. axillary, 1-flowered, usually bibracteolate near the calyx.*

1. G. AUREA. *Muhl. Golden Hedge Hyssop.*

Smooth; *lvs.* oblong-lanceolate, subentire, half-clasping ; *sterile fil.* 2, minute.—A small, perennial herb, 6—8' high, frequenting the borders of muddy ponds and other moist places, Mass.! N. Y.! to Flor. Stem declining and rooting at the base, quadrangular, simple or branching. Leaves opposite, sessile, a little clasping, smooth, punctate, acute or nearly so, often with a few teeth near the end. Flowers yellow, axillary, alternate, on slender stalks, as long as the leaves. Filaments 4, adhering to the corolla, 2 of them minute, without anthers. Aug.

2. G. VIRGINICA. (G. aurea. *Ph.* G. Missouriana. *Beck.* G. officinalis. *Mx.*)—*St.* ascending, branched; *lvs.* lanceolate, subacute ; *ped.* as long or longer than the leaves; *sterile fil.* none. -2⊥ U. S. and Can. Stem 4—8' high, more or less pubescent, round, declining and branching at base. Leaves 1—2' long, and ⅓ as wide, smooth, lanceolate, sessile, dentate or nearly entire near the ends, subconnate or amplexicaul. Corolla white or pale-yellow, pubescent within, twice as long as the calyx, and on long, pubescent stalks. Calyx with 5 equal segments, and 2 bracts which are linear-lanceolate and rather longer than the sepals. July.

3. G. SPHÆROCARPA. Ell. (G. Caroliniensis. *Le Conte.*) *Round-fruited Hedge Hyssop.*—Glabrous, ascending, branched; *lvs.* lanceolate-ovate, attenuate to the base, sparingly toothed ; *ped.* scarcely longer than the calyx.— Low grounds, Western States! to Ga. Plant a few inches high, differing from the last chiefly in the short peduncles, round capsules, broader leaves, &c. Flowers whitish, 5—6" long. June.

14. ILYSANTHUS. Rafinesque.

Gr. ιλυς, mud, ανθος; flower; from the habitat of the plants.

Calyx 5-parted ; cor. upper lip short, erect, bifid, lower lip larger, spreading, trifid ; sta. 2 fertile, 2 sterile fil. forked, one of the divisions glandular, obtuse, the other acute, or rarely with half an anther ; caps. ovate or oblong, about equaling the calyx.—① *with opposite lvs., and axillary, 1-flowered ped., resembling Gratiola in habit.*

1. G. GRATIOLÖIDES. Benth. (Lindernia dilatata. *Ell.* L. attenuata. *Muhl.* L. pyxidaria. *Ph.* Gratiola anagalloidea. *Michx.*)—Glabrous, ascending, much branched; *lvs.* ovate or oblong, obtusish, subdentate, lower attenuated to a petiole; *cor.* erect, twice longer than the calyx, on bractless peduncles; *sterile fil.* bearing the glabrous, acute lobe below the middle.—In wet places, Can., N Y.! Ohio.! Ia., Ill.! to Tex., frequent. A low, inconspicuous plant, 3—6 or 9' high. Leaves 5—8" long, sometimes mostly sessile, commonly the lower distinctly petiolate. Corolla bluish-white, much exserted, 5" long. July, Aug. —Varies with the leaves somewhat dilated at base and sessile, and the peduncles longer or shorter, being sometimes a little longer than the leaves.

15. HEMIANTHUS. Nutt.

Gr. ημι, half, ανθος, flower; alluding to the absence of the upper lip.

Calyx 4-toothed ; cor. upper lip very short or obsolete, lower 3 cleft, the middle segment long, spreading ; cells of the 2 anthers divaricate ; sterile fil. 0 ; caps. 1-celled, 2-valved.—① *Minute glaurous. creeping. Lvs. opposite.*

H. MICRANTHEMÖIDES. Nutt. (Herpestis micrantha. *Ell.*)
Inundated banks, Del., *Nuttall.* Stems a few inches in length, dichotomously branched. Leaves roundish-ovate, opposite, crowded, sessile, obscurely 3-veined. Flowers axillary, solitary, minute. Aug. Sept.

TRIBE 6. SIBTHORPEÆ.

16. LIMOSELLA.

Lat. limus, mud; the plant grows by the edge of puddles and muddy places.

Calyx 5-cleft; corolla shortly campanulate, 5-cleft, equal; stamens approximating in pairs; capsule partly 2-celled, 2-valved, many-seeded.—*Minute, aquatic herbs. Scape 1-flowered.*

L. TENUIFOLIA. Nutt. (L. subulata. *Ives* and 1*st edit.*) *Mudwort.*
Acaulescent; *lvs.* linear, scarcely distinct from the petiole; *scape* as long as the leaves; *cor. segments* oval-oblong, shorter than the calyx.—① R. I.!
Mass.! N. Y., Penn. A minute plant, an inch in height, growing on the muddy banks of rivers. Leaves and flower-stalks radical. Flowers very small, blue and white. Aug.

TRIBE 7. DIGITALEÆ.

Inflorescence centripetal, racemose. Leaves all alternate, the lower crowded, petiolate. *Benth.*

17. SYNTHῨRIS. Benth.

Calyx 4-parted; cor. subcampanulate, segments 4, erect-spreading or 0; sta. 2, inserted into the tube of the corolla, exserted; anth cells parallel, distinct; caps. compressed, obtuse or emarginate, loculicidal, seeds plano-convex.—♃ *N. American, with a thick root. Radical lvs. petiolate, cauline bract-like, on the scape-like stem, alternate. Fls racemed or spicate.*

S. HOUGHTONIANA. Benth. (Gymnandra Houghtoniana. *Torr. & Gray in edit.*)—Hirsute; *radical lvs.* ovate, subcordate at base, crenulate, obtuse *scape* erect, clothed with foliaceous bracts, dense-flowered above; *cor.* as long as the calyx, upper segment longer than the other very short ones.—Dry hills. Wis., *Lapham!* Leaves 2—3′ by 1½—2′, on petioles about 1′ long, some of the leaves often suborbicular. Bracts much smaller, ovate and ovate-lanceolate, clasping. Scape 9—12′ high. Spike elongating in fruit.

18. DIGITĀLIS.

Lat. digitabulum, a thimble; from the form of the flowers.

Calyx 5-parted; corolla campanulate. ventricose, in 5 subequal lobes; capsule ovate, 2-celled, 2-valved, with a double dissepiment.—*Herbs or shrubs of Europe and Asia. Lower lvs. crowded, petiolate, upper alternate. Fls. in showy racemes.*

1. D. PURPUREA. *Purple Foxglove.*—*Lvs.* oblong, rugose, crenate; *cal. segments* ovate-oblong; *cor.* obtuse, upper lip entire; *ped.* as long as the calyx.-
Native of Europe. A well known, showy border flower of easy culture. It is a biennial plant 2—3f high, with large, rough, downy leaves. Flowers numerous, in a long, simple spike, large, crimson, often white, with beautiful eye-like spots within. The whole plant is a violent and dangerous poison when taken in considerable quantities, producing delirium, convulsions and death. But in the hand of the judicious physician it becomes a valuable medicine, acting as a sedative and diuretic. July. ‡†

2. D. FERRUGINEA. *Iron-colored Foxglove.*—*Lvs.* oblong-lanceolate, very smooth; *rac.* many-flowered; *cal. segments* oval-elliptical, obtuse; *cor. limb* subglobose, woolly, lower segment ovate.—♃ in Greece, Armenia and Circassia. Corolla *rust-colored*, 16″ long, lower lip longest, densely bearded. †

3. D. grandiflòra. Allioni. *Great Yellow Foxglove.—Lvs.* ovate or oblong-lanceolate, veiny, serrulate, amplexicaul; *rac.* tomentose, lax; *cal. segments* lanceolate, acute; *cor.* ventricose-campanulate, segments broader than long, lowest twice broader than the lateral.—♃ in Europe. Plant 2—3f high. Flowers 1½′ long, yellow, varying to brownish or orange. †

4. D. lùtea. *Small Yellow Foxglove.—*Very smooth; *lvs.* oblong or lanceolate, denticulate; *rac.* secund, many-flowered; *cal. segments* lanceolate, acute; *cor.* glabrous, tube subventricose, lower segment half as long again as the rest.—♃ Europe. Stem 3f high. Flowers 8—10″ long, yellow, varying to white. †

5. D. orientàlis. Lam. *Oriental Spotted Foxglove.—St.* and *lance-linear lvs.* glabrous; *spike* interrupted, glandular-villose; *pedicels* very short; *cal. segments* ovate-lanceolate, acute; *cor.* pubescent, lower segments oblong, obtuse.—♃ Bythinia. Height 3f. Corolla purplish, spotted.

Obs.—Several other species are sometimes seen in gardens, among which are D. Thapsi, with mullein-like leaves all radical and flat on the ground; D. leucophæa, with very large, dense, leafy racemes of downy white fls. Numerous hybrids also occur in gardens, produced between the above species which are often difficult to distinguish.

Tribe 8. VERONICEÆ.

19. VERONICA.

Etymology doubtful; perhaps named for St. Veronica.

Calyx 4-parted; corolla subrotate, deeply 4-cleft, lower segments mostly narrow; sta. 2, inserted into the tube, exserted; sterile fil. 0; caps. compressed, 2-sulcate, often obcordate, 2-celled, few-seeded. —*Herbs or shrubs (the following species herbs). Lvs. opposite. Fls. solitary, axillary or in racemes, blue, flesh-colored or white.*

§ 1. *Erect, tall. Lvs. verticillate; racemes dense, terminal, often panicled; corolla tube longer than the limb.* Leptandra. Benth.

1. V. Virginica. (Leptandra Virginica. *Nutt.) Culver's Physic.*
Erect, tall, glabrous; *lvs.* briefly petiolate, in 4s, 5s or 6s, lance-ovate to lance-linear; *spikes* mostly several, paniculate.—Woods, thickets and barrens, Can. to Ga., W. to the Miss.! A conspicuous plant arising 2—5f. Stem simple, straight, smooth, with whorls of lanceolate, acuminate, finely serrate leaves which are subpetiolate and glaucous beneath, and 4—6 in a whorl. Flowers numerous, nearly sessile, in long, terminal and verticillate, subterminal spikes. Corolla white, tubular, pubescent inside. Stamens and style twice as long as the corolla. Jl.

§ 2. *Leaves opposite. Corolla tube very short.*
* *Racemes axillary.*

2. V. Anagallis. (V. tenerrima. *Schmidt.*)
Glabrous, erect; *lvs.* sessile, clasping and subcordate, lanceolate, acutish, entire or serrulate; *rac.* in opposite or alternate axils; *caps.* orbicular, slightly notched.—♃ A smooth, succulent plant, frequenting the borders of brooks and pools, Can. and U. S.! not common. Stem fleshy, 12—20′ high. Leaves about 2—3′ by 5—7″. Racemes longer than the leaves, loose, pedicels (2—3″) scarcely longer than the bracts. Flowers bluish-purple, small. Jn. Jl.

3. V. Americàna. Schweinitz. (V. Beccabunga *Am. authors.) Brooklime.*
Glabrous, decumbent at base, erect above; *lvs.* ovate or ovate-oblong, acute or obtusish, serrate, petiolate, abrupt at base; *rac.* opposite, loose; *caps.* roundish, turgid, emarginate.—♃ in brooks and clear waters, Can. and U. S. Plant rather fleshy, very smooth, 12—18′ long, more or less decumbent and rooting at base. Leaves 1—2′ long, ⅓—½ as wide, petioles 1—3″ long, margined. Racemes longer than the leaves. Pedicels (3—5″) twice longer than the bracts. Flowers blue or bluish-purple. Jn. Jl.—This plant is variable, some of its species approaching V. Anagallis, others V. Beccabunga, of Europe.
a. Lvs. ovate, acute, acutely serrate, truncate or subcordate at base.—Frequent!
β. Lvs. ovate-lanceolate, serrulate, rounded at base, petiolate.—Common!

4. V. scutellāta. *Skull-cap or Marsh Speedwell.*

Glabrous, ascending, weak; *lvs.* linear or lance-linear, sessile, acute, remotely denticulate; *rac.* in alternate axils, very loose; *pedicels* divaricate; *caps.* compressed, broadly obcordate.—♃ slender and weak, in swamps and marshes, N. Eng. and Western States, and Brit. Am., common. Stem 10—16′ high. Leaves (2—3′ by 2—3″) much longer than the internodes. Peduncles and pedicels filiform, the latter (6—9″) six times longer than the bracts. Flowers rather large, flesh-color, with purple lines. Jn.—Aug.

5. V. officinālis. *Officinal Speedwell.*

Roughish-pubescent; *st.* prostrate, branched; *lvs.* briefly petiolate, and subsessile, obovate-elliptic or oblong, obtuse, serrate, mostly narrowed to the base; *rac.* dense, many-flowered; *pedicels* shorter than the calyx; *caps.* puberulent, obovate-triangular, slightly emarginate.—♃ in dry woods and open fields, Can. to Ga. Plant trailing, 6—12′ long, with ascending branches. The leaves vary from ovate to obovate, but are generally elliptical, 1—1¼′ in length. The flowers are pale blue, forming rather long, axillary, erect, pedunculate spikes. Found in dry woods and open fields. May—Jl. §

* * *Raceme terminal.*

6. V. serpyllifolia. *Thyme-leaved or Smooth Speedwell.*

Subglabrous, much branched below; *sts.* ascending; *lvs.* oval, subcrenate, obtuse, lower roundish and petiolate, upper sessile, passing abruptly into oblong, entire, alternate bracts; *ped.* as long as the ovate sepals; *caps.* obcordate, broader than long.—♃ Meadows and mountain valleys, in grass, &c., U. S. and Can. Plant varying in height from 3′ to 12′. Leaves rather fleshy, 3-veined, orbicular and oval and ovate, 4—12″ long, petioles 0—2″. Racemes bracted, rather close in flower, elongating in fruit to 2—5′. Corolla scarcely exceeding the calyx, blue and white, beautifully penciled with purple lines. May-August.

* ** *Annual. Flowers axillary, solitary, scarcely racemed.*

7. V. peregrīna. (V. Marilandica. *Willd.*) *Purslane Speedwell.*

Ascending, subglabrous; *lowest lvs.* petiolate, oval-oblong, dentate-serrate, obtuse, *upper* sessile, oblong, obtuse, serrate or entire, *floral ones* oblong-linear, entire, longer than the subsessile flowers; *caps.* suborbicular, slightly notched, the lobes rounded.—① Throughout N. America, in fields or clayey soils. Plant often branched from the base, 4—10′ high. Leaves rather fleshy, the upper cauline 6—11″ long, floral much smaller. Sepals oblong, longer than the pale blue or white corolla. Capsule hardly broader than long. May, June.

8. V. arvensis. *Field Veronica. Corn Speedwell.*

Puberulent-pilose, simple or branched, erect or assurgent; *lvs.* cordateovate, incisely crenate, lower ones petiolate, upper and floral alternate, lanceolate, crenate, sessile.—Frequent in dry fields, N. H. to Car. W. to the Miss. A small, pubescent, pale green plant, 2—6′ high. Stem nearly erect, branching from the base, the leaves assurgent. Flowers on short peduncles, corolla shorter than the calyx, pale blue, beautifully penciled with purple lines. May, June. §

β. ? (V. reniformis. *Raf.*) Lvs. sessile, reniform, entire.

9. V. agrestis. *Neckweed. Field Speedwell.*

St. procumbent, diffusely branching; *lvs.* cordate-ovate, petiolate, deeply serrate, floral ones lanceolate; *ped.* as long as the leaves.—① In cultivated fields, Can. to Ga. and La. A small, pilose plant, 2—8′ long, with a round, leafy, hairy stem, branching mostly at the base. The leaves are roundishovate, shorter than their petioles, the upper alternate. Flowers small, light blue, veined, their stalks recurved in fruit. Segments of the calyx fringed, ovate, equal. Seeds concave beneath. May—Sept.

10. V. hederæfolia. *Ivy-leaved Speedwell.*

Prostrate, pilose; *lvs.* petiolate, cordate, roundish, mostly 3—5-toothed or lobed; *ped.* scarcely longer than the leaves; *sep.* triangular, subcordate, acute, at length erect.—Dry or rocky soils, L. I. to Del. Stem diffusely branched. Leaves rather fleshy, the lower smaller, opposite, upper cauline broadly cordate or truncate at base, alternate as well as the floral. Calyx somewhat 4-

angled in fruit, segments ciliate at edge. Corolla smaller than the calyx, blue, caducous. Capsule turgid, broader than long, 4-seeded. Mar.—May.

11. V. SPICĀTA. *Spiked Speedwell.*—Erect, tall; *lvs.* petiolate, ovate-oblong or lanceolate, lower ones obtuse, crenate, upper acute, crenate-serrate, entire at apex; *rac.* mostly solitary; *pedicels* much shorter than the sepals; *cal.* mostly hoary-pubescent.—♃ Europe and Asia. A beautiful garden species with numerous varieties. Flowers blue, roseate, &c. †

Tribe 9. BUCHNEREÆ.

20. BUCHNĒRA.

Named by Linnæus, in honor of J. G. Buchner, a German botanist, 1745.

Calyx 5-toothed; corolla tube slender, limb flat, in 5 cordate, sub-equal lobes; capsule 2-valved.—*Herbs with the lower lvs. opposite, the upper alternate. Flowers in a terminal spike (sta. 4, included).*

B. AMERICĀNA. *Blue-hearts.*

Lvs. ovate-lanceolate, denticulate, scabrous, 3-veined, sessile; *fls.* remotely spiked.—♃ In low grounds, N. Y. to Mo. and Ga. Stem 1—2f high, simple or slightly branched, slender and terete, ending in a long, loose and somewhat virgate spike of purple flowers. Leaves 1—2′ long, very rough, appressed to the stem. Flowers axillary and sessile. Stamens inserted, 2 in the throat of the corolla, and 2 in the middle of the tube. Calyx half as long as tube of corolla. Aug.

Tribe 10. GERARDIEÆ.

Inflorescence centripetal, racemose. Leaves, at least the lower, opposite. Corolla tube dilated, limbs spreading, lobes flat, subregular. Stamens approximating in pairs.

21. SEYMERIA. Pursh.

Calyx deeply 5-cleft; cor. tube short, dilated, 5-lobed, lobes ovate or oblong, entire, equaling or longer than the tube; sta. 4, subequal; valves of the capsule loculicidal, entire; seeds 00.—*Herbs erect, branching. Cauline lvs. mostly opposite and incised. Fls. yellow.*

S. MACROPHYLLA. Nutt. *Large-leaved Seymeria.*

Erect, tall, sparingly pubescent; *lvs.* large, the lower deeply pinnatifid, segments lance-oblong, incised, terminal one the largest, upper lanceolate, se-rate or entire; *cor. tube* incurved, scarcely longer than the limb; *sty.* short, dilated and slightly bifid at apex; *caps.* ovate-acuminate.—♃ In woods, White River Valley, Ia.! Ohio, *Clark!* to Ark. The plant has much the aspect of Dasystoma. Height, 4—6f. Lower leaves (5—7′ by 2—3′) lance-ovate in outline, floral (2—3′) mostly opposite. Corolla ½′ long, very woolly within. Capsule a little shorter and broader. July.

22. GERARDIA.

Named by Linnæus, in honor of John Gerard, an English botanist of the 16th century.

Calyx campanulate, briefly or narrowly 5-toothed; cor. tubular, ventricose or subcampanulate, tube longer than the 5 broad, entire lobes; sta. didynamous, in pairs, shorter than the corolla, anth. all equal; caps. obtuse, or briefly acuminate; seeds 00.—*American herbs, rarely suffruticose. Lvs. opposite. Fls. axillary, solitary, purple or rose-color.*

1. G. PURPŪREA. *Large-flowered Purple Gerardia.*

St. angular, much branched; *lvs.* linear, narrow, acute; *fls.* subsessile, scattered; *cal. segments* subulate.—① Found in wet pastures and swamps, N. Eng. to Ill. and Ga. Stem slender, branching, erect, smooth, obtuse-angled, 1—2f high. Leaves entire, roughish, 8—15″ long, and about 2″ wide, coiled up in drying. Flowers large, axillary, often opposite, purple, on very short stalks. Aug.

2. G. MARITIMA. Raf. *Marsh Gerardia.*

St. angular; *lvs.* linear, fleshy, short, rather obtuse; *fls.* stalked; *cal.* truncate; *upper segments of the corolla* fringed.—① Native of salt marshes, Mass. to N. J., also shores of L. Mich. *Houghton.* This species resembles the foregoing, of which Pursh describes it as a variety. It is a smaller plant 6—12′ in height, and with smaller flowers. The leaves are shorter and thicker. The calyx segments are cut square off, not acute as in the preceding. Corolla purple. Flower-stalks axillary and terminal. July—Sept.

3. G. TENUIFOLIA. Vahl. *Small-flowered Purple Gerardia.*

Branching; *lvs.* linear; *ped.* axillary, longer than the flowers; *cal. teeth* short, acute.—① A slender and delicate species, usually very branching, but often simple, in fields and woods, U. S. Stem 6—12′ high. Leaves about an inch long, very narrow (scarcely a line in width), entire, rough, obtuse, often coiled. Flowers opposite, axillary, on slender stalks an inch or more in length. Corolla purple, spotted within, the border much spreading, smooth and nearly equal. Calyx teeth short and acute. Aug. Sept.

4. G. ASPÉRA. Doug. *Rough Gerardia.*

St. a little branched; *lvs.* long and narrowly linear, floral ones exceeding the calyces; *ped.* twice longer than the calyx; *calyx teeth* lanceolate, acute, about as long as the tube; *cor.* glabrous.—① Illinois, Buckley in DC., Prod. x. 518. Peduncles sometimes but little exceeding the calyx, sometimes twice as long. Flowers as large as those of G. purpurea, to which species this is perhaps too nearly allied.

5. G. SETACEA. Walt.

Branches slender, roughish; *lvs.* setaceous, roughish; *fls.* few; *ped.* alternate and opposite, very long; *cal. teeth* short and setaceous; *caps.* ovate, larger than the calyx.—① Penn. ? to Car. *Nuttall?* Scarcely darkens in drying.

6. G. SKINNERIANA. Wood. (G. aphylla. *Benth. in part, not Nutt.*)

Scabrous, pale green; *st.* erect, sparingly branched, slender, 4 angles margined; *lvs.* remote, linear, acute at each end, the floral ones 2 or 3 times shorter than the very long peduncles; *cal. teeth* very short, glandular-acute; *cor.* infundibuliform-campanulate, lobes short, spreading; *caps.* roundish-ovoid, scarcely exceeding the calyx.—① Barrens, Ia.! Plant 12—18′ high, the stem and few branches quite slender and rough on the slightly winged angles. Leaves (8—12″ by ¼—⅓″) much shorter than the internodes, margin slightly revolute. Peduncles 1—1½′ long, erect. Corolla (5—6″) glabrous, light purple or rose-color. Capsule loculicidal, about 30-seeded. Jl. Aug.

Obs.—I detected this delicate species in July, 1846, in Greene Co., Ia., on land belonging to Dr. A. G. Skinner, whose zeal in botanical pursuits deserves more than this slight notice. It does not turn black in drying.

23. OTOPHYLLA. Benth.

Gr. ους (oτος), ear, φυλλον, leaf; alluding to the auriculate leaves.

Calyx deeply 5-parted, sepals leafy, unequal; cor. tube enlarged upwards, lobes broad, entire; sta. didynamous, the upper pair with smaller abortive anthers! caps. subglobose, many-seeded.—*Erect, hairy herbs with opposite leaves. Fls. axillary, solitary, subsessile.*

O. MICHAUXII. Benth. (Gerardia auriculata. *Michx.*) *Michaux's Gerardia.* Scabrous-hirsute, subsimple; *lrs.* ovate-lanceolate, lower entire, upper mostly auriculate-lobed; *fls.* sessile.—① Penn. to Ill., *Rev. E. Jenny!* in prairies and low grounds. A rough, rigid plant, 9—18′ high. Leaves (1—1½′ by ¼—⅓′) entire on the margin, sessile; floral ones with an oblong-lanceolate lobe each side at base. Flowers alternate or mostly opposite, calyx deeply cleft, corolla purple or rarely white, pubescent, dilated at the mouth, 9—12″ long. Aug. Sept.

24. DASYSTŌMA. Raf.

Gr. δασυς, hairy, στομα, mouth; alluding to the character of the corolla.

Calyx campanulate, half 5-cleft, imbricate in æstivation; cor. tube dilated, longer than the 5 entire lobes; sta. included, didynamous; anth. all equal, awned at base; caps. ovate, acute, 2 valves bearing the septum in the middle, seeds 00.—♃ erect, *N. American. Lower lvs. opposite, upper generally alternate and entire. Cor. large, yellow, villous within as well as the stamens.*

1. D. QUERCIFOLIA. Benth. (Gerardia quercifolia. *Ph.* G. flava. *Linn.* G. glauca. *Eddy.*) *Oak-leaved Dasystoma.*—Glabrous; *st.* paniculate-branched; *lvs.* paler beneath, petiolate, lower ample, bipinnatifid, upper oblong-lanceolate, pinnatifid or entire. *fls.* pedunculate; *cal. lobes* lanceolate, acute, longer than the tube.—Woods and barrens, Northern and Western States! frequent. It has a tall, smooth, glaucous, branching stem 3—5f high. Leaves sinuate-pinnatifid; the upper ones only cut-dentate, all acute at each end, stalked, paler beneath. Flowers large and of a brilliant yellow, opposite and axillary, near the top of the stem, forming a loose spike. Corolla trumpet-shaped. The flowers resemble in form those of the foxglove, while the leaves may be likened to those of the oak. The whole plant turns black in drying making but a shabby appearance in the herbarium. Aug.

β. integrifolia. Benth. *Lvs.* lanceolate, entire; *cor.* smaller.—Ohio.

1. D. PUBESCENS. Benth. (Gerardia flava. *Ph.*) *Downy Dasystoma.* (Fig. 50.

Pubescent; *st.* subsimple; *lvs.* sessile, oblong-lanceolate, entire or sinuate-lobed; *cal. segments* oblong, obtuse, shorter than the tube.—In woods throughout the U. S. A tall and very showy plant. Stem 2—3f high, erect, pubescent. Lower leaves variously pinnatifid, or cut and toothed; upper ones very entire or toothed, obtuse; all opposite and sessile. Flowers large, yellow, opposite, axillary, trumpet-shaped. This also with the next species, turns black in drying. Aug. Sept.

3. D. PEDICULARIA. Benth. (Gerardia pedicularia. *Linn.*) *Lousewort Dasystoma.—St.* panicled, pubescent; *lvs.* oblong, pinnatifid, the segments serrate; *cal. seg.* leafy, cut-dentate.—One of the most elegant species, found in woods and mountains, Can. to Ga. and Ky. Stem tall and bushy, 2—3f high, covered with a scattered, woolly pubescence. Leaves numerous, pinnatifid with serrate lobes, opposite, on short, hairy stalks. Flowers large. Corolla trumpet-shaped, yellow, with roundish, spreading, leaf-like segments. The leaves have the general appearance of those of the lousewort, or some of the ferns. Aug.

TRIBE 11. EUPHRASIEÆ.

Inflorescence centripetal, racemose. Corolla upper lip galeate or concave, erect or incurved. Stamens ascending beneath the upper lip.

25. CASTILLÉJA.
Named for one Castillejo, a Spanish botanist.

Calyx tubular, 2—4-cleft; cor. galea (upper lip) linear. very long, crenate-concave. lower short, 3-lobed; sta. beneath the galea, didynamous; anth. oblong-linear, with unequal lobes, cohering in the form of an oblong disk, the exterior fixed by the middle, interior pendulous.—*Herbaceous or suffruticose. Lrs. alternate, the floral often colored at the apex. Fls. subsessile, in terminal, leafy bracts.*

1. C. COCCINEA. Spreng. (Euchroma *Nutt.* Bartsia. *Linn.*) *Painted-cup. Lvs.* sessile, pinnatifid, with linear and divaricate segments; *bracts* about 3-cleft and colored at the summit, longer than the corolla; *cal.* 2-cleft, nearly equal with the corolla, segments retuse and emarginate.—♃ Wet meadows, Can. and U. S., rare in N. Eng. A very beautiful plant, remarkable for its large, bright scarlet bracts. Stem angular, simple, 8—12' high. Leaves alternate, sessile, with about 2, long, linear segments on each side. Bracts crowded near the summit of the stem, in 3 segments, the middle one larger than the linear lateral ones. Flowers one in the axil of each bract. Calyx and corolla tubular, dull yellow, the former tinged with scarlet towards the tip. May, Ju-

2. C. sessiliflòra. Ph. (Euchroma grandiflora. *Nutt.*) *Great-flowered Painted-cup.*—Pilose-pubescent; *lvs.* sessile, clasping, oblong-linear, mostly cuneate-trifid, lobes divaricate; *cal.* sessile, elongated; *spikes* dense; *cor.* long, exserted, arched, segments of the lower lip acuminate.—A plant of curious appearance, prairies, Wis., *Lapham!* to the plains of the upper Missouri. Stem 8—14' high, several from the same root, simple, leafy. Leaves grayish, 2—3' long. Flowers crowded, pubescent. Corolla tube slender, 2—3' in length, greenish-white, with a slight tinge of purple. Style and stamens enfolded by the upper lip, and a little exserted. May.

3. C. septentrionàlis. Lindl. (Bartsia pallida. *Ph.* not ? *of Linn.*)

Lvs. alternate, linear, undivided, the upper ones lanceolate, the floral ones subovate, subdentate at the end, all 3-veined; *cal.* with acute teeth.—♃ This hardy plant inhabits the alpine regions of the White Mts. in N. H.! particularly the heights of Mt. Clinton, where it may be found in blossom in Aug. It is also a native of Siberia and Hudson's Bay. Stem about a foot high, furrowed, simple. Leaves alternate, smoothish, the lower ones linear, becoming broader towards the upper part of the stem, where they are lanceolate and all usually with but 3 veins. Tuft of flowers at top of the stem. Bracts broader and shorter than the leaves, 5—7-veined, with about 3 teeth at the end, of a pale straw-color, tipped with purple. Flowers straw-colored, nearly concealed by the bracts.

26. SCHWALBEA.
Named by Linnæus in honor of Schwalbe, a German botanist.

Calyx tubular, inflated, obliquely 4-cleft, upper division small, lower division large, emarginate or 2-toothed ; corolla ringent, upper lip entire, arched, lower 3-lobed ; seeds many, chaffy.—♃ *with alternate leaves and flowers in a terminal spike.*

S. Americàna. *Chaff-seed.*

In sandy barrens and marshes, N. Y. to Flor. Stem 1—2f high, pubescent, square, simple. Leaves sessile, ovate-lanceolate or oblong, 3-veined, with a ciliate margin. Bracts ovate, acuminate, diminishing upwards. Flowers on simple, alternate, very short pedicels, in a long spike. Corolla dull purple or brownish-yellow, twice as long (1—1½') as the calyx. Jn.

27. RHINANTHUS.
Gr. ῥιν, nose, ανθος ; alluding to the singular appearance of the compressed galea.

Calyx 4-toothed, ventricose ; cor. tube cylindrical, as long as the calyx, limb ringent, galea appendaged, compressed, lip broader, deeply divided into 3 obtuse segments ; caps. 2-valved, compressed obtuse.—① *erect, with opposite lvs.*

R. minor. Ehrh. (R. Crista-galli. *Linn.*) *Yellow Rattle.*

Mostly glabrous; *lvs.* oblong or lanceolate; *cor.* scarcely a third longer than the calyx ; *appendages of the galea* transversely ovate, broader than long. —① Meadows and pastures, Mass., N. Y. to Arc. Am. Stem a foot high, smooth, branching. Leaves opposite, nearly sessile, cordate-lanceolate, acutely serrate, rough. Flowers axillary, crowded into a leafy spike. Calyx inflated, contracted at the mouth, with 4 nearly equal teeth, and much shorter than the yellow, ringent corolla, but becoming very large after flowering. July.

28. PEDICULÀRIS.
Lat. *pediculus,* a louse ; probably from its efficacy in destroying that insect.

Calyx ventricose, 2—5-cleft, the segments leafy. or sometimes obliquely truncate ; corolla vaulted, upper lip compressed, emarginate ; lower lip spreading, 3-lobed ; capsule 2-celled, oblique, mucronate ; *seeds* angular.—*Herbs. Lrs. alternate, rarely sub-opposite, often pinnatifid. Fls. spicate.*

1. P. CANADENSIS. (P. gladiata. *Michx.*) *Lousewort.*

Hirsute; *st.* simple; *lvs.* alternate, petiolate, lance-oblong, pinnatifid lobes oblong-ovate, crenate-dentate; *spike* short, dense, leafy; *cal.* truncate downwards, *cor. galea* abruptly incurved, with 2 setaceous teeth; *caps.* acuminate.—⅃ Pastures and low grounds, U. S.! & Can. Stem erect, a foot high Leaves 3—6′ by 1—2′, chiefly radical. Spike short, hairy, with a few small leaves at the base. Calyx truncated in an oblique direction downwards. Corolla yellowish and purple; the upper lip long, erect, forming a galea or helme' cut square off at the end, with a bristle-like tooth at each corner. Capsule prolonged into a lanceolate point ½′ long. May—July.

β. *gladiata* (P. gladiata *Mx.*). *Caps.* prolongated into an ensiform point which is ¼—1′ in length. Plant rather taller.

2. P. LANCEOLATA. Michx. (P. pallida. *Ph.*) *Branching Lousewort.*

Nearly glabrous; *st.* branched; *lvs.* subopposite, briefly petiolate or sessile, oblong-lanceolate, doubly incised-crenate; *spike* rather dense; *cal.* 2-lobed; *cor. galea* as long as the lip, incurved at apex, ending in a short, conical beak.— ⅃ In alluvial woods, &c. N. Y.! to Wis. *Lapham!* S. to Va. Stem 1—2′ in height, smooth, with pubescent lines, nearly opposite leaves and a few axillary branches. Leaves 3—5′ by 1—1½′. Spikes 1—2′ in length, with ovate-lanceolate bracts. Calyx and corolla smooth, the latter greenish-yellow, an inch in length, with the galea somewhat emarginate at the end. Capsule short, broadly ovoid. Sept.

29. MELAMPYRUM.

Gr. μελας, black, πυρος, wheat; the seeds blacken the flour of wheat if ground with it.

Calyx 4-cleft; upper lip of the corolla compressed, the margis folded back; lower lip grooved, trifid; capsule 2-celled, oblique, opening laterally, cells 2-seeded; seeds cylindric-oblong, smooth.— *Herbs with opposite lvs. Fls. solitary in the upper axils.*

M. PRATENSE. β. *Americanum.* Benth. (M. Americanum. *Mx.*) *Cow Wheat.*—*Lvs.* linear and lanceolate, petiolate, glabrous, the upper ones toothed at base; *fls.* axillary, distinct.—① Inhabits woods, Can. to Ga., W. to Ky. Stem with opposite branches, 8—10′ high, round, erect. Leaves opposite, 1 1½′ by 3—5″, the floral ones broader, with setaceous teeth at base and tapering to an obtuse point. Flowers in the axils of the upper leaves, yellowish, slender, the corolla twice the length of the calyx. Capsules acute, declined, 4-seeded. Jl.

ORDER XCII. VERBENACEÆ.—VERVAINS.

Trees and shrubs, sometimes Herbs. *Lvs.* generally opposite, simple or compound, exstipulate.
Fls. in axillary corymbs or dense heads, or alternate-spicate.
Cal. tubular, 4—5-toothed, inferior, persistent.
Cor. tubular, the limb bilabiate or irregularly 4—5-cleft, deciduous.
Sta. 4, didynamous, seldom equal, occasionally only 2.
Ova. 2—4-celled; ovules erect or pendulous, solitary or twin. *Style* 1.
Fr. drupaceous, baccate or dry, dividing into 2 or 4 1-seeded portions.
Seeds with little or no albumen.

Genera 55, species 616, the herbs chiefly natives of temperate regions and the shrubs and trees of the tropics, where they are in some instances very large. The teak-wood (Tectona grandis), native of India, justly styled the "Oak of the East," is a timber tree of immense size and great durability, often attaining the height of 100 feet. The wood contains silex. The medicinal properties of the tribe are little known or unimportant.

Conspectus of the Genera.

Flowers { in spikes or spiked corymbs. { Corolla subequal, funnel-form. . . . . . Verbena. 1
{ Corolla bilabiate, nodding in fruit. . . . Phryma. 2
{ in small peduncticulate, axillary heads. . . . . Lippia. 2

1. VERBENA.

Celtic *ferfaen,* to remove stone, hence Eng. *vervain* and Lat. *verbena.*

Calyx 5-toothed, with one of the teeth often truncate; corolla funnel-form, limb 5-cleft, nearly equal; stamens 4 (rarely but 2); seeds 2—4, enclosed in a thin, evanescent pericarp.—*Herbs with opposite lvs. Fls. mostly alternately spicate, rarely capitate or corymbed.*
35*

1. V. HASTĀTA. *Vervain.* *Simpler's Joy.*

Erect; *lrs.* lanceolate, acuminate, incisely serrate, petiolate, the lower ones lobed or hastate; *spikes* erect, slender, panicled; *fls.* tetrandrous.—♃ An erect, ta'l and elegant plant, frequent by roadsides and in low grounds, mostly throughout the U. S. and Can. Stem 3—4f high, with paniculate, opposite branches above. Leaves rough in appearance and to the touch, opposite, lower ones often somewhat hastate. Flowers small, blue, arranged in long, close, imbricated spikes, which are somewhat fascicled at the summit of the stem, erect and parallel to each other. Seeds 4. July—Sept.

β. pinnatifida. *Lvs.* incisely pinnatifid and coarsely dentate.—Western States! common.

γ. oblongifolia. Nutt. (V. paniculata. *Lam.*) *Lrs.* lance-ovate or lance-oblong, sharply serrate; *spikes* filiform, loosely paniculate; *fls.* smaller.—Penn. to Ia.! and Mo. I have frequently observed this tall (4—6f) variety, and many others, on the sandy prairies of Indiana. They appear to be *hybrids* between V. hastata and V. urticæfolia.

2. V. URTICÆFOLIA. *Nettle-leaved Vervain.*

Erect, subpubescent; *lrs.* ovate and ovate-lanceolate, serrate, acute, petiolate; *spikes* axillary and terminal, loose, filiform; *fls.* tetrandrous.—♃ About roadsides and rubbish. A weed of uninviting appearance, 2—3f high, with leaves resembling those of the nettle. It has long, slender, weak, green, divergent spikes, remotely filled with small, white, distinct flowers. Seeds 4. Jl. Aug.

3. V. SPURIA. *Spurious or Jagged-leaved Vervain.*

St. decumbent at base, divaricately branching, hairy; *lrs.* ovate-lanceolate, petiolate, laciniately lobed and toothed; *spikes* slender, loose; *bracts* a little longer than the calyx.—Conn. *Eaton,* Md.! to Ga. and Western States. An unsightly plant, with a square stem 1—2f high, half erect, di- and trichotomous above. Spikes 3—6' long, the bracts and flowers minute. Calyx 1' in length. Corolla blue. Aug. Sept.—This plant appears to be constantly though slightly different from V. officinalis of Europe.

4. V. BRACTEOSA. Michx. (Zapania. *Lam.*) *Prostrate Verbena.*

St. decumbent, branched, divaricate, pilose; *lrs.* laciniate, hirsute, rugose; *spikes* terminal, thick, many-flowered; *bracts* linear, squarrose, much longer than the calyx.—Dry fields and roadsides, Middle and Western States! Whole plant hairy, 8—16' long, remarkable for its squarrose-bracteate spikes. Leaves 1—2' long. Flowers small, blue. Capsule 4-celled, 4-seeded. Seeds bony. June—Sept.

5. V. STRICTA. Vent. *Mullein-leaved Verbena.*

Hirsute and hairy; *st.* thick, rigidly erect, branched above; *lvs.* ovate, oval or obovate, unequally dentate, sessile, acute, rugose; *spikes* erect, strict, imbricate, subfalcate.—An erect, rigid, and rather handsome species, in dry fields, Western States! common. Very hirsute, 1—3f high. Leaves 2—3' by 1—2', numerous, veiny and whitish beneath. Corolla blue, thrice larger than in V. hastata. July.

6. V. ANGUSTIFOLIA. Michx. (V. rugosa. *Willd.*) *Narrow-leaved Vervain.*

Erect, mostly simple; *lrs.* lanceolate-linear, tapering to the base, remotely serrate, with furrowed veins; *spikes* filiform, solitary, axillary and terminal.—A small, hairy species, found on rocky hills and in other dry soils, N. Y. to Va.! W. to the Miss. Stem not more than a foot high, with narrow (3' by 5"), rough leaves and slender spikes of deep blue flowers. July.

7. V. AUBLETIA. *Garden Verbena.*—*St.* weak, assurgent; *spikes* solitary, imbricate, long-pedunculate; divisions of the *cor.* emarginate; *lrs.* oval, deeply serrate and divided, petiolate.—Native at the South. A slender and delicate plant of the green-house, producing numerous, successive clusters of rose-colored or scarlet flowers. Stem square, viscidly pubescent, 1—2f high, with opposite branches and leaves. Leaves deeply cut and toothed, rhombic-oval, on short stalks. Flowers larger than others of the genus, in corymbose spikes. Bracts nearly as long as the calyx, narrow, permanent, downy as well as the calyx. May.

2. LIPPIA.

In honor of Augustus Lippi, a French physician.

Flowers in dense, pedunculate heads; calyx 2-parted, compressed erect, membranaceous, shorter than the tube of the corolla; corfunnel-shaped, limb subbilabiate, upper lip entire or emarginate, lower 3-lobed; sta. 4, didynamous, included; drupe dry, thin, enclosed in the calyx, 2-seeded.—*Shrubs or prostrate herbs, with opposite leaves. Heads on axillary peduncles.*

L. NUDIFLÓRA, Michx. (Zapania nudiflora. *Ph.* and *authors.*) *Fog-fruit.* Glabrous, procumbent; *st.* 4-angled, geniculate, simple; *lvs.* lanceolate or linear-lanceolate, acute, serrate, cuneate at base, petiolate, shorter than the peduncles.—♃ On river banks, Penn. to Ia.! Ill. and La. Stems 1f or more long. Leaves with conspicuous veins, 1—2′ long, ⅓—½ as wide, petioles 3—6″. Peduncles 2—3′. Heads ovoid or roundish. Flowers small, purplish-white. July, Aug.
β. *lanceolata.* (Lippia lanceolata. *Mx.*) *Lvs.* linear-lanceolate.

3. PHRYMA.

Calyx cylindric, bilabiate, upper lip longer, 3-cleft, lower lip 2-toothed; corolla bilabiate, upper lip emarginate, much smaller than the 3-lobed lower one; seed solitary.—*Herbs with opposite lvs. Fls opposite, spicate, deflexed in fruit.*

P. LEPTOSTACHYA. (Priva. *Lindl.*) *Slender-spiked Phryma or Lopseed.*
Lvs. stalked, ovate, serrate; *spikes* long and slender; *cal.* in fruit reflexed. —♃ Found in rocky woods, Can. and U. S. Stem 2—3f high. Leaves large, (3- 6′ long), thin and coarsely toothed, on short stalks. Flowers small, opposite, light-purple, in very long and slender spikes, of which one is terminal, the rest opposite and axillary, each often with a pair of bracts below. After flowering the calyx closes upon the fruit and becomes reflexed backwards close to the stem. Hence the common name *lopseed.* The specific name refers to the *slender spikes.* Seeds solitary, rather large, invested with a thin, membranous capsule, and enclosed in the matured calyx. July.

ORDER XCIII. LABIATÆ.—LABIATE PLANTS.

Herbs or undershrubs with 4-cornered stems and opposite branches.
Lvs. opposite, without stipules, replete with receptacles of aromatic oil.
Fls. in axillary, subsessile verticillasters, sometimes as if in whorls, spikes or heads.
Color almost always of the cyanic series, blue, purple, red, white, &c.
Cal. tubular, regularly 5-toothed or cleft, or bilabiate, persistent.   [3-cleft ons.
Cor. bilabiate (rarely regular, 5-toothed), the upper lip bifid or entire, overlapping in æstivation the lower
    Sta. 4, didynamous, or sometimes only 2, the upper pair being abortive or wanting, situated on the corolla
    tube. *Anth.* mostly 2-celled.
Ova. free, deeply 4-lobed, the single style arising from the base of the lobes.
Fr. 1—4 hard nuts or achenia. *Embryo* erect. *Cotyledons* flat.
Sds. erect, with little or no albumen. *Embryo* erect. *Cotyledons* flat.

Genera 135, species 2350, chiefly natives of temperate regions, being most abundant between latitudes 40° and 50° of the northern hemisphere.

Properties.—This well-known family is universally pervaded by an aromatic, volatile oil, and a bitter principle; the former rendering them eminently tonic, cordial and stomachic; the latter, where it prevails, febrifugal. The *pennyroyal, lavender, sage, hoarhound, thyme, spearmint, peppermint, horse mint, rosemary, &c., &c.,* plants whose qualities are too well known to require particular mention here, are all members of this useful family. Not one species is poisonous or even suspicious.

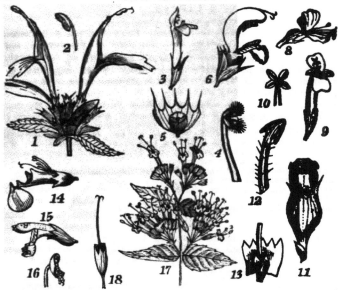

FIG. 51.—1. Monarda didyma. 2. An anther enlarged. 3. Flower of Galiopsis Tetrahit. 4. One of its stamens, much enlarged. 5. The calyx opened, showing the 4 achenia. 6. Flower of Salvia Sclarea. 8. Flower of Ocimum basilicum. 9. Flower of Nepeta Glechoma. 10. A pair of the anthers forming a cross. 11. Flower of Physostegia Virginica seen from beneath. 12. One of its stamens. 13. The ovaries with the rudimentary filament. 14. Flower of Teucrium Canadense. 15. Flower Nepeta Cataria. 16. One of its anthers. 17. Cunila Mariana. 18. A calyx and style.

Conspectus of the Genera.   **• Flowers diandrous.**

				{ Bracts whitish.	*Blephilia.* 8
			{ Corolla not yellow.	{ Bracts green.	*Hedeoma.* 18
		{ Connectile erect, short.	{ Corolla yellow.		*Collinsonia.* 16
	bilabiate.	{ Connectile transverse, long ; anthers dimidiate.			*Salvia.* 9
				{ Herbs.	*Cunila.* 17
			{ Upper lip emarginate.	{ Shrubs.	*Rosmarinus.* 16
		{ Corolla bilabiate.	{ Upper lip linear, embracing the filament.		*Monarda.* 7
Calyx	subequal.	{ Corolla subregular, 4-lobed.			*Lycopus.* 6

• • Flowers didynamous.

				{ Lips of calyx { Filaments simple.	*Melissa.* 20
				toothed . . . { Filaments forked.	*Prunella.* 22
		{ Tube of corolla	{ Stamens ascending.	Lips of the calyx entire.	*Scutellaria.* 21
Calyx		exserted. . . .	{ Stamens very long, arching the 5-cleft limb of corolla.		*Trichostemma.* 25
bilabiate.				{ Stamens scarcely exsert.	*Thymus.* 13
	{ Tube of the corolla scarcely longer than calyx.			{ Stamens exserted.	*Origanum.* 12
				{ equal.	*Hyssopus.* 15
			{ Cal. 15-nerved. { oblique.		*Lophanthus.* 24
			{ Flowers		*Dracocephalum.* 23
			{ Bracts { spicate. { Cal. 10-nerved. .	*Satureja.* 14
			green. { Flowers capitate, involucrate.		*Pycnanthemum.* 11
		{ Corolla bilabiate. { Bracts colored.			*Origanum.* 12
	{ Stamens erect	{ Corolla limb of 4 subequal lobes, one of them emarginate.			*Mentha.* 4
	or divergent.	{ Corolla limb of 5 equal lobes.			*Isanthus.* 5
				{ Calyx 5-ribbed.	*Lamium.* 28
		{ Leaves crenate. { Calyx 15-ribbed.			*Nepeta.* 29
				{ Plants glabrous.	*Physostegia.* 26
		{ included beneath { Leaves sharply serrate. { Plants hairy.			*Ballota.* 30
	{ Stamens as-	the upper lip. . . { Leaves entire, lance-linear and elliptic.			*Micromeria.* 19
	cending and	{ exserted through the fissure of the upper lip of corolla.			*Teucrium.* 10
			{ Corolla not spurred.	{ Upper lip of the corolla 2-lobed.	*Lavandula.* 2
		{ Stamens declinate.	{ Corolla tube spurred.	{ Upper lip of the corolla 4-cleft.	*Ocimum.* 1
					Plectranthus. 3
		{ Leaves	{ Lower lip of the corolla with 2 lateral teeth.		*Galeopsis.* 30
		undivided.	{ Lower lip of the cor. with lateral lobes reflexed.		*Stachys.* 31
	the teeth spinescent.	{ Leaves mostly 3-lobed.			*Leonurus.* 32
Calyx 4-toothed.	Upper pair of anthers cohering, half empty.				*Synandra.* 27
Calyx subentire, very large, campanulate.					*Melucella.* 34
Calyx 10-toothed, the alternate teeth shorter.					*Marrubium.* 10

TRIBE 1. OCIMOIDEÆ.—Corolla subbilabiate, the 4 upper lobes nearly equal, the lower one declinate, flat or concave, carinate or saccate. Stamens 4, declined.

1. OCIMUM.

Gr. οζα, to smell; on account of the powerful scent of the plants.

Upper lip of calyx orbicular, lower 4-fid; corolla resupinate, one lip 4-cleft, the other undivided; exterior filaments with a process at their base.

O. BASILICUM. *Royal Ocimum. Sweet Basil.*—*Lvs.* smooth, ovate-oblong, sublentate, petiolate; *cal.* ciliate.—① An exotic from Persia, cultivated. Plant about a foot high, with peculiarly smooth and soft leaves variously colored, exhaling a delightful odor. Stem retrorsely pubescent above, branched. Flowers white, in simple, terminal racemes. Jl., Aug. ‡

2. LAVANDULA.

Lat. *lavera*, to wash. The use of the distilled water of this plant is well known.

Calyx ovoid-cylindric, with 5 short teeth, the upper one often largest; corolla upper lip 2-lobed, lower 3-lobed, lobes all nearly equal; tube exserted; stamens included.

L. SPICA. *Lavender.*—*Lvs.* linear-oblanceolate, tapering to the base, sessile, revolute at the edge, the upper ones linear-lanceolate, the highest shorter than the calyx; *spike* interrupted; *bracts* subulate.—♃ Plant 12—18′ high, suffruticose, branching from the base. Leaves crowded at the base of the branches, clothed with a whitish down. Calyx villose. Corolla much exserted and of a lilac color. The plant is well known as an aromatic of a delightful fragrance. It is stimulant and tonic, and the oil extracted by distillation enters into many compositions in medicine. Jl. ‡

3. PLECTRANTHUS. L'Her.

Gr. πληκτρον, a cockspur, ανθος; from the spur-like appendage of the corolla.

Calyx upper lip largest; cor. resupinate, ringent, with the tube gibbous or spurred on the upper side at base.—*Half shrubby plants, with purple flowers. Natives of hot climates.*

P. PARVIFLORUS. "*Sage Geranium.*"—*St.* suffruticose, smoothish, branched; *rac.* compound, terminal, leafless; *peduncles* 1-flowered, verticillate; *corolla* gibbous.—S. America. Sometimes seen in house cultivation and called *Sage Geranium!* It is a large, coarse plant, 2—3f high, with large, soft, ovate, crenate leaves, and terminal racemes of very delicate bluish-purple flowers.

TRIBE 2. MENTHOIDEÆ.—Corolla somewhat campanulate or funnelform; tube scarcely exserted, limb subequally 4—5-lobed. Stamens 4, sometimes 2, distant, straight, diverging.

4. MENTHA.

Mintha, the daughter of Cocytus, is fabled to have been changed into one of these plants.

Calyx equally 5-toothed; cor. nearly regular, 4-cleft, the broadest segment emarginate; sta. 4, straight, distant; anth. cells parallel; filaments naked.

1. M. CANADENSIS. (M. borealis. *Michx.*) *Horsemint.*

Ascending, pubescent; *lvs.* petiolate, lanceolate, serrate, acute at each end; *fls.* in axillary cymes; *sta.* generally exserted.—♃ Can. to Ky. and Penn. An herbaceous, grayish plant, 1—2f high, growing in muddy situations. The stem is square and usually branched, the angles beset with reversed hairs. Leaves serrate, on opposite, downy footstalks, and punctate with resinous dots at each end. Flowers apparently in whorls, pale purple, usually distinguished by the projecting stamens which are sometimes twice as long as the corollas. Calyx hairy. Aromatic like the other species. June, July

β. *stamens* equaling the corolla

2. M. viridis. (M. tenuis. Michx. M. gracilis. Muhl.) Spearmint.

Lvs. subsessile, oblong-lanceolate, acute, incisely serrate; bracts setaceous, and, with the teeth of the calyx, somewhat hairy; spikes slender, interrupted, attenuate above.—24 Can. and U. S. A well known plant, highly esteemed for its agreeable, aromatic properties. It grows in wet soils, rapidly spreading by its creeping roots, with erect, branching, 4-angled stalks, 1—2f high. The spikes are somewhat panicled, long, composed of distinct, axillary cymes, apparently whorled, a little remote from each other. Peduncles smooth, round, shining. Corollas pale purple. Styles much exserted.

3. M. piperita. Smith. Peppermint.

Lvs. smooth, ovate-lanceolate, serrate, petiolate; bracts lanceolate; cal. quite smooth at base, punctate.—24 This species, introduced from Europe, has become naturalized in wet places, and cultivated in gardens. It has a more penetrating taste and stronger smell than the other species, pungent to the tongue followed by a sensation of coldness. The essence of peppermint is a well known medicine, acting as a cordial, used in flatulency, nausea, &c. It has a purplish stem, 2—3f high, with scattered, deflexed hairs. Leaves sharply serrate, dark green. Corolla purplish. July.

4. M. arvensis. Corn Mint. Field Mint.

St. ascending, much branched; lvs. ovate, serrate, petiolate, acute, hirsute; verticils axillary; pedicels smooth; cal. hirsute.—Naturalized in Penn. Ohio, &c., native in Europe. Stem stout, often erect, about 1f in height. Leaves varying to oblong or ovate-lanceolate, sometimes nearly smooth, about twice longer (1—2') than wide, several times longer than the petioles. Flowers small, numerous, pale purple. Stamens exserted. The plant smells like decayed cheese. July.

5. ISANTHUS. DC.

Gr. ισος, equal, ανθος, the flowers being regular, a character very rare among the labiatæ.

Calyx subcampanulate ; corolla 5-parted, tube straight and narrow, segments of the border ovate and equal; stamens subequal; stigma linear, recurved.

1. cœruleus. Blue Gentian.

Viscid, hairy; lvs. oval-lanceolate, acute at both ends, 3-veined; ped. 1—2-flowered.—24 A branching, leafy herb, in dry fields, Northern and Western States! with the aspect of the pennyroyal. Stem rounded, slender, 12—18' high with branches and leaves opposite. Leaves an inch or less in length, and a fourth as wide, distinctly triple-veined. Flowers numerous, blue, with included stamens. Calyx leaves lanceolate, longer than the tube. July.

6. LYCOPUS.

Gr. λυκος, a wolf. πους, a foot; a fanciful name.

Calyx tubular, 4—5-cleft; cor. subregular, 4-cleft, the tube as long as the calyx, upper segment broadest, emarginate; sta. 2, distant, diverging, simple; sty. straight, as long as the stamens; ach. 4, obliquely truncate at apex.—24

1. L. sinuatus. (L. Europæus. Michx. L. Americanus. Muhl.) Water Hoarhound.—Lvs. oblong-lanceolate, sinuate-dentate, lower ones incised; teeth of the calyx acuminate-spinescent; st. square, 1—2f high; fls. small, white, many in a whorl.—A perennial plant, widely diffused throughout the U. States, growing in damp grounds. In habits and general appearance it resembles the Menthæ, but is sufficiently distinguished by the number of stamens, form of the flattened triquetrous achenia, and its being inodorous. Stem sharply 4-angled, the sides concave, 1—2f high. Lower leaves deeply and pinnatifidly toothed. Verticils dense. Calyx teeth longer than the achenia. It dyes a permanent black. Aug.

2. L. Virginicus. Virginian Water Hoarhound. Bugle-weed.

Lvs. broad-lanceolate, serrate, tapering and entire at the base; calyx teeth spineless, usually 4, shorter than the achenia.—A plant as widely diffused as the preceding, growing in wet soils. Stem smooth, obtusely 4-angled, with the sides concave, 12—18' high, usually simple, bearing small whorls of minute, purplish flowers. Leaves with coarse, tooth-like serratures, sessile. The whole plant often changes to purple. It is reputed a remedy for blood-spitting. July, Aug.

TRIBE 3. **MONARDEÆ.**—Corolla bilabiate; tube exserted. Stamens 2, fertile, ascending, the upper pair abortive; anthers linear with the 2 cells contiguous, or halved with the 2 cells widely separated on opposite ends of a long, transverse connectile.

7. MONARDA.

Name in honor of Monardus, a Spanish botanist of the 16th century.

Calyx elongated, cylindric, striate, subequally 5-toothed; cor. ringent, tubular, upper lip linear, lower lip reflexed, 3-lobed, the middle lobe narrowest; sta. 2, fertile, ascending beneath the upper lip, and mostly exserted; anth. cells divaricate at base, connate at apex.—♃

1. M. DIDÝMA. (M. purpurea. *Lam.* M. coccinea. *Michx.*) *Mountain Mint.* St. acutely 4-angled; *lvs.* broadly ovate, acuminate, somewhat rough and villous, on short petioles, veins and veinlets hairy beneath; *fls.* in terminal often proliferous heads; *bracts* colored.—A handsome, fragrant plant, 2—3f high, with crimson or scarlet flowers. Stem mostly branching. Leaves 2—5' long, very broad at base, often cordate, serrate, with scattered hairs above, and prominent, hairy veins beneath. Flowers in heads which are often proliferous, with large, ovate-lanceolate bracts tinged with tne same color as the corollas. Calyx colored. Corollas large and showy. Styles 4, 2 of which are minute and abortive; hence the specific name. A beautiful plant in cultivation Swamps, Can. to Ga.

2. M. FISTULÒSA. (M. oblongata, rugosa, clinopodia, allophylla, mollis, purpurea, *of authors.*) *Horsemint. Wild Bergamot.*—*Lvs.* ovate-lanceolate, acute or acuminate, petiolate, more or less pubescent; *hds. of fls.* terminal, few, but many-flowered; *bracts* sessile; *cal.* slightly curved with the throat hirsute.— A handsome, variable plant, growing in hedges, thickets, rocky banks, Mass. to Ga. W. to the Saskatchawan. The stem is 2—4f high, quadrangular with the sides somewhat concave, hollow in various degrees, nearly smooth or pubescent above, simple or with a few opposite branches. Leaves obtuse at base, ovate or oblong-lanceolate, mostly acuminate, acutely serrate, nearly smooth, 2—4' long and on petioles ½ their length. Flowers in involucrate, terminal heads, 20—50 in a head. Outer bracts leafy, often partially colored. Calyx slender, ½' in length. Corolla much exserted, varying from greenish-white and pale purple to blue, the upper lip long and linear, enfolding the 2 stamens, which, with the style are somewhat exserted at its end. Jl. Aug.

α. St. simple, hollow; *hds.* simple or proliferous; *cor.* pale yellow.

β. (M. allophylla. *Mx.*) St. branched, hollow, or solid with pith; *lvs.* oblong-lanceolate; *hds.* simple; *bracts* partially colored; *cor.* pale blue.

γ. (M. clinopodia. *L.*) S'. solid; *lvs.* tapering at base, remotely serrate; *hds.* simple; *cal.* short; *cor.* pale purple.

ε. *purpurea.* St. tall (3—5f), glabrous, dark purple.— la.! common.

ζ. *mollis.* Lvs. softly pubescent; *upper lip of cor.* densely bearded.

3. M. PUNCTÀTA. (M. lutea. *Michx.*) *Horsemint.* Nearly glabrous; st. obtusely angled, hoary-pubescent; *lvs.* oblong-lanceolate to oblong, remotely and obscurely serrate; *verticils* axillary, dense; *bracts* lanceolate, colored, longer than the verticils; *al.* teeth unequal. Pine barrens, N. J., common, to Car. and Western States. Stem 2—3f high, branched. Leaves punctate. Corolla yellow, with brown spots, upper lip villous at the apex. Bracts large, yellow and red. Sept.—It contains an essential oil which is valuable in medicine.

4. M. Bradburiàna.

St. simple, glabrous; *lvs.* ovate- or oblong-lanceolate, subsessile, rounded at base, hirsute-pubescent both sides, margin subdentate, apex acute; *cal.* pilose, densely bearded at throat, segments subulate-spinose; *hds.* large, terminal, outer bracts broad-lanceolate, ciliate, colored.—Ohio to Ill. *Mead!* Stem slender, about 3f high. Leaves sometimes slightly petiolate, 2—3' long. Bracts purple. Corolla purple. Jl.

8. BLEPHILIA. Raf.

Gr. βλεφαρις, the eyelash; probably referring to the ciliate bracts

Calyx 13-ribbed, bilabiate, upper lip 3-toothed, lower lip shorter, 2-toothed, the teeth setaceous; corolla bilabiate, upper lip short, erect, oblong, obtuse, entire; lower lip of 3 unequal, spreading lobes, the lateral ones orbicular; stamens 2, fertile, ascending, exserted.

1. B. hirsùta. (Monarda hirsuta. *Ph.* M. ciliata. *Michx.*) *Hairy Blephilia.* Whole plant hirsute; *lvs.* ovate-lanceolate, acuminate, serrate, petiolate; *fls.* in axillary verticillasters and terminal heads; *bracts* colored, shorter than the flowers, oblong, acuminate.—♃ In damp woods, rare N. Eng., common in the W. States. Stem 1—2f high, diffusely branching, roughly pubescent. Petiole ⅓—½' long, leaves 3 or 4 times as long, somewhat rounded at base. Flowers small, forming several dense whorls near the ends of each branch. Corolla scarcely ⅓' long, pale purple with spots of a deeper hue. Style longer than stamens or corolla. Jn. Jl.

2. B. ciliàta. (Monarda ciliata. *Linn.* not *Michx.*) *St.* hirsute, simple, acutely 4-angled; *lvs.* few, ovate-lanceolate, tapering to an obtuse point, subsessile, serrate, minutely pubescent; *fls.* in dense, approximate, involucrate, terminal and subterminal verticils; *bracts* ovate, veiny, glabrous, ciliate, as long as the calyx.—Fields, barrens, Penn. to the Miss., very abundant in the Western States! Plant 2—4f high, generally simple, rarely with 1 or 2 branches. Leaves 1—2½' long, ½—1' wide. Flowers small, numerous. Verticils subglobose. Outer bracts 5'' by 3—4'', whitish. Calyx subbilabiate. Jn.—Aug.

9. SALVIA.

Lat. *salveo,* to be in health; probably from its salutary qualities.

Calyx striate, bilabiate, upper lip 2—3-toothed or entire, lower lip divided; corolla ringent; stamens 2; connectile transversely articulated to the filament, supporting at each end a cell of the dimidiate anther; achenia 4.—*A large genus of which but few species are native. The transverse connectile constitutes the essential character.*

1. S. lyràta. *Wild or Meadow Sage. Cancer-weed. Radical lvs.* lyrate, erosely dentate; *upper lip of the cor.* very short, straight. —♃ in shady woods, Can. to Ga. Stem erect, quadrangular, nearly leafless, 1—2f high, branching above and covered with hairs pointing downwards. Radical leaves oblong, lyrate or sinuate-pinnatifid, petiolate. Cauline leaves but 1—2 pairs, just below the raceme. Flowers in whorls of about 6, distant, constituting a long, interrupted raceme. Corolla blue, the tube much exserted. Native of shady woods. May, June.

2. S. officinàlis. *Common Sage.*—*Lvs.* ovate-lanceolate, crenulate, rugose; *whorls* few-flowered; *cal.* mucronate; *upper lip of the cor.* as long as the lower and somewhat vaulted.—A well known garden plant, with a shrubby stem, rugose leaves of a dull green color and an aromatic fragrance. Flowers in whorls forming a spike. Corolla ringent, blue, with a lengthened tube and viscid calyx, somewhat brown. Native in the south of Europe. Very useful in domestic economy and medicine. July. ‡

3. S. Sclarea. *Clary.*—*Lvs.* oblong, heart-shaped, rugose, villous, serrate; *bracts* colored, concave, longer than the calyx.—♀ A strong-scented exotic, 1—3f nigh, with viscid leaves as large as the hand. The flowers and bracts are

variegated with pale purple and yellowish-white, in whorled spikes. Ca.yx with spinous teeth. Native in Italy. ‡

4. S. MEXICÀNA. Willd.? *Mexican Salvia.*—*St.* branching at base, weak, ascending, pubescent; *lvs.* long-petiolate, ovate-lanceolate, subcordate, crenate-dentate, smooth above, pubescent beneath; *fls.* opposite, in terminal racemes; *bracts* deciduous; *cal.* slightly colored, upper lip truncate, subentire; *cor.* fimbriate-ciliate, 3—4 times longer than the calyx; *sty.* exserted.—♃ Mexico. A beautiful and popular house plant. Flowers bright crimson or scarlet, near 9' long. There are several varieties. †

5. S. SPLENDENS. Ker.?—*St.* erect, glabrous; *lvs.* broad-ovate and ovate, petio'ate, rounded or acute at base, glabrous both sides, dent-serrate, acuminate; *fls.* opposite, racemose; *bracts* deciduous; *cal.* scarlet, and, with the corolla, pubescent, upper lip entire, acuminate, lower lip 2-toothed.—♃ Mexico. Gardens. Plant 9—4f high, branched. Flowers large, scarlet. After flowering the calyces enlarge, and become as showy as the corollas. †

10. ROSMARÍNUS.

An ancient Latin name, compounded of *ros*, dew, and *marinus*, of the sea.

Corolla bilabiate, upper lip 2-parted, lower lip reflexed, in 3 divisions of which the middle is the largest; fil. 2 fertile, elongated, ascending towards the upper lip, having a tooth on the side.

R. OFFICINALIS. *Rosemary.*—*Lvs.* sessile, linear, margins revolute.—An erect, evergreen shrub, 4f high, much branched. Leaves opposite, obtuse, linear-oblong, entire, smooth. dark green and shining above, downy and sometimes whitish beneath. Flowers axillary and terminal, of a bright blue color, having, like the leaves, a strong aromatic fragrance like camphor. It yields by distillation a large proportion of fragrant oil. ‡

TRIBE 4. **SATUREÍNEÆ.**—Calyx 5-toothed and equal, or bilabiate with the upper lip trifid and the lower bifid. Corolla subbilabiate, upper lip erect, flat, entire or bifid, lower spreading, trifid, lobes subequal; tube about as long as the calyx. Stamens 2—4, distant, straight, diverging.

11. PYCNANTHÉMUM. Benth.

Gr. πυκνος, dense, ανθος, alluding to the dense, capitate inflorescence

Calyx tubular, striate, 5-toothed; upper lip of corolla nearly entire, lower lip trifid, middle lobe longest; sta. distant; anth. with parallel cells.

1. P. INCÀNUM. Michx. (Clinopodium. *Willd.*) *Mountain Mint. Wild Basil.*—*Lvs.* oblong-ovate, acute, subserrate, rounded at base, with short petioles and hoary tomentum; *hds. of fls.* compound, terminal and lateral, pedunculate; *bracts* subulate.—♃ Grows in rocky woods and nills, Can., N., Mid. and W. States. Stem 2—4f high, obtusely 4-angled, erect, and, like the rest of the plant, covered with soft, whitish down. Leaves whitish beneath Flowers pale red with purple spots, on white, tomentose peduncles, in dense heads, and with numerous bracts, of which the inner ones are setaceous, bearded at the end; the outer ones are lanceolate. Plant aromatic. Jl. Aug.

β. *St.* taller, branched; *upper lvs.* with white blotches.—Ind.!

2. P. ARISTÀTUM. Michx. (Nepeta Virginica, *Linn.?*) *Wild Basil.*

St. hirsute-pubescent, brachiate-corymbose; *lvs.* ovate-lanceolate, briefly petiolate, acute at base, subserrate, pubescent, acuminate; *verticils* terminal, capitate and subterminal, hirsute; *bracts* lance-subulate, and with the calyx terminated by awns.—Woods and barrens, N. Y., Mass. to Miss.! Stem 1—3f high, subsimple or much branched. Leaves 1—2½' by ½—¾', generally with small, remote serratures and ciliate on the margin. Each branch terminates in a small (½' diam.) head with one or two dense whorls just below it. It is a more spreading and hairy plant than P. muticum. Jl. Aug.

3. P. PILÒSUM. Nutt.　*Hairy Pycnanthemum.*

St. and *lvs.* beneath pilose; *st.* subsimple; *lvs.* lanceolate, nearly entire, sessile; *fls.* in large, terminal, sessile heads; *bracts* lanceolate, and, with the calyx, canescently villous and awnless; *cor.* pubescent; *sta.* exserted.—♃ Low grounds, Ill. *Mead*, to Tenn. *Nuttall.* I am unacquainted with this species, but it is probably distinct, closely allied to the last. "Stem a little branched at the summit. Bracts shorter than the calyx, acute but not awned. Calyx teeth minute. Corolla white, without spots."

4. P. MUTICUM.　(Brachystemum muticum. *Michx.*)　*Awnless Pycnanthemum.*—*St.* pubescent, paniculate-branching above; *lvs.* ovate-lanceolate, subdentate, sessile, nearly smooth; *hds.* terminal; *bracts* ovate-lanceolate, acuminate, awnless; *sta.* included.—♃ Found in woods and dry hills, Penn. to Ga. and W. States. Stem 2f high, square, with larger, opposite leaves and white flowers. Leaves large, the width a third of the length, entire or denticulate. Heads mostly terminal, and with the bracts and upper leaves, whitish pubescent. Corolla tinged with purple, with spots of a deeper hue.　Aug.

5. P. LANCEOLÀTUM. Pursh.　(P. verticillatum. *Pers.* Brachystemum Virginicum. *Mx.?*)—*St.* straight, corymbosely branched, pubescent on the angles; *lvs.* subsessile, ovate-lanceolate and linear-lanceolate, feather-veined, entire; *verticils* sessile, fasciculate-corymbed; *bracts* linear-lanceolate, acuminate; *sta.* exserted.—♃ This species much resembles the next, but is distinct in several important characters. Grows in dry woods and hills, abundant W. and Mid. States. Stem 2 or more feet high, square, with obtuse angles, somewhat scabrous. Branches corymbed, downy above. Leaves varying in width from one-sixth to one-half of their length. Flowers collected in dense, canescent heads. Corolla purplish-white with darker spots.　Aug.

6. P. LINIFÒLIUM. Pursh.　*Flax-leaved Pycnanthemum.*

St. straight, smooth; *branches* trichotomous, fastigiate; *lvs.* linear, very entire, 3-veined, smooth; *hds.* terminal, dense, in a fasciculate corymb.—♃ An erect plant with fastigiate branches, 1½f high, growing in exsiccated swamps, Penn., N. Y. to the Miss. and S. States. Stem often purplish, slightly 4-angled, corymbose at the summit. Leaves very narrow, entire, smooth and punctate, with fascicles of smaller ones in the axils. Flowers small, white, in numerous, small, roundish heads, mostly terminal, and with imbricated bracts. Aug.—I have generally found this species nearly destitute of the mint-like flavor of the foregoing. The statement in a former edition was made on the authority of others.

12. ORIGÀNUM.
Gr. ορος, a mountain. and γανος, joy.

Flowers collected into dense clusters, imbricated with bracts; upper lip of the corolla erect, flat. emarginate, lower lip with 3 nearly equal segments.

1. O. VULGÀRE.　*Wild Marjoram.*

Lvs. ovate, entire, petiolate; *spikes* roundish, panicled, fasciculate, smooth, erect; *bracts* ovate, longer than the calyx, colored.—♃ grows in fields and thickets. Stem 12—18′ high, purple, leafy, branching above. Leaves a very little serrate, opposite, hairy, sprinkled with resinous dots, paler beneath. Petioles hairy one-fourth as long as the leaves. Bracts tinged with purple. Flowers purplish-white. The plant has a highly aromatic taste. Jl. Aug. ‡

2. O. MAJORÀNA. Ph.　(Majorana hortensis. *Mœnch.*)　*Sweet Marjoram.*—*Lvs.* oval or obovate, obtuse, entire, petiolate, hoary-pubescent; *spikes* roundish, compact, pedunculate, clustered at the end of the branches; *bracts* roundish.—♃ Native of Portugal, cultivated in gardens. It has a pleasant aromatic flavor and is employed in various ways as a *seasoning*. Plant soft-downy, a foot high. Flowers pink-colored. Jl. Aug. ‡

13. THYMUS.
Gr. Sυμος, courage : on account of its invigorating smell.

Flowers capitate or verticillate; calyx subcampanulate, bilabiate,

10-ribbed, the throat closed with hairs; upper lip of the corolla flat, emarginate, shorter than the lower.

1. T. VULGÀRIS. *Garden Thyme.*—*St.* procumbent at base, erect; *lvs.* revolute at the sides, oblong-ovate and lanceolate; *verticils* in terminal, leafy spikes. —♄ Native of S. Europe and cultivated for culinary purposes. Stems suffruticose, numerous, branched, 6—10′ high. It is highly aromatic, as well as the other species, and is peculiarly attractive to bees. Blossoms in summer. ‡

2. T. SERPYLLUS. *Wild Thyme. Mother of Thyme.*
St. decumbent; *lvs.* flat, elliptical, obtuse, ciliate at base; *fls.* capitate.— ♃ Mass., N. Y. and Penn. An aromatic plant, similar to the preceding, but milder and rather more pleasantly flavored. Stems suffruticose, wiry, slender and wavy, with leafy, downy and ascending branches, each terminating in a small, dense, oblong head of purple flowers, much frequented by bees. Leaves entire, petiolate, punctate, smoothish, ciliate. Corolla purple, spotted. June. Cultivated and naturalized.

14. SATURÈJA.

Arabic *satur* the general name for labiate plants.

Calyx tubular, 10-ribbed; segments of the bilabiate corolla nearly equal; stamens diverging, scarcely exserted.

S. HORTÈNSIS. *Summer Savory.*—*St.* branching; *lvs.* linear-oblong, entire, acute at the end; *ped.* axillary, cymose.—♃ Native of Italy. Cultivated as a culinary aromatic. Stem branching and bushy, 1½f high, woody at base, frequently changing to purple. Leaves numerous, small and narrow, with axillary cymes of pink-colored flowers. Calyx about as long as the corolla. Jl. Aug. ‡

15. HYSSÒPUS.

Hebrew *ezob;* Arabic *azzof;* Eng. hyssop.

Upper lip of the corolla erect, flat, emarginate, lower lip 3-parted, the middle segment largest, tube about as long as the calyx; stamens exserted, diverging.

H. OFFICINÀLIS. *Hyssop.*—*Lvs.* linear-lanceolate, acute, entire, sessile; *cal. teeth* erect; *fls.* in racemose, secund verticils, middle division of the corolla 2-lobed, entire.—♃ The common hyssop is a native of S. Europe, often met with in our gardens, being cultivated for its reputed medicinal properties. It is a handsome plant, growing in tufts, 2f high, with delicate foliage and bright blue flowers. July. ‡

16. COLLINSONIA.

Name in honor of John Collinson, an English botanist.

Corolla exserted, campanulate-ringent, upper lip in 4 subequal lobes, lower lip longer, declined, fimbriate; stamens 2, (rarely 3) much exserted, divergent.

C. CANADENSIS. *Horse Balm.*
Lvs. ovate, acuminate, coarsely serrate, petiolate, glabrous; *teeth of the cal.* subulate, shorter than the tube; *rac.* paniculate, terminal.—♃ A tall herb with large leaves and yellow flowers, woods and fields, Can. to Ky. and Car. Stem 4-sided, 3—4f high, smooth or a little pubescent. Leaves thin, 6—8′ long and 3—4′ wide. Flowers in a large, compound raceme, with opposite branches and pedicels. Corolla greenish-yellow, the lower lip elongated and fringed. Style and stamens very long. Flowering in summer.

17. CUNÌLA.

The ancient Roman name for pennyroyal.

Calyx 10-ribbed, equally 5-toothed, throat densely villose; upper lip of corolla flat, emarginate; stamens 2, erect, exserted, distant.

C. MARIÀNA. (Ziziphora, *R. & S.*) *Dittany.*
Lvs. ovate, serrate, subsessile; *cymes* pedunculate, corymbose, axillary and

terminal.—♃ Grows on rocks and in dry woods, N. Y. to Ga. and Ark. Stem 4-angled, mostly purple, branching, smoothish, 1—2f high. Leaves small, nearly smooth, roundish or subcordate at base, tapering to a point and punctate with pellucid dots. Flowers with subulate bracts at the base of the 3-forked pedicels. Calyx punctate. Corolla nearly twice as long as the calyx, pubescent, pale red. Stamens and style much exserted, of the same hue as the corolla. The herb is delightfully fragrant, and used in febrifugal infusions. Jl. Aug.

TRIBE 5. **MELISSINEÆ.**—Calyx bilabiate; corolla bilabiate; upper lip straight, lower lip spreading, cleft into 3 flat lobes, of which the middle one is often broadest. Stamens 4, sometimes 2, ascending.

18. HEDEOMA. Pers.
Gr. ἡδεια, sweet or agreeable, οσμη, smell; on account of the fragrance.

Calyx gibbous beneath at base, 13-ribbed, throat hairy; upper lip of corolla erect, flat, lower lip spreading, 3-lobed; stamens 2, fertile, ascending.

H. PULEGIÖIDES. Pers. (Cunila. *Linn.* Ziziphora. *R. & S.*) *Pennyroyal.* Lvs. oblong, few-toothed; fls. axillary, whorled.—① A small, strong-scented herb, held in high repute in the domestic *materia medica.* Stem erect, branching, half a foot high. Leaves opposite, with 1—2 teeth each side, on very short petioles, smooth on the upper surface, roughish beneath. Calyx ciliate, 2 lower divisions spined. Abundant in dry pastures, N. Eng., Can. to Ga. and Ark. Flowering all summer.

19. MICROMERIA. Benth.
Gr. μικρος, small, μερος, division; on account of the slightly 2-lipped calyx and corolla.

Calyx 13 (rarely 15)-ribbed, 5-toothed, nearly equal; corolla subbilate, tube exserted, upper lip bifidly emarginate, lower subequally 3-lobed; stamens ascending, the upper pair shorter.

M. GLABELLA. Benth. (Cunila. *Michx.* Hedeoma glabra. *Nutt.*) Glabrous; st. branching above, and mostly surculose at base; lvs. entire, those of the suckers elliptic-ovate, of the stem linear-oblong, obtuse; *verticillasters* about 6-flowered.—♃ A delicate little herb nearly or quite smooth, growing on lime-stone rocks near the base of Niagara Falls! W. to the Falls of St. Anthony. Rare. It has the general aspect and fragrance of pennyroyal. Stem erect (prostrate at base), 4-angled, slender, 6—10' high. Suckers at the base often numerous and several inches in length, with leaves about 3" by 2", brownish-purple beneath. Stem leaves 9—12" long, very narrow, the lowest sometimes with a few teeth. Flowers somewhat regular, on pedicels ¼' long, with linear bracts at the base. Corolla pale purple. Stamens 4, the upper pair much the shortest, all antheriferous. July, Aug.

20. MELISSA. Benth.
Gr. name of the bee, from μελ, honey, which is sought in these flowers by bees with avidity.

Calyx 13-ribbed, flattish above, the upper lip 3-toothed, lower bifid, upper lip of the corolla erect, flattish, lower lip spreading, 3-lobed, the middle lobe mostly broadest; stamens ascending.

1. M. OFFICINÀLIS. *Balm.* Pubescent; st. erect, branching; fls. in dimidiate verticils, subsessile; lvs. ovate, acute, coarsely crenate-serrate, rugose; bracts few, ovate-lanceolate, petiolate.—♃ N. Eng. ! to Ia. ! found in the deepest forests! A well known garden plant. Stem 1—2f high. Flowers white or yellowish. The plant is a stomachic and diuretic, generally administered in the form of tea. For medicinal use it should be cut before flowering, which occurs in June and after. ‡ ¶ †

2. M. CLINOPODIUM. Benth. (Clinopodium vulgare. *Linn.*) *Wild Basil.* Villose; lvs. ovate, subserrate; verticils many-flowered, hairy; bracts numerous, subulate.—♃ Low woods, Northern and Western States. A common plant, 1—2f high. Stem square, simple or sparingly branched, and, as well

as the whole plant, clothed with a whitish wool. Leaves petiolate, tapering to an obtuse point, pale, with whitish down beneath. Flowers purplish, in very hairy, mostly terminal whorls or heads. Peduncles cymosely branched, short. Involucre of narrow and bristle-like bracts, about equaling the hairy, subulate calyx teeth. July.

TRIBE 6. **SCUTELLARINEÆ.**—Calyx bilabiate, upper lip truncate. Corolla bilabiate, upper lip vaulted, tube ascending, exserted. Stamens 4, ascending beneath the upper lip of the corolla.

21. SCUTELLARIA.

Lat. *scutella*, a small vessel; from the resemblance of the calyx with its appendages.

Calyx campanulate, bilabiate, lips entire, upper one appendaged on the back and closed after flowering; cor. bilabiate, upper lip vaulted, lower dilated, convex, tube much exserted, ascending; sta. ascending beneath the upper lip; anth. approximate in pairs.

* *Flowers axillary, solitary.*

1. **S. GALERICULÀTA.** *Common Scull-cap.*
St. erect, simple, or branched; lvs. lanceolate-cordate, remotely crenate-serrate; fls. axillary, solitary.—⚥ Meadows and ditches, Can. to Penn. Abundant. The whole plant glabrous. Stem square, 12—18' high. Leaves truncate-cordate at base and acutish at apex, scarcely petiolate, 1½' by ½'. Flowers much larger than the preceding, rarely more than 1 from the same axil, with a vizor-like calyx like that of the other species. Cor. an inch in length, blue. Aug.

2. **S. NERVÒSA.** Pursh. (S. gracilis. *Nutt.*)
St. slender, erect, subsimple, stoloniferous; lvs. broad-ovate, subcordate, crenate-serrate, sessile, glabrous, 3—5-veined, lower roundish-ovate, upper ovate and lance-ovate, slightly petioled; fls. axillary, solitary.—Rocky shades, along streams, Penn.! to Ill.! and La. Roots creeping, often sending out long, filiform stolons. Stem 8—15' high, weak, often with a few filiform branches. Leaves 2—15" by 5—12", the middle pairs largest, acute or obtusish. Floral leaves entire, small. Flowers few, sometimes on the slender branches only. Corolla pale blue, 4—6" long. May—Jl.

3. **S. PARVÙLA.** Michx. (S. ambigua. *Nutt.*)
St. simple or branching at base, square, puberulent; lower lvs. suborbicular, petiolate, upper oblong-ovate, obtuse, entire, sessile, axillary, opposite.—Pastures, Mid.! and Western States! Plant 3—5' high. Root generally (not always ?) with tuberous internodes, and fibrous at each joint. Leaves 3—6" long, ½ as wide, lower 3" diam. Flowers 4—6" long, rather numerous, longer than the leaves, blue. June.

* * *Flowers in axillary and terminal racemes.*

4. **S. LATERIFLÒRA.** *Mad-dog Scull-cap.*
St. branching, nearly glabrous; lvs. ovate-lanceolate, acuminate, serrate, petiolate; rac. lateral, axillary, leafy.—Meadows and ditches, lat. 38° to Arc. Am. Stem square, 1—2f high, very branching. Leaves opposite, rounded at base, acuminate or acute, coarsely serrate, on petioles an inch in length. Racemes opposite, axillary, somewhat one-sided, on long stalks, and consisting of numerous small, blue flowers intermixed with small leaves. The English name is due to the singular form of the ca'yx, which after flowering, closes upon the seeds like a cap or vizor. July, Aug.

5. **S. PILÒSA.** Michx. Darl. (S. ovalifolia. *Bart.*)
St. erect, mostly simple, hirsute-pubescent; lvs. pubescent, rhomboid ovate or oval, crenate-serrate, petiolate, in remote pairs; rac. terminal, rather short; bracts elliptic-ovate.—Open woodlands, Penn. to Car. Stem 1—1½f high, purplish. Leaves few, 1—2½ by ½—1½', cuneately narrowed to the petiole, rather obtuse. Raceme generally simple and few-flowered, with opposite, elliptical bracts. Pedicels and calyx hairy. Corolla tube nearly white below blue at summit, 6—9" long. June—Aug.

6. S. RUGOSA. Wood. (*Nov. sp.*)

St. decumbent at base, diffusely branched, pubescent; *lvs.* oval and ovate, rugose, pubescent, petiolate, obtuse at each end, subcordate, crenate-serrate; *rac.* simple, elongated, terminal on the stem and branches; *bracts* broad-ovate, petiolate, subcordate, as long as the calyx.—At Harper's Ferry, on the rocky shores of the Shenandoah! A rough, diffuse plant, about 1f high. Stem with the angles obtuse and the sides grooved. Leaves rather numerous, 12—18″ by 9—13″, scarcely longer than the petioles, the bracts 2—3″ diam. Racemes 5—8′ long, rather dense-flowered. Corolla 8″ long. July—Sept.

7. S. INTEGRIFOLIA. (S. hyssopifolia. *Pers.* S. Caroliniana. *Ph.*)

St. erect, nearly simple, and, with the whole plant, densely pubescent; *lvs.* ovate-lanceolate, and linear-lanceolate, tapering to the base, subacute, entire, subsessile; *rac.* loose, leafy; *bracts* lanceolate; *fls.* large.—2f Mid. States to Ark., on dry hills. Stem 1—2f high, with large, blue flowers in terminal racemes. The leaves (1—3′ long) vary in breadth and margin, the lowest being sometimes ovate and crenate. Corolla bright blue at the summit, nearly white at base, 8—9″ long. June, July.—The plant is intensely bitter.

8. S. CANESCENS. Nutt. (S. serrata and S. incana. *Spr.* fide *Hook.*)

St. erect, tall, pubescent; *lvs.* petiolate, oblong-ovate or ovate, rounded or attenuate at base, minutely pubescent both sides, paler beneath, margin crenate, apex acute, the lower cordate; *rac.* terminal and axillary, pedunculate, paniculate; *bracts* lanceolate and lance-linear; *fls.* canescent.—Dry grounds, Middle and Western States, abundant. Stem usually purple, 1—3f high. Leaves 2—3′ long, ½ as wide, often with a purple margin and purplish spots. Flowers rather numerous, large and showy. Corolla 10″ long, tube white, lips blue.

9. S. CORDIFOLIA. Muhl. (S. versicolor? *Nutt.*)

Stout, branching, clothed with a soft, glandular pubescence in all its parts; *lrs.* broadly cordate, large, obtusely dentate, nearly smooth; *petioles* very long; *rac.* ternate, terminal; *bracts* ovate; *fls.* smaller. *Nuttall.*—Open woods and prairies, Western States. I have specimens essentially agreeing with the above, in which the leaves are 3—4′ long, 2—3′ wide. Flowers in a large, diffuse panicle, less showy than in the last species. Bracts broad-ovate and nearly sessile, viscidly pubescent. Corolla 8″ long, upper lip blue, lower white.

22. PRUNELLA.

Calyx about 10-ribbed, upper lip dilated, truncate, with 3 short teeth, lower lip with 2 lanceolate teeth; filaments forked, one point of the fork bearing the anther.

P. VULGARIS. *Self-heal. Blue-curls.*

St. ascending, simple; *lvs.* oblong-ovate, toothed, petiolate; *upper lip of cor.* truncate, with 3 awns.—2f A very common plant, in meadows and low grounds, N. Am., lat. 33° to the Arc. Sea. The stem is nearly a foot high, (2f in Ia.! and Ill.!) obtusely 4-angled, hairy, simple or slightly branched. Leaves few, opposite, slightly toothed, the stalks gradually becoming shorter from the lower to the upper pair which are sessile. Flowers blue, in a large ovate spike of dense verticils. Bracts imbricated, reniform, 2 beneath each vertici.. Flowering all summer.

TRIBE 7 NEPETEÆ.—Calyx oblique, upper teeth longer. Corolla bilabiate, upper lip vaulted, lower spreading, throat mostly inflated. Stamens ascending or diverging, the upper pair longer.

23. NEPÈTA.

Said to be from Nepet, a town in Tuscany.

Calyx arid, striate, upper lip of the corolla emarginate, lower 3-lobed, the middle lobe largest and crenate, margin of the orifice reflected; stamens approximate.

1. N. CATARIA. *Catnep.* *Catmint.* (Fig. 51.)

Erect, tall, hoary-tomentose; *lvs.* petiolate, cordate, coarsely crenate-serrate; *fls.* spiked, the whorls slightly pedunculated.—♃ This common plant is naturalized everywhere about old buildings and fences. Cats are very fond of it and will often devour it with the greatest avidity. Stem square, pubescent, branching, 2—3f high. Leaves very evenly bordered by tooth-like or crenate serratures, and as well as the whole plant, covered with a soft, hoary down, paler beneath. Flowers many, white or purplish, the lower lip dotted with crimson. July.

2. N. GLECHOMA. Benth. (Glechoma hederacea. *Linn.*) *Gill-over-the-ground.* *Ground Ivy.* (Fig. 51.)—*Lvs.* reniform, crenate; *cor.* about 3 times as long as the calyx.—♃ A creeping plant, naturalized about hedges, walls, &c. Stems prostrate, radicating at base, square, varying in length from a few inches to 1—2f. Leaves petiolate, opposite, roundish, cordate-reniform, hairy and glaucous. Flowers axillary, about 3 together. Corolla bluish-purple, with a variegated throat. The 2 anthers of each pair of stamens meet with their 2 divaricate cells, forming the appearance of a cross. The plant is aromatic, and was formerly used in ale, also in medicine. May.

24. LOPHANTHUS. Benth.

Gr. λοφος, a crest, ανθος; flowers in dense, terminal spikes.

Calyx 15-ribbed, oblique, 5-cleft, upper segments longer; corolla bilabiate, upper lip bifidly emarginate, lower lip 3-lobed, the middle lobe broader and crenate; stamens diverging.

1. L. NEPETOIDES. Benth. (Hyssopus. *Linn.*)

St. smooth, quadrangular, with the angles acute and slightly winged; *lvs.* ovate and ovate-lanceolate, acutely serrate; *petioles* smoothish.—♃ Middle! and Western States! A tall, branching, pale green herb, common about fences and dry hedges. Stem 3—6f high, the sides somewhat concave, and the angles prominent. Leaves acuminate, about 4′ by 2′. Flowers in crowded, axillary verticils, forming a terminal, green spike, which is nearly continuous above. Corolla greenish-yellow. Stamens exserted. July, Aug.

2. L. SCROPHULARIFOLIUS. Benth. (Hyssopus. *Linn.*)

St. pubescent, quadrangular, with the angles obtuse; *lvs.* cordate-ovate, crenate-serrate; *petioles* ciliate-pubescent.—♃ Tall, stout and branching, with the general aspect of the former species, and found in similar situations. The herbage is often changed to dark purple. Stem 2—4f high, purple. Leaves about 5′ by 3′, coarsely serrate, acuminate. Flowers in crowded, axillary verticils, forming a long, dense, terminal spike. Corolla pale purple, more conspicuous than in the first. Stamens and style exserted. July, Aug.

25. DRACOCEPHALUM.

Gr. δρακω, dragon, κεφαλος, head; from the resemblance of the flowers.

Calyx subequal, oblique, 5-cleft, upper segments larger; cor. bilabiate, upper lip vaulted, emarginate, throat inflated, lower lip spreading, 3-cleft, middle lobe much larger, rounded or subdivided; sta. distinct, ascending, the upper pair longer than the lower.—*Flowers axillary and terminal, usually with large, conspicuous bracts*

1. D. CORDATUM. Nutt. *Cordate-leaved Dragonhead.*

Stoloniferous; *st.* and elongated petioles pubescent; *lvs.* cordate, obtusely crenate, sparingly hirsute above; *spike* unilateral; *bracts* broad-ovate, entire, nearly as long as the calyx; *ped.* bi′racteolate, mostly 1-flowered; *cal.* segments acute, almost pungent.—♃ Islands of the Ohio, 40 miles below Pittsburg, *Nuttall.* Stem about 1f high, quadrangular. Leaves 3 or 4 pairs, obtusely cordate, almost as broad as long, petiole about as long as the lamina (1′), upper pairs subsessile. Flowers secund. Corolla pale blue, about 1′ long, orifice much dilated. June.

2. D. PARVIFLÓRUM. Nutt. *Small-flowered Dragonhead.*

Subpubescent; *lvs.* ovate-lanceolate, deeply serrate, petiolate; *bracts* leafy ovate, ciliate, mucronate-serrate; *cal.* upper segment much the largest; *fls* small, verticillate, subcapitate, corolla scarcely exceeding the calyx.—② Woods, Watertown, N., Y. *Vasey.* Mo., *Nuttall.* Saskatchawan River, *Richardson.* Very rare. Flowers whitish, very small, the verticils involucrate and almost spicate. Calyx dry and membranaceous. Upper lip of the corolla arched, emarginate, central lobe of lower lip crenate. July.

TRIBE 8. **STACHYDEÆ.**—Calyx oblique or rarely subbilabiate, 3—10-toothed. Corolla bilabiate, upper lip galeate or flat, lower lip unequally 3-lobed. Stamens ascending, upper pair shorter.

26. PHYSOSTEGIA. Benth.

Gr. φυσα, a bladder, and στηγη, a covering ; from the inflated corolla.

Calyx campanulate, subequally 5-toothed ; corolla tube much exserted, throat inflated, upper lip concave, middle division of lower lip largest, roundish, emarginate ; sta. 4, unconnected, ascending beneath the upper lip, the two lower rather longer.—*Flowers opposite, in a terminal, bracteate, 4-rowed spike.*

P. VIRGINIÀNA. Benth. (Dracocephalum Virg., denticulatum, variegatum and obovatum of *auth.* fide *Benth.*) *Lion's Heart.*—*Calyx teeth* acute, subequal.—♃ A beautiful plant, native in Penn., S. and W. States ! often adorning our gardens, where it spreads rapidly. It is 2—3f high, very smooth, dark green. Stem square, thick, rigid. Leaves opposite, closely sessile, 4—5' by 1', with remote and shallow teeth, of a shining dark green. Flowers in 4-rowed spikes, numerous, dense, or often subremote. Bracts subulate. Corolla pale purple, about an inch long, spotted inside. Aug. Sept.

27. SYNANDRA. Nutt.

Gr. συν, together, αιδρας ; in allusion to the coherence & æ anthers.

Calyx 4-cleft, segments unequal, subulate, converging to one side ; upper lip of cor. entire, vaulted, the lower obtusely and unequally 3-lobed ; throat inflated ; upper pair of anthers cohering, having the contiguous cells empty.

S. GRANDIFLÒRA. Nutt. *Large-flowering Synandra.*
St. subsimple, nearly smooth, subterete ; *lvs.* cordate-ovate, acuminate, obtusely dentate, often dilated at base, upper ones smaller, cauline sessile, lower subpetiolate ; *fls.* solitary and sessile ; *cal. seg.* ovate, setaceously acuminate, two upper larger than the two lower ; *cor.* tube somewhat funnel-form, mouth much inflated, upper lip entire, vaulted, lower 3-lobed, lobes entire.—Banks of the Ohio, Cincinnati, *Nutt.* Woods, near Vermillion River, Ill.! Stem about 1f high. Corolla about 1' long, yellowish-white, lower lip elegantly striated with purple lines. June.

β. ? *petiolata.* Wood. *St.* quadrangular, hirsute ; *lvs.* hirsute, deltoid, truncate-cordate, all long-petiolate, lower petioles 4—6' long, upper 1— 4'. Calyx hirsute, almost hispid.—Cincinnati, *Clark!* This is, perhaps, a distinct species.

28. LAMIUM.

Lat. *lamia,* the name of a sea monster, to which the grotesque flowers may be likened.

Upper lip of the corolla vaulted, galeate, nearly entire, lower lip broad, emarginate, lateral lobes truncate, often toothed on each side near the margin of the dilated throat.

L. AMPLEXICAULE. *Henbit.*
Lvs. roundish, incisely crenate, floral ones broadly cordate, obtuse, sessile, amplexicaul, lower ones petiolate.—① A small, slender herb, found in cultivated grounds. Stems ascending, several from the same root, 6—10' high, with

opposite, short, broad, hairy leaves deeply crenate or cut; lower ones on stalks an inch or more in length. Flowers in dense verticils, closely sessile in the axils of the upper leaves. Calyx hairy. Corolla purple, downy, the tube much exserted, the lower lip spotted with white. May—Nov. ♃

29. LEONÛRUS.

Gr. λεων, a lion, ουρα, tail; from the appearance of the spikes of flowers.

Calyx teeth subspinescent; upper lip of the corolla entire, hairy, concave, erect, lower lip 3-lobed, the middle lobe obcordate; anthers sprinkled with shining dots.

1. L. CARDĪACA. *Motherwort.*

Lower stem lvs. palmate-lobed, uppermost lanceolate, often trifid, all of them toothed, cuneiform at base; *cor.* longer than the calyx, the tube with a hairy ring within.—♃ Tartary, whence it was first introduced into Europe and thence to America, ever following the footsteps of civilized man. It is common about rubbish, stone walls and waste places. Stem 3—5f high, downy, square, large, purplish, bearing its opposite, stalked, rough caves arranged in 4 vertical rows. Flowers in many whorls. Calyx rigid and bristly. Corolla purplish, hairy without, variegated within. July.—It has a strong and pungent smell, and has considerable reputation as an ingredient in herb drinks for colds, coughs, &c.

2. L. MARRUBIASTRUM. *Hoarhound Leonurus.*

Cauline lvs. oblong-ovate, incisely and coarsely serrate, the floral lanceolate, tapering at each end, incisely dentate; *cor.* shorter than the calyx teeth, tube naked within, upper lip somewhat vaulted, pubescent.—Naturalized in Penn. *Darlington!* and Ia.! A plant of vigorous growth, 2—4f high, with opposite, ascending branches. Leaves 2—3′ in length, the lowest on long petioles. Verticils many-flowered, remote but numerous, forming an interrupted, leafy spike. Corolla reddish-white. July, Aug.

30. GALEOPSIS.

Gr. γαλη, a weasel. οψις, appearance; its grotesque flowers are likened to that animal.

Calyx 5-cleft, spinescent; upper lip of the corolla vaulted, subcrenate, lower lip with 3 unequal lobes, having 2 teeth on its upper side, middle lobe largest, cleft and crenate; sta. ascending beneath the upper lip.

1. G. TETRÂHIT. *Hemp Nettle.* (Fig. 51.)

St. hispid, the internodes thickened upwards; *lvs.* ovate, hispid, serrate; *cor.* twice as long as the calyx, the upper lip nearly straight, concave.—① A common weed, naturalized in waste and cultivated grounds, 1—2f high. Stem obtusely 4-angled, remarkably swelled below the joint, and covered with prickly, deflexed bristles. Leaves hairy on both sides, ovate, acute, serrate. Flowers in dense verticils. Calyx with 5 acute, bristly teeth. Corolla variegated with white and purple, upper lip concave, purple, longer than the 3-lobed lower one. June, July.—This plant is so prevalent in many parts of N. Eng. as to nearly ruin some farms.

2. G. LADANUM. *Red Hemp Nettle.*

St. hairy, internodes equal; *lvs.* lanceolate, subserrate, hairy; *upper lip of the cor.* slightly crenate.—① A smaller species, naturalized, growing among rubbish in gravely soils, &c. Stem about a foot high, not swollen below the joints, with opposite branches. Flowers in dense, remote whorls. Corollas usually rose-colored, often white or variegated, spotted with crimson. Aug. Sept.

31. STACHYS.

Gr. σταχυς, a spike; this being the inflorescence of all the species.

Calyx tube angular, 5 or 10-ribbed, 5-toothed, upper teeth often larger; cor. bilabiate, upper lip erect, spreading or somewhat vaulted, lower lip spreading, 3-lobed, middle lobe largest; sta. ascending lower ones longer; anth. approximated in pairs.

1. S. ASPERA. Michx. (S. hispida *Ph?*) *Hedge Nettle. Wound-wort.*

St. erect, the angles retrorsely hispid or rough; *lvs.* subpetiolate, oblong lanceolate, acutely serrate, smooth or nearly so; *verticils of the spike* about 6 flowered; *cal.* smoothish, with spreading teeth, at length spinescent.—♃ Fields and roadsides. About a foot high and rather slender. Stem erect, square, generally hispid backward on the angles, sometimes nearly smooth. Leaves smooth, membranaceous, generally rounded at base and acute at apex. Spike terminal, leafy, composed of verticils, each with 4—8 pale purple flowers. Variable in pubescence. July.

2. S. SYLVATICA. (S. aspera. *Muhl. & Bw.* S. hispida. *Nutt.*) *Wood Stachys.*—St. very hispid on the angles; *lvs.* on short petioles, ovate-lanceolate, sometimes cordate at base, acuminate, crenate-serrate, hirsute; *verticils* 3—6 flowered; *floral lvs.* very small, lanceolate-linear, hispid-ciliate; *cal.* hirsute, with ciliate, spinescent teeth.—♃ A very rough and hairy herb, in low woods and on shady banks. Stem erect, 12—18' high. Leaves 3—4' long and ⅓ as wide, with rounded or uncinate serratures, the upper surface with short, bristly hairs. Petioles 0—½' long, and with the veins beneath, hispid. Spike long, slender, terminal, very rough and hairy. Corolla exserted, purple, spotted. July., Aug.

3. S. HYSSOPIFOLIA. Michx. (S. palustris. *Walt.*)

St. scarcely pubescent, slender, erect; *lvs.* sessile, linear-lanceolate, slightly dentate; *verticils* about 4-flowered; *cal.* subspinescent.—♃ A slender species 6—12' high, in meadows, N. Y. and Ms., rare, West to La. where it is frequent! Leaves very narrow, often linear, 2—3' by 4—10", with minute teeth or finely serrulate. Flowers sessile. Corolla a little hairy, purple. July. A smoother and elegant plant compared with the last.

4. S. INTERMEDIA. Ait.

St. somewhat villose; *lrs.* oblong, subcordate, crenate; *verticils* many-flowered; *cal.* somewhat spinescent.—Ill. Jn., Jl.

32. MARRUBIUM.

Calyx tubular, 5—10-ribbed, with 5 or 10 subequal teeth; cor. bilabiate, upper lip erect, flattish or concave, entire or bifid, lower lip spreading, 3-lobed, middle lobe broadest, emarginate, tube included; sta. included beneath the upper lip.

M. VULGÀRE. *Hoarhound.*

St. ascending, hoary-pubescent; *lvs.* roundish-ovate, crenate-dentate, downy-canescent beneath; *cal.* of 10 setaceous, uncinate teeth.—♃ Introduced into fields and roadsides. Stem 1—2f high, branching at base, or several from the same root, covered with a white, downy pubescence. Leaves petiolate, 1—2' in diameter, whitish and rough-veined above, very woolly beneath, rounded and toothed. Flowers white, in sessile, axillary, dense, hairy verticils. Calyx woolly, the teeth spreading and alternately shorter. The hoarhound is an aromatic and bitter herb, well known as an ingredient in cough candy. It is tonic and diuretic, and much used in pulmonary affections. ‡

33. BALLOTA.

Gr. βαλλω, to reject; on account of its offensive odor.

Calyx hypocrateriform, 10-ribbed, 5-toothed; cor. bilabiate, tube cylindrical, as long as the calyx, upper lip concave, crenate, lower lip 3-cleft, middle segment largest, emarginate; ach. ovoid-triangular.

B. NIGRA. *Black* or *Fœtid Hoarhound.*

Lrs. ovate, subcordate, undivided, serrate; *cal.* somewhat truncate, throat dilated, teeth spreading, acuminate.—♃ Said to have been introduced, but is frequently met with about hedges, &c., in Ms. and Conn. Stem 2—3f high, pubescent as well as the opposite, broad leaves. Flowers purple or white, in axillary verticils. This plant has the general appearance of hoarhound (Marrubium) but not its fragrance. July.

34. MOLUCCELLA.
Brought from the Molucca Islands, &c.

Calyx campanulate, very large, the margin expanding, often repand-spinose; cor. much smaller, included within it.

M. LÆVIS. *Molucca Balm. Shell Flower.*—*St.* ascending, subsimple, glabrous; *lvs.* petiolate, roundish-ovate, dentate; *fls.* in a terminal, leafy raceme; cal. campanulate, equally 5-toothed, nearly twice longer than the corolla, teeth awnless.—① Syria. A curious plant in gardens, smooth in all its parts and of a glaucous green, 1—2f high. It is chiefly remarkable for its ample, bell-shaped calyx, in the bottom of which is seated the yellowish-green flower.

TRIBE 9. AJUGOIDEÆ.—Corolla upper lip very short, or split to the base, or rarely erect and vaulted, lower lip longer. Stamens ascending, much exserted. Achenia reticulately rugose.

35. TRICHOSTEMA.
Gr. θριξ, τριχος, hair, στημα, a stamen; for its long, hair-like stamens.

Calyx resupinate, oblique, unequally 5-cleft; upper lip (lower by the twisting of the peduncle) of 2 short, acute teeth, lower (at length the upper) twice as long, 3-toothed; cor. tube slender, very short, unequally 5-cleft, lobes oblong, declined; sta. much exserted, lower ones longer.

T. DICHOTOMA. *Blue-curls.*
Lvs. oblong-lanceolate, attenuate at base, obtuse, entire, pubescent; *fls.* resupinate; *sta.* very long, exserted.—① Found on dry or rocky hills and in sandy soils, Mass.! to Md.! Ga., La., Ill. An interesting plant, a foot high. Stem obtusely 4-angled, hairy, bushy. Branches opposite, divided, the upper pair generally forming a dichotomous division of the stem. Leaves petiolate, of a rhombic, ovate-lanceolate form. Flowers axillary and terminal, becoming inverted by the twisting of the petiole. Corolla purple. Stamens slender, of a delicate purplish hue, gracefully bending from the lower lip of the corolla to the upper, forming a beautiful arch. Aug.

36. TEUCRIUM.
Teucer, the founder of Troy, is said by Pliny to have first employed it medicinally.

Calyx subcampanulate, and subregular, in 5 acute segments; corolla with the 4 upper lobes nearly equal, the lowest largest, roundish; stamens exserted from the cleft in the upper side of the tube.

T. CANADENSE. *Wild Germander.*
Plant erect, hoary-pubescent; *lvs.* lanceolate, acute, serrate, petiolate; *bracts* linear-lanceolate, longer than the calyx; *spike* long, of many crowded verticils of flowers; *upper teeth of cal.* broader.—♃ Can. and U. S., fields and roadsides. Stem about 2f high, simple or branched, square, with concave sides. Leaves 3 times as long as wide, somewhat rounded at base, green above, hoary with down beneath. Bracts longer than the calyx. Flowers disposed in axillary verticils, each of 4—6. Calyx with 5 broad, nearly equal segments, the 2 lower ones narrower. Corolla purplish, apparently without the upper lip, instead of which is a fissure through which the stamens are exserted. July.

β. Virginicum. Upper lvs. ovate-oblong, nearly sessile; *bracts* about as long as the calyx.—Habits similar to the last.

ORDER XCIV. BORRAGINACEÆ.—BORRAGEWORTS.

Herbs, shrubs or trees. with round stems and branches.
Lvs. alternate, often rough with stiff hairs which are scale-like and indurated at base. (flowers expand.
Stip. none. Spikes, racemes or corymbs generally secund, and circinate before flowering, evolving as the
Fls. of the genus series, white, blue, red, &c., rarely yellow.
Cal.—Sepals 5, regular, more or less united at base, persistent.
Cor.—Petals 5, regular (very rarely irregular), united at base, hypogynous, imbricate in æstivation.

Sta. 5, inserted into the corolla and alternate with its lobes.
Ova. deeply 4-lobed. the style arising from the base of the lobes.
Fr.—Nuts or achenia 4, distinct, 1-seeded. *Seeds* without albumen.
Embryo with a superior radicle. *Cotyledons* plano-convex.

Genera 88, species 600? very abundant in the south of Europe and middle of Asia, becoming rare as we approach the arctic circle. All our native species are herbaceous.

Properties.—Mucilaginous and emollient plants, never poisonous.

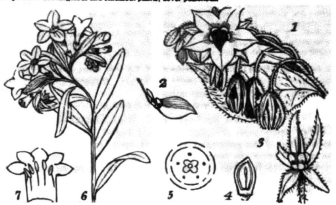

FIG. 92.—1. *Borrago officinalis.* 2. A petal with its appendage at base, and anthers produced at apex. 3. Calyx with the 4 achenia and style. 4. Vertical section of one of the achenia, showing the seed, embryo and albumen. 5. Plan of the flower. 6. Lithospermum (Batschia) canescens. 7. Corolla laid open, showing the stamens inserted on the tube.

Conspectus of the Genera.

irregular.				Echium.	1
			Corolla rotate, blue. . .	Borrago.	2
			Corolla campanulate. . .	Symphytum.	3
		excavated	Cor. funnel or (tube straight.	Anchusa.	4
		at base.	salver-form. (tube twice bent.	Lycopsis.	5
			(Lvs. rough.	Onosmodium.	6
			dilated. (Lvs. smooth.	Mertensia.	9
			Cor. throat (nar- (lobes erect. .	Pulmonaria.	10
		not ex-	open and (row. (lobes spreading.	Lithospermum.	7
	Ach. free,	cavate	Cor. closed (yellow, large. .	Pentalophus.	8
	unarmed.	at base.	at throat. (white, small. .	Myosotis.	11
Ovary			(Corolla salver-form. .	Echinospermum.	12
deeply			Corolla salver-form.	Cynoglossum.	13
4-parted.	Achenia echinate, cohering. . orolla funnel-form.				
Corolla regular.	Ovary entire (partible in fruit) bearing the style at the top. .			Heliotropium.	14

TRIBE 1. **BORRAGEÆ.**—Ovary consisting of two bipartible (rarely 2-celled) carpels. Style arising from the base between the segments of the ovary. Fruit deeply 4-(rarely 2-)parted. Seeds without albumen.

1. ECHIUM. Buek.

Gr. εχις, a viper; from the spotted stem of some species.

Calyx 5-parted, segments subulate, erect; corolla campanulate, obliquely and unequally lobed. with a short tube and naked orifice; stigma cleft; achenia tuberculate, imperforate —*Herbs or shrubs. Fls irregular, in spicate, panicled racemes. Cor. cyanic.*

E. VULGÀRE.   *Viper's Bugloss.*
St. herbaceous, rough with bristles and tubercles; *cauline lvs.* lanceolate, and rough with bristles; *spikes* lateral, hairy, deflected.—① A rough plant, with large, handsome, violet-colored flowers, found in fields and waste grounds, N. States. Stem 18—20' high, round, with entire, dull green leaves, which are 2—6' long, and ¼ as wide, lower ones petiolate, upper ones amplexicaul. Flowers in numerous, crowded, axillary, recurved spikes, appearing in June and July. ⚥

2. BORRÅGO. Tourn.

Calyx 5-parted; corolla rotate, with acute segments; orifice crowned; filaments converging; achenia rounded, imperforate at base, inserted lengthwise into an excavated receptacle.—*European herbs.*

1. B. officinÅlis.   *Common Borrage.*—*Lvs.* ovate, alternate, the lower ones petiolate; *cal.* spreading; *ped.* terminal, many-flowered.—① Native of England, and with us a common inhabitant of the garden. The whole plant is rough with short, bristly hairs, erect, 2f high, with terminal clusters of handsome, sky-blue flowers during summer. It was formerly in high repute as a cordial. The young leaves form a good salad and pot-herb. ‡

2. B. orientÅlis.   (Psilostemon. *DC.*)   *Oriental Borrage.*—*Lvs.* cordate, petiolate; *ped.* many-flowered; *sta.* exserted, villous.—① An ornamental garden plant, native of Turkey. Stem and leaves hairy. Flowers blue, appearing in the spring months. ‡

3. SYMPHŸTUM.

Gr. συμφυσις, a joining or healing; from its reputation for healing wounds.

Calyx 5-parted; corolla tubular-campanulate, orifice closed with 5, subulate scales, converging into a cone; achenia gibbous, imperforate.—♃ *Oriental herbs.   Flowers cyanic.*

S. officinÅle.   *Comfrey.*
Pilose; *st.* branching above; *lvs.* extensively decurrent, the lower and radical petiolate, ovate-lanceolate, upper and floral lanceolate; *sep.* lanceolate, acuminate; *cor. limb* with 5 recurved teeth.—A large, coarse-looking but showy exotic, in our gardens and shrubberies, also naturalized in low grounds, Middle States. Whole plant rough with dense hairs. Stem 3—4f high, winged by the decurrent leaves, bearing terminal, revolute racemes. Corollas white, pink and red, appearing all summer. Root perennial. It abounds with mucilage and has long been regarded as an efficient vulnerary. ‡ §

4. ANCHŪSA.

Gr. αγχουσα, paint; the root of one species was once used for staining the features.

Calyx 5-parted; corolla infundibuliform, vaulted; tube straight, orifice closed with 5 prominent scales; achenia perforate at the base and their surfaces generally rugose; stamens included; stigma emarginate.—*Handsome herbs, mostly European.   Fls. cyanic.*

A. officinÅlis.   *Bugloss, or Ox-tongue.*—*Lvs.* lanceolate, strigose; *spikes* one-sided, imbricated; *cal.* as long as the tube of the corolla.—♃ A rough garden plant, native of Britain. The English name, Bugloss, comes from the Greek, signifying *ox-tongue*, on account of the long, rough leaves. Stem 2f high, rough with bristly hairs. Bracts ovate. Flowers purple, with a melliferous corolla very attractive to bees. The leaves are juicy, and the root mucilaginous, used in medicine to promote the eruption of the small-pox. Blossoms all summer. †

5. LYCOPSIS.

Gr. λυκος, a wolf, and οψ, the eye; name suggested by the small blue flowers.

Calyx 5-cleft; corolla funnel-form, tube incurved, orifice closed with ovate, converging scales; achenia perforated at base, ovoid, angular. —① *Distinguished from Anchusa only by the curved corolla tube.*

L. arvensis.   *Wild Bugloss.*
Plant hispid; *lvs.* lanceolate, repand-denticulate; *rac.* leafy; *fls.* sessile; *cal.* shorter than the tube of the corolla.—A very hispid, almost bristly plant, found in fields and roadsides, Northern States, probably introduced. Stem erect, branching, roundish, about a foot high. Leaves 5 or 6 times as long as wide, the margin irregularly and slightly toothed. Flowers small. Calyx erect. Corolla sky-blue with white scales within. June, July. §

6. ONOSMODIUM. Michx.

From Onosma, another genus of this order, and εἶδος, appearance or resemblance.

Calyx deeply 5-parted, with linear segments; corolla subcampanu late, having a ventricose, half 5-cleft limb, with the segments con verging and the orifice open; anthers sessile, sagittate, included; style much exserted; achenia imperforate, shining.—♃ *North American. Rac. terminal, subspicate, one-sided. Fls. white.*

1. O. VIRGINIÀNUM. Alph. DC. (O. hispidum. *Michx.* Lithospermum Virg. *Linn.*) *False Gromwell.*—*St.* with appressed hairs; *lvs.* oblong-lanceolate, minutely strigose; *cal. lobes* lanceolate, pilose both sides, half as long as the corolla; *cor. lobes* lance-subulate, clothed externally with long, hispid hairs. —N. Y. to Flor., in dry, hilly grounds. A very rough, erect plant about 1½f high. Leaves 1½—2½′ by ½—⅔′, 3—5-veined, often oval and even ovate-lanceolate. Flowers greenish-white, in leafy racemes which are recurved at first but finally erect. Styles (6—7″) twice longer than the corolla.

2. O. CAROLINIÀNUM. DC. (O. molle. *Michx.* Lithospermum Carol. *Lam.* L. molle. *Muhl.*)—Villose-canescent; *lvs.* oblong-oval, rather obtuse, each side whitish with scattered hairs; *bracts* ovate-lanceolate; *cal. segments* lanceolate, half as long as the glabrous corolla; *cor. segments* ovate, acute.—Rocky hills, N. Y. to Car. and Tenn. Plant a foot or more high, clothed with a soft, white pubescence. Anthers silky-pubescent, as long as the glabrous filaments, its lobes scarcely diverging. Jl. Aug.

3. O. STRIGÒSUM. G. Don. *St.* erect, simple, pilose-hispid, very leafy; *lrs.* lance-linear, sessile, very long, 3-veined, with appressed hairs; *bracts* lance-linear, silky; *cal. lobes* linear, acute, silky with appressed hairs both sides, very long; *cor.* cylindrical, a third longer than the calyx, silky-puberulent outside; *sta.* included; *sty.* exserted.— Ill., (*Mead,*) in wet prairies and woods. Leaves 3′ long, ½′ wide, nearly glabrous beneath the veins. Corolla yellowish-white. Fruit smooth and shining.

7. LITHOSPERMUM.

Gr. λίθος, a stone, and σπερμ', seed; the seeds being hard and shining like little pebbles.

Calyx 5-parted, persistent; corolla funnel-form or salver-form; limb 5-lobed, orifice open; stamens included; stigma obtuse, bifid; achenia bony, rugose or smooth, imperforate at base.—*Herbaceous or suffruticose, generally with a thick, reddish root. Fls. spiked or racemed, bracted, white or yellow.*

* *Flowers white.*

1. L. OFFICINÀLE. *Officinal Gromwell.* *St.* herbaceous, erect, very branching above; *lvs.* lanceolate, acute, veiny; *cal.* nearly equal to the tube of the corolla; *ach.* smooth.—♃ A rough, weed-like plant, introduced from Europe. Grows in dry, gravely soils. Stems much branched, clustered, arising 1—2f from a white, fusiform root. Leaves grayish-green, rough on the upper side, hairy beneath, rather acute, entire, 2—3′ by ½—⅔′. Flowers small, white, axillary, solitary, pedicellate, in recurved, leafy spikes. Achenia ovate, white or grayish, polished, stony, usually but 1 or 2 perfected. Jl. ⚥

2. L. LATIFOLIUM. Michx. *Broad-leaved Gromwell.* Herbaceous, erect, sub-simple, scabrous; *lrs.* ovate-lanceolate, acute at each end, veined, scabrous; *rac.* leafy, few-flowered; *sep.* lance-linear, longer than the corolla and spreading in fruit; *ach.* punctate, shining-white, ovoid-turgid.—Woods and thickets, N. Y., *Torrey,* to Ohio and Ill., *Mead!* and Va., *Pursh.* Differs from the first in the less branching stem, much broader leaves, longer calyx and larger fruit "punctate with minute impressions." Leaves 2—4′ by 1—2′, strongly veined. Nuts generally but 2, half as long as the calyx. Flowers small, white.

3. L. ANGUSTIFOLIUM. Michx. *St.* herbaceous, procumbent; *lvs.* linear, strigose with an appressed pubes-

cence; *fls.* scattered, lateral, axillary; *ach.* turgid, ovoid, shining, impressed-punctate.—Banks of streams, sandy prairies, Ohio river, *Michaux*, Ill. *Mead.?* An obscure species, wholly unknown to me.

4. L. ᴀʀᴠᴇɴsᴇ. *Corn Gromwell. Wheat-thief.*

Lvs. linear-lanceolate, obtuse, hairy; *cal.* nearly equal to the corolla, with spreading segments; *ach.* rugose.—① A rough, pilose weed, introduced into our fields and waste grounds, much to the annoyance of the farmers. The stem is branching, erect, 12—15' high, from a fusiform root with reddish bark. Leaves bright green, rough, sessile, 1—2' in length, with only the central vein; the lower ones obtuse and narrowed to the base; upper ones subacute. Flowers small, white, subsessile, solitary, in the axils of the upper leaves. May, Jn.

* * *Flowers yellow.*

5. L. ᴄᴀɴᴇsᴄᴇɴs. Lehmann. (Batschia canescens. *Michx.*)   *Puccoon.*

St. erect, subsimple, softly villose; *lvs.* oblong, obtuse, silky-canescent above, villous beneath; *fls.* axillary; *tube of the corolla* thrice as long as the very short calyx.—♃ A handsome plant, with bright yellow flowers, found in prairies, fields and dry hills, Can., N. Y.! to Ill.! and Southern States. Stem 8—12' high, erect, simple, rarely a little branched above, hoary-villose. Leaves sessile, 2—3" wide and 4 times as long, 1-veined. Flowers crowded near the summit of the stem. Calyx segments lanceolate, acute. Corolla bright orange-yellow, including the subsessile stamens and short style. Jn., Jl. The root is used to dye red by the Indians.

6. L. ʜɪʀᴛᴜᴍ. Lehm. (Anchusa. *Muhl.* Batschia Carolinensis. *Gmel.*)

Herbaceous, erect, simple, hairy above; *lvs.* sessile, linear-lanceolate, obtuse, ciliate-hirsute both sides, floral ovate-lanceolate; *cal. lobes* linear, hirsute, a little shorter than the tube of the corolla; *cor.* segments spreading, obovate, entire, tube hispid inside at base; *ach.* ovoid, shining.—♃ Can., Penn., to La. Stems 8—12' high, clustered. Flowers crowded, racemed. Corolla orange-yellow, large.

7. L. ᴀᴘᴜʟᴜᴍ. Vahl. (Myosotis. *Linn.* M. lutea. *Lam.*)

St. herbaceous, erect, rough with hairs; *lvs.* linear-lanceolate, acute, erect, rough and hairy; *spikes* hispid; *bracts* foliaceous; *cor.* longer than the calyx; *ach.* muricate.—① Dry woods, Ohio. *Pursh.* Stem 2—6' high, generally simple at base, branched above. Corollas small, yellow, in the axils of lanceolate bracts.

8. PENTALÓPHUS. Alph. DC.

Gr. πεντε, five, λοφος, crest; from the character.

Calyx 5-parted, with linear segments; cor. hypocrateriform, tube cylindric, much longer than the calyx, throat closed with 5 glabrous valves alternating with the stamens, segments ovate, spreading, much shorter than the tube; achenia solitary by abortion, ovoid, white, smooth.—*American herbs.*

P. ʟᴏɴɢɪғʟᴏʀᴜs. Alph. DC. (Lithospermum. *Spreng.* Batschia. *Nutt.*)

Erect, strigose with a cinerous pubescence; *lower lvs.* lance-linear, attenuated to the base, obtuse, *upper ones* linear, acutish; *rac.* leafy, terminal; *cal. seg.* linear, much longer than the pedicel; *cor. tube* 4 times longer than the calyx, a little dilated upwards.—♃ Prairie du chien to Mo.! Stem 10—15' high, slender, branched near the top. Leaves 1—1½' by 2—3", the floral ones about as long as the flowers. Corolla yellow, the tube 8—10" long, lobes crenulate. Style scarcely exserted. Fruit much shorter than the calyx, smooth, white. Jl.

9. MERTENSIA. Roth.

Calyx short, 5-cleft; cor. tube cylindric, twice longer than the calyx, limb subcampanulate, 5-cleft, throat naked, or oftener with 5 folds or ridges between the insertion of the stamens; sta. inserted at top of the tube; anth. subsagittate; ach. smooth or reticulated.—

♃ *St. and leaves usually glabrous and pellucid-punctate, the radical ones many-veined, cauline sessile.   Rac. terminal.*

1. M. Virginica. DC. (Pulmonaria. *Linn.* Lithospermum pulchrum. *Lehm.*) *Virginian Lungwort.—Plant* erect, smooth; *cal.* much shorter than the tube of the corolla, limb longer than the tube; *radical lvs.* (large) obovate-elliptical, obtuse; *cauline ones* long-lanceolate.—A smooth, erect, and elegant plant, about 20′ high, native in N. Y. to Ga. and W. States, sometimes cultivated. The leaves of the stem are sessile, much narrower than those of the root, whose width is ⅓ of their length. Flowers in terminal clusters. Corolla blue, funnel-form, sitting upon a short, 5-toothed calyx. Stamens and style included. May.

2. M. maritima. G. Don. (Pulmonaria. *Linn.* Lithospermum. *Lehm.*) Glabrous; *sts.* and *branches* procumbent or ascending; *lvs.* ovate, obtuse, fleshy, glaucous, the radical petiolate, cauline sessile; *rac.* leafy; *cal.* deeply cleft, scarcely half as long as the glabrous corolla.—Sea shore, Northern States, *Pursh,* N. to Greenland. Stem diffusely branched. Flowers purplish-blue, limb longer than the tube. Jl.

3. M. denticulata. G. Don. (Pulmonaria. *Roem.* Lithospermum. *Lehm.*)- *St.* erect; *lvs.* glaucous, rather fleshy, acute-mucronate, ciliate-denticulate, radical ovate, petiolate, cauline oblong, sessile, 3-veined at base; *cal. segments* acute; *pedicels* as long as the flower; *sty.* finally exserted.—N. Y. *Muhl. Torrey.* Stems 6—12′ high, clustered. Peduncles many-flowered. Corolla pale purple.

10. PULMONARIA.
Probably named from its having been used in lung complaints.

Calyx prismatic, 5-angled, 5-toothed; cor. infundibuliform, with a cylindric tube, orifice hairy in 5 lines alternating with the stamens; ach. imperforate.—♃ *European herbs.*

P. officinalis. *Common Lungwort.—Plant* rough; *cal.* the length of the tube of the corolla; *radical lvs.* ovate, cordate, scabrous, *cauline ones* ovate, sessile.—Native of England, but naturalized and cultivated in our gardens. Flowers blue, in terminal clusters. Stem a foot high. This as well as other foreign species of this genus, is a rough-leaved plant, while the several American species are uniformly smooth. May. ‡

11. MYOSOTIS. Dill.
Gr. μυος, a rat, and (ους) ωτος, an ear; from the form of the leaves.

Calyx 5-cleft; corolla hypocrateriform, the 5 lobes slightly emarginate, orifice closed with short, concave scales; achenia ovate, smooth, with a small cavity at base.—*Herbs, slightly villous.   Rac. at length elongated.*

1. M. cæspitosa. Schultz. *a. laxa.* DC. (M. palustris. *Roth.* M. scorpoides. *Willd.*) *Marsh Scorpion Grass.*—Nearly smooth, somewhat branched, erect; *lvs.* linear-oblong, obtuse, with short, scattered hairs; *rac.* without bracts; *pedicels* divaricate in fruit, twice as long as the short, spreading, smooth segments of the calyx. ♃ Grows about ditches and marshes, Can. and U. S., often called *water-mouse-ear* from the leaves, which are roughish with appressed hairs Stem about a foot high, with scattered hairs, ascending from long, creeping roots. Leaves scattered, sessile, 1—3′ long, ⅛ as wide. Racemes terminal, or often one of them supra-axillary, one-sided. Flowers small, blue, on pedicels ½′ long. May—Aug.

2. M. stricta. Link. (M. arvensis. *Rich.* and *1st edit.* M. inflexa. *Engelm.*) *Forget-me-not.—St.* branching; *lvs.* oval-lanceolate, hairy; *rac.* long; *pedicels* in fruit suberect, about as long as the calyx; *cal. segments* oval, acuminate, hairy, closed, about the length of the corolla.—① Found in sandy woods, N. Eng. *! to* Ill. *!* Whole plant of a grayish hue from its dense pubescence. Stem 4—10′ high, at length much branched. Leaves ½—1′ in length, sessile, acutish,

the lower ones oblanceolate, obtuse, tapering to a short petiole. Racemes revolute at the end, not secund, short at first, but arising at length 6′, 8′, or even 12′. Flowers very small, white. June.

12. ECHINOSPERMUM. Swartz.

Gr. εχινος, the sea-urchin, σπερμα, seed ; from the character.

Calyx 5-parted ; corolla hypocrateriform, orifice closed with concave scales ; seeds echinate, compressed or angular, fixed to a central column.—*Herbs erect. Lvs. oblong or linear. Rac. bracted. Pedicels short, erect.*

E. LAPPŮLA. Lehm. (Rochelia. *Roem.* Cynoglossum. *Scop.*) *Burr-seed.*
St. branched above ; *lvs.* lanceolate or linear-lanceolate, hairy ; *cor.* longer than the calyx, the border erect-spreading ; *ach.* each with 2 rows of hooked prickles on the margin.—① An erect herb, in dry soils, roadsides, N. States to Arc. Am. Stem having a dry, grayish aspect from its dense hairs, about a foot high, undivided except at the top where it branches into a kind of panicle. Leaves 1′ by 1—2′, sessile. Flowers very small, blue. Jl.

13. CYNOGLOSSUM.

Gr. κυων, a dog. γλωσσα, tongue ; from the form of the long, soft leaves.

Calyx 5-parted ; corolla short, infundibuliform, vaulted ; orifice closed by 5 converging. convex scales ; achenia depressed, fixed laterally to the style.— *Cor. blue, purple or white.*

1. C. OFFICINÁLE. *Hound's-tongue.*
Silky-pubescent ; *lvs.* lanceolate, acute, radical ones alternate at the base, petiolate, cauline ones sessile ; *sta.* shorter than corolla.—♃ An erect, downy plant, of a dull green color, 18—20′ high, and emitting a disagreeable smell, which several distinguished botanists have compared to the smell of young mice ! Grows in waste grounds and roadsides. Stem erect, hairy, 1—2f high. Leaves hoary with soft down on both sides, entire, upper ones clasping, with broad bases, lower or es 6—10′ by 1—2′, tapering into a long, attenuated base (winged petiole ?), pointed at apex. Clusters terminal, panicled, recurved at the end. Flowers with a downy calyx and a dull red corolla. Calyx leaf-like in fruit. Seeds rough, with hooked prickles. Jl. ♃

2. C. VIRGINICUM. (C. amplexicaule. *Mx.* and *1st edit.*)
Hirsute-pilose ; *lvs.* oblong-oval, acute, upper ones clasping, cordate at base ; *corymb* terminal, leafless, on a long peduncle.—♃ Inhabiting woods and thickets, Vt.! to Md.! W. to Ill.! rare in N. Eng. A very hairy plant, 2f high, simple, bearing at the top of its leafless summit, a small, panicled corymb of pale purple flowers. Radical leaves 5—6′ long and half as wide. Calyx and pedicels very hairy. Jn.

3. C. MORRISÓNI. DC. (Echinospermum Virginicum. *Lehm.* and *1st edit.*)
Virginian Mouse-ear.—*St.* much branched ; *lvs.* oblong-lanceolate, acuminate, scabrous above ; *rac.* divaricate, dichotomous ; *fruit* densely covered with hooked prickles.—① An erect, hairy weed, in rocky grounds and rubbish, Can. to Flor. Stem furrowed, 2—3f high, with many slender, remote, wide-spread branches, each terminating in a centrifugal, racemose inflorescence. Leaves entire, remote, large (3—4′ long), tapering to each end, the lower ones petioled. Flowers very small, white, the pedicels nodding in fruit. Jl.

TRIBE 2. HELIOTROPEÆ.—Ovary bearing a simple, terminal style. Fruit dryish, drupaceous, partible. Seeds without albumen.

14. HELIOTROPIUM. Tourn.

Gr. ηλιος, the sun. τρεπω, to turn ; the flowers were said to be always turned towards the sun.

Calyx 5-parted ; corolla hypocrateriform, orifice naked, limb 5-cleft, with the sinuses plaited ; stamens included ; stigma peltate ; achenia
37*

cohering without a common receptacle, at length separable.—*Herbs or shrubs. Spikes unilateral. Flowers never yellow.*

1. H. EUROPÆUM. *Heliotrope.*—Herbaceous; *lvs.* ovate, entire, rugose and tomentose; *spikes* in pairs.—A delicate annual, native of Europe and still growing on the banks of the Shenandoah at Harper's Ferry! where *Nuttall* saw it in 1818. Cultivated among stove-plants. Stem 8—12' high. Leaves 1—2' by 8—15''. Flowers white, mostly in 2, terminal, long, scorpoid racemes. ⚥

2. H. PERUVIANUM. *Peruvian Heliotrope.*—Shrubby; *lvs.* ovate-lanceolate; *fls.* in numerous, aggregated spikes.—Native of Peru. A small, elegant greenhouse shrub, 1—2f high. Leaves rough, serrulate, twice as long as wide, on short petioles. Flowers small but numerous, very fragrant, white or tinged with purple.

ORDER XCV. **HYDROPHYLLACEÆ.**—HYDROPHYLLS.

*Herbs shrubs or small trees, often hispid. Lvs. alternate, or the lower opposite, often lobed.
Fls. in circinate racemes or unilateral spikes, rarely axillary and solitary.
Cal. 5 cleft, the sinuses usually with reflexed appendages, persistent.
Cor. 5-lobed, regular, with 10 melliferous scales near the base.
Sta. 5, inserted into the base of the corolla and alternate with the lobes.
Anth. 2-celled, versatile.   [from the base of the cavity.
Ova. free, simple, 1-celled. Style single, terminal, bifid. Stigmas 2. Placenta 2, parietal or on stalks
Fr.—Capsule invested with the permanent calyx.
Sds. few, crustaceous. Embryo conical, in abundant, cartilaginous albumen.*

Genera 16, species 75, chiefly American. Of no known use.

Conspectus of the Genera.

		Placenta large, fleshy, free.	*Hydrophyllum* 1
appendag'd inside with	Sta. exserted.	Placenta on the middle of the valves.	*Phacelia.* 4
5 melliferous grooves.	Stamens included. Sepals very large.		*Ellisia.* 2
Corolla destitute of grooves or scales inside. Stamens as long as the corolla.			*Cosmanthus.* 3

1. **HYDROPHYLLUM.** Tourn.

Gr. ὕδωρ, water, and φύλλον, a leaf; the leaves in Spring hold each a quantity of water.

Sepals slightly united at base; corolla campanulate, with 5 longitudinal, margined, nectariferous grooves inside; stamens exserted; capsule globose, 2-celled, 2-valved, 4-seeded, 3 of the seeds mostly abortive; placenta fleshy, free.—*North American herbs. Radical lvs. on long petioles, pinnately or palmately veined, cauline alternate. Cymes scorpoid, bractless.*

* *Calyx appendaged between the sepals at base. Stamens as long as the corolla.*

1. H. APPENDICULATUM. Michx. (Nemophila paniculata. *Spreng.*)

Lvs. hairy, lower ones pinnatifid, cauline palmately 5-lobed, dentate, lobes diverging, and with the long *petioles, ped.* and *cal.* hispid; *sep.* lance-subulate, the appendages at the base ovate, acute, 4 times shorter; *cor.* glabrous except the minute appendages inside; *sta.* included.—⚥ Mich. to Ia., *Plummer!* and Ill. to Va., moist woods and bottoms. Stems 1—1½f high, branched. Petiole: 1—4' long. The leaves are of a singular form, roundish in outline, the broad acute lobes diverging in a stellate manner; teeth mucronate. Calyx 4—5'' long appendages deflexed, 1' long. Corolla blue, on long peduncles. May.

* * *Calyx not appendaged. Filaments much exserted.*

2. H. VIRGINICUM. *Virginian Water-leaf.*

Plant nearly smooth; *lvs.* pinnatifid and pinnate, the segments oval-lanceolate, incisely serrate; *fascicles* conglomerate; *ped.* as long as the petioles.—⚥ An inhabitant of wet or moist woods, Can. to Car. and Western States. Stem a foot high, bearing large, roundish tufts of flowers peculiarly distinguished by their exserted stamens and style, which are twice the length of the bell-shaped corollas. Leaves few, on long, clasping petioles, with about 5 distinct leaflets, the upper 3 more or less confluent at base, all irregularly toothed. Corollas varying from white to sky-blue. June.

3. H. CANADENSE. *Canadian Water-leaf. Burr-flower.*

Lvs. smoothish, palmate, roundish, with 5—7 shallow lobes, unequally dentate, teeth obtuse-mucronate; *fls.* in crowded fascicles; *peds.* shorter than the petioles.—♃ Quite different in aspect from the last. Found in alpine woods, Can. to Car. W. to Ohio! Stem 12—18' high, with large, roughish leaves divided into 5—7 lobes. Fascicles of flowers dense, axillary and terminal. Corollas white or variously tinged with purple. Stamens and style much exserted, as in the last. Jn., Jl.

4. H. MACROPHYLLUM. Nutt. (H. hispidum. *Riddell.*)
Whole plant reversely hispid with white hairs; *lvs.* oblong-oval in outline, pinnatifid, lower segments distinct, upper confluent, all incised into rounded, mucronate teeth, cauline solitary or few, much smaller; *cymes* terminal, long-pedunculate, dense-flowered; *cor.* glabrous except the grooves inside.—♃ Ohio, *Locke!* to the Alleghany Mts. Stem a foot high, almost leafless, with a terminal, globose cyme of white flowers. Radical leaves 8—12' (including the petiole 3—4') by 3—5', the segments ovate-oblong. Corolla twice longer than the ovate, acute sepals, half as long as the capillary filaments which are 9" in length. Jn.

2. ELLISIA.

In honor of Joseph Ellis, F. R. S., an English naturalist, correspondent of Linnæus.

Calyx 5-parted, equaling the tubular-campanulate, caducous corolla; tube with 10 minute appendages within, limb 5-lobed; sta. included; nectary annular, 5-toothed; sty. bifid, with linear lobes; caps. ovoid-globose, 2-valved; seeds 4.—① *North American herbs, with pinnatifid lvs.   Cor. white.*

E. NYCTELÆA.
Ascending, branching, with few, scattered hairs; *lvs.* pinnatifid, petiolate, upper ones alternate, segments linear-oblong, nearly distinct, sparingly dentate; *ped.* 1-flowered, opposite the leaves, about as long as the sepals; *cal. seg.* triangular-acuminate, broad at base, longer than the tube of the corolla.—① Ill. *Mead!* to Va., woods and river banks. Stem 4—10' long. Leaves 1—2' long ⅔ as wide. Calyx at length remarkably large for the size of the plant, nearly an inch in diam. Corolla lobes obtuse, emarginate, with purple spots at base inside.

3. COSMANTHUS. Nolte.

Gr. κοσμος, elegance, ανϑος, a flower.

Calyx 5-parted; cor. broadly campanulate, caducous, 5-cleft, tube without appendages; sta. 5, about equaling the corolla; nectary minute; ova. hairy except at base, 1-celled; sty. bifid; caps. 2-valved, valves septiferous in the middle; seeds 4—10, rugulose.—① *N. American herbs, with alternate lvs.   Rac. long, bractless.   Fls. small, white or pale blue.*

1. C. PARVIFLORUS. Alph. DC. (Phacelia. *Ph.* Eutoca. *Br.*)
Diffuse, pubescent; *lvs.* subsessile, pinnatifid or trifid, segments oblong x ovate, sparingly lobed or entire; *rac.* solitary; *pedicels* in flower longer than the calyx; *sta.* subequal to the corolla; *placentæ* 6—8-ovuled.—Penn. to Va. Stem 6—8' high. Flowers pale blue, 4" diam., the lobes rounded and entire Capsule ovoid, shorter than the calyx. May.

2. C PURSHII. Wood. (C. fimbriatus. *Nolte.* Phacelia fimbriata. *Ph.* not *Mx.* P. Purshii. *Buckley.*) *Miami Mist.*—Nearly glabrous; *lower lvs.* petiolate, pinnatifid, segments few, entire, ovate, terminal one largest; *upper lvs.* sessile, pectinately pinnatifid, with oblong, acute lobes; *rac.* terminal, simple, 5—10-flowered; *pedicels* longer than the lance-linear sepals; *lobes of the cor.* fimbriate.—Fields and river bottoms, Penn. to Ga., W. to Ia.! and Ky. Plant 8—12' high, slender and with slender branches. Radical leaves with obtuse lobes, mostly shorter than the petiole. Flowers light blue, 4—5" broad, spreading. May, Jn.

4. PHACELIA.

Gr. φακελος, a bundle or fascicle ; alluding to the fasciculate racemes.

Calyx 5 parted ; cor. tubular-campanulate, caduccus, 5-lobed, tube within furnished with 5 melliferous grooves ; sta. 5, mostly exserted ; ova. 1-celled, hispid ; sty. bifid ; caps. ovoid, 2-valved, valves placentiferous in the middle ; seeds 4.—*American herbs, hispid, with alternate lvs., and loose or dense, one-sided racemes.*

P. BIPINNATIFIDA. Michx.

Hairy, suberect; *lvs.* incisely pinnatifid, long-petivate, lateral segments 3—4, incisely lobed and toothed, terminal trifid ; *rac.* elongated, bifid or subpaniculate ; *cor. lobes* entire, twice longer than the calyx, shorter than the stamens.—♃ or ♀ Woods and hill sides, Penn. to Ohio! and Ia.! Plant sometimes nearly smooth, 1—2f high, bearing several leafless racemes at top. Leaves 3—6' long, including the petiole. Corolla 6" broad, blue, the grooves bordered with narrow, pubescent margins. May, Jn.

β. *Plummeri. St.* ferruginous-hirsute; *ls.* much smaller; *cor.* scarcely longer than the calyx.—Richmond, Ia. *Plummer!*

ORDER XCVI . POLEMONIACEÆ.—PHLOXWORTS.

Herbs, with opposite, occasionally alternate, compound or simple leaves.
Cal. 5 united sepals, inferior, persistent, sometimes irregular.
Cor. 5 united petals, regular, the lobes imbricate or twisted in æstivation.
Sta. 5, inserted into the midst of the corolla tube and alternate with its lobes.
Ova. 3-celled, free. *Styles* united into 1. *Stigma* trifid.
Caps. 3-celled, 3-valved, loculicidal, valves also separating from the 3-cornered axis.
Sds. few or many. *Albumen* horny. *Embryo* foliaceous.

Genera 17, species 104, chiefly North American. They are valued only in cultivation as ornamental plants.

Conspectus of the Genera.

	hypocrateriform.	Phlox.	1
	infundibuliform.	Gilia.	2
Corolla	campanulate.	Polemonium.	3

1. PHLOX.

Gr. φλοξ, a flame ; from the color and profusion of the flowers.

Calyx prismatic, deeply 5-cleft ; corolla hypocrateriform, the tube more or less curved ; stamens very unequal, inserted in the tube of the corolla above the middle ; capsule 3-celled, cells 1-seeded.—*A highly ornamental, North American genus. Lvs. mostly opposite, sessile, simple, entire. Fls. in terminal corymbs or panicles.*

1. P. PANICULÂTA. *Panicled Phlox or Lychnidea.*

St. glabrous. erect; *lvs.* oblong or ovate-lanceolate, acuminate at each end, rough-edged, flat; *corymbs* paniculate, subpyramidal, many-flowered ; *cal. teeth* setaceous-acuminate, shorter than the tube; *pet.* roundish, obovate, entire.—♃ This well-known favorite of gardens is found native in woods and on river banks, Western States! to Penn. and Car. It flourishes in rich, moist soil, or in leaf-mould or peat. Stem 2—3f high, ending in a large, oblong-pyramidal panicle of innumerable pink-colored, scentless flowers. Leaves 3—5' by 9—16", lower ones distinctly petioled, the highest sometimes subcordate at base. July—Sept. †

2. P. ACUMINÂTA. Ph. *Acuminate Lychnidea.*

St. erect, paniculate, branching above; *lvs.* oblong or ovate-lanceolate, acuminate, the upper ones subcordate at base, all pubescent beneath, as well as the stem; *panicle* pyramidal-corymbose, many-flowered; *cal. teeth* briefly setaceous-acuminate.—♃ This species, whose numerous varieties are common in gardens, is a native of Mo., Ky. and Ill. Although distinguishable a sight from P. paniculata, being stoute and rougher, its chief technical distinction is *its pubescence* and shorter calyx teeth. Petals rounded at the end, light purple varying to deep purple and red. June—Aug. †

3. P. MACULÀTA. (P. pyramidalis. *Sm.* P. latifolia. *Michx.*)

St. erect, subsimple, scabrous or nearly smooth; *lower lvs.* lanceolate, *the highest* ovate, cordate at base, all subcoriaceous, roughish or smooth; *panicle* oblong or subpyramidal; *cal. teeth* lanceolate, acute; *pet.* orbicular.—♃ Moist fields, Penn. to Car. and Western States. Stem 2—3' high, mostly punctate, with purple spots. Lower branches of the panicle shorter than the leaves, or often elongated. Corolla tube more or less curved, smooth. Petals obtuse or retuse. purple, varying in gardens from white to crimson. †

β. suaveolens. (P. suaveolens. *Ait.*) *Fls.* white, fragrant. †

4. P. CAROLÌNA. (P. revoluta. *Aiken.* P. nitida. *Ph.*) *Carolina Lychnidea.*

Glabrous, erect or ascending; *lvs.* linear-lanceolate, attenuated to the apex, subcoriaceous, margins revolute, lower attenuated to the base, upper rounded at base; *panicle* corymbose, puberulent or smooth; *cal. teeth* short-acuminate; *cor. tube* awned, segments obovate, entire.—♃ A very smooth species, fields and barrens, Ohio, Ia.! common; also Md. to Car. Stem slender, 1—2' high. Leaves 2—4' long, mostly quite narrow (3—4''), thick and shining. Panicle few (15—25)-flowered. Corolla purple. May—Aug.

β. ovata. Benth. (P. ovata. *Linn.*) *Lrs.* ovate and lance-ovate.—Car. †

5. P. GLABERRÌMA. *Ohio Lychnidea.*

Glabrous; *st.* branching at base, the branches subsimple; *lvs.* lance-linear or lance-oblong, rather obtuse, thin, sessile, flat, upper ones lance-ovate, often rounded at base; *panicle* corymbose, few-flowered, glabrous; *cal. teeth* lanceolate, subacuminate, half as long as the corolla tube; *pet.* obovate, entire.—♃ Pine barrens, Ohio! Very distinct from the former by its shorter, obtuse, never acuminate or revolute leaves and its much larger calyx. Stem 2' high. Leaves 1½—2' by 3—5''. Flowers purple. June, July.

6. P. REPTANS. Michx. *Creeping Lychnidea.*

Stolons creeping; *sts.* low, assurgent; *lrs.* ovate, obovate or oblong; *cor.* few-flowered; *cal.* puberulent, segments linear-subulate; *pet.* obovate, entire.—♃ Hill-sides and mountains, Ia. *Plummer!* to S. Car. Flowering-stems 6' high, with small (4—9'' by 2—4'') and remote leaves. Stolons with leaves 2—3 times larger, somewhat crowded at the end. Flowers 3—8. Corolla bluish-purple, tube scarcely twice longer than the calyx. June.

7. P. DIVARICÀTA. *Early-flowering Lychnidea.*

Low, diffuse, pubescent; *lrs.* lanceolate, ovate or oblong; *panicle* corymbose, loose; *cal.* roughish-puberulent, segments linear-subulate; *cor. segments* emarginate-bifid.—♃ Can., N. Y.! to Va. Stems loosely branched, a foot or more long, flaccid. Leaves 1—2' long, acute, the lower tapering to the base, the upper broad and clasping at base, the floral linear-setaceous. Pedicels diverging, longer than the calyx which is half as long as the corolla tube. Corolla of a peculiar light but brilliant grayish-blue. May.

β? Laphami. *Lrs.* ovate; *pet.* obtuse, entire.—Wis. *Lapham!* Western Reserve. *Cowles!*—Intermediate between P. divaricata and P. glaberrima, and may prove distinct from both.

8. P. PILÒSA. (P. aristata. *Michx. Benth.*)

Pilose-puberulent, erect or decumbent at base; *lvs.* lance-linear, margin subrevolute, subamplexicaul; *panicle* corymbose; *cal.* hirsute, segments setaceous-acuminate; *pet.* obovate, entire.—♃ Penn. to Ohio, *Clark!* Wis. *Lapham!* and Southern States. Stem 12—18' high, rigid. Leaves 1½ —3' by 2—4'', attenuated to the apex. Corolla pale red or bluish-white, the tube a third longer than the long, slender, awn-like teeth of the calyx. May, June.

β Lvs. shorter, broadest at base, sessile; *fls.* smaller.—Indiana!

9. P. BÌFIDA. Beck. *Beck's Lychnidea.*

Low, assurgent, diffusely branched, puberulent; *lvs.* amplexicaul, subrevolute on the margin, acutish, lower lance-ovate, upper lance-linear; corymbs very loose, 2—5-flowered; *cal. segments* linear, acute; *cor. tube* curved, segments deeply bifid.—♃ A very distinct species, and very rare, in Mo. *Beck,* Cass Co., Ill., *Mead!* Stem brownish-purple, slender, 6' high. Leaves 12—15'' by 1—2'', lower much shorter. Pedicels 1' long. Tube of corolla much curved, segments cleft nearly half way down, purple. Apr.

10. P. Drummondii. *Drummond's Lychnidea.*

Erect, dichotomously branched, glandular-pilose; *lvs.* oblong or lanceolate, scabrous; *corymb* dense-flowered; *cal.* hairy, segments lanceolate, setaceous, elongated, revolute; *cor. tube* pilose, segments obovate, entire.—One of the handsomest species of the genus, common in cultivation. Whole plant glandular-scabrous, 8—12' high. Flowers very showy, all shades from white to dark purple. †

11. P. subulata (and P. setacea. *Linn.*) *Moss Pink.*

Procumbent, cæspitose, much branched, pubescent; *lvs.* rigid, subulate or linear-subulate, ciliate, fascicled in the axils; *cal. teeth* linear-subulate, very acute; *cor. lobes* cuneate, emarginate.—Rocky hills and mountains, Penn. to Ga. and Ky., abundant in its localities, in dense, turfy masses, sprangled over in May with rose-colored flowers. Flowering branches, 2—4' long, numerous and fascicled. Corymb 3—6-flowered. Corolla white or pink, deeper purple in the centre. May. †

2. GILIA. Ruiz & Pavon.

Calyx 5-cleft, segments acute; cor. tube long or short, limb regularly 5-lobed; sta. 5, equally inserted at top of the tube; disk cupform; caps. oblong or ovoid, few or many-seeded.—*Herbs with alternate, pinnatifid lvs. Fls. paniculate, capitate or scattered, generally bractless.*

§ 1. *Corolla subrevolute, tube included in the calyx.*

1. G. tricolor. Benth. *Tri-colored Gilia.*—*St.* erect, nearly smooth; *lvs.* twice or thrice pinnatifid, with narrow, linear segments; *cymes* paniculate, 3—6-flowered; *cor.* tricolored, 2 or 3 times longer than the calyx, tube very short.—① An elegant little garden plant, from California, 1f high. Flowers numerous, limb pale lilac-blue, throat purple and tube yellow. †

§ 2. Ipomopsis. *Corolla infundibuliform, tube much exserted.*

2. G. (ipomopsis) coronopifolia. Pers.

Erect, tall; *st.* strict, hairy; *lvs.* crowded, pinnatifid with subulate divisions; *thyrse* elongated, with very short branches; *cor.* elongated, segments oval-oblong, erect-spreading; *sta.* exserted.—② Southern States! A splendid herb, 2—4f high, bearing at top a long (1f) thyrse of scarlet-red flowers. Corollas 1½' long. †

3. POLEMONIUM.

Gr. πολεμος, war; Pliny relates that two kings fought for the merit of its discovery.

Calyx campanulate, 5-cleft; corolla rotate-campanulate, limb 5-lobed, erect, tube short, closed at the base by 5 stameniferous valves; capsule 3-celled, 3-valved, cells many-seeded.—*Herbs with alternate, pinnately-divided lvs. Fls. terminal.*

1. P. reptans. *American Greek-Valerian.*

St. smooth, branching, erect; *lvs.* pinnately 7—11-foliate, leaflets oval-lanceolate, acute; *fls.* terminal, nodding; *cells of caps.* 2—3-seeded.—♃ A handsome plant of woods and damp grounds in N. Y. to Ill.! and sometimes cultivated. Stem 12—18' high, weak, fleshy. Leaflets mostly 7, subopposite, smooth, entire, sessile, an inch long and half as wide. Flowers numerous, rather large, on short petioles. Segments of the calyx lanceolate-acute, persistent, much shorter than the tube of the corolla. Corolla blue, lobes short, rounded at the ends. Anthers introrse. Root *creeping.*

2. P. cœruleum. *Greek Valerian.*—*S'.* smooth, simple, erect; *lvs.* pinnately 11—17-foliate, segments acuminate; *fls.* erect; *cal.* equaling the tube of the corolla; *cells of caps.* 6—10-seeded.—② A handsome, cultivated plant, native in England. Stems clustered, several from the same root, about 2f high, hollow, stout, each dividing at top into a corymbose panicle. Leaves mostly radical, on long, grooved petioles; leaflets all sessile, ovate-lanceolate, subopposite, oblique, odd one lanceolate. Fls. terminal, suberect. Cor. blue, about ½' diam.

Order XCVII. DIAPENSIACEÆ.

Undt shrubs, prostrate, with crowded, heath-like leaves and solitary, terminal flowers.
Cal.—Sepals 5, much imbricated, surrounded at base with imbricated scales.
Cor.—Petals 5, united, regular, imbricated in æstivation.
Sta. 5, equal, the filaments petaloid and inserted on the corolla tube.
Anth. 2-celled, transversely valved.
Ova. 2-celled, free. Styles united into 1. Stigma 3-lobed.
Fr.—Capsule 3-valved, loculicidal. Seeds many, small, albuminous.

Genus 1, or 2 according to many authors, species 2, natives of the north of Europe and the northern parts of North America.

DIAPENSIA.

Calyx 5-parted, calyculate with 3 bracts at base; corolla hypocrateriform, limb 5-cleft, flat; stamens 5, from the summit of the tube; stigmas 3; capsule 3-celled, 3-valved, many-seeded.—*Low, evergreen undershrubs.*

§ 1. *Anthers awnless.*

1. D. Lapponica. *Northern Diapensia.*

Cæspitose; *lvs.* dense, spatulate, fleshy, evergreen, obtuse and entire; *fls.* pedunculated.—♃ A little, leafy plant, 2—3' high, growing on the summits of the White Mts. in N. Hampshire, forming dense tufts among the rocks. Leaves crowded, pale beneath, fleshy, 5—8'' by 1'' with a revolute margin, clasping base, and broadly obtuse point. Flowers on short, terminal, solitary peduncles, which are an inch long in fruit. Calyx of 5, obtuse leaves, longer than the leafy bracts at its base. Corolla white, with 5, flat segments. July.

§ 2. Pyxidanthera. *Anthers with the lower valves beaked.*

2. D. barbulata. Ell. (Pyxidanthera barbulata. *Michx.* P. cuneifolia. *Ph.*)—Branches short, ascending; *lvs.* lance-cuneiform, acute, pubescent at base; *fls.* terminal, sessile; *lower valve of the anther* beaked or awned at base.—A prostrate, creeping plant, abundant in pine barrens, N. J. to Car., forming dense beds. It has also been found by Dr. Peck on the White Mts. (*fide Torr.*) Stems 3—6' long, subhispid. Leaves 1—2'' by ⅓—1''. Flowers white, 3'' diam. Sepals denticulate, as long as the corolla tube. May, Jn.—The beak of the anther appears to be quite variable, sometimes reduced to an acute point.

Order XCVIII. CONVOLVULACEÆ.—Bindweeds.

Herbs or shrubs, with a milky juice, mostly twining, sometimes erect.
Lvs. alternate, without stipules, sometimes wholly wanting. Fls. showy.
Cal.—Sepals 5, much imbricated, usually united at base, persistent.
Cor. regular, limb 5-lobed or entire, plaited and twisted in æstivation.
Sta. 5, inserted into the base of the corolla and alternate with its lobes.
Ova. 2—4-celled, free. Styles united into 1.
Fr.—Capsule 2—4-celled, valves with septifragal dehiscence.
Sds. few, large, with thin mucilaginous albumen. Cotyledons foliaceous, or wanting.

Genera 43, species 650, very abundant in tropical climates, rare in cold.

Properties.—The roots abound in an acrid, milky juice which is strongly purgative. *Jalap* of the shops is the product of the root of Exogonium purga of Mexico, and other species. *Scammony* of Convolvulus Scammonia, native of Levant. The drastic qualities of both depend upon the presence of a peculiar resin. The *sweet potato*, a valuable article of food, is the product of C. Batatas, native of the South.

Conspectus of the Genera.

		Style simple.				Convolvulus. 1
	Stamens included.	Styles 3 or 2.				Stylisma. 2
Calyx naked.	Stamens exserted.	Corolla scarlet.				Quamoclit. 3
leafy, green.	Calyx enclosed in two large, leaf-like bracts.					Calystegia. 4
Plants { leafless, parasitic, orange-colored.						Cuscuta. 5

Suborder 1.—CONVOLVULEÆ.

Embryo with cotyledons. Carpels united. Fruit capsular, dehiscent.

1. CONVOLVULUS.

Lat. *convolvere*, to entwine; from the habit of most of the plants.

Calyx 5-parted, naked or with 2 small bracts near the base; cor. campanulate, or funnel-form, limb 5-plaited; sta. shorter than the limb, rarely a little longer; ova. 2—4-celled, cells 1—2-ovuled; sty.

simple; stig. simple or 2-lobed; caps. valvate, 2—4-celled, 4—6-seeded.—*A large genus of twining or prostrate herbs, rarely shrubby or arborescent.*

Obs.—The generic distinctions adopted by Choisy in the Prodromus of De Candolle, Vol. ix., appear to me to be too indefinite to be generally useful in a work like the present. I have adopted them merely as sections of the present genus.

§ 1. *Stigmas 2, linear-cylindric, often revolute. Capsule 2-celled.*

1. C. ARVENSIS. *Small Bindweed.*

St. striate, angular, generally prostrate; *lvs.* sagittate, somewhat auriculate; *ped.* mostly 1-flowered, bibracteate near the apex; *sep.* roundish-ovate; *caps.* smooth.—♃ A twining plant, growing in fields and pastures, Maine to Car., not common. Stems several feet long, climbing or prostrate, a little hairy. Leaves 1—2′ long, the lower ones obtuse. Flowers small, white, often with a tinge of red. The small, acute bracts are near the middle of the peduncle. Jn.

2. C. TRICOLOR. *Tricolored Bindweed.—St.* ascending, villose; *lvs.* lance-obovate, subspatulate, sessile, ciliate at base; *ped.* 1-flowered, bracteate, longer than the leaves; *sep.* ovate-lanceolate, acute; *cor.* tricolored; *capsule* villose.—ⓘ About the Mediterranean. Stem weak, 1—3f long. Corollas yellowish in the centre, white in the middle, and of a fine sky-blue on the upper part of the border. July.

§ 2. IPOMŒA. *Stigma capitate, entire or 2-lobed. Capsule 2-celled, 4-seeded.*

3. C. PANDURÀTUS. (Ipomœa. *Meyer.*) *Wild Potato. Man-of-the-earth.*

St. twining; *lvs.* broad-cordate or panduriform; *ped.* long, 1—4-flowered; *cal.* smooth; *cor.* tubular-campanulate.—♃ In sandy fields, N. Y. to Ga. Stems several from the same root, 4—8f long, slender, smooth. Leaves 2—3′ long and of about the same width, acute or obtuse, with rounded lobes at the base, sometimes lobed and hollowed on the sides and becoming fiddle-shaped. Petioles 2—3′ long. Peduncles axillary, longer than the petioles, generally branching at the top, and bearing several large flowers. Corolla 2′ long, purple and white. July, Aug.

4. C. LACUNÒSUS. (C. micranthus. *Riddell.*) *Small-flowered Bindweed.*

Minutely pubescent; *st.* twining; *lvs.* cordate, acuminate, angular-lobed or entire, on long petioles; *ped.* 1—3-flowered, half as long as the petioles; *sep.* oblong-lanceolate, acute, half as long as the corolla, ciliate, lobes acute; *caps.* pilose.—ⓘ Penn., Md.! to Flor., W. to Ohio and Ill. A small, prostrate species, 2—6f long, in dry fields and hills. Leaves 2′ by 1½′, deeply cordate, often deeply 3-lobed! petioles 1—3′ long. Flowers 8″ diam., 9″ long, white with a purplish rim. Aug. Sept.

§ 3. PHARBITIS. *Stigma capitate, granulate. Ovary 3- rarely 4-celled, cells 2-seeded.*

5. C. PURPUREUS. (Ipomœa. *Ph.* Pharbitis hispida. *Choisy.*) *Common Morning Glory.—St.* climbing and twining, retrorsely pilose; *lvs.* cordate, entire; *fl.* nodding; *ped.* 2—5-flowered; *pedicels* thick; *cal.* hispid.—ⓘ In fields, Mid. and W. States. Stems climbing many feet. Leaves roundish, heart-shaped. Flowers large, beautiful, generally of a dark purple, sometimes blue, flesh-colored, striped, &c. A well known and favorite climber and free flower of the easiest culture. Jn. § †

6. C. NIL. (Pharbitis. *Choisy.*) *Morning Glory.*

Lvs. cordate, 3-lobed; *fls.* half 5-cleft; *ped.* shorter than the petioles, 1—3-flowered.—A very beautiful twining plant, found wild, Penn. to Flor., but best known as a garden annual. Stem and leaves somewhat hairy. Calyx very hairy, the segments long-acuminate. Flowers large, the tube white and the border of a clear blue color (whence its specific name, Anil or Nil, indigo). It is of the easiest culture, and raised from the seed. Blossoms from July to September. †

§ 4. BATATAS. *Stig. capitate, 2-lobed. Ova.* 4-, *or by abortion,* 3—2-celled.

7. C. JALÀPA. (C. macrorhizus. *Ell.* Batatas Jalapa. *Choisy.*)

St. creeping or twining; *lvs.* cordate, entire, sinuate or lobed, tomentose-pubescent beneath ; *ped.* scarcely equaling the long petiol*t*, 1—3-flowered ; *sep.* roundish-ovate, pubescent; *seeds* villose, with long hairs.—♃ We-tern States! frequent, to Flor. and La. Stems 2—6f in length, on the ground, or trailing over fences, &c. Leaves 2—3′ long, the petioles 2—5′. Corolla large *(2½—3′* diam.) and showy, white or rose-colored, purple at base. July, Aug.—The root is tuberous and mildly purgative.

8. C. BATĀTAS. (Batatas edulis. *Choisy.*) *Sweet Potato.* — *St.* creeping, rarely twining ; *lvs.* cordate, hastate, angular, 5-veined, smoothish; *ped.* long; *fls.* fascicled ; *sep.* lanceolate, acuminate.—① The sweet potato is native of both Indies and cultivated in all tropical climates. Not only the tubers, but the leaves and tender shoots are boiled and eaten. The tubers are sweet and considered nutritive. This is the *potato* of the old English botanists, of Shakespeare, and their cotemporaries, the *Solanum tuberosum* then being unknown. The stem is round, hispid, prostrate, creeping, sending out scattered, oblong tubers which are purplish without. Flowers large, purple or white. ‡

2. QUAMÓCLIT. Tourn.

Gr. κναμος, a bean, κλιτος, dwarf; resembles the climbing bean, but smaller.

Sepals 5, mostly mucronate ; cor. tubular-cylindric ; sta. exserted ; sty. 1 , stig. capitate, 2-lobed ; ovary 4-celled, cells 1-seeded.—*Twining herbs, mostly American.*

1. Q. VULGĀRIS. Choisy. (Ipomæa. *Linn.* Convolvulus. 1*st. edit.*) *Jasmine.* *Bindweed.   Cypress Vine.*—*Lvs.* pinnatifid to the midvein, segments linear, parallel, acute ; *ped.* 1-flowered ; *sep.* ovate-lanceolate.—① An exceedingly delicate vine, Penn. *Eaton,* Southern States! generally cultivated. Stems glabrous, very slender, twining and climbing to the height of 5—10f. Flowers much smaller than those of the common morning glory, scarlet, varying to crimson and rose-color. Trained upon twine it forms a most delicate and beautiful awning. July, Aug.

2. Q. COCCINEA. Mœnch. (Ipomæa. *Linn.* Convolvulus. *Spreng.*) *Lvs.* cordate, acuminate, entire or angular at base ; *ped.* elongated, about 5-flowered ; *cal.* awned.—① Southern States, naturalized in the Western, occasionally cultivated. Flowers varying from yellow to scarlet. †

3. STYLISMA. Raf.

The name has reference to the plurality of the styles.

Sepals 5, equal ; cor. campanulate ; ovary 2-celled; styles 2, rarely 3, stigmas thick ; sta. included.—♃ *Slender, creeping.*

S. TENELLUS. Wood. (S. evolvuloides. *Choisy.* Convolvulus tenellus *Lam.* C. Sherardi. *Ph.*)—Dry, sandy or rocky soils, Ohio, to Flor. Stem long, prostrate, branching, pubescent. Leaves lance-linear or linear, obtuse, 1′ in length, with short petioles. Peduncles longer than the leaves, 1—5-flowered. Sepals ovate-lanceolate, 3″ long. Corolla twice longer, hairy outside.

4. CALYSTEGIA. Br.

Gr. καλυξ, calyx, στεγη, a covering; alluding to the conspicuous calycine bracts.

Calyx 5-parted, included in 2 large, foliaceous bracts ; cor. campanulate, 5-plicate ; sta. subequal, shorter than the limb; ova. half bilocular, 4-ovuled ; sty. simple ; stig. 2, obtuse ; caps. 1-celled, 4-seeded.—*Herbs twining or prostrate. Ped.* 1-flowered, *solitary.*

1. C. SPITHAMÆUS. Br. (Convolvulus. *Linn.* C. stans. *Michx.*) *Erect Bindweed.—St.* erect or assurgent; *lvs.* oblong-lanceolate, subcordate, hoary-pubescent; *ped.* 1-flowered, generally longer than the leaves.—♃ An erect, downy species, (*a span*) 8—10′ high, found in fields and hilly pastures, Can. to Penn., W. to Ill. Stem branching, leafy, bearing one, often two or more large, white flowers, on peduncles 2—4′ long, issuing from near the root. Leaves 2—3′ long, ⅓ as wide, oval, with an abrupt, cordate base, and on petioles ⅓ as long. Bracts concealing the calyx. June.

38

2. C. SEPIUM. Br. (Convolvulus. *Linn.*) *Hedge Calystegia. Rutland Beauty.*—*St.* twining; *lvs.* sagittate, the lobes being truncate and the apex generally acute; *ped.* quadrangular, 1-flowered; *bracts* cordate, much longer than the calyx.—♃ A vigorous climber, in hedges and low grounds, Can. to Car., W. to Ill. Stems 5—8f in length. Leaves cordate-sagittate, 2—4 long, ½ as wide. Flowers numerous, large, white, with a reddish tinge, appearing in long succession. The bracts are so close to the corolla as to appear like the calyx which they entirely conceal. It is cultivated as a shade for windows, arbors, &c. June, July.

SUBORDER 2. **C U S C U T E Æ.**

Embryo without cotyledons. Leafless, parasitic herbs.

5. CUSCÜTA. Tourn.

Calyx 5 (rarely 4)-cleft; corolla globose-campanulate, 4—5-cleft, marescent; stamens 4—5, inserted upon the corolla at the clefts, stigmas, 2; capsule 2-celled, circumscissile at the base; cells 2-seeded. —*Herbs without verdure, germinating in the soil, at length withering at the root, and deriving their nourishment from other plants about which they twine from right to left. Stem yellowish or reddish. Leaves none, or minute scales instead. Fls. variously aggregated.*

1. C. GRONOVII. Willd. (C. Americana. *Linn.* C. vulgivaga and saururi. *Eng.*)—*St.* filiform, thick; *fls.* densely glomerate, in paniculate spikes, sessile; *sep.* broad-ovate, obtusish; *cor.* 5-cleft, segments short, spreading or reflexed, withering at the base of the capsule; *scales* oblong, fimbriate; *sty.* diverging; *stig.* capitate.—⓵ An extremely delicate vine, found in damp places, by rivulets, Can. and U. S. The stem is smooth, slender, 3—5f long. springing from the soil at first, but after having twined itself about the low plants in its way, and becoming fixed upon them by its lateral radicles, it withers away at base, and is henceforth disconnected with the soil. It is of a light orange color, wholly destitute of green, furnished with a few minute scales, branching, always turning from right to left, or hanging in festoons. Flowers nearly globose, about a line long, and on peduncles of about the same length. Calyx segments round-obtuse. Corolla twice as long, yellowish-white. Aug.

2. C. LEPIDANCHE. Wood. (C. glomerata. *Choisy.* Lepidanche compositarum. *Engelm.*)—*St.* filiform; *fls.* in compact masses surrounding the stem, sessile, with scarious bracts intermixed; *cal.* 5-sepaled, scarious; *cor.* tubular-campanulate, 5-lobed, longer than the calyx, lobes lanceolate, acute, spreading or reflexed; *anth.* elongated; *scales* fimbriate.—⓵ Abundant in Mo., Ill.! and Ia.! on the Labiates, composites, &c. Flowers about 2″ long, forming compact, cylindrical masses while the stems decay, appearing as if springing from the stems of other plants. Corolla white and scarious. Anthers partly exserted. July.

β. adpressa. Chois. (Lepidanche adpressa. *Eng.*) Bracteate; *sep.* obtuse or orbicular-ovate.—Ill.

3. C. EPILINUM. Weih. (C. Europæa. *Darl. & others.*) *Flax Dodder.*— *Fls.* sessile, in small, dense, remote heads; *cal.* 5-parted, segments rather obtuse; *cor.* globose-cylindric, scarcely longer than the calyx, withering around the capsule; scales minute, crenate-dentate.—⓵ Europe, introduced into the Mid. States, growing on flax. *Torr. Darl.* Stems reddish-orange. Flowers yellowish-white. Calyx thickish; stamens included. Stigmas acute. Capsule depressed-globose, surrounded with the withering corolla. June.

ORDER XCIX. SOLANACEÆ.—NIGHTSHADES.

Plants herbaceous or shrubby, with a colorless juice. *Lvs.* alternate, the floral ones sometimes collateral. *Inflorescence* often supra-axillary; pedicels bractless.
Cal.—Sepals 4—6. more or less united, mostly persistent.
Cor. regular, limb 4—5-cleft, plaited in æstivation, deciduous.
Sta. 4—5 (sometimes 1 abortive), inserted on the corolla, alternate with its segments.

Anth. bursting longitudinally, rarely by terminal pores.
Ova. free (superior), 2-celled, (4-celled in Datura) with the placenta in the axis.
Styles and *stigmas* united into 1.
Fr. a capsule or berry. *Seeds* numerous. *Embryo* curved, lying in fleshy albumen.

Genera 60, species 550, diffused throughout the world, except the frigid zones, but most abundant in the torrid

Properties. These are highly important. A large portion of the genera are pervaded by a narcotic principle, rendering the herbage and fruit dangerously poisonous, yet furnishing some of the most active medicines; as the *henbane* (Hyoscyamus), *belladonna* (Atropa), *stramonium* (Datura), *tobacco* (Nicotiana), &c. At the same time several species of Solanum afford wholesome and nutritious food, not because they are free from the narcotic principle, but because it is expelled in the process of cooking or ripening in the sun. Such are the tubers of the invaluable potato, the fruit of the *tomato* and *egg plant*. The genus Capsicum is entirely free from narcotine, and produces the well-known stimulant fruit, *Cayenne pepper*.

Conspectus of the Genera.

		rotate, with a	Fruit { sitting on the calyx.		.	.	.	Solanum	5
		very short tube.	baccate, { enclosed in the calyx.		.	.	.	Physalis	6
			Fruit capsular, dry.	.	.	.	.	Capsicum	7
		campanulate.	Sepals lanceolate	.	.	.	.	Atropa	2
			Sepals leafy, sagittate.	.	.	.	.	Nicandra	3
			{ Capsule spinose	.	.	.	.	Datura	8
	regular,	funnel-form. .	Herbs. { Capsule smooth,	.	.	.	.	Nicotiana	2
			Trailing shrubs.	.	.	.	.	Lycium	10
Corolla { irregular.	salver-form, lower segments larger.			.	.	.	.	Petunia	1
	funnel-form, upper segments larger.			.	.	.	.	Hyoscyamus	4

1. PETUNIA. Juss.

The Brazilian name is *petun*, Latinized, *petunia*.

Calyx tube short, the limb 5-cleft, foliaceous; corolla hypocrateriform, the tube cylindric, limb in 5, unequal, flat, plicate lobes; stamens 5, unequal, included, arising from the middle of the corolla tube; capsule 2-valved.—*Herbs with simple lvs. and axillary, solitary, showy fls.*

1. P. VIOLACEA.—*St.* weak, viscid-pilose; *lvs.* acute, on short petioles; *cor.* ventricose, cleft into rounded, acute lobes.—① ② A pretty, trailing or climbing plant, becoming quite popular in cultivation, native of Brazil. Whole plant clothed with clammy hairs. Stems simple, several from the same root, 2—8f long. Leaves 1—2' long, nearly as broad, tapering at base into a winged petiole, fleshy, nearly smooth beneath. Sepals obtuse. Peduncles as long as the leaves, and scarcely longer than the corolla tube. Limb of the corolla bright purple, an inch or more broad, upper segment smallest. Capsule furnished with a tube each side of the sutures.

2. P. ALBA.—*St.* weak, viscid-pilose; *lvs.* ovate, acute, upper ones sessile; *cor. tube* cylindric, scarcely dilated above, 2 or 3 times longer than the obtuse, spatulate sepals, limb flat, spreading, greenish-white.—① ② Brazil. Usually regarded as a variety of the first, and perhaps it may have originated from that species by cultivation. It is usually a stouter plant, with larger leaves and flowers, the latter constantly yellowish or greenish-white, with a long, slender tube.

2. NICOTIÄNA. Tourn.

In honor? of John Nicot, of Languedoc, who seems to have introduced it into Europe.

Calyx urceolate, 5-cleft; corolla infundibuliform, regular, limb 5-lobed; stamens 5; stigma emarginate; capsule 2-celled, 2—4 valved.—① *Coarse, narcotic herbs, with simple lvs. and terminal fls. Cor. white, tinged with green or purple.*

1. N. RUSTICA. *Common Tobacco.*
Viscid-pubescent; *lvs.* petioled, ovate, entire; *tube of the cor.* cylindric, longer than the calyx, segmt, ts round, obtuse.—For the purposes of *tobacco* this plant is considered inferior to the Virginian. Stem 12—18' high. Flowers greenish-yellow, in a terminal panicle or raceme. In western N. Y., &c., said to have been introduced by the Indians. Aug. ⚥

2. N. TABACUM. *Virginian Tobacco.*
Viscid-pubescent; *lvs.* lanceolate, sessile, decurrent; *cor. tube* inflated at the throat, lobes acute.—Native of Central America, particularly the Island of Tobago, and the Province of Tabasco in Mexico, whence it was first exported

to Europe, 1586. It is extensively cultivated in the Middle and Western States, and is exported in vast quantities. Stem 4—6f high, paniculate above. Leaves 1—2f by ½—1f entire. Flowers rose-color, not inelegant. July.

Obs.—Sir Walter Raleigh has the honor of first introducing the practice of smoking into England, more than 280 years ago, and in his house at Islington is still to be seen a shield bearing his arms, with a tobacco plant at the top. Loudon. The use of this nauseous weed has become almost universal, and furnishes a striking illustration of the force of habit. Its first use, whether smoked or chewed, produces a deadly sickness; and it is only by repeated and painful trials that it can be tolerated. At length, however, it becomes so necessary to the comfort of its victim, that, at all times and places, its precious smoke or extract must be flowing continually from his mouth. Taken into the stomach, it is a powerful narcotic poison. July.

3. DATÚRA.
An alteration of the Arabic name *Tatorah.*

Calyx large. tubular, ventricose. 5-angled, deciduous, with a persistent, orbicular, peltate base; corolla infundibuliform, tube cylindric, long, limb 5-angled and plaited; stamens 5; stigma obtuse, bilamellate; capsule 2-celled, 4-valved; cells 2—3-parted.—① *herbs, with bluish-white or purple, solitary, axillary flowers.*

1. D. STRAMONIUM. *Thorn Apple.*
St. dichotomous; *lrs.* ovate, smooth, angular-dentate; *caps.* spiny, erect. A well-known poisonous plant, growing among rubbish in waste places. Stem about 3f high, smooth, hollow. Leaves large, situated at the base of the dichotomous branches, their sides unequal, with large, irregular teeth and sinuses. Flowers solitary, axillary; corolla funnel-shaped, with a long tube and a plaited, 5-toothed border, the color white with a slight tinge of purple. Fruit egg-shaped, the size of a small apple, covered with spines. Aug.—Every part is poisonous, but, when used with certain restrictions, is a useful medicine for asthma, &c. ⚘
β. *Tatula. St.* and *fls.* purple.—This variety has advanced along the national road to Ia., *Plummer !* and Ill., *Mead.*

2. D. METEL.
Lvs. cordate, nearly entire, pubescent; *fr.* prickly, globose, nodding. · Banks of the Ohio, *Locke.* Doubtless introduced, having escaped from gardens. Plant 2f high. Flowers white. ⚘ †

4. HYOSCYÄMUS. Tourn.
Gr. ὑς, ὑος, a pig. and κυαμος, bean; the fruit is said to be not poisonous to swine.

Calyx tubular, 5-cleft; corolla infundibuliform, irregular; one of the 5, obtuse lobes larger; stamens 5, declinate; stigma capitate; capsule ovoid, 2-celled, opening with a lid near the summit.—*Coarse, weed-like herbs, native in eastern countries.*

H. NIGER. *Common Henbane.*
St. branching, erect, very leafy; *lrs.* sinuate, clasping; *fls.* sessile.—⊛ A tall, well known, fœtid weed, growing about the rubbish of old houses, roadsides, &c. The whole plant is hairy, viscid, and of a sea-green hue, emitting a fœtid odor. Stem 2f high, round. Leaves large, oblong, cut into acute, sinuate lobes. Flowers in terminal, one-sided spikes; the corolla straw-color, finely reticulated with dark purple veins. The whole plant is reputed poisonous, but has long been regarded as an excellent medicine in nervous diseases, coughs, convulsions, &c. Jl. ⚘

5. NICANDRA. Adans.
In honor of Nicander, a Greek physician, who lived about 50 years B. C.

Calyx 5-cleft, 5-angled, the angles compressed, sepals sagittate; corolla campanulate; stamens 5, incurved; berry 3—5 celled, enveloped in the persistent calyx.—① *Peruvian herbs.*

N. PHYSALÖIDES. Adans. (Atropa physaloides. *Linn.*) *Apple of Peru.*—*St.* herbaceous; *lrs.* glabrous, sinuate, angular; *fls.* solitary, axillary, on short peduncles; *cal.* closed with the angles very acute.—Native of Peru, cultivated in gardens, from whence it has in a few instances strayed into the neighboring fields. It is a large, coarse herb, 2—5' high, very branching. Leaves large, oblong, decurrent. Corolla slightly lobed, pale blue, white and with 5 blue spots in the centre. July—Sept. §

6. PHYSÄLIS.

Gr. φυσις, a bladder; the inflated calyx enclosing the fruit.

Calyx 5-cleft, persistent, at length ventricose; corolla campanulate-rotate, tube very short, limb obscurely 5-lobed; stamens 5, connivent; berry globose, enclosed within the inflated, 5-angled, colored calyx.—*Herbs, rarely shrubs, with axillary or supra-axillary flowers.*

1. P. viscósa. Aikin. (P. viscosa, obscura, pubescens, Pennsylvanica and Philadelphica, *of authors.*) *Yellow Henbane. Ground Cherry.*—Pubescent; *st.* decumbent, herbaceous; *branches* somewhat dichotomous and angular; *lvs.* solitary or in pairs, ovate, more or less cordate, repand-toothed or entire; *fls.* solitary, axillary, pendulous.—Dry fields, roadsides, &c. Stem more or less decumbent, about a foot high, often viscid as well as the whole plant. Leaves very variable in the same plant, 1—4' long, of ½, ⅓, or even of equal breadth, acute, acuminate, or often obtuse at the apex, often abrupt at base, sometimes nearly or quite entire on the margin, twice as long as the petioles; when in pairs one of them is much smaller. Corolla twice as long as the calyx, greenish-yellow, with 5 brownish spots at base inside. Fruit yellow or orange-colored, not unpleasant to the taste, enclosed in the enlarged, inflated, angular calyx. Jl.

α. *Lvs.* somewhat viscid, oval, subcordate, geminate.
β. (P. Pennsylvanica. *Linn.*) *Lvs.* ovate and lance-ovate, subentire, nearly smooth, geminate.
γ. (P. obscura. *Michx.*) *Lvs.* pubescent, broad-ovate, subcordate, subsolitary.

Obs.—Many other varieties have been noticed as species, but having examined specimens in numerous localities, I am but confirmed in concurring with Dr. Aikin in the above view.

2. P. lanceoláta. Michx. *Lance-leaved Physalis.*
St. herbaceous, dichotomously branched, densely pubescent; *lvs.* mostly in pairs, ovate-lanceolate, acuminate, entire, unequal at base; *fls.* solitary, nodding; *cal.* villose.—4 Penn., Western States, S. to Ga. Stem 1—2f high. Leaves 3—6' by 1½—3', often very unequal at base. Flowers nodding. Calyx half-cleft, with lanceolate, acuminate segments. Corolla pale greenish-yellow, with dark spots at base. Jl.- *Darl.* Fl. Cest., p. 139.—I strongly suspect this to be only another variety of the preceding.

P. Alkekengi. *Winter Cherry.—St.* somewhat branching below; *lvs.* in pairs, entire, acute; *cal.* of the fruit red or reddish.—Native of S. Europe, cultivated for ornament. Plant about a foot high. Flowers white. Berries acid and somewhat bitter. †

7. CAPSÏCUM. Tourn.

Gr. καπτω, to bite; from the acridity of the fruit.

Calyx erect, 5-cleft, persistent; cor. rotate, tube very short, limb plaited, 5-lobed; anth. connivent; fr. capsular, dry, inflated, 2—3-celled; seeds flat, very acrid.—*A large genus of herbaceous or shrubby plants, pervaded by a heating, acrid principle. Lvs. often in pairs. Ped. axillary, solitary.*

1. C. annuum. *Red Pepper. Cayenne Pepper.—St.* herbaceous, angular, branching above; *lvs.* ovate, acuminate, entire, petiolate, glabrous; *ped.* smooth, axillary; *cal.* angular, with short, acute lobes; *cor. lobes* spreading, longer than the stamens; *berry* oblong or subglobose, red.—① India. Cultivated for its fruit, whose stimulant properties are well known.—There are in gardens several varieties in respect to the fruit. *a.* The long, or Cayenne, *β.* the depressed-globose or squash pepper, best for pickling, *γ.* the cherry pepper, used for pepper-sauce and in seasoning meat, *δ.* the sweet Spanish pepper, used as a salad.—Sown in March in hot-beds, transplanted in May. *Kenrick,* Am. Orch., p. 374.

8. SOLANUM.

Calyx 5—10 parted, persistent; cor. rotate, subcampanulate, tube very short, limb plicate, 5—10-lobed; anth. erect, slightly cohering or connivent, opening at the top by 2 pores; berry 2—6-celled, subglobose or depressed, often torose; seeds 00.—*Herbs or shrubs, unarmed or prickly. Lvs. sometimes geminate, pinnatifid or undivided. Ped. solitary or several, 1—00-flowered.*

§ 1. *Berry 2-celled. Stem and leaves unarmed.*

1. S. DULCAMARA. *Bittersweet. Woody Nightshade.*
St. shrubby, flexuous, thornless; lvs. ovate-cordate, upper ones hastate; clusters cymose.—A well-known, shrubby climber, with blue flowers and red berries, N. Eng. to Ark. Stem branching, several feet in length, climbing about hedges and thickets in low grounds. Lower leaves entire; the upper ones becoming auriculate or hastate. Flowers drooping, on branching peduncles from the side of the stem. Corolla of 5 reflexed segments, purple, with 2 green spots at the base of each segment. Berries bright red. The root being chewed, gives at first a sensation of bitterness, then of sweetness. The berries are poisonous. The leaves and twigs have been used medicinally with good effect. July.

2. S. NIGRUM. β. *Virginicum. Black Nightshade.*
St. herbaceous, thornless; lvs. ovate, toothed and waved; umbels lateral, drooping.—② A weed-like plant without beauty and of suspicious aspect, about rubbish, in old fields, N. and W. States. Stem erect, branching, angular, a foot high. Leaves almost always with the lamina perforated and the margin erose as if gnawed by insects. Peduncles branching into a sort of umbel, from the side of the stem, generally remote from the leaves. Flowers white, anthers yellow. Berries globose, black. It is reputed poisonous, but is used medicinally. Flowers in summer.

3. S. TUBEROSUM. *Common Potato.*—Rt. tuberous; st. herbaceous; segments of the lvs. unequal, the alternate ones minute; fls. subcorymbed; cor. 5-angled.— ② This most valuable plant is supposed to be a native of S. America, where it still grows wild. Although it now constitutes so large a portion of the food of civilized man, it was scarcely known until the 17th century, and was not extensively cultivated before the middle of the 18th. The varieties of the potato are very numerous, differing in their time of ripening, quality, color, form, size, &c. New varieties are readily procured by sowing the seeds, which, with care, will produce good tubers the third year. Potatoes thus reared, are now thought to be less liable to the "potato rot." ‡

4. S. PSEUDO-CAPSICUM. *Jerusalem Cherry.*—St. shrubby; lvs. oblong-lanceolate, subrepand; ped. 1-flowered, opposite the leaves.—♄ A small, ornamental shrub, native of Madeira, cultivated. Stem 2—4f high, branching into a symmetrical summit. Leaves dark evergreen, smooth and shining, about 2' long. Flowers white, with orange anthers, drooping, succeeded by a few scarlet, globose berries of the size of small cherries. †

§ 2. *Berry 2-celled. Stems and leaves prickly.*

5. S. CAROLINENSE. *Horse Nettle.*
St. and petioles aculeate; lvs. oblong-ovate, petiolate, strigose, angular-lobate, acute, midvein beneath with a few spines; rac. loose, supra-axillary, few-flowered; berries globose.—♃ Roadsides, &c., Penn. to Car.! W. to Ia. and Ill.! A rough weed, 1—2f high, armed with straw-colored, scattered prickles. Leaves 4—6' by 2—3', usually in unequal pairs with a few large, repand lobes or teeth. Flowers white, lateral and terminal. Corolla white, 12—15" diam. Berries yellowish. June.

6. S. MELONGENA. (S. insanum. L.) *Egg Plant.*—St. prickly; lvs. ovate, subsinuate, downy, prickly; fls. many-parted.—① An herbaceous, branching plant, about 2f high. The fruit, with which it is heavily laden, consists of egg-shaped berries, from the size of an egg to that of an ordinary water melon,

smooth, and of a glossy purple. It is prepared for food in various ways, and considered wholesome and delicious eating. Like the tomato, it is cultivated from the seed sown early in warm, dry and mellow soil. ‡

β. *Fr.* smaller, white.—Cultivated for the curiosity of the fruit, which when ripe can scarcely be distinguished by its appearance from a hen's egg.

§ 3. *Berries 3—6-celled, often torose.*

7. S. LYCOPERSICUM. *Tomato.*—Hairy; *st.* herbaceous, weak; *lvs.* unequally pinnatifid, segments cut, glaucous beneath; *fr.* torulose, furrowed, smooth.— ① This plant resembles the potato in its general aspect. It grows 3—4f high, with jagged leaves, greenish-yellow flowers, and an unpleasant odor. The fruit is large and abundant, with acute furrows, at first green, becoming when ripe of a beautiful red. This plant has come into high repute, and its cultivation is rapidly extending. The fruit is prepared in various ways, for sauces, stews, &c., having an agreeable acid taste. ‡

Obs.—Cultivation has produced numerous varieties. One has large, torulose, bright-yellow fruit; another has small, globose, golden-yellow fruit, not torulose; the fruit of a third is small, pear-shaped, less juicy, &c.

9. ATRÖPA.

Name of one of the three Fates in Grecian mythology, whose office was to cut the thread of human life.

Calyx persistent, 5-cleft; corolla campanulate; stamens 5, distant; berry globose, 2-cellod, sitting on the calyx.—*Herbs, shrubs or trees, natives of the Old World.*

A. BELLADONNA. *Deadly Nightshade.*—*St.* herbaceous; *lvs.* ovate, entire; berries black.—This foreigner is far less repulsive in its appearance than most others of its order. The lurid, pale purple of the flower, indeed, looks suspicious, but not its smell; nor is there any warning of its deadly nature given by the aspect, taste or smell of the berries, which are larger than cherries, round, green, at length of a fine glossy black, full of a purple juice. Stem 5f high, branching below, and with the large leaves, inclines more or less to a purplish hue. Every part of the plant, especially the berries, is poisonous. †

10. LYCIUM.

Named from Lycia, the native country of the original species.

Calyx 2—5-cleft, short; corolla tubular, limb mostly 5-lobed, spreading, orifice closed by the beard of the filaments; stamens 4— 5, exserted; berry 2-celled; seeds several, reniform.—*Shrubs, the branches ending in a spinose point, and often having axillary spines. Fls. axillary, solitary, or in pairs.*

L. BARBARUM. *Matrimony Vine.*—*St.* angular; *branches* long, pendulous, somewhat spiny; *lvs.* often fasciculate, lanceolate; *cal.* mostly 3-cleft.—Native of Barbary, cultivated and nearly naturalized. It is a shrub, with long, slender, trailing or hanging branches which overspread walls, &c., with a thick, tangled mass. Leaves smooth, 3 times as long as wide, often broadest above, acute or obtuse, tapering into a petiole. Flowers greenish-purple. Berries orange-red. †

ORDER C. GENTIANACEÆ

Plants herbaceous, rarely shrubby, generally smooth, sometimes twining. *Juice colorless.*
Lvs. opposite, entire and smooth. *Stip.* 0.
Fls. conspicuous, terminal or axillary, regular or sometimes irregular.
Cal.—Sepals 4—5—10, united at base, persistent.
Cor. usually regular, limb divided into as many lobes as there are sepals, mostly twisted a æstivation.

Sta. issuing from the tube of the corolla as many as its lobes and alternate with them.
Ova. 1-celled, sometimes rendered apparently 2-celled by the introflexed placentæ.
Sty. united into 1, or wanting. *Stig.* 1—2.
Fr. Capsule many-seeded. *Seeds* small. *Embryo* straight, with fleshy albumen.

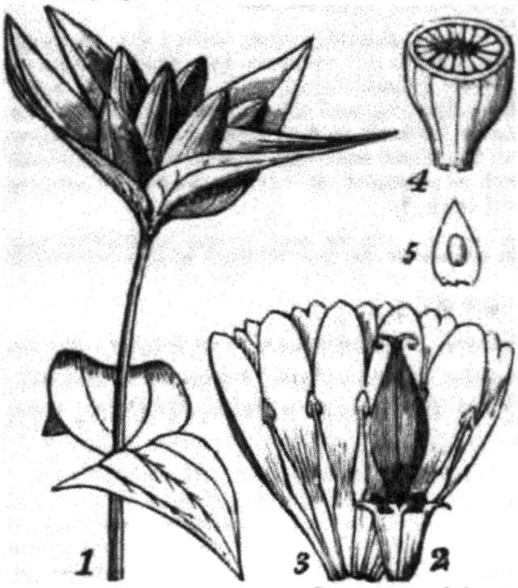

Genera 60, species 450, found in every part of the world.

Properties.—An intensely bitter principle, called *gentianine*, pervades the whole order without exception, residing in every part, rendering them tonic and febrifugal. The *gentian* of the shops is most commonly the product of Gentiana lutea, but almost any of our species may be substituted for it. (Dr. Gray.) In the other genera of the order, the buck-bean (Menyanthes trifoliata), Limnanthemum nymphoides, Sabbatia angularis, Frasera Caroliniensis, &c., are valued in medicine for the same properties. Many are cultivated for ornament.

FIG. 53.—1. Gentiana Saponaria. 2. The calyx and capsule. 3. The corolla laid open, showing the folds (2-lobed) between the proper petals, and the stamens attached at base. 4. Capsule cut across. 5. Seed magnified, with its large, loose testa.

Conspectus of the Genera.

			Petals 5, or 4 and fringed. .	*Gentiana.*	2
		Anthers { Pet. 4. { Sep. 4, subulate. .	*Cicendia.*	3
		straight. { entire. { Sep. 2, leaf-like. .	*Obolaria.*	4
	tubular. { Anthers spirally twisted. . . .	*Erythræa.*	1
Corolla with- {	rotate. { Petals with no gland. . . .	*Sabbatia.*	1
out horns... {	{ Petals with a glandular pit. .	*Swertia.*	5
opposite. { Corolla without 4 horns at base.	*Halenia.*	7
verticillate in whorls of 4s, 5s and 6s.	*Frasera.*	8
none, or reduced to small opposite scales. . . .	*Centaurella.*	6
{ simple, floating in water. . . .	*Limnanthemum.*	10
Leaves { alternate and radical, { trifoliate.	*Menyanthes.*	11

TRIBE 1. Corolla imbricate from right to left in æstivation. Testa of the seed membranaceous. Terrestrial herbs with opposite leaves.

1. SABBATIA. Adans.
In honor of Sabbati, a distinguished Italian botanical author.

Calyx 5—12-parted; corolla rotate, limb 5—12-parted; stamens 5(—12); anthers erect, at length recurved, 2-celled, cells distinct; stigma 2-parted, with spiral divisions; capsule 1-celled, the valves a little introflexed.—② *Slender herbs, with pedicellate, mostly roseate fls. Native of the temperate regions of North America.*

1. S. GRACILIS. Salisb. (S. campanulata. *Torr.* Chironia. *Linn.*)
St. slightly angular, internodes twice longer than the leaves; *branches* alternate, spreading; *lvs.* linear and lance-linear, the lowest lance-ovate; *panicle* few-flowered; *cal.* segments linear-setaceous, about equaling the corolla; *cor.* 5-parted, lobes elliptic-oblong, obtuse.—Wet meadows, Penn. to Flor., W. to Ky. Stem a foot high, with long, diverging branches. Flowers terminal, subsolitary, purple, on long peduncles. July, Aug.

2. S. CORYMBOSA. Baldwin. (S. paniculata. *s. Ph.* Chironia. *Walter.* Swertia. *Linn.*)—*St.* slightly 4-angled, internodes twice longer than the leaves; *branches* opposite; *lvs.* ovate-lanceolate, 3-veined, acutish, upper ones

anceolate; *cyme* fastigiate, terminal; *sep.* linear, 3 times shorter than the corolla; *cor.* 5—6-parted, white, lobes obovate-oblong, obtuse.—Pine barrens, N. J. to Ga. Stem a foot high, branching near the summit. Leaves an inch in length, closely sessile. Flowers few, generally pentamerous.

3. S. CONCINNA. Wood. (*Nov. sp.*) *Elegant Star Flower.*

St. slender, subquadrangular, internodes 2—4 times longer than the leaves: *branches* opposite, suberect; *lvs.* linear and lance-linear, lower ones ovate, all acutish, sessile; *panicle* oblong; *cal.* segments linear, twice longer than the tube, twice shorter than the corolla; *cor.* 5-parted, segments oblong-obovate, obtuse, light purple.—Dry, grassy prairies, Ia.! abundant. Stem a foot high, few or many-flowered. Leaves 9—12″ by 1—3″, Flowers 15″ diam., of a delicate blush-purple, the star in the centre yellow, bordered with green. Jl. Aug.

4. S. STELLARIS. Ph. (S. gracilis. *Ell.* Chironia amœna. *Raf.*)

St. erect, terete; *branches* dichotomous, elongated, 1-flowered; *lvs.* lanceolate, acute, *seg. of the cal.* subulate, half as long as the corolla; *seg. of the cor.* obovate.—Frequent in salt marshes, N. Y. to Flor. Stems somewhat angular, 12—18′ high, with many forked divisions, forming a sort of loose corymb. Leaves somewhat fleshy, 1—2′ long, sessile. Flowers rose-color, with a yellow star in the centre bordered with a purple ring. Aug.

5. S. CALYCÒSA. Ph. (Chironia dichotoma. *Walter.*)

St. erect, leafy, few-flowered; *lvs.* oblong, 3-veined, obtuse; *fls.* solitary, 7—9-parted; *cal.* leafy, longer than the corolla; *pet.* oblanceolate.—Fields and meadows, N. Y. to Ga. Stem a foot high, subangular, with a few axillary, spreading branches. Leaves 1—2′ long, sessile, mostly obtuse, oval, thin. Flowers large, terminal, often solitary, variable in the number of its parts, but mostly in 7s. Corolla pink-colored. Sepals acute. This species is quite variable.

6. S. ANGULÀRIS. Ph. (Chironia. *Linn.*) *Angular-stemmed Star Fl.*

St. quadrangular, with winged angles; *lvs.* ovate, amplexicaul, 5-veined; *panicle* corymbose; *ped.* elongated; *sep.* lance-linear, half as long as the corolla, distinct almost to the base; *cor.* segments obovate, obtuse.—Wet meadows and prairies, Can. to Car. and Ark. Stem 10—18′ high, much branched, branches opposite. Leaves closely embracing the stem, 1—2′ by ½—1½′, as long as the internodes or often shorter. Flowers numerous, 1½—1½′ diam., deep rose color, the star in the centre greenish. July, Aug.

7. S. CHLOROÏDES. Ph. (Chironia dodecandra. *Linn.*)

St. slender, weak, angular; *lvs.* lanceolate, erect; *branches* few, 1-flowered; *fls.* 7—12-parted; *sep.* linear, shorter than the corolla.—An elegant plant, with large, showy flowers, in wet grounds, Mass., R. I.! to Car. The stem is 2—3f high, somewhat angular, with few, opposite, spreading branches. Leaves 1—1½′ long, opposite, entire, smooth, closely sessile, acute, veinless. Flowers solitary, terminal. Corolla nearly 2′ diam., much larger than the calyx, bright purple, with a yellow base, segments spatulate, rounded at end, varying in number with the other parts of the flower. June.

Obs. The species of this genus are very ornamental, some of them perhaps among the most beautiful of our native plants.

2. E R Y T H R Æ A. Renealm.

Gr. ερυθρος red; from the color of the flowers.

Calyx 5, rarely 4-parted; cor. infundibuliform, twisted and withering above the capsule, tube cylindric, limb 5—4-parted; sta. 5—4, inserted near the top of the tube; anth. exserted, spirally twisted; sty. 1; stig. bilamellate or capitate; caps. 2-valved, 1 or partly 2-celled.—① *St. subangular. Lrs. connate at base. Fls. cymose, roseate, white or yellow.*

1. E. MUHLENBERGII. Griseb. (E. Centaurium. *Beck?* E. pulchella. Hook. Chironia dubia. *Willd.*)—*St.* simple below, dichotomously branched above; *lvs.* ovate-oblong, obtusish; *cymes* loose, dichotomous; *fls.* pedicellate; *cor.* tube a little longer than the calyx, segments oblong-lanceolate, acutish.—N. Y., Penn. Very rare. Stem 3—8′ high, 1—3 times forked, sometimes with oppo

site or brachiate branches. Leaves 4—7'' by 1—3'', closely sessile. Flowers lateral and terminal and central, the pedicels in the forks near ½' long, the others shorter. Corolla bright purple, tube yellowish-green, slender, persistent and withering on the capsule. July—Sept.

2. E. PICKERINGII. Oakes. *Pickering's Erythræa.*
St. dichotomously branched, erect; *lvs.* clasping and slightly decurrent, lower ones oval, obtuse, upper lanceolate, acute; *fls.* sessile, mostly lateral on the long branches; *sep.* linear, acute, erect; *cor.* tube slender, contracted at the neck, lobes spreading, obtuse; *anth.* linear-oblong, finally twisting outwards.— ① Coast of Maryland. *Dr. Pickering.* Sandy margins of the sea-shore, Nantucket, *Mr. Oakes.* Whole plant very smooth and intensely bitter, 6—12' high. Leaves 1' long, fleshy, pale green. Corolla 8'' long, rose or nearly white.

3. CICENDIA. Adans.

Calyx 4—5-parted; cor. infundibuliform, marescent, tube cylindric, limb 4—5-parted; sta. 4—5, inserted into the throat of the corolla; anth. erect, roundish, scarcely exserted; sty. deciduous; stig. capitate; caps. 1 or partly 2-celled, 2-valved.—① *Low herbs, with yellow or rose-colored flowers.*

C.? PULCHELLA. Griseb. (Exacum. *Ph.*)
Lower lvs. suborbicular, *upper* subulate; *panicle* corymbose; *ped.* filiform; *cal.* 4-parted, segments subulate.—Sea coast, N. J. This plant appears not to have been detected by any botanist since Pursh, and from his brief description it is impossible to ascertain its true position in the order.

4. OBOLARIA.

Gr. οβολος, a small coin, with which the leaves of these plants are compared.

Calyx of 2 cuneate-oblong sepals or bracts; corolla tubular-campanulate, marescent, 4-cleft, lobes entire or crenulate; stam. inserted on the corolla at the clefts; stigma subcapitate, bifid; capsule 1-celled, 2-valved; seeds 00, very minute.—♃ *Leaves opposite. Fls axillary and terminal, sessile, with leaf-like sepals.*

O. VIRGINICA. *Penny-wort.*
Penn. to Car., W. to Ky., in woods. Stem 4—8' high, often in clusters, subsimple or with a few opposite branches above. Leaves cuneate-obovate or roundish-rhomboidal, sessile and decurrent at base, fleshy, obtuse or truncate at apex, lower ones small and remote, upper crowded, glaucous-purple, sepals or bracts similar. Corolla pale purple or whitish, longer than the stamens. Capsule ovoid, obtuse, surrounded by the withered corolla. Apr., May.—There is some doubt in respect to the order of this genus. It has more recently been referred to Orobanchaceæ.

5. GENTIANA. Tourn.

From Gentius, king of Illyria, who discovered the tonic virtues of this genus.

Calyx 5—4-parted or cleft; cor. marescent, tubular at base, limb 4—5-parted, segments either spreading, erect or convergent, often furnished with intermediate, plicate folds; sta. 5—4, inserted in the corolla tube; stig. 2, revolute or erect; sty. short or 0; caps. 2-valved, 1-celled, many-seeded.—*Herbs of various habit. Lvs. opposite. Fls. terminal or cymose.*

§ 1. *Flowers 5—10-merous.*

1. G. PNEUMONANTHE. (G. saponaria. *Griseb.*)
St. ascending; *lvs.* linear-lanceolate, or the lower oblanceolate, the margin smooth! *cyme* terminal; *fls.* aggregated, or in a racemose cyme, sessile; *cal.* 5-cleft, the lobes linear, equaling the tube; *cor.* clavate, connivent at apex, *twice longer* than the calyx, lobes ovate-obtuse, much longer than the interme-

diate folds.—Can. and U. S., rare in N. Eng. A fine plant, with large, showy flowers. Stem simple, 8—15' high, often purple. Leaves 1—2' by 2—3'', thick, entire, sessile, acute. Flowers 2' in length, 2—4 together at top of the stem and a few solitary ones in the axils of the upper leaves. Corolla blue. Calyx segments 6—9'' long, acute, distant. Aug., Sept.

β. *rubricaulis.* (G. rubricaulis. *Schw.*) *Lvs.* ovate-lanceolate; *folds of the cor.* more or less cleft.

2. G. SAPONARIA. Linn. *in part.* (G. Andrewsii. *Griseb.*) *Soapwort Gentian.*—*Lvs.* oval-lanceolate, 3-veined, acute; *fls.* in whorled heads, sessile; *cor.* ventricose, clavate-campanulate, closed at top, 10-cleft, the inner segments plicate and fringed, equaling the exterior.—4 Brit. Am. to Car. A handsome plant, conspicuous in meadows and by brook-sides. Stem 12—18' high, simple, erect, smooth, with opposite, smooth leaves, scabrous on the margin, resembling those of the common soapwort. Flowers large, bright-blue, erect, 1½' long, subsessile, in bunches at the top of the stem, and often solitary in the upper axils. The inflated corollas are so nearly closed at the top as to be easily mistaken for buds; and the young botanist waits in vain to see them expand. Calyx of 5 ovate segments, shorter than the tube. Sept., Oct.

3. G. OCHROLEUCA. Frœl. *Ochroleucous* or *Straw-colored Gentian.*
St. ascending; *lvs.* ovate-lanceolate or lanceolate, margins slightly scabrous; *cymes* terminal, aggregated; *cal.* 5-cleft, lobes unequal, as long as the tube; *cor.* clavate, apex connivent or slightly expanding, lobes ovate, obtuse, the folds entire, acute, short; *anth.* free.—Can., Western States! to Flor. Stems 1—1½' high, stout. Leaves amplexicaul or sessile, 2—4' by ½—1½', acute, or slightly acuminate. Flowers 2' in length, ½' thick. Corolla open at top, ochroleucous or straw-color. Seeds smooth, wingless. Aug., Sept.

4. G. ANGUSTIFOLIA. Michx. (G. purpurea. *Walt.*) *Narrow-leaved Gentian.*—*St.* erect, slender, 1-flowered; *fl.* pedunculate; *lvs.* linear-obtuse, smooth, the lower ones subcuneate; *cor.* funnel-form, narrow, open, 5-cleft, twice as long as the calyx, lobes ovate-oblong, obtuse, twice as long as the lacerate folds.—4 N. J. to Car., in sandy fields. Stem a foot high. Flower large, sky-blue. Calyx deeply cleft, with linear segments.

5. G. QUINQUEFLORA. *Fire-flowered Gentian.*
St. 4-angled, branching; *lvs.* ovate-lanceolate, acute, 3-veined; *fls.* terminal and axillary, about in 5s, pedicellate; *cor.* tubular-campanulate, in 5, lanceolate, setaceously acuminate segments; *cal.* very short.—② Woods and pastures. Stem a foot high, smooth, generally branched. Leaves 3—5-veined, half-clasping, acute, smooth. Flowers small, on pedicels half an inch in length. Corolla pale blue, 4 times as long as the subulate sepals. Sept., Oct.

β. *parviflora.* Raf. *Cal.* enlarged, lobes foliaceous, lance-linear, half as long as the smallish corolla.—This variety prevails in Ohio! Ia.! and Ky.

6. G. LUTEA. (Swertia. *Vent.*) *Yellow Gentian.*
St. tall, strict; *lvs.* oval and ovate, margin smooth; *cymes* umbellate, dense-flowered, pedunculate, axillary and terminal; *cor.* yellow, rotate, segments oblong-linear, acuminate, spreading, without folds.—4 In alpine and hilly lands, N. Y. and N. J. *Eaton.* Doubtless not native. It is a handsome plant, often cultivated, both for ornament and for the sake of its powerfully tonic virtues. † ‡

§ 2. *Corolla 4-cleft, segments fimbriate.* CROSSOPETALUM.

7. G. CRINITA. Frœlich. *Blue Fringed Gentian.*
St. terete, erect; *lvs.* lanceolate, acute; *fls.* tetramerous; *seg. of the cor.* cut-ciliate.—4 The Fringed Gentian is among our most beautiful and interesting native plants; not uncommon in cool, low grounds, Can. to Car. The stem is about 1f high, round and smooth. The branches are 'ong, and, with a slight curve at base, become perfectly erect and straight, each bearing 2 leaves at the middle, and a single, large, erect flower at the top. Leaves broadest at base, tapering to the apex, 1—2' long and ½ as wide. Calyx square, segmens acuminate, equaling the tube of the corolla. Cor. of a bright bluish-purple, the segments obovate, finely fringed at the margin, and expanded in the sunshine. Aug.

8. G. detonsa. *Shorn or Slightly-fringed Gentian.*

St. nearly strict, simple or branched; *lvs.* oblong-lanceolate or linear, the lowest rosulate, spatulate; *ped.* 1-flowered, very long, subsolitary; *cal.* 4(—5)-cleft, lobes ovate and lanceolate, nearly equaling the corolla; *cor.* lobes round-ish-obtuse, ciliate and crenate-ciliate, erect-spreading; stig. distinct.—N. Y., Wis. *Lapham!* N. to Hudson's Bay. A fine species, with large blue flowers. Stem a foot high. Leaves 1—2½' by 1—3", tapering to an acute point. Peduncles 4—7' long, each with a single large, erect, showy flower. Cor. 15—20" long, obconic or bell-shaped, blue.

6. CENTAURELLA.
Name a Latin diminutive of Centaurea.

Flowers tetramerous; sepals appressed; corolla subcampanulate; petals slightly united, nearly erect, stigma thick, glandulous, somewhat bifid; capsule 1-celled, 2-valved, invested by the permanent calyx and corolla; seeds many.—①N. *American, slender, erect herbs with scale-like leaves and small white flowers.*

1. C. autumnalis. Pursh. (Bartonia tenella. *Muhl.* C. paniculata. *Mx.*) *Screw-stem.*—*St.* smooth, branching above, branches subdivided; *lvs.* subulate, minute; *panicle* erect, many-flowered; *cal.* equaling the corolla; *sty.* much shorter than the ovary.—A slender and nearly naked plant, 5—8' high, of a yellowish-green color. Native in wet grounds, Mass. to Ga. Stem square, often twisted, with very minute, bract-like leaves, which are mostly opposite. Peduncles opposite or terminal, simple or branched. Pedicels bracteate at base, ½' in length. Calyx segments linear-lanceolate, acute, shorter than the oval segments of the corolla. Cor. white, small, bearing the stamens at its clefts. Aug.
β. *æstivalis.* (C. æstivalis. *Ph.*) *Cor.* segments oblong, obtusish, twice longer than the calyx; *ova.* attenuated at apex.—In Penn. *Pursh.*

2. C. Moséri. Stendel. *Moser's Centaurella or Screw-stem.*
Cyme racemiform; *branches* alternate, spreading, racemiform; *cor.* segments ovate, acute, twice longer than the calyx segments; *ovary* oblong, obtuse; *sty.* 0; *lvs.* mostly alternate.—In Penn., *Moser*, in DC. Prod. 9; 121.

7. HALENIA. Borkh.

Flowers tetramerous; corolla short campanulate, petals spurred at base, with glands at the base of the spur within; stigmas 2, terminating the acuminate ovary; capsule 1-celled; seeds indefinite, fixed to the sutures of the valves.—*Erect, branching.*

H. deflexa. Griseb. (Swertia corniculata. *Mx.* S. deflexa. *Sm.*) *Fel-wort.*—*St.* erect, leafy; *lvs.* 3—5-veined, radical ones oblong-spatulate, tapering into a petiole, cauline ones oblong-lanceolate, acute, sessile; *spurs* cylindric, obtuse, deflexed, half as long as the corolla.—② Swamps, Can., Bangor, Maine, *Miss Towle!* N. Y., rare. Stem about 18' high, obtusely 4-angled, smooth, with few branches above. Leaves opposite, 1½—2' long, ⅓ as wide, smooth. Flowers greenish-yellow, in terminal fascicles. Sepals linear-lanceolate, half as long as the petals. Corolla persistent, with 4, spreading horns or spurs descending betwen the sepals. Seeds numerous, obtuse, yellow. Aug

8. SWERTIA.

Calyx 5—4-parted; cor. marescent, rotate, 5—4-parted, without crown or folds, each segment with a glandular pit fringed at the margin; sta. 5—4, inserted in the throat; anth. nodding; ova. 1-celled; sty. 0; caps. 1-celled, many-seeded.—⚴

S. pusilla. Ph.
St. simple, 1-flowered; *lvs.* oblong; *cal.* segments obtuse, half as long as the large corolla; *cor.* rotate, segments oblong-acuminate.—White Hills, N. H., *Pursh.* Stem an inch or two high, with one or two pairs of small leaves and a

single large, terminal, blue flower. Jn.—I understand that this interesting plant
has been rediscovered on the White Mts. by *Mr. Abel Storrs.*

9. FRASÈRA. Walter.

In honor of John Fraser, an American cultivator of exotics.

Flowers mostly tetramerous; petals united at base, oval, spreading, deciduous, each with 1 or 2 bearded, orbicular glands in the middle; style 1; stigmas 2, distinct; capsule compressed, 1-celled; seeds few, imbricate, large, elliptic, margined.—⁴ *Showy and tall, with opposite or verticillate leaves.*

F. CAROLINENSIS. Walt. (F. Walteri. *Mx.* and *1st edit.*) *Columbo.*
St. tall, erect, glabrous, branched above; *lvs.* verticillate, oblong-lanceolate, acutish, sessile, feather-veined, entire or wavy; *panicle* compound, pyramidal, leafy, verticillate; *cal.* segments acute, shorter than the oblong, obtusish petals; gland solitary, oval-orbicular.—A tall and showy plant, in moist woods, Western N. Y. to Car. W. to Ohio! Ia.! Ky., Ill. Stem dark purple, 4—7—9f! high, perfectly straight, 1—2' thick at base. Leaves smooth, subcarnose, 3—12' by 1—3', in whorls of 4—6, rarely opposite. Branches of the panicle also whorled, spreading-erect. Flowers tetramerous, 1½' diam. Petals greenish with blue dots and a large purple gland near the base. June, July.—Highly valued as a tonic.

TRIBE 2. **MENYANTHEÆ.**—Æstivation of the corolla induplicate. Aquatic or marsh herbs. Sheaths of the leaves alternate.

10. LIMNANTHÈMUM. Gmel.

Gr. λιμνη, a lake, ανθος, a flower; from its aquatic habitat.

Calyx 5-parted; corolla subcampanulate, with a short tube and spreading, 5-lobed limb, deciduous, segments obtuse; stamens 5, alternating with 5 glands; capsule 1-celled, without valves; placentæ fleshy, many-seeded.—⁴ *submersed, generally in stagnant water. Lvs. floating, on long petioles.*

L. LACUNÒSA. Gmel. (Villarsia. *Ph.* V. trachysperma. *Mx.*) *Lake-flower.*
Floating; *lrs.* reniform, subpeltate, scabrous above, spongy and lacunose beneath; *fs.* umbellate, from the summit of the stem (petiole?); *cor.* smooth; *glands* from the base of the petals, stipitate; *cal.* shorter than the capsule; *seeds* muricate.—A curious aquatic, in ponds and lakes, N. Y.! to Car. The stems are 1, 2 or 3f long, according to the depth of the water, bearing at the top three kinds of organs; the summit is prolonged into a petiole bearing a leaf about an inch in diameter, resembling that of Nymphæa; on the upper side is an umbel of small white flowers, blossoming successively at the surface of the water, and beneath, a cluster of short, simple, tuberous radicles, each of which is capable of producing a new plant. July.

11. MENYANTHES. Tourn.

Gr. μην, a month; ανθος; in allusion to its supposed properties as an emenagogue

Calyx 5-parted; corolla funnel-form, limb spreading. 5-lobed, villous within; stamens 5; style 1; stigma bifid; capsule 1-celled.

M. TRIFOLIÀTA. *Buck Bean.*
Lvs. trifoliate.—Grows in swamps, margins of ponds, &c., N. Am. N. of latitude 38°. This fine plant arises from large, black roots descending deep into the boggy earth. Stem 8—12' high, round. Leaves on long, round footstalks stipuled at base. Leaflets obovate. Peduncle long, naked, terminal, bearing a pyramidal raceme of flesh-colored flowers. Pedicels thick, bracteate at base. Sepals obtuse, about a third as long as the corolla. Petals acute, about as long as the stamens, remarkably and beautifully distinguished by the soft, fringe-like hairs at the base and in the throat of the tube. Bitter herb, actively medicinal, sometimes substituted for hops. May.

ORDER CI. APOCYNACEÆ.—DOGBANES.

Trees, shrubs and *herbs* with a milky juice. *Lvs.* opposite, entire, without stipules.
Cal.—Sepals 5, united at base, persistent.
Cor. 5-lobed, regular, twisted in æstivation, deciduous.
Sta. 5, arising from the corolla and alternate with its segments.

Fil. distinct. *Anthers* 2-celled, opening lengthwise, sometimes slightly connected.
Pollen granular, globose or 3-lobed, immediately applied to the stigma.
Ova. 2, distinct or rarely united. *Styles* distinct or united. *Stigmas* united into 1 which is common to both styles.
Fr.—Follicles 2, rarely 1 of them abortive.
Sds. numerous, pendulous, with or without a coma, albuminous. *Embryo* foliaceous.

Genera 100, species 666, chiefly natives of the torrid zone.

Properties.—These plants possess active and often suspicious qualities residing in the white juice with which the order is pervaded, and in the seeds which are often deadly poisons. The alkaloid *strychnine* or *strichnia*, one of the most violent of poisons, is the active principle of the seeds of the Strychnos Nux-vomica of India. It is sometimes administered as a medicine, but with doubtful success. S. Tieute of Java is one kind of Upas. Cerbera Tanghin, a tree of Madagascar, is powerfully poisonous, a single seed being sufficient to destroy twenty persons. The Apocyneæ are emetic, and becoming highly valued in hydrocephalus, &c. The juice contains caoutchouc in small quantities, but in Sumatra this is obtained largely from the juice of Urceola elastica.

FIG. 54.—1. Apocynum androsæmifolium. 2. A flower, somewhat enlarged. 3. The flower cut open, showing the stamens with distinct filaments and united anthers. 4. The ovaries and stigma. 5. Plan of the flower. 6. Matured follicles. 7. A seed with the long, silky coma.

Conspectus of the Genera.

Herbs {	with white or flesh-colored flowers.	*Apocynum.* 1
	with blue flowers.	*Amsonia.* 2
	with opposite leaves and bluish-purple flowers. . . .	*Vinca.* 3
Shrubs {	with ternately ver cillate leaves and rose-colored flowers. .	*Nerium.* 4

1. APOCYNUM.

Gr. απο, away, κυων, dog; Pliny says this plant is fatal to dogs.

Calyx very small; corolla campanulate, lobes short; stamens included; filaments short, arising from the base of the corolla, and alternate with 5 glandular teeth; anthers sagittate, connivent, cohering to the stigma by the middle; ova. 2; stigmas connate; follicles long, sublinear, distinct.—*Herbs, suffrutescent, erect, with opposite, entire, mucronate lvs. Cymes terminal and axillary. Pedicels not longer than the pale flowers.*

1. **A. ANDROSÆMIFOLIUM.** *Dog's-bane.*
Smooth; *lvs.* ovate; *cymes* lateral and terminal; *limb of cor.* spreading, the *tube* longer than the calyx.—A smooth, elegant plant, 3f high, in hedges and borders of fields. Stem reddened by the sun, erect, branching above. Leaves dark green above, paler beneath, opposite, rounded at base and acute at apex, 2—3′ long and ½ as wide, on petioles ¼′ long. Cymes paniculate, at the top of the branches and in the axils of the upper leaves. Pedicels ¼′ long. Calyx much shorter than the corolla. Corolla as long as the pedicels, bell-shaped, white, striped with red, with 5, acute, spreading segments. Medicinal. U. S. and Brit. Am. June, July.

β incanum. *Lvs.* hoary pubescent beneath.

2. **A. CANNABINUM.** *Indian Hemp.*
Lvs. oblong, obtuse at each end, mucronate; *cymes* paniculate, many-flowered, terminal and lateral; *cal. seg.* lanceolate, equaling the tube of the corolla; *cor. seg.* erect.—A species with smaller leaves and erect flowers. It is

in low shades and hedges, Can. to Ga. and Ark. Stem 2—4f high, generally
dividing above into long, slender branches. Leaves 1½—2½' by 8—11", opposite, on petioles 2" long, and, when young, downy beneath. Cymes terminal,
with linear bracts. Flowers about half as long as those of the last species.
Sepals lanceolate, acute. Corolla white, with straight, obtuse segments. The
fibres of the bark are strong and pliable, said to be used by the Indians in
various ways as hemp. July, Aug.

β. *pubescens. Lvs.* beneath and cymes pubescent.

3. A. HYPERICIFOLIUM. *St. John's-wort. Dog's-bane.*

Smooth; *lvs.* oblong, on very short petioles, obtuse or subcordate at base,
mucronate; *cymes* terminal, shorter than the leaves ; *cal.* nearly as long as the
tube of the corolla.—Gravely banks of streams. Stem erect, 2f high, with
opposite branches. Leaves 2—4' long, ⅓ as wide, lower ones often sessile and
cordate, smooth both sides but paler beneath. Flowers very small, in dense
cymes at the ends of the stems and branches. Sepals lance-linear, about as
long as the tube of the greenish-white, erect corolla. Aug.

2. AMSONIA. Walt.

Calyx 5-cleft, segments acuminate ; cor. 5-cleft, tube narrowly
funnel-form, bearded inside, hispid at throat ; segments linear, convolute in æstivation ; sta. 5 ; sty. 1 ; ova. 2, connate at base ; follicles 2. erect, slender, fusiform ; seeds in one row, cylindric, truncate
at each end.—*Lvs. alternate, entire, subsessile. Cymes terminal, corymbose. Fls. blue.*

A. TABERNÆMONTANA. Walt. (A. latifolia. *Michx.*)

Erect ; *lvs.* ovate-lanceolate, acuminate, acute at base, briefly petiolate,
margin slightly revolute; *sep.* glabrous, lanceolate, acuminate; *cor.* pilose outside near the top of the tube.—A plant of singular appearance, in prairies and
damp grounds, Western ! and Southern States ! Stem terete, smoothish, 2f
high, branched above. Leaves numerous, 3—4' by 1—1½', conspicuously
veined beneath. Flowers pale blue, in several terminal, cymose clusters.
Corolla 8" diam., very hairy at top of tube. Follicles in pairs, 2—3' long,
about 6-seeded. May, June.

3. VINCA.

Lat. *vinculum*, a band; from the long twining branches.

Corolla hypocrateriform, contorted, border 5-cleft, with the lobes
oblique. orifice 5-angled ; 2 glands at the base of the ovary ; capsule
follicular, erect, fusiform ; seed oblong.—*Trailing shrubs. Lvs.
evergreen.*

1. V. MINOR. *Lesser Periwinkle.—Sts.* procumbent; *lvs.* elliptic-lanceolate,
smooth at the margins ; *fls.* pedunculate ; *seps.* lanceolate.—Native in Europe.
A handsome evergreen, flowering in May. Stems several feet in length, round,
smooth and leafy. Leaves opposite, smooth and shining, about an inch long.
Flowers solitary, axillary, alternate, violet, inodorous.

2. V. MAJOR. *Greater Periwinkle.—Sts.* nearly erect; *lvs.* ovate, ciliate; *fls.*
pedunculate ; *sep.* setaceous, elongated.—Native in Europe. Shrub with numerous, slender straggling branches, very leafy, forming light masses of evergreen foliage flourishing best beneath the shade of other plants. Leaves 1—2'
in length, rounded or somewhat cordate at base. Flowers blue, appearing in
May and June.

4. NERIUM.

Calyx with 5 teeth at the base outside of the corolla ; corolla hypocrateriform. segments contorted. orifice with a corona consisting of
5, laciniate leaflets ; filaments inserted into the middle of the tube ,
anthers sagittate, adhering to the stigma by the middle.—*Oriental
shrubs. Lvs. evergreen.*

N. Oleander. *Rose Bay-tree. Oleander.*—*Lvs.* linear-lanceolate; *asp.* squarrose; *corona* flat, its segments 3-toothed. Native in S. Europe and the Levant. Stem branched. Leaves 3 together, on short stalks, smooth, very entire, coriaceous, with prominent, transverse veins beneath. Flowers terminal, corymbose, large and beautiful, rose-colored. One variety has white flowers, another variegated, and a third, double. This splendid shrub is common in Palestine, '*Rev. S. Hebard!*) growing by rivulets, &c. It is commonly supposed by travellers to be the plant to which the Psalmists alludes, Ps. i. 3, and xxxvii. 35.

Order CII. ASCLEPIADACEÆ.—Asclepiads.

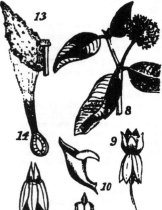

Herbs or *shrubs*, with a milky juice. *Lvs.* almost constantly opposite, entire, exstipulate. *Fls.* somewhat umbeled, fascicled or racemose. Sep. 5, slightly united, persistent.

Cor. petals 5, united at base, regular, deciduous, twisted-imbricate in æstivation.

Sta. 5, inserted into the base of the corolla and alternate with its segments.

Fil. connate. Anth. 2-celled, cells sometimes nearly divided by partial septa.

Pol. when the anther bursts cohering in masses which are as many as the cells, or confluent into pairs and adhering to the 5 processes of the stigma either by 2s, by 4s or singly.

Ova. 2, styles 2, approximate, often very short. Stigmas united into 1, which is common to both styles, and with 5 glandular angles.

Fr.—Follicles 2, one of them sometimes abortive.

Sds. numerous, pendulous, almost always comose at the hilum. Albumen thin.

Embryo straight. Cotyledons foliaceous. Radicle superior.

Genera 111, species 910, chiefly natives of tropical regions, and especially abundant in S. Africa, S. India and New Holland, but are not uncommon in temperate regions.

Properties.—similar to those of the Apocynaceæ, but far less active. The juice is acrid and stimulating, and generally to be, at least, suspected. A few of the species are medicinal, but none of much consequence.

FIG 55 —6 Asclepias cornuti. 9. A flower, the petals and sepals reflexed, and the corona erect. 10. One of the segments of the corona with the horn bent inwardly. 11. A pair of pollen masses suspended from the glands at an angle of the antheridium. 12. The two ovaries. 13. A mature follicle. 14. A seed with its long silky coma.

Conspectus of the Genera.

erect. Segments of the corona 5. distinct, { each with a horn.	. . . .	Asclepias.
{ without horns Fls. greenish.	. .	Acerates. 2
Corona leaflets 5, distinct. each 2-lobed. Fls. white.	. .	Enslenia. 3
Corona annular, undulate. 0-awned. Fls. purple.	.	Gonolobus. 4
Herbs { twining and climbing. { Corona urceolate, 5-cleft. 5 awned. Fls. purple.	.	Periploca. 5

1. ASCLEPIAS.

The Gr. name, from Esculapius, the fabulous god of medicine and physicians.

Calyx deeply 5-parted; cor. deeply 5-parted, valvate in æstivation, finally reflexed; staminal corona 5-leaved, leaflets cucullate, with an averted, horn-like process from the base, curved towards the stigma; antheridium (connate mass of anthers) 5-angled, truncate, opening by 5 longitudinal fissures; pollinia (masses of pollen) 5 distinct pairs, fixed by the attenuated apex, pendulous; follicles 2, ventricose; seeds comose.—♃ *Mostly North American, with opposite, verti cillate, rarely alternate leaves.* *Umbels between the petioles.*

* *Leaves opposite.*

1. A. cornuti. Decaisne. (A. Syrica. *Linn.* and *1st edit.*) *Common Silk weed.*—*St.* simple; *lvs.* oblong-lanceolate, petiolate, gradually acute, tomentose beneath; *umbels* nodding; *seg. of the corona* bidentate; *follicles* muricate—*A coarse*, very lactescent plant, common by roadsides, and in sandy

fields. Stem 3—4f high, seldom branched. Leaves 5—8' by 2—3', tapering at both ends. Umbels several, axillary, subterminal, dense, globose, each of 20 or more sweet-scented flowers. Calyx segments lanceolate. Corolla pale purple, reflexed, leaving the corona, which is of nearly the same hue, quite conspicuous. But few of the flowers prove fertile, producing oblong, pointed, rough pods or follicles, which contain a mass of long, silky fibres with seeds attached. July.

2. A. PHYTOLACCOÏDES. Ph. *Poke-leaved Silkweed.*

St. simple, erect, puberulent; *lvs.* broadly ovate, attenuated at base and apex, acute, smoothish both sides, glaucous beneath; *ped.* terminal, subequaling the leaves, whitish-puberulent, many-flowered; *pedicels* slender, loose; antheridium stipitate; *seg. of the corona* truncate, bidentate; horns exserted.— A tall and elegant species, found in low, shady grounds, Can. to Ga. and Ark. Stem 4—5f high, smooth and slender. Leaves acuminate at each end, 6—9' long and nearly half as wide. Umbels near the top on lateral peduncles, 4—6' long and consisting each of 10—20 large flowers, on pedicels about 2' in length. Petals green. Corona flesh-colored, each segment truncate, with its inner margin 2-toothed, and with a long, slender, incurved horn. June.

3. A. OBTUSIFOLIA. Michx. (A. cordata. *Walt.*) *Blunt-leaved Silkweed.*

St. simple, erect; *lvs.* oblong-ovate or oval, obtuse, mucronate, sessile, cordate and subamplexicaul, undulate, very smooth both sides; *umbels* terminal, many-flowered, glabrous, long-pedunculate; *corona* horns arcuate, falcate, inflexed.—In shady grounds, prairies, Mid.! W.! and S. States. Stem 2—3f high, bearing a single (rarely 2) terminal umbel of 30—40 large, reddish-green flowers. Leaves much waved on the margin, 4—5' long, ¾ as wide, with a broad, rounded, mucronate apex. Corolla light purple. Corona nearly white its segments large, slightly 2-toothed. July.

4. A. PURPURASCENS. *Purple Silkweed.*

St. simple, erect, puberulent; *lvs.* elliptical, ovate-elliptical or ovate, mucronate, narrowed at base into a short petiole, smooth above, tomentose-pubescent and paler beneath; *corona* segments oblong or lance-ovate, obtuse; *horns* falcate, acute, resupinate.—In hedges and thickets, N. H.! Mass.! to Ia.! Stem 3f or more high, simple or slightly branched at top. Leaves paler and downy beneath, the midvein purple, smooth above. Flowers in terminal, erect umbels. Calyx small, green. Corolla dark purple, with reflexed segments Corona purple, twice as long as the antheridium, its horns abruptly bent inwards to a horizontal position and lying close upon it. July.

5. A. INCARNATA. (A. amœna. *Michx.*) *Rose-colored Silkweed.*

St. erect, branching above; *lvs.* lanceolate, on short petioles, slightly tomentose; *umbels* numerous, erect, mostly terminal, often in opposite pairs; *seg. of the corona* entire, horns exsert.—A handsome species, found in wet places, Can. and U. S. Stem 3—4f high, with 2 hairy lines. Leaves 4—7' by ¾—1½', rather abrupt at base, tapering to a very acute point, on petioles ¼' long. Umbels close, 2—6 together at the top of the stem or branches, each an inch or more in diameter, on a peduncle 2' long, and consisting of 10—20 small flowers. Corolla deep purple, corona paler. Horns subulate, curving inwards over the summit of the antheridium. July.

β. *pulchra.* *St.* and *lvs.* densely tomentose, the latter elliptical-lanceolate, 3—4 times as long as wide, sessile or on very short, hairy stalks.—Stem 4—5f high. A remarkable variety.

6. A. QUADRIFOLIA. Pursh.

St. erect, simple, smooth; *lvs.* smooth, thin, petiolate, ovate, acuminate, mostly in 4s; *umbels* few, lax, on long, terminal or axillary peduncles; *corona* long, segments 2-toothed, horns short.—An elegant species, in dry woods, Can. and U. S. Stem about 2f high, slender, often with 1—2 hairy lines. Leaves opposite, the middle or upper pairs near together so as to appear in 4s, 2—3' long, ¾ as wide, acute or acuminate, on petioles 2—4' long. Flowers small, white, on filiform stalks with a pubescent line. Corona twice as long as antheridium. July.

β. *lancrolata*. Decaisne. Lvs. lanceolate, acuminate, acute at base and narrowed into the petiole.—Ohio, Ia., *Dr. Plummer!*

7. A. PARVIFLÒRA. Ph. (A. debilis. *Mx.*) *Small-flowered Asclepias.*
Suffrutescent and branched at base; *sts.* ascending, terete, smooth; *lvs.* lanceolate, attenuate at base and apex, petiolate, smooth both sides, thin ; *ped.* shorter than the leaves, umbellate, many-flowered; *umbels* small, pubescent; *corolla* 3 or 4 times shorter than the pedicels ; *horns* filiform, acute.—A very delicate species, with small, white flowers. Woods, along rivers, N. Y. to Ga., W. to Ia.! Stems often clustered, 1½—3f high, very leafy. Leaves 4—6' (including the 1' petiole) by 1—1½'. Umbels several, 1' diam., 15—20-floweres July, Aug.

8. A. VARIEGÀTA. (A. hybrida. *Michx.*)
St. simple, erect, pubescent; *lvs.* ovate or obovate, mucronate, glabrous, glaucous beneath ; *ped.* lateral or terminal, one-third as long as the leaves, umbellate, many-flowered; *cor.* segments ovate ; *corona* segments rounded at apex ; *horns* broad-falcate, with the apex horizontal or suberect ; *follicles* oblong, with a long and slender point, minutely puberulent.—Woods, N. Y. to Carolina. Stem 3—4f high. Leaves with a slight acumination, at length slightly undulate. Umbels about 2, 20—30-flowered. Corolla white, slightly tinged with purple.

9. A. PAUPERCÙLA. Michx. (A. laurifolia. *Ph.* not *Mx.* A. lanceolata. *Walt.*)—*St.* virgate, erect, glabrous ; *lvs.* linear and linear-oblong, margin narrowly revolute, both sides glabrous, tapering into a short petiole ; *ped.* 1 or 2 at top of the stem, umbel puberulent, few-flowered ; *corona* segments ovate, dilated above ; *horns* short, included.—N. J. to La. in wet woods. Stem 3—4f high, very smooth. Leaves green on both sides, rough on the edges, mostly very narrow. Flowers greenish-red, petals linear-oblong, half as long as the pedicel.

10. A. RÙBRA. (A. laurifolia. *Mx.* A. acuminata. *Ph.* A. periplocæfolia. *Nutt.*)
St. simple, erect ; *lvs.* ovate-acuminate, very acute, subcordate or rounded at base, subsessile, glabrous ; *umbels* on long, mostly terminal peduncles ; *corona* segments acute, rather longer than the suberect horns.—A small and elegant species, in Penn., N. J. to Car., not common. Stem 1—2f high, with a pubescent line on one side. Leaves 3—5' by 1—2', in remote pairs, the upper sometimes alternate. Peduncles 1—5, 2—3' long, pedicels about 1'. Flowers greenish-purple Follicles ventricose-acuminate, smoothish. July, Aug.

* *Leaves alternate.*

11. A. TUBERÒSA. *Tuber-root Asclepias.* *Butterfly Weed.*
St. ascending, hairy, with spreading branches at top; *lvs.* alternate, oblong-lanceolate, sessile; *umbels* numerous, forming large, terminal corymbs.—Found in sandy fields, Can. and U. S., rare in N. Eng. Root large, fleshy, sending up numerous stems. These are about 2f high, leafy, erect or ascending, hairy and colored. Leaves hairy, scattered, only the upper ones quite sessile, lanceolate, acute or acuminate, obtuse at base, 2—4' by ⅓—1'. The corymb consists of numerous, bright orange-colored flowers. Petals 5, oblong, reflexed, concealing the small calyx. Pods or follicles lanceolate, pointed, and like the other species, containing long, silky down, uniting the flat, ovate seeds to the placentæ. Aug.—Medicinal.

* * *Leaves verticillate.*

12. A. VERTICILLÀTA. *Whorl-leaved Asclepias.*
St. erect, simple, marked with pubescent lines; *lvs.* generally verticillate, very narrowly linear, revolute ; *segs. of the corona* short, 2-toothed, horn falcate, exsert.—A slender and delicate species, 2f high, in swamps or moist meadows, Can. and U. S. Leaves in whorls of 4—6, 3—5' long and a line in width. Flowers small, greenish-white, in small, lateral umbels. Peduncles half as .ong as the leaves. July.

2. ACERATES. Ell.

Gr. a, privative, κερατος, horns; the corona being destitute of these processes

Calyx 5-parted ; cor. deeply 5-parted limb spreading or reflexed , staminal corona 5-leaved, leaflets without horns, concave, appressed to the angles of the antheridium; pollinia 5 pairs, pendulous; follicles smooth or muricate.—♃ *Herbs differing from Asclepias only in the absence of the horn-like processes of the corona. Lvs. mostly opposite.*

1. A. VIRIDIFLÒRA. Ell. (and A. lanceolata. *Ives*, and *1st. edit.* Asclepias. *Pursh.*)—Suffruticose at base, pubescent above; *lvs.* elliptical, varying to oblong and lanceolate, briefly petiolate, scabrous above and on the margin, tomentose-pubescent beneath; *umbels* sessile, globose, many-flowered; *pedicels* pubescent; *pet.* ovate, reflexed; segments of the corona erect and adnate to the antheridium.—Can., Conn. to Ark., in dry, stony soils. Stem 2—3f high, ascending and often branched at base, clothed with dense close hairs. Leaves 3—6 or 8' long, ⅓—½ as wide, obtuse or acute, or even acuminate, scarcely petiolate. Flowers green, small, inelegant, in 2 or 3 small, subglobose umbels. July.—The plant varies greatly in respect to its leaves.

2. A. LONGIFÒLIA. Ell. *Long-leaved Acerates.*
Scabrous-puberulent; *st.* ascending, simple; *lvs.* alternate, numerous, linear and lance-linear, subsessile, acute; *umbels* half as long as the leaves, numerous, many-flowered, pubescent, axillary, pedunculate; *corona* scales shorter than the antheridium.—Mich. to Ia.! and Miss., in meadows and prairies. Stem stout, 2—3f high. Leaves 3—5' (including the 1—2" petiole) by 3—5". Flowers very numerous in each umbel, green, peduncle and pedicels about 1' long. July, Aug.

3. ENSLENIA. Nutt.

In memory of Mr. Aloysius Enslen, who collected many plants in the Southern States.

Calyx small, 5-parted; cor. 5-parted, segments erect; corona 5-leaved, leaflets membranaceous, free, truncate, each terminated by 2 filiform, flexuous lobes; pollinia oblong. obtuse at base and apex, pendulous; stig. 5-angled, conical; follicles cylindraceous, smooth.— ♃ *A twining herb, with opposite, cordate-ovate, acuminate lvs. Ped. racemose-umbellate, many-flowered. Fls. white.*

E. ALBIDA. Nutt.
Ohio, *Clarke!* to Va. and Ark. Stems slender, with an alternate, pubescent line. Leaves thin, glabrous, with rounded, auriculate lobes at base, 2—3' diam., margin entire. Peduncles axillary, as long as the petioles. Flowers ochroleucous, sweet-scented. July, Aug.

4. GONOLÒBUS. Michx.

Calyx much smaller than the corolla; cor. rotate, deeply 5-parted; corona small, shield-form, undulate, 5-lobed; antheridium depressed, discoid, pentangular, terminated by a membrane; pollinia transverse, 5-pairs; follicles 2; seeds comose.—♃ *St. climbing. Lvs. opposite, cordate. Ped. interpetiolar, racemed or corymbose.*

G. MACROPHYLLUS. (and G. hirsutus. *Michx.* Cynanchium obliquum. *Jacq.* C. macrophyllus. *Muhl.*)—*St.* tomentose-pubescent and with soft, scattered hairs; *lvs.* broad, ovate or oval, cordate, acuminate, tomentose-pubescent; *ped.* shorter than the petioles, 2—5-flowered, with linear bracts at summit; *pet.* linear or linear-oblong, obtuse, smooth above, minutely puberulent beneath.—Thickets, along streams, Penn. to Car., W. to Ohio! Ky. Vine trailing or climbing, 3—5f. Leaves 3—5' by 2—4', the lobes at base rounded, and often nearly or quite closed, with a short acumination at apex. Flowers dark purple. Petals 5—7" by 1". June, July.

5. PERIPLÒCA.

Gr. περι, around. πλοκη, a binding or twining; from the habit of the plant.

Calyx minute; corolla rotate, flat, 5-parted, orifice surrounded by a 5-cleft, urceolate corona, terminating in 5 filiform awns; filaments distinct, anthers cohering, bearded on the back; pollinia solitary, 4-lobed; follicles 2, smooth, divaricate; seeds comose.—*Twining shrubs. Fls. in umbels or cymes*

P. GRÆCA.

Lvs. ovate, acuminate; *corymbs* axillary; *cor.* villous within.—A climbing shrub, 10—15′ long, sparingly naturalized in Western N. Y., also cultivated in gardens. Leaves opposite, 3—4′ long, ¼ as wide, and on petioles ½′ long. Flowers in long, branching, axillary peduncles. Sepals minute, lanceolate, acute. Petals very hairy within, linear, obtuse, dark purple. Follicles about 2′ long. Aug.

ORDER CIII. JASMINACEÆ.—JASMINWORTS.

Shrubs, often with twining stems. *Lvs.* opposite or alternate, mostly compound.
Fls. opposite, in corymbs, white or yellow, mostly fragrant.
Cal. divided or toothed, persistent.
Cor. regular, hypocrateriform, limb in 5—8 divisions, twisted-imbricate in æstivation.
Sta. 2, arising from the corolla and included within its tube.
Ova. free, 2-celled, each cell with 1 erect ovule. *Sty.* 1. *Stig.* 2-lobed.
Fr. either a double berry, or a capsule separable into 2. *Sds.* 2.

Genera 5, species 100, ornamental shrubs abounding in tropical India. The essential oil which pervades the order, residing chiefly in the flowers, is exquisitely fragrant. On this account, as well as for their beauty, many of these plants are cultivated.

JASMINUM.

Gr. ιασμη, perfume; from the fragrance of the flowers.

Calyx tubular, 5—10-cleft; corolla hypocrateriform, tube long, limb flat, 5—10-cleft; berry double; seeds 2, solitary, ariled.—*Shrubs bushy or climbing. Lvs. opposite, compound. Petioles articulated. Fls. paniculate.*

1. J. FRUTICANS. *Yellow Jasmine.*—Glabrous, erect; *branches* angular; *lvs.* alternate, trifoliate, rarely simple; *lfts.* curved; *fls.* few, subterminal; *cal.* segments subulate; *cor.* tube twice longer than the calyx, limb of 5, obtuse lobes.—S. Europe. Stem 3f high. Flowers yellow, inodorous, tube about 6″ long. Propagated by layers. †

2. J. OFFICINALE. *White Jasmine.*—Smooth, scarcely climbing; *branches* subangulate; *lfts.* 3—7, lanceolate, acuminate; *panicles* terminal, few-flowered, corymbose; *cor.* tube twice longer than the calyx.—Asia. Stem several feet in length. Flowers white. Both species are beautiful and much cultivated. The deliciously fragrant *oil of Jasmine* of the shops is extracted from this plant. †

ORDER CIV. OLEACEÆ.—OLIVES.

Trees and *shrubs*, with opposite, simple, sometimes pinnate leaves.
Fls. perfect (sometimes diœcious). Sepals united at base, persistent (vation; rarely 8.
Cor.—Petals 4. united below, sometimes distinct but connected in pairs by the filaments, valvate in æsti-
Sta. 2, alternate with the petals. *Anth.* 2-celled, bursting longitudinally.
Ova. free, 2-celled. *Ovules* in pairs, pendulous. *Style* 1 or 0. *Stigma* entire or bifid.
Fr. drupaceous, baccate or samaræ, usually 1-seeded by abortion.
Sds.—Albumen dense, fleshy, abundant, twice as long as the straight *embryo.*

Genera 24, species 130, natives of temperate climates. The ash is very abundant in N. America. The Phillyreas and the Syringas are all Oriental.

Properties.—*Olive oil* is expressed from the pericarp of the *olive* (Olea Europæa). The bark of this tree, and also of the ash, is bitter, astringent and febrifugal. *Manna,* a sweet, gentle purgative, is the concrete discharge of several species of the Fraxinus, particularly of the European F. Ornus. The species of the ash are well known for their useful timber.

Conspectus of the Genera.

Corolla tube long, including the short stamens.	*Syringa.*	1
{ colored. { limb of spreading, ovate segments. . . .	*Ligustrum.*	2
{ Corolla tube short, { limb of long, linear segments. . .	*Chionanthus.*	3
Flowers { green. Fruit a simple samara. Trees with pinnate leaves, ...	*Fraxinus.*	4

1. SYRINGA.

Gr. συριγξ, a shepherd's pipe; from the use once made of its branches.

Calyx small, teeth erect; corolla hypocrateriform, tube several times longer than the calyx, limb cleft into deep, obtuse, spreading segments; stamens short, included within the tube. Capsule 2-celled, 2 valved.—*Oriental, flowering shrubs, with simple, entire leaves.*

1. S. VULGÀRIS. *Common Lilac.*—Lvs. cordate-ovate, entire, glabrous, green both sides; *inflorescence* thyrsoid; *limb of cor.* subconcave.—Hungary. There are many varieties of this beautiful shrub. *a.* Corolla lilac-purple, in a dense thyrse. *β. cærulea.* Fls. purplish-blue. *γ. alba. Cor.* white, thyrse subcompound.—One of the most popular shrubs, beautiful in foliage and flowers.

2. S. PERSICA. *Persian Lilac.*—Lvs. lanceolate, acute, smooth, both sides green, sometimes pinnatifid; *limb of the cor.* flattish.—Persia. A smaller shrub than the first, with smaller thyrses of white or lilac-blue flowers. The leaves vary from entire to pinnatifid, small at flowering time. Apr. May.

2. LIGUSTRUM.

Lat. *ligo,* to bind, from the use made of its shoots.

Calyx minutely toothed; cor. tube short, limb with spreading, ovate lobes; sta. 2; sty. very short; berry 2-celled, 2—4-seeded, seeds convex on one side, angular on the other.—*Shrubs with simple lvs. Fls. in terminal panicles, tetramerous.*

L. VULGÀRE. *Privet. Prim.*

Lvs. lanceolate and obovate, acute or obtuse, on short petioles; *panicle* dense, terminal.—A smooth shrub, 5—6f high, in woods and thickets, N. Y.! to Va. W. to the Miss. Branches wand-like with opposite, entire, smooth, dark green leaves which are 1—2' long, ½ as wide, varying from obovate to elliptical, with a rounded, obtuse or acute point. Flowers small, numerous, white. Anthers large, exserted. Berries black, in conical bunches, bitter. It is said to have been introduced from England where it is used for hedges. May, June.

3. CHIONANTHUS.

Gr. χιων, snow, ανθος; alluding to the whiteness of the flowers.

Calyx 4-parted, short; cor. tube very short, limb 4-parted, lobes linear, elongated; sta. 2, inserted into the cor. tube, included; sty. very short; drupe fleshy; putamen bony, 1-celled, 1-seeded.—*Trees with opposite leaves. Branchlets compressed. Racemes terminal and axillary.*

C. VIRGINICA. *Virginian Fringe-Tree.*

Lvs. oval and oblong-lanceolate; *pedicels* long, 1-flowered; *cal.* glabrous; *cor.* segments linear, acute, flaccid.—An ornamental shrub or small tree, 8—25f high, Penn. to Tenn.! on mountains. Leaves coriaceous, smooth. Flowers in rather dense, pendulous panicles. Petals snow-white, 8—10'' in length. Drupes oval, purple. May, Jn. †

β. maritima. Pursh. Lvs. obovate-lanceolate, membranaceous, pubescent; *panicles* very loose.—In low, maritime woods, N. J. †

4. FRAXINUS.

Gr. φραξις, a separation; in allusion to the easy separation of its annual layers into laminæ.

Flowers diœciously polygamous. *Staminate fls.* (often perfect).—Calyx 0 or 3—4-parted; corolla 0 or 4-petaled; stamens 2. *Pistillate fls.*—Calyx and corolla as the perfect; samara 2-celled, by abortion 1-seeded.—*Trees. Lvs. unequally pinnate. Fls. paniculate, the staminate ones densely so.*

1. F. AMERICÀNA. (F. acuminata. *Lam.*) *White Ash.*

Lfts. petiolate, oblong, shining, entire or slightly toothed, acuminate, glaucous beneath; *fls.* calyculate.—The white ash is one of the most desirable tenants of our forests. It is chiefly confined to the northern parts of the U. States and Canada. Few trees exceed it in the beauty and magnitude of its groves-

tious. The trunk arises often more than 40f without a branch and then expands into a regular summit of an equal additional height. The leaves are a foot or more in length, opposite, pinnate, consisting of about 7 leaflets. Flowers in loose panicles, the fertile ones with a calyx and the barren ones without. The wood is light, firm, elastic and durable, furnishing a most excellent timber for carriage frames, agricultural implements, pins, handspikes, bars, &c.   May.

2. F. juglandifolia. Lam.   (F. viridis. *Michx.*)

Lfts. 3—4 pairs, petiolulate, ovate, serrate, opaque, green above and with the branchlets, glabrous and glaucous beneath, pubescent in the axils of the veins; *fls.* calyculate; *samara* cuneate-lanceolate, obtuse.—A small tree, 15—25f high, in wet woods from the banks of the Ohio to Car., not common. Leaves 10—15′ long, consisting of 7—15 leaflets which are green both sides, with a glaucous hue beneath, margin denticulate. Flowers greenish. Fruit much smaller than in the other species.   May.

3. F. pubescens. Walt.   (F. tomentosa. *Michx.*)   *Red Ash.*

Lfts. petiolate, elliptical-ovate, acuminate, serrate, covered with a dense tomentum beneath, as well as the *petioles* and *branchlets;* *fls.* calyculate.—The red ash is abundant in Penn. and the southern parts of N. England, resembling the last species, so as often to be confounded with it. It arises 60f, with a straight trunk covered with bark of a deep brown color. Leaves of about 7 leaflets, which become reddish underneath. The wood is similar to that of the white ash, and is valuable for about the same diversified uses.   May.

4. F. sambucifolia. Lam.   *Black Ash.   Water Ash.*

Lfts. sessile, ovate-lanceolate, serrate, rugose and shining, round-oblique at the base; *axils of the veins* villous beneath; *fls.* naked.—This species is common in the northern U. S. and the British Provinces, where it is almost universally known as *black ash.* It prefers moist woods and even swamps, which it sometimes almost exclusively occupies. It grows to the height of 60—70f, with a diameter of 2f. The bark is of a darker hue than that of the white ash and less deeply furrowed. Buds of a deep blue, not yellow as in the former. Leaves 9—16′ in length, of about 7 sessile leaflets, which are smooth above and red-downy on the veins beneath. The wood is purplish, very tough and elastic, but less durable than the white ash. The young saplings are in great requisition for hoops, and the mature trunks for baskets.   May.

5. F. quadrangulata. Michx.   *Blue Ash.*

Lfts. 3—4 pairs, sessile, elliptic-lanceolate, serrate, pubescent beneath; *branchlets* glabrous, quadrangular-winged, at length subterete; *buds* velvety; *samara* oblong, obtuse at each end, apex slightly emarginate.—A tall tree, in rich woods, Ohio to Tenn., E. to the Alleghanies. Trunk often 60—80f high, 16 - 20′ diam. at base. Leaves 12—18′ long, consisting of 7—11 leaflets. Leaflets vary from oval to lanceolate, distinctly serrate, 3—4′ long. The branchlets are furnished with 4 membranous margins 2 or 3″ wide, which disappear when the twig is older. Samara slightly narrowed towards the base. Sterile panicles compound, much shorter than the leaves. May.—The wood possesses the same strength and elasticity that characterize the other species.

SUBCLASS III. APETALÆ.

Corolla none; the floral envelops consisting of a single series of organs (calyx) only, or sometimes wholly wanting.

ORDER CV. ARISTOLOCHIACEÆ.—BIRTHWORTS.

Plants herbaceous or shrubby, the latter often climbing. Wood without concentric layers.
Lvs. alternate, simple, petiolate, often with a stipule opposite the leaf, or exstipulate.
Fls. perfect, axillary, solitary, brown or of some dull color.
Cal.—Tube adherent to the ovary, segments 3, valvate in æstivation.
Sta. 6—12. epigynous or adhering to the base of the short and thick styles.
Ova. 3—4-celled. *Stig.* radiate, as many as the cells of the ovary.
Fr. capsule or berry, 3—6-celled many-seeded.
Embryo minute, in the base of fleshy albumen.

Genera 8, species 130, most abundant in the tropical countries of S. America, and thinly diffused throughout the northern hemisphere.

Properties.—Tonics and stimulants. Both the following genera are successfully employed in medicine.

Conspectus of the Genera.

Calyx limb { equal. *Asarum.* 1
{ unequal. *Aristolochia.* 2

1. ASĂRUM.

Said to be from the Gr. *a*, not, and *σειρω*, to bind; because not used in garlands.

Calyx campanulate; stamens 12, placed upon the ovary; anthers adnate to the middle of the filaments; style very short; stigma 6-rayed; capsule 6-celled, crowned with the calyx.—*Herbs with creeping rhizomas and* 1—2 *lvs. on each branch. Fls. solitary.*

1. A. CANADENSE. *Wild Ginger. Asarabacca.*

Lvs. 2, broad-reniform; *cal.* woolly, deeply 3-cleft, the segments reflected. —♃ A small, acaulescent plant, growing in rich, shady soil, Can. to Ga. and W. States. The leaves are radical, large, 2—4′ by 3—5′, with a deep sinus at base, on long, hairy stalks, and having a soft, velvet-like surface. The flower grows from between the bases of the leaf-stalks, solitary, on a nodding peduncle, and is close to the ground, sometimes even buried just beneath the surface. Calyx purplish, of 3, broad, long-pointed divisions abruptly spreading. The 12 filaments bear the anthers on their sides just below the extremity. The root or rhizoma is aromatic, and has been considered useful in whooping-cough. May—July.

2. A. VIRGINICUM. Michx. *Sweet-scented Asarabacca.*

Lvs. solitary, orbicular-ovate, glabrous, coriaceous, cordate, entire, obtuse; *fl.* subsessile; *cal.* short, subcampanulate, glabrous externally.—Grows in light soils among rocks, N. J. to Ga. A low, stemless plant, very similar in habit to the preceding. Each branch of the rhizoma bears a terminal leaf and a flower. Leaf 3—4′ diam., very smooth, clouded with spost, the petiole 2 or 3 times longer, lobes at base rounded and nearly closed. Flower many times shorter than the petiole. Calyx segments obtuse, of a dusky purple, greenish outside. Apr

2. ARISTOLOCHIA.

Gr. αριστος, excellent, λοχος, pertaining to parturition; alluding to the medicinal properties.

Calyx ligulate, with an inflated base and an unequal border: anthers 6, subsessile upon the style; stigma 6-cleft; capsule 6-celled, many-seeded.—*St. erect or twining.*

1. A. SERPENTARIA. *Virginia Snake-root.*

St. erect, flexuous; *lvs.* oblong, cordate, acuminate; *ped.* radical; *lip of the cal.* lanceolate.—A curious vegetable of low growth, in hedges and thickets, Penn. to Ill. and La. Stem 8—13′ high, subsimple, jointed, herbaceous. Leaves 2—4′ by 1—2′, rarely larger, petioles 3—9″ in length. Flowers few, near the base of the stem. Calyx dull purple, of a leathery texture, tubular, bent almost double, gibbous at the base and at the angle, limb 2-lipped, upper lip 2-lobed. Capsule obovate, 6-angled, 6—9″ long, with numerous small seeds. June.— The dried root is a valuable stimulant, diaphoretic and tonic, containing camphor. It has a warm, bitter, pungent taste.

2. A. Sipho. L'Her. *Dutchman's pipe.*

St. twining, shrubby; *lvs.* ample, suborbicular, cordate, entire, acute, petiolate; *ped.* 1-flowered, furnished with a single, ovate bract; *cal.* tube bent, ascending, limb 3-cleft, equal.—A vigorous climber in mountainous woods, Western Penn. to Ky. and S. States. St. woody, twining, and ascending trees 30 or 40f. Leaves 6—12′ diam., alternate, sprinkled with soft hairs. Flowers solitary, the tube long and bent at nearly a right angle, in the form of a (siphon or) tobacco pipe, and of a dull brown color. It is highly ornamental in cultivation, for arbors. June. †

Order CVI. CHENOPODIACEÆ.—Chenopods.

Herbs or undershrubs, with alternate (rarely opposite) leaves without stipules.
Fls. inconspicuous, generally perfect, often diœcious or polygamous.
Cal. deeply divided, often tubular at base, imbricate in æstivation.
Sta. from the base of the calyx, as many as its lobes or fewer, and opposite to them.
Ova. 1, with 1 ovule attached to its base within. *Styles* 2—4, rarely 1.
Fr. a utricle. *Embryo* usually curved around fleshy albumen.

Genera 63, species 360, often maritime plants, and more generally weeds, abounding in the northern temperate zone.

Properties.—Some are useful for food, as the *beet, mangel-wurtzel, orache, spinach, &c.* Others contain an essential oil, which renders them tonic, antispasmodic and anthelmintic; as *Chenopodium botrys, C. ambrosioides, C. anthelminticum;* the latter yields the officinal *worm-seed oil.* *Salsoli, Salicornia* and other sea-side species yield soda from their ashes in great abundance.

FIG. 92.—1. Flower of Chenopodium album. 2. Calyx, &c., removed, showing the ovary and two (hypogynous) stamens. 3. Cross section of the seed, showing the coiled embryo. 4. Branch of Salicornia herbacea. 5. Two joints magnified. 6. Ovary of a flower. 7. Flower of Blitum capitatum, with the fleshy calyx. 8. Vertical section of the ovary. 9. Flower of Beta vulgaris.

Conspectus of the Genera.

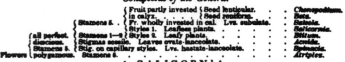

		Fruit partly invested	Seed lenticular.		Chenopodium.	7
		in calyx.	Seed reniform.		Beta.	9
	Stamens 5.	Fr. wholly invested in cal. Lvs. subulate.			Salsola.	2
		Styles 1. Leafless plants.			Salicornia.	1
all perfect.	Stamens 1—2.	Styles 2. Leafy plants.			Blitum.	6
diœcious.		Stigmas sessile. Leaves ovate-lanceolate.			Acnida.	4
Stamens 5.		Stig. on capillary styles. Lvs. hastate-lanceolate.			Spinacia.	3
Flowers polygamous.	Stamens 5.				Atriplex.	5

1. SALICORNIA.

Lat. *sal,* salt, *cornu,* horn; in allusion both to its locality and appearance.

Calyx turbinate, fleshy, closed, entire; sta. 1—2; style 1, bifid; utricle invested in the calyx, 1-seeded.—*Salt marsh herbs, rarely shrubby, destitute of leaves.*

1. S. herbacea. (S. mucronata? Bw.) *Herbaceous Samphire or Saltwort.*

St. erect, herbaceous, spreading; *joints* compressed; *internodes* dilated upwards, truncated; *branches* numerous, opposite, light green, jointed, succulent, smooth, terminating in a spike; *spikes* lateral and terminal, tapering upwards; *fls.* small, sessile, about three on each side of the base of every joint. -A leafless plant with succulent and jointed branches, about a foot high, growing abundantly on sea shores and salt marshes, N. Eng.! to Ga.; also at Salina, N. York. This and other species are said to make a good pickle for the table. When burned, its ashes yield soda. Aug.

2. S. ambigua. Michx. *Dubious or Prostrate Samphire.*

Perennial, procumbent, branching; *joints* small, crescent-shaped; *spikes* opposite and alternate; *cal.* truncate; *anth.* purplish-yellow; *stig.* 2.—A small species, found in the vicinity of New Bedford, Ms. *Dewey.*

2. SALSÖLA.
Latin *salsus,* salt.

Calyx 5-parted, persistent, embracing the fruit with its base, and crowning it with its enlarged limb; styles 2; seed horizontal; embryo spiral.—*Salt-marsh herbs, with linear or subulate leaves.*

1. S. Kali.

Herbaceous, decumbent; *lvs.* subulate, channeled, spinose, smooth; *cal.* margined, axillary.—A rigid, prickly and very branching plant, of th⁰ sea-coast, N. Eng.! to Ga. Stem 1—2f high, diffuse. Leaves about an inch long, sessile, ending with a spine. Flowers green, succulent, sessile, bracteate. Seed cochleate, enclosed in the calyx. Aug. Used in the preparation of soda.
β. *Caroliniana.* (*Nutt.* S. Caroliniana. *Walt.*); *lvs.* dilated; *cal.* with a broader margin; *st.* smooth.

2. S. soda.

Herbaceous, glabrous; branches ascending; *lvs.* semiterete, rather acute; *cal.* in fruit transversely connate, somewhat membranaceous.—In N. Y. *Muhlenberg,* who also attributes S. Tragus, another European species, to the shores of N. J. But this is very uncertain. July. ♃

3. SPINACIA.
Lat. *spina,* a prickle; on account of the spiny processes of the fruit.

Flowers ♂ ♀.—♂ Calyx 5-parted; stamens 5. ♀ Calyx 2—4-cleft; styles 4, capillary; utricle contained within the indurated and sometimes muricated calyx.

S. oleracea. *Spinage.*—*Lvs.* hastate-lanceolate, petiolate; *fr.* sessile, prickly or unarmed.—① Native country unknown, but it has long been a common plant in gardens, and in some esteem as an esculent. Stem 1—2f high, with leaves between hastate and sagittate, 2—3' long, and nearly half as wide, tapering at base into a long petiole. Flowers greenish, the sterile ones in a terminal-panicle, the fertile ones in dense, sessile, axillary racemes. June, July.

4. ACNIDA. Mitchell.
Gr. *a,* not, *κνιδη,* the nettle; a nettle-like plant which does not sting.

Flowers ♂ ♀.—♂ Calyx 5-parted; stamens 5. ♀ Calyx 3-parted; stig. 3—5, sessile; utricle 1-seeded.—*Herbaceous weeds, mostly aquatic.*

1. A. Cannabina. *Water Hemp.*

Lvs. ovate-lanceolate; *caps.* smooth, acute-angled.—① In salt marshes and inland swamps, Can. and U. S. Stem furrowed, smooth, 2—4f high. Leaves alternate, petiolate, 2—5' long, tapering to a long point. Panicles axillary and terminal, with numerous small, green flowers. Aug.

2. A. ruscocarpa. Michx.

Lvs. oval-lanceolate; *caps.* obtusely angled, rugose.—① Can. and U. S. A tall, branching, unsightly plant in similar situations with the last. Stem 6—8f high, angular. Flowers greenish-white, in terminal and axillary spikes. Jl.

5. ATRIPLEX

Fls. monœciously polygamous. ♀ Calyx 5-parted; *sta.* 5; style bipartite: utricle depressed, invested in the calyx. ♀ Calyx 2-leaved; *sta.* 0; style and fruit as above.—*Mostly* ①, *rarely shrubby plants. Lvs. alternate. Fls. glomerate, paniculate, ♂ and ♀ on the same plant.*

1. A. hortensis. *Garden Orache. Golden Orache.*

St. erect, herbaceous; *lvs.* triangular, toothed, of a uniform color both sides; *cal. of the fruit* ovate, reticulate, entire.—① Sparingly naturalized in cultivated grounds. Stem 3f or more high, with thick leaves of variable shape, and 2—3' long. Flowers green, in terminal, interrupted racemes or spikes. It is sometimes cultivated as a pot-herb. July. ♃

2. A. PATÚLA. *Spreading Orache.*
St. spreading, herbaceous; *lvs.* triangular-hastate, acuminate; *cal. of the fruit* submuricate on the disk.—① A very branching plant, 1—2f long, found in salt marshes, N. Y. to Ga. Stem and leaves dull green, somewhat glaucous. Lower leaves hastate, with coarse teeth, upper ones lanceolate and nearly entire. Flowers in long, terminal and axillary, interrupted racemes. Sepals of the fertile flowers triangular, studded with tubercles in the midst. Aug.

3. A. ARENARIA. Nutt. *Sand Orache.*
St. spreading, herbaceous; *lvs.* entire, oblong-ovate, subsessile, silvery-white beneath, upper ones acute or acuminate; *fls.* aggregated, axillary; *cal. of the fertile fls.* muricate, dentate, retuse.—① Grows on sandy shores, N. Y. to Car. Stem about a foot high, reddish, angular, very branching. Lower leaves often wedge-shaped. Flowers monœcious, the sterile ones in short, dense spikes at the ends of the branches, the fertile in axillary clusters. July.

4. A. LACINIÁTA. *Frosted Sea Orache.*
Stem spreading, herbaceous; *lrs.* triangular-ovate, deeply toothed, hoary pubescent beneath, lower ones opposite; ♀ tetrandrous; *cal. of the fruit,* rhomboid, 3-veined, denticulate.—① In salt marshes, N. Y. to Car. Stem a foot long, mostly procumbent, mealy, alternately branched. Leaves stalked, entire at base, covered with small, grayish scales. The perfect flowers in terminal, sessile clusters, their ovaries about 5. Pistillate flowers axillary, 2—3 together. Jl.

5. A. HALÌMUS. *Common Orache.*
St. shrubby; *lvs.* often subopposite, oblong-rhomboidal, entire, attenuated into the margined petiole.—A tall, shrubby species, 6f high, said by *Dr. Muhlenberg* to be naturalized on the sea-coasts of N. J. It is ornamental on account of its silvery foliage, and sometimes cultivated. ♃

6. BLITUM.

Gr. βλητον, insipid; in allusion to the fair but insipid berries.

Fls. ♀. Calyx 3-cleft, segments ovate, equal; sta. 1, exserted; sty. 2; ova. ovoid, acuminate; seed 1, contained in the calyx which becomes a berry.—① *Herbaceous weeds. Fls. and fr. in capitate clus ters terminal and axillary.*

1. B. CAPITÁTUM. *Strawberry Blite.*
Lvs. triangular-hastate, toothed; *hds.* in a terminal, leafless spike; *fr.* consisting of the reddened flowers, appearing like strawberries, full of a purple juice, taste insipid.—Va. to Arc. Circle. A weed-like plant, about a foot in height, branching, growing in fields, and sometimes cultivated for borders in the flower garden. Heads of flowers sessile, near together, on the branches and summit of the stem. June. †

2. B. VIRGÁTUM. *Slender Blite.*
St. with spreading branches; *lrs.* triangular-hastate, sinuate-dentate; *glomerules* scattered, lateral.—Fields and waste places. Stem 2f in length, spreading or procumbent. Leaves 2—3' by 1—2' coarsely toothed, on petioles half as long. Flowers always in axillary clusters, never terminal. Calyx finally fleshy and red in fruit. Jn. ♃

3. B. MARITÌMUM.
Lrs. lanceolate, attenuate at each extremity, incisely dentate; *hds.* axillary, sessile, spiked; *cal.* membranaceous.—① A coarse, unsightly plant, in salt marshes, N. Y. Stem 1—2f high, very branching. Leaves fleshy, with 2 or more large teeth each side. Flowers very numerous and minute, not becoming red in fruit. Aug.

7. CHENOPODIUM

Gr. χην, a goose; πους, a foot; from the resemblance of the leaves.

Calyx 5-parted, obtusely 5-angled, free, partially enveloping the fruit; stamens 5; styles 2; utricle membranaceous; seed lenticular, vertically depressed.—*Mostly* ☉ *weeds with alternate leaves. Fls. glomerate, paniculate.*

1. C. ALBUM (and C. viride. *Linn.*) *White Goose-foot. Hogweed.*
Lvs. rhomboid-ovate, crowded, entire at base, the upper ones oblong, very entire; *rac.* branched, leafy, smooth.—☉ A weed, common in cultivated grounds, Can. and N. States. Stem 2—3f high, furrowed, smooth, branching, leaves more or less mealy as also the whole plant. Flowers numerous, small, green, in irregular, terminal, erect racemes. July—Sept.

2. C. RUBRUM. *Red Goose-foot.*
Lvs. rhomboid-triangular, deeply toothed and sinuate; *rac.* erect, compound, leafy.—☉ A weed in waste grounds, rubbish, &c., N. Eng. and Brit. Am. Stem reddish, 1½—2f high, with short branches, very compound, and with compact clusters of small, reddish-green flowers. Leaves dark green, the upper ones small, and intermixed with the flowers. July.

3. C. HYBRIDUM. *Tall Goose-foot.*
Lvs. cordate, angular-toothed, acuminate; *rac.* branched, subcymose, divaricate, leafless.—A weed in waste places, &c., N. Eng.! to Ky.! rather taller than the foregoing. Stem slender, 2—3f high, bearing a loose, branching cluster of unsightly and ill-scented flowers, remote from the leaves. Leaves bright green, with large, remote teeth and a tapering point. July. ♀

4. C. RHOMBIFOLIUM. Muhl.
Lvs. rhombic-triangular, acute, sinuate-dentate, *upper ones* lanceolate, cuneate at base; *rac.* axillary, erect, mostly leafless; *bracts* minute, inflexed.—Penn. and Ohio. Plant yellowish-green, 2—3f high. Stem branching, angular with stripes of a deeper green. Leaves 2—3' by 1—1½', widest near the base, with a few acute dentures, petioles nearly half as long. Flowering branches shorter than the leaves, axillary, nearly leafless, with several roundish, dense clusters of green flowers.

5. C. AMBROSIOÏDES. *Ambrosia Goose-foot.*
Lvs. lanceolate, remotely dentate; *rac.* simple, axillary, leafy.—Fields and roadsides, N. Eng. to Ill. Plant rather fragrant. Stem 1—2f high, much branched, angular, slightly pubescent. Leaves acute at each end (the upper ones nearly linear), about 4 times as long as wide, the petioles 0—½' long. Flowers green, in sessile clusters on short, erect, slender, leafy branches. Stamens exsert. Aug. Sept.

6. C. BOTRYS. *Oak-of-Jerusalem.*
Lvs. oblong, sinuate; *rac.* much divided.—Sandy fields, &c., N. Eng. to Ill. This plant is sometimes cultivated both on account of its fragrance, and the remarkable appearance of its compound clusters of innumerable flowers. Plant 1—3f high, viscid-pubescent. Leaves petiolate, the sinuses deep giving them some resemblance to oak leaves. The branches put forth numerous leaves and short, axillary clusters on every side, forming long, leafy, cylindric, green, compound racemes, of which the central one is much the tallest. June †

7. C. ANTHELMINTICUM. *Worm-seed.*
Lvs. oblong-lanceolate, toothed; *spike* simple, slender, interrupted, leafless.—Maine! to Ill. A strong-scented species, said to be a good vermifuge, as both its specific and common name would imply. Stem 1½—2f high, its branches ending in long spikes of green, inconspicuous flowers. Aug.

8. C. GLAUCUM. *Sea-green Goose-foot.*
Lvs. oblong and ovate-oblong, repand-sinuate on the margin, glaucous beneath; *spike* simple, leafless, glomerate, axillary and terminal.—N. Y. *Muhl.*

9. C. MARITIMUM.
Lvs. linear, subulate, fleshy, semi-cylindrical; *fls.* in sessile, axillary clusters; *sta.* shorter than the sepals.—A fleshy plant growing in salt meadows, Can. to Flor. Stem 1½—2f high, branching. Leaves numerous, very acute, ½—1' long. Flowers in axillary glomerules, green. Utricle thin and semi-transparent, containing a black and shining seed. Aug. Sept.

8. BETA.

Celtic *bett*, red; the usual color of the beet.

Calyx 5-sepaled; sta. 5; styles 2, very short, erect, with acute stigmas; seed reniform, imbedded in the fleshy calyx.—② *Stems furrowed.
Lvs. alternate. Fls. glomerate, green, in spikes or paniculate racemes.*

1. B. VULGARIS. *Common Beet.*—Fls. in dense, sessile, axillary clusters; lower
lvs. ovate; *rt.* fleshy.—This useful culinary is said to grow wild in S. Europe
Besides its use in salads, pickles, soups, &c., the beet yields sugar equal to that
of the cane. There are several varieties, of which the purple-leaved is the
most esteemed for the kitchen, and the green-leaved for extracting sugar. Aug.

2. B. CICLA. *Scarcity.*—Lvs. with very thick veins; *fls.* 3 together; *rts.*
scarcely any.—Native of Portugal. Root leaves stalked, those of the stem sessile. Flowers green, numerous, in very long spikes. A culinary plant, with
very large leaves, used as a salad, &c. Aug.

β. *Mangel-Wurtzel. Rt.* very large.—Cultivated as food for cattle, for which
purpose it is highly prized by many farmers.

ORDER CVII. SCLERANTHACEÆ.—SCLERANTHS.

Herbs small, inconspicuous, with opposite leaves, no stipules and minute, axillary, sessile flowers.
Cal.—Tube urceolate, limb 4—5-toothed.
Sta. inserted on the calyx tube, and usually twice as many as its lobes.
Ova. 1, free, 1-seeded. *Styles* 2 or 1. *Fr.* a utricle, in the hardened calyx.
Seed pendulous from the apex of a funiculus which arises from base of cell.
Embryo curved around farinaceous albumen.

Genera 4, species 14, natives of the northern hemisphere.

SCLERANTHUS.

Gr. σκληρος, hard, ανθος, when in fruit, the floral envelope appears hard and dry.

Calyx persistent, 5-cleft, the tube contracted at the orifice, sta.
10, rarely 5 or 2; styles 2; utricle very smooth, enclosed in the calyx.

S. ANNUUS. *Common Knawel.*
St. spreading, slightly pubescent; *sta.* 10; *cal. of the fr.* with acute, spreading segments.—① weed in dry fields and roadsides, N. Eng. and Mid. States,
Stems numerous, branching, decumbent, short, ending with leafy clusters of
sessile, green flowers. The leaves are numerous, linear, acute, short, opposite
partially united at their basis. Fls. very small, green, in axillary fascicles. Jl.

ORDER CVIII. AMARANTHACEÆ.—AMARANTHS.

Herbs or *shrubs*, with opposite or alternate leaves, without stipules.
Fls. in heads or spikes usually colored, generally perfect.
Cal.—Sepals 3—5, dry and scarious, persistent, generally with dry, colored bracts.
Sta. 5 or some multiple of 5, distinct or monadelphous, hypogynous.
Ova. 1, free, 1 or few-ovuled. *Style* 1 or none. *Stigma.* simple or compound.
Fr. a utricle. *Seeds* pendulous. *Embryo* curved around farinaceous albumen.

Genera 38, species 388, most abundant in warm latitudes. A few of them are cultivated for their richly
colored, imperishable flowers. Others are mere weeds.

Conspectus of the Genera.

Leaves {alternate.	{ Rachis terete or furrowed.		.	.	.	.	.	.	*Amaranthus*	1
	{ Rachis broad, flat and crest-form.			.	.	.	.	.	*Celosia.*	5
		{ Flowers paniculate.		.	.	.	.		*Iresine.*	2
	{ Bracts whitish or greenish.	{ Flowers spicate.	.	.	.	.	.		*Oplotheca.*	3
{opposite.	{ Bracts crimson.	Flowers capitate.		.	.	.	.	.	*Gomphrena.*	4

1. AMARANTHUS.

Gr. α, not, μαραινω, to wither; the flowers of some of the species are imperishable.

Fls. ♂ ♀; calyx deeply 3—5-parted, mostly colored, persisten*;
segments lanceolate, acute. ♂ Stamens 3—5. ♀ Styles 2—3; capsule 1-celled, circumscissile; seed 1.—①*Herbs with alternate leaves
Fls. in axillary and terminal clusters, rarely scattered.*

* *Flowers triandrous.*

1. '. ALBUS. *White Cock's-comb.*
St. btusely angular; *lvs.* obovate, retuse; *fls.* triandrous, in small, axillary

lusters.—A common garden weed, 1—2f high, simple or branched. Leaves entire, varying from oval to obovate, emarginate, with a mucronate point, tapering to a petiole which is nearly as long as the blade, those upon the branches very small. Flowers inconspicuous, pale green, accompanied with numerous, setaceous-pointed bracts. July. §

2. A. BLITUM. *Low Amaranth.*
St. diffusely branched and spreading; *lvs.* ovate, long-petiolate, obtuse or abruptly decurrent at base, entire, lower ones retuse, upper obtuse or acute; *fls.* in short, axillary, spicate clusters, shorter than the petioles.—A weed naturalized in waste places. Stem mostly prostrate and spreading. Leaves as long as the petioles, ½—⅓ as wide. §

3. A. OLERACEUS. *Pot Amaranth.*
Lvs. rugose, oblong, very obtuse, emarginate; *clusters* axillary, branching; *fls.* sometimes pentandrous.—Fields and waste places, Mass. to Penn. Stem 18—24' high. July.

* * *Flowers pentandrous.*

4. A. HYBRIDUS. *Hybrid Amaranth.*
St. furrowed, erect; *lvs.* ovate-lanceolate; *rac.* pentandrous, decompound crowded, erect.—A common weed in waste and cultivated grounds, N. Eng. to Miss. Stem 3f high, or more, leaves alternate, long-stalked, mostly entire, obtuse, emarginate, mucronate, the lowest ones retuse. Flowers minute, in large, green, oblong spikes becoming at length a dull red, axillary and terminal.

5. A. PUMILUS. Nutt. *Dwarf Amaranth.*
Lvs. ovate, obtuse, smooth and fleshy, often retuse; *clusters* axillary; *fls.* pentandrous; *cal.* 5-leaved, concave.—Sandy sea shores, N. Y. Stem 8—12' high, often decumbent. Flowers green, obscure.

6. A. RETROFLEXUS.
Lvs. ovate, undulate; *branches* downy; *rac.* pentandrous, triply compound, compact, erect.—Waste grounds, among rubbish, N. Y. to Va. Stem 2f high. Aug.

7. A. SPINOSUS. *Spiny Amaranth.*
St. glabrous, much branched; *lvs.* ovate-lanceolate, with two stipular spines at base of the petioles; *fls.* in compound, terminal and axillary spikes.—In cultivated grounds and roadsides. Middle States. A vile weed, 2f or more high, readily known by its stipular spines. Aug. §

8. A. HYPOCHONDRIACUS. *Prince's Feather.*
Lvs. oblong, lanceolate, mucronate; *rac.* pentandrous, compound, compact, erect.—This species is native in the Middle States, and cultivated often as a garden annual. The whole plant is dark red, 3—4f high, with long, plume-like clusters. †

9. A. MELANCHOLICUS. *Love-lies-bleeding.*—Lvs. ovate-lanceolate, colored; *rac.* axillary, peduncled, roundish.—① From India. The whole plant purple, 18' high. † Several other species are rarely cultivated.

2. IRESINE.

Gr. ειρεσιωνη. an olive branch bound with tufts of wool, borne by suppliants; from the resemblance.

Flowers ♂ ♀—♂ Calyx deeply 5—7-parted, subtended by 2 bracts; scales or nectaries (petals?) 5 or 7. ♀ Stigmas 2, sessile; capsule at length tomentose, many-seeded.—*Mostly* ♃. *Lvs. opposite, entire, Fls. paniculate, axillary and terminal.*

1. CELOSIOÏDES.
St. erect, furrowed, paniculate above; *lvs.* scabrous, punctate, lower oblong, acuminate, upper ovate-lanceolate; panicle compound, large, rather dense.—A tall, handsome annual, 3—4f high, on river banks, Ohio, near Cincinnati, &c. Leaves tapering to the base into a winged petiole, 3—6' by 2—4'. Panicle of whitish flowers large, with opposite branches, branchlets and pedicels, nearly or quite leafless. Sept. Oct.

3. OPLOTHÉCA. Nutt.

Gr. οπλος, armor, θηκη, sheath; alluding to the armed cover of the fruit.

Fls. ☿—Calyx scarious, tubular, 5-toothed, densely tomentose, subtended at base by 2 truncated bracts; sta. 5, filaments united into a sheath below; stig. simple; utricle 1-seeded, enclosed in the indurated, muricate calyx.—① *Lvs. opposite, entire. Spikes opposite, sessile.*

O. FLORIDÀNA. Nutt.
St. simple, erect, arachnoid-pubescent; *lvs.* linear, tapering to the base, obtusish at apex; *fls.* imbricated, in short, dense, cottony spikes.—On sandy river banks, Ill. *Mead !* Plant 1—2f high, with a terminal, virgate inflorescence 6—10′ long. Leaves 1—2′ by 3—5″. Spikes remote, ½—1½′ long. Calyx white-scarious, persistent, contracted above, enclosing the utricle.

4. GOMPHRÉNA.

Bracts 5, colored, the 3 outer ones connivent, carinate; sepals 5, villous, disk (nectary) cylindric, 5-toothed; utricle circumscissile, 1-seeded.—*Herbs and shrubs with opposite leaves. None of the species native.*

1. G. GLOBÒSA. *Globe Amaranth.*—*St.* erect, hairy; *lvs.* oblong, pubescent; *hds.* globose, solitary, 2-leaved; *keels of the bracts* winged.—A tender annual from India, valued for its heads of flowers, which, if gathered before too far advanced, will retain their beauty several years. Height 18′. Branches opposite, axillary. Flowers purple.

2. G. PERENNIS. *Perennial Globe Amaranth.*—*Lvs.* lanceolate; *hds.* 2-leaved; *fls.* distinguished by a peculiar calyx.—♃ Plant about 2f high, native of S. America. The heads 2-leaved and terminal, resemble heads of clover. The crowded, purple perianths are chiefly conspicuous. Gathered like the former species, its flowers are equally fadeless and durable. July—Oct.

5. CELOSIA.

Gr. κηλεος, burnt; some of the species appear as if singed.

Sepals 3—5, colored; stamens united at base by a plicate disk (nectary); style 2—3-cleft; utricle circumscissile.—*A genus of ornamental foreign herbs. Lvs. mostly alternate.*

C. CRISTÀTA. *Cock's-comb.*—*Lvs.* ovate, acuminate; *stip.* falcate; *common ped.* striated; *spike* oblong, compressed.—This curious annual is said to have come from Japan, where the flowers or crests are a foot in diameter, and of an intense, purplish-red. Height 2f. June—Sept.

ORDER CIX. NYCTAGINACEÆ.—NYCTAGOS.

Herbs or shrubs. Lvs. opposite, one of each pair smaller than the other.
Cal. colored, tubular, the upper part resembling a corolla with a plaited limb, falling off from the lower part which becomes indurated in fruit.
Sta. hypogynous, definite. *Anth.* 2-celled.
Ova. free, with a single, erect ovule. *Style* 1. *Stigma* 1.
Fr. a thin utricle, enclosed within the enlarged and persistent calyx.
Seed with its testa coherent with. he utricle. *Cotyledons* leafy.

Genera 14, species 100, natives of warm latitudes. Nearly all, except the following beautiful genus, are obscure weeds. Roots purgative

MIRABÌLIS.

Calyx funnel-form, tube contracted, free from the ovary, limb plaited, entire, deciduous; sta. 5; stig. globose.

1. M. JALÀPA. *Four-o'clock.*—*Lvs.* smooth; *fls.* in clusters, stalked.—♃ This well-known and much admired plant is from the W. Indies. Root large, tuberous, and is one of the substances which furnish the Jalap of the shops. Stem 2f high. Leaves opposite, cordate, acuminate. Flowers *large, very fragrant,* in axillary and terminal clusters; border wide-spreading, *opening* at about 4 o'clock, P. M. Calyx bright purple. By cultivation it

sports into many pleasing varieties with yellow and white, red and white, .ed and yellow flowers. June—Sept.

2. M. dichotoma. *Mexican Four-o'clock.*—*Fls.* sessile, erect, axillary, solitary.—♃ From Mexico. Stem 2f high, dichotomous, with yellow flowers, opening at 4 o'clock like the former. July, Aug.

3. M. longiflora. *Long-flowered Four-o'clock.*—*Lrs.* pubescent; *fl.* crowded; *tube of the cal.* very long.—♃ Native of Mexico. Stem 2f high. Tube of the calyx slender, hairy, twice as long as the leaves. Flowers white. Jn.—Sep.

Order CX. POLYGONACEÆ.—Buckwheats.

Herbs, rarely shrubs, with alternate leaves.   *base of the leaf-stalks; occasionally a.*
Stip. of that remarkable kind called ochreæ, cohering round the stem in the form of a sheath above the
Fls. mostly perfect, and in racemes.
Cal.—Sepals united at base, imbricate in æstivation.
Sta. definite, inserted on the calyx near the base.
Ova. free, with a single erect ovule.   *Styles or stigmas* several.
Fr.—Achenium usually triangular.
Sds.—Embryo generally on one side of farinaceous albumen.

Genera 38, species 632, widely diffused throughout the world.

Properties.—The roots of these plants are nauseous and purgative. *Rhubarb* of the shops is the root of some unknown species of this order, native of Tartary. But the *leaves* and *stalks* of sorrel, the garden rhubarb, &c., are agreeably tart, and contain oxalic acid; the petioles of the latter, together with the farinaceous seeds of the buck-wheat, are well-known articles of food.

Conspectus of the Genera.

	(**6.** Stigmas 3, multifid. 3 outer sepals smaller.				*Rumex.* 4
(d. Stamens	(**9.** Stigmas multifid.				*Rheum.* 1
	mostly 5, united at base. persistent and enclosing the fruit.				*Polygonum.* 3
Sepals (4. Radical leaves reniform, on long petioles.					*Oxyria.* 2

1. RHEUM.

Rha, the river Volga, on whose banks the plants are said to be native.

Calyx colored, 6-sepaled, persistent; stamens 9; styles 3; stigmas multifid, reflexed; achenia 3-angled, the angles margined.—♃ *Fls. fasciculate, in racemose panicles.*

R. Rhaponticum. *Garden Rhubarb or Pie-plant.*—*Lrs.* smooth, coriate-ovate, obtuse; *petioles* channeled above, rounded at the edges.—Native of Asia, cultivated in gardens for the sake of the juicy, acid petioles. These are taken in a green state, in the spring of the year, and made into tarts and pies, whose excellence is well known to every one. Stem stout and fleshy, 3—4f high, hollow, with large, sheathing stipules at the joints. Leaves very large, 1—2f long, ½ as wide, on petioles of nearly the same length. Panicle terminal, at first enclosed in a white, membranous bract which at length bursts, disclosing innumerable greenish-white flowers. May.

2. OXYRIA. R. Br.

Gr. οξυς, acid; in allusion to the qualities of its leaves.

Cal. 4-sepaled, 2 inner sepals largest; achenium 3-sided, with a broad, membranaceous margin; sta. 2—6; styles 2, stigmas large, plumose.—♃ *Lrs. mostly radical, petiolate. Stem nearly leafless, paniculate-racemose.*

O. reniformis. Hook. (Rumex digynus. *Linn.*) *Mountain Sorrel.*
Radical lrs. reniform, on long petioles; *outer sepals* oblong, half as long as the inner, valvular sepals; *stamens* 2; *styles* 2.—Found on the summits of the White Mts., in moist ravines, and N. to the Arc. Sea. The plant is acid to the taste, like Rumex acetosus. Stem 3—4' in height. June.

3. POLYGONUM.

Gr. πολυς, many, γονυ, knee; that is, plants with many joints.

Sepals 4—6, mostly 5, connected at base. colored or corolla-like, persistent; sta. 5—9, mostly 8; sty. 2—3, mostly 3, short, filiform; ach. mostly triangular. usually covered by the persistent calyx.—*Herbaceous. Sts. jointed. Fls. in axillary and terminal fascicles and spikes or paniculate racemes.*

§ 1. AVICULARIA. *Flowers axillary. Stamens 5—8. Stigmas 3*

1. P. AVICULARE. *Bird Polygonum or Knot-grass.*

St. procumbent; *lvs.* elliptical-lanceolate, rough-edged, acute, sessile, *flowers* subsessile.—① A common weed in fields, highways and door-yards, U. S. and Brit. Am. Stems slender, ½—1½f long, striate, smooth, branching, with short, white, torn, remotely veined stipules at the joints. Leaves smooth except the edges, ½—1′ long and ½ as wide. Flowers reddish, small, 2 or 3 together in the axils of the leaves, appearing all summer.

β. *glaucum.* (P. glaucum. *Nutt.*) *Lvs.* fleshy, glaucous, revolute on the margin.—Grows on the sea-coast, Long Island.

2. P. ERECTUM. (P. aviculare. β. latifolium. *Michx.*) *Erect Knot-grass.* *St.* mostly erect, branched; *lvs.* smooth, broadly oval, petiolate; *fls.* pedicellate; *sta.* mostly 5.—♃ Western and Mid. States and Brit. Am. A species remarkably distinct in appearance from the last, in similar situations, but seldom growing with it. Stem 1—2½f high, branched, smooth. Leaves 1—2′ long and about ½ as wide, rather obtuse, the petioles 0—½′ long. Flowers 2—3 together, pedicellate, in the axils of the leaves, yellowish. Jn.—Sept.

3. P. TENUE. Michx. (P. linifolium. *Muhl.*) *Slender Knot-grass.* *St.* slender, erect, branched, acute-angled; *lvs.* linear-lanceolate, erect, acuminate; *stip.* tubular, villose at top; *fls.* alternate, subsolitary, axillary.—① A small, slender plant, on rocky soils, N. Eng., Mid. States. Stem ½—1f high. Leaves 1—1½′ long, 1—2″ wide, 3-veined, sessile. Flowers white. Jl., Aug.

§ 2. PERSICARIA. *Spikes terminal or axillary.*

4. P. PUNCTATUM. Ell. (P. Hydropiper. *Michx.* not of *Linn.*) *Water Pepper.*—*St.* branched, often decumbent at base; *lvs.* lanceolate, punctate with pellucid dots, waved and scabrous on the margin; *spike* loose, interrupted, drooping; *sta.* 6—8; *sty.* 2, united half way up.—① Can. to Flor. A plant well known for its acrid taste, growing in ditches, low grounds, among rubbish, &c. Stem smooth, swelling above the joints, 2f high, and, like the leaves, sprinkled with glandular dots, in which the acrimony is said to reside. Leaves 2—3′ long and not more than ½ as wide. Flowers green, tinged with purple and white. Aug., Sept.

5. P. MITE. Pers. (P. hydropiperoides. *Michx.* P. barbatum. *Walt.*) *Mild or Tasteless Knot-grass.*—*St.* mostly decumbent at base, erect and hairy above; *lvs.* narrow, lanceolate, subhirsute; *stip.* hirsute, long-ciliate; *spikes* linear; *bracts* ciliate, subimbricate; *sta.* 8; *sty.* 3.—① Ditches and ponds, Can. to Car. and Tenn. Stem a foot or more high. Leaves 2—4′ long, ½ as wide, sessile. Spikes several, crowded near the summit of the stem, composed of small fascicles of reddish flowers. Jl., Aug. *See also Addenda,* p. 638.

6. P. PENNSYLVANICUM. *Pennsylvanian Knot-grass.* *St.* smooth, tumid at the joints; *lvs.* lanceolate, petiolate; *stip.* glabrous, not ciliate; *spikes* oblong, crowded; *ped.* hispid; *sta.* 8; *sty.* 2 or 1.—① Margins of ponds and ditches, N. H. to Car. Stem geniculate, branched above, 2—4f high. Leaves 3—5′ long, ½ as wide, slightly scabrous. Spikes short and dense, large, and somewhat nodding. Flowers large, rose-colored, pedicellate. Jl.

7. P. LAPATHIFOLIUM. (P. incarnatum. *Ell.*) *St.* geniculate, smooth; *lvs.* ovate-lanceolate, petiolate, often hoary beneath; *spikes* numerous, rather crowded, erect, on scabrous peduncles; *sta.* 6; *sty.* 2.—① A rare species in swamps and ditches, N. Y. to Ga. Stem 2—4f high. Leaves 3—5′ long, ½—¾ as wide. Petioles ½—½′ long. Flowers small, white, or tinged with red, in numerous, panicled spikes. Aug.

8. P. PERSICARIA. *Ladies' Thumb. Spotted Knot-weed.* *St.* erect; *lvs.* lanceolate, the upper surface spotted; *stip.* fringed; *spikes* dense, oblong, erect; *ped.* smooth; *sta.* 6; *sty.* 2, half united.—① A common species about buildings, fences, wet grounds, &c. Stem smooth, branched, leafy, 1—2f high, often colored. Leaves 2—4′ long, ½ as wide, entire, short-stalked, acuminate, generally marked with a brownish, heart-shaped spot near the middle. Flowers rose-colored, in short, dense, terminal spikes. Jn., Aug. ⸹

9. P. ORIENTÀLE. *Oriental Knot-grass. Prince's Feather.*

St. erect, paniculately branched; *lvs.* large, with hairy, salver-form stipules; *sta.* 7; *sty.* 2.—① Native of the East, naturalized in fields and roadsides, throughout the U. S. A tall, showy plant, often cultivated for ornament. Stem 5—8f high. Leaves 8—12′ long, ½ as wide, ovate, acuminate. Spikes numerous, large, red, plume-like, terminal. Aug. ♦ †

10. P. AMPHIBIUM. (P. coccinium. *Muhl.*) *Amphibious Knot-weed.*

St. assurgent, prostrate or decumbent at base, rooting at the lower joints; *lvs.* oblong-lanceolate and oblong, acute, or rounded or cordate at base, petiolate, smooth, acute or acuminate at apex; *spike* terminal, ovoid or oblong, dense; *sta.* 5; *sty.* 2-cleft.—Marshes, ponds, N. Eng. to Ill. A very variable species, with large leaves and a terminal, dense spike of bright red flowers. Stems smooth, furrowed, short-jointed, often very long and creeping or floating and rooting. Stipules large, sheathing, mostly lacerated. Leaves 5—7 by 1—2′, often shining. Spikes 1—2′ long, the shorter mostly thicker. Aug.—The principal varieties are as follows:

a. coccineum. *St.* thick, suberect, 1—3f long; *lvs.* oblong, acute, shining both sides; *stipules* truncate; *spikes* ovoid. (N. Eng.!)

β. natans. *St.* very long, thick, rooting, prostrate, with erect branches; *lvs.* lance-oblong, subcordate, acuminate, *stip.* lacerate, long; *spike* long, slender (Indiana! &c.)

γ. (P. fluitans. Ea.?) *Lvs.* lance-linear, tapering to each end; *ochreæ* long, hirsute, with a leafy, spreading summit; *spike* oblong.—Lancaster, N. H *Rickard!* Wisconsin, *Lapham!*—I am by no means *certain* that these three varieties are not distinct.

11. P. VIRGINIÀNUM.

St. simple, pilose above; *lvs.* broadly lanceolate, ciliate-serrulate, acuminate, smooth, on short petioles; *rac.* long, slender, few-flowered; *fls.* tetramerous, unequal, remote; *sta.* 5; *sty.* 2, unequal.—♃ Shades, Can. to Flor., W. to Miss. Stem 2—4f high. Leaves 3—6′ long, ½ as wide, petioles 1—5′′ long. Stipules hairy. Spike or raceme terminal, simple or with one or two branchlets, leafless, 1—2f long. Flowers small, white, Jl., Aug.

§ 3. BISTORTA. *Fls. in terminal, solitary spikes. Sta. 7—9. Stig. 3.*

12. P. VIVIPÀRUM. *Viviparous Bistort.*

St. simple; *lvs.* linear-lanceolate, revolute at the margin; *spike* linear, solitary.—Stem erect, leafy, ½f high, bearing a single spike of white flowers which are often transformed into bulblets while on the stem.—♃ White Mts. N. to Arc. Am. Leaves 1—1½′ by 2—3′′, with entire, obtuse, smooth stipules. Jl.

13. P. HIRSÙTUM. Walt. *Hairy Knot-grass.*

St. and *stip.* very hirsute; *lvs.* lanceolate, hirsute, punctate with pellucid dots; *spikes* filiform; *sta.* 7; *sty.* 3-parted.—① On river banks, Ohio and Southern States. The whole plant is clothed with soft hairs. Stem 2f high. Flowers white. July.

§ 4. POLYGONELLA. *Flowers in paniculate spikes. Stamens 8.*

14. P. ARTICULÀTUM. *Jointed Polygonum.*

St. erect; *lvs.* linear; *spikes* panicled, filiform; *fls.* solitary, pedunculated, with imbricate, truncated bracts; *sta.* 8; *sty.* 3.—① N. Y., Mich., found in dry, hilly pastures. Stem slender, branching, straight, with numerous, racemed spikes, and imbricate, sheathing bracts. Leaves ½—1′ by 1—2′′, obtuse. Flowers flesh-colored, on nodding, hair-like peduncles issuing from above the bracts. Achenia acutely triangular. Aug.

§ 5. FAGOPYRUM. *Fls. in racemose panicles. Lvs. subcordate or sagittate.*

15. P. SAGITTÀTUM. *Scratchgrass.*

St. prostrate, rough-angled; *lvs.* sagittate; *fls.* capitate; *sta.* 8; *sty.* 3.—① Wet grounds, N. Eng. to Flor. and W. States. A rough, climbing species, often several feet in length. Stem square, the angles very rough with prickles pointing downwards. Leaves acute, 1—3′ long, ½ as wide, on petioles ½—1′ long, with smooth stipules. Flowers in small, terminal heads, whitish. Jn.

16. P. ARIFOLIUM. *Hastate Knotgrass.*

St. aculeate with reversed prickles; lvs. hastate; spikes few-flowered; fls. distinct; sta. 6; sty. 2.—① Wet grounds, Can. to Ga. and W. States. Distinguished from the last chiefly by its larger, halbert-shaped leaves, which are 3—4' long and ¼ as wide. Petioles ¼—1' long. Clusters racemose, slender, loose, few-flowered, at the ends of the branches. June, July.

17. P. CONVOLVŮLUS. *Knot Bindweed.*

St. twining, angular; lvs. cordate-hastate; seg. of the cal. obtusely keeled; sts. 8; sty. 3.—① A common climbing species, in fields and waste grounds, Can. to Ky. and Car. Stem 2—6f long, roughish, angled, with axillary branches. Leaves 1—2' long, ¼ as wide, on petioles ¼—1' long, with somewhat spreading and acute lobes at base. Flowers whitish, in terminal, interrupted spikes. June— Sept.

β. cilinode. (P. cilinode. Mx.) Plant minutely pubescent; stip. fringed with reflexed hairs at base.

18. P. SCANDENS. *Climbing Knotgrass.*

St. twining, smooth; lvs. cordate, acuminate; seg. of the cal. winged; sta. 8; sty. 3.—24 N. Eng. to Ark. Stem 3—7f long, climbing, often colored and with axillary branches. Leaves heart-shaped, with distinct, rounded lobes. Flowers in long, interrupted racemes. Calyx and fruit conspicuously 3-winged, the wings decurrent on the slender, jointed pedicels. Aug.

19. P. FAGOPŸRUM. *Buckwheat.—*St. erect, smooth; lvs. cordate-sagittate; rac. panicled; sta. 8; sty. 3; angles of the ach. equal.—① The name from the Lat. fagus, beech, and pyrum, a pear; the fruit resembles in shape a beech-nut. Native of Asia, but here naturalized. A valuable grain cultivated for the flour which is made into pan-cakes and eaten warm. Stems 2—4f high. Leaves 2—4' long, ¼ as wide. Flowers numerous, white, very grateful to bees. ‡ §

4. RUMEX.

Calyx persistent, of 6 colored sepals, the 3 inner (valves) larger, sta. 6; sty. 3, spreading; stigmas many-cleft; achenium 3-cornered, covered by the 3 valve-like inner sepals.—*Herbs with the flowers in dense, fasciculate panicles.*

§ 1. LAPATHUM. *Flowers all ☿ . Inner sepals (valves) granuliferous.*

* Inner sepals entire.

1. R. CRISPUS. *Yellow Dock.*

Lvs. lanceolate, waved, acute; valves (inner sepals) of the cal. entire, ovate each bearing a tubercle.—24 Can. and U. S. A weed so common as hardly to need description, growing in cultivated grounds, about rubbish, &c., much to the annoyance of the farmer. Stem 2—3f high, smooth, channeled, from a yellow, fusiform root. Flowers numerous, in a large panicle, consisting of many racemes of half-whorls, interspersed with leaves. Calyx-valves 3, enclosing the seed, each with a grain on the back. The root is used in medicine for cutaneous diseases. June. §

2 R. SANGUINEUS. *Bloody-veined Dock.*

Lvs. petiolate, cordate, lanceolate; valves of the cal. entire, one of them principally bearing the granule.—24 Can. and N. States. Stem of a reddish color, branching, leafy, 2—3f high. Leaves smooth, radical ones large, mostly with red veins, somewhat cordate, slightly curled at the edges. Flowers in small, distant whorls. Grows in waste and shady places. July. §

3. R. BRITTANICUS. *British Water Dock.*

Lvs. broad-lanceolate; joints with nearly obsolete, torn sheaths; fls. polygamous; valves entire, all bearing granules.—24 Can., U. S. Aquatic, growing in muddy places. Root yellow internally, large. Stem 3f high, furrowed, angular and branched. Leaves large, petiolate, acute. Flowers in verticillate fascicles, collected into a large, terminal panicle. Pedicels nodding in fruit. Calyx valves large, cordate. July.

[redacted]

4. R. verticillatus. *Naked-spiked Dock.*

Lvs. oblong-lanceolate; *valves* entire, all bearing granules; *spikes* leafless with flowers in half-whorls.—♃ Northern States. An aquatic species of muddy situations. Stem 2f high, with long, tubular sheaths and few branches. Leaves long, narrow, acute, flat. Whorls few-flowered. Pedicels ½—1′ long. June.

5. R. aquaticus. *Great Water Dock.*

Lvs. lanceolate, acute, lower ones cordate, on long petioles; *valves* ovate, entire, all of them bearing granules.—♃ Northern U. S. Ponds and ditches. Root large, very astringent. Stem 3—4f high. Leaves somewhat glaucous, the lower ones distinctly cordate at base. Flowers verticillate, in a terminal, leafy panicle. Pedicels nodding.

6. R. altissimus. Wood. (R. verticillata, *Mead*?) *Peach-leaved Dock.*

Glabrous, tall, erect; *lvs.* linear-elliptic, entire, petiolate, tapering to each end; *rac.* slender, paniculate, somewhat secund, leafless or the lowest verticil axillary; *fls.* all ☿; *inner sep.* broad-cordate, one graniferous, one abortively so, and the third naked.—♃ Marshy prairies and borders of streams, Indiana! A very showy Rumex, 3—6f high, slightly branched above. Leaves 3—5′ by ½—1′, somewhat acuminate, broadest in the middle. Verticils approximate, pedicels 2—3″ long. Achenium exactly resembling a beach-nut in form and color. June. (*Nov. sp.*)

* * *Inner sepals toothed.*

7. R. acutus. *Dock.*

Lvs. oblong-cordate, acuminate; *whorls* leafy; *valves* oblong, subdentate, all of them bearing granules.—♃ N. States. Ditches and waste places. Stem 2—3f high. Leaves large, the lower ones distinctly cordate, on long petioles. Racemes paniculate, composed of dense, leafy, dimidiate verticils. Granules large, red, one upon the back of each valve. May. §

8. P. pallidus. Bw. *White Dock.*

Lvs. linear-lanceolate, acute; *spikes* slender; *valves* ovate, entire, hardly larger than the granule.—♃ Found in salt marshes, Ms. Stems numerous, ascending. Leaves smooth, acute, petioled, wavy on the margin. Flowers crowded, on short pedicels. Granule large, white, nearly covering the back of each valve. June.

9. R. maritimus. *Golden Dock.*

Lvs. linear, very long, entire, acute at each end; *fls.* in leafy racemes; *inner sepals* dentate, all graniferous.—♃ "Borders of brackish ponds in Martha's Vineyard and Nantucket." *Oakes.* Plant a foot or more high. Flowers yellowish-green, in crowded half-whorls on the branches and main stem above. §

10. R. obtusifolius. *Broad-leaved Dock.*

St. roughish; *radical lvs.* ovate, obtuse; *valves* toothed, one of them principally bearing a granule on the back.—♃ N. States. A weed as troublesome as the first, growing about houses and fields wherever it is least welcome. Stem 2—3f high, furrowed, branching, leafy. Leaves oblong, cordate, obtuse, crenate-wavy; upper ones narrower and more acute; root leaves very large, oblong, heart-shaped, often with stalk and veins red. Flowers in long, nearly naked racemes. July. §

§ 2. Acetosa. *Flowers ♂ ♀. Inner sepals without granules.*

11. R. acetosella. *Field Sorrel. Sheep Sorrel.*

Lvs. lanceolate-hastate; *fls.* ♀ ♂.—♃ A common weed, growing in pastures and waste grounds throughout the U. S., preferring dry, hard soils. Stem ½—1f high, leafy. Leaves halbert-shaped, very acid, but pleasant to the taste Flowering all summer. Flowers small, red or reddish, collected in panicled racemes, the valves destitute of granules. The stamens and styles are on separate plants.

Order CXI. PHYTOLACCACEÆ.—Phytolaccads.

Herbs or *undershrubs*, with alternate, exstipulate leaves.
Fls. racemose, perfect. Sepals 4—5, petaloid.
Sta. 4—5 and alternate with the sepals, or indefinite.
Ova. 1—several-celled. *Styles* and *stigmas* equal in number to the cells.

Fr. baccate or dry. *Seeds* solitary, ascending.
Embryo cylindric, curved around fleshy albumen.

Genera 9, species 90, chiefly natives of the tropics. *Phytolacca decandria* possesses active properties, but they appear to be yet little understood and of doubtful application.

PHYTOLACCA. Tourn.

Gr. φυτον, a plant. Lat. lacca, lac; from the purple juice of the berries.

Calyx 5-sepaled, resembling a corolla; stamens 7—20; styles 5—10; berry superior, 10-celled, 10-seeded.—*Herbaceous. Racemes often opposite the leaves.*

P. DECANDRIA. *Poke. Garget. Jalap.*
Lvs. ovate, acute at both ends; *fls.* with 10 stamens and 10 styles.—♃ A common, well-known plant, growing about roadsides, hedges, U. S. and Can. Root very large and branching. The stem, with the diameter of an inch, is 5 or 6 high, round, smooth, branching, and when mature, of a fine, deep purple. Leaves 5' by 2—3', smooth, of a rich green color, entire and petioled. Racemes cylindric, long, at first terminal, becoming finally opposite to the leaves. Flowers greenish-white, consisting of 5 ovate, concave sepals, 10 stamens with white 2-lobed anthers, and 10 short, recurved styles. The fruit is a dark purple berry, of a round, depressed form. The juice of the berries stains paper and linen a beautiful purple color, which, however, is not durable. In Spain, it is said they are used to color wine. July—Sept.

ORDER CXII. LAURACEÆ.—LAURELS.

Trees and *shrubs*, with alternate, exstipulate lvs. and umbelled or panicled fls.
Cal.—Sepals 4—6, somewhat united, free from the ovary, imbricated in 2 series.
Sta. definite, perigynous, usually twice as many as the sepals, the 3 inner sterile or wanting.
Anth. adnate, 2—4-celled, opening by recurved valves, from the base to the apex.
Ova.—*Style* and *stigma* single.
Fr.—Berry or drupe on a thickened pedicel. *Seed* large, without albumen. *Plumula* conspicuous, 2-leaved.

Genera 66, species 450, chiefly natives of the tropics, but few being sufficiently hardy to endure our climate.

Properties.—The species of this highly important order are throughout pervaded by a warm and stimulant aromatic oil. *Cinnamon* is the dried bark of Cinnamomum Zeylanicum, of Ceylon, &c. *Camphor* is obtained from many trees of this order, but chiefly from Camphora Officinarum, of Japan, China, &c. *Cassia Bark*, from Cinnamomum aromaticum, of China. Persea gratissima, a tree of the W. Indies, yields a delicious fruit called the *Avocado pear*. Both of the following species are also moderately medicinal.

Genera.

Anthers opening by { 2 valves.   *Benzoin.* 1   { 4 valves.   *Sassafras.* 2

1. BENZOIN. Nees.

Flowers ☿, rarely by abortion ♂ ♀. Calyx deeply 6-parted; sta. 9, the inner row each with a pair of opposite, pedicellate glands at base; anth. opening by 2 valves, detached below and reflected upward; sty. short, thick; drupe 1-seeded.—*Lvs. deciduous, entire. Fls. preceding the leaves, lateral.*

B. ODORIFERUM. Nees. (Laurus Benzoin. *Linn.*) *Fever-bush. Spice Wood.*
Lvs. obovate-lanceolate, veinless, entire, deciduous; *fls.* in clustered umbels, often diœcious; *buds* and *pedicels* smooth.—A shrub 6—12f in height, in moist woods, U. S. and Can. It has an aromatic flavor resembling gum Benzoin, and the bark an agreeable, spicy taste. Leaves oval or obovate, cuneiform and acute at base, 2—4' long and ⅓ as wide, paler beneath. Flowers pedicellate, in small, sessile umbels, greenish, appearing in advance of the leaves. Calyx 6-cleft, with oblong segments. Berries red. May.

2. SASSAFRAS. Nees.

Diœciously-polygamous; calyx colored 6-parted. ♂ Sta. 9, fertile, *the* inner row with each a pair of opposite, pedicellate glands at base; anth. opening by 4 valves. ♀ Sta. 6, short, abortive; sty. filiform;

drupe superior, 1-seeded.—*Lvs. deciduous, lobed. Fls. and lvs. co-
emporary and from the same bud.*

S. OFFICINÀLE. (Laurus Sassafras. *Linn.*) *Sassafras.*
Lvs. of two forms, ovate and entire, or 3-lobed and cuneate at base; *fls*
in terminal and axillary, corymbose racemes, with linear bracts.—Not uncommon in N. Eng., very abundant in the forests and barrens of the other states.
It varies in height from 10—40f. Leaves alternate, petioled, those of the young
shoots ovate-lanceolate, others with 3 large lobes. Flowers greenish-yellow,
appearing in May and June, in clustered racemes at the end of the last year's
twigs, and after the leaves have expanded. Every part of the tree has a pleasant fragrance, and a sweetish, aromatic taste, which is strongest in the bark of
the root. These qualities depend upon an essential oil which may be obtained
by distillation, and which has been highly valued in medicine. The young
shoots are a common ingredient in *small beer*, imparting to it a grateful flavor.

ORDER CXIII. SANTALACEÆ.—SANDALWORTS.

Trees, shrubs and herbs, with alternate, undivided leaves.
Fls. small, in spikes, rarely in umbels or solitary.
Cal. tube adherent to the ovary, limb 4—5-cleft, valvate in æstivation.
Sta. as many as the sepals, inserted at their base and opposite to them.
Ova. 1-celled, 1—4-ovuled. Style 1. Stigma often lobed.
Fr. hard, dry and drupaceous, 1-seeded, crowned with the persistent calyx.

Genera 18, species 110, natives of Europe, America, Australasia, &c. The fragrant sandal-wood is the
product of Santalum album, &c., of India.

Conspectus of the Genera.

Herbs.	Comandra.	1
Shrubs.	Pyrularia.	2
Trees.	Nyssa.	3

1. COMANDRA. Nutt.

Gr. κομη, *hair,* ανδρος, *stamens; stamens connected to the petals by a tuft of hairs.*

Calyx somewhat urceolate, tube adherent, limb 4—5-parted; sta
mens 4—5, opposite the sepals, inserted into the top of the tube;
disk perigynous, 5-lobed, the lobes alternating with the stamens.
Fruit drupaceous, 1-seeded, crowned with the limb of the calyx.—
*Very smooth, suffruticose plants of N. America. Ped. axillary and ter-
minal. Fls. small, umbellate.*

C. UMBELLÀTA. Nutt. (Thesium umbellatum. *Linn.*) *Bastard Toad-flax.*
Erect; *lvs.* oval-lanceolate; *fls.* subcorymbed, terminal.—♃ Plant about a
foot high, in rocky woods, U. S. and Brit. Am. Stem slender, striate, generally branching at top. Leaves entire, alternate, acutish, 1—1½' long, and ⅓ as
wide, tapering to a very short petiole. Flowers small, white, in little umbels
of about 3. Each umbel is furnished with a deciduous involucre of about 4
small leaflets, the whole constituting a kind of corymb. Each anther is attached to its opposite sepal by several hair-like, yellow filaments. June.

2. PYRULARIA. L. C. Rich.

Flowers ♀ ☿ ♂. Calyx 5-cleft, subcampanulate; disk 5-toothed,
glandular, half adherent to the ovary; sty. 1; stig. 2 or 3, sub-
lenticular; drupe pyriform, 1-seeded, enclosed in the adhering base of
the calyx.—*Shrubs with the habit of Celastrus. Lvs. alternate, entire.
Rac. terminal.*

P. OLEIFÈRA. L. C. Rich. (P. pubera. *Michx.* Hamiltonia. *Muhl.*) *Oilnut.*
Margins of mountain streams, Penn. to Ga. Shrub 4—6f high, hirsute-
pubescent. Root fœtid. Leaves 2—3' by 1—1½', oblong-ovate, entire, acuminate, petiolate, veins prominent beneath. Flowers small, in a terminal raceme,
greenish-yellow. Calyx tube short, nearly filled with the glandular disk in the
♂ flowers, the segments reflexed in the ♀. Stamens alternate with the lobes
of the disk, opposite to those of the calyx. Drupe globose-depressed. May.

June.—The whole plant is more or less oily, and is greedily devoured by cattle. Sometimes called *Buffalo Oil.*.

3. NYSSA.

The name of a nymph, according to Linnæus.

Flowers diœcious-polygamous. ♂ Calyx 5-parted; stamens 5—10, inserted around a glandular disk; pistil 0. ♀ ♀ Calyx 5-parted; stamens 5 or 0; pistil 1; drupe inferior; nut 1-seeded.—*Trees.*

1. N. MULTIFLORA. Walt. (N. villosa. *Michx.*) *Pepperidge. Tulepo. Gum Tree.—Lvs.* oblong, obovate, very entire, acute at each end, the petiole, midvein and margin villous; *fertile ped.* 3—6-flowered; *nut* short, obovate, obtuse, striate.—This tree is disseminated throughout the U. S.; it is found 30—70ft high, the trunk 1—3ft in diameter, with horizontal branches forming a pyramidal summit. The bark is light gray, similar to that of the white oak, and, like the next species, broken into hexagons. Leaves tough and firm, 2—5' long, and ½ as wide, often with 1 or more blunt teeth. Flowers small, greenish, in small clusters on a long, branching peduncle, the fertile ones succeeded by a few deep blue, oblong drupes. The wood is white, fine-grained, rather soft, the texture consisting of interwoven bundles of fibres, rendering it very difficult to split. It is therefore useful for beetles, naves of wheels, hatters' blocks, &c. Jn.

2. N. AQUATICA. (N. biflora. *Walt.*) *Water Tulepo or Gum Tree.*
" *Lvs.* oblong-ovate, entire, acute at both ends, smooth; *fertile ped.* 2-flowered; *drupe* short, obovate; *nut* striate.—This tree grows in swamps, in certain sections of the Northern and Middle States. The trunk when full grown is 30—50f high and 15—20' in diameter, the bark divided by deep furrows into hexagons. Leaves alternate, smooth, 2—4' in length. Flowers small and obscure, the fertile ones producing a fruit of a deep blue color, growing in pairs on a common stalk which is shorter than the leaves. The wood is dark brown, similar in quality and uses to the last. Jn." *Michx. abr.*

ORDER CXIV. THYMELACEÆ.—DAPHNADS.

Shrubs with a very tenacious bark, alternate or opposite, entire leaves and perfect flowers.
Cal. free, tubular, colored, limb 4 (rarely 5)-cleft, imbricated in æstivation.   *Sta.* many.
Sta. definite, inserted into the calyx and opposite to its lobes when equal to them in number; often twice
Ova. solitary, with 1 ovule. *Style* 1. *Stigma* undivided.
Fr. hard, dry, drupaceous. *Albumen* 0 or thin.

Genera 38, species 300, very abundant in Australasia and S. Africa, sparingly disseminated in Europe and Asia. The only North American genus is that which follows.

Properties.—The bark is acrid and caustic, raising blisters upon the skin. It is composed of interlaced fibres which are extremely tough, but easily separable. The lace-bark tree (Lagetta) of Jamaica is particularly remarkable for this property.

Genera.

| | obsolete. Stamens exserted | Dirca. | 1 |
| Corolla limb | spreading. Stamens included | Daphne | 2 |

1. DIRCA.

Gr. διρκα, a fountain; the shrub grows wet places.

Calyx colored, tubular, with an obsolete limb; stamens 8, unequal, longer than, and inserted into the tube; style 1; berry 1-seeded.— *Lvs. alternate, simple. Fls. expanding before the leaves.*

D. PALUSTRIS. *Leather-wood.*
Lvs. oblong-ovate or obovate; *fls.* axillary, 2 3 in a hairy, bud-like involucre.—A shrub, 5f in height, when full grown, U. S. and Can. The flowers appear in April and May, much earlier than the leaves. They are small, yellow, funnel-shaped, about 3 together, issuing from the same bud. Leaves entire, on short petioles, pale underneath. Stamens much longer than the sepals, alternately a long and a short one. Berry oval, small, red. Every part of this shrub is very tough. The twigs furnish " rods for the fool's back," the bark is used for ropes, baskets, &c.

2. DAPHNE.

For the nymph Daphne, who, it is said, was changed into a laurel which this resembles.

Calyx 4-cleft, marescent, limb spreading; stamens 8, included in calyx tube; sty. 1; drupe 1-seeded.—*Mostly evergreen shrubs, of much beauty and fragrance. Lvs. generally from the terminal buds, fls. from the lateral. None are American.*

1. D. MEZEREUM.—*Lvs.* deciduous, lanceolate, ix terminal tufts, entire, sessile; *fls.* sessile, about three from each lateral bud; *cal.* hypocrateriform, segments ovate, spreading; *sta.* inserted in 2 rows near the top of the tube; *fil.* very short; *stig.* sessile.—A beautiful, early-flowering, European shrub. The whole plant, especially the bark, is very acrid. Flowers pink-colored in one variety, white in another, clothing nearly the whole plant in March. Height 2–3f. †

2. D. LAUREŎLA. *Spurge Laurel.*—*Lvs.* evergreen, lanceolate, glabrous, acute, entire, subsessile; *fls.* 5 together in each axillary raceme. -A taller shrub than the preceding, from Europe. Stem 4—6f high, with ample and beautiful foliage. Flowers fragrant, greenish-yellow. Mar., Apr. †

3. D. COLLINA.—*Lvs.* evergreen, oblanceolate, obtuse, tapering to a short petiole, very smooth above, villous beneath; *fls.* fasciculate, terminal. Apr.—Jn. +

ORDER CXV. ELÆAGNACEÆ.—OLEASTERS.

Trees and shrubs usually covered with a leprous scurf. Lvs. alternate or opposite, entire. Stip 0
Fls. axillary, diœcious, rarely perfect.
Sterile fl.   Cal.—4-parted.   Sta. 3, 4 or 8, sessile.   Anth. 2-celled.
Fertile fl.   Cal.—Free, tubular, persistent, limb entire or 2—4-toothed.
Ova. simple, 1-celled.   Ovule solitary, stipitate.   Stig. simple, subulate, glandular.
Fr.-Achenium crustaceous, enclosed within the calyx, which becomes succulent and baccate.
Sds. ascending.   Embryo straight, in thin, fleshy albumen.
Genera 4, species 28, thinly dispersed throughout the whole northern hemisphere.

Genera.

Stamens { 2.   *Shepherdia.* 1
{ 4.   *Elæagnus.* 2

1. SHEPHERDIA. Nutt.

In honor of John Shepherd, curator of the botanic garden of Liverpool.

Flowers ♀ ♂.—♂ Calyx 4-cleft; sta. 8, alternating with 8 glands. ♀ Calyx tube closely investing the ovary, but not adhering to it, limb 4-lobed; sty. 1; stig. oblique; berry globose, composed of the fleshy calyx.—*North American shrubs, with spinescent branches, and opposite, deciduous leaves. Fls. aggregated.*

1. S. CANADENSIS. Nutt. (Hippophæ. *Willd.*)
Lvs. elliptic-ovate, nearly smooth above, clothed beneath with stellate hairs and ferruginous, deciduous scales.—A shrub 6—8f high, found in Vt., N. Y. and W. to Wis., *Laplam!* by streams and on river banks. Leaves obtuse at each end, the upper surface green, with few, scattered, stellate hairs, lower surface white, with rust-colored spots, densely tomentose, margin entire; petioles 2—4" long, lamina 1—2' by ½—1'. Flowers minute, in small, lateral, nearly sessile clusters. Berries oval, scaly, consisting of the fleshy calyx enclosing the achenia in its tube, sweetish to the taste. Jl.—A curious and ornamental shrub.

2. S. ARGENTEA. Nutt. (Hippophæ. *Ph.*)—*Lvs.* oblong-ovate, obtuse, both surfaces smooth and equally covered with silvery scales.—A small tree, 12—18f high, with thorny branches. Leaves 1—2' by 4—9". Petioles ⅓' long, margin entire, the surface of a light, silvery hue, sprinkled with rust-colored spots. Fruit the size of a currant, scarlet, well-flavored. †

2. ELÆAGNUS.

Gr. ἐλαια, the olive; the trees having a resemblance to the olive.

Calyx 4-cleft, campanulate, colored on the inner side; sta. 4, alter nate with the calyx lobes; anth. subsessile; sty. short; fruit baccate, consisting of the achenium enclosed in the dry, farinaceous calyx tube marked with 8 furrows.—*Trees or shrubs, cultivated for the silvery foliage. Leaves alternate.*

1. E. ARGENTEA. *Silver-leaved Oleaster.*
Lvs. broad-ovate or oval, wavy, acutish at each end, both surfaces, particularly the lower, silvery and shining with ferruginous scales.—Missouri, &c. A beautiful shrub, with reddish branches and small, roundish-ovate, cartilaginous drupes. †

2. E. ANGUSTIFOLIA. *Narrow-leaved Oleaster.*—Lvs. narrow-lanceolate, acute at each end, entire, alternate, smooth, canescent; fls. axillary, aggregate.—A tree of middle size from Europe, cultivated for its beautiful foliage and fine fruit, which, "when dried, resembles an oblong plum with a reddish skin and a flavor similar to that of a date." ‡

3. E. LATIFOLIA. *Broad-leaved Oleaster.*—Lvs. ovate, evergreen.—E. India. †

ORDER CXVI. ULMACEÆ.—ELMWORTS.

Trees and *shrubs* with alternate, simple, deciduous leaves.
Stip. 2 at the base of each leaf, deciduous.
Fls. sometimes by abortion diœcious, in loose clusters, never in catkins.
Cal. free from the ovary, campanulate, 4–5-cleft, imbricate in æstivation.
Sta. inserted on the base of the calyx, as many as its lobes and opposite to them.
Ova. 1–2-celled. *Ovules* solitary. *Stigmas* 2, distinct.
Fr. indehiscent, either a samara or drupe, 1-celled; 1-seeded.
Seed pendulous, without albumen. *Cotyledons* foliaceous.

Genera 9, species 60, natives of the northern temperate zone. The mucilaginous bark of the *slippery elm* (Ulmus fulva) is the only important medicinal product. Several of the elms afford excellent timber.

Genera.

Flowers all perfect. Calyx 4–5-cleft. Fruit samara. . . . . Ulmus. 1
Flowers polygamous. Calyx of sterile fl 4-parted. Fruit a drupe. . . . . Celtis. 2

1. ULMUS.

From *elm*, its original name in Anglo-Saxon, Teutonic, Gothic, and other Celtic dialects.

Flowers ☿. Calyx campanulate, 4—5-cleft; stamens 5—8; styles 2; samara compressed, with a broad, membranaceous border.—*Trees, rarely shrubs. Lvs. retrorsely scabrous, often abrupt at base. Fls. fasciculate, appearing before the leaves.*

1. U. AMERICANA. *American Elm. White Elm.*
Lvs. ovate, acuminate, serrate, unequal at the base; fls. pedicelled; fr. fimbriate.—This majestic tree is common in the Northern, Middle and Western States. It is a native of the forest, but often grows spontaneously in open fields, where it is readily distinguished by its long, pendulous branches. The trunk, with a diameter of 3—5f, towers to the height of 30, 50, and even 70f, perfectly straight and naked, when it divides into 2 or more primary branches. These ascend, gradually spreading, and repeatedly dividing into other long, flexible limbs bending in broad, graceful curves. It is a great favorite as a shade tree, and is frequently seen rearing its stately form and casting its deep shade over the "sweet homes" of New England. Leaves short-stalked, oval-acuminate, doubly denticulate, and 4—5' long. The veins are quite regular and prominent. Flowers small, purplish, collected into small, terminal clusters, and appearing in April, before the leaves. Fruit flat, fringed with a dense down. The wood is tough and strong, but not easily wrought; used for the naves of wheels, &c.

2. U. FULVA. *Slippery Elm. Red Elm.*
Branches rough; lvs. oblong-ovate, acuminate, nearly equal at base, unequally serrate, pubescent both sides, very rough; buds covered with fulvous down; fls. sessile.—Woods and low grounds, N. Eng. to Car. The slippery elm is much sought on account of the mucilage of the inner bark. Its diameter is 1—2f, and height 20—10. The leaves are larger, thicker and rougher than those of the white elm, and exhale a pleasant odor. Flowers collected at

the ends of the young shoots. Calyx downy, sessile. Stamens short, reddish, 7 in number. April.

3. U. RACEMÓSA. Thomas. *Cork Elm.*

Lvs. ovate, acuminate, auriculate on one side; *fls.* in racemes; *pedicels* in distinct fascicles, united at their base.—A tree found in Meriden, N. H.! to N. Y.! The twigs and branches are remarkably distinguished by their numerous, corky, wing-like excrescences. Leaves 3—4' long, ½ as wide, produced into a rounded auricle on one side, doubly serrate, smooth above, veins and under surface minutely pubescent. Flowers pedicellate, 2—4 in each of the fascicles which are arranged in racemes. Calyx 7—8-cleft. Stamens 7—10. Stigmas recurved. Samara ovate, pubescent, the margin doubly fringed.

4. U. CAMPESTRIS. *English Elm.*—*Lvs.* (small) ovate, doubly serrate, unequal at base; *fls.* subsessile, densely clustered; *sta.* 5—8; *cal.* segments rounded, obtuse; *samara* suborbicular, the border destitute of a fringe of hairs.—From Europe. Introduced and naturalized in the Eastern States to a small extent. It is a majestic tree, 50—70f high. The main trunk is 2—4f diam. at base, usually prolonged to the summit. The bark is more rough and broken than that of the American elm, the branches more rigid and thrown off at a larger angle, the foliage more dense, consisting of smaller and darker-colored leaves. In form it favors the oak more than our own native elms. Many trees of this kind, in the vicinity of Boston, are particularly mentioned in Emerson's Report, pp. 301, 302.

5. U. MONTÀNA. *Scotch Elm. Wych Elm.*—*Lvs.* doubly serrate, unequal at base; *fls.* pedicellate; *sta.* 6—8; *samara* with the margin fringed.—Another European elm, often attaining an enormous size. The timber is highly valuable. In form it more nearly resembles the slippery elm than the white elm.

6. U. NEMORÀLIS. *River Elm.*

Lvs. oblong, nearly smooth, equally serrate, nearly equal at the base; *fls.* sessile.—Banks of rivers, N. Eng. to Virginia, according to *Pursh,* Fl. p. 200, but unknown to any subsequent botanist.

2. CELTIS.

Celtis was the ancient name for the lotus, which this genus is said to resemble.

Flowers monœcious-polygamous. ♂ Calyx 6-parted; stamens 6; ♀ Calyx 5-parted; stamens 5; styles 2; stigmas subulate, elongated, spreading; drupe 1-seeded.—*Trees or large shrubs. Lvs. mostly oblique at base. Fls. subsolitary, axillary.*

1. C. OCCIDENTÀLIS. *American Nettle Tree. Hoop Ash. Beaver Wood.*

Lvs. ovate, acuminate, entire and unequal at base, serrate, rough above, and rough-hairy beneath; *fr.* solitary.—This species is some 30f high in New England, where it is rarely found, but is much more abundant at the South and West. The trunk has a rough, but unbroken bark, with numerous slender, horizontal branches. Leaves with a very long acumination, and remarkably unequal at the base. Flowers axillary, solitary, small and white, succeeded by a small, round, dull red drupe. The wood is tough, and is used for making hoops, &c.

2. C. CRASSIFOLIA. Lam. *Hackberry.*

Lvs. ovate, acuminate, serrate, unequally cordate at base, subcoriaceous, rough both sides; *ped.* about 2-flowered.—This tree is thinly disseminated in the northern parts of the U. S. In woods it is distinguished by its straight, slender trunk, undivided to a great height, covered with an unbroken bark. The leaves are of a thick and firm texture, very large, heart-shaped at base. Flowers small, white, succeeded by a round, black drupe about the size of the whortleberry. The wood is white and close-grained, but neither strong nor durable. May.

3. C. PÙMILA. Pursh. (C. occidentalis. β. *Muhl.*)

Lvs. broadly ovate, acuminate, equally serrate, unequal at the base, smooth on both surfaces, the younger only pubescent; *ped.* mostly 3-flowered; *fruit* solitary. A small shrub, on the banks of rivers, Md., Va. *Pursh.* Leaves

41*

sometimes cordate, nearly as broad as long, with a very short acumination
Berries brown and glaucous.

ORDER CXVII. SAURACEÆ.—SAURURADS.

Herbs aquatic, with jointed stems. Leaves alternate, with stipules.
Fls. in spikes, perfect, achlamydeous, each in the axil of a bract or scale.
Sta. definite, persistent; anth. 2-celled, connectile thick, continuous with the slender filament.
Ova.—Carpels 3—4, each few-ovuled.
Fr.—Capsule or berry 3—5-celled, few-seeded.
Sds. usually solitary in the cells. Embryo minute, cordate, outside of hard, farinaceous albumen.
  Genera 4, species 7, natives of China and North America, growing in marshes and pools. Properties unimportant.

1. SAURURUS.
Gr. σαυρα, a lizard, ουρα, a tail; alluding to the form of the inflorescence.

Inflorescence an ament or spike of 1-flowered scales; stamens 6, 7,
8 or more; anthers adnate to the filaments; ovaries 4; berries 4,
1-seeded.

S. CERNUUS. Willd. (Anonymus aquatica. *Walt.*) *Lizzard Tail.*
St. angular; *lvs.* cordate, acuminate, petiolate.—♃ An aquatic plant, with
neat foliage, and yellowish, drooping spikes of flowers, Can. to Car. and
W. States! Stem 1½—2f high, weak, furrowed. Leaves 4—6' long, and ½ as
wide, smooth and glaucous, with prominent veins beneath and on petioles 1—2
long. Spikes slender, drooping at summit, longer than the leaf. Scales tubular, cleft above, white. Flowers very small and numerous, sessile, consisting
only of the long stamens, and the ovaries with their recurved stigmas. Aug.

ORDER CXVIII. CERATOPHYLLACEÆ.—HORNWORTS.

Herbs submersed, with dichotomous, cellular, verticillate leaves.
Fls. monœcious. Calyx many-parted.
Sterile.—Sta. indefinite (16—20). Anth. tricuspidate, sessile, 2-celled.
Fertile.—Ova. free, 1-celled, with a suspended, solitary ovule. Style filiform, oblique, sessile.
Fr.—Achenium beaked with the indurated stigma.
Seed orthotropous, suspended, exalbuminous, and containing 4 cotyledons.
  Herbs, floating. Leaves cellular, many-cleft, verticillate.
  This order consists of the following genus only, with 6? species. They grow in ditches and pools
Europe, Northern Asia, Barbary, and North America.

CERATOPHYLLUM.
Gr. κερας, a horn, φυλλον, a leaf; alluding to the horn-like divisions of the leaves.

Character the same as that of the order.

C. DEMERSUM. *Hornwort.*
Lvs. 6—8 in a whorl, doubly dichotomous, dentate-spinescent on the back;
fls. axillary; *fr.* 3-spined.—♃ An aquatic weed, in ditches, &c., N. Y. to Va.
W. to Ill. Stem floating or prostrate, 8—16' long, filiform, with numerous
whorls of leaves. They are dichotomously divided into 2 or more filiform segments. Flowers minute, axillary, sessile, with sessile anthers. Fruit an
oblong, beaked capsule, with 1 seed. Jl.—Sept.

ORDER CXIX. CALLITRICHACEÆ.—STARWORTS.

Herbs aquatic, small, with opposite, simple, entire leaves.
Fls. axillary, solitary, very minute, monœcious, achlamydeous, with 2 colored bracts.
Sta. 1, rarely 2; filaments slender; anther 1-celled, 2-valved, reniform.
Ova. 4-celled, 4-lobed; ovules solitary. Styles 2; stigmas simple points.
Fr. 1-celled, 4-seeded, indehiscent. Seeds peltate, albuminous.
  Genus 1, species 6, growing in stagnant waters both of Europe and America.

CALLITRICHE.
Gr. καλος, beautiful, θριξ, τριχος, hair; alluding to the slender stems.

Character the same as that of the order.—♃.

C. VERNA. (C. intermedia. *Willd.* C. aquatica. *Bw.*)
Floating; *upper lvs.* oblong-spatulate, lower ones linear, obtuse or emar-

ginate; *caps.* with the lobes obtusely margined.—① This littl? polymorphous plant is common in shallow streams and muddy places, U. S. Stem floating, 1—2f long, composed of 2 tubes, simple or branched. Leaves 2 at each node, becoming crowded above into a star-like tuft upon the surface of the water, the lower ones becoming gradually narrower, and the lowest quite linear. Flowers white, axillary, 1 or 2 together, often monœcious. Anther a little exserted, yellow. May—Sept.

β. *autumnalis.* Darl. (C. autumnalis. *L.* C. linearis. *Ph.*) Floating; *lvs.* nearly all linear, 1-veined, truncate or emarginate, the upper ones a little dilated towards the end.—In similar situations with, and generally accompanying the former.

γ. *terrestris.* Darl. (C. terrestris. *Raf.* C. brevifolia. *Ph.*) Procumbent and diffuse; *lvs.* all oblong, obtuse, crowded, fleshy.—Grows on soft mud, evers. spreading the surface. This form evidently depends on the locality.

ORDER CXX. PODOSTEMACEÆ.—THREADFOOTS.

Herbs aquatic, without spiral vessels, of the habit of the liverworts or mosses.
Lvs. capillary or linear, not articulated with the stem.
Fs. monœcious, achlamydeous, bursting through an irregularly lacerated spathe.
Sta. often reduced to 1 or 2, and monadelphous.
Ova. free, 2—3-celled. *Sty.* 2 or 0. *Stig.* 2—3.
Fr.—Capsule subpedicellate, opening by 2 valves.
Sds. numerous, without albumen, attached to a central column.

Genera 5, species 25, natives of N. America and Asia.

PODOSTEMUM. L. C. Rich.

Gr. πους, ποδος, a foot, στημον; the stamens being apparently on a common foot-stalk.

Stamens 2, with the filaments united below; ovary oblong-ovoid; stigmas 2, sessile, recurved; capsule 2-celled; seeds minute.—*Small, submersed herbs, adhering to stones and pebbles.*

P. CERATOPHYLLUM. Michx. (Lacis ceratophylla. *Bongard.*) *Threadfoot.*
Lvs. dichotomously dissected; *fls.* solitary, axillary.—Middle! Western and Southern States. Stem a few inches long, usually destitute of roots and attached to stones by lateral, fleshy processes. Leaves numerous, alternate, coriaceous, divided into many long, linear-setaceous segments. Flowers on short, thick peduncles, the 2 stamens and styles at length bursting through the lacerated calyx. In shallow, running streams. July.

ORDER CXXI. EUPHORBIACEÆ.—SPURGEWORTS.

Trees, shrubs or herbs, often abounding in an acrid milk.
Lvs. opposite or alternate, simple, rarely compound, often furnished with stipules.
Fls. monœcious or diœcious, axillary or terminal. *Cal.* inferior, lobed or wanting.
Cor.—Petals or scales equal in number to the sepals, or wanting.
Sterile fl.—*Sta.* definite or indefinite, distinct or monadelphous. *Anth.* 2-celled.
Fertile fl.—*Ova.* free, of 2—9 more or less united carpels, coherent to a central prolongation of the axis.
Styles distinct, often 2-cleft.
Fr.—Capsule of 2 dehiscent carpels which open elastically.
Sd. with a large embryo in fleshy albumen.

Genera 191, species 2500? chiefly natives of S. America, not more than 50 species being found in N. America, north of Mexico.

Properties.—An acrid, stimulant and poisonous principle, residing chiefly in the milky juice, pervades the whole order This principle varies in activity from mild stimulants to the most active poisons: but it is volatile and easily expelled by heat. *Tapioca* is a starch-like accumulation formed in the roots of the Jatropha Manihot. When fresh, this root is a violent poison, but loses its deleterious properties by washing and exposure to heat. *Castor-oil* is expressed from the seeds of Ricinus communis. *Croton-oil* from the seeds of Croton Tiglium. Caoutchouc is yielded in abundance by several S. American species.

FIG. 63.—1. Head or capitulum of Euphorbia cordata, with the corolla-like involucre, and pedicellate pistillate flower. 2. The involucre tube cut open, showing the monandrous, staminate flowers surrounding the pistillate. 3. One of the flowers, with a toothed bract at base. 4. Cross section of the ovary, showing the 3 one-seeded cells or carpels.

Conspectus of the Genera.

9—16-androus. { Fruit in a toothed, leafy involucre.				Acalypha. 2
{ Fruit not involucrate.				Croton. 5
polyandrous. Leaves large, glaucous, peltate.				Ricinus. 3
tetrandrous. Leaves evergreen, or te.				Buxus. 7
pentandrous. Leaves clothed with shining scales.				Crotonopsis. 4
triandrous. Leaves very small.				Phyllanthus. 6
Sterile fls. { monandrous, several in a corolla-like involucre.				Euphorbia 1

1. EUPHORBIA.

Eupherbus, physician to Juba, king of Mauritiana, first used these plants in medicine.

Flowers ♂, mostly achlamydeous; involucre monophyllous, sub-campanulate, with 4—5 petaloid segments alternating with as many external, gland-like teeth. ♂ 12 or more; stamen 1; filament articulated in the middle. ♀ solitary, central; ovary pedicellate; styles 3, bifid; capsule 3-lobed, 3-celled; cells 1-seeded.—*Herbs or shrubs, with a-milky juice. Lvs. generally opposite, sometimes wanting, often stipulate. Invol. axillary or subumbellate.*

* *Heads of flowers in involucrate umbels.* † *Cauline leaves alternate.*

1. E. COROLLÀTA. *Flowering Spurge.*
Erect; *cauline* and *floral lvs.* oblong, narrow, obtuse; *inner segments of the invol.* obovate, petaloid; *umbel* 5-rayed, rays 2 or 3 times di- or trichotomous.— ♃ In dry fields, &c., Can. and U. S. Stem slender, erect, 1—2' high, generally simple and smooth. Leaves 1—2' long, often quite linear, very entire, scattered on the stem, verticillate and opposite in the umbel. The umbel, as in other species, consists of about 5 verticillate branches from the summit of the stem, each of which is subdivided into about 3, and finally into 2 peduncles. Corolla-like involucre large, white, showy. July Aug.—The central head is 2 or 3 weeks earliest.

2. E. HELIOSCOPIA. *Sun Spurge.*
Erect; *floral lvs.* obovate, *cauline* wedge-form, serrate, smooth; *umbel* 5-rayed, then 3-rayed and forked.—① A milky weed in cultivated grounds, N. H.! to Niagara ! S. to Car., remarkable for the symmetry of its vegetation. Stem smooth, erect, 8—16' high. Leaves scattered, ½—1½' long, ⅔ as broad at the rounded or retuse apex, finely and sharply serrate, entire, and tapering to the base. Umbels subtended by a large involucre of 5 obovate leaves. Each of the 5 rays is pilose with scattered hairs and subdivided into an umbellet of 3 rays with a 3-leaved involucel, and these finally into 2 or more pedicellate fascicles. Capsules smooth. June, July.

3. E. NEMORÀLIS. Darl. (E. pilosa. *Pursh.*) *Hairy Spurge.*
Leaves oblong-lanceolate and oblanceolate, acute, narrowed to the base, subsessile, pilose beneath, those of the involucre ovate; *umbel* 5—8-rayed, rays ones or twice divided; *seg. of the floral invol.* colored, entire, subreniform; *ova.* verrucose.—♃ Moist woods, Can. to Va. Stem 2—3f high, smooth, rarely branched below the umbel. Leaves 3—4' by 1', entire or slightly serrulate above, those of the stem alternate, those of the branches opposite and nearly as broad as long. Floral involucre purplish-brown within. Capsule at length nearly smooth. May, June.

4. E. OBTUSÀTA. Pursh. *Obtuse-leaved Spurge.*
Erect; *lvs.* alternate, sessile, spatulate, serrulate, smooth; *umbel* 3-rayed, rays twice dichotomous; *floral lrs.* ovate, subcordate, somewhat obtuse; *caps.* muricate.—♃ Grows in waste grounds, Can. to Va. Stem 12—18' high

5. E. PEPLUS.
St. erect or ascending, branched below; *umbels* 3-rayed, rays mostly dichotomous; *involucrate lvs.* ovate, acute.—① A small species, in cultivated grounds, Penn. to Va. Floral leaves large. Flowers conspicuous. *Torr.*

† *Cauline leaves opposite.*

6. E. MERCURIALINA. Muhl.
St. weak and slender, simply 3-cleft; *lvs.* opposite and ternately verti-

cillate, subsessile, oval and entire; *ped.* terminal, solita.;, 1-lowered.—♃ In rocky shades, Penn. *Pursh.* Jl. Aug.

7. E. Lathyris. *Mole-tree., Caper Spurge.*
St. erect, stout, smooth; *lvs.* lance-linear, rather acute, entire, glabrous, sessile; *umbel* mostly 4-rayed, rays dichotomous; *glands of the invol.* lurate, 2-horned, the horns dilated and obtuse.—♁ Cultivated grounds and gardens. Stem 2—3f high. Leaves 2—4′ by 3—9″, numerous and arranged in 4 rows on the stem. Umbel of 4 verticillate branches with a central subsessile head. Jl.—Sept. §

8. E. Herronii. Riddell. *Herron's Euphorbia.*
Erect, pilose, with opposite branches; *lower lvs.* rhombic-ovate, middle ones lanceolate, upper lance-linear, denticulate, obtuse at apex, acute at base; *umbels* small, few flowered.—♁ Ohio. Plant hairy in all its parts, 1—2f high. Petiole ciliate, half as long as the leaves. Aug. Sept.

9. E. angustifolia. Haworth?—*St.* much branched at base, ascending and with barren branches above; *lvs.* crowded, linear, obtuse; *umbel* many-rayed, proliferous; *floral invol.* 2-leaved, leaflets broad-cordate.—♃ A heath-like plant of the gardens, native of ——, 1f high. Leaves 1′ by 1″, very numerous, spi-rally arranged, apparently verticillate. †

10. E. marginata. Pursh.—*Lvs.* oblong-lanceolate, subcordate, sessile, acute, mucronate, entire on the margin, glabrous; *umbel* 3-rayed, once or twice dicho-tomous; *involucrate lvs.* oblong, cordate, colored and membranaceous at the margin; inner segments of the floral involucre roundish; *caps.* hoary-pubes-cent.—♁ Gardens. A handsome species, remarkable for the variegated leaves of the involucre. †

* * *Heads axillary or fasciculate. Leaves opposite.*

11. E. hypericifolia. *Spurge. Eye-bright.*
St. smooth, branching, nearly erect, branches divaricate-spreading; *lvs.* opposite, oval-oblong, serrate, sub-falcate; *corymbs* terminal.—♁ A slender and branching plant, found in dry and rich soils. Stem 10—20′ high, usually pur-ple, very smooth, the branches often pubescent, U. S. and Can. Leaves tripli-veined, marked with oblong dots and blotches, ciliate, 6—12″ long, and ⅓ as wide, oblique, on very short petioles. Corymbs of small white heads, terminal and axillary. July, Aug.

12. E. maculata. (E. depressa. *Ell.* E. thymifolia. *Linn.*) *Spotted Spurge.*
Procumbent; *branches* spreading; *lvs.* serrate, oblong, hairy; *fls.* axillary. —♁ A prostrate plant, spreading flat u;n the ground, in sandy fields, Mass.! to Ind.! and S. States. Stem 6—12′ in length, much branched, hairy. Leaves opposite, 3—6″ l-ng and ⅓ as wide, oblong, obtuse, serrulate, smooth above, often spotted with dark purple, the margin ciliate, pale and hairy beneath, on short stalks. Heads of flowers small, crowded near the summit, involucre minute, white. Jl.—Sept.

13. E. polygonifolia. *Knot-grass Spurge.*
Procumbent; *lvs.* entire, lanceolate and oblong, obtuse at base; *fls.* in the axils of the branches, solitary.—♁ Sea shores, R. I.! to Flor. A very smooth, succulent, prostrate plant, with milky juice. Stems 6—10′ long, dichotomous, procumbent. Leaves oblong and linear-lanceolate, rarely cordate at base, 3—5″ by 1″, petioles about 1″. Stipules subulate and simple. Heads small, in the forks of the purple stem. June, July.

14. E. Ipecacuanhæ. *Ipecac Spurge.*
Procumbent or suberect, small, smooth; *lvs.* opposite, obovate and oblan-ceolate; *ped.* elongated, axillary, 1-flowered.—♃ Sandy soil, Middle! and S. States. Root perennial, very long. Stem rather thick and succulent, 3—6′ long. Leaves 1½—2′ by 3—6″, sessile, varying from obovate to linear. Heads solitary. Peduncles as long as the leaves. June.

15. E. dentata. Michx. *Toothed-leaf Spurge.*
Hairy; *lvs.* opposite, oval, dentate; *fls.* crowded at the summit of the stem.—♁ Shady rocks, Penn., *Torrey,* to Tenn., *Michaux.* Upper leaves spotted.—Probably a variety of E. hypericifolia. Jl. Aug.

16. E. PUNICEA. *Scarlet or Splendid Euphorbia.*—*St.* suffruticose, fleshy, armed with rigid, sharp thorns; *lvs.* ovate, tapering to the base, glabrous, entire, acute, mucronate; *ped.* axillary. 2 or 3 times dichotomous; *involucrate bracts* scarlet. ·—A singular and showy garden plant.

2. ACALŸPHA.

The Greek name for the nettle, which this plant resembles.

Fls. δ.—δ Calyx 3—4-parted; sta. 8—16, united at base.—φ Calyx 3-parted, segments connivent, persistent; styles 3, elongated, 2—3-parted; caps. 3-celled, cells 1-seeded.—*Herbaceous or shrubby Lvs. alternate.*

A. VIRGINICA. *Three-seeded Mercury.*
Pubescent, branched; *lvs.* petiolate, oblong-lanceolate, serrate; *invol.* of the fertile flowers cordate, broad-ovate, acuminate, veined and toothed.—① In dry and gravely soils, U. S. and Can., rare in N. Eng. Stem erect or ascending at base, 10—18' high. Leaves 3-veined, 1—2½' long, ½ as wide, hairy, obtusish. Pistillate flower at the base of the peduncle of the staminate spike. Involucrum of the fruit axillary, on a short stalk, shorter than the leaves, its margin cut half way down into long, acute segments. Aug.
β. (A. Caroliniana. *Walt.*) *Lvs.* rhombic-ovate, on long petioles.—Penn. to Ind. !

3. RICÌNUS.

Lat. *ricinus*, an insect, which the fruit of these plants resemble.

Flowers δ.—δ Calyx 5-parted; sta. many. φ Calyx 3-parted; sty. 3, 2-cleft; caps. echinate, 3-celled, 3-seeded.—*Herbs and shrubs with peltate, palmate lvs.*

R. COMMÙNIS. *Castor-oil Bean. Palma Christi.*—*St.* frosted or glaucous, white, herbaceous; *lvs.* peltate, palmate, lobes lanceolate, serrate; *caps.* prickly.—Native of the E. Indies, where it becomes a tree, although an herbaceous annual with us. In our gardens it is a tall, smooth plant of a light bluish-green color. Leaves 4—12" diam., on long petioles. From its seeds is expressed the well known castor oil of the shops. For this purpose it is extensively cultivated in the U. S. July, Aug. ‡

4. CROTONOPSIS. Michx.

Named from its resemblance ($o\psi\iota s$) to the next genus below.

Flowers δ.—δ Calyx 5-parted; cor. of 5 petals; stamens 5. φ Calyx 5-parted; cor. 0; stig. 3, twice bifid; caps. 1-seeded, indehiscent.—① *Lvs. alternate, stellately pubescent and shining. Fls. aggregate, the upper ones sterile.*

C. LINEÀRIS. Michx.
St. erect, dichotomously branched; *lvs.* clothed with a stellate pubescence above, with hairs and shining scales beneath.—In the sandy swamps of N. J. to Car. and Ill. *Nuttall.* Stem 12—18' high, and like the leaves sprinkled with silvery, shining scales. Leaves on short petioles, linear-lanceolate or lance-ovate. Flowers very minute, in terminal and axillary spikes. June.

5. CROTON.

A Greek name, synonymous with the Lat. *ricinus.* See genus No. 3.

Flowers δ.—δ Calyx cylindrical, 5-toothed; cor. of 5 petals or 0; sta. 10—15. φ Calyx 5—many-sepaled; cor. 0; styles 3 or 6, bifid; capsule of 3, coherent, 1-seeded carpels.—*A large genus, mostly tropical, and inconspicuous weeds.*

1. C. CAPITÀTUM. Michx.
St. woolly, tomentose; *lvs.* val-oblong, obtuse, rounded and entire at the base, clothed with soft tomentum on both surfaces; *fertile fls.* at the base of the

spikelets; *sty.* 6, twice bifid; *staminate fls.* capitate, crowded.—① Grows in sandy prairies, Ill. to the sources of the Missouri.

2. C. ELLIPTICUM. Nutt. (Crotonopsis elliptica. *Willd.*)

Plant clothed with a stellate pubescence; *lvs.* elliptical-ovate, the older ones obtuse at apex, smoothish and green on the upper surface; *fls.* glomerate; *sty.* 3, bifid; *caps.* angular, 2-seeded.—① Ill. *Mead,* and Mo.

3. C. GLANDULÒSUM.

St. trichotomous; *lvs.* oblong, serrate, hairy beneath, nearly entire, and bearing 2 glands at the base; *spikes* of flowers situated in the division of the stem.—① Ill. (*Mead*), river bottoms.

6. PHYLLANTHUS.
Gr. φυλλον, ανθος; the leaves of the original species bear flowers at the edges.

Flowers ♂.—♂ Calyx persistent, with 6 spreading, colored segments; stamens 3, very short, filaments united at base, anthers didymous. ♀ Calyx as in the ♂; styles 3, bifid; capsule 3-celled; cells 2-valved, 1—2-seeded.—*Herbs or shrubs with alternate, stipulate leaves and minute, axillary flowers.*

P. CAROLINENSIS. Walt. (P. obovatus. *Willd.*)

St. erect, herbaceous, with alternate branches; *lvs.* simple, entire, glabrous, oval and obovate, obtuse, slightly petioled; *fls.* few, subsolitary, axillary. —① A small-leaved, delicate plant, Penn.! to Ill.! Stem 6—10′ high, slender, the branches filiform. Leaves of the stem 6—8″ by 4—5″, of the branches twice, and of the branchlets four times smaller. Flowers 1—3 in each axil, the ♂ with the ♀, ½—1″ diam., whitish. July, Aug.

7. BUXUS.
The Greek name of this plant was πυξος.

Flowers ♂.—♂ Calyx 3-leaved; petals 2; sta. 4, with the rudiment of an ovary. ♀ Cal. 4-sepaled; pet. 3; sty. 3; caps. with 3 beaks and 3 cells; seeds 2.—*Shrubs. Lvs. evergreen, opposite.*

B. SEMPERVIRENS. *Box.—Lvs.* ovate; *petioles* hairy at edge; *anth.* ovate, saggittate.—Var. *angustifolia* has narrow, lanceolate leaves. Var. *suffruticosa,* the dwarf box has obovate leaves and a stem scarcely woody, highly esteemed for edgings in gardens.—The box with its varieties is native of Europe.

ORDER CXXII. EMPETRACEÆ.—CROWBERRIES.

Shrubs small, evergreen, heath-like, with exstipulate leaves and minute, axillary flowers.
Fls. dioecious. *Cal.* consisting of hypogynous, imbricated scales.
Sta. equal in number to the inner sepals and alternate with them.
Ova. 3—9-celled, with a single erect ovule in each cell.
Styles short or 0. *Stigmas* lobed and often lacerated.
Fr.—Drupe seated in the persistent calyx, containing 3—9 bony nucules.
Sds. solitary, ascending, albuminous. *Radicle* inferior.

Genera 4. species 4. natives of Europe, North America and the Straits of Magellan. They are acrid The berries are used for food in Greenland.

Genera.

Drupe { 6—9-seeded. Empetrum. 1
       { 3-seeded.   Oakesia. 2

1. EMPÈTRUM.
Gr. εν, upon, πετρος, a stone; from the places of its natural growth.

Flowers ♀ ♂. Perianth consisting of 2 series of sepaloid scales ♂ Stamens 3, anthers pendulous on long filaments. ♀ Styles 6—9, very short, erect, or 0; stigmas oblong, radiate-spreading; drupe globose, 1-celled; seeds 6—9.—*Low, alpine shrubs.*

E. NIGRUM *Crowberry.*

Procumbent; *branches* smooth; *lvs.* imbricated, linear-oblong, obtuse at each end, nearly smooth, with a revolute margin.—A small, prostrate, alpine shrub, found on the granite rocks of the White Mts. of N. H., and the calcareous mountains of Vt. The stem is 1 to 3 or 4f long, much branched and closely covered all around with evergreen leaves, which are ⅓—½' long and a line wide. Flowers very small, reddish, crowded in the axils of the upper leaves. Berries black, not ill-flavored. May, Jn.

2. O A K E S I A. Tuckerman.

Dedicated to William Oakes, Esq., of Ipswich, Mass., to whom N. Eng. botany is greatly indebted.

Flowers ♂ ☿ ♀.—♂ Stamens 3, enclosed in 3—6 membranaceous, sepaloid scales; fil. filiform, exserted, distinct; anth. 2-lobed, opening by lateral clefts. ♀ or ☿ Calyx of 3 equal, membranaceous scales in the axis of a larger, ovate, ciliate scale; stam. 3 or 0; sty. trifid; ova. 3-celled; drupe 3-seeded.

O. CONRÁDI. (Empetrum. Torr. Tuckermania. Klotzch.) *Plymouth Crowberry.*—A low, bushy, tufted shrub, about 1f high, in Plymouth, Mass., *Emerson*, pine barrens, N. J., *Torrey*, N. Y., *Vasey*. Stems slender, with a reddish-ash-colored bark, with short, verticillate branches. Leaves evergreen, numerous, spiral or imperfectly verticillate, ¼' long, linear, revolute. Flowers in terminal clusters of 10—15, with brownish scales and purple stamens and styles.—Plants with ☿ are less common than those with ♂ or ♀. March, Apr.

ORDER CXXIII. JUGLANDACEÆ.—WALNUTS.

Trees, with alternate and unequally pinnate leaves and no stipules.
Fls. green, inconspicuous, monœcious. *Sterile* in aments. Coro a 0.
Cal membranaceous, oblique, irregular. *Stg.* indefinite (5—20).
Fertile in small clusters. *Corolla* 0 or sometimes present and 3—5-petaled.
Cl.—Tube adherent, limb 3—5-parted. [dilated.
Ova. 1-celled (partially 2—4-celled). *Ovule* solitary, erect. *Styles* 0—2, very short. *Stig.* 1—2, much
Fl. drupaceous, 1-celled, with 2—4 imperfect partitions; endocarp bony.
fil. 2—4-lobed, without albumen, oily.
Genera 4, species 27, mostly North American.
Properties:—The well known fruit of the *butternut, walnut, pecan-nut, &c.,* is sweet and wholesome, abounding in a rich, drying oil. The epicarp, and even the integument of the kernel, are very astringent. The timber is highly valuable.

Genera.

corolla of the fertile flowers { 4-cleft. Leaves 15—21-foliate.	. . . . . .	*Juglans.* 1
{ none. Leaves 5—9-foliate.	. . . . . .	*Carya.* 2

1. JUGLANS.

Lat. *Jovis glans;* i. e. the nut of Jove; a name given it by way of eminence.

♂ in an imbricated, simple ament; calyx scale 5—6-parted, somewhat bracteate at base; stamens about 20. ♀ Calyx 4-cleft, superior; corolla 4-parted; stigmas 2; fruit drupaceous, epicarp spongy, indehiscent, endocarp rugose and irregularly furrowed.—*Trees of large size, with alternate, unequally pinnate leaves. Leaflets numerous. Sterile aments axillary. Fertile flowers terminal.*

1. J. CINEREA. (J. cathartica. Michx.) *Butternut. White Walnut.*

Lfts. numerous (15—17), lanceolate, serrate, rounded at the base, softpubescent beneath; *petioles* villous; *fr.* oblong-ovate, with a terminal, obtuse point, viscid, hairy; *nucleus* oblong, acuminate, deeply and irregularly furrowed.—The butternut is found throughout the N. England, Middle and Western States, and Canada, growing on elevated river-banks and on cold, uneven, rocky soils. It is 40—50f high, with a large, but short trunk. The branches are horizontal, and unusually wide-spreading, forming a very large head. Leaves 12—20' long, consisting of 7 or 8 pairs of leaflets, with an odd one. Barren flowers in long aments; fertile in short spikes. The kernel is oily, pleasant-flavored, and well-known in N. England. The wood is of a redd sh hue, light, and is considerably used in panneling and ornamental work. From the bark is extracted an excellent cathartic. April, May

2. J. NIGRA. *Black Walnut.*

Lfts. numerous (15—21), ovate-lanceolate, serrate, subcordate, tapering above; *petioles* and under side of the leaves subpubescent; *fr.* globose, with scabrous punctures.—The black walnut is a common and stately forest tree in the Middle and Western States, but sparingly found in the Northern. It arises 60—90f! high with a diameter of 3—6. In open lands it spreads widely into a spacious head. The duramen of the wood is compact and heavy, of a deep violet color, surrounded with a white alburnum. It is used extensively, west of the Alleghanies, for building, and every where for cabinet work. April, May.

2. CARYA.

Gr. καρυα, *the walnut, from* καρα, *roundish; in allusion to the shape of the nut.*

♂ Aments imbricated, slender and mostly 3-parted or trichotomous; scales 3-parted; stamens 4—6; anthers hairy. ♀ Calyx 4-cleft, superior; corolla 0; styles 0; stigma divided, 2-lobed, the lobes bifid; epicarp 4-valved; nucleus subquadrangular, even.— *Large trees. Pubescence stellate. Lfts. few. ♂ aments branched.*

1. **C. ALBA.** Nutt. (Juglans squamosa. *Mx. f.*) *Shag-bark Walnut or Hickory.*—*Lfts.* 5—7, on long petioles, oblong-lanceolate, acuminate, sharply serrate, villous beneath, the odd one sessile; *aments* filiform, smooth; *fr.* somewhat quadrangular, smooth.—Native throughout the Atlantic States, and abundant farther west. In forests it is very tall and slender, with rough and shaggy bark consisting externally of long, narrow plates loosely adhering by the middle. Like other hickories, the wood is strong and elastic, compact and heavy, and is much used where these qualities are required, as in making hoops, whip stalks, axe handles, the keels of vessels, &c. It is considered superior to all other wood for fuel. The fruit is covered with a very thick epicarp separating into 4 parts, and containing a thin-shelled, richly-flavored kernel. April, May.

2. **C. TOMENTOSA.** Michx. (Juglans alba. *Linn.*) *White Walnut. Mockernut Hickory.*—*Lfts.* 7 or 9, oblong-lanceolate, acuminate, slightly serrate, pubescent beneath, odd one subpetiolate; *aments* filiform, tomentose; *fr.* subglobose, smooth, with a very thick pericarp; *nut* with a hard, thick shell and a small, but agreeably-flavored kernel.—This tree is found in all the Atlantic States, growing 50f high in woods. The bark is thick and rugged, but never scaly. In winter it may be known by its large, greyish-white and very hard buds. The drupes, which are very various in size, have a small kernel difficult to extract, on which account they are less sought than the *shag-barks.* April, May.

3. **C. PORCINA.** Nutt. (Juglans glabra. *Willd.*) *Hog Walnut.*
Lfts. mostly 7, lanceolate, acuminate, serrate, smooth both sides, odd one subsessile; *fr.* and *nucleus* obcordate or oblong.—Found in woods throughout the U. S., growing to the height of 50—70f. The drupes afford small, bitter kernels. The wood possesses the general properties of the hickories in a superior degree, and is used wherever great strength is required. It makes excellent fuel. May.

4. **C. AMARA.** Nutt. (Juglans amara. *Michx.*) *Bitternut Hickory.*
Lfts. about 9, ovate-oblong, acuminate, sharply serrate, smooth both sides except the pubescent veins and midvein, odd one short, petiolate, the rest sessile; *fr.* subglobose, with the sutures prominent above; *drupe* smooth, subglobose; *kernel* bitter.—Grows in most of the U. S., but attains its greatest size in Penn. and along the Ohio valley. The drupe has a thin shell which may be broken by the fingers and contains a kernel so bitter that animals will scarcely touch it.

5. **C. SULCATA.** Nutt. (Juglans. *Willd.* J. laciniosa. *Mx.* Thick Shell-bark.
Lfts. 7—9, obovate-lanceolate, acuminate, serrate, pubescent beneath, terminal one subsessile and attenuate to the base; *fr.* roundish, 4-angled, smooth, nut oblong, slightly compressed, conspicuously mucronate.—N. Y. to Car. and

42

Western States, generally growing in the vicinity of rivers. Rare east of the Alleghanies. It more nearly resembles C. alba than any other species. It is a large tree, 50—70f high. The bark is divided into long strips which at length are attached only by the middle, narrower and of a lighter color than C. alba. Leaves 10—20′ in length, composed of 7, or more frequently 9 leaflets. Sterile aments 3-parted, very long. Nut with a very thick, 4-parted pericarp, and nearly twice larger than in C. alba.

6. C. MICROCARPA. Nutt. *Small-fruited Hickory.*

Lfts. 5- 7, oblong-lanceolate, glabrous, glandular beneath, serrate, conspicuously acuminate; *aments* glabrous; *fr.* roundish-ovoid, pericarp thin; nut small, slightly quadrangular.—A large tree, 60—80f high, in moist woodlands, Penn. *Darlington.* Trunk 1½—2f diam., with an even bark. Leaflets mostly 5, often 7, 4—8′ by 2—3′, the under surface tufted in the axils of the veinlets and sprinkled with dark, glandular dots. Aments long and slender. Pistillate flowers 2 or 3 together, terminal, on a common peduncle, with conspicuous sepals. Fruit about the size of a nutmeg. Nut with a thin shell, not mucronate.

ORDER. CXXIV. CUPULIFERÆ.—MASTWORTS.

Trees and shrubs. Lvs. stipulate, alternate, simple, straight-veined, that is with the veinlets proceeding straight from the midvein to the margin.
Fls. generally monœcious. *Sterile* in aments, *fertile* solitary, or 2 or 3 together, or in fascicles.
Cal.—Sepals regular and membranous, or scale-like.
Sta. 1—3 times as many as the sepals, inserted into their bases.   (each
Ova. adherent, seated within a coriaceous involucrum (cupule), with several cells and several ovules in each
Stig. several, subsessile, distinct.
Fr. A bony or coriaceous nut, more or less enclosed in the cupule.
Sds. 1, 2 or 3 (most of the ovules being abortive), pendulous. *Albumen* 0.
Embryo large. *Cotyledons* fleshy, plano-convex. *Radicle* minute, superior.

Genera 8, species 285, constituting a large portion of the forests of the northern temperate regions, and of mountainous tracts within the tropics.

Properties.—The bark of the oak and other genera is well known for its astringent qualities. The edible fruit of the *hazel-nut, chestnut, beechnut, &c.*, are too well known to require description. Cork is the bark of Quercus Suber. *Nut-galls* are produced from the petioles of Q. infectoria of Asia Minor, being caused by wounds made by insects.

Conspectus of the Genera.

in an echinate, valvate cupule, and ovoid-compressed.	. . . .	Castanea. 2
in a muricate, valvate cupule, and sharply 3-angled	. . . .	Fagus. 3
in a hairy, { coriaceous, involucrate cupule. . .	. . . .	Corylus. 4
enveloped { inflated . . { membranous, closed cupule. . . .	. . . .	Ostrya. 5
(acorn) partly immersed in a scaly cupule. . . .	. . . .	Quercus. 1
Nut { naked, concealed in the axil of a foliaceous bract . . .	. . . .	Carpinus. 6

1. QUERCUS.

Celtic *quer,* fine, and *cuez,* a tree; so called emphatically, because the sacred mistletoe grows upon it. The more common Celtic name was *derw;* hence *druid.*

♂ in a loose ament; calyx mostly 5-cleft; stamens 5—10. ♀ Cupule cup-shaped, scaly; calyx incorporated with the ovary, 6-lobed; ovary 3-celled, 2 of the cells abortive; style 1; stigmas 3; nut (acorn) coriaceous, 1-celled, 1-seeded, surrounded at the base by the enlarged, cup-shaped, scaly cupule.—*A noble genus of trees, rarely shrubs. Aments axillary, pendulous, filiform, with the flowers distinct.*

§ 1. *Fructification annual. Fruit pedunculate. Leaves not mucronate.*

* *Leaves lobed.*

1. Q. ALBA. *White Oak.*

Lvs. oblong, pinnatifid-sinuate, smooth, lobes linear-oblong, obtuse, entire dilated upwards; *fr.* pedunculate, cup deep, warty, acorn ovate.—The white oak grows in woods throughout the U. S. and Can., and for grandeur, strength and usefulness, stands preeminent among the sons of the forest. With a diameter of 5—6f, it attains the height of 70—80, but its magnitude varies with the soil and climate. Leaves obliquely divided into rounded, obtuse and entire lobes, not terminated by mucronate points, pubescent beneath when young *Fruit rather large.* Bark white, often with dark spots. The trunk yields timber of great value for strength and durability. It is extensively employed in

QUERCUS. CXXIV. CUPULIFERÆ. 503

ship-building, in coopering, in carriage-making, in ploughs, mills, &c. The bark is useful in tanning, and in medicine. May.

2. Q. MACROCARPA. *Overcup White Oak.*
Lvs. tomentose beneath, deeply and lyrately sinuate-lobed (most deeply in the middle), lobes obtuse, repand, upper ones dilated; *cupule* deep, with the upper scales setose; *acorn* ovate, turgid.—Most common in the Western States! frequenting limestone hills, but is occasionally met with in N. Eng. and N. Y. It is 60—70f in height, clothing itself with dark green, luxuriant foliage. Leaves 10—15' long, being larger than those of any other species here described. The acorns are also of extraordinary size, enclosed ⅔ of their length in the cup which is usually bordered with hair-like filaments. May.

3. Q. STELLÂTA. (Q. obtusiloba. *Michx.*) *Iron Oak.*
Lvs. deeply sinuate, cuneiform at the base, pubescent beneath, lobes very obtuse, the 3 upper ones dilated, 2-lobed; *cal.* hemispherical; *acorn* oval.—The iron oak, called also *post oak, box white oak, turkey oak, &c.*, is common in the Western! Middle and Southern States, rare in N. Eng. It is a tree of moderate size, with widely spreading and very crooked branches. The bark is greyish-white. Leaves thick, strongly tomentose beneath, in 4 or 5 lobes which are sometimes so arranged as to appear cruciform or stellate. Acorns very sweet. The timber is finer grained, stronger and more durable than white oak, hence it is useful for posts, staves, carriages, &c. The crooked branches afford *knees* for ship-building.

4. Q. OLIVÆFORMIS. *Mossy-cup Oak.*
Lvs. oblong, smooth, glaucous beneath, deeply and unequally sinuate-pinnatifid; *cup.* deeply bowl-shaped, fringed on the edge; *acorn* oval-ovate.—This species of oak is confined to a few districts of N. Y. and Penn. It is a large and majestic tree, chiefly remarkable for its smaller branches always inclining downwards. The leaves are so irregularly cut and lobed that scarcely two can be found alike. May.

** Leaves dentate, not lobed.*

5. Q. PRINUS. Willd. (Q. Prinus palustris. *Michx.*) *Swamp Chestnut Oak.*
Lvs. on long petioles, obovate, acute, pubescent beneath, with coarse, unequal, dilated teeth, callous at the tip; *cup* deep, attenuate at base; *acorn* ovate. —This oak is seldom met with in N. England, but abounds in the rest of the U. S. It is one of the loftiest trees of the forest, arising to the height of 50f with its undivided, straight and uniform trunk, and thence with its expansive top to the height of 80—90f. Flowers appear in May, succeeded by large and sweet acorns. The timber is valuable in the arts, and makes excellent fuel.

6. Q. BICÔLOR. Willd. (Q. Prinus discolor. *Michx.*) *Swamp White Oak.*
Lvs. oblong-ovate, downy, white underneath, coarsely toothed, entire at the base, the teeth unequal, dilated, rather acute, callous at the tip; *petioles* short; *fr.* on long peduncles, in pairs, cup hemispherical, acorn oblong-ovate.— The swamp oak is diffused throughout most of the U. S., growing in low, swampy woods. It is a beautiful tree, attaining in favorable situations the height of 70f. Foliage rich and luxuriant, leaves smooth and green above and white downy beneath, 5—7' by 2½—4'. The trunk is covered with a grayish-white bark which divides into large, flat scales. It affords excellent fuel and timber.

7. C. MONTÂNA. Willd. (Q. Prinus monticola. *Mx.*) *Mountain Oak.*
Lvs. broad-ovate, oblong, white downy beneath, shining above, coarsely toothed, obtuse and unequal at the base; *teeth* obtuse (or rarely acute!) subequal; *fr.* in pairs, on short peduncles, cup hemispherical, with rugose and tubercula: scales, acorn ovate. This oak, sometimes called rock chestnut oak, is native of the Northern and Middle States, growing in woods and mountain sides. Its height seldom exceeds 60f, and is generally much less. In open situations its top spreads widely and symmetrically. The petiole is yellow, rather short. Timber valuable in ship-building, &c.

8. Q. CASTÂNEA Willd. (Q. Prinus acuminata. *Michx.*) *Yellow Oak.*
Chestnut Oak.- *Lvs.* on long petioles, oblong-lanceolate, obtuse at base, acuminate, downy beneath, with coarse, subequal, dilated, obtuse teeth, cup

hemispherical; *acorn* roundish ovate.—The yellow oak abounds in the Middle and Western States, in rocky and mountainous woods. It is a large tree, 60f in height. Bark whitish, slightly furrowed. Leaves regularly toothed, light green above, whitish beneath. Flowers in May, succeeded by acorns usually sweet-flavored. Timber little used.

9. Q. Chinquápin. Michx. (Q. prinoides. *Willd.*) *Dwarf Chestnut Oak.*
Lvs. on short petioles, obovate, acute at the base, glaucous beneath, with coarse, subequal, dilated teeth, callous at the tip; *cup* hemispherical; *acorn* ovate.—This is one of the most diminutive of all the oaks, never exceeding 3—4f in height. It is native of the Northern and Middle States, in barren woods, but not common. The flowers appear in May, followed by acorns of middle size, very sweet and so abundant as often to weigh the shrub prostrate on the ground.

§ 2 *Fructification biennial. Fr. subsessile. Lvs. setaceously mucronate.*
    * *Leaves sinuate-lobed.*

10. Q. rubra. *Red Oak.*
Lvs. on long petioles, smooth, obtusely sinuate, lobes rather acute, dentate; *cup* shallow and flat, smoothish; *acorn* subovate.—The red oak is the most common species in the Northern States and in Canada. It is a lofty, wide-spreading tree, 70f in height with a diameter of 3 or 4. Leaves 6—10' long, smooth on both sides, with deep and rounded sinuses between the narrow, mucronate lobes. The flowers appear in May, succeeded by very large acorns contained in cups so shallow as rather to resemble saucers than cups, and are greedily devoured by wild and domesticated animals. The bark is extensively used in tanning. The wood is reddish, coarse-grained, of little value as timber, but excellent for fuel.

11. R. tinctoria. Bartram. *Black Oak. Yellow-bark Oak.*
Lvs. obovate, oblong, sinuate, pubescent beneath, finally glabrous, lobes oblong, obtuse, mucronate; *cup* flat; *acorn* depressed-globose.—This oak is found throughout the U. S. It is one of the loftiest trees of the forest, 80—90f in height and 4—5 in diameter. Bark deeply furrowed, black or deep brown. Leaves 6—8' long, broadest towards the end, quite variable. Acorns brown, nearly sessile, about half covered with the thick, scaly cup. From the bark of this species, *quercitron*, used in dyeing, is obtained, hence it is called quercitron oak. The bark is used in tanning.

12. Q. coccinea. Wang. *Scarlet Oak.*
Lvs. on long petioles, oblong, deeply sinuate, smooth, lobes divaricate, dentate, acute; *cup* turbinate, scaly; *acorn* short, ovate.—The scarlet oak is most abundant in the Middle and Southern States, but is often met with in the more southern parts of N. England. It is a large tree, 80f in height, with a diameter of 3 or 4. Leaves of a bright, shining green, with about 4 deep sinuses, remarkably rounded and broad at the base. By the frosts of autumn they are changed to *scarlet*, unlike those of the red oak which become dull red or brown. Acorns large, similarly rounded at both ends, half immersed in the cup. Bark very thick, used in tanning. The wood is little valued for timber or fuel.

13. Q. palustris. Michx. *Pin Oak. Water Oak.*
Lvs. on long petioles, oblong, deeply sinuate, smooth; *axils of the veins* tufted-villous beneath, lobes divaricate, rather narrow, dentate, acute; *cup* flat, smooth; *acorn* subglobose.—The pin oak is most luxuriant in the W. States and the adjacent districts of other States, rare in New England, growing in swamps and wet woods. Height 60—90f, with a diameter of 2—4. It is remarkable for its unusual number of secondary branches which die as the tree advances, giving the trunk the appearance of having *pins* or tree nails driven into it; also for its light and open foliage. Bark smooth. Wood coarse-grained, little esteemed as timber. Acorns small, round, in shallow cups. May.

14. Q. elongáta. (Q. rubra. *Wall.* Q. falcata. *Michx.*) *Spanish Oak.*
Lvs. on long petioles, 3-lobed or sinuate, tomentose beneath, lobes somewhat falcate, setaceously mucronate, the terminal one elongated; *cup* shallow, somewhat turbinate; *acorn* globose.—Sandy soils, N. J. to Ga. Trunk 70—80f high, 4—5f diam. in the Southern States, not half these dimensions in N. J

Bark blackish and deeply furrowed. Wood coarse-grained, reddish and porous. Lobes of the leaves often not at all falcate in the smaller trees but always clothed with a thick tomentum beneath. Acorns small, round, on peduncles 1—2′ in length. May.—The bark is highly esteemed in tanning.

* * *Leaves dentate or slightly lobed.*

15. Q. ILICIFOLIA. Willd. (Q. Bannisteri. *Michx.*) *Shrub or Scrub Oak.*

Bear Oak.—Lvs. on long petioles, obovate-cuneate, 3—5-lobed, entire on the margin, whitish-downy beneath; *cup* subturbinate; *acorn* subglobose.—A shrub, common throughout the U. S., growing only on gravely hills and barrens which it occupies exclusively in large tracts. Stem 3—4f high, divided into numerous, straggling branches. Acorns small and abundant, and said to be greedily eaten by bears, deer and swine. May.

16. Q. NIGRA. Willd. (Q. ferruginea. *Michx.*) *Barren-Oak. Black-Jack.*

Iron Oak.—Lvs. coriaceous, cuneiform, obtuse or subcordate at base, 3-lobed at apex, lobes nearly equal, entire or retuse, mucronate when young, at length wholly awnless, smooth and shining above, ferruginous-pulverulent beneath, villose in the axils of the veins; *fr.* with a turbinate cup and roundish ovoid acorn; *scales of the cup* obtuse, scarious.—A small, gnarled tree, with dark, massy foliage, in sandy soils, N. J. to Ill.! and S. States. Trunk 20—30f high, with a thick, black, broken bark. The leaves are very firm in texture, 3—5′ by 2½—4′, broadest near the apex, middle lobe scarcely as wide and but little longer than the other two. Petioles 3—6″ long. May.—The wood is very valuable for fuel.

17. Q. TRILÖBA. *Downy Black Oak.*

Lvs. oblong-cuneiform, acute at the base, somewhat 3-lobed at the end, tomentose beneath, lobes equal, mucronate with setaceous awns, middle one longer; *fruit* with a flat cup and a depressed-globose acorn.—A tree of rapid growth, 25—40f high, in the pine barrens of N. J. to Flor.

18. Q. HETEROPHYLLA. Pursh. (Q. Leana. *Clark.*)

Lvs. on long petioles, coriaceous, oblong or oblong-ovate, acute or rounded or subcordate at base, margin with a few shallow, tooth-like lobes, or often only wavy or entire; *lobes* setaceous-acuminate; *acorn* subglobose, in a hemispherical cup; *scales of the cup* oblong-ovate, obtuse.—Ohio! I have specimens of the leaves and fruit of this remarkable and long lost species from *Mr. J. Clark,* re-discovered in Ohio, by the late *Mr. T. G. Lea.* The leaves are exceedingly variable, usually 4—6′ by 1½—2′, smooth and shining above, tomentose along the veins beneath, generally broad and abrupt at base. Fruit ½′ diam.

* * * *Leaves entire.*

19. Q. PHELLOS. *Willow Oak.*

Lvs. deciduous, linear-lanceolate, tapering to each end, very entire, glabrous, mucronate at apex; *acorn* subglobose, in shallow cups.—A tree 30—60f high, borders of swamps, N. J. to Flor. and Western States. Trunk straight, 10—20′ diam., covered with a smooth, thick bark. The leaves, which bear considerable resemblance to those of the willow, are of a light green color, dentate when young, 3—5′ in length. Acorn ½′ diam. May.—The timber is of little value.

20. Q. IMBRICARIA. *Laurel Oak. Shingle Oak.*

Lvs. deciduous, lance-oblong, acute at each end, briefly petiolate, very entire, shining-glabrous above, subpubescent beneath, mucronate at apex; *acorn* subglobose, in a shallow cup; *scales of the cup* broad-ovate.—A beautiful tree, very abundant in the Western States, also common along rivers, Penn. to Ga. Trunk 40—50f high, 1—2f diam., with a smooth unbroken bark, and a large head of coarse, irregular branches. The leaves are dark green, thick and firm in texture, 3—5′ by 1—1½′, forming a dense, heavy foliage. June.—The timber makes miserable shingles. In Indiana it is called *Jack Oak.*

2. CASTANEA. Tourn.

From *Castanea,* a town in Thessaly, where this tree still grows to magnificent dimensions.

♂ in a long, cylindric ament; cal. 6-cleft; sta. 10—12. ♀ 3, within a 4-lobed, densely muricated involucre; cal. 5—6-lobed; sta. 10—12.

abortive rudiments; sty. 6; nut mostly 1-seeded, invested with the enlarged, echinate involucre or cupule.—*Trees and shrubs. Lvs. mostly deciduous, alternate, acuminate. Sterile aments axillary, pendulous. Fruit enclosed in very prickly 4-lobed burrs.*

1. C. Vesca. Gært. *β. Americana.* Michx. (Fagus Castanea. Linn.) *Chestnut.*—Lvs. oblong-lanceolate, acuminate, mucronately serrate, smooth both sides.—Abundant in particular districts throughout the U. S. It is a lofty tree, with a large, straight trunk. Leaves quite large (6—9' long and ¼ as wide), with large, uniform teeth, mucronate with the prolonged, straight veins. Aments as long as the leaves and so numerous as to impart their yellowish hue to the whole tree when in blossom. The nuts are about 3 together, of a peculiar brown, villous above, enclosed in the enlarged cupule or burr which is beset on all sides with strong, compound, acute spines. Timber coarse-grained, strong, elastic, light and very durable, hence much used for posts, &c. July.— The nuts are smaller, but sweeter than those of the European variety (the Spanish chestnut.)

2. C. pumila. Michx. *Dwarf Chestnut. Chinquapin.*

Lvs. oblong, ovate or obovate, mucronate-serrate, hoary-tomentose beneath; nut solitary.—Sterile places, N. J., Penn. to Ga. and Tenn.! Shrub 6—12f high, much branched. Leaves 3—5' by 1¼—2', smooth above, generally obtuse at base, acute at apex, margins mucronate with the projecting, straight veinlets; petioles 6" long; under surface nearly white. Aments axillary, the lower staminate, 6—10' long, upper fertile with remote, pistillate flowers. Involucre of fruit bristly and prickly, 4-lobed. Nut (by abortion) solitary, small, ovoid, sweet. *Fl.* Jn. *Fr.* Oct.

3. FAGUS.

Gr. φηγος, the beech; it also signifies something eatable.

♂ in a globose ament; cal. 6-cleft, campanulate; sta. 5—12. ♀ 2, within a 4-lobed, prickly involucre; cal. single, with 4—5 minute lobes; sty. 3; nut 1-seeded, enclosed within the enlarged, spiny involucre or capsule.—*Lofty trees, with smooth, ash-colored bark. Lvs. alternate, plicate in vernation. ♂ aments on long, pendulous peduncles.*

F. sylvatica. · Linn. *β. Americana.* Nutt. (F. sylvestris. Michx. F. ferruginia. Ait.) *Beech.*—Lvs. broadly ovate-lanceolate, briefly petiolate, obtuse at base, ciliate with soft white hairs when young, at length nearly glabrous, margin with small, remote teeth, apex acuminate; *buds* lanceolate-cylindric, imbricated with brown scales, developing both leaves and flowers; *nuts* ovoid-triangular, obtuse-mucronate.—A common forest tree, abundant in N. Eng., frequent in the Western States and British provinces. The trunk is tall and straight in forests, 50—80f high, but lower and with an expansive head in open situations, always known by the light gray, unbroken bark. Leaves with very regular and straight veinlets, 4—6' long, ¼ as wide, often persistent through the winter. ♂ aments pubescent, peduncles 2' long. Nut small, 2 together in the 4-lobed burr, oily, sweet and nutritious. Timber fine-grained, with reddish duramen and white alburnum. May.

Obs.—The Red Beech is now regarded only as a variety; with the wood softer, and of more easy cleavage, and perhaps a slight difference in foliage. There are several beautiful varieties in cultivation, with purple foliage, silver foliage, &c. (See garden catalogues.)

4. CORYLUS.

Gr. κορυς, a bonnet; to which the cupule enwrapping the nut may well be compared.

♂ in a cylindric ament; cal. scale 3-cleft; sta. 8; anth. 1-celled. ♀ Calyx obsolete; ova. several; stig. 2; nut ovoid, surrounded with the enlarged, coriaceous, lacerated involucre (capsule).—*Shrubs. Aments and capitate fertile clusters subterminal.*

1. C. Americana. Hazel.

Lvs. rout ish, cordate, acuminate; *invol.* roundish-campanulate, mu a

:arger than the roundish nut, its border dilated and coarsely serrate.—Shrub 5—6f high, growing in thickets and borders of fields, U. S. Leaves 3—6' long and ⅓ as wide. From the ends of the branches hang the long, pendulous aments of barren flowers in April. The nuts are remarkably distinguished by the large, bell-shaped involucre in which each one is enveloped. They are a well-flavored fruit, though somewhat inferior to the European hazel or *filbert.*

2. C. ROSTRÀTA. Ait. *Beaked Hazel.*

Lvs. oblong-ovate, acuminate; *stip.* linear-lanceolate; *invol.* campanulate-tubular, longer than the nut, 2-parted, with dentate segments.—This species is found in the same localities as the former, is a rather smaller shrub, and chiefly differs from it in the involucre, which is covered with short, stiff hairs, and contracted at the top into a long (1—1½') narrow neck, like a bottle. Nuts as in C. Americana. May.

5. OSTRYA.

Gr. οστρεω, a scale; in allusion to the conspicuous sacs (not scales) of the fertile aments.

♂ in a cylindric ament; cal. scale roundish-ovate, ciliate, 1-flowered; anth. conspicuously bearded at the summit. ♀ geminate, in a loose, linear ament; cal. 0; fls. enclosed each in an inflated membranous sac, which, at length, enlarged, contains the matured nut.— *Small trees.*

O. VIRGINÌCA. *Hop Hornbeam. Iron-wood. Lever-wood.*

Lvs. ovate, acuminate, serrate; *fertile ament* oblong, pendulous; *buds* rather acute.—A small tree disseminated throughout the U. S., 25—30f in height Its bark is remarkable for its fine, narrow, longitudinal divisions. Leaves about twice as long as wide. The fruit is similar in appearance to hops, suspended from the ends of the branches, consisting of membranous, imbricated sacs, (cups?) containing each a flower. The wood is very white, hard and strong, much used for levers, &c. Apr. May.

6. CARPÌNUS.

Celtic *car,* wood, and *pino,* the head; alluding to its use in making yokes for cattle.

♂ in a long, cylindric ament; cal. scale roundish, ciliate; sta. 8—14, slightly bearded at summit. ♀ in a loose ament; scale large, oblong, 3-lobed, 1—3-flowered; cal. 6-toothed; stig. 2; nut long, ovoid, furrowed, 1-seeded.—*Small trees. Scales of the ♀ aments persistent and becoming foliaceous*

C. AMERICÀNA. *Hornbeam.*

Lvs. oblong-ovate, acuminate, unequally serrate; *scales of the fertile ament* 3-parted, the middle segment much the largest, oblique, with a lateral tooth.— A small tree (12—20f high), common in woods throughout the U. S. The wood is very fine-grained, compact and white, covered with a light gray or ash-colored bark. Leaves 2—4' long, ⅓ as wide, petiolate. From the ends of the branches hang the long, loose, pale green, leafy aments, consisting of alternate pairs of enlarged scales, with a dark-colored nut at the base of each. Apr. May

ORDER CXXV. BETULACEÆ.—BIRCHWORTS

Trees or *shrubs,* with deciduous stipules.
Lvs. alternate, simple, with the veinlets running straight from the midvein to the margin.
Fls. monœcious, amentaceous, mostly achlamydeous, ternate in the axil of a 3-lobed bract.
Perianth wanting or of several small scales, sometimes resembling a real calyx.
Stem's.—Sta. definite, distinct. *Anth.* 2-celled.
Fertile.—Ova. 2-celled, 2-ovuled. *Styles* or *stigmas* 2, distinct.
Fr. 1-celled and 1-seeded (by abortion), membranous and indehiscent.
Seed pendulous, without albumen.

Genera 2,? species 65, chiefly natives of the cool parts of the northern hemisphere. Properties generally astringent. The branches are often fine timber trees.

Genera.

§ 10—12. Scales of the fertile aments 3 flowered. . . . . Betula. 1
Aments { 1. Scales of the fertile aments 1-flowered. . . . . Alnus. 2

1. BETÜLA. Tourn.

Betu is the Celtic name for the birch.

♂ in a cylindric ament; bracts deeply 3-parted, peltate; calyx 0; stamens 10—12. ♀ Ament oblong-ovoid; scales subtrilobate; calyx 0; nut compressed, with a membranaceous margin.—*Trees and shrubs mostly with the outer bark laminated and horizontally fibrous. Lvs ovate, serrate, alternate.*

* Trees.

1. B. LENTA. *Black Birch. Sweet Birch. Mahogany Birch.*
Lvs. cordate-ovate, acuminate, acutely serrate, veins beneath and petioles hairy; *fertile aments* erect.—This noble species is common in the Eastern and Middle States, often exceeding 60f in height, with a diameter of 2 to 3f. The trunk is invested with a dark brown or reddish bark, which becomes rough in old trees, and is remarkable for its agreeably aromatic fragrance and flavor. Leaves 3—4' long, about ½ as wide. Sterile aments 2—3' long, fertile much shorter and thicker. In spring the cambium affords the boys a delicious morsel. The wood is of a reddish color, strong, compact, and takes a good polish. It is much used in cabinet work. April, May.

2. B. EXCELSA. Ait. *Lofty or Yellow Birch.*
Lvs. ovate, acute, serrate, on pubescent petioles, shorter than the peduncles; *barren aments* ovate, erect; *scales* with rounded, lateral lobes.—A common forest tree in N. England, arising in woods to the height of 60—80f, with a trunk 2—3f diam., invested with a thin, yellowish cuticle. Barren aments 2—4' long, cylindric, clustered, and pendulous at the ends of the branches. The wood is chiefly valuable as fuel. April, May.

3. B. NIGRA. Ait. (B. rubra. *Michx.*) *Red Birch.*
Lvs. rhombic-ovate, acute at each end, doubly serrate, glaucous beneath; *fertile ament* sessile, erect, ovoid, scales villous, the segments linear, equal.—A tree 30—50f high, growing on the banks of streams, Methuen, Mass., *Emerson*, to Car. W. to Ia.! and Ill., *Mead*. Trunk covered with a reddish or chocolate-colored bark, which at length becomes very loose and torn, hanging in shreds, and finally rough like that of the black cherry. Branches arched and slender; branchlets almost filiform, often clothing the trunk to the base. Leaves dark green above, about 3' by 2', often smaller, petioles 6—8" long, pubescent. May.

4. B. POPULIFOLIA. Ait. *Poplar-leaved Birch. White Birch.*
Lvs. deltoid, long-acuminate, unequally serrate, very smooth, on smooth petioles; *fertile aments* pedunculate; *scales* with roundish, lateral lobes.—This species, like the preceding, is distinguished for the white cuticle with which the trunk is invested. It is common in the rocky and mountainous woods of N. England, where it seldom exceeds 30—40f in height. The branches are covered with a reddish-brown bark, very slender, and throw out, in May, long, pendulous aments.

5. B. PAPYRACEA. Ait. *Paper Birch. Canoe Birch.*
Lvs. ovate, acuminate, doubly serrate, the veins hairy beneath; *fertile aments* nodding, pedunculate; *lateral lobes of the calyx* short, roundish.—This birch is abundant in the hillside woods of N. England, &c. It sometimes attains the height of 60—70f, but is generally smaller. The trunk, which is 1 — 2f in diameter, is covered with a tough cuticle consisting of numerous laminæ, the outer of which is snow white. Of this the Indians construct their light canoes. The bark upon the branches is dark brown. Leaves 2—3' long, ½ as wide. Sterile aments 1—2' long. The wood is of a fine, compact texture, but not durable, and is used in turnery and furniture work. May, June.

β. *minor.* Tuckerman. *Lvs.* smaller, ovate, glabrous, acute some of them roundish-obtuse.—White Mts. Shrubs 6—9f high.

* Shrubs.

6. B. PUMILA. *Dwarf Birch.*
Low, shrubby; *young branches* pubescent, without glandular dots; *lvs.* or-bicular-obovate, petioles densely pubescent beneath; *fertile ament* cylindrical

Shrub 2—3f high, mountains, N. Y. and Penn. *Pursh.* "In several low places towards the hills" Penn. *Bartram.* Cedar swamps, Columbus, Ohio, *Sullivant* (fide *Tuckerman.*) A very obscure species, unless it be the following.

7. B. GLANDULÒSA. Michx.   *Glandular Dwarf Birch.*

Low; *branches* glandular-punctate, glabrous; *lvs.* obovate. entire at base, obtusely serrate, glabrous; *fertile aments* oblong; *scales* half 3-cleft, lobes ovate-oblong, middle one rather longest; *nut* orbicular, with a narrow margin.—A beautiful shrub, inhabiting the mountainous districts of the N. and N. W. States! N. to Hudson's Bay. Height 2—4f. Leaves about 9" by 6 or 7", very regularly toothed.—If this shrub be distinct from .he preceding, it may be known by its glandular-dotted branches and its want of pubescence—scarcely by its variable leaves.

8. B. LITTELLIÀNA. Tuckerman.

Low, glabrous; *branches* resinous-punctate; *lvs.* suborbicular, coarse.y serrate; *fertile aments* oblong-cylindric, scales trifid, lobes oblong-obovate, middle one longest.—White Mts., *Tuckerman.* Shrub somewhat erect. Leaves 2 —4 times larger than those of B. nana.

9. B. NANA.   *Tiny Birch.*

Low, smooth; *lvs.* orbicular, crenate, reticulated beneath; *scales of the ament* deeply 3-parted; *seeds* orbicular, nearly wingless.—This miniature tree is found on the summits of Mt. Clinton, Mt. Franklin, &c., of the White Mts.! It is scarcely more than a foot in height, often but a few inches, the branches few and straggling, the leaves ½—¾' in diameter, smooth both sides, pale and distinctly reticulate beneath, and on petioles 1—2" long.

2. ALNUS. Tourn.

♂ Ament long, cylindric, composed of cuneate, truncate, 3-lobed, 3-flowered bracts; cal. 4-parted; sta. 4. ♀ Ament ovoid; bracts 2-flowered; 3-fid; cal. 0; nut wingless, compressed.—*Shrubs, arising from large and strong roots. Buds pedunculate. Lvs. plicate in vernation, alternate, simple, deciduous.*

1. A. INCÀNA. Willd. (A. glauca. *Michx.* Betula incana. *Linn.*)

Lvs. submembranaceous, oblong, acutish, obtuse at base or cordate, margin somewhat lobed, sharply serrate, glaucous-pubescent beneath; *veins* hirsute, their axils naked; *stip.* oblong-lanceolate; *fertile aments* oval.—Not uncommon in N. Eng. and Mid. States. A tall shrub or small tree, readily distinguishable by the form and pubescence of the leaves.

2. A. RUBRA. Marsh. (A. serrulata. *Willd* and 1*st edit.*)   *Common Alder.*

Lvs. obovate, acuminate, doubly serrulate, the veins and their axils hairy beneath; *stip.* elliptical, obtuse.—A well known shrub growing in clumps, and forming thickets on the borders of ponds and rivers, and in swamps. Stems numerous, rather straight, 10—15f in height. Leaves 2—4' long and ¾ as wide, strongly veined; petioles ¼—½' long. Aments 2—3' long, slender, pendulous, fascicled at the ends of the branches; fertile ones short, thick, dark brown, persistent, several together a little below the sterile ones. March, April.

3. A. CRISPA. Michx. (Betula crispa. *Ait.*)

Lvs. oval, acute, obtusish at base, doubly serrate, clothed with a soft viscid pubescence, or subglabrous, villous on the veins and axils beneath; *stip.* broadly ovate; *fertile aments* on long peduncles, oval.—White Mts., *Tuckrrman*, Green Mts., *Robbins*, Can., *Michaux.* An elegant shrub, 3—4f high. Leaves varying to broad-ovate, rarely cordate, nearly smooth in the alpine state, otherwise softly pubescent and sprinkled with resinous particles. Apr.

ORDER CXXVI. MYRICACEÆ.—GALEWORTS.

Shrubs or small trees. aromatic, covered with resinous glands or dots. *Lvs.* alternate, simple. *Fls.* monœcious or diœcious. amentaceous, each axillary to a bract. *Sterile.—Sta.* 2—6. *Anth.* 2—4-celled, opening longitudinally. *Fertile.—Ova.* 1-celled, 1-ovuled, surrounded by several hypogynous scales.

Stig. ♀, subulate, or dilated and petaloid.
Fr. drupaceous or dry. Seed solitary, erect, without albumen.

Genera 3, species about 20, found in the temperate parts of North America, in India and South Africa one species in Europe. Sweet Fern is highly aromatic and astringent. The fruit of the bayberry bush yields wax in abundance.

Genera.

Leaves { cuneate-lanceolate, serrate. Flowers diœcious . . : : : : : : : **Myrica.** 1
{ sinuate-pinnatifid. Flowers monœcious . . . . **Comptonia.** 2

1. MYRICA.

(Gr. μυρω, to flow; because some of the species are native of river banks and inundated places.

Flowers ♂ ♀. Aments ovate-oblong; scales loosely imbricate, lunate.
♂ Stamens 4—6, short, erect; anth. large, 4-valved. ♀ Ovary 1,
superior; sty. 2, spreading; stig. 2, acute; drupe 1-celled, 1-seeded.
—Stipules very fugacious or 0.

1. M. GALE. Sweet Gale. Dutch Myrtle.
Lvs. glabrous, cuneate-lanceolate, obtuse and serrate above, margin very
entire and slightly revolute below, tapering to a very short petiole; sterile aments
of ovate, cordate, acuminate, ciliate scales; fr. in an oblong, dense, amentaceous head.—A branching shrub, 3—4f high, on the inundated borders of ponds
and mountain lakes, Northern States and Can. Leaves dark green, paler beneath with a strong midvein, 9—18″ by 4—6″, entire ½ the length. ♂ and ♀
aments on separate plants, the former terminal, about 1′ in length, the latter
axillary and much shorter. Fruit and leaves, when crushed, with a pungent,
spicy odor. May.

2. M. CERIFÉRA. Bayberry. Wax Myrtle.
Lvs. glabrous, cuneate-oblong, rather acute or obtuse, distinctly petiolate,
margin entire or remotely dentate above, paler and with distinct veinlets beneath; aments cotemporary with the leaves, lateral, naked, the ♂ larger, with
lax, roundish scales; fr. spherical, distinct, clustered, naked, covered with wax.
—This interesting and useful shrub is found in dry woods or in open fields,
Nova Scotia to Flor. W. to Lake Erie. It varies in height from 2—8f, covered
with a grayish bark. It has a very branching top, numerous dry-looking leaves
1½—2½′ by ½—¾′. The ♂ and ♀ aments on separate plants, below the leaves,
½—¾′ long, the former much thicker. The fruit consists of a globular stone enclosing a kernel, and covered with a coating of whitish wax, which, being separated by boiling water, constitutes the bayberry tallow of commerce. May.

2. COMPTONIA.

In honor of Henry Compton, Lord Bishop of London, who made an extensive collection of plants.

Flowers ♂. ♂ Ament cylindric; bract reniform-cordate, acuminate; cal. scale 2-parted; sta. 3, forked; anth. 6. ♀ Ament ovate;
cal. scales 6, longer than the bract; sty. 2; nut ovoid, 1-celled.—
Low shrubs. Lvs. long and narrow, pinnatifid-lobed, with small stipules.

C. ASPLENIFOLIA. Ait. (Liquidambar. Linn.) Sweet Fern.
Lvs. long, linear-lanceolate, alternately sinuate-pinnatifid.—A well known,
handsome, aromatic shrub, 2f high, common in dry woods and hills. The main
stem is covered with a rusty, brown bark, which becomes reddish in the branches,
and white downy in the young shoots. Leaves numerous, on short peduncles,
3—4′ by ½′, divided nearly to the midvein into numerous, rounded lobes so as
to resemble those of the spleenwort. Stipules in pairs, acuminate. Barren
flowers in erect, cylindric catkins, terminal and lateral. Fertile flowers in a
dense, rounded burr or head, situated below the barren ones. Fruit a small,
ovate, brown, 1-celled nut. May.

ORDER CXXVII. SALICACEÆ.—WILLOWORTS.

Trees or shrubs, with alternate, simple leaves and deciduous or persistent stipules.
Fls. diœcious, amentaceous, achlamydeous, axillary to 1-flowered bracts.
Sterile.—Sta. 2—several, distinct or monadelphous. Anth. 2-celled.
Fertile.—Ovs. 1 2-celled. Ovules numerous, erect. Styles or stigmas 2.

Fr. coriaceous, 1-celled, 2-valved.
Sds. numerous, ascending, furnished with a silky coma. *Albumen* 0.

Genera 2, species 220, chiefly natives of the northern temperate and frigid zones, one species, *Salix arctica*, extending farther north than any other known woody plant.

Properties.—The bark is astringent and tonic, possessing the febrifugal properties of the sulphate of quinia. The wood is employed for various economical purposes.

Genera.

Stamens { 2—4. Capsule 1-celled. *Salix.* 1 }
{ 5—20. Capsule 2-celled. *Populus.* 2 }

1. SALIX.*

Celtic *sal*, near, and *lis*, water; alluding to their usual locality.

Aments cylindric, bracts imbricated, 1-flowered, each with a neo-tariferous gland at base. ♂ Calyx 0; sta. 2—7. ♀ Calyx 0; ova. ovoid-lanceolate, acuminate; stig. 2, mostly bifid; caps. 1-celled, 2-valved, valves acuminate, finally revolute at summit; seeds numerous, minute, comose.—*Trees, shrubs and undershrubs. Lvs. usually narrow and elongated, each with 2 conspicuous stipules. Aments terminal and lateral.*

§ 1. CINEREÆ. Borrer. *Upland, grayish shrubs. Leaves obovate-lanceolate, mostly entire, rugose, canescent-pilose, margins often revolute. Aments oval or oval-cylindric, expanding before the leaves, with centrifugal inflorescence. Stamens 2. Scales red, finally black. ♀ aments recurved when young. Ovary pedicellate; stigma red or pale yellow, finally green.*—Barratt.

1. S. TRISTIS. Ait. (S. longirostris. *Michx.*) *Sage Willow.*

Lvs. linear-lanceolate or oblanceolate, cuneate at base, entire or remotely undulate-toothed, margin subrevolute, apex acute or obtusish; *stip.* minute, narrow-lanceolate, caducous; *aments* very small; *scales* orbicular-oblong, hairy at the margin; *ova.* with grayish, silky pubescence; *sty.* short.—Sandy or dry fields, borders of woods, pastures, N. Eng.! to Ind.! and Ill. A small, downy shrub, with a profusion of aments in spring, appearing before the leaves. Leaves at length numerous, often crowded and rosulate at the ends of the branches, 1—2′ long, tapering from above the middle to a very short petiole, the margin often revolute, under surface glaucous, often pubescent, upper generally smoothish. In starved specimens the whole plant is grayish-white, with very small leaves.

β. (S. Muhlenbergiana. *Ph.* and *1st edit.*) Shrub larger in all its parts. Bark green on the stem, yellowish and downy on the new branches. Leaves 2—3′ long, often abrupt at base.

γ. *Monadelphia.* Barratt. *Sta.* united half their length.—Conn.

2. S. MUHLENBERGIÀNA. Barratt. (S. conifera. *Muhl.* and *1st edit.*) *Muhlenberg's Willow.*—*Lvs.* oblanceolate, remotely serrate, glabrous above, pubescent and not rugose beneath; *young branches* smooth; *stip.* lunate, subdentate; *aments* precocious, diandrous; *scales* lanceolate, obtuse, villous; *ova.* pedicellate, lanceolate, silky; *sty.* long, bifid; *stig.* 2-lobed. –A shrub in dry soils, Northern States, 4—8f high, with brown twigs. On the ends of these, cone-like excrescences are often produced by the punctures of insects. Aments covered with very hairy scales, appearing before the leaves in April.

3. S. CANDIDA. *Willd. White Willow.*

Lvs. lanceolate or linear-lanceolate, very long, obscurely serrulate at the summit, pubescent above, hoary-tomentose beneath, revolute on the margin; *stip.* lanceolate, as long as the petioles; *aments* cylindric; *scales* obovate, obtuse, very long, hairy; *stig.* 2-lobed.—A beautiful species in shady woods. Stems 4—6f high. Leaves 3—12′ by 1—2′. Catkins dense, white with dense wool. Styles and stigmas dark red, ¼′ in length. April, May.

§ 2. DISCOLORES. Borrer. *Trees or shrubs blossoming in early spring. Leaves serrate or denticulate, smooth and shining above, glaucous and pubescent beneath. Aments oval or cylindrical, preceding the leaves, smooth, silky or woolly, without floral*

* Arranged according to Dr. Joseph Barratt. See preface.

leaves, centrifugal. Scales turning black. Stamens 2, free or united. Ova. stipitate, subpubescent. Barratt.

4. S. DISCÓLOR. Muhl.   *Two-colored or Bog Willow.*

Lvs. oblong, rather acute, glabrous above, entire at the end, glaucous beneath; *stip.* lanceolate, serrate, deciduous; *aments* cotemporary with the leaves, oblong, downy, diandrous; *scales* oblong, acute, black, hairy; *ova.* sessile, downy; *stig.* 2-parted.—A shrub, 8—10f high, in swampy grounds, Can. to Car., with tough brown twigs, and white, glossy catkins. Leaves 1—3' long, finely serrate except at the end. Sterile aments about ¾' long, fertile 1'. Filaments white, anthers at length yellow. April.

β. *Monadelphia.* Barratt. Tree 10—15f high. Filaments 3—4, half-united.

5. ERIOCEPHÀLA. Michx.   *Woolly-headed Swamp Willow.*

Branchlets very pubescent, brown or purplish; *lvs.* lanceolate-elliptic or oblong, cuneate at base, entire or remotely serrulate above, under surface glaucous or ferruginous, both surfaces pubescent when young, at length the upper surface green and nearly smooth; *stip.* semicordate, with sharp serratures; *aments* oval-oblong, densely villous.—A small tree, putting forth its large and exceedingly woolly catkins in April. Grows in swamps, N. Eng.

6. S. PRINÖÏDES. Ph.   *Prinos-leaved Willow.*

Branchlets puberulent when young, at length glabrous and dark brown; *lvs.* oval-oblong or lance-oblong, glabrous, glaucous beneath, cuneate at base, remotely serrulate, acute or abruptly acuminate; *stip.* semicordate, incisely serrate; *aments* preceding the leaves, hairy; *ova.* ovoid, acuminate, silky; *sty.* long; *stig.* bifid.—Shrub 6—8f high, N. Eng. to Penn., W. to Mich. Catkins appearing in April, 1—2' long Ovaries distinctly stipitate, tapering at apex into the long, exserted style.

7. S. CRASSA. Barratt.   *Dense-flowered Early Willow.*

Lvs. elliptic-lanceolate, rather remotely serrate, entire towards the base, glabrous and dull green above, veiny and clothed with short, ferruginous hairs beneath, adult subcoriaceous; *stip.* small, lanceolate, serrate or often wanting; ♂ *aments* ovate, sessile, densely clothed with yellowish-white, silky hairs; *scales* obovate.—A hairy and beautiful willow, rare in N. Eng. Tree about 15f high, with rough, ash-colored bark. Branches irregular and knotty, with thick, densely flowered twigs in April. Leaves 3½' by 1'. ♂ catkins 1—2' long.

8. S. SENSITIVA. Barratt.   *Frost or Tender Willow.*

Lvs. ovate-lanceolate, acuminate, cuneate and entire at base, finely serrate at the apex, and more distantly and strongly serrate towards the base, glabrous and rather thin; *stip.* subfalcate, serrate; ♂ *aments* rather lax; *scales* rather lax, lightly clothed with grayish-black hairs.—A small tree, about 15f high, found in various parts of N. Eng., &c. The aments and twigs are frequently destroyed by frost at flowering-time, being thinly protected with hairs. Leaves smooth, 3—5' by 1½—2'. Aments 1½' long. Apr.—This and the three next preceding species are very closely allied, as suggested by *Mr. Emerson, Rep., p.* 262, and it is possible that they may hereafter be united under one species.

‡ 3. GRISEÆ. Borrer. *Shrubs with branches brittle at base, and an intensely bitter bark. Leaves lanceolate, serrate, grayish-silky beneath, turning black in drying. Aments cylindrical, rather short, preceding the leaves, with 2 or 3 minute leaves at base. Stamens 2, beginning to appear from the middle of the ament. Ovaries grayish-silky.* Barratt.

9. S. GRISEA.   *Gray Willow.*

Lvs. lanceolate, serrulate, acuminate, smooth above, silky beneath; *stip.* ovate-oblong, denticulate, deflected, deciduous; *scales* oblong, hairy, black at the tip; *ova.* oblong, pedicellate, silky; *stig.* sessile, obtuse. A shrub 6—8f high, in inundated meadows. Branches purplish, long and slender, very tough, except at the base, where they are very brittle. Leaves 2—4' by ¼—1' ♀ aments very abundant, ¼' long. Apr.

10. S. PETIOLÀRIS. Smith.   *Long-stalked Green Osier.*

Lvs. lanceolate, serrate, smooth, glaucous beneath, silky at base, mostly unequal, stipules lunate, dentate; *aments* appearing before the leaves; *scales* lax, obovate, obtuse, hairy, black; *ova.* on long pedicels, ovate, silky; *stig.* sessile two-lobed.—Low grounds, banks of streams, Conn., N. Y., Car.—A small tree, with long, slender, smoot , purplish or yellowish-green, tough and elastic branches, which may be useful in basket-making.

11. S. MYRICOÏDES. Muhl. *Gale-leaved Willow.*

Les. oblong-lanceolate, acute, with 2 glands at base, obtusely serrate, smooth, glaucous beneath; *stip.* ovate, acute, glandular-serrate; *aments* villous, black; *ova.* on long pedicels, glabrous; *sty.* bifid; *stig.* bifid.—Swamps, N. Eng. to Va. A small shrub, with green branches, the branchlets purple, smooth. Leaves a length thick and coriaceous, the serratures each tipped with a gland. Apr.

12. S. FUSCATA. Pursh. *Leaden-flowered Willow.*

Les. lanceolate-obovate, acute, glabrous, subserrate, glaucous beneath, in the young state pubescent; *stip.* very narrow; *aments* nodding; *scales* obtuse, scarcely hairy within; *ova.* short, pedicellate, ovoid, silky; *stig.* sessile, 2-lobed. —Grows in pools, swamps and on wet banks, seldom exceeding 3 or 4f in height, distinguished by the leaden hue of its aments. It furnishes excellent twigs for basket-work, and is well adapted for embankments and mill-dams.

§ 4. VIMINALES. Borrer.

13. S. VIMINALIS. *Basket Osier.*

Les. linear-lanceolate, very long, acuminate, subentire, silky-canescent beneath; *stip.* minute; *branches* virgate; *aments* precocious (appearing before the leaves); *scales* roundish, very hairy; *ova.* sessile, ovoid; *sty.* filiform; *stig.* undivided, acute.—This beautiful willow was probably introduced from Europe. Wet meadows and margins of rivers. Stems 10—12f high, with long, straight, slender and flexible branches. Leaves often a foot in length, narrow, covered with a snow-white pubescence beneath. Aments very hairy. May.

§ 5. FRAGILES. *Trees. Leaves lanceolate or lance-falcate, serrate, denticulate or entire. Aments pedunculate, cylindric, loose, acuminate, cotemporary with the leaves; scales greenish-yellow, pubescent or smooth. Stamens 2—5, expanding first from the base of the ament. Ovary smooth. Barratt.*

14. S. FRAGILIS. *Crack Willow.*

Les. ovate-lanceolate, glabrous, whole margin serrate, acuminate, petioles glandular; *stip.* semicordate, pointed, dentate; *ova.* on short pedicels, oblong-ovoid, glabrous; *sty.* short; *stig.* bifid, longer than the styles; *scales* oblong, about equaling the ovaries, pubescent, ciliate; ♂ with an abortive ovary.—A tall tree, 60 or 80f high, native in Great Britain. It has a bushy head, with numerous oblique, irregular branches. The twigs break off at base by a slight touch. The wood is of a salmon-color.

15. S. DECIPIENS. Hoffm.

Branches smooth, highly polished; *lvs.* lanceolate, glabrous, serrate, acuminate, floral ones often obovate and recurved, petioles somewhat glandular; *stip.* small, semi-ovate, acute, dentate, often 0; *ova.* pedicellate, glabrous, acuminate; *sty.* longer than the 2-cleft stigma.—A small, elegant tree, remarkable for the polished, light reddish-brown twigs, appearing as if varnished. The young twigs stained with crimson. It is often set in rows for ornament and shade.

16. S. RUSSELLIANA. Sm. *Bedford Willow.*

Lvs. glabrous, lanceolate, tapering to each end, whole margin serrate, very pale beneath, petioles glandular or margined; *stip.* semi-cordate, strongly serrate, acuminate; *ova.* glabrous, pedicellate, longer than the scales; *sty.* as long as the bifid stigmas; *scales* narrow-lanceolate, slightly ciliate.—A large tree, native of Britain, often propagated in this country. It has long, green shoots, long, bright green, serrated leaves. Apr. May. §

17. S. PAMEACHIANA. Barratt. *Pameachy Willow.*

Lvs. long-lanceolate, acuminate, with fine cartilaginous serratures glaucous beneath; *stip.* small, lanceolate, often 0; ♂ aments cylindrical, lax and somewhat recurved; *scales* obtuse, yellowish; *s'a.* ♀; *bark* of the twigs smooth, yellowish and variegated, in flowering time red at the ends.—Grows on the banks of Pameachy river, Middletown, Conn. A tree of small size, intermediate between S. decipiens and S. vitellina, but "certainly distinct from either."

18. S. LUCÎDA. Muhl.

Lvs. ovate-lanceolate, long-pointed, rounded at base, smooth and shining, *stip.* oblong, serrate; *aments* triandrous; *scales* lanceolate, obtuse, serrate and smooth at the tip, hairy at the base; *ovaries* lanceolate-subulate, smooth; *style* bifid *stigmas* obtuse.—A small and beautiful tree, common in N. Eng., Middle States, Mich. and British Am. Trunk 10—15f high, 3—4' diam. Branches smooth, dark, shining green. Leaves broad and glossy, dark green above, tapering to a long point. May.

19. S. NÎGRA. Marshall. *Black Willow.*

Lvs. lanceolate, acute at each end, serrulate, smooth and green on both sides, petiole and midvein above tomentose; *stip.* dentate; *aments* erect, cylindric, villous; *scales* oblong, very villous; *fil.* 3—6 (generally 5), bearded at base; *ova.* pedicellate, ovoid, smooth; *sty.* very short; *stig.* bifid.—A small tree, on the banks of rivers, chiefly in N. Y. and Penn. Branches very brittle at base, pale yellow. The trunk has a blackish bark. Sterile aments 3' long. May.

20. S. PURSHIÄNA. Spreng. (S. falcata. *Hook.*) *Pursh's Willow.*

Lvs. very long, lance-linear, often falcate, gradually attenuate to the apex, acute at base, finely dentate-serrate, smooth on both sides, silky pubescent when young; *stip.* somewhat lunate or obliquely reniform-cordate, dentate, reflexed; *ova.* glabrous, pedicellate; *sty.* short.—Readily recognized by its very long, falcate leaves and the broad persistent stipules. Grows in swamps and margins of ponds, Middle States, N. Eng., Western States, Can. It is a shrub or small tree, with a slender trunk sometimes 30 or 40f high. Leaves green both sides, 6—8' long. Aments 2' long.

§ 6. ALBÆ. Borrer. *Trees of the largest size, with lanceolate, serrate leaves, the serratures glandular, lower surface clothed with long, appressed, silky hairs, often the upper also, giving to the foliage a whitish or bluish hue. Aments lax. Stamens* 2. *Ovaries glabrous.*

21. S. ALBA. *White Willow.*

Lvs. elliptic-lanceolate, regularly glandular-serrate, silky beneath, often above, acute at apex; *ova.* ovoid, acuminate, glabrous, subsessile; *stig.* short, recurved, deeply cleft; *sta.* 2, with hairy filaments; *scales* short, pubescent at the margin.—A large tree of rapid growth, native of Europe, introduced in Mass. (*Emerson*) and probably in other states.

β. *cærulea.* (*Blue Willow.*) *Lvs.* of a bluish hue, less silky beneath.—A tree of rapid growth, completely naturalized in Mass. *Emerson*

22. S. VITELLÎNA. *Yellow Willow. Golden Osier.*

Lvs. lanceolate, acuminate, with thickened serratures, smooth above, paler and somewhat silky beneath; *stip.* 0; *aments* cylindric; *scales* ovate-lanceolate, pubescent outside; *ova.* sessile, ovate-lanceolate, smooth; *stig.* subsessile, 2-lobed. —This willow was probably introduced, but is now very common by roadsides, &c. It is a tree of moderate height, with shining yellow branches. May.

23 S. BABYLONÎCA. *Babylonian or Weeping Willow.*

Branches pendulous; *lvs.* lanceolate, acuminate, smooth, glaucous beneath, *stip.* roundish, oblique, acuminate; *ova.* sessile, ovate, smooth.—This elegant species has been introduced from the East, and cultivated until nearly naturalized. The long, slender, drooping branchlets very naturally indicate the English name of the tree, and give it a place in the church-yard to "weep" over the remains of the departed. The Latin name was happily suggested to Linnæus by the 137th Psalm:

> "By the rivers of *Babylon* there we sat down :
> Yea, we wept, when we remembered Zion.
> We hanged our harps upon the *willows* in the midst thereof."

§ 7. FULVÆ. Borrer.

24. S. ROSTRÂTA. Richardson. *Beaked Willow.*

Branches erect, straight, pubescent, at length smooth; *lvs.* broadly or obovate-lanceolate, acute, subentire. at length coriaceous, smooth above, glaucous-pubescent beneath; *stip.* semicordate, dentate; *aments* short, cylindric, dense, the fertile ones becoming very long and loose; *scales* oblong, membranous, hairy at the apex; *ova.* narrow-lanceolate, silky, long-acuminate, on very long pedicels, *sty.* very short; *stig.* lobed, the lobes bifid or entire.—Shrub or small tree 8—10f high. Bark of the trunk dark-colored, of the branches yellow.

§ 8. CORDÀTÆ. *Tall shrubs with dichotomous, flexuous, smooth branches. Leaves cordate or attenuate at base, glabrous. Stipules semi-cordata, serrate. Aments slightly pedunculate, ovoid-cylindrical, cotemporary; scales subciliate, red or yellowish. Stamens 2 or 3. Ovaries pedicellate, glabrous.* Barratt.

25. S. CORDÀTA.   *Heart-leaved Willow.*
Lvs. oblong-lanceolate, acuminate, cordate at base, smooth; *stip.* large, roundish-ovate, serrate; *aments* triandrous; *scales* lanceolate, woolly, black; *ova.* pedicellate, lanceolate, smooth; *sty.* very short; *stig.* bifid.—An elegant shrub, 6—8f high, in swamps throughout the Middle States. Branches green and smooth, with light green leaves an inch wide and 3′ long. Aments an inch long, accompanying the leaves in April and May.

26. S. RIGIDA.   *Stiff-leaved Willow.*
Lvs. oblong-lanceolate, acuminate, subcordate, rigid, smooth, coarsely serrate, the lowest serratures elongated, petioles villous; *stip.* large, cordate, obtuse, glandular-serrate; *aments* triandrous; *scales* lanceolate, woolly, black; *ova.* on long pedicels, lanceolate, smooth; *sty.* very short; *stig.* 2-parted.—A small tree, 10—15f high, growing in swamps. Branches green, red towards the end, the younger ones pubescent. Much used in basket-making. April, May.

27. S. TORREYÀNA.   Barratt.   *Torrey's Willow.*
Lvs. cordate-ovate, sharply pointed, margin wavy, finely serrate, paler beneath; *stip.* large, semicordate; ♂ *aments* slender, scales lanceolate, blackish, ciliate; *sta.* 2; *fil.* rather short; *ova.* on short pedicels, smooth, deltoid-lanceolate; *stig.* 4-parted, flesh-colored; *caps.* green.—A fine, shrubby willow, 6—10f high, river banks, N. Eng., recognized by its broad, heart-shaped, glossy leaves with a wavy margin, sharp point, and very large stipules. Branches of a light gray color, branchlets yellowish-green. Apr.

28. S. LONGIFOLIA.   Muhl.   *Long-leaved Willow.*
Lvs. linear, acuminate at each end, elongated, remotely toothed, smooth, nearly of the same color on both sides; *stip.* lanceolate, dentate; *ament* tomentose, pedunculate; *sta.* 2; *scales* flat, retuse; *fil.* bearded at base, twice longer than the scales.—River banks from the Conn. and Ohio to Oregon and Brit. Am. It possesses a remarkable power of rooting, extending itself and binding the loose sands together. Stems about 2f high, with brown branches and white branchlets.

29. S. ANGUSTÀTA.   Ph.   *Narrow-leaved Heart Willow.*
Lvs. lanceolate, acute, very long, gradually attenuated at base, very glabrous, serrulate, nearly the same color both sides; *stip.* semi-cordate; *aments* erect, somewhat glabrous; *ova.* pedicellate, ovoid, glabrous; *sty.* bifid; *stig.* 2-lobed.—Banks of streams from the Conn. to the Miss. An excellent osier, with very long and slender twigs, long and narrow leaves.

§ 9. ARBUSCULÆ. *Small shrubs, inhabiting arctic or alpine regions. Aments cotemporary with the leaves.*

30. S. HERBACEA.   *Herb Willow.*   Arctic Willow.
Dwarf; *lvs.* orbicular, serrate, glabrous, veiny; *aments* few-flowered, sessile; *scales* small, glabrous; *ovaries* sessile, lanceolate, glabrous; *style* short; *stig.* lobes bifid.—On the alpine regions of the White Mountains! N. to Lab. and the Arc. Islands. An interesting little shrub, the smallest of its tribe. Stem ascending, 1—2′ high. Leaves about 3″ diameter, smooth and shining on both sides. Stipules wanting. Roots long, creeping, branching. Jn. Jl.

31. S. MYRTILLÒIDES.   (S. pedicellaris. *Ph.* and *auct. Am.*)
Lvs. oblong-elliptic, acute or obtuse, rather obtuse at base, entire, both sides glabrous, beneath glaucous and reticulate-veined; *aments* pedunculate; *caps.* ovate-conic, glabrous, long-pedicellate; *scales* short, obtuse, a little hairy; *sty.* very short; *lobes* of the stigma cleft.—Swamps, N. Eng. and N. Y. A low and elegant shrub, with rather a virgate habit, remarkable for its entire smooth-

ness. Leaves elliptical, revolute at edge, obtuse at base, somewhat glaucous beneath.

32. S. AMBIGUA. Ehrh. *Dubious Willow.*
Lvs. elliptical, obovate or lanceolate, recurved at the point, entire or remotely denticulate, rugose-veiny beneath, silky-villous, at length glabrous; *stip.* semi-ovate, straight; *aments* sessile, briefly pedunculate in fruit; *caps.* tomentose, long-pedicellate; *sty.* short; *stig.* emarginate.—White Mts. *Tuckerman.* A prostrate shrub, with leaves about 1½′ by ⅓′.

33. S. PHYLICIFOLIA.
Lvs. ovate or lanceolate, remotely repand-serrate, glabrous, glaucous beneath; *stip.* semicordate, oblique at apex; *aments* bracteate, ♂ sessile; *caps.* pedicellate, conical-elongated, somewhat silky; *sty.* long.—White Mts. *Tuckerman.* A handsome, low shrub, spreading, with broad-elliptical, very smooth leaves, the margins repand-serrate.

34. S. CUTLERI. Tuckerman. (S. Uva-ursi *of 1st. edit., etc.?*) *Cutler's Willow.*—*Lvs.* elliptical, acute or obovate, obtuse at base, glandular-denticulate, smooth above, glaucous-smoothish beneath, silky-villous when young; *aments* pedunculate, cylindric, dense; *caps.* ovate-conic, briefly pedicellate, glabrous; *scales* obovate, black, silky; *stig.* bifid, lobes at length cleft.—White Mts. *Tuckerman.* A low or prostrate shrub.

2. POPULUS.

Lat. *populus,* the people; being often planted by the public ways.

Aments cylindric; bracts lacerately fringed; calycine scales turbinate, oblique, entire. ♂ Stamens 8—30. ♀ Ova. superior; style very short, bifid; stigma large, 2-lobed; caps. 2-valved, 2-celled.—*Trees of large dimensions. Wood soft and light. Lvs. broad, petioles long, often compressed vertically and glandular. Aments lateral, preceding the leaves.*

1. P. TREMULOÏDES. *American Aspen. White Poplar.*
Lvs. orbicular-cordate, abruptly acuminate, dentate-serrate, pubescent at the margin.—Abundant in N. England and in the Middle States, growing in woods and open lands. Stem 25—40f in height, with a diameter of 8—12′. Bark greenish, smooth except on the trunk of the oldest trees. Leaves small (2—2½′ long, and 1½ as wide, dark green, on petioles which are 2—3′ long and laterally compressed, so that they can scarcely remain at rest in any position, and are thrown into excessive agitation by the slightest breeze. The trembling of the "aspen leaf" is proverbial. Aments plumed with silken hairs, about 2′ long, pendulous, appearing in April, long before the leaves. The wood is white, soft and light, of little value.

2. P. GRANDIDENTATA. Michx. *Large Poplar.*
Lvs. roundish-ovate, acute, with large, unequal, sinuate teeth, smooth, villous when young.—Woods and groves in the northern parts of the U. S., less common than the preceding species. Stem 40f high, with a diameter of 1f, straight, covered with a smooth, greenish bark. Branches distant, coarse and crooked, clothed with leaves only at their extremities. Leaves 3—5′ long and nearly as wide, clothed with thick, white down in spring, but becoming perfectly smooth. The wood is white, soft, and quite durable. May.

3. P. BETULIFOLIA. Ph. (P. Hudsonica. *Michx.*) *Birch-leaved or Black Poplar.*—*Lvs.* rhomboidal, long-acuminate, dentate, smooth; *young branches* pilose.—This poplar is found chiefly in the valleys of the Hudson and Connecticut. It is a tree of middle size, with grayish-white twigs, and dark brown buds. Leaves 3½′ long and 2 broad. Aments 4—5′ long, without hairs. April.

4. P. BALSAMIFERA. *Balsam Poplar.*
Lvs. ovate-acuminate, with close-pressed serratures, white and reticulate-veined beneath; *buds* resinous. — The balsam poplar, though nowhere abun

ᴗ at, is found in woods and fields, disseminated throughout N. England and Canada. With a trunk 18′ diam., it arises 60—70′. The buds of this species, as well as of most of the poplars, are covered with an aromatic resin, which may be separated in boiling water. April.

5. P. MONILIFÈRA. Ait. *Necklace Poplar.*

Lvs. subcordate-deltoid, smooth, glandular at base, with cartilaginous, hairy, hooked serratures; *veins* spreading; *petioles* compressed above; *older branches* terete; *fertile aments* long and pendulous.—Banks of the Hudson, near Troy, N. Y., apparently native. *Beck.* A tree 60—70′ high, with a cylindric trunk. Leaves 2½—4′ diam., on long petioles. April.

6. P. HETEROPHYLLA. *Various-leaved Poplar.*

Lvs. roundish-ovate, obtuse, uncinately toothed, cordate and somewhat auricled at base, the sinus small, tomentose when young.--A tree 60—70′ high, found in swamps. Branches cylindric. Leaves with auriculate lobes at base, which often conceal the insertion of the petiole. May.

7. P. CANDICANS. Ait. *Balm-of-Gilead.*

Lvs. ovate-cordate, acuminate, obtusely and unequally serrate, whitish beneath, reticulate-veined, somewhat 3-veined; *petioles* hirsute; *buds* resinous; *branches* terete.—This tree is sometimes met with in New England, growing about houses as a shade tree. It is 40—50′ high, and 18—30′ in diameter. Bark smooth, greenish. Foliage copious, dark green. Apr.

8. P. LÆVIGÀTA. Ait. (P. Canadensis. *Michx.* and 1st. *edit.*) *River Poplar.*

Cotton Tree.—*Lvs.* roundish-ovate, deltoid, acuminate, subcordate, unequally serrate, shining, smooth, glandular; *petioles* compressed; *younger branches* angled.—The cotton-tree grows 70—80′ high in N. Y. and Vt. The fertile aments are 6—8′ long, and pendulous. The seeds are clothed with a white, cotton-like down which gives name to the tree. Buds sealed against the frosts and rains with resin. April.

9. P. ANGULÀTA. *Water Poplar. Western Cotton Tree.*

Lvs. ovate-deltoid, subcordate, uncinate-serrate, acuminate, glabrous, younger ones broadly cordate; *branches* winged, angular.—A tree of noble dimensions, growing along the rivers of the Southern and Western States. Trunk 40—80′ high, 1—2′ diam., bearing a broad summit, with coarse branches and branchlets. Leaves on adult trees 2—3′ long, about the same width, truncate at base; on younger trees they are 2 or 3 times larger, with a cordate base. Petioles longer than the leaves, compressed near the base of the lamina. Branchlets remarkably thick, greenish, spotted with white, striate. Buds short-ovoid, green, not coated with resin. Timber not valuable. March, April.

10. P. DILATÀTA. *Lombardy Poplar.*—*Lvs.* smooth, acuminate, deltoid, serrate, the breadth equaling or exceeding the length; *trunk* lobed and sulcate.— This tree is native in Italy as its name imports. It was early brought to this country, and has been planted about many a dwelling and in village streets. Its rapid growth is the only commendable quality it possesses, while the huge worms by which it is often infested render it a nuisance.

11. P. ALBA. *Abele or Silver-leaf Poplar.*—*Lvs.* cordate, broad-ovate, lobed and toothed, acuminate, dark green and smooth above, very white-downy beneath; *fertile aments* ovate; *stigmas* 4.—A highly ornamental tree, native of Europe. Nothing can be more striking than the contrast between the upper and lower surface of the leaves.

ORDER CXXVIII. BALSAMIFLUÆ.—LIQUIDAMBARS.

Trees with alternate, simple or lobed leaves, with glandular serratures and deciduous stipules.
Aments monœcious, roundish, with achlamydeous flowers.
Sterile.—Anth. numerous, oblong, subsessile, with scales intermixed.
Fertile.— Ova. ⚭-celled, collected into a globe, each surrounded by a few scales.
Styles 2, long. *Fr* a kind of strobile, composed of the indurated scales and capsules.
Caps. 2-beaked, 2-celled, opening between the beaks. *Sds.* several, winged.

Genus 1, species 2, natives of India, Levant and North America. The fragrant resin, *liquid storax*, is the product of some of the species.

LIQUIDAMBAR.

Lat. *liquidam*, fluid, *amber*; a resin resembling amber flows from the trees.

Character of the genus the same as that of the order.

L. STYRACIFLUA. *Sweet Gum.*

Lvs. palmate, with acuminate, serrate lobes; *veins* villous at their bases.
—The sweet gum or gum-tree is thinly disseminated throughout the U. S.
With a diameter of 5f it arises to the height of 60. The trunk is covered
with a deeply furrowed bark. The young twigs are yellowish, putting forth
leaves of a rich green, which are deeply divided into 5 lobes more regularly
formed than those of the rock maple. The fruit is in a globular, compact
ball, suspended by a slender pedicel, consisting of numerous capsules, each containing 1 or 2 seeds. When wounded in summer, a gum of an agreeable odor
is distilled from the trunk. May.

ORDER CXXIX. PLATANACEÆ.—SYCAMORES.

Trees and *shrubs*, with alternate, palmately lobed leaves and sheathing, scarious stipules.
Aments monoecious, globose, with achlamydeous flowers.
Sterile.—Sta. single, with only small scales intermixed. *Anth.* 2-celled, linear.
Fertile.—Ova. terminated by a thick style with one side stigmatic.
Fr.—Nut clavate, tipped with the persistent, recurved style. Seed, solitary albuminous.

Genera 1, species 6? Trees of the largest dimensions, natives of Barbary, Levant and N. America.

PLATÁNUS.

Gr. πλατυς, broad; in reference to the ample foliage.

Character of the genus the same as that of the order.

P. OCCIDENTALIS. *Plane Tree. Button-wood. Sycamore.*

Lvs. lobed, angular; *branches* whitish.—The plane-tree is a native of all
the U. S., and is by far the largest (though not the loftiest) tree of the American
forest. On the margins of the great rivers of the West, trees are found whose
trunks measure from 40—50f in circumference, or more than 13f in diameter!
In N. England it also grows to magnificent dimensions. It flourishes in any
soil, but is most frequently met with on the stony borders and beds of streams.
Leaves very large, tomentose beneath when young. Flowers in globular aments
or balls, which hang upon the tree on long pedicels most of the winter. The
bark is yearly detached from the trunk in large scales leaving a white surface
beneath. May.

ORDER CXXX. URTICACEÆ.—NETTLEWORTS.

Trees and *shrubs*, with a milky juice, or herbs with a watery juice.
Lvs. alternate or opposite, rough or covered with stinging hairs, often stipulate.
Fls. monoecious, dioecious, or polygamous, in panicles, aments or dense heads.
Cal. membranous, lobed, persistent.
Sta. definite, distinct, inserted into the base of the calyx and opposite its lobes.
Ova. free, simple, 1-ovuled. *Style* 1.
Fr. achenium or utricle, surrounded by the membranous or fleshy calyx.

Genera 56, species 540, widely diffused throughout the world.

Properties.—The juice is almost always deleterious, sometimes in a high degree. It contains caoutchouc. The celebrated *Bohon Upas*, the most deadly of all poisons, is the concrete juice of *Antiaris toxi-
caria* of the Indian Archipelago. Its poisonous property is said to be due to the presence of *strychnia.*
Meanwhile the famous *cow tree* of S. America yields a copious supply of milk which is rich and wholesome. *Gum lac* is obtained abundantly from Ficus Indica. The renowned *Banyan tree* is Ficus reli-
giosa. In this order are also found many excellent fruits. *Figs* are the fruit of Ficus Carica, &c. *Bread
fruit* is the compound fruit of Artocarpus; *mulberries* of Morus nigra. *Fustic*, a yellow dye, is the wood
of M. tinctoria of S. America. The use of *hemp* in the manufacture of cordage is well known, as are
likewise the uses of the *hop.* The nettles are remarkable for their stinging, venomous hairs.

This order is composed of four principal suborders, viz. *Artocarpeæ*, *Moreæ*, *Urticeæ*, and *Cannabineæ*,
of which the three last are represented in the following genera.

Conspectus of the Genera.

Trees and shrubs. { Fruit a compound, fleshy, purple berry.	Morus.	1
Fruit simple, fleshy, dark red, small.	Broussonetia.	2
Fruit a large, compound, yellow globe.	Machura.	3
Fruit a fig!	Ficus.	4
{ Fls. spicate or paniculate. { Fertile cal. 2-sepaled.	Urtica.	5
{ Fertile calyx 0.	Bœhmeria.	6
Herbs { erect. { Lvs. simple. Sta. 4. { Flowers capitate, involucrate.	Parietaria.	7
{ Leaves palmately 5—7-foliate. Stamens 5.	Cannabis.	8
{ climbing dioecious. Stamens 5. Fertile flowers in aments.	Humulus.	9

SECTION I. MOREÆ.

Shrubs or trees with a milky juice. Fruit fleshy, composed of the fleshy calyx or receptacle.

1. MORUS.

Celtic *mor*, black; the color of the fruit of some of the species.

Flowers ♂, rarely ♂ ♀—♂ in loose spikes; calyx 4-parted. ♀ in dense spikes; calyx 4-parted; styles 2; achenium compressed, enclosed within the baccate calyx; spike constituting a compound berry.—*Trees with alternate, generally lobed leaves. Fls. inconspicuous.*

1. M. RUBRA. *Red Mulberry.*
Lvs. scabrous, pubescent beneath, rounded or subcordate at base, equally serrate, acuminate, either ovate or 3-lobed; *fertile spikes* cylindric; *fr.* dark red.— This tree varies greatly in height according to its situation. In New England, where it is not very common, it is but a shrub 15—20f high. In the Middle and Western States, it attains the elevation of 50—60f, with a diameter of 2f. Trunk covered with a grayish bark, much broken and furrowed. Wood fine-grained, strong and durable. Leaves 4—6' long, ⅔ as wide, entire or divided into lobes, thick, dark green. Flowers small. Berries of a deep red color, compounded of a great number of small ones, of an agreeable acid flavor. May.

2. M. ALBA. *White Mulberry.*—*Lvs.* nearly glabrous, cordate and oblique at base, unequally serrate, either undivided or lobed; *fr.* whitish.—Native of China. Cultivated for the sake of its leaves as the food of silk worms. A tree of humble growth. Leaves 2—4' long, ⅔ as wide, acute, petiolate. Flowers green, in small, roundish spikes or heads. Fruit of a yellowish-white, insipid.
β. *multicaulis.* (*Chinese Mulberry.*) *Lvs.* large (4—7' long, ⅔ as broad).—Shrub.

3. M. NIGRA. *Black Mulberry.*—*Lvs.* scabrous, cordate, ovate or lobed, obtuse, unequally serrate; *fertile spikes* oval.—Native of Persia, cultivated for ornament and shade, in this as well as in many other countries. Fruit dark red or blackish, of an aromatic, acid flavor.

2. BROUSSONETIA. L'Her.

In honor of P. N. V. Broussonet, a distinguished French naturalist.

Flowers ♂♀.—♂ Ament cylindric; cal. 4-parted. ♀ Ament globose; receptacle cylindric-clavate, compound; cal. 3-4-toothed, tubular; ovaries becoming fleshy, clavate, prominent; sty. lateral; seed 1, covered by the calyx.—*Trees, from Japan.*

B. PAPYRIFERA. *Paper Mulberry.*—*Lvs.* of the younger tree, roundish-ovate, acuminate, mostly undivided, of the adult tree 3-lobed; *fr.* hispid.—A fine, hardy tree, occasionally cultivated. It is a low, bushy headed tree, with large, light green, downy leaves, and dark red fruit a little larger than peas, with long, purple hairs.

3. MACLURA. Nutt.

Dedicated to William Maclure, Esq., of the U. S., a distinguished geologist.

Flowers ♂♀.—♂ in aments. Calyx 0; ova. numerous, coalescing into a compound, globose fruit, of 1-seeded, compressed, angular, cuneiform carpels; sty. 1, filiform, villous.—*A lactescent tree, with decid uous, alternate, entire, ex-stipulate leaves, and axillary spines.*

M. AURANTIACA. Nutt. *Osage Orange.*
A beautiful tree, native on the banks of the Arkansas, &c. Leaves 4—5' by 1½—2½', glabrous and shining above, strongly veined and paler beneath, on short petioles, ovate or ovate-oblong, margin obscurely denticulate, apex subacuminate, rather coriaceous. The fruit is about the size of an orange, golden-yellow when ripe, suspended by an axillary peduncle amid the dark glossy

leaves. No tree cultivated in our climate can surpass this in richness and beauty.

4. FICUS

Gr. συκη. Lat., ficus. Celtic, figueron. Teutonic, feige. Anglo-Saxon, fic. Eng. ch. fig.

Flowers ♂, fixed upon the inside of a turbinate, fleshy, closed re ceptacle. ♂ Calyx 3-parted; stam. 3. ♀ Calyx 5-parted; ovary 1; seed 1.—*A large genus of trees and shrubs, none North American.*

F. CARICA. Willd. *Common Fig.—Lvs.* cordate, 3-5-lobed, repand-dentate, lobes obtuse, scabrous above, pubescent beneath.—Supposed to be a native of Caria, Asia, although cultivated for its fruit in all tropical climes. With us it is reared only in sheltered locations as a curiosity. The delicious fruit is we l- known. Leaves very variable.

SECTION 2. URTICEÆ.

Herbs (in cool climates), with a watery juice. Flowers spicate or paniculate, with a membranaceous calyx.

5. URTICA.

Lat., uro, to burn; in reference to the stinging species.

Flowers ♂, sometimes ♂♀.—♂ Calyx 4-sepaled, with a cup shaped, central rudiment of an ovary; sta. 4. ♀ Calyx 2-leaved, persistent, at length surrounding the shining, compressed achenium; sty. 1.— *Herbs often with stinging hairs. Lvs. accompanied with stipules. Flowers green, in axillary or subterminal clusters.*

1. U. CANADENSIS (and divaricata. *Linn.*) *Common Nettle.*
Hispid and stinging; *lvs.* on long petioles, broad-ovate, rounded or sub- cordate at base, serrate, acuminate; *panicles* axillary, solitary or in pairs, di- varicate, mostly shorter than the petioles, lower sterile, upper fertile and sub- terminal, elongated in fruit.—Damp places, U. S. and Can. Stem 2—6f high, mostly simple, flexuous at top. Leaves alternate, large (3—5' by 2—3') more or less hispid both sides, sometimes nearly smooth. Lower petioles 3' long. Flowers minute, in panicles which vary from 1—4' in length, the fertile pani- cles about 2, nearly terminal and erect, enlarged in fruit. Aug.

2. U. DIOICA. *Diœcious or Stinging Nettle.*
Hispid and stinging; *lvs.* cordate, lance-ovate, conspicuously acuminate, coarsely and acutely serrate, the point entire, petioles thrice shorter; *fls.* ♂ or ♂ ♀, in branching, clustered, axillary, interrupted spikes longer than the petioles.— 4 Waste places, common. Stem 2—4f high, branching, obtusely 4-angled, with opposite, short-stalked leaves which are 3—4' long, and about ½ as wide. Flowers small, green, in axillary clusters, of mean aspect, corresponding with the insidious character of the plant. " Its power of stinging resides in minute, tubular hairs or prickles, which transmit a venomous fluid when pressed." *Bigelow.* July, Aug.

3. U. PROCERA. Willd. *Tall Nettle.*
St. tall, simple, 4-sided, slightly hispid; *lvs.* lanceolate, rough, hispid, prominently 5-veined, acutely serrate; *panicles* axillary, very branching, nu- merous, interruptedly spicate, lower ones sterile, upper fertile.- Borders of fields, waste places, N. H., &c. Stem 3—5f high, with a tough bark. Leaves densely strigose-hispid, serratures incurved, acute at each end, or somewhat acuminate at apex, 3 times longer than the petioles. Flowers small, green, in glomerate panicles, on the upper part of the stem. July.—Does not sting.

4. U. PUMILA. *Richweed. Stingless Nettle.*
St. ascending, often branched, weak and succulent; *lvs.* on long petioles, rhombic-ovate, crenate-serrate, membranaceous and glabrous; *fls.* ♂, triandrous, in corymbed heads shorter than the petioles.—4 In waste places, about build- ings, and in woods, U. S. and Can. Stem fleshy, semi-transparent when grow-

ing in shades, smooth and shining. Leaves on long petioles, especially the lower ones, smoothish, about 2′ long and ⅓ as wide. Flowers in short heads or corymbs, axillary. A species without stings. Aug., Sept.

5. U. URENS. *Burning or Dwarf Nettle.*

Lvs. broadly elliptic, about 5-veined, acutely serrate; *clusters* glomerate, by pairs.—① Weed, in cultivated grounds. Stem 12—20′ high, hispid with venomous stings, branching. Leaves 1—2′ long, ⅓ as broad, on short petioles and with large serratures. Stipules small, lanceolate, reflexed. Flowers in drooping, pedunculate clusters about as long as the petioles, both the sterile and fertile in the same axil. Rare. June, July. ⚥

6. S. GRACILIS. Ait. *Slender Nettle.*

St. erect, strict, sparingly hispid; *lvs.* ovate-lanceolate, subacuminate, coarsely and somewhat doubly serrate, 3-veined, smoothish above, hispid beneath on the veins; *spikes* elongated, pinnately branched, a little shorter than the leaves; *fls.* glomerate.—⅔ Northern and Western States, and Brit. Am. Stem 2—3′ high. Flowers minute, green. July, Aug.

6. BŒHMERIA. Willd.
Named for G. F. Bœhmer, a German botanist.

Flowers ☿ or ♂ ♀.—♂ Calyx 4-parted, with lanceolate, acute segments; stamens 4. ♀ achlamydeous; ovary and style 1, in the axil of a bract; achenium compressed, margined.—*Herbs or shrubs, nearly allied to Urtica. Lvs. opposite or alternate. Fls. clustered.*

B. CYLINDRICA. (Urtica cylindrica and capitata. *Linn.*) *False Nettle.*

Herbaceous; *lvs.* opposite, ovate-lanceolate, acuminate, dentate, smooth; *fls.* ♂ ♀; *sterile spikes* glomerate, interrupted, *fertile* cylindric.—A coarse, nettle-like plant, in swamps and bottoms, Mid. and Western States! Stem slender, obtusely 4-angled, channeled on each side, 2—3′ high. Leaves 3-veined, 3—5′ long, ⅓ as wide, on long petioles. Flowers minute, the fertile ones in axillary, cylindric spikes, 1—2′ in length, the barren spikes rather longer and more slender. July, Aug.

β *Spike* shorter, subcapitate; *petioles* somewhat shorter.

γ. (B lateriflora. *Muhl.*) *Lvs.* roughish; *spikes* longer and much interrupted.

7. PARIETARIA.
Lat. *paries*, a wall; some of the species prefer to grow on old walls, &c.

Flowers monœcious-polygamous, in clusters surrounded by a many-cleft involucre; calyx 4-parted; stamens 4, at first incurved, then expanding with an elastic force; ovary and style 1; achenium polished, enclosed within the persistent calyx.—*Herbs with usually alternate leaves. Clusters of green flowers axillary.*

P. PENNSYLVANICA. *Pellitory.*

Lvs. oblong-lanceolate, veiny, tapering to an obtuse point, punctate with opaque dots; *invol.* longer than the flowers.—① A rough, pubescent herb, found in damp, rocky places, Vt., N. Y., W. to Wisc.! &c. Stem erect, simple or sparingly branched, 6—12′ high. Leaves alternate, entire, hairy and rough, about ½′ wide and 3 or 4 times as long, petiolate, and ending with an obtuse acumination. Segments of the involucre about 3, lance-linear. Flowers dense, greenish and reddish-white. Rare. June.

SECTION 3. CANNABINEÆ.

Herbs, erect or twining, with a watery juice. ♂ racemose or paniculate, ♀ in a cone-like ament. Albumen 0.

8. CANNABIS.
Arabic *ganab*, hemp.

Flowers ♂ ♀.—♂ Calyx 5-parted. ♀ Calyx entire, oblong-acuminate, opening longitudinally at the side; sty. 2; ach. ? 2-valved, enclosed within the persistent calyx.—① *Lvs. opposite, digitate. Fls, axillary, ♂ in cymose panicles, ♀ in sessile spikes.*

C. SATIVA. *Hemp.*
Lvs. palmately 5—7-foliate.—The hemp was introduced originally from India, but it springs up spontaneously in our hedges and waste grounds. It is a tall, erect plant, with handsome petiolate leaves. Leaflets lanceolate, serrate, 3—5′ long, ⅓ as wide, the middle one the largest. Flowers small, green, solitary and axillary in the barren plants, spiked in the fertile ones. It is cultivated in many countries for the sake of its fibre, which is stronger than that of flax, and is the best of all materials for cordage and sail-cloth. The seeds are nutritious, but the leaves are stimulant and narcotic, producing intoxication. June. ⚥

9. HUMULUS.

Lat. *humus*, moist earth; the hop grows only in rich soils.

Flowers ♂ ♀.—♂ Calyx 5-sepaled; stamens 5; anthers with 2 pores at the summit. ♀ Bracts imbricate, large, entire, concave, persistent, 1-flowered; calyx membranous, entire, persistent; styles 2; achenium invested by the thin calyx.—♃ *twining with the sun. Lvs. opposite. Fls. in axillary panicles and strobile-like aments.*

H. LUPULUS. *Common Hop.*
The hop vine is found wild in hedges, &c., throughout this country, and is, as every one knows, extensively cultivated for the sake of its fertile aments, which are chiefly used as a preservative in beer. It has a long, annual stem of rapid growth, always twining with the sun, rough backwards with reflexed prickles. Leaves very rough, generally 3-lobed, deeply cordate at base, on long stalks. Flowers of the barren plants extremely numerous, panicled, greenish; those of the fertile, in aments with large scales. In the cultivation of the hop it has been found *profitable* to plant a few layers of the barren vines among the fertile ones, as the produce is thus increased in weight through the fertilization of the seeds. Aug.

CLASS II. GYMNOSPERMS.

OVULES not enclosed in an ovary, fertilized by the pollen without the intervention of a pistil, and becoming truly NAKED SEEDS, the carpel being represented by a flat open scale or entirely wanting. EMBRYO with 2 opposite, or several whorled cotyledons.

ORDER CXXXI. CONIFERÆ.—CONIFERS.

Trees or evergreen shrubs, with branching trunks, abounding in a resinous juice.
Lvs. scattered or fascicled, linear or acerose (rarely lanceolate), parallel-veined, rigid, generally evergreen.
Fls. monœcious or diœcious, destitute of calyx or corolla.
Sterile, monandrous or monadelphous, collected in a kind of loose ament.
Anth. 2 or many-lobed, often tipped with a crest. *Pollen* large, usually compound.
Fertile, in aments composed of open, scale-like carpels, or solitary and without a carpel.
Ovary, *style* and *stigma* wanting. *Ovules* 1, 2 or many, erect or inverted.
Fr.—A strobile (cone), or a solitary seed. *Integuments* hard and crustaceous.
Embryo in the axis of oily albumen.

Genera 29, species 180, natives of all climates, but most abundant in the temperate zones, those of the southern, however, very different from the pines, spruces, larches and cedars of the northern.

Properties.—Few orders can be named, which are of more importance to mankind, whether in reference to their invaluable timber or their resinous secretions. *Turpentine, tar, pitch* and *resin* are the product of the pines. *Burgundy pitch is* yielded by Pinus sylvestris of Europe; *Venetian turpentine* by the Larix; *oil of Savin* by Juniperus Sabina of Europe, &c.

FIG. 54.—1. Branch of Thuja occidentalis, with strobiles. 2. A magnified branchlet with a cone of abundant flowers. 3. A carpellary scale with the two winged seeds. 4. A vertical transverse section of one of the seeds, showing the embryo, &c. 5. The immature, erect ovules. 6. One of the ovules enlarged, showing the micropyle at top. 7. Branch of Abies Americana. 8. Scale, with the bract. 9. Scale with immature ovules. 10. Scale with ripe seeds. 11. A pair of leaves of Pinus resinosa. 12. Anther of Pinus sylvestris. 13. Scale of the cone, with the ovules turned downwards. 14. Staminate ovule of Cupressus, with pollen. 15. Fertile scale, with many erect ovules.

Conspectus of the Genera.

{ Leaves linear or acerose. . .	{ Scales thick and blunt at edge.	.	*Pinus.*
	{ Scales thin and even at edge.	.	*Abies.* 2
{ a woody cone. { Leaves scale-like, imbricate.	{ Fertile scales 4—8-ovuled.	.	*Cupressus.* 3
	{ Fertile scales 2-ovuled.	.	*Thuja.* 4
{ a fleshy berry with 3 bony seeds. Leaves mostly acerose.	.	.	*Juniperus.* 5
Fruit { a fleshy drupe with a single seed. Leaves linear, 2-ranked.	.	.	*Taxus.* 6

Tribe 1. ABIETINEÆ.

Flowers ♂ ☿ ♀.—♂ aments numerous, deciduous. Scales peltate, each bearing 2 sessile, 1-celled anthers. ♀ Strobile ovoid; carpellary scales closely imbricated, each bearing a pair of ovules adhering to the base inside, and subtended by a bract outside; fruit a woody strobile or cone; seeds winged, cotyledons 2—15.

1. PINUS.

Celtic *pin* or *pen*, a rock or mountain ; many species of this noble genus prefer such situations.

Strobile large, conical; carpellary scales thickened at the summit, becoming strong and woody in fruit; cotyledons 4—8.—*Trees, often of the loftiest dimensions. Branches often verticillate. Leaves evergreen, acerose, in fascicles of 2—5, each fascicle invested with a membranous sheath at base.*

1. P. RESINÓSA. Ait. (P. rubra. *Michx.*) *Norway Pine. Red Pine.*

Lvs. in pairs, channeled, elongated, with elongated sheaths; *cones* ovoid conic, rounded at the base, subsolitary, about half as long as the leaves; *scales* unarmed, dilated in the middle.—It abounds in the northern parts of the U. S. and in Canada, attaining the height of 80f, with a trunk 2f in diameter, very straight and uniform. Bark smoother, and of a clearer red than other pines. Leaves chiefly collected towards the ends of the branches, always in pairs, 5—8' in length, the sheaths ½—1½'. This pine affords a fine-grained, resinous timber of much strength and durability, and highly valued in architecture. May.

2. P. BANKSIÀNA. Lambert. (P. rupestris. *Michx.*) *Scrub Pine.*

Lvs. in pairs, rigid, curved, acute, terete upon the back and channeled above, margins somewhat scabrous; *cones* ovate-acuminate, recurved, tortuous; *scales* unarmed, obtuse, smooth.—A small tree, with long, spreading, flexible branches, abounding in barrens, in Me. and British America. Leaves about an inch in length. Cones nearly twice as long as the leaves, usually in pairs. April, May.

3. P. INOPS. Ait. *Jersey* or *Scrub Pine.*

Lvs. in pairs, rather short, obtuse, rigid, channeled above, terete beneath, margins obscurely serrulate; *cones* recurved, ovoid-oblong, as long as the leaves; *scales of the cone* compact, obtuse at base, with a straight, subulate point.—A tree 15—25f high, on barrens in the Middle States. Branches straggling, and, with the trunk, covered with a rough, blackish bark. Leaves 1—2' long. The wood abounds in resin. May.

4. P. VARIÁBILIS. Lamb. (P. mitis. *Michx. f.*) *Yellow Pine. Spruce Pine.*—*Lvs.* 2—3 together, channeled on the inner surface; *cones* ovoid, subsolitary; *scales* armed with short, incurved spines.—Widely diffused throughout the country, attaining the height of 50—60f. Leaves dark green, 5—6' long, covering the branchlets. Cones 2—3' long, rugged with the projecting point of the scales. It furnishes close-grained and moderately resinous timber, which is used in immense quantities for all kinds of architecture. May.

5. P. RÍGIDA. *Pitch Pine.*

Lvs. in 3s, with short sheaths; *cones* pyramidal-ovoid, clustered; *scales* with reflexed spines.—Common in barren, sandy plains, which it often exclusively occupies. It is of moderate height at the north (25—30f), but attains a great height in the Southern States. The trunk, which is seldom straight, is covered with a very thick and rough bark cleft with deep furrows. Leaves 4—6' long. Cones usually several together, 2—3' long. The wood is heavy with resin, little used in architecture except for floors, but makes excellent fuel. May.

6. P. palustris. Lamb. (P. australis. ?) *Long-leaved* or *Broom Pine.*

Lvs. in 3s, very long, conglomerate at the ends of the branches; *cone* subcylindrical, muricate, with small, recurved spines; *stip.* pinnatifid, ragged, persistent.—Found in the Middle, Southern and Western States. The trunk is 15—30' diam., arising with a slight diminution 40 or 50f to the branches, thence 20—40f to the summit. Bark slightly furrowed. Leaves a foot in length. Buds very long, whitish. Sterile aments violet-colored, 2' long. Cone 6—10' long. Seeds with a thin, white testa. Timber strong, compact and durable, used at the South in vast quantities.

7. P. strobus. *White Pine. Weymouth Pine.*

Lvs. in 5s, slender, with very short sheaths; *cones* solitary, cylindric, loose, pendant longer than the leaves.—This pine is one of the most majestic and the most useful forest trees of this, or of any other country. The trunk is perfectly straight, covered with a comparatively smooth bark, and, in some instances, 5—7f in diameter, and 100f in height without a limb; then, sending out a few branches, it forms a tufted head far above the surrounding forest. The branches are given off in whorls which are very observable in young trees. The leaves are about 4' long, numerous, slender, of a bluish green, forming an extremely soft and delicate foliage. The wood is soft, fine-grained, easily wrought, very durable, and is used in immense quantities in various kinds of architecture. The large trunks are in particular sought for the masts of ships. May.

2. ABIES. Juss.

Name probably derived from the Celtic abies.

Strobile smaller, roundish-oblong; carpellary scales attenuated to a thin, even edge; cotyledons 3—9.—*Trees or shrubs. Lvs. evergreen or deciduous, linear and solitary, or acerose and fasciculate, never sheathed at base.*

§ 1. *Leaves evergreen, solitary, linear.*

1. A. canadensis. (Pinus. Linn.) *Hemlock.*

Lvs. linear, flat, obscurely denticulate, glaucous beneath, in 2 rows; *cones* ovoid, terminal, scarcely longer than the leaves; *scales* rounded, entire.—A well known evergreen inhabitant of the rocky, mountainous woods of the Northern States, and Brit. Am., commonly attaining the height of 70—80f. The trunk is large in proportion, straight, covered with a rough bark. Branches brittle and nearly horizontal, with pubescent twigs. Leaves 6—8" in length, less than 1" wide, arranged in 2 opposite rows. Cones very small. The wood of the hemlock is soft, elastic, of a coarse, loose texture, not much valued for timber, but is sometimes substituted for pine. The bark is extensively used in tanning. May.

2. A. nigra. Michx. (Pinus. Linn.) *Black or Double Spruce.*

Lvs. 4-cornered, scattered, straight, erect; *cones* ovoid, pendulous; *scales* elliptical-obovate, erosely dentate at the edge, erect.—This fine tree abounds in the northern parts of the U. S., where dark, mountain forests, are often wholly composed of it. It is a large tree, 70—80f high, with a straight trunk and a lofty pyramidal head. The leaves thickly cover the branches, are of a dark green color, little more than ½' in length. Cones 1—2' long. The timber is light, strong and elastic, and, although inferior to the white pine, is much used in architecture. That salutary beverage, *spruce beer*, is made from the young branches. May.

3. A. alba. Michx. (Pinus. Ait.) *White or Single Spruce.*

Lvs. 4-sided, incurved; *cones* lax, pendulous, subcylindric, with entire, broadly obovate, somewhat 2-lobed scales.—Very abundant in the northern sections of the U. States, preferring humid and rocky woods. Height 50f. Trunk 1—2f in diameter at the base, regularly dimin ing upward. Lower branches longest, the others becoming gradually shorter upwards. Leaves ½—¾' in length, placed on all sides of the branches. Cones small. The timber is useful in the frames of buildings, &c. May.

44

§ 2. *Leaves solitary, evergreen. Bark smooth, with reservoirs of balsam. Cones long, erect.*

4. A. BALSAMEA. Willd. (Pinus. *Linn.* Picea. *Michx.*) *Fir Balsam.*

Balsam Spruce.—Lvs. linear, flat, obtuse, glaucous, with a grooved line above and an elevated one beneath; *cones* cylindric, erect, reflexed on the margin; *scales* broad, compact; *bracts* obovate, shorter than the scale.—A beautiful evergreen, common in humid forests of the northern part of the U. States. Its branches are nearly horizontal, gradually becoming shorter upwards, forming a regularly pyramidal head. The leaves are little larger than those of the hemlock (8—10" long) growing upon the sides and top of the branches, of a bright green above, and silvery-white beneath. Cones 2—3' in length. Bark smooth, abounding in reservoirs filled with a resin or balsam which is considered a valuable medicine. May.

5. A. FRASÈRI. Pursh. *Fraser's* or *Double Balsam Fir.*

Lvs. flat, glaucous beneath, linear, often emarginate, subsecund, erect above; *strobile* ovoid-oblong, erect, very small; *bracts* elongated, reflexed, oblong-cuneate, emarginate, briefly mucronate, incisely toothed.—A smaller tree than the last, much resembling it in habit, found on mountains, from N. Eng.! to Car. Leaves 3' long, and much crowded. Cones about 1—2' long when mature; singularly distinguished by the long-pointed, violet-colored, reflexed bracts. Sterile aments terminal. May.—A highly ornamental shade tree.

§ 3. *Leaves deciduous, collected in fascicles of 20—40.*

6. A. (LARIX) AMERICÀNA. Michx. (Pinus pendula and microcarpa of *authors.*) *American Larch.—Lvs.* short, in dense fascicles, without sheaths, very slender; *cones* oblong, inclining upwards; even when the branches are pendulous; *scales* thin and inflexed on the margin; *bracts* elliptical, often hollowed at the sides, abruptly acuminate with a slender point.—A beautiful tree, often seen in our shrubberies, and thinly interspersed, in forests, throughout N. England. It is remarkably distinguished from the pines by its deciduous leaves, the branches being bare nearly half the year. The tree arises 80—100f, with a straight and slender trunk and horizontal branches. Leaves 1—2' long, collected in bunches of 12—20 on the sides of the branches. Cones deep purple, ½—1' long. The wood is considered most valuable of all the pines or spruces, being very heavy, strong, and durable. Apr., May.

β. *pendula. Branches* slender and drooping.—A most beautiful variety.

TRIBE 2.   CUPRESSINEÆ.

Carpellary scales not bracteate, each with 1—8 erect ovules at base inside, becoming concreted and fleshy in a drupe-like fruit. Anthers of several cells.

3. CUPRESSUS.

From the Isle of Cyprus, where the cypress is very abundant.

Flowers ♂.—♂ in an ovoid ament; scales peltate; anthers 4. sessile. ♀ in a strobile; scales peltate. bearing 4—8, erect (orthotropous) ovules at base inside; seed angular, compressed; integuments membranous; cotyledons 2 or more.—*Trees, with evergreen, flat, squamose, imbricated leaves. Fertile aments becoming indurated cones.*

1. C. THYÒIDES. Michx. *White Cedar.*

Branchlets compressed; *lvs.* imbricate in 4 rows, ovate, tuberculate at base *cones* spherical.—This tree is thinly disseminated in N. England, but quite common in the Middle States. It usually occurs in swamps, which it densely and exclusively occupies. Height 40—60f. The leaves consist of short, minute, evergreen scales, covering the finely divided branchlets, in 4 imbricated rows, and each one furnished with a minute gland or tubercle on the back. The wood is white, fine-grained, and wonderfully light, soft and durable. Used in the manufacture of shingles, pails, fences, &c. Posts made of this cedar it is said will last 50 years. May.

2 C. DISTYCHA. (Taxodium distychum. *L. C. Rich.*) Cypress.
Lvs. in 2 rows (distychous), deciduous, flat; *sterile aments* paniculate, leaf-less, pendulous; cone oblong-globose.—One of the largest trees of the forest, native of N. J. to Mexico. It grows in wet soils, forming what is called the cypress or cedar swamps of the Southern States. The trunk arises to the height of 125f with a circumference of 25—40f above the conical base. The roots produce large, conical excrescences, which, being hollow, are sometimes used for beehives. The head is wide-spread, and often depressed. Foliage light green and open. Cones 1′ diam., composed of the indurated, combined scales. Timber light, fine-grained and durable.

4. THUJA.

Gr. θυω, to sacrifice; the wood is fragrant in burning and was used in sacrifices.

Flowers ♂.—♂ in an imbricated ament; anthers 4, sessile. ♀ in a strobile, each scale bearing 2 erect ovules at the base inside; seed winged; integument membranous; cotyledons 2 or more.—*Trees or shrubs. Lvs. evergreen, squamose, imbricate.*

T. OCCIDENTALIS. *Arbor Vitæ.*
Branchlets ancipital; *lvs.* imbricate in 4 rows, rhomboid-ovate, appressed, tuberculate; *cones* oblong, the inner scales truncated and gibbous below the tip.—This tree is often called *white cedar*, and from its resemblance might easily be mistaken for the *Cupressus thyoides*. It abounds in the British Provinces and in the northern parts of the U. S. on the rocky borders of streams and lakes, and in swamps. It has a crooked trunk, rapidly diminishing in size upwards, throwing out branches from base to summit. The evergreen foliage consists of branchlets much more flat and broad than those of the white cedar. Cones terminal, consisting of a few long, loose scales, unlike the round, compact cones of that tree. The wood is very light, soft and durable. Its most important use is for fences. May.

5. JUNIPERUS.

Celtic Juneprus. rough or rude.

Flowers ♂ ♀, rarely ♂.—♂ ament ovate; scales verticillate, peltate; anthers 4—8, 1-celled. ♀ ament globose; scales few, united at base, concave; ovules 1 at the base of each scale; berry formed of the enlarged, fleshy scales containing 2—3 bony seeds; cotyledons 2. —*Trees or shrubs. Lvs. evergreen, mostly acerose, opposite or in whorls of 3.*

1. J. COMMUNIS. *Common Juniper.*
Lvs. ternate, spreading, subulate, mucronate, longer than the berry.—A shrub, with numerous, prostrate branches, growing in dry woods and hills, often arising in a slender pyramid, 6—8f high (rarely arboreous, *Dr. Robbins!*) Leaves arranged in whorls of 3, 5—8″ long, acerose-lanceolate, ending in a sharp, bristly point, channeled and glaucous on the midvein above, keeled and green below. Barren flowers in small, axillary aments or cones; fertile ones on a distinct shrub, small, axillary, sessile. Berries roundish, oblong, dark blue, ripening the second year from the flower. They are then sweetish, with a taste of turpentine. In medicine they are diuretic and cordial. May.

2. J. VIRGINIANA. (J. Sabina. *Hook.*) *Red Cedar.*
Upper lvs. imbricate in 4 rows, ovate-lanceolate, pungently acute, appressed, older ones acerose, cuspidate, spreading; *trunk* arboreous.—Found throughout the U. S., but chiefly in the maritime parts, growing in dry, rocky situations. It is a tree of middle size, sending out numerous, horizontal branches. Leaves dark green, the younger ones small, ovate, acute, scale-like, overlaying each other in 4 rows, upon the subdivided branchlets; the older ones ¼′ long. Flowers inconspicuous, the staminate in oblong, terminal aments, ¼′ long; the fertile on separate trees, producing small, bluish berries covered with a white powder. The wood is fine-grained and compact, of a reddish hue, very light

and durable. It is used for fences, aqueducts, tubs and pails, and in the manufacture of drawing pencils. April, May.

β. *prostrata.* *Lvs.* ovate, submucronate, glandular in the middle, appressed,
berries tubercular; *st.* prostrate, creeping.—A shrub, on gravely shores, with
creeping branches 4—6f long.

TRIBE 3. TAXINEÆ.

Fertile flowers solitary, terminal, consisting of a naked ovule maturing into a
kind of drupe.

6. TAXUS.

Gr. τaξ̃oυ, an arrow; arrows were formerly poisoned with the juice of the yew tree.

Flowers ♂♀ or ⚥, surrounded with numerous scales. ♂ Stamens
8—10, monadelphous; anthers peltate, 6—8-celled, cells dehiscent
beneath. ♀ solitary, consisting of a single ovule, becoming in fruit a
fleshy, 1-seeded drupe.—*Trees or shrubs, with evergreen, linear, alternate leaves.*

T. CANADENSIS. *Dwarf Yew. Ground Hemlock.*

Lvs. linear, mucronate, 2-ranked, revolute on the margin; *sterile receptacles* globose.—A small, evergreen shrub, with the general aspect of a dwarf
hemlock spruce (*Pinus Canadensis*). It grows on thin, rocky soils in shady
places, 2—3f long, Can. to Penn. and Ky. Leaves nearly an inch long, arranged in 2 opposite rows, on the sides of the branchlets. Staminate flowers in
small, roundish, axillary heads. Drupes oval, concave or open at the summit,
red and juicy when mature. May.

SUBDIVISION SECOND.

ENDOGENS, OR MONOCOTYLEDONOUS PLANTS.

STEM not distinguishable into bark, pith and concentric zones or layers of wood. GROWTH by irregular, internal accretions, consisting of bundles of woody fibre and vessels, successively descending from the leaves above, through the cellular tissue already formed. LEAVES mostly with simple, parallel veins, alternate, entire, frequently sheathing at base, and seldom falling off by an articulation. SEPALS and PETALS, when present, commonly in 3s. OVULES produced within an ovary. EMBRYO with one cotyledon, rarely with two, the second being much smaller than, and alternate with, the first.

CLASS III. AGLUMACEOUS ENDOGENS.

Flowers without glumes. Organs developed on the usual and normal plan, consisting of stamens and pistils, either or both, surrounded by verticillate, floral envelops; or the latter are wanting, and the stamens and pistils are achlamydeous.

ORDER CXXXIV. ARACEÆ.—ARADS.

Herbs or tropical shrubs, with a fleshy rhizoma or cormus.
Lvs. sheathing at base, often with branching veins, and sometimes compound.
Fls. mostly monœcious and achlamydeous, arranged upon a naked or spathaceous spadix.
Perianth, when present, consisting of 4—6 parts.
Sta. definite or indefinite, hypogynous, very short. *Anth.* ovate, extrorse.
Ova. free, 1—several-celled. *Stigma* sessile.
Fr.—Berry succulent or dry. *Seeds* solitary or several, with fleshy albumen.

Genera 28, species 179, abundant in tropical regions, more rare in temperate, one only, Calla palustris, extending to the northern frigid zone.

Properties.—An acrid, volatile principle pervades the order, which is, in some instances, so concentrated as to become poisonous. The corms and rhizomes abound also in starch, which in some cases, when the volatile acridity is expelled in drying or cooking, is edible and nutritious.

Conspectus of the Genera.

		and covered	cylindric.	Berry 1-seeded.	*Peltandra.*	2
		with flowers,	oval, preceding the leaves.	Berry many-seeded.	*Calla.*	3
	Spadix in a spathe	and naked above. Stem a corm.			*Symplocarpus.*	6
broad.	Spadix naked, yellow, on a clavate scape.				*Arum.*	1
Leaves	linear-ensiform. Scape leaf-like. Spadix lateral.				*Orontium.*	4
					Acorus.	5

1. ARUM.

Coptic *aron,* the name of the Egyptian species, A. colocasia.

Flowers sometimes ♀ ♂. Spathe cucullate, convolute at base; perianth 0; spadix cylindric, naked above, staminate below the middle and pistillate at the base; berry 1-celled, many-seeded.—♃

1. A. TRIPHYLLUM. *Dragon-Root. Jack-in-the-Pulpit.*

Acaulescent; lvs. trifoliate, mostly in pairs, leaflets oval, acuminate; spadix clavate; spathe ovate, acuminate, flat and deflected above.—A curious and well known inhabitant of wet woodlands, Can. to Car. W. to the Miss. The stem is a rugose, fleshy, subterraneous corm giving off radicles in a circle from

44*

the edge. Scape 8—12' high, erect, round, embraced at the base by the long sheaths of the petioles. Leaves 2, on long stalks, each consisting of 3 smooth leaflets, 2—7' long. ½ as wide. Spathe green without, usually variegated within with stripes of dark purple alternating with pale green. Spadix much shorter than the spathe varying from green to dark purple. Fruit a bunch of bright scarlet berries. The corm loses its fiercely acrid principle by drying, and is then valued as a carminative, &c. May, June.

β. *atrorubens.* Dewey. (A. atrorubens. L.) *Spathe* sessile, spreading horizontally above, dark brown.—Plant rather smaller, and with a disagreeable odor.

2. A. Dracontium. *Green Dragon.*

Acaulescent; *lf.* mostly solitary, pedate; *lfts.* oblong-lanceolate; *spadix* subulate, longer than the convolute, oblong spathe.—Less common in N. Eng. than the former species, found in wet places, banks of streams, U. S. Stem a fleshy, subterraneous corm. Scape slender, 6—12' high. Leaf on an erect, sheathing petiole, which is dichotomous above, each half bearing 2—4 leaflets with an odd one at the fork. Leaflets rather smaller than in *A. triphyllum.* Spathe green, 1—2' long, rolled into a tube at base. Spadix slender, with its long, tapering point much exserted. Fruit a bunch of red berries. June, July.

2. PELTANDRA. Raf.

Gr. πελτη, a shield or target, ανδρες, stamens ; from the character.

Spathe convolute; spadix covered with flowers, staminate above, pistillate below; perianth 0; stamens peltate; berry 1-celled, 1-seeded.—♃.

P. Virginica. Raf. (Arum. *Linn.* Calla. *Bw.* Lecontia. *Cooper.* Rensselaeria. *Beck.* Caladium. *Lindl.*)—Acaulescent; *lvs.* oblong, hastate-cordate, acute at apex, the lobes obtuse; *spathe* elongated, incurved; *spadix* covered with staminate flowers the greater part of its length.—A smooth, dark green plant, in wet grounds, N. Y. and Ms. to Car. Leaves radical, numerous, 8—12' long, ½ as wide, on petioles as long as the scapes. Scapes many from the same root, 8—15' long. Spathe closely involving the spadix, green, 2—3' long, lanceolate, wavy on the margin. Spadix slender, acuminate, shorter than the spathe, bearing the ovaries and finally the berries in a dense cluster at its base. June. It is to be hoped that this persecuted plant will soon find, if it has not here found, a permanent abode. Jl.

3. CALLA.

Gr. καλλος, beautiful ; a term well applied.

Spathe ovate, spreading; spadix covered with flowers, staminate intermixed with, or above the pistillate; perianth 0; berry many-seeded.—♃ *Aquatic herbs.*

1. C. palustris. *Northern Calla.*

Lvs. cordate; *spathe* ovate, flat; *spadix* covered with ovaries intermixed with stamens.—A fine plant, growing in shallow water, Mid. States, N. Eng. N. to Arc. Am. Rhizoma creeping, rooting at the joints. Leaves 2—3' long, ½ as wide, on long stalks, involute at the acuminate point, smooth and entire. Scape smooth, green, roundish, thick, 4—6' high. Spathe clasping at the base, spreading, recurved, with an involute point, greenish-yellow without, white and soft within. Spadix 1' in length. The root-stock is acrid, but Linnæus tells us that the Laplanders extract a wholesome bread stuff from it. July.

2. C. Æthiopica. *Ethiopian Calla.*—*Lvs.* sagittate-cordate; *spathe* cucullate; *spadix* with the sterile flowers above the fertile.—A magnificent plant from Cape Good Hope, often met with in green-houses and parlors. The leaves are very large, smooth and entire, on long, sheathing, radical footstalks. Scape smooth, round, arising a little above the leaves, 3—5f high. Spathe very large, white, involute at base, reflexed and terminating abruptly in a long *acumination.* Spadix yellowish-white, about half the length of the spathe. *Flowers from* Jan. to May.

3. ORONTIUM.

The ancient Gr. name of a plant, so called from its growing by the Orontes, a river in Asia Minor.

Spadix cylindric, covered with flowers; perianth 4—6-sepaled; stamens 4—6; ovary free; stigma sessile; fruit a dry berry or utricle.—♃ *acaulescent, aquatic. Fls. yellow, at the summit of the scape. Spathe radical.*

O. AQUATICUM. *Golden Club.*
Lvs. ovate-lanceolate; *spike* or *spadix* cylindric, on a clavate scape.—This interesting plant is a native of inundated banks and pools, U. S., but not very common. The leaves are large (often becoming 10—12′ long and ½—1 as wide), smooth, of a deep green, velvet-like surface above, paler beneath, on long, radical petioles. Scape thick and terete, about a foot in length, closely invested by the short spathe at base, and ending in a spadix of a rich yellow color, covered with small, perfect, yellow flowers of an offensive odor,—the upper ones often tetramerous. May.

4. ACORUS.

Gr. a, privative, and κορη, the pupil of the eye; supposed to cure maladies of the eye.

Spadix cylindric, covered with flowers; perianth 6-sepaled; ovary free; stigma sessile, minute; fruit dry, 3-celled, many-seeded.— ♃ *herbs, with a fleshy rhizoma. Lvs. radical, ensiform. Scape foliaceous.*

A. CALAMUS. *Sweet Flag.*
Summit of the scape above the spadix very long and leaf-like.—Grows in wet soils throughout the U. States. The thick, prostrate, creeping rhizoma is highly valued for its aromatic flavor, its warm and pungent taste. The long, sword-shaped leaves are readily distinguished by the ridge running their whole length. The cylindrical spadix is about 3′ long and 3″ diam., covered with small, green flowers, and bursting from the side of the leaf-like scape in June and July.

5. SYMPLOCARPUS. Salisb.

Gr. συμπλοκη, connection, καρπος, fruit.

Spathe ventricose; spadix oval, covered with perfect flowers; perianth deeply 4-parted, segments cucullate, cuneate, truncate, persistent, becoming thick and spongy; berries globose, 2-seeded, imbedded in the spadix.—♃ *Aquatic, acaulescent herbs.*

S. FŒTIDUS. Nutt. (Pothos fœtida. *Michx.* Ictodes. *Bw.*) *Skunk Cabbage.*
Lvs. cordate-oval, acute; *spadix* subglobose, preceding the leaves.—A common plant, Can., N. Eng., Mid. and W. States, growing in swamps, meadows and ditches, renowned for its odor, which is scarcely less offensive than that of the animal whose name it bears. Early in spring, the swelling spathe is seen emerging first from the ground or water, more or less covered with purplish spots, its edges partly infolded, and its point incurved. It encloses the spadix, which is oval, covered with flowers of a dull purple. The leaves, which arise fter the flowers, are of a bright green, numerous, becoming very large (often 20′ by 12′.)

ORDER CXXXV. LEMNACEÆ—DUCKMEATS.

Plants mostly floating, cellular, annual, consisting of a frond (stem and leaf in one).
Fls. bursting from the margin of the fronds, 2 or 3, achlamydeous, enclosed in a spathe.
Sterile fls.—Sta. definite, often monadelphous.
Fertile fls.—Ova. 1-celled, with 1 or more erect ovules. Sty. short. Stig. simple.
Fr.—Utricle 1—several-seeded. Sds. with a fungous testa.

Genera 6, species 23, widely diffused. They are almost entirely destitute of spiral vessels. Some of them may be regarded as the simplest of Phænogamous plants.

LEMNA.

Gr. λεμμα, a scale or husk; from the resemblance of the frond.

Sterile and fertile flowers in the same spathe, the former of 2. cal-lateral stamens, tne latter of a simple, carinate ovary, with a style and stigma.—① *Herbs, consisting of a frond (stem and leaf confounded) sending down from the under surface, roots which hang loosely in the water, and producing from the margins the spathaceous flowers.*

1. L. TRISULCA. *Ivy-leaved Duck-meat.*
Fronds elliptic-lanceolate, thin, serrate at one extremity and caudate at the other; *roots* solitary.—Floating in ponds and pools of clear water. Fronds nearly ½' in length, diaphanous, with a tail-like appendage at base, obtuse at apex, the new ones issuing in a cruciate manner from lateral fissures in the margin of the old. Root a solitary fibre, ending in a sheath. Flowers very minute. Utricle sitting on the upper surface of the frond. June—Sept.

2. L. MINOR. *Lesser Duck-meat.*
Fronds nearly ovate, compressed; *root* solitary.—This little floating plant occurs in dense patches on the surface of stagnant waters. The leaves, properly fronds, adhere 2—3 together, 1' in length, rather thick, and convex below. Root undivided, sheathed at the end. Flowers minute from a cleft in the margin of the fronds, near the base. Jn.—Sept.

3. L. GIBBA. *Gibbous Duck-meat.*
Fronds obovate, hemispherical beneath, nearly plain above; root solitary. —Floating on the surface of stagnant waters, N. York. Fronds about a line in length, pellucid and reticulated beneath. June—Sept.

4. L. POLYRHIZA. (Spirodela. *Schleiden.*)
Fronds broad-ovate, a little convex beneath; *roots* numerous.—Floating in stagnant waters. Fronds resembling flax-seed, but larger (2—4" long), scattered on the surface of the water, of a firm, but succulent texture, becoming purplish. Roots in thick bundles of 8—10 black fibres from the under surface of the fronds. All these species are eaten by ducks and other aquatic birds. June—Sept.

ORDER CXXXVI. TYPHACEÆ.—TYPHADS.

Herbs, growing in marshes or ditches. Stems without joints.
Lvs. rigid, ensiform, with parallel veins.
Fls. monœcious, arranged upon a spadix with no spathe.
Cal.—Sepals 3 or 6. *Corolla* 0.
Sta. 3—6. *Filaments* long and slender. *Anthers* cuneiform, erect.
Ovs. 1, free, 1-celled, with a solitary, pendulous *ovule. Styles* short. *Stig.* 1—2.
Fr.—Utricle with an albuminous seed.

Genera 2, species 12, in ditches and marshes throughout the world.

Genera.

Spadix of flowers { long and cylindrical . . . . Typha. 1
{ globose . . . . Sparganium. 2

1. TYPHA.

Gr. τυφος, a marsh; where all the species grow.

Spadix of flowers long, cylindric, dense. ♂ Stamens about 3 together, united into a common filament. ♀ flowers below the sterile; ovary pedicellate, surrounded at base by a hair-like pappus.—*Root* ♃. *Spadix terminal. Fls. very numerous.*

1. T. LATIFOLIA. (and angustifolia. *Linn.*) *Cat-tail. Reed Mace.*
Lvs. ensiform, concave within near the base; *sterile* and *fertile spikes* close together, or a little remote.—A common, smooth, tall inhabitant of the water in muddy pools and ditches, U. S., Can. The stem arises from 3 to 5f, round and smooth, leafy below, terminated by the large cylindric spikes. Spikes of a brown color, 6—10' in length, composed of slender, downy flowers so compact, particularly the fertile ones, as to be of considerable hardness. The upper portion is smaller, composed of the sterile flowers. Leaves somewhat sword-shaped, erect, 2—4f long and nearly 1' wide. They are called flags, and made useful for weaving the seats of chairs, &c. July.

β. *angustifolia* Sterile and *fertile spikes* a little remote (½—2').—Found in the same situations with the former. A well marked variety, but differing only in the more slender habit, and less complete development of its parts.

2. SPARGANIUM.

Gr. σπαργανον, a band or fillet; in reference to the long, ribbon-like leaves.

Spadix of flowers globose. ♂ Calyx 3—6-sepaled. ♀ Calyx 3—6-sepaled; utricle turbinate, acuminate, 1—2-seeded.—*Root* ♃. *Fls. collected in several dense, roundish heads, the sterile heads above the fertile.*

1. S. ERECTUM. (S. ramosum. *Smith.*) *Burr Reed.*

Lvs. triangular at base, their sides concave; *common flower-stalks* branched; *stig.* 2, linear.—Grows in pools and ditches, where it is conspicuous among other reedy plants for its globular burrs of flowers. Stem 1—2f high, flexuous, round, with a few branches above. Leaves ½—2f long, 4—8″ wide, linear, arising above the stem, triangular towards the base, and sword-form upwards, tapering, but obtuse. Heads of flowers light green; fertile ones 2—5, the lowest generally raised on a short, axillary stalk; sterile ones above, more numerous, smaller, sessile. Aug.

2. S. SIMPLEX. Smith. (S. Americanum. *Nutt.*)

Lower lvs. equal with, or exceeding the stem, which is nearly simple, *floral ones* concave at base and erect; *stig.* always simple, ovate oblong, oblique, scarcely more than half the length of the style.—Ponds and lakes. Stem 1—2f high, simple or divided at base. Leaves mostly radical, 1—2½f by 3″, carinate at base. Fertile heads sessile, generally 3, below the several barren ones, with the simple styles conspicuous. Aug.

3. S. NATANS. Michx. *Floating Burr Reed.*

Lvs. floating, flat; *common flower-stalk* simple; *stig.* ovate, very short; *head of sterile fls.* subsolitary.—Lakes and pools, U. S. and Brit. Am. Stem long and slender, and, with the leaves, floating upon the surface of the water. Leaves thin and pellucid. Heads of fertile flowers axillary, generally 2, mostly sessile. Sterile cluster terminal. Aug.

Order CXXXVII. NAIADACEÆ.—Naiads.

Water plants, with cellular leaves and inconspicuous flowers. Fls. perfect or monoecious. *Calyx* 2—4-sepaled or 0. *Sta.* definite. *Ovaries* 1, or 2—4, free, 1-ovuled. *Stigma* simple, often sessile. *Fr.* dry, indehiscent, 1-celled, 1-seeded. *Seed* pendulous. *Albumen* 0.

Genera 5, species 14, in waters and marshes, salt and fresh, in nearly all countries.

Conspectus of the Genera.

(solitary, diclinous, monandrous.						Najas. 2
2. Flowers axillary, { 2 together, a sterile and fertile one.						Zannichellia. 3
short, 2-flowered, borne on a long, tortuous peduncle.						Ruppia. 4
linear, bearing the monoecious flowers in a double row.						Zostera. 1
Spike { cylindric, covered with perfect, tetramerous flowers.						Potamogeton. 5

1. ZOSTĒRA.

Gr. ζωστηρ, a girdle; alluding to its ribbon-like leaves.

Spadix linear, bearing the separated flowers in 2 rows on one side; perianth 0. ♂ Anther ovoid, sessile, parallel to the ovary. ♀ Ovaries 2, ovoid; style bifid; utricle 1-seeded.

Z. MARINA. *Sea Wrack-grass.*

St. trailing, throwing out tufts of fibrous roots at the joints; *branches* floating, simple; *lvs.* alternate, linear, entire, sheathing at base, 1—several feet in length; *receptacle* or *spadix* linear, flat, pale green, 2′ long, issuing from a cleft in the base of the leaf, covered in front with a double series of naked flowers.—♃ Habits aquatic, growing in the sea on sandy banks and shallows (Maine to Ga.), and is thence washed upon the shore by the waves. Like other sea-weeds, it is gathered for manure. Aug.

2. NAJAS.

Gr. ναω, to flow; hence Nais, or Naides, Nymph of the waters; from the habitat.

Flowers often ♂ ♀. ♂ Calyx cylindric, 2-cleft; stamen 1 (rarely

t slender, often elongated; anther 4-valved, valves
Perianth 0; style filiform; stigma 2—3-fid; capsule
axillary.

ssis. Michx. (Caulinia flexilis. *Willd.* Fluvialis flexilis
ter Nymph.—St. filiform, cæspitose, dichotomously branching;
ciculate in 3s, 4s or 6s, at the nodes, linear, obscurely denticu-
veined.—A slender, flexible, rather erect, submersed aquatic
. J. and W. States, consisting of tufts of thread-like, knotted
. Leaves ¼—1' long, ½" wide, sessile and sheathing at base.
sessile, axillary, very small, the fertile ones consisting of an
ed with a filiform style, with 2—3 stigmas at summit. Aug.
ragilis. *Willd.*) St. and lvs. rather rigid, the latter mostly
rved.

3. ZANNICHELLIA.

In honor of Zannichelli, an eminent botanist of Venice.

—♂ Stamen 1; filament elongated. ♀ Calyx mono-
lla 0; ovaries 4 or more, each with a single style and
oming in fruit an oblong, incurved, subsessile achenium.

s. *Horn Pond-weed.*
floating; lvs. opposite, linear; anth. 4-celled; stig. entire; ach.
ack.—In pools and ditches, N. States. Stem round, smooth,
hing, leafy. Leaves grass-like, 2—3' long, sessile. Flowers
lary bracts, small, 2 together, a sterile and fertile, the former
ngle, naked, erect, yellowish-brown stamen, the latter of 4—6
e free from the inflated, one-sided, 2—3-toothed calyx. Jl. Aug

4. RUPPIA.

In honor of Ruppi, a German botanist.

together on a spadix or spike arising from the sheath-

1—3' long according to the depth of the water, branched. Upper leaves 2—4' by 8—16", petioles 2—8', submersed about ⅓ as wide. Spike 1—2' long, 20—40-flowered. Jl. Aug.—Varies with the lower leaves all reduced to petioles. In the Wisconsin plant the leaves are all subcordate, fruit acutely carinate but not lunate.

2. P. CLAYTONII. Tuckerman. (P. fluitans? *Ph.*, *Bw.*, *Torr.*, *&c.*)
Floating; *lvs.* lanceolate or oblong, tapering to long petioles (sometimes on short petioles, *E. T.*), scarcely coriaceous, submersed leaves long, narrowly linear, membranaceous, acute, 1-veined, slightly tapering to the sessile base; *spikes* rather loose, on long peduncles; *fr.* compressed, suborbicular.—Ponds and slow waters, frequent. Stems round, slender or filiform, often branched. Lower leaves 3—6' by 1½", remote, upper about 2—3' by ⅓'. Spikes 1' long, peduncles 2—4'

β? (P. heterophyllus. *Torr.*) *Lower lvs.* approximate, lowest slightly petiolate; *ped.* shorter.—Uxbridge, Mass., *Rickard*! Probably common. A beautiful variety. Fruit not seen.

3. P. DIVERSIFOLIUS. Bart. (P. setaceum. *Ph.*) *Setaceous-leaved Pond-weed.*
Upper lvs. lanceolate, opposite, 5-veined, on short petioles, *lower ones* submerged, sessile, filiform, alternate, dense, axillary.—Common in pools and ditches. A very slender and delicate species, only the upper leaves arising to the surface. These are 6—10" by 2—4", acute at each end, on hair-like petioles 5—6" long. Spadices dense, short, 5—6 flowered. July.

β. Submersed leaves few, not fascicled in the axils.

§ 2. *Leaves all submersed, uniform.*

4. P. LUCENS. *Shining Pond-weed.*
Lvs. lanceolate, flat, large, the short petioles continuing in a thick midvein; *spikes* long, cylindric, many-flowered.—♃ Can., N. Eng. &c. Rivers and lakes. Distinguished for its large leaves which are very pellucid, and, when dry, shining above, beautifully veined, 3—5' long, acuminate, ½—1' wide, each with a lanceolate bract above its base. Spadix 2' long, of numerous, green flowers, on a peduncle 2 or 3 times as long, thick and enlarged upwards. June.

5. P. OBRUTUS. Wood. *Lyndon Pond-weed.*
Lvs. glossy, linear-lanceolate, sessile, rather acute, only the midvein conspicuous, alternate, approximate, the lower stipules wanting; *spikes* long-pedunculate; *achenia* inflated, subhemispherical, margined on the back, beak incurved both sides, conspicuously umbilicate.—Passumpsic river, Lyndon, Vt.! A remarkable species, differing widely from any other with which I am acquainted. Stem round, slender, simple. Leaves uniform, 3—4' by ⅓', tapering to the slightly clasping base, the two upper opposite. Spike dense, 1½' long, peduncle 3' in length. Fruit with 2 little pits.

6. P. PRÆLONGUS. Wolff. *Long-stalked Pond-weed.*
Lvs. oblong or ovate, obtuse, many-veined, with 3 stronger veins, all reticulately connected, base amplexicaul; *ped.* very long; *spike* cylindrical, many-flowered; *fr.* ventricose, lunate, acutely carinate on the back.—Ponds and rivers, Northern States and Can. The plant is wholly submersed, sending up its spike to the surface on a very long stalk. I have gathered it in Niagara river, growing in depths of 6 or 8f. July, Aug.

7. P. PERFOLIATUS. *Perfoliate or Clasping Pond-weed.*
Lvs. cordate, clasping the stem, uniform, all immersed; *spikes* terminal; *fls.* alternate.—A common species growing in ponds and slow waters, wholly below the surface except the purplish flowers. Stem dichotomous, very leafy, 6—10' long. Leaves alternate, apparently perfoliate near the base, 1½' long, ½ as wide, obtuse, pellucid. Spadix on a short peduncle (1—2'), few flowered. Jl.

8. P. ROBBINSII. Oakes. *Robbins' Pond-weed.*
Lvs. lance-linear, approximate, sheathing the stem with the adnate stipules, lamina auriculate at base, margin minutely ciliate-serrulate; spikes oblong, small and few-flowered; *ped.* shorter than the leaves.—First discovered by Dr. *Robbins!* in Poodicberry Pond, Jefferson, N. H. Since found in many other

ones subcordate; *scape* simple, few-flowered; *lower ped.* elongated.—In water,
Penn. (*Muhlenberg*) to Car. Scape mostly erect, 3—6' long. Leaves 1—3' long.
Flowers few, small, the upper sterile. *Elliot.*

8. S. LANCIFOLIA. Willd. (S. falcata. *Ph.*) *Lance-leaved Sagittaria.*

Lvs. broad-lanceolate or ovate, acute at each end, glabrous, coriaceous,
and somewhat perennial; *scape* simple; *ach.* compressed, subfalcate.—This remarkable species (but unknown to me) has been found along the shores of the
Connecticut river, Mass. and Ct. Stem 2—3f high.

SUBORDER. JUNCAGINEÆ.

Sepals and petals both herbaceous (green), or 0. Stamens 6. Ovaries
3 or 6, coherent, ovules 1—2 in each carpel. Seeds erect, with
the embryo straight. Herbaceous, bog plants. Leaves ensiform.
Flowers in spikes or racemes.

3. TRIGLOCHIN.

Gr. τρις, three, γλωχις, a corner; on account of the 3-angled fruit.

Sepals and petals concave, deciduous, the former inserted a little
below the latter; stamens 6, very short; anthers large, extrorse;
ovaries 1-ovuled; stigmas adnate; fruit clavate, composed of 3—6
united, indehiscent, 1-seeded carpels.—♃ *Lvs. grass-like, all radical.*

1. T. MARITIMUM. (T. elatum. *Nutt.*) *Sea Arrow-grass.*

Fruit ovate-oblong, grooved, of 6 united carpels; *scape* longer than the
leaves.—A rush-like plant in salt marshes and ditches on the sea-coast, and at
Salina, N. Y., also lake shores, Wisc.! Leaves linear, semicylindric, smooth,
thick, 6—12' long, less than a line wide. Scape obtusely angled, simple, 9—18'
long, bearing a long raceme of 30—40 green flowers on pedicels 1—2" long.
Fruit separating into 6 linear carpels, each containing a linear seed. The
plant has a sweetish taste, and cattle are fond of it. July.

2. T. PALUSTRE. *Marsh Arrow-grass.*

Fruit nearly linear, of 3 united carpels; *scape* scarcely longer than the leaves.
—In marshes, Salina, N. Y.! N. to Arc. Am. Leaves very numerous, fleshy;
smooth, very narrow. Scape 6—12' high, ending in a raceme with rather remote
very small, green flowers on pedicels 2—3" long. The slender fruit is attenu
ated at base, obtuse at apex, grooved and margined, consisting of 3 very slen
der carpels. July.

4. SCHEUCHZERIA.

In honor of the Scheuchzers, two brothers, distinguished botanists.

Sepals and petals oblong, acute, persistent; sta. 6, with linear anthers; stigmas sessile, lateral; ovaries 1—2-ovuled; capsules inflated,
compressed, 2-valved, 1—2-seeded.—♃ *Lvs. cauline, linear, sheathing
at base.*

S. PALUSTRIS.

A rush-like plant, in swamps, Vt.! to Penn. Rare. Root-stock horizontal,
fleshy. Stem about a foot high, simple, angular. Leaves semicylindric, 4—6'
long, in the barren shoots much longer, sheathing at base. Raceme terminal,
5—8-flowered. Flowers yellowish-green, on short pedicels, each axillary to a
bract. Stamens large, exserted, erect. July.

ORDER CXXXIX. HYDROCHARIDACEÆ.—HYDROCHARADS.

Plants aquatic, floating, with parallel-veined leaves.
Fls. diœcious or perfect, issuing from a spathe.
Perianth.—Sepals 3, herbaceous. Petals 3, colored.
Sta. definite or indefinite, epigynous.
Ova. adherent to the perianth, single. *Stigmas* 3—6. *Ovules* indefinite.
Fr. dry or succulent, indehiscent, 1 or more celled. *Seeds* without albumen.

Genera 12, species 20, native of Europe, N. America, and the East Indies. Of no important use.

Conspectus of the Genera.

Leaves { cordate, petiolate. { short or | verticillate, sessile. Flowers axillary, solitary. . . . . . . Hydrocharis. 1 { long, linear, radical. Flowers elevated on long scapes. . . . . Udora. 2 Vallisneria. 3

1. HYDROCHÁRIS.

Gr. ὕδωρ, water, χαίρω, to rejoice; as water is its own element.

Flowers ♂ ♀.—Spathe 2-leaved; calyx 3-parted; corolla of 3 petals; styles 3, abortive in the ♂; sta. 10—12, infertile in ♀; stig. 6—8, bifid, with several barren filaments and 3 nectariferous glands; caps. inferior, 6-celled, ∞-seeded.—*Floating aquatics.*

H. spongiósa. Bosc. (H. cordifolia. *Nutt.*) *Frogbit.*

Lvs. roundish-cordate; ♂ *fls.* on long peduncles, ♀ on short ones; *anth.* .inear; *stig.* 6—7, 2-parted nearly to the base; *segments* lanceolate-subulate.—2⟂ Braddock's bay, L. Ontario, *Sartwell.* Stems jointed, producing offsets from which new plants arise. Root somewhat fibrous. Leaves petiolate, 5-veined, purplish beneath. ♂ peduncles about 3′ long. Sepals oval, green. ♀ peduncles 1—1½′ long, thick. Ovary oblong. Seeds hirsute. Petals white. Aug.

2. UDÓRA. Nutt.

Gr. ὕδωρ, water; from its aquatic habits.

Flowers ♂ ♀; spathe bifid; spadix 1-flowered. ♂ Stamens 9, 3 of them interior. ♀ Tube of the perianth very long; abortive filaments 3; capsule ventricose, 3-seeded.—*Creeping. Lvs. verticillate.*

U. Canadensis. Nutt. (Elodea. *Michx.* Sepicula verticillata. *Muhl.*)

Ditch Moss.—*Lvs.* verticillate, in 3s and 4s, lanceolate, oblong or linear, serrulate; *tube of the perianth* filiform.—Resembling a coarse moss, in still waters. Stem filiform, diffusely dichotomous, very leafy, submersed. Leaves 3—6″ by (less than) 1″, thin and diaphanous, sessile, obtuse. Flowers axillary, solitary, minute, of a dingy white, the slender, hair-like tube 2—3′ long (4—6′ according to *Nuttall*), the lower part (pedicel?) 2-margined. Stigmas recurved between the segments, crested with glandular hairs. Aug.

3. VALLISNERIA.

In honor of Anthony Vallisner, a French botanist.

Flowers ♂ ♀; spathe ovate, 2—4-parted. ♂ Spadix covered with minute flowers; corolla 0. ♀ Spathe bifid, 1-flowered; perianth elongated; sepals linear; stigmas 3, ovate, bifid; capsule 1-celled, many-seeded.—2⟂ *Submersed. Lvs. all radical. Scape spiral, very long.*

V. Americána. Michx. *Tape Grass.*

Lvs. linear, obtuse, serrulate at the end, tapering at the base, floating; *peduncle of the fertile flower* long, *of the sterile* short, erect.—A curious plant, in slow moving or stagnant waters, U. S. Leaves linear, 1—2f long, about ½′ wide, the edges thinner than the middle. Scapes several, of the sterile plants short, of the fertile plants very tortuous, 2—4f long when extended, thread-like, thickened at top, bearing each a single, white flower at or near the surface. Sepals and petals crowning the (1′) long, narrow, incurved ovary, which is half concealed in the spathe. July, Aug.

Order CXL.   ORCHIDACEÆ.—Orchids.

Herbs perennial, often acaulescent, with fleshy corms, or tuberous, flocculated roots.

Lvs. simple, parallel-veined, entire. *Fls.* in terminal or radical racemes, spikes or panicles, rarely solitary.

Fls.—Very irregular, with an adherent, ringent perianth of 6 parts.

Cal.—Sepals 3, usually colored, odd one uppermost by the twisting of the ovary.

Cor.—Petals 3, usually colored, odd one lowest by the twisting of the ovary.

Lip. (labellum, or lowest petal.) diverse in form, often lobed, frequently spurred at base.

Sta. 3, united into a central column, the 2 lateral ones generally abortive, and the central one perfect; more rarely the central abortive and lateral perfect.

Anth. 2, 4 or 8-celled, persistent or deciduous, often overrulate.

Pol. either powdery, or cohering in waxy masses (pollinia), which are either constantly adhering to a gland, or breaking loose in their cells.

1

Ova. 1-celled, with 3 parietal placentæ. *Ovules* indefinite.
Sty. consolidated with the stamens. *Stig.* a viscid cavity in front of the column.
Fr.—Capsule 3-ribbed, 3-valved. *Sds.* many, without albumen.

Genera 394, species 3000) They are among the most interesting and curious of plants, almost always remarkable for the grotesque form of their tortuous roots and stems, and the fragrance, brilliancy and odd structure of the flowers.

The Orchidaceæ are natives of every part of the world. In the tropics multitudes of them are epiphytes, growing on living trees or decaying timber.

This order is remarkable for those qualities only which please the eye. Many of its species are cultivated for ornament, but few of them possess either active or useful properties. The *salep* of commerce is a nutritive mucilaginous substance afforded by the roots of some Asiatic Orchis. The aromatic *vanilla*, used to flavor chocolate, &c., is the fruit of the West Indian Vanilla clavicalata.

Conspectus of the Genera.

			cauline.	*Microstylis.* 1
		ovate,	radical.	*Aplectrum.* 4
	Leaf solitary.	ensiform (rarely 2).		*Calopogon.* 10
		near base of stem.		*Liparis.* 2
	Leaves 2 only,	near middle of stem.		*Listera.* 13
		radical,	ringent.	*Goodyera.* 12
		Sepals	erect.	*Spiranthes.* 11
Flowers several	Leaves several,	cauline. Fls. 3 or 4.		*Calopogon.* 10
Spur 0. Flower solitary, lip bearded within.				*Pogonia.* 7
			Pollinia 2. Fls. bracteate	*Orchis.* 7
Leaves 1 Fertile		flattish.	Pollinia 4. Fls. bractless.	*Tipularia.* 8
or more. anth. 1. Lip spurred at base,	ventricose. Flower and leaf solitary.			*Calypso.* 5
Anthers 2 fertile, mid.'le one sterile, petaloid. Lip saccate.				*Cypripedium.* 14
Plants green. Flowers solitary. Lip bearded within.				*Arethusa.* 9
Leaves 0. Plants destitute of green herbage. Flowers racemose.				*Corallorhiza.* 3

SECTION 1. Pollen cohering in grains which finally become waxy, and are definite in number.

1. MICROSTÝLIS. Nutt.

Gr. μικρος, little, στυλος, style; alluding to the slender column.

Segments of the perianth distinct, petals filiform; lip sessile, concave, erect, truncate and bidentate at summit; column minute; pollinia 4, loose.

1. M. OPHIOGLOSSOÏDES. Nutt. (Malaxis orph. *Willd.* M. unifolia. *Michx.*)—*Lf.* solitary, ovate, amplexicaul; *st.* 5-angled; *rac.* short, obtuse, capitate; *pedicels* much longer than the flowers.—A small plant, in woods, &c., Can. and N. States. Stem 5—9′ high, with a single leaf a little below the middle. This leaf is rather acute, smooth, ovate or oval, about 2½′ in length, 1 in width. At the base of the stem is an abrupt sheath. Flowers whitish, minute, numerous, in a terminal raceme an inch or more in length, dense at top. Bracts minute. Pedicels about 4″ long. June

2. M. MONOPHYLLOS. Lindl. (M. brachypoda. *Gray.* Malaxis mon. *Willd.*)—*Leaf* solitary, ovate, sheathing at base; *rac.* elongated, with numerous flowers on short pedicels; bracts minute; *sep.* acute, spreading; late. ral *pet.* reflexed, linear; *lip* triangular-hastate, cucullate, acuminate with a recurved point.—In shady swamps, N. Y., rare. *Prof. Hadley, Dr. Gray.* Stem 2—6′ high, 3-angled, with a subspicate raceme of 20—40 small, greenish flowers. July.

2. LIPÁRIS. Rich.

Gr. λιπαρος, elegant, shining; a term characteristic of these plants.

Segments of the perianth distinct, sublinear. spreading or deflexed; lip spreading, flat, ascending, often exterior; column winged; pollinia 4, parallel with each other, without pedicels or glands.

1. L. LILIFOLIA. Rich. (Malaxis lilifolia. *Sw.*) *Tway-blade.* *Lvs.* 2, ovate-lanceolate; *scape* triangular; *inner pet.* filiform, reflexed; *lip* concave, obovate, acute at the tip.—In wet woods, Can. to Car., Ohio. Leaves radical, 3—6′ long, ½—¾ as wide, rather acute, tapering into a sheathing base. Scape about 6′ high. Flowers 10—20, in a terminal, rather showy raceme. Pedicels near an inch in length. The 3 sepals greenish-white, linear. 2 upper petals capillary, yellowish-white. Lip much larger than the other petals, white. June.

2. L. Lœsælii. Rich. (L. Correana. *Spr.* Malaxis Lœselii. *Sw.*)

Lvs. 2, ovate-oblong, obtuse, plicate, shorter than the few-flowered racemes; *scape* angular; *lip* ovate, entire; *sep. and pet.* linear, subequal.—About half as large as the preceding, in moist meadows and fields, Can., Ohio, N. Eng. and Mid. States. Leaves 2—3' long, about 1' wide, obtuse or acute, sheathing at base. Scape 3—5' high. Flowers about 6, appressed to the rachis, in a thin raceme. Pedicels about 2'' in length. Sepals and petals greenish-white. Ovaries clavate, as long as the pedicels. Jn.

3. CORALLORHIZA. Brown.

Gr. κοραλλιον, coral, ριζα, root : its branched roots much resemble coral.

Segments of the perianth nearly equal, converging ; lip produced behind ; spur short and adnate to the ovary ; column free ; pollinia 4, oblique (not parallel).—*Plants leafless.*

1. C. odontorhiza. Nutt. (C. verna. *Nutt.* C. innata. *Br.*) *Dragon's-claw. Coral-root.*—*Lip* undivided, oval, obtuse, crenulate, spotted ; *spur* obsolete, adnate to the ovary ; *capsule* oblong or subglobose.—A singular plant, with no leaves or green herbage, inhabiting old woods, Can. to Car. and Ky. The root is a collection of small, fleshy tubers, articulated and branched much like coral. Scape 9—14' high, rather fleshy, striate, smooth, invested with a few long, purplish-brown sheaths. Flowers 10—20, in a long spike, of a brownish-green. Lip white, generally with purple spots. Capsules large, reflexed, strongly ribbed. July, Aug.

β. verna.—*Lip* white, without spots, minutely toothed each side.—More delicate and slender than the variety *α.* Flowers fewer (7—10).

2. C. multiflora. Nutt. *Flowering Coral-root.*

Scape many-flowered ; *lip* cuneate-oval, spotted, 3-parted ; the middle lobe recurved, lateral ones short and tooth-like ; *spur* conspicuous, adnate ; *cap.* elliptic-obovoid.—In woods, growing on the roots of trees, N. Eng. and Mid. States. Root coralline. Scape 10—15' high, leafless, brownish-purple, sheathed with a few bracts. Flowers larger than in the other species, 15—20, erect-spreading, in a long raceme. Lip showy, 3—4'' long, white, sprinkled with purple spots. Spur yellowish, conspicuous, but short and adnate to the ovary. Jl.

4. APLECTRUM. Nutt.

Gr. α, privative, πληκτρον, a spur : the lip being without a spur.

Segments of the perianth distinct, nearly equal, converging ; lip unguiculate, not produced at base ; column free ; anther a little below the apex ; pollinia 4, oblique, lenticular.

A. hyemale. (Cymbidium hyemale. *Willd.*) *Adam-and-Eve. Putty-root.*—*Lf.* solitary, radical, petiolate, ovate, striate; *lip.* trifid, obtuse, with the palate ridged.—A fine plant, in woods, Can., Ohio! to N. Eng. (rare) and Flor. Root bearing large, roundish, mucilaginous tubers. Leaf rather elliptic than ovate, 4—5' long, ½—⅔ as wide, twice as long as the petiole, which arises from the summit of the tuber a short distance from the scape. Scape arising from beneath the tuber, about 1f high, invested with 2—3 sheaths. Flowers resembling those of Corallorhiza, brownish-purple, erect, in a terminal raceme. Lip dilated near the end. Capsule large, smooth, nodding. May, Jn.

5. CALYPSO. Salisb.

Named for the goddess Calypso (*Gr.* καλυπτω, to conceal).

Segments of the perianth ascending, secund ; lip ventricose spurred beneath near the end ; column petaloid ; pollinia 4.

C. bulbosa. Salisb. (C Americana. *Br.* C. borealis. *Ph.* Cypripedium. *Linn.*)—*Lf.* solitary, radical, broad-ovate, veined ; *l.p* narrowed and subunguiculate at base ; *spur* bifid, longer than the lip, with acute teeth ; *ped.* longer than the ovary.—This rare and beautiful plant is found in Vt., (Carey) Nova Scotia, Mich., W. to Oreg. Scape 6—8' high, sheathed, bearing a single, purplish flower at top, as large as that of a Cypripedium. Leaf petiolate, 2—3′ in diameter, subcordate at base.

6. TIPULARIA. Nutt.

Tipula, the crane-fly ; from the fancied resemblance of the flowers.

Segments of the perianth spatulate, spreading ; lip entire, sessile, conspicuously spurred at base ; column wingless, free ; anther operculate, persistent ; pollinia 4, parallel.

T. DISCÔLOR. Nutt. (Orchis. *Ph.* Limodorum uniflorum. *Muhl.)*
A slender, green-flowered plant, resembling a Corallorhiza, growing in pine woods, Vt., Mid. States to Car. Root bulbous. Leaf solitary, petiolate, ovate, plaited, smooth, and longitudinally veined. Flowers minute, greenish-white, nodding, in a terminal raceme destitute of bracts. Jl.

SECTION 2. Pollen cohering in waxy masses, which are pedicellate, with glands at the base of the pedicels. Anthers of 2 distinct, vertical cells.

7. ORCHIS.

The Greek name of these plants.

Perianth ringent, the upper sepal vaulted ; lip entire or lobed, produced at base into a spur which is distinct from the ovary ; anther terminal ; pollinia 2, adnate, pedicellate.—*Fls. racemose or spicate.*

§ 1. ORCHIS *vera. Lip broad, entire. Glands of the pedicels of the pollinia enclosed.*

1. O. SPECTABILIS. *Showy Orchis.*
Lvs. about 2, nearly as long as the scape ; *lip* obovate, undivided, crenate, retuse ; *segments of the perianth* straight, the lateral ones longer ; *spur* clavate, shorter than the ovary, *bracts* longer than the flower.—This pretty little plant is found in shady woods and thickets, among rocks, &c., U. S. and Can. Root fasciculate. Leaves few, radical, ovate, 3—6′ long, ½—⅓ as wide. Scape 4—6′ nigh, acutely-angled, with a lanceolate, acute bract and 3—5 large, showy flowers. Segments of the perianth purple, ovate-lanceolate. Lip and spur white or whitish, each about 8″ long. May, Jn.

§ 2. PLATANTHERA. *Lip narrow, entire. Cells of the anther widely separated at base by the broad, interposed stigma. Glands of the pedicels of the pollinia naked.*

2. O. ORBICULÀTA. Pursh. (Platanthera. *Lindl.* Habenaria. *Torr.*)
Lvs. 2, radical, suborbicular, rather fleshy ; *scape* bracteate ; *upper sep.* orbicular, *lateral ones* ovate ; *lip* linear-subspatulate, nearly twice as long as the sepals ; *spur* arcuate, compressed, clavate, twice as long as the ovary.—A remarkable plant, not uncommon in old woods and in thickets, Penn. to Can. at d W. States. Leaves lying flat upon the ground, 3—6′ diam., rather inclining to oval or ovate with the apex acute. Scape 1—2f high, sheathed with a few bracts, bearing a raceme of numerous, greenish-white flowers. Lip ½—1′ by ½—1″. Spur 1½—2′ long. Jl.

3. O. HOOKERIÀNA. Wood. (Platanthera. *Lindl.* Habenaria. *Torr.*)
Lvs. 2, radical, suborbicular or suboval, fleshy ; *scape* naked ; *bracts* lanceolate, nearly as long as the flowers ; *upper sepal* ovate, erect, *lateral ones* deflexed and meeting behind ; *pet.* acute, lip lanceolate, projecting, acuminate, a little longer than the ·sepals ; *spur* subulate, arcuate, about twice longer than the ovary.—Woods, Can., N. Eng. ! to Wis., *Lapham !* rare. Resembles O. orbiculata, but is very distinct. Scape 8—12′ high, without a bract below the flowers. Leaves 4—5′ long, nearly or quite as wide. Flowers 12—18, in a straight raceme, yellowish-green, the spur 9—12″ in length. Jn. Jl.

4. O. HYPERBOREA. Willd. (Habenaria hyp. & Huronensis. *Spr.* Platanth. *Lindl.)*—*St.* leafy ; *lvs.* very erect, acute ; *spike* somewhat second

bracts linear-lanceolate, acute, longer than the flower; *sep.* deflexed; *pet.* and *lip* linear, obtuse, subequal, (the latter dilated at base!) and about as long as the pendulous, obtuse spur.—A tall, leafy, variable species, found in mountainous woods and open meadows, N. Y. to Mich. and Can. Stems thick, 2 or 3, or even 4 feet high. Leaves lanceolate, 4—7′ by 1—1½′. Flowers greenish in shades, nearly white in open situations, forming a long, more or less dense spike. July.

β. *Huronensis.* More slender; *lvs.* lance-oblong and lance-linear, obtuse or acute; *spike* rather loose, often long.—Vt., *Dr. Phelps!* W. to Mich.

5. O. DILATĀTA. Pursh. (Habenaria. *Hook.* Platanth. *Lindl.*)
St. slender; *lvs.* lanceolate and linear, acute; *spike* loose; *bracts* lance-linear, about as long as the flowers; *upper sepal* ovate, obtuse, the lateral narrower and spreading; *lip* linear, entire, obtuse, dilated at the base, about equaling the petals and a little shorter than the obtuse, incurved spur, which is longer than the ovary.—Swamps, Northern States! (rare) and Can. It is a slender and delicate species, with pure white flowers. Stem 10—15′ high. Leaves often narrow and grass-like, the lower lanceolate. Flowers 10—90, spur about 4″ long. July.

6. O. OBTUSĀTA. Pursh. (Platanthera. *Lindl.* Habenaria. *Rich.*)
Lf. solitary, oblong-obovate, obtuse; *st.* bearing the leaf near its base; *spike* loose; *upper sep.* broadest; *pet.* subtriangular; *lip* linear, entire, with 2 tubercles at base, as long as the arcuate, acute spur.—Found in muddy ponds and ditches, N. H., *Storrs!* N. to Lab. Stem slender, angular, 6—8′ high, terminating in a thin spike of about a dozen small, greenish-white flowers. Leaf tapering at base, and usually obtuse at the summit, 2—3′ in length, and 1 in breadth, issuing with the stem from 2—3 radical, sheathing bracts. July.

7. O. INTĒGRA. Nutt. (Habenaria. *Spreng.*)
St. leafy; *lvs.* lanceolate and lance-linear; *bracts* shorter than the flowers, *lip* oblong, entire, longer than the petals; *spur* subulate, longer than the ovary. —Swamps, N. J., *Nuttall.* A species very nearly allied to O. ciliaris, apparently differing only in the flowers being smaller, and with the lip entire, not fringed. Flowers orange-yellow. Jl.

§ 3. HABENARIA. *Lip dilated, variously divided. Glands of the pedicels of the pollinia naked, distinct.*

* *Lip toothed or 3-parted, not fimbriate.*

8. O. FLAVA. (Habenaria herbiola. *Br.* Platanthera. *Lindl.*)
St. leafy; *lower lvs.* oblong, acute, *upper* lanceolate, acuminate; *spike* rather dense, cylindric; *bracts* longer than the flowers; *lip* oblong, obtuse, dentate at base; *palate* with 1 tuberculate tooth; *spur* filiform, rather shorter than the sessile ovary.—A small-flowered orchis found in alluvial soil. Stem flexuous, 12—18′ high. Leaves about 3, with long sheaths, 3—6 or 7′ by ½—2′, tapering to an acute summit. Flowers in a long, thin spike. Sepals short, ovate, green. Petals yellowish. Upper bracts about as long as the flowers, lower ones 2 or 3 times as long. The tubercle of the lip is a remarkable character. June.

9. O. VIRIDIS. Swtz. (O. bracteata. *Muhl.* Peristylus bracteatus. *Lindl.*)
St. leafy; *lvs.* oblong, obtuse, upper ones acute; *spike* lax; *bracts* 2—3 times as long as the flowers; *sep.* connivent, ovate; *pet.* linear, erect; *lip* linear-cuneate, truncate, 3-toothed at the end, the middle tooth small or obsolete; *spur* short, inflated, obtuse.—A small, green-flowered orchis, in shades. Stem 6—9′ high. Leaves about 3, 18—30″ by 6—12″, upper bracts as short as the flower. Spikes 2—3′ long. Flowers yellowish-green. Lip as long as the ovary, 3 times as long as the spur. Can. to Va. W. to Ill. July, Aug.

10. O. TRIDENTĀTA. Willd. (Habenaria. *Hook.* Gymnadenia. *Lindl.*)
Radical lf. solitary, oblong, obtuse, *cauline* 2—3, much smaller; *sep.* campanulate, obtuse, converging; *lip* lanceolate, 3-toothed, at the extremity; *spur* filiform, curved, clavate, longer than the ovary.—Grows in woods and swamps, Can., Mich., Penn., Va. Stem slender, 1-2′ high, with small, greenish-white flowers in a short and rather loose spike, appearing in July

11. O. ROTUNDIFOLIA. Pursh. (Habenaria. *Rich.* Platanthera. *Lindl.*) *Lf.* solitary, roundish-ovate; *scape* naked; *spike* few-flowered; *bracts* obtuse, shorter than the ovary; *sep.* and *pet.* obtuse; *lip* 3-lobed, lateral, lobes subfalcate, middle one obcordate; *spur* as long as the lip.—Ct., Penn., *Eaton*, Can. Scape about a foot high, slender, without a bract. Leaf 2—4' long, ⅔ as wide, spotted, sheathing at base. Flowers about a dozen, of a greenish-white, remarkable for their broad, 3 (almost 4)-lobed, pendant lip.

** *Lip fimbriate. Stem leafy.*

12. O. CILIARIS. (Habenaria. *Rich.* Platanthera. *Lindl.*) *Yellow Fringed Orchis.*—*Lower lvs.* linear-lanceolate; *spike* oblong, dense; *bracts* shorter than the ovary; *lip* oblong-lanceolate, pinnate-ciliate, twice as long as the petals; *spur* longer than the ovary.—A delicately beautiful orchis, with bright orange-colored flowers, in swamps, Can. to Ga. and Ky., rare. Stem about 2f high. Leaves sheathing at base; lower ones 3—5' long, rapidly diminishing upwards. Sepals roundish, obtuse, concave. 2 petals linear, very small, incised at the summit; the lip narrow, lanceolate, conspicuously fringed, 4'' long. Spur 1' in length. July, Aug.

13. O. CRISTATA. Michx. (Habenaria. *Brown.* Platanthera. *Lindl.*) *Lvs.* lanceolate and lance-linear; *spike* somewhat crowded, many-flowered; *segments of the perianth* rounded, the two lateral petals toothed, lip oblong, pinnately ciliate, *spur* shorter than the ovary.—Swamps, N. J., Penn. to Car. A small species, distinguished from the foregoing by smaller and more crowded flowers which are of a bright orange-yellow, and by its shorter spur, &c. Jn. Jl.

14. O. BLEPHARIGLOTTIS. Willd. (Habenaria. *Rich.* Platanthera. *Lindl.*) *White Fringed Orchis.*—*Lower lvs.* lanceolate, channeled; *spike* oblong, dense; *bracts* linear, acuminate, shorter than the flowers; *lip* lanceolate, ciliate, as long as the upper sepal; *spur* much longer than the long-beaked ovary.—A delicate orchis, in swamps, (N. Y.! to Car.,) resembling the last species, but distinguished at least, by the color of its flowers which are of a pure white. Stem 1—2f high. Flowers fewer than in the last. Sepals roundish-oblong, lateral reflexed. Petals spatulate, dentate. Lip fringed in the middle, 2'' long. June, July.

15. O. FISSA. Willd. (Habenaria. *Br.*) *St.* tall, leafy; *lvs.* lanceolate and lance-linear; *bracts* nearly equaling the ovary; *sep.* roundish-ovate; *lateral petals* denticulate; *lip* 3-parted, divisions cuneiform, dentate, middle one 2-lobed; *spur* filiform, clavate at end, curved, longer than the ovary.—Wet grounds and marshes, Penn. to Va., W. to Ind. ! A truly beautiful species 2—4f high. Stem slightly winged. Leaves 4—6' long. Flowers violet-purple, large, 20—50, in a terminal spike. Ovary 1', and spur 1½' long. June, July.

16. O. LACÈRA. Michx. (O. Psycodes. *Spr.* Habenaria. *Br.* Platanthera. *Lindl.*) *Ragged Orchis.*—*Lower lvs.* oblong, obtuse, *upper ones* narrow, acuminate; *bracts* longer than the flowers; *sep.* retuse; *pet.* emarginate; *lip* 3-parted, *segments* cuneate, capillaceous-multifid; *spur* filiform, clavate, as long as the ovary.—Swamps and meadows, Can. to Car. Stem 1—2f high, smooth, slender. Leaves few, 3—6' by ¼—1', mostly acute. Flowers numerous, in a long, loose spike, of a greenish-white, not showy. Sepals ovate. Petals oblong-linear, entire, lip reflexed, very deeply laciniate. Readily distinguished from the following by its more slender habit, greenish flowers, and the entire (not fringed) petals. July.

17. O. PSYCÒDES. (O. fimbriata. *Br.* Habenaria. *Rich.* Platanthera. *Lindl.*) *Purple Fringed Orchis.*—*Lower lvs.* lanceolate, diminishing upwards; *lip.* 3-parted, scarcely longer than the petals, the segments cuneiform, ciliate-fimbriate; *lateral pet.* ovate, fimbriate-dentate; *spur* filiform, clavate, longer than the ovary.—A beautiful plant, common in meadows, Can., N. Eng. Mid. and W. States. Stem 1½—2½f high, smooth, slender. Leaves 3—6' long. Flowers showy, numerous, in a terminal, cylindric spike, light purple. Lip of ovary somewhat longer than the petals, its 3, fan-like, spreading segments, the petals, beautifully fringed. Spur an inch in length. July.

β. (O. incisa. *Willd.*) *Lateral petals* subdentate, terminal one incisely dentate, *spur* subulate.

18. O. leucophæa. Nutt. *White-flowered or Prairie Orchis.*

St. leafy; *lvs.* lanceolate, tapering to a narrow, obtuse point, channeled; *bracts* shorter than the ovaries; *rac.* oblong; *sep.* roundish-oblong, acutish; *lateral petals* obovate, denticulate; *lip* 3-parted, flabelliform, segments deeply fimbriate; *spur* subulate-clavate, curved, twice as long as the ovary.—Wet prairies, Ohio, Ind.! Ill. Stem 1—3f high. Leaves 2—6′ long. Raceme abou: 12-flowered Sepals and spur yellowish, petals white. Ovary curved, 1′ long.

19. O. grandiflóra. Bw. (Habenaria. *Torr.* Platanthera fimbriata. *Lindl.*) *Large-flowering Orchis.*—*Lower lvs.* oblong, oval, obtuse, upper ones very narrow; *bracts* shorter than the ovary; *rac.* oblong; *lip* dependent, twice as long as the petals, 3-parted, the segments cuneiform and fimbriate, the middle one largest, with connivent fimbriæ; *lateral pet.* fimbriate; *spur* ascending, clavate, longer than the ovary.—A superb plant, considered the most beautiful of the genus, in wet meadows, N. H.! Can. Stem 2—3f high, thick, hollow, with several sheathing bracts at base. Leaves 2 or 3 principal ones, 4—7′ by 1—2′, upper ones linear, an inch or two long. Flowers very large, purple, in a terminal raceme, 3—6′ long. Middle seg. of the lip nearly semicircular, twice as long as the lateral ones. June.

SECTION 3.   Pollen powdery, or consisting of loosely cohering granules.

8. ARETHÚSA. Swtz.

Arethusa was a fabulous nymph of Diana, who was transformed into a fountain.

Perianth with its segments cohering at base; lip spurless, adnate to the column at base, deflected at the end and bearded inside; pollen angular.—*Small plants, inhabiting wet places.*

A. bulbósa. *Bulbous Arethusa.*

Leafless; *rt.* producing a globular tuber; *scape* sheathed, 1-flowered.—This beautiful and interesting plant is found only in wet meadows and swamps, Can. to Va. Stem 6—12′ high, invested with about 3, long, loose sheathes with lanceolate points, the upper ones rarely at length produced into a short, linear-spatulate leaf. At the top is a single, large, fragrant flower of a rich purple color. A little below the base of the flower is a small spathe of 2 unequal bracts. June.

9. POGONIA. Brown.

Gr. πωγων, beard; in allusion to the bearded lip.

Perianth with its leaflets distinct; lip sessile or unguiculate, cucullate, bearded inside; pollen farinaceous.

1. P. ophioglossóides. Br. (Arethusa. *Linn.*)
Rt. fibrous; *st.* furnished with an oval-lanceolate leaf and a foliaceous bract near the flower; *lip* fimbriate.—An interesting plant, much taller than the bulbous Arethusa, found in swamps and muddy shores, Can., N. Eng. to Car. and Ky. The stem is very slender, 9—16′ high, with 2 remote leaves, the one placed about midway, 2—3′ long, lanceolate, acute, sheathing at the base; the other (a bract?) much smaller, situated near the flower. Flower large, nodding, pale purple. Lip long as petals and sepals (¾′). June.

2. P. verticilláta. Nutt. (Arethusa. *Willd.*)
Lvs. 5, oblong-lanceolate, verticillate; *fls.* solitary, the 3 outer petals very long, linear, inner ones shorter, lanceolate, obtuse; *lip* 3-lobed, the middle lobes undulate.—Swamps. Stem 8—12′ high, with a whorl of leaves near the top and a flower 1—2′ above it. Leaves 1½′ long, ½ as wide, abruptly acuminate. The flower is remarkable for its sepals being about 2′ long, very narrow, and of a greenish-brown color. Lip crested in the middle. July.

3. P. pendúla. Lindl. (Triphora. *Nutt.* Arethusa. *Willd.*)
Rt. tuberous; *st.* leafy, about 3-flowered at the top; *lvs.* clasping, ovate, alternate; *fls.* axillary, nodding; *lip* entire, scabrous, not bearded; *fr.* pendu

lous.—A small, delicate plant, in swamps, Mid. W. ! and S. States. Stem scarcely 6' high, slightly angled, with about 3 flowers, which from their singular form suggest the common name. Leaves 3—6, ¼—½' long, ½ as wide, purplish. Flowers white or greenish, the segments of the perianth equal, converging, and rather longer than the lip. Aug.

10. CALOPOGON. Brown.

Gr. καλος, beautiful, πωγων, beard ; in allusion to the bearded lip.

Segments of the perianth distinct ; lip on the upper side of the flowers, unguiculate, bearded ; column free, winged at the summit ; pollen angular.

C. PULCHELLUS. Br. (Cymbidium. *Willd.*) *Grass Pink.*
Rt. tuberous; *lf.* radical, ensiform, veined; *scape* few-flowered; *lip* erect, narrowed at base, with an expanded border, and a concave, hairy disk.—A truly beautiful plant, in swamps and moist meadows, U. S. and Can. Scape slender, 10—90' high, furnished with a single, long leaf (8—12' by ¼'), sheathing its base. Flowers 3—8, large, purple, remarkable for their inverted position; lip expanded at the end and fimbriate on the upper side of the flower, while the column is below. Petals and sepals expanded. July.

11. SPIRANTHES. Rich.

Gr. σπειρα, a cord; in reference to the twisted spike.

Flowers in a spiral spike ; petals connivent ; lip unguiculate, parallel with the column, entire, with 2 callous processes at base ; column free, clavate, bidentate at summit ; ovary oblique ; stigma rostrate.

1. S. GRACILIS. Beck. (Neottia. *Bw.*) *Ladies' Tresses.*
Lvs. radical, ovate, caducous; *scape* sheathed, *fls.* in a spiral row; *lip* obovate, curled.—A very delicate plant, not uncommon in old woods, N. Eng., Can. Scape leafless, with several remote, sheathing scales, very slender, and 8—12' high. Leaves 3—4, close to the ground, 1—2' long, ½ as wide, on short petioles, mostly withering and falling away before the flowers expand. Flowers small, white, arranged in a row which winds once or twice around the stem. July.

2. S. TORTILIS. Sw. (Ophrys. *Mx.* S. æstivalis. *Mx.*)
Radical lvs. linear; *scape* sheathed; *fls.* spirally secund; *lip* somewhat 3-lobed, the middle lobe larger, crenulate.—A plant mostly similar to the last, in woods and meadows, N. Eng. to Car. Leaves 3—6' by 2—4'' commonly disappearing before flowering. Scape slender, a foot or more high, with a spiral row of oblique, small, white flowers, forming a twisted spike 2—4' long. July.

3. S. CERNUA. Rich. (Neottia. *Willd.*)
Lvs. radical, linear-lanceolate, veined; *stem* sheathed; *spike* dense ; *fls.* recurved, drooping ; *sep.* and *pet.* cohering ; *lip* oblong, entire or crenulate, dilated at the apex.—In moist grounds, N. Eng. to Ga. Scape ½—1½f high, rather stout, pubescent above, with a dense, twisted spike at summit 1—2' long. Leaves 3 or more, nearly or quite radical, 3—8 or 10' long, ½—1' wide. Bracts ovate, acuminate, as long as the greenish flowers. Aug. Oct.

12. GOODYERA. Brown.

Named for John Goodyer, an obscure English botanist.

Perianth ringent ; calyx herbaceous, upper sepal vaulted, the 2 lower ones beneath the saccate and entire lip ; column free, pollen angular ; stigma prominent, roundish.

G. PUBESCENS. Br. *Rattlesnake Plantain.*
Lvs. radical, ovate, petiolate, reticulate; *scape* sheathed, and with the flowers, pubescent; *lip* ovate, acuminate ; *pet.* ovate.—A plant found in woods, Can. and U. S. remarkably distinguished for its leaves which are all radical.

and of a dark green, reticulated above with white veins. They are ovate, 1—2 in length, contracted at base into winged petioles scarcely half as long. Scaps erect, 6—12' high. Flowers white, in a terminal, oblong, cylindric spike. Lip roundish, saccate, inflated. July, Aug.

β. *repens*. (G. repens. *Br.*) *Lvs.* less conspicuously reticulated; *spike* somewhat unilateral.—A reduced form of G. pubescens, certainly unworthy of being exalted into a species. Stem 6—8' high. Flowers in one row, which is more or less spiral.

13. LISTÈRA. Brown.

Named for Dr. Martin Lister, an English naturalist, died 1711.

Lip 2-lobed, pendant, with no callous processes; column wingless, minute; anther fixed by its base, persistent.

1. L. CORDÁTA. Br. (Ophris. *Mx.*) *Tway-blade.*
St. 2-leaved, the leaves opposite, deltoid-subcordate, acute; *rac.* few-flowered; *lip* linear, 2-toothed at base, deeply bifid, with divaricate and acute segments; *column* very short.—Root fibrous. Stem 4—8' high, furrowed. Leaves ½—1' diam., sessile, about half way up the stem. Flowers minute, greenish-purple, 10—15, in a short raceme. A delicate little plant, in woods and sphagnous swamps, among mountains, &c., N. States and Brit. Am. July, Aug.

2. L. CONVALLARIÖÍDES. Nutt. (Epipactis. *Sw.*)
St. 2-leaved, the leaves opposite, roundish-ovate; *rac.* few-flowered, loose, pubescent; *sep.* ovate-lanceolate; *lip* oblong, 2-toothed at base, with 2 roundish lobes and an intermediate minute one at the apex; *column* elongated.—Car. to Arc. Am. Root fibrous. Stem very slender, 5—10' high, sheathed with a few bracts bearing the 2 leaves near the middle. Leaves near an inch long, ¾ as wide. Flowers small, the broad, obcordate lip twice as long as the sepals. May.

SECTION 4. Lateral anthers fertile; the middle one sterile and petaloid.

14. CYPRIPEDIUM.

Gr. Κυπρις, Venus, ποδιον, a slipper; from the slipper-like form of the lip.

The 2 lower sepals united into 1 segment, or rarely, distinct; lip ventricose, inflated, saccate, obtuse; column terminated by a petaloid lobe (barren stamen).—*Fls.* large, very showy, distinguished for the large, inflated lower petal or lip.

1. C. CANDIDUM. Willd. *White-flowered Ladies'-slipper.*
St. leafy; *lvs.* oblong-lanceolate; *fl.* terminal, solitary; *sep.* elliptic-lanceolate, acuminate, lower scarcely bifid at apex; *pet.* lance-linear, longer than the compressed lip; *lobe* of the style lanceolate, rather obtuse.—Border of woods, prairies, Penn. to Ind. *Plummer!* Resembles the next in foliage, bt remarkably distinguished by the white flower. Stem about 1f high, simple Leaves 3—6' by 1—1½', sheathing the stem, acute. Ovary pedicellate. Lip 1 in length. Petals and sepals nearly 2'. May.

2. C. ACAULE. Ait. (C. humile. Sw. ?) *Acaulescent Ladies'-slipper.*
Scape leafless, 1-flowered; *lvs.* 2, radical, elliptic-oblong, rather acute; *lobe of the column* roundish-rhomboidal, acuminate, deflexed; *pet.* lanceolate; *lip* longer than the petals, cleft before.—A beautiful plant, in dark woods, Car. to Arc. Am. Leaves large, plaited and downy. Scape 10—14' high, with a single lanceolate bract at the base of the large, solitary flower. Sepals 1' lkng, the two lower completely united into a broad lanceolate one beneath the lip. Petals lateral, wavy. Lip 2' by 1', purple, forming the most showy part of the flower. May, June.

3. C. PARVIFLÓRUM. Salisb. *Yellow Ladies'-slipper.*
St. Leafy; *lvs.* broad-lanceolate, acuminate; *lobe of the column* triangular-oblong, acute; *sep.* ovate, oblong, acuminate; *pet.* long, linear, contorted; *lip* shorter than the petals, compressed.—Woods and meadows, Newf. to Car Stems usually several from the same root, about a foot high. Leaves 3 6

vy 2—3', veined, alternate, clasping, pubescent. Flower mostly solitary. Segments 4, greenish, with purple stripes and spots, the lower one bifid, composed of 2 united sepals, the two lateral ones 2—3' by ¼', waved and twisted. Lip a large, inflated sac, bright yellow, spotted inside, with a roundish aperture above. May, June.

β. *pubescens. Leaves* lanceolate; *lobe of the column* obtuse. The flowers somewhat smaller. Meriden, N. H.

4. C. SPECTABILE. Sw. *Showy Ladies'-slipper.*

St. leafy; *lvs.* ovate-lanceolate, acuminate; *lobe of the column* elliptic-cordate, obtuse; *sep.* broad-ovate, obtuse; *lip* longer than the petals, cleft before.—A tall, superb species, found in swamps, Can. to Ky. and Car. Stem thick, 2 feet or more high, hairy. Leaves 6—10' by 2—4', veined, plaited, hairy. Flowers 2—3 on each plant, very large. Lip white, striped with purple, 2' long, 1½ broad; upper segment largest, lower one smaller, composed of 2 sepals completely united. July.

5. C. ARIETINUM. Ait. (Arietinum Americanum. *Beck.*) *Ram's-head.*

St. leafy; *lvs.* elliptical, striate-veined; *sep.* 3, distinct (the 2 lower not united), linear-lanceolate, the upper oblong-ovate, acuminate; 2 *lateral pet.* linear; *lip* as long as the petal, saccate, obconic.—In damp woods, Can., Maine, Vt. Stems usually clustered, flexuous, 8—12' high, lower part sheathed. Leaves 3—5, 2—3' by ½—1', sessile, amplexicaul. Flower mostly solitary, with a leafy bract at base. Segments about equal in length, the upper one as broad as the other 4 together. The singular form of the lip readily suggests the name of this curious plant. May.

ORDER CXLVII. AMARYLLIDACEÆ.—AMARYLLIDS.

Herbs, perennial. *Leaves* parallel-veined.
Fls. showy, almost always either yellow or white, often on scapes and with spathaceous bracts.
Perianth mostly regular, adherent to the ovary, colored, consisting of 3 sepals and 3 petals.
Sta. 6, arising from the perianth segments. *Anthers* introrse.
Ova. 3-celled, the cells many-ovuled (sometimes 1—3-ovuled). *Style* 1. *Stig.* 3-lobed.
Pr. a 3-celled capsule or berry. *Seeds* with fleshy albumen.

Genera 65, species 400, chiefly tropical plants, most abundant in Brazil and S. Africa. Very few are found in our climate.

Properties.—A few of the Amaryllidaceæ possess poisonous properties, which is very rare among the Endogens. The Hottentots are said to poison their arrows by dipping them in the viscid juice of the bulbs of Hæmanthus toxicarius. The bulbs of Narcissus poeticus, and other species, are emetic. The fermented juice of the Agave forms the intoxicating *pulque* of the Mexicans. Many are highly ornamental, and are therefore cultivated.

Conspectus of the Genera.

	Scape tall, with numerous flowers.	*Agave.* 4
Corona 0. { Scape bearing 3—5 small, yellow flowers.	*Hypoxis.* 5
Corona of 2 emarginate segments.	*Galanthus.* 3
Flowers regular. { Corona monophyllous.	*Narcissus.* 2
Flowers irregular. Stamens declined.	*Amaryllis.* 1

1. AMARYLLIS.

Lat. Amaryllis, the name of a nymph, from αμαρνσσω, to shine with splendor.

Perianth irregular, funnel-shaped, nodding; filaments declined, arising from the orifice, unequal in proportion or direction; seeds flat, numerous.—*A splendid genus, with a few native and many foreign species.*

1. A. ATAMASCO. *Atamasco Lily.*

Spathe 2-cleft, acute; *fl.* pedicelled; *cor.* campanulate, with nearly equal petals, suberect.—A pretty species found in Penn., south to Carolina, sometimes cultivated. Leaves linear, a foot long. Scape round, 6' high. Spathe a little colored, bifid at the summit. Flower large, solitary, white and pink. Sepals acuminate. June.

2. A. FORMOSISSIMA. *Jacobea Lily.—Lvs.* radical; *fls.* nodding, very ringent, tube fringed; *sta.* included in the involute lower segments.—A splendid flower, from Mexico, often grown with us in large pots of light, loamy soil. Root bulbous. Leaves thick, oblong, narrow. Scape a foot high. Spathe red, disclosing a single large flower of a fine dark red color. Jn.—Aug.

2. NARCISSUS.

Gr. ναρκη, *stupor; from the effects produced by the smell of some of the species which are poisonous.*

Perianth regular; corona monophyllous, funnel-form, consisting of a whorl of united sterile stamens, within which the fertile ones are inserted.

Obs.—A well known, popular genus, whose species are easily cultivated, many of them very fragrant and beautiful. They have bulbous roots, ensiform leaves, and usually yellow flowers, with a long, compressed spathe, opening on one side, and deciduous.

1. N. JONQUILLA.—*Scape* 1—3-flowered; *segments* reflexed, spatulate; *cup* (corona) much shorter than the segments, saucer-shaped, spreading, crenate.—Native of Spain. Scape a foot high, round, slender, bearing at the summit a few flowers of a rich yellow, and very fragrant. May, Jn.

2. N. POETICUS. *Poet's Narcissus.*—*Scape* 1-flowered; *segments* imbricate at base, reflexed; *corona* expanded, flat, rotate, crenulate; 3 *anth.* shorter than the tube.—Native of S. Europe. Scape about a foot high, leaves of the same length. It bears a single flower, which is mostly white, but having the crown singularly adorned with circles of crimson, white and yellow. Jn.

3. N. PSEUDO-NARCISSUS. *Daffodil.*—*Scape* 2-edged, straight, striated; *segments* sulphur color; *corona* with a serrate-crenate orifice.—Native of England. Root bulbous. Leaves linear, a foot long, striate, veined. Scape a foot high, bearing at the top a single, very large flower, with a very long cup or corona. April, May.

4. N. TAZETTA.—*Spathe* many-flowered; *corona* campanulate, truncate, shorter than the petals; *lvs.* flat.—Native of Spain. Root a large bulb. Leaves smooth, sword-shaped. Scape naked, striate, a foot high, with 10—12 flowers. Corolla white, cup a strong yellow, not fragrant. April, May.

3. GALANTHUS.

Gr. γαλα, *milk,* ανθος; *on account of the delicate whiteness of the flower.*

Flowers spathaceous; sepals 3, concave; corona formed of 3 small, emarginate segments; stigma simple.—*Ornamental, bulbous exotics.*

G. NIVALIS. *Snow-drop.*—*Lvs.* linear, radical, keeled, acute; *scape* 1-flowered. —Native of the Alps, well known in gardens, flowering early in Spring. It 's a small plant, half a foot high, arising from a perennial bulb, bearing a single large, nodding flower white as snow. Stem usually furnished with 2 long, narrow leaves towards the top.

4. AGAVE.

Gr. αγαυος, *admirable; a term eminently applicable.*

Perianth tubular, funnel-form, adherent to the ovary, 6-parted; stamens 6, exserted; anthers versatile; capsule ovate, attenuate at each end, obtusely triangular, 3-celled, many-seeded.—*A splendid American genus. Root sometimes ligneous. Stem herbaceous. Lvs. mostly radical, rigid, channeled, often spiny. Panicle large, pyramidal.*

1. A. VIRGINICA. *False Aloe.*

Acaulescent, herbaceous; *lvs.* linear-lanceolate, fleshy, glabrous, with cartilaginous serratures on the margin; *scape* simple, glabrous, with leaf-like scales and sessile flowers.—Rocky banks, Penn. to Ga. Root premorse, tuberous. Scape 6f high, terete, glabrous, loosely spicate above. Radical leaves long, acute. Flowers greenish-yellow, very fragrant, tube longer than the acute segments. Capsule roundish, obscurely 3-angled, 3-furrowed, 3-valved, 3-celled. Sept.

2. A. AMERICANA. *American Aloe. Century Plant.*—Acaulescent; *lvs.* spinose-dentate, lanceolate, coriaceous and fleshy; *scape* branched, lofty and arborescent; *cor. tube* contracted in the middle; *pedicel* as long as the corolla.—The largest of all herbaceous plants, native of tropical America, often cultivated. It is a popular notion that it flowers but once in a hundred years, but it is known to flower much oftener, according to the culture it receives. Leaves radical, thick, 3—6 or 8f long, 4—12' wide. The scape arises from the centre of the

46

mass of leaves, to the height of 15—25f, bearing a pyramidal panicle. Flowers yellow. There is a variety with striped leaves.

5. HYPOXIS.

Gr. ὑπο, under, ὀξυς, sharp; on account of the pointed base of the fruit.

Spathe 2-leaved; perianth persistent; capsule elongated, narrowed at the base; seeds numerous, roundish, with a black, crustaceous integument.—*Small, bulbous, grass-like plants, with yellow flowers. Lvs. radical, linear.*

H. ERECTA. *Star-grass.*
Pilose; *scape* about 4-flowered, shorter than the linear-lanceolate leaves —In woods and meadows, Can. and U. S. Leaves all radical, 6—12' by 3—5", very acute. The slender, hairy scapes, several from the same root, arise 6—8', divided at top into a sort of umbel with 3—5 peduncles having each a minute, subulate spathe at the base. Perianth hairy and greenish without, yellow within; segments oval, rather obtuse. June.

ORDER CXLVIII. HÆMODORACEÆ.—BLOODROOTS.

Herbs with fibrous, perennial roots and perfect flowers.
Lvs. permanent, ensiform, equitant, usually in 2 ranks.   [cylindrical tube
Perianth more or less woolly. adherent. the sepals and petals often indistinguishable, and united into a
Sta. arising from the perianth, either 3 and opposite the sepals, or 6.
Ova. 1 or 3-celled, cells 1, 2, or many-ovuled. *Sty.* and *stig.* simple.
Fr.—Capsule covered with the withered perianth, valvular, seldom indehiscent.
Sds. definite or indefinite. *Embryo* short, straight, in cartilaginous albumen.

Genera 13, species 56, sparingly occurring in N. America, S. Africa, New Holland, &c. The root of Lacnanthes tinctoria abounds in a red coloring matter. One of the most intense bitters known is Aletris farinosa.

Conspectus of the Genera.

Stamens { 3. { Perianth woolly, 6-parted.   *Lacnanthes.* 1
        {   { Perianth woolly, 6-parted.   *Lophiola.* 2
        { 6. { Perianth mealy, 6-toothed.   *Aletris.* 3

1. LACNANTHES. Elliot.

Gr. λαχνη, soft hair, ανθος; in allusion to the woolly corollas.

Perianth woolly, tube adherent; calyx lobes exterior, of 3 linear sepals, as long as the 3 lance-oblong petals; stam. 3, equaling the sepals, and opposite to them; stig. minutely 2-lobed; caps. 3-celled, truncated, many-seeded.—*Lvs. lanceolate-linear. Fls. corymbose.*

L. TINCTORIA. Ell. (Dilatris. *Pursh.*) *Red-root.*
Swamps and borders of ponds, R. I., *Olney!* N. J. to Flor. An interesting plant, with rush-like leaves. Stem erect, strict, 18—24' high, clothed with white wool above. Leaves mostly radical, fleshy, 3—4" wide and nearly as high as the stem. Cauline leaves remote and bract-like. Corymb terminal, close, 15—30-flowered. Flowers densely clothed with white wool outside, glabrous and yellow within. Anthers bright yellow. Jl.—The root is said to be employed in dyeing.

2. LOPHIÖLA. Ker.

Gr. λοψος, a crest; alluding to the crested petals.

Perianth half superior, 6-cleft, persistent, woolly; petals narrower than the sepals, somewhat interior; sta. 6; filament naked; anthers erect; style conical, 3-partible; stigma simple; capsule opening at the summit, 3-celled, 3-valved, many-seeded.—*Lvs. ensiform. Fls. corymbose.*

L. AMERICÀNA. Wood. (L. aurea. *Ker.* Conostylos. *Br.*) *Golden Crest-flower.*—Sandy swamps, pine barrens, N. J. Stem 1—2f high, erect, hoary-tomentose when young. Leaves glaucous, narrowly linear, two-edged, glabrous, the lower and radical long, cauline 2 or 3, shorter. Corymb finally much expanded, many-flowered. Corolla woolly and yellow within, segments reflexed, about as long as the stamens. Capsule ovate, dissepiments arising from the centre of each valve. Seeds white. July, Aug.

3. ALÉTRIS.

Gr. (αλευρ) αλειτρος, meal; from the powdery dust with which the plant is covered.

Perianth 6-cleft, tubular, rugose, persistent; stamens issuing at the top of the tube, style 3-sided, 3-partible; capsule opening at top, many-seeded.—*Lvs. radical, rosulate. Scape many-flowered.*

1. A. FARINÓSA. (A. alba. *Mz.*) *Star-grass. Colic Root.*
Lvs. broad-lanceolate; *fls.* oblong-tubular, pedicelled; *perianth* in fruit rugose or mealy in appearance.—Grows in low grounds, in most of the States. Root premorse, intensely bitter. Scape 20—30' high, with remote scales or bracts, and surrounded at base with a circle of lanceolate, sessile leaves. These are 3—4' long, ½ as wide, and lie flat upon the ground. Flowers in a long, thin raceme. Perianth white, ½' long, on very short pedicels, rugose without when old. Medicinal. July.

2. A. AÚREA. Walt. *Yellow Aletris.*
Lvs. lanceolate; *fls.* subsessile; *perianth* short, tubular-campanulate, yellow, finally rugose and very scabrous.—In the pine barrens of N. J. to Car., abundant. Scarcely distinct from the preceding. *Torrey.* Scape 2—3f high, with few, yellow flowers in the spicate raceme. Leaves all radical. Jl., Aug.

ORDER CXLIX.   IRIDACEÆ.—IRIDS.

Herbs perennial, arising from bulbs, corms or rhizomas, rarely from fibrous roots.
Lvs. equitant, mostly distichous. *Fls.* with spathaceous bracts.
Per.—Tube adherent to the ovary, limb 6-parted, colored, in 2 often unequal series.
Sta. 3, alternate with the 3 petals. *Anthers* 2-celled, extrorse.
Ova. 3-celled, many-ovuled. *Style* 1. *Stigmas* 3, dilated or petaloid.   (albumen.
Fr.—Capsule 3-celled, 3-valved, with loculicidal dehiscence. *Seeds* numerous, with hardened, fleshy

Genera 52, species 500, chiefly natives of the Cape of Good Hope, or of the middle of Europe or N. Amer.

Properties—More remarkable for their beauty than their utility. Some of them are cathartic, as Iris tuberosa. The aromatic *orris root* is the dried rhizoma of Iris florentina of S. Europa. *Saffron consists* of the dried orange-colored stigmas of Crocus sativus.

Conspectus of the Genera.

	｛Petals smaller than the sepals.	｛Stamens distinct.			*Iris.*	1
		｛Stamens united.			*Tigridia.*	4
			｛Tube short.		*Ixia.*	2
		｛subsect.	｛Tube very long.		*Crocus.*	6
Flowers ｛regular.	｛Petals and sepals subequal, ｛spreading, flat.	Tube short.			*Sisyrinchium.*	5
｛irregular.	Seeds winged.				*Gladiolus.*	3

1. IRIS.

Named from the Greek, signifying rainbow; on account of the varied color of the flowers.

Sepals 3, reflexed, larger than the 3 erect petals; stamens distinct; style short or 0; stigmas petaloid, covering the stamens.—*Lvs. mostly ensiform.*

1. I. VERSICÓLOR. *Blue Flag.*
St. terete, flexuous; *lvs.* ensiform; *fls.* beardless; *ova.* triangular, with concave sides and roundish angles.—Grows in wet grounds, (U. S. and Can.,) where its large, blue flowers are conspicuous among the grass. Rhizoma large, horizontal, acrid. Stem 2—3f high, acute on one side, often branched and bearing several flowers. Leaves a foot long, ⅓—1' wide, erect, sheathing at base. Sepals spatulate, purple, the claw variegated with green, yellow and white, with purple lines. Petals erect, paler, a little shorter than the stigmas. Style short, bearing 3 petaloid stigmas which are reflexed and bifid at the end, purple or violet, concealing the stamens beneath. Anther oblong; seeds flat. June.
β. *sulcata* (*Torr.*) *St.* nearly straight; *pet.* longer than the stigmas; *angles of the ovary* sulcate.

2. I. PRISMÁTICA. Pursh. (I. Virginica. *Torr.*) *Boston Iris.*
St. round, slender, few-flowered; *lvs.* linear, long; *fls.* beardless; *ova.* triangular, the side doubly grooved.—In similar situations with the last, reaching

distinguished by its very slender habit. Mass. ! to N. J. Rhizoma fleshy. Stem smooth, 1—2″ in diam., 1—2f high, branching at top and bearing 2—6 flowers. Bracts at the base of the branches withering. Leaves few, alternate, grass-like, 6—10′ long, amplexicaul. Sepals narrow, yellow, edged with purple. Petals linear-lanceolate. June.

3. I. LACUSTRIS. Nutt.
Lvs. ensiform, longer than the low, compressed, 1-flowered scape; *seg.* of the perianth nearly equal, obtuse, emarginate, the sepals scarcely crested; *caps.* turbinate, 3-sided, margined.—Islands of Lake Huron, near Mackinaw, *Nuttall.* Roots extensively creeping. Leaves 2—5′ by 3—4″, those of the scape bract-like. Scape 1—2′ high. Flowers pale blue, the sepals rather broader. Jl.

4. I. SAMBUCINA. *Flower-de-Luce.* Fr., *Fleur-de-lis.*—*St.* many-flowered, longer than the leaves; *segments of the perianth* emarginate, outer ones flat; *lvs.* bent inwards at the point; *spathe* membranaceous at the apex; *fil.* beardless, lower ones pedunculate; *stig.* with acute, serrate divisions.—Native of the South of Europe. Common in gardens. The prevailing color of the flower is light blue, often fading to white. May. †

5. I. PUMILA. *Dwarf Iris.*—*Scape* very short (3—6′), 1-flowered; *spathe* shorter than the tube; *sep.* reflexed, narrower than the erect petals.—A small species from Hungary, cultivated in the edgings of walks. Leaves numerous, broad-ensiform, suberect. Flowers large, deep purple, appearing in early spring.†

6. I. OCHROLEUCA.—Beardless; *lvs.* ensiform, depressed, striate; *scape* subterete; *ova.* 6-angled.—A tall species from Levant. Stem 3—4f high. Flowers ochroleucous or sulphur-yellow. July. †

7. I. CHINENSIS. *China Iris.*—*Scape* compressed, many-flowered; *stigma* is lacerated. A small species from China. Flowers elegantly striped. A few other species are rarely cultivated in gardens. †

2. IXIA.
Gr. ιξος, sticky ; alluding to the glutinous juice of some species.

Spathe of 2 or 3, ovate, short bracts ; petals and sepals distinct or slightly united, similar, regular; sta. 3 ; filaments filiform; stigma straight or incurved, fixed by the base, subfiliform.

1. I. CHINENSIS.—*St.* terete, flexuous, leafy; *lvs.* ensiform, vertical, sheathing; *panicle* somewhat dichotomous and corymbose; *perianth* campanulate, segments 6, lance-linear.—Native of S. Africa, frequent in gardens, Western and Southern States. Whole plant smooth, 3—5f high. Leaves erect, tapering to an acute point. Flowers yellow, spotted with red inside, about 1′ long. Jl., Aug.†

3. GLADIOLUS.
Lat. *gladius,* a sword ; in reference to the form of the leaves.

Spathe 2-leaved ; perianth irregular ; stamens distinct, ascending; stigmas 3 ; seeds winged.—*A large genus of bulbous plants, with large and showy flowers.* None native.

G. COMMUNIS.—*Spike* unilateral; *upper segment of the perianth* covered by the lateral ones, 3 lower marked by a white, linear-lanceolate spot, lowest very large.—A fine border flower, from S. Europe. Stem 2—3f high, with the flowers arranged in a long, somewhat spiral row upon it. Perianth large, deep red, variegated with white. Its colors are liable to considerable variation. Jl. †

4. TIGRIDIA.
Name in reference to the large spotted flowers.

Spathe 2-leaved ; the 3 sepals larger than the 3 petals ; stamens monadelphous ; filaments united into a long tube.

T. PAVONIA. *Tiger Flower.*—*S.* simple, flexuous; *lvs.* ensiform, veined; *segments* flat; *pet.* panduriform.—A superb, bulbous plant, with large, beautiful flowers, native of Mexico and Peru. Stem 2f high or more, erect, round, leafy, somewhat branched. Leaves erect, smooth, a foot long. Flower inodorous,

5—6' broad, yellow, variegated with scarlet, crimson and purple. It is very evanescent, lasting but a few hours, but a new one appears daily for several weeks. It ripens seeds, from which, or from offsets, it may be increased. Jl.—Sept.†

5. SISYRINCHIUM.

Gr. ους, a hog. and ρυγχος, a snout; alluding to the singular spathe.

Spathe 2-leaved; segments of the perianth flat, equal; stamens monadelphous; stigma 3-cleft.—♃ *Grass-like plants, with compressed, ancipital scapes.*

1. S. ANCEPS. (S. gramineum. *Lam.*) *Blue-eyed Grass.*
Scape simple, winged; *valves of the spathe* unequal, the longer scarcely equaling the flowers; *pet.* mucronate.—A delicate little plant, with blue flowers, common in low grass-lands, Can. and U. S. Stem or scape 10—12' high, so winged as to resemble the leaves, smooth and mostly simple. Leaves linear, about as long as the scape, sheathing at base. Spathe 2—5-flowered, the longer valve acuminate. Flowers purple or blue, on filiform pedicels. Sepals a little broader than the petals, spreading. Capsules globose. Jn. Jl.

2. S. MUCRONATUM. Michx. *Blue-eyed Grass.*
Scape simple, subsetaceous; *spathe* colored, outer valve longer than the flowers, ending in a long, mucronate point.—Middle States, W. to Ky. Found in wet meadows, where the grass is not luxuriant. Leaves radical, a line wide. Scape 6—10' high, narrowly winged, setaceously slender. Spathe of 2 very unequal valves, 3—4-flowered, tinged with purple. Flowers smaller than in the preceding, of a fine blue color. Jn.

6. CROCUS.

Named from the youth *Crocus*, who, according to Grecian mythology, was changed into this flower.

Perianth funnel-form, the segments united at base into a long and slender tube; stigma 3-cleft, convolute, crested.—*Spathe radical*, 1—2-leaved, thin, transparent. *The long tube of the flower nearly or quite sessile upon the bulb. After flowering, the ovary arises from the ground by the growth of the scape, to ripen its seeds in the sun.*

1. C. SATIVUS. *Saffron.*—*Lvs.* linear, revolute at the margins; *stig.* 3-parted, as long as the corolla, reflexed.—From Asia. Stem bulbous. Leaves radical, with a longitudinal, white furrow above. Flower nearly or quite sessile on the bulb, with a long, white tube, and purple, elliptical segments. Stigmas long, emarginate, exsert, of a deep orange-color. Its virtues, both medicinal and coloring, reside chiefly in the large stigmas. Sept.—A variety, perhaps the most common, has yellow perianths. ‡

2. C. VERNUS. *Spring Crocus.*—*Stig.* included within the flower, with 3, short, wedge-shaped segments.—Native of the Alps. Stem bulbous. Scape an inch or two high, 3-sided. Flowers vary in color, generally purple, often yellow or white; tube very long, slender, gradually enlarged upwards, closed at the mouth with a circle of hairs, limb campanulate, much shorter than the tube. Anthers yellow, sagittate. Flowers in March or April. The Crocus is propagated in gardens chiefly by bulbs. †

ORDER CL. DIOSCOREACEÆ.—YAM ROOTS.

Shrubs twining. *Lvs.* usually alternate and reticulate-veined.
Fl. dioecous. *Perianth* tube adherent to the ovary; segments of limb 6, in 2 series.
Steril s.—*Sta.* 6, inserted into the base of the sepals and petals.
Fertil s.—*Ova.* 3-celled, cells 1—2-ovuled. *Styles* and *stigmas* nearly distinct.
Fr.—*Capsule* 3-winged, compressed, 2 of the cells sometimes abortive.
Sds. flat, compressed. *Embryo* small, in cartilaginous albumen.

Genera 6. Species 110. The only remarkable or useful product of this order is *yams*, an important article of food in all tropical countries. They are the large, mucilaginous, sweetish tubers of *Dioscorea sativa*, &c.

DIOSCOREA.

In honor of Pedacius Dioscorides, a Greek Physician and florist of about the reign of Nero.

Flowers ♂ ♀; styles of the fertile flowers 3; cells of the capsule

2-seeded; seeds membranaceously margined. — *Slender, shrubby climbers, twining with the sun. Lvs. simple and palmately veined or palmately divided. Fls. green, inconspicuous, in axillary spikes or panicles.*

D. VILLÒSA. (D. quaternata. *Ph.*) *Yam Root.*

Lvs. broad-ovate, cordate, acuminate, 9—11-veined, the margin entire or wavy, lower surface villose with short, soft hairs, upper surface glabrous; *petioles* elongated, the lowest somewhat verticillate in 4s, the next subopposite, the middle and upper alternate; ♂ plant with the spikes paniculate, ♀ with the spikes simple.—A delicate twining vine, in thickets and hedges, U. S. and Can., rare in N. Eng. Stem woolly, smooth, reddish-brown, 1—2″ diam., 5—10—15f long, running over bushes and fences. Leaves 2—4′ long, ⅘ as wide, distinctly cordate and acuminate. The two outside veins in some of the leaves meet a little above the base, in others at the base—a character of no value. Petioles 2—4′ long. Peduncles axillary. Ovaries at first elliptic, finally almost as broad as long. June, July.

β. *læviuscula.* Wood. (D. quaternata. *Pursh.*) *Lvs.* smooth both sides, margin slightly wavy; otherwise indistinguishable from variety *a.*

ORDER CLI. SMILACEÆ.—SARSAPARILLAS.

Herbs or shrubs, often climbing. *Lvs.* reticulate-veined.
Fls. diœcious or perfect. *Perianth* free from the ovary, 6-parted, regular.
Sta. 6. inserted into the base of the segments.
Ova. 3-celled; cells 1 or many-seeded.
Fr.—Berry roundish, few or many-seeded. *Seeds* with cartilaginous albumen.

Genera 2, species 120, thinly disseminated through most countries. The diuretic and demulcent *areas parillæ* are the roots of several chiefly S. American species of Smilax.

SMILAX.

Gr. σμιλη, a grater; the stems of some species are rough with prickles.

Flowers ♂ ♀; perianth broad-campanulate. ♂ Anthers adnate. ♀ Style minute; stigmas 3; berry 3-celled, 1—3-seeded.—⁴ *or shrubby, mostly climbing by stipular tendrils, often prickly. Lvs. entire, petiolate, palmately veined. Fls. umbellate.*

* *Stem shrubby and prickly.*

1. S. ROTUNDIFOLIA (and caduca. *Linn.* S. quadrangularis. *Ph.*)

St. terete or sub-4-sided, flexuous, aculeate, ligneous, climbing; *lrs.* short-petiolate, roundish-ovate, acuminate, subcordate, a little paler and glaucous beneath, 5—7-veined, glabrous; *ped.* axillary, solitary, many-flowered, longer than the petioles; *berries* black, glaucous.—A strong, thorny vine, extending 10—40f in hedges and thickets, U. S. and Can. Stem woody, smooth, except the scattered thorns which proceed from the wood. Branches 4-angled. Leaves 2—3′ by 1½—3′, cordate or tapering at base. Tendrils strong, from the top of the wings of the petioles. Flowers small, greenish, in small, axillary umbels. Berries round, black. June.

β. *caduca.* Smaller, with ovate leaves shorter than the peduncles.

γ. *quadrangularis.* *St.* unarmed above, obtusely 4-angled; *lrs.* ovate.

2. S. SARSAPARILLA. *Medicinal Sarsaparilla.*

"*St.* slightly 4-angled, aculeate above; *lrs.* unarmed, elliptical-ovate, cuspidate, abrupt, 3-veined, glaucous beneath; *ped.* longer than the short petioles."—In swampy thickets, Penn. to Car. *Pursh.* Root long, slender, with a thick bark brown externally, white within. Stem stout, somewhat flexuous, armed with a few scattered, hooked prickles. Leaves finally nearly orbicular, 2- 3′ diam., abruptly contracted at each end, with 3 strong veins and 2 lateral smaller secondary ones. Petioles short, margined, with 2 tendrils. Flowers in small, thin umbels, yellowish-white. Berries (red, *Woodville*, black, *Pursh.*) 3-seeded.

3. S. LAURIFOLIA. *Laurel-leaved Green Brier.*

St. aculeate, terete, branches unarmed; *lvs.* coriaceous, oval-lanceolate, slightly acuminate, 3-veined; *umbels* on very short peduncles, which are arranged alternately on a common rachis.—N. J. to Ga. A vigorous, evergreen climber, ascending trees to a great height. Stem with few scattered prickles. Leaves numerous, very thick and smooth. Raceme of umbels longer than the leaves, peduncles shorter than the ultimate pedicels. Jn.—Aug.

4. S. PANDURATUS. Pursh. (S. tamnoides. Ell.)

St. branched, terete, aculeate; *lvs.* ovate, somewhat panduriform, acuminate, cordate, 3-veined; *ped.* twice as long as the petioles; *umbel* many-flowered; *berries* black.—In sandy woods, N. J. to Car. Stem twining, 6—12f. Leaves smooth and shining both sides, with shallow depressions or sinuses on each edge. July.

* *Stems shrubby, unarmed.*

5. S. PSEUDO-CHINA.

St. terete, unarmed; *cauline lvs.* ovate, cordate, *ramial* ovate-oblong, all 5-veined, on short petioles; *ped.* very long.—Sandy woods, N. J. to Car., W. to Ohio. Root large, tuberous. Stem purplish-brown, very smooth, branching and climbing by tendrils which arise from the base of the petioles. Leaves 2—4' by 1—2', slightly hispid on the veinlets beneath. Peduncles longer than the leaves. May, Jn.

* * * *Stem herbaceous.*

6. S. HERBACEA. (S. pedunculata. *Muhl.*)

St. herbaceous, unarmed, angular, erect, or inclining; *lvs.* ovate, 7—9-veined, cuspidate; *umbels* on long peduncles. A coarse, smooth, ill-scented plant, in thickets and low grounds, N. Eng. to Ky. and Wisc.! Stem slightly angled, 3—6 or 8f high, usually nodding with its slender summit and few small branches, and leaning on other plants or on each other. Leaves 2—4' by 1½—3, often roundish, paler beneath, the petioles winged at base and producing a long, slender tendril from the top of each wing. Fertile umbels simple, about 40-flowered, on peduncles 6—9' long, those of the sterile umbels shorter. Flowers yellowish-green, diffusing about the plant an intolerably offensive and sickening odor. Berries dark blue. Jn.

β. *St.* more generally climbing by its tendrils; *lvs.* broadly ovate, subcordate.

7. S. LASIONEURON. Hook.

St. terete, climbing, subsimple, unarmed; *lvs.* oblong, broadly ovate, cordate, rounded and mucronate at apex, 7-veined, glaucous and hispid-pubescent on the veinlets beneath, glabrous and green above; *ped.* a little longer than the petioles, many-flowered; *tendrils* from the base of the petioles.—Green Co., Ind.! also Can. Stem 3—6f high. Peduncle much shorter than the leaves, which are often about 4½' by 3'.

ORDER CLII. TRILLIACEÆ.—TRILLIADS.

Herbs with simple stems, tuberous roots and verticillate, net-veined leaves.
Fls. large, terminal, solitary, perfect, trimerous, rarely tetramerous.
Cal.—Sepals 3, green or herbaceous.　*Cor.*—Petals 3, larger than the sepals, colored or herbaceous.
Sta. 6—10. *Fil.* subulate. *Anth.* linear, with cells on their edges and the connective extended.
Ova. free, 3—5-celled. *Sty.* distinct. *Stig.* small. *Ovules* ∞, in 2 rows, ascending.
Fr. succulent, 3—5-celled. *Sds.* ∞. *Embryo* minute, in fleshy albumen.

Genera 4, species 30, in woodlands, temperate parts of Europe, Asia and N. America. The roots of some species are emetic.

Genera.

Leaves { in one whorl. *Trillium.* 1
　　　{ in two whorls. *Medeola.* 2

TRILLIUM. Miller.

Lat. *trilix*, triple; because the sepals, petals, carpels, cells, stigmas and leaves are in 3s.

Perianth deeply 6-parted, in 2 distinct series, outer of 3 sepals, inner of 3 colored petals; stamens nearly equal; stigmas sessile, distinct or approximate; berry 3-celled; cells many-seeded.—⅜ *Stem simple. Lvs.* 3, *whorled at the top of the stem, reticulate-palmate veined. Fls. solitary, terminal.*

1. T. sessile. *Sessile-flowered Trillium.*
Lvs. broad-ovate or suborbicular, rather acute, sessile; *fl.* closely sessile, erect; *sep.* erect, ovate-lanceolate or lanceolate, acute; *pet.* linear-lanceolate, purple, a third longer than the sepals; *anth.* long, erect.—A small species in fertile soils, Middle, Western (*Clark! Plummer!*) and Southern States. Rhizoma horizontal, thick. Stem 6—8' high, slender. Leaves rather thick, 1½—2½' by 1—2', smooth and entire. Sepals green, about 8" by 3", the petals narrower and much longer, dark purple. Apr. May.

2. T. recurvatum. Beck.
Lvs. ovate or obovate, attenuated to a petiole, acute; *fl.* closely sessile; *pet.* lanceolate-ovate, very acute, attenuate at base, erect, as long as the recurved sepals.—A small Trillium quite distinct, although allied to the last, in shady woods, Wis., *Lapham!* Ill., *Jenney!* Mo., *Beck.* Stem 8—10' high, rather thick. Leaves 2—2½' by 1½—2', with distinct, short petioles. Petals purple, and with the green, reflexed sepals about 1' long. May.

3. T. erythrocarpum. Michx. (T. pictum. *Ph.*) *Smiling Wake-robin.*
Lvs. ovate, acuminate, rounded at base, abruptly petioled; *ped.* erect; *pet.* lanceolate-ovate, recurved, twice as long as the sepals.—Can. to Ga. A beautiful flower, adorning our woods in May and June. Stem 8—12' high, with a whorl of 3 broad-ovate leaves at top. These are 3-veined, rounded at base, long acuminate, 3—4' long, ⅔ as wide, petiole 2—3" long. Flower nearly erect. Petals wavy at the edges, white, finely radiated with purple lines at base. The root is considered medicinal.
β. Cleavelandicum. Wood. (T. Cleavelandicum. *Swallow!*) *Sep.* developed into leaves, which are but little smaller than the true leaves; *pet.* 6, the 3 outer but partly colored. Otherwise as in *α.*—Brunswick, Me.! This is probably a metamorphosis; but Mr. S. has gathered it three years in succession, and also finds it thus far unaltered when cultivated from the root. Its claims to the rank of a species must be tested by plants reared from the seeds. (*Dr. T. Rickard* comm.)

4. T. pusillum. Michx. (T. pumilum. *Ph.*)
Lvs. oval-oblong, obtuse, sessile; *ped.* erect; *pet.* scarcely longer than the calyx.—Penn., *Muhlenberg.* A very small species. Petals flesh-colored. This plant appears to be lost to the later botanists.

5. T. nivale. Riddell. *Snowy Trillium.*
Rt. tuberous, premorse; *st.* low; *lvs.* ovate or oval, rather obtuse, distinctly and abruptly petiolate; *fl.* short, pedunculate, erect; *pet.* spatulate-obovate, obtuse, white, one-third longer than the calyx.—The smallest species here described, in stony or dry fields, Ohio, *Clark!* Wis., *Lapham!* Stem 2—4' high, from a thick, tuberous root. Leaves 8—18" by 5—12", petioles 2—4", about equaling the peduncle. Sepals green, much narrower than the snowy petals which are about 8" by 4". Mar. Apr.

6. T. pendulum. Muhl. (T. cernuum. *Bart.*, *Ph.*, *&c.*) *Drooping Trillium.*—*Lvs.* suborbicular-rhomboidal, abruptly acuminate, shortly petiolate; *fl.* cernuous, on a recurved peduncle.—A large species, with a small flower, Mid. and W. States! Stem slender, 10—15' high. Leaves 3—5' diam., nearly round, on petioles 1" long. Flower white, pendulous beneath the leaves. Peduncle 1—2½' long. Sepals green, oblong-lanceolate, acuminate, 1' long. Petals oblong-ovate, acute, 1½' by ½', white. Stigmas erect, recurved at top, lower part styloid (or styles 3, erect, with recurved stigmas!). May, Jn.

7. T. erectum. (T. atropurpureum. *Curt.*) *Bath Flower.*
Lvs. rhomboidal, acuminate, sessile; *ped.* inclining; *fl.* nodding; *pet.* scarcely longer, but much broader than the sepals.—A conspicuous plant in woods, of fine appearance, but of an intolerably offensive odor. At the top of the stem, which is a foot high, is a whorl of 3 leaves which are 3-veined, 3—5' long, of equal width, and a single, nodding flower, on a nearly *erect* peduncle. Petals broad-ovate, an inch long, twice as wide as the sepals and of a dusky purple, greenish outside. May.

β. *Fls.* white and much smaller.—N. Y., &c.
γ. *Fls.* white and somewhat larger.—Western States !

8. T. GRANDIFLÓRUM. Salisb. *Large-flowering Trillium.*
Lvs. broadly rhomboid-ovate, sessile, abruptly acuminate; *ped.* inclined; *fl.* suberect; *pet.* much longer than the calyx, spatulate-oblanceolate, conniveni at base.—Damp, rocky woods, Mid., S. and W. States, abundant. Stem 9—12′ high. Leaves 3—5′ diam. Flower larger than in any of the preceding species. Petals 1½—2′ in length, broadest near the apex, with a short, abrupt acumination, white, varying to rose-colored. May.

2. MEDEÓLA.

Named after the fabulous sorceress, *Medea*, for its supposed medicinal virtues.

Perianth deeply parted into 6 petaloid, revolute segments; stigmas 3, divaricate, united at base; berry 3-celled; cells 3—6-seeded. —*Stem simple.*

M. VIRGÍNICA. (Gyromia. *Nutt.*) *Cucumber Root.*
Lvs. verticillate in the middle of the stem, 3 at the top.—None can but admire the symmetry of its form. Rhizoma white, fleshy, tuberous, thought to resemble the cucumber in flavor. Stem erect, 1—2′ high, invested with loose, cottony wool. Leaves in two whorls, one just above the middle of the stem, consisting of 6—8 wedge-lanceolate leaves (3—4′ by 9—12″); the other at the top, of about 3 ovate, shorter ones. Flowers in the upper whorl, 1, 2 or 3, pendulous, with greenish, revolute segments. The stigmas are very long, reflexed, dark red. July.

ORDER CLIII. LILIACEÆ.—LILYWORTS.

Herbs with parallel-veined leaves. Stems often bulbous or tuberous at base.
Fls. perfect, regular, generally large and richly colored.
Perianth free from ovary, of 6 segments (rarely 4), colored.
Sta. 6 (rarely 4), inserted into the sepals and petals. *Anthers* introrse.
Ova. 3-celled, many-ovuled. *Styles* united into 1. *Stigmas* often 3-lobed.
Fr. capsular or fleshy, with several or many seeds in each cell.
Sds.—Albumen fleshy.

Genera 133, species 1200, chiefly natives of temperate regions. The flowers of most are beautiful, of many brilliant, and of some truly splendid.

Properties.—The order abounds in a bitter, stimulant principle and also in mucilage. Some of the bulbous species yield a nutritious diet, as the *asparagus*, *onion*, *garlic*. The well known active medicine, *squills*, is the bulb of *Scilla maritima*, of S. Europe. The various kinds of officinal *aloes*, are the product of several species of Aloe. The powerful astringent, *dragon's blood*, is the concentrated juice of *Dracaena Draco* of the Canary Isles.

Conspectus of the Genera.

* Segments of the perianth scarcely cohering in a tube.

				{ erect.	.	.	*Tulipa.*	1
			solitary. { drooping.	.	.	*Erythronium.*	4
		on a scape, .	{ umbellate, &c.	.	.	*Convallaria.*	12
			{ Petals with a grooved line.	.	*Lilium.*	2
Fls. campanulate,	{ on a leafy stem.	{ Petals with a honey cavity at base.	*Fritillaria.*	3
			{ Leaves broad, cauline.	*Convallaria.*	12
			Valves 6, { Leaves linear, radical.	*Phalangium.*	11
		racemose. . { Valves 6, bearing the stamens.	*Asphodelus.*	10
Branches or leafless. Fls. subrotate,	corymbose, pedicels bracted at base.	*Ornithogalum.*	9
	{ umbellate, from a leafy spathe.	*Allium.*	7
	Sepals and petals recurved. Seeds few.	*Streptopus.*	15
Branches leafy.	{ Leaves ovate, &c. { Sepals and petals erect. Seeds many.	*Uvularia.*	14
	{ Leaves filiform, fasciculate. . . .	*Asparagus.*	16

* * Segments of the perianth cohering into a tube at base.

Flowers on a scape.	{ Stamens inserted at the top of the incurved tube.	.	.	*Polyanthes.*	6
	{ Stamens inserted in the middle of the tube.	.	.	*Hyacinthus.*	5
	{ Stamens inserted at base of tube, declined.	.	.	*Hemerocallis.*	8
Flowers axillary on a leafy stem.	.	.	.	*Polygonatum.*	13

TRIBE 1. TULIPACEÆ.—Bulbous. Sepals and petals scarcely adhering in a tube. Integuments of the seed soft and pale.

1. TULIPA.

Persian *thouliban*, a turban; alluding to the form of these magnificent flowers.

Perianth campanulate; stamens short, subulate; anthers 4-angled; stigma thick; capsule oblong, triangular.—4 *Lvs. radical*, *Flower* scape. *The species are chiefly oriental.* Only 2 are ...ated.

...ANA. *Common Tulip.*—*Scape* 1-flowered, smooth; *lvs.* ovate-...rect, segments obtuse, smooth.—Named for *Gesner*, a Zurich ...ears to have been introduced into Europe from Persia in 1559 ...endless, and may be produced by first planting the seed in a ...terwards transplanting the bulbs into a poorer soil. After a ...owers become broken or variegated with colors in that exqui-...uch admired. In catalogues there are enumerated and describ-...) varieties. May.

...NS, vulgò, *Van Thol*, differs from T. Gesneriana, in having a ...and fragrant flowers. It is moreover much smaller, and blos

2. LILIUM.

. λειρον, Celtic, *li*, white; one species is the emblem of purity.

...npanulate, segments mostly recurved, each with a lon-...ve within, from the middle to the base; *stam.* shorter ...; capsule subtriangular, the valves connected with lat-...4 *Herbs, with bulbous and leafy stems. Lvs. sessile,* ...ticillate. *Fls. terminal.*

...DENSE. *Yellow Lily.*

...verticillate, lanceolate, the veins hairy beneath; *ped.* terminal, ...y by 3s; *fl.* nodding, the segments spreading, never revolute.—...A plant of much beauty, frequently adorning our meadows in ...root affords a fine example of the scaly bulb. Stem round, ...ounded by several remote whorls, each consisting of 4—6 ...n a few scattered ones at base. These are 2—3' by ¼—1'. ...ometimes 7—20, pendulous, yellow or orange-colored, spotted ...inside. July.

6. L. JAPONICUM.—*Lvs.* scattered, lanceolate; *fls.* cernuous, campanulate.—Native of China. A noble species, requiring careful management. Its flower is large, nodding, terminal, white, on a stem 2f high. †

7. L. TIGRINUM. *Tiger-spotted Lily.*—*Lvs.* scattered, sessile, 3-veined, the upper cordate-ovate; *perianth* revolute, papillose inside.—Native of China, very common in cultivation. Stem 6f high, with a pyramid of dark orange colored, spotted flowers. Axils of leaves bulbiferous. Aug. †

3. FRITILLARIA.

Lat. *fritillus*, a chess-board; alluding to the checkered petals.

Perianth campanulate, with a broad base and nectariferous cavity above the claw of each segment; stamens as long as the petals.—4

1. F. IMPERIÀLIS. *Crown Imperial.*—*Rac.* comose, naked below; *lvs.* entire.—Native of Persia. A fine, showy flower of easy culture. Stem thick, striate, 3f high, the lower part invested with the long, narrow, entire leaves; the upper part is naked, bearing at the top a raceme of several large, red or yellow, nodding flowers, beneath a crown formed by the pairs of small, narrow leaves at the base of each pedicel. May. †

2. F. MALEÀGRIS. *Checkered Fritillary.*—*Lvs.* alternate, linear, channeled; *st.* 1-flowered.—Native of Britain. Stem a foot high, with alternate, long, very narrow leaves. The flower, which is usually solitary, is large, nodding, and beautifully checkered with purple and pale red or yellow. May. †

4. ERYTHRONIUM.

Gr. ερυθρος, red; in allusion to the color of the flower and leaves of some species.

Perianth campanulate, segments recurved, the 3 inner ones (petals) usually with a tubercle attached to each side at base, and a groove in the middle; capsule somewhat stipitate, seeds ovate.—4 *Leaves 2, subradical. Scape 1-flowered. Fls. nodding, liliaceous.*

1. E. AMERICÀNUM. Smith. (E. Dens-canis. *Michx.*) *Yellow Erythronium.*—*Scape* naked; *lvs.* spotted, lanceolate and involute at the point; *segments* oblong-lanceolate, obtuse, inner ones bidentate near the base; *sty.* clavate; *stig.* undivided.—A beautiful little plant, among the earliest of our vernal flowers, found in rich, open grounds, or in thin woods, U. S. and Can. The bulb is deep in the ground. Scape slender, 3—4' high. The 2 leaves are of equal length (5'), one of them nearly twice as wide as the other, both clouded with brown spots. Flower drooping, yellow, revolute in the sunshine. May.

2. E. ALBIDUM. Nutt. *White Erythronium.*—*Lvs.* elliptic-lanceolate; *segments of perianth* linear-lanceolate, rather obtuse, inner ones without dentures at base, subunguiculate; *stig.* 3-cleft, lobes reflexed.—About the size of the last, in wet meadows, near Albany, N. Y., *Storrs!* to Wisc., *Lapham!* Leaves without an acumination, tapering to the base, of equal length including the petiole (4—5'), one of them twice as wide as the other. Scape a little longer than the leaves, bearing a single, white, nodding flower. Segments 1½' long. April, May.

3. E. BRACTEÀTUM. Bw. *Bracted Erythronium.*—*Scape* bracted; *lvs.* lanceolate, very unequal.—An alpine species, found in Vt. *Boott.* It is a smaller plant, distinguishable by the inequality of the leaves, one of which is 3 or 4 times as large as the other. Scape shorter than the leaves with a narrow, lanceolate bract 1½' long, a little below the flower. Flower greenish-yellow. Segments about 9'' long, gibbous at base. Jn.

TRIBE 2. HEMEROCALLIDEÆ.—Bulbous. Sepals and petals united into a tube. Integuments of the seed soft and pale.

5. HEMEROCALLIS.

Gr. ἡμερα, the day, and καλλος, beautiful; flowers beautiful but lasting only a day.

Perianth campanulate, with a cylindric tube; stamens declined.

stigma simple, villous, small.—♃ *An ornamental genus, natives of the old world. Lvs. radical. Scape corymbose.*

1. H. FULVA. *Common Day Lily.*—*Lvs.* linear-lanceolate, carinate; *pet.* obtuse, wavy; *veins of sep.* branched.—Native of the Levant, naturalised in some parts of this country. A well known, showy, border flower. Leaves very numerous, mostly radical, an inch wide and a foot or more long. Scape round, thick, naked, smooth, branching, 3f high. Flowers very large, liliaceous, of a tawny red. Style striate. July. †

2. H. FLAVA. *Yellow Day Lily.*—*Lvs.* broad-linear, carinate; *segments* flat, acute; *veins of the sepals* undivided.—Native of Siberia. A foot high. Flowers a bright yellow, much smaller than those of H. fulva. Scape branching. Jl. †

3. H. JAPONICA. *White Day Lily.*—*Lvs.* cordate, ovate, acuminate; *fls.* infundibuliform.—A fine species from Japan. Leaves as large as the hand, very smooth, on long, radical petioles. Flowers large, white, on a scape a foot high. June. †

6. POLYANTHES.

Gr. πολυς, many. ανθος; the flowers of the plant being numerous.

Perianth funnel-form, incurved; filaments inserted into the throat; ovary at the bottom of the tube.—♃

P. TUBEROSA. *Tuberose.*—*Lvs.* linear-lanceolate; *pet.* oblong.—A fine parlor plant, native of Ceylon. Stems bulbous at base with tuberous branches. Scape scaly, 2—3f high, with alternate, large, white, regular flowers of a delicious fragrance which is most powerful at evening. Aug. Sept. †

TRIBE 3. SCILLEÆ.—Bulbous. Flowers usually smaller than in the preceding. Integument of the seed black and brittle.

7. ALLIUM.

Celtic *all*, hot or burning.

Flowers in a dense umbel, with a membranous, 2-leaved spathe; perianth deeply 6-parted, segments mostly spreading, ovate, the 3 inner somewhat smaller; ovary angular; stigma acute; capsule 3-lobed.—*Strong-scented, bulbous plants. Lvs. mostly radical. Umbel on a scape.*

1. A. TRICOCCUM. Ait. *Lance-leaved Garlic.*
Scape terete; *lvs.* lanceolate, oblong, flat, smooth; *umbel* globose; *seed* solitary in each cell of the 3-celled capsule.—♃ A strong-scented plant, common in damp woods, N. H. to Va. and to Ill. Bulb oblong, acuminate. Leaves 5—8 long, an inch or more wide, acute, tapering into a petiole, all withering and disappearing before the opening of the flowers. Scape a foot or more high, bearing a thin, 2-leaved, deciduous spathe at top, with an umbel of 10—12 white flowers. June, July.

2. A. CANADENSE. *Canadian Garlic.*
Scape terete; *lvs.* linear; *umbel* capitate, bulbiferous.—♃ In woods. Leaves radical, ½ as long as the scape, smooth, nearly flat above. Scape 12—18' high, round, smooth, bearing a spathe of 2 ovate, acute bracts at the top, with a head of bulbs and flowers. The bulbs are sessile, each furnished with a bract beneath, and among them are a few whitish flowers on slender pedicels. June.

3. A. VINEALE. *Crow Garlic.*
St. slender, with a few leaves; *cauline lvs.* terete, fistulous, *umbel* bulbiferous; *sta.* exsert; *fil.* alternately tricuspidate, the middle point bearing the anther.—♃ Meadows, Mid. and W. States. Leaves 6—12' long. Scape 1—2f high, bearing a spathe of 2 small bracts at top, and an umbel of flowers with which bulbs are sometimes intermixed. Perianth purple. June, July. §

4 A. CERNUUM. Roth. *Nodding Garlic.*
Scape angular; *lvs.* linear, flat, very long; *umbel* cernuous, *sta.* simple.

Banks of Seneca lake, N. Y., W. to Ohio, *Leck!* and Wis., *Lapham!* Bulb 6—8" diam. Scape mostly 4-angled, smooth, slender, 15—24' high, rec urved at top. Umbel 12—20-flowered. Pedicels 7—8" long. Flowers rose-colored. Ovary 6-toothed, becoming a roundish, 3-seeded capsule. July.

5. A. TRIFLÒRUM. Pursh. *Few-flowered Leek. Mountain Leek.*

Scape naked, terete, shorter than the leaves; *lvs.* lanceolate, veined; *um bel* few-flowered.—In shady woods on the high mountains of Pennsylvania, *Pursh.* May, June.

6. A. SCHÆNOPRÀSUM. *Cives.—Scape* equaling the round, subulate leaves.— (*Gr.* σχοινος, a rush, and πρασον, a leek. The leaves resemble rush-leaves. Ja.

7. A. ASCALONICUM. *Shallot.—Scape* terete; *lvs.* subulate; *umbel* globose · *sta.* tricuspidate.—Native about Ascalon, Palestine. It has a soboliferous bulb small, fistulous leaves, and seldom flowers. July.

8. A. PORRUM. *Leek.—St.* compressed, leafy; *lvs.* sheathing at base; *sta.* tricuspidate.—Native of Switzerland. Root bearing a scaly, cylindrical bulb Stem 2f high, bearing long, linear, alternate, sheathing leaves, and at the top a large umbel of small, white flowers. July.

9. A. SATIVUM. *Common Garlic.—Bulb* compound; *st.* leafy, bulbiferous; *sta.* tricuspidate.—Native of Sicily. The bulb is composed of several smaller ones surrounded by a common membrane, acrid and very strong-scented. Stem ₁'f high. Flowers small, white. Used as seasoning and sometimes in medicine. July.

10. A. PROLIFÈRUM.—*Scape* fistulous, twisted; *lvs.* fistulous; *umbels* bulbifer ous and proliferous; *sta.* tricuspidate, the middle point antheriferous.—A curi ous species, native of the W. Indies. Scape 2—3f high, producing several bulbs among, or instead of, the white flowers. July.

11. A. CÈPA. *Common Onion.—Scape* fistulous, swelling towards the base, longer than the terete, fistulous leaves.—② *Cep,* in the Celtic, signifies a head. Native of Hungary. The root bears a tunicated bulb, compressed, or round, or oblong in figure. The scape, which appears the second year, is 3—4f high, straight, smooth, stout, bearing at top a large, round umbel of greenish-white flowers. Universally cultivated for the kitchen, and its peculiar merits as a pot-herb are, no doubt, well known to our readers. Culture has produced numerous varieties.

8. ORNITHOGÀLUM.

Gr. ορνιθος, a bird, γαλα, milk ; why so called is not obvious.

Perianth deeply 6-parted, spreading above the middle; filaments dilated at the base ; capsule roundish, angular.—*Lvs. radical. Scape naked, racemose or corymbose.*

O. UMBELLÀTUM. *Star-of-Bethlehem.—Fls.* corymbose; *ped.* longer than bracts; *fil.* subulate.—24 From England, but naturalized in many parts of this country. Leaves linear and narrow, emarginate, as long as the scape. Scape near a foot high. Flowers few, in a kind of loose corymb. Petals and sepals white, beautifully marked with a longitudinal green stripe on the outside. May.

9. HYACINTHUS.

H acinthus of Grecian fable, was killed by Zephyrus, and transformed into this flower.

Perianth subglobose or campanulate, regular, 6-cleft: 3 nectariferous pores at the top of the ovary : stamens issuing from the middle of the segments , cells of the capsule about 2-seeded.—*Natives of the Levant.*

H. ORIENTÀLIS.—*Perianth* funnel-form, half 6-cleft, ventricose at the base.—24 The hyacinth is a well known, splendid flower, long prized and cultivated. Leaves thick, linear-lanceolate, 3—5' long. Scape twice as long as the leaves, thick, bearing a dense, thyrsoid raceme of numerous blue flowers. A plant peculiarly adapted to parlor cultivation in bulb glasses.

Other ornamental species sometimes cultivated are H. botryoides, *grape hyacinth*, with globose flowers; H. comosus, *purple grape hyacinth*, with prismatic flowers; and H. racemosus, *hare-bell hyacinth*, with ovoid flowers.

Tribe 4. **ANTHERICEÆ.**—Stem subterraneous, or if developed, erect. Root fasciculate or fibrous. Leaves never coriaceous nor permanent.

10. ASPHODĒLUS.

Gr. a, privative, σφαλλω, to surpass; a flower not surpassed in beauty.

Perianth 6-parted, spreading, with 6 valves, covering the ovary; sta. issuing from the valves.—*Fine garden plants, native of S. Europe.*

1. A. luteus. *King's Spear. Yellow Asphodel.*—St. leafy; lvs. 3-cornered —♃ Native of Sicily. A plant of easy culture and rapid increase. Stem 3f high, thickly invested with 3-cornered, hollow leaves. Flowers yellow, in a long spike, reaching from the top almost to the base of stem. June.

2. A. ramōsus. *Branching Asphodel.*—St. naked, branched; ped. alternate, longer than bract; lvs. ensiform, carinate, smooth.—♃ Native of S. Europe. Not so tall as the preceding, but with larger, white flowers. June.

11. PHALANGIUM. Tourn.

Gr. φαλαγξ, tarantula, a venomous species of spider, whose bite it was supposed to cure.

Perianth 6-parted, petals and sepals similar, spreading; filaments 6. smooth; caps. free from the perianth, ovoid; seeds angular.—♃ *Lvs. flat, linear. Fls. small, white or bluish.*

P. esculentum. Nutt. *Esculent Phalangium. Quamash.*
St. bulbous; lvs. all radical, linear, carinate at base; sta. subexserted; stig. minutely 3-cleft.—Wet prairies, along rivers, lakes, Wis. *Lapham!* Ill. *Jenney!* Ind. *Skinner!* &c. An interesting little plant, usually in thick grass. Bulb 1—1½' diam., resembling a small onion. Scape 1—2f high. Leaves nearly as long as the scape, grass-like, 3" wide, smooth. Raceme short, pedicels longer than the flowers, each with a subulate bract. Petals and sepals pale blue, about 3" long. Anthers oblong, small, yellow. Seeds black. May.

Tribe 5. **CONVALLARINEÆ.**—Stem arising from a horizontal rhizoma or tuber.

12. CONVALLARIA.

Lat. convallis, a valley; the locality of some species.

Perianth 4—6-parted, segments spreading; stamens 4—6, divergent, arising from the base of the segments; berry globose, 2—3-celled.—*Plants somewhat various in habit, with simple stems and alternate leaves. Fls. in terminal racemes or umbels, reddish or greenish-white.*

§ 1. Majanthemum. *Perianth 4-parted. Stamens 4. Berry 2-celled.*

1. C. bifolia. (Smilacina. *Desf.* Styrandra. *Raf.*) *Two-leaved Solomon's Seal.*—Lvs. 2—3, cordate, ovate; fls. in a terminal raceme; lfts. of the perianth spreading.—♃ A small plant, frequent along the edges of woodlands, Can., N. Eng., W. to Wis.! Stem angular, about 6' high. Leaves 2, rarely 3, about 2' long, ½ as wide, ovate, distinctly cordate, sessile, or the lowest on a petiole. Raceme terminal, erect, an inch long, consisting of 12—20 white flowers. Berry small, round, and when mature pale red, speckled with deep red. May.

§ 2. Smilacina. *Perianth 6-parted. Stamens 6. Berry 3-celled.*

2. C. trifoliāta. (Smilacina. *Desf.*) *Three-leaved Solomon's Seal.*
Lvs. 3—4, oval-lanceolate, tapering to both ends, amplexicaul; rac. terminal, simple.—♃ A delicate little species in mountain swamps, Can. N. Eng (rare), W. to Wis.! Stem 3—5' high, pubescent, angular. Leaves 2—3'

long, ⅓ as wide, acuminate, smooth. Flowers 4—6, white, 6-parted, the segments spreading. May.

3. C. STELLÀTA. (Smilacina. *Desf.*) *Star-flowered Solomon's Seal.*

St. erect; *lvs.* numerous, 3-veined, lanceolate, amplexicaul, acute; *fls.* few, in a simple, terminal raceme.—4 Along rivers, Can. and Northern States, W. to the Miss. Stem 10—20' high, round and smooth. Leaves 8—10, smooth, glaucous beneath, 4—6' by 9—12", tapering to the apex. Flowers white, about 8, stellate, rather larger than in the next. May, Jn.

4. C. RACEMÒSA. (Smilacina. *Desf.*) *Clustered Solomon's Seal.*

St. recurved; *lvs.* oval, acuminate, subsessile; *rac.* compound.—4 A larger species than the preceding. Rhizoma thick, sweet to the taste. Stem 1¼—2f high, downy, gracefully recurved at top. Leaves 4—6' long, about ⅓ as wide, contracted into a long acumination, veined, minutely pubescent. Petioles 0—2' long. Flowers very numerous, small, white, on white pedicels, and with white, exserted, tapering filaments, constituting a large, compound, terminal raceme.

§ 3 CLINTONIA. *Perianth subcampanulate, 6-parted. Berry 2-celled.*

5. C. BOREÀLIS. (C. umbellulata. *Michx.* Dracæna. *Ait.* Clintonia. *Raf.*) *Wild Lily of the Valley.*—Scape umbellate; *lvs.* broad-oval-lanceolate; *fls.* cernuous; *berries* blue.—4 Mountainous or hilly woods, Can., N. Eng. to Car., W. to the Miss. Rhizoma creeping to some extent. Leaves 4—7' long, ⅓ as wide, petiolate, radical or nearly so, smooth and glossy, fringed with scattered hairs. Scape erect, round, 8—18' high, bearing at top a beautiful umbel of 3—6 yellowish-green, nodding flowers. Perianth liliaceous, of 6 oblanceolate, erect-spreading segments. Berries of a rich amethystine blue. Jn.

6. C. MAJÀLIS. *Lily of the Valley.*—Scape naked, smooth, semi-cylindric; *lvs.* nearly radical, ovate; *rac.* simple, 1-sided.—4 An elegant, sweet-scented plant, native of woods at the South, and is, or deserves to be, a frequent inhabitant of our gardens. Leaves 2, seldom 3, ovate-elliptical. Scape 6' high, with white flowers depending from its upper half in a single rank. May.

13. POLYGONÀTUM.

Gr. πολυς, many, γονυ, knee; from the many-jointed rhizoma.

Perianth tubular, cylindrical, 6-cleft; stamens inserted near the summit of the tube; berry globose, 3-celled, cells 2-seeded.—*St. simple. Lvs. alternate. Fls. axillary.*

P. MULTIFLÒRUM. Desf. (P. latifolium, angustifolium, biflorum, pubescens and canaliculatum, of *Ph.*, &c.)—*St* recurved, smooth; *lvs.* distichous, lanceolate, amplexicaul, smooth above; *peduncles* axillary, 1—4-flowered. —4 In woods, free States and Can. Stem 1—3f high, most recurved in the tallest plants. Leaves more or less clasping at base, or only sessile in the smallest plants, 2½—6' by 1—2½', veined, smooth and glossy above, paler and generally pubescent beneath. Peduncles filiform, branching, scarcely a filth as long as the leaves. Flowers 5—8" long, pendulous, greenish, sub-cylindric. Berries dark blue or blackish when ripe. Jn.

α. *Lvs.* very amplexicaul, smooth both sides, distinctly veined; *lower peduncles* 4-flowered.—Plant 2—3f high. In rich, damp soils.

β. *pubescens. Lvs.* pubescent beneath, slightly clasping; *st.* 1—2f high.— This variety is most common in New England.

γ. *biflora. Lvs.* smooth both sides, ⅓ as wide as long, sessile; *fls.* greenish-white, 4—5" long; *st.* round, 1—1½f high.

δ. *canaliculata. St.* channeled on the upper side.

ε. *latifolia. Lvs.* ovate, acuminate, sessile, glabrous.—Stem angled, 4—5f high—Middle States.

14. UVULARIA.

Perianth deeply 6-parted; segments linear-oblong, acute, erect, with a nectariferous cavity at the base of each; filaments very short;

anthers linear, half as long as the petals; style trifid; capsule 3-celled, many-seeded, seeds with an aril.—*Lvs. alternate. Fls. solitary, terminal and axillary.*

1. U. sessilifolia. *Bellwort. Wild Oats.*

Lvs. sessile, lance-oval, glaucous beneath; *caps.* stiped, ovate.—2l Can. and U. S. A common species, found in woods and in grass lands. Stem smooth, slender, 6—10' high, dividing at the top into 2 branches, one bearing leaves only, the other, leaves and a flower. Leaves smooth and delicate, dark green above, paler beneath, 1—1½' long. The flower is cylindric, near an inch long, yellowish-white, of 6, long, linear petals. May.

2. U. perfoliàta. (U. flava. *Smith.*) *Mealy Bellwort.*

Lvs. perfoliate, elliptical, subacute; *perianth* subcampanulate, tuberculate-scabrous within; *anths.* cuspidate; *caps.* truncate.—2l Can. and U. S. A handsome, smooth plant, in woods. Stem 10—14' high, passing through the perfoliate leaves near their bases, and dividing into 2 branches at top. Leaves 1—3' by ½—1', rounded at the base, acute at apex. Flower pale yellow, pendulous from the end of one of the branches. Segments linear-lanceolate, 1½' long, twisted, covered within with shining grains. Anthers ⅔' long. May.

3. U. grandiflòra. *Large-flowered Bellwort.*

Lvs. perfoliate, elliptic-oblong, acute; *fl.* terminal, solitary, pendulous; *segments* acuminate, smooth within and without; *anth.* obtuse.—2l Can. and U. S. Larger than either of the foregoing. In woods. Stem 12—15 inches high, passing through the perfoliate leaves near their bases, dividing into 2 branches at top, one of which bears the large, yellow, pendulous flower. Leaves almost acuminate, rounded at base. Anthers ⅔' long. May.—Readily distinguished by the smooth petals.

4. U. lanuginòsa. Pers. (Streptotus. *Michx.*)

Lvs. ovate, acuminate, sessile, without dots, ciliate, the lower amplexicaul, pubescent beneath when young; *ped.* terminal, pubescent; *fls.* in pairs; *perianth* acute at base, *lfts.* linear-lanceolate; *sty.* glabrous.—2l Mountains, Penn. to Car. and L. Winipeg to Oreg. Leaves with an abrupt and long acumination. Flowers greenish. Berry with the cells by abortion 1—2-seeded. May.

15. STREPTÓPUS. Michx.

Gr. στρεφω, to turn, πους, a foot; a twisted footstalk or peduncle.

Perianth 6-parted, campanulate; segments with a nectariferous pore at the base of each; anthers longer than the filaments; stigma very short; berry roundish, 3-celled; seeds few, hilum without an aril.—*St. branched. Fls. axillary, solitary, generally with the peduncle distorted.*

1. S. ròseus. Michx. (Uvularia. *Linn.*) *Rose Twist-foot.*

Smooth; *lvs.* oblong-ovate, clasping, margin serrulate-ciliate, under surface green like the upper; *pedicels* short, generally distorted in the middle *segments* spreading at apex; *anth.* short, 2-horned; *stig.* trifid.—2l Can. to Car and Tenn. A common species, native of woods. Stem a foot or more high, round, dichotomously branching. Leaves 2—4' long, ½ as wide, ending in a slender point, smooth, but conspicuously edged with minute, rough hairs. Flowers reddish, spotted, suspended beneath the branches, one under each leaf. Jn.

2. S. amplexifolius. DC. (S. distortus. *Michx.* Uvularia. *Linn.*)

Smooth; *lvs.* oblong-ovate, clasping, smooth and entire on the margin, glaucous beneath; *pedicels* solitary, geniculate and distorted in the middle; *sep.* long-acuminate, reflexed; *an'h.* very acute, entire; *stig.* truncate.—2l Can. and Mid. States. Native of woods. Stem round, dichotomous, 2f high. Leaves 2—3' long, ½ as wide, very smooth. Peduncles opposite the leaf, twisted and bent downwards each with a bell-form, drooping flower gibbous at base, of a pale straw-color. Anthers sagittate, attenuate at the apex into a long, subulate point. Fruit oblong, red, many-seeded. June.

TRIBE 6. **ASPARAGEÆ.**—Stem usually fully developed, or if not, the leaves are coriaceous and permanent.

16. ASPARAGUS.

Gr. σπαρασσω, to tear; some of the species are armed with strong prickles.

Perianth 6-parted, erect; ovary turbinate; stamens erect; style very short; stigmas 3; berry 3-celled, cells 2-seeded.

A. OFFICINÀLIS. *Asparagus.*
St. herbaceous, unarmed, very branching, erect; *lvs.* setaceous, flexible, fasciculate.—♃ Native of England, and other parts of Europe, naturalized on rocky shores. Stem 2—4f high. Leaves filiform, ½—1¼′ long, pale pea-green. Flowers axillary, solitary or in pairs. Berries globose, red. It is one of the oldest and most delicate of culinary vegetables, was no less praised in ancient Rome, by Pliny, Cato and other writers, than a the present day. Diuretic. Jl.

ORDER CLIV. PONTEDERACEÆ.—PONTEDERADS.

Plants aquatic or marsh. *Lvs.* sheathing, parallel-veined, mostly cordate or dilated at base.
Inflorescence various, often spathaceous.
Perianth tubular, colored, 6-parted, often irregular, circinate in æstivation.
Sta. 3 or 6, unequal, perigynous.
Ova. free or sometimes adherent to the perianth at base, 3-cel ed. *Style* 1. *Stig.* simple.
Fr.—Capsule 3 (sometimes 1)-celled, 3-valved, with loculicide dehiscence.
Seeds numerous (sometimes solitary), attached to a central axis. Albumen farinaceous.

Genera 5, species 20, found exclusively in America, E. Indies and Tropical Africa. They are of no known use.

Conspectus of the Genera.

Flowers { equal. { solitary.	*Leptanthus.* 2
{ 2—4 together in a spathe.	*Heteranthera.* 3
{ unequal, in a terminal spike.	*Pontederia.* 1

1. PONTEDERIA.

In honor of Julius Pontedera, a botanic author and professor, of Padua, about 1720.

Perianth bilabiate, tubular at base, under side of the tube perforated with 3 longitudinal clefts, the lower part persistent; stamens unequally inserted, 3 near the base and 3 at the summit of the tube; utricle 1-seeded.—*Fls. blue, mostly spicate.*

P. CORDÀTA. *Pickerel-weed.*
Lvs. subradical, cordate-oblong; *fls.* spiked.—♃ Can. and U. S. A fine conspicuous plant, native of the borders of muddy lakes, &c., growing in patches extending from the shores to deep water. Stem thick, round, erect, arising 1—2f above the water, bearing a single leaf Leaves 4—7′ by 1½—3′, very smooth and glossy, almost sagittate, with veins beautifully arranged to conform to the margin. Flowers in a spike, arising above the spathe, very irregular. Perianth 2-lipped, each lip 3-cleft, always blue, appearing in July.
β. angustifolia. Torr. *Lvs.* narrow, truncate and subcordate at base.

2. LEPTANTHUS. Michx.

Gr. λεπτος, slender, ανθος; in reference to the long tube of the perianth.

Spathe 1-flowered; tube of the perianth very long and slender, limb 6-parted, equal; anthers of 2 forms; capsule 1-celled, many-seeded.—*Lvs. alternate, sheathing at base.*

L. GRAMÍNEA. Vahl. (Schollera graminea. *Schreb.*)
St. floating, rooting at the lower joints; *lvs.* linear.—A grass-like aquatic, in flowing water, Northern States. Stem slender, dichotomous, 1—2f long. Leaves 3—6′ long, 1—2″ wide, obtuse at apex, slightly sheathing at base. Flower solitary, issuing from a short (1′) spathe. Tube 1½′ long, limb in 6, linear-lanceolate segments, yellow. Stamens 3 (4, *authors*); filaments broad, one of them abortive, the other 2 with linear anthers longer than the thick style. Jl. Aug.

47*

3. HETERANTHÉRA. R. & P.

Gr. ἱτεψs, otherwise, ανηρ; the anthers being dissimilar in the same flower.

Spathe several-flowered; tube of the perianth long and slender, limb 6-parted, equal; stamens 3; anthers of 2 forms; capsule 3-celled, many-seeded; dissepiment contrary.

H. reniformis. R. & P. (Leptanthus. *Michx.*)
St. prostrate or floating; *lvs.* suborbicular, reniform or auriculate at base; *spathe* acuminate, few-flowered.—On muddy or inundated banks, Mid. and W. States. Stem 4′ to a foot or more in length. Leaves ⅓′ by ⅓′, on petioles 1—2′ long, with a broad sinus at base and a short, abrupt acumination. Spathe closely enveloping the 2 or 3 very evanescent, white flowers. Tube of the perianth ½′ long, limb in 6 oblong segments. Filaments inserted at the orifice, 2 of 'he anthers small, round, yellow, the other oblong, greenish. Jl. Aug.

Order CLV. MELANTHACEÆ.—Melanths.

Herbs perennial, with bulbs, rhizomas, corms or fasciculated roots.
St simple, often scapiform. *Lvs.* parallel-veined. *Fls.* perfect, or (by abortion) polygamous.
Perianth regular, in 2 series, each of 2 segments which are distinct or united at base, generally involute *Sta.* 6, with extrorse anthers. [in æstivation.
Ova. 3-celled, 9—many-ovuled. *Styles* distinct or 0. *Stigmas* undivided.
Fr.—Capsule or berry 3-celled, generally with septicidal dehiscence.
Seeds with a membranous testa, and dense, fleshy albumen.

Genera 30, species 130, rather generally diffused in northern countries.

Properties.—The order is generally pervaded by drastic, narcotic and poisonous qualities, most powerful in Veratrum and Colchicum. The corms and seeds of the latter are the most important medicinal products of the order. Their virtue is due to an *alkaline* principle called *veratria*, which is found in this genus, as well as in most of the others.

Conspectus of the Genera.

		Sepals and petals sessile.	*Zigadenus.*	1
with 2 glands at base.	Sepals and petals unguiculate.	*Melanthium.*	2
		Perianth calyculate.	*Tofieldia.*	7
		Leaves linear. Perianth naked.	*Amianthium.*	4
	Flowers perfect	Leaves setaceous, dry.	*Xerophyllum.*	5
Sepals and	or diœcious.	Leaves dilated.	*Helonias.*	6
petals each without glands.	Flowers polygamous. Racemes compound.	*Veratrum.*	3

1. ZIGADÉNUS. Michx.

Gr. ζευγος, a pair, αδην, a gland; alluding to the glands of the segment.

Perianth deeply 6-parted, spreading, colored, each segment with 2 glands above its contracted base; stamens inserted in contact with the ovary; capsule membranaceous, 3 celled, many-seeded.

1. Z. glaberrimus. Michx. *Zigadene.*
Rt. bulbous; *st.* leafy; *lvs.* linear, channeled, recurved; *bracts* ovate, acuminate; *segments of the perianth* acuminate.—Wet meadows, N. York. Found near Rochester. *Eaton.* Southern States. Stem 2—3f high. Lower leaves about 10′ long; upper ones gradually diminishing, all concave and spreading. Panicle terminal, loose, consisting of several greenish-white flowers. Sepals ovate-lanceolate, free from the stamens, with the 2 glands at the base of each distinct and conspicuous. June.

2. Z. glaucus. (Z. cloranthus. *Rich.* Melanthium. *Nutt.*)
St. bulbous, nearly naked; *lvs.* shorter than the stem, linear, rather obtuse; *rac.* subsimple; *bracts* lanceolate, shorter than the pedicels; *sep.* and *pet.* oval or obovate, obtuse, each with an obcordate gland.—Sandy shores, Can. to Ark., Niagara, Lake Erie, *Nuttall.* Stem 10—15′ high. Leaves glaucous, upper gradually reduced to bracts. Raceme subsimple, sometimes a little compound at base. Flowers few (10—20), greenish-white, on pedicels 1′ long, the segments with the 2 glands united. Capsule oblong-ovoid, carpels divergent at apex, 6—8-seeded. July, Aug.

2. MELANTHIUM.

Gr. μελας, black, ανθος; alluding to the dark color of the flower.

Flowers monœcious-polygamous; perianth rotate, 6-parted, seg ments unguiculate, with 2 glands at base, the claws bearing the stamens; ovary often abortive; capsule exserted, subovoid, summit trifid and tipped with the 3 persistent styles; seeds margined.—*St. erect, puberulent above. Lvs. alternate, narrow. Panicle terminal.*

1. M. VIRGINICUM. (Leimanthium. *Willd.* Veratrum. *Ait.* Helonias.)
Lvs. linear-lanceolate; *panicle* pyramidal; *segments of the perianth* suborticular, hastate or auriculate at base.—Native of wet meadows and margins of swamps, N. Y. to Flor. Stem 3—4f high, leafy. Leaves about a foot long, and an inch wide, sessile on a contracted and subclasping base. Flowers greenish-yellow, becoming brown, on short pedicels, arranged in simple, alternate racemes, and together constituting a pyramidal panicle 10—15' in length. Lower flowers generally sterile. July, Aug.

2. M. HYBRIDUM. Walt. (Leimanthium. *Roem. & Sch.*)
Lvs. long-linear-lanceolate, upper ones few and short; *panicle* long, of simple racemes, pedicels filiform, much longer than the flowers; *segments of the perianth* narrowly unguiculate, roundish-rhomboidal; *glands* connivent; *claws* channeled, stameniferous below the middle.—Woods, Penn. to Ga.! Stem 3— 4f high, somewhat leafy. Leaves varying from lanceolate-linear to lanceolate, the lowest contracted to the base or subpetiolate, shorter than the stem. Perianth very open, yellowish-green, segments acuminate, the long claws adhering to and involving the filaments. June, July.

β. *robustior.* Gray. *Lower lvs.* lanceolate-oval; *lower branches* paniculate, compound.

3. VERÁTRUM.

Lat. *vere, atrum,* truly black: alluding to the dark color of the flowers or root.

Flowers by abortion ♂ ♀ ♀; segments of the perianth united at base, petaloid, spreading, sessile and without glands; sta. 6, shorter than the perianth and inserted on its base; ovaries 3, united at base, often abortive; styles short; capsule 3-lobed, 3-partible, ∞-seeded. —*Lvs. alternate, broad and plicate, or narrow and grass-like. Fls. paniculate.*

1. V. VIRIDE. (V. album. *Michx.*) *Poir. White Hellebore.*
Lvs. broad-oval, acuminate; *panicle* compound, racemose; *bracts* oblong-lanceolate, bracteoles longer than the downy pedicels.—Can. to Ga.—A large-leaved, coarse-looking plant, of our meadows and swamps. Root large, fleshy with numerous long fibres. Stem 2—4f high, striate and pubescent. Leaves strongly veined and plaited, the lowest near a foot long and half as wide, sheathing at the base. Flowers numerous, green, in many axillary (or bracted) racemes, which together form a very large, pyramidal, terminal panicle. July. The root is emetic and stimulant, but poisonous, and should be used with caution. When powdered it causes violent sneezing.

2. V. WOODII. Robbins. (Nov. sp.) *Indiana Veratrum.*
Lvs. mostly radical, lanceolate and linear-lanceolate, glabrous, veined and plicate, acute, tapering to a long, winged, sheathing petiole; *st.* or *scape* terete, tall, erect, with remote, lance-linear bracts; *panicle* simple, slender, pyramidal, many-flowered; *fls.* ♂ ♀, subsessile; *segments* of perianth oblanceolate, dark brownish-purple within.—Woods, Linton, Green Co., Ia.! Root fasciculate. Leaves 10—16' long (including the 4—8' petiole), 2—4' wide. Bracts 1—3' long. Scape 3—6f high, paniculate ⅓ its length. Flowers ⅓' diam., nearly black, with red stamens, upper and lower sterile. Ovary oblong, crowned with 3 spreading styles half its length. Seeds compressed, winged with the broad, loose, membranous testa. July.

3. V. ANGUSTIFOLIUM. Pursh. *Grass-leaved Veratrum.*
Lvs. narrowly linear, flat, very long, lowest obtuse, upper ones diminishing to subulate bracts; *fls.* in a slender panicle of racemes, those of the terminal raceme (except a few of the highest) perfect and fertile, those of the lateral racemes mostly sterile; *segments* narrowly lanceolate, subulate, acuminate.—A

very slender, grass-like species, in woods, Western States. Stem 3′ high, solid erect, with a virgate, thin panicle of greenish-white flowers. Leaves 1—2f by 2—3″, half-clasping. Panicle 1½f long, the lateral racemes 1—3′, the terminal one much longer; pedicels shorter than the flowers, each with a very minute bract. June, July.

4. AMIANTHIUM. A. Gr.

Gr. αμιαντος, pure, immaculate; ανθος; alluding to the white flowers.

Flowers ☿; perianth segments scarcely united at base, petaloid, spreading, sessile and without glands; stam. 6, inserted with the segments; anthers reniform; ovaries 3, more or less united; caps. 3-lobed, 3-partible; carpels follicular, 1—4-seeded; testa of the seeds loose, at length fleshy.—*Herbs with scapiform stems, grass-like leaves and numerous white flowers.*

A. MUSCITOXICUM. Gray. (Melanthium. *Walt.* Helonias erythrosperma. *Michx.*) *Fly-poison.*—*St.* bulbous; *lvs.* flat, lower broad-linear, obtuse, upper reduced to bracts; *rac.* simple; *segments* oblong, obtuse; *pedicels* filiform; *carpels* distinct above; *sty.* divergent; seeds ovoid, red.—Shady swamps, N. J., Penn. and Southern States. Stem 1—2f high. Leaves mostly radical, about 1f long. Raceme 3—9′ long, dense-flowered, pedicels 6—9″ long. Perianth and stamens white, the latter rather the longest. Carpels united only at base, the summits horn-like and diverging. Seeds rather large, scarlet-red when ripe. June, July.

2. A. LEIMANTHÖIDES. Gray. *Rt.* fibrous; *lvs.* linear, flat; *panicle* simple, terminal raceme elongated; *segments* of the perianth broad-oval, longer than the linear styles; *sds.* winged at the apex, lanceolate, compressed.—N. J. to La. Stem roundish, 2—4f high, the lower leaves about half as long, pale green, acute. Flowers white, on filiform pedicels, finally recurved. Segments of the perianth obtuse, a little shorter than the capillary filaments. July.

5. XEROPHYLLUM.

Gr. ξηρος, dry. φυλλον, leaf.

Flowers ☿; leaflets of the perianth oval, spreading, petaloid, sessile and without glands; stam. 6; filaments dilated and contiguous at base; ovary subglobose; styles 3, linear, revolute; caps. subglobose, 3-lobed, 3-celled, cells 2-seeded.—*Herbs with numerous dry, setaceous leaves. Racemes simple, with white, showy flowers.*

X. ASPHODELÖIDES. Nutt. (X. setifolium. *Michx.* Helonias asph. *Linn.*) *Lvs.* radical and cauline, rigid, diminishing above; *pedicels* with 2 alternate bracteoles, bractless at base; *filaments* at length equaling the segments of the perianth.—Sandy plains, N. J. to Car. Stem 3—5f high, very leafy. Radical leaves 1f long, very narrow, crowded and cæspitose. Flowers in a long, terminal raceme, numerous, small. Sepals and petals obtuse, the latter a little longer. June.

6. HELONIAS.

Gr. ελος, a marsh; where some species grow.

Perianth 6-parted, spreading, petaloid, the segments sessile, and without glands; styles 3, distinct; capsule 3-celled, 3-horned; cells many-seeded.—*Lvs. mostly radical, narrow, often gramineous, sheathing at base. Fls. in a terminal, simple raceme.*

1. H. BULLÀTA. (H. latifolia. *Ph.*) *Lvs.* crowded, mostly radical, linear-spatulate, mucronate; *scape* simple, hollow, with few remote bracts, or naked; *rac.* spicate, ovoid-cylindric, dense; *suth.* blue.—N. J., Penn. to Va. Scape 10—18′ high, rather thick and fleshy. Leaves about as long as the scape, 1—1½′ wide. Racemes short. Pedicels as long as the flowers, colored. Flowers purple, segments obtuse. May.

2. H. DIOICA. Ph. (Veratrum luteum. *Linn.*) *Unicorn Root.*

St. leafy; *lvs.* lanceolate, radical ones oblanceolate; *rac.* spiked, nodding, diœcious; *ped.* short, without bracts; *sta.* exserted; *segments* linear.—In low grounds, Can. to Ga. and La. Root premorse. Stem or scape 12—30' high, furrowed. Radical leaves 4—8' by ½—1', in a sort of whorl at the base of the scape. Flowers small, very numerous, greenish-white, in long, terminal, spicate racemes which are more slender and weak on ne barren plants. Ovaries as long as the linear petals, subtriangular. Capsule 3-furrowed, oblong, tapering to the base, opening at the top. The fertile plants are taller, more erect, but with fewer flowers. June.

7. TOFIELDIA. Hudson.

In honor of Mr. Tofield, a Scotch gentleman, residing near Doncaster.

Flowers ♀, calyculate, with 3 remotish, united bracts; lfts. of the perianth petaloid, spreading, sessile and without bracts; sta. 6; anth. roundish-cordate, introrse; ovaries 3, united; styles distinct, short; caps. 3-lobed, 3-partible; capsule 00-seeded.—*Lvs. equitant, subradical. Scape not bulbous. Fls. spicate or racemose.*

T. GLUTINOSA. Nutt.

St. leafy below, glandular-scabrous, simple; *lvs.* shorter than the stem, linear-ensiform, glabrous, obtuse; *rac.* oblong, few-flowered, close, composed of 3-flowered, alternate fascicles; *caps.* longer than the perianth.—Woods, Ohio, *Sullivant!* to Wis. *Lapham!* N. to Arctic Am. A plant remarkable for its glutinous-glandular stem. Stem slender, scape-like, 1—1½f high, dotted with its dark-colored glands. Leaves 3—6' by 3—6", conduplicate. Spicate raceme 1—1½' long, 9—18-flowered. Pedicels nearly as long as the flowers. Involucre truncate, 3-toothed, a little below the perianth. Petals and sepals subequal, oblanceolate, less than 2" long. Capsule of 3, half-united, inflated carpels, twice longer than the perianth.

ORDER CLVI. JUNCACEÆ.—RUSHES.

Plants herbaceous, generally grass-like, often leafless, with small, dry, green flowers.
Lvs. fistular, or flat and channeled, with veins parallel. *Inflorescence* cymous, capitate or fascicled.
Perianth more or less glume-like, regular, 6-leaved, in 2 series (sepals and petals)
Sta. 6, rarely 3, hypogynous. *Anth.* 2-celled.
Ova. 3-carpeled, 3 (or by the dissepiments not reaching the centre 1)-celled.
Styles united into 1. *Stigmas* 3
Fr.—Capsule 3-valved, with the dissepiments from the middle of the valves.
Seeds few or many, with a fleshy albumen.

Genera 12, species 200, chiefly natives of the cool parts of the earth. Properties unimportant.

Conspectus of the Genera.

Perianth { green. {	Capsule mostly 3-celled. Seeds numerous.	.	.	.	*Juncus.* 1
	Capsule 1-celled. Seeds 3, fixed to the bottom of cell.	.	.	.	*Luzula.* 2
colored, yellow.	. . . .	.	.	.	*Narthecium.* 3

1. JUNCUS.

Lat. *jungo*, to join; because ropes were anciently made of these plants.

Perianth persistent; stamens 6; capsule mostly 3-celled; seeds numerous, attached to the inner edge of the dissepiments.

* *Leaves none. Cymes apparently lateral.*

1. J. BALTICUS. Willd. *Baltic Rush.*

Rhizoma creeping, prostrate, rooting; *scapes* numerous, sheathed at base, opaque, terete, rigid, slender, pungently acute; *panicle* small, short, lateral; *perianth segments* subequal, ovate-lanceolate, very acute, equaling the elliptical, mucronate capsule.—Sandy shores, Milwaukie, Wis., *Lapham!* N. to the Saskatchawan and Labrador. Scape leafless, 12—18' high, hard, tough, closely arranged along the scaly rootstock, the sheaths 3'—3' long. Panicle 2—3' below the apex of the scape, 1' long. Flowers 20—40, reddish-brown.

2. J. ACUTUS.

Cæspitose; *scapes* numerous, tall, rigid, terete, sheathed at base; *panicle* with an involucre o: two unequal, pungent bracts; *perianth segments* lanceolate, acute, half as long as the roundish-obovate, mucronate capsule. Sandy sea-coasts, N. J, to Ga. Scapes fascicled on the rhizomas, forming dense tufts, 2—2½ high. Panicle 2—3′ long. Involucre with one of the bracts longer, the other shorter than the panicle. Perianth brown, the 3 sepals longest, acuminate, reflexed at apex. July.

3. J. EFFÙSUS. *Soft Rush. Bull-rush.*

Scape straight, not rigid; *panicle* lateral, loose, decompound, *caps.* obovate, obtuse.—Very common in ditches and moist lands, forming tufts, Can. and U. S. Scape solid, with a spongy pith, soft, striate, 2—3f high, bearing a loose, spreading panicle, which protrudes from a fissure opening in the side of the stem about half way up. Flowers small, green, numerous, with 3 white anthers and yellowish seeds. June, July.

4. J. SETÀCEUS. Rostkow. (J. filiformis *Michx.*) *Bristly Rush.*

Scape filiform, striate; *umbel* lateral, compound, few-flowered; *ped.* compressed, several-flowered; *perianth segments* very acute.—Swamps, Can. and U. S. A very slender species, growing in tufts about 2f high. Scapes sheathed at base. Panicle small, 20—30-flowered, bursting from the side of the scape some distance below the summit. July.

5. J. FILIFORMIS. Linn. (not *Michx.*) *Thread Rush*

Creeping, leafless; *scape* slender, filiform, minutely striate, flaccid; *panicle* subsimple, lateral, near the middle of the scape; *sep.* pale, nearly equal, lanceolate, a little longer than the pale, shining, obovate, mucronate capsule. White Hills, N. H., *Green.* Scape a few inches in length.

* * *Leaves nodose-articulate, subterete.*

6. J. MILITÀRIS. Bw. *Bayonet Rush.*

Lf. solitary, jointed, longer than the stem; *panicle* terminal, proliferous, *hds.* about 5-flowered; *st.* thick, round, smooth, 2, 3 or 4f high.—Ponds, N. Eng. Leaf jointed, cylindrical, loosely cellular within, 2—3f long, inserted below the middle of the stem, but rising above its summit. Panicle erect, terminal, composed of several pedunculate heads, each with 4—6 sessile flowers.

7. J. NODÒSUS. (J. polycephalus. *Michx.* J. echinatus. *Muhl.*)

St. leaves subcompressed; *panicle* terminal, decompound; *hds.* globose, dense, 10—15-flowered; *sep.* subequal, lanceolate, rigid, very acute, about equaling the oblong-lanceolate, acute, triquetrous, shining capsule; *segments* 3—6. In boggy meadows, U. S. and Can. Stem 1½—2f high. Leaves thick, jointed by internal, transverse partitions. Heads resembling small burrs, some sessile, others pedunculate. Leaflets of the perianth produced into a short cusp or awn. Aug. Sept.

α. major. *St.* and *lvs.* thick, the latter longer, compressed; *hds.* few, 6″ diam.
β. altior. *St.* (2f high) and *lvs.* terete, very slender; *hds.* 5—9, 4″ diam.
γ. minor. *Lvs.* almost filiform; *hds.* 1—3, as large as in *β.*

8. J. ACUMINÀTUS. Michx. (J. acutiflorus. *Hook?*)

St. erect; *lvs.* terete; *panicle* terminal, compound; *hds.* 3—6-flowered, both pedunculate and sessile; *leaflets of the perianth* linear-lanceolate, mucronate, shorter than the acute capsule.—Very common in boggy meadows, U. S. Stem 12—18′ high, tough and wiry. Leaves few and short, with knot-like joints. Panicle erect. Aug.

9. J. SUBVERTICILLÀTUS. Willd. (J. fluitans. *Michx.*)

St. few-leaved, compressed; *lvs.* compressed, fistulous, articulate; *panicle* subcorymbose, elongated; *fls.* in dense, capitate fascicles; *hds.* many-flowered, pedunculate or sessile; *perianth* shorter than the triquetrous, acuminate capsules; *sep.* linear-lanceolate, keeled, striate, cuspidate. Swamps and shores of ponds, Penn., *Dr. Darlington,* to Wis., *Lapham!* Stem 18′—2f high, slender, Leaves much shorter. Panicle 4—8′ in length, the branches subverticillate, diverging, very unequal. Sepals with scarious, white margins.

10. J. CONRÀDI. Tuckerman, in Torr., N. Y. State Fl. II. p. 328, *inedit.*

St. leafy; *vs.* few, terete, subfiliform, obscurely articulate, shorter than the stem; *fls.* single, scattered, central and unilateral on the slender branches of the terminal, di-*tricho.omous* panicle; *perianth segments* lanceolate, margins scarious, rather shorter than the acuminate capsule. R. l. *Olney!* to N. Y., *Torrey.* Root fibrous. Stems erect, 6—9' high, wiry. Bracts much shorter than the rather diffuse, thin panicle.

*** *Leaves not articulate, radical.*

11. J. GREENII. Oakes & Tuckerman. (J. squarrosus. *Muhl.?*)
Scape tall, subterete, striate; *lvs.* filiform-setaceous, subterete, scarcely channeled, shorter than the scape, with sheathing bases; *panicle* subumbellate. 5-rayed; *bracts* setaceous, one of them very long; *fls.* single, approximate; *sep.* and *pet.* ovate, acute, twice shorter than the triangular-acute, shining capsule.— Wet grounds, R. Isl. and Mass., *Dr. Truman Rickard!* The handsomest of the rushes, about 2f high, rigid, strict. Leaves all radical, 1f or more high. Panicle 2—3' long, one of the bracts twice longer, the other twice shorter. Capsule 2'' long, of a glossy mahogany color. Seeds very minute, linear-oblong

12. J. TENUIS. Willd. *Slender Rush.*
St. scape-like, slender, erect; *lvs.* subradical, linear-setaceous, shorter than the stem; *bracts* 2—3, much longer than the panicle; *fls.* single, approximate, subsessile; *perianth segments* acuminate, longer than the subglobose-triangular capsule.—A very common rush, about foot-paths and roadsides, and in fields and meadows, U. S. and Can. Stems wiry, 6—24' high. Leaves very narrow, 3—8' long. Panicle subfasciculate, 5—10-flowered, varying to subumbellate and 20—30-flowered, the rays very unequal. June, July.

**** *Leaves flattish, channeled, cauline and radical.* ·

13. J. BUFONIUS. *Toad Rush.*
St. dichotomous above; *lvs.* grooved, subsetaceous; *fls.* oblong, subsolitary, sessile, unilateral.—A small, cæspitose species, common in wet grounds, ditches, &c. Stems numerous, 3—8' high, with a large, few-flowered panicle at top. Leaves few, 2—3' long. Perianth segments twice as long as the ovary. July, Aug.

14. J. BULBOSUS. (J. uliginosus. *Sibthorp.*)
St. leafy, very slender, compressed; *lvs.* mostly radical, linear-setaceous, shorter than the stem; *panicle* small, few-flowered, subtrichotomous, longer than the bracts; *fls.* about in 3s; *sep.* and *pet.* equal, acute, incurved, rather shorter than the subglobose, obtuse capsule.—A common rush, in salt marshes, N. J. to the Arctic Sea, usually with dark green foliage and brown capsules. Stems tufted, erect or decumbent and stoloniferous, about 1f in length, tough and wiry. Leaves 3—8' long. Bracts 6—12''. Flowers 12 or more, at length brown or blackish. July, Aug.—It makes good hay.

15. J. TRIFIDUS. *Three-leaved Rush.*
St. sheathed at base; *lf.* solitary, linear-setaceous near the top; *sheaths* ciliate; *bracts* foliaceous, long, grooved.—Heads about 3-flowered, terminal. White Hills, N. H., *Bw.* Stems crowded, thread-like, 1f high. Radical leaves 1—2, very short. The cauline leaf resembles the 2 bracts, apparently forming with them a foliaceous, 3-bracted involucre. July.

16 J. MARGINATUS. Rostkow.
St. compressed; *lvs.* flat, smooth, gramineous; *panicle* corymbose, simple, proliferous; *fls.* in capitate clusters, triandrous; *perianth* about as long as the obtuse capsule, the sepals and bracts somewhat awned.—In low grounds, Mass.! N. Y. to Car. Stems 1—3f high. Radical leaves numerous, sheathing; cauline 1 or 2. Panicle consisting of several globose, 3—6-flowered heads both pedunculate and sessile, longer than the erect bracts at base. Sepals edged with dark purple, unequal. Aug.

2. LUZULA. DC.

Italian *lucciola,* a glow-worm; from the dew glistening upon its flowers.

Perianth persistent, bibracteate at base; stamens 6; capsule 1-celled, 3-seeded; seeds fixed to the bottom.—Stem jointed, leafy. *Lvs. flat, grass-like, generally pilose.* Fls. terminal.

1. L. CAMPESTRIS. Willd. (Juncus. *Linn.*) *Field Rush.*
Lvs. hairy ; *spikes* terminal, with or without peduncles; *lfts. of the perianth*
lanceolate, acuminate, awned, longer than the obtuse capsule.—In meadows,
U. S. and Can. Stem simple, straight, 3—12' high, according to the moisture.
Leaves grass-like, 2—6' long, very hairy at the margins. Heads in a sort of
umbel, with an involucre of 2 or 3 short, unequal leaves. Perianth dark brown.
An early species, flowering in May.

2. L. PILÓSA. Willd. (Juncus. *Linn.*) *Hairy Wood Rush.*
Lvs. pilose; *panicle* cymose, spreading ; *fls.* solitary ; *caps.* obtuse.—Common in woods and groves, Free States. Stem 4—10' high. Radical leaves
numerous, 2—4' long, linear-lanceolate, veined, fringed with long, white hairs.
Panicle 8—12-flowered, with a leafy bract. Pedicels 5—10'' long, finally deflexed. Perianth brown, with 2 green bracteoles. May.

3. L. MELANOCARPA. Desv. (Juncus. *Michx.*)
St. elongated; lvs. sublanceolate, glabrous; *corymb* decompound; *ped.*
elongated, the branches with 3—5 pedicellate flowers; *sep.* ovate, acuminate,
longer than the oval-triangular, obtuse-mucronate capsule.—Native of the
White Hills, N. H., *Bw.* Stem 12—18' high. Radical leaves 8—10' by 3—5'',
those of the stem much shorter, all very smooth. Panicle large, nodding, many-
flowered. Capsule black. June.

4. L. SPICÀTA. DC. (Juncus. *Willd.*)
Lvs. linear, hairy at the base; *spike* cernuous, compound ; *sep.* acuminate-
awned, about equal in length to the subglobose capsule.—White Hills, N. H.,
Boott. Stem 8—10' high, slender, simple. Leaves 2—3' long, a line wide,
smooth except at the base. Spike an inch long. Aug.

3. NARTHECIUM. Moehr.

Gr. ναρθηξ, a rod or wand ; in allusion to the slender inflorescence.

Perianth 6-parted, colored, spreading, persistent; stam. 6; filaments hairy; caps. prismatic, 3-celled ; seeds ∞, ovate-oblong, appendaged at each extremity.—♃ *Root fibrous.* *Lvs. ensiform. Scape
nearly naked. Fls. yellow.*

N. AMERICÀNUM. Ker. (Phalangium ossifragum. *Muhl.*)
Lvs. radical, striate, narrow-ensiform ; *scape* simple, bracted ; *rac.* lax, interrupted; *pedicels* with a bract at base, and a setaceous bracteole near the
flower.—An interesting little plant, in pine barrens and sandy swamps, Middle
States. Scapes 10—15' high, terete, with 2 or 3 subulate bracts. Leaves numerous, much shorter than the scape. Pedicels 3—7'' long. Perianth greenish
externally, yellow within, about half as long as the yellowish, mature capsule.
Aug.

ORDER CLVII. COMMELYNACEÆ.—SPIDERWORTS.

Herbs with flat, narrow leaves which are usually sheathing at base.
Perianth in 2 series, the outer (calyx) of 3 herbaceous sepals, the inner (corolla) of 3 colored petals.
Sta. 6, some of them usually deformed or abortive, hypogynous.
Ova. 2—3-celled, cells few-ovuled. Styles and stigmas united into one.
Fr.—Capsule 2—3-celled, 2—3-valved ; cells often but 2-seeded, with loculicidal dehiscence.
Seeds few, with dense, fleshy albumen. Embryo opposite the hilum.

Genera 16, species 250, chiefly natives of the Indies, Australasia and Africa, —a few of N. America.
They are of little importance to man.

Genera.

Stamens 6, { 2 or 4 of them sterile.   *Commelyna.* 1
         { all perfect . . . *Tradescantia.* 2

1. COMMELÝNA. Dill.

In honor of the brothers Commelyn, two German botanists.

Sepals herbaceous ; petals colored; stam. 6, 3—4 of them sterile
and furnished with cruciform glands ; caps. 3-celled, 3-valved, one of
the valves abortive.—Lvs. lance-linear, with sheaths at base. Fls. enfolded in a conduplicate, persistent, spathaceous bract.

1. C. ANGUSTIFOLIA. Michx.? (C. erecta. *Willd.*) *Day Flower.*
St. assurgent, branching, subgeniculate; *lvs.* lanceolate, subpetiolate, sheaths split to the base; *spathe* broad-cordate, distinct and open at base, enfolding 2 peduncles and several flowers; *pedicels* contorted; *pet.* unequal, the lower one much smaller, unguiculate; *sta.* 2, perfect.—Dry soils, Middle! Southern and Western States! Plant nearly smooth, 12—18' high, glabrous. Leaves 3—5' by 8—14'', varying from lance-linear to lance-ovate. Spathe veiny, 3—5-flowered. Petals deep blue. July, Aug.

2. C. VIRGINICA. Linn.? (C. longifolia. *Michx.*)
St. erect, branched at base, ciliate-pubescent; *lvs.* lanceolate, subpetiolate, sheaths entire, elongated, ciliate-pilose; *bracts* deltoid-falcate, united and entire at base as if peltate, about 2-flowered; *pet.* nearly equal; *sta.* 3, perfect.—Rocky woods, thickets, Penn. (*Muhl.*) Harper's Ferry! to Ga. A more slender, but erect species, 1—2f high. Leaves 3—5' by 6—12'', usually narrow-lanceolate, pilose-scabrous, the sheaths near 1' long. Spathe broadly funnel-shaped. Petals blue. July, Aug.—Neither of these plants agrees with the descriptions in the books.

2. TRADESCANTIA.
Named in honor of John Tradescant, gardener to Charles I.

Sepals persistent; petals large, suborbicular, spreading; filaments clothed with jointed hairs; anthers reniform.—4 *Fls. in terminal, close umbels, subtended by 2 or 3 long, leafy bracts.*

1. T. VIRGINICA. *Spider-wort.*
St. erect, simple or branched; *lvs.* lance-linear, channeled above, sessile, glabrous; *fls.* in a terminal, subumbellate cluster, pedicels finally elongated and reflexed; *cal.* pubescent.—Moist meadows, prairies, &c., Middle! and Western States! common. Stem thick, round, jointed, 2—3f high. Leaves numerous, subpilose, 12—18' by 6—12'', the bracts similar. Petals large, suborbicular, of a deep, rich blue, soon fading. May—Aug.—The juice of the plant is viscid and spins into thread; hence the common name.
β.? (T. subaspera. *Sims.?*) *Lvs.* lanceolate, narrowed to the base, pilose both sides, sheaths entire, ciliate with long, white hairs; *umbels* both axillary and terminal; *fls.* small, rose-colored.—Shady river banks, Ia.!

2. T. ROSEA. Michx.
St. erect, simple; *lvs.* linear, glabrous, channeled, amplexicaul; *ped.* elongated; *cal.* glabrous.—Penn. to Ga., in moist woods. Stem 8—12' high. Leaves 6—8' by 2—3'. Umbel terminal, subtended by 2 or 3 subulate bracts. Pedicels nearly 1' long. Flowers much smaller than in the preceding species. Petals rose-colored, twice longer than the smooth calyx. May.

ORDER CLVIII. XYRIDACEÆ.—XYRIDS.

Herbs, sedge-like, with linear or ensiform leaves. Fls. capitate at the top of a simple scape.
Perianth 6-parted, in 2 series, sepals 3, glumaceous, petals 3, unguiculate. [inante.
Sta. 6, 3 of them with extrorse anthers and inserted on the claw of the petals, the other 3 abortive fila-
Ova. single. Style trifid. Stigmas obtuse, lobed.
Fr.—Capsule 3-valved, 1 celled, with parietal placentæ, or 3-celled.
Seeds numerous, albuminous.

Genera 5, species 70, natives of tropical Asia, Africa and America, a few species of Xyris extending into the United States. Of no important use.

XYRIS.
Gr. ξυρος, acute-pointed; in allusion to the form of the leaves.

Heads of flowers ovoid-cylindric; sepals cartilaginous; petals equal, ovate, crenate, with narrow claws as long as the sepals; capsule 1-celled, with parietal placentæ.—*Lvs. narrowly linear, rigid, radical, sheathing the base of the scape. Fls. in a terminal, dense head, petals yellow.*

1. X. CAROLINIANA. Lam. (X. Jupacai. Mz. X. flexuosa. Ell.) *Yellow-eyed Grass.*—Scape slender, compressed and ancipitous above; *lvs.*

48

linear-ensiform, more or less twisted, acute, rigid, erect, shorter than the scape; *head* oblong-globose, obtuse; *scales* coriaceous, imbricated, obtuse, concave, scarious and yellowish at edge; lower ones empty; *sep.* not exserted; claw ♀ petals as long as the scales.—♃ Meadows, swamps and prairies, U. S. not uncommon. Scape firm and wiry, often twisted or flexuous, 1—2f high, 1″ diam Leaves few, flat, 6—12′ by 2—3,″ clasping or equitant at base. Petals yellow, limb spreading, retuse. July, Aug.

β. *brevifolia.* Wood. (X. brevifolia. *Mx.*) *Lvs.* linear-subulate, short, much twisted.—Evidently a variety of this polymorphous species.

γ. ? *Olneyi.* Wood. *Lvs.* larger, nearly as long as the scape; strongly equitant-clasping at base; scales rather loosely imbricated; *sep.* a little exserted; filaments hairy.—Cumberland, R. I., *Olney!* Scape 18—24′ high. Leaves 3—4″ wide.—Perhaps distinct, but its claims cannot now be satisfactorily determined.

2. X. FIMBRIATUS. Ell. *Fringed Xyris.*

Scape tall, erect; *lvs.* linear-ensiform, about equaling the scape; *head* oblong, with the scales loosely imbricated; *sepals* much exserted, fimbriate.—A large species, found in N. J. (*Darlington*) S. to Ga., *Elliott.* Scape 2—3f high ✓ I have never seen this species.

ORDER CLIX. ERICAULONACEÆ.—PIPEWORTS.

Herbs perennial, aquatic, with linear, spongy, cellular leaves sheathing at base. *Fls.* monœcious or diœcious, in a dense head. *Perianth* 2—6-parted, or wanting. *Sta.* 6, some of them generally abortive. *Anthers* mostly 1-celled. (*Ova.* 1 or more-celled, cells 1-seeded. *Seeds* pendulous.

Genera 2, species 200, chiefly South American. They are of no known use.

ERIOCAULON.

Gr. εριον, wool. καυλος, stem; the stem being sometimes woolly or tomentose.

Flowers ♂, collected into an imbricated head; involucre many-leaved. ♂ in the disk; perianth single, 3-cleft, the 2 inner segments united nearly to their summit; stamens 4—6. ♀ in the margin; perianth single, deeply 4-parted; style 1; stigmas 2 or 3; capsule 2—3-celled, 2—3-lobed; cells 1-seeded.

1. E. SEPTANGULARE. With. (E. pellucidum. *Michx.*) *Pipewort.*

Smooth; *scape* slender, about 7-furrowed; *lvs.* linear-subulate, pellucid, channeled, 5-veined; *hd.* small, globose; *scales of the involucre* obtuse.—A small plant of simple structure, in water, only the scape arising above the surface. Leaves radical, submersed, in a small tuft at the bottom, 1—3′ by 1—2′, tapering to a point, transparent at base. Stem simple, erect, 4—12′ high, with a small, terminal, hemispherical head of close, white flowers. Jn.

2. E. DECANGULARE. Michx. *Tall Pipewort.*

Scape 10-furrowed; *lvs.* ensiform, glabrous; *hds.* large, depressed-globose: *invol. scales* oval, acute, those of the receptacle mucronate.—Ponds, N. J. ti Car. Scape 2—3f high. Flowers very white. Aug.

3. E. GNAPHALOIDES. Michx. (E. decangulare. *Walt.*)

Scape somewhat compressed, with 10 furrows; *lvs.* short, subulate, ensiform, glabrous; *hds.* hemispheric-convex; *invol.* of shining, scarious, ovan round-obtuse scales.—In still waters. Scape 10—14′ high. July.

CLASS IV. GLUMACEOUS ENDOGENS.

FLOWERS with glumes; or floral organs enclosed in imbricated bracts, and arranged in spikelets, having no proper perianth (calyx or corolla). OVARY with one cell containing a solitary ovule and becoming a one-seeded fruit (achenium or caryopsis).

ORDER CLX. CYPERACEÆ—SEDGES.

Herbs mostly perennial, coarse, grass-like, cæspitose. *Root* fibrous or rarely tuberous. *Stems* (culms) usually solid with pith, generally without joints or nodes, and triangular. *Lvs.* with their sheaths entire. *Inflorescence* mostly capitate or spicate. *Fls.* perfect or monœcious, solitary in the axil of each bract (glume, scale). *Pri.* wanting, or represented by a few hypogynous bristles, or a cup-shaped or saccate *perigynium.* *Sta.* definite (1—12), mostly 3. *Anthers* fixed by their base, entire, 2-celled. *Ova.* generally either surrounded by bristles, or invested in the saccate *perigynium.* *Ovule* erect. *Sty.* 2 or 3, more or less united. *Stigmas* undivided, rarely bifid. *Fr.* an achenium. *Embryo* in the end of the albumen next the hilum.

Genera 112, species 2000. The sedges abound in almost all countries and climes of the globe, and in all localities, but are more common in the meadows, marshes and swamps of the temperate zones. About 25 genera and 340 species are known in North America.

Properties.—They are in general little used for food or in the arts. Their coarse herbage is often eaten by cattle, but they are nearly destitute of the sweet and nutritious properties of the grasses. The leaves of some of the larger species are used in Italy to bind flasks, and in weaving the bottoms of chairs. Yet, although of so little apparent value, their vast numbers authorise the belief that they subserve many highly important ends in the economy of nature.

Conspectus of the Genera.

distichously imbricated.	Flowers in axillary, racemose spikelets.					Dulichium.	1
	Fls. all terminal.	Spikelets linear, many-flowered, distinct, loose.				Cyperus.	2
				Ach. triangular.		Mariscus.	3
		Spikelets 1—2-flowered, in hds.	Ach. lenticular.			Kyllingia.	4
				Ach. triangular.		Putrena.	5
	Perigynium of 3 ovate, unguiculate petals, with 3 bristles.						6
			Leaves 0.	Spike solitary, terminal.		Eleocharis.	6
		Fls. all termin.	Leaves present.	Bristles 3—6.		Scirpus.	7
				Bristles numerous, long.		Eriophorum.	8
				Bristles 0.	Style 2-cleft.	Fimbristylis.	9
					Style 3-cleft.	Trichelostylis.	10
		Flowers all lateral.		Bristles 3—6. Sty. deciduous.		Scirpus.	7
				Bristles 0. Leaves radical.		Hemicarpha.	11
	spirally or irregularly imbricated.	Flowers both axil. and termin.	Brist. 3—6.	Style short, bulbous at base.		Rhynchospora.	12
			Bristles 0.	Style very long, horn-like.		Ceratoschœnus.	14
				Spikelets 20—30-flowered.		Psilocarya.	13
		Ach. not enclosed.		Spikelets	Ach. corky, brown.	Cladium.	15
				few flowrd.	Ach. bony, white.	Scleria.	16
		Achenium enclosed in a saccate perigynium. Fls. diœcious.				Carex.	17

TRIBE 1. CYPEREÆ.—Flowers ☿. Spikelets distichously imbricated.

1. DULICHIUM.

Gr. δύο, two, λειχην, scale; alluding to the glumes in two rows?

Spikelets linear-lanceolate, subcompressed; glumes sheathing, closely imbricated in 2 rows; style long, bifid, the persistent base crowning the compressed achenium; ovary invested with setæ.— *Stem leafy. Spikes axillary, racemose.*

D. SPATHACEUM. Pers. (Cyperus. *Linn.* Scirpus. *Michx.*)
St. round, leafy and somewhat 3-sided above, thick, sheathed below; *lvs.* alternate, pointing 3 ways, 2—4' by 3"; *sheaths* tubular, shorter than the internodes; *spikes* axillary from within the sheaths and terminal, each consisting of 8—10 linear-lanceolate, alternate spikelets in 2 rows; *spikelets* 5—7-flowered, nearly an inch in length; *glumes* linear-lanceolate.—♃ Marshes, borders of streams, U. S. and Can. Aug.

2. CYPERUS.

Spikelets compressed, distinct, many-flowered; glumes imbricated in two, opposite rows, nearly all with a flower enclosed; ovary generally without setæ.—*Mostly* ♃. *Stem simple, leafy at base, mostly triangular, bearing an involucrate, simple or compound umbel at top.*

§. 1. *Style 2-cleft. Achenium compressed-lenticular.*

1. C. FLAVESCENS. *Yellow Sedge.*
St. leafy, triquetrous; *spikelets* linear-lanceolate, 15—20-flowered, in fascicles of 3 or 4; *invol.* of 3, unequal leaves, longer than the spikes; *glumes* ovate, obtuse; *style* 2-cleft; *ach.* mucronate, somewhat rugose, dark brown.—♃ Marshy grounds U. S., not common. Stems and leaves about 8' high, the former with yellowish-green spikes in a terminal umbel with unequal rays. Aug., Sept.

2. C. DIANDRUS. TORR. *Diandrous Sedge*
St. slender, reclining; *umbels* sessile, 1—2-rayed; *rays* unequal; *invol.* 3-leaved, the 2 outer leaves very long; *spikelets* oblong-lanceolate, flat, 14—16-flowered, collected into capitate fascicles; *glumes* acute; *sta.* 2; *sty.* 2-cleft; *ach.* compressed.—Marshes, N. Eng.! Mid. and W. States. Stem 6—12' long. Umbel somewhat paniculate. Glumes chestnut-colored. A delicately beautiful sedge.
β. 1 *castaneus.* TORR. (C. castaneus. *Bw.*) *Rays* very short; *glumes* close, subcoriaceous, green on the back.—N. Eng.! Stems 4—8' high.

3. C. NUTTALLII. TORR. *Nuttall's Sedge.*
St. acutely triquetrous, leafy at base; *lvs.* narrow-linear, nearly as high as the stem; *umbel* loose, subsessile, about 3-rayed; *rays* short; *invol.* 4-leaved, the 2 outer leaves very long; *spikelets* very acute, linear-lanceolate, fasciculate, brownish; *ach.* oblong-obtuse.—♃ Salt marshes, N. Eng. to La. Stems 6—12' high, in dense tufts. Scales minutely 3-toothed. Stamens 2.

4. C. TENELLUS. Linn.? TORR. *Slender Sedge.*
St. setaceous, very slender, 3—5' high; *lvs.* setaceous; *spikelets* solitary, lance-linear; *invol.* mostly of one erect, setaceous leaf 1' in length; *scales* rather loose, 3-veined on the keel; *sta.* 1; *sty.* 2-cleft; *ach.* oblong-obovate, much compressed, brown.—Monmouth Co., N. J., *Dr. Cleaver*, (TORR. Cyn., p. 258.)

§ 2. *Style 3-cleft. Achenia triangular.*

* *Spikelets alternate, in pinnatiform, subdistichous spikes.*

5. C. STRIGOSUS. *Bristle-spiked Galingale.*
St. triquetrous, leafy only at base; *lvs.* broad-linear, rough-margined, about as long as the stem; *umbel* with elongated rays and oblong, loose spikes; *involucels* 0 or setaceous; *spikelets* numerous, linear-subulate, spreading horizontally, 8—10-flowered, 7—9' long; *invol.* of about 6 leaves, the 2 outer ones very long.—Wet grounds, U. S., frequent. Stem 1—2f high, bulbous at base. Umbel yellowish. Sept.
β. *Rays* with setaceous involucels 1' long; *spikelets* very numerous, subulate, 3—4'' long.—Ia.!

6. C. MICHAUXIANA. Schultes. (C. eruthrorhizos. *Torr.*)
St. acutely triangular; *umbel* compound, with short rays; *spikelets* 6—9-flowered, the lower ones compound; *rachis* very broad, easily separating at the joints; *ova.* ovoid-triangular, enfolded by the interior, adnate scales.—① Brackish swamps, generally near the sea, Middle and Southern States. Stem 12—15'' high, reddened at the base, longer than the leaves. Spikelets 9'' long, 7—9-flowered.

7. C. REPENS. Ell. (C. phymatodes. *Muhl.*)
Rt. creeping, bearing small, round tubers at the extremities; *st.* 1—2f high, 3-angled, striate; *lvs.* subradical, as long as the stem; *umbel* 4—6-rayed; *rays* often branched, bearing 12—20 linear, obtuse spikelets somewhat in 2 rows, *sheaths* obliquely truncate, *involucels* 0; *spikelets* 12—20-flowered, 6—8'' long, the lowest generally fasciculate; glumes yellowish.—♃ Moist fields, N. Y. to Ia.! and S. States. Aug.

8. C. SPECIOSUS. Vahl. *Showy Sedge.*
St. acutely angled, about 3f high; *lvs.* deeply channeled, half as long as the stem; *umbels* compound, about 8-rayed, rays alternate, 1—3' long; *ochreæ* (sheaths) deeply 2-parted; *partial umbels* with numerous, compound spikes, involucellate with setaceous bracts; *spikelets* umbellate. 5—8-flowered; *sta.* 3.—Wet places, Columbus, Ohio, *Sullivant.*

* * *Spikelets irregularly inserted all around the rachis.*

9. C. SCHWEINITZII. Torr. *Schweinitz's Sedge.*

St. 8—12′ high, triquetrous, rough on the angles; *lvs.* shorter than the stem, about a line wide; *umbel* simple, erect, 4—6-rayed, rays elongated, unequal; *sheaths* truncate, entire; *invol.* 3—5-leaved, longer than the leaves, scabrous on the margin; *spikelets* 6—7, alternate, approximate, 6—8-flowered, with a small, setaceous bract at the base of each; *scales* membranaceous on the margin; *sta.* 3; *sty.* 3-cleft, scarcely longer than the smooth achenia.—Shore of L. Ontario, *Sartwell*, of Lake Erie, *Sullivant!* to Ark.

10. C. ERUTHROERIZOS. Muhl. *Red-root Sedge.*

St. 2—3f high, obtusely triquetrous, longer than the leaves; *umbel* compound; *rays* 5—9, 3—4′ long, each with 3—4 sessile spikes; *sheaths* entire; *spikelets* very numerous, 6″ long, crowded and spreading in the spikes, a little flattened, about 13-flowered; *outer glumes* mucronate, closely imbricated, chestnut-brown, veinless and shining, the inner ones entirely free from the rachis; *sta.* 3; *ach.* smooth and shining, much shorter than the glume.—Wet grounds. Penn. and Southern States.

* * * *Spikelets more or less aggregated in terminal masses.*

11. C. FILICULMIS. Vahl. (C. mariscoides. *Ell.*)

St. slender, almost filiform, tuberous at base, 8—12′ long, leafy only at base; *lvs.* mostly radical, carinate; *umbel* simple and sessile, or with 1 or 2 rays; *spikelets* linear-lanceolate, 3—8-flowered, flattened when old, collected into globose heads; *glumes* remote, loose, ovate, yellowish.—Dry, rocky hills, N. Eng.! to Flor., W. to Ill.! Aug.

12. C. GRAYII. Torr.

St. 8—12′ high, filiform, obtusely triangular, erect, tuberous at base; *lvs.* radical, channeled, about ½″ wide; *umbel* 4—6-rayed, capillary, erect, spreading; *sheaths* truncate; *hds.* loose, of 6—8 spikelets; *spikelets* linear, compressed, 6—7-flowered; *scales* ovate, veined, obtuse, imbricated, interior ones lanceolate; *sta.* 3; *sty.* 3-cleft; *ach.* obovate-triquetrous, ¼ the length of the scale, gray, dotted.—Sandy fields, Mass., *Oakes*, L. I., *Kneiskern*, N. J., *Torrey.* Sept. (*Dr. Sartwell* comm.)

13. C. DENTATUS. Torr.

St. about 1f high, leafy at base, triquetrous; *lvs.* a little shorter than the stem, strongly keeled; *umbel* compound, 6—10-rayed; *invol.* of 3 unequal leaves, one of them longer than the umbel; *spikelets* 3 on each peduncle, 3—7′ long, ovate, flat, 8-flowered; *glumes* acute, spreading at the points, giving the spikelets a serrated appearance; *sty.* 3-cleft; *ach.* triangular.—♃ Swamps, N. Eng.! and Middle States.

14. C. INFLEXUS. *Muhl.*

St. setaceous, leafy at base, 2—3′ high; *lvs.* equaling the stem; *umbel* 2—3-rayed, or conglomerate and simple; *invol.* of 3 long leaves; *spikelets* oblong, 8—12-flowered, 10—20 together, densely crowded into the ovoid heads; *glumes* yellowish, veined, squarrose-uncinate at tip· *sta.* 1.—Banks of streams, Free States! and British Provinces. Aug., Sept.

15. C. ACUMINATUS. Torr. & Hook.

St. 5—10′ (3—4′ *Torr.*) high, slender, obtusely triquetrous; *lvs.* erect, radical, as long as the stem; *umbel* 1—6 (1—2 *Torr.*)-rayed; *invol.* 3—4-leaved, very long; *rays* unequal, each with a globose head of 15—40 spikelets; *spikelets* 3—11″, oblong-linear, obtuse, 15—25-flowered; *fls.* very regularly imbricated in 2 rows; *glumes* acute, with the point recurved; *sta.* 1; *ach.* dull-grayish.—Ill. *Mead!* Mo. *Drummond.*

3. MARISCUS. Vahl.

Celtic *mar*, a marsh; alluding to the place where some species grow.

Spikelets subterete, clustered in heads, rarely with but 1 or 2 fertile flowers; glumes imbricated somewhat in 2 rows, the lower ones short and empty; rachis margined with the adnate, persistent, in-

terior glumes; stamens 3; style 3-cleft; achenium triangular.—♃
Habit of Cyperus.

1. M. OVULÀRIS. Vahl. (Cyperus. *Gron. Torr.* Scirpus echinatus. *Linn.*)
St. triquetrous, nearly naked, 1—2f high; *lvs.* shorter than the stem,
nearly smooth; *umbel* simple; *rays* 3″—2′ long; *hds.* 1—5, globose, 1 sessile,
the rest on the spreading rays; *spikelets* linear-subulate, 6″ long, 50—100 in
each head; *fls.* 2—4, 1—2 fertile; *invol.* 3—4-leaved, outer leaves very long.—
Bogs and low grounds, Middle and Western States, common. Aug., Sept.
δ. tenellus. Torr. Slender and cæspitose; *hds.* ovate, small.
γ. cylindricus. Torr. *Hds.* oblong, cylindrical.

2. M. RETROFACTUS. Vahl. (Scirpus. *Linn.*)
St. obtusely triangular, nearly leafless, pubescent, 2—3f high; *lvs.* pubescent, 3—4″ wide, about half as long as the stem; *umbels* simple; *rays*
unequal, long, 6—8; *invol.* 3—5-leaved; *bracts* unequal, not longer than the
rays; *spikelets* 70—100, subulate, 1-flowered, finally retrorsely imbricate into
obovate heads; 2 *lower glumes* empty.—A rare species, Middle and Western
States. Aug., Sept.

4. KYLLINGIA.
In honor of Peter Kylling, a Danish botanist.

Spikelets compressed; scales about 4, the two lowest short and
empty, the third only usually with a fertile flower; stam. 1—3;
style long, 2-cleft; achenia lenticular.—*Stems triangular. Hds. sessile, solitary or aggregated, involucrate.*

1. K. PUMILA. Michx.
Cæspitose; *st.* 2—12′ high, slender; *lvs.* mostly radical, shorter than the
stem, smooth; *hds.* generally solitary, sometimes triple, closely sessile, oval or
oblong; *invol.* 3-leaved, 1—2′ long; *spikelets* 1-flowered, very numerous, about
2″ long; the lowest glume or glumes very small; *sta.* always 2; *ach.* lens-
shaped, fulvous.—Wet banks, Columbus, Ohio (*Sullivant!*) Ky. (*Short*) and
Southern States.

2. K. MONOCEPHÀLA.
St. slender, 10—15′ high, leafy at base; *lvs.* much shorter than the stem,
1″ wide; *hd.* simple, globose, dense, inclining; *invol.* 3-leaved, the highest
leaflet erect; *spikelets* numerous, 1-flowered; two lower glumes minute, two
upper striate, subequal, serrulate, ciliate on the keel.—Mass., N. Y. to Ga.

TRIBE 2. HYPOLYTREÆ.

5. FUIRÈNA. Rottböll.
In honor of George Fuiren, a Dutch botanist.

Glumes awned, imbricated on all sides into a spike; petaloid
scales 3, cordate, awned, unguiculate, investing the achenium.—♃
St. angular, leafy. Spikes subumbeled, axillary and terminal.

1. F. SQUARRÒSA. Michx.
St. 1—2f high, obtusely triangular, sulcate; *lvs.* ciliate, shorter than the
stem; *sheaths* hispid-pilose; *spikes* clustered, ovate, mostly terminal, 6—12;
awns nearly as long as the glumes; *petaloid scales* ovate, cuspidate with a short
bristle; *ach.* twice the length of the stipe.—Bogs and swamps, N. J. to Car.
Sept.

2. F. PUMÌLA. Spr. (F. squarrosa. *β. Torr.*)
St. pubescent above, 3—6′ high; *lvs.* linear-lanceolate, flat, striate, as
long as the stem; *spikes* 1, 2 or 3, thick, subtended by 2—3 long, involucrate,
unequal leaves; *glumes* ovate-lanceolate, with short awns; *petaloid scales* ovate-
lanceolate; *ach.* pedicellate, with retrorsely hispid seta. Wet, sandy places. Aug

TRIBE 3. **SCIRPEÆ.**—Flowers ☿. Glumes of the spikelet imbricated all
around. Perigynium none, or setaceous.

6. ELEOCHÂRIS. R. Br.

Gr. ελος, a marsh, χαιρω, to rejoice ; plants delighting in marshy grounds.

Spikes terete ; bristles of the perigynium mostly 6 (3—12), rigid,
persistent ; styles 2—3-cleft, articulated to the ovary ; achenium
crowned with a tubercle which is the persistent, bulbous base of the
style.—*Stem simple, leafless.   Spike solitary, terminal.*

§ *Spikes cylindrical, length more than three times the diameter.*

1. E. EQUISETÖIDES. Torr. (Scirpus. *Elliott.*)   *Horse-tail Rush.*
St. about 2f high, papillose, terete, 2—3" diam., with about 90 joints, produced by internal, transverse partitions ; *sheath* radical, obtuse, membranous ;
spike oblong-cylindrical, about 1' in length, acute and slightly contracted at
base ; *glumes* roundish-ovate, cartilaginous, obtuse ; bristles 6, as long as the
achenium ; *sty.* 3-cleft ; *ach.* brown, shining.—Bogs, Cumberland, R. I., *Olney!*
Del. to Ga. It strikingly resembles Equisetum hyemale.

2. E. QUADRANGULÂTA. R. Br. (Scirpus. *Michx.*)
St. 2—4f high, acutely and unequally quadrangular, the broadest side
convex, the others concave ; *sheaths* radical, purplish ; *spike* 1' or more in length ;
glumes roundish-ovate, obtuse, coriaceous ; *bristles* 6 ; *ach.* obovate, of a dull
white.—Penn., Md., *Dr. Robbins,* to Ga. and La. In swamps and inundated
banks.

3. E. ROBBINSII. Oakes.   *Robbins' Club Rush.*
Sts. clustered, 9—26' high, rigid, sharply triangular, pale green, several
of them fruitless ; *sheath* truncate ; *spike* 1—1½" long, scarcely thicker than the
stem, placed 2—5" below its apex ! *glumes* 3—9, linear-lanceolate, acute, finally
brownish ; *bristles* 6, twice longer than the achenium ; *ach.* 1" long, pale brown
tubercle closely sessile.—Ponds and ditches, N. H. and Mass., *Rickard!* Very
distinct. In water a part of the stems are floating and as fine as hairs. July

§ § *Spike ovate, length less than three times the diameter: * Stems terete*

4. E. PALUSTRIS. R. Br. (Scirpus. *Linn.*)   *Marsh Club Rush.*
St. leafless, round, inflated ; *spikelets* smooth and shining, lance-oblong,
acute, often oblique, terminal ; *glumes* subacute, the lower ones larger, sometimes empty.—Low grounds, U. S. and Brit. Am. Root creeping. Stems numerous, 1—2½f high, each with an obtuse sheath at the base. Achenium roundish-obovoid, rugose, punctate, surrounded with 3 or 4 scabrous bristles, and
crowned with a tubercle. July.

5. E. OBTÚSA. Schultes. (Scirpus obtusus. *Willd.* Scirpus capitatus. *Walt.*)
St. sulcate, subterete, 6—15' high ; *spikelet* ovoid, very obtuse, often nearly globose ; *glumes* round, dark brown, with whitish margins ; *ach.* obovate,
compressed, smooth, brown, invested with 6 setæ as long as the glumes.—Shallow waters, Can. and U. S., common. July.

6. E. TUBERCULÔSA. R. Br. (Scirpus. *Michx.*)
St. columnar, striate, 12' high, leafless, sheathed at base ; *spikelet* ovate-
lanceolate ; *glumes* very obtuse, loose ; *ach.* somewhat triquetrous, smaller than
the sagittate tubercle with which it is crowned ; *bristles* 6, as long as the tubercle.—Sandy swamps, N. Eng. ! to Flor. Remarkable for its large tubercle. Jl

* * *Stems compressed or angular.*

7. E. OLIVACEA. Torr.
Sts. cæspitose, 2—4' high, slender, compressed, sulcate, soft ; *spike* ovate,
acutish 2—3" long, 20—30-flowered ; *glumes* ovate, obtuse, reddish-brown, with
scaricus edges and a green midvein, the lowest largest ; *bristles* 6 ; *sty.* 2-cleft ;
ach. broadly obovate, smooth, of a dull, blackish-olive color when ripe.—Sands,
generally partly submersed, Providence, R. I., *Olney!* Mass. to N. J.

8. E. INTERMEDIA. Schultes. (Scirpus. *Muhl.*)   *Turf Club Rush.*
St. cæspitose, setaceous, diffuse, compressed, furrowed, hard and wiry,

e acute, 2—3" long, 7—9-flowered; *glumes* ance-ovate, acute,
with a green midvein; *bristles* 6, white, longer than the ache-
it; *ach.* obovate, attenuated to the base, striate, of a light brown
ing water, forming a strong, dense turf, N. H.! to Ga., W. to

ULÃRIS. R. Br. (Scirpus. *Linn.*) *Hair Club Rush.*
, setaceous, quadrangular, very slender, 3—6' high; *spikelets*
ute, 4—8-flowered; *glumes* obtusish, the lowest one larger and
oid, triangular, verrucose, yellow and shining.—Edges of ponds,
mersed, U. S. and Brit. Am. Very delicate. June, July.

uis. Schultes. (Scirpus. *Willd.*) *Slender Club Rush.*
, almost filiform, quadrangular, the sides sulcate, 8—15' high,
ple sheath at base; *spike* terminal, elliptic-oval, acute at each
k purple, ovate, obtuse, the lower ones larger and empty; *ova.*
ng below, invested with 2 or 3 or 0 setæ.—Common in wet places,
id. States. June, July.

LANOCARPA. Torr. (Scirpus. *Baldwin.*) *Black-fruited Club Rush.*
essed, furrowed, slender, almost filiform, wiry, 12—18' high;
; *spike* lance-oblong, rather acute, 4—6" in length, 20—40-flow-
ate, obtuse, brownish, with scarious margins and a prominent,
ein; *bristles* 3, purple; *ach.* obovate-turbinate, blackish; *tubercle*
ted in the centre.—Providence, R. I., *Olney!*

MÆA. Torr.
high, setaceous, compressed, sulcate; *spikes* ovate, compressed,
mostly empty; *bristles* 6, longer than the achenium, slender,
ards; *ach.* ovate, acute, triangular, whitish and shining; *tuber-*
a coast, Mass., *Oakes.* Sept. (*Dr. Sartwell*, comm.)

TELLÃTA. Torr. in N. Y. Fl. ined. (Scirpus. *Torr.* Cyp., p. 318.)
' high, clustered, angular and sulcate, slender, almost filiform,

almost capillary, 3—4f long; *spike* solitary, somewhat terminal (the stem being continued above it in the form of a bract), lanceolate; *style* 2-cleft: *bristles* 6.— Streams, &c., Mass. ! to N. Y. Aug.

3. S. CÆSPITÒSUS. *Bracted Mountain Rush.*

Sts. cæspitose, round, sheathed at base with numerous rudiments of leaves; *spikes* compressed, terminal; 2 *lower glumes* involucre-like, as long as the spike; *ach.* with 6 bristles.—Grows in dense tufts, 4—12' high. Spike 4—5-flowered, reddish-brown. On the White Mts., N. H., *Bw.* July.

•• *Spikes many, lateral.*

4. S. TORREYI. Olney. (S. mucronatus. *Ph.? Torr.*)

St. 2f high, 3-angled, with concave sides, rather slender, leafy at the base; *lvs.* 2 or 3, 1f or more long, slender; *spikes* 2-4 (rarely 1), sessile, distinct, acute, ovate-oblong; scales ovate, mucronate, smooth; *sty.* 3-cleft; *ach.* obovate, acuminate, unequally 3-sided, shorter than the bristles.—Borders of ponds, N. Eng. to Mich. The stem, here and in the following, is prolonged above the spikes, in the form of an involucral leaf. Jl. Aug.

5. S. DEBILIS. Pursh. *Weak-stemmed Rush.*

St. cæspitose, roundish, deeply striate, 9—16' high, with a few subulate leaves at base; *spikelets* about 3, short-ovoid, sessile, crowded, subterminal; *glumes* ovate, obtuse, carinate, pale green; *ach.* obovate, mucronate; *bristles* 4— 5.—Borders of ponds and rivulets, N. Eng. to Car. Aug.

6. S. TRIQUÈTER. Michx. (S. Americanus. *Pers.*) *Three-cornered Rush*

St. nearly naked, 3-angled, corners acute and two of the sides concave, about 3f high and ending in a sharp point; *lvs.* few and short, from the top of the sheath; *spikes* lateral, 1—5, ovate, crowded and sessile, at various distances below the point; *glumes* round-ovate, mucronate; *bristles* 6.—Ponds and marshes, fresh and salt, throughout N. America.

7. S. LACUSTRIS. (S. acutus. *Muhl.*) *Lake Bulrush.*

Scape smooth, leafless, filled with a porous pith, 5—8f high, cylindric, tapering above the panicle, and abruptly ending in a short cusp; *panicle* cymose near the top; *ped.* rough, twice compound; *spikelets* ovoid, closely imbricate; *scales* ovate, mucronate, pubescent; *bracts* shorter than the panicle.—The largest species of bullrush, frequenting the muddy margins of rivers and ponds U. S. to Arc. Am. July.

8. S. OLNEYI. A. Gr. *Olney's Rush.*

St. triquetrous-winged, leafless, 2—7f high; *sheath* radical, tipped with a short (1—2') leaf; *spikes* 6—12, sessile, aggregated, 2—3" long, placed 9—12" below the triangular apex of the stem; *glumes* roundish-ovate, mucronate; *bristles* 6—12; *ach.* obovate, plano-convex, gibbous at apex.—Salt marshes, Sekonk river, R. I., *Olney!* Tom river, N. Y., *Kneiskern.* Remarkably distinguished by its 3-winged stem. July. *See also Addenda, p.* 638.

••• *Spikes terminal.*

9. S. MARITÌMUS. *Sea Bulrush.*

St. acutely 3-angled, leafy, 2—3f high; *lvs.* broad-linear, rough-edged, carinate, taller than the stem; *spikes* conglomerate, 6—10, nearly an inch long, corymbose; *invol.* of about 3 very long leaves; *glumes* ovate, 3-cleft, the middle segment subulate and reflexed; *style* 3-cleft; *bristles* 3—4, much shorter than the broad-obovate, lenticular, dark brown, polished achenium.—Salt marshes, N. Eng.! to Flor. Aug.

β. *fluviatilis.* Torr. *Umbel* somewhat compound; *bristles* 6, as long as the obovate, triangular, dull grayish achenium.—Fresh water swamps and lakes, Western N. Y. to Wis., *Lapham!* and Mo.

10. S. ATROVIRENS. Muhl.

St. obtusely triangular, leafy, 2f high; *cyme* compound, proliferous; *invol.* of 3 leaf-like bracts, longer than the cyme; *spikes* ovate, acute, crowded, 10—20 in a globose head; *hds.* numerous, 4' diam., dark green; *glumes* ovate, mucronate; *ach.* white, smooth; *bristles* 4.—Common in meadows, Middle and Western States. June, July.

11. S. **BRUNNEUS.** Muhl.
St. obtusely triangular, leafy, 2—3f high; *cyme* decompound, its principal branches about 5, unequal, with truncate sheaths at base; *spikelets* clustered in heads of 3—6; *glumes* obtuse, reddish-brown; *ach.* smooth, yellowish-white, shorter than the 4 or 5 tortuous bristles.—Much resembles the last species. Margins of waters, N. Eng. to Penn, W. to Ohio, rare.

§ 2. TRICOPHORUM. *Bristles 6, much longer than the achenium, tortuous smoothish. Stem leafy. Umbel decompound.*

12. S. ERIOPHORUM. Michx. (Tricophorum cyperinum. *Pers.*)
St. obtusely triangular, leafy, 3—5f high; *lvs.* 2f long, rough-edged; *lvs.* terminal, decompound, large and loose; *spikes* mostly pedicellate; *bristles* 6, capillary, curled, very conspicuous, being 5 or 6 times as long as the white achenium.—A common, stiff, rank meadow sedge, which cattle do not eat, U. S. and Can. Spikes numerous, 2—3" long, ovoid, obtuse, in small clusters, in a large, showy panicle. Involucre 4-leaved. Aug.

13. S. LINEATUS. Michx. (Tricophorum. *Pers.*)
St. triangular, very leafy, 2—3f high; *umbels* terminal and axillary, decompound, at length nodding; *invol.* of 1—2 bracts, longer than the leaves, *spikes* ovoid, pedunculate, solitary; *glumes* lanceolate, ferruginous; *bristles* 6, as long as the glumes.—Swamps, in most of the States. Aug.

8. ERIOPHÖRUM.

Gr. ερεον, wool, φερω, to bear; alluding to the copious bristles of the perigynum.

Glumes imbricated all around into a spike; achenium invested in very long, dense, woolly or cottony hairs.—*Stem generally leafy Spikelets mostly in umbels, finally clothed with the long, silky hairs.*

* *Spikelet solitary.*

1. E. ALPÍNUM.
St. very slender, acutely 3-angled, naked, somewhat scabrous, 8—16' high, with 3—4 radical sheaths; *radical lvs.* very short, subulate; *spike* oblong, terminal, about 2" in length; *hairs* 6 to each flower, woolly, white, crisped, 4 times as long as the spike.—Bog meadows, often alpine, N. H.! to N. Y. and Penn. Jl.

2. E. VAGINÃTUM. *Sheathed Cotton Grass.*
Sts. densely cæspitose, obtusely triangular, slender, smooth and rigid, 1—2f high; *uppermost sheaths* inflated; *spikelet* ovate, oblong, 6—8" long, of a blackish color, with scarious glumes; *hairs* 30—40 to each flower, straight, white and glossy, twice as long as the spikelet, conspicuous, as well as in other species, even at a distance among the meadow grass.—N. Eng. to Mich., N. to Arc. Am.   June, July.

* * *Spikelets numerous.*

3. E. CONFERTISSÌMUM. Wood. *Dense-headed Eriophorum.*
St. strictly erect, firm but slender, 2—3f high; *lvs.* narrowly linear, channeled, rigid, triangular-subulate above, 8—12' long; *sheaths* close; *invol.* 2-leaved, one leaf twice, the other 4—5 times longer than the spikelets; *spikelets* 5—8, crowded, erect, on very short (2—4"), slightly scabrous peduncles, 20—30. flcwered; *glumes* obovate, carinate, very obtuse; *ach.* compressed, oblanceolate, 1½" long, flat on one side, carinate on the other; *bristles* 100—200, white, ½' long, straight and silky.—Bogs, Meriden, N. H.! Distinguished for its very large and dense heads.

4. E. POLYSTÃCHYUM. *Many-spiked Cotton Grass.*
St. somewhat triangular, smooth, 1—2f high; *cauline lvs.* 2—3, broad-linear, flattened below, triquetrous at the end; *spikelets* about 10, on rough peduncles which are long and drooping and sometimes branched; *setæ* 30—40 to each flower, reddish-white, 6—8" long.—Very conspicuous in meadows and swamps, U. S. and Brit. Am.

5. E. ANGUSTIPOLIUM. Rich. *Narrow-leaved Cotton Grass.*
St. slender, leafy, smooth, 10—15' high; *cauline lvs.* narrow, 3-cornered with concave sides, 1—3' long; *invol.* of one bract, with a loosely sheathing

base; *spikelets* 2—4, on short peduncles, nodding; *setæ* 40—50 to a flower, long, white and cottony.—Swamps, N. States and Brit. Am. July.

6. E. VIRGINICUM. *Virginian Cotton grass.*

St. nearly round, leafy, smooth, 2—3f high; *lvs.* flat, few, long, with scabrous margins; *invol.* 2—4-leaved, outer leaves much longer; *spikelets* in a sort of umbel, erect, nearly sessile; *glumes* ovate, brown at the sides; *keel* green; *hairs* 50—60, reddish-white, long and cottony.—Wet grounds, U. S. and Can.

9. FIMBRISTŸLIS. Vahl.

Lat. *fimbria*, a fringe, *stylus*, style ; from the ciliate style.

Glumes imbricated on all sides ; bristles 0 ; style compressed, 2-cleft, bulbous at base, deciduous, often ciliate on the margin.—♃ *with the habit of Scirpus.*

1. F. BALDWINIÀNA. Torr. (Scirpus Baldwinianus. *Schult.* S. ferrugineus. *Darl.*)—*St.* 2—12' high, compressed, deeply striate, leafy at base; *umbel* mostly simple, 3—4-rayed, central spikelets sessile; *invol.* subulate, 2-leaved, as long as the umbel; *spikelets* ovoid, acute; *glumes* ovate, brown; *sty.* bifid, ciliate; *ach.* white, longitudinally furrowed.—Swamps and damp places, Middle, Southern and Western States. July.

2. F. SPADICEA. Vahl. (Scirpus spad. *Linn.* S. castaneus. *Michx.*)

S'. 1—2f high, hard and rigid, compressed, nearly naked; *lvs.* 5—6 high, filiform, channeled inside, semi-terete outside, lower ones rust-colored; *umbel* of few rays, rather exceeding the 2—3 subulate, involucre leaves; *spikes* ovate-oblong, 3—6" long; *glumes* broad-ovate, mucronate, finally of a dark, shining, chestnut brown; *sty.* conspicuously fimbriate; *ach.* whitish.—Marshes, N. J. to La. Aug.

10. TRICHELOSTŸLIS. Lestiboudois.

Gr. τριχηλος, three-fold, στυλος ; from the character.

Glumes in 4—8 ranks, carinate ; bristles 0 ; style 3-cleft, deciduous below the bulb at the base ; achenium triangular.—♃ *Stems leafy at the base. Spikes usually in terminal umbels.*

1. T. MUCRONULÀTUS. Torr. (Scirpus muc. *Michx.* Fimbristylis autumnalis. *R. & S.*)—*St.* compressed, 2-edged, cæspitose, leafy at base, 3—10' high; *lvs.* flat, linear, shorter than the stem; *umbel* compound; *invol.* 2-leaved, *spikelets* lanceolate, acute, somewhat 4-sided, 2—3 together; *glumes* brown, mucronate; *ach.* white.—Wet places, along rivers, &c., N. Eng.! to Ga., W to Mo. July.

2. T. CAPILLÀRIS. Wood. (Scirpus. *Linn.* Isolepis. *R. & S.*)

St. cæspitose, nearly naked, 3-angled, capillary, 4—8' high; *lvs.* subradical, setaceous, shorter than the stem; *spikelets* ovoid, 2—4, pedunculate, inner one sessile; *glumes* oblong, ferruginous, margin pubescent; *ach.* white.—Ir sandy fields, Mass. to Car., W. to Ky. and Ohio. Aug.

11. HEMICARPHA. Nees.

Gr. ημισυς, half, καρφα, straw or chaff?

Glumes imbricated all around ; bristles 0 ; stam. 1 ; style 2-cleft, not bulbous at base, deciduous ; achenium compressed, oblong, subterete.—♃ *Spikes glomerate.*

H. SQUARRÒSA. Nees. (Isolepis subsquarrosa. *Schrad.* Scirpus subsq. *Muhl.* S. minimus. *Ph.*)—*Scape* setaceous, compressed, sulcate, recurved, 2—3' high; *lvs.* setaceous, shorter than the scape; *spikes* 2—3, terminal (apparently lateral), subsessile, ovoid, nearly 2'' long; *invol.* of 2 bracts, one appearing like a continuation of the scape, thrice longer than the other; *glumes* 00, with a short, recurved or squarrose point, finally brown; *ach.* minute, of a dull, brownish-white.—Sandy banks, N. Eng.! to Penn. and Ky.

12. PSYLOCARYA. Torr.

Flowers ♀. Glumes 00, imbricated all around, all fertile; perigynium 0; stam. 2; filaments long, persistent; style 2-cleft, dilated or tuberculate at base; achenium bioonvex, crowned with the persistent style.—*Stems leafy. Spikes lateral and terminal, cymose.*

P. SCIRPOIDES. Torr.

St. slender, leafy, smooth, 3-sided, 5—9′ high; *lvs.* linear, smooth, 3—5′ by 1·′, cauline about 2; *cymes* terminal, and one from the sheath of each cauline leaf; *spikes* about 3″ long, oblong-ovate, in small, loose clusters, 20—30-flowered; *glumes* chestnut-colored, thin, ovate, acute; *ach.* tumid, dark brown, crowned with the long style, which is much dilated at base.—Borders of ponds, Smithfield, R. I., *Olney!* Mass., *Greene.* Rare.

13. RHYNCHOSPÖRA. Vahl.

Gr. ρυγχος, a beak, σπορα, seed; from the character.

Flowers ♀ or ♂ ♀ ♀, few in each spikelet; glumes loosely imbricated, the lowest small and empty; perigynium of 6—12 bristles; stam. 3; style bifid; achenium lens-shaped or subglobose, crowned with the distinct, bulbous base of the style.—♃ *Stem leafy, 3-sided. Inflorescence terminal and axillary.*

* *Achenium smooth.*

1. R. ALBA. Vahl. (Schœnus albus. *Linn.*) *White Bog-Rush.*
   St. triangular above, very slender, leafy, smooth, 10—16′ high; *lvs.* setaceous, channeled; *corymbose fascicles* pedunculate, both terminal and from the axils of the sheaths, with setaceous bracts; *spikelets* lanceolate, acute at each end, with crowded, lanceolate, white glumes.—In wet, shady grounds; common. July—Sept.

2. R. CAPILLACEA. Torr. (Schœnus. *Muhl.*)
   St. 6—12′ high, filiform, glabrous, triangular; *lvs.* setaceous, much shorter than the stem; *spikelets* 3—6, mostly terminal, oblong, each with a setaceous bract; *glumes* chestnut-colored, with scarious edges; *bristles* 6, much longer than the oblong, substipitate achenium; *tubercle* about half the length of the achenium.—Swamps, N. Y., *Sartwell,* Penn., *Muhl.*

3. R. FUSCA. Rœm. & Schult. (Schœnus fuscus. *Linn.*)
   St. 3-angled, about 2f high; *lvs.* linear, carinate, smooth; *fascicles* alternate, pedunculate; *bracts* setaceous, longer than the ovoid spikes; *glumes* brown, ovate; *ach.* brown, rugose, with an acute, black tubercle as long as the hispid bristles.—Wet places, Mass. to N. J. Rare.

4. R. GRACILENTA. A. Gr.
   St. 1—2f high, very slender or filiform, smooth; *lvs.* linear-setaceous, much shorter than the stem; *corymbs* small, fasciculate, the lateral on slender peduncles exserted from the sheaths; *spikelets* ovoid; *glumes* ovate, acute, dark brown; *bristles* 6, a third longer than the roundish-ovoid achenium; *tubercle* flat, dilated at base.—Dry grounds, N. Y. to Car.

5. R. GLOMERÄTA. Vahl. (Schœnus. *Linn.*)
   St. slender, smooth, leafy, a foot or more high; *lvs.* flat, carinate, rough-edged; *corymbed fascicles* very remote, in pairs, axillary and terminal; *spikelets* lanceolate; *glumes* keeled, mucronate, brown; *ach.* obovoid or cuneiform, very smooth, as long as the tubercle; *setæ* 6, rough, backwards.—In bogs, Can. to Flor. July, Aug.

6. R. CEPHALANTHA. A. Gr.
   St. 2—3f high, triangular, stout; *lvs.* linear, very narrow, the lower and radical nearly as long as the stem; *hds.* roundish, axillary and terminal, solitary or rarely two together; *spikelets* lance-oblong; *glumes* ovate-oblong, dark brown; *bristles* 6, twice longer than the achenium; *ach.* roundish-ovoid, a little compressed, very obtuse.—N. J. pine barrens.

* * *Achenium rugose.*

7. R. CYMÒSA. Nutt. (Schœnus. *Willd.*)

St. 1—2f high, triangular, angles acute; *radica. lvs.* shorter than the ~ent. *cauline* rising above the stem; *corymbs* 3—4, the terminal largest; *spikelets* ovoid, in close fascicles of about 5; *glumes* broad-ovate, dark brown; *bristles* 6, as long as the broad-ovate, transversely rugose achenium: *tubercle* depressed, much shorter than the ach.—N. J. to La.　Jl., Aug.

8. R. TORREYÀNA. A. Gr.

St. 2f high, slender, cæspitose, striate; *lvs.* setaceous, radical, 6—10" long, *cauline* much shorter; *corymbs* diffuse, the lateral, if any, on short peduncles; *spikelets* ovoid, pedicellate or sessile; *glumes* ovate, mucronate, brown; *bristles* 6, scarcely half as long as the oblong-ovate achenium; *tubercle* short, nearly as broad at base as the achenium.—N. J. *Torrey.* Jl., Aug.

15. CERATOSCHŒNUS. Nees.

Gr. κερας,—ατος, a horn, σχοινος, rush; alluding to the long, persistent style of the achenium.

Spikelets 2—5-flowered, one flower ☿, the rest ♂; glumes loosely imbricated, somewhat in 2 rows, lower ones empty; perig. of 5 or 6 rigid, hispid or scabrous bristles; stam. 3; style simple, very long, persistent on the smooth, compressed achenium.—♃ *Stems leafy. Corymbs compound.*

1. C. LONGIROSTRIS. A. Gr. (Schœnus longirostris. *Michx.* 8. corniculatus. *Lam.* Rhyncospora cornic. *A. Gr.*)—Glabrous and glaucous; *st.* 3—4f high, triangular; *lvs.* 12—16' by 4—6", flat, rough-edged; *fls.* in very large, terminal and axillary corymbs, terminal one the largest; *spikelets* loosely fascicled in 4s or 5s on the long peduncles; *glumes* brown, ovate; *bristles* shorter than the achenium, which is 2" long, and crowned with the (7") long, subulate, horny style.—Ohio! to Flor.　Common in wet places　Aug.

2. C. MACROSTACHYA. Torr. (Rhyncospora *ejusd.*)

Glabrous; *st.* 2—3f high, triangular; *lvs.* 1—2f by 2—4", rough-edged; *axillary corymbs* subsimple, *terminal ones* compound; *upper spikelets* densely fascicled; *ach.* ovate, smooth; *bristles* erectly hispid, twice as long as the achenium; *style* persistent, nearly 4 times as long as the achenium.—Mass. *Robbins.!*

16. CLADIUM. Browne.

Flowers ♂ ☿ ♀; glumes imbricated somewhat in 3 rows, lower ones empty; bristles 0; stam. 2; style 2—3-cleft, deciduous; achenium subglobose, the pericarp hard, thickened and corky above.— ♃ *Stem leafy.　Corymbs or panicles terminal and axillary.*

C. MARISCÖIDES. Torr. (Schœnus. *Muhl.*)　*Bog Rush.*

St. terete, leafy, 20—30' high, hard and rigid; *lvs.* narrowly linear, channeled above, rounded beneath, much shorter than the stems; *bracts* short; *umbels* 1—3, erect, the lateral on long exserted peduncles; *rays* 3—7, some of them very short; *spikelets* aggregated in heads of 4—8, lance-ovate, 3" long; *glumes* tawny-brown, about 6, the upper usually ☿, the next ♂, and the rest empty *ach.* ovoid, short-beaked with the remains of the 3-cleft style.—Bogs, Can. to Penn.　July.

TRIBE 3. **SCLEREÆ.**—Flowers monœcious or diclinous.　Achenium naked (without a perigynium), more or less hard and bony.

17. SCLERIA.

Gr. σκληρος, hard; alluding to the indurated shell of the fruit.

Flowers diclinous, fertile spikelets 1-flowered, glumes fasciculate perigynium cup-shaped or 0; achenium globose, ovoid or triangular with a thick, bony pericarp; style 3-cleft, deciduous.—♃ *Stems leafy Spikelets in spikes, fascicles or panicles.*

49

Perigynium cup-shaped, lobed, repand or annular.

CULĀRIS. Michx.

high, triangular, rather slender; *lvs.* 1″ wide, channeled, radical
uline few, much shorter; *fascicles* 3—5, lateral and terminal,
:owered, subsessile; *spikelets* somewhat in pairs, the ♂ many-
base of the ♀; *glume* light brown, ovate, acuminate; *sta.* 3;
ach. globose, of a dead white, ½″ diam., conspicuously reticulated
:d.—Borders of ponds, R. I., *Olney!* to Flor. The achenium is
:eautiful object.

A. Torr. (S. reticularis. *Muhl.*)

high, weak, diffuse, acutely triangular, slender; *lvs.* flat, 2′
fascicles about 3, one terminal, the others lateral and very re-
;″ long, compressed, slender, often recurved; *spikelets* distant, in
e at the base of the ♀; *sta.* 2; *perig.* deeply 3-lobed; *ach.* about
se, whitish, marked with brown, hairy, transverse ridges and
sea coast, N. J. to Flor. Sept.

UIFLŌRA. Muhl.

6′ high, triangular, slender, smoothish; *lvs.* narrow, nearly
: pubescent; *fascicles* 2—3, lateral and terminal, few-flowered,
on long peduncles; *bracts* foliaceous, ciliate; *spikelets* in pairs;
naceous, acute; ♀ *spikelet* of 3 ovate, mucronate glumes, some-
the keel; *sty.* 3-cleft; *ach.* globose, rough, white and shining;
ring upon which are 6 roundish, minute, tubercles.—N. Y.
I. (*Carey*), to Car. Aug.
olitary, terminal, (apparently lateral) sometimes with a small
se of the stem.—Ohio, *Sullivant!*

:LOMERĀTA. Michx. *Whip-grass.*

acutely triangular, rough, leafy, 3—4f high; *lvs.* linear-lanceo-

A. *Spike single.* 2. *Diœcious.*

2. C. Davalliána. Smith.
Spike oblong, rather loose-flowered; *perig.* ovate-lanceolate, attenuate, convex, terete, recurved, longer than the ovate glume; *st.* and *lvs.* are usually serrulate.—Wayne Co., N. Y., *Sartwell.*

3. C. exilis. Dewey.
Fertile spike ♂ below, ovate, rather densely-flowered; *perig.* ovate-lanceolate, convex on both sides, diverging, serrulate on the margin, a little longer than the ovate, acute glume; *lvs.* setaceous; *st.* 12—20' high.—Grows in Danvers and Ipswitch, Ms., *Oakes,* in N. Y. and N. J. May.

β. squamacea. Dew. *Spike* often an inch long, having many ♂ glumes at the base and few *perig.* at the summit.—Longer than the other, and grows with it in Ipswich, Mass., *Oakes.*

B. *Spikes several, androgynous.*

1. *Stamens variously situated—above, below, or in the middle; sometimes diœcious.*

4. C. sterîlis. Willd.
Spike compound, ♂ below, often diœcious; *spikelets* 4—6, ovate, subapproximate; *perig.* ovate, acuminate or subrostrate, bifid, compressed, triquetrous, scabrous on the margin, equaling the ovate, acutish glume; *st.* 2f high, erect and stiff.—Wet places, common.

5. C. bromoîdes. Schk.
Spikelets numerous, alternate, ♂ below, sometimes all ♀; *perig.* lanceolate, erect, acuminate, scabrous, nerved, bifid, twice longer than the ovate-lanceolate glume.—Common in small bogs, in wet places.

6. C. siccàta. Dewey.
Spikelets numerous, ♂ above, often wholly ♂, ovate, close or approximate; *fr.* ovate, lanceolate, acuminate, compressed, nerved, bifid, scabrous on the margin, equaling the ovate and lanceolate glume.—Sandy plains, Westfield, Mass., *Davis;* Ipswitch, Mass., *Oakes;* widely spread over the country, but not abundant.

7. C. Sartwellii. Dewey.
Spikelets 12—20, ovate, sessile, compact, bracteate, lower ones especially fructiferous; upper often ♂; *perig.* ovate, lanceolate, convexo-concave, subulate, short, 2-toothed, a little longer than the ovate and acute glume; *lvs.* flat, linear, shorter than the stem.—Junius, Seneca Co., N. Y., *Sartwell.*

2. *Stamens at the summit of the spikelets.*
a. *Cephalous, or fruit in heads.*

8. C. cephalophôra. Willd.
Spikelets ovate, densely aggregated into an ovate head, bracteate, about 5; *perig.* ovate, acuminate, compressed, bifid, scabrous on the margin, with a short, ovate, and scabro-cuspidate glume which equals it; *st.* 8—16' high.—Borders of fields and woods, common, but not abundant.

9. C. vulpinoidea. Michx. (C. vulpinæformis. *Tuckerman.* C. multiflora. *Muhl.*)—*Spikelets* ovate-oblong, obtuse; *spike* decompound, bracteate, conglomerate; *perig.* ovate, acuminate, densely imbricate, bifid, triplinerved, diverging, a little shorter than the ovate-cuspidate glume; *st.* obtusely triangular, round and leafy towards the base.—Common in fields.

β. microsperma. Dew. (C. microsperma. *Wahl.*) *Spikelets* closely aggregated, whole spike less compact; *perig.* more convex, shorter, less acuminated into a beak, very abundant.—Grows with the other, in dry and moist situations.

10. C. setacea. Dewey.
Spikelets ovate, alternate, obtuse, conglomerate, bracteate; *perig.* ovato-lanceolate, acuminate, compressed, bifid, some diverging, about equal to the ovate-lanceolate, awned glume; *st.* 2f high, acutely triangular, scabrous above and striate.—Wet places—not abundant.

11. C. Muhlenbergii. Schk.
Spikelets alternate, obtuse, approximate, with a long bract at the lower

one; *perig.* ovate, convex above, very smooth, nerved, bifid, scabrous on the margin, some diverging, a little shorter than the ovate and mucronate glume; *st.* 12—18′ high.—In fields, not very common, readily distinguished from the three preceding and following.

12. C. CHORDORRHIZA.

Spikelets 3—5, aggregated into a head, ovate, sessile; *perig.* ovate, acuminate, subrostrate, convex above, equaling the broad, ovate and acute glume *st.* branching towards the base and sending out roots at the joints; *spikes* rarely bearing only stamens.—Marshes, New York; common, *Sartwell.* Michigan. *Cooley.*

13. C. PRAIREA. Dew.

Spike below branched; *spikelets* ovate, sessile, 5—7 on a branch; *perig.* ovate-lanceolate, convex both sides, scabrous on the margin, slightly bifid, equaling the ovate-lanceolate glume; *st.* 2—3f high, leafy towards the base.— Abundant in the prairies of Michigan, and sparingly found in N. England and N. Y. Resembles *C. paniculata. L.;* which has a much broader ovate glume, shorter than the perigynium and is far more paniculate, and for which this has been taken.

14. C. TERETIUSCULA. Good.

Spikelets ovate, acute, sessile, decompound, brownish, lower one bracteate; *perig.* ovate, acute, convex and gibbous, scabrous on the edge, spreading, longer than the ovate and acute glume; *fr.* brown; *st.* 18—36′ high, leafy towards the root.—Wet places, common, in tufts.

15. C. DECOMPOSITA. Muhl.

Spike decompound or paniculate; *spikelets* very many, ovate, alternate, *perig.* ovate, convex on both sides, triangular, acutish or short-rostrate, short, brownish, glabrous, about equal to the ovate and acuminate, whitish glume; *st.* 18—30′ high.—Found in swamps, Michigan, and in Yates Co., N. York, *Sartwell.*

16. C. PANICULATA.

Spike paniculate, often ♂ ♀, long and spreading; *spikelets* ovate, sessile, 6—18 on a branch below, short bracteate; *perig.* ovate, acute, gibbous, nerved, 2-toothed, brownish or tawny, 2-toothed, serrulate on the margin, a little shorter than the broad-ovate, short-acute glume; *st.* 2f high.—Found in Northern America, and hardly known in the United States.

b. *Perigynia radiating.*

17. C. ROSEA. Schk.

Spikelets 3—5, subremote, sessile, alternate, stellate, even before maturity, lowest long bracteate; *perig.* oblong-lanceolate, 5—12, convex above, scabrous on the margin, 2-toothed, very diverging or even reflexed, twice as long as the ovate-obtuse glume; *st.* 8—16′ high.

β. *radiata.* Dew. *Spikelets* distant, about 3-flowered, with setaceous bracts, *perig.* oblong, acute; *st.* 4—8′ high, flaccid or lax, setaceous, with very narrow leaves.—Common in pastures and moist woods; the variety is about woods, o. open places in woods.

18. C. RETROFLEXA. Muhl.

Spikelets about 4, ovate, alternate, subapproximate, sessile, bracteate and stellate in maturity; *perig.* ovate, acutish, 2-toothed, subscabrous or smooth on the margin, reflexed and spreading, about equal to the ovate and acute glume; *st.* about a foot high.—Readily distinguished from the preceding. Woods and pastures, not abundant.

19. C. STIPATA. Muhl.

Spike often decompound; *spikelets* oblong, aggregated, numerous, bracteate; *perig.* ovate-lanceolate, round at the base, plano-convex, nerved, bifid, subscabrous on the margin, diverging, twice longer than the ovate-lanceolate glume; *st.* thick, acutely triquetrous, concave on the sides.—Wet places and marshes, abundant.

20. C. ALOPECOIDA. Tuckerman. (C. cephalophora, β. maxima. Dew.) *Spike* compound, rather loose; *spikelets* 8—10, aggregated into an oblong

head, bracteate, sessile; *perig.* ovate, plano-convex, scarcely nerved, acuminate, serrulate on the edge, bifid, subrostrate, a little longer than the ovate and acuminate glume; *st.* triquetrous, scabrous on the edges.—Moist woods, Penn. and N. York, *Sartwell.*

21. C. CEPHALOIDEA. Dew.

Spikelets 4—6, ovate, aggregated closely, sessile and bracteate; *perig.* ovate, obtusish, bifid, scabrous on the margin, plano-convex, very diverging in maturity, about twice as long as the short, ovate, obtusish glume.—Dry fields— not abundant, but common over New England and New York. In hedges it is often four feet long, and subrostrate, leafy towards the base.

22. C. SPARGANÖIDES. Muhl.

Spikelets 7—10, ovate, rather distant, bracteate, sessile; *perig.* ovate, acute, compressed, diverging, acuminate, 2-toothed, scabrous on the margin, nearly twice the length of the ovate, acute, or mucronate glume; *st.* about 2f high, with long, striate leaves.

β. ramea. Dew. has one branch or more at the base, with several spikelets in the place of the lower spikelet, and is the *C. divulsa* of Pursh.—About cultivated and moist fields, common.

23. C. MURICÀTA.

Spikelets about 5, ovate, sessile, approximate, bracteate, lower ones sometimes remotish; *perig.* ovate-lanceolate, plano-convex, 2-toothed, horizontal, scabrous on the margin, sometimes longer than the ovate-lanceolate glume.— Fields near Boston, *B. D. Greene,* and common in Arctic America; Charlestown, Mass., *M. A. Curtis.*

e. Perigynia few.

24. C. DISPERMA. Dew.

Spikelets 3—4, erect, subapproximate, lowest bracteate; *perig.* ovate, obtuse, about two, nerved, plano-convex, short-beaked, glabrous, twice longer than the ovate, acute, submucronate glume; *st.* slender, 6—18' high, with narrow and linear leaves.—*Perigynia* 1—2, sometimes 3. Wet woods, N. England, N. York, Michigan and Wisconsin Territory.

3. Androgynous; stamens at the base of the spikelets.

d. Perigynia radiating.

25. C. STELLULÀTA. Good.

Spikelets 4—6, ovate, remotish, sessile; *perig.* broad ovate, contracted into a short beak, compressed, slightly bifid, scabrous on the edge, diverging and reflexed, a little longer than the ovate, obtusish glume; *st.* erect, stiff, leafy below, 8—24' high.—Common in wet places over the Northern States.

26. C. SCIRPÖIDES. Schk.

Spikelets about 4, ovate, approximate, sessile, obtuse, lowest bracteate; *perig.* ovate, cordate, compressed, lanceolate or rostrate, scabrous on the margin, diverging or horizontal, longer than the ovate-lanceolate, acute glume; *st.* 6—16' high, leafy towards the base.—Wet places in the country. The more lanceolate fruit and glume, and more flexible stem, separate it from the preceding. *C. scirpoides* has the stamens chiefly below the upper spikelet.

27. C. CURTA. Good.

Spikelets 4—7, ovate-oblong, upper subapproximate, lower often remote; *perig.* round-ovate, acutish, obtusish, diverging, convexo-concave, 2-toothed, slightly scabrous, longer than the ovate, white, hyaline glume; *st.* 1—2f high, usually light green, with silvery or hoary spikelets.—Moist places over the country.

28. C. SPHÆROSTACHYA. Dew. (C. canescens, β. sphærostachya. *Tuck.*)

Spikelets 3—4, ovate, roundish, remote, sessile, few fruited, 2—6; *perig.* ovate-lanceolate or roundish, rostrate, longer than the ovate and hyaline, white glume; *st.* 1—2f high, slender, flaccid, subrostrate, and with the leaves, green. —Common in N. England and N. York, in wet places.

e. Ovate-lanceolate spikelets; few-fruited.

29. C. DEWEYÀNA. Schk.

Spikelets about 3, sessile, ovate-lanceolate, alternate, subremote, highest

49*

bracteate; *perig.* ol .ong-lanceolate, rostrate, acuminate, bifurcate, plano-convex, slightly scabrous on the margin, a little longer than the ovate-lanceolate, awned, hyaline glume; *st.* 1—4' long, subprocumbent, with radical leaves; whole plant yellowish-green.—Common in open woods or on the borders of woods.

30. C. TRISPERMA. Dew.
Spikelets about 3, remote, sessile, alternate, highest ebracteate; *perig.* ovate-oblong, acute or short-rostrate, plano-convex, at the orifice entire, nerved, subscabrous on the edges, somewhat diverging, longer than the oblong, acute and hyaline glume; *st.* 10—24' high, prostrate or recurved, filiform, slender, longer than the leaves.—In tufts in marshes or wet woods; common in N. England and N. York.

ζ. Spikelets oval.

31. C. SCOPARIA. Schk.
Spikelets 5—10, usually 5—7, ovate, sessile, approximate, the lowest with a long deciduous bract; *perig.* ovate, lanceolate, nerved, erect, slightly margined, glabrous, longer than the lanceolate, acuminate glume; *st.* 18—24' high, leafy towards the root.—Moist places, very common.
β. aggregata. Dew. *Spikelets* aggregated into a head, somewhat spiral.

32. C. LAGOPODIŌIDES. Schk.
Spikelets 8—20, cylindric, ovate, rather near, alternate and sessile; *perig.* lanceolate, tapering at both ends, concavo-convex, nerved, bidentate, scabrous on the margin, nearly twice as long as the ovate-lanceolate glume; *st.* nearly 2' high, leafy; the whole light green.—Common.

33. C. STRAMINEA. Wahl.
Spike compound, erect; *spikelets* about 6, ovate, short-oblong, alternate, sessile, subapproximate; *perig.* broad, roundish-ovate, compressed, ciliate-serrate on the margin, beaked, 2-toothed, widely winged, commonly shorter than the ovate-lanceolate glume; *st.* 12—20' high, longer than the leaves; *spikelets* whitish or tawny.—Common in woods and fields.
α. brevior. Dew. *Spikelets* 3—5, often closely approximate, and more nearly round; *perig.* shorter-ovate, and shorter-rostrate, scarcely longer than the ovate-lanceolate glume.—This is the plant originally described by Willdenow.
β. minor. Dew. *Spikelets* small, 5—6, globose or obovate, less approximate; *perig.* small, ovate, acuminate, less winged, serrulate, about equaling the ovate, acute glume.

34. C. TENÈRA. Dew. (C. adusta. *Bott.*)
Spike compound, recurved; *spikelets* about 5, obovate, remotish, alternate, sessile, brownish, attenuated below, the lowest bracteate; *fr.* ovate, compressed, somewhat winged, rostrate, nerved, ciliate-serrate, longer than the oblong-lanceolate scale; *st.* 15—30' high, small and slender, erect, with a nodding spike, longer than the leaves.—Light green. Common.

35. C. FESTUCACEA. Schk.
Spikelets 5—8, obovate and clubform, sessile and alternate, approximate, lower one bracteate; *perig.* roundish-ovate, rostrate, winged, striate, 2-toothed, scabrous on the margin, longer than the ovate, lanceolate glume; *st.* 15—30' high, erect and stiff, leafy below.—Plant pale green. Spikelets greenish to brown. Common in fields, but not abundant. The *club-form* spikelets from the decurrent scales of the ♂ flowers, especially mark this species.

36. C. MIRABILIS. Dew.
Spikelets 7—11, ovate-globose, alternate, sessile, often closely-aggregated, and stiff-form, bracteate below; *perig.* ovate, sublanceolate, scabrous on the margin, concavo-convex, rostrate, 2-toothed, subdiverging, scarcely twice longer than the ovate. lanceolate glume; *st.* 18—36' high, erect, stiff, rough above, rather slender; plant light green.—Common about fences and hedges, and has a specially rigid appearance.

37. C. CRISTÀTA. Schw.
Spikelets 6—14, globose, sessile, closely aggregated into a head of a crested form, bracteate; *perig.* ovate, oblong, compressed, winged, rostrate-acuminate,

bifid, concavo-convex, scabrous on the margin, longer than the oblong, lanceolate glume; *st.* 1—3f high, acutely triangular.—Plant yellowish-green. Common in fields and meadows on colder soils.

38. C. TENUIFLÔRA. Wahl.

Spikelets 2—3, ovate, clustered, sessile, alternate, lower one bracteate; *perig.* ovate-oblong, acutish, plano-convex, equaling the oblong-ovate, hyaline or white glume; *st.* a foot or more high, slender, subprostrate, longer than the flat andnarrow leaves.—Light green. Spikelets whitish. Burlington and Salem, Vt., in swamps, *Robbins,* Oriskany and Ogdensburg, N. Y., *Kneiskern,* Southampton, Mass., *Chapman.*

39. C. CYPERÔIDES.

Spikelets ovate, closely aggregated into a head, with long and leafy bracts; *perig.* ovate, long-lanceolate, or drawn into a long awn scabrous on its edges, slightly stipitate, 2-toothed, a little longer than the lanceolate and cuspidate glume; *plant* very pale green.—Jefferson Co., N. Y.—first found in our country last summer, by Dr. Crawe.

40. C. MUSKINGUMENSIS. Schw.

Spikelets oval-oblong, 5—10, somewhat tapering at both ends, large and approximate, close-flowered, dry and chaff-like; *perig.* lanceolate, compressed, thin, distinctly winged, bidentate, nerved, acuminate, twice longer than the ovate-lanceolate glume; *plant* light green in all its parts.—Common in Ohio and Mich., 18—36' high.

41. C. LIDDÔNI. Boott.

Spikelets 5—7, oblong-ovate, closely aggregated; *perig.* ovate, lanceolate, acuminate, oblique at the orifice, glabrous, on the margin serrulate, scarcely longer than the ovate-lanceolate glume, which is acute and hyaline on the edges; *perig.* and *glumes* rather chestnut brown; *plant* yellowish-green.—Arctic Am., *Boott,* Mich., *Dr. Cooley.*

C. *Stamens and Stigmas on separate spikes.*

1. *Staminate spike single.*

42. C. AUREA. Nutt. (C. pyriformis. Schw.)

♂ *Spike* short, cylindric, pedunculate; ♀ *spikes* 3, oblong, loose-flowered, subpendulous, exsertly pedunculate, subapproximate, bracteate; *perig.* globose, obovate or pear-form, obtuse, nerved, entire at the mouth, longer than the ovate acute or short-mucronate glume; *st.* 3—10' high, slender, often subprocumbent –Plant glabrous, green. Common in wet grounds.

43. C. SAXATILIS.

♂ *Spike* oblong, thick; ♀ *spikes* 2 or 3, oblong, obtuse, sessile, lower pedunculate; *perig.* elliptic, plano-convex, obtuse, short-rostrate, about equaling the oblong and obtuse glume; *st.* 6—10' high, erect, with long and leafy sheath and bracts.—Spikes nearly black. White Mts., N. H., *Barratt;* woods, Vt., *Pursh.*

44. C. CONCÔLOR. R. Br.

♂ *Spike* erect, cylindric; ♀ *spikes* 2—3, erect, subsessile, cylindric; *perig.* oval, entire, smooth, mucronate, about equal to the oblong and obtuse glume; *st.* 10—15' high, smooth, leafy below; *bracts* auriculate; ♂ spike sometimes pistillate above.—White Mts., N. H., *Boott.* Closely related to *C. cæspitosa,* L., but has a smooth stem; scales of light color.

2. *Staminate spikes one or more, and the upper part of the pistillate sometimes staminate.*

45. C. RIGIDA. Good.

♂ *Spike* oblong, cylindric, rarely 2; ♀ *spikes* 2—3, oblong, cylindric, densely-flowered, short and thick, approximate, lower one subpedicellate, with a bract surpassing the stem; *perig.* ovate, obtusish, entire at the orifice; *glume* nearly twice longer than the mature fruit and subequal before; *st.* 3—6' high, thick and stiff, often recurved; *lvs.* stiff and glaucous.—Ipswich, Ms, *Oakes.* Has been confounded with *C. cæspitosa.*

46. C. CÆSPITOSA.

♂ *Spike* single, oblong, cylindric, sometimes 2, with oblong, black scales; ♀ *spike* 2—3, cylindric, obtuse, rather thick, remotish, bracteate, lowest one short-pedunculate; *perig.* ovate, obtuse, glabrous, entire at the orifice, scarcely rostrate, a little longer than the oblong, obtuse, black glume; *st.* 6—14' high, scabrous on the edge, leafy towards the base; *lvs.* flat.—Wet places, Ipswich, Mass., *Oakes;* N. Y. and Michigan.

47. C. STRICTIOR. Dew.

♂ *Spikes* 1—2, with oblong and blackish, acutish glumes; ♀ *spikes* 2—3, cylindric, ♂ above, and hence acutish, lowest short-pedunculate; *perig.* ovate, compressed, acute, glabrous, entire at the orifice, early falling off, glabrous, a little longer than the oblong and acute glume; *st.* a foot and more high, trique-trous and rough on the angles, with reticulated filaments connecting the leaves towards the base; *lvs.* erect, close; whole plant glaucous except the spikes.— Wet places, common.

48. C. STRICTA.

♂ *Spikes* 1—2, cylindric, lower one sessile, and the scale rusty brown and obtuse; ♀ *spikes* 2—3, long-cylindric, upper half ♂, lower longer, short-peduncu-ate, loosely-flowered below; *perig.* ovate-acuminate or elliptic, compressed, at the orifice entire or slightly emarginate and its glume strongly ferruginous, the lower ones acute-lanceolate, the upper linear and obtuse, commonly longer and narrower than the perigynia; *st.* 2f high, with reticulated filaments connecting the leaves, *Boott.*—Wet places, as bogs, common.

49. C. ACUTA.

Spikes long and slender; ♂ 2—3; ♀ 3—4, long, slender, cylindric, short-pedunculate, nodding towards maturity, remotish, bracteate; *perig.* oval or ob-long, obtuse, orifice protended, or very short-rostrate, about equaling the oblong, acute glume; *st.* acute, triquetrous, lax; the stamens at the summit of the pistillate spikes render them acute.—Common.

β. *erecta.* Dew. (*Schk.* fig. 85, *c.*) *Spikes* shorter, 2 of each; ♀ nearly erect, oblong, close-flowered; *perig.* shorter than the ovate-lanceolate glume.—Evi-dently misplaced by *Schkuhr.*

γ. *sparsiflora.* Dew. (*Schk.* fig. 92, *b.*) ♀ *Spikes* very long, recurved, very sparsely flowered below.—Common.

50. C. AQUATILIS. Wahl.

♂ *Spikes* 1—4, erect, cylindric, lowest bracteate, the glume oblong, ob-tusish; ♀ *spikes* often 3, cylindric, thick and thickened above, 1—2' long, sub-erect, short-pedunculate, densely-flowered; *perig.* elliptic, lenticular, rather small, entire, glabrous, protruded at the orifice, about equal to the ovate, acutish glume; *st.* 20—30' high, rather obtuse-angled and scarcely scabrous.-n marshes and wet places, common.

51. C. CRINITA. Lam.

♂ *Spikes* one or more, lax, oblong, sometimes with a few ♀ flowers; ♀ *spikes* about 3, oblong, cylindric, pedicellate, nodding, attenuated below, and more loosely flowered, often ♂ at summit; *perig.* ovate, sub-inflated, short-ros-trate, entire at the orifice, glabrous, about ⅓ as long as the oblong, obtusish, scabrous-awned glume; *st.* 12—24' high, rough, triquetrous.—Common in w.t places.

β. *gynandra.* Dew. (C. gynandra. *Schw.*) ♀ *Spikes* pendulous, thicker in the midst; *glumes* about twice as long as the perigynia.

52. C. PALEACEA. Schreb. Schk., fig. 125.

♀ *Spikes* about 4, long-cylindric, densely-flowered, recurved, with a long, reclined peduncle; *perig.* ovate, suborbicular, obtusish, emarginate at the ori-fice, convex both sides; *glumes* terminated by a long, serrate point more than thrice the length of the perigynia; *st.* 20—42' high, recurved, rough-edged, pa'e green.—Common in dry grounds.

II. Stigmas three.

D. *Spikes androgynous. Monœcious.*

1. *Stamens at the summit.*

a. *Spikes single.*

53. C. POLYTRICHÖIDES. Muhl. (C. microstachya. Mx.)
Spike oblong, terminal; *perig.* 3—8, oblong, alternate, subtriquetrous, glabrous, emarginate, twice longer than the ovate and obtuse, and rarely mucronate glume; *st.* 4—12' high, very slender, with setaceous and subradical leaves.—Common in wet and cold grounds.

54. C. LENEOGLÖCHIN. Ehrh. (C. pauciflora. *Lightfoot*.)
Spike about 4-flowered, with 1 or 2 ♂ flowers at the apex; *perig.* lanceolate, subtriquetrous and tapering, much reflexed, twice longer than the oblong-lanceolate glume; *st.* 3—8' high, with subradical and linear leaves.—In Ashfield and Hawley, Mass., in a marsh, *Porter.*

b. *One or more radical peduncles with a single spike.*

55. C. PEDUNCULATA. Muhl.
Spikes about 5, 3-sided, distant, long, recurved, pedunculate; *perig.* obovate, triquetrous, recurved at the apex, commonly glabrous, a little longer than the oblong or obovate, mucronate glume; *st.* 4—12' high, triangular, rather procumbent; *sta.* sometimes removed a little from the ♀ spike.—Common in woods. Flowers early in the spring.

56. C. WILLDENOWII. Schk.
Sts. or *radical ped.* 1—3; *spike* commonly single, stameniferous above, or the stamens removed a little; *perig.* 3—6, alternate, loose, oblong and inflated a little, tapering at the base and conic-rostrate above; ♀ *glumes* ovate and acute, the lower ones long and leaf-like, much surpassing the stem.—On dry grounds, common throughout the U. S.—One variety has the ♂ spike distinct; another is destitute of the long and leafy scales, and is frequent at the North as well as in Flor.

57. C. STENDELII. Kth.
Sts. or *radical ped.* 1—8' long; *spike* commonly single, stameniferous above; *perig.* 1—4, subglobose or ellipsoid and inflated, alternate, stipitate, terete and conic-rostrate, with an oblique orifice; ♀ *glumes* usually long and leafy; *lvs.* smooth, soft, narrow, longer far than the stems.—Jefferson Co., N. Y., and in Ohio and the Western States.

58. C. BACKII. Boott.
Ped. radical, 1—4f high, stiff, thick or large; *spike* single, commonly stameniferous above, short; *perig.* ovate, globose, smooth, conic-rostrate, entire at the orifice, when mature pear-shaped, the beak articulated to the fruit; ♀ *glumes* usually long and leaf-like, enclosing the fruit; *lvs.* radical, flat, thick, rough or scabrous and short.—Jefferson Co., N. Y. and Arctic Am.—The three preceding species are closely related, and yet look very different.

2. *Spikes staminate at the base.*

c. *Spikes one, often more.*

59. C. SQUARROSA.
Spikes 1—4, oblong, cylindric, obtuse, upper one attenuated below at first by the decurrent, ♂ flowers, all very densely flowered; *perig.* ovate, subglobose, long-rostrate, 2-toothed, horizontal, glabrous and subsquarrose, longer than the lanceolate glume; *st.* 1—2f high, slender for the large spike or spikes; *lower spikes* pedunculate.—Large and fine. It is *C. typhirea Mx.* when only *one spike* is present.
β. (*C. typhinoides.* Schw.) *Spikes* 2, the lower on a very long peduncle, and both longer and smaller.

E. *Spikes diœcious.*

60. C. SCIRPÖIDEA. Mx.
Spike oblong, cylindric, acutish; ♂ *glume* oblong, obtusish; *perig.* ovate, (oval), subrostrate, pubescent, longer than the ovate, acutish glume; scarious on the edge; *st.* 4—10' high, erect; *lvs.* flat and long.—White Mts., N. H., Oakh

F. *Terminal spike androgynous, pistillate at the summit ; the other pis-tillate.*

61. C. virescens. Muhl.
Spikes 2—4, oblong, erect, alternate, *the lower* subsessile, bracteate ; *upper spike* very rarely wholly ♂ ; *perig.* ovate, obtuse, costate, pubescent, longer than the ovate, pubescent and mucronate glume, or about equal to it ; *st.* 1—2f high, rather slender ; *lvs.* towards the base.—Whole plant pubescent and light green.
β. *costata.* Schw. *Perig.* strongly costate, *outer sheaths* purplish-brown ; *lvs* numerous and larger.—Both are common in open woods and hedges.

62. C. hirsúta.
Spikes 3, short-oblong, thick, alternate, erect, the lower subsessile and long-bracteate, all approximate and densely flowered ; *perig.* ovate, triquetrous, nerved, obtuse, entire at the orifice, glabrous in maturity, about equal to the ovate, acuminate, glabrous glume ; *st.* 12—20′ high ; *lvs.* and *sheaths* retrorsely pubescent ; *upper spike* very rarely all ♂.—Moist upland meadows. Common.
β. *pedunculata.* Torr. *Spikes* oblong-cylindric, pedunculate ; *lvs.* slightly pubescent.—Common. C. triceps (*Mx.*) much resembles this,—is not pubes-cent but glabrous.

63. Buxbaumii. Wahl.
Spikes about 4, cylindric, thick, upper one sometimes wholly ♂, and sometimes ♂ above and below ; *pistilliferous* oblong, subremote, subsessile, bracteate ; *perig.* ovate-oblong, acutish, or obovate, obtuse, subtriquetrous, entire at the orifice, nerved and glabrous, scarcely equal to the oblong and mu-cronate glume ; *st.* 10—18′ high, leafy towards the base.—Common in wet grounds. It is described as sometimes having 2 stigmas in Europe, but placed by Schk., Wahl., &c., in the division having 3.

64. C. gracillima. Schw.
Spikes 3—4, long, graceful, sub-loose-flowered, distant, long-pedicellate, recurved in maturity, bracteate, upper one rarely all ♂ ; *perig.* oblong, trique-trous, obtuse, oblique at the orifice, slightly 2-lobed, longer than the oblong and obtuse and short-awned glume ; *st.* often 2f high, reddish towards the base, leafy and subprocumbent, pale green.—Common in damp meadows.

65. C. formósa. Dew.
Spikes 3—4, oblong, short and thick, distant, 1-sided, on a long and slen-der peduncle, recurved ; *perig.* oblong, triquetrous, subinflated, acutish at either end, nearly entire or 2-lobed at the orifice, twice longer than the ovate and acute glume ; *st.* 1—2f high, 3-sided, dark brown towards the base, yellowish bright green.—Common in wet meadows.

66. C. Davisii. Torr. (C. Torreyana. *Dew.*)
Spikes 4, oblong, cylindric, subsparsely flowered, remote, pedicellate, pen-dulous in maturity ; *perig.* oblong-conic, subinflated, subtriquetrous, nerved, acutish, short-rostrate, 2-lobed at the orifice, glabrous towards maturity, about equaling the oblong, scabrous-awned glume ; *st.* 1—2f high, triquetrous, sca-brous above, with leaves equaling it ; *lvs.* and *sheaths* pubescent, sometimes but very little, light green.—First found on the alluvial meadows of the Housa-toric in Mass., *Dewey.* Sometimes nearly pubescent.

G. *Staminate spike single.*

1. *Pistillate spikes short and sessile or nearly sessile. Peryginia radiating or diverging.*

67. C. varia. Muhl.
♂ *Spike* erect, short or subelongated ; ♀ *spikes* 3, ovate, sessile, rather near, bracteate, few-flowered ; *perig.* ovate or sub-globose, subtriquetrous, acu-minate-rostrate, bifid, scabro-pubescent, about equal to the ovate, acuminate glume ; *st.* 6—15′ high, erect, slender, purple towards the base. Pale green.—Dry woods and hedges ; common.
β. *pedicella'a.* Dew. has pistillate spikes ovate-oblong, short-pedicellate erect, loose-flowered ; *perig.* more numerous.—Grows in the same situations.

68. C. PENNSYLVANICA. Lam. (C. marginata. *Muhl.*)

♂ *Spike* erect, pedunculate, subtriquetrous, with an obtuse glume; ♀ *spikes* 1--3, ovate, subsessile, subapproximate, few-flowered; *perig.* ovate-globose, tomentose, short-rostrate, slightly 2-toothed, about equal to the ovate-acuminate, or oblong-acuminate, deep reddish glume; *st.* 4--12' high, erect, stiff, with short leaves.—Open woods and hedges, common—much resembles the preceding, but larger in all its parts, and readily distinguished by its different aspect and its deep reddish-brown scales.

69. C. EMMONSII. Dew.

♂ *Spike* sessile, short; ♀ *spikes* 2--3, approximate, sessile, few-flowered, often one long radical peduncle; *perig.* globose-triquetrous, attenuated at the base, rostrate, pubescent, at the orifice oblique, about equal to the ovate glume; *st.* decumbent, 6--10' high, leafy at the base, pale ash-green.—On dry fields and hills; common.

70. C. NOVÆ-ANGLIÆ. Schw.

♂ *Spike* short, slender, oblong; ♀ *spikes* 2--3, ovate, alternate, sessile, remotish, few-flowered, bracteate; *perig.* 3--6, oval-triquetrous, rostrate, costate, slightly pubescent, a little longer than the ovate, mucronate glume; *st.* 4--8' high, slender, subdecumbent, longer than the leaves.—Pale green. Open woods in high grounds.

β. *collecta*. Dew. (C. collecta. *Dew.*) *St.* 10--16' high, very slender erect; ♀ *spikes* 2--4, lower short-pedunculate; *perig.* more tapering into a beak, slightly bidentate.—High lands of Mass.; not abundant.

71. C. UMBELLATA. Schk.

♂ *Spike* short, erect; ♀ *spikes* several, each on its radical peduncle, ovate, subumbellate; *perig.* ovate or globose, 5--8, acutish at either end, rostrate, short-bidentate, pubescent, equaling the ovate-lanceolate glume; *st.* ½--4' high, with very long leaves.

β. *ricina*. Dew. 1 or 2 ♀ spikes close to the ♂, sessile; *the other* ♀ *spikes* on their own stems or radical peduncles.—In small tufts on dry hills. Both varieties grow on the same root, but Schk. saw and figured only the first.

72. C. PRÆCOX. Jacq.

♂ *Spike* erect, subclavate; ♀ *spikes* 1--3, ovate, bracteate, approximate, lower one short-pedunculate; *perig.* 6--12, ovate and subglobose, triquetrous, pubescent, short-rostrate, equal to the ovate, acute, or mucronate glume; *st.* 2--6' high, leafy at the base.—On rocky hills, Salem, Mass., *Pickering*, Ipswich, Mass., *Oakes*.

2. Pistillate spikes with nearly inclosed peduncles.

73. C. VESTITA. Willd.

♂ *Spike* single, rarely 2, cylindric, oblong; ♀ *spikes* 2, ovate-oblong, sessile, subapproximate, bracteate, often with stamens above; *perig.* ovate, oblong, subtriquetrous, nerved, short-rostrate, bifid, pubescent, a little longer than the ovate-oblong, acutish, submucronate glume; *st.* 18--30' high, acutely triangular and leafy below.—Common in wet places over the country.

74. C. PUBESCENS. Muhl.

♀ *Spikes* 2--3, oblong, rather loose-flowered, erect, bracteate, the lowest pedunculate; *perig.* ovate-triquetrous, rostrate, nearly entire at mouth, pubescent, a little longer than the ovate-oblong, carinate, mucronate glume; *st.* 10--20' high, and with the leaves, pubescent.—Moist woods and meadows; common.

75. C. FLAVA. L.

♀ *Spikes* 2--4, ovate-oblong, approximate, sometimes androgynous; *perig.* ovate, closely imbricate, costate, bidentate, reflexed with a long, curved beak, longer than the ovate-lanceolate glume; *st.* 10--20' high, rather obtusely angled or triquetrous· glabrous; yellowish-green.—Wet and cold soils; common.

76. C. LEPIDOCARPA. Tausch.

♀ *Spikes* 1--3, short and round-ovate, often aggregated, sessile, dense-flowered, the lowest sometimes remote and pedunculate; *perig.* ovate trique trous, inflated, nerved, rostrate, and at last recurved, 2-toothed, diverging, twice longer than the ovate and obtuse glumes ; *plant* yellowish-green.—Mass., N. Y., Mich.—Formerly confounded with C. flava.

77. C. ÆDÉRI.
Spikes sometimes androgynous; ♀ about 4, clustered, nearly sessile, short-oblong, sometimes ♂ above or below, bracteate ; *perig.* rather obovate, subinflated, nerved, bidentate, diverging with a subulate beak, a little longer than the ovate glume ; *st.* 2—10′ high, leafy.—Pale yellow. Mass. and N. Y. —abundant in Pittsfield, Mass., and at Niagara Falls.

78. C. TENTACULÀTA. Muhl.
♀ *Spikes* 2—4, oblong, cylindric, bracteate, upper one sessile, the rest nearly sessile, densely flowered ; *perig.* ovate, inflated, long-rostrate, bidentate, nerved, diverging, glabrous, twice longer than the ovate and small scabro-mucronate glume ; *st.* 1—2f high, often large, triquetrous ; *lvs.* linear-lanceo-ate, longer than the stem.—In clusters in wet or marshy places ; common.

79. C. ROSTRÀTA. Michx.
♂ *Spike* short and small; ♀ *spikes* 2—3, sub-globose, or capitate, bracte ate ; *perig.* aggregated into a head, small, erect, or subdiverging, oblong-conic, very long-rostrate, slightly inflated at the base, twice longer than the ovate-oblong, acutish glume ; *st.* 8—16′ high, few-leaved, erect, stiff.—Pale yellow. At the base of the White Mts., N. H., *Oakes;* also in Canada, where Mx. found it. Has been called a variety of *C. Xanthophysa Wahl.*)

80. C. INTUMÉSCENS. Rudge. (C. folliculata. *Schk.* fig. 52.)
♂ *Spike* oblong, pedunculate; ♀ *spikes* 1—3, few-flowered, approximate, bracteate, erect, nearly sessile, the lower one sometimes remote and exsertly pedunculate ; *perig.* ovate-conic, large and much inflated, acuminate-rostrate, bidentate, nerved, diverging, very glabrous, thrice longer than the ovate-cuspi-date glume ; *st.* a foot or more high, erect, stiff, leafy, dark green and very glabrous.—Wet grounds, in open woods or marshes ; common.
β. *globularis.* Gray. ♀ *spikes* large, globular, many-fruited.—Grows in the same situations.

81. C. FOLLICULÀTA. (C. Xanthophysa. *Wahl.*)
♀ *Spikes* 2—4, ovate or capitate, densely flowered, distant, the peduncles sometimes projecting far beyond the sheaths, often ♂ at the apex, long brac-teate ; *perig.* oblong-conic, much inflated, diverging or horizontal, long-rostrate, twice longer than the oblong-ovate, acute glume ; *st.* 2—5f high, leafy ; *lvs* linear-lanceolate, long and flat.—Pale yellow. In wet or marshy places ; com. mon.

82. C. LUPULINA. Muhl. (C. lurida. *Wahl.*)
♂ *Spike* erect, slender, subsessile; ♀ *spikes* 2—4, ovate-oblong, large and thick, or oblong-cylindric, short-pedunculate, erect, densely flowered, approximate, the lowest sometimes long-pedunculate and distant ; *perig.* ovate-conic, ventricose, long, conic-rostrate, bicuspidate, nerved, glabrous, about thrice longer than the ovate-lanceolate, acuminate glume ; *st.* 1—3f high, tri-quetrous, leafy ; *lvs.* and *bracts* long, flat, wide, striate, scabrous on the edge.— Bright green. Finely named from its *hop-like* spikes. Marshes and about ponds, common.
β. *polystachya.* Torr. ♀ *Spikes* about 5, very long-cylindric, the lowest re mote and very long-pedunculate ; *perig.* less inflated.—Swamps, in Phillips town, N. Y., on the Highlands, *Barrati.*

3. *Pistillate spikes exsertly pedunculate.*

83. C. PLANTAGINEA. Lam. Schk., fig. 70. (C. latifolia. *Wahl.*;
♂ *Spike* erect, large, subclavate, with oblong and acute glume.; ♀ *spikes* 3—5, oblong, erect, remote, sparse-flowered, 2 upper nearly inclosed-peduncu iate, the lower ones exsertly-pedunculate, with subulate bracts; *perig.* oblong, triquetrous-elliptic or cuneiform, tapering at either end, recurved at the apex, and entire at the orifice, longer than the ovate-cuspidate glume ; *s′.* 8—18′ high, erect, triquetrous, with dark brown sheaths; *lvs.* radical, broad, ensiform, strongly

3-nerved.—Bright green. Hedges and open woods, common and one of the first appearing species in the spring.

84. C. CAREYÀNA.

♂ *Spike* erect, oblong, with oblong and obtuse glumes; ♀ *spikes* 2—3, ovate, loose and few-flowered, distant, upper subsessile, all bracteate; *perig.* ovate, triquetrous, subinflated, nerved, acuminate, tapering at the base, smooth and glabrous, entire at the orifice, twice longer than the ovate, mucronate glume; *st.* 1—2′ high, erect, smooth, leafy towards the base; *lvs.* linear-lanceolate.— Pale green. Woods, Auburn, N. Y., *Carey*, and in various places in Ohio; closely related to C. *plantaginea*, and to C. *Fraseri* of the Southern States.

85. C. ANCEPS. Schk. (C. plantaginea. *Muhl.*)

♀ *Spikes* 2—4, subfiliform, erect, attenuate, sparse-flowered, remote, with a 2-edged peduncle, leafy-bracteate, upper one subsessile; *perig.* oval-triquetrous, tapering at both ends, short-rostrate, attenuate, glabrous, striate, excurved at the apex, a little longer than the oblong-mucronate or ovate-acute glume; *st.* 6—12′ high, acutely triquetrous; *lvs.* radical, of medium width.—Glaucous or light green. Woods and hedges, common.

β. *patulifolia.* Dew. (C. anceps. *Schk.*, fig. 195.) *Lvs.* radical, broad, many-veined, narrower at the base; *sheaths* with long and leafy bracts; *perig.* longer-rostrate.

γ. *angustifolia.* Dew. (*Schk.* fig. 198.) *St.* a foot high; *lvs.* narrow, striate, long; *perig.* short-rostrate and much recurved.

86. C. BLANDA. Dew. (C. conoidea. *Muhl.*)

♀ *Spikes* 2—4, oblong, cylindric, subsparse-flowered, alternate, approximate, bracteate, highest subsessile, the lowest on a long, 2-edged peduncle *perig.* obovate, subtriquetrous, nerved, recurved at the apex, entire at the orifice, little longer than the ovate, scabro-mucronate glume; *st.* 8—12′ high, triquetrous, leafy towards the base; *lvs.* long as the stem.—Pale green or glaucous. Meadows and dry, open woods, common.

87. C. CONOIDEA. Schk. (C. granularioides. *Schw.*)

♂ *Spikes* 2—3, oblong, or ovate-oblong, remote, erect, rather dense-flowered, bracteate; *perig.* oblong-conic, obtusish, glabrous, nerved, subdiverging, entire at the mouth, a little longer than the ovate-subulate glume; *st.* 8—12′ high; *lvs.* towards the base, shorter than the stem.—Bright green. Moist, upland meadows, common.

88. C. TETANICA. Schk., fig. 207.

♀ *Spikes* 2—3, oblong, loose-flowered, remote; *perig.* obovate, recurved at the apex, entire at the orifice, with an ovate glume, obtusish at the upper and mucronate at the lower part of the spike; *st.* 6—10′ high, triquetrous, longer than the flat and linear-lanceolate leaves.—Light green. Upland meadows, rare.

89. C. DIGITÀLIS. Willd.

♀ *Spikes* about 3, 4—10-flowered, oblong, distant, loose-flowered, lax and recurved; *perig.* ovate, triquetrous, alternate, nerved, glabrous, short and obtuse, entire at the orifice, longer than the ovate-lanceolate glume; *st.* 4—12′ high, triquetrous, shorter than the long, decumbent leaves.—Pale green.

β. *Van Weckii.* Dew. Smaller; *perig.* more remote and smaller.—Open, moist woods, common. Has been mistaken for C. *oligocarpa*, *Schk. & Muhl.*

90. C. RETROCURVA. Dew.

♀ *Spikes* 2—4, on long, filiform and recurved peduncles, bracteate, sub-dense-flowered, short and thick, oblong; *perig.* ovate, triquetrous, nerved, obtusish, equaling the ovate, cuspidate glume; *st.* 6—12′ high, prostrate; *lvs.* radical and wide.—Glaucous. Open woods, rare. Has been considered C. *digitalis*, *Willd.*, but is different.

91. C. OLIGOCARPA. Schk.

♀ *Spikes* 2—3, erect, 3—4-flowered, bracteate; *perig.* obovate, roundish-triquetrous, short-rostrate, entire at the mouth, longer than the oblong-mucronate glume; *st.* 6—12′ high; *lvs.* flat and shorter towards the base; *plant* light green.—Open woods or hedges, rare. Differs from the following species in its fruit and pubescence.

92. C. Hitchcockiana. Dew.

♂ *Spike* erect, pedunculate; ♀ *spikes* 2—3, erect, few-flowered, lowest distant; *perig.* oval-triquetrous, tapering at both ends, inflated, alternate, bent at the apex, striate, with a short, truncated and open beak, about equaling or shorter than the oblong or ovate, mucronate glume; *st.* 10—24' high, erect, stiff, scabrous above, with long and leafy bracts; *st. lvs.* and *bracts* scabrous and subpubescent.—Borders of woods. Cannot be the *C. oligocarpa* figured by Schkuhr.

93. C. laxiflora. Lam.

♂ *Spike* oblong, slender; ♀ *spike* 2—4, oblong, lax-flowered, few-flowered, erect, remote; *perig.* ovate or oblong-ovate, obtusish, glabrous, ventricose, nerved, subtriquetrous, entire at the mouth, a little longer than the ovate, scabro-mucronate glume; *st.* 10—18' high, triquetrous, leafy.—Bright to pale green. Woods, hedges and meadows, common.

94. C. granularis. Muhl.

♀ *Spikes* 2—4, cylindric, oblong, dense-flowered, suberect; *perig.* roundish-ovate, nerved, very short-beaked and recurved, entire at the orifice, nearly twice as long as the ovate-acuminate glume; *st.* 8—16' high, erect or subdecumbent, smooth, leafy.—Glaucous green except the mature, yellow spikes. Moist soils in meadows and hedges, along brooks, abundant.

95. C. panicea.

♀ *Spikes* 2—3, loose-flowered, remotish, lowest long-pedunculate; *perig.* subglobose, obtuse, entire at the mouth, a little greater than the ovate, subacute glume; *st.* a foot high, triquetrous, leafy at the base; *lvs.* shorter than the stem. —Light green. Near Boston, *Pickering.*

96. C. binervis. Smith.

♀ *Spikes* 3, oblong, cylindric, subdense-flowered; *perig.* ovate, round, short-rostrate, bicuspidate, smooth, binerved, twice longer than the ovate, subacute glume; *st.* a foot high or more, triquetrous, leafy towards the base.— Pale green. Near Boston, *B. D. Greene.*

97. C. Greeniana. Dew.

♂ *Spike* one and erect, sometimes 2; ♀ *spikes* 2—3, oblong, bracteate, pedunculate; *perig.* ovate-lanceolate, triquetrous, nerved, rostrate, bifurcate, subdense-flowered, about equal to the ovate, cuspidate glume; *st.* 1—2f high, scabrous above, leafy towards the base.—Light green. Resembles *C. fulva, Good,* but differs in its fruit and glume. Near Boston, *B. D. Greene.* Rare.

98. C. Grayana. Dew.

♂ *Spike* oblong; ♀ *spikes* 2—3, oblong-cylindric, subloose-flowered; *perig.* ovate-oblong, subtriquetrous, subinflated, obtuse or acutish, entire at the orifice, longer than the obtuse, oblong glume; *st.* 6—16' high, erect, triquetrous, striate, with leaves about its own length.—Glaucous green. Sphagnous swamp, near Utica, N. Y., *Gray;* cedar swamp, N. J., *Torrey.* Has been supposed to be *C. lirida, Wahl.,* from which it differs in several respects.

99. C. Halseyana. Dew.

♂ *Spike* oblong, erect, sessile, often 2, approximate; ♀ *spikes* 1—2, oblong-cylindric, e.ect, loose-flowered, sometimes ♂ above; *perig.* ovate, short-rostrate, subtriquetrous, inflated, glabrous, oblique at the orifice, a little longer than the ovate, subacute glume; *st.* 1—2f high, acutely triquetrous; *lvs.* linear-lanceolate, shorter towards the base.—Dark green.—Upland meadows, Westfield, Ms., *Davis;* plains of N. J., *Kneiskern.*

100. C. capillaris.

♂ *Spike* small; ♀ *spikes* 2—3, ovate, oblong, about 6-flowered, loose-flowered, long and recurved pedunculate; *perig.* oval, short-rostrate, oblong, oblique at the orifice, longer than the ovate, obtuse glume; *st.* 2—7' high, leafy at the base; *lvs.* narrow, long.—Grows in tufts. Pale green. Alpine regions of the White Mts., *Robbins.*

101. C. eburnea. Boott. (C. alba. β. setifolia. Dew.)

♀ *Spikes* 2—3, erect, 3—6-flowered, ovate, with white, leafless sheaths, and the upper higher than the ♂ spike; *perig.* ovate-globose, rostrate or slightly obovate, glabrous and brown in maturity, twice longer than the white, ovate,

hyaline glume; *st.* 4—10′ high, erect, with subradical and bristle-form leaves.—
l'ale green, common. Abundant along the banks of the Genesee.

102. C. DEBILIS. Michx. (C. flexuosa. *Schk.*)

♂ *Spike* erect, filiform; ♀ *spikes* 3—4, filiform, loose-flowered, flexuous,
nodding, remotish, 1—2′ long; *perig.* oblong-lanceolate, subtriquetrous, alternate, rostrate, bifid, glabrous, nerved, nearly twice longer than the ovate-lanceolate glume; *st.* 1—2′ high, triquetrous and scabrous above, leafy towards the
base.—Bright green. Moist woods and meadows, common.

103. C. ARCTÁTA. Boott. (C. sylvatica. *Dew.*)

♀ *Spikes* 3—4, long and slender, loose-flowered, nodding and remote ,
perig. ovate, triquetrous, lanceolate or long-rostrate, subventricose, bifid, glabrous, little surpassing the ovate, membranaceous, mucronate glume; *st.* 10—
20′ high, scabrous above and leafy below.—Pale green In the same situations
as the preceding, common.

104. C. FLEXILIS. Rudge. (C. castanea. *Wahl.* C. blephoriphora. *Gray.*`

♀ *Spikes* 2—4, ovate-oblong, cylindric, nodding; *perig.* ovate, subconic
rostrate, bidentate, scarcely shorter than the ovate, obtusish, oblong glume; *st*
12—18′ high, erect, striate; *lvs.* short, and shorter below; *lvs.* and *bracts* ciliate
—Bright green. Oneida Co., N. Y., *Gray.*

105. C. WASHINGTONIÀNA. Dew.

♂ *Spike* erect, with oblong and obtuse black glumes; ♀ *spikes* 2—5,
oblong, cylindric, subremote, erect, loose-flowered, black or dark brown, subdistant, upper sessile ; *perig.* oval, acutish at both ends, glabrous, short-rostrate
entire at the orifice, about equaling the ovate-oblong, subacute, blackish glume
with a white edge; *st.* a foot or more high, triquetrous, subscabrous above.—
Light green. Seed distinctly triquetrous. Near summit of Mt. Washington
N. H., *Barratt.* Is distinct from *C. saxatilis, L.,* already described as foun.
on the White Mts.

106. C. SULLIVANTII.

♀ *Spikes* 3, oblong, erect, cylindric, rather loose-flowered, bracteate, am.
the lowest long-pedunculate and sparsely flowered below; *perig.* ovate, acute
and subrostrate, subtriquetrous and 2-toothed, equaling the ovate-oblong and
mucronate glume; *plant* light green.—Ohio.

107. C. KNIEISKERNII. Dew.

♀ *Spikes* 3, long-cylindric, rather distant, sublax-flowered, with recurved
peduncles; *perig.* ovate, oblong, subtriquetrous, terete-conic, rostrate, short-2-
toothed, a little longer than the ovate and oblong glume, which is obtusish and
short-mucronate.

108. C. WOODII. Dew.

♀ *Spikes* 1—3, ovate-oblong, loose-flowered, erect, lower long-pedunculate, recurved; *perig.* obovate, obtuse, subtriquetrous, closed at the orifice,
tapering below, twice longer than the ovate and acutish glume; *lvs.* narrow
and linear, and with the stem closely and slightly pubescent.—Jefferson Co.,
N. Y.

4. *Pistillate spikes scarcely sheathed.*

109. C. PALLESCENS. L.

♀ *Spikes* 2—3, oblong, short, cylindric, distant, nodding towards maturity;
perig. oval, obtuse, round, about equal to, or a little shorter than, the ovate
glume; *st.* 6—16′ high, hardly erect; *bracts* sometimes transversely rugose.—
Plan' often subpubescent, and of a light green In dry meadows. Common.

110. C. UNDULÀTA. Kunze.

♀ *Spikes* 2, erect, ovate-oblong; *perig.* oblong, round, triquetrous, obtuse,
striate, very short-beaked, bidentate, longer than the oblong, cuspidate, mucronate glume; *st.* 12—18′ high, erect, triquetrous, scabrous; *lower bract* transversely waved-plicate; *lvs.* pubescent.—In the same situation as the preceding,
and scarcely to be distinguished from it.

111. C. TORREYI. Tuckerman.

♂ *Spike* oblong, short pedunculate; ♀ *spikes* 2 3, short, oblong, subses sile, erect; *perig.* oblong, obovate, very obtuse, glabrous, subtriquetrous, entire at the orifice, subrostrate, twice longer than the acute glume; *st.* 12—18′ high, erect, triquetrous, with subradical and pubescent leaves.—Pale green. N. Y *Tuckerman.*

112. C. miliacea. Muhl.

♂ *Spikes* erect, slender; ♀ *spikes* 2—3, long-cylindric, slender, loose-flowered below, nodding; *perig.* ovate, triquetrous, glabrous, subrostrate, entire at the orifice, longer than the oblong, emarginate or obcordate, awned glume; *st.* 12—24′ high, slender, scabrous; *lvs.* linear-lanceolate.—Yellowish-green. Wet meadows; common.

113. C. limosa.

♀ *Spikes* 1—3, ovate or oblong, long-pedunculate, subloose-flowered, smoothish, pendulous; *perig.* elliptic, compressed, very short-rostrate, entire at the orifice, about equal to the oblong and obtuse, or ovate, cuspidate glume; *st.* 8—16′ high, ascending, obtusely triquetrous, with subradical, flat and narrow leaves.—Glaucous green. Marshes; common.

114. C. rariflora. Smith. (C. limosa. β. rariflora. *Wahl.*)

♀ *Spikes* about 2, linear, quite loose-flowered, long-pedunculate, nodding; *perig.* ovate-oblong, triquetrous, depressed, equaling the ovate, subcircinate, brown glume; *st.* 10′ high.—Glaucous. White Mountains, N. H., *Barratt.*

115. C. irrigua. Smith. (C. limosa. β. irrigua. *Wahl.*)

♀ *Spikes* 2—3, ovate-oblong, thickish, nodding; *perig.* roundish-ovate, short-rostrate, subcompressed, shorter than the ovate-lanceolate, red-brown glume; *st.* near a foot high, longer than the flat, subrecurved leaves; glaucous.—♂ Spike rarely ♀ at the summit, or ♀ spikes with stamens at the base. Marsh. Bridgewater, N. Y., *Gray;* also in marshes in Mass. and Mich., *Cooley.* Rare.

116. C. hystericina. Willd.

♂ *Spike* rarely pistillate at the summit; ♀ *spikes* 2—4, oblong, cylindric, attenuate, subdistant, long-bracteate, nodding, rarely sheathed; *perig.* ovate, inflated, subtriquetrous, nerved, bifid, glabrous, twice longer than the oblong, emarginate, submucronate glume; *st.* 12—24′ high, scabrous above, with long, linear-lanceolate leaves.—Yellowish green. Wet places; very common.

117. C. Pseudo-cyperus.

♂ *Spike* cylindric and elongated; ♀ *spikes* 3—4, cylindric, long-peduncu-late, rather remote, recurved-pendulous, with long and leafy bracts; *perig.* ovate, lanceolate, bidentate, reflexed, and a little shorter than the ovate-lance-olate or setaceous glume.—Common about ponds and ditches. It is smaller in all its parts than C. comosa, *Boott;* and, besides, the fruit of the latter is deeply and widely bifurcate, and its glume is hispid or ciliate. The two have been confounded in our country, though long known.

118. C. comosa. (C. furcata. *Ell.* C. Pseudo-cyperus. 1*st edit.*)

♂ *Spike* long and slender, rarely pistillate above; ♀ *spikes* 2—5, long-cylindric, pendulous, thick, dense-flowered, with very long and leafy bracts; *perig.* ovate-lanceolate, acuminate, rostrate, 2-forked, reflexed, triquetrous, gla-brous, generally longer than the lanceolate, mucronate, setaceous glume; *st.* 18—30′ high, large, rough, with long and wide, rough leaves and bracts. Plant very glabrous and yellowish-green. Wet places about ponds and ditches; common.

119. C. Cooleyi. Dew.

♂ *Spike* short and small, with oblong-lancelate glumes; ♀ *spikes* 2—4, cylindric, oblong, or ovate and short, rather dense-flowered, upper sessile, lower on very long, recurved peduncles; *perig.* ovate-rostrate or oblong-lanceolate, bifurcate, nerved, about equal to the ovate, awned, scabrous glume; *st.* filiform and scabrous, subrostrate, a foot or more high, much shorter than the subradi-cal, narrow leaves.—Light green. Marsh in Macomb Co., Mich., *Cooley.*

120. C. scabrata. Schw.

♀ *Spikes* 3—6, cylindric, subrecurved, remotish, long-pedunculate; *perig.* ovate-oblong, subinflated, subbifid, rostrate, quite scabrous, longer than the ovate-lanceolate, acuminate, short-bidentate, ciliate glume; *st.* 1—2f high, acutely triquetrous, rough above, longer than the leaves towards the base.—Bright green. Along brooks and streams; common.

121. C. CRAWEI. Dew.

♀ *Spikes* 3—6, cylindric, short and thick, densely flowered, sometimes aggregated, sometimes remote, the lowest often subradical and long-pedunculate; *perig.* ovate, terete, scarcely rostrate, diverging, entire at the orifice, twice longer than the ovate and obtusish glume; ♂ *spike* with one or two small ones at its base.

122. C. POLYMORPHA. Muhl.

♂ *Spikes* long cylindric, upper one pedunculate, with oblong and ovate scales; ♀ *spikes* 2, oblong, cylindric, close-fruited, erect, nearly sessile, upper one staminate at the apex, with nearly enclosed peduncles; *perig.* ovate, acuminate, bilobate, scarcely rostrate, striate, longer than the ovate glume.—N. J. and the Southern States. Differs from *D. Halseyana* in its ♂ spikes and obtuse glume, in its acuminate, not rostrate fruit, and its larger, fertile spikes.

H. *Staminate spikes usually two or more.*

123. C. SCHWENITZII. Dew.

♂ *Spikes* 2, rarely 1, *upper* long and slender, *lower* with a few perigynia at the base; ♀ *spikes* 2—4, oblong, cylindric, subapproximate, subrecurved, subloose-flowered, lowest often long-pedunculate; *perig.* ovate-oblong, tapering above, rostrate, inflated, nerved, glabrous, bifurcate, longer than the lanceolate, subulate, subsetaceous glume; *st.* 6—12' high, scabrous above, very leafy.—Pale yellowish-green. Wet sandy grounds. Not abundant.

124. C. RETRORSA. Schw.

♂ *Spikes* about 3, rarely 1, often with a few perigynia at the base; ♀ *spikes* 4—6, oblong-cylindric, approximate, dense-flowered, with long and leafy bracts, the lowest often remote and long-pedunculate; *perig.* ovate-inflated, subglobose, rostrate, bifurcate, nerved, reflexed, twice longer than the lanceolate glume; *st.* 15—30' high, scabrous above, large, stiff and leafy.—Bright green. In clusters, about pools of water, common. The lower spikes sometimes have 1 or 2 smaller spikes attached to them.

125. C. ARISTATA. R. Br.

♀ *Spikes* 2—4, cylindric, distant, close-flowered, erect; *perig.* ovate, oblong, nerved, deeply bifid, very glabrous, long-rostrate, longer than the oblong, awned glume; *lvs.* and *sheaths* villose on the under side; *st.* a foot or more high.—Bright green. Watertown, N. Y., *Torr. & Gray.* Is not this very closely related to the following species?

126. C. TRICHOCARPA. Muhl.

♂ *Spikes* about 3, erect, rarely 1, or ♀ above, cylindric, *lower* shorter; ♀ *spikes* 2—4, erect, long-cylindric, smoothish, rather loose-flowered; *perig.* ovate, conic, inflated, nerved, rostrate, densely pubescent, about twice longer than the ovate-lanceolate glume; *st.* 15—30' high, scabrous above, and with pubescent leaves and sheaths.—Light green. In wet and marshy places; common.

β. *turbinata.* Dew. ♀ *Spikes* ovate, or short-oblong, thick, remote, dense-flowered; *perig.* subdiverging, ovate and conic, rostrate, longer than the ovate-oblong, mucronate glume; *st.* 2—3f high.—Glaucous green. In a pond in Beckman, N. Y., there abundant.

127. C. LONGIROSTRIS. Torr.

♂ *Spikes* 3, short; ♀ *spikes* 2—3, cylindric, quite loose-flowered, pendulous, subdistant, with filiform peduncles; *perig.* ovate, globose, inflated, glabrous, long-rostrate, hispid, a little longer than the lanceolate or ovate, cuspidate glume; *st.* 15—30' high, rather slender, stiff, leafy below.—Bright green. On light soil of hedges in N. England and N. York. Common.

128. C. LANUGINOSA. Michx. (C. pellita. *Muhl.*)

♂ *Spikes* 2, oblong, slender, erect; ♀ *spikes* 2—3, cylindric, erect, dense-flowered, sometimes short-oblong and thick, subrostrate; *perig.* ovate, short-rostrate, bicuspitate, subtriquetrous, thick, pubescent and woolly, about equaling the ovate-lanceolate, awned glume; *st.* 12—24' high, nearly round below, with flat, linear-lanceolate leaves and bracts.—Glabrous and yellowish-green. Wet places and marshes. Common.

129. C. FILIFORMIS. Gooden.
♂ *Spikes* 2—3, with oblong glumes; ♀ *spikes* 2—3, ovate, oblong, short-cylindric, close-flowered, remotish, erect; *perig.* ovate, villose, short-rostrate, bifurcate, about equaling the ovate, acute glume; *st.* 20—30' high, erect, slender stiff, with convolute leaves and bracts.—Pale green. Marshes. Common.

130. C. LACUSTRIS. Willd.
♂ *Spikes* 3—4, erect, sessile; ♀ *spikes* 2—3, erect, oblong, cylindric, short-pedunculate; *perig.* ovate-oblong, tapering or lanceolate, bifurcate, glabrous, a little longer than the oblong, mucronate glume; *st.* 2—3f high, scabrous above, erect and large, with long and large leaves and bracts.—Light green. Marshes. Common.

131. C. RIPARIA. Gooden.
♂ *Spike* 3—5, oblong, thick, erect, sessile; ♀ *spikes* 2—3, erect, oblong, often long-cylindric; *perig.* ovate-elliptic, contracted into a short, bifurcate beak, glabrous, about equaling or shorter than the ovate, mucronate, or oblong-lanceolate glume; *st.* 2—3f high, scabrous above, leafy below.—Bright green.

132. C. OLIGOSPERMA. Michx. (Oakesiana. *Dew.*)
♂ *Spikes* several, sometimes one, erect, slender, long-cylindric, with an oblong obtusish glume; ♀ *spikes* 1—3, ovate, globular, sessile, distant; *perig.* few, ovate, inflated, acute, nerved, short-rostrate, entire at the orifice, glabrous, a little longer than the ovate-lanceolate glume; *st.* 1—2f high, scabrous above, leafy below; *lvs.* involute and rush-like.—Light green. About the lakes of N. Eng. and N. Y. Abundant in the marshes of Micu.

133. C. VESICARIA.
♂ *Spikes* about 3, erect, oblong; ♀ *spikes* 2—3, cylindric, erect, dense-flowered, alternate, long-bracteate; *perig.* ovate, oblong-conic, terete, inflated, rostrate, nerved, diverging, glabrous, bicuspidate, nearly twice longer than the oblong-lanceolate glume; *st.* about 2f high, shorter than the leaves.—Bright green. Marshes. Not common.
β. *utriculata.* Dew. (C. utriculata. *Boott.*) *Perig.* oblong-elliptic, nerved, cylindric-rostrate, bicuspidate, more or less longer than the lanceolate, scabrous-awned glume.—Marshes, with the other.

134. C. AMPULLACEA. *Gooden.*
♂ *Spikes* 2—4, oblong, cylindric, erect; ♀ *spikes* 2—3, long-cylindric, erect, close-flowered, short-pedunculate, sometimes ♂ above; *perig.* subglobose, inflated, diverging, nerved, glabrous, setaceous, rostrate, bifurcate, little longer than the lanceolate glume; *st.* 2—3f high, obtusely triquetrous, leafy.—Light green. Marshes. Common.

135. C. MONILE. *Tuckerman.*
♂ *Spikes* 2—4, long, slender, cylindric, with a long, lanceolate glume; ♀ *spikes* 2, long, cylindric, short-pedunculate, subloose-flowered, erect; *perig.* ovate, long-conic, subtriquetrous, inflated, rostrate, bicuspidate, more than twice longer than the oblong-lanceolate glume; *st.* 15—30' high, erect, with long leaves and bracts.—Bright green. Marshes. Not common. More loose-flowered and fruit longer than that of C. *visicaria.* L.

136. C. BULLATA. Schk.
♂ *Spikes* 3, erect, slender, cylindric, with oblong-lanceolate glumes; ♀ *spikes* 2—3, rather long, cylindric, nearly erect; *perig.* ovoid-globose, inflated, glabrous, costate, with a long, scabrous beak, bifurcate, longer than the lanceolate glume; *st.* 20—30' high, rather slender, triquetrous, scabrous above, leafy and shorter than the leaves.—Glabrous, light green. In wet meadows. Common.

137. C. TUCKERMANI. Dew
♂ *Spikes* 2—3, cylindric, lower ones sessile and short, with an oblong, acutish glume; ♀ *spikes* 2—3, oblong, cylindric, thick and large, pedunculate, subloose-flowered; *perig.* inflated, ovate, large, conic, costate, bifurcate, gla-

brous, nerved, twice longer than the ovate-lanceolate glume; *st.* about 2f high, erect, scarce.y scabrous; *bracts* and *lrs.* long, not wide; light green.—Wei places in meadows, common, and has been ranked under *C. bullata.*

138. C. MIRATA. Dew. (C. arista. *Dew.* not of *R. Br.*)

♂ *Spikes* 2 or more, long-cylindric; ♀ *spikes* about 2, long-cylindric, pedunculate, subdense-flowered, suberect; *perig.* ovate, conic, long rostrate, costate, bifurcate, glabrous, subinflated at the base, about equaling the ovate, long-setaceous or long-awned glume; *st.* about 2f high, rough; *lvs.* and *bracts* longer than the stem; light green.—Shores of lake Ontario, N. Y. *Sartwell.* Also found in the State of Georgia.

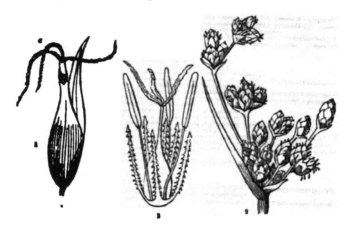

FIG. 85.—1. Carex; a single, fertile flower; *a,* the glume; *b,* the perigynium, containing the ovary with (*c*) the three stigmas. 2. Scirpus lacustris; the inflorescence. 3. A single (magnified) flower, showing the 6 hypogynous bristles of the perigynium, ovary with three stigmas, and the three stamens.

ORDER CLXI. GRAMINEÆ.—GRASSES.

Herbs perennial, with fibrous or bulbous rhizomas, or often annual or biennial,
Stems (*culms*) cylindrical, fistular, closed at the nodes, covered with a coat of silex, often solid.
Lvs. narrow and undivided, parallel-veined, alternate, with a sheath split down to the nodes, and a membranous ligule or stipule at the juncture of the blade and sheath.
Inflorescence arranged in spikes, racemes or panicles.
Flo. generally perfect, in little spikelets composed of bracts imbricated in 2 rows.
Glumes.—Outer bracts (*calyx, Linn.*) generally 2 and unequal, sometimes 1 only.
Palea.—Inner bracts (*corolla, Linn.*) 2, alternate, the lower (exterior) one simple, the upper (interior) often doubly carinate, being composed of 2 pieces united by their edges.
Scales.—Innermost bracts (*nectary, Linn.* rudimentary petals) 1—3, distinct or united, membranous, hyaline.
Sta. 1—6, commonly 3. *Anthers* versatile. [pogynous
Ova. simple, with 2 *styles* and 2 feathery *stigmas.* *Fruit* a caryopsis.
Seed with the embryo situated on the outside of farinaceous albumen, at the base, next the hilum.

Genera 291, species about 3800, universally diffused throughout the world, having no other limits than those that bound vegetation in general. But the species and their characters are widely different in different climes. In temperate zones the grasses clothe a large portion of the earth's surface with a compact, soft, green, carpet-like turf; but in tropical regions this beautiful grassy turf disappears and the grasses become larger, more isolated like other plants, fewer in the number of individuals, with broader leaves and more showy flowers.

Properties.—This family doubtless contributes more to the sustenance of man and beast than all others combined. Its sweet and nutritious properties reside both in the farinaceous albumen of the seed and in the herbage. No poisonous or even suspicious herb is found among them, with the single exception of Lolium temulentum. The poisonous and medicinal *ergot* or *spurred rye* is only a parasitic fungus, and therefore forms no exception to this remark. The stems of many grasses contain sugar, as the *maize* and *sugar cane.* Silex is also a frequent ingredient. To this order belong the common *grains,* maize, wheat, rye, rice, barley, oats, &c. The most important of the cultivated *grasses* are Phleum or *Timothy grass,* several kinds of Poa. Agrostis, Alopecurus, Festuca, Aira, Panicum, Cinna, Briza, &c.

CLXI. GRAMINEÆ.

Conspectus of the Genera.

Spikelets { 1-fruited, . . { in panicles, { with awns. . . 1
{ without awns. 2
{ in spikes, { with awns. . . 3
{ without awns. 4
{ 2—∞-fruited, { in panicles, { with awns. . . 5
{ without awn. 6
{ in spikes. . 7

§ 1. { Glumes large, { 1-flowered. { Palea 2. { the lower with 2 awns, often twisted. . . . . . . Fr. white. *Oryzopsis.* . 4
{ cartilaginous, { seeds. Fr. black. *Piptatherum.* 3
{ stipitate. Fr. black *Stipa.* . 2
{ naked. Stam. 1. *Oryza.* 10
{ herbaceous, { hairy. Stam. 2. *Calamagrostis.* 13
{ membranaceous, 1-keeled. . *Agrostis.* 5
{ with 1 awn, { Aristida.* 1
{ Palea 2, upper one with an awn. Flowers polygamous. *Sorghum.* 57
{ 2-flowered (one flower abortive), longer than the palea. *Holcus.* 26
{ unequal, one of them hardly perceptible. . . *Muhlenbergia.* 9
{ Glumes minute, { subequal, one of them awned. . . . . *Trichochloa.* 11
{ Glumes 0. Aquatic grasses. . . . . . . *Zizania.* 50

§ 2. { Spikelets 1-flowered. { Glumes 2. { Palea 2, { Stig. 2. { Fr. coated. *Agrostis.* . 5
{ Fr. naked. *Sporobolus.* . 6
{ membran'ous. { Stigmas 3. *Pramna.* 12
{ coriaceous. *Milium.* . 9
{ Palea 1. Panicle mostly capillary. *Trichodium.* 8
{ Glumes 0. Leaves retrorsely scabrous. . *Leersia.* 56
{ Spikelets 2-flowered. { Glumes 2, very unequal, resembling the outer abortive palea. *Panicum.* 21
{ Glumes 2, equal, longer than the palea. *Phalaris.* 24
{ Spikelets 3-flowered, 2 of the flowers staminate, the other perfect. *Hierochloa.* 25

§ 3. { Spike solitary, { Involucre 0. { Palea with awns 1—3 times their length. *Alopecurus.* 14
{ Palea with awns 5 times their length. *Hordeum.* 51
{ Spikelets with an involucre of 2 or more bristles. *Setaria.* 22
{ subsimple. { Spikelets with a burr-like involucre. *Cenchrus.* 23
{ Both glumes and palea awned. *Polypogon.* 7
{ paniculate or lobed. { Glumes awnless, palea awned. *Anthoxanthum.* 27
{ Spikes ∞, not bearded, unilateral, panicled. Palea cartilaginous. *Optismenus.* 20
{ Spikes 2—∞, polygamous. Sterile flowers plumosely bearded. *Andropogon.* 58

§ 4. { Fls. perfect or polygamous. { Spikes cylindrical, { solitary, terminal. *Phleum.* 16
{ several, terminal and lateral. *Crypsis.* 15
{ 1—∞; spikelets suborbicular. *Paspalum.* 17
{ digitate or verticillate, linear. *Digitaria.* 19
{ Spikes unilateral, { pedunculate in a 2-sided panicle *Spartina.* 49
{ sessile in a unilateral panicle. *Lepturus.* 60
{ Spikes monœcious, . . { all terminal, sterile above, fertile at base. *Tripsacum.* 61
{ fertile ones lateral, sterile ones terminal, panicled. *Zea.* 62

§ 5. { Lower palea awned on the back { near the base. { Apex bifid. Awn bent. *Avena.* 30
{ Apex multifid. *Aira.* 29
{ which has two bristly teeth. *Trisetum.* 28
{ near the apex { which is merely bifid. *Bromus.* 39
{ which is entire. *Gymnopogon.* 45
{ Lower palea awned at the apex { which is entire. Fruit coated. *Festuca.* 36
{ with a cusp each side of it. *Uralepis.* 32
{ between the two teeth: awn twisted. *Danthonia.* 31
{ Palea entire, not mucronate, bristly at base. . *Arundo.* 33
{ Palea bidentate, outer one mucronate. *Ceratochloa.* 35
{ Palea entire, outer one mucronate. *Festuca.* 36
{ Palea cartilaginous, outer cuspidate, coating fruit. *Diarrhena.* 37
{ Outer palea bifid and tricuspidate. *Tricuspis.* 39
{ Outer palea erose-denticulate. Spikelets terete. *Glyceria.* 40
{ Terminal fl. { Spikelets { not cordate. { Palea not mucronate, inner one bifid. *Poa.* 41
{ Outer palea truncate-mucronate, inner bifid. *Uniola.* 42
{ perfect. . . { Spikelets cordate at base, turgid, pendulous. *Briza.* 43
{ Panicle contracted. *Kœleria.* 38

§ 6. { Terminal flower abortive or a mere pedicel. { Panicle large, diffuse. *Melica.* 44
{ one tin the top spikelet 2). *Lolium.* 54
{ Spikes 2-ranked, { Glumes broad, . . { two in each spikelet. *Triticum.* 51
{ collateral: spikelets in 2s, &c. *Elymus.* 55
{ Glumes subulate. { opposite: spikelets solitary. *Secale.* 52
{ Spikelets in two rows. *Eleusine.* 48
{ digitate (rarely solitary). { Spikelets in one row. *Cynodon.* 47
§ 7. { Spikes unilateral, { remote, short, forming a long, slender raceme. *Atheropogon.* 60
{ conglomerate or paniculate. . . *Dactylis.* 46

FIG. 98.—1 *Agrostis alba*; a 1-flowered spikelet; *a*, the two glumes. 2. A flower, with the two palea, three stamens and two plumose stigmas. 3. *Leersia oryzoides*; a flower removed from its glumes, showing its 2 hypogynous scales, three stamens and ovary with the two stigmas. 4. *Phleum pratense*; a 1-flowered spikelet; *a*, glumes; *b*, truncate palea; &c. 5. *Polypogon*; a 1-flowered spikelet; glumes and lower palea awned. 6. *Holcus lanatus*; a two-flowered spikelet; *a*, glumes; *b*, the two flowers (upper staminate). 7. *Poa pratensis*; a 4-flowered spikelet; *a*, the two glumes; *b*, a single flower, with two palea, &c. 8. *Festuca duriuscula*; a 5-flowered spikelet; *a*, two glumes; *b*, a single flower. 9. The caryopsis of *Hordeum*, showing the embryo at the base of the copious albumen.

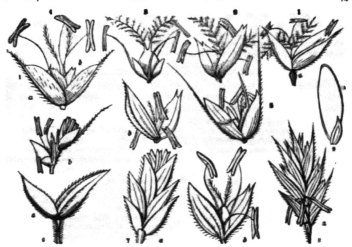

Tribe 1. **STIPACEÆ.**—Inflorescence panicled. Spikelets solitary, 1-flower-ed. Glumes membranaceous. Paleæ mostly two, lower one coriaceous, involute, awned.

1. ARISTĬDA.

Lat. *arista*, an awn ; characteristic of the genus.

Panicle contracted or racemose; glumes 2, unequal; paleæ pedicellate, lower one with 3 long awns at the tip, upper one very minute or obsolete.

1. A. DICHOTŎMA. Michx. *Poverty Grass.*

Cæspitose; *st.* dichotomously branching; *panicle* contracted-racemose, *lateral awns* very short, the intermediate one nearly as long as the paleæ, contorted.—A slender grass, in sandy soils, U. S., common. Stems 8—12' high, branching at each joint. Leaves very narrow, with very short, open sheaths, and a very short stipule. Spikelets slender, on clavate peduncles. Aug.

2. A. PURPURASCENS. Poir.

St. erect, simple, filiform, 2—3f high; *lvs.* very narrow, flat, erect, a foot in length, with short, open sheaths; *panicle* long, loosely spicate; *spikelets* on short, clavate, appressed pedicels; *awns* nearly equal, divaricate, twice the length of the paleæ; *paleæ* often dark purple.—⅔ Sandy woods, Northern States. Sept.

3. A. GRACĬLIS. Ell.

St. very slender, a foot or more high; *lvs.* setaceous, erect, with short sheaths, pilose at the throat; *panicle* very slender; *spikelets* somewhat remote, appressed; *lateral awns* short, erect, intermediate one longer, spreading. ⅔ Mass. and S. States. A grass of little value, as well as the other species of this genus.

4. A. TUBERCULŎSA. Nutt. *Long-awned Poverty Grass.*

St. erect (declinate at base), 8—20' high, rigid, with small tubercles in the axils of the numerous branches; *nodes* tumid; *lvs.* long and narrow-linear; *panicle* large, loose, simple; *spikelets* pedicellate; *glumes* nearly 1' long, linear, awned; *upper paleæ* involute, the awns 2' long, hispid upwards, twisted together to near the middle, thence finally horizontally divaricate.—⅔ A very singular species, in dry prairies, Ill., *Mead!* July, Aug.

5. A. **stricta**. Michx. *Upright Aristida.*
St. strictly erect, cæspitose, branched, 2—3f high; *lvs* straight, erect, pubescent, linear, convolute above; *panicle* long, loosely racemose; *spikelets* appressed; *glumes* unequal, very acute; *lower paleæ* hairy at base; *awns* twice as long as the paleæ, spreading, the middle one the longest.—♃ Penn. to Car. W. to Mich

2. STIPA.
Lat. *stipa*, a foot-stalk; alluding to the stipitate fruit of some of the species.

, Glumes 2; paleæ mostly 2, shorter than the glumes, the lower with a long awn at the apex, the upper entire; awn jointed at the base, deciduous; caryopsis striate.

1. S. **avenacea**. *Feather Grass.*
St. naked above; 2—3f high; *lvs.* smooth, striate, setaceous, chiefly radical; *panicle* spreading, somewhat 1-sided, 4—6' long, at length diffuse, branches capillary, solitary and in pairs; *glumes* nearly equal, mucronate, as long as the dark brown, cylindric fruit; *scales* 2, lanceolate; *awn* twisted, 2—3' in length.— ♃ N. Y. to Car.

2. S. **juncea**. Pursh. *Rush-leaved Feather Grass.*
St. 2—3f high; *lvs.* convolute-filiform, smooth inside, long; *panicle* loose; *glumes* loose, filiformly acuminated to more than twice the length of the fruit; *fr.* attenuated at base into a stipe which is a third of its length, stipe acute, pubescent; *paleæ* obtuse, distinctly articulated to the awn, which is smooth and slender, scarcely contorted and 4—6' in length.— ♃ Prairies, Ill., Mo. When in fruit, the pungent stipe adheres to everything that comes in its way. Aug.

3. S. **canadensis**. Lam.
Lvs. setaceous; *panicle* small; *glumes* smooth, ovate-obtuse, as long as the pubescent fruit; *awn* thick and short.— ♃ Amherst, Mass. *Dewey.* Neither this nor the preceding species is common or of much value in agriculture.

3. PIPTATHËRUM. Palis.
Gr. πιπτω, to fall. 3ερος, harvest.

Panicle racemed; glumes membranaceous, longer than the elliptical, cartilaginous paleæ; *lower palea* awned at the tip; *scales* ovate, entire; *caryopsis* coated.

P. **nigrum**. Torr. (Oryzopsis melanocarpa. *Muhl.*) *Black-seeded Millet.*
St. erect, simple, leafy, 18—24' high; *panicle* simple, flexuous, few-flowered; *spikelets* racemose, ovoid-lanceolate; *glumes* acuminate, mucronate, 5—6" in length, smooth; *paleæ* hairy, nearly black when ripe, the lower one tipped with an awn an inch in length; *fruit* black.— ♃ Rocky hills, N. Eng. to Ky., frequent. Aug.

4. ORYZOPSIS. Rich.
Named for its resemblance (οψις) to the genus Oryza.

Panicle racemed; glumes 2, subequal, loose, obovate, awnless; paleæ 2, cylindric-ovate, hairy at base; scales linear-elongated.

O. **asperifolia**. Michx. *Mountain Rice.*
St. nearly naked, purple at base, 10—20' high; *lvs.* subradical, erect rigid, pungent at the point, nearly as long as the stem, cauline ones few and very short; *spikelets* in a racemose, simple, flexuous panicle, 1—2 upon each branch; *glumes* abruptly acuminate; *paleæ* white, the lower one with a long bent awn.—Woods, Free States, N. to Subarctic Am. Leaves green through the winter. Caryopsis white, about as large as rice, farinaceous. May.

Tribe 2. AGROSTIDEÆ.—Inflorescence panicled, rarely spiked. Spikelets solitary, 1-flowered. Glumes and paleæ of nearly similar texture, usually carinate.

5. AGROSTIS.
Gr. αγρος, a field, it being eminently an occupant of fields and pastures.

Inflorescence paniculate; glumes 2, acute, subequal, the lower one larger, sometimes longer, often shorter than the paleæ; paleæ 2, unequal, lower one larger, awnless or awned, larger than the glumes, coating the caryopsis.

§ 1. *Glumes longer than the paleæ.*

1. A. VULGARIS. Smith. (A. polymorpha. *Gray.*) *Red-top. Bent Grass.*
St. erect, 1—2f high; *panicle* spreading, with the branches finally divaricate; *lvs.* linear-lanceolate, veined, scabrous, with smooth, striate sheaths, and short, truncate stipules; lower *palea* twice as large as the upper, and nearly as long as the lanceolate, acute glumes.—♃ U. S. A common and very valuable grass, spread over hills, vales, and meadows, forming a soft, dense turf. Flowers very numerous, purplish. July.

2. A. ALBA. (A. decumbens. *Muhl.*) *White-top. Florin Grass.*
St. decumbent, geniculate, rooting at the lower joints, sending out stolons; *lvs.* linear-lanceolate, smooth, those of the stolons erect and subulate; *sheaths* smooth, with a long, membranaceous stipule; *panicle* dense, narrow, at length spreading, whitish, sometimes purplish; *lower palea* 5-veined, rarely awned.—Ⅰ N. Eng. to Ohio, in meadows, or in dry soils; hence its characters are variable, being often nearly erect. June. §

3. A. STRICTA. Willd. *Bent Grass.*
St. erect, smooth, with black nodes; *lvs.* linear-lanceolate, scabrous on the margin, with cleft, white stipules; *panicle* elongated, strict, the branches about 5, flexuous, scabrous, erect; *glumes* equal, lanceolate; *palea* unequal, smaller than the glumes, with an awn at the base of the outer one twice longer than the flower.—♃ Fields, N. Eng., N. Y. June.

4. A. CANINA. *Dog's Bent Grass.*
St. prostrate, somewhat branched, rooting at the lower nodes, about 2f long; *panicle* at length spreading, with angular, rough branches; *glumes* elongated; *lower palea* furnished with an incurved awn upon the beak twice its length.—♃ Introduced and common in wet meadows. July. §

5. A. PICKERINGII. Tuckm. (A. canina, *β. alpina. Oakes.*)
St. erect; *lvs.* flat, linear; *pan.* ovate, diffuse, branches verticillate, rather erect, scabrous; *gl.* subequal, keel of the lower mucronate at tip, upper acute, smoothish; *lower palea* ovate-lanceolate, acute or erose, veined, upper ovate, veinless; awn from the middle of the back, contorted, twice longer than the fls.—White Mts.
β. *rupicola.* Tuckm. Smaller; *pan.* contracted, smoothish, often purplish.—Mountains, Vermont.

§ 2. VILFA. *Glumes not longer than the subequal, awnless paleæ.*

6. A. VIRGINICA. (Vilfa vaginiflora. *Gray.*)
Sts. numerous, assurgent, procumbent and hairy at base, nearly simple, about a foot long; *lvs.* somewhat 2-rowed, involute, rigid, erect, 2—3' long with smooth sheaths which are hairy at the throat and swollen with the enclosed panicles; *panicles* spike-form, terminal and lateral, the lateral ones concealed; *glumes* nearly equal, about as long as the subequal paleæ.—Ⅰ Sandy cils, Middle States. Sept., Oct.

7. A. COMPRESSA. Torr. (Vilfa compressa. *Trinius.*) *Flat-stemmed*
Agrostis.—Glabrous; *st.* erect, compressed, simple, leafy, branched at base, 1—2f high; *lvs.* narrowly linear, compressed, scarcely shorter than the stem; *keel* prolonged into the open sheath; *stip.* very short; *panicle* purple, subsimple, contracted, the branches few and erect; *glumes* equal, acute, shorter than the paleæ, the upper emarginate, rarely mucronate; *palea* ovate, obtuse, smooth, sometimes deeply cleft; *stig.* purple.—Sandy swamps, N. J. Sept.

8. A. SEROTINA. Torr. (Vilfa serotina. *ejusdem.*)
St. 12—18' high, filiform, compressed, growing in patches, smooth, often viviparous at the nodes; *lvs.* 2—3' by ¼'', keeled, smooth; *sheaths* open; *stip.* ovate, short; *panicle* 3—10' long, capillary, diffuse, branches flexuous, alternate; *spikelets* elliptical, scarcely ¼'' long; *glume* ovate, 1-veined, unequal, bais

the length of the paleæ; *paleæ* smooth, the lower one shorter; *sta.* 3.—Long Island, *Kneiskern.* July.

9. A HETEROLĒPIS. Wood. (Vilfa heterolepis. *Gray.*)
St. 1—2f high, smooth; *lvs.* setaceous, somewhat convolute, scabrous on the margins; lower *sheaths* pubescent, upper ones smooth; *panicle* spreading, pyramidal, few-flowered; *glumes* purplish, outer one subuliform, inner one ovate, cuspidate, membranaceous in texture, 1-veined; *valves* of the perianth oblong, obtuse, thin, a little shorter than the superior glume, inferior valve veined, apiculate, superior valve 2-veined, shorter than the outer one; *sta.* 3; *anth.* linear, reddish; *stig.* 2; *sty.* short; *fr.* roundish, smooth.—Watertown, N. Y. *Crawe.* Aug., Sept.

10. A. JUNCEA. Michx. (A. Indica. *Mühl.*)
Glaucous; *st.* erect, 1—2f high, terete, slender; *lvs.* erect, 2—6" by 1", concave, convolute when dry, margin scabrous; *sheaths* much shorter than the internodes; *stip.* short; *pan.* oblong-pyramidal, branches verticillate, about in 6s; *glumes* purple, lanceolate, acute, upper as long as the paleæ, the lower twice shorter; *paleæ* subequal; *anth.* and *sty.* whitish.—2↓ Penn. to Flor., barrens. Oct.

§ 3. MUHLENBERGIOIDEÆ.   *Glumes shorter than the paleæ.*

11. A. MEXICĀNA. (A. lateriflora. *Mx.* Muhlenbergia. Mex. *Trin.*)
St. erect or ascending, with swelling nodes, much branched and leafy above, often nearly leafless below, 1½—3f high; *lvs.* lanceolate, scabrous, with half-clasping sheaths; *panicles* numerous, terminal and lateral, narrow and dense-flowered, lateral ones partly enclosed in the sheath; *glumes* narrow, acuminate, mostly shorter than the subequal, pubescent paleæ.—2↓ Wet shades, N. Eng. to Ohio! and Ill., common.

12. A. SOBOLIFĒRA. Muhl. (Tricochloa. *Trin.* Muhlenbergia. *Gray.*)
St. erect, slender, producing shoots at base, sparingly branched, 18—30 high; *branches* erect and filiform; *nodes* not swelling; *lvs.* linear-lanceolate, with open sheaths; *panicle* simple, filiform, with appressed branches, and crowded spikelets; *paleæ* equal, longer than the acute glumes.—2↓ Rocky hills, New Eng. to Ill., frequent. Aug.

13. A. SYLVATĪCA. Torr. (Muhlenbergia. *Gray.*)
St. ascending, 2—3f long, much branched, diffuse, smooth, with swelling nodes; *lvs.* lanceolate, scabrous, veined, 4—6' long, with smooth, open sheaths; *panicles* slender, rather dense; *glumes* nearly equal, acuminate, a little shorter than the paleæ; *awn* several times longer than the spikelet.—2↓ Rocky shades N. Y. to Ill., N. J., Penn. Sept.

14. A. WILLDENOWII. Trin. (A. tenuiflora. *Willd.*)
St. erect, subsimple, pubescent at the nodes, with a few appressed branches; *lvs.* 6—9' by 2—3", lanceolate, veined, scabrous, spreading, with pubescent sheaths; *panicle* contracted, very slender and long, with remote, filiform branches; *glumes* subequal, acuminate, half as long as the paleæ; *awn* 3—4 times the length of the spikelet.—2↓ Rocky woods, Can. and U. S. July, August.

15. A CRYPTANDRIA. Torr.
Panicle pyramidal, with spreading, generally alternate branches, hairy at the axils; *fls.* subracemed; *lower glume* very short, upper one as long as the nearly equal paleæ; *stems* 3f high; *sheaths* bearded at the throat. Very abundant at Buffalo. *Aikin.*

16? A. LONGIFOLIA. Torr. (A. aspera. *Michx.?*)
S. erect, simple, 2—4f high; *lvs.* 2f long, filiform at the end, with smooth, closed sheaths and bearded stipules; *panicles* terminal and lateral, contracted into a spiked form, generally concealed in the swelling sheaths; *glumes* dusky-purple, much shorter than the subequal, smooth, spotless paleæ.—Sandy fields, Northern States. Sept., Oct.—Perhaps a Sporobolus.

6. SPOROBÓLUS. Brown.

Gr. σπορα, a seed; βαλλω, to cast forth; its fruit is loose, and easily falls out.

Panicles contracted; glumes 2, glabrous, awnless, unequal, one or both much shorter than the paleæ; paleæ 2, concave, nearly equal, beardless; fruit loose, free, not enclosed in the paleæ.

S. ᴀꜱᴘᴇʀ. Sullivant?

Rt. long, white, fibrous; *st.* stout, glabrous, geniculate at base, 2f high; *lvs.* rigid, glabrous, 2—8′ by 1—3″, tapering to a long, pungent point; *branches* with short leaves, barren, also ending in a long, pungent point; *sheaths* ciliate at edge and bearing dense tufts of long, white hairs at top; *panicles* small, terminal and lateral, half enclosed in the long sheaths; *spikelets* blackish-green; lower *glume* very short, upper nearly as long as the paleæ; *fr.* compressed, obovate, ¼″ in length,—♃ Ohio, *Sullivant!*

7. POLYPÓGON. Desf.

Gr. πολυς, many, πωγων, beard; a characteristic term.

Inflorescence contracted into a spike; glumes 2, nearly equal, obtuse, with long awns; paleæ shorter than the glumes, lower one entire, with a short, straight, tender awn (sometimes awnless), upper one bifid, toothed.

P. ʀᴀᴄᴇᴍóꜱᴜꜱ. Nutt. (P. glomeratus. *Willd.* Agrostis racemosus. *Mx.* Muhlenbergia glomerata. *Trin.*)—Glaucous; *st.* compressed, erect, smooth, with appressed branches or subsimple, 1½—4f high; *lvs.* somewhat 2-rowed, erect, flat, rough, 3—5′ long, with closed sheaths; *panicle* spicate, dense, conglomerated, interrupted, 2—3′ long, many-flowered; *glumes* linear, ½ the length of their awns; *lower paleæ* mucronate.—♃ Bog meadows, also on rocky mountains N. Eng. to Mo.! Aug., Sept.

8. TRICHODIUM.

Gr. θριξ, τριχος, hair; from its capillary inflorescence.

Inflorescence a capillary panicle; glumes 2, subequal, narrow and acute; palea 1 (or 2, the upper very minute), awnless, shorter than the glumes, loosely enclosing the caryopsis.

1. T. ʟᴀxɪꜰʟóʀᴜᴍ. Michx. (Agrostis Michauxii. *Trin.*) *Thin grass.*
St. erect, smooth, very slender, 1—2f high; *lvs.* 3—6′ long, linear-lanceolate, scabrous, lower ones involute, upper ones shorter and flat; *sheath* rather open; *panicle* large and very diffuse, with long, capillary, verticillate branches, trichotomously divided near the end; *spikelets* in terminal clusters, purple; *glumes* linear-lanceolate.—♃ Pastures and roadsides, U. S. and Brit. Am. Jn.

2. T. ᴍᴏɴᴛᴀɴᴜᴍ. Torr. (Agrostis Torreyi. *Tuckm.*) *Mountain Hair Grass.*
Sts. cæspitose, erect, filiform, simple, in small tufts, 8—12′ high; *radical lvs.* 2—3′ long, involute-filiform, cauline rather longer; *stip.* bifid, serrate; *pan.* ovate, branches spreading, finally divaricate, capillary, hispid; *spikelets* fasciculate at the ends of the branches; *glumes* equal; *paleæ* with a short, twisted awn at the back.—♃ Mts. and rocky woods, N. H. and N. Y.

3. T. ꜱᴄᴀʙʀᴜᴍ. Muhl. (Agrostis scabra. *Willd.*) *Rough Hair Grass.*
St. geniculate at base, assurgent, branched, 1—2f high; *lvs.* rough, striate, linear-lanceolate, 4—6′ long, with the sheaths commonly closed and smooth; *panicle* long, with verticillate, divaricate, dichotomously divided branches which are much shorter than in *T. laxiflorum*; *spikelets* pale green, not clustered.—♃ Common in dry soils, N. Eng. to Ill. July.

4. T. ᴀʟᴛɪꜱꜱɪᴍᴜᴍ. Michx. (T. elatum. *Ph.* Cornucopiæ alt. *Walt.*)
St. erect, rigid, simple, slender, 3f high, leafy; *lvs.* broadly linear, scabrous flat, 6—8′ long; *sheaths* scarcely smooth; *panicle* purple, exserted, contracted, branches in whorls of 4—6s, erect, rather rigid, and dense-flowered at the ends; *glumes* subequal, lanceolate, acuminate, scabrous on the keel, about ¼′

long; *paleæ* 5-veined, a little shorter than the upper glume; *sta.* 3; *stig.* plumose.—♃ N. J. to Car. Aug.

β. *laxa.* Panicle more loose, with long, green branches.—White Mts.

5. T. concinna. Wood. (Agrostis concinna. *Tuckm.*)
St. low, erect; *lvs.* filiform-setaceous; *pan.* ovate, spreading, glabrous; *gl.* unequal, lower acute-mucronate, roughish above, upper acute, glabrous; *paleæ* glabrous, awn from below the middle, contorted, scabrous, exceeding the flower, with a few hairs at its base.—White Mts. *Tuckerman.*—The species are not valuable in agriculture.

9. MUHLENBERGIA. Schreb.

In honor of the late Henry Muhlenberg, D.D., a well known, eminent botanist.

Panicle nearly simple; glumes 2, very minute, unequal, fringed; paleæ many times longer than the glumes, linear-lanceolate, veined, hairy at base; the lower one terminating in a long awn.

1. M. diffúsa. Schreb. *Dorp-seed Grass.*
St. decumbent, diffuse, branching, slender, compressed; *branches* assurgent; *lvs.* 2—3′ by 2″, linear-lanceolate, rough, with smooth, striate, open sheaths; *panicles* terminal and lateral, with remote, appressed, rough branches; *spikelets* pedicellate, often purple; *awn* about as long as the paleæ.—♃ Borders of woods and fields, N. Eng. to Car. and Ky. Aug.

2. M. erecta. Roth. (Brachyelytrum aristatum. *Palis.*)
St. erect, simple, retrorsely pubescent at the nodes, 2—3′ high; *lvs.* lanceolate, scabrous, ciliate on the margin, 4—6′ long, 3″ or more wide, with somewhat open sheaths; *panicle* terminal, simple racemose, contracted; *spikelets* pedicellate, large; *glumes* very unequal, upper one subulate; *lower paleæ* half as long as its awn, *upper paleæ* with a short awn at base lodged in the dorsal groove.—♃ Rocky hills, Can. and U. S., frequent. July.

10. CINNA.

Glumes 2, subequal, compressed, without involucre or awns, upper one 3-veined; paleæ 2, naked at base, on short stipes, lower one larger, enclosing the upper, with a short awn a little below the tip; stamen 1.

C. arundinacea. Willd. (Agrostis Cinna. *Lam.* A. monandra. *Hornem.*)
St. simple, erect, smooth, 3—5′ high; *lvs.* linear-lanceolate, 12—18′ by 3—5′ pale green, rough-edged, with smooth, striate sheaths; *stip.* long, lacerated; *panicle* near a foot in length, rather attenuated above and nodding, with the branches capillary, drooping, and arranged somewhat in 4s; *glumes* linear-lanceolate; *lower paleæ* with a short, straight awn a little below the tip.—♃ A beautiful grass, sought by cattle, in rich, shady soils, U. S. and Can. Aug.

11. TRICHOCHLOA. DC.

Gr. τριξ, τριχος, hair, χλοα, grass; from the capillary inflorescence.

Glumes 2, very minute; paleæ many times longer than the glumes; naked at base, lower one convolute at base, terminating in a long, unarticulated awn.

T. capilláris. DC. (Stipa. *Lam.* Agrostis. *Muhl.* Muhlenbergia. *Lindl.*)
Cæspitose; *sts.* erect, very slender and smooth, 18—24′ high; *lvs.* erect, becoming filiform towards the end. 1—1½f long; *panicle* diffuse, with the branches 1—4′ long, in pairs, and as fine as hairs; *spikelets* purple; *lower paleæ* produced into an awn 3 or 4 times its length.—♃ An exceedingly delicate grass, with large, purple, glossy and almost gossamer-like panicles, waving in the breeze. Sandy soils. Jn. Jl.

12. PSAMMA. Palis.

Gr. ψαμμος, sand; in which this grass grows on the sea shore.

Panicle spicate; glumes 2, awnless; paleæ 2, shorter than the glumes, surrounded with hairs at base; scales linear-lanceolate, longer than caryopsis; styles 3-parted; stig. 3.

P. ARENARIA. Palis. (Arundo. *Linn.* Ammophila. *Lindl.*) *Mat Grass.*
Rt. creeping extensively; *st.* erect, rigid, 2—4f high; *lvs.* involute, if by *t*', smooth and glaucous, pungently acute; *sheaths* smooth; *stip.* oblong; *panicle* dense, with erect, appressed branches, 6—10' long, and an inch thick; *spikelets* compressed, greenish-white; *lower palea* longer than the upper. 2l On sandy sea-coasts, Can. to N. J. At Dorchester, Mass., this grass is extensively manufactured into paper. Aug.

13. CALAMAGROSTIS. Adans.

Name compounded of *Calamus* and *Agrostis.*

Panicle contracted; glumes 2, subequal, acute or acuminate, paleæ 2, mostly shorter than the glumes, surrounded with hairs at base, lower one mucronate, mostly awned below the tip, the upper one often with a stipitate pappus at base.

1. C. CANADENSIS. Palis. (C. Mexicana. *Nutt.* C. agrostoides. *Ph.* Arundo. *Michx.*) *Reed Grass. Blue-joint.—St.* smooth, erect, rigid, 3—5f high; *lvs.* linear-lanceolate, striate, with smooth, veined sheaths; *panicle* erect, rather loose, oblong, the branches capillary, aggregated in 4s and 5s; *glumes* very acute, smoothish, much longer than the paleæ; *lower palea* bifid at the apex, with an exserted awn arising from below the middle of the back. 2l Wet grounds, N. Eng. W. to Mich. Makes good hay. Common. Aug.

2. C. COARCTÁTA. Torr. (Agrostis glauca. *Muhl.* Arundo stricta. *Spr.*)
Glaucous; *st.* erect, 2—4f high; *lvs.* linear-lanceolate, scabrous. with the veins and keel white; *sheaths* striate; *stip.* oblong, obtuse; *panicle* condensed and spike-form, the branches rigidly erect, short and aggregated; *glumes* acuminate, lanceolate, lower 1-veined, upper 3-veined; *lower palea* 5-veined, bifid at the apex, with a short, straight awn a little below the tip.—2l Bogs, Free States and Brit. Am. July, Aug.

3. C. BREVIPILIS. Torr. (A. Epigeios. *Muhl.*)
St. terete, 3—4f high; *lvs.* broad-linear, the sheaths glabrous; *stip.* hairy; *panicle* pyramidal, loose, with the diffuse, capillary branches solitary or in pairs, *glumes* unequal, bearded at base, acute, 1-veined, shorter than the equal, obtuse, awnless paleæ; *pappus* very short, not half the length of the paleæ.—2l In sandy swamps, N. J., *Torrey.*

4. C. PURPURASCENS. Brown. (C. sylvatica. *Trin.*)
Panicle spicate; *glumes* scabrous; *palea* 2, the lower scabrous, 4-toothed at the apex, awned upon the back; abortive *rudiment* plumose, twice longer than the hairs at its base.—White Mts., N. H., *Tuckerman.* Rocky Mts., *Richardson.*—Very rare and unimportant.

5. C. INEXPANSA. Gray.
St. 2—5f high, erect, simple; *lvs.* 2—3" wide, smooth; *panicle* 4—8' long, slender, contracted, branches short, appressed, 4 or 5 together; *glumes* oblong-lanceolate, 2½" long, rough on the keel and sides, acute; *palea* nearly equal, acute, oblong, as long as the glumes, lower one rough, 3-veined, notched at tip, with a short awn inserted below the middle, nearly as long as the flower.—Penn Yan, N. Y. *Sartwell.* Aug.

TRIBE 3. **PHLEOIDEÆ.**—Inflorescence in dense, cylindric or unilateral spikes. Spikelets 1-flowered. Glumes 2, of nearly similar texture with the paleæ.

14. ALOPECURUS.

Glumes subequal, connate, distinct; paleæ united into an inflated glume, cleft on one side below the middle, generally awned; styles often connate.

TENSIS. *Fox-tail Grass.*
smooth, leafy, about 2f high, bearing an erect, dense, many-
ric, obtuse, compound spike, about 2' long; *lvs.* flat, smooth,
sheaths and ovate stipules; *glumes* ciliate, connate below the
as the *paleæ*; *awn* twisted, scabrous, twice the length of the
ds and pastures, Northern States. An excellent grass. Jn., Jl. §

ICULÀTUS. *Bent Fox-tail Grass.*
ling, geniculate, rooting below, sparingly branched, 1—2f high;
, about 2' long; *lvs.* linear-lanceolate, smooth, flat, acute, a few
, with slightly inflated sheaths, and long, entire stipules; *glumes*
at base, hairy outside; *paleæ* truncate, smooth, half as long as
wn.—♃ Wet meadows, N. Eng.! Mid. States and Brit. Am. Jn.
Torr. (A. aristulatus. *Mx.*) *Awns* very short.

15. CRYPSIS. Ait.

υψις, concealment; from the flowers being concealed in the sheaths.

ce an oblong spike; glumes 2, unequal, compressed, 1-
eæ 2, unequal, longer than the glumes; sta. 2—3; ca-
covered by the paleæ.

ICA. Nutt.
nbent and geniculate, 6—12' long, much branched from the base;
ute, divaricate, short, rigid and pungent, subpilose above; *spikes*
al, thick and lobed, more or less enclosed in the inflated sheaths
he terminal one about 1' long, lateral shorter and subcapitate;
ed on the keel, the upper a little longer.—About Philadelphia,
Oct.

16. PHLEUM.

Gr. φλεος; used by the ancients probably for a different plant.

very remote ,oints; *lvs.* lance-linear, 3—7' by 2—3", ciliate and hairy; *sheaths* pubescent, upper one very long; *spike* generally solitary, often 2, on a long, very slender peduncle, sometimes with another scarcely exserted from the sheaths; *spikelets* plano-convex, with the flat side out, ½" diam., about 2 on each very short pedicel, appearing 2—3-rowed in the 1-sided spike.—Dry fields, Mass.! to Car. W. to Ky. Aug.

2. P. LÆVE (and precox. *Michx.*)
St. erect, rather firm, 18'—3f high, glabrous; *lvs.* generally smooth, pilose only at the base, broadly linear; lower *sheaths* sometimes hairy; *spikes* 2—6, alternate, spreading, with a few long, white hairs at the base; *spikelets* in 2 rows; *rachis* flexuous, flat on the back; *pedicels* undivided, with one spikelet; *spikelets* twice as large (1½" diam.) as in the preceding; *glumes* orbicular-ovate, 1-veined.—Grassy banks of rivers, Penn. to Ky. and Ga. Aug.

3. P. STOLONIFÉRUM. Bosc.
St. about 2f long, procumbent at base, geniculate, *stoloniferous* and branched; *lvs.* short, subcordate; *spikes* very numerous (30—50), subverticillate, spreading, in elongated, terminal and lateral racemes; common *rachis* 4—5' long, angular, smooth, partial ones 3—15" long; *spikelets* ovate, alternate.—Cedar swamps, N. J., *Pursh.* July, Aug.

18. DIGITARIA. Haller.
Lat. *digitus*, a finger; alluding to the digitate form of the inflorescence.

Inflorescence digitate or fasciculate; spikes linear, unilateral; spikelets in pairs, on short, bifid pedicels, 2-flowered; glumes 2, the lower very small, sometimes wanting; lower flower abortive, with a single, membranaceous palea; upper flower ♀, with 2 cartilaginous, subequal paleæ; caryopsis striate.

1. D. SANGUINÀLIS. Scop. (Panicum sanguinale. *Linn.*) *Purple Finger Grass. Crab Grass.*—*Sts.* decumbent at base, radiating and branching at the lower joints, 1—2f long; *lvs.* linear-lanceolate, on long, loose sheaths, softly pilose, the sheaths strigosely hairy; *spikes* 3—5' long, fasciculate at the top of the stem, 5—9 together; *spikelets* in pairs, oblong-lanceolate, closely appressed to the flexuous rachis, in 2 rows.—① Common in cultivated grounds, N. Eng., W. Ind.! Aug.—Oct.

2. D. GLABRA. Roem. & Schultz. (Panicum. *Jand.*)
St. generally decumbent, rarely rooting at the joints, a foot long; *lvs.* short, flat, nearly glabrous; *spikes* digitate, spreading, 3—4; *spikelets* crowded, ovoid; *glume* equaling the abortive flower, both hairy.—① Sandy fields, N. Y., Penn. to Ohio! Spikes rather more slender than in the foregoing.

3. D. SEROTINA. Michx. (D. villosa. *Ell.*)
Rt. creeping; *st.* decumbent, 12—18' long, terete, hairy at the joints, forming a dense carpet where it grows; *lvs.* linear-lanceolate, thin, and with the sheaths, very pubescent with long hairs; *spikes* numerous, setaceous, 2—3' long; *spikelets* all pedicellate; lower *glume* very minute, the margin ciliate.—① N. Y.

4. D. FILIFORMIS. Ell. (Panicum. *Willd.*)
St. erect, filiform, simple, 12—18' high; *lvs.* short, nearly smooth, narrow-lanceolate; *lower sheaths* very hairy, *upper* glabrous; *spikes* 2—4, filiform, erect; *rachis* flexuous; *spikelets* in 3s, all pedicellate; *glume* solitary, as long as the abortive flower.—① Dry, gravely soils, N. Y. to Ky. Aug.

19. MILIUM.
Celtic *mil*, a pebble; alluding to its hard, turgid fruit.

Inflorescence paniculate; spikelets 1-flowered; glumes 2, without involucre or awns; paleæ 2, shorter than the glumes, awnless, oblong, concave, persistent and cartilaginous, coating the caryopsis.

1. M. EFFÙSUM. *Spreading Millet Grass.*
St. erect, simple, smooth, 5—8f high, bearing a compound, diffuse panicle; *lvs.* flat, 8—12' by ¼—1', on smooth, striate sheaths; *branches of the panicle* clus-
51*

tered. horizontal, 1—6' long; *spikelets* ovate, few and scattered; *paleæ* smooth and polished.—♃ In woods, Penn. to Can. Plant pale green. Summer.

2. M. PUNGENS. TORR. *Dwarf Millet Grass.*

St. erect, simple, rigid, 12—18' high; *lvs.* lanceolate, *cauline* very short, pungent, at length involute, *radical* 6—8' long, erect, acute and pungent; *sheaths* striate, rough, tumid; *panicle* contracted, few-flowered; *ped.* bifid; *glumes* awn-ess; *paleæ* hairy, about equaling the glumes; *sty.* 2-parted.—♃ Rocky hills, Northern States, rare. May.

3. M. AMPHICARPON. Pursh. (M. ciliatum. *Muhl.*)

Sts. numerous, assurgent, 18—24' high, somewhat branched and geniculate; *lvs.* 2—3' by 2—4'', lance-linear, hairy and ciliate; *sheath.* striate, the upper ones leafless; *panicle* s.mple, 2—3' long, its branches few, erect, appressed, racemose, bearing ♀ flowers; *spikelets* oblong, purplish; radical *peduncles* clustered, 1—3' high, sheathed, each bearing a single ♀ spikelet; *caryopsis* brown. --N J. Aug.

20. OPLISMÉNUS. Kunth.

Gr. οπλισμα, armament, μενος, courage; alluding to the stout awns

Panicle compounded of alternate, dense racemes; glumes 2, unequal, echinate, 2-flowered, lower short, upper as long as the ♀, acuminate; abortive flowers with 2 paleæ, the lower terminating in a long awn; ♀ paleæ cartilaginous, shining, coating the caryopsis.

1. O. CRUS-GALLI. Kunth. (Panicum. *Linn.*) *Cock's-foot Grass.*

St. terete, smooth, 3—4f high; *lvs.* linear-lanceolate, flat, serrulate, with smooth, striate sheaths and no stipule; *panicle* simple or apparently so, branches spike-form, compound, alternate and in pairs; *rachis* hairy and rough; *glumes* hispid with bristles; *lower abortive palea* ending in a rough awn, nearly 1' in length.—① A coarse, weedy grass, introduced into cultivated grounds, barn yards, &c., common. Aug., Sept. ♀

2. O. HISPIDUM. Wood. (O. muricatus. *Kunth.* Panicum. *Muhl.*)

St. thick, 3—4f high; *lvs.* broad, flat; *panicle* compound, nodding, dense, 4—6' long, with alternate racemes; *fls.* always awned.—① Salt marshes, N. Y. to Car. Best distinguished from the preceding by its hispid sheaths. Sept., Oct.

21. PANICUM.

Lat. *panicula*, the mode of flowering, or *panis*, bread, which some species afford.

Glumes 2, unequal, the lower mostly very small; flowers 2, dissimilar, the lower abortive or sterile, with 1—2 paleæ, the upper palea membranaceous; the upper ♀, with the paleæ cartilaginous, equal, concave, awnless, coating the caryopsis.

* *Spikelets in racemose panicles.*

1. P AGROSTOIDES. Muhl. (P. elongatum. *Ph.* P. fusco-rubens ? *Nutt.*)

St. compressed, glabrous, 2—3f high, often geniculate at base; *lvs.* long and numerous, cauline linear-lanceolate, carinate, rough-edged, on short, striate sheaths. *panicles* terminal and lateral, pyramidal, composed of racemed, spreading or deflexed branches; *spikelets* 1'' long, purple, lanceolate, acute, crowded and appressed; *upper glume* 5-veined; *paleæ* of the neutral flower nearly equal.—♃ ? Meadows, frequent. July.

2. P. ANCEPS. Michx. (P. rostratum. *Muhl.*)

St. compressed, 2—3f high; *lvs.* linear, carinate, very long; *sheaths* ancipital, pilose on the throat and margin; *panicle* erect, contracted, with nearly simple branches; *spikelets* interruptedly racemose, very acuminate; *upper palea* of the neutral flower oblong, obtuse or emarginate.—♃ Fields and meadows Common, Mid States, N. Eng. July.

3. P. PROLIFERUM. Lam. (P. dichotomiflorum. *Mx.* P. geniculatum *Muhl.*) *St.* assurgent, geniculate at base, very smooth, thick and succu-*ent; lvs.* linear-lanceolate, 4—6'' wide, 10—15' long, on tumid sheaths, hairy

at throat; *panicles* large, pyramidal, terminal and axillary, smooth; *spikelets* racemose; *abortive flower* with one palea.—♃ ? Wet meadows, Mid and S. States. Sept.

4. P. PEDUNCULATUM. Torr.
St. dichotomously branched above, round, 3—4f high; *lvs.* ½' wide, tapering to the point; *sheaths* hispid and papillose; *panicle* compound, smooth, on a long peduncle, branches in pairs, racemed; *spikelets* ovate, smooth; *upper palea* of the abortive flower half as long as the lower.—♃ Moist woods, N. Y. Jl.

5. P. RECTUM. Roem. & Schultz. (P. involutum. *Torr.* P. de pauperatum. *Muhl.*)—*St.* cæspitose, mostly simple, hairy at the joints, erect, 10—15' high; *lvs.* lance-linear, erect, involute at the end, forming a long, slender, pungent point, rather rough and hairy, sometimes smooth, upper ones longer than the lower; *sheaths* scabrous, hairy; *panicle* erect, rather few-flowered, the branches tortuous, in pairs, one longer with 2 spikelets; *spikelets* rather large, pedunculate; *glumes* veined, lower one short. broad-ovate, obtuse; *paleæ* hard, whitish, shining.—N. Eng. ? and Mid. States ! May, June.

6. P. XANTHOPHYSUM. Gray.
St. 12—15' high, glabrous, generally simple; *lvs.* lanceolate, 3—6' by 5—7", acute, veined, nearly smooth, ciliate at base; *sheaths* pilose, shorter than the joints; *ped.* elongated; *panicles* simple, few-flowered; *spikelets* obovate, *glumes* pubescent, the inferior one acute, 3-veined, half the length of the many-veined, superior one; ♂ 2-valved, ♀ cartilaginous, obtuse, smooth and shining, about equal to the superior glume.—Near Oneida Lake, *Kneiskern*, Conn. River, N. H.! to Conn. ! June, July.

* * Spikelets in loose panicles.

7. P. CLANDESTINUM.
St. with short, axillary, appressed branches, 2—3f high, rigid, leafy; *lvs.* 3—6' by 1', lanceolate, subcordate at base; *sheaths* hispid, enclosing the short, lateral panicles; *upper palea* of the neutral flower obtuse.—♃ Moist woods Mass. and Mid. States. July, Aug.

8. P. LATIFOLIUM. (P. scoparium. *Lam.* ?)
St. nearly simple, with the nodes retrorsely pilose; *lvs.* lance-ovate, clasping, somewhat pubescent; *panicle* terminal, a little exserted from the sheath, simple, pubescent; *spikelets* rather large, oblong-ovate; *abortive flower* staminate.—♃ Common in ditches, woods, &c., U. S. Readily known by broad, short leaves. June, July.

9. P. NERVOSUM. Muhl.
St. simple, smooth at the nodes, 3—4f high; *lvs.* oblong-lanceolate, 2—3' long, smooth, a little ciliate on the margin, cordate at base, an inch wide, with short sheaths; *panicle* much branched, smooth, many-flowered, pedunculate or sessile, branches flexuous, somewhat spreading; *spikelets* oblong; *abortive flower* staminate.—Bogs, N. Y. to Car. W. to Ill.—Perhaps not distinct from the last. July.

10. P. MACROCARPON. Torr.
St. erect, simple, straight, 2—3f high; *lvs.* linear-lanceolate, erect, subpilose beneath, 3—6' long; *sheaths* hispid, villous on the margin, with no stipules; *panicle* rather compound, smooth, with few, spreading, flexuous, subsimple branches; *spikelets* ovoid-globose; *abortive flower* neutral.—Banks of Connecticut river, Mass. July.

11. P. PUBESCENS. Lam. (P. nitidum, β. villosum. *Gray.*)
St. 8—30' high, simple or branched, erect, and with the sheaths, covered with a dense, villous, deflexed pubescence; *sheaths* bearded at the throat; *lvs.* remote, linear-lanceolate, short, upper ones pubescent, lower villose; *panicle* terminal, rather crowded, compound, sometimes rather loose, branches subverticillate, pubescent; *spikelets* pubescent, rather small, at the extremity of the branches; lower *glume* small, upper one 5-veined; *abortive flower* neutral, ♀ longer than the upper glume, smooth.—Penn Yan, N. Y., *Sartwell*, Penn. to Ohio, *Sullivant!* June.

12. P. dichotòmum.

St. much branched and dichotomous above, erect or decumbent, 1 ⚤ high, branches fasciculate; *lvs.* linear-lanceolate, smooth, very numerous, 1—2' long, ciliate at base; *panicles* lateral and terminal, simple, capillary, with loose, spreading branches; *spikelets* minute, ovoid-obtuse.—Varies with the stem very tall, or low, and decumbent; rather rigid or very slender. N. Eng. to Ky. July.

13. P. nitìdum. Lam. (P. discolor. *Muhl.*)

St. erect, slender, 8—24' high, the nodes with a ring of dense, deflexed hairs; *lvs.* lance-linear, narrow, acuminate, rather remote, 2—4' by 2—4"; *sheaths* as long as the leaves, hairy at the throat and often all over; *stip.* 0; *panicle* rather small, exserted, roundish-ovoid, diffuse, nearly glabrous, branches spreading; *spikelets* purplish, numerous, small (½" long), oblong, obtuse; lower *glume* minute; *abortive fl.* neutral, its outer palea equaling the upper glume; ♀ white, polished.—♃ ↑ Woods and fields, U. S. June.

a. Tall, simple, smooth, except the densely bearded nodes.—Middle and Western States.

β. St. with short branches; *lvs.* and *sheaths* pubescent.—N. Eng. and Middle States, common.

γ. Low, branched, very hairy, purplish.—Dry fields, N. Eng., common.—These are the extreme forms, between which there are many intermediate ones.

14. P. microcarpon. Muhl. Darl.

St. 18—30' high, erect, simple, glabrous; *joints* glabrous; *lvs.* lanceolate, veined, ciliate at base, undulate and scabrous on the margin, scabrous above, smooth beneath, 6—10" wide; *sheaths* deeply striate, smooth; *stip.* 0; *panicle* large, much branched, nearly smooth; *spikelets* small, numerous, scarce.y pubescent; ♀ *flower* as long as the upper glume; *fr.* shining, bluish-white.—♃ Woods and low grounds, Penn. *Darlington.*

15. P. capillàre. *Annual Field Panic.*

St. nearly simple, assurgent and thick at base, 1—2f high; *lvs.* hairy, broad-linear, acuminate, 4—6' long; *sheaths* covered with hispid hairs; *panicle* large, pyramidal, capillary, loose, expanding; *spikelets* small, purple, lanceolate, acuminate, smooth, on long, hispid peduncles; *abortive fl.* of 1 palea.—① Fields and roadsides, U. S. Aug.

β. sylvaticum. Torr. *St.* very slender, branched at the base.—Woods.

16. P. verrucòsum. Muhl. (P. debilis. *Ell.*)

St. slender, decumbent and geniculate, branching from the base, 1—2f high; *lvs.* linear, a few inches long, spreading, smooth; *panicle* much expanded, few-flowered, flowers verrucose; *abortive flower* of one palea, and neuter.—①? Swamps and thickets, Mid. and S. States. Panicle terminal and lateral, loose and capillary. Aug.

17. P. virgàtum. *Salt-lick Panic.*

Glabrous and often purple; *st.* 3—5f high; *lvs.* long, linear-lanceolate, hairy at base; *sheaths* striate; *stip.* with long, white ciliæ; *panicle* virgate, at length spreading, diffuse, very large; *fls.* acuminate, and with the glumes, divaricate, paleæ of the abortive flower nearly equal, enfolding the purple stamens.—♃ Salt-lick prairies, fields, &c., N. Y. to Ind.! Aug.—None of these species are of much value in agriculture.

22. SETARIA. Palis.

Lat. *seta,* a bristle; from the bristly involucres of the spikelets.

Inflorescence a compound, cylindrical spike; spikelets 2-flowered, invested with an involucre of 2 or more bristles; glumes, flowers, paleæ and fruit as in the genus *Panicum.*

1. S. virìdis. Palis. *Wild Timothy.*

St. smooth, 2—3f high; *lvs.* lanceolate, flat, minutely serrulate; *sheaths* striate, hairy on the margin, and with a setose stipule; *spike* cylindric, compound, terminal, green; *involucre* of 4—10 fasciculate bristles, much longer than the spikelets; *paleæ of the perfect flower* longitudinally striate, puncta'e.—① Common in cultivated grounds, Free States. July, Aug.

2. S. glauca. Palis. *Bottle Grass.*

St. 2—3f high; *lvs.* lanceolate, carinate, rough, hairy at base; *sheaths* striate, smooth; *stip.* setose; *spike* cylindric, yellowish-green, 2—4' long; *invol.* of 6—10 fascicled, scabrous bristles much longer than the spikelets; *palea of the perfect fl.* transversely rugose.—① Fields and roadsides, N. Eng. to Ohio. Jl. Aug. *β.* (S. purpurascens. *R. & S.*) *Sheaths* and *spikelets* pilose.—Penn.

3' S. verticillata. Palis.   —

St. smooth, about 2f high; *lvs.* lanceolate, rough-edged; *sheaths* smooth, hairy on the margin; *spicate panicle* composed of short, divided branches in interrupted verticils, 2—3' long; *bristles of the invol.* in pairs, rough backwards; *palea of the* ♀ roughish punctate.—① Sandy fields, N. Eng. to Ohio. July. ¶

4. S. Italica. Palis.

St. somewhat compressed, about 4f high; *lvs.* lanceolate, 1—2f long, an inch wide; *sheaths* roughish, pilose at the throat; *spike* compound, interrupted at the base, nodding, 6—8' long; *spikelets* conglomerate; *invol.* of 2 bristles, several times longer than the flower.—① Ditches, Mid. and S. States. July.

5. S. Germanica. Palis. *Millet. Bengal Grass.*

St. 2—4f high, simple, leafy; *lvs.* lance-linear, flat, acuminate, serrulate on the margin; *sheaths* striate, close, pubescent; *stip.* bearded; spike compressed, yellowish, oblong-cylindric; *rachis* densely hirsute; involucrate *bristles* 4—8, as long as, or longer than the spikelets, yellowish; *glumes* unequal, ovate; ♀ *palea* smooth, obscurely 3-veined.—① In fields, not often cultivated. ¶

23. CENCHRUS.

Gr. κεγχρος, millet; this grass bearing some resemblance to the millet.

Flowers racemose or spicate; involucre burr-form, laciniate, echinate, persistent, including 1—3 spikelets; glumes 2, 2-flowered, outer smaller; flowers dissimilar, the lower sterile, the upper perfect: scales 0.

C. tribulöides. (Also C. echinatus Linn. ?) *Burr Grass.*

St. 1—2f long, erect or procumbent and geniculate at base; *lvs.* lance-linear, conduplicate, gradually acuminate, 3—6' by 2—3''; *sheaths* open, about as long as the colored joints; *spike* with the burr-like involucres approximate; *invol.* cartilaginous, beset externally with many sharp, retrorsely hispid spines as long as itself and containing 2—3 spikelets; *glumes* acuminate-mucronate, about 3'' long, producing but 1 caryopsis.—① Sandy soils, N. J.

Tribe 5. **PHALAREÆ.**—Inflorescence a contracted panicle. Spikelets solitary, with 1 perfect flower and 1—2 imperfect ones. Lower palea awned or mucronate, upper with two keels.

24. PHALÃRIS.

Gr. φαλαρος, brilliant; on account of its smooth, shining seeds.

Spikelets 1-flowered; glumes 2, subequal, carinate; palea 2, coriaceous, awnless, shorter than the glumes, coating the caryopsis, each with an external, accessory palea or abortive rudiment at base.

1. P. arundinacea. (P. Americana. *Torr.*)

St. erect, sparingly branched or simple, 2—5f high; *lvs.* spreading, lanceolate, veined, rough-edged, on smooth, striate sheaths; *panicle* oblong, spicate, somewhat secund, 3—4' long, *glumes* 3-veined, whitish, scabrous; *rudiments* pilose.—♃ Common in ditches and swamps, Can. to Car. and Ky. A large showy grass, but not valuable. July, Aug.

β picta is the well-known striped or ribbon grass, with beautifully variegated leaves longitudinally striped in endless diversity. †

2. P Canariensis. *Canary Grass.*

St. erect, or geniculate at the lower joints, round, striate, leafy; *lvs.* lanceolate; *panicle* spicate, ovoid, erect; *glumes* whitish, with green veins; *rudiments* smooth.—① Fields and pastures, not common. Jl. ¶

25. HOLCUS.

Spikelets 2-flowered; glumes herbaceous, boat-shaped, mucronate; flowers pedicellate, the lower one perfect and awnless, upper one ♂ or neuter, awned on the back.

H. LANÀTUS. *Soft Grass.*
Hoary pubescent; *st.* 1½—2f high; *lvs.* lanceolate, 2—5' long; *sheaths* striate; *panicle* oblong, dense, whitish with a purple tinge; *fls.* shorter than the glumes,; sterile one with a recurved, included awn.—♃ Common in wet meadows, N. Eng. | Mid. and W. States. Very soft with whitish down. Jl.

26. HIEROCHLOA. Gmel.

Spikelets 3-flowered; glumes 2, scarious; lateral flowers staminate, triandrous; central flower ♀, diandrous (rarely triandrous).

1. H. BOREÀLIS. Roem. & Schultz. *Seneca Grass.*
Smooth, glossy; *st.* simple, erect, 15—20' high; *radical lvs.* as long as the stem, cauline 2—4' long, lanceolate, mucronate; *panicle* rather 1-sided and spreading, pyramidal, few-flowered, 2—3' long; *branchlets* flexuous; *spikelets* broad, subcordate, colored, unarmed; *lower palea* ciliate.—♃ Wet meadows, Virg. to Arc. Am. Very fragrant. May.

2. H. ALPÌNA. Roem. & Schultz.
Smooth; *st.* erect, stout, 6—8' high; *lvs.* linear-lanceolate, acute; *sheaths* tumid, longer than the internodes; *panicle* ovoid, 1½—2' long, with the branches in pairs; *spikelets* purple, compressed, large, longer than the branches; *glumes* lanceolate; *lower fl.* with an awn about as long as the palese.—♃ Summits of the White Mts., *Bigelow.* Jn.

27. ANTHOXANTHUM.

Gr. av$os, a flower, avos, yellow; from the color of its spikes.

Spikelets 3-flowered, the central one ♀, the 2 lateral ones neuter, each consisting of one bearded palea; glumes 2, unequal, the upper one larger, enclosing the flowers; paleæ of the ♀ 2, short, awnless; stamens 2.

A. ODORÀTUM. *Sweet-scented Vernal Grass.*
St. slender, erect, 10—18' high; *lvs.* short, striate, pale green; *panicle* spicate, oblong-ovoid; *spikelets* pubescent, on short peduncles; *palea of the lateral fls.* linear-oblong, ciliate on the margin, one of them with a bent awn from near the base, the other with a straight awn from the back near the summit.—An early-flowering, deliciously fragrant grass, in most of the States, and Can. May, June. ⚥

TRIBE 6. **AVENEÆ.**—Inflorescence paniculate. Spikelets solitary, fewflowered. Glumes and paleæ of similar texture. Upper flowers generally pedicellate, with awn-like processes or abortive rudiments between the upper and the lower ones. Upper palea with two keels.

28. AÏRA.

Gr. αιρα, a deadly weapon; originally applied to a poisonous plant.

Spikelets 2-flowered, without abortive rudiments; glumes 2, membranaceous and shining, subequal; one of the flowers pedicellate; paleæ subequal, pilose at base, the lower one lacerate at apex and awned on the back.

1. A FLEXUÒSA.
St. smooth, 1—2f high, nearly naked; *lvs.* setaceous, smooth, with striate sheaths and truncate stipules; *panicle* loose, spreading, trichotomous, with long, flexuous branches; *awns geniculate*, longer than the palese.—♃ Vales and hills, U. S. and Brit. Am., common. An erect, elegant grass, growing in tufts. Jn.

2. A. CÆSPITOSA. (A. aristulata. Torr.)

Cæspitose, glabrous; *st.* 18—30' high; *lvs.* narrow-linear, scabrous above, smooth beneath, flat; *panicle* pyramidal, capillary, oblong, finally diffuse; *awns* traight, about as long as the paleæ which are longer than the bluish glumes.— ♃ Swamps, Free States and Can. May.

3. A. PUMILA. *Pursh.*

Sts. scarcely 1' high, erect, growing in tufts, scarcely longer than the eaves; *lvs.* flat, smooth; *panicle* small, fastigiate, few-flowered; *pedicels* short; *alcæ* awnless, obtuse, twice the length of the glumes; *glumes* with a membranaceous margin.—♃ In barren, clayey soils, near brickyards, Penn. *Pursh.* Jn.

29. TRISÉTUM.

Lat. *tris*, three, *setum*, a bristle; a characteristic term.

Spikelet 2—5-flowered; glumes 2, as long as the flowers; lower palea with 2 bristles at the apex and a soft, flexuous awn from above the middle of the back; scales ovate; fruit coated, furrowed.

1. T. PALUSTRE. Torr. (Avena. *Mx.* Aira pallens. *Muhl.*)

St. erect, contracted at the nodes, slender, smooth, about 2f high; *lvs.* lance-linear, about 3' long, roughish, on smooth, striate sheaths; *panicle* oblong, contracted, nodding, yellowish-green; *spikelets* 2—3 flowered, middle flower abortive, upper one pedicellate, its lower palea ending in 2 setose teeth, and awned below the tip, lower one mostly awnless.—♃ Wet meadows, Mass., N. Y. to Flor. May—July.

2. T. PURPURASCENS. Torr. (Avena striata. *Michx.* ?)

St. leafy, 2f high; *lvs.* narrow-linear, keeled, 4—6' long, and with the sheaths smooth; *panicle* very simple, almost a raceme, few-flowered, 4—6' long; *glumes* 3-flowered, very unequal, entire; *spikelets* lanceolate, terete, often purple, smooth; *lower palea* 7-veined, 2-cleft at the extremity; *awn* geniculate.—♃ Mountain bogs, N. Eng., N. Y., Can. June.

3. T. SUBSPICATUM. Brown. (T. aroides. *Palis.* Aira. *Linn.*)

St. a foot high; *lvs.* narrow, 2—4' long; *panicle* contracted into a spike 2 ong; *awn* at length deflexed, longer than the glume.—♃ Rocks and mountains, Little Falls, N. Y. *Gray.* White Mts., N. H. *Pickering.* Jn.

30. AVÉNA.

Spikelets 2—5-flowered; glumes 2, loose and membranaceous, subequal, longer than the flowers; paleæ 2, mostly hairy at base, the lower one bifid, with a twisted or bent awn at the back.

1. A. ELATIOR. (Arrhenatherum. *Palis.*)

St. 2—4f high, geniculate, smooth; *lvs.* lance-linear, rough on the margin and upper surface; *panicle* loose, equal, nodding, branches in pairs or ternate; *spikelets* 2-flowered; *awn* twice as long as the palea; *upper flower* ♀, mostly awnless.—A tall grass, introduced and naturalized in cultivated grounds May, June.

2. A. PENNSYLVANICA. (Arrhenatherum. *Torr.*)

St. erect, smooth; *lvs.* linear-lanceolate; *panicle* slender, with short, alternate branches; *awn* twice as long as the flowers, geniculate, from the base of the lower palea; *upper flower* awnless.—① Fields and open woods, N. Eng. to Car. and Ill., rare. July.

3. A. PRÆCOX. Palis. (Aira. *Linn.*)

Cæspitose; *st.* erect, a few inches high; *lvs.* ½—1' long, rough; *sheaths* deeply striate; *panicle* dense, racemose; *spikelets* ovate, 2-flowered, *glumes* as long as the flowers; *lower palea* with a bent awn from the lower part of the back twice its length.—① N. Y. to Virg. Jn.

4. A. SATIVA. *Common Oat.*—*St.* smooth, 2—4f high; *lvs.* linear-lanceolate, veined, rough, with loose, striate sheaths; *stip.* lacerate; *panicle* loose; *spikelets* pedunculate, pendulous, 2-flowered, both flowers perfect, the lower one mostly awned; *paleæ* somewhat cartilaginous, closely embracing the caryopsis.

—① A highly important grain, one of the staple productions of the soil; said to have been first discovered in the Island of Juan Fernandez.

β. *nigra.* *Black Oats.*—*Paleæ* dark brown, almost black, awnless.

γ *secunda.* *Horse-mane Oats.*—*Panicle* 1-sided; *awns* short.

5. A. STERILIS. *Animated Oat.*—*St.* 3—4f high, and with the leaves smooth, the latter long, acute, flat; *spikelets* 5-flowered, outer flowers and awns hairy, inner flowers awnless.—① From Barbary. Cultivated as a curiosity The awns are 2′ long, geniculate, and twisted more or less according to the state of the atmosphere. Hence the tumbling motion of these spikelets in the moist and warm hands, like a grotesque insect. †

31. DANTHONIA.
In honor of M. Danthoine, a French botanist.

Spikelets 2—7-flowered; glumes 2, subequal, longer than the flowers, cuspidate; paleæ hairy at the base, lower one bidentate at the apex, with a twisted awn between the teeth, the upper one obtuse, entire.

D. SPICÀTA. Palis. (Avena. *Linn.*)

St. slender, nearly erect, 12—18′ high, *lower lvs.* numerous, 4—6′ long, flat, hairy above, *cauline lvs.* much shorter, subulate, erect, on very short sheaths; *panicle* simple, spicate, short, erect; *spikelets* 3—8 or 10, about 7-flowered; *glumes* a little longer than the flowers; *lower pales* hairy, about half as long as its spirally twisted awn.—Pastures and open woods, Free States. June—Aug.

32. URALÈPIS. Nutt.
Gr. ουρα, tail, λεπις a scale or palea; a characteristic term.

Spikelets 2—3-flowered; glumes 2, shorter than the flowers; flowers stipitate and distinct; paleæ 2, very unequal, distinctly villous on the margins, the lower one tricuspidate, the central cusp setose; upper palea concave on the back; fruit gibbous, coated.

U. ARISTULÀTA. Nutt.

Cæspitose; *sts.* procumbent at base, bearded at the nodes, 10—18′ high; *lvs.* subulate, the upper ones shorter than the sheaths, hairy beneath; *panicles* simple, racemose, terminal and lateral, concealed in the sheaths of the leaves, the upper one partly exsert; *spikelet* 3-flowered; *awn of the palea* as long as the lateral cusps. ① Sea-coast and sandy fields, Mid. States. Aug.

33. ARUNDO.
Lat. *arundo,* a reed; Celtic *aru,* water; from its place of growth.

Spikelets many-flowered; glumes 2, awnless, lanceolate, unequal; lower flower ♀ and naked at the base, the others perfect, pedicellate; paleæ unequal, the lower one mucronate, acuminate or slightly awned.

A. PHRAGMÌTES. (Phragmites communis. *Trin.*)

St. smooth, stout, erect, 6—12f high, often an inch in diameter at base; *lvs.* lanceolate, 1—2f by 1—2′, rough-edged, smooth and glaucous; *panicle* large and loosely branched, branches in half whorls, rather erect, slender; *spikelets* 3—5-flowered, very slender, erect; *glumes* shorter than the flowers which are of a dark hue, with tufts of white, silky hairs, about as long as the paleæ.—♃ Swamps and about ponds, Mass. to Ind.! July.

TRIBE 7. FESTUCACEÆ.—Inflorescence panicled. Spikelets many-flowered, oblong. Flowers sessile, closely arranged in 2 rows on the rachis. Paleæ of similar texture with the glumes, the upper one with 2 keels.

34. BROMUS.
Gr. βρωμα, food; this name was formerly applied to a species of wild oats.

Spikelets 3—20-flowered; glumes 2, shorter than the flowers;

lower palea cordate, bifid at the apex, usually awned a little below the tip, upper palea conduplicate, ciliate on the margin; scales ovate, smooth.

1. B. secalinus. *Cheat. Chess.*
St. smooth, erect, 3f high; *lvs.* flat, rough at the edge and above; *sheaths* veined, smooth; *stip.* laciniate; *panicle* spreading, the branches mostly simple, each bearing 1—3 spikelets; *spikelets* ovate, compressed, about 10-flowered, large, 2-ranked, oval, appearing not unlike short heads of wheat. Flowers distinct, a *wn* very short.—① A handsome grass in fields, often among wheat. June. §

2. B. arvensis. (B. mollis *of Am. auth.?*) *Soft Brome Grass.*
Root ①; *st.* erect, mostly pubescent, 1—2f high; *lvs.* and *sheaths* downy pubescent; *panicle* erect, close, compound, 3—4' long; *spikelets* oblong-ovate, slightly compressed, tomentose, 5—10-flowered; *fls.* elliptical; *lower palea* oblong-lanceolate, 7-veined, with a straight awn nearly its length. A coarse grass, in fields and roadsides. June, July. §

3. B. purgans, (Canadensis, ciliatus and pubescens, *of Linn., Muhl., &c.*)
St. terete, rather slender, simple, erect, 2—4f high, nodes blackish; *lvs.* broadly linear, flat, rough-edged, more or less pubescent, striate, 6—12' by 3—5''; *sheaths* more or less pilose with deflexed hairs; *panicle* large, erect, 5—8' long, finally nodding, branches in 2s—4s, compound, scabrous-pubescent; *spikelets* numerous, lance-ovoid, subterete, 7—11-flowered, pedicellate, 9—13'' long, acute at each end; *fls.* imbricated, lower palea pubescent, longer than its straight awn upper green at edge and beautifully ciliate.—⁊ Fields and woods, U. S. and Brit. Am. Jn., Jl.—Varies in size, pubescence, &c., but its forms can scarcely be characterized even as varieties.—A coarse, showy grass, of little value.

35. CERATOCHLOA. Pais.
Gr. κερας, κερατος, a horn, χλοα, grass.

Panicle simple; spikelets lanceolate, compressed, many-flowered; glumes shorter than the flowers, paleæ bifid-toothed, the lower one mucronate between the teeth; fruit coated, furrowed, 3-horned.

C. unioïdes. Palis.
St. 1—2f high; *lvs.* lance-linear, pubescent, veined; *sheaths* bearded at the throat; *stip.* ovate; *panicle* small, nodding, spreading, branches in pairs; *spikelets* oblong-lanceolate, pedicels hairy; *glumes* nearly equal, acuminate, striate; *palea* unequal, lanceolate, acuminate, margined; *sta.* 3.—River bottoms, Penn., Car.

36. FESTUCA.
A Latin name for the shoot or stalk of a plant.

Spikelets oblong, acute at each end, subterete; glumes 2, unequal, shorter than the flowers; paleæ lanceolate, lower one sharply acuminate or awned at the extremity; caryopsis coated.

1. F. tenella. Willd. *Slender Fescue Grass*
St. filiform, wiry, often growing in tufts and geniculate at base, 6—12' high; *lvs.* erect, linear-setaceous, 2—3' long; *sheaths* subpubescent, with lacerated stipules; *panicle* simple, contracted, rather secund, branches alone or in pairs; *spikelets* 5—7-flowered, with subulate, subequal glumes, at length brownish; *fls.* subulate, longer than their awns.—① Sandy fields, N. Eng to Ill., S. to Car. June.

2. F. elatior. *Tall Fescue Grass.*
St. smooth, 3—4f high; *lvs.* lanceolate, smooth, rough-edged, a foot long, on smooth, loose sheaths; *panicle* drooping, very branching, loosely spreading, branches in pairs; *spikelets* lance-ovate, acute, 4—6-flowered, 6—8'' long, racemose on the branches; *lower glume* shorter; *lower palea* acuminate or mucronate.—A fine grass, in meadows, U. S. and Can. June.

3. F. pratensis. Huds. *Meadow Fescue Grass.*
St. smooth, 2—3f high; *lvs.* lance-linear, veined, smooth, rough-edged,

52

about 8' long; *sheaths* veined, smooth with obsolete stipules; *panicle* branched, spreading, somewhat 1-sided, branches subsolitary; *spikelets* lance-linear, 7—9-flowered, about 8'' long; *lower glume* smaller; *lower palea* acuminate.—♃ Introduced in fields and meadows. June, July.

4. F. DURIUSCÙLA. *Hard Fescue Grass.*
    St. smooth, 12—18' high; *lvs.* linear, very acute, a little scabrous; *stipules* membranaceous, lacerate; *panicle* oblong, spreading, inclining to one side branches in pairs; *spikelets* nearly terete, 5—7-flowered; *lower glume* smaller, *upper one* 3-veined; *palea* unequal, *lower* with short awns.—♃ Fields and pastures. A fine grass, common, Car. to Can. June, July.

5. F. NUTANS. Willd.
    St. erect, slender, smooth, with black nodes, about 3f high; *lvs.* narrow-linear, a foot long, veined; *panicle* slender, diffuse, at length nodding, branches in pairs; *spikelets* lance-ovate, 3—5-flowered; *fls.* smooth, awnless and nearly veinless.—♃ Open woodlands, in most of the States. June.

6. F. FASCICULÀRIS. Willd.
    St. much branched from the base, with short internodes, procumbent, geniculate, 12—18' long; *lvs.* linear, very long, 5-veined, scabrous, on long, loose sheaths; *panicle* erect, inclining to one side, with strict, spike-form branches; *spikelets* appressed, secund, 8—10-flowered; *glumes* 1-veined, lower one very short; *lower palea* tipped with awns of their own length.—① Wet meadows, Mid. States. Aug.

7. F. OVINA. *Sheep's Fescue.*
    St. erect, ascending at base, 6—10' high; *lvs.* very narrow, rough, radical ones very numerous, 2—4' long, cauline few, short, erect; *panicle* few-flowered, simple, contracted; *spikelets* ovate, 4-flowered; *palea* roundish.—♃ ? A valuable grass, recently introduced. June.

8. F. MYÙRUS.
    St. 6—12' long, erect, geniculate near the base; *lvs.* 2—3' long, subulate, concave; *stip.* bifid or retuse; *panicle* slender, crowded; *spikelets* 4—7-flowered; *glumes* minute, equal; *fls.* subulate, hairy; *lower palea* with an awn twice its length; *sta.* 1; *stig.* plumose, white.—① Sandy fields, N. J. to Car.

37. DIARRHÈNA. Palis.
Gr. δις, two, αρρην, rough; from the two scabrous keels of the upper palea.

Panicle racemose or simple; glumes 2, very unequal, 2—5-flowered, rigid, acuminate, mucronate; paleæ cartilaginous, lower cuspidate, upper much smaller, emarginate; caryopsis coated, as long as the upper palea: scales ovate, ciliate.

D. DIANDRA. Wood. (D. Americana. *Palis.* Festuca diandra. *Mx.*)
    St. erect, nearly leafless, slender, rigid, 15—30' high; *lvs.* few, subradical, broadly linear, flat, rough-edged, 10—16' by 5—7'', nearly glabrous; *sheaths* close; *stip.* obsolete; *panicle* very simple and slender, branches erect, few; *spikelets* 2-flowered; *glumes* broad-ovate, upper twice larger, 5-veined; *palea* much longer than the glumes, the upper with 2 roughish, green keels, and conspicuously mucronate; *sta.* 2?—River banks, Ohio to Ill. !

38. KŒLERIA. Pers.
In honor of M. Kœler, a German botanist.

Spikelets compressed, 2—3-flowered; glumes 2, unequal, shorter than the flowers; upper flower pedicellate, with a short, awn-like rudiment at the base of the upper palea; paleæ 2, the lower awnless, or awned beneath the tip.

1. K. CRISTÀTA. Smith. (K. nitida. *Nutt.* K. tuberosa. *Pers.* ?)
    St. 20—30' high, smooth, leafy to one-half its height, rigidly erect; *lvs.* flat, erect, pubescent, 2—3' by 1—2'', shorter than their pubescent sheaths; *stip.* short, lacerate; *panicle* spicate, narrow, 3—5' long, 6—8'' diam., branches very short; *spikelets* 2'' long, silvery and shining, compressed, about 2-flowered, with

an abortive pedicel; *glume* linear-oblong, acute, serrulate on the keel, upper one
longer.—♃ Ohio, *W. S. Sullivant !*
  β. *Nuttalii. St.* 8' high.—Mich.

  2. K. ᴏʙᴛᴜsᴀᴛᴀ. Torr. (Aira obtusata. *Michx. ?*)
  St. erect, geniculate below, leafy, 18—24' high; *nodes* pubescent, blackish,
contracted; *lvs.* 3—6' by 2'', scabrous, acuminate, shorter than the sheaths; *stip.*
lacerate; *panicle* contracted, 3—5' long, 6—12'' diam., dense, *branches* fascicled,
short, appressed; *spikelets* 1½'' long, 2-flowered, tumid; lower glume linear-oblong, upper larger, obovate, obtuse, puberulent; *palea* equal, awnless, obtuse,
scarious at summit, a little exserted.—♃ Ohio, *Sullivant !*

  3. K. ᴛʀᴜɴᴄᴀ̄ᴛᴀ. Torr. (Aira truncata. *Muhl.* Holcus striatus. *Linn.*)
  St. slender, 2f high; *lvs.* smooth, narrow, 4—6' long; *panicle* oblong,
loose, racemose; *spikelets* 2-flowered, in clustered racemes, on short, suberect
branches; *glumes* subequal, the lower one linear, upper one much broader, very
obtuse or truncate; *paleæ* awnless.—♃ Fields and open woods, Can. to Ky. Jn.
  β. *major. Lvs.* broad-linear, very long; *panicle* large, spreading.

  4. K. ᴘᴇɴɴsʏʟᴠᴀɴɪᴄᴀ. DC. (Aira mollis. *Muhl.*)
  St. smooth, 2—3f high; *nodes* black; *lvs.* 1—2' long, narrow, flat, lower
ones soft pubescent; *panicle* very slender, loose, 4—8' long; *spikelets* about 3-
flowered, shining; *lower glume* linear, *upper one* much broader, oblanceolate, 3-
veined; *paleæ* awnless.—♃ Rocky woods, N. England (rare) to Ky.! Ill.
May, June.

  5. K. ᴘᴀɴɪᴄᴜʟᴀ̄ᴛᴀ. Nutt.
  St. tall (2—3f high); *lvs.* elongated (4—6'), on long sheaths; *panicle* oblong, glabrous; *spikelets* 2—3-flowered, shining; *glumes* awnless, very unequal,
the larger one truncately obtuse.—♃ Michigan. Also Florida.

39. TRICUSPIS. Palis.

Lat. tres, three, *cuspis,* a point; alluding to the structure of the flowers.

Spikelets terete, tumid, about 5-flowered; glumes 2, unequal, cari
nate, shorter than the flowers; lower palea bifid-toothed, tricuspidat
by the projecting keel and two lateral veins, upper one truncate, al
most emarginate; caryopsis 2-horned.

  T. sᴇsʟᴇʀɪᴏ̈ɪᴅᴇs. Torr. (Poa. *Michx.* Winsoria pœformis. *Nutt.*)
  False Red-top.—St. very hard and smooth, erect, 4—5f high; *lvs.* smooth
beneath, lance-linear, veined, 12—18' long; *lower sheaths* often hairy; *stip.* 0;
panicle loose, expanding, branches flexuous, smooth, long; *spikelets* ovate-
lanceolate, purple, shining, 4—5-flowered; *glumes* unequal, mucronate, *lower*
palea with 3 projecting veins.—♃ A splendid grass in sandy fields, N. Eng. to
Ill. and S. States.

40. GLYCERIA. Brown.

Gr. γλυκυς, sweet; on account of the sweet taste of the seeds.

Spikelets slender, many-flowered; glumes 2, unequal, veinless,
truncate, shorter than the flowers; lower palea herbaceous, embracing the upper, bidentate one; scales connate, truncate.

  1. G. ᴘʟᴜ̈ɪᴛᴀɴs. Brown. (Festuca fluitans. *Linn.*)
  St. compressed or ancipitous, ascending at base, 3—5f high; *lvs.* lance-
linear, smooth beneath, about a foot long; *sheaths* veined, smooth, with a very
large stipule; *panicle* secund, long, slender, slightly branched; *spikelets* 1—1½'
long, linear, appressed, about 10-flowered; *fls.* obtuse; *lower palea* 7-veined, denticulate.—♃ Aquatic. N. and N. W. States. June, July.

  2. G. ᴀᴄᴜᴛɪғʟᴏ̄ʀᴀ. Torr. (Festuca brevifolia. *Muhl.*)
  St. somewhat compressed, 1—2f high; *lvs.* narrow, attenuated above, half
as long as the stem; *panicle* simple, long, appressed; *spikelets* linear, 4—6-flow-
ered; *fls.* very slender, acute, indistinctly veined.—♃ Inundated meadows, N.
Eng., N. Y. June.

41. POA.

Gr. **ποα**, *the general name for grasses or herbage.*

Spikelets compressed, ovate, oblong or linear, many-flowered (3—20); glumes 2, shorter than the lower flowers; paleæ subequal, awnless, often with an arachnoid web at base, bifid-toothed, the lower one herbaceous, scarious on the margin; scales ovate, acute, smooth.

* *Flowers webbed at base.*

1. P. **pratensis**. *Spear Grass.*
St. terete, smooth, 1—2f high; *lvs.* carinate, linear, abruptly acute, radical ones very long and numerous, cauline shorter than the veined, smooth sheaths; *stip.* short, truncate; *panicle* diffuse, branches 3—5 together in half whorls; *spikelets* ovate, acute, with about 4, acute flowers; *glumes* lanceolate, rather acuminate.—♃ An excellent grass both for hay and pasturage, very abundant. May, June. Varies much in luxuriance according to the soil.

2. P. **trivialis**. (P. stolonifera. *Muhl.*) *Roughish Meadow Grass.*
Sts. sometimes stoloniferous at base, roughish backwards, 2—3f high; *sts.* lance-linear, acute, rough-edged, lower ones very long, cauline as long as the roughish sheaths, with long, acuminate stipules; *panicle* diffuse, expanding, scabrous, branches 4—5 together in half-whorls; *spikelets* oblong-ovate, 2—3 flowered.—♃ A grass equally common and valuable with the last, N. States. June, July.

3. P. **compressa**. *Blue Grass.*
St. decumbent and rooting at base, much compressed, smooth, striate, 12—18' high; *lvs.* linear, carinate, veined, smooth, short, bluish-green; *sheaths* smooth, rather loose, with a short, obtuse stipule; *panicle* contracted, somewhat secund, branches scabrous, in 2s and 3s; *spikelets* ovate-oblong, 3—6-flowered, subsessile.—♃ Less abundant than the last, forming tufts in moist places, Free States. June.

4. P. **serotina**. Erhr. (P. palustris. *Muhl.*) *Meadow Red-top.*
St. erect, 2—3f high; *lvs.* flat, narrow-linear, smooth, 10—15' long; *stip* long, lacerated; *panicle* diffuse, somewhat secund, 6—10' long, branches in half-whorls; *spikelets* ovate-lanceolate, 2—3-flowered; *fls.* but little webbed at base, yellow at the tip, obscurely 5-veined.—♃ Common in wet meadows, Free States. June.

5. P. **pungens**. Torr. (P. flexuosa. *Muhl.*)
St. compressed, 1—2f high; *lvs.* of the stem about 2, flat, oblong, lanceolate, cuspidate and pungent, lower about 1' long, upper minute, root lvs. long and narrow, all erect, keeled and pungent at the point; *stip.* truncate, lacerate; *sheaths* nearly as long as the nodes; *panicle* small, racemose at apex, branches in half-whorls, capillary; *spikelets* ovate, 3—4-flowered; *fls.* rather obtuse, webbed; *glumes* smooth, upper acute; *sty.* doubly plumose, white.—♃ Middle and Western States. April.

6. P. **laxa**. Hœnke. (P. alpina. *Torr.*)
St. cæspitose, 6—8' high; *lvs.* linear, acute, smooth; *stip.* lanceolate; *pan.* 1—2' long, contracted, nodding, branches mostly in pairs, smooth, flexuous; *spikelets* 2¼" long, ovate, 3-flowered; *fls.* often purple, acute, hairy, somewhat webbed at base; *glume* lance-ovate, slightly scabrous on the keel; *lower palea* hairy below, upper rough-edged; *anth.* violet.—♃ Mountains and woods, N. Eng.! and Mid. States.

7. P. **nemoralis**. *Wood Spear Grass.*
St. slender, 2—3f high; *lvs.* narrow-linear, pale green, smooth as well as the sheaths; *stip.* scarcely any; *panicle* 6—10' long, slender, nodding when in fruit, branches capillary, flexuous, in 2s or 3s; *spikelets* ovate, about 3-flowered, the flowers spreading and at length remote, slightly webbed at base.—♃ A tall, rank grass, in wet, open woods, N. Eng.! and Mid. States. July

* * *Flowers free, or not webbed at base.*

8. P. **annua**. *Annual Spear Grass.*
Sts. decumbent and rooting at the base, smooth, compressed, 3—8' long ·

lvs. lance-linear, short, smooth, carinate, on loose, glabrous sheaths; *stip.* oblong, dentate; *panicle* spreading, the branches generally solitary, at length horizontal; *spikelets* ovate-oblong, rather numerous, containing about 5, loose flowers.—① A small, abundant, annual grass, N. Eng. to Ohio, forming a dense, soft and beautiful turf. May—Sept.

9. P. **nervata.** Willd. (P. striata. *Michx.* Briza Canadensis. *Nutt.*) *Fowl Meadow.*—*St.* smooth, 3—4f high; *lvs.* lance-linear, striate, rough above, about a foot long, on striate, roughish sheaths; *stip.* lacerate; *panicle* large, loose, diffuse, equal, branches weak, pendulous in fruit, long and capillary, in 2s or 3s; *spikelets* ovate-oblong, containing about 5, obtuse, conspicuously 7-veined flowers.—♃ A beautiful and valuable grass in wet meadows, N. Eng. to Mich. June.

10. P. **elongàta.** Torr.
St. round, erect, smooth, 3f high; *lvs.* narrow-linear, smooth, 8—15' long; *sheaths* striate, smooth; *stip.* very short; *panicle* (8—10') elongated, racemose, nodding, branches solitary or in 2s, appressed; *spikelets* ovate-obtuse, tumid, containing about 3, obtuse, 5-veined flowers.—♃ Wet meadows N. Eng. to Ill. July.

11. P. **obtùsa.** Muhl.
St. smooth, firm, 2—3f high; *lvs.* dark green, linear, often surpassing the stem, and with the sheaths smooth; *panicle* dense, ovate, many-flowered, 3—4' long; *spikelets* ovate, tumid, thick, containing 5—7, smooth, ovate, obtuse flowers; *lower palea* 7 veined.—♃ Swamps, N. Eng. to Penn. Aug., Sept.

12. P. **conferta.** Ell. (P. glomerata. *Walt.*)
St. erect, geniculate, 2—3f high; *lvs.* glabrous, flat, serrulate on the margin; *panicles* terminal and axillary, 4—8' long, erect, compressed, with the spikelets densely clustered; *spikelets* 8-flowered, glabrous.—♃ Penn., *Schweinitz* (fide *Beck*), S. to Car.

13. P. **modesta.** Tuckerman.
St. short, geniculate at base, branched, compressed, glabrous; *lvs.* 3—4' by ¼″, rather rigid; *sheaths* striate, smoothish; *stip.* conspicuous, truncate, erose and laciniate; *panicle* strict, 6—9' long, branches solitary, filiform, scabrous; *spikelets* scattered, briefly pedicellate; *glumes* unequal, obtuse, erose, glabrous; *lower fl.* larger, sessile, veinless; *caryopsis* ovate, fuscous.—Brooksides, Cambridge, Mass. *E. T.*

14. P. **Canadénsis.** Torr. (Briza Canadensis. *Michx.*)
St. round, smooth, erect, 3—4f high; *lvs.* broad-linear, rough, glaucous, on smooth sheaths; *stip.* lacerate, ovate-obtuse; *panicle* large, 6—9' long, branches flexuous, in half-whorls, much spreading or pendulous in fruit; *spikelets* short, ovate, tumid, 6—8-flowered; *glumes* much shorter than the lower flower; *upper palea* very obtuse, *lower* about 7-veined; *sta.* 2.—♃ A large and beautiful grass, in shady grounds, Free States, Can. July, Aug.

15. P. **capillàris.**
St. much branched at base, smooth, a foot high; *lvs.* linear, attenuated above, flat, smooth; *sheaths* striate, with long hairs about the throat and margin; *stip.* short; *panicle* very large (near a foot long), with diffusely spreading, capillary branches, axils smooth, *spikelets* ovate, acute, about 3-flowered, on long pedicels; *palea* scabrous.—① Dry grounds, U. S. Aug.

16. P. **hirsùta.** Michx.
St. subsimple, compressed, erect, 1—2f high; *lvs.* lance-linear, attenuate at end, surpassing the stem, hairy at base; *sheaths* loose, longer than the internodes, lower ones hairy, upper ones smooth; *stip.* fringed; *panicle* very large, capillary, branches spreading, reflexed in fruit, hirsute in the axils; *spikelets* oblong, about 5-flowered; *palea* ciliate.—♃ Sandy fields. July, Aug.
β. *spectabilis.* (*Torr.* P. spectabilis. *P.*) *Spikelets* linear, 10—15-flowered *st.* taller.

17. P. **marítima.** Huds.
St. somewhat geniculate, round, about a foot high; *lvs.* somewhat glaucous, rough-edged, involute; *panicle* erect, dense, branches in pairs, scabrous,

spikelets terete, linear, purplish, about 5-flowered; *fls.* obtuse, indistinctly 5-veined.—24 Salt marshes, Ms., *Bigelow.* June.

18. P. AQUATICA. *β. Americana.* Torr. (P. aquatica. *Ph.*)

Smooth; *st.* stout, leafy, 4—5f high; *lvs.* broad-linear, flat, thin; *panicle* erect, diffuse, branches at length spreading, flexuous, 3—5 together, in half whorls; *spikelets* linear, purple, with 6—8 ovate-obtuse flowers.—24 Wet meadows, Free States and Can. A very large, handsome poa. Aug.

19. P. DENTATA. Torr.

Smooth; *st.* erect, round, 3f high; *lvs.* flat, linear, 10—16' long, glaucous beneath; *stip.* elongated; *panicle* large, loose, few-flowered, branches capillary spreading; *spikelets* lanceolate, about 5-flowered; *lower glume* 3-veined; *lower palea* 5-veined, 5-toothed at the apex when old.—24 Swamps, Mass., N. H.! Penn. Not very common. June, July.

20. P. FASCICULATA. Torr.

Very smooth; *st.* firm and leafy, oblique, round, branched at base, 1—2r high; *lvs.* flat, lance-linear; *panicle* spreading, branches fasciculate, crowded, straight; *spikelets* oblong, somewhat racemed, sessile, crowded, about 3-flowered; *glumes* minute, unequal.—24 Salt marshes, N. Y.

21. P. PECTINACEA. Michx. (P. pilosa. *Muhl.* P. tenella. *Ph.*)

St. cæspitose, oblique, geniculate at base, 8—12' high; *lvs.* flat, smooth, pilose at base, 5-veined, 2—4' long; *sheaths* bearded at the throat; *panicle* large, loose, capillary, purplish, hairy in the axils, branches subverticillate; *spikelets* linear, with 5—9 acute flowers; *upper palea* persistent on the rachis which thus is made finally to appear pectinate.—① In sandy fields, Mid. and S. States. July, Aug.

22. P. REPTANS.

♂ ♀; *st.* branched, creeping, rooting at the joints, 6—12' long; *lvs.* subulate, flat, 2—3' long; *sheaths* open, pilose on the margin and throat; *panicle* 1—2' long, branches short, simple, in fascicles, few-flowered; *spikelets* linear-lanceolate, with 12—20 acuminate flowers.—① Swamps, N. Y. to Ky.! Jl, Aug.

23. P. ERAGROSTIS. (P. obtusa. *Nutt.* Briza eragrostis. *Muhl.*)

St. oblique or decumbent, geniculate, 1—2f long; *lvs.* lanceolate, attenuate at end, scabrous on the margin and above; *sheaths* pilose at the throat; *stip.* short, bearded; *panicle* expanding, branches subdivided, flexuous, subpilose in the axils; *spikelets* ovate-oblong, 12—20-flowered; *glumes* nearly equal.—① A beautiful grass, introduced into fields and roadsides, N. Eng. to Ill.! It has a strong, peculiar odor. Aug. ⑨

42. BRIZA.

Gr. βριζω, to nod, or hang down: alluding to the pendulous spikelets.

Spikelets cordate-ovate, 6—9-flowered; glumes 2, shorter than the lower flowers; paleæ ventricose, lower one cordate at base, embracing the upper which is suborbicular and much shorter; caryopsis beaked.

B. MEDIA.

S'. naked above, 1—2f high; *lvs.* flat, smooth, lance-linear; *stip.* short obtuse; *panicle* erect, few-flowered, branches wide-spreading, capillary, purplish, bearing the ovate or cordate, tumid, pendant and tremulous spikelets at the ends, these are about 7-flowered, greenish-purple; *paleæ* veinless.—24 Naturalized in the vicinity of Boston, *Bigelow.* May.

43. UNIÖLA.

Lat. *unus*, one; on account of the aggregation of many flowers into one spikelet.

Spikelets compressed, 3—20-flowered; lower flower abortive; glumes 2, shorter than the lower flower; lower paleæ boat-shaped at the end, truncate and mucronate between the lobes, upper subulate, somewhat bifid; scales emarginate; caryopsis with 2 horns.

1. U. LATIFOLIA. Michx. *Broad-leaved Uniola.*

St. 2—4f high, smooth, subsimple; *lvs.* 8—18' by 6—12'', lance-linear, glabrous, rough edged; *sheaths* longer than the internodes; *panicle* loose, 8—19' long, nodding; *spikelets* all on long peduncles, about 10" long, ovate, flat, about 10 flowered; *glumes* unequal, near twice shorter than the flowers; *fs.* subfalcate, 6" long, lower one abortive; *sta.* 1.- ♃ Dry woods, middle and Western States. Singularly elegant and showy.   Aug.

2. U. GRACILIS. Michx.   (Holcus laxus. *Linn.*)
St. slender, leafy, 3—4f high; *lrs.* broad-linear, tapering to a slender point, flat, 12—18' long; *sheaths* shorter than the joints; *panicle* long, racemose, branches solitary, short, remote, erect; *spikelets* 3-flowered; *fs.* spreading, straight, monandrous; *glumes* rigid, acute.—♃ Sea coasts, N. Y. to Ga.   Aug.

3. U. SPICÀTA.   (Festuca distachophylla. *Michx.*)
St. smooth, round, branched at base, 1—2f high; *cauline lvs.* numerous, 3—6' long, involute, rigid and acute; *sheaths* longer than the joints, close, upper ones hairy at throat; *stip.* inconspicuous; *panicle* densely spicate, consisting of short, fasciculate branches with sessile spikelets; *spikelets* oblong, 5—9-flowered; *fs.* triandrous.—♃ Salt marshes, N. Y. to Car.   July.

44. MELICA.

Panicle simple or compound; glumes 2, unequal, membranous, 2—5-flowered; fls. a little longer than the glumes, the upper incomplete and abortive; scales truncate, fimbriate; caryopsis free, not furrowed.

M. GLABRA. Walt.   (M. speciosa. *Muhl.*)
St. 3—4f high, glabrous; *lrs.* linear, flat, pubescent beneath; *stip.* lacerate; *panicle* glabrous, loose, few-flowered, erect or a little nodding, branches simple, solitary; *spikelets* 6—8" long; lower *glume* shorter, very smooth: *paleæ* veined; upper *fl.* neuter, pedicellate, consisting of very short, roundish paleæ.— ♃ Mountains, Penn. to Car.

45. DACTYLIS.

Spikelets aggregated, compressed, 3—5-flowered; glumes unequal, the larger one carinate, shorter than the flowers; paleæ subequal, lanceolate, acuminate, the lower one emarginate, carinate, mucronate, upper bifid at apex; scales dentate.

D. GLOMERÀTA.   *Orchard Grass.*
St. roughish, 2—4f high; *lrs.* linear-lanceolate, carinate, a little scabrous, glaucous; *sheaths* striate, *stip.* lacerate; *panicle* remotely branched, rather secund; *spikelets* about 4-flowered, in dense, glomerate, unilateral, terminal clusters; *glumes* very unequal; *anth.* large, yellow.—♃ A fine, well known grass, of rapid growth, introduced in shady fields, as orchards, &c.   June.

TRIBE 8. **CHLOREÆ.**—Inflorescence spiked. Spikelets solitary, few-flowered, the terminal flower often abortive. Glumes carinate, not opposite. Upper palea with two keels.

46. ELEUSINE.

Spikes digitate, unilateral; spikelets 5—7-flowered; glumes obtuse, unequal, lower one smaller; paleæ unequal, upper one bifid toothed; scales truncate, fimbriate; caryopsis triangular, ovate, enclosed in a separate membrane or perigynium.

E. INDÌCA.   *Wire Grass.*
St. oblique, compressed, procumbent and branching at base, 12—16' long; *lrs.* linear, somewhat hairy, on smooth, loose sheaths hairy at the throat; *spikes* 2—4, rarely more or less, linear, straight, divaricate, 2—4' long; *spikelets*

closely imbricate, smooth; *upper glume* 5-veined; *fr.* dark brown.—① Common about houses, foot-paths, &c., Mid. and W. States. Aug.

47. CYNŎDON. Rich.
Gr. κυων, a dog, οδος, tooth; alluding to the singular, one-sided spikelets.

Spikes digitate or fasciculate; spikelets unilateral, in a single row; glumes membranaceous, shorter than the flowers, persistent; ♀ upper palea bifid-toothed; rudiment minute, pedicellate, in a groove of the upper palea; scales truncate.

C. DACTŸLON. Pers. (Digitaria. *Ell.* Panicum. *Linn.*) *Bermuda Grass.*
Rt. creeping extensively; *st.* creeping, stoloniferous at base, 1—2f long; *lvs.* hairy on the margin and towards the base, narrow-linear; *sheaths* hairy; *spikes* 4—5, digitate, spreading, 2—3' long, serrated with the uneven spikelets; *glumes* scabrous on the keel, lanceolate, acute; *palea* subequal, the lower broader, enfolding the upper.—♃ A vigorous creeper, in sands and hard soils, Penn. to Ga.

48. GYMNOPŎGON. Palis.
Gr. γυμνος, naked, πωγων, beard; alluding to the long awn of the palea.

Spikes setaceous, paniculate; glumes 2, keeled, subequal, the lower with a straight awn from a little below the tip; rudiment aristiform.

G. RACEMŌSUM. Palis. (Anthropogon lepturoides. *Nutt.*)
St. ascending, 18—24' high, with short internodes; *lvs.* ovate-lanceolate, 1—2' by 4—8'', glabrous, flat, spreading, in 2 rows; *sheaths* hairy at the throat; *stip.* obsolete; *panicle* large, pyramidal, branches simple, rigid, verticillate, spreading, 3—5' long; *spikelets* sessile, appressed; *glumes* linear, pungent; lower *palea* with an awn at its back 3—4 times its length, upper bifid.—♃ Sandy fields, N. J. to Ga. Aug.

49. SPARTÎNA.

Spikelets imbricated in a double row in unilateral, paniculate spikes; glumes 2, unequal, compressed; paleæ 2, subequal, compressed, awnless; style long, bifid.

1. S. CYNOSURŌIDES. Willd. (Limnetis. *Pers.*)
St. slender, smooth, 3—5f high; *lrs.* 2—3f long, sublinear, convolute and filiform at the end; *sheaths* striate, glabrous; *panicle* loose, slender, composed of 20 or more alternate, one-sided, pedunculate spikes 2—3' long; *spikelets* arranged on 2 sides of a triangular rachis; *glumes* acuminate, one of them with a short awn; *paleæ* white and awnless.—♃ Marshes, Free States and Can. A coarse, sedge-like grass. Aug.

2. S. JUNCEA. Willd. (Limnetis. *Pers.*)
Rt. creeping extensively; *st.* erect, rigid, round, smooth, 1—2f high; *lvs.* convolute at the edges, spreading, in 2 rows; *spikes* 3—5; *ped.* smooth; *rachis* compressed; *lower glume* 3 times as long as the upper; *palea* obtuse, lower one shorter; *sty.* 2.—♃ Marshes and river banks, Free States and Can. Jl., Aug.

3. S. GLABRA. Muhl.
St. smooth, succulent, terete, 3—5f high; *lvs.* concave, erect, about 2f long, ½' wide at base, tapering to a long acumination; *spikes* 10—15, erect and appressed, alternate and sessile upon a triangular rachis; *spikelets* in a dense, double row.—♃ Marshes, Indiana! Aug., Sept.

50. ATHEROPŎGON. Muhl.
Gr. αθηρ, chaff, πωγων, beard; a characteristic term.

Spikes in a thin, simple raceme; glumes 2, membranaceous, 2-flowered, lower one setiform; ♀ paleæ 2, lower one 3-toothed or 3-bristled, upper bifid; abortive flower pedicellate, paleæ 2—3-bristled

A. APLULÖIDES. Muhl.

St. 1—2f high, geniculate at base, ascending, terete; *lvs.* linear-lanceolate, smoothish beneath, pilose above; *stip.* short, truncate; *spikes* 4—6″ long, 20—40, on short, flat peduncles, thinly arranged in 2 opposite rows, each with 4—8 spikelets; *spikelets* 2-flowered, arranged in 2 rows on the under side of the flat, partial rachis; *glumes* unequal, the lower awn-like and slightly adhering to the rachis; *anth.* 3, bright red; *fr.* oblong; *abortive fl.* pedicellate, empty.— ♃ Middle and Western States. Guilford, Conn., *Robbins!*

TRIBE 9. **HORDEÆ.**—Inflorescence spiked. Spikelets solitary, in pairs, or several together, one, few or many-flowered. Glumes mostly two, equal and opposite, rarely unequal and alternate. Lower palea awned or awnless, upper one with 2 keels.

51. TRITICUM.

Lat. *tritum,* rubbed or ground; alluding to the manner of its preparation for food.

Spikelets imbricated in 2 rows, sessile on the teeth of the rachis, about 5-flowered, with the upper flowers abortive; glumes 2, equal, opposite, ovate, concave, mucronate; paleæ 2, lower awned or mucronate; scales 2, collateral.

1. T. **SATIVUM.** *β. hybernum. Winter Wheat.*

St. round, smooth, the internodes somewhat inflated, 3—5f high; *lvs.* lance-linear, veined, roughish above; *stip.* truncate; *spike* parallel, somewhat 4-sided; *spikelets* crowded, broad-ovate, about 4-flowered; *glumes* ventricose; *awns of the upper palea* generally longer than the flowers.—① and ② This is without doubt the most valuable plant of the order; is universally cultivated, and may be regarded as naturalized. Many varieties are known to farmers, of which the most important are

γ. *æstivum. Summer Wheat.* Glumes always awned.—①
δ. *compositum. Egyptian Wheat.* Spike compound; *spikelets* awned.

2. T. **REPENS.** (Agropyron. *Palis.*)

St. trailing at the lower joints, about 2f high; *lvs.* lance-linear, rough above and somewhat hairy; *stip.* short, truncate; *spike* compressed, about 3′ in length; *spikelets* remote, alternate, lance-oblong, 5—6-flowered; *glumes* lanceolate, 5-veined, acuminate.—♃ A vile weed, in fields and gardens, extremely difficult to eradicate. June—Aug. ♦

3. T. **CANINUM.** R. & S. *Dog's Couch Grass.*

St. 2—3f high, erect or oblique; *lvs.* flat, smooth; *stip.* almost wanting; *spikelets* about 5-flowered; *glumes* 3-veined, and with the outer palea, terminating in a straight, scabrous bristle, longer than the flowers.—Delaware, *Muhlenberg.* Probably ♦.

4. T. **CRISTATUM.** Schreb. (Bromus cristatus. *Linn.*)

St. erect, glabrous; *spike* oblong, compressed, imbricated in 2 rows, about 5-flowered, smoothish, spreading; *palea* subulate-awned.—Penn. *Schweinitz (Beck, bot., p. 416).*

52. SECALE.

Celtic *segal,* from *sega,* a sickle.

Spikelets solitary on the teeth of the rachis, 2—3-flowered, the 2 lower flowers fertile, sessile, opposite, the upper one abortive; glumes subulate, opposite, shorter than the flowers; lower palea with a very long awn, upper often bifid at apex; scales abortive, hairy.

S. **CEREALE.** *Rye.*

St. hairy beneath the spike, 4—6f high; *lvs.* lance-linear, rough-edged and rough above, glaucous; *spike* about 5′ long, linear, compressed; *palea* smooth, lower ciliate on the keel and margin; *awns* scabrous-ciliate, long, straight, erect.—① or ② The native country of this highly valuable grain unknown. It has long been cultivated, and like the wheat, may be considered naturalized. June, July.

53. HORDEUM.

Spikelets 3 at each joint of the rachis, 1-flowered, the lateral ones sometimes abortive ; glumes 2, subulate, nearly equal, awned ; paleæ 2, lower lance-ovate, long awned, upper obtusely acuminate ; caryopsis adhering to the paleæ.

1. H. vulgàre. *Barley.*
St. smooth, 2—3f high ; *lvs.* lance-linear, carinate, nearly smooth ; *sheaths* auriculate at the throat ; *spike* thick, about 3′ long ; *spikelets* all fertile, 1-flowered, with an awn-like rudiment at the base of the upper palea ; *glumes* collateral, shorter than the flowers ; *fr.* arranged in 4 rows.—① Extensively ultivated. May. §

2. H. distìchum. *Two-rowed Barley.*
St. 2—3f high ; *lvs.* lance-linear, scabrous above ; *sheaths* auriculate at the throat ; *spike* 3—4′ long, linear, compressed ; *lateral spikelets* abortive, awnless ; *fr.* arranged in 2 rows.—① More common, and is generally preferred for malting to the former species. June. §

3. H. jubàtum. *Squirrel-tail Grass.*
St. slender, round, smooth, simple, about 2f high ; *lvs.* broad-linear, 4—6′ long, rough-edged, otherwise smooth as well as the sheaths ; *spikes* 2—3′ long ; *spikelets* with the lateral flowers neuter ; *glumes* and *paleæ* produced into fine, smooth awns, 6 times as long as the flowers ; *abortive flowers* on short pedicels.—② Marshes, N. Eng. to Mo., N. to Subarc. Am. June.

4. H. pusìllum. Nutt.
St. 4—6′ high, decumbent or geniculate at the base ; *lvs.* about 1½′ long, rather obtuse, glaucous, striate ; upper *sheath* tumid, embracing the spike ; *spike* linear, about 1½′ long ; *glumes* by 3s, collateral, imbricated, lateral ; abortive *fls.* awnless ; *awn* of the central sessile, ☿ as long as those of the involucre ; *glumes* all awned, the inner setaceous from the base ; *awns* 1′ or more long.—Ohio ! to Ill. and Mo.

54. LOLIUM.
Celtic *loloa* ; a name applied to one of the species.

Spikelets many-flowered, sessile, remote, with the edge to the rachis ; glume to the lower spikelet single, to the terminal one 2 ; paleæ herbaceous, subequal, lower one short-awned or mucronate, upper bifid-toothed.

1. L. perenne. *Darnel Grass.*
Smooth ; *st.* terete, 1—2f high ; *lvs.* lance-linear, shining-green, on striate sheaths with truncate stipules ; *rachis* flexuous, grooved, 5—6′ long ; *spikelets* about 16, longer than the glumes, 7—9-flowered, alternate, in two opposite rows ; *lower palea* 5-veined, upper with 2, prominent, rough keels.—♃ N. turalized in meadows, cultivated grounds, &c. May, June.

2. L. temulentum. *Poisonous Darnel.*
St. terete, smooth, 2f high ; *lvs.* lance-linear, rough-edged, and with the sheaths, smooth on the surface ; *stip.* truncate ; *rachis* flexuous, 4—6′ long ; *spikelets* much compressed, 5—7-flowered, longer than the glumes ; *lower palea* 5-veined, produced into an awn twice its length.—① Remarkably distinguished from all other grasses by its poisonous seeds. N. Eng. to Penn. July.

55. ELÝMUS.
Gr. ἐλύω, to fold up ; the spike is enveloped in the sheaths in some of the species.

Spikelets 2 or more at each joint of the rachis, 2—6-flowered ; glumes 2, collateral, subequal, subulate ; paleæ lanceolate, lower one entire, mucronate or awned ; scales ciliate.

1. E. virgìnicus. *Lime Grass. Wild Rye.*
St. erect, smooth, 3—4′ high ; *lvs.* lance-linear, flat, scabrous, deep green, ′ broad ; *sheaths* veined ; *stip.* very short ; *spike* erect, thick, 3—5′ long ; *spike-*

lets in pairs about 3-flowered; *glumes* both in front, lance-linear, slightly connate at base, produced into a scabrous awn; *fls.* smooth; *lower palea* awned.— 2↓ Banks of streams, N. Eng. to Ill., S. to Va. Aug.

2. E. CANADENSIS. (E. glaucifolius. *Willd.*)

St. erect, smooth, stout, 3—5f high; *lrs.* lance-linear, flat, smooth, dark green or often glaucous; *spike* rather spreading, 4—8' long, generally nodding at the summit; *rachis* hairy; *spikelets* 2—5-flowered; *glumes* 5—7-veined, short-awned, hairy; *lower palea* hairy, awned.—2↓ A tall, showy grass, with long, recurved, waving spikes. River banks, &c., Free States and Brit. Am. Aug.

3. E. VILLOSUS. Muhl. *Rye Grass.*

St. slender, striate, smooth, 2—3f high; *lrs.* rough-edged, pubescent above, ½' broad; *sheaths* hairy, especially the lower ones; *spike* 2½—3½' long, a little nodding and spreading; *rachis* and *flowers* hispid, pilose; *spikelets* 1—3-flowered; *glumes* linear; *lower palea* with a long, straight awn.—2↓ Dry grounds, Free States. July.

4. E. HYSTRIX. *Hedgehog Grass.*

St. round, smooth, 2—4f high; *lvs.* lance-linear, carinate, scabrous, generally glaucous and with the sheaths striate; *spike* 4—6' long, erect; *rachis* nearly smooth, flexuous; *spikelets* remote, diverging, almost horizontal, 2—3-flowered; *glumes* 0, rarely 1 or 2; *fls.* smoothish; *lower palea* terminating in a very long awn.—2↓ An odd-looking grass, in moist woods, Free States, common. July.

5. E. STRIATUS. Willd. *Striated Lime Grass.*

St. slender, erect, 8—12' high; *lrs.* and *sheaths* smooth, the former lance-linear, acuminate, scabrous on the upper surface; *spike* erect, 2—3' long; *rachis* 4-leaved, strongly veined, 2-flowered; one flower commonly abortive; *spikelets* in pairs, somewhat spreading, hispid, each 2-flowered; *awns* 3 or 4 times as long as the paleæ.—2↓ Mass., *Bigelow*, to Penn., W. to Ohio, rare. A small and slender species. July.

TRIBE 10. **ANDROPOGONEÆ.**—Inflorescence panicled or spiked. Spikelets generally in pairs, one sessile and perfect, the other mostly pedicellate and imperfect. Glumes of stouter texture than the paleæ. Paleæ delicate and membranaceous, the lower commonly awned.

56. ANDROPÓGON.

Gr. ανδρος, of a man. πωγων, beard; in allusion to the hairy flowers.

Spikelets in pairs, polygamous, the lower one incomplete, on a plumosely bearded pedicel, upper one 1-flowered, perfect; glumes subcoriaceous, awnless; paleæ shorter than the glumes, one generally awned.

1. A. FURCATUS. Muhl. *Forked Spike.*

St. semiterete above, 4—7f high; *lvs.* lance-linear, rough-edged, radical ones very long; *spikes* digitate or fasciculate, in 2s—5s, 3—5' long, purple; *spikelets* appressed, abortive one on a plumose pedicel, ♂ with 2 paleæ, awnless, perfect one with 2 unequal glumes; *lower palea* bifid, awned between the divisions.—2↓ Meadows and low grounds, Free States and Can. Aug.

2. A. SCOPARIUS. Michx. (A. purpurascens. *Muhl.*) *Broom Grass.*

St. slender, paniculate, 3f high, branched, one side furrowed, branches fasciculate, erect; *lrs.* lance-linear, somewhat hairy and glaucous; *spikes* simple, lateral and terminal, on long peduncles, 2—3 from each sheath, purple; *spikelets* remote, abortive one neuter, mostly with 2 paleæ, awned. 2↓ Woods, U. S. Aug.

3. A. VIRGINICUS.

Cæspitose; *st.* subcompressed, 3f high, branches few and short, half concealed; *lrs.* linear, lower ones a foot or more long, rough-edged and hairy; *sheaths* smooth; *spikes* short, in slender, half concealed fascicles of 2 or 3, lateral and terminal; *abortive spikelet* a mere pedicel, without paleæ; ☿ monandrous, with a straight awn.—2↓ Swamps, meadows, &c., N. Eng. to Ky. Sept.

4. A. MACROURUS. Michx. *Indian Grass.*
St. sulcate on one side, much branched above, 2—3f high; *lvs.* linear, rough, lower ones very long, upper ones erect; *sheaths* hairy; *spikes* conjugate, ½—1′ long, in dense lateral and terminal, fastigiate panicles, partly concealed; *abortive spikelet* without paleæ; ♀ monandrous, with a straight awn.—♃ Swamps, Mid. States to Car. Sept.

5. A. NUTANS. *Beard Grass.*
Glabrous; *st.* terete, simple, 3—5f high; *lvs.* glaucous, lance-linear, rough, ½′ broad; *panicle* oblong, branched, nodding, 6—10′ long; *abortive spikelet* without paleæ; *glumes of the* ♀ hairy, ferruginous, shining; *awn* contorted. —♃ Sandy fields, Ü. S. and Can. Oct.

57. SORGHUM.
Formed from *sorght*; the Asiatic name of one of the species.

Spikelets in 2s or 3s, abortive ones pedicellate, awnless, with 2 paleæ, the perfect, sessile, 1-flowered; glumes 2, coriaceous; paleæ 3, the upper one awned.

1. S. SACCHARĀTUM. *Broom Corn.*—St. thick, solid with pith, 6—10f high; *lvs.* lanceolate, acuminate, pubescent at base; *panicle* large, diffuse, with long, verticillate, at length nodding branches; *glumes of the perfect spikelet* hairy, persistent.—① From the E. Indies. The uses of this fine, cultivated plant are doubtless well known to our readers. ‡

2. S. VULGĀRE. *Indian Millet.*—St. erect, round, solid with pith, 6—10f high; *lvs.* carinate, lanceolate; *panicle* compact, oval, erect until mature; *fls.* pubescent; *paleæ* caducous; *fr.* naked.—① From the E. Indies. Rarely cultivated as a curiosity, or for the seed as food for poultry. ‡

TRIBE 11. ORYZEÆ.—Inflorescence panicled or spiked. Spikelets, solitary 1—3-flowered. Flowers perfect or diclinous. Stamens 1—6.

58. LEERSIA.
In honor of John Daniel Leers, a German botanist.

Spikelets 1-flowered, compressed; glumes 0; paleæ 2, compressed, carinate, awnless; scales 2, membranaceous.

1. L. ORYZŌIDES. Swartz. *Cut Grass.*
St. retrorsely scabrous, 3—5f high; *lvs.* lanceolate, carinate, the margin very rough backwards; *sheaths* also very rough with retrorse prickles; *panicle* much branched, diffuse, sheathed at the base; *spikelets* spreading; *paleæ* ciliate on the keel, white, compressed and closed; *sta.* 3.—♃ A very rough grass, common in swamps, by streams, &c., U. S. and Can. Aug.

2. L. VIRGINICA. Willd. *White Grass.*
St. slender, branched, geniculate or decumbent at base, 2—3f long, nodes retrorsely hairy; *lvs.* lance-linear, roughish; *sheaths* roughish backwards, striate; *panicle* simple, at length much exserted, the lower branches diffuse; *fls.* pedicellate, in short, appressed, flexuous racemes; *lower palea* boat-shaped, mucronate; *sta.* 1—2.—♃ Damp woods, U. S. and Can. Aug.

3. L. LENTICULĀRIS. Michx. *Catch-fly Grass.*
St. erect, 2—4f high; *panicle* erect; *fls.* large, roundish, imbricated; *sta.* 2; *paleæ* with the keel and veins ciliate.—♃ Wet places, Ohio, *Frank,* Ct., *Eaton.*

59. ZIZANIA.

♂ Glumes 0; spikelets 1-flowered; paleæ 2, herbaceous. ♂ Paleæ subequal, awnless; stamens 6. ♀ Spikelets subulate; paleæ unequal, linear, lower one with a straight awn; styles 2; caryopsis enveloped in the plicate paleæ.

1. Z. AQUATICA. Lamb. (Z. clavulosa. *Michx.) Indian Rice.*
St. ½′ in diameter, fistular, smooth, 6f high; *lvs.* lance-linear, 2—3f long, an inch wide, smooth, serrulate; *panicle* a foot or more long, pyramidal, the

[redacted]

lower branches divaricate and sterile, the upper spicate and fertile; *spikelets* on clavate pedicels; *awns* long, hispid; *fr.* slender, ½′ long, blackish, deciduous, farinaceous.—♃ Inundated shores of ponds and rivers, U. S. and Can. The fruit, which is very abundant, affords sustenance to wild geese, ducks, and other water fowls. Aug.

2. Z. MILIACEA. Michx.

St. erect, 6—10f high; *lvs.* very long, narrow, glaucous; *panicle* terminal, large, diffuse, pyramidal; *glumes* with short awns; ♂ and ♀ *fls.* intermixed; *sty.* 1; *fr.* ovate, glabrous. - ♃ Penn. to Car., W. to Ohio, growing in water. Aug.

3. Z.? FLUITANS. Michx. (Hydrocochloa. *Palis.* Hydropyrum. *Kunth.*)

St. long, slender, branching, floating in the water; *lvs.* linear, flat; *spike* solitary, axillary, setaceous about 4-flowered; *paleæ* awnless; *stig.* 2, very long; *fr.* reniform.—♃ Can and N. States? July.

60. LEPTŪRUS. R. Br.

Gr. λεπτος, slender, ουρα, tail; from the long, slender, cylindrical spike.

Flowers ♂ ☿ ♀, spicate; rachis filiform, jointed, joints with one spikelet; glumes 1 or 2, rigid, connate with the rachis, simple or 2-parted.

L. PANICULĀTUS. Nutt.

St. scarcely 1f high, compressed; *lvs.* short, rigid, sheathing the base of the panicle; *panicle* or naked rachis incurved, acutely triangular, rigid, bearing 6—10 compressed, subulate spikes on one side, each 1—2′ long; *spikelets* remote, on one side the rachis; *glumes* rigidly fixed, unequal, parallel; *paleæ* 2, the outer of the same texture as the glumes, inner membranaceous.—Ill., *Mead*, Mo., *Nuttall.*

61. TRIPSACUM.

Gr. τριβω, to grind; application not obvious.

☿ Spikes digitate; glumes 2, coriaceous; paleæ 2, membranaceous. ♂ Spikelets 2-flowered, outer flower staminate, inner neuter. ♀ Spikelets 3-flowered, the 2 lateral flowers abortive; outer glume enclosing the flowers in a cavity of the rachis, with an aperture each side at base.

T. DACTYLŌIDES. *Sesame Grass.*

St. slightly compressed, smooth, solid with pith, brown at the nodes, 4—6f high; *lvs.* near an inch broad, long, lance-linear, smooth beneath, roughish above; *spikes* 5—8′ long, usually 2—3 together, digitate, terminal, ♂ flowers above, ♀ below, without awns.—♃ River banks and sea shores, Mid., W. and S. States. A large, coarse and very singular grass. Jl.

β. monostachyon. *Spike* single.

62. ZEA.

Gr. ζαω, to live; the seeds contributing eminently to the support of life.

☿—♂ in terminal, paniculate racemes; spikelets 2-flowered; glumes 2, herbaceous, obtuse, subequal; paleæ membranaceous, awnless, obtuse. ♀ lateral, axillary, on a spadix enclosed in a spathe of numerous bracts spikelets 2-flowered, one flower abortive; glumes 2, very obtuse, paleæ awnless; style 1, filiform, very long, pendulous; caryopsis compressed.

Z. MAYS. *Maize. Indian Corn.*

Rt. fibrous; *st.* erect, 5—10f (in some varieties 15—20f) high, channeled on one side, leafy; *lvs.* lance-linear, entire, 2—3f long.—① The varieties of this noble plant are numerous, produced by climate and culture. It is a native of the warm latitudes of America, but how widely it has been cultivated on both continents, and how important it is to man, it is unnecessary here to state. Jl. §

SECOND GRAND DIVISION,
CRYPTOGAMIA, OR FLOWERLESS PLANTS.

*Plants chiefly composed of cellular tissue, without spiral vessels, des-
titute of true flowers, and producing* SPORES
instead of seeds.

CLASS V. ACROGENS.

Flowerless plants with a proper STEM or AXIS, often with a vascular system composed chiefly of annular ducts, usually furnished with
leaves. GROWTH by the extension of the apex, without subsequent
increase in diameter. SPORES with a proper integument, and contained in a vessel analogous to an ovary, called THECA or SPORANGIUM

ORDER CLXII. EQUISETACEÆ.—HORSETAILS.

Plant leafless, simple or with verticillate branches.
Stem striate-sulcate, jointed, fistular between, and separable at, the joints.
Sheaths dentate, crowning the summit of each internode.
Inflorescence (by analogy) a dense, cylindric, terminal spike or strobile.
Scales of the strobile peltate, hexagonal, subverticillate.
Thecæ 4—7. attached to the under surface of the scales, with lateral dehiscence.
Spores numerous, globose, surrounded by minute granules.   [manner.
Elaters, bodies of unknown use, consisting of 4 elastic, clavate filaments involving the spores in a spiral

An order consisting at present of a single genus, growing in wet grounds, on river banks, and borders
of woods, throughout most countries. The Equisetaceæ abound in the fossil remains of coal measures
with other Cryptogamia, as Lycopodineæ and Filices, indicating that these plants were once of gigantic
dimensions, and formed a large part of the original flora of our globe. Species about 10.

Properties—They abound in silex, and hence are used by cabinet-makers, comb-makers, &c., in polish-
ing their work.

EQUISETUM.
Lat. *equus,* a horse, *seta,* hair; perhaps alluding to the general resemblance

Character the same as that of the order.

1. E. HYEMÀLE. *Scouring Rush.*
Sts. all simple, erect, very rough, each bearing a terminal, ovoid spike;
sheath cinerous white, black at the base and summit, short, with subulate,
awned and deciduous teeth.—Very noticeable in wet, shady grounds, and by
brooksides. Stems about 2' high, often 2 or more united at base from the same
root. Sheaths 2—3'' long, 1—2½' apart, the white ring much broader than the
black, at length entire from the falling off of the teeth. The roughness of the
cuticle is owing to the silex in its composition. June.

2. E. ARVENSE. *Field Horsetail.*
Fertile s'ts. erect, simple, *sterile* with simple, quadrangular branches, decumbent at base.—Low grounds, Free States and Brit. Am. Fertile stems first
appearing, 6—8' high, with 3—5 joints surmounted by large, inflated sheaths cut
into long, dark brown teeth. Spike oblong, ½—2' long. Sterile stems rather
taller than the fertile, remaining through the season, after these have decayed.
At each joint is a whorl of simple, rough, ascending branches, issuing from the
base of the sheaths, their joints also sheathed. April.

3. E. SYLVATICUM. *Wood Horsetail.*
Sterile and *fertile sts.* with compound, rough, deflexed, angular branches.—
Grows in woods and low grounds, Free States and Brit. Am. Stems 9—16

high; the fertile with 4—5 whorls of branches from the base of the sheaths which are 2—3' apart, and cleft into several large, tawny red teeth or segments; the sterile taller and more slender, with more numerous whorls of branches. The branches are all subdivided and curved downwards. Spike oval-cylindric, pedicellate. May.

4. E. VARIEGÀTUM. Smith. (E. scorpoides. Mx.)

Cæspitose; sts. branching at base, filiform, scabrous; spike blackish; sheaths 3-toothed, blackish, teeth membranaceous, whitish, deciduous at the tips.—Hilly woods, Free States and Brit. Am. Stems numerous, 3—6' long, 6-furrowed (5-furrowed. Beck), sheaths very short, 1—2' apart. Spikes small, ovoid, terminal. Not common. July.

5. E. LIMÒSUM. (E. uliginosum. Willd.) Pipes.

S's. somewhat branched, erect, striate-sulcate; branches from the middle joints, simple, short, 5-sided, smooth; spike oblong-ovoid; sheaths appressed.—Borders of ponds and swamps, frequent. Stems 2—3f high, slender, rarely simple, generally with 2—6 whorls of branches about the middle. Branches very irregular in length and position. Sheaths 3—4" long, white at the summit, tipped with as many black, subulate teeth as there are furrows (15—20). This species is greedily devoured by cattle. July.

6. E. PALÙSTRE. Marsh Horsetail.

Sts. branched, smooth, sulcate; branches simple, pentagonal, curved upwards; sheath somewhat appressed, remote, 10-toothed at the apex; spike oblong, dark brown.—Marshes, common. Stems 1—2f high, deeply furrowed. Branches short and like the other species produced in whorls from the bases of the sheaths, at first horizontal, finally bending to an upright position. Spike an inch long. May, June.

ORDER CLXIII. LYCOPODIACEÆ.—CLUB MOSSES.

Stems creeping or erect, branching, rarely simple, abounding in ducts.
Leaves small, numerous, crowded, entire, lanceolate or subulate, 1-nerved.
Inflorescence axillary, or crowded into a sort of ament or spike.
Theca of two kinds in the same plant, sessile, 1, often 2-celled.
Spores few, rather large in some of the thecæ, other thecæ containing minute grains, appearing like fine powder.

Like the Equisetaceæ, these plants appear to have been very abundant in the first ages of the world, and to have attained a gigantic size, though at present but a few feet in length. Properties unimportant. Some are emetic. The powder contained in the theca is highly inflammable, and is used in the manufacture of fire-works. Genera 5, species 200.

Genera.

Leaves cauline, on erect or creeping stems.   .   .   .   . Lycopodium. 1
Leaves or fronds radical, long, linear-subulate.   .   .   .   . Isoëtis.   2

1. LYCOPODIUM.

Gr. λυκος, a wolf, πους, a foot; from some fancied resemblance.

Thecæ axillary, sessile, 1-celled, some of them 2-valved, filled with minute, farinaceous grains, others 3-valved, containing several larger globular spores.

* *Inflorescence in pedunculate spikes.*

1. L. CLAVÀTUM. (L. tristachium. Nutt.) Common Club Moss.

St. creeping; branches ascending; lvs. scattered, incurved, capillaceous-acuminate; spikes in pairs, rarely in 3s, cylindrical, pedunculate; bracts of the spike ovate, acuminate, erosely denticulate.—A well known evergreen, trailing upon the ground in shady pastures and woods, common. Stem and branches clothed with numerous linear-lanceolate leaves which are entire or serrulate, and end in a pellucid, curved bristle. Spikes perfectly straight, parallel, erect, and upon an erect peduncle. July.

2. L. COMPLANÀTUM. Ground Pine.

St. trailing; branches dichotomous; lvs. 4-ranked, unequal, the marginal ones connate, diverging at apex, the superficial ones solitary, appressed; ped. elongated, supporting 4—6, cylindric spikes.—A trailing evergreen, common in woods and shady grounds. Stem round, creeping among the moss and leaves, often 10f in length. Branches numerously subdivided, compressed, somewhat resembling the branchlets of the cedar. Leaves minute, very acute. July.

3. L. CAROLINIĀNUM.
St. creeping; *lvs.* somewhat 2-ranked, spreading, lanceolate, entire; *ped.* erect, solitary, elongated, bearing a single spike; *bracts* sublanceolate, entire.—In muddy grounds. Both the stem and its branches are prostrate, with erect, slender peduncles 3—6' high. July.

4. L. SABINÆFOLIUM. Willd. (L. alpinum. *Mx.*)
St. elongated, creeping; *branches* erect, short, dichotomous, with fastigiate divisions; *lvs.* imbricated on all sides, erect, terete-subulate, aristate-acuminate; *spikes* pedunceled by the attenuated and slightly leafy summits of the branches, cylindric, solitary, with cordate, acuminate bracts.—White Mts.; extensively creeping among the rocks, with erect, numerously divided branches, a few of the divisions terminating in spikes an inch in length. July.

* * *Spikes sessile* † *Leaves surrounding the stem.*

5. L. DENDROIDEUM. Michx.   *Tree Club Moss.*
St. erect; *branches* alternate, crowded, dichotomous, erect; *lvs.* linear-lanceolate, in 6 equal rows, spreading; *spikes* numerous, solitary.—An elegant little plant, common in woods, readily distinguished by its upright, tree-like form. Plant about 8' high, with branches more or less diverging. These are subdivided into numerous, forked branchlets, radiant, so as together to represent a spiral arrangement. Spikes 2—6, an inch long. July.
β obscurum. Torr. (L. obscurum. *L.*) *Branches* spreading; *spike* mostly solitary.

6. L. RUPESTRE.   *Rock Club Moss.*
St. creeping; *branches* ascending, subdivided; *lvs.* scattered, imbricate, linear-lanceolate, capillaceous-acuminate, ciliate; *spike* solitary, quadrangular.
—A very small species, creeping on rocks, &c. Stem a few inches in length, with numerous branches, which are ½—1' long, clothed with grayish-green leaves. Spike ½' long, 4-rowed, seeming a mere continuation of the branch. Jl.

7. L. ALOPECURÖIDES.   *Fox-tail Club Moss.*
St. creeping, subramose; *branches* simple, long, ascending, bearing a single sessile spike at top; *lvs.* linear-subulate, ciliate-dentate at base, spreading; *spike* leafy.—Swamps. Stem extensively creeping. Branches 6—8' high, rarely subdivided, densely clothed with a fine, soft foliage. Spike 1—2' long, very leafy. Aug.

8. L. ANNOTĪNUM.   *Interrupted Club Moss.*
St. creeping; *branches* twice dichotomous, ascending; *lvs.* in 5 rows, linear-lanceolate, mucronate, spreading and serrulate near the tip; *spike* oblong, solitary.—In mountain woods, not common. Branches subdivided near their base, branchlets simple, 4 or more, 6—8' high. Leaves at length reflexed at end. Spike rather cylindric, an inch in length, distinct from the branches. Jl.

9. L. INUNDĀTUM.   *Marsh Club Moss.*
St. creeping, often submersed; *branches* simple, solitary, erect, with a single leafy spike at top; *lvs.* linear. scattered, acute, entire, curved upwards.—In swamps, Mass., N. Y., Can., &c. Spikes ½—1' long, at the summit of branches which are 5—7' long, arising from the base of the stem. Bracts of the spikes leaf-like, dilated at base, spreading at the end, larger than the stem leaves which are 1—2" long. July.

10. L. SELAGINÖIDES.   *Savin-leaved Club Moss.*
St. filiform, creeping; *branches* nearly erect, the flowering ones simple; *lvs.* scattered, lanceolate, a little spreading, ciliate-denticulate; *spike* solitary, leafy.—In moist woods, N. States and Can. Spikes yellowish-green, about ½ long, the bracts foliaceous and twice larger than the true leaves, which are about a line in length. Branches 3—6' high, the sterile ones much divided. Jl.

* * *Spikes sessile.* †† *Leaves 2-ranked.*

11. L. APŎDUM. (L. albidulum. *Muhl.*)
St. branching, prostrate and rooting near the base; *lvs.* orbicular-ovate, acute, membranaceous, alternate, amplexicaul, in 2 rows, with minute, acuminate, superficial ones in a third row on the upper side; *spikes* subsolitary.—A

small, creeping, moss-like species, in wet, rocky shades, U.S., not common. Stem a few inches long, filiform. Leaves less than a line in length. Spikes leafy, scarcely distinguishable from the branches. July, Aug.

*** *Spikes indistinguishable from the branches.*

12. L. LUCIDŬLUM. Michx. *Shining Club Moss.*

St. ascending, dichotomously divided; *lvs.* in 8 rows, linear-lanceolate, denticulate, shining, spreading, or a little reflexed; *thecæ* in the axils of leaves not changed nor crowded into a spike.—In wet woods, U.S. and Can. The foliage of this species is dark green and shining, more ample than is common to the genus. Stems 8—16' long, nearly erect. Leaves 3—5" long, distinctly serrate. Thecæ hemispherical or reniform, in the axils of the leaves near the top of the stem. Jl.

13. L. SELĬGO. (L. recurvum. *Willd.*) *Fir Club Moss.*

St. erect, dichotomously and fast' giately branched; *lvs.* scattered, imbricate, lance-linear, entire, rigid and pungent, but awnless.—A smaller species than the last, found on the summits of the White Mts. Stems 4—8' high, densely clothed with stiff, shining, spreading leaves arranged somewhat in 8 rows and 2—3" in length. Thecæ axillary. Aug.

ISOÊTES.

Gr. ισος, equal, ετος, the year : from its being evergreen.

Thecæ membranaceous, oblong, cordate, 1-celled, immersed in the dilated base of the frond; spores subglobose, slightly angular, attached to numerous filiform receptacles.

1. LACUSTRIS? (I. riparia. *Engelman.*) *Quill-wort.*

Leaves cæspitose, subulate, semiterete, dilated and imbricated at base.—A curious aquatic, in water at or near the margin of ponds and rivers, N. Engl and Mid. States, often wholly submersed! Leaves or fronds numerous, tufted and simple, 2—10' long, somewhat spreading, containing numerous cells divided both by longitudinal and transverse partitions. Thecæ whitish, imbedded in corresponding cavities in the bases of the fronds, traversed within by many threads to which the numerous, small, white, granular spores are attached. Aug.—Our plant differs slightly from the European (with which I have compared it), but I think not specifically.

ORDER CLXIV. FILICES.—FERNS.

Stem a perennial, creeping, horizontal rhizoma. or sometimes erect and arborescent. [nation.
Fronds (fruit-bearing leaves) variously divided. rarely entire, with forked veins, mostly circinate in ver-
Inflorescence occupying the back or margin of the fronds (leaves) arising from the veins.
Thecæ or *sporangia* of one kind only in the same plant, 1-celled, dehiscing irregularly.
Sori, somewhat regular collections of thecæ ; or the thecæ are isolated and scattered.
Indusium. a scale investing each sorus ; or the sori are covered with the revolute margin of the frond, or they are naked.

Genera 192. species 2040. A large and interesting order of flowerless plants, distinguished for their elegant, plume-like foliage. They are usually a few inches to a few feet high, but some of the tropical species, as the Cyatheæ of both Indies, are 15—25 feet high, vieing with the palms in size and beauty.

Properties.—Generally mucilaginous and mildly astringent, hence considered pectoral. Aspidium and Pteris are anthelmintic. Osmunda regalis has been successfully administered for the rickets.

Observation.—The fructification of the ferns, with its various appendages, is too minute to be well observed by the naked eye ; but an examination of it with the aid of a good lens, cannot fail to be interesting and satisfactory. In regard to the localities of the ferns, it should here be remarked, that the species respectively are more generally disseminated throughout the States represented by this flora, and also British America, than are the species of the Phænogamia. They are, however, far more common in the hard, mountainous or rocky soils of the East than in the fertile regions of the West. It seems, therefore, unnecessary to make particular mention of the geographical range of the more common species.

Conspectus of the Genera.

* Fertile leaflets or fronds contracted into the form of a panicle or spike.

		Stipes angular.		Onoclea.	11
	the fert. fronds	smoothish, deeply grooved within		Struthiopteris.	12
	distinct.	Stipes clothed with reddish wool.		Osmunda.	12
Fronds ma-	divided.	each one partly fertile.		Osmunda.	13
ny, radical,	entire and narrowly linear.			Schizæa.	15
		entire Scape spicate.		Ophioglossum.	16
Fern erect,	Frond solitary, on a scape.	divided. Scape paniculate.		Botrychium.	17
Fern climbing, stem long and slender.				Lygodium.	14

53*

sometimes contracted, but never paniculate or spicate.

of 1 scale, opening outwardly.	*Aspidium.*	9
of 2 scales, 1 of them marginal.	*Dicksonia.*	10
superficial, of several scales, opening inwardly.	*Cheilanthes.*	6
Indusium beneath the sorus, fimbriate.	*Woodsia.*	4
Indusium 0.	*Polypodium.*	1
transverse, on the veinlets.	*Asplenium.*	3
not marginal, parallel with the midvein.	*Woodwardia.*	5
or oblong, closely marginal, continuous.	*Pteris.*	7
g-acuminate, striking root in the ground.	*Asplenium.*	3
ate, not rooting. Sori linear, scattered.	*Scolopendrium.*	6
ck, polished, bifurcate stipe.	*Adiantum.*	8

rPODIACEÆ.—Thecæ furnished with a vertical, jointed, nally incomplete ring, and bursting transversely and irregu-

1. POLYPODIUM.

r. πολυς, many, πους, foot; from the multitude of its roots.

h, scattered on various parts of the under surface of usium none.

ım. Pursh. (P. ceteraccinum. *Mx.* Acrostichum polypodoi- —*Fronds* deeply pinnatifid; *segments* alternate, linear, very aly beneath, the upper ones gradually smaller; *stipe* scaly, : segments near the apex; *sori* solitary and distinct.—A para- igh, growing on the inclined, moss-clad trunks of living trees, e huge sycamores, to the height of 10—20 feet. In the damp rs, Western States! and also Southern.

RE. (P. Virginianum. *Willd.*) *Common Polypod.* ly pinnatifid, smooth; *segments* linear-oblong, obtuse, crenu- nes gradually smaller; *sori* large, distinct.—Rather common ınd in woods, forming tangled patches with their roots which

* Frond pinnate.

1. A. ACROSTICHŌIDES. Willd. (Nephrodium. *Michx.*)

Leaflets of the frond distinct, alternate, subsessile, falcate-lanceolate, auriculate on the upper side at base, ciliate-serrulate, only the upper ones fertile; *sori* at length confluent; *stipe* chaffy.—Common in rocky shades. Frond 15—18' high, of a narrow-lanceolate outline. Stipe with loose, chaffy scales. Leaflets numerous, slightly curving upwards, 1—2' in length, the terminal ones, which alone are fruitful, are contracted in size, the under side becoming overspread with the sori. June—Aug.

β. incisum. Gray. (A. Schwenitzii. *Beck.*) Pinnæ irregularly and incisely dentate; *sori* mostly distinct.—N. Y., N. J. and Penn. Passes insensibly into a

* * Frond pinnate with pinnatifid leaflets.

2. A. THELYPTÉRIS. Willd. (A. Noveboracense. *ejusd.*) *Shield-Fern.*

Leaflets of the frond linear-lanceolate, deeply pinnatifid, distinct, subsessile; *segments* oblong, obtuse or acute, subentire, ciliate; *sori* marginal, small, at length confluent; *stipe* smooth and naked.—Damp woods. Fronds pale green, thin and delicate, about a foot long and ⅓ as wide, acuminate at apex. Leaflets acuminate, becoming entire above, sometimes crossing (decussating) at base. Rachis pubescent. Stipe slender, channeled on the upper side. Sori in 2 marginal rows on each segment, finally nearly covering their under surface. July.

β. Noveboracense. Rather paler and more delicate in texture; *leaflets* more narrow and remote.

3 A. CRISTÁTUM. *Willd.* (Nephrodium. *Mx.*) *Crested Shield-Fern.*

Frond nearly bipinnate, lanceolate-ovate; *leaflets* subcordate, oblong-pinnatifid, segments oblong, obtuse, ciliate-serrate; *stipe* scaly.—Moist woods and meadows, N. Eng., Mid. States, rather. rare. Frond 12—18' high, pale green, remarkable for its broad, ovate-lanceolate outline. Segments of the leaflets sometimes almost distinct, broad and obtuse, with sharp teeth. Sori large, in double rows, tawny when mature, chiefly on the upper half of the frond. July.

4. A. LANCASTRIENSE. Spreng. (A. cristatum. *β.* 1*st. edit.*) *Lancaster Shield-Fern.*—*Stipe* with a few large, oblong, torn scales, chiefly at base; *frond* narrowly lanceolate; *leaflets* subopposite, remote, short-petiolulate, broadest at base, the lower triangular-ovate; *sori* large, in a single row each side the midvein of each dentate segment; *indusium* fixed near one side.—Woods, Meriden, N. H., *Rickard!* to N. Y. and N. J. A beautiful fern, quite distinct from the preceding, 24—30' high. Frond dark green, 15—18' by 5—8'. Leaflets gradually narrowing from base to apex. Segments nearly distinct, more or less distinctly serrate-dentate, each with 1—25 dark-brown sori (lower leaflets fruitless). July.

5. A. GOLDIÀNUM. Hook. (A. Filix mas. *Pursh.*) *Goldie's Fern.*

Leaflets of the frond lanceolate, acuminate, deeply pinnatifid; *segments* oblong, subacute, subfalcate, mucronate-serrate; *sori* in 2 rows, each side of the vein of each segment. A tall species, 2—3f high, in rocky woods. Fronds numerous, bright green, scaly upon the stipe and rachis, 5—10' wide. Leaflets 3—6' long, not widening at base, with elongated, narrow segments. July.

* * * Frond bipinnate.

6. A. MARGINÀLE. Sw. (Nephrodium. *Mx.*) *Marginal Shield-Fern.*

Segments of the leaflets oblong, obtuse, decurrent, crenate-sinuate, repand at base, lower ones almost pinnatifid; *sori* marginal; *stipe* chaffy.—A large, handsome fern, in rocky woods, common. Frond 12—18' high, very smooth (rachis a little chaffy), its divisions nearly opposite. Segments of the leaflets distinct, near an inch long, ⅓ as wide. contracted at base, then decurrent, forming a narrow margin along the rachis. Fruit in round dots, in regular rows along the margins of the segments. Indusium large, orbicular, with a lateral sinus. July.

7. A. TENUE. Sw. (A. fragile. *Willd.* Cistopteris. *Bernh.*) *Brittle Shield-Fern.*—*Segments of the leaflets* oblong, obtuse or acute, incisely serrate, approaching to pinnatifid, its serratures subentire; *rachis* winged by the

decurrent leaflets; *stipe* chaffy at base.—A delicate fern, on moist rocks, frequent. Fronds 6—12′ high, dark green, its divisions rather remote, and with the subdivisions, considerably variable in form. Sori large and numerous, near the margins of the segments. June, July.

8. A. ACULEATUM. Sw. *Prickly Shield-Fern*.
Segments of the leaflets ovate, subfalcate, acute, aculeate-serrate, upper ones truncate at base, lower cuneate at base; only the *upper leaflets* fertile; *stipe* and *rachis* chaffy.—Mansfield Mt., Vt., and Mts. in Essex Co., N. Y., *Macrae*. Fronds dark green, in tufts 1—2′ high. Segments of the leaflets on very short petioles, somewhat dilated at base on the upper side, deeply serrate, each serrature tipped with a short spinose bristle. Sori brown, in single rows, distinct. Indusium reniform. Aug.

9. A. DILATATUM. Sw. (A. spinulosum. *Willd.* ?) *Broad Shield-Fern*.
Leaflets oblong-lanceolate, distinct ; *segments* distinct, oblong, obtuse, incisely pinnatifid ; *ultimate segments* mucronate-serrate; *stipe* chaffy ; *indusium* umbilicate.—Woods and shady pastures. Fronds 1—2′ high, nearly tripinnate, the foliage about twice as long as wide, acuminate at apex, abrupt at base. Leaflets also acuminate, but the segments rather obtuse, all distinct at base, except those near the summit, serratures with short, soft bristles. Stipe with large, tawny scales. Sori rather large, somewhat in 2 rows. Jl.

10. A. BULBIFERUM. Sw. (Cistopteris. *Bernh.*) *Bulbiferous Shield-Fern*.
Frond bipinnate, oblong-lanceolate, segments opposite, oblong, serrate, the lower one pinnatifid ; *rachis* bulbiferous ; *sori* roundish, the indusium attached to one side.—In damp woods, frequent. Frond 12—18′ high, remarkable for the little bulbs produced in the axils of the rachis, which, falling to the ground, take root. Foliage narrow, tapering to an acuminate summit. Stipe smooth. Jl.

3. ASPLENIUM.

Gr. a, privation, σπλην, the spleen; from its supposed medicinal virtues.

Sori linear, oblique, or somewhat transverse, scattered ; indusium arising from the lateral veins and opening longitudinally, usually towards the midvein.

1. A. RHIZOPHYLLUM. Willd. *Walking Fern*.
Frond mostly undivided, lanceolate, stipitate, subcrenate, cordate-auriculate at base, the apex attenuated into a long, slender acumination, rooting at the point.—This singular fern grows in rocky woods, not very common. The frond is 4—8′ long ; the long, slender, linear point bending over backwards, reaches the earth, and there strikes root, giving rise to a new plant. Though usually with slightly crenate margins, the plant varies by imperceptible degrees, becoming sometimes so deeply crenate as to form a well-marked variety with pinnatifid fronds. July.

2. EBENEUM. Willd. *Ebony Spleenwort*.
Frond pinnate ; *lfts.* lanceolate, subfalcate, serrate, auriculate at base on the upper side ; *stipe* smooth and polished.—A beautiful fern, in dry woods, hills, &c. Fronds 8—14′ high, on a slender stipe of a shining brown or black color. Foliage 5—9′ long, 1—1½′ wide, linear-lanceolate in outline. Leaflets near an inch in length, rather acuminate and curved at apex, dilated at base on the upper side, and sometimes on the lower. Fruit arranged in short lines on each side the midrib. July.

3. A. ANGUSTIFOLIUM. Michx. *Swamp Spleenwort*.
Frond pinnate ; *lfts.* alternate, upper ones subopposite, linear-lanceolate, serrate towards the apex, somewhat repand, the base truncate on the upper side and rounded on the lower.—In low woods, frequent. Fronds 1—2′ high, in tufts, the outer ones barren, inner fertile. Sori large, diverging from the midrib, parallel with the veins, at length confluent. July.

4. A. TRICHOMANES. (A. melanocaulon. *Willd.*) *Dwarf Spleenwort*.
Frond pinnate ; *lfts.* roundish, small roundish-obovate, obtusely cuneate and entire at base, crenate above ; *stipe* black and polished.—A small and delicate fern, forming tufts on shady rocks. Frond 3—6′ high, lance-linear

in outline, with 8—12 pairs of roundish, sessile leaflets, 3—4″ long. Fruit in several linear-oblong, finally roundish sori on each leaflet, placed oblique to the midvein. July.

5. A. THELIPTERÖIDES. Michx. *Silvery Spleenwort.*

Frond bipinnatifid; *lfts.* pinnatifid, oblong-lanceolate, acuminate; *segments* oblong, obtuse, serrate-crenate; *sori* in parallel, oblique lines.—A fine, large fern, on shady banks of streams. Fronds 1½—3f high, of an ovate-acuminate outline, on a slightly chaffy, pale stipe. Leaflets distinct and rather remote, narrow, 4—6′ long. Segments rounded at the end, near ½′ long. Sori arranged in 2 rows on each segment, one on each side the midvein, convergent below, with shining, silvery indusia when young. July.

6 A. FILIX-FŒMINA. Bernh. (Aspidium Filix-fœmina and asplenoides. *Sw.* A. angustum. *W.*)—*Frond* bipinnate; *lfts.* lanceolate, acuminate; *seg.* oblong-lanceolate, deeply cut-pinnatifid; *ultimate seg.* 2—3-toothed; *sori* reniform or lunate, arranged near the veins; *stipe* smooth.—A delicate, finely-divided fern in moist woods. Fronds 1—2f high, with subopposite divisions. These are subdivided into distinct, obtuse segments, which are themselves cut into oblong, deep serratures, and lastly, the serratures are mostly with 2—3 teeth at the summit. Sori large, at first in linear curves, finally confluent, giving the whole frond a dark brown hue. July.

7. A. RUTA-MURARIA. *Wall-rue Spleenwort.*

Frond bipinnate at base, simply pinnate above; *lfts.* small, petiolate, cuneate, obtusely dentate above.—An extremely small and delicate fern, in dry, rocky places. Frond 2—3′ high, ½ as wide, smooth, growing in tufts, somewhat coriaceous. Segments usually 3 on each leaflet, less than ½′ long. Stipe flat and smooth. Sori linear-oblong, slightly oblique, of a rusty-brown color, finally confluent. July.

8. A. MONTÀNUM. Willd. (A. Adiantum-nigrum. *Michx.*)

Frond glabrous, bipinnate; *lfts.* oblong-ovate, pinnatifid; *seg.* 2—3-toothed at the apex; *sori* linear, finally confluent.—Mountain rocks, Bethlehem, Penn. *Schweinitz* (fide *Beck*), S. to Car., W. o Ky. Fronds growing in tufts, 4—8′ high, narrowly oblong-lanceolate in outline, mostly bipinnate, but more or less divided according to the size. Segments more obtuse than in the foreign A. Adiantum-nigrum. July.

4. WOODSIA. Brown.

In honor of Joseph Woods, an excellent English botanist.

Sori roundish, scattered; indusium beneath the sorus, open, with a multifid or fringed margin, including the pedicellate thecæ, like a calyx.

1. W. ILVENSIS. Br. (Polypodium. *Willd.*)

Frond pinnate, leaflets pinnatifid, lanceolate; *segments* ovate-oblong, obtuse; *sori* near the margin, at length confluent; *rachis* and *stipe* chaffy.—Growing in tufts, on rocks and in dry woods. Fronds 5 or 6′ high, on chaffy and woolly stipes, most chaffy at base. Foliage 3 or 4′ long, ½ as wide, oblong-lanceolate in outline, woolly or chaffy beneath, with opposite and alternate leaflets about an inch in length. The lower leaflets are pinnatifid, upper ones wavy on the margin or entire. June.

2. W. PERRINIÀNA. Hook. & Grev. (Hypopeltis obtusa. *Torr.*)

Frond subbipinnate, minutely glandular-pilose; *segments of the leaflets* pinnatifid; *ultimate segments* roundish-oblong, obtuse, bidentate; *sori* submarginal; *stipe* somewhat chaffy.—About a foot high, among and on rocks. Fronds lance oblong in outline, 3 times as long as wide. Segments of the leaflets crenate-serrate, the lower ones distinct, upper confluent. Sori orbicular, becoming nearly confluent, each subtended by a half round indusium notched into little teeth on the margin. July.

3. W. HYPERBOREA. Br. (Polypodium. *Willd.*) *Flower-cup Fern.*

Frond pinnate; *lfts.* suborbicular, subcordate, 3-parted or incisely pinnatifid, cuneate at base, rough pilose beneath.—A very small species, much resem-

bling the last forming tufts on rocks. Plant 2—4' high. Fronds lance-linear
in outline, on very scaly stipes. Leaflets 8 or 10 pairs, suboppposite, nearly
round, 2 or 3" in diam., the margins only crenate above, deeply pinnatifid in
the lowest pairs. July.

4. W. RUFIDULA. Beck. (W. ilvensis and Aspidium rufidulum. *Pursh.*)
Frond bipinnate; *segments of the leaflets* hairy, oblong, obtuse, pinnatifid,
with obtuse, ultimate segments; *sori* at length confluent; *stipe* and *rachis* hairy.
—Grows on rocks. Fern 6—8' high. Stipe dark brown, densely clothed with
woolly hairs. Frond hairy both sides, its leaflets 4—8" long, lower ones distinctly pinnate, upper pinnatifid. July.

5. WOODWARDIA. Smith.

In honor of Thomas J. Woodward, a distinguished English botanist.

Sori oblong, straight, parallel with the ribs on either side of them;
indusium superficial, arched or vaulted, opening inwardly.

1. W. ONOCLEOIDES. Willd. (W. angustifolia. *Smith.*)
Sterile fronds pinnatifid; *lfts.* lanceolate, repand, slightly serrulate; *fertile
fronds* pinnate, the leaflets entire, linear, acute.—In swamps, not common.
Fern about a foot high, growing in tufts. Barren fronds numerous, of a narrow-
lanceolate, acuminate outline. Leaflets with decurrent or confluent bases.
Fertile fronds fewer, with linear segments nearly covered on the back with the
fruit in oblong, longitudinal sori ⅓' in length. Aug.

2. W. VIRGINICA. Willd.
Frond pinnate, very smooth, the leaflets pinnatifid, lanceolate, sessile;
sori in interrupted lines near the midvein of the leaflets and segments.—In low
woods and swamps. Frond about 2f high, on a smooth stipe, lanceolate in outline, and pale green. Leaflets alternate, deeply pinnatifid, with numerous,
spreading, obtuse and slightly crenate lobes. Fruit arranged in lines along each
side of the midveins, both of the segments and leaflets. July, Aug.

6. SCOLOPENDRIUM. Smith.

Gr. σκολοπενδρα, the centipede; from the number of its roots?

Sori linear, transverse, scattered; indusium double, occupying both
sides of the sorus, superficial, finally opening lengthwise.

S. OFFICINARUM. Willd. (Asplenium Scolopendrium. *Linn.*) *Hart's-tongue.*
Frond simple, ligulate, acute, entire, cordate at base.—Shady rocks, Chi-
tenango, N. Y., *Sartwell.* Stipe rather short (3—5' long), chaffy, bearing the
frond suberect, 8—15' high, 2—3' wide, bright green, paler beneath. Sori oblique
to the midvein, 6—9" in length. Rhizoma large, creeping. July.—This curious fern appears to be confined to the vicinity above mentioned, where it was
first detected by *Pursh*, unless the true plant has also been found in Ky. by
M'Murt.

7. PTERIS.

Gr. πτερον, or πτερυξ, a wing; from the general resemblance of the frond.

Sori in a continuous, marginal line; involucre formed of the inflected margin of the frond, opening inwardly.

1. P. AQUILINA. *Common Brake.*
Frond 3-parted; *branches* bipinnate; *lfts.* linear-lanceolate, lower ones
pinnatifid, upper ones entire; *segments* oblong, obtuse.—Abundant in woods,
pastures and waste grounds. Fern 2—5f in height, upon a smooth, dark purple,
erect stipe. Frond broad-triangular in outline, consisting of 3 primary divisions, which are again subdivided into obtusely pointed, sessile leaflets. These
are entire above, becoming gradually indented towards the base of each subdivision. Sori covered by the folding back of the margins of the segments.
July, Aug.

2. P. ATROPURPUREA. *Rock Brake.*
Frond pinnate; *lower lfts.* ternate or pinnate, segments lanceolate, obtuse,
obliquely truncate or subcordate at base.—Fern 6—10' high, growing on rocks

Frond twice as long as wide, of a grayish hue, the two lower divisions consisting of 1—3 pairs of leaflets with a large, terminal segment. All the segments lance-linear, distinct, with margins conspicuously revolute. Stipe and rachis dark purple, with dense, paleaceous hairs at base. June—Aug.

3. P. GRACILIS. Michx. (Cheilanthes. *Spreng.*)
Frond slender, lanceolate, sterile ones pinnate, leaflets pinnatifid, segments broad-ovate, obtuse; *fertile* bipinnate, leaflets linear-oblong, crenate; *stipe* dark brown.—A delicate species, growing on rocks. Fern 4—6′ high, smooth and shining in all its parts. Aug.

8. CHEILANTHES. Swartz.
Gr. χειλος, lip, ανθος; from the form of its indusium.

Sori roundish, distinct, situated at the margin of the fronds; in dusium of membranous, distinct, inflexed scales, opening inwardly, sometimes continuous with the frond.

C. VESTITA. Swartz. (Nephrodium lanosum. *Mx.*) *Hairy Cheilanthes.*
Stipe and *rachis* hairy; *frond* bipinnate, oblong-ovate in outline, hairy on both sides; *leaflets* alternate; *segments* oblong, alternate, sessile, distinct, crenately pinnatifid, the ultimate segment very entire; *sori* finally continuous along the margin.—Rocky banks, Mid. and W. States, frequent. Stipe slender, rigid, 2—3′ long, dark brown. Fronds 3—6′ by 1—2′. Leaflets lance-ovate in out line, 6—12″ long. Sori marginal, distinct when young, finally crowded. July.

9. ADIANTUM.
Gr. α, privation, διαινω, to moisten; as the rain slides off without wetting it.

Sori oblong or roundish, marginal; indusia membranaceous, arising from the reflexed margins of distinct portions of the frond and opening inwardly.

A. PEDATUM. *Maidenhair.*
Frond pedate; *divisions* pinnate; *segments* oblong-rhomboid, incisely lobed on the upper side, obtuse at apex; *sori* oblong, subulate.—This is, doubtless, the most beautiful of all our ferns, abounding in damp, rocky woods. Stipe 8—14′ high, slender, of a deep, glossy purple approaching to a jet-black. At top it divides equally into 2 compound branches, each of which gives off, at regular intervals, 6—8 simply pinnate leaflets from the outer side, giving the whole frond the form of the crescent. Ultimate segments dimidiate, the lower margin being bounded by the midvein and the veinlets all unilateral. July.

10. DICKSONIA. L'Her.
In honor of James Dickson, a distinguished English cryptogamist.

Sori marginal, roundish, distinct; indusium double, one superficial, opening outwards, the other marginal and opening inwards.

D. PILOSIUSCULA. Willd. *Fine-haired Mountain Fern.*
Frond bipinnate; *leaflets* lanceolate, sessile; *segments* pinnatifid, decurrent, oblong-ovate, ultimate segments toothed; *stipe* a little hairy.—A large and delicate fern, in pastures, roadsides, among rocks and stones. Fronds 2—3f high, in tufts, and remarkable for their numerous divisions and subdivisions. Stipe and rachis smooth, with the exception of a few, soft, scattered hairs. Leaflets alternate, approximate; segments deeply divided into 4-toothed, ultimate segments. Sori minute, solitary, on the upper margin of the segments. July.

11. ONOCLEA.
Gr. ονος, a kind of vessel, κλειω, to close; alluding to the contracted thecæ.

Thecæ covering the whole lower surface of the frond; indusia formed of the segments of the frond, whose margins are revolute and contracted into the form of a berry, opening, but not expanding.

O. SENSIBILIS. *Sensitive fern.*
Sterile fronds pinnate, *leaflets* lanceolate, acute, laciniate, the upper ones

nds bipinnate, with recurved and globular contracted seg-
in low grounds. Fronds about a foot high, the barren ones
at triangular in outline, composed of broad, oblong, sinuate
r ones smaller, nearly entire, becoming united at base. The
y dissimilar in its form to the others, resembling a compound
e fruit in the globular segments of its short divisions. Color
/.

orr. (O. obtusilobata. *Schk.*) *Lfts.* opposite; *segm.* rounded.

12. STRUTHIOPTÉRIS. Willd.

Sos, the ostrich, πτερος, wing. or plume; from the resemblance.

ly covering the back of the frond; indusia scaly,
g internally.

cA. Willd. (S. Pennsylvanica. *Willd.*) *Ostrich Fern.*
pinnate; *leaflets* pinnatifid, sessile; *segments* entire, rather
nes somewhat elongated.—A fern of noble size and appear-
ow woods and swamps. The sterile fronds are often 5 or 6f
bout 3, numerous, in circular clumps. Stipes smooth, chan-
natifid, with numerous segments, the lower of which are the
acute, all more or less connected at base. Fertile fronds few,
e sterile, much smaller, the leaflets with numerous, brown-
ments, densely covered by the fruit beneath. Aug.

*N*DIACEÆ.—Thecæ destitute of a ring, reticulated, stri-
at the apex, opening lengthwise and usually externally.

13. OSMUNDA.

bose. pedicellate, radiate-striate, half-2-valved, collect-
surface of the frond or a portion of it, which is more
d into the form of a panicle.

long, rather acute, entire; some of the intermediate leaflets fertile.—A large fern in low grounds. Fronds 2—3f high, light green., interrupted near the middle by 2—4 pairs of fertile leaflets, which are so much contracted in size as to resemble dense, compound racemes, and densely covered with small reddish-brown thecæ. Stipe channeled, smooth above, chaffy at base. June.

14. LYGODIUM. Swartz.

Gr. λυγωδες, flexible, slender; from the slender, climbing habit.

Thecæ sessile, arranged in 2-ranked spikelets issuing from the margin of the frond, opening on the inner side from the base to the summit; indusium a scale-like veil covering each theca.

L. PALMATUM. Sw. *Climbing Fern.*
Stem flexuous, climbing; *fronds* conjugate, palmate, 5-lobed, lobes entire, obtuse, *spikelets* oblong-linear, from the upper fronds, which are divided and contracted into a compound spike.—This is one of the few ferns with climbing stems, and the only one found in the U. S. Plant of a slender and delicate structure, smooth. Stem 3—4f long. Stipes alternate on the stem, forked, supporting a pair of fronds which are palmately divided into 5—9 segments. Fertile fronds terminal, numerously subdivided into linear-oblong segments or spikelets, with the fruit in 2 rows on the back. Mass.! to Penn., &c. July.

15. SCHIZÆA. Smith.

Gr. σχιζω, to cut, cleave; alluding to the many-cleft spikes.

Spikes unilateral, flabelliform, aggregate; thecæ roundish, radiate at top. sessile, bursting laterally; indusium continuous, formed of the inflexed margins of the spikes.

S. PUSILLA. Pursh. (S. tortuosa. *Muhl.*)
Frond simple, linear, tortuous; *spikes* few, crowded at the top of a long, slender stipe or scape.—A very delicate fern, found in the pine barrens, Quaker Bridge, N. J., also in Western N. Y. by *Mr. Timothy Westmore!* Fronds numerous, cæspitose, 2—3' long, ½—1'' wide. Fertile stipes several, 3—6' high, filiform, with a few short, unilateral spikelets at-top arranged in 2 rows. Thecæ somewhat turbinate, in 2 rows on the inner side of each spikelet. Aug.

TRIBE 3. **OPHIOGLOSSEÆ.**—Thecæ 1-celled, adnate at base, subglobose, coriaceous, opaque, half-2-valved, not cellular, and destitute of a ring.

16. OPHIOGLOSSUM.

Gr. οφις, a serpent, γλωσσα, tongue; from the resemblance.

Thecæ roundish, opening transversely, connate, arranged in a 2-ranked, articulated spike.

O. VULGATUM. *Adder's Tongue.*
Frond simple, oblong-ovate, obtuse, closely reticulated; *spike* cauline.—A curious little plant in low grounds. Fronds solitary, 2—3' long, ½ as wide, amplexicaul, entire, smooth, without a midvein, situated upon the stem or stipe a little below the middle. Stipe 6—10' high, terminating in a lance-linear, compressed spike, 1—2' long, with the thecæ arranged in 2, close, marginal ranks. Thecæ opening outwards and horizontally, becoming lunate, distinct, straw-colored. Vernation straight, not circinate. June.

17. BOTRYCHIUM. Swartz.

Gr. βοτρυς, a cluster of grapes; from the resemblance of the fructification.

Thecæ subglobose, 1-celled, 2-valved, distinct, coriaceous, smooth, adnate to the compound rachis of a racemose panicle; valves opening transversely.

1. B. NEGLECTUM. Wood. *Meriden Botrychium.*
Stipe bearing the frond near the top; *frond* simply pinnate, very short with subentire or subpinnatifid segments; *spikes* paniculate.—A singular species

growing in woods, Meriden, N. H.! It bears a general resemblance to B. Lunaria of Europe, but is quite distinct from that species. Height 5—8'. Frond 9—20' long, ⅓ as wide; segments 3—4 pairs, oblong, obtuse, erosely or incisely dentate. Panicle subsimple, often larger than the fronds. Thecæ 2-valved. Jl

B. **simplex**. Hitchcock.
Stipe bearing the frond above; *frond* ternate, pinnatifid; *segments* cuneate-obovate, incised, unequal; *spike* subcompound, unilateral, interrupted.—Grows in dry, hilly pastures, Ms.! Frond sheathed at base, with a lacerate membrane, nearly simple, divided into 3 or more segments which are 1—2' long, often much dissected. Stipe or scape 3—6' high. Thecæ sessile. Jn.

3. B. Virginicum. (B. gracile. *P.*) *Rattlesnake Fern.*
Stipe with a single frond in the middle; *frond* twice and thrice pinnate, tne lowest pair of pinnæ springing from the base; *ultimate segments* obtuse, somewhat 3-toothed; *spikes* decompound; *plant* subpilose.—A beautiful fern, the largest of its genus, in low woods. Stipe or scape 1—2' high, bearing the frond about half-way up. This is apparently ternate, the lower pair of divisions arising from the base. It is almost tripinnate, the ultimate segments being decurrent and more or less confluent at base, with 3—5 cut serratures. Panicle terminal, 3—6' long, reddish-tawny. June, July.

4. B. fumarioïdes. Willd. (B. obliquum. *Muhl.*)
Scape bearing the frond near the base; *frond* in 3 bipinnatifid divisions; *segments* obliquely lanceolate, crenulate; *spikes* bipinnate. –Native of shady woods and pastures. Frond almost radical, of a triangular outline, 3—5' long and wide, of a stouter texture than the last, distinctly petiolate. Scape thick, 8—12' high, bearing a tawny, compound panicle 2—4' in length, composed of numerous little 2-ranked spikes. Aug.
β. dissectum. Oakes. (B. dissectum. *Willd.*) *Frond* near the base of the scape, more numerously dissected, almost tripinnatifid.

Order CLXV. MARSILEACEÆ.—Pepperworts.

Plants stemless, creeping or floating.
Leaves veiny, usually petiolate, often sessile and scale-like, sometimes destitute of a lamina.
Reproductive organs of two kinds; the one compound, the other simple, oval, radical bodies separate from or mixed with the first, with many cells.

Genera 4, species 20? inhabiting ditches and inundated places in nearly all countries, but chiefly in temperate latitudes.

1. SALVINIA. Micheli
In honor of Salvini, professor of Greek at Florence.

Stamens? jointed hairs on the stalks of the ovary; ovary nearly sessile, among the roots, hairy, 1—5, opening at top; fruit capsular, covered with bristly hairs, containing reproductive bodies of two kinds, one kind globular, the other oblong.

S. natans. Willd. (Marsilea natans. *Linn.*)
Leaves opposite, arranged in two rows upon the rhizoma, elliptical, entire, sub-ordate at base, obtuse at apex, clothed with fascicles of hairs above; *ovaries* or *fruit* nearly globose, aggregated in subsessile clusters on the rhizoma, submersed.—ⓛ Floating, like a Lemna, in lakes and other still waters, Western N. Y. and Can. Leaves nearly an inch long, of fine green. The plant is quite rare.

2. AZOLLA. Lam.
Gr. οζω, to make dry, ολλυμι, to kill; as the plants speedily die when taken from the water.

Stamens? capillary; reproductive organs in pairs or numerous, of two kinds, the one of two transverse cells, the upper containing several angular, stalked bodies, the other stipitate, numerous, globose, 1-celled, enclosed in an ovate, close involucre, and containing several angular spores.

A. CAROLINIANA. Willd.

Leares arranged in two rows upon the rhizoma, imbricated, oblong-subu-
.ate, obtusish, spreading, fleshy, the floating ones reddish beneath.—① A small
plant, resembling some of the mosses, floating in still or sluggish waters, Northern and Western States. It has been found in Lake Ontario, and in Cayuga
Marsh (*Eaton*), also in the Ohio river at Louisville, Ky. (*M'Murt*,) and other
western rivers.

ORDER CLXVI. CHARACEÆ.—CHARADS.

Plant aquatic, submersed; *axis* consisting of parallel, tubular cells, either transparent, or encrusted
with carbonate of lime, furnished with leaves or branches consisting of verticillate tubes.
   Organs of reproduction consisting of round, succulent *globules*, containing filaments and a fluid; and
axillary *nucules* formed of a few short tubes twisted spirally around a centre, endowed with the power
of germination.
   These are remarkable for the distinct current, readily observable with a microscope, in the fluid of each
tube of which the plant is composed. The currents instantly cease when the plant is injured.

CHARA.

Gr. χαιρω, to rejoice; because it delights in the water?

Globules minute, round, reddish, dehiscent, filled with a mass of
elastic filaments; nucules (thecæ?) sessile, oval, solitary, membranaceous, spirally striated, the summit indistinctly cleft into 5 valves,
the interior filled with minute spores.

1. C. VULGÁRIS. *Feather-beds.*
Sts. and *branches* naked at base; *branches* terete, leafy at the joints; *lvs.* (or
branchlets?) oblong-subulate; *bracts* shorter than the fruit.—A slender, flexile
plant of a dull green color, found in ponds and ditches generally stagnant. It
appears in dense tufts, like a soft bed, undulating with the motion of the water.
When taken out, it has an offensive odor. Stems slender, a foot or more long,
with a verticil of about 8 filiform branchlets at each joint. June.

2. C. FLEXÍLIS.
St. translucent, naked; *branches* jointless, leafless, compressed; *nucules*
lateral, naked.—Found in ponds, Stockbridge, Ms., in company with Najas.
Resembles the last, but the stems are shorter and more erect, nearly destitute of
the verticils of branchlets. It is annual, as are also all the other species. Aug.

3. C. FOLIÒSA. Willd. (C. squamosa. *Desf.*)
St. solitary, 8—10' high, calcarious and brittle, scabrous, striate, with 2
or 3 branches, the younger parts bearing numerous whorls of minute, leaf-
like scales; *branchlets* about 13 in a whorl, 6—9'' long, involucrate at their origin by a dense whorl of subulate scales; each branchlet with 4—7 whorls of
nearly obsolete scales; *thecæ* 2—4, on the inner side and lower half of each
branchlet.—① Rivers, Ohio, *Riddell*, Ind.! Aug.

4. C. HUMILIS. Riddell.
St. incrusted, solitary and branching, scabrous, not sulcate, 2—3' high;
branchlets 8—10 in a whorl, without an involucre, generally equaling the internodes (6—9''); *thecæ* 1—2 on each of the branchlets in the upper whorls.—
① In shallow water, rivers, Ohio, *Riddell.* Color deep green. Resembles the
last. Aug.

5. C. SABULÒSA. Riddell. *S'one-wort.*
St. 18' high, thickly encrusted and very brittle, with a few erect branches,
subsulcate, papillose; *branchlets* generally 10, sometimes 8, in each whorl, without an involucre, nearly 6'' long, (about half as long as the internodes,) each
with 3—4 verticils of scales; *thecæ* on the inner side of the branchlets, in the
axils of the scales.—① Pools of clear water, 2 or 3 feet deep, Ohio, Ind.! Color
.ight pea-green. Aug.

  β. spiralis. Riddell. *Stems* spirally sulcate; *branchlets* longer.

INDEX.

419	English Ivy,	295	Hart's-tongue,	633	Leaf-cup,			363
304	English Moss,	277	Hawkweed,	368, 369	Leather-flower,			139
352	Eternal Flower,	355	Hawthorn,	243	Leather wood,			496
250	Eyebright,	487	Hazel,	496, 497	Leek,		550, 551	
8, 233	False Flax,	163	Heartsense,	178	Lemon Tree,			204
—627	False Mermaid,	200	Hedge Hyssop,	403	Lettuce,		358, 361	
9, 570	False Red-top,	613	Hedge Mustard,	168	Lever-wood,			497
472	False Wall Flower,	169	Hedgehog,	228	Life-everlasting,			351
604	Feather-beds,	637	Hedgehog Grass,	624	Lilac,			463
237	Feather Grass,	596	Heliotrope,	436	Lily,		548	549
541	Felwort,	454	Hellebore,	144, 557	Lily of the Valley,			553
272	Fennel,	590	Hemlock, 286, 515,	516	Lime Tree,		209	210
318	Fennel Flower,	149	Hemp,	515	Lime Grass,		42?	62?
145	Fern,	629—636	Hemp Nettle,	427	Linden Tree,			209
455	Fescue Grass, 611,	612	Henbane,	446, 447	Lion's-heart,			428
431	Fever-bush,	478	Henbit,	428	Lion's foot,			361
7, 338	Feverfew,	343	Herb Robert,	196	Live-forever,			277
531	Feverwort,	300	Herd's Grass,	602	Liverwort,			141
294	Field Mint,	416	Hickory,	491, 492	Lizzard's-tail,			484
191	Fig,	510	High Cranberry,	302	Locust,		294,	237
416	Figwort,	389	High-water Shrub,	153	Loosestrife, 280, 285,			387
309	Finger Grass,	605	Hoarhound, 315, 416,	128	Lopseed,			413
188	Fir Balsam,	516	Hobble-bush,	302	Lousewort,			411
296	Fire-weed,	352	Hogweed,	363, 469	Lovage,			289
207	Flax,	195	Holly,	381	Love-lies-bleeding,			471
355	Fleabane, 225,	327	Hollyhock,	208	Lungwort,			434
507	Florin Grass,	597	Honesty,	164	Lupine,			233
2, 573	Flower-of-an-hour,	209	Honewort,	287, 289	Lychnidea,	428, 429,		440
6, 291	Fly-poison,	556	Honeysuckle, 298, 299,	300	Madwort,			163
145	Fly-trap,	192	Hoop Ash,	485	Maidenhair,			633
411	Fog fruit,	413	Hop,	512	Maize,			623
470	Fool's Parsley,	289	Hornbeam,	497	Mallow,		206—208	
603	Forbed Spike,	621	Horn Poppy,	156	Mangel-Wurtzel,			470
2, 370	Forked Chickweed,	185	Hornwort,	484	Man-of-the-Earth,			442
6, 197	Forget-me-not,	434	Horse Balm,	621	Maple,		212, 213,	214
9, 490	Foul Meadow,	615	Horse Chestnut,	214	Mare's tail,			36?
2, 14,	Four-o'clock, 472,	473	Horsemint,	415, 417	Marigold,		142,	341
347	Foxglove, 404,	405	Horse Nettle,	448	Marjoram,			420
549	Fox tail Grass,	60?	Horse Radish,	162	Marsh Rosemary,			386
1, 272	Franklinia,	205	Horsetail,	624, 625	Marvel of Peru,			473
547	Fraxinella,	204	Hound's tongue,	425	Mat Grass,			601
150	Fringe Tree,	463	Houseleek,	277	Matrimony Vine,			442
352	Fritillary,	548	Huckleberry,	368	May Apple,			132
405	Frogbit,	529	Hyacinth,	551	May Flower,			374
356	Frost-plant,	191	Hydrangea,	281, 282	May weed,			342

INDEX

OF THE

NATURAL ORDERS AND GENERA.

°°° The names of the Orders are in Capitals.—Suborders are marked with the (*), and Sections with the (°).—The Tribes are in Roman spaced,—the Genera in Roman,—and the Synonyms in Italic. The figures after Syn. refer to the Synonyms of the Genus next above.

Abies,	515	Amphicarpæa,	396	Arundo,	616	Brachyelemum,	489
Abietineæ,	514	Amphieris,	396		Syn. 601	Brasenia,	180
Abutilon,	260	Amsonia,	457	Asarum,	486	Brassica,	171
Acalypha,	499	Amsyralen,	349	ASCLEPIADACEÆ,	455	Briza,	616
ACANTHACEÆ,	395	Amygdalus,	363	Asclepias,	456		Syn. 615, 616
Acer,	213		Syn. 363		Syn. 461	Bromus,	611
	Syn. 214	ANACARDIACEÆ,	206	Ascyrum,	182		Syn. 616
ACERACEÆ,	212	Amaryllis,	567	Asparageæ,	555	Broussonetia,	505
Acerates,	455	Anchusa,	431	Asparagus,	545	Bryophyllum,	277
Actæa,	177		Syn. 432	Asphodelus,	554	Buchnera,	407
Achillea,	342	Andromeda,	371	Aspidium,	631	Buchnereæ,	407
Achyranthes,	155		Syn. 373		Syn. 631	Bunias,	171
Acnida,	487	Andropogon,	621	Asplenium,	630	Bupleurum,	330
Aconitum,	146	Andropogoneæ,	621		Syn. 630	Buxus,	502
Acorus,	521	Anemone,	139	Aster,	312—315	CABOMBACEÆ,	180
Acrostichum,	629		Syn. 141, 148		Syn. 295, 323, 327, 329	Cacalia,	342
Actæa,	146	Anethum,	330	Astereideæ,	312	CACTACEÆ,	274
	Syn. 147	Angelica,	325, 331	Astragalus,	229	Cactus,	275
Actinomeris,	344	Avena,	151	Atheropogon,	619	° Cæsalpina,	238
	Syn. 347	ANONACEÆ,	151	Atragene,	139	Cakile,	171
Adenarium,	199	Anonymos,	484	Atriplex,	487	Caladium,	520
Adiantum,	633	Antennaria,	351	Atropa,	441	Calamagrostis,	601
Adlumia,	156	Anthemis,	342		Syn. 440	Calendula,	353
Adoxa,	141		Syn. 342	AURANTIACEÆ,	204	Calla,	520
Aeschynomene,	230	Anthericeæ,	553	Avena,	609	Calliopsis,	345, 346
Æsculus,	214	Anthoxanthum,	598		Syn. 610	Callistephus,	327
Æthusa,	331	Anthropogon,	621	Aveneæ,	609	CALLITRICHA-	
Agathyrsus,	360	Antirrhinum,	404	Avicularia,	474	CEÆ,	484
Agave,	560	Antirrhineæ,	404	Amies,	374	Callitriche,	484
Agrimonia,	296		Syn. 405		Syn. 375, 376, 377	Calopogon,	533
Agropyrum,	612	Anychia,	155	Azalia,	186	Caltha,	145
Agrostemma,	191, 192	Apargia,	359	Baccharis,	324	CALYCANTHA-	
Agrostideæ,	595	Apios,	231		Syn. 324	CEÆ,	284
Agrostis,	598	Apium,	331	Ballota,	422	Calycanthus,	285
	Syn. 598, 599, 601		Syn. 332	BALSAMIFLUÆ,	397	Calypso,	531
Ailanthus,	208	Aplectrum,	531	BALSAMINACEÆ,	219	Calystegia,	445
Aira,	608	APOCYNACEÆ,	456	Baptisia,	236	Camelina,	168
	Syn. 608, 613	Apocynum,	456	Barbarea,	165	Camellia,	182
Ajugoideæ,	430	AQUIFOLIACEÆ,	381	Berberis,	390, 454	Campanula,	365
Alcea,	261	Aquilegia,	142	Barosia,	460, 430		Syn. 365, 366
Aletris,	561	Arabis,	164	Batatas,	442, 443	CAMPANULACEÆ,	365
Alphenettia,	557		Syn. 163, 164	Batschia,	430	° Campylospermæ,	325
Alisma,	523	ARACEÆ,	519	Begonia,	270	° Cannabineæ,	511
ALISMACEÆ,	522	Aralia,	304	BEGONIACEÆ,	270	Cannabis,	511
Allium,	543	ARALIACEÆ,	304	Bellis,	327	CAPPARIDACEÆ,	172
Alnus,	498	Arbutus,	371	Bessera,	473	Capraria,	408
Alopecurus,	597	Archangelica,	331	BERBERIDACEÆ,	154	CAPRIFOLIACEÆ,	305
Alpinia,	560	Archemora,	187	Berberis,	391	Caprifolium,	306, 309
Alsineæ,	187	Arctium,	357	Beta,	487	Capsella,	161
Althæa,	261	Arctostaphylos,	371	Betula,	498	Capsicum,	447
Alyssum,	163	Arenaria,	196		Syn. 498	Cardamine,	167
	Syn. 168		Syn. 198, 199	BETULACEÆ,	497		Syn. 163
AMARANTHA-		Arethusa,	532	Bidens,	345	Cardiospermum,	216
CEÆ,	470		Syn. 532	Bigelovia,	324	Carduus,	356, 357
Amaranthus,	470	Argemone,	158	Bignonia,	394, 395	Carex,	575
AMARYLLIADA-		Aristida,	605	BIGNONIACEÆ,	393	Carices,	575
CEÆ,	567	Aristolochia,	486	Blitoria,	485	Carpinus,	497
Amaryllis,	567	ARISTOLOCHIA-		Blephilia,	418	Carthamus,	354
Amberboa,	354	CEÆ,	485	Blitum,	485	Carum,	331
Ambrosia,	333	Aritoncum,	530	Barbera,	528	Carya,	347
Amelanchier,	298	Armeniaca,	362	Behmeria,	512	Caryophylla-	411
Amellus,	323	Armeria,	390	Boltonia,	329	CEÆ,	188
Amianthum,	551	Arnica,	336	Bootria,	552	Cassandra,	372
Ammania,	290	Arenas,	245, 246	Borragoæ,	431	Cassia,	238
Ammi,	331	Arrhenatherum,	609	BORRAGINACEÆ,	429	Cassiope,	371
Ammophila,	601	Artemisia,	340	Borrago,	431	Castanea,	495
Amorpha,	228	Artocarpus,	519	Botrychium,	628	Castilleja,	409
Ampelopsis,	212		Syn. 519	Brachyelytrum,	598	Catalpa,	393

Catananche.	258	Commelyna.	562
Caulinia.	544	COMMELYNACEÆ.	562
Caulophyllum.	155	COMPOSITÆ.	310
Ceanothus.	217	Comptonia.	500
Celastrus.	215	CONIFERÆ.	513
CELASTRACEÆ.	215	Conioselinum.	290
Celosia.	472	Conium.	293
Celtis.	483	Conobea.	402
Cenchrus.	607	Conoclineum.	318
Centaurea.	354	Conostylea.	540
Syn.	354	Convallaria.	552
Centaurella.	454	Convallarineæ.	552
Centunculus.	397	CONVOLVULA-	
Cephalanthus.	307	CEÆ.	441
Cerastium.	187	† Convolvuleæ.	441
Cerasus.	240	Convolvulus.	441
Syn.	241	Syn. 443,	444
Ceratochloa.	611	Conyza.	395, 334
CERATOPHYLLA-		Coptis.	144
CEÆ.	484	Corallorhiza.	531
Ceratophyllum.	484	Corchorus.	207
Ceratochœnus.	575	Coreopsis.	344
Cercis.	235	Syn. 344,	347
Cereus.	275	Coriandrum.	294
Chærophyllum.	292	CORNACEÆ.	298
Chara.	637	Cornucopiæ.	599
CHARACEÆ.	637	Cornus.	298
Cheilanthes.	633	Coronilla.	230
Syn.	633	Corydalis.	156
Cheiranthus.	169	Syn.	158
Syn.	170	Corylus.	496
Chelidonium.	156	Cosmanthus.	437
Syn.	156	Crantzia.	285
Chelone.	400	CRASSULACEÆ.	278
Syn. 400,	401	Cratægus.	243
Chelonem.	399	Crocus.	543
CHENOPODIACEÆ.	466	Cronopeta'um.	453
Chenopodium.	468	Crotalaria.	224
Chimaphila.	379	Croton.	488
Chiogenes.	570	Crotonopsis.	488
Chionanthus.	463	Syn.	489
Chironia.	450, 451	CRUCIFERÆ.	159
Chlorea.	617	Crypsis.	602
Chondrilla.	352	Crypta.	195
Chrysanthemum.	343	Cryptotænia.	247
Syn. 328,	343	Cucubalus.	190
Chrysis.	157	Cucumis.	271
Chrysocoma.	334	Cucurbita.	272
Chrysopsis.	333	Syn.	272
Syn.	334	CUCURBITACEÆ.	270
Chrysosplenium.	281	Cunila.	421
Chrysostemma.	345	Syn.	422
Cicendia.	452	Cuphea.	261
Cichoraceæ.	357	Cupressus.	516
Cichorium.	357	Cupressineæ.	516
Cicuta.	286	CUPULIFERÆ.	492
Cimicifuga.	147	Cuscuta.	444
Syn.	147	Syn.	147
† Cinchoneæ.	305	† Cuscuteæ.	444
Cinna.	600	Cydonia.	245
Circæa.	286	Cymbidium.	531, 536
Circæa.	286	Cynanchum.	457
Cirsium.	356	Cynara.	355
CISTACEÆ.	190	Cynareæ.	353
Clateria.	629, 630	Cynodon.	618
Cistus.	191	Cynoglossum.	435
Citrus.	204	Syn.	435
Cladium.	575	Cynthia.	358
Clarkia.	264	Syn.	358
Claytonia.	194	CYPERACEÆ.	565
Clematis.	139	Cypereæ.	565
Cleome.	172	Cyperus.	565
Syn. 172,	173	Syn. 566,	568
Clethra.	373	Cypripedium.	531
Clinopodium.	419, 422	Syn.	531
Clintonia.	565	Dactylis.	617
Syn.	553	Dahlia.	327
Clitoria.	223	Dalea.	226
Cnicus.	355	Syn.	226
Syn. 356,	357	Daliborda.	251
Cnidium.	290	Syn.	253
Cochlearia.	163	Danthonia.	610
Cœlestina.	317	Daphne.	481
† Cœlospermæ.	293	Darlingtonia.	234
Collinsia.	399	Dasystoma.	408
Collinsonia.	431	Datura.	446
Colutea.	224	Daucus.	292
Comandra.	479	Decodon.	261
Comaropsis.	253	Delphinium.	145
Comarum.	252	Dentaria.	167
		Desmanthus.	238

Desmodium.	231	† Escallonea.	280
Dianthera.	562	Eschscholtzia.	157
Dianthus.	192	Euchroma.	409, 410
Diapensia.	441	Eupatoriaceæ.	313
DIAPENSIACEÆ.	441	Eupatorium.	314
Diarrhena.	612	Syn.	317
Dicksonia.	633	Euphorbia.	485
Diclytra.	386	EUPHORBIACEÆ.	485
Dictamnus.	204	Euphrasia.	409
Dielytra.	156	Eubotrya.	216, 319
Diervilla.	300	Eutoca.	437
Digitaleæ.	404	Exacum.	452
Digitalis.	404	Fagopyrum.	475
Digitaria.	618	Fœd.	176
Syn.	618	Fagus.	494
Dilatris.	540	Fedia.	309
Diodia.	307	Festucaceæ.	610
Dionæa.	190	Festuca.	611
Dioscorea.	543	Syn. 612, 613,	617
DIOSCOREACEÆ.	542	Ficus.	516
Diospyros.	383	Filago.	382
Diploclinium.	270	FILICES.	627
Diplopappus.	335	Fimbristylis.	573
Syn.	333	Syn.	573
DIPSACEÆ.	308	Florkea.	200
Dipsacus.	308	Fluviales.	544
Dirca.	480	Foeniculum.	290
Discopleura.	286	Fragaria.	252
Dodecatheon.	396	Franklinia.	206
Draba.	163	Frasera.	455
Dracæna.	553	Fraxinus.	463
Dracocephalum.	425	Fritillaria.	549
Syn.	426	Fuchsia.	265
Drosera.	179	Fuirena.	568
DROSERACEÆ.	179	Fumaria.	156
Dryas.	257	Syn. 156,	159
Dulichium.	568	FUMARIACEÆ.	157
Dysodia.	347	Galactia.	223
EBENACEÆ.	383	Galanthus.	539
Echinacea.	338	Galega.	225
Echinocystis.	270	Galeopsis.	427
Echinospermum.	435	Galium.	304
Echium.	430	Gaultheria.	373
Eclipta.	333	Syn.	371
ELATINACEÆ.	194	Gaura.	265
Elatine.	195	Genista.	274
Elatine'la.	308	Gentiana.	452
ELÆAGNACEÆ.	479	GENTIANACEÆ.	450
Eleagnus.	481	GERANIACEÆ.	196
Eleocharis.	568	Geranium.	196
Eleophantopus.	313	Syn.	197
Eleusine.	617	Gerardia.	407
Ellisia.	437	Syn. 408,	410
Elodea.	184	Geum.	253
Syn.	529	Gilia.	440
Elymus.	620	Gillenia.	257
Emilia.	349	Gladiolus.	542
EMPETRACEÆ.	489	Glaux.	396
Empetrum.	489	Glechoma.	425
Enandromeda.	372	Gleditschia.	237
Enemion.	148	Glyceria.	613
Enonymes.	215	Glycine.	222, 223
Enonymus.	216	Glycyphylla.	371
Ensiena.	541	Gnaphalium.	380
Epigæa.	373	Syn. 381,	382
Epilobium.	262	Gomphrena.	472
Epipactis.	525	Gonolobus.	461
Euphegus.	382	Goodyera.	526
EQUISETACEÆ.	624	Gossypium.	207
Equisetum.	624	GRAMINEÆ.	588
Erechtites.	352	Gratiola.	403
Erica.	377	Syn.	403
ERICACEÆ.	371	Gratioleæ.	401
† Ericineæ.	371	GROSSULACEÆ.	273
Erigeron.	336	Gymnadenia.	532
Eriocaulon.	544	Gymnandra.	404
ERICAULONA-		Gymnocladus.	237
CEÆ.	544	Gymnopogon.	618
Eriophila.	162	Gynandropsis.	172
Eriophorum.	481	Gurrania.	547
Erodium.	197	Hæmaria.	525, 526
Ervum.	221	HÆMODORACEÆ.	540
Erycenia.	293	Halenia.	454
Erysimum.	169	Haloua.	385
Syn. 168,	169	† Halonæa.	387
Erythræa.	451	HAMAMELACEÆ.	282
Erythronium.	549	Hamamelis.	282

Bmeditonia, . . 479	IRIDACEÆ, . . 541	Liquidambar, . . 508	Mitella, . . 288
Harpalyce, . . 361, 362	Iris, . . . 541	Syn. 508	Moluccella, . . 429
Hedeoma, . . . 422	Isanthus, . . 416	Liriodendron, . . 150	Mollugo, . . . 189
Syn. 422	Isoetes, . . . 164	Listera, . . . 537	Momordica, . . 271
Hedera, . . . 295	Innardia, . . 298	Lithospermum, . . 432	Syn. 271
Hedyotis, . . . 306	Isoetis, . . . 677	Syn. 432, 433, 434	Monarda, . . . 417
Hedysarum, . . 220	Isoiris, . . . 573	LOASACEÆ, . . 280	Syn. 418
Syn. 236—235	Isopyrum, . . 145	Lobelia, . . . 363	Monardeæ, . . 417
Helenium, . . 343	Itea, . . . 281	LOBELIACEÆ, . . 362	Moneses, . . . 379
Syn. 344	Iva, . . . 353	Loise euria, . . 373	Monnerie, . . 440
Helianthemum, . 181	Ixia, . . . 542	Lolium, . . . 629	MONOPETALÆ, . 662
Helianthus, . . 330	JASMINACEÆ, . 462	* Lomentaceæ, . . 171	Monotropa, . . 383
Syn. 337	Jasminum, . . 462	Lonicera, . . 290	Syn. 383
Heliastrum, . . 324	Jeffersonia, . . 152	Syn. 301	*Monotropeæ, . 383
Heliopsis, . . 327	JUGLANDACEÆ, . 440	Loniceræ, . . . 290	*Moreæ, . . . 508
Heliotropea, . . 435	Juglans, . . . 490	Lonicereæ,. . . 290	Morea, . . . 542
Heliotropium, . 435	Syn. 491	Lophanthus, . . 425	Morus, . . . 509
Helleborus, . . 144	JUNCACEÆ, . . 559	Lophiola, . . . 540	Mulgedium, . . 359
Helonias, . . 558	* Juncagineæ, . . 552	LORANTHACEÆ, . 297	Muhlenbergia, . 600
Syn. 557, 558	Juncus, . . . 559	Ludwigia, . . 263	Syn. 595, 600
Hemerocallideæ, 549		Lunaria, . . . 164	Myagrum, . . 164
Hemerocallis, . 549	Juniperus, . . 517	Lupinus, . . . 205	Myosotis, . . . 434
Hemianthus, . . 401	Justicia, . . . 395	Luzula, . . . 560	Syn. 434
Hemicarpha, . . 573	Kalmia, . . . 374	Lychnis, . . . 191	Myosurus, . . 146
Hepatica, . . 141	Kerria, . . . 257	Lycium, . . . 410	Myrica, . . . 499
Heracleum, . . 291	Kœleria, . . . 612	LYCOPODIACEÆ, 625	MYRICACEÆ, . 499
Herpestis, . . 402	Krigia, . . . 357	Lycopodium, . . 625	Myriophyllum, . 287
Syn. 404	Syn. 358	Lycopsis, . . . 431	MYRTACEÆ, . . 260
Hesperis, . . . 168	Kuhnia, . . . 315	Lycopus, . . . 416	Myrtus, . . . 260
Syn. 168	Kyllingia, . . 568	Lygodium, . . 636	Nabalus, . . . 361
Heteranthera, . 556	LABIATÆ, . . 413	Lyonia, . . . 372	NAIADACEÆ, . 555
Heteromeris, . . 181	Laburnum, . . 205	Lysimachia, . . 413	Najas, . . . 555
Heuchera, . . 279	Lacis, . . . 493	Syn. 385	Napæa, . . 208, 210
Hibiscus, . . . 205	Lachnanthes, . . 546	LYTHRACEÆ, . . 259	Narcissus, . . 550
Hieracium, . . 359	Lactuca, . . . 359	Lythrum, . . . 260	Nasturtium, . . 310
Hierochloa, . . 608	Lagenaria, . . 272	Syn. 260	Narthecium, . . 552
HIPPOCASTANA-	Lamium, . . . 426	Maclura, . . . 509	Nasturtium, . . 164
CEÆ, . . 214	Lapathum, . . 476	Macrotys, . . . 147	Naumburgia, . . 380
Hippophae, . . 481	Lappa, . . . 357	Magnolia, . . . 150	Negundo, . . . 213
Hippuris, . . 285	Lathyrus, . . 219	MAGNOLIACEÆ, . 149	NELUMBIACEÆ, 153
Holcus, . . . 608	Lavandula, . . 420	Majorana, . . 420	Nelumbium, . . 153
Syn. 613, 617	LAURACEÆ, . . 478	Malaxis, . . 530, 531	Nemopanthus, . 201
Honckenya, . . 189	Laurus, . . 478, 479	Malope, . . . 209	Nemophila, . . 438
Hordeæ, . . . 619	Lavatera, . . 207	Malva, . . . 206	Neottia, . . . 537
Hordeum, . . 620	Lechea, . . . 180	MALVACEÆ, . . 205	Nereta, . . . 424
Hottonia, . . 384	Leontia, . . . 152	Malva, . . . 520	Syn. 419
Houstonia, . . 306	Ledum, . . . 377	Marrubium, . . 429	Nepeteæ, . . . 424
Hudsonia, . . 181	Leersia, . . . 622	Marsilea, . . . 636	Nephrodium, . . 633
Humulus, . . 512	LEGUMINOSÆ, . 217	MARSILEACEÆ, . 636	Syn. 633
Hyacinthus, . . 551	Leimanthium, . . 557	Martynia, . . 394	Nerium, . . . 457
Hydrangea, . . 281	Leiophyllum, . . 374	Maruta, . . . 342	Nesæa, . . . 260
Hydrastis, . . 146	Lemna, . . . 521	Matthiola, . . 170	Nicandra, . . 445
HYDROCHARIDA-	LEMNACEÆ, . . 521	Meconopsis, . . 156	Nicotiana, . . 444
CEÆ, . . 528	LENTIBULACEÆ, . 390	Medeola, . . . 547	Nigella, . . . 145
Hydrocharis, . . 529	Leontice, . . . 152	Medicago, . . . 226	Nuphar, . . . 155
Hydrocochlos, . 629	Leontodon, . . 358	Melampyrum, . . 411	NYCTAGINACEÆ, 472
Hydrocotyle, . . 284	Syn. 362	MELANTHACEÆ, . 556	Nymphæa, . . 154
Syn. 285, 294	Leonurus, . . 427	Melanthium, . . 556	Syn. 154, 155
Hydropeltis, . . 153	Lepachys, . . 338	Syn. 556, 557	NYMPHEACEÆ, . 154
HYDROPHYLLA-	Lepidanche, . . 444	MELASTOMACEÆ, 259	Nyssa, . . . 508
CEÆ, . . 438	Lepidium, . . 161	Melica, . . . 617	Oakesia, . . . 548
Hydrophyllum, . 438	Leptandra, . . 405	Melilotus, . . 229	Obolaria, . . 450
Hydropyrum, . . 622	Leptanthus, . . 555	Syn. 229	Ocimeæ, . . 412
Hymenopappus, . 346	Syn. 556	Melilineæ, . . . 236	Ocimoideæ, . 416
Hyoseramus, . . 444	Leptopoda, . . 344	Melocactus, . . 273	Ocimum, . . 416
Hyoseris, . . . 356	Lepturus, . . . 623	Melothria, . . 271	Œnanthe, . . 291
HYPERICACEÆ, . 187	Lespedeza, . . 233	MENISPERMA-	Œnothera, . . 261
Hypericum, . . 182	Leucanthemum, . 343	CEÆ, . . . 151	OLEACEÆ, . . 461
Syn. 184	Leucospora, . . 402	Menispermum, . 151	ONAGRACEÆ, . 261
Hypobrichia, . . 261	Leucothoe, . . 372	Mentha, . . . 415	Onagreæ, . . 262
Hypolytreæ, . . 568	Liatris, . . . 317	Menthoideæ, . . 415	Onoclea, . . 630
Hypopitis, . . 631	*Ligulifloræ, . . 357	Mentzelia, . . 280	Onosmodium, . 433
Hypopitys, . . 380	Ligusticum, . . 289	Menyantheæ, . . 458	Onosma, . . 433
Hypoporum, . . 576	Syn. 289	Menyanthes, . . 458	Ophioglossæ, . 635
Hypoxis, . . . 547	Ligustrum, . . 462	Menziesia, . . 373	Ophioglossum, . 636
Hyssopus, . . 421	Liliaceæ, . . 548	Mertensia, . . 433	Ophrys, . . 536
Syn. 425	Lilium, . . . 548	MESEMBRYACEÆ, 276	Syn. 537
	LIMNANTHA-	Mesembryanthemum, 276	Oplismenus, . 604
Iberis, . . . 164	CEÆ, . . 280	Mespilus, . . 243, 244, 245	Oplotheca, . . 472
Ictodes, . . 521	Limnanthemum, . 455	Micromeria, . . 422	Opuntia, . . 274
Ilex, . . . 201	Limnetis, . . . 618	Micropetalon, . . 197	ORCHIDACEÆ, . 529
Syn. 201	Limodorum, . . 532	Microstylis, . . 530	Orchis, . . . 532
ILLICEBRACEÆ, . 183	Limosella, . . 404	Mikania, . . . 316	Syn. 530
Ilysanthus, . . 403	Lindernia, . . 403	Milium, . . . 603	Origanum, . . 420
Impatiens, . . 199	LINACEÆ, . . 195	Mimosa, . . . 235	Ornithogalum, . 551
Inula, . . . 304	Linnæa, . . . 301	Syn. 235	OROBANCHACEÆ, 392
Syn. 323	Linaria, . . . 396	*Mimoseæ, . . 235	Orobanche, . . 392
Ionæa, . . 442, 443	Linum, . . . 195	Mimulus, . . . 401	Orontium, . . 521
Isnopeæ, . . 440	Liparis, . . . 530	Mirabilis, . . 472	*Orthospermeæ, . 284
Ionius, . . . 671	Lippia, . . . 413	Mitchella, . . 305	Oryzeæ, . . 593

Orumpais,	658	Podophyllum, 182	Rubus, 218	Sisymbrium, 169
Syn. 806	PODOSTEMACEÆ, 485	Rudbeckia, 337	Syn. 168, 166
Osmorhiza, 283	Podostemum, 485	Syn. 328	Sisyrinchium, 543
Osmunda, 634	Pogonia, 535	Ruellia, 395	Sium, 287
Osm. undiaceæ, 634	Polanisia, 172	Rumex, 473	SMILACEÆ, 476
Ostrya, 497	POLEMONIACEÆ, 435	Syn. 473	Smilacina, 532, 533
Otophylla, 408	Polemonium, 446	Ruppia, 524	Smilax, 644
OXALIDACEÆ, 200	Polyanthes, 550	Ruta, 204	Smyrnium, 287, 288, 289
Oxalis, 200	Polygala, 173	RUTACEÆ, 204	SOLANACEÆ, 444
Oxycoccus, 370	POLYGALACEÆ, 173	Sabbatia, 450	Solanum, 445
Oxydendron, 372	POLYGONACEÆ, 473	Sagina, 189	Soles, 178
Oxyria, 473	Polygonatum, 563	Sagittaria, 508	Solidago, 329—332
Pæonia, 149	Polygonella, 475	SALICACEÆ, 497	Sonchus, 362
Panax, 296	Polygonum, 473	Salicornia, 465	Syn. 366, 361
Panicea, 602	Polymnia, 336	Salix, 497	Sophora, 228
Panicum, 604	Polypodiaceæ, 628	Salpiglossen, 467	Sorbus, 247
Syn. 600, 604, 616	Polypodium, 628	Salvia, 419	Sorghum, 622
Papaver, 157	Syn. 631	Salvinia, 638	Sparganium, 512
PAPAVERACEÆ, 156	Polypogon, 600	Sambucea, 301	Sparganophorus, 314
Papilionaceæ, 219	Polytænia, 294	Sambucus, 301	Spartina, 615
Parietaria, 511	*Pomeæ, 243	Samolus, 368	Specularia, 368
Parnassia, 180	PONTEDERACEÆ, 556	Sanguinaria, 155	Spergula, 195
Paronychia, 195	Pontederia, 556	Sanguisorba, 255	Syn. 189
Parthenium, 336	Populus, 500	Sanicula, 285	Spermacoce, 308
Paspalum, 602	Portulaca, 198	SANTALACEÆ, 479	Syn. 307
Passiflora, 233	PORTULACACEÆ, 193	SAPINDACEÆ, 215	Spigelia, 307
PASSIFLORACEÆ, 233	Potamogeton, 564	*Spireliæ, 192	Spireliæ, 307
Pastinaca, 291	Syn. 567	Saponaria, 192	Spinacia, 467
Paulownia, 389	Potentilla, 251	Scrothra, 184	Spiræa, 255
Pavia, 214	Poterium, 255	Sarracenia, 155	Syn. 257
PEDALIACEÆ, 304	Pothos, 521	SARRACENIA-		Spiranthes, 538
Pedicularis, 410	Prenanthes, 361, 362	CEÆ, 155	Spirodela, 522
Pelargonium, 197	Primula, 364	Sassafras, 478	Sporobolus, 600
Paktandra, 520	PRIMULACEÆ, 363	Satureja, 421	Stachydeæ, 426
Pentalophus, 433	Prinos, 361	Satureineæ, 409	Stachys, 427
Penthorum, 278	Priva, 413	SAURACEÆ, 484	Staphylea, 215
Pentstemon, 400	Proserpinaca, 267	Saururus, 484	Staphyleæ, 215
Poplu, 195, 291	Prunella, 424	Saxifraga, 278	Statice, 390
Periploca, 461	Prunus, 241	SAXIFRAGACEÆ, 278	Syn. 390
Persica, 242	Syn. 240, 242	*Saxifrageæ, 278	Stellaria, 187
Peristylus, 533	Psamma, 600	Scabiosa, 310	*Stellatæ, 304
Persicaria, 474	Pellostemon, 431	Scandix, 292	Stipa, 596
Petalostemon, 226	Psoralea, 225	Scheuchzeria, 526	Syn. 600
Syn. 226	Psylocarya, 574	Schizæa, 635	Stipaceæ, 596
Petroselinum, 288	Ptelea, 201	Schizanthus, 397	Streptopus, 564
Petunia, 445	Pteris, 632	Schoenus, 574, 575	Syn. 564
Phaca, 223	Pterospora, 380	Scholleræ, 555	Struthiopteris, 634
Pharelia, 433	Pulmonaria, 434	Schrankia, 228	Stylipus, 264
Syn. 437	Syn. 434	Schwalbea, 410	Stylisma, 443
Phalangium, 552	Punica, 562	Scilleæ, 550	Stylophorum, 156
Syn. 562	Pycnanthemum, 419	Scirpeæ, 569	Stylosanthes, 223
Phalareæ, 607	Pyrethrum, 343	Scirpus, 570	STYRACACEÆ, 382
Phalaris, 607	Syn. 343	Syn. 565, 569, 568, 570, 573	Styrandra, 552
Phalerocarpus, 371	Pyrola, 379	SCLERANTHA-		Subularia, 163
Pharbitis, 442	Syn. 379	CEÆ, 470	Sullivantia, 279
Phaseolus, 221	*Pyroleæ, 379	Scleranthus, 470	Swertia, 454
*Philadelpheæ, 282	Pyrularia, 479	Sclereæ, 573	Syn. 450, 453, 454
Philadelphus, 282	Pyrus, 244	Scleria, 575	Symphoria, 300, 301
Phleoideæ, 601	Syn. 243, 245	Sclerolepis, 313	Symphoricarpus, 300
Phleum, 602	Pysidanthera, 441	Scolopendrium, 632	Symphitum, 431
Phlox, 438	Quamoclit, 443	Scrophularia, 399	Symplocarpus, 521
Phragmites, 616	Quercus, 492	SCROPHULARIA-		Synandra, 430
Phryma, 413	Queria, 195	CEÆ, 396	Synthyrus, 403
Phyllanthus, 489	RANUNCULA-		Scutellaria, 423	Syringa, 483
Phyllodoce, 373	CEÆ, 137	Scutellarineæ, 423	Tagetes, 341
Physalis, 447	Ranunculus, 141	Secale, 619	Talinum, 194
Physostegia, 426	Raphanus, 171	Sedum, 278	Tanacetum, 380
Phytolacca, 479	Reseda, 173	Selinum, 290	Syn. 347
PHYTOLACCA-		RESEDACEÆ, 173	Sempervivum, 277	Taraxacum, 362
CEÆ, 477	Rennelaria, 590	Senecio, 347	Taxus, 515
Pimpinella, 288	RHAMNACEÆ, 916	Senecionideæ, 335	Taxineæ, 515
Pinguicula, 390	Rhamnus, 216	Syn.	Tecoma, 389
Pinus, 514	Rheum, 473	Sepicula, 529	Tephrosia, 224
Syn. 515, 516	Rhexia, 259	Sericocarpus, 325	TERNSTROEMIA
Piptatherum, 596	Rhinanthus, 410	Sesamum, 305	CEÆ, 205
Pisum, 221	Rhododendron, 375	Setaria, 603	Teurium, 429
Syn. 219	Syn. 375	Seymeria, 407	Thalictrum, 147
PLANTAGINA-		Rhodora, 375	Shepherdia, 481	Thaspia, 289
CEÆ, 388	Rhus, 202	Sibbaldia, 257	Thaspium, 289
Plantago, 388	Rhyncospora, 574	Sibthorpeæ, 404	Syn. 287, 289
PLATANACEÆ, 504	Syn. 575	Sicyos, 270	Thesium, 479
Platanus, 504	Ribes, 271	Syn. 271	Thlaspi, 160
Platanthera, 532—535	Ricinus, 488	Sida, 209	Syn. 161
Plectranthus, 415	Robinia, 224	Silene, 190	Thuja, 517
Plushea, 334	Rochelia, 435	Sileneæ, 190	THYMELACEÆ, 480
PLUMBAGINA-		Ræn, 246	*Siliculosæ, 160	Thymus, 420
CEÆ, 390	ROSACEÆ, 238	Silphium, 335	Tinrella, 240
Poa, 614	*Rosaceæ proper, 243	Sinapis, 170	Tigridia, 542
Syn. 613	Rosmarinus, 419	Sison, 287, 294	Tilia, 210
Podalyria, 228	RUBIACEÆ, 303	Sium,	TILIACEÆ, 208

Tilia, 276	Tropæolum, 160	Vaccinium, 333	VITACEÆ, 211
Tipularia, 532	Troximon, 362	Syn. 270, 271	Vitis, 21
Tofieldia, 550	Tubularia, 313	Valeriana, 308	Waldsteinia, 255
Trachysperma, 455, 492	Tuckermania, 490	VALERIANACEÆ, 308	Wimoria, 613
Tradescantia, 363	Tulipa, 547	Valerianella, 309	Wistaria, 222
Tragopogon, 356	Tulipaceæ, 547	Vallisneria, 529	Woodsia, 631
Syn. 358	Syn. 105	Veratrum, 557	Woodwardia, 632
Trautvetteria, 147	Turritis, 106	Syn. 557, 560	Xanthium, 326
Tricochloa, 600	Syn. 108	Verbascum, 397	Xeranthemum, 345
Trichodium, 599	Tussilago, 318	Verbena, 411	Xerophyllum, 556
Tricophorum, 572	Syn. 318	VERBENACEÆ, 411	Xylosteum, 301
Trichostema, 420	Typha, 522	Verbesina, 329	XYRIDACEÆ, 562
Trichostylis, 573	TYPHACEÆ, 522	Vernonia, 313	Xyris, 562
Tricuspis, 613	Udora, 529	Vernoniaceæ, 313	Zannichellia, 504
Trientalis, 366	ULMACEÆ, 482	Veronica, 405	ZANTHOXYLA-
Trifolium, 228	Ulmus, 482	Veronicæ, 405	CEÆ, 201
Syn. 228, 230	UMBELLIFERÆ, 283	Viburnum, 301	Zanthoxylum, 201
Triglochin, 548	Uniola, 616	Vicia, 229	Zapania, 412, 413
TRILLIACEÆ, 545	Uralepis, 610	Vilfa, 597, 598	Zea, 623
Trillium, 545	Urospermum, 361	Villarsia, 455	Zenobia, 272
Triosteum, 300	Urtica, 510	Vinca, 457	Zizadenus, 556
Triphora, 536	Syn. 511	Viola, 175	Zinnia, 332
Tripsacum, 623	URTICACEÆ, 509	Syn. 178	Zizania, 622
Trisetum, 609	Urticeæ, 510	Zizia, 297
Triticum, 619	Utricularia, 391	VIOLACEÆ, 175	Zizia, 297
Trollius, 144	Uvaria, 151	Viscum, 297	Ziziphora, 421, 422
TROPÆOLACEÆ, 160	Uvularia, 553	Syn. 305	Zostera, 520
	Vaccineæ, 333		

ENGLISH INDEX.

Adam-and-Eve, 531	Basil, 419, 632	Blue-curls, 404, 419	Cabbage, 171
Adder's-tongue, 635	Basswood, 218	Blue-eyed Grass, 542	Calabash, 278
Agrimony, 265	Bath Flower, 546	Blue Grass, 614	Cale, 171
Albany Beech-drops, 390	Bayberry, 509	Blue Flag, 541	Calico-bush, 274
Alder, 501, 492	Beech Plum, 241	Blue-hearts, 407	Campion, 189
Alexanders, 287	Beech Pea, 219	Blue Tangles, 370	Canary Grass, 607
Allspice, 252	Bean, 220, 221, 222	Bog Rush, 574, 575	Cancer-weed, 418
Almond, 242, 243	Bear-berry, 371	Boneset, 316	Candytuft, 164
Aloe, 530	Beard-tongue, 400, 401	Borrage, 531	Caraway, 290, 348
Alum-root, 280	Beard Grass, 622	Bottle Grass, 607	Cardinal Flower, 363, 364
Amaranth, 471, 472	Beaver-wood, 483	Bowman's Root, 267	Cardoon, 355
American Cowslip, 366	Bedstraw, 304	Box, 489	Carnation, 189
Angelica, 290, 291	Beech, 498	Boxberry, 371, 372	Carrot, 288
Angelica Tree, 285	Beech-drops, 306, 362	Brake, 632	Carpet-weed, 196
Anise, 290	Beet, 470	Broccoli, 171	Castor Oil Bean, 498
Annual Spear Grass, 614	Bell-flower, 365, 366	Broom Grass, 611, 621	Catch-fly, 188, 191
Apple, 244	Bell-wort, 554	Brooklime, 405	Cat-gut, 233
Apple of Peru, 448	Bengal Grass, 607	Broomrape, 390	Catnep, 420
Apricot, 242	Bent Grass, 597	Broom Corn, 622	Cattail, 522
Arbor Vitæ, 517	Berberry, 152	Buck-bean, 455	Cauliflower, 171
Arrow-grass, 548	Bilberry, 369	Buck-eye, 214	Cedar, 516, 517
Arrow-head, 527	Bindweed, 442, 443, 178	Buck-thorn, 216, 217	Celandine, 156
Arrow-wood, 303	Bird's-nest, 290	Buck-wheat, 478	Celery, 286
Artichoke, 341, 355	Birch, 496, 499	Bugle-weed, 416	Century Plant, 530
Asarabacca, 465	Bishop's Cap, 280	Burdock, 340	Chaff-seed, 410
Ash, 453, 454, 463	Bishop-weed, 286	Bullrush, 573	Chamomile, 342
Asparagus, 558	Bistort, 478	Burdock, 340	Cheat, 611
Asphodel, 559	Bitterwect, 448	Burnet, 265	Checkerberry, 371
Atamasco Lily, 539	Blackberry, 259	Burning Bush, 216	Cherry, 242
Austrian Eglantine, 267	Black Haw, 303	Burr Flower, 425	Chess, 611
Avens, 263, 264	Black Snake-root, 167	Burr Grass, 607	Chestnut, 497
Awlwort, 163	Black Thorn, 241	Burr Marigold, 346	Chickweed, 185, 187, 193
Bachelor's Button, 354	Bladder-nut, 215	Burr-reed, 523	Chickweed Winter-
Balloon-vine, 215	Bladder Senna, 234	Burr-seed, 436	green, 366
Balm, 422, 429	Bladderwort, 391, 392	Buttercups, 142	China Aster, 357
Balm of Gilead, 507	Blazing Star, 317	Butterfly-weed, 460	Chinquapin, 498
Balsam Apple, 271	Blessed Thistle, 356	Butternut, 490	Choke Berry, 246
Balsam Spruce, 516	Blood-root, 156	Butterwort, 390	Cinquefoil, 261, 262
Baneberry, 146, 147	Blueberry, 369, 370	Burton-bush, 307	Citron Tree, 219
Barley, 618	Blue-bottle, 354	Button-wood, 504	Chives, 564

Clarry, 418	Hart's-tongue, 636	Leaf-cup, 505
Cleavers, 364	Hawkweed, 352, 356	Leather-flower, 139
Clotweed, 366	Hawthorn, 243	Leather-wood, 497
Cloudberry 360	Hazel, 496, 497	Leek, 523, 551
Clover, 227, 228, 233	Heartsease, 175	Lemon Tree, 204
Club Moss, 626—627	Hedge Hyssop, 403	Lettuce, 359, 361
Club Rush, 569, 570	Hedge Mustard, 164	Life-everlasting, 351
Cock's-comb, 472	Hedgehog, 473	Lilac, 463
Cock's-foot Grass, 604	Hedgehog Grass, 621	Lily, 548 549
Coffee Tree, 227	Heliotrope, 436	Lily-of-the-Valley, 553
Colic Root, 541	Hellebore, 144, 557	Lime Tree, 209 210
Colocynth, 272	Hemlock, 286, 515, 516	Lime Grass, 621 622
Colt's-foot, 318	Hemp, 515	Linden Tree, 210
Columbine, 145	Hemp Nettle, 427	Lion's-heart, 426
Columbo, 466	Henbane, 446, 447	Lion's-foot, 361
Comfrey, 431	Henbit, 426	Live-forever, 277
Cone-flower 307, 329	Herb Robert, 196	Liverwort, 141
Coral-root, 531	Herd's Grass, 602	Lizzard's-tail, 484
Coriander, 204	Hickory, 491, 496	Locust, 226, 227
Corn Cockle, 191	High Cranberry, 302	Loosestrife, 290, 295, 312, 347
Corn Mint, 416	High-water Shrub, 753	Lopseed, 413
Corn Salad, 369	Hoarhound, 315, 416, 126	Lousewort, 411
Corn Spurry, 166	Hobble-bush, 302	Lovage, 290
Cornel, 296	Hogweed, 363, 469	Love-lies-bleeding, 471
Cotton Plant, 307	Holly, 301	Lungwort, 434
Cotton Thistle, 356	Hollyhock, 208	Lupine, 226
Cotton Tree, 507	Honesty, 164	Lychnidea, 439, 439, 440
Cotton Grass, 572, 573	Honewort, 287, 298	Madwort, 165
Cowbane, 296, 291	Honeysuckle, 298, 299, 300	Maidenhair, 633
Cowslips, 143	Hoop Ash, 463	Maize, 623
Cow Wheat, 411	Hop, 512	Mallow, 264—280
Coxcomb, 470	Hornbeam, 497	Mangel-Wurtzel, 470
Crab Grass, 603	Horn Poppy, 156	Man-of-the-Earth, 442
Cranberry, 202, 370	Hornwort, 484	Maple, 212, 213, 214
Crane's-bill, 196, 197	Horse Balm, 421	Mare's-tail, 288
Crowberry, 489, 490	Horse Chestnut, 214	Marigold, 143, 341
Crowfoot, 141, 142, 143	Horsemint, 415, 417	Marjoram, 420
Crownbeard, 347	Horse Nettle, 448	Marsh Rosemary, 399
Crown Imperial, 549	Horse Radish, 163	Marvel of Peru, 472
Cucumber, 270, 271, 272	Horse-tail, 624, 625	Mat Grass, 601
Cucumber-root, 547	Hound's tongue, 435	Matrimony Vine, 448
Cucumber Tree 150	Houseleek, 277	May Apple, 152
Cudweed, 352	Huckleberry, 365	May Flower, 374
Culver's Physic 405	Hyacinth, 551	May weed, 342
Cup plant, 346	Hydrangea, 261, 262	Meadow Beauty, 259
Currant, 273	Hyssop, 415	Meadow Grass, 614
Currant-leaf, 340	Ice plant, 278	Meadow Rue, 147, 148
Cut Grass, 622	Indian Cress, 200	Meadow-sweet, 256, 257
Cypress, 517	Indian Grass, 602	Medick, 229
Cypress Vine, 443	Indian Hemp, 456	Melilot, 228
Daffodil, 529	Indian Millet, 622	Mercury, 408
Daisy, 326, 327, 343	Indian Physic, 257	Mermaid-weed, 287
Dandelion, 359, 362	Indian Pipe, 360	Miami Mist, 437
Darnel Grass, 620	Indian Rice, 622	Mignonette, 173
Day flower, 563	Indian Tobacco, 364	Milful, 267, 288, 312
Day Lily, 550	Ink-berry, 389	Milk-weed, 350
Deadly Nightshade, 449	Innocence, 308	Milkwort, 174
Deerberry, 369	Iron weed, 313	Millet, 596, 607
Deer Grass, 259	Iron-wood, 497	Millet Grass, 602, 604
Devil's-bit, 310	Jack-in-the-Pulpit, 519	Mistletoe, 297
Dewberry, 349	Jacobea, 276	Mitrewort, 260
Ditch Moss, 529	Jacobea Lily, 538	Modesty, 286
Dittany, 421	Jalap, 473	Moneywort, 287
Dock, 476, 477	Japan Rose, 266	Monkey Flower, 401, 656
Dockmackie, 301	Jasmine, 443, 463	Monk's-hood, 146
Dodder, 444	Jersey Tea, 217	Moonseed, 151
Dog's bane, 456, 457	Jerusalem Cherry, 448	Morning Glory, 442
Dog's-couch Grass, 619	Jewel-weed, 199	Moss Pink, 440
Dogwood 203, 296, 297	Judas Tree, 236	Motherwort, 427
Dorp weed Grass, 600	July Flower, 170	Mountain Ash, 245
Dragon's-claw, 531	June Berry, 245	Mountain Fringe, 158
Dragon head, 425, 426	Juniper, 517	Mountain Heath, 373
Dragon-root, 519	King's Spear, 552	Mountain Laurel, 374
Dropwort, 291	Knapweed, 354	Mountain Mint, 417, 419
Duckmeat, 519	Knawell, 184	Mountain Rice, 602
Dutchman's Breeches, 156	Knot Grass, 474, 475, 476	Mourning Bride, 310
Dutchman's Pipe, 488	Knot-weed, 474, 475	Mouse-ear, 435
Dwarf Cassia, 337	Labrador Tea, 377	Mouse-tail, 148, 596
Dwarf Pink, 308	Ladies' Ear drop, 265	Mudwort, 404
Dyer's broom, 231	Ladies' Slipper, 537, 538	Mugwort, 350
Dyer's weed, 173	Ladies' Thumb, 474	Mulberry, 249, 250, 509
Egg plant, 448	Ladies' Tresses, 536	Mullein, 397
Eglantine, 247	Lake Flower, 455	Mullein Pink, 192
Egyptian Wheat, 619	Lamb Lettuce, 369	Muskmelon, 271
Elder, 295, 301, 302	Larch, 516	Mustard, 166—170
Elecampane, 354	Larkspur, 145, 146	Myrtle, 258, 366
Elm, 492, 493	Laurel, 374, 401	Narcissus, 529
Endive, 357	Laurestine, 303	Nasturtion, 200
Enchanter's Night-ade, 364	Lavender, 415	Neckweed, 405
	Lead plant, 225	

(English Ivy, 206; English Moss, 277; Eternal Flower, 366; Eyebright, 487; False Flax, 163; False Mermaid, 200; False Red-top, 613; False Wall Flower, 169; Feather-beds, 637; Feather Grass, 506; Felwort, 454; Fennel, 280; Fennel Flower, 149; Fern, 629—636; Fescue Grass, 611, 612; Fever-bush, 478; Feverfew, 343; Feverwort, 300; Field Mint, 416; Fig, 510; Pigwort, 300; Finger Grass, 603; Fir Balsam, 516; Fire-weed, 362; Flax, 185; Fleabane, 325, 347; Florin Grass, 597; Flower-of-an-hour, 209; Fly-poison, 558; Fly-trap, 192; Fog fruit, 413; Fool's Parsley, 289; Forbed Spike, 621; Forked Chickweed, 185; Forget-me-not, 434; Foul Meadow, 615; Four-o'clock, 472, 473; Foxglove, 404, 465; Fox-tail Grass, 602; Franklinia, 265; Fraxinella, 204; Fringe Tree, 463; Fritillary, 549; Frostbit, 529; Frost-plant, 191; Fumitory, 159; Galangale, 566; Garget, 475; Garlic, 550, 551; Gay Feather, 317; Gem-fruit, 280; Gentian, 416, 453, 454; Geranium, 196—198; Gill over the-ground, 425; Ginseng, 286; Globe Flower, 144, 257; Goat's Beard, 356; Goat's Rue, 225; Golden Chain, 275; Golden Club, 521; Golden Cress flower, 540; Goldenrod, 328—333; Gold of pleasure, 163; Goldthread, 144; Gooseberry, 273, 274; Goose grass, 252, 304; Goose-foot, 469; Gourd, 272; Grape, 211; Grass, 580—624; Grass of Parnassus, 180; Grass Pink, 536; Grass poly, 290; Greek Valerian, 440; Green Brier, 541; Green Dragon, 620; Green head, 506; Gromwell, 432, 433; Ground Cherry, 447; Ground Ivy, 425; Ground Nut, 222, 295; Ground Pine, 625; Groundsel, 347; Groundsel Tree, 335; Guelder Rose, 302; Gum Tree, 496; Huckleberry, 482; Hair Bell, 365; Hair Grass, 599; Hardhack, 256)

Nectarine, 242
Nettle, 429, 510, 511
Nettle Tree, 483
Nightshade, 444
Nine-bark, 256
Noosuch, 226
Nutmeg Flower, 149
Oak, 492-495
Oak-of-Jerusalem, 469
Oat, 609, 610
Oilnut, 479
Oily-seed, 395
Okra, 209
Oleander, 454
Okaster, 482
Onion, 551
Orache, 467, 468
Orange Tree, 305
Orchard Grass, 617
Orpine, 277
Osage Orange, 509
Osier, 504
Ostrich Fern, 634
Ox-eye, 337
Ox-tongue, 431
Pæony, 148
Painted-cup, 409, 410
Palma Christi, 488
Panic, 608
Pansey, 178
Parsley, 289
Parsnep, 287, 291
Partridge-berry, 305
Passion Flower, 269
Pawnee Lettuce, 309
Pawpaw, 151
Pea, 219, 220, 221
Peach, 242
Pear, 244
Pearlwort, 189
Pea Vine, 223
Pellitory, 511
Pencil Flower, 217
Penny Cress, 160
Pennyroyal, 422
Pennywort, 284, 452
Pepper, 447
Peppergrass, 161
Pepperidge, 480
Peppermint, 416
Pepper-root, 168
Periwinkle, 457
Persimmon Tree, 382
Pettymorrel, 294
Pheasant's Eye, 141, 193
Pickerel-weed, 585
Pigmy-weed, 276
Pimpernell, 387, 388
Pine, 514, 515
Pine Sap, 380
Pink, 192, 193, 400
Pink-root, 308
Pinweed, 141
Pines, 625
Pipewort, 564
Pipsissewa, 379
Plantain, 388, 389, 390, 536
Plum, 241
Poison Hemlock, 282
Poison Ivy, 300
Poison Oak, 300
Poisonous Darnel, 620
Poke, 478, 557
Polar-plant, 336
Polypod, 628
Pomegranate, 259
Pond Lily, 154, 155
Pond weed, 524-526
Poor-man's Weather-glass, 387
Poplar, 150, 506, 507
Poppone root, 153
Poppy, 156, 157
Potato, 445
Pot Marigold, 354
Poverty Grass, 595
Prairie Burdock, 336
Prickly Ash, 301
Prickly Pear, 275
Pride-of-the-Meadow, 257
Prim, 463

Primrose, 293, 294, 384, 385
Prince's Feather, 471, 475
Prince's Pine, 379
Privet, 463
Puccoon, 433
Pumpkin, 272
Pumpkin-wood, 216
Purslain, 193, 194, 195, 266
Putty-root, 531
Quamash, 552
Queen-of-the-Meadow, 256
Quillwort, 627
Quince, 245
Radish, 172
Ragged Robin, 192
Ram's-head, 536
Raspberry, 249, 250
Rattle-box, 234
Red-bud, 236
Red Osier, 256
Red-root, 217, 540
Red-top, 597, 612, 614
Reed Grass, 601
Reed Mace, 522
Resin-weed, 336
Rhubarb, 473
Ribwort, 388
Rice, 604
Robin's Plaintain, 326
Rock Brake, 632
Rock Rose, 181
Rocket, 168, 171
Rose, 246-248, 250
Rose Acacia, 224
Rose-bay, 282, 376
Rose bay Tree, 458
Rosemary, 372, 419
Rosin-weed, 336
Rue, 204
Rush, 559-562, 571
Rutland Beauty, 444
Rye, 619, 620
Rye Grass, 621
Saffron, 365, 543
Sage, 418
Sage Geranium, 415
Salt-rheum-weed, 400
Saltwort, 398, 466
Samphire, 466, 467
Sandwort, 189, 190
Sanicle, 285
Sarsaparilla, 294, 544
Satin Flower, 164
Saxifrage, 255, 279, 279
Scabish, 279
Scarcity, 470
Scorpion Grass, 434
Scorpion Senna, 230
Scratch Grass, 475
Screw-stem, 454
Scurvy Grass, 167
Sea Wrack Grass, 521
Sedge, 566, 567
Seed-box, 265
Self-heal, 424
Seneca Grass, 606
Senna, 226
Sensitive Brier, 238
Sensitive Plant, 237, 238
Sensitive Pea, 337
Sesame Grass, 623
Shad-bush, 245
Shaddock Tree, 306
Shallot, 551
Sheep-poison, 374
Shell Flower, 429
Shepherd's Purse, 161
Shield Fern, 629, 630
Short-styled Cicely, 293
Sicklepod, 166
Side-saddle Flower, 155
Silkweed, 456, 459
Silver-weed, 251
Simpler's Joy, 412
Skull-cap, 416, 423
Shamrock, 227
Skunk Cabbage, 521
Slender Blite, 468
Sloe, 241
Snail, 279

Snake-head, 400
Snake Cactus, 275
Snake-root, 147, 174, 285, 465
Snap-dragon, 398, 399
Sneezewort, 342, 343
Snowball, 302
Snow-berry, 306
Snowdrop, 539
Snowdrop Tree, 383
Soapwort, 192
Soft Grass, 608
Solomon's Seal, 552, 553
Sorrel, 200, 201, 473, 477
Sorrel Tree, 372
Southernwood, 350
Sow Thistle, 363
Spanish Needles, 345
Spear Grass, 614
Spearmint, 416
Spearwort, 141
Speedwell, 406, 407
Spicewood, 478
Spiderwort, 172, 563
Spikenard, 294
Spinage, 467
Spindle Tree, 216
Spleenwort, 275, 630, 631
Spring Beauty, 194
Spruce, 515
Spurge, 486, 487
Squash, 272
Squirrel-tail Grass, 620
Staff Tree, 215
Star Flower, 451
Star-Grass, 540, 541
Star-of-Bethlem, 551
St. John's-wort, 183, 184
St. Peter's-wort, 182
St. Peter's Wreath, 256
Stone-crop, 277, 279
Stonewort, 637
Strawberry, 253
Strawberry Blite, 468
Succory, 357
Sumach, 292, 293
Summer Savory, 421
Sundew, 179
Sunflower, 339, 340, 341, 343
Swamp Pink, 375, 376
Sweet Basil, 415
Sweet Brier, 247
Sweet Cicely, 293
Sweet Fern, 500
Sweet Flag, 521
Sweet Gum, 506
Sweet-pepper Bush, 372
Sweet Potato, 443
Sweet-scented Grass, 606
Sweet-scented Shrub, 258
Sweet Sultan, 354
Sweet William, 192, 193
Sycamore, 213, 508
Syringa, 282
Tansey, 350
Tape Grass, 520
Tares, 223
Tassel Flower, 349
Tea Plant, 206
Teasel, 309
Ten-weeks' Stock, 170
Thimble-berry, 250
Thistle, 275, 356, 363
Thorn, 243, 244
Thorn Apple, 446
Thoroughwax, 286
Thoroughwort, 316
Threadfoot, 485
Thrift, 390
Thyme, 421
Tickseed, 344, 346
Tiger Flower, 542
Timothy Grass, 602
Toad Flax, 398, 479
Tobacco, 445
Tomato, 419
Touch-me-not, 199
Trailing Arbutus, 374
Tree-of-Heaven, 301
Trefoil, 202, 227, 228, 234
Trumpet Flower, 284

Trumpet-weed, 314
Tuberose, 536
Tulepo, 430
Tulip, 547
Tulip Tree, 130
Turk's Cap, 175, 548
Turmeric-root, 146
Turnip, 171
Tway-blade, 536, 537
Twin flower, 301
Twist-foot, 564
Umbrella Tree, 150
Unicorn Plant, 394
Unicorn-root, 550
Valerian, 308
Vegetable Oyster, 360
Venus' Fly-trap, 180
Venus' Looking-glass, 387
Vernal Grass, 606
Vervain, 412
Vetch, 2, 6, 221, 229
Violet, 175-178, 353
Viper's Bugloss, 436
Virgin's Bower, 130
Wake Robin, 546
Wall Flower, 170
Wall Pepper, 277
Walnut, 490, 491
Water Carpet, 281
Water Cress, 166
Water Feather, 394
Water Hemp, 467
Water leaf, 438
Water Lily, 154
Water melon, 272
Water Nymph, 524
Water Pepper, 474
Water-target, 153
Wayfaring Tree, 302
Wheat, 619
Wheat thief, 433
Whip Grass, 576
Whistle-wood, 213
White Bay, 150
White Grass, 622
White top, 597
White-weed, 338, 343
White-wood, 150
Whitlow Grass, 162
Whortleberry, 368, 369
Wild Basil, 419, 422
Wild Bergamot, 417
Wild Bullace Tree, 241
Wild Germander, 420
Wild Ginger, 465
Wild Indigo, 225
Wild Mandrake, 152
Wild Oats, 564
Wild Peppergrass, 161
Wild Potato, 443
Wild Service-berry, 245
Wild Timothy, 597
Willow, 501-506
Willow-herb, 268
Winter Berry, 381, 382
Winter Cherry, 447
Winter Cress, 166
Wintergreen, 373, 378, 379
Wire Grass, 617
Witch Hazel, 282
Withe Rod, 302
Woad, 164
Wolf-berry, 306
Wolf's-bane, 146
Wood Sorrel, 206, 201
Wood-waxen, 224
Worm Grass, 308
Worm-seed, 469
Wormwood, 349, 350
Woundwort, 420
Wrack Grass, 602
Yam-root, 544
Yarrow, 342
Yellow-eyed Grass, 564
Yellow Phlox, 180
Yellow Rattle, 410
Yellow-seed, 161
Yellow-weed, 162
Yew, 516
Zigadene,

Lightning Source UK Ltd.
Milton Keynes UK
UKHW020336280219
338009UK00006B/517/P